This book is dedicated to
Florence Haseltine, with thanks
for having taught me how
to do my first experiment.

G.M.C.

Brief Contents

PART I   **INTRODUCTION**   1

   1   *An Overview of Cells and Cell Research*   3
   2   *The Composition of Cells*   43
   3   *Cell Metabolism*   73
   4   *Fundamentals of Molecular Biology*   103

PART II   **THE FLOW OF GENETIC INFORMATION**   153

   5   *The Organization and Sequences of Cellular Genomes*   155
   6   *Replication, Maintenance, and Rearrangements of Genomic DNA*   201
   7   *RNA Synthesis and Processing*   253
   8   *Protein Synthesis, Processing, and Regulation*   309

PART III   **CELL STRUCTURE AND FUNCTION**   353

   9   *The Nucleus*   355
   10   *Protein Sorting and Transport:* The Endoplasmic Reticulum, Golgi Apparatus, and Lysosomes   385
   11   *Bioenergetics and Metabolism:* Mitochondria, Chloroplasts, and Peroxisomes   433
   12   *The Cytoskeleton and Cell Movement*   473
   13   *The Plasma Membrane*   529
   14   *Cell Walls, the Extracellular Matrix, and Cell Interactions*   569

PART IV   **CELL REGULATION**   597

   15   *Cell Signaling*   599
   16   *The Cell Cycle*   649
   17   *Cell Death and Cell Renewal*   689
   18   *Cancer*   719

Contents

Preface xiv
Organization and Features of The Cell xvi
Media and Supplements to Accompany The Cell xviii

PART I **Introduction**

1

An Overview of Cells and Cell Research 3

The Origin and Evolution of Cells 4

The First Cell 4

The Evolution of Metabolism 7

Present-Day Prokaryotes 8

Eukaryotic Cells 9

The Origin of Eukaryotes 10

The Development of Multicellular
  Organisms 12

Cells as Experimental Models 16

E. coli 16

Yeasts 17

Caenorhabditis elegans 17

Drosophila melanogaster 18

Arabidopsis thaliana 19

Vertebrates 19

Tools of Cell Biology 21

Light Microscopy 21

Electron Microscopy 27

Subcellular Fractionation 30

Growth of Animal Cells in Culture 33

Culture of Plant Cells 36

Viruses 36

■ **KEY EXPERIMENT** *Animal Cell
  Culture 35*

■ **MOLECULAR MEDICINE** *Viruses
  and Cancer 37*

Summary and Key Terms 39

Questions 41

References and Further Reading 41

2

The Composition of Cells 43

The Molecules of Cells 43

Carbohydrates 44

Lipids 46

Nucleic Acids 50

Proteins 52

Cell Membranes 58

Membrane Lipids 60

Membrane Proteins 61

Transport across Cell Membranes 63

**Proteomics: Large-Scale Analysis
  of Cell Proteins 65**

Identification of Cell Proteins 65

Global Analysis of Protein Localization 68

Protein Interactions 69

■ **KEY EXPERIMENT** *The Folding of
  Polypeptide Chains 54*

■ **KEY EXPERIMENT** *The Structure
  of Cell Membranes 59*

Summary and Key Terms 70

Questions 71

References and Further Reading 71

3

Cell Metabolism 73

**The Central Role of Enzymes as
  Biological Catalysts 73**

The Catalytic Activity of Enzymes 73

Mechanisms of Enzymatic Catalysis 74

Coenzymes 78

Regulation of Enzyme Activity 79

Metabolic Energy 81

Free Energy and ATP 81

The Generation of ATP from Glucose 84

The Derivation of Energy from Other
  Organic Molecules 89

Photosynthesis 90

The Biosynthesis of Cell Constituents 91

Carbohydrates 92

Lipids 93

Proteins 94

Nucleic Acids 97

■ KEY EXPERIMENT *Antimetabolites and Chemotherapy 98*

■ MOLECULAR MEDICINE *Phenylketonuria 96*

Summary and Key Terms 100

Questions 101

References and Further Reading 101

4

Fundamentals of Molecular Biology 103

Heredity, Genes, and DNA 103

Genes and Chromosomes 104

Genes and Enzymes 107

Identification of DNA as the Genetic Material 107

The Structure of DNA 108

Replication of DNA 110

Expression of Genetic Information 111

Colinearity of Genes and Proteins 111

The Role of Messenger RNA 112

The Genetic Code 113

RNA Viruses and Reverse Transcription 115

Recombinant DNA 118

Restriction Endonucleases 118

Generation of Recombinant DNA Molecules 121

Vectors for Recombinant DNA 122

DNA Sequencing 125

Expression of Cloned Genes 126

Detection of Nucleic Acids and Proteins 129

Amplification of DNA by the Polymerase Chain Reaction 129

Nucleic Acid Hybridization 131

Antibodies as Probes for Proteins 134

Gene Function in Eukaryotes 137

Genetic Analysis in Yeasts 137

Gene Transfer in Plants and Animals 139

Mutagenesis of Cloned DNAs 142

Introducing Mutations into Cellular Genes 142

Interfering with Cellular Gene Expression 145

■ KEY EXPERIMENT *The DNA Provirus Hypothesis 116*

■ MOLECULAR MEDICINE *HIV and AIDS 120*

Summary and Key Terms 147

Questions 149

References and Further Reading 150

PART II *The Flow of Genetic Information*

5

The Organization and Sequences of Cellular Genomes 155

The Complexity of Eukaryotic Genomes 155

Introns and Exons 157

Repetitive DNA Sequences 161

Gene Duplication and Pseudogenes 164

The Composition of Higher Eukaryotic Genomes 165

Chromosomes and Chromatin 166

Chromatin 166

Centromeres 171

Telomeres 175

The Sequences of Complete Genomes 176

Prokaryotic Genomes 177

The Yeast Genome 178

The Genomes of *Caenorhabditis elegans* and *Drosophila melanogaster* 180

Plant Genomes 183

The Human Genome 185

The Genomes of Other Vertebrates 190

Bioinformatics and Systems Biology 192

Systematic Screens of Gene Function 193

Regulation of Gene Expression 194

Variation among Individuals and Genomic Medicine 195

■ KEY EXPERIMENT *The Discovery of Introns 158*

■ KEY EXPERIMENT *The Human Genome 188*

Summary and Key Terms 196

Questions 198

References and Further Reading 199

6

Replication, Maintenance, and Rearrangements of Genomic DNA 201

DNA Replication 202
DNA Polymerases 202
The Replication Fork 202
The Fidelity of Replication 209
Origins and the Initiation of
   Replication 211
Telomeres and Telomerase: Maintaining
   the Ends of Chromosomes 214

DNA Repair 216
Direct Reversal of DNA Damage 216
Excision Repair 219
Translesion DNA Synthesis 223
Recombinational Repair 225

**Recombination between Homologous
   DNA Sequences 227**
Models of Homologous
   Recombination 228
Enzymes Involved in Homologous
   Recombination 229

DNA Rearrangements 233
Site-Specific Recombination 233
Transposition via DNA Intermediates 239
Transposition via RNA Intermediates 242
Gene Amplification 247
■ **KEY EXPERIMENT** *Rearrangement of
   Immunoglobulin Genes 240*
■ **MOLECULAR MEDICINE** *Colon
   Cancer and DNA Repair 224*

Summary and Key Terms 249
Questions 250
References and Further Reading 251

7

RNA Synthesis and Processing 253

Transcription in Prokaryotes 254
RNA Polymerase and Transcription 254
Repressors and Negative Control of
   Transcription 258
Positive Control of Transcription 261

**Eukaryotic RNA Polymerases and
   General Transcription Factors 262**
Eukaryotic RNA Polymerases 262
General Transcription Factors and
   Initiation of Transcription by RNA
   Polymerase II 262
Transcription by RNA Polymerases I
   and III 266

**Regulation of Transcription in
   Eukaryotes 268**
cis-Acting Regulatory Sequences:
   Promoters and Enhancers 269
Transcription Factor Binding Sites 272
Transcriptional Regulatory Proteins 273
Structure and Function of Transcriptional
   Activators 277
Eukaryotic Repressors 280
Relationship of Chromatin Structure to
   Transcription 281
Regulation of Transcription by
   Noncoding RNAs 285
DNA Methylation 286

RNA Processing and Turnover 287
Processing of Ribosomal and Transfer
   RNAs 287
Processing of mRNA in Eukaryotes 290
Splicing Mechanisms 292
Alternative Splicing 299
RNA Editing 300
RNA Degradation 301
■ **KEY EXPERIMENT** *Isolation of a
   Eukaryotic Transcription Factor 276*
■ **KEY EXPERIMENT** *The Discovery
   of snRNPs 294*

Summary and Key Terms 303
Questions 306
References and Further Reading 306

8

Protein Synthesis, Processing, and Regulation 309

Translation of mRNA 309
Transfer RNAs 310
The Ribosome 311
The Organization of mRNAs and the
   Initiation of Translation 317
The Process of Translation 319
Regulation of Translation 325

Protein Folding and Processing 329
Chaperones and Protein Folding 330
Enzymes that Catalyze Protein
   Folding 332
Protein Cleavage 333
Glycosylation 335
Attachment of Lipids 337

Regulation of Protein Function 339
Regulation by Small Molecules 340
Protein Phosphorylation 341
Protein-Protein Interactions 344

Protein Degradation 344
The Ubiquitin-Proteasome Pathway 344
Lysosomal Proteolysis 347
■ **KEY EXPERIMENT** *Catalytic Role of
   Ribosomal RNA 316*
■ **MOLECULAR MEDICINE** *Antibiotic
   Resistance and the Ribosome 320*

Summary and Key Terms 348
Questions 350
References and Further Reading 350

PART III Cell Structure and Function

9

The Nucleus 355

The Nuclear Envelope and Traffic between the Nucleus and the Cytoplasm 355

Structure of the Nuclear Envelope 356

The Nuclear Pore Complex 361

Selective Transport of Proteins to and from the Nucleus 362

Regulation of Nuclear Protein Import 368

Transport of RNAs 369

Internal Organization of the Nucleus 371

Chromosomes and Higher-Order Chromatin Structure 371

Sub-Compartments within the Nucleus 374

The Nucleolus and rRNA Processing 375

Ribosomal RNA Genes and the Organization of the Nucleolus 376

Transcription and Processing of rRNA 377

Ribosome Assembly 379

■ **KEY EXPERIMENT** *Nuclear Lamina Diseases 359*

■ **MOLECULAR MEDICINE** *Identification of Nuclear Localization Signals 364*

Summary and Key Terms 380

Questions 382

References and Further Reading 382

10

Protein Sorting and Transport: The Endoplasmic Reticulum, Golgi Apparatus, and Lysosomes 385

The Endoplasmic Reticulum 386

The Endoplasmic Reticulum and Protein Secretion 386

Targeting Proteins to the Endoplasmic Reticulum 387

Insertion of Proteins into the ER Membrane 393

Protein Folding and Processing in the ER 398

Quality Control in the ER 400

The Smooth ER and Lipid Synthesis 403

Export of Proteins and Lipids from the ER 406

The Golgi Apparatus 408

Organization of the Golgi 409

Protein Glycosylation within the Golgi 410

Lipid and Polysaccharide Metabolism in the Golgi 413

Protein Sorting and Export from the Golgi Apparatus 414

The Mechanism of Vesicular Transport 417

Experimental Approaches to Understanding Vesicular Transport 417

Cargo Selection, Coat Proteins, and Vesicle Budding 419

Vesicle Fusion 420

Lysosomes 424

Lysosomal Acid Hydrolases 424

Endocytosis and Lysosome Formation 424

Phagocytosis and Autophagy 428

■ **KEY EXPERIMENT** *The Signal Hypothesis 390*

■ **MOLECULAR MEDICINE** *Gaucher Disease 426*

Summary and Key Terms 431

Questions 431

References and Further Reading 431

11

Bioenergetics and Metabolism: Mitochondria, Chloroplasts, and Peroxisomes 433

Mitochondria 434

Organization and Function of Mitochondria 434

The Genetic System of Mitochondria 435

Protein Import and Mitochondrial Assembly 437

The Mechanism of Oxidative Phosphorylation 443

The Electron Transport Chain 444

Chemiosmotic Coupling 445

Transport of Metabolites across the Inner Membrane 450

Chloroplasts and Other Plastids 451

The Structure and Function of Chloroplasts 451

The Chloroplast Genome 452

Import and Sorting of Chloroplast Proteins 454

Other Plastids 456

Photosynthesis 458

Electron Flow through Photosystems I and II 459

Cyclic Electron Flow 461

ATP Synthesis 462

Peroxisomes 462

Functions of Peroxisomes 463

Peroxisome Assembly 465

■ **KEY EXPERIMENT** *The Chemiosmotic Theory 448*

■ **MOLECULAR MEDICINE** *Diseases of Mitochondria: Leber's Hereditary Optic Neuropathy 438*

Summary and Key Terms 467

Questions 469

References and Further Reading 470

12

The Cytoskeleton and Cell Movement 473

Structure and Organization of Actin Filaments 473

Assembly and Disassembly of Actin Filaments 474

Organization of Actin Filaments 480

Association of Actin Filaments with the Plasma Membrane 482

Protrusions of the Cell Surface 485

Actin, Myosin, and Cell Movement 486

Muscle Contraction 487

Contractile Assemblies of Actin and Myosin in Nonmuscle Cells 491

Nonmuscle Myosins 493

Formation of Protrusions and Cell Movement 495

Intermediate Filaments 497

Intermediate Filament Proteins 497

Assembly of Intermediate Filaments 498

Intracellular Organization of Intermediate Filaments 499

Functions of Keratins and Neurofilaments: Diseases of the Skin and Nervous System 502

Microtubules 505

Structure and Dynamic Organization of Microtubules 505

Assembly of Microtubules 507

Organization of Microtubules within Cells 510

Microtubule Motors and Movement 511

Identification of Microtubule Motor Proteins 511

Cargo Transport and Intracellular Organization 514

Cilia and Flagella 517

Reorganization of Microtubules during Mitosis 520

Chromosome Movement 521

■ **KEY EXPERIMENT** *Expression of Mutant Keratin Causes Abnormal Skin Development 502*

■ **KEY EXPERIMENT** *The Isolation of Kinesin 514*

Summary and Key Terms 523

Questions 526

References and Further Reading 526

13

The Plasma Membrane 529

Structure of the Plasma Membrane 529

The Phospholipid Bilayer 530

Membrane Proteins 532

Mobility of Membrane Proteins 537

The Glycocalyx 540

Transport of Small Molecules 540

Passive Diffusion 541

Facilitated Diffusion and Carrier Proteins 542

Ion Channels 543

Active Transport Driven by ATP Hydrolysis 550

Active Transport Driven by Ion Gradients 555

Endocytosis 556

Phagocytosis 557

Receptor-Mediated Endocytosis 558

Protein Trafficking in Endocytosis 563

■ **KEY EXPERIMENT** *The LDL Receptor 559*

■ **MOLECULAR MEDICINE** *Cystic Fibrosis 554*

Summary and Key Terms 566

Questions 567

References and Further Reading 567

14

Cell Walls, the Extracellular Matrix, and Cell Interactions 569

Cell Walls 569

Bacterial Cell Walls 570

Eukaryotic Cell Walls 570

The Extracellular Matrix and Cell-Matrix Interactions 575

Matrix Structural Proteins 575

Matrix Polysaccharides 578

Matrix Adhesion Proteins 579

Cell-Matrix Interactions 580

Cell-Cell Interactions 584

Adhesion Junctions 584

Tight Junctions 588

Gap Junctions 589

Plasmodesmata 592

■ **KEY EXPERIMENT** *The Characterization of Integrin 582*

■ **MOLECULAR MEDICINE** *Gap Junction Diseases 591*

Summary and Key Terms 593

Questions 594

References and Further Reading 595

PART **IV** *Cell Regulation*

15

Cell Signaling 599

Signaling Molecules and Their Receptors 600

Modes of Cell-Cell Signaling 600

Steroid Hormones and the Nuclear Receptor Superfamily 601

Nitric Oxide and Carbon Monoxide 603

Neurotransmitters 604

Peptide Hormones and Growth Factors 605

Eicosanoids 607

Plant Hormones 608

Functions of Cell Surface Receptors 609

G Protein-Coupled Receptors 610

Receptor Protein-Tyrosine Kinases 612

Cytokine Receptors and Nonreceptor Protein-Tyrosine Kinases 616

Receptors Linked to Other Enzymatic Activities 616

Pathways of Intracellular Signal Transduction 617

The cAMP Pathway: Second Messengers and Protein Phosphorylation 618

Cyclic GMP 620

Phospholipids and Ca²⁺ 621

The PI 3-Kinase/Akt and mTOR Pathways 624

MAP Kinase Pathways 627

The JAK/STAT and TGF-β/Smad Pathways 633

NF-κB Signaling 634

The Hedgehog, Wnt, and Notch Pathways 634

Signal Transduction and the Cytoskeleton 637

Integrins and Signal Transduction 637

Regulation of the Actin Cytoskeleton 638

Signaling Networks 640

Feedback and Crosstalk 640

Networks of Cellular Signal Transduction 641

■ **KEY EXPERIMENT** *The Src Protein-Tyrosine Kinase 612*

■ **MOLECULAR MEDICINE** *Cancer: Signal Transduction and the* ras *Oncogenes 629*

Summary and Key Terms 643

Questions 646

References and Further Reading 646

16

The Cell Cycle 649

The Eukaryotic Cell Cycle 650

Phases of the Cell Cycle 650

Regulation of the Cell Cycle by Cell Growth and Extracellular Signals 652

Cell Cycle Checkpoints 654

Restricting DNA Replication to Once per Cell Cycle 656

Regulators of Cell Cycle Progression 657

Protein Kinases and Cell Cycle Regulation 657

Families of Cyclins and Cyclin-Dependent Kinases 661

Growth Factors and the Regulation of G₁ Cdk's 665

DNA Damage Checkpoints 667

The Events of M Phase 669

Stages of Mitosis 669

Cdk1/Cyclin B and Progression to Metaphase 672

The Spindle Assembly Checkpoint and Progression to Anaphase 675

Cytokinesis 677

Meiosis and Fertilization 678

The Process of Meiosis 678

Regulation of Oocyte Meiosis 681

Fertilization 684

■ **KEY EXPERIMENT** *The Discovery of MPF 658*

■ **KEY EXPERIMENT** *The Identification of Cyclin 662*

Summary and Key Terms 685

Questions 687

References and Further Reading 687

17

Cell Death and Cell Renewal 689

Programmed Cell Death 690

The Events of Apoptosis 690

Caspases: The Executioners of Apoptosis 692

Central Regulators of Apoptosis: The Bcl-2 Family 695

Signaling Pathways that Regulate Apoptosis 698

Stem Cells and the Maintenance of Adult Tissues 700

Proliferation of Differentiated Cells 700

Stem Cells 703

Medical Applications of Adult Stem Cells 708

Embryonic Stem Cells and Therapeutic Cloning 709

Embryonic Stem Cells 710

Somatic Cell Nuclear Transfer 713

■ **KEY EXPERIMENT** *Identification of Genes Required for Programmed Cell Death 694*

■ **KEY EXPERIMENT** *Culture of Embryonic Stem Cells* 710

Summary and Key Terms 716
Questions 717
References and Further Reading 717

18

Cancer 719

The Development and Causes of Cancer 719

Types of Cancer 720
The Development of Cancer 721
Causes of Cancer 723
Properties of Cancer Cells 724
Transformation of Cells in Culture 728

Tumor Viruses 729

Hepatitis B and C Viruses 729
SV40 and Polyomavirus 730
Papillomaviruses 731
Adenoviruses 731
Herpesviruses 732
Retroviruses 732

Oncogenes 733

Retroviral Oncogenes 734
Proto-Oncogenes 735
Oncogenes in Human Cancer 738
Functions of Oncogene Products 741

Tumor Suppressor Genes 746

Identification of Tumor Suppressor Genes 746
Functions of Tumor Suppressor Gene Products 750
Roles of Oncogenes and Tumor Suppressor Genes in Tumor Development 753

Molecular Approaches to Cancer Treatment 755

Prevention and Early Detection 755
Molecular Diagnosis 756
Treatment 756
■ **KEY EXPERIMENT** *The Discovery of Proto-Oncogenes* 737
■ **MOLECULAR MEDICINE** *STI-571: Cancer Treatment Targeted against the* bcr/abl *Oncogene* 759

Summary and Key Terms 761
Questions 763
References and Further Reading 763

Answers to Questions 767
Glossary 779
Index 799

Preface

The three years since publication of the Third Edition of *The Cell* in 2003 have been marked not only by rapid progress in many areas of cell biology, but also by revolutionary changes in some of the ways that molecular and cellular biologists are approaching their science. The human genome project introduced large scale experimentation to cell and molecular biology, together with the need for sophisticated computational approaches to analyze the vast amounts of data generated by genome sequencing. The field of bioinformatics developed in response to this need and has now become an integral part of our efforts to understand the molecular basis of cell behavior. The success of genomics has also spawned global experimental approaches in other areas and given rise to the new field of systems biology, which combines large-scale experimentation with quantitative analysis and modeling.

These changes in experimental approaches have been coupled with striking advances in several areas. Perhaps foremost among these, it has become apparent that microRNAs play widespread roles in regulating gene expression, the full biological significance of which remains to be discovered. Our understanding of stem cells has progressed enormously, together with the potential for using these cells to treat a variety of devastating diseases. Likewise, the last few years have seen major progress in the development of anticancer drugs that specifically target the abnormal proteins responsible for the uncontrolled growth of cancer cells.

The Fourth Edition of *The Cell* has been revised and updated to incorporate these advances and to highlight the exciting opportunities that mark the future of molecular and cellular biology. This edition not only provides an up-to-date review of genomics, but also includes major new sections dealing with the current challenges of bioinformatics, proteomics, and systems biology. The chapter on Cell Signaling has similarly been expanded to include a new section on signaling networks and quantitative analysis of signaling pathways within the cell. A new chapter on Cell Death and Cell Renewal provides expanded treatment of both programmed cell death and stem cells, including their medical applications. And the chapter on Cancer has been updated to cover the most recent developments in designer drugs.

While the text has been updated, the goals and approaches of *The Cell* remain the same as in previous editions. *The Cell* continues to be an accessible and readable text that can be appreciated and mastered by undergraduate students who are taking a first course in cell and molecular biology. Without sacrificing accessibility, we have also sought to ensure that *The Cell* is an intellectually satisfying and rigorous text that will engage students and enable them to appreciate the outstanding unanswered questions in our field. With this goal in mind, we have emphasized experimental approaches and the frontiers of current knowledge throughout the book. Each chapter contains one or two Key Experiment essays that describe a seminal paper and its background in detail, with the intent of giving students a sense of "doing science." Most chapters also contain Molecular Medicine essays that relate basic science to clinical applications. The Fourth Edition of *The Cell* also includes a new feature: each chapter contains several short "sidebars" designed to stimulate students by highlighting additional areas of interest and medical relevance. Finally, we have provided an updated and expanded set of questions at the end of each chapter, with answers to all questions at the back of the book. These questions are designed not only to facilitate review of the chapter, but also to stimulate students to think about the design of experiments and interpretation of experimental results.

The overarching theme of this and previous editions of *The Cell* has been to convey not only the facts, but also a sense of the excitement and challenges of research in contemporary molecular and cellular biology. We hope that *The Cell* stimulates our students to meet these challenges and contribute to the research upon which future texts will be based.

Acknowledgments

The Fourth Edition of *The Cell* has benefited, as always, from the comments of instructors and students who used the previous edition. Many readers have offered helpful suggestions that have been incorporated into the current edition, and we are grateful for their advice.

We are also grateful to the following colleagues who reviewed the new edition and whose comments and criticisms have improved the text:

David G. Bear, *University of New Mexico Health Science Center*
Wade E. Bell, *Virginia Military Institute*
James O. Deshler, *Boston University*
Larry A. Feig, *Tufts University School of Medicine*
Ulla Hansen, *Boston University*
Jennifer Lippincott-Schwartz, *National Institutes of Health*
Robert Mackin, *Creighton University School of Medicine*
Ross N. Nazar, *University of Guelph*
Karen E. Nelson, *The Institute for Genomic Research*
Carol S. Newlon, *University of Medicine and Dentistry of New Jersey*
Kirsten Prufer, *Louisiana State University*
Joel D. Richter, *University of Massachusetts Medical School, Worcester*
Howard Riezman, *Universite de Geneve*
Mendell Rimer, *University of Texas at Austin*
Michael S. Risley, *Albert Einstein College of Medicine*
Teri Shors, *University of Wisconsin, Oshkosh*
Nahum Sonenberg, *University of McGill*
Karsten Weis, *University of California, Berkeley*

We are especially pleased to thank Utsav Saxena, one of our graduate students, for his contributions to this new edition. Based on his experience in teaching the course, Utsav wrote the sidebars and extensively revised the end-of-chapter questions. We are most grateful for his contribution.

We are also grateful to our publishers and editors. Andy Sinauer and Dean Scudder at Sinauer Associates and Jeff Holtmeier at ASM Press have been continuing sources of support and help. Christopher Small, Jefferson Johnson, and Janice Holabird did an exceptional job crafting the book. Suzanne Lain did an outstanding job of copyediting the manuscript, and Chelsea Holabird once again coordinated the production of *The Cell* with skill, patience, and good humor.

GEOFFREY M. COOPER • ROBERT E. HAUSMAN • APRIL 2006

Organization and Features of
The Cell A Molecular Approach

The Cell has been designed to be an approachable and teachable text that can be covered in a single semester while allowing students to master the material in the entire book. It is assumed that most students will have had introductory biology and general chemistry courses, but will not have had previous courses in organic chemistry, biochemistry, or molecular biology. Several aspects of the organization and features of the book will help students to approach and understand its subject matter.

Organization

The Cell is divided into four parts, each of which is self-contained, so that the order and emphasis of topics can be easily varied according to the needs of individual courses.

Part I provides background chapters on the evolution of cells, methods for studying cells, the chemistry of cells, and the fundamentals of modern molecular biology. For those students who have a strong background from either a comprehensive introductory biology course or a previous course in molecular biology, various parts of these chapters can be skipped or used for review.

Part II focuses on the molecular biology of cells and contains chapters dealing with genome organization and sequences; DNA replication, repair, and recombination; transcription and RNA processing; and the synthesis, processing, and regulation of proteins. The order of chapters follows the flow of genetic information (DNA → RNA → protein) and provides a concise but up-to-date overview of these topics.

Part III contains the core block of chapters on cell structure and function, including chapters on the nucleus, cytoplasmic organelles, the cytoskeleton, the plasma membrane, and the extracellular matrix. This part of the book starts with coverage of the nucleus, which puts the molecular biology of Part II within the context of the eukaryotic cell, and then works outward through cytoplasmic organelles and the cytoskeleton to the plasma membrane and the exterior of the cell. These chapters are relatively self-contained, however, and could be used in a different order should that be more appropriate for a particular course.

Finally, Part IV focuses on the exciting and fast-moving area of cell regulation, including coverage of topics such as cell signaling, the cell cycle, programmed cell death, and stem cells. This part of the book concludes with a chapter on cancer, which synthesizes the consequences of defects in basic cell regulatory mechanisms.

Features

Several pedagogical features have been incorporated into *The Cell* in order to help students master and integrate its contents. These features are reviewed below as a guide to students studying from this book.

Chapter organization. Each chapter is divided into three to five major sections, which are further divided into a similar number of subsections. An outline listing the major sections at the beginning of each chapter provides a brief overview of its contents.

Key Terms and Glossary. Key terms are identified as boldfaced words when they are introduced in each chapter. These key terms are reiterated in the chapter summary and defined in the glossary at the end of the book.

Illustrations and micrographs. An illustration program of full-color art and micrographs has been carefully developed to complement and visually reinforce the text.

Key Experiment and Molecular Medicine Essays. Each chapter contains either two Key Experiment essays or one Key Experiment and one Molecular Medicine essay. These features are designed to provide the student with a sense of both the experimental basis of cell and molecular biology and its applications to modern medicine. We have also found these essays to be a useful basis for student discussion sections, which can be accompanied with a review of the original paper upon which the Key Experiments are based.

Sidebars. Each chapter contains several sidebars that provide brief descriptive highlights of points of interest related to material covered in the text. The sidebars supplement the text and provide starting points for class discussion.

Chapter Summaries. Chapter summaries are organized in outline form corresponding to the major sections and subsections of each chapter. This section-by-section format is coupled with a list of the key terms introduced in each section, providing a succinct but comprehensive review of the material.

Questions and Answers. An expanded set of questions at the end of each chapter (with answers in the back of the book) are designed to further facilitate review of the material presented in the chapter and to encourage students to use this material to predict or interpret experimental results.

References. Comprehensive lists of references at the end of each chapter provide access to both reviews and selected papers from the primary literature. In order to help the student identify articles of interest, the references are organized according to chapter sections. Review articles and primary papers are distinguished by [R] and [P] designations, respectively.

Companion Website icons. New icons in the margin direct students to the website's animations, videos, quizzes, problems, and other review material.

Media & Supplements to Accompany
The Cell, Fourth Edition

For the Student
Companion Website

(*www.sinauer.com/cooper*)
The Fourth Edition of *The Cell* features a new companion website, available free of charge. This robust site features a wealth of study and review material coupled with rich multimedia resources including detailed animations, video microscopy, and micrographs. The companion site includes online quizzes, problem sets, animations, videos, chapter summaries, Web links, and a glossary. (See the inside front cover for details.)

For the Instructor*
Instructor's Resource Library

The Fourth Edition Instructor's Resource Library includes an expanded collection of digital resources to aid in planning your course, presenting lectures, and assessing your students. Included are:

- A convenient browser interface
- All textbook figures (art and photos) and tables in both high- and low-resolution JPEG formats
- PowerPoint® presentations of all figures, tables, supplemental photos, videos, and animations
- Complete, ready-to-use lecture PowerPoint® presentations
- An expanded collection of video microscopy
- The entire collection of animations from the student website, for use in lecture
- Over 100 supplemental micrographs
- Problems and quizzes from the student website, with answers
- Textbook end-of-chapter questions, with answers
- The complete Test File, in Microsoft® Word® format
- The computerized Test File (includes Brownstone Diploma® software)
- Chapter outlines

Test File By Dennis Goode

Expanded for the Fourth Edition, the Test File includes a collection of over 1300 multiple-choice, true/false, and short answer questions covering the full range of content covered in every chapter. Questions are organized by chapter heading, making it easy for the instructor to find questions on specific topics. The Test File also includes the quiz questions and problems from the companion website, as well as the textbook end-of-chapter discussion questions. The Test File is included in the Instructor's Resource Library and is provided both as Microsoft Word® files and in the Brownstone Diploma® exam-creation software (software included). Diploma® enables the instructor to easily create exams from the bank of questions provided, as well as add their own questions, publish secure Internet exams, and more.

Online Quizzing

New for the Fourth Edition, online quizzes are available for student use on the companion website. These quizzes are linked to a new online grade book that makes it easy to track student performance. In addition, instructors can add their own questions to the online quizzes, control quiz availability, and more.

Assessment via Course Management Systems

Instructors using course management systems such as WebCT®, Blackboard®, and Angel® can easily create and export quizzes and exams (or the entire test bank) for integration into their online course. The entire test file is provided in WebCT® and Blackboard® formats, and other formats can be easily generated from the Brownstone Diploma® software (included).

Overhead Transparencies

A set of 150 full-color figures from the textbook is available as overhead transparencies. These have been formatted and color-adjusted for optimal projection in the classroom.

Other Media Resources

The following videos and CD-ROMs are available to qualified adopters of the text:

- Fink, *CELLebration* (VHS, ISBN 0-87893-166-X)
- Pickett-Heaps and Pickett-Heaps, *Diatoms: Life in Glass Houses* (DVD, ISBN 0-9586081-7-2)
- Pickett-Heaps and Pickett-Heaps, *The Dynamics and Mechanics of Mitosis* (VHS, ISBN 0-9586081-5-6)
- Pickett-Heaps and Pickett-Heaps, *Living Cells: Structure and Diversity* (VHS, ISBN 0-646-29291-9)
- Pickett-Heaps and Pickett-Heaps, *Remarkable Plants: The Oedogoniales (Green Algae)* (DVD, ISBN 0-9586081-8-0)
- Sardet, Larsonneur, and Koch, *Voyage Inside the Cell* (DVD, ISBN 0-87893-755-2; VHS, ISBN 0-87893-763-3)

*Available to qualified adopters

PART 1

Introduction

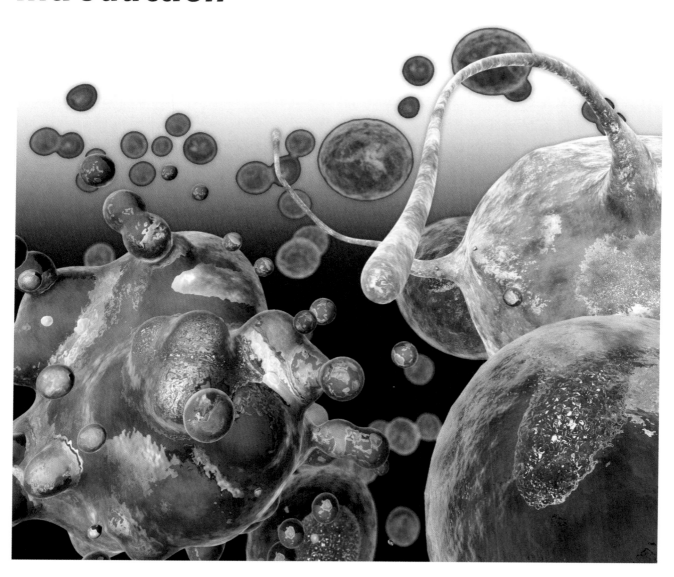

CHAPTER 1 ■ An Overview of Cells and Cell Research

CHAPTER 2 ■ The Composition of Cells

CHAPTER 3 ■ Cell Metabolism

CHAPTER 4 ■ Fundamentals of Molecular Biology

An Overview of Cells and Cell Research

■ **The Origin and Evolution of Cells 4**

■ **Cells as Experimental Models 16**

■ **Tools of Cell Biology 21**

■ **KEY EXPERIMENT:** Animal Cell Culture 35

■ **MOLECULAR MEDICINE:** Viruses and Cancer 37

UNDERSTANDING THE MOLECULAR BIOLOGY OF CELLS is an active area of research that is fundamental to all of the biological sciences. This is true not only from the standpoint of basic science, but also with respect to a growing number of applications in medicine, agriculture, biotechnology, and biomedical engineering. Especially with the completion of the sequence of the human genome, progress in cell and molecular biology is opening new horizons in the practice of medicine. Striking examples include the development of new drugs specifically targeted to interfere with the growth of cancer cells and the potential use of stem cells to replace damaged tissues and treat patients suffering from conditions like diabetes, Parkinson's disease, Alzheimer's disease, spinal cord injuries, and heart disease.

Because cell and molecular biology is a rapidly growing field of research, it is important to understand its experimental basis as well as the current state of our knowledge. This chapter will therefore focus on how cells are studied, as well as reviewing some of their basic properties. Appreciating the similarities and differences between cells is particularly important to understanding cell biology. The first section of this chapter therefore discusses both the unity and the diversity of present-day cells in terms of their evolution from a common ancestor. On the one hand, all cells share common fundamental properties that have been conserved throughout evolution. For example, all cells employ DNA as their genetic material, are surrounded by plasma membranes, and use the same basic mechanisms for energy metabolism. On the other hand, present-day cells have evolved a variety of different lifestyles. Many organisms, such as bacteria, amoebas, and yeasts, consist of single cells that are capable of independent self-replication. More complex organisms are composed of collections of cells that function in a coordinated manner, with different cells specialized to perform particular tasks. The human body, for example, is composed of more than 200 different kinds of cells, each specialized for such distinctive functions as memory, sight, movement, and digestion. The diversity exhibited by the many different kinds of cells is striking; for example, consider the differences between bacteria and the cells of the human brain.

The fundamental similarities between different types of cells provide a unifying theme to cell biology, allowing the basic principles learned from experiments with one kind of cell to be extrapolated and generalized to other cell types. Several kinds of cells and organisms are widely used to study different aspects of cell and molecular biology; the second section of this chapter discusses some of the properties of these cells that make them particularly valuable as experimental models. Finally, it is important to recognize that progress in cell biology depends heavily on the availability of experimental tools that allow scientists to make new observations or conduct novel kinds of experiments. This introductory chapter therefore concludes with a discussion of some of the experimental approaches used to study cells, as well as a review of some of the major historical developments that have led to our current understanding of cell structure and function.

The Origin and Evolution of Cells

Cells are divided into two main classes, initially defined by whether they contain a nucleus. **Prokaryotic cells** (bacteria) lack a nuclear envelope; **eukaryotic cells** have a nucleus in which the genetic material is separated from the cytoplasm. Prokaryotic cells are generally smaller and simpler than eukaryotic cells; in addition to the absence of a nucleus, their genomes are less complex and they do not contain cytoplasmic organelles or a cytoskeleton (Table 1.1). In spite of these differences, the same basic molecular mechanisms govern the lives of both prokaryotes and eukaryotes, indicating that all present-day cells are descended from a single primordial ancestor. How did this first cell develop? And how did the complexity and diversity exhibited by present-day cells evolve?

The First Cell

It appears that life first emerged at least 3.8 billion years ago, approximately 750 million years after Earth was formed. How life originated and how the first cell came into being are matters of speculation, since these events cannot be reproduced in the laboratory. Nonetheless, several types of experiments provide important evidence bearing on some steps of the process.

TABLE 1.1 Prokaryotic and Eukaryotic Cells

Characteristic	Prokaryote	Eukaryote
Nucleus	Absent	Present
Diameter of a typical cell	$\approx 1\ \mu m$	$10–100\ \mu m$
Cytoskeleton	Absent	Present
Cytoplasmic organelles	Absent	Present
DNA content (base pairs)	1×10^6 to 5×10^6	1.5×10^7 to 5×10^9
Chromosomes	Single circular DNA molecule	Multiple linear DNA molecules

FIGURE 1.1 Spontaneous formation of organic molecules Water vapor was refluxed through an atmosphere consisting of CH_4, NH_3, and H_2, into which electric sparks were discharged. Analysis of the reaction products revealed the formation of a variety of organic molecules, including the amino acids alanine, aspartic acid, glutamic acid, and glycine.

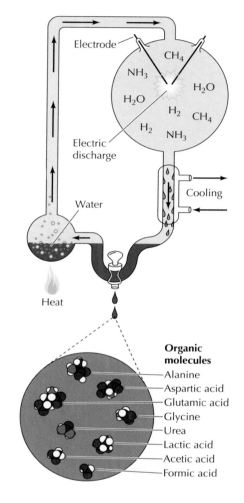

It was first suggested in the 1920s that simple organic molecules could form and spontaneously polymerize into macromolecules under the conditions thought to exist in primitive Earth's atmosphere. At the time life arose, the atmosphere of Earth is thought to have contained little or no free oxygen, instead consisting principally of CO_2 and N_2 in addition to smaller amounts of gases such as H_2, H_2S, and CO. Such an atmosphere provides reducing conditions in which organic molecules, given a source of energy such as sunlight or electrical discharge, can form spontaneously. The spontaneous formation of organic molecules was first demonstrated experimentally in the 1950s, when Stanley Miller (then a graduate student) showed that the discharge of electric sparks into a mixture of H_2, CH_4, and NH_3, in the presence of water, leads to the formation of a variety of organic molecules, including several amino acids (Figure 1.1). Although Miller's experiments did not precisely reproduce the conditions of primitive Earth, they clearly demonstrated the plausibility of the spontaneous synthesis of organic molecules, providing the basic materials from which the first living organisms arose.

The next step in evolution was the formation of macromolecules. The monomeric building blocks of macromolecules have been demonstrated to polymerize spontaneously under plausible prebiotic conditions. Heating dry mixtures of amino acids, for example, results in their polymerization to form polypeptides. But the critical characteristic of the macromolecule from which life evolved must have been the ability to replicate itself. Only a macromolecule capable of directing the synthesis of new copies of itself would have been capable of reproduction and further evolution.

Of the two major classes of informational macromolecules in present-day cells (nucleic acids and proteins), only the nucleic acids are capable of directing their own self-replication. Nucleic acids can serve as templates for their own synthesis as a result of specific base pairing between complementary nucleotides (Figure 1.2). A critical step in understanding molecular evo-

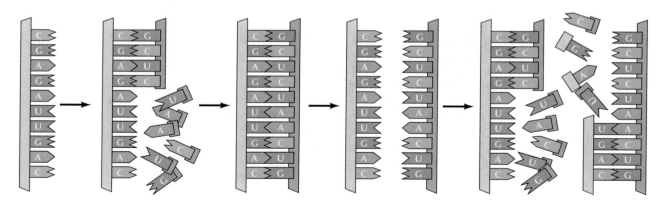

FIGURE 1.2 Self-replication of RNA Complementary pairing between nucleotides (adenine [A] with uracil [U] and guanine [G] with cytosine [C]) allows one strand of RNA to serve as a template for the synthesis of a new strand with the complementary sequence.

lution was thus reached in the early 1980s, when it was discovered in the laboratories of Sid Altman and Tom Cech that RNA is capable of catalyzing a number of chemical reactions, including the polymerization of nucleotides. Further studies have extended the known catalytic activities of RNA, including the description of RNA molecules that direct the synthesis of a new RNA strand from an RNA template. RNA is thus uniquely able to both serve as a template for and to catalyze its own replication. Consequently, RNA is generally believed to have been the initial genetic system, and an early stage of chemical evolution is thought to have been based on self-replicating RNA molecules—a period of evolution known as the **RNA world**. Ordered interactions between RNA and amino acids then evolved into the present-day genetic code, and DNA eventually replaced RNA as the genetic material.

The first cell is presumed to have arisen by the enclosure of self-replicating RNA in a membrane composed of **phospholipids** (Figure 1.3). As discussed in detail in the next chapter, phospholipids are the basic components of all present-day biological membranes, including the plasma membranes of both prokaryotic and eukaryotic cells. The key characteristic of the phospholipids that form membranes is that they are **amphipathic** molecules, meaning that one portion of the molecule is soluble in water and another portion is not. Phospholipids have long, water-insoluble (**hydrophobic**) hydrocarbon chains joined to water-soluble (**hydrophilic**) head groups that contain phosphate. When placed in water, phospholipids spontaneously aggregate into a bilayer with their phosphate-containing head groups on the outside in contact with water and their hydrocarbon tails in the interior in contact with each other. Such a phospholipid bilayer forms a stable barrier between two aqueous compartments—for example, separating the interior of the cell from its external environment.

The enclosure of self-replicating RNA and associated molecules in a phospholipid membrane would thus have maintained them as a unit, capa-

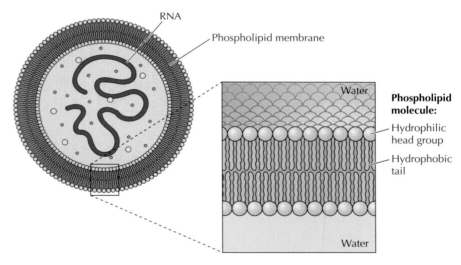

FIGURE 1.3 Enclosure of self-replicating RNA in a phospholipid membrane
The first cell is thought to have arisen by the enclosure of self-replicating RNA and associated molecules in a membrane composed of phospholipids. Each phospholipid molecule has two long hydrophobic tails attached to a hydrophilic head group. The hydrophobic tails are buried in the lipid bilayer; the hydrophilic heads are exposed to water on both sides of the membrane.

ble of self-reproduction and further evolution. RNA-directed protein synthesis may already have evolved by this time, in which case the first cell would have consisted of self-replicating RNA and its encoded proteins.

The Evolution of Metabolism

Because cells originated in a sea of organic molecules, they were able to obtain food and energy directly from their environment. But such a situation is self-limiting, so cells needed to evolve their own mechanisms for generating energy and synthesizing the molecules necessary for their replication. The generation and controlled utilization of metabolic energy is central to all cell activities, and the principal pathways of energy metabolism (discussed in detail in Chapter 3) are highly conserved in present-day cells. All cells use **adenosine 5′-triphosphate (ATP)** as their source of metabolic energy to drive the synthesis of cell constituents and carry out other energy-requiring activities, such as movement (e.g., muscle contraction). The mechanisms used by cells for the generation of ATP are thought to have evolved in three stages, corresponding to the evolution of glycolysis, photosynthesis, and oxidative metabolism (Figure 1.4). The development of these metabolic pathways changed Earth's atmosphere, thereby altering the course of further evolution.

In the initially anaerobic atmosphere of Earth, the first energy-generating reactions presumably involved the breakdown of organic molecules in the absence of oxygen. These reactions are likely to have been a form of present-day **glycolysis**—the anaerobic breakdown of glucose to lactic acid, with the net energy gain of two molecules of ATP. In addition to using ATP as their source of intracellular chemical energy, all present-day cells carry out glycolysis, consistent with the notion that these reactions arose very early in evolution.

Glycolysis provided a mechanism by which the energy in preformed organic molecules (e.g., glucose) could be converted to ATP, which could then be used as a source of energy to drive other metabolic reactions. The

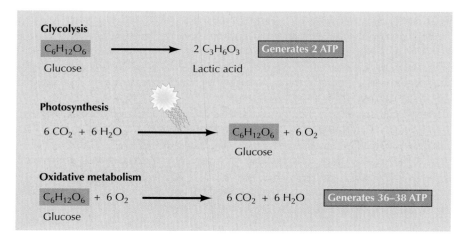

Glycolysis

$C_6H_{12}O_6$ $\longrightarrow$ $2\ C_3H_6O_3$ Generates 2 ATP

Glucose   Lactic acid

Photosynthesis

$6\ CO_2 + 6\ H_2O$ $\longrightarrow$ $C_6H_{12}O_6 + 6\ O_2$

                                        Glucose

Oxidative metabolism

$C_6H_{12}O_6 + 6\ O_2$ $\longrightarrow$ $6\ CO_2 + 6\ H_2O$ Generates 36–38 ATP

Glucose

FIGURE 1.4 Generation of metabolic energy   Glycolysis is the anaerobic breakdown of glucose to lactic acid. Photosynthesis utilizes energy from sunlight to drive the synthesis of glucose from CO_2 and H_2O, with the release of O_2 as a by-product. The O_2 released by photosynthesis is used in oxidative metabolism, in which glucose is broken down to CO_2 and H_2O, releasing much more energy than is obtained from glycolysis.

development of **photosynthesis** is generally thought to have been the next major evolutionary step, which allowed the cell to harness energy from sunlight and provided independence from the utilization of preformed organic molecules. The first photosynthetic bacteria, which evolved more than 3 billion years ago, probably utilized H_2S to convert CO_2 to organic molecules—a pathway of photosynthesis still used by some bacteria. The use of H_2O as a donor of electrons and hydrogen for the conversion of CO_2 to organic compounds evolved later and had the important consequence of changing Earth's atmosphere. The use of H_2O in photosynthetic reactions produces the by-product free O_2; this mechanism is thought to have been responsible for making O_2 abundant in Earth's atmosphere.

The release of O_2 as a consequence of photosynthesis changed the environment in which cells evolved and is commonly thought to have led to the development of **oxidative metabolism**. Alternatively, oxidative metabolism may have evolved before photosynthesis, with the increase in atmospheric O_2 then providing a strong selective advantage for organisms capable of using O_2 in energy-producing reactions. In either case, O_2 is a highly reactive molecule, and oxidative metabolism, utilizing this reactivity, has provided a mechanism for generating energy from organic molecules that is much more efficient than anaerobic glycolysis. For example, the complete oxidative breakdown of glucose to CO_2 and H_2O yields energy equivalent to that of 36 to 38 molecules of ATP, in contrast to the 2 ATP molecules formed by anaerobic glycolysis. With few exceptions, present-day cells use oxidative reactions as their principal source of energy.

Present-Day Prokaryotes

Present-day prokaryotes, which include all the various types of bacteria, are divided into two groups—the **archaebacteria** and the **eubacteria**—which diverged early in evolution. Some archaebacteria live in extreme environments, which are unusual today but may have been prevalent in primitive Earth. For example, thermoacidophiles live in hot sulfur springs with temperatures as high as 80°C and pH values as low as 2. The eubacteria include the common forms of present-day bacteria—a large group of organisms that live in a wide range of environments, including soil, water, and other organisms (e.g., human pathogens).

Most bacterial cells are spherical, rod-shaped, or spiral, with diameters of 1 to 10 μm. Their DNA contents range from about 0.6 million to 5 million base pairs, an amount sufficient to encode about 5000 different proteins. The largest and most complex prokaryotes are the **cyanobacteria**, bacteria in which photosynthesis evolved.

The structure of a typical prokaryotic cell is illustrated by *Escherichia coli (E. coli)*, a common inhabitant of the human intestinal tract (Figure 1.5). The cell is rod-shaped, about 1 μm in diameter and about 2 μm long. Like most other prokaryotes, *E. coli* is surrounded by a rigid **cell wall** composed of polysaccharides and peptides. Beneath the cell wall is the **plasma membrane**, which is a bilayer of phospholipids and associated proteins. Whereas the cell wall is porous and readily penetrated by a variety of molecules, the plasma membrane provides the functional separation between the inside of the cell and its external environment. The DNA of *E. coli* is a single circular molecule in the nucleoid, which, in contrast to the nucleus of eukaryotes, is not surrounded by a membrane separating it from the cytoplasm. The cytoplasm contains approximately 30,000 **ribosomes** (the sites of protein synthesis), which account for its granular appearance.

■ **Existence of organisms in extreme conditions has led to the hypothesis that life could exist in similar environments elsewhere in the solar system. The field of astrobiology (or exobiology) seeks to find signs of this extraterrestrial life.**

FIGURE 1.5 Electron micrograph of *E. coli*   The cell is surrounded by a cell wall, beneath which is the plasma membrane. DNA is located in the nucleoid. (Menge and Wurtz/Biozentrum, University of Basel/Science Photo Library/Photo Researchers, Inc.)

Plasma membrane

Cell wall

Nucleoid

0.5 μm

Eukaryotic Cells

Like prokaryotic cells, all eukaryotic cells are surrounded by a plasma membrane and contain ribosomes. However, eukaryotic cells are much more complex and contain a nucleus, a variety of cytoplasmic organelles, and a cytoskeleton (Figure 1.6). The largest and most prominent organelle of eukaryotic cells is the **nucleus**, with a diameter of approximately 5 μm. The nucleus contains the genetic information of the cell, which in eukaryotes is organized as linear rather than circular DNA molecules. The nucleus is the site of DNA replication and of RNA synthesis; the translation of RNA into proteins takes place on ribosomes in the cytoplasm.

In addition to a nucleus, eukaryotic cells contain a variety of membrane-enclosed organelles within their cytoplasm. These organelles provide compartments in which different metabolic activities are localized. Eukaryotic cells are generally much larger than prokaryotic cells, frequently having a cell volume at least a thousandfold greater. The compartmentalization provided by cytoplasmic organelles is what allows eukaryotic cells to function efficiently. Two of these organelles, **mitochondria** and **chloroplasts**, play critical roles in energy metabolism. Mitochondria, which are found in almost all eukaryotic cells, are the sites of oxidative metabolism and are thus responsible for generating most of the ATP derived from the breakdown of organic molecules. Chloroplasts are the sites of photosynthesis and are found only in the cells of plants and green algae. **Lysosomes** and **peroxisomes** also provide specialized metabolic compartments for the digestion of macromolecules and for various oxidative reactions, respectively. In addition, most plant cells contain large **vacuoles** that perform a variety of functions, including the digestion of macromolecules and the storage of both waste products and nutrients.

Because of the size and complexity of eukaryotic cells, the transport of proteins to their correct destinations within the cell is a formidable task. Two cytoplasmic organelles, the **endoplasmic reticulum** and the **Golgi apparatus**, are specifically devoted to the sorting and transport of proteins destined for secretion, incorporation into the plasma membrane, and incorporation into lysosomes. The endoplasmic reticulum is an extensive network of intracellular membranes, extending from the nuclear membrane throughout the cytoplasm. It functions not only in the processing and transport of proteins, but also in the synthesis of lipids. From the endoplasmic reticulum, proteins are transported within small membrane vesicles to the Golgi apparatus, where they are further processed and sorted for transport to their final destinations. In addition to this role in protein transport, the Golgi apparatus serves as a site of lipid synthesis and (in plant cells) as the site of synthesis of some of the polysaccharides that compose the cell wall.

Eukaryotic cells have another level of internal organization: the **cytoskeleton**, a network of protein filaments extending throughout the cytoplasm. The cytoskeleton provides the structural framework of the cell, determining cell shape and the general organization of the cytoplasm. In

Animal cell

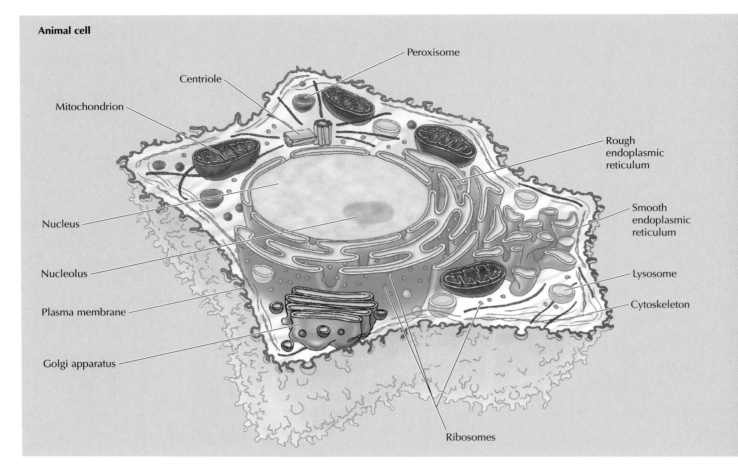

FIGURE 1.6 Structures of animal and plant cells Both animal and plant cells are surrounded by a plasma membrane and contain a nucleus, a cytoskeleton, and many cytoplasmic organelles in common. Plant cells are also surrounded by a cell wall and contain chloroplasts and large vacuoles.

addition, the cytoskeleton is responsible for the movements of entire cells (e.g., the contraction of muscle cells) and for the intracellular transport and positioning of organelles and other structures, including the movements of chromosomes during cell division.

The Origin of Eukaryotes

A critical step in the evolution of eukaryotic cells was the acquisition of membrane-enclosed subcellular organelles, allowing the development of the complexity characteristic of these cells. The organelles of eukaryotes are thought to have arisen by **endosymbiosis**—one cell living inside another. In particular, eukaryotic organelles are thought to have evolved from prokaryotic cells living inside the ancestors of eukaryotes.

The hypothesis that eukaryotic cells evolved by endosymbiosis is particularly well supported by studies of mitochondria and chloroplasts, which are thought to have evolved from eubacteria living in larger cells. Both mitochondria and chloroplasts are similar to bacteria in size, and like bacteria, they reproduce by dividing in two. Most important, both mitochondria and chloroplasts contain their own DNA, which encodes some of their components. The mitochondrial and chloroplast DNAs are replicated each time the organelle divides, and the genes they encode are transcribed within the organelle and translated on organelle ribosomes. Mitochondria and chloroplasts thus contain their own genetic systems, which are distinct from the nuclear genome of the cell. Furthermore, the ribosomes and ribosomal

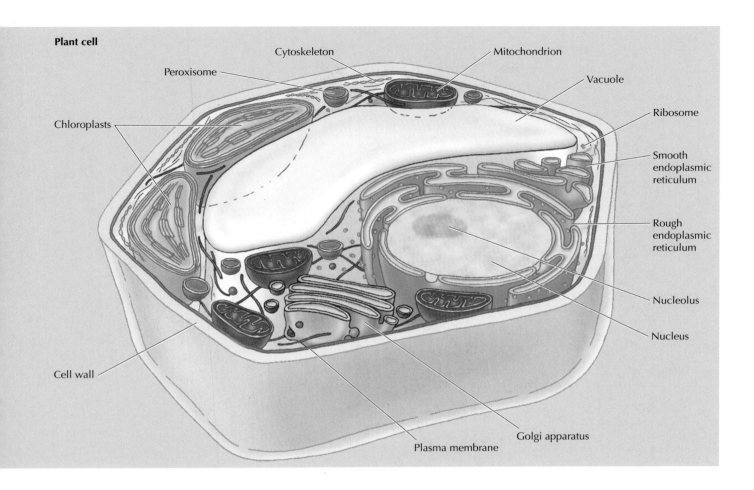

Plant cell

Peroxisome

Cytoskeleton

Mitochondrion

Vacuole

Ribosome

Chloroplasts

Smooth endoplasmic reticulum

Rough endoplasmic reticulum

Nucleolus

Nucleus

Cell wall

Plasma membrane

Golgi apparatus

RNAs of these organelles are more closely related to those of bacteria than to those encoded by the nuclear genomes of eukaryotes.

An endosymbiotic origin for these organelles is now generally accepted, with mitochondria thought to have evolved from aerobic eubacteria and chloroplasts from photosynthetic eubacteria, such as the cyanobacteria. The acquisition of aerobic bacteria would have provided an anaerobic cell with the ability to carry out oxidative metabolism. The acquisition of photosynthetic bacteria would have provided the nutritional independence afforded by the ability to perform photosynthesis. Thus, these endosymbiotic associations were highly advantageous to their partners and were selected for in the course of evolution. Through time, most of the genes originally present in these bacteria apparently became incorporated into the nuclear genome of the cell, so only a few components of mitochondria and chloroplasts are still encoded by the organelle genomes.

The precise origin of eukaryotic cells remains an unsettled issue in our understanding of early evolution. Studies of their DNA sequences indicate that the archaebacteria and eubacteria are as different from each other as either is from present-day eukaryotes. Therefore, a very early event in evolution appears to have been the divergence of two lines of descent from a common prokaryotic ancestor, giving rise to present-day archaebacteria and eubacteria. However, it has proven difficult to determine whether eukaryotes evolved from eubacteria or from archaebacteria. Surprisingly, some eukaryotic genes are more similar to eubacterial genes whereas others

■ **Certain present day marine protists engulf algae to serve as endosymbionts that carry out photosynthesis for their hosts.**

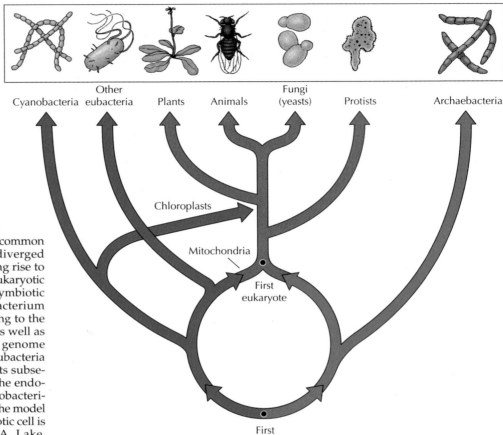

FIGURE 1.7 Evolution of cells
Present-day cells evolved from a common prokaryotic ancestor which diverged along two lines of descent, giving rise to archaebacteria and eubacteria. Eukaryotic cells may have arisen by endosymbiotic association of an aerobic eubacterium with an archaebacterium, leading to the development of mitochondria as well as the formation of a eukaryotic genome with genes derived from both eubacteria and archaebacteria. Chloroplasts subsequently evolved as a result of the endosymbiotic association of a cyanobacterium with the ancestor of plants. The model for formation of the first eukaryotic cell is based on M. C. Rivera and J. A. Lake, 2004. *Nature* 431: 152.

are more similar to archaebacterial genes. The genome of eukaryotes thus appears to consist of some genes derived from eubacteria and others from archaebacteria, rather than reflecting the genome of either a eubacterial or archaebacterial ancestor.

A recent hypothesis explains the mosaic nature of eukaryotic genomes by proposing that the genome of eukaryotes arose from a fusion of archaebacterial and eubacterial genomes (Figure 1.7). According to this proposal, an endosymbiotic association between a eubacterium and an archaebacterium was followed by fusion of the two prokaryotic genomes, giving rise to an ancestral eukaryotic genome with contributions from both eubacteria and archaebacteria. The simplest version of this hypothesis is that an initial endosymbiotic relationship of a eubacterium living inside an archaebacterium gave rise not only to mitochondria but also to the genome of eukaryotic cells, containing genes derived from both prokaryotic ancestors.

The Development of Multicellular Organisms

Many eukaryotes are unicellular organisms that, like bacteria, consist of only single cells capable of self-replication. The simplest eukaryotes are the **yeasts**. Yeasts are more complex than bacteria, but much smaller and simpler than the cells of animals or plants. For example, the commonly studied yeast *Saccharomyces cerevisiae* is about 6 μm in diameter and contains 12 million base pairs of DNA (Figure 1.8). Other unicellular eukaryotes, how-

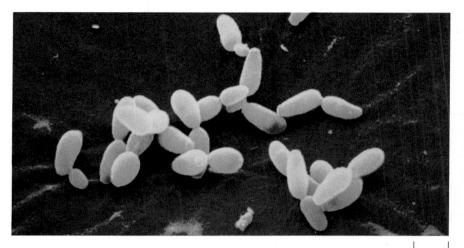

5 µm

FIGURE 1.8 Scanning electron micrograph of *Saccharomyces cerevisiae*
Artificial color has been added to the micrograph. (Andrew Syed/Science Photo
Library/Photo Researchers, Inc.)

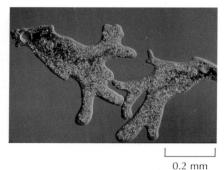
0.2 mm

**FIGURE 1.9 Light micrograph of
*Amoeba proteus***   (M. I. Walker/
Photo Researchers, Inc.)

ever, are far more complex cells, some containing as much DNA as human
cells have. They include organisms specialized to perform a variety of tasks,
including photosynthesis, movement, and the capture and ingestion of
other organisms as food. *Amoeba proteus*, for example, is a large, complex
cell. Its volume is more than 100,000 times that of *E. coli*, and its length can
exceed 1 mm when the cell is fully extended (Figure 1.9). Amoebas are highly
mobile organisms that use cytoplasmic extensions, called **pseudopodia**, to
move and to engulf other organisms, including bacteria and yeasts, as food.
Other unicellular eukaryotes (the green algae) contain chloroplasts and are
able to carry out photosynthesis.

Multicellular organisms evolved from unicellular eukaryotes more than
1 billion years ago. Some unicellular eukaryotes form multicellular aggregates that appear to represent an evolutionary transition from single cells to
multicellular organisms. For instance, the cells of many algae (e.g., the
green alga *Volvox*) associate with each other to form multicellular colonies
(Figure 1.10), which are thought to have been the evolutionary precursors of
present-day plants. Increasing cell specialization then led to the transition
from colonial aggregates to truly multicellular organisms. Continuing cell
specialization and division of labor among the cells of an organism have led
to the complexity and diversity observed in the many types of cells that
make up present-day plants and animals, including human beings.

Plants are composed of fewer cell types than are animals, but each different kind of plant cell is specialized to perform specific tasks required by the
organism as a whole (Figure 1.11). The cells of plants are organized into three
main tissue systems: ground tissue, dermal tissue, and vascular tissue. The
ground tissue contains parenchyma cells, which carry out most of the metabolic reactions of the plant, including photosynthesis. Ground tissue also
contains two specialized cell types (collenchyma cells and sclerenchyma
cells) that are characterized by thick cell walls and provide structural support to the plant. Dermal tissue covers the surface of the plant and is composed of epidermal cells, which form a protective coat and allow the
absorption of nutrients. Finally, several types of elongated cells form the

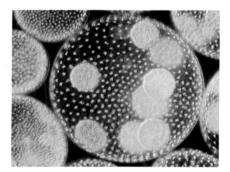

FIGURE 1.10 Colonial green algae
Individual cells of *Volvox* form colonies
consisting of hollow balls in which hundreds or thousands of cells are embedded in a gelatinous matrix. (Cabisco/
Visuals Unlimited.)

FIGURE 1.11 Light micrographs of representative plant cells
(A) Parenchyma cells, which are responsible for photosynthesis and other metabolic reactions. (B) Collenchyma cells, which are specialized for support and have thickened cell walls. (C) Epidermal cells on the surface of a leaf. Tiny pores (stomata) are flanked by specialized cells called guard cells. (D) Vessel elements and tracheids are elongated cells that are arranged end to end to form vessels of the xylem. (A, Jack M. Bastsack/Visuals Unlimited; B, A. J. Karpoff/Visuals Unlimited; C, Alfred Owczarzak/Biological Photo Service; D, Biophoto Associates/Science Source/Photo Researchers Inc.)

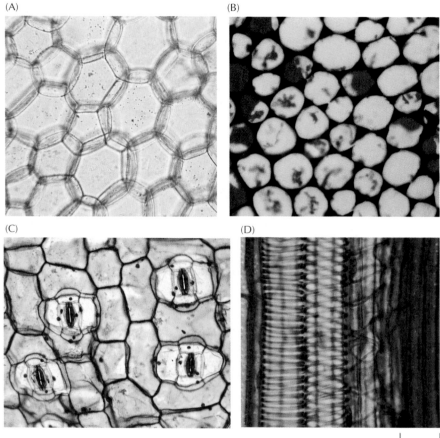

50 μm

vascular system (the xylem and phloem), which is responsible for the transport of water and nutrients throughout the plant.

The cells found in animals are considerably more diverse than those of plants. The human body, for example, is composed of more than 200 different kinds of cells, which are generally considered to be components of five main types of tissues: epithelial tissue, connective tissue, blood, nervous tissue, and muscle (Figure 1.12). **Epithelial cells** form sheets that cover the surface of the body and line the internal organs. There are many different types of epithelial cells, each specialized for a specific function, including protection (the skin), absorption (e.g., the cells lining the small intestine), and secretion (e.g., cells of the salivary gland). Connective tissues include bone, cartilage, and adipose tissue, each of which is formed by different types of cells (osteoblasts, chondrocytes, and adipocytes, respectively). The loose connective tissue that underlies epithelial layers and fills the spaces between organs and tissues in the body is formed by another cell type, the **fibroblast**. Blood contains several different types of cells: red blood cells (**erythrocytes**) function in oxygen transport and white blood cells (**granulocytes, monocytes, macrophages,** and **lymphocytes**) function in inflammatory reactions and the immune response. Nervous tissue is composed of supporting cells and nerve cells, or **neurons**, which are highly specialized to transmit signals throughout the body. Various types of sensory cells, such as cells of the eye and ear, are further specialized to receive external signals from the environment. Finally, several different types of muscle cells are responsible for the production of force and movement.

The evolution of animals clearly involved the development of considerable diversity and specialization at the cellular level. Understanding the mechanisms that control the growth and differentiation of such a complex array of specialized cells, starting from a single fertilized egg, is one of the major challenges facing contemporary cell and molecular biology.

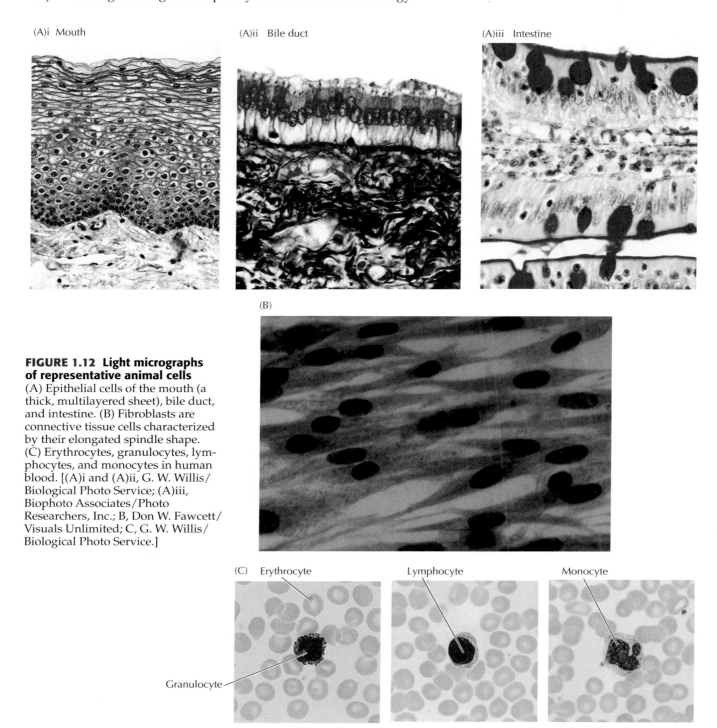

(A)i Mouth

(A)ii Bile duct

(A)iii Intestine

(B)

(C) Erythrocyte   Lymphocyte   Monocyte

Granulocyte

FIGURE 1.12 Light micrographs of representative animal cells (A) Epithelial cells of the mouth (a thick, multilayered sheet), bile duct, and intestine. (B) Fibroblasts are connective tissue cells characterized by their elongated spindle shape. (C) Erythrocytes, granulocytes, lymphocytes, and monocytes in human blood. [(A)i and (A)ii, G. W. Willis/ Biological Photo Service; (A)iii, Biophoto Associates/Photo Researchers, Inc.; B, Don W. Fawcett/ Visuals Unlimited; C, G. W. Willis/ Biological Photo Service.]

Cells as Experimental Models

The evolution of present-day cells from a common ancestor has important implications for cell and molecular biology as an experimental science. Because the fundamental properties of all cells have been conserved during evolution, the basic principles learned from experiments performed with one type of cell are generally applicable to other cells. On the other hand, because of the diversity of present-day cells, many kinds of experiments can be more readily undertaken with one type of cell than with another. Several different kinds of cells and organisms are commonly used as experimental models to study various aspects of cell and molecular biology. The features of some of these cells that make them particularly advantageous as experimental models are discussed in the sections that follow. In many cases, the availability of complete genome sequences further enhances the value of these organisms as model systems in understanding the molecular biology of cells.

E. coli

Because of their comparative simplicity, prokaryotic cells (bacteria) are ideal models for studying many fundamental aspects of biochemistry and molecular biology. The most thoroughly studied species of bacteria is *E. coli*, which has long been the favored organism for investigation of the basic mechanisms of molecular genetics. Most of our present concepts of molecular biology—including our understanding of DNA replication, the genetic code, gene expression, and protein synthesis—derive from studies of this humble bacterium.

E. coli has been especially useful to molecular biologists because of both its relative simplicity and the ease with which it can be propagated and studied in the laboratory. The genome of *E. coli*, for example, consists of approximately 4.6 million base pairs and contains about 4300 genes. The human genome is nearly a thousand times larger (approximately 3 billion base pairs) and is thought to contain 20,000 to 25,000 genes (Table 1.2). The small size of the *E. coli* genome (which was completely sequenced in 1997) provides obvious advantages for genetic analysis.

Molecular genetic experiments are further facilitated by the rapid growth of *E. coli* under well-defined laboratory conditions. Under optimal culture conditions, *E. coli* divide every 20 minutes. Moreover, a clonal population of *E. coli*, in which all cells are derived by division of a single cell of origin, can be readily isolated as a colony grown on semisolid agar-containing medium (Figure 1.13). Because bacterial colonies containing as many as 10^8 cells can develop overnight, selecting genetic variants of an *E. coli* strain—for example, mutants that are resistant to an antibiotic, such as penicillin—is easy and rapid. The ease with which such mutants can be selected and analyzed was critical to the success of

TABLE 1.2 DNA Content of Cells

Organism	Haploid DNA content (millions of base pairs)	Number of genes
Bacteria		
Mycoplasma	0.6	470
E. coli	4.6	4300
Unicellular eukaryotes		
Saccharomyces cerevisiae (yeast)	12	6000
Dictyostelium discoideum	70	Not known
Euglena	3000	Not known
Plants		
Arabidopsis thaliana	125	26,000
Zea mays (corn)	5000	Not known
Animals		
Caenorhabditis elegans (nematode)	97	19,000
Drosophila melanogaster (fruit fly)	180	14,000
Chicken	1200	20–23,000
Zebrafish	1700	Not known
Mouse	3000	20–25,000
Human	3000	20–25,000

experiments that defined the basic principles of molecular genetics, discussed in Chapter 4.

The nutrient mixtures in which *E. coli* divide most rapidly include glucose, salts, and various organic compounds, such as amino acids, vitamins, and nucleic acid precursors. However, *E. coli* can also grow in much simpler media consisting only of salts, a source of nitrogen (such as ammonia), and a source of carbon and energy (such as glucose). In such a medium, the bacteria grow a little more slowly (with a division time of about 40 minutes) because they must synthesize all their own amino acids, nucleotides, and other organic compounds. The ability of *E. coli* to carry out these biosynthetic reactions in simple defined media has made them extremely useful in elucidating the biochemical pathways involved. Thus, the rapid growth and simple nutritional requirements of *E. coli* have greatly facilitated fundamental experiments in both molecular biology and biochemistry.

Yeasts

Although bacteria have been an invaluable model for studies of many conserved properties of cells, they obviously cannot be used to study aspects of cell structure and function that are unique to eukaryotes. Yeasts, the simplest eukaryotes, have a number of experimental advantages similar to those of *E. coli*. Consequently, yeasts have provided a crucial model for studies of many fundamental aspects of eukaryotic cell biology.

The genome of the most frequently studied yeast, *Saccharomyces cerevisiae*, consists of 12 million base pairs of DNA and contains about 6000 genes. Although the yeast genome is approximately three times larger than that of *E. coli*, it is far more manageable than the genomes of more complex eukaryotes, such as humans. Yet, even in its simplicity, the yeast cell exhibits the typical features of eukaryotic cells (Figure 1.14): It contains a distinct nucleus surrounded by a nuclear membrane, its genomic DNA is organized as 16 linear chromosomes, and its cytoplasm contains a cytoskeleton and subcellular organelles.

Yeasts can be readily grown in the laboratory and can be studied by many of the same molecular genetic approaches that have proved so successful with *E. coli*. Although yeasts do not replicate as rapidly as bacteria, they still divide as frequently as every 2 hours and they can easily be grown as colonies from a single cell. Consequently, yeasts can be used for a variety of genetic manipulations similar to those that can be performed using bacteria.

These features have made yeast cells the most approachable eukaryotic cells from the standpoint of molecular biology. Yeast mutants have been important in understanding many fundamental processes in eukaryotes, including DNA replication, transcription, RNA processing, protein sorting, and the regulation of cell division, as will be discussed in subsequent chapters. The unity of molecular cell biology is made abundantly clear by the fact that the general principles of cell structure and function revealed by studies of yeasts apply to all eukaryotic cells.

Caenorhabditis elegans

The unicellular yeasts are important models for studies of eukaryotic cells, but understanding the development of multicellular organisms requires the experimental analysis of plants and animals, organisms that are more complex. The nematode *Caenorhabditis elegans* (Figure 1.15) possesses several notable features that make it one of the most widely used models for studies of animal development and cell differentiation.

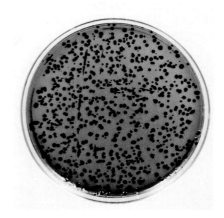

FIGURE 1.13 Bacterial colonies
Photograph of colonies of *E. coli* growing on the surface of an agar-containing medium. (A. M. Siegelman/Visuals Unlimited.)

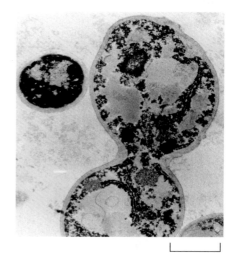

2 μm

FIGURE 1.14 Electron micrograph of *Saccharomyces cerevisiae*
(David Scharf/Peter Arnold, Inc.)

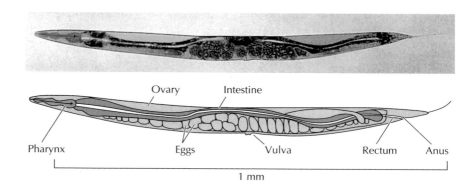

FIGURE 1.15 *Caenorhabditis elegans* (From J. E. Sulston and H. R. Horvitz, 1977. *Dev. Biol.* 56: 110.)

Ovary Intestine

Pharynx Eggs Vulva Rectum Anus

1 mm

Although the genome of *C. elegans* (approximately 100 million base pairs) is larger than those of unicellular eukaryotes, it is simpler and more manageable than the genomes of most animals. Its complete sequence has been determined, revealing that the genome of *C. elegans* contains approximately 19,000 genes—more than three times the number of genes in yeast, and nearly the same number of genes in humans. Biologically, *C. elegans* is a relatively simple multicellular organism: Adult worms consist of only 959 somatic cells, plus 1000 to 2000 germ cells. In addition, *C. elegans* can be easily grown and subjected to genetic manipulations in the laboratory.

The simplicity of *C. elegans* has enabled the course of its development to be studied in detail by microscopic observation. Such analyses have successfully traced the embryonic origin and lineage of all the cells in the adult worm. Genetic studies have also identified many of the mutations responsible for developmental abnormalities, leading to the isolation and characterization of critical genes that control nematode development and differentiation. Importantly, similar genes have also been found to function in complex animals (including humans), making *C. elegans* an important model for studies of animal development.

Drosophila melanogaster

Like *C. elegans*, the fruit fly ***Drosophila melanogaster*** (Figure 1.16) has been a crucial model organism in developmental biology. The genome of *Drosophila* is 180 million base pairs, larger than that of *C. elegans*, but the *Drosophila* genome only contains about 14,000 genes. Furthermore, *Drosophila* can be easily maintained and bred in the laboratory, and the short reproductive cycle of *Drosophila* (about 2 weeks) makes it a very useful organism for genetic experiments. Many fundamental concepts of genetics—such as the relationship between genes and chromosomes—were derived from studies of *Drosophila* early in the twentieth century (see Chapter 4).

Extensive genetic analysis of *Drosophila* has uncovered many genes that control development and differentiation, and current methods of molecular biology have allowed the functions of these genes to be analyzed in detail. Consequently, studies of *Drosophila* have led to striking advances in understanding the molecular mechanisms that govern animal development, particularly with respect to formation of the body plan of complex multicellular organisms. As with *C. elegans*, similar genes and mechanisms exist in vertebrates, validating the use of *Drosophila* as a major experimental model in contemporary developmental biology.

FIGURE 1.16 *Drosophila melanogaster* (Darwin Dale/ Photo Researchers, Inc.)

Arabidopsis thaliana

The study of plant molecular biology and development is an active and expanding field of considerable economic importance as well as intellectual interest. Since the genomes of plants cover a range of complexity comparable to that of animal genomes (see Table 1.2), an optimal model for studies of plant development would be a relatively simple organism with some of the advantageous properties of *C. elegans* and *Drosophila*. The small flowering plant ***Arabidopsis thaliana*** (Figure 1.17) meets these criteria and is therefore widely used as a model to study the molecular biology of plants.

Arabidopsis is notable for its genome of only about 125 million base pairs. Although *Arabidopsis* contains a total of about 26,000 genes, many of these are repeated, so the number of unique genes in *Arabidopsis* is approximately 15,000—a complexity similar to that of *C. elegans* and *Drosophila*. In addition, *Arabidopsis* is relatively easy to grow in the laboratory, and methods for molecular genetic manipulations of this plant have been developed. These studies have led to the identification of genes involved in various aspects of plant development, such as the development of flowers. Analysis of these genes points to many similarities, but also to striking differences, between the mechanisms that control the development of plants and animals.

Vertebrates

The most complex animals are the vertebrates, including humans and other mammals. The human genome is approximately 3 billion base pairs—about 20–30 times larger than the genomes of *C. elegans*, *Drosophila*, or *Arabidopsis* – and contains 20,000 to 25,000 genes. Moreover, the human body is composed of more than 200 different kinds of specialized cell types. This complexity makes the vertebrates difficult to study from the standpoint of cell and molecular biology, but much of the interest in biological sciences nonetheless stems from the desire to understand the human organism. Moreover, an understanding of many questions of immediate practical importance (e.g., in medicine) must be based directly on studies of human (or closely related) cell types.

One important approach to studying human and other mammalian cells is to grow isolated cells in culture, where they can be manipulated under controlled laboratory conditions. The use of cultured cells has allowed studies of many aspects of mammalian cell biology, including experiments that have elucidated the mechanisms of DNA replication, gene expression, protein synthesis and processing, and cell division. Moreover, the ability to culture cells in chemically defined media has allowed studies of the signaling mechanisms that normally control cell growth and differentiation within the intact organism.

The specialized properties of some highly differentiated cell types have made them important models for studies of particular aspects of cell biology. Muscle cells, for example, are highly specialized to undergo contraction, producing force and movement. Because of this specialization, muscle cells are a crucial model for studying cell movement at the molecular level. Another example is provided by nerve cells (neurons), which are specialized to conduct electrochemical signals over long distances. In humans, nerve cell axons may be more than a meter long, and some invertebrates, such as the squid, have giant neurons with axons as large as 1 mm in diameter. Because of their highly specialized structure and function, these giant neurons have provided important models for studies of ion transport across

FIGURE 1.17 *Arabidopsis thaliana*
(Jeremy Burgess/Photo Researchers, Inc.)

FIGURE 1.18 **Eggs of the frog *Xenopus laevis***
(Courtesy of Michael Danilchik and Kimberly Ray.)

the plasma membrane, and of the role of the cytoskeleton in the transport of cytoplasmic organelles.

The frog ***Xenopus laevis*** is an important model for studies of early vertebrate development. *Xenopus* eggs are unusually large cells, with a diameter of approximately 1 mm (Figure 1.18). Because those eggs develop outside of the mother, all stages of development from egg to tadpole can be readily studied in the laboratory. In addition, *Xenopus* eggs can be obtained in large numbers, facilitating biochemical analysis. Because of these technical advantages, *Xenopus* has been widely used in studies of developmental biology and has provided important insights into the molecular mechanisms that control development, differentiation, and embryonic cell division.

The **zebrafish** (Figure 1.19) possesses a number of advantages for genetic studies of vertebrate development. These small fish are easy to maintain in the laboratory and they reproduce rapidly. In addition, the embryos develop outside of the mother and are transparent, so that early stages of development can be easily observed. Powerful methods have been developed to facilitate the isolation of mutations affecting zebrafish development, and several thousand such mutations have now been identified. Because the zebrafish is an easily studied vertebrate, it promises to bridge the gap between humans and the simpler invertebrate systems, such as *C. elegans* and *Drosophila*. Its utility as a model organism will be further enhanced with the completion of its genome sequence.

Among mammals, the mouse is the most suitable for genetic analysis, which is facilitated by the availability of its complete genome sequence. Although the technical difficulties in studying mouse genetics (compared, for example, to the genetics of yeasts or *Drosophila*) are formidable, many mutations affecting mouse development have been identified. Most important, recent advances in molecular biology have enabled the production of genetically engineered mice in which specific mutant genes have been introduced into the mouse germ line, allowing the functions of these genes to be studied in the context of the whole animal. The suitability of the mouse as a model for human development is indicated not only by the similarity of the mouse and human genomes but also by the fact that mutations

(A)

(B)

FIGURE 1.19 **Zebrafish**   (A) A 24-hour-old embryo. (B) An adult fish. (A, courtesy of Charles Kimmel, University of Oregon; B, courtesy of S. Kondo.)

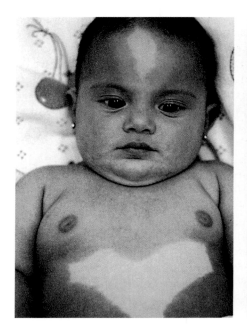

FIGURE 1.20 The mouse as a model for human development A child and a mouse show similar defects in pigmentation (piebaldism) as a result of mutations in a gene required for normal migration of melanocytes (the cells responsible for skin pigmentation) during embryonic development. (Courtesy of R. A. Fleischman, Markey Cancer Center, University of Kentucky.)

in homologous genes result in similar developmental defects in both species; piebaldism is a striking example (Figure 1.20).

Tools of Cell Biology

As in all experimental sciences, research in cell biology depends on the laboratory methods that can be used to study cell structure and function. Many important advances in understanding cells have directly followed the development of new methods that have opened novel avenues of investigation. An appreciation of the experimental tools available to the cell biologist is thus critical to understanding both the current status and future directions of this rapidly moving area of science. Some of the important general methods of cell biology are described in the sections that follow. Other experimental approaches, including the methods of biochemistry and molecular biology, will be discussed in later chapters.

Light Microscopy

Because most cells are too small to be seen by the naked eye, the study of cells has depended heavily on the use of microscopes. Indeed, the very discovery of cells arose from the development of the microscope: Robert Hooke first coined the term "cell" following his observations of a piece of cork with a simple light microscope in 1665 (Figure 1.21). Using a microscope that magnified objects up to about 300 times their actual size, Antony van Leeuwenhoek, in the 1670s, was able to observe a variety of different types of cells, including sperm, red blood cells, and bacteria. The proposal of the cell theory by Matthias Schleiden and Theodor Schwann in 1838 may be seen as the birth of contemporary cell biology. Microscopic studies of plant tissues by Schleiden and of animal tissues by Schwann led to the same conclusion: All organisms are composed of cells. Shortly thereafter, it was recognized that cells are not formed *de novo* but arise only from division of pre-existing cells. Thus, the cell achieved its current recognition as the

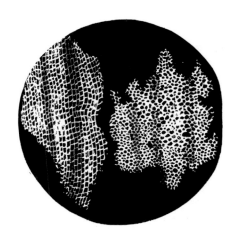

FIGURE 1.21 The cellular structure of cork A reproduction of Robert Hooke's drawing of a thin slice of cork examined with a light microscope. The "cells" that Hooke observed were actually only the cell walls remaining from cells that had long since died.

fundamental unit of all living organisms because of observations made with the light microscope.

The light microscope remains a basic tool of cell biologists, with technical improvements allowing the visualization of ever-increasing details of cell structure. Contemporary light microscopes are able to magnify objects up to about a thousand times. Since most cells are between 1 and 100 μm in diameter, they can be observed by light microscopy, as can some of the larger subcellular organelles, such as nuclei, chloroplasts, and mitochondria. However, the light microscope is not sufficiently powerful to reveal fine details of cell structure, for which **resolution**—the ability of a microscope to distinguish objects separated by small distances—is even more important than magnification. Images can be magnified as much as desired (for example, by projection onto a large screen), but such magnification does not increase the level of detail that can be observed.

The limit of resolution of the light microscope is approximately 0.2 μm; two objects separated by less than this distance appear as a single image, rather than being distinguished from one another. This theoretical limitation of light microscopy is determined by two factors—the wavelength (λ) of visible light and the light-gathering power of the microscope lens (numerical aperture, NA)—according to the following equation:

$$\text{Resolution} = \frac{0.61\lambda}{NA}$$

The wavelength of visible light is 0.4 to 0.7 μm, so the value of λ is fixed at approximately 0.5 μm for the light microscope. The numerical aperture can be envisioned as the size of the cone of light that enters the microscope lens after passing through the specimen (Figure 1.22). It is given by the equation

$$NA = \eta \, \sin \alpha$$

where η is the refractive index of the medium through which light travels between the specimen and the lens. The value of η for air is 1.0, but it can be

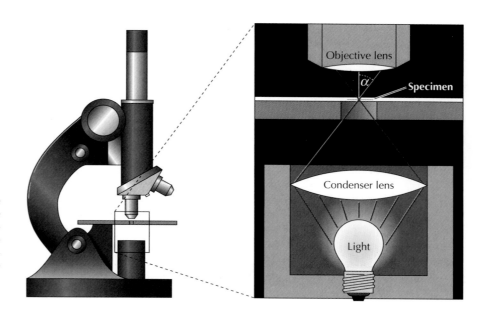

FIGURE 1.22 Numerical aperture
Light is focused on the specimen by the condenser lens and then collected by the objective lens of the microscope. The numerical aperture is determined by the angle of the cone of light entering the objective lens (α) and by the refractive index of the medium (usually air or oil) between the lens and the specimen.

increased to a maximum of approximately 1.4 by using an oil-immersion lens to view the specimen through a drop of oil. The angle α corresponds to half the width of the cone of light collected by the lens. The maximum value of α is 90°, at which sin α = 1, so the highest possible value for the numerical aperture is 1.4.

The theoretical limit of resolution of the light microscope can therefore be calculated as follows:

$$\text{Resolution} = \frac{0.61 \times 0.5}{1.4} = 0.22 \ \mu m$$

Microscopes capable of achieving this level of resolution had already been made by the end of the nineteenth century; further improvements in this aspect of light microscopy cannot be expected.

Several different types of light microscopy are routinely used to study various aspects of cell structure. The simplest is **bright-field microscopy**, in which light passes directly through the cell and the ability to distinguish different parts of the cell depends on contrast resulting from the absorption of visible light by cell components. In many cases, cells are stained with dyes that react with proteins or nucleic acids in order to enhance the contrast between different parts of the cell. Prior to staining, specimens are usually treated with fixatives (such as alcohol, acetic acid, or formaldehyde) to stabilize and preserve their structures. The examination of fixed and stained tissues by bright-field microscopy is the standard approach for the analysis of tissue specimens in histology laboratories (Figure 1.23). Such staining procedures kill the cells, however, and therefore are not suitable for many experiments in which the observation of living cells is desired.

Without staining, the direct passage of light does not provide sufficient contrast to distinguish many parts of the cell, limiting the usefulness of bright-field microscopy. However, optical variations of the light microscope can be used to enhance the contrast between light waves passing through regions of the cell with different densities. The two most common methods for visualizing living cells are **phase-contrast microscopy** and **differential interference-contrast microscopy** (Figure 1.24). Both kinds of microscopy use optical systems that convert variations in density or thickness between different parts of the cell to differences in contrast that can be seen in the final image. In bright-field microscopy, transparent structures (such as the nucleus) have little contrast because they absorb light poorly. However, light is slowed down as it passes through these structures so that its phase is altered compared to light that has passed through the surrounding cytoplasm. Phase-contrast and differential interference-contrast microscopy convert these differences in phase to differences in contrast, thereby yielding improved images of live, unstained cells.

The power of the light microscope has been considerably expanded by the use of video cameras and computers for image analysis and processing. Such electronic image-processing systems can substantially enhance the

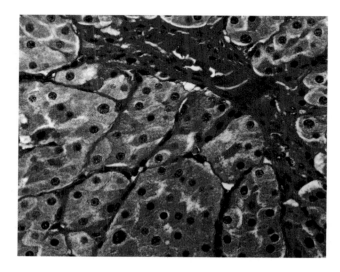

FIGURE 1.23 Bright-field micrograph of stained tissue Section of a benign kidney tumor. (G. W. Willis/Biological Photo Service.)

(A)

(B)

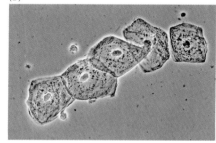

(C)

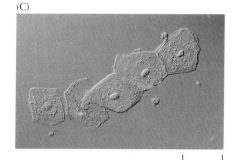

50 μm

FIGURE 1.24 Microscopic observation of living cells Photomicrographs of human cheek cells obtained with (A) bright-field, (B) phase-contrast, and (C) differential interference-contrast microscopy. (Courtesy of Mort Abramowitz, Olympus America, Inc.)

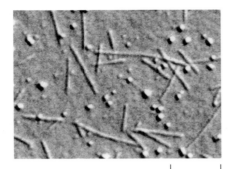

2.5 µm

FIGURE 1.25 Video-enhanced differential interference-contrast microscopy Electronic image processing allows the visualization of single microtubules. (Courtesy of E. D. Salmon, University of North Carolina, Chapel Hill.)

contrast of images obtained with the light microscope, allowing the visualization of small objects that otherwise could not be detected. For example, **video-enhanced differential interference-contrast microscopy** has allowed visualization of the movement of organelles along microtubules, which are cytoskeletal protein filaments with a diameter of only 0.025 µm (Figure 1.25). However, this enhancement does not overcome the theoretical limit of resolution of the light microscope, approximately 0.2 µm. Thus, although video enhancement allows the visualization of microtubules, the microtubules appear as blurred images at least 0.2 µm in diameter and an individual microtubule cannot be distinguished from a bundle of adjacent structures.

Light microscopy has been brought to the level of molecular analysis by methods for labeling specific molecules so that they can be visualized within cells. Specific genes or RNA transcripts can be detected by hybridization with nucleic acid probes of complementary sequence, and proteins can be detected using appropriate antibodies (see Chapter 4). Both nucleic acid probes and antibodies can be labeled with a variety of tags that allow their visualization in the light microscope, making it possible to determine the location of specific molecules within individual cells.

Fluorescence microscopy is a widely used and very sensitive method for studying the intracellular distribution of molecules (Figure 1.26). A fluorescent dye is used to label the molecule of interest within either fixed or living cells. The fluorescent dye is a molecule that absorbs light at one wavelength and emits light at a second wavelength. This fluorescence is detected by illuminating the specimen with a wavelength of light that excites the fluorescent dye and then using appropriate filters to detect the specific wavelength of light that the dye emits. Fluorescence microscopy can be used to

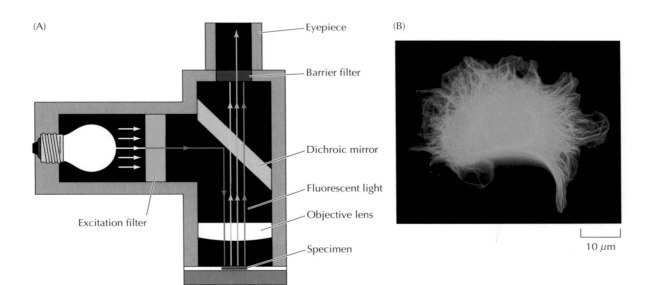

FIGURE 1.26 Fluorescence microscopy (A) Light passes through an excitation filter to select light of the wavelength (e.g., blue) that excites the fluorescent dye. A dichroic mirror then deflects the excitation light down to the specimen. The fluorescent light emitted by the specimen (e.g., green) then passes through the dichroic mirror and a second filter (the barrier filter) to select light of the wavelength emitted by the dye. (B) Fluorescence micrograph of a newt lung cell in which the DNA is stained blue and microtubules in the cytoplasm are stained green. (Conly S. Rieder/Biological Photo Service.)

study a variety of molecules within cells. One frequent application is to label antibodies directed against a specific protein with fluorescent dyes, so that the intracellular distribution of the protein can be determined.

An important recent advance in fluorescence microscopy has been the use of the **green fluorescent protein (GFP)** of jellyfish to visualize proteins within living cells. GFP can be fused to any protein of interest using standard methods of recombinant DNA, and the GFP-tagged protein can then be expressed in cells and detected by fluorescence microscopy, without the need to fix and stain the cell as would be required for the detection of proteins with antibodies. Because of its versatility, the use of GFP has become extremely widespread in cell biology, and been used to study the localization of a wide variety of proteins within living cells (Figure 1.27). Several related fluorescent proteins with blue, yellow or red emissions are also available, further expanding the utility of this technique.

A variety of methods have been developed to follow the movement and interactions of GFP-labeled proteins within living cells. One widely used method for studying the movements of GFP-labeled proteins is **fluorescence recovery after photobleaching (FRAP)** (Figure 1.28). In this technique, a region of interest in a cell expressing a GFP-labeled protein is bleached by exposure to high-intensity light. Fluorescence recovers over time due to the movement of unbleached GFP-labeled molecules into the bleached region, allowing the rate at which the protein moves within the cell to be determined.

The interactions of two proteins with one another within a cell can be analyzed by a technique called **fluorescence resonance energy transfer (FRET)** (Figure 1.29). In FRET experiments, the two proteins of interest are coupled to different fluorescent dyes, such as two variants of GFP. The GFP variants are chosen to absorb and emit distinct wavelengths of light, such that the light emitted by one GFP variant excites the second. Interaction between the two proteins can then be detected by illuminating the cell with

■ Semiconducting nanocrystals (termed quantum dots) are increasingly being use in place of fluorescent dyes for many applications in fluorescence microscopy. Quantum dots fluoresce brighter and are more stable than traditional fluorescent dyes.

■ GFP is derived from the Pacific jellyfish *Aequoria victoria*. Proteins that fluoresce in different colors have been isolated from other marine organisms.

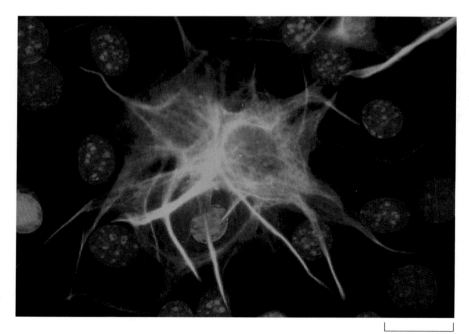

5 μm

FIGURE 1.27 Fluorescence microscopy of a protein labeled with GFP   A microtubule-associated protein fused to GFP was introduced into mouse neurons in culture and visualized by fluorescence microscopy. Nuclei are stained blue. (From A. Cariboni, 2004. *Nature Cell Biol.* 6:929.)

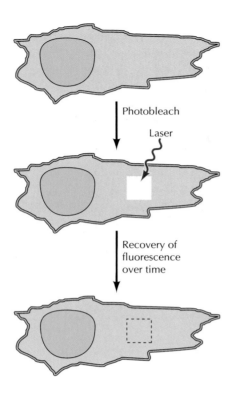

Photobleach

Laser

Recovery of
fluorescence
over time

FIGURE 1.28 Fluorescence recovery after photobleaching (FRAP) Region of a cell expressing a GFP-labeled protein is bleached with a laser. Fluorescence recovers over time as unbleached GFP-labeled molecules diffuse into the bleached region. The rate of recovery of fluorescence therefore provides a measurement of the rate of protein movement within the cell.

a wavelength of light that excites the first GFP variant and analyzing the wavelength of emitted light. If the proteins coupled to these GFP variants interact within the cell, the fluorescent molecules will be brought close together and the light emitted by the first GFP variant will excite the second, resulting in emission of light at the wavelength characteristic of the second variant of GFP.

The images obtained by conventional fluorescence microscopy are blurred as a result of out-of-focus fluorescence. These images can be improved by a computational approach called image deconvolution, in which a computer analyzes images obtained from different depths of focus and generates a sharper image as would have been expected from a single focal point. Alternatively, **confocal microscopy** allows images of increased contrast and detail to be obtained by analyzing fluorescence from only a single point in the specimen. A small point of light, usually supplied by a laser, is focused on the specimen at a particular depth. The emitted fluorescent light is then collected using a detector, such as a video camera. Before the emitted light reaches the detector, however, it must pass through a pinhole aperture (called a confocal aperture) placed at precisely the point

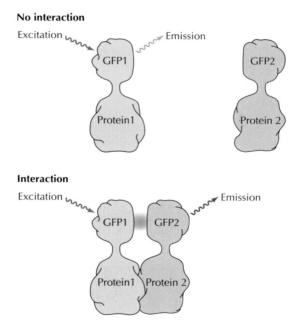

FIGURE 1.29 Fluorescence resonance energy transfer (FRET) Two proteins are fused to different variants of GFP (GFP1 and GFP2) with distinct wavelengths for excitation and emission, chosen such that the light emitted by GFP1 excites GFP2. Cells are then illuminated with a wavelength of light that excites GFP1. If the proteins do not interact, light emitted by GFP1 will be detected. However, if the proteins do interact, GFP1 will excite GFP2, and light emitted by GFP2 will be detected.

FIGURE 1.30 Confocal microscopy A pinpoint of light is focused on the specimen at a particular depth, and emitted fluorescent light is collected by a detector. Before reaching the detector, the fluorescent light emitted by the specimen must pass through a confocal aperture placed at the point where light emitted from the chosen depth of the specimen comes into focus. As a result, only in-focus light emitted from the chosen depth of the specimen is detected.

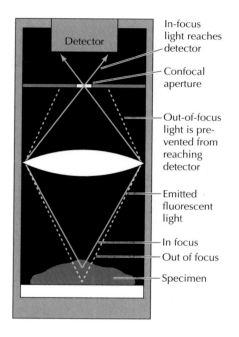

where light emitted from the chosen depth of the specimen comes to a focus (Figure 1.30). Consequently, only light emitted from the plane of focus is able to reach the detector. Scanning across the specimen generates a two-dimensional image of the plane of focus, a much sharper image than that obtained with standard fluorescence microscopy (Figure 1.31). Moreover, a series of images obtained at different depths can be used to reconstruct a three-dimensional image of the sample.

Multi-photon excitation microscopy is an alternative to confocal microscopy that can be applied to living cells. The specimen is illuminated with a wavelength of light such that excitation of the fluorescent dye requires the simultaneous absorption of two or more photons (Figure 1.32). The probability of two photons simultaneously exciting the fluorescent dye is only significant at the point in the specimen upon which the input laser beam is focused, so fluorescence is only emitted from the plane of focus of the input light. This highly localized excitation automatically provides three-dimensional resolution, without the need for passing the emitted light through a pinhole aperture, as in confocal microscopy. Moreover, the localization of excitation minimizes damage to the specimen, allowing three-dimensional imaging of living cells.

Electron Microscopy

Because of the limited resolution of the light microscope, analysis of the details of cell structure has required the use of more powerful microscopic techniques—namely electron microscopy, which was developed in the 1930s and first applied to biological specimens by Albert Claude, Keith Porter, and George Palade in the 1940s and 1950s. The electron microscope can achieve a much greater resolution than that obtained with the light microscope because the wavelength of electrons is shorter than that of light. The wavelength of electrons in an electron microscope can be as short as 0.004 nm—about 100,000 times shorter than the wavelength of visible light. Theoretically, this wavelength could yield resolution of 0.002 nm, but such a resolution cannot be obtained in practice, because resolution is determined not only by wavelength, but also by the numerical aperture of the microscope lens. Numerical aperture is a limiting factor for electron microscopy because inherent properties of electromagnetic lenses limit their aperture angles to about 0.5 degrees, corresponding to numerical apertures of only about 0.01. Thus, under optimal conditions, the resolving power of the electron microscope is approximately

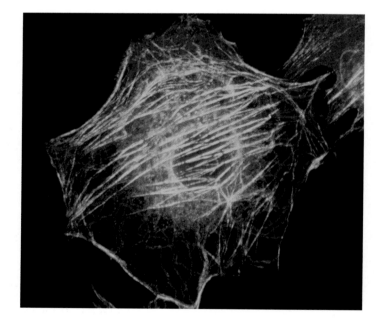

FIGURE 1.31 Confocal micrograph of human cells Microtubules and actin filaments are stained with red and green fluorescent dyes, respectively. (K. G. Murti/Visuals Unlimited.)

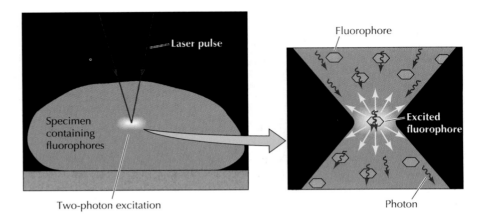

FIGURE 1.32 Two-photon excitation microscopy   Simultaneous absorption of two photons is required to excite the fluorescent dye. This only occurs at the point in the specimen upon which the input light is focused, so fluorescent light is only emitted from the chosen depth of the specimen.

0.2 nm. Moreover, the resolution that can be obtained with biological specimens is further limited by their lack of inherent contrast. Consequently, for biological samples the practical limit of resolution of the electron microscope is 1 to 2 nm. Although this resolution is much less than that predicted simply from the wavelength of electrons, it represents more than a hundredfold improvement over the resolving power of the light microscope.

Two types of electron microscopy—transmission and scanning—are widely used to study cells. In principle, **transmission electron microscopy** is similar to the observation of stained cells with the bright-field light microscope. Specimens are fixed and stained with salts of heavy metals, which provide contrast by scattering electrons. A beam of electrons is then passed through the specimen and focused to form an image on a fluorescent screen. Electrons that encounter a heavy metal ion as they pass through the sample are deflected and do not contribute to the final image, so stained areas of the specimen appear dark.

Specimens to be examined by transmission electron microscopy can be prepared by either positive or negative staining. In positive staining, tissue specimens are cut into thin sections and stained with heavy metal salts (such as osmium tetroxide, uranyl acetate, and lead citrate) that react with lipids, proteins, and nucleic acids. These heavy metal ions bind to a variety of cell structures, which consequently appear dark in the final image (Figure 1.33). Alternative positive-staining procedures can also be used to identify specific macromolecules within cells. For example, antibodies labeled with electron-dense heavy metals (such as gold particles) are frequently used to determine the subcellular location of specific proteins in the electron microscope. This method is similar to the use of antibodies labeled with fluorescent dyes in fluorescence microscopy. Three-dimensional views of structures with resolutions of 2–10 nm can also be obtained using the technique of **electron tomography**, which generates three-dimensional images by computer analysis of multiple two-dimensional images obtained over a range of viewing directions.

Negative staining is useful for the visualization of intact biological structures, such as bacteria, isolated subcellular organelles, and macromolecules (Figure 1.34). In this method, the biological specimen is deposited on a sup-

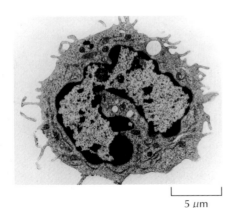

|—————— 5 μm ——————|

FIGURE 1.33 Positive staining
Transmission electron micrograph of a positively stained white blood cell. (Don W. Fawcett/Visuals Unlimited.)

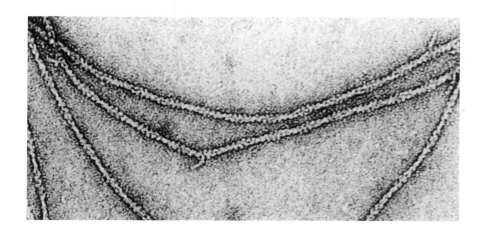

FIGURE 1.34 Negative staining
Transmission electron micrograph of negatively stained actin filaments. (Courtesy of Roger Craig, University of Massachusetts Medical Center.)

porting film, and a heavy metal stain is allowed to dry around its surface. The unstained specimen is then surrounded by a film of electron-dense stain, producing an image in which the specimen appears light against a stained dark background.

Metal shadowing is another technique used to visualize the surface of isolated subcellular structures or macromolecules in the transmission electron microscope (Figure 1.35). The specimen is coated with a thin layer of evaporated metal, such as platinum. The metal is sprayed onto the specimen from an angle so that surfaces of the specimen that face the source of evaporated metal molecules are coated more heavily than others. This differential coating creates a shadow effect, giving the specimen a three-dimensional appearance in electron micrographs.

The preparation of samples by **freeze fracture**, in combination with metal shadowing, has been particularly important in studies of membrane structure. Specimens are frozen in liquid nitrogen (at −196°C) and then fractured with a knife blade. This process frequently splits the lipid bilayer, revealing the interior faces of a cell membrane (Figure 1.36). The specimen is then shadowed with platinum, and the biological material is dissolved with acid, producing a metal replica of the surface of the sample. Examination of such replicas in the electron microscope reveals many surface bumps, corresponding to proteins that span the lipid bilayer. A variation of freeze frac-

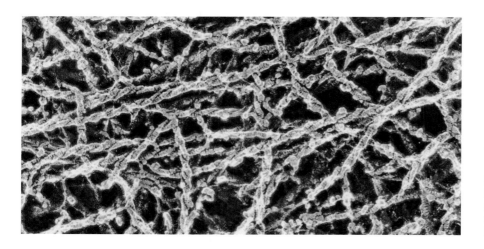

FIGURE 1.35 Metal shadowing
Electron micrograph of actin/myosin filaments of the cytoskeleton prepared by metal shadowing. (Don W. Fawcett, J. Heuser/Photo Researchers, Inc.)

(A)

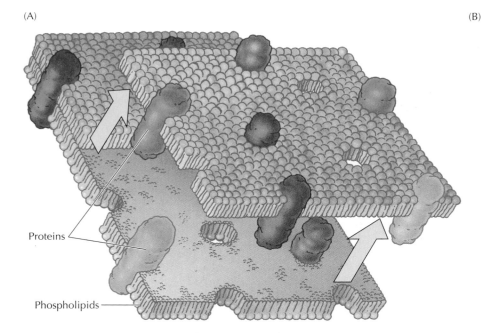

Proteins

Phospholipids

(B)

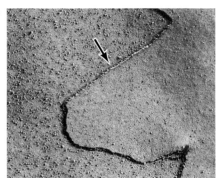

FIGURE 1.36 Freeze fracture
(A) Freeze fracture splits the lipid bilayer, leaving proteins embedded in the membrane associated with one of the two membrane halves. (B) Micrograph of freeze-fractured plasma membranes of two adjacent cells. Proteins that span the bilayer appear as intramembranous particles (arrow). (Don W. Fawcett/ Photo Researchers, Inc.)

1.1 WEBSITE ANIMATION

Cell Fractionation
After cells are broken apart by sonication, the subcellular constituents are fractionated using different types of centrifugation.

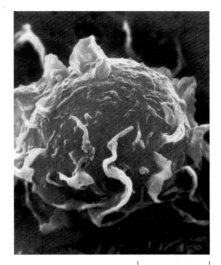

5 μm

FIGURE 1.37 Scanning electron microscopy Scanning electron micrograph of a macrophage. (David Phillips/Visuals Unlimited.)

ture called freeze etching allows visualization of the external surfaces of cell membranes in addition to their interior faces.

The second type of electron microscopy, **scanning electron microscopy**, is used to provide a three-dimensional image of cells (Figure 1.37). In scanning electron microscopy the electron beam does not pass through the specimen. Instead, the surface of the cell is coated with a heavy metal, and a beam of electrons is used to scan across the specimen. Electrons that are scattered or emitted from the sample surface are collected to generate a three-dimensional image as the electron beam moves across the cell. Because the resolution of scanning electron microscopy is only about 10 nm, its use is generally restricted to studying whole cells rather than subcellular organelles or macromolecules.

Subcellular Fractionation

Although the electron microscope has allowed detailed visualization of cell structure, microscopy alone is not sufficient to define the functions of the various components of eukaryotic cells. To address many of the questions concerning the function of subcellular organelles, it has proven necessary to isolate the organelles of eukaryotic cells in a form that can be used for biochemical studies. This is usually accomplished by **differential centrifugation**—a method developed largely by Albert Claude, Christian de Duve, and their colleagues in the 1940s and 1950s to separate the components of cells on the basis of their size and density.

The first step in subcellular fractionation is the disruption of the plasma membrane under conditions that do not destroy the internal components of the cell. Several methods are used, including sonication (exposure to high-frequency sound), grinding in a mechanical homogenizer, or treatment with a high-speed blender. All these procedures break the plasma membrane and the endoplasmic reticulum into small fragments while leaving other components of the cell (such as nuclei, lysosomes, peroxisomes, mitochondria, and chloroplasts) intact.

FIGURE 1.38 Subcellular fractionation
Cells are lysed and subcellular components are separated by a series of centrifugations at increasing speeds. Following each centrifugation, the organelles that have sedimented to the bottom of the tube are recovered in the pellet. The supernatant is then recentrifuged at a higher speed to sediment the next-largest organelles.

The suspension of broken cells (called a lysate or homogenate) is then fractionated into its components by a series of centrifugations in an **ultracentrifuge**, which rotates samples at very high speeds (over 100,000 rpm) to produce forces up to 500,000 times greater than gravity. This force causes cell components to move toward the bottom of the centrifuge tube and form a pellet (a process called sedimentation) at a rate that depends on their size and density, with the largest and heaviest structures sedimenting most rapidly (Figure 1.38). Usually the cell homogenate is first centrifuged at a low speed, which sediments only unbroken cells and the largest subcellular structures—the nuclei. Thus, an enriched fraction of nuclei can be recovered from the pellet of such a low-speed centrifugation while the other cell components remain suspended in the supernatant (the remaining solution). The supernatant is then centrifuged at a higher speed to sediment mitochondria, chloroplasts, lysosomes, and peroxisomes. Recentrifugation of the supernatant at an even higher speed sediments fragments of the plasma membrane and the endoplasmic reticulum. A fourth centrifugation at a still higher speed sediments ribosomes, leaving only the soluble portion of the cytoplasm (the cytosol) in the supernatant.

The fractions obtained from differential centrifugation correspond to enriched, but still not pure, organelle preparations. A greater degree of purification can be achieved by **density-gradient centrifugation**, in which organelles are separated by sedimentation through a gradient of a dense substance, such as sucrose. In **velocity centrifuga-**

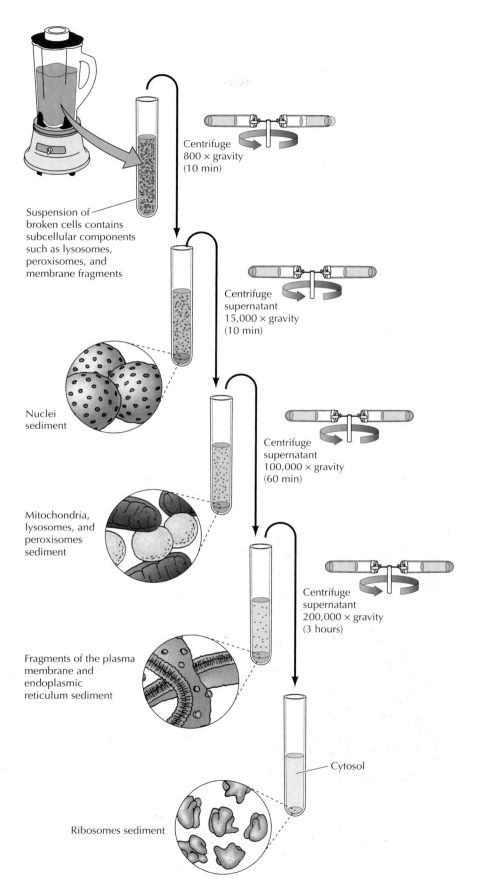

Suspension of broken cells contains subcellular components such as lysosomes, peroxisomes, and membrane fragments

Centrifuge
800 × gravity
(10 min)

Centrifuge
supernatant
15,000 × gravity
(10 min)

Nuclei
sediment

Centrifuge
supernatant
100,000 × gravity
(60 min)

Mitochondria, lysosomes, and peroxisomes sediment

Centrifuge
supernatant
200,000 × gravity
(3 hours)

Fragments of the plasma membrane and endoplasmic reticulum sediment

Cytosol

Ribosomes sediment

FIGURE 1.39 Velocity centrifugation in a density gradient The sample is layered on top of a gradient of sucrose, and particles of different sizes sediment through the gradient as discrete bands. The separated particles can then be collected in individual fractions of the gradient, which can be obtained simply by puncturing the bottom of the centrifuge tube and collecting drops.

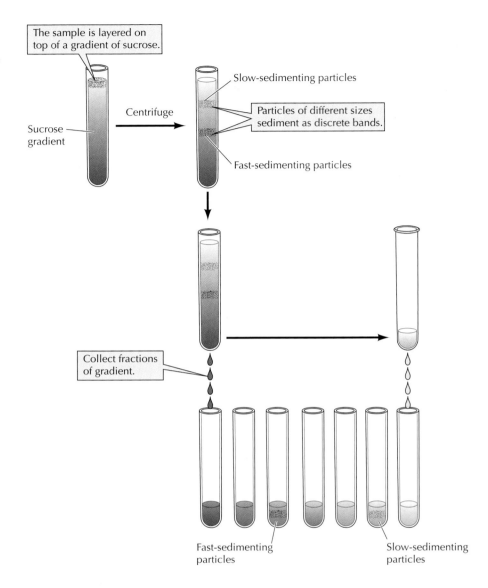

The sample is layered on top of a gradient of sucrose.

Sucrose gradient

Centrifuge

Slow-sedimenting particles

Particles of different sizes sediment as discrete bands.

Fast-sedimenting particles

Collect fractions of gradient.

Fast-sedimenting particles

Slow-sedimenting particles

tion, the starting material is layered on top of the sucrose gradient (Figure 1.39). Particles of different sizes sediment through the gradient at different rates, moving as discrete bands. Following centrifugation, the collection of individual fractions of the gradient provides sufficient resolution to separate organelles of similar size, such as mitochondria, lysosomes, and peroxisomes.

Equilibrium centrifugation in density gradients can be used to separate subcellular components on the basis of their buoyant density, independent of their size and shape. In this procedure, the sample is centrifuged in a gradient containing a high concentration of sucrose or cesium chloride. Rather than being separated on the basis of their sedimentation velocity, the sample particles are centrifuged until they reach an equilibrium position at which their buoyant density is equal to that of the surrounding sucrose or cesium chloride solution. Such equilibrium centrifugations are useful in separating different types of membranes from one another and are sufficiently sensitive to separate macromolecules that are labeled with different isotopes. A classic example, discussed in Chapter 4, is the analysis of DNA replication by separating DNA molecules containing heavy and light isotopes of nitrogen (^{15}N and ^{14}N) by equilibrium centrifugation in cesium chloride gradients.

Growth of Animal Cells in Culture

The ability to study cells depends largely on how readily they can be grown and manipulated in the laboratory. Although the process is technically far more difficult than the culture of bacteria or yeasts, a wide variety of animal and plant cells can be grown and manipulated in culture. Such *in vitro* cell culture systems have enabled scientists to study cell growth and differentiation, as well as to perform genetic manipulations required to understand gene structure and function.

Animal cell cultures are initiated by the dispersion of a piece of tissue into a suspension of its component cells, which is then added to a culture dish containing nutrient media. Most animal cell types, such as fibroblasts and epithelial cells, attach and grow on the plastic surface of dishes used for cell culture (Figure 1.40). Because they contain rapidly growing cells, embryos or tumors are frequently used as starting material. Embryo fibroblasts grow particularly well in culture and consequently are one of the most widely studied types of animal cells. Under appropriate conditions, however, many specialized cell types can also be grown in culture, allowing their differentiated properties to be studied in a controlled experimental environment. **Embryonic stem cells** are a particularly notable example. These cells are established in culture from early embryos and maintain their ability to differentiate into all of the cell types present in adult organisms. Consequently, embryonic stem cells have played an important role in studying the functions of a variety of genes in mouse development, as well as offering the possibility of contributing to the treatment of human diseases by providing a source of tissue for transplantation therapies.

The culture media required for the propagation of animal cells are much more complex than the minimal media sufficient to support the growth of bacteria and yeasts. Early studies of cell culture utilized media consisting of undefined components, such as plasma, serum, and embryo extracts. A major advance was thus made in 1955, when Harry Eagle described the first defined media that supported the growth of animal cells. In addition to salts and glucose, the media used for animal cell cultures contain various amino acids and vitamins, which the cells cannot make for themselves. The growth media for most animal cells in culture also include serum, which serves as a source of polypeptide growth factors that are required to stimulate cell division. Several such growth factors have been identified. They serve as critical regulators of cell growth and differentiation in multicellular organisms, providing signals by which different cells communicate with each other. For example, an important function of skin fibroblasts in the intact animal is to proliferate when needed to repair damage resulting from a cut or wound. Their division is triggered by a growth factor released from platelets during blood clotting, thereby stimulating proliferation of fibroblasts in the neighborhood of the damaged tissue. The identification of individual growth factors has made possible the culture of a variety of cells in serum-free media (media in which serum has been replaced by the specific growth factors required for proliferation of the cells in question).

The initial cell cultures established from a tissue are called **primary cultures** (Figure 1.41). The cells in a primary culture usually grow until they cover the culture dish surface. They can then be removed from the dish and replated at a lower density to form secondary cultures. This process can be

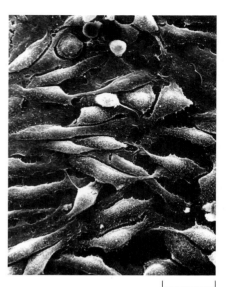

FIGURE 1.40 Animal cells in culture
Scanning electron micrograph of human fibroblasts attached to the surface of a culture dish. (David M. Phillips/Visuals Unlimited.)

10 μm

FIGURE 1.41 Culture of animal cells

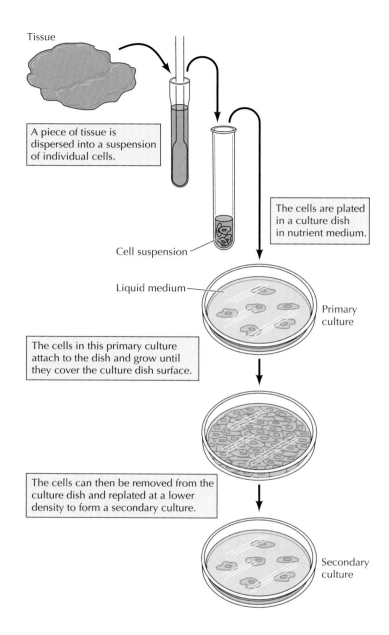

Tissue

A piece of tissue is dispersed into a suspension of individual cells.

Cell suspension

The cells are plated in a culture dish in nutrient medium.

Liquid medium

Primary culture

The cells in this primary culture attach to the dish and grow until they cover the culture dish surface.

The cells can then be removed from the culture dish and replated at a lower density to form a secondary culture.

Secondary culture

repeated many times, but most normal cells cannot be grown in culture indefinitely. For example, normal human fibroblasts can usually be cultured for 50 to 100 population doublings, after which they stop growing and die. In contrast, embryonic stem cells and cells derived from tumors frequently proliferate indefinitely in culture and are referred to as permanent or immortal **cell lines**. In addition, a number of immortalized rodent cell lines have been isolated from cultures of normal fibroblasts. Instead of dying as most of their counterparts do, a few cells in these cultures continue proliferating indefinitely, forming cell lines like those derived from tumors. Such permanent cell lines have been particularly useful for many types of experiments because they provide a continuous and uniform source of cells that can be manipulated, cloned, and indefinitely propagated in the laboratory.

Even under optimal conditions, the division time of most actively growing animal cells is on the order of 20 hours—ten times longer than the divi-

KEY EXPERIMENT

Animal Cell Culture

Nutrition Needs of Mammalian Cells in Tissue Culture
Harry Eagle
National Institutes of Health, Bethesda, MD
Science, Volume 122, 1955, pages 501–504

Harry Eagle

The Context

The earliest cell cultures involved the growth of cells from fragments of tissue that were embedded in clots of plasma—a culture system that was far from suitable for experimental analysis. In the late 1940s, a major advance was the establishment of cell lines that grew from isolated cells attached to the surface of culture dishes. But these cells were still grown in undefined media consisting of varying combinations of serum and embryo extracts. For example, a widely used human cancer cell line (called HeLa cells) was initially established in 1952 by growth in a medium consisting of chicken plasma, bovine embryo extract, and human placental cord serum. The use of such complex and undefined culture media made analysis of the specific growth requirements of animal cells impossible. Harry Eagle was the first to solve this problem, by carrying out a systematic analysis of the nutrients needed to support the growth of animal cells in culture.

The Experiments

Eagle studied the growth of two established cell lines: HeLa cells and a mouse fibroblast line called L cells. He was able to grow these cells in a medium consisting of a mixture of salts, carbohydrates, amino acids, and vitamins, supplemented with serum protein. By systematically varying the components of this medium, Eagle was able to determine the specific nutrients required for cell growth. In addition to salts and glucose, these nutrients included 13 amino acids and several vitamins. A small amount of serum protein was also required.

The basal medium developed by Eagle is described in the accompanying table, reprinted from his 1955 paper.

The Impact

The medium developed by Eagle is still the basic medium used for animal cell culture. Its use has enabled scientists to grow a wide variety of cells under defined experimental conditions, which has been critical to studies of animal cell growth and differentiation, including identification of the growth factors present in serum—now known to include polypeptides that control the behavior of individual cells within intact animals.

Table 4. Basal media for cultivation of the HeLa cell and mouse fibroblast (*10*)

L-Amino acids* (mM)			Vitamins‡ (mM)		Miscellaneous	
Arginine	0.1		Biotin	10^{-3}	Glucose	5mM§
Cystine	0.05	(0.02)†	Choline	10^{-3}	Penicillin	0.005%#
Glutamine	2.0	(1.0)‖	Folic acid	10^{-3}	Streptomycin	0.005%#
Histidine	0.05	(0.02)†	Nicotinamide	10^{-3}	Phenol red	0.0005%#
Isoleucine	0.2		Pantothenic acid	10^{-3}		
Leucine	0.2	(0.1)†	Pyridoxal	10^{-3}		
Lysine	0.2	(0.1)†	Thiamine	10^{-3}	For studies of cell nutrition	
Methionine	0.05		Riboflavin	10^{-4}	Dialyzed horse serum, 1%†	
Phenylalanine	0.1	(0.05)†			Dialyzed human serum, 5%	
Threonine	0.2	(0.1)†	Salts§ (mM)			
Tryptophan	0.02	(0.01)†			For stock cultures	
Tyrosine	0.1				Whole horse serum, 5%†	
Valine	0.2	(0.1)†	NaCl	100	Whole human serum, 10%	
			KCl	5		
			NaH$_2$PO$_4$ · H$_2$O	1		
			NaHCO$_3$	20		
			CaCl$_2$	1		
			MgCl$_2$	0.5		

* Conveniently stored in the refrigerator as a single stock solution containing 20 times the indicated concentration of each amino acid.
† For mouse fibroblast.
‡ Conveniently stored as a single stock solution containing 100 or 1000 times the indicated concentration of each vitamin; kept frozen.
§ Conveniently stored in the refrigerator in two stock solutions, one containing NaCl, KCl, NaH$_2$PO$_4$, NaHCO$_3$, and glucose at 10 times the indicated concentration of each, and the second containing CaCl$_2$ and MgCl$_2$ at 20 times the indicated concentration.
‖ Conveniently stored as a 100mM stock solution; frozen when not in use.
Conveniently stored as a single stock solution containing 100 times the indicated concentrations of penicillin, streptomycin, and phenol red.

FIGURE 1.42 Plant cells in culture
An undifferentiated mass of plant cells (a callus) growing on a solid medium. (John N. A. Lott/Biological Photo Service.)

sion time of yeasts. Consequently, experiments with cultured animal cells are more difficult and take much longer than those with bacteria or yeasts. For example, the growth of a visible colony of animal cells from a single cell takes a week or more, whereas colonies of *E. coli* or yeast develop from single cells overnight. Nonetheless, genetic manipulations of animal cells in culture have been indispensable to our understanding of cell structure and function.

Culture of Plant Cells

Plant cells can also be cultured in nutrient media containing appropriate growth regulatory molecules. In contrast to the polypeptide growth factors that regulate the proliferation of most animal cells, the growth regulators of plant cells are small molecules that can pass through the plant cell wall. When provided with appropriate mixtures of these growth regulatory molecules, many types of plant cells proliferate in culture, producing a mass of undifferentiated cells called a **callus** (Figure 1.42).

It is noteworthy that many plant cells are capable of forming any of the different cell types and tissues ultimately needed to regenerate an entire plant. Consequently, by appropriate manipulation of nutrients and growth regulatory molecules, undifferentiated plant cells in culture can be induced to form a variety of plant tissues, including roots, stems, and leaves. In many cases, even an entire plant can be regenerated from a single cultured cell. In addition to its theoretical interest, the ability to produce a new plant from a single cell that has been manipulated in culture makes it easy to introduce genetic alterations into plants, opening important possibilities for agricultural genetic engineering.

Viruses

Viruses are intracellular parasites that cannot replicate on their own. They reproduce by infecting host cells and usurping the cellular machinery to produce more virus particles. In their simplest forms, viruses consist only of genomic nucleic acid (either DNA or RNA) surrounded by a protein coat (Figure 1.43). Viruses are important in molecular and cellular biology because

(A)

(B)

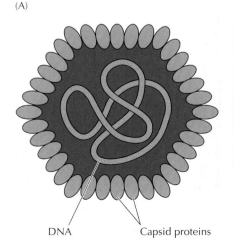

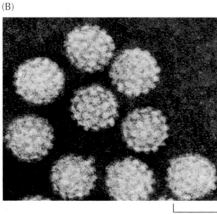

50 nm

FIGURE 1.43 Structure of an animal virus   (A) Papillomavirus particles contain a small circular DNA molecule enclosed in a protein coat (the capsid). (B) Electron micrograph of human papillomavirus particles. Artificial color has been added. (B, Linda Stannard/Science Photo Library/Photo Researchers, Inc.)

DNA   Capsid proteins

they provide simple systems that can be used to investigate the functions of cells. Because virus replication depends on the metabolism of the infected cells, studies of viruses have revealed many fundamental aspects of cell biology. Studies of bacterial viruses contributed substantially to our under-

---

MOLECULAR MEDICINE

Viruses and Cancer

The Disease

Cancer is a family of diseases characterized by uncontrolled cell proliferation. The growth of normal animal cells is carefully regulated to meet the needs of the complete organism. In contrast, cancer cells grow in an unregulated manner, ultimately invading and interfering with the function of normal tissues and organs. Cancer is the second most common cause of death (next to heart disease) in the United States. Approximately one out of every three Americans will develop cancer at some point in life and, in spite of major advances in treatment, nearly one out of every four Americans ultimately die of this disease. Understanding the causes of cancer and developing more effective methods of cancer treatment therefore represent major goals of medical research.

Molecular and Cellular Basis

Cancer is now known to result from mutations in the genes that normally control cell proliferation. The major insights leading to identification of these genes came from studies of viruses that cause cancer in animals, the prototype of which was isolated by Peyton Rous in 1911. Rous found that sarcomas (cancers of connective tissues) in chickens could be transmitted by a virus, now known as Rous sarcoma virus, or RSV. Because RSV is a retrovirus with a genome of only 10,000 base pairs, it can be subjected to molecular analysis much more readily than the complex genomes of chickens or other animal cells can. Such studies eventually led to identification of a specific cancer-causing

gene (oncogene) carried by the virus, and to the discovery of related genes in normal cells of all vertebrate species, including humans. Some cancers in humans are now known to be caused by viruses; others result from mutations in normal cell genes similar to the oncogene first identified in RSV.

Prevention and Treatment

The human cancers that are caused by viruses include cervical and other anogenital cancers (papilloma viruses), liver cancer (hepatitis B and C viruses), and some types of lymphomas (Epstein-Barr virus and human T-cell lymphotropic virus). Together, these virus-induced cancers account for about 20% of worldwide cancer incidence. In principle, these cancers could be prevented by vaccination against the responsible viruses, and considerable progress in this area has been made by the development of an effective vaccine against hepatitis B virus.

Other human cancers are caused by mutations in normal cell genes, most of which occur during the lifetime of the individual rather than being inherited. Studies of cancer-causing viruses have led to the identification of many of the genes responsible for non-virus-induced cancers, and to an understanding of the molecular mechanisms responsible for cancer development. Major efforts are now under way to use these insights into the molecular and cellular biology of cancer to develop new approaches to cancer treatment. Indeed, the first designer drug effective in treating a human cancer (the drug STI-571 or Gleevec, discussed in chapter 18) was developed against a gene very similar to the oncogene of RSV.

Reference

Rous, P. 1911. A sarcoma of the fowl transmissible by an agent separable from the tumor cells. *J. Exp. Med.* 13: 397–411.

The transplantable tumor from which Rous sarcoma virus was isolated.

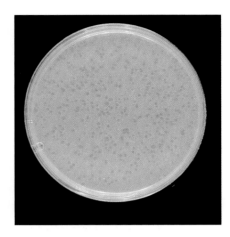

FIGURE 1.44 Bacteriophage plaques
T4 plaques are visible on a lawn of *E. coli*. Each plaque arises by the replication of a single virus particle. (E. C. S. Chen/Visuals Unlimited.)

■ Animal viruses are often used in gene therapy as carriers of genes to be introduced into cells.

standing of the basic mechanisms of molecular genetics, and experiments with a plant virus (tobacco mosaic virus) first demonstrated the genetic potential of RNA. Animal viruses have provided particularly sensitive probes for investigations of various activities of eukaryotic cells.

The rapid growth and small genome size of bacteria make them excellent subjects for experiments in molecular biology, and bacterial viruses (**bacteriophages**) have simplified the study of bacterial genetics even further. One of the most important bacteriophages is T4, which infects and replicates in *E. coli*. Infection with a single particle of T4 leads to the formation of approximately 200 progeny virus particles in 20 to 30 minutes. The initially infected cell then bursts (lyses), releasing progeny virus particles into the medium, where they can infect new cells. In a culture of bacteria growing on agar medium, the replication of T4 leads to the formation of a clear area of lysed cells (a plaque) in the lawn of bacteria (Figure 1.44). Just as infectious virus particles are easy to grow and assay, viral mutants—for example, viruses that will grow in one strain of *E. coli* but not another—are easy to isolate. Thus, T4 is manipulated even more readily than *E. coli* for studies of molecular genetics. Moreover, the genome of T4 is 23 times smaller than that of *E. coli*—approximately 0.2 million base pairs—further facilitating genetic analysis. Some other bacteriophages have even smaller genomes—the simplest consisting of RNA molecules of only about 3600 nucleotides. Bacterial viruses have thus provided extremely facile experimental systems for molecular genetics. Studies of these viruses are largely what have led to the elucidation of many fundamental principles of molecular biology.

Because of the increased complexity of the animal cell genome, viruses have been even more important in studies of animal cells than in studies of bacteria. Many animal viruses replicate and can be assayed by plaque formation in cell cultures, much as bacteriophages can. Moreover, the genomes of animal viruses are similar in complexity to those of bacterial viruses (ranging from approximately 3000 to 300,000 base pairs), so animal viruses are far more manageable than are their host cells.

There are many diverse animal viruses, each containing either DNA or RNA as their genetic material (Table 1.3). Most viruses with RNA genomes replicate by synthesizing new RNA copies of their genomes from RNA tem-

TABLE 1.3 Examples of Animal Viruses

Virus family	Representative member	Genome size (thousands of base pairs)
RNA genomes		
Picornaviruses	Poliovirus	7–8
Togaviruses	Rubella virus	12
Flaviviruses	Yellow fever virus	10
Paramyxoviruses	Measles virus	16–20
Orthomyxoviruses	Influenza virus	14
Retroviruses	Human immunodeficiency virus	9
DNA genomes		
Hepadnaviruses	Hepatitis B virus	3.2
Papovaviruses	Human papillomavirus	5–8
Adenoviruses	Adenovirus	36
Herpesviruses	Herpes simplex virus	120–200
Poxviruses	Vaccinia virus	130–280

plates in infected cells. However, one family of animal viruses—the **retroviruses**—contain RNA genomes in their virus particles but synthesize a DNA copy of their genome in infected cells. These viruses provide a good example of the importance of viruses as models, because studies of the retroviruses are what first demonstrated the synthesis of DNA from RNA templates—a fundamental mode of genetic information transfer now known to occur in all eukaryotic cells. Other examples in which animal viruses have provided important models for investigations of their host cells include studies of DNA replication, transcription, RNA processing, and protein transport and secretion.

It is particularly noteworthy that infection by some animal viruses, rather than killing the host cell, converts a normal cell into a cancer cell. Studies of such cancer-causing viruses, first described by Peyton Rous in 1911, not only have provided the basis for our current understanding of cancer at the level of cell and molecular biology, but also have led to the elucidation of many of the molecular mechanisms that control animal cell growth and differentiation.

COMPANION WEBSITE

Visit the website that accompanies **The Cell** (www.sinauer.com/cooper) for animations, videos, quizzes, problems, and other review material.

SUMMARY

KEY TERMS

THE ORIGIN AND EVOLUTION OF CELLS

The First Cell: All present-day cells, both prokaryotes and eukaryotes, are descended from a single ancestor. The first cell is thought to have arisen at least 3.8 billion years ago as a result of enclosure of self-replicating RNA in a phospholipid membrane.

prokaryotic cell, eukaryotic cell, RNA world, phospholipid, amphipathic, hydrophobic, hydrophilic

The Evolution of Metabolism: The earliest reactions for the generation of metabolic energy were a form of anaerobic glycolysis. Photosynthesis then evolved, followed by oxidative metabolism.

adenosine 5′-triphosphate (ATP), glycolysis, photosynthesis, oxidative metabolism

Present-Day Prokaryotes: Present-day prokaryotes are divided into two groups, the archaebacteria and the eubacteria, which diverged early in evolution.

archaebacteria, eubacteria, cyanobacteria, *Escherichia coli* (*E. coli*), cell wall, plasma membrane, ribosome

Eukaryotic Cells: Eukaryotic cells, which are larger and more complex than prokaryotic cells, contain a nucleus, cytoplasmic organelles, and a cytoskeleton.

nucleus, mitochondria, chloroplasts, lysosome, peroxisome, vacuole, endoplasmic reticulum, Golgi apparatus, cytoskeleton

The Origin of Eukaryotes: Eukaryotic cells are thought to have evolved from symbiotic associations of prokaryotes. The genome of eukaryotes may have arisen from a fusion of eubacterial and archaebacterial genomes.

endosymbiosis

The Development of Multicellular Organisms: The simplest eukaryotes are unicellular organisms, such as yeasts and amoebas. Multicellular organisms evolved from associations between such unicellular eukaryotes, and division of labor led to the development of the many kinds of specialized cells that make up present-day plants and animals.

yeast, *Saccharomyces cerevisiae*, pseudopodia, epithelial cell, fibroblast, erythrocyte, granulocyte, monocyte, macrophage, lymphocyte, neuron

KEY TERMS

SUMMARY

CELLS AS EXPERIMENTAL MODELS

E. coli: Because of their genetic simplicity and ease of study, bacteria such as *E. coli* are particularly useful for investigation of fundamental aspects of biochemistry and molecular biology.

Yeasts: As the simplest eukaryotic cells, yeasts are an important model for studying various aspects of eukaryotic cell biology.

Caenorhabditis elegans

Caenorhabditis elegans: The nematode *C. elegans* is a simple multicellular organism that serves as an important model in developmental biology.

Drosophila melanogaster

Drosophila melanogaster: Because of extensive genetic analysis, studies of the fruit fly *Drosophila* have led to major advances in understanding animal development.

Arabidopsis thaliana

Arabidopsis thaliana: The small flowering plant *Arabidopsis* is widely used as a model for studies of plant molecular biology and development.

Xenopus laevis,
zebrafish

Vertebrates: Many kinds of vertebrate cells can be grown in culture, where they can be studied under controlled laboratory conditions. Specialized cell types, such as neurons and muscle cells, provide useful models for investigating particular aspects of cell biology. The frog *Xenopus laevis* and zebrafish are important models for studies of early vertebrate development, and the mouse is a mammalian species suitable for genetic analysis.

TOOLS OF CELL BIOLOGY

resolution, bright-field microscopy, phase-contrast microscopy, differential interference-contrast microscopy, video-enhanced differential interference-contrast microscopy, fluorescence microscopy, green fluorescent protein (GFP), fluorescence recovery after photobleaching (FRAP), fluorescence resonance energy transfer (FRET), confocal microscopy, multi-photon excitation microscopy

Light Microscopy: A variety of methods are used to visualize cells and subcellular structures and to determine the intracellular localization of specific molecules using the light microscope.

transmission electron microscopy, electron tomography, metal shadowing, freeze fracture, scanning electron microscopy

Electron Microscopy: Electron microscopy, with a resolution that is approximately a hundredfold greater than that of light microscopy, is used to analyze details of cell structure.

differential centrifugation, ultracentrifuge, density-gradient centrifugation, velocity centrifugation, equilibrium centrifugation

Subcellular Fractionation: The organelles of eukaryotic cells can be isolated for biochemical analysis by differential centrifugation.

SUMMARY

Growth of Animal Cells in Culture: The propagation of animal cells in culture has allowed studies of the mechanisms that control cell growth and differentiation.

Culture of Plant Cells: Cultured plant cells can differentiate to form specialized cell types and, in some cases, can regenerate entire plants.

Viruses: Viruses provide simple models for studies of cell function.

KEY TERMS

embryonic stem cell, primary cultures, cell line

callus

bacteriophage, retrovirus

Questions

1. What did Stanley Miller's experiments show about the formation of organic molecules?

2. What kind of macromolecule is capable of directing its own replication?

3. Discuss the evidence that mitochondria and chloroplasts originated from bacteria that were engulfed by the precursor of eukaryotic cells.

4. Why is the evolution of photosynthesis thought to have favored the subsequent evolution of oxidative metabolism?

5. Given that the diameter of a *S. aureus* bacterial cell is 1 μm and the diameter of a human macrophage is 50 μm, how many *S. aureus* cells can fit inside a single human macrophage? Assume that the cells are spherical.

6. Which model organism provides the simplest system for studying eukaryotic DNA replication?

7. You are studying a gene involved in mammalian embryonic development. Which model organism would be best suited for your studies?

8. What resolution can be obtained with a light microscope if the specimen is viewed through air rather than through oil? Assume that the wavelength of visible light is 0.5 μm.

9. You are about to purchase a research microscope for your laboratory, and you have a choice between two different objective lenses. One has a magnification of 100× and a numerical aperture (NA) of 1.1. The other has a magnification of 60× and a numerical aperture of 1.3. Assuming

that price is not a concern, which of these objectives would you choose?

10. What advantage does the use of green fluorescent protein (GFP) have over the use of fluorescent-labeled antibodies for studying the location and movement of a protein in cells?

11. Identify the different characteristics or properties of organelles that allow separation by velocity centrifugation as compared to equilibrium centrifugation in a sucrose gradient.

12. Why is serum usually required in the media used to grow animal cells in culture?

13. Distinguish between primary cell cultures and immortal cell lines.

14. Why is the ability to culture embryonic stem cells important?

References and Further Reading

The Origin and Evolution of Cells

Andersson, S. G. E., A. Zomorodipour, J. O. Andersson, T. Sicheritz-Ponten, U. C. M. Alsmark, R. M. Podowski, A. K. Naslund, A.-S. Eriksson, H. H. Winkler and C. G. Kurland. 1998. The genome sequence of *Rickettsia prowazekii* and the origin of mitochondria. *Nature* 396: 133–140. [P]

Cech, T. R. 1986. A model for the RNA-catalyzed replication of RNA. *Proc. Natl. Acad. Sci. USA* 83: 4360–4363. [P]

Crick, F. H. C. 1968. The origin of the genetic code. *J. Mol. Biol.* 38: 367–379. [P]

Darnell, J. E. and W. F. Doolittle. 1986. Speculations on the early course of evolution. *Proc. Natl. Acad. Sci. USA* 83: 1271–1275. [P]

Dyall, S. D., M. T. Brown and P. J. Johnson. 2004. Ancient invasions: From endosymbionts to organelles. *Science* 304: 253–257. [R]

Gesteland, R. F., T. R. Cech and J. F. Atkins (eds.). 1999. *The RNA World.* 2nd ed. Plainview, NY: Cold Spring Harbor Laboratory Press.

Gilbert, W. 1986. The RNA world. *Nature* 319: 618. [R]

Johnston, W. K., P. J. Unrau, M. S. Lawrence, M. E. Glasner and D. P. Bartel. 2001. RNA-catalyzed RNA polymerization: Accurate and general RNA-templated primer extension. *Science* 292: 1319–1325. [P]

Joyce, G. F. 1989. RNA evolution and the origins of life. *Nature* 338: 217–224. [R]

Kasting, J. F. and J. L. Siefert. 2002. Life and the evolution of Earth's atmosphere. *Science* 296: 1066–1068. [R]

Margulis, L. 1992. *Symbiosis in Cell Evolution.* 2nd ed. New York: W. H. Freeman.

Martin, W. and T. M. Embley. 2004. Early evolution comes full circle. *Nature* 431: 134–137. [R]

Miller, S. L. 1953. A production of amino acids under possible primitive earth conditions. *Science* 117: 528–529. [P]

Pace, N. R. 1997. A molecular view of microbial diversity and the biosphere. *Science* 276: 734–740. [R]

Rivera, M. C. and J. A. Lake. 2004. The ring of life provides evidence for a genome fusion origin of eukaryotes. *Nature* 431: 152–155. [P]

Cells as Experimental Models

Adams, M. D. and 194 others. 2000. The genome sequence of *Drosophila melanogaster*. *Science* 287: 2185–2195. [P]

Blattner, F. R., G. Plunkett III., C. A. Bloch, N. T. Perna, V. Burland, M. Riley, J. Collado-Vides, J. D. Glasner, C. K. Rode, G. F. Mayhew, J. Gregor, N. W. Davis, H. A. Kirkpatrick, M. A. Goeden, D. J. Rose, B. Mau and Y. Shao. 1997. The complete genome sequence of *Escherichia coli* K–12. *Science* 277: 1453–1462. [P]

Botstein, D., S. A. Chervitz and J. M. Cherry. 1997. Yeast as a model organism. *Science* 277: 1259–1260. [R]

Goffeau, A. and 15 others. 1996. Life with 6000 genes. *Science* 274: 546–567. [P]

International Human Genome Sequencing Consortium. 2001. Initial sequencing and analysis of the human genome. *Nature* 409: 860–921. [P]

International Human Genome Sequencing Consortium. 2004. Finishing the euchromatic sequence of the human genome. *Nature* 431: 931–945. [P]

Meyerowitz, E. M. 2002. Plants compared to animals: The broadest comparative study of development. *Science* 295: 1482–1485. [R]

Mouse Genome Sequencing Consortium. 2002. Initial sequence and comparative analysis of the mouse genome. *Nature* 420: 520–562. [P]

Shin, J. T. and M. C. Fishman. 2002. From zebrafish to human: Molecular medical models. *Ann. Rev. Genomics Hum. Genet.* 3: 311–340. [R]

The *Arabidopsis* Genome Initiative. 2000. Analysis of the genome sequence of the flowering plant *Arabidopsis thaliana*. *Nature* 408: 796–815. [P]

The *C. elegans* Sequencing Consortium. 1998. Genome sequence of the nematode *C. elegans*: A platform for investigating biology. *Science* 282: 2012–2018. [P]

Venter, J. C. and 273 others. 2001. The sequence of the human genome. *Science* 291: 1304–1351. [P]

Tools of Cell Biology

Bowers, W. E. 1998. Christian de Duve and the discovery of lysosomes and peroxisomes. *Trends Cell Biol.* 8: 330–333 [R]

Cairns, J., G. S. Stent and J. D. Watson (eds.). 1992. *Phage and the Origins of Molecular Biology*. Plainview, NY: Cold Spring Harbor Laboratory Press.

Chudakov, D. M., S. Lukyanov and K. A. Lukyanov. 2005. Fluorescent proteins as a toolkit for *in vivo* imaging. *Trends Biotechnol.* 23: 605–613. [R]

Claude, A. 1975. The coming of age of the cell. *Science* 189: 433–435. [R]

De Duve, C. 1975. Exploring cells with a centrifuge. *Science* 189: 186–194. [R]

Eagle, H. 1955. Nutrition needs of mammalian cells in tissue culture. *Science* 235: 442–447. [P]

Flint, S. J., L. W. Enquist, V. R. Racaniello and A. M. Skalka. 2003. *Principles of Virology: Molecular Biology, Pathogenesis, and Control of Animal Viruses*. 2nd ed. Washington, DC: ASM Press.

Kam, Z., E. Zamir and B. Geiger. 2001. Probing molecular processes in live cells by quantitative multidimensional microscopy. *Trends Cell Biol.* 11: 329–334. [R]

Koster, A. J. and J. Klumperman. 2003. Electron microscopy in cell biology: Integrating structure and function. *Nature Rev. Molec. Cell Biol.* 4: SS6–SS10. [R]

Lippincott-Schwartz, J., N. Altan-Bonnet and G. H. Patterson. 2003. Photobleaching and photoactivation: Following protein dynamics in living cells. *Nature Cell Biol.* 5: S7–S14. [R]

McIntosh, R., D. Nicastro and D. Mastronarde. 2005. New views of cells in 3D: An introduction to electron tomography. *Trends Cell Biol.* 15: 43–51. [R]

Miyawaki, A., A. Sawano and T. Kogure. 2003. Lighting up cells: Labelling proteins with fluorophores. *Nature Cell Biol.* 5: S1–S7. [R]

Palade, G. 1975. Intracellular aspects of the process of protein synthesis. *Science* 189: 347–358. [R]

Piston, D. W. 1999. Imaging living cells and tissues by two-photon excitation microscopy. *Trends Cell Biol.* 9: 66–69 [R]

Porter, K. R., A. Claude and E. F. Fullam. 1945. A study of tissue culture cells by electron microscopy. *J. Exp. Med.* 81: 233–246. [P]

Rous, P. 1911. A sarcoma of the fowl transmissible by an agent separable from the tumor cells. *J. Exp. Med.* 13: 397–411. [P]

Salmon, E. D. 1995. VE-DIC light microscopy and the discovery of kinesin. *Trends Cell Biol.* 5: 154–158. [R]

Spector, D. L., R. Goldman and L. Leinwand. 1998. *Cells: A Laboratory Manual*. Plainview, NY: Cold Spring Harbor Laboratory Press.

CHAPTER 2

The Composition of Cells

■ *The Molecules of Cells 43*

■ *Cell Membranes 58*

■ *Proteomics: Large-Scale Analysis of Cell Proteins 65*

■ *KEY EXPERIMENT:*
The Folding of Polypeptide Chains 54

■ *KEY EXPERIMENT:*
The Structure of Cell Membranes 59

CELLS ARE INCREDIBLY COMPLEX AND DIVERSE STRUCTURES, capable not only of self-replication—the very essence of life—but also of performing a wide range of specialized tasks in multicellular organisms. Yet cells obey the same laws of chemistry and physics that determine the behavior of nonliving systems. Consequently, modern cell biology seeks to understand cellular processes in terms of chemical and physical reactions.

This chapter considers the chemical composition of cells and the properties of the molecules that are ultimately responsible for all cellular activities. Proteins are given particular emphasis because of their diverse roles within the cell, including acting as enzymes that catalyze almost all biological reactions and serving as key components of cell membranes. In recent years, a variety of new technologies have been developed that allow the analysis of cell proteins on a large scale. This emerging field of proteomics is expected to make a major impact on our understanding of the expression and interactions of proteins within cells, much like the impact that large scale genome sequencing has made on our understanding of the genetic content of cells and organisms.

The Molecules of Cells

Cells are composed of water, inorganic ions, and carbon-containing (organic) molecules. Water is the most abundant molecule in cells, accounting for 70% or more of total cell mass. Consequently, the interactions between water and the other constituents of cells are of central importance in biological chemistry. The critical property of water in this respect is that it is a polar molecule, in which the hydrogen atoms have a slight positive charge and the oxygen has a slight negative charge (Figure 2.1). Because of their polar nature, water molecules can form hydrogen bonds with each other or with other polar molecules, as well as interact

2.1 WEBSITE ANIMATION

Bond Formation The polymerization of sugars, amino acids, and nucleotides to form polysaccharides, polypeptides, and nucleic acids, respectively, occurs through covalent bonding.

(A)

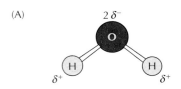

FIGURE 2.1 Characteristics of water   (A) Water is a polar molecule, with a slight negative charge (δ^-) on the oxygen atom and a slight positive charge (δ^+) on the hydrogen atoms. Because of this polarity, water molecules can form hydrogen bonds (dashed lines) either with each other or with other polar molecules (B), in addition to interacting with charged ions (C).

(B)

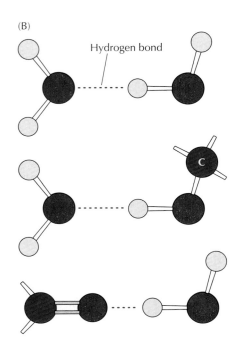

Hydrogen bond

(C)

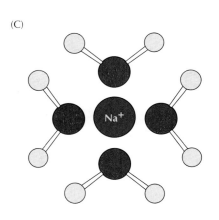

■ The artificial sweetener sucralose (Splenda) is a synthetic derivative of common sugar in which some of the hydroxyl groups of the sugar are replaced with chlorine.

with positively or negatively charged ions. As a result of these interactions, ions and polar molecules are readily soluble in water (hydrophilic). In contrast, nonpolar molecules, which cannot interact with water, are poorly soluble in an aqueous environment (hydrophobic). Consequently, nonpolar molecules tend to minimize their contact with water by associating closely with each other instead. As discussed later in this chapter, such interactions of polar and nonpolar molecules with water and with each other play crucial roles in the formation of biological structures, such as cell membranes. The inorganic ions of the cell, including sodium (Na^+), potassium (K^+), magnesium (Mg^{2+}), calcium (Ca^{2+}), phosphate (HPO_4^{2-}), chloride (Cl^-), and bicarbonate (HCO_3^-), constitute 1% or less of the cell mass. These ions are involved in a number of aspects of cell metabolism, and thus play critical roles in cell function.

It is, however, the organic molecules that are the unique constituents of cells. Most of these organic compounds belong to one of four classes of molecules: carbohydrates, lipids, proteins, and nucleic acids. Proteins, nucleic acids, and most carbohydrates (the polysaccharides) are macromolecules formed by the joining (polymerization) of hundreds or thousands of low-molecular-weight precursors: amino acids, nucleotides, and simple sugars, respectively. Such macromolecules constitute 80 to 90% of the dry weight of most cells. Lipids are the other major constituent of cells. The remainder of the cell mass is composed of a variety of small organic molecules, including macromolecular precursors. The basic chemistry of cells can thus be understood in terms of the structures and functions of four major classes of organic molecules.

Carbohydrates

The **carbohydrates** include simple sugars as well as polysaccharides. These simple sugars, such as glucose, are the major nutrients of cells. As discussed in Chapter 3, their breakdown provides both a source of cellular energy and the starting material for the synthesis of other cell constituents. Polysaccharides are storage forms of sugars and form structural components of the cell. In addition, polysaccharides and shorter polymers of sugars act as markers for a variety of cell recognition processes, including the adhesion of cells to their neighbors and the transport of proteins to appropriate intracellular destinations.

The structures of representative simple sugars (**monosaccharides**) are illustrated in Figure 2.2. The basic formula for these molecules is $(CH_2O)_n$, from which the name carbohydrate is derived (C = "carbo" and H_2O = "hydrate"). The six-carbon ($n = 6$) sugar glucose ($C_6H_{12}O_6$) is especially important in cells, since it provides the principal source of cellular energy. Other simple sugars have between three and seven carbons, with three- and five-carbon sugars being the most common. Sugars containing five or more carbons can cyclize to form ring structures, which are the predominant forms of these molecules within cells. As illustrated in Figure 2.2, the cyclized sugars exist in two alternative forms (called α or β), depending on the configuration of carbon 1.

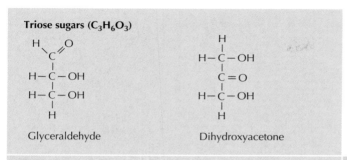

FIGURE 2.2 Structure of simple sugars   Representative sugars containing three, five, and six carbons (triose, pentose, and hexose sugars, respectively) are illustrated. Sugars with five or more carbons can cyclize to form rings, which exist in two alternative forms (α and β), depending on the configuration of carbon 1.

Monosaccharides can be joined together by dehydration reactions in which H_2O is removed and the sugars are linked by a **glycosidic bond** between two of their carbons (Figure 2.3). If only a few sugars are joined together, the resulting polymer is called an **oligosaccharide**. If a large number (hundreds or thousands) of sugars are involved, the resulting polymers are macromolecules called **polysaccharides**.

Two common polysaccharides—**glycogen** and **starch**—are the storage forms of carbohydrates in animal and plant cells, respectively. Both glycogen and starch are composed entirely of glucose molecules in the α configuration (Figure 2.4). The principal linkage is between carbon 1 of one glucose and carbon 4 of a second. In addition, both glycogen and one form of starch (amylopectin) contain occasional $\alpha(1{\rightarrow}6)$ linkages, in which carbon 1 of one glucose is joined to carbon 6 of a second. As illustrated in Figure 2.4, these linkages lead to the formation of branches resulting from the joining of two separate $\alpha(1{\rightarrow}4)$ linked chains. Such branches are present in glycogen and amylopectin, although another form of starch (amylose) is an unbranched molecule.

The structures of glycogen and starch are thus basically similar, as is their function: to store glucose. **Cellulose**, in contrast, has a quite distinct function as the principal structural component of the plant cell wall. Perhaps surprisingly, then, cellulose

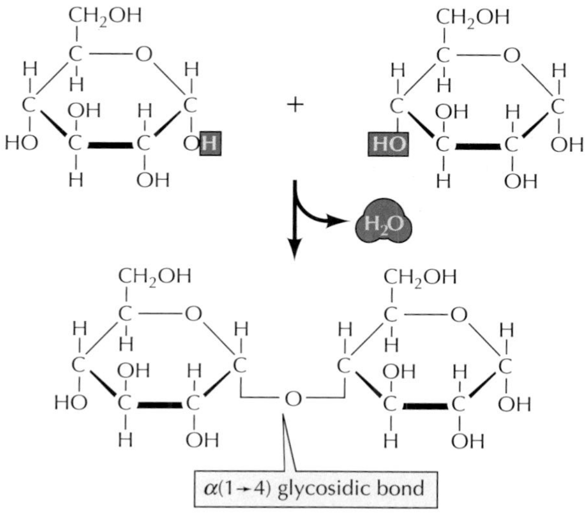

FIGURE 2.3 Formation of a glycosidic bond
Two simple sugars are joined by a dehydration reaction (a reaction in which water is removed). In the example shown, two glucose molecules in the α configuration are joined by a bond between carbons 1 and 4, which is therefore called an $\alpha(1{\rightarrow}4)$ glycosidic bond.

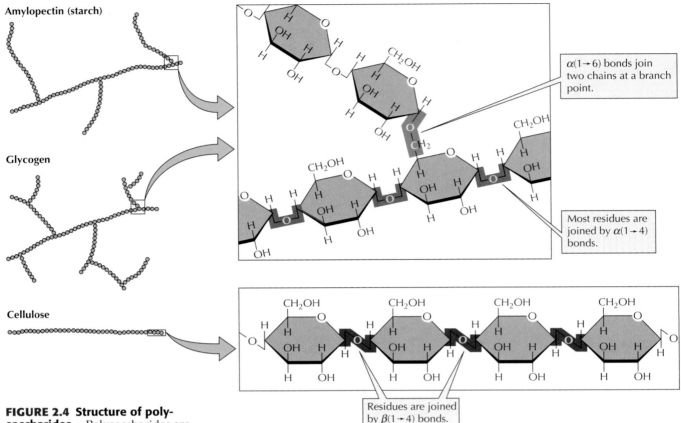

Amylopectin (starch)

Glycogen

Cellulose

$\alpha(1\rightarrow6)$ bonds join two chains at a branch point.

Most residues are joined by $\alpha(1\rightarrow4)$ bonds.

Residues are joined by $\beta(1\rightarrow4)$ bonds.

FIGURE 2.4 Structure of poly-saccharides Polysaccharides are macromolecules consisting of hundreds or thousands of simple sugars. Glycogen, starch, and cellulose are all composed entirely of glucose residues, which are joined by $\alpha(1\rightarrow4)$ glycosidic bonds in glycogen and starch but by $\beta(1\rightarrow4)$ bonds in cellulose. Glycogen and one form of starch (amylopectin) also contain occasional $\alpha(1\rightarrow6)$ bonds, which serve as branch points by joining two separate $\alpha(1\rightarrow4)$ chains.

is also composed entirely of glucose molecules. The glucose residues in cellulose, however, are in the β rather than the α configuration, and cellulose is an unbranched polysaccharide (see Figure 2.4). The linkage of glucose residues by $\beta(1\rightarrow4)$ rather than $\alpha(1\rightarrow4)$ bonds causes cellulose to form long extended chains that pack side by side to form fibers of great mechanical strength.

In addition to their roles in energy storage and cell structure, oligosaccharides and polysaccharides are important in a variety of informational processes. For example, oligosaccharides are frequently linked to proteins, where they play important roles in protein folding and serve as markers to target proteins for transport to the cell surface or incorporation into different subcellular organelles. Oligosaccharides and polysaccharides also serve as markers on the surface of cells, playing important roles in cell recognition and the interactions between cells in tissues of multicellular organisms.

Lipids

Lipids have three major roles in cells. First, they provide an important form of energy storage. Second, and of great importance in cell biology, lipids are the major components of cell membranes. Third, lipids play important roles in cell signaling, both as steroid hormones (e.g., estrogen and testosterone) and as messenger molecules that convey signals from cell surface receptors to targets within the cell.

The simplest lipids are **fatty acids**, which consist of long hydrocarbon chains, most frequently containing 16 or 18 carbon atoms, with a carboxyl group (COO–) at one end (Figure 2.5). Unsaturated fatty acids contain one or

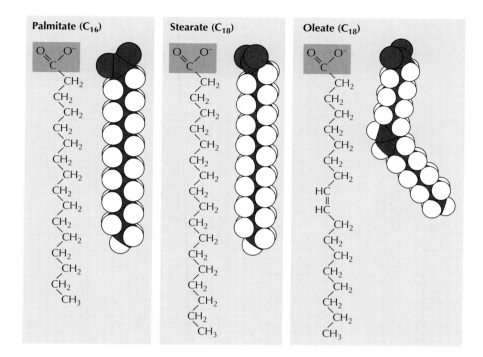

Palmitate (C$_{16}$)   Stearate (C$_{18}$)   Oleate (C$_{18}$)

FIGURE 2.5 Structure of fatty acids
Fatty acids consist of long hydrocarbon chains terminating in a carboxyl group (COO$^-$). Palmitate and stearate are saturated fatty acids consisting of 16 and 18 carbons, respectively. Oleate is an unsaturated 18-carbon fatty acid containing a double bond between carbons 9 and 10. Note that the double bond introduces a kink in the hydrocarbon chain.

more double bonds between carbon atoms; in saturated fatty acids all of the carbon atoms are bonded to the maximum number of hydrogen atoms. The long hydrocarbon chains of fatty acids contain only nonpolar C—H bonds, which are unable to interact with water. The hydrophobic nature of these fatty acid chains is responsible for much of the behavior of complex lipids, particularly in the formation of biological membranes.

Fatty acids are stored in the form of **triacylglycerols,** or **fats,** which consist of three fatty acids linked to a glycerol molecule (Figure 2.6). Triacylglycerols are insoluble in water and therefore accumulate as fat droplets in the cytoplasm. When required, they can be broken down for use in energy-yielding reactions (see Chapter 3). It is noteworthy that fats are a more efficient form of energy storage than carbohydrates, yielding more than twice as much energy per weight of material broken down. Fats therefore allow energy to be stored in less than half the body weight that would be required to store the same amount of energy in carbohydrates—a particularly important consideration for animals because of their mobility.

Phospholipids, the principal components of cell membranes, consist of two fatty acids joined to a polar head group (Figure 2.7). In the **glycerol phospholipids,** the two fatty acids are bound to carbon atoms in glycerol, as in triacylglycerols. The third carbon of glycerol, however, is bound to a phosphate group, which is in turn frequently attached to another small polar molecule, such as choline, serine, inositol, or ethanolamine. **Sphingomyelin,** the only nonglycerol phospholipid in cell membranes, contains two hydrocarbon chains linked to a polar head group formed from serine rather than from glycerol. All phospholipids have hydrophobic tails, con-

Glycerol

Fatty acids

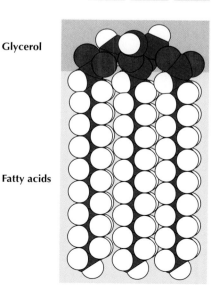

Glycerol

Fatty acids

FIGURE 2.6 Structure of triacylglycerols   Triacylglycerols (fats) contain three fatty acids joined to glycerol. In this example, all three fatty acids are palmitate, but triacylglycerols often contain a mixture of different fatty acids.

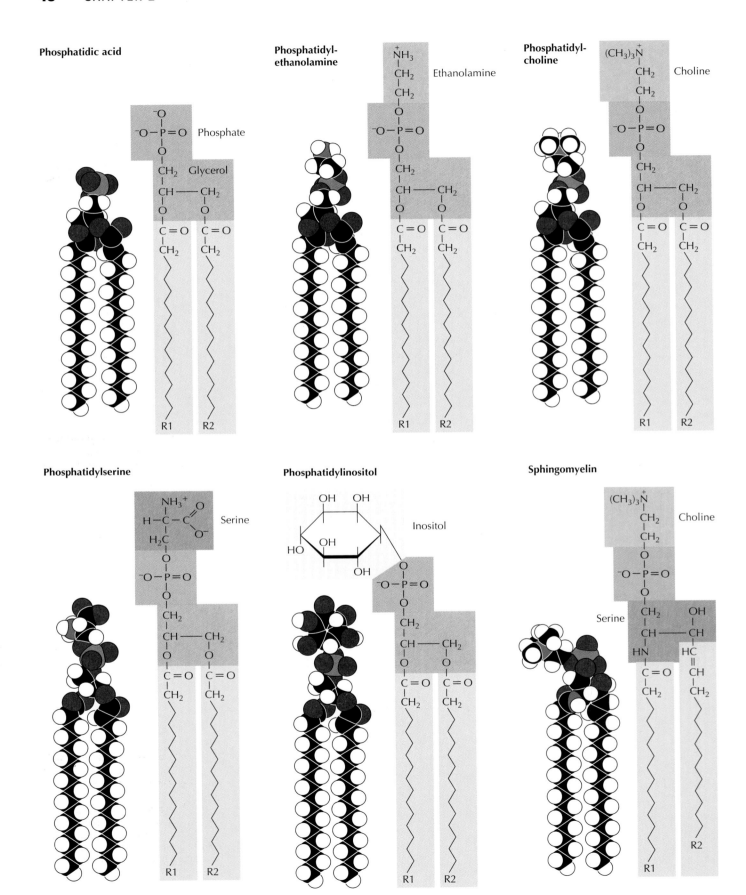

Phosphatidic acid

Phosphate

Glycerol

R1 R2

Phosphatidyl-ethanolamine

Ethanolamine

R1 R2

Phosphatidyl-choline

Choline

R1 R2

Phosphatidylserine

Serine

R1 R2

Phosphatidylinositol

Inositol

R1 R2

Sphingomyelin

Choline

Serine

R1 R2

◀ **FIGURE 2.7 Structure of phospholipids**   Glycerol phospholipids contain two fatty acids joined to glycerol. The fatty acids may be different from each other and are designated R1 and R2. The third carbon of glycerol is joined to a phosphate group (forming phosphatidic acid), which in turn is frequently joined to another small polar molecule (forming phosphatidylethanolamine, phosphatidylcholine, phosphatidylserine, or phosphatidylinositol). In sphingomyelin, two hydrocarbon chains are bound to a polar head group formed from serine instead of glycerol.

sisting of the two hydrocarbon chains, and hydrophilic head groups, consisting of the phosphate group and its polar attachments. Consequently, phospholipids are **amphipathic** molecules, part water-soluble and part water-insoluble. This property of phospholipids is the basis for the formation of biological membranes, as discussed later in this chapter.

In addition to phospholipids, many cell membranes contain **glycolipids** and **cholesterol**. Glycolipids consist of two hydrocarbon chains linked to polar head groups that contain carbohydrates (Figure 2.8). They are thus similar to the phospholipids in their general organization as amphipathic molecules. Cholesterol, in contrast, consists of four hydrocarbon rings rather than linear hydrocarbon chains (Figure 2.9). The hydrocarbon rings are strongly hydrophobic, but the hydroxyl (OH) group attached to one end of cholesterol is weakly hydrophilic, so cholesterol is also amphipathic.

In addition to their roles as components of cell membranes, lipids function as signaling molecules, both within and between cells. The **steroid hormones** (such as estrogens and testosterone) are derivatives of cholesterol (see Figure 2.9). These hormones are a diverse group of chemical messengers, all of which contain four hydrocarbon rings to which distinct functional groups are attached. Derivatives of phospholipids also serve as messenger molecules within cells, acting to convey signals from cell surface receptors to intracellular targets that regulate a wide range of cellular processes, including cell proliferation, movement, survival, and differentiation (see Chapter 15).

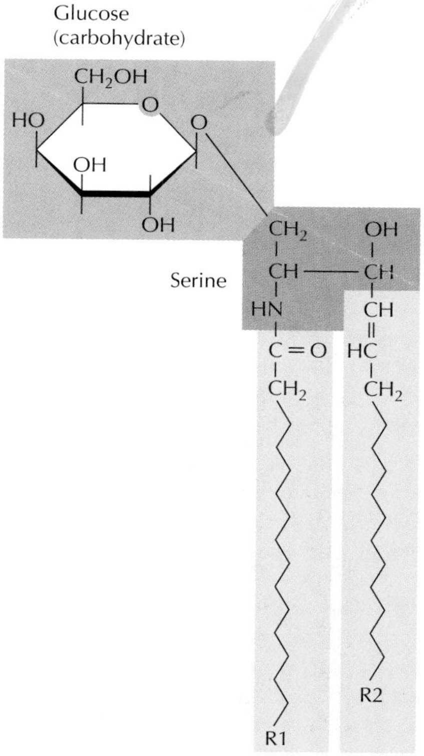

FIGURE 2.8 Structure of glycolipids
Two hydrocarbon chains are joined to a polar head group formed from serine and containing carbohydrates (e.g., glucose).

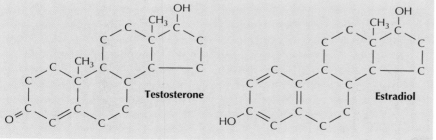

FIGURE 2.9 Cholesterol and steroid hormones   Cholesterol, an important component of cell membranes, is an amphipathic molecule because of its polar hydroxyl group. Cholesterol is also a precursor to the steroid hormones, such as testosterone and estradiol (a form of estrogen). The hydrogen atoms bonded to the ring carbons are not shown in this figure.

Nucleic Acids

The nucleic acids—DNA and RNA—are the principal informational molecules of the cell. **Deoxyribonucleic acid (DNA)** has a unique role as the genetic material, which in eukaryotic cells is located in the nucleus. Different types of **ribonucleic acid (RNA)** participate in a number of cellular activities. **Messenger RNA (mRNA)** carries information from DNA to the ribosomes, where it serves as a template for protein synthesis. Two other types of RNA (**ribosomal RNA** and **transfer RNA**) are involved in protein synthesis. Still other kinds of RNAs are involved in the processing and transport of both RNAs and proteins. In addition to acting as an informational molecule, RNA is also capable of catalyzing a number of chemical reactions. In present-day cells, these include reactions involved in both protein synthesis and RNA processing.

DNA and RNA are polymers of **nucleotides**, which consist of **purine** and **pyrimidine** bases linked to phosphorylated sugars (Figure 2.10). DNA con-

FIGURE 2.10 Components of nucleic acids Nucleic acids contain purine and pyrimidine bases linked to phosphorylated sugars. A nucleic acid base linked to a sugar alone is a nucleoside. Nucleotides additionally contain one or more phosphate groups.

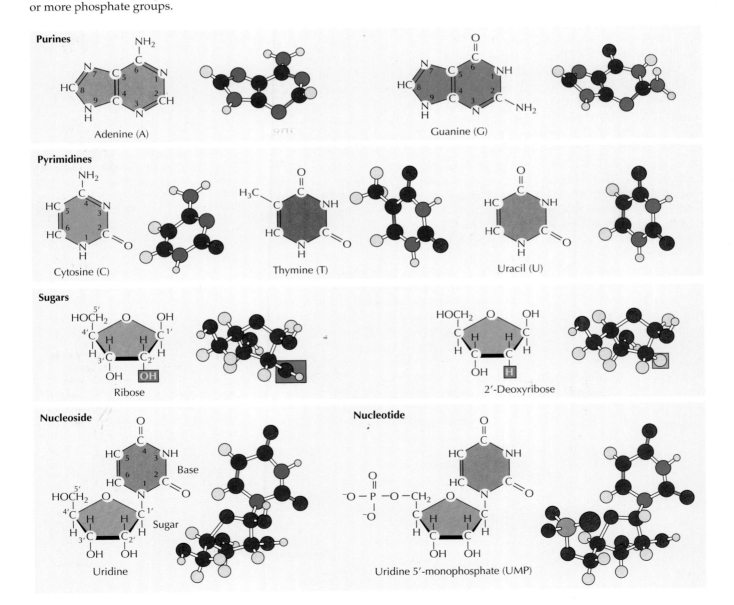

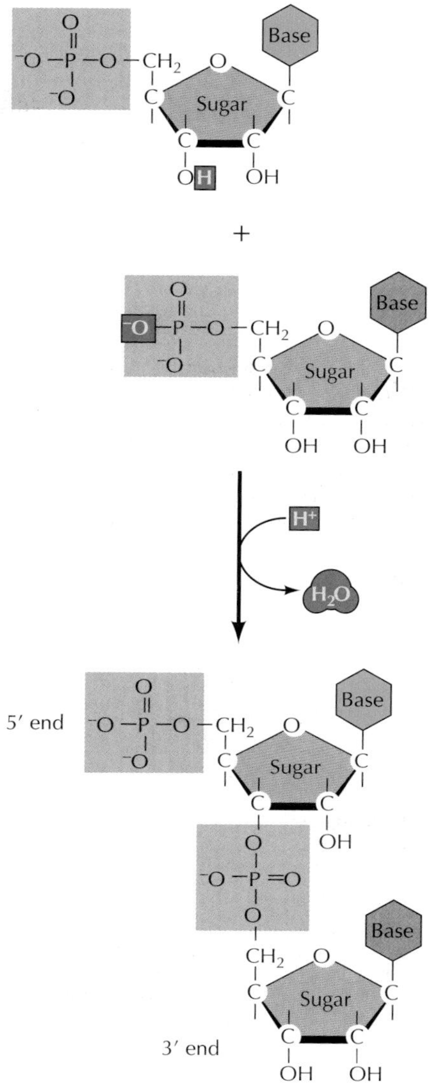

FIGURE 2.11 Polymerization of nucleotides A phosphodiester bond is formed between the 3' hydroxyl group of one nucleotide and the 5' phosphate group of another. A polynucleotide chain has a sense of direction, one end terminating in a 5' phosphate group (the 5' end) and the other end terminating in a 3' hydroxyl group (the 3' end).

tains two purines (**adenine** and **guanine**) and two pyrimidines (**cytosine** and **thymine**). Adenine, guanine, and cytosine are also present in RNA, but RNA contains **uracil** in place of thymine. The bases are linked to sugars (**2'-deoxyribose** in DNA, or **ribose** in RNA) to form **nucleosides**. Nucleotides additionally contain one or more phosphate groups linked to the 5' carbon of nucleoside sugars.

The polymerization of nucleotides to form nucleic acids involves the formation of **phosphodiester bonds** between the 5' phosphate of one nucleotide and the 3' hydroxyl of another (Figure 2.11). **Oligonucleotides** are small polymers containing only a few nucleotides; the large **polynucleotides** that make up cellular RNA and DNA may contain thousands or millions of nucleotides, respectively. It is important to note that a polynucleotide chain has a sense of direction, with one end of the chain terminating in a 5' phosphate group and the other in a 3' hydroxyl group. Polynucleotides are always synthesized in the 5' to 3' direction, with a free nucleotide being added to the 3' OH group of a growing chain. By convention, the sequence of bases in DNA or RNA is also written in the 5' to 3' direction.

The information in DNA and RNA is conveyed by the order of the bases in the polynucleotide chain. DNA is a double-stranded molecule consisting of two polynucleotide chains running in opposite directions (see Chapter 4). The bases are on the inside of the molecule, and the two chains are joined by hydrogen bonds between complementary base pairs—adenine pairing with thymine and guanine with cytosine (Figure 2.12). The important consequence of such complementary base pairing is that one strand of DNA (or RNA) can act as a template to direct the synthesis of a complementary strand. Nucleic acids are thus uniquely capable of directing their own self-replication, allowing them to function as the fundamental informational molecules of the cell. The information carried by DNA and RNA directs the synthesis of specific proteins, which control most cellular activities.

Nucleotides are not only important as the building blocks of nucleic acids; they also play critical roles in other cell processes. Perhaps the most prominent example is adenosine 5'-triphosphate (ATP), which is the principal form of chemical energy within cells. Other nucleotides similarly function as carriers of either energy or reactive chemical groups in a wide variety of metabolic reactions. In addition, some nucleotides (e.g., cyclic AMP) are important signaling molecules within cells (see Chapter 15).

FIGURE 2.12 Complementary pairing between nucleic acid bases The formation of hydrogen bonds between bases on opposite strands of DNA leads to the specific pairing of guanine (G) with cytosine (C) and adenine (A) with thymine (T).

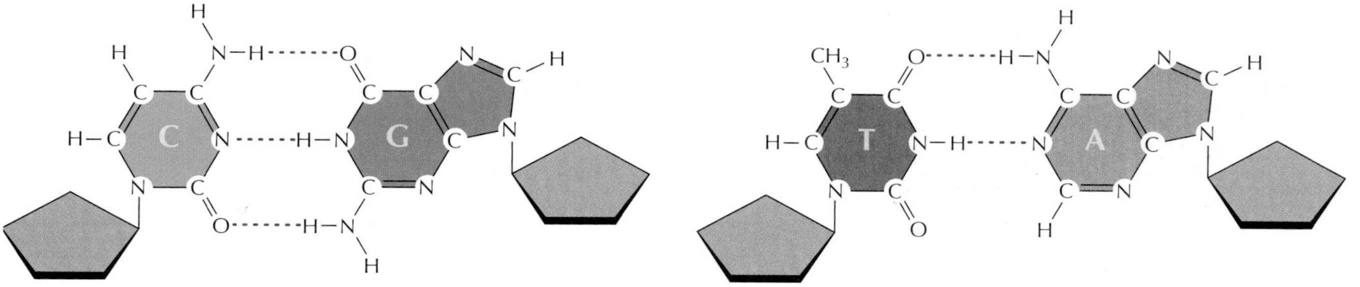

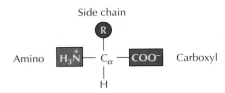

FIGURE 2.13 Structure of amino acids   Each amino acid consists of a central carbon atom (the α carbon) bonded to a hydrogen atom, a carboxyl group, an amino group, and a specific side chain (designated R). At physiological pH, both the carboxyl and amino groups are ionized, as shown.

◼ **Besides being the building blocks of proteins, amino acids serve another important biological function as molecules used by cells to communicate. The amino acids glycine and glutamate, along with several other amino acids that are not found in proteins, act as neurotransmitters (see Chapter 15).**

Proteins

While nucleic acids carry the genetic information of the cell, the primary responsibility of **proteins** is to execute the tasks directed by that information. Proteins are the most diverse of all macromolecules, and each cell contains several thousand different proteins, which perform a wide variety of functions. The roles of proteins include serving as structural components of cells and tissues, acting in the transport and storage of small molecules (e.g., the transport of oxygen by hemoglobin), transmitting information between cells (e.g., protein hormones), and providing a defense against infection (e.g., antibodies). The most fundamental property of proteins, however, is their ability to act as enzymes, which, as discussed in the next chapter, catalyze nearly all the chemical reactions in biological systems. Thus, proteins direct virtually all activities of the cell. The central importance of proteins in biological chemistry is indicated by their name, which is derived from the Greek word *proteios*, meaning "of the first rank."

Proteins are polymers of 20 different **amino acids**. Each amino acid consists of a carbon atom (called the α carbon) bonded to a carboxyl group (COO–), an amino group (NH_3^+), a hydrogen atom, and a distinctive side chain (Figure 2.13). The specific chemical properties of the different amino acid side chains determine the roles of each amino acid in protein structure and function.

The amino acids can be grouped into four broad categories according to the properties of their side chains (Figure 2.14). Ten amino acids have nonpolar side chains that do not interact with water. Glycine is the simplest amino acid, with a side chain consisting of only a hydrogen atom. Alanine, valine, leucine, and isoleucine have hydrocarbon side chains consisting of up to four carbon atoms. The side chains of these amino acids are hydrophobic and therefore tend to be located in the interior of proteins, where they are not in contact with water. Proline similarly has a hydrocarbon side chain, but it is unique in that its side chain is bonded to the nitrogen of the amino group as well as to the α carbon, forming a cyclic structure. The side chains of two amino acids, cysteine and methionine, contain sulfur atoms. Methionine is quite hydrophobic, but cysteine is less so because of its sulfhydryl (SH) group. As discussed later, the sulfhydryl group of cysteine plays an important role in protein structure because disulfide bonds can form between the side chains of different cysteine residues. Finally, two nonpolar amino acids, phenylalanine and tryptophan, have side chains containing very hydrophobic aromatic rings.

Five amino acids have uncharged but polar side chains. These include serine, threonine, and tyrosine, which have hydroxyl groups on their side chains, as well as asparagine and glutamine, which have polar amide (O=C—NH_2) groups. Because the polar side chains of these amino acids can form hydrogen bonds with water, these amino acids are hydrophilic and tend to be located on the outside of proteins.

The amino acids lysine, arginine, and histidine have side chains with charged basic groups. Lysine and arginine are very basic amino acids, and their side chains are positively charged in the cell. Consequently, they are very hydrophilic and are found in contact with water on the surface of proteins. Histidine can be either uncharged or positively charged at physiological pH, so it frequently plays an active role in enzymatic reactions involving the exchange of hydrogen ions, as discussed in Chapter 3.

Finally, two amino acids, aspartic acid and glutamic acid, have acidic side chains terminating in carboxyl groups. These amino acids are negatively charged within the cell and are therefore frequently referred to as aspartate

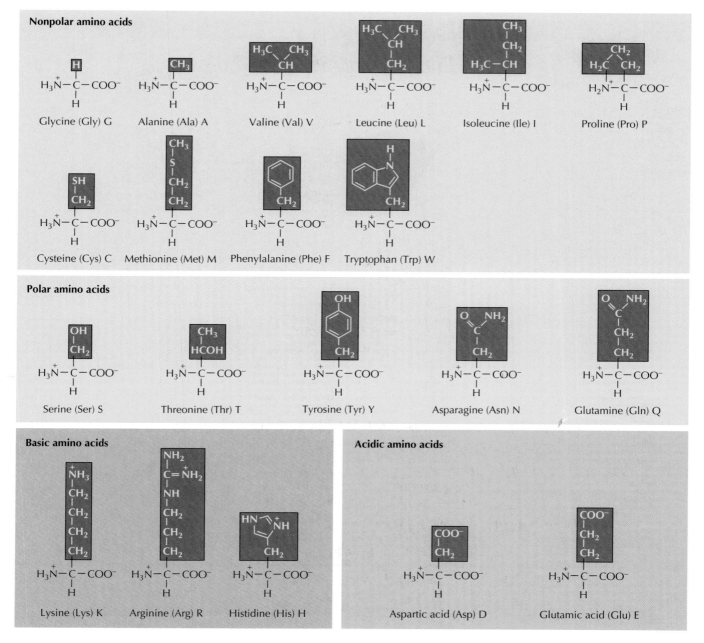

FIGURE 2.14 The amino acids
The three-letter and one-letter abbreviations for each amino acid are indicated. The amino acids are grouped into four categories according to the properties of their side chains: nonpolar, polar, basic, and acidic.

and glutamate. Like the basic amino acids, these acidic amino acids are very hydrophilic and are usually located on the surface of proteins.

Amino acids are joined together by **peptide bonds** between the α amino group of one amino acid and the α carboxyl group of a second (Figure 2.15). **Polypeptides** are linear chains of amino acids, usually hundreds or thousands of amino acids in length. Each polypeptide chain has two distinct ends, one terminating in an α amino group (the amino, or N, terminus) and the other in an α carboxyl group (the carboxy, or C, terminus). Polypeptides are synthesized from the amino to the carboxy terminus, and the sequence of amino acids in a polypeptide is written (by convention) in the same order.

KEY EXPERIMENT

The Folding of Polypeptide Chains

Reductive Cleavage of Disulfide Bridges in Ribonuclease
Michael Sela, Frederick H. White, Jr., and Christian B. Anfinsen
National Institutes of Health, Bethesda, MD
Science, Volume 125, 1957, pages 691–692

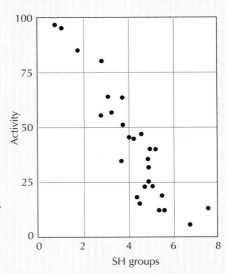

A summary of results of the renaturation experiments. The enzymatic activity of ribonuclease is plotted as a function of the number of sulfhydryl groups present after various treatments. Activity is expressed as percent activity of the native enzyme.

The Context

Functional proteins are structurally far more complex than linear chains of amino acids. The formation of active enzymes or other proteins requires the folding of polypeptide chains into precise three-dimensional conformations. This difference between proteins and polypeptide chains raises critical questions with respect to understanding protein structure and function. How is the correct protein conformation chosen from the many possible conformations that could be adopted by a polypeptide, and what is the nature of the information that directs protein folding?

Classic experiments done by Christian Anfinsen and his colleagues provided the answers to these questions. Studying the enzyme ribonuclease, Anfinsen and his collaborators were able to demonstrate that denatured proteins can spontaneously refold into an active conformation. Therefore, all the information required to specify the correct three-dimensional conformation of a protein is contained in its primary amino acid sequence. A series of such experiments led Anfinsen to the conclusion that the native three-dimensional structure of a protein corresponds to its thermodynamically most stable conformation, as determined by interactions between its constituent amino acids. The original observations leading to the formulation of this critical principle were reported in this 1957 paper by Christian Anfinsen, Michael Sela, and Fred White.

The Experiments

The protein studied by Sela, White, and Anfinsen was bovine ribonuclease, a small protein of 124 amino acids that contains four disulfide (S—S) bonds between side chains of cysteine residues. The enzymatic activity of ribonuclease could be

---

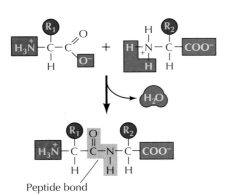

FIGURE 2.15 Formation of a peptide bond The carboxyl group of one amino acid is linked to the amino group of a second.

The defining characteristic of proteins is that they are polypeptides with specific amino acid sequences. In 1953 Frederick Sanger was the first to determine the complete amino acid sequence of a protein, the hormone insulin. Insulin was found to consist of two polypeptide chains, joined by disulfide bonds between cysteine residues (Figure 2.16). Most important, Sanger's experiment revealed that each protein consists of a specific amino acid sequence. Protein sequences are now usually deduced from the sequences of mRNAs, and the complete amino acid sequences of over 100,000 proteins have been established. Each consists of a unique sequence of amino acids, determined by the order of nucleotides in a gene (see Chapter 4).

The amino acid sequence of a protein is only the first element of its structure. Rather than being extended chains of amino acids, proteins adopt distinct three-dimensional conformations that are critical to their function. These three-dimensional conformations of proteins are the result of interactions between their constituent amino acids, so the shapes of proteins are determined by their amino acid sequences. This was first demonstrated by

KEY EXPERIMENT

determined by its ability to degrade RNA to nucleotides, providing a ready assay for function of the native protein. This enzymatic activity was completely lost when ribonuclease was subjected to treatments that both disrupted the noncovalent bonds (e.g., hydrogen bonds) and cleaved the disulfide bonds by reducing them to sulfhydryl (SH) groups. The denatured protein thus appeared to be in a random inactive conformation.

Sela, White, and Anfinsen made the crucial observation that enzymatic activity reappears if the denatured protein is incubated under conditions that allow the polypeptide chain to refold and disulfide bonds to re-form. In these experiments the denaturing agents were removed, and the inactive enzyme was incubated at room temperature in a physiological buffer in the presence of O_2. This procedure led to oxidation of the sulfhydryl groups and the re-formation of disulfide bonds. During this process, the enzyme regained its catalytic activity, indicating that it had refolded to its native conformation. Because no other cellular components were present, all the information required for proper

protein folding appeared to be contained in the primary amino acid sequence of the polypeptide chain.

The Impact

Further experiments defined conditions under which denatured ribonuclease completely regains its native structure and enzymatic activity, establishing the "thermodynamic hypothesis" of protein folding—that is, the native three-dimensional structure of a protein corresponds to the thermodynamically most stable state under physiological conditions. Thermodynamic stability is governed by interactions of the constituent amino acids, so the three-dimensional conformations of proteins are determined directly by their primary amino acid sequences. Since the order of nucleotides in DNA specifies the amino acid sequence of a polypeptide, it follows that the nucleotide sequence of a gene contains all the information needed to determine the three-dimensional structure of its protein product.

Although Anfinsen's work established the thermodynamic basis of protein folding, understanding the

Christian B. Anfinsen

mechanism of this process remains an active area of research. Protein folding is extremely complex, and we are still unable to deduce the three-dimensional structure of a protein directly from its amino acid sequence. It is also important to note that the spontaneous folding of proteins *in vitro* is much slower than protein folding within the cell, which is assisted by enzymes (see Chapter 8). The protein folding problem thus remains a central challenge in biological chemistry.

FIGURE 2.16 Amino acid sequence of insulin
Insulin consists of two polypeptide chains, one of 21 and the other of 30 amino acids (indicated here by their one-letter codes). The side chains of three pairs of cysteine residues are joined by disulfide bonds, two of which connect the polypeptide chains.

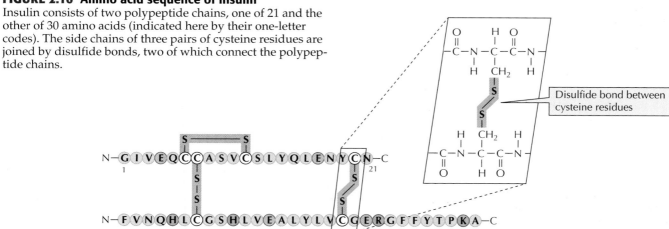

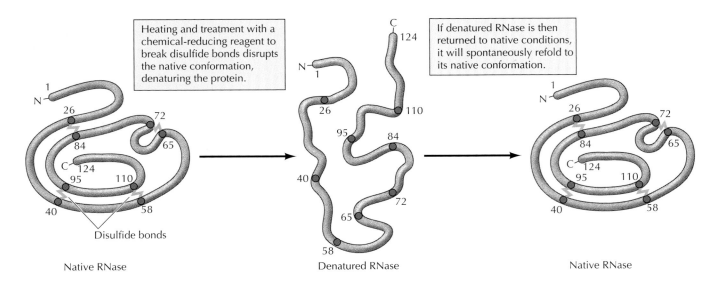

FIGURE 2.17 Protein denaturation and refolding   Ribonuclease (RNase) is a protein of 124 amino acids (indicated by numbers). The protein is normally folded into its native conformation, which contains four disulfide bonds (indicated as paired circles representing the cysteine residues).

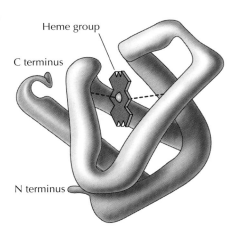

FIGURE 2.18 Three-dimensional structure of myoglobin   Myoglobin is a protein of 153 amino acids that is involved in oxygen transport. The polypeptide chain is folded around a heme group that serves as the oxygen-binding site.

experiments of Christian Anfinsen in which he disrupted the three-dimensional structures of proteins by treatments, such as heating, that break noncovalent bonds—a process called denaturation (Figure 2.17). Following incubation under milder conditions, such denatured proteins often spontaneously returned to their native conformations, indicating that these conformations were directly determined by the amino acid sequence.

The three-dimensional structure of proteins is most frequently analyzed by **X-ray crystallography,** a high-resolution technique that can determine the arrangement of individual atoms within a molecule. A beam of X-rays is directed at crystals of the protein to be analyzed, and the pattern of X-rays that pass through the protein crystal is detected on X-ray film. As the X-rays strike the crystal, they are scattered in characteristic patterns determined by the arrangement of atoms in the molecule. The structure of the molecule can therefore be deduced from the pattern of scattered X-rays (the diffraction pattern).

In 1958 John Kendrew was the first person to determine the three-dimensional structure of a protein, myoglobin—a relatively simple protein of 153 amino acids (Figure 2.18). Since then, the three-dimensional structures of several thousand proteins have been analyzed. Most, like myoglobin, are globular proteins with polypeptide chains folded into compact structures, although some (such as the structural proteins of connective tissues) are long fibrous molecules. Analysis of the three-dimensional structures of these proteins has revealed several basic principles that govern protein folding, although protein structure is so complex that predicting the three-dimensional structure of a protein directly from its amino acid sequence is impossible.

Protein structure is generally described as having four levels. The **primary structure** of a protein is the sequence of amino acids in its polypeptide chain. The **secondary structure** is the regular arrangement of amino acids within localized regions of the polypeptide. Two types of secondary struc-

α helix

Hydrogen bond

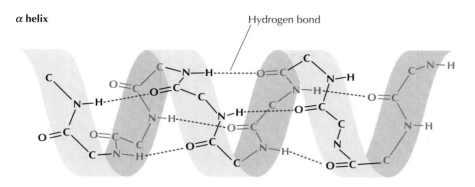

β sheet

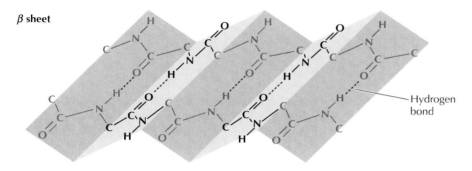

Hydrogen bond

FIGURE 2.19 Secondary structure of proteins The most common types of secondary structure are the α helix and the β sheet. In an α helix, hydrogen bonds form between CO and NH groups of peptide bonds separated by four amino acid residues. In a β sheet, hydrogen bonds connect two parts of a polypeptide chain lying side by side. The amino acid side chains are not shown.

ture, which were first proposed by Linus Pauling and Robert Corey in 1951, are particularly common: the **α helix** and the **β sheet**. Both of these secondary structures are held together by hydrogen bonds between the CO and NH groups of peptide bonds. An α helix is formed when a region of a polypeptide chain coils around itself, with the CO group of one peptide bond forming a hydrogen bond with the NH group of a peptide bond located four residues downstream in the linear polypeptide chain (Figure 2.19). In contrast, a β sheet is formed when two parts of a polypeptide chain lie side by side with hydrogen bonds between them. Such β sheets can be formed between several polypeptide strands, which can be oriented either parallel or antiparallel to each other.

Tertiary structure is the folding of the polypeptide chain as a result of interactions between the side chains of amino acids that lie in different regions of the primary sequence (Figure 2.20). In most proteins, combinations of α helices and β sheets, connected by loop regions of the polypeptide chain, fold into compact globular structures called **domains**,

N terminus

Loop region

α helix

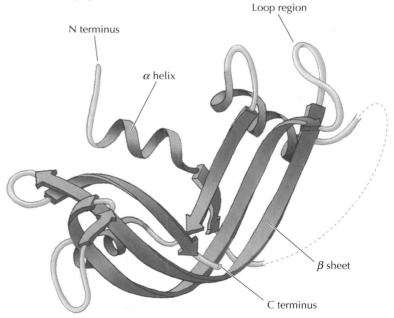

β sheet

C terminus

FIGURE 2.20 Tertiary structure of ribonuclease Regions of α-helix and β-sheet secondary structures, connected by loop regions, are folded into the native conformation of the protein. In this schematic representation of the polypeptide chain as a ribbon model, α helices are represented as spirals and β sheets as wide arrows.

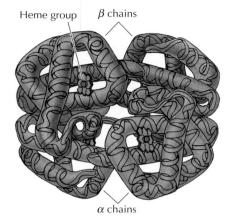

Heme group β chains

α chains

FIGURE 2.21 Quaternary structure of hemoglobin Hemoglobin is composed of four polypeptide chains, each of which is bound to a heme group. The two α chains and the two β chains are identical.

which are the basic units of tertiary structure. Small proteins, such as ribonuclease or myoglobin, contain only a single domain; larger proteins may contain a number of different domains, which are frequently associated with distinct functions.

A critical determinant of tertiary structure is the localization of hydrophobic amino acids in the interior of the protein and of hydrophilic amino acids on the surface, where they interact with water. The interiors of folded proteins thus consist mainly of hydrophobic amino acids arranged in α helices and β sheets; these secondary structures are found in the hydrophobic cores of proteins because hydrogen bonding neutralizes the polar character of the CO and NH groups of the polypeptide backbone. The loop regions connecting the elements of secondary structure are found on the surface of folded proteins, where the polar components of the peptide bonds form hydrogen bonds with water or with the polar side chains of hydrophilic amino acids. Interactions between polar amino acid side chains (hydrogen bonds and ionic bonds) on the protein surface are also important determinants of tertiary structure. In addition, the covalent disulfide bonds between the sulfhydryl groups of cysteine residues stabilize the folded structures of many cell-surface or secreted proteins.

The fourth level of protein structure, **quaternary structure**, consists of the interactions between different polypeptide chains in proteins composed of more than one polypeptide. Hemoglobin, for example, is composed of four polypeptide chains held together by the same types of interactions that maintain tertiary structure (Figure 2.21).

The distinct chemical characteristics of the 20 different amino acids thus lead to considerable variation in the three-dimensional conformations of folded proteins. Consequently, proteins constitute an extremely complex and diverse group of macromolecules, suited to the wide variety of tasks they perform in cell biology.

■ **Proteins exhibit incredible diversity. Some unusual proteins include:**

Cement proteins that are used by barnacles to attach to underwater surfaces. Scientists are trying to develop these proteins for industrial uses as glues or anticorrosive coatings.

Silk from the Golden Orb spider is extremely light and flexible but almost as strong as Kevlar, which is used to make bullet-proof vests.

Cell Membranes

The structure and function of cells are critically dependent on membranes, which not only separate the interior of the cell from its environment but also define the internal compartments of eukaryotic cells, including the nucleus and cytoplasmic organelles. The formation of biological membranes is based on the properties of lipids, and all cell membranes share a common structural organization: bilayers of phospholipids with associated proteins. These membrane proteins are responsible for many specialized functions; some act as receptors that allow the cell to respond to external signals, some are responsible for the selective transport of molecules across the membrane, and others participate in electron transport and oxidative phosphorylation. In addition, membrane proteins control the interactions between cells of multicellular organisms. The common structural organization of membranes thus underlies a variety of biological processes and specialized membrane functions, which will be discussed in detail in later chapters.

KEY EXPERIMENT

The Structure of Cell Membranes

The Fluid Mosaic Model of the Structure of Cell Membranes

S. J. Singer and Garth L. Nicolson

University of California at San Diego and the Salk Institute for Biological Sciences, La Jolla, CA

Science, Volume 175, 1972, pages 720–731

S. J. Singer

Garth L. Nicolson

The Context

Cell membranes play central roles in virtually all aspects of cell biology, serving as boundaries of the subcellular organelles of eukaryotic cells as well as defining the cell itself by separating its contents from the external environment. Understanding the structure and organization of cell membranes was thus a critical step in unraveling the molecular basis of cell behavior. In the 1960s, it was known that cell membranes were composed of proteins and lipids, but it was not clear how these molecules were organized. Moreover, there was a large amount of variation in the protein and lipid compositions of different cell membranes, leading some investigators to believe that different membranes might not share a common structural organization.

In contrast, Jonathan Singer and Garth Nicolson approached the problem of understanding membrane structure with the assumption that the same general principles would describe the organization of lipids and proteins in all cell membranes. They applied the principles of thermodynamics as well as integrating a variety of experimental results to develop the fluid mosaic model—a model that has stood the test of time and provided the structural basis for understanding the diverse functions of membranes in cell biology.

The Experiments

Considerations of thermodynamics were basic to Singer and Nicolson's development of the fluid mosaic model. In particular, they reasoned that the overall structure of membranes would be determined by a combination of hydrophobic and hydrophilic interactions. These interactions were recognized as determinants of the organization of phospholipid bilayers, and Singer and Nicolson assumed that similar interactions would govern the organization of membrane proteins. They therefore proposed that the nonpolar amino acid residues of membrane proteins would be sequestered within the membrane, like the fatty acid chains of the phospholipids, whereas polar groups of proteins would be exposed to the aqueous environment. These considerations argued strongly against an earlier model of membrane structure in which it was proposed that membranes had a three-layer structure consisting of an inner lipid bilayer sandwiched between two protein layers.

Singer and Nicolson instead proposed a mosaic structure, in which globular integral membrane proteins were inserted into a phospholipid bilayer (see figure). The proteins, like the phospholipids, were postulated to be amphipathic, with nonpolar regions inserted into the lipid bilayer and polar regions exposed to the aqueous environment. The lipid bilayer was thought to constitute the structural matrix of the membrane in which the proteins were embedded.

(*Continued on next page*)

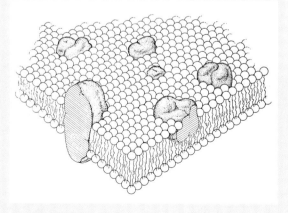

The lipid-globular protein mosaic model. Globular proteins are distributed in a matrix of phospholipids.

Several lines of experimental evidence were cited in support of the fluid mosaic model. First, electron microscopy of membranes using the freeze fracture technique provided direct evidence suggesting that proteins were embedded in the lipid bilayer (see Figure 1.36). Also as predicted by the fluid mosaic model, proteins were found to be randomly distributed in the plane of the membrane when detected by staining the surface of cells. And finally, as discussed in Chapter 13, experiments of Larry Frye and Michael Edidin demonstrated

that proteins could move laterally through the membrane. Multiple experimental approaches thus indicated that membrane proteins were inserted into and able to move laterally through the lipid bilayer, providing strong support for the fluid mosaic model of membrane structure.

The Impact

Singer and Nicolson's fluid mosaic model has been abundantly supported by further experimentation, and continues to represent our basic understanding of the structure of cell

membranes. Given the diverse roles of membranes in cell biology, the fluid mosaic model is thus fundamental to understanding virtually all aspects of cell behavior, including phenomena as diverse as energy metabolism, hormone action, and the transmission of information between nerve cells at a synapse. Initially based on thermodynamic considerations of the properties of lipids and proteins, Singer and Nicolson developed a model of membrane structure that has impacted all aspects of cell biology.

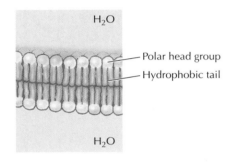

FIGURE 2.22 A phospholipid bilayer Phospholipids spontaneously form stable bilayers, with their polar head groups exposed to water and their hydrophobic tails buried in the interior of the membrane.

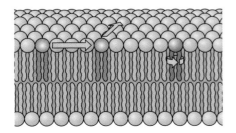

FIGURE 2.23 Mobility of phospholipids in a membrane Individual phospholipids can rotate and move laterally within a bilayer.

Membrane Lipids

The fundamental building blocks of all cell membranes are phospholipids, which are amphipathic molecules, consisting of two hydrophobic fatty acid chains linked to a phosphate-containing hydrophilic head group (see Figure 2.7). Because their fatty acid tails are poorly soluble in water, phospholipids spontaneously form bilayers in aqueous solutions, with the hydrophobic tails buried in the interior of the membrane and the polar head groups exposed on both sides, in contact with water (Figure 2.22). Such **phospholipid bilayers** form a stable barrier between two aqueous compartments and represent the basic structure of all biological membranes.

Lipids constitute approximately 50% of the mass of most cell membranes, although this proportion varies depending on the type of membrane. Plasma membranes, for example, are approximately 50% lipid and 50% protein. The inner membrane of mitochondria, on the other hand, contains an unusually high fraction (about 75%) of protein, reflecting the abundance of protein complexes involved in electron transport and oxidative phosphorylation. The lipid composition of different cell membranes also varies (Table 2.1). The plasma membrane of *E. coli* consists predominantly of phosphatidylethanolamine, which constitutes 80% of total membrane lipid. Mammalian plasma membranes are more complex, containing four major phospholipids—phosphatidylcholine, phosphatidylserine, phosphatidylethanolamine, and sphingomyelin—which together constitute 50 to 60% of total membrane lipid. In addition to the phospholipids, the plasma membranes of animal cells contain glycolipids and cholesterol, which generally correspond to about 40% of the total membrane lipid molecules.

An important property of lipid bilayers is that they behave as two-dimensional fluids in which individual molecules (both lipids and proteins) are free to rotate and move in lateral directions (Figure 2.23). Such fluidity is a critical property of membranes and is determined by both temperature and lipid composition. For example, the interactions between shorter fatty acid chains are weaker than those between longer chains, so membranes containing shorter fatty acid chains are less rigid and remain fluid at lower temperatures. Lipids containing unsaturated fatty acids similarly increase

TABLE 2.1 Lipid Composition of Cell Membranes[a]

Lipid	Plasma membrane		Rough endoplasmic reticulum	Outer mitochondrial membranes
	E. coli	Erythrocyte		
Phosphatidylcholine	0	17	55	50
Phosphatidylserine	0	6	3	2
Phosphatidyl-ethanolamine	80	16	16	23
Sphingomyelin	0	17	3	5
Glycolipids	0	2	0	0
Cholesterol	0	45	6	<5

Source: Data from P. L. Yeagle, 1993. *The Membranes of Cells*, 2nd ed. San Diego, CA: Academic Press.
[a] Membrane compositions are indicated as the mole percentages of major lipid constituents.

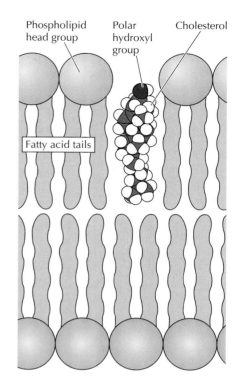

FIGURE 2.24 Insertion of cholesterol in a membrane Cholesterol inserts into the membrane with its polar hydroxyl group close to the polar head groups of the phospholipids.

■ **How do plants sense cold?** Evidence suggests that plants detect drops in temperature in part by sensing changes in the fluidity of their membranes.

membrane fluidity because the presence of double bonds introduces kinks in the fatty acid chains, making them more difficult to pack together.

Because of its hydrocarbon ring structure (see Figure 2.9), cholesterol plays a distinct role in determining membrane fluidity. Cholesterol molecules insert into the bilayer with their polar hydroxyl groups close to the hydrophilic head groups of the phospholipids (Figure 2.24). The rigid hydrocarbon rings of cholesterol therefore interact with the regions of the fatty acid chains that are adjacent to the phospholipid head groups. This interaction decreases the mobility of the outer portions of the fatty acid chains, making this part of the membrane more rigid. On the other hand, insertion of cholesterol interferes with interactions between fatty acid chains, thereby maintaining membrane fluidity at lower temperatures.

Membrane Proteins

Proteins are the other major constituent of cell membranes, constituting 25 to 75% of the mass of the various membranes of the cell. The current model of membrane structure, proposed by Jonathan Singer and Garth Nicolson in 1972, views the membrane as a **fluid mosaic** in which proteins are inserted into a lipid bilayer (Figure 2.25). While phospholipids provide the basic structural organization of membranes, membrane proteins carry out the specific functions of the different membranes of the cell. These proteins are divided into two general classes, based on the nature of their association with the membrane. **Integral membrane proteins** are embedded directly within the lipid bilayer. **Peripheral membrane proteins** are not inserted into the lipid bilayer but are associated with the membrane indirectly, generally by interactions with integral membrane proteins.

Most integral membrane proteins (called **transmembrane proteins**) span the lipid bilayer, with portions exposed on both sides of the membrane. The membrane-spanning portions of these proteins are usually α-helical regions of 20 to 25 nonpolar amino acids. The hydrophobic side chains of these amino acids interact with the fatty acid chains of membrane lipids, and the formation of an α helix neutralizes the polar character of the peptide bonds, as discussed earlier in this chapter with respect to protein folding. The only other protein structure known to span lipid bilayers is the β-**barrel**, formed by the folding of β sheets into a barrel-like structure, which is found in some transmembrane proteins of bacteria, chloroplasts, and mitochondria (Figure 2.26). Like the phospholipids, transmembrane proteins are amphipathic molecules,

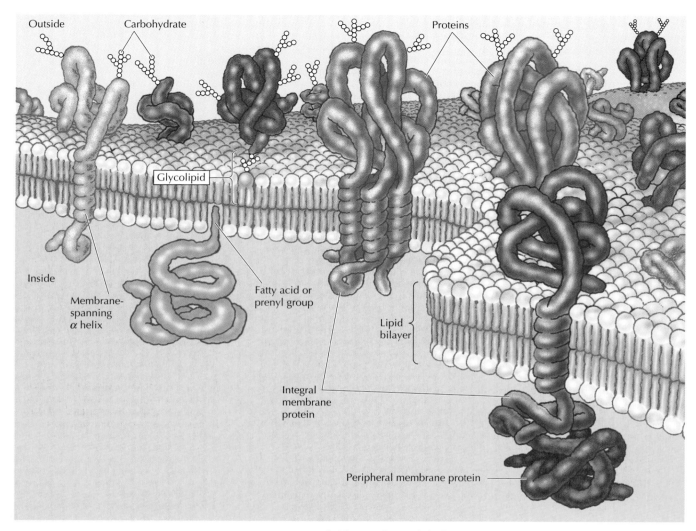

Outside · Carbohydrate · Proteins · Glycolipid · Inside · Membrane-spanning α helix · Fatty acid or prenyl group · Lipid bilayer · Integral membrane protein · Peripheral membrane protein

FIGURE 2.25 Fluid mosaic model of membrane structure   Biological membranes consist of proteins inserted into a lipid bilayer. Integral membrane proteins are embedded in the membrane, usually via α-helical regions of 20 to 25 hydrophobic amino acids. Some transmembrane proteins span the membrane only once; others have multiple membrane-spanning regions. In addition, some proteins are anchored in the membrane by lipids that are covalently attached to the polypeptide chain. These proteins can be anchored to the extracellular face of the plasma membrane by glycolipids and to the cytosolic face by fatty acids or prenyl groups (see Chapter 8 for structures). Peripheral membrane proteins are not inserted in the membrane but are attached via interactions with integral membrane proteins.

with their hydrophilic portions exposed to the aqueous environment on both sides of the membrane. Some transmembrane proteins span the membrane only once; others have multiple membrane-spanning regions. Most transmembrane proteins of eukaryotic plasma membranes have been modified by the addition of carbohydrates, which are exposed on the surface of the cell and may participate in cell-cell interactions.

Proteins can also be anchored in membranes by lipids that are covalently attached to the polypeptide chain (see Chapter 8). Distinct lipid modifications anchor proteins to the cytosolic and extracellular faces of the plasma membrane. Proteins can be anchored to the cytosolic face of the membrane either by the addition of a 14-carbon fatty acid (myristic acid) to their amino terminus or by the addition of either a 16-carbon fatty acid (palmitic acid) or 15- or 20-carbon prenyl groups to the side chains of cysteine residues. Alternatively, proteins are anchored to the extracellular face of the plasma membrane by the addition of glycolipids to their carboxy terminus.

Transport across Cell Membranes

The selective permeability of biological membranes to small molecules allows the cell to control and maintain its internal composition. Only small uncharged molecules can diffuse freely through phospholipid bilayers (Figure 2.27). Small nonpolar molecules, such as O_2 and CO_2, are soluble in the lipid bilayer and therefore can readily cross cell membranes. Small uncharged polar molecules, such as H_2O, also can diffuse through membranes, but larger uncharged polar molecules, such as glucose, cannot. Charged molecules, such as ions, are unable to diffuse through a phospholipid bilayer regardless of size; even H^+ ions cannot cross a lipid bilayer by free diffusion.

Although ions and most polar molecules cannot diffuse across a lipid bilayer, many such molecules (such as glucose) are able to cross cell membranes. These molecules pass across membranes via the action of specific transmembrane proteins, which act as transporters. Such transport proteins determine the selective permeability of cell membranes and thus play a critical role in membrane function. They contain multiple membrane-spanning regions that form a passage through the lipid bilayer, allowing polar or charged molecules to cross the membrane through a protein pore without interacting with the hydrophobic fatty acid chains of the membrane phospholipids.

As discussed in detail in Chapter 13, there are two general classes of membrane transport proteins (Figure 2.28). **Channel proteins** form open pores through the membrane, allowing the free passage of any molecule of the appropriate size. Ion channels, for example, allow the passage of inor-

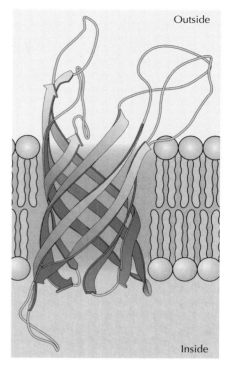

FIGURE 2.26 Structure of a β-barrel Some transmembrane proteins span the phospholipid bilayer as β-sheets folded into a barrel-like structure.

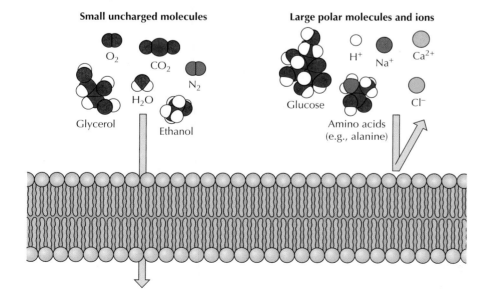

FIGURE 2.27 Permeability of phospholipid bilayers Small uncharged molecules can diffuse freely through a phospholipid bilayer. However, the bilayer is impermeable to larger polar molecules (such as glucose and amino acids) and to ions.

(A)

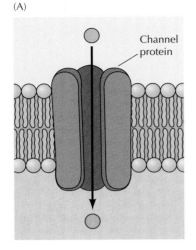

Channel
protein

(B)

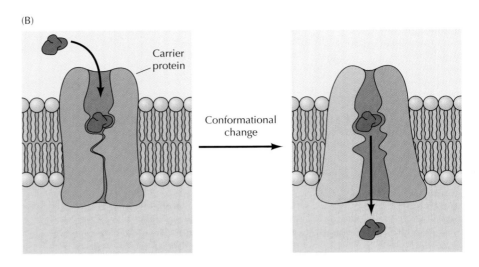

Carrier
protein

Conformational
change

FIGURE 2.28 Channel and carrier proteins   (A) Channel proteins form open pores through which molecules of the appropriate size (e.g., ions) can cross the membrane. (B) Carrier proteins selectively bind the small molecule to be transported and then undergo a conformational change to release the molecule on the other side of the membrane.

2.2 **WEBSITE ANIMATION**

Passive transport Simple diffusion and facilitated diffusion are types of passive transport by which certain molecules cross biological membranes.

ganic ions such as Na^+, K^+, Ca^{2+}, and Cl^- across the plasma membrane. Once open, channel proteins form small pores through which ions of the appropriate size and charge can cross the membrane by free diffusion. The pores formed by these channel proteins are not permanently open; rather, they can be selectively opened and closed in response to extracellular signals, allowing the cell to control the movement of ions across the membrane. Such regulated ion channels have been particularly well studied in nerve and muscle cells, where they mediate the transmission of electrochemical signals.

In contrast to channel proteins, **carrier proteins** selectively bind and transport specific small molecules, such as glucose. Rather than forming open channels, carrier proteins act like enzymes to facilitate the passage of specific molecules across membranes. In particular, carrier proteins bind specific molecules and then undergo conformational changes that open channels through which the molecule to be transported can pass across the membrane and be released on the other side.

As described so far, molecules transported by either channel or carrier proteins cross membranes in the energetically favorable direction, as determined by concentration and electrochemical gradients—a process known as **passive transport**. However, carrier proteins also provide a mechanism through which the energy changes associated with transporting molecules across a membrane can be coupled to the use or production of other forms of metabolic energy, just as enzymatic reactions can be coupled to the hydrolysis or synthesis of ATP. For example, molecules can be transported in an energetically unfavorable direction across a membrane (e.g., against a concentration gradient) if their transport in that direction is coupled to ATP hydrolysis as a source of energy—a process called **active transport** (Figure 2.29). Membrane proteins can thus use the free energy stored as ATP to control the internal composition of the cell.

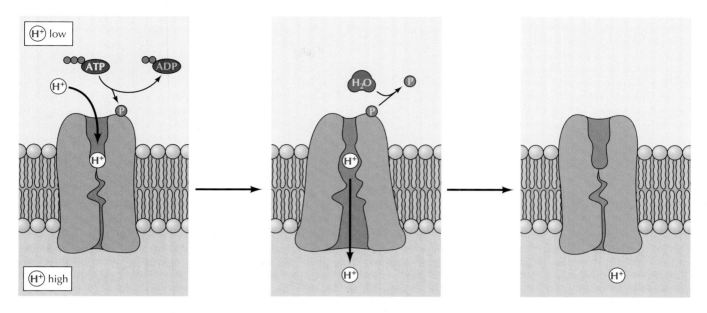

FIGURE 2.29 Model of active transport   Energy derived from the hydrolysis of ATP is used to transport H$^+$ against the electrochemical gradient (from low to high H$^+$ concentration). Binding of H$^+$ is accompanied by phosphorylation of the carrier protein, which induces a conformational change that drives H$^+$ transport against the electrochemical gradient. Release of H$^+$ and hydrolysis of the bound phosphate group then restores the carrier to its original conformation.

Proteomics: Large-Scale Analysis of Cell Proteins

The last few years have seen a number of major changes in the way scientists approach cell and molecular biology, with large-scale experimental approaches being applied to understand the complexities of biological systems. Most dramatic have been the genome sequencing projects, which have yielded the complete sequence of the human genome, as well as the genome sequences of a number of model organisms. As will be discussed in later chapters, this systematic analysis of cell genomes (**genomics**) is making a major impact on our understanding of cell function.

However, the analysis of cell genomes is only the first step in understanding the workings of a cell. Since proteins are directly responsible for carrying out almost all cell activities, it is necessary to understand not only what proteins can be encoded by a cell's genome but also what proteins are actually expressed and how they function within the cell. A complete understanding of cell function therefore requires not only the sequence of its genome, but also a systematic analysis of its protein complement. This large-scale analysis of cell proteins (**proteomics**) has the goal of identifying and quantifying all of the proteins expressed in a given cell (the **proteome**), as well as establishing the localization of these proteins to different subcellular organelles and elucidating the networks of interactions between proteins that govern cell activities.

Identification of Cell Proteins

The number of distinct species of proteins in eukaryotic cells is typically far greater than the number of genes. This arises because many genes can be expressed to yield several distinct mRNAs, which encode different polypeptides as a result of alternative splicing (discussed in Chapter 5). In addition, proteins can be modified in a variety of different ways, including the addition of phosphate groups, carbohydrates, and lipid molecules. The human genome, for example, contains approximately 20,000 to 25,000 different genes, and the number of genes expressed in any given cell is thought to be around 10,000. However, because of alternative splicing and protein modifications, it is estimated that these genes give rise to more than 100,000 different proteins. In addition, these proteins can be expressed at a wide range of different levels. Characterization of the complete protein complement of a cell, the goal of proteomics, thus represents a considerable challenge. Although major progress has been made in the last several years, substantial technological hurdles remain to be overcome before a complete characterization of cell proteomes can be achieved.

The first technology developed for the large scale separation of cell proteins was **two-dimensional gel electrophoresis** (Figure 2.30), which remains a well-established strategy for proteomic experiments. A mixture of cell proteins is first subjected to electrophoresis in a tube with a pH gradient running from end to end. Each protein migrates according to charge until it reaches a pH at which the charge of the protein is neutralized, as determined by the protein's content of acidic and basic amino acids. The proteins are then subjected to electrophoresis in a second dimension under conditions where they separate according to mass, with lower molecular weight proteins moving more rapidly through a gel. This approach is capable of resolving several thousand protein species from a cell extract. However, this is much less than the total number of proteins expressed in mammalian cells, and it is important to note that two-dimensional gel electrophoresis is biased towards the detection of the most abundant cell proteins.

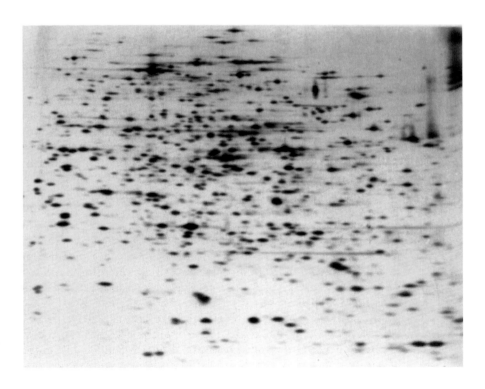

FIGURE 2.30 Two-dimensional gel electrophoresis A mixture of cell proteins is first subjected to electrophoresis in a pH gradient so that they separate according to charge (horizontal axis). The proteins then undergo another electrophoresis in a second dimension under conditions where they separate by size (vertical axis). The gel is stained to reveal spots corresponding to distinct protein species. The example shown is a gel of proteins from *E. coli*. (From P. H. O'Farrell. 1975. *J. Biol. Chem.* 250: 4007; courtesy of Patrick H. O'Farrell.)

FIGURE 2.31 Identification of proteins by mass spectrometry
A protein sample (for example, a spot excised from a two-dimensional gel) is digested with a protease that cleaves the protein into small peptides. The peptides are then ionized and analyzed in a mass spectrometer, which determines the mass-to-charge ratio of each peptide. The results are displayed as a mass spectrum, which is compared to a database of theoretical mass spectra of all known proteins for protein identification.

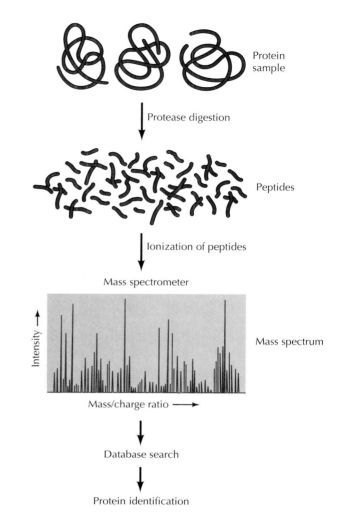

Proteins detected as spots on two-dimensional gels can be excised from the gels and identified by **mass spectrometry**, which was developed in the 1990s as a powerful tool for protein identification (Figure 2.31). The protein spot excised from the gel is digested with a protease to cleave the protein into small fragments (peptides) in the range of approximately 20 amino acid residues long. A commonly used protease is trypsin, which cleaves proteins at the carboxy-terminal side of lysine and arginine residues. The peptides are then ionized by irradiation with a laser or by passage through a field of high electrical potential and introduced into a mass spectrometer, which measures the mass-to-charge ratio of each peptide. This generates a mass spectrum in which individual tryptic peptides are indicated by a peak corresponding to their mass-to-charge ratio. Computer algorithms can then be used to compare the experimentally determined mass spectrum with a database of theoretical mass spectra representing tryptic peptides of all known proteins, allowing identification of the unknown protein. Alternatively, more detailed sequence information can be obtained by tandem mass spectrometry. In this technique, individual peptides from the initial mass spectrum are automatically selected to enter a "collision cell" in which they are partially degraded by random breakage of peptide bonds. A second mass spectrum of the partial degradation products of each peptide is then determined, and the amino acid sequence of the peptide can be deduced. Protein modifications, such as phosphorylation, can also be identified because they alter the mass of the modified amino acid.

Although powerful, the two-dimensional gel/mass spectrometry approach is limited. As noted above, two-dimensional gels favor detection of the most abundant cell proteins, and membrane proteins are characteristically difficult to resolve by this approach. Because of these limitations, it appears that two-dimensional gels are capable of resolving proteins corresponding to only several hundred genes, representing less than 10% of all cell proteins. An alternative "shotgun" approach has therefore been developed to eliminate the initial separation of proteins by two-dimensional gel electrophoresis. In this approach, an unfractionated mixture of cell proteins is digested with a protease (e.g., trypsin), and the complex mixture of peptides is subjected to sequencing by tandem mass spectrometry. The sequences of individual peptides are then used for database searching to identify the proteins present in the starting mixture. Additional methods have been developed to compare the amounts of proteins in two different samples, allowing a quantitative analysis of protein levels in different types

■ Mass spectrometry has many uses, including testing the blood samples of athletes for the presence of certain performance-enhancing drugs.

of cells or in cells that have been subjected to different treatments. Although several problems with the sensitivity and accuracy of these methods remain to be solved, the analysis of complex mixtures of proteins by shotgun mass spectrometry provides a powerful approach to the systematic analysis of cell proteins.

Global Analysis of Protein Localization

Understanding the function of eukaryotic cells will require not only the identification of the proteins expressed in a given cell type, but also characterization of the locations of those proteins within the cell. As reviewed in Chapter 1, eukaryotic cells contain a nucleus and a variety of subcellular organelles. Systematic analysis of the proteins present in these organelles has become an important goal of proteomic approaches to cell biology.

One approach to global analysis of protein localization is to isolate the organelle of interest by subcellular fractionation, as discussed in Chapter 1. The proteins present in isolated organelles can then be analyzed by mass spectrometry, which is simplified since any given organelle contains fewer proteins than would be present in total cell extracts. Mitochondria, for example, are estimated to contain about 1,000 different proteins—more than 600 of these have been identified by mass spectrometry of isolated mitochondria. Similar analysis of the proteome of a variety of subcellular organelles is currently a major research effort in a number of laboratories and is expected to provide a database of the protein constituents of these compartments of eukaryotic cells.

An alternative approach has recently been applied to a systematic analysis of protein localization in yeast. In these experiments, investigators constructed a collection of yeast strains in which each protein encoded in the yeast genome was expressed as a fusion with green fluorescent protein (GFP), allowing the subcellular localization of the GFP-tagged protein to be determined by fluorescence microscopy (see Figure 1.27). Using this approach, the subcellular localizations of over 4000 proteins was deter-

FIGURE 2.32 Subcellular localization of yeast proteins
The protein-coding regions of over 4000 yeast genes were expressed in a collection of yeast strains as fusions with green fluorescent protein (GFP). The subcellular localization of each fusion protein was then determined by fluorescence microscopy. The micrographs shown illustrate the indicated localization of representative fusion proteins. (After W. -K. Huh, J. V. Falvo, L. C. Gerke, A. S. Carroll, R. W. Howson, J. S. Weissman and E. K. O'Shea, 2003. *Nature* 425: 686. Images courtesy of James Falvo.)

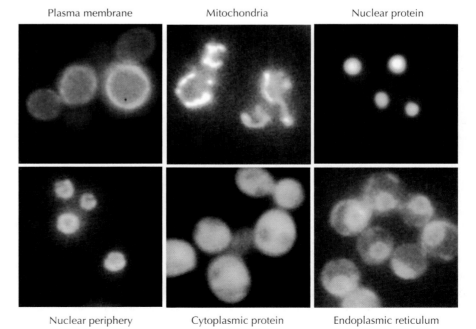

Plasma membrane   Mitochondria   Nuclear protein

Nuclear periphery   Cytoplasmic protein   Endoplasmic reticulum

mined (Figure 2.32). This represents 75% of the total number of yeast proteins, including many proteins whose subcellular localization was previously unknown, so this global analysis of yeast protein localization provides a comprehensive catalog of protein localization in a model eukaryotic cell.

Protein Interactions

Proteins almost never act alone within the cell. Instead, they generally function by interacting with other proteins in protein complexes and networks. Elucidating the interactions between proteins therefore provides important clues as to the function of novel proteins, as well as helping to understand the complex networks of protein interactions that govern cell behavior. Along with global studies of subcellular localization, the systematic analysis of protein complexes and interactions has therefore become an important goal of proteomics.

One approach to analysis of protein complexes is to isolate a protein from cells under gentle conditions, so that it remains associated with the proteins it normally interacts with inside the cell. The isolated protein complexes can then be analyzed by mass spectrometry to identify all of the proteins present in the complex. This approach to analysis of protein interactions has been used in large-scale investigations of protein complexes in yeast, and has identified numerous interactions between proteins involved in processes such as cell signaling or gene expression.

Alternative approaches to systematic analyses of protein complexes include screens for protein interactions *in vitro* as well as genetic screens that detect interactions between pairs of proteins that are introduced into yeast cells. The latter approach (called the yeast two-hybrid method) was initially used for large scale studies of interactions between yeast proteins, but has more recently been applied to systematic screens of interactions between proteins of higher eukaryotes, such as *Drosophila* and *C. elegans*. The screens have identified thousands of protein-protein interactions, which can be presented as maps that depict an extensive network of interacting proteins within the cell (Figure 2.33). Continuing elucidation of these protein networks is expected to extend our understanding of the complexities of cell regulation, as well as illuminating the functions of many so-far unidentified proteins.

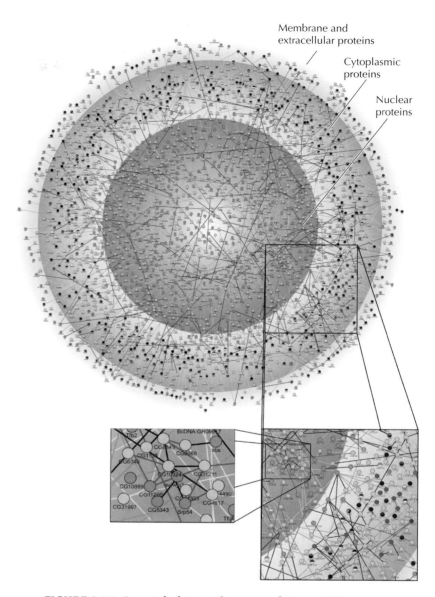

FIGURE 2.33 A protein interaction map of *Drosophila melanogaster*   Interactions among 2346 proteins are depicted, with each protein represented as a circle placed according to its subcellular localization. (From L. Glot et al., 2003. *Science* 302: 1727).

COMPANION WEBSITE

Visit the website that accompanies **The Cell** (www.sinauer.com/cooper) for animations, videos, quizzes, problems, and other review material.

KEY TERMS

carbohydrate, monosaccharide, glycosidic bond, oligosaccharide, polysaccharide, glycogen, starch, cellulose

lipid, fatty acid, triacylglycerol, fat, phospholipid, glycerol phospholipid, sphingomyelin, amphipathic, glycolipid, cholesterol, steroid hormone

deoxyribonucleic acid (DNA), ribonucleic acid (RNA), messenger RNA (mRNA), ribosomal RNA, transfer RNA, nucleotide, purine, pyrimidine, adenine, guanine, cytosine, thymine, uracil, 2′-deoxyribose, ribose, nucleoside, phosphodiester bond, oligonucleotide, polynucleotide

protein, amino acid, peptide bond, polypeptide, X-ray crystallography, primary structure, secondary structure, α helix, β sheet, tertiary structure, domain, quaternary structure

phospholipid bilayer

fluid mosaic, integral membrane protein, peripheral membrane protein, transmembrane protein, β-barrel

channel protein, carrier protein, passive transport, active transport

genomics, proteomics, proteome

SUMMARY

THE MOLECULES OF CELLS

Carbohydrates: Carbohydrates include simple sugars and polysaccharides. Polysaccharides serve as storage forms of sugars, structural components of cells, and markers for cell recognition processes.

Lipids: Lipids are the principal components of cell membranes, and they serve as energy storage and signaling molecules. Phospholipids consist of two hydrophobic fatty acid chains linked to a hydrophilic phosphate-containing head group.

Nucleic Acids: Nucleic acids are the principal informational molecules of the cell. Both DNA and RNA are polymers of purine and pyrimidine nucleotides. Hydrogen bonding between complementary base pairs allows nucleic acids to direct their self-replication.

Proteins: Proteins are polymers of 20 different amino acids, each of which has a distinct side chain with specific chemical properties. Each protein has a unique amino acid sequence, which determines its three-dimensional structure. In most proteins, combinations of α helices and β sheets fold into globular domains with hydrophobic amino acids in the interior and hydrophilic amino acids on the surface.

CELL MEMBRANES

Membrane Lipids: The basic structure of all cell membranes is a phospholipid bilayer. Membranes of animal cells also contain glycolipids and cholesterol.

Membrane Proteins: Proteins can either be inserted into the lipid bilayer or associated with the membrane indirectly, by protein-protein interactions. Some proteins span the lipid bilayer; others are anchored to one side of the membrane.

Transport across Cell Membranes: Lipid bilayers are permeable only to small uncharged molecules. Ions and most polar molecules are transported across cell membranes by specific transport proteins, the action of which can be coupled to the hydrolysis or synthesis of ATP.

PROTEOMICS: LARGE-SCALE ANALYSIS OF CELL PROTEINS

Identification of Cell Proteins: Characterization of the complete protein complement of cells is a major goal of proteomics. One well-established method for proteomic studies is two-dimensional gel electrophoresis, which can be used to separate thousands of cell proteins. Mass spectrometry provides a powerful tool for protein identification, which can

SUMMARY

be used to analyze protein samples separated by two-dimensional gel electrophoresis or to identify proteins in unfractionated cell extracts.

Global Analysis of Protein Localization: Isolated subcellular organelles can be analyzed by mass spectrometry to determine their protein constituents. Alternatively, a large fraction of yeast proteins have been tagged with green fluorescent protein for global studies of protein localization.

Protein Interactions: A variety of large-scale approaches are being applied to systematically identify protein-protein interactions and complexes with the goal of elucidating the complex networks of protein interactions that regulate cell behavior.

KEY TERMS

two-dimensional gel electrophoresis, mass spectrometry

Questions

1. What characteristics of water contribute to its function as the most abundant molecule in cells?

2. Although glycogen and cellulose are both composed of glucose monomers, they are chemically distinct. In what way?

3. What are the major functions of fats and phospholipids in cells?

4. In addition to serving as the building blocks for nucleic acids, what other important functions do nucleotides have in cells?

5. If you removed all the bases from an mRNA molecule, could it still function as an information-carrying molecule?

6. What was the experimental evidence that demonstrated that the information necessary for the folding of proteins is contained in the sequence of their amino acids?

7. How does the side chain of cysteine help in the folding of certain proteins?

8. What are the biological roles of cholesterol?

9. Why are the membrane-spanning regions of transmembrane proteins frequently α helical? What other protein structure is capable of spanning membranes?

10. Which of the following amino acids would most likely be found in a membrane-spanning α helix?

a. Lysine
b. Glutamine
c. Aspartic acid
d. Alanine
e. Arginine

11. What properties of proteins are exploited for the separation of individual proteins from a complex mixture by two-dimensional electrophoresis?

12. List and briefly explain the large-scale proteomic approaches used to determine the subcellular location of proteins.

References and Further Reading

The Molecules of Cells

Anfinsen, C. B. 1973. Principles that govern the folding of protein chains. *Science* 181: 223–230. [P]

Berg, J. M., J. L. Tymoczko and L. Stryer. 2002. *Biochemistry.* 5th ed. New York: W. H. Freeman.

Branden, C. and J. Tooze. 1999. *Introduction to Protein Structure.* 2nd ed. New York: Garland.

Kendrew, J. C. 1961. The three-dimensional structure of a protein molecule. *Sci. Am.* 205(6): 96–111. [R]

Mathews, C. K., K. E. van Holde and K. G. Ahern. 2000. *Biochemistry.* 3rd ed. Redwood City, CA: Benjamin Cummings.

Nelson, D. L. and M. M. Cox. 2005. *Lehninger Principles of Biochemistry.* 4th ed. New York: W. H. Freeman.

Pauling, L., R. B. Corey, and H. R. Branson. 1951. The structure of proteins: Two hydrogen bonded configurations of the polypeptide chain. *Proc. Natl. Acad. Sci. USA* 37: 205–211. [P]

Petsko, G. A. and D. Ringe. 2003. *Protein Structure and Function.* Sunderland, MA: Sinauer.

Sanger, F. 1988. Sequences, sequences, and sequences. *Ann. Rev. Biochem.* 57: 1–28. [R]

Tanford, C. 1978. The hydrophobic effect and the organization of living matter. *Science* 200: 1012–1018. [R]

Cell Membranes

Bretscher, M. 1985. The molecules of the cell membrane. *Sci. Am.* 253(4): 100–108. [R]

Dowham, W. 1997. Molecular basis for membrane phospholipid diversity: Why are there so many lipids? *Ann. Rev. Biochem.* 66: 199–212. [R]

Englund, P. T. 1993. The structure and biosynthesis of glycosyl phosphatidylinositol protein anchors. *Ann. Rev. Biochem.* 62: 121–138. [R]

Farazi, T. A., G. Waksman and J. I. Gordon. 2001. The biology and enzymology of protein N-myristoylation. *Ann. Rev. Biochem.* 276: 39501–39504. [R]

Petty, H. R. 1993. *Molecular Biology of Membranes: Structure and Function.* New York: Plenum Press.

Singer, S. J. 1990. The structure and insertion of integral proteins in membranes. *Ann. Rev. Cell Biol.* 6: 247–296. [R]

Singer, S. J. and G. L. Nicolson. 1972. The fluid mosaic model of the structure of cell membranes. *Science* 175: 720–731. [P]

Tamm, L. K., A. Arora and J. H. Kleinschmidt. 2001. Structure and assembly of β-barrel membrane proteins. *J. Biol. Chem.* 276: 32399–32402. [R]

Towler, D. A., J. I. Gordon, S. P. Adams and L. Glaser. 1988. The biology and enzymology of eukaryotic protein acylation. *Ann. Rev. Biochem.* 57: 69–99. [R]

White, S. H., A. S. Ladokhin, S. Jayasinghe and K. Hristova. 2001. How membranes shape protein structure. *J. Biol. Chem.* 276: 32395–32398. [R]

Yeagle, P. L. 1993. *The Membranes of Cells.* 2nd ed. San Diego, CA: Academic Press.

Zhang, F. L. and P. J. Casey. 1996. Protein prenylation: Molecular mechanisms and functional consequences. *Ann. Rev. Biochem.* 65: 241–269. [R]

Proteomics: Large-Scale Analysis of Cell Proteins

Brunet, S., P. Thibault, E. Gagnon, P. Kearney, J. J. M. Bergeron and M. Desjardins. 2003. Organelle proteomics: Looking at less to see more. *Trends Cell Biol.* 13: 629–638. [R]

De Hoog, C. L. and M. Mann. 2004. Proteomics. *Ann. Rev. Genomics Hum. Genet.* 5: 267–293. [R]

Giot, L. and 48 others. 2003. A protein interaction map of *Drosophila melanogaster. Science* 302: 1727–1736. [P]

Gavin, A.-C. and 37 others. 2002. Functional organization of the yeast proteome by systematic analysis of protein complexes. *Nature* 415: 141–147. [P]

Ho, Y. and 45 others. 2002. Systematic identification of protein complexes in *Saccharomyces cerevisiae* by mass spectrometry. *Nature* 415: 180–183. [P]

Huh, W.-K., J. V. Falvo, L. C. Gerke, A. S. Carroll, R. W. Howson, J. S. Weissman and E. K. O'Shea. 2003. Global analysis of protein localization in budding yeast. *Nature* 425: 686–691. [P]

Li, S. and 47 others. 2004. A map of the interactome network of the metazoan *C. elegans. Science* 303: 540–543. [P]

Resing, K. A. and N. G. Ahn. 2005. Proteomics strategies for protein identification. *FEBS Letters* 579: 885–889. [R]

Steen, H. and M. Mann. 2004. The ABC's (and XYZ's) of peptide sequencing. *Nature Rev. Mol. Cell. Biol.* 5: 699–711. [R]

Zhu, H., M. Bilgin and M. Snyder. 2003. Proteomics. *Ann. Rev. Biochem.* 72: 783–812. [R]

3

Cell Metabolism

■ **The Central Role of Enzymes as Biological Catalysts** 73

■ **Metabolic Energy** 81

■ **The Biosynthesis of Cell Constituents** 91

■ **KEY EXPERIMENT:** Antimetabolites and Chemotherapy 98

■ **MOLECULAR MEDICINE:** Phenylketonuria 96

CELLS CARRY OUT A VAST ARRAY OF CHEMICAL REACTIONS, which are responsible both for the breakdown of molecules that serve as food and for the synthesis of cell constituents. This chapter presents an overview of this complex network of chemical reactions, which constitute the metabolism of the cell. It is intended neither to be a comprehensive discussion of biochemistry nor to chart all the metabolic reactions within cells. Rather, the chapter will focus on three major topics: the role of proteins as biological catalysts, the generation and utilization of metabolic energy, and the biosynthesis of major cell constituents. An appreciation of these key aspects of cell metabolism is basic for understanding the diverse aspects of cell structure and function that will be discussed throughout the rest of this text.

The Central Role of Enzymes as Biological Catalysts

A fundamental task of proteins is to act as enzymes—catalysts that increase the rate of virtually all the chemical reactions within cells. Although RNAs are capable of catalyzing some reactions, most biological reactions are catalyzed by proteins. In the absence of enzymatic catalysis, most biochemical reactions are so slow that they would not occur under the mild conditions of temperature and pressure that are compatible with life. Enzymes accelerate the rates of such reactions by well over a million-fold, so reactions that would take years in the absence of catalysis can occur in fractions of seconds if catalyzed by the appropriate enzyme. Cells contain thousands of different enzymes, and their activities determine which of the many possible chemical reactions actually take place within the cell.

■ Enzymes are remarkable catalysts. Some enzymes are capable of increasing the rate of a reaction by as much as 10^{18}-fold as compared to an uncatalyzed reaction, such that a process that would normally take billions of years can be carried out in less than a second.

The Catalytic Activity of Enzymes

Like all other catalysts, **enzymes** are characterized by two fundamental properties. First, they increase the rate of chemical reactions without themselves being consumed or permanently altered by the reaction. Second, they increase reaction rates without altering the chemical equilibrium between reactants and products.

These principles of enzymatic catalysis are illustrated in the following example in which a molecule acted upon by an enzyme (referred to as a **substrate** [S]) is converted to a **product** (P) as a result of the reaction. In the absence of the enzyme, the reaction can be written as follows:

$$S \rightleftharpoons P$$

The chemical equilibrium between S and P is determined by the laws of thermodynamics (as discussed further in the next section of this chapter) and is represented by the ratio of the forward and reverse reaction rates ($S{\rightarrow}P$ and $P{\rightarrow}S$, respectively). In the presence of the appropriate enzyme, the conversion of S to P is accelerated, but the equilibrium between S and P is unaltered. Therefore, the enzyme must accelerate both the forward and reverse reactions equally. The reaction can be written as follows:

$$S \overset{E}{\rightleftharpoons} P$$

Note that the enzyme (E) is not altered by the reaction, so the chemical equilibrium remains unchanged, determined solely by the thermodynamic properties of S and P.

The effect of the enzyme on such a reaction is best illustrated by the energy changes that must occur during the conversion of S to P (**Figure 3.1**). The equilibrium of the reaction is determined by the final energy states of S and P, which are unaffected by enzymatic catalysis. In order for the reaction to proceed, however, the substrate must first be converted to a higher energy state, called the **transition state**. The energy required to reach the transition state (the **activation energy**) constitutes a barrier to the progress of the reaction, limiting the rate of the reaction. Enzymes (and other catalysts) act by reducing the activation energy, thereby increasing the rate of reaction. The increased rate is the same in both the forward and reverse directions, since both must pass through the same transition state.

The catalytic activity of enzymes involves the binding of their substrates to form an enzyme-substrate complex (ES). The substrate binds to a specific region of the enzyme, called the **active site**. While bound to the active site, the substrate is converted into the product of the reaction, which is then released from the enzyme. The enzyme-catalyzed reaction can thus be written as follows:

$$S + E \rightleftharpoons ES \rightleftharpoons E + P$$

Note that E appears unaltered on both sides of the equation, so the equilibrium is unaffected. However, the enzyme provides a surface upon which the reactions converting S to P can occur more readily. This is a result of interactions between the enzyme and substrate that lower the energy of activation and favor formation of the transition state.

Mechanisms of Enzymatic Catalysis

The binding of a substrate to the active site of an enzyme is a very specific interaction. Active sites are clefts or grooves on the surface of an enzyme, usually composed of amino acids from different parts of the polypeptide chain that are brought together in the tertiary structure of the folded protein. Substrates initially bind to the active site by noncovalent interactions, including hydrogen bonds, ionic bonds, and hydrophobic interactions.

3.1 WEBSITE ANIMATION

Catalysts and Activation Energy In a chemical reaction, an enzyme lowers the amount of energy required to reach the transition state, allowing the reaction to proceed at an accelerated rate.

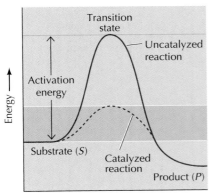

FIGURE 3.1 Energy diagrams for catalyzed and uncatalyzed reactions
The reaction illustrated is the simple conversion of a substrate S to a product P. Because the final energy state of P is lower than that of S, the reaction proceeds from left to right. For the reaction to occur, however, S must first pass through a higher energy transition state. The energy required to reach this transition state (the activation energy) represents a barrier to the progress of the reaction and thereby determines the rate at which the reaction proceeds. In the presence of a catalyst (e.g., an enzyme), the activation energy is lowered and the reaction proceeds at an accelerated rate.

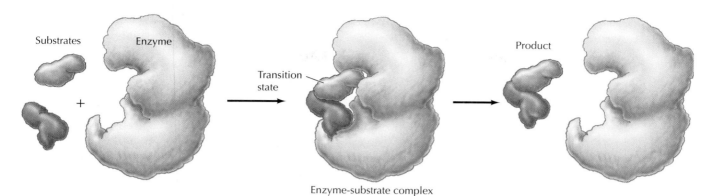

Substrates   Enzyme

+

Transition
state

Enzyme-substrate complex

Product

Once a substrate is bound to the active site of an enzyme, multiple mechanisms can accelerate its conversion to the product of the reaction.

Although the simple example discussed in the previous section involved only a single substrate molecule, most biochemical reactions involve interactions between two or more different substrates. For example, the formation of a peptide bond involves the joining of two amino acids. For such reactions, the binding of two or more substrates to the active site in the proper position and orientation accelerates the reaction (Figure 3.2). The enzyme provides a template upon which the reactants are brought together and properly oriented to favor the formation of the transition state in which they interact.

Enzymes accelerate reactions also by altering the conformation of their substrates to approach that of the transition state. The simplest model of enzyme-substrate interaction is the **lock-and-key model** in which the substrate fits precisely into the active site (Figure 3.3). In many cases, however, the configurations of both the enzyme and substrate are modified by substrate binding—a process called **induced fit**. In such cases the conformation of the substrate is altered so that it more closely resembles that of the transition state. The stress produced by such distortion of the substrate can further facilitate its conversion to the transition state by weakening critical bonds. Moreover, the transition state is stabilized by its tight binding to the enzyme, thereby lowering the required energy of activation.

In addition to bringing multiple substrates together and distorting the conformation of substrates to approach the transition state, many enzymes participate directly in the catalytic process. In such cases, specific amino acid side chains in the active site may react with the substrate and form bonds with reaction intermediates. The acidic and basic amino acids are often involved in these catalytic mechanisms, as illustrated in the following discussion of chymotrypsin as an example of enzymatic catalysis.

FIGURE 3.2 Enzymatic catalysis of a reaction between two substrates The enzyme provides a template upon which the two substrates are brought together in the proper position and orientation to react with each other.

3.2   WEBSITE ANIMATION

Enzyme-Catalyzed Reactions In this example, an enzyme provides a template upon which two substrates are brought together in the proper position and orientation to react with each other.

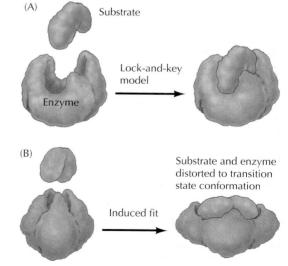

(A)   Substrate

Lock-and-key
model

Enzyme

(B)

Induced fit

Substrate and enzyme
distorted to transition
state conformation

FIGURE 3.3 Models of enzyme-substrate interaction   (A) In the lock-and-key model, the substrate fits precisely into the active site of the enzyme. (B) In the induced-fit model, substrate binding distorts the conformations of both substrate and enzyme. This distortion brings the substrate closer to the conformation of the transition state, thereby accelerating the reaction.

Chymotrypsin is a member of a family of enzymes (serine proteases) that digests proteins by catalyzing the hydrolysis of peptide bonds. The reaction can be written as follows:

$$\text{Protein} + H_2O \rightarrow \text{Peptide}_1 + \text{Peptide}_2$$

The different members of the serine protease family (including chymotrypsin, trypsin, elastase, and thrombin) have distinct substrate specificities; they preferentially cleave peptide bonds adjacent to different amino acids. For example, whereas chymotrypsin digests bonds adjacent to hydrophobic amino acids, such as tryptophan and phenylalanine, trypsin digests bonds next to basic amino acids, such as lysine and arginine. All the serine proteases, however, are similar in structure and use the same mechanism of catalysis. The active sites of these enzymes contain three critical amino acids—serine, histidine, and aspartate—that drive hydrolysis of the peptide bond. Indeed, these enzymes are called serine proteases because of the central role of the serine residue.

Substrates bind to the serine proteases by insertion of the amino acid adjacent to the cleavage site into a pocket at the active site of the enzyme (Figure 3.4). The nature of this pocket determines the substrate specificity of the different members of the serine protease family. For example, the binding pocket of chymotrypsin contains hydrophobic amino acids that interact with the hydrophobic side chains of its preferred substrates. In contrast, the binding pocket of trypsin contains a negatively charged acidic amino acid (aspartate), which is able to form an ionic bond with the lysine or arginine residues of its substrates.

Substrate binding positions the peptide bond to be cleaved adjacent to the active site serine (Figure 3.5). The proton of this serine is then transferred to the active site histidine. The conformation of the active site favors this proton transfer because the histidine interacts with the negatively charged aspartate residue. The serine reacts with the substrate, forming a tetrahedral transition state. The peptide bond is then cleaved, and the C-terminal portion of the substrate is released from the enzyme. However, the N-terminal

■ Proteases play a variety of roles in cells and viruses. Some HIV proteins are synthesized as precursors that need to be cleaved by a viral protease to make infectious viral particles. Inhibition of the protease blocks HIV replication, and drugs designed to inhibit the protease are currently used in the treatment of AIDS.

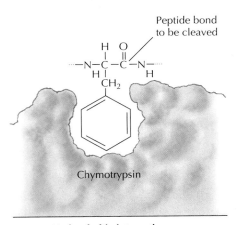
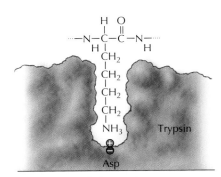

Hydrophobic interaction   **Ionic interaction**

FIGURE 3.4 Substrate binding by serine proteases   The amino acid adjacent to the peptide bond to be cleaved is inserted into a pocket at the active site of the enzyme. In chymotrypsin, the pocket binds hydrophobic amino acids; the binding pocket of trypsin contains a negatively charged aspartate residue that binds basic amino acids via an ionic interaction.

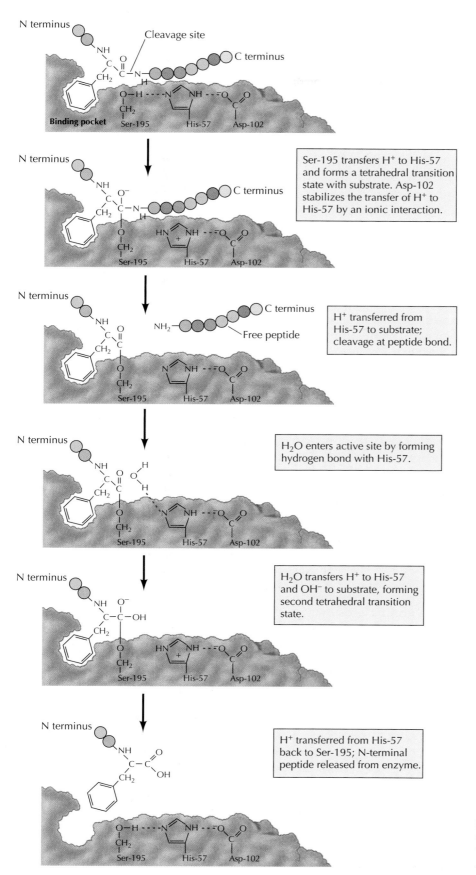

Ser-195 transfers H+ to His-57 and forms a tetrahedral transition state with substrate. Asp-102 stabilizes the transfer of H+ to His-57 by an ionic interaction.

H+ transferred from His-57 to substrate; cleavage at peptide bond.

H₂O enters active site by forming hydrogen bond with His-57.

H₂O transfers H+ to His-57 and OH⁻ to substrate, forming second tetrahedral transition state.

H+ transferred from His-57 back to Ser-195; N-terminal peptide released from enzyme.

FIGURE 3.5 Catalytic mechanism of chymotrypsin   Three amino acids at the active site (Ser-195, His-57, and Asp-102) play critical roles in catalysis.

peptide remains bound to serine. This situation is resolved when a water molecule (the second substrate) enters the active site and reverses the preceding reactions. The proton of the water molecule is transferred to histidine, and its hydroxyl group is transferred to the peptide, forming a second tetrahedral transition state. The proton is then transferred from histidine back to serine, and the peptide is released from the enzyme, completing the reaction.

This example illustrates several features of enzymatic catalysis: the specificity of enzyme-substrate interactions, the positioning of different substrate molecules in the active site, and the involvement of active-site residues in the formation and stabilization of the transition state. Although the thousands of enzymes in cells catalyze many different types of chemical reactions, the same basic principles apply to their operation.

Coenzymes

In addition to binding their substrates, the active sites of many enzymes bind other small molecules that participate in catalysis. **Prosthetic groups** are small molecules bound to proteins in which they play critical functional roles. For example, the oxygen carried by myoglobin and hemoglobin is bound to heme, a prosthetic group of these proteins. In many cases metal ions (such as zinc or iron) are bound to enzymes and play central roles in the catalytic process. In addition, various low-molecular-weight organic molecules participate in specific types of enzymatic reactions. These molecules are called **coenzymes** because they work together with enzymes to enhance reaction rates. In contrast to substrates, coenzymes are not irreversibly altered by the reactions in which they are involved. Rather, they are recycled and can participate in multiple enzymatic reactions.

Coenzymes serve as carriers of several types of chemical groups. A prominent example of a coenzyme is **nicotinamide adenine dinucleotide (NAD⁺)**, which functions as a carrier of electrons in oxidation–reduction reactions (Figure 3.6). NAD^+ can accept a hydrogen ion (H^+) and two electrons (e^-) from one substrate, forming NADH. NADH can then donate these electrons to a second substrate, re-forming NAD^+. Thus, NAD^+ transfers electrons from the first substrate (which becomes oxidized) to the second substrate (which becomes reduced).

Several other coenzymes also act as electron carriers, and still others are involved in the transfer of a variety of additional chemical groups (e.g., carboxyl groups and acyl groups; Table 3.1). The same coenzymes function together with a variety of different enzymes to catalyze the transfer of spe-

TABLE 3.1 Examples of Coenzymes and Vitamins

Coenzyme	Related vitamin	Chemical reaction
NAD⁺, NADP⁺	Niacin	Oxidation–reduction
FAD	Riboflavin (B₂)	Oxidation–reduction
Thiamine pyrophosphate	Thiamine (B₁)	Aldehyde group transfer
Coenzyme A	Pantothenate	Acyl group transfer
Tetrahydrofolate	Folate	Transfer of one-carbon groups
Biotin	Biotin	Carboxylation
Pyridoxal phosphate	Pyridoxal (B₆)	Transamination

(A)

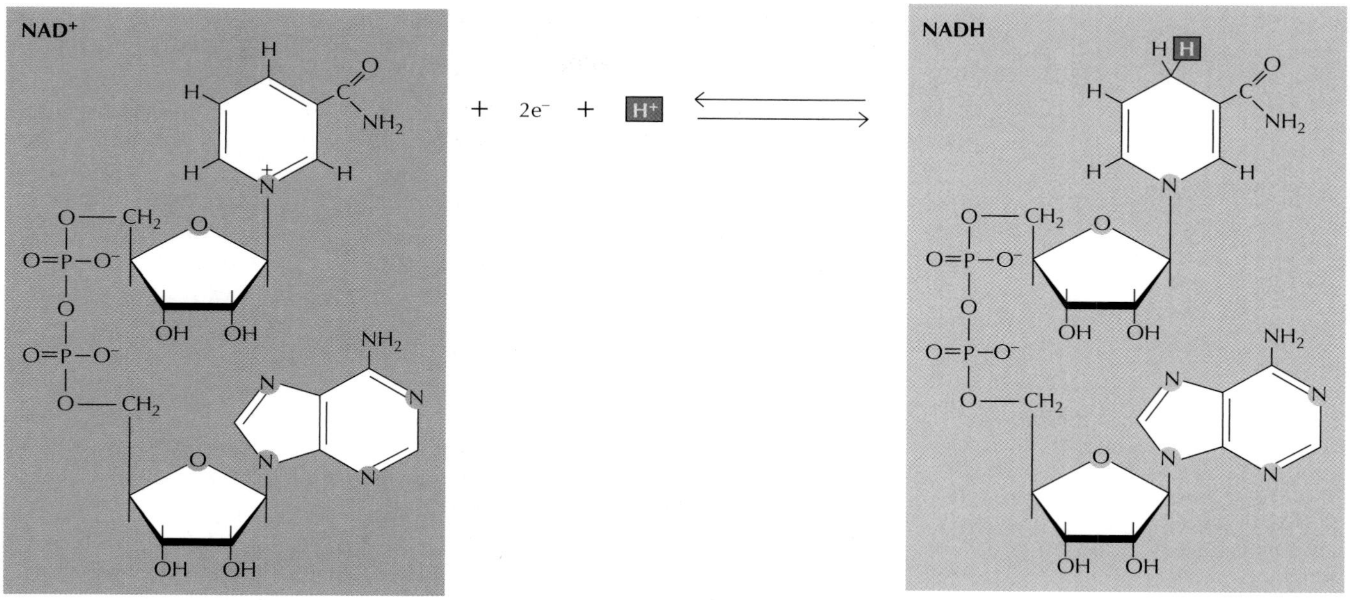

(B)

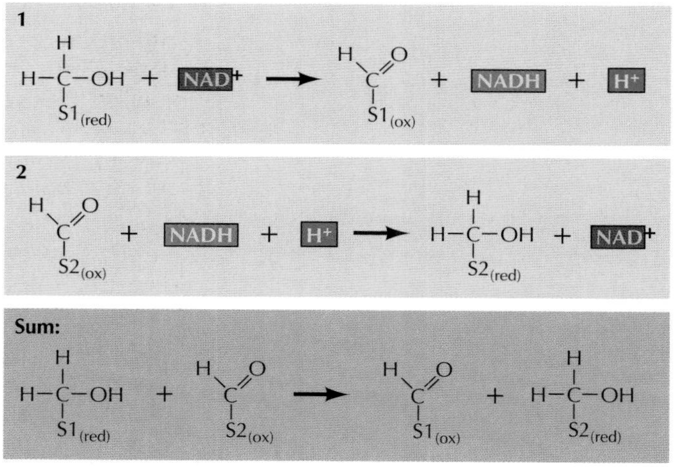

FIGURE 3.6 Role of NAD⁺ in oxidation–reduction reactions (A) Nicotinamide adenine dinucleotide (NAD^+) acts as a carrier of electrons in oxidation–reduction reactions by accepting electrons (e^-) to form NADH. (B) For example, NAD^+ can accept electrons from one substrate (S1), yielding oxidized S1 plus NADH. The NADH formed in this reaction can then transfer its electrons to a second substrate (S2), yielding reduced S2 and regenerating NAD^+. The net effect is the transfer of electrons (carried by NADH) from S1 (which becomes oxidized) to S2 (which becomes reduced).

cific chemical groups between a wide range of substrates. Many coenzymes are closely related to vitamins, which contribute part or all of the structure of the coenzyme. Vitamins are not required by bacteria such as *E. coli* but are necessary components of the diets of humans and other higher animals, which have lost the ability to synthesize these compounds.

Regulation of Enzyme Activity

An important feature of most enzymes is that their activities are not constant but instead can be modulated. That is, the activities of enzymes can be regulated so that they function appropriately to meet the varied physiological needs that may arise during the life of the cell.

One common type of enzyme regulation is **feedback inhibition** in which the product of a metabolic pathway inhibits the activity of an enzyme involved in its synthesis. For example, the amino acid isoleucine is synthe-

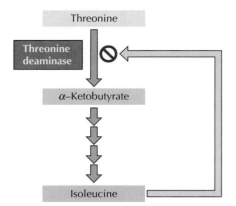

FIGURE 3.7 Feedback inhibition
The first step in the conversion of threonine to isoleucine is catalyzed by the enzyme threonine deaminase. The activity of this enzyme is inhibited by isoleucine, the end product of the pathway.

sized by a series of reactions starting from the amino acid threonine (Figure 3.7). The first step in the pathway is catalyzed by the enzyme threonine deaminase, which is inhibited by isoleucine, the end product of the pathway. Thus, an adequate amount of isoleucine in the cell inhibits threonine deaminase, blocking further synthesis of isoleucine. If the concentration of isoleucine decreases, feedback inhibition is relieved, threonine deaminase is no longer inhibited, and additional isoleucine is synthesized. By regulating the activity of threonine deaminase, the cell synthesizes the necessary amount of isoleucine but avoids wasting energy on the synthesis of more isoleucine than is needed.

Feedback inhibition is one example of **allosteric regulation** in which enzyme activity is controlled by the binding of small molecules to regulatory sites on the enzyme (Figure 3.8). The term "allosteric regulation" derives from the fact that the regulatory molecules bind not to the catalytic site but to a distinct site on the protein (*allo* = "other" and *steric* = "site"). Binding of the regulatory molecule changes the conformation of the protein, this in turn alters the shape of the active site and the catalytic activity of the enzyme. In the case of threonine deaminase, binding of the regulatory molecule (isoleucine) inhibits enzymatic activity. In other cases, regulatory molecules serve as activators, stimulating rather than inhibiting their target enzymes.

The activities of enzymes can also be regulated by their interactions with other proteins and by covalent modifications, such as the addition of phosphate groups to serine, threonine, or tyrosine residues. **Phosphorylation** is a particularly common mechanism for regulating enzyme activity; the addition of phosphate groups either stimulates or inhibits the activities of many different enzymes (Figure 3.9). For example, muscle cells respond to epinephrine (adrenaline) by breaking down glycogen into glucose, thereby providing a source of energy for increased muscular activity. The breakdown of glycogen is catalyzed by the enzyme glycogen phosphorylase, which is activated by phosphorylation in response to the binding of epinephrine to a receptor on the surface of the muscle cell. Protein phosphorylation plays a central role in controlling not only metabolic reactions but also many other cellular functions, including cell growth and differentiation.

FIGURE 3.8 Allosteric regulation
In this example, enzyme activity is inhibited by the binding of a regulatory molecule to an allosteric site. In the absence of an inhibitor, the substrate binds to the active site of the enzyme and the reaction proceeds. The binding of an inhibitor to the allosteric site induces a conformational change in the enzyme and prevents substrate binding. Most allosteric enzymes consist of multiple subunits.

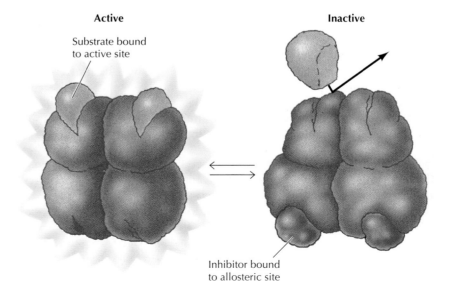

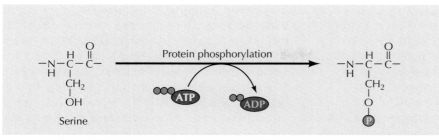

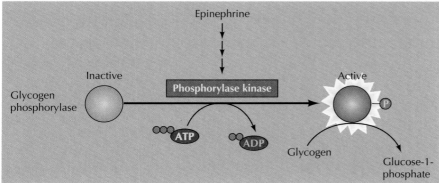

FIGURE 3.9 Protein phosphorylation
Some enzymes are regulated by the addition of phosphate groups to the side-chain OH groups of serine (as shown here), threonine, or tyrosine residues. For example, the enzyme glycogen phosphorylase, which catalyzes the conversion of glycogen into glucose-1-phosphate, is activated by phosphorylation in response to the binding of epinephrine to muscle cells.

Metabolic Energy

Many tasks that a cell must perform, such as movement and the synthesis of macromolecules, require energy. A large portion of the cell's activities is therefore devoted to obtaining energy from the environment and using that energy to drive energy-requiring reactions. Although enzymes control the rates of virtually all chemical reactions within cells, the equilibrium position of chemical reactions is not affected by enzymatic catalysis. The laws of thermodynamics govern chemical equilibria and determine the energetically favorable direction of all chemical reactions. Many of the reactions that must take place within cells are energetically unfavorable, and are therefore able to proceed only at the cost of additional energy input. Consequently, cells must constantly expend energy derived from the environment. The generation and utilization of metabolic energy is thus fundamental to all of cell biology.

Free Energy and ATP

The energetics of biochemical reactions are best described in terms of the thermodynamic function called **Gibbs free energy (G)**, named for Josiah Willard Gibbs. The change in free energy (ΔG) of a reaction combines the effects of changes in enthalpy (the heat that is released or absorbed during a chemical reaction) and entropy (the degree of disorder resulting from a reaction) to predict whether or not a reaction is energetically favorable. All chemical reactions spontaneously proceed in the energetically favorable direction, accompanied by a decrease in free energy ($\Delta G < 0$). For example, consider a hypothetical reaction in which A is converted to B:

$$A \rightleftharpoons B$$

If $\Delta G < 0$, this reaction will proceed in the forward direction, as written. If $\Delta G > 0$, however, the reaction will proceed in the reverse direction, and B will be converted to A.

The ΔG of a reaction is determined not only by the intrinsic properties of reactants and products, but also by their concentrations and other reaction conditions (e.g., temperature). It is thus useful to define the free-energy change of a reaction under standard conditions. (Standard conditions are considered to be a 1-M concentration of all reactants and products, and 1 atm of pressure). The standard free-energy change ($\Delta G°$) of a reaction is directly related to its equilibrium position because the actual ΔG is a function of both $\Delta G°$ and the concentrations of reactants and products. For example, consider the reaction

$$A \rightleftharpoons B$$

The free-energy change can be written as follows:

$$\Delta G = \Delta G° + RT \ln [B]/[A]$$

where R is the gas constant and T is the absolute temperature.

At equilibrium, $\Delta G = 0$ and the reaction does not proceed in either direction. The equilibrium constant for the reaction ($K = [B]/[A]$ at equilibrium) is thus directly related to $\Delta G°$ by the above equation, which can be expressed as follows:

$$0 = \Delta G° + RT \ln K$$

or

$$\Delta G° = - RT \ln K$$

If the actual ratio $[B]/[A]$ is greater than the equilibrium ratio (K), $\Delta G > 0$ and the reaction proceeds in the reverse direction (conversion of B to A). On the other hand, if the ratio $[B]/[A]$ is less than the equilibrium ratio, $\Delta G < 0$ and A is converted to B.

The standard free-energy change ($\Delta G°$) of a reaction therefore determines its chemical equilibrium and predicts in which direction the reaction will proceed under any given set of conditions. For biochemical reactions, the standard free-energy change is usually expressed as $\Delta G°'$, which is the standard free-energy change of a reaction in aqueous solution at pH = 7, approximately the conditions within a cell.

Many biological reactions (such as the synthesis of macromolecules) are thermodynamically unfavorable ($\Delta G > 0$) under cellular conditions. In order for such reactions to proceed, an additional source of energy is required. For example, consider the reaction

$$A \rightleftharpoons B \qquad \Delta G = +10 \text{ kcal/mol}$$

The conversion of A to B is energetically unfavorable, so the reaction proceeds in the reverse rather than the forward direction. However, the reaction can be driven in the forward direction by coupling the conversion of A to B with an energetically favorable reaction, such as:

$$C \rightleftharpoons D \qquad \Delta G = -20 \text{ kcal/mol}$$

If these two reactions are combined, the coupled reaction can be written as follows:

$$A + C \rightleftharpoons B + D \qquad \Delta G = -10 \text{ kcal/mol}$$

The ΔG of the combined reaction is the sum of the free-energy changes of its individual components, so the coupled reaction is energetically favorable and will proceed as written. Thus, the energetically unfavorable conversion of A to B is driven by coupling it to a second reaction associated with a large

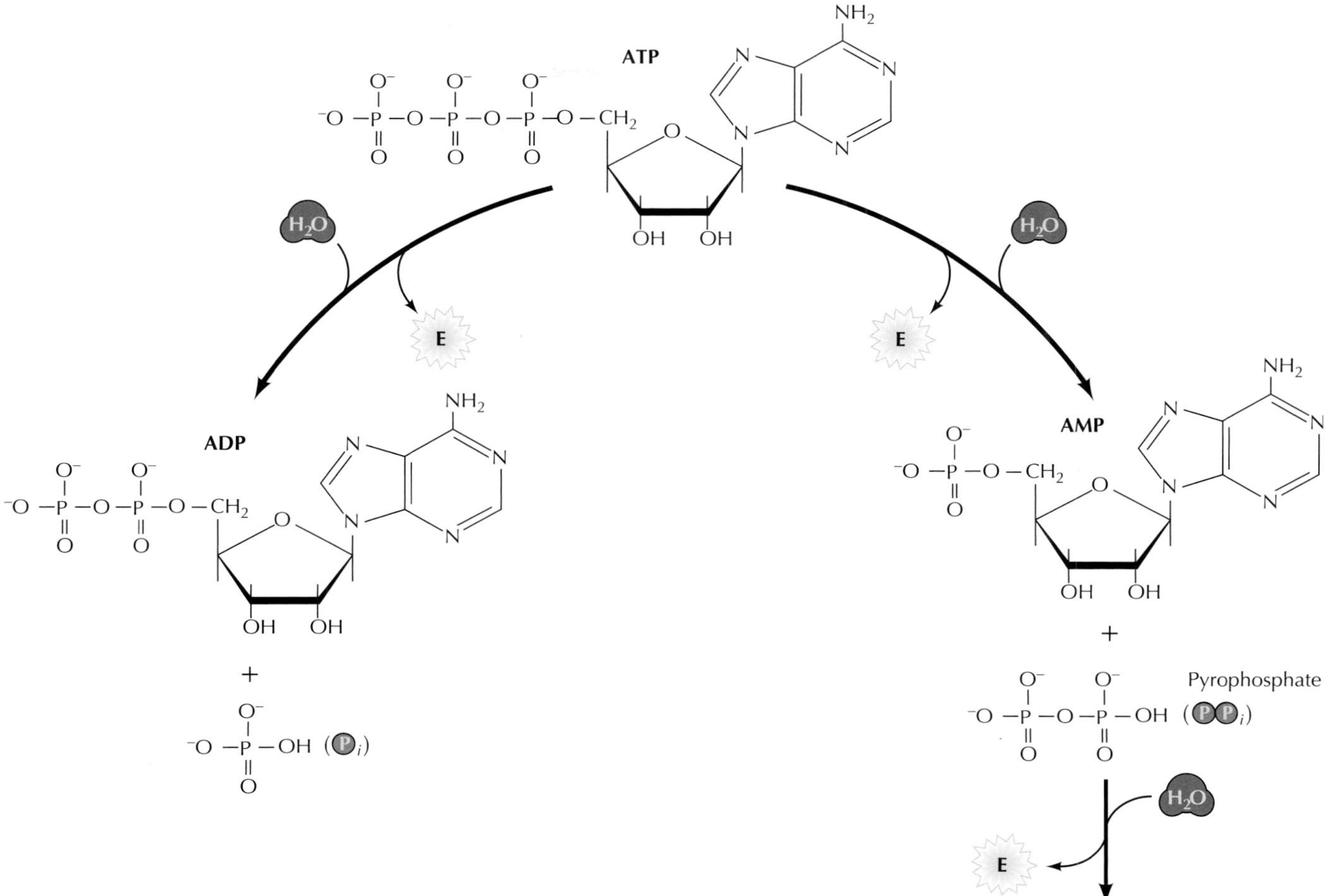

FIGURE 3.10 ATP as a store of free energy The bonds between the phosphate groups of ATP are called high-energy bonds because their hydrolysis results in a large decrease in free energy. ATP can be hydrolyzed either to ADP plus a phosphate group (HPO_4^{2-}) or to AMP plus pyrophosphate. In the latter case, pyrophosphate is itself rapidly hydrolyzed, releasing additional free energy.

decrease in free energy. Enzymes are responsible for carrying out such coupled reactions in a coordinated manner.

The cell uses this basic mechanism to drive the many energetically unfavorable reactions that must take place in biological systems. **Adenosine 5′-triphosphate (ATP)** plays a central role in this process by acting as a store of free energy within the cell (Figure 3.10). The bonds between the phosphates in ATP are known as **high-energy bonds** because their hydrolysis is accompanied by a relatively large decrease in free energy. There is nothing special about the chemical bonds themselves; they are called high-energy bonds only because a large amount of free energy is released when they are hydrolyzed within the cell. In the hydrolysis of ATP to ADP plus phosphate (P_i), $\Delta G°′ = -7.3$ kcal/mol. Recall, however, that $\Delta G°′$ refers to "standard conditions" in which the concentrations of all products and reactants are 1 M. Actual intracellular concentrations of P_i are approximately $10^{-2}\,M$, and intracellular concentrations of ATP are higher than those of ADP. These differences between intracellular concentrations and those of the standard state favor ATP hydrolysis, so for ATP hydrolysis within a cell, ΔG is approximately -12 kcal/mol.

Alternatively, ATP can be hydrolyzed to AMP plus pyrophosphate (PP_i). This reaction yields about the same amount of free energy as the hydrolysis of ATP to ADP. However, the pyrophosphate produced by this reaction is then itself rapidly hydrolyzed, with a ΔG similar to that of ATP hydrolysis. Thus, the total free-energy change resulting from the hydrolysis of ATP to AMP is approximately twice that obtained by the hydrolysis of ATP to ADP.

For comparison, the bond between the sugar and phosphate group of AMP, rather than having high energy, is typical of covalent bonds; for the hydrolysis of AMP, $\Delta G°' = -3.3$ kcal/mol.

Because of the accompanying decrease in free energy, the hydrolysis of ATP can be used to drive other energy-requiring reactions within the cell. For example, the first reaction in glycolysis (discussed in the next section) is the conversion of glucose to glucose-6-phosphate. The reaction can be written as follows:

$$\text{Glucose} + \text{Phosphate (HPO}_4^{2-}) \rightarrow \text{Glucose-6-phosphate} + \text{H}_2\text{O}$$

Because this reaction is energetically unfavorable as written ($\Delta G°' = +3.3$ kcal/mol), it must be driven in the forward direction by being coupled to ATP hydrolysis ($\Delta G°' = -7.3$ kcal/mol):

$$\text{ATP} + \text{H}_2\text{O} \rightarrow \text{ADP} + \text{HPO}_4^{2-}$$

The combined reaction can be written as follows:

$$\text{Glucose} + \text{ATP} \rightarrow \text{Glucose-6-phosphate} + \text{ADP}$$

The free-energy change for this reaction is the sum of the free-energy changes for the individual reactions, so for the coupled reaction $\Delta G°' = -4.0$ kcal/mol, favoring glucose-6-phosphate formation.

Other molecules, including other nucleoside triphosphates (e.g., GTP), also have high-energy bonds and can be used as ATP is used to drive energy-requiring reactions. For most reactions, however, ATP provides the free energy. The energy-yielding reactions within the cell are therefore coupled to ATP synthesis, while the energy-requiring reactions are coupled to ATP hydrolysis. The high-energy bonds of ATP thus play a central role in cell metabolism by serving as a usable storage form of free energy.

The Generation of ATP from Glucose

The breakdown of carbohydrates, particularly glucose, is a major source of cellular energy. The complete oxidative breakdown of glucose to CO_2 and H_2O can be written as follows:

$$\text{C}_6\text{H}_{12}\text{O}_6 + 6\ \text{O}_2 \rightarrow 6\ \text{CO}_2 + 6\ \text{H}_2\text{O}$$

The reaction yields a large amount of free energy: $\Delta G°' = -686$ kcal/mol. To harness this free energy in usable form, glucose is oxidized within cells in a series of steps coupled to the synthesis of ATP.

Glycolysis, the initial stage in the breakdown of glucose, is common to virtually all cells. Glycolysis occurs in the absence of oxygen and can provide all the metabolic energy of anaerobic organisms. In aerobic cells, however, glycolysis is only the first stage in glucose degradation.

The reactions of glycolysis result in the breakdown of glucose into pyruvate, with the net gain of two molecules of ATP (Figure 3.11). The initial reactions in the pathway actually consume energy, using ATP to phosphorylate glucose to glucose-6-phosphate and then fructose-6-phosphate to fructose-1,6-bisphosphate. The enzymes that catalyze these two reactions—hexokinase and phosphofructokinase, respectively—are important regulatory points of the glycolytic pathway. The key control element is phosphofructokinase, which is inhibited by high levels of ATP. Inhibition of phosphofructokinase results in an accumulation of glucose-6-phosphate, which in turn inhibits hexokinase. Thus, when the cell has an adequate supply of metabolic energy available in the form of ATP, the breakdown of glucose is inhibited.

3.3   WEBSITE ANIMATION

Glycolysis Glycolysis is the initial stage in the breakdown of glucose, resulting in two molecules of pyruvate and a net gain of two molecules of ATP.

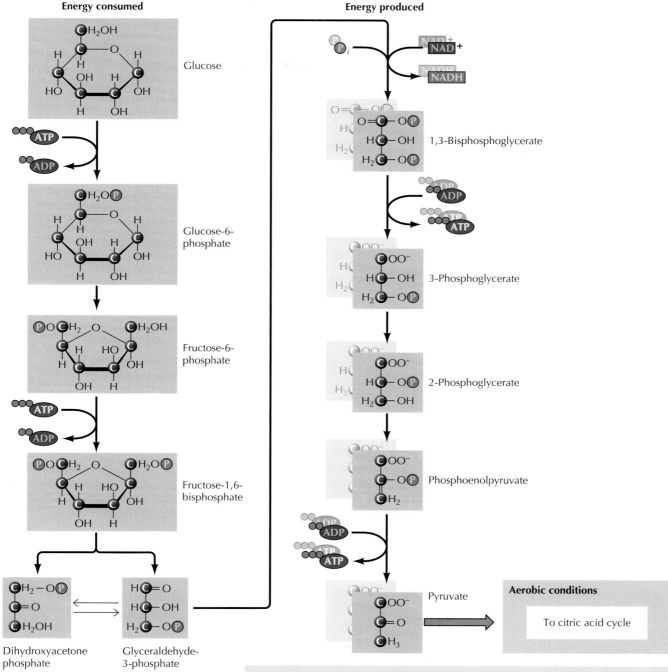

Energy consumed

Glucose

Glucose-6-phosphate

Fructose-6-phosphate

Fructose-1,6-bisphosphate

Dihydroxyacetone phosphate

Glyceraldehyde-3-phosphate

Energy produced

1,3-Bisphosphoglycerate

3-Phosphoglycerate

2-Phosphoglycerate

Phosphoenolpyruvate

Pyruvate

Aerobic conditions

To citric acid cycle

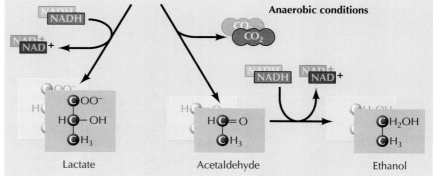

Anaerobic conditions

Lactate

Acetaldehyde

Ethanol

FIGURE 3.11 Reactions of glycolysis
Glucose is broken down to pyruvate, with the net formation of two molecules each of ATP and NADH. Under anaerobic conditions, the NADH is reoxidized by the conversion of pyruvate to ethanol or lactate. Under aerobic conditions, pyruvate is further metabolized by the citric acid cycle. Note that a single molecule of glucose yields two molecules each (shadow boxes) of the energy-producing three-carbon derivatives.

The reactions following the formation of fructose-1,6-bisphosphate constitute the energy-producing part of the glycolytic pathway. Cleavage of fructose-1,6-bisphosphate yields two molecules of the three-carbon sugar glyceraldehyde-3-phosphate, which is oxidized to 1,3-bisphosphoglycerate. The phosphate group of this compound has a very high free energy of hydrolysis ($\Delta G°' = -11.5$ kcal/mol), so it is used in the next reaction of glycolysis to drive the synthesis of ATP from ADP. The product of this reaction, 3-phosphoglycerate, is then converted to phosphoenolpyruvate, the second high-energy intermediate in glycolysis. In the hydrolysis of the high-energy phosphate of phosphoenolpyruvate, $\Delta G°' = -14.6$ kcal/mol, its conversion to pyruvate is coupled to the synthesis of ATP. Each molecule of glyceraldehyde-3-phosphate converted to pyruvate is thus coupled to the generation of two molecules of ATP; in total, four ATPs are synthesized from each starting molecule of glucose. Since two ATPs were required to prime the initial reactions, the net gain from glycolysis is two ATP molecules.

In addition to producing ATP, glycolysis converts two molecules of the coenzyme NAD^+ to NADH. In this reaction, NAD^+ acts as an oxidizing agent that accepts electrons from glyceraldehyde-3-phosphate. The NADH formed as a product must be recycled by serving as a donor of electrons for other oxidation–reduction reactions within the cell. In anaerobic conditions, the NADH formed during glycolysis is reoxidized to NAD^+ by the conversion of pyruvate to lactate or ethanol. In aerobic organisms, however, the NADH serves as an additional source of energy by donating its electrons to the electron transport chain, where they are ultimately used to reduce O_2 to H_2O, coupled to the generation of additional ATP.

In eukaryotic cells, glycolysis takes place in the cytosol. Pyruvate is then transported into mitochondria, where its complete oxidation to CO_2 and H_2O yields most of the ATP derived from glucose breakdown. The next step in the metabolism of pyruvate is its oxidative decarboxylation in the presence of **coenzyme A (CoA)**, which serves as a carrier of acyl groups in various metabolic reactions (Figure 3.12). One carbon of pyruvate is released as CO_2, and the remaining two carbons are added to CoA to form acetyl CoA. In the process, one molecule of NAD^+ is reduced to NADH.

The acetyl CoA formed by this reaction enters the **citric acid cycle** or **Krebs cycle** (Figure 3.13), which is the central pathway in oxidative metabo-

3.4 WEBSITE ANIMATION

The Citric Acid Cycle
The citric acid cycle is the central pathway in oxidative metabolism and completes the oxidation of glucose to six molecules of carbon dioxide.

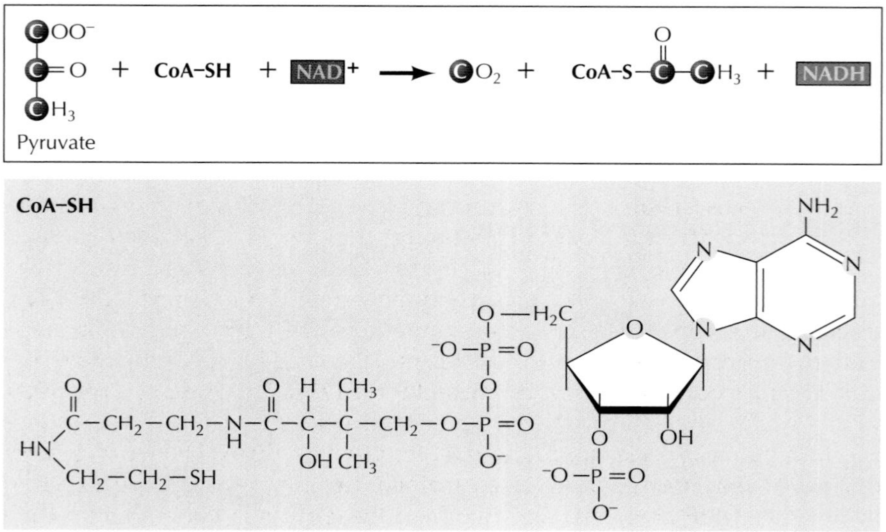

FIGURE 3.12 Oxidative decarboxylation of pyruvate
Pyruvate is converted to CO_2 and acetyl CoA, and one molecule of NADH is produced in the process. Coenzyme A (CoA–SH) is a general carrier of activated acyl groups in a variety of reactions.

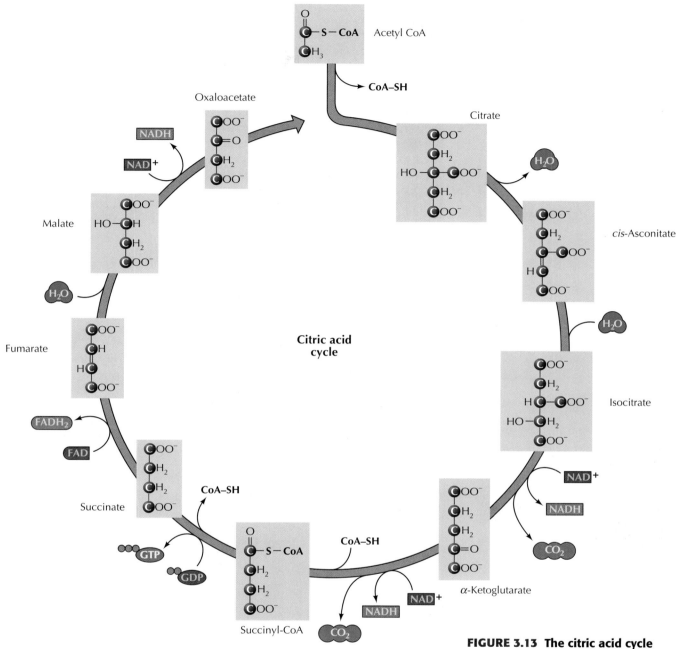

FIGURE 3.13 The citric acid cycle
A two-carbon acetyl group is transferred from acetyl CoA to oxaloacetate, forming citrate. Two carbons of citrate are then oxidized to CO_2 and oxaloacetate is regenerated. Each turn of the cycle yields one molecule of GTP, three molecules of NADH, and one molecule of $FADH_2$.

lism. The two-carbon acetyl group combines with oxaloacetate (four carbons) to yield citrate (six carbons). Through eight further reactions, two carbons of citrate are completely oxidized to CO_2 and oxaloacetate is regenerated. During the cycle, one high-energy phosphate bond is formed in GTP, which is used directly to drive the synthesis of one ATP molecule. In addition, each turn of the cycle yields three molecules of NADH and one molecule of reduced **flavin adenine dinucleotide ($FADH_2$)**, which is another carrier of electrons in oxidation–reduction reactions.

The citric acid cycle completes the oxidation of glucose to six molecules of CO_2. Four molecules of ATP are obtained directly from each glucose molecule—two from glycolysis and two from the citric acid cycle (one for each molecule of pyruvate). In addition, ten molecules of NADH (two from glycolysis, two from the conversion of pyruvate to acetyl CoA, and six from the citric acid cycle) and two molecules of $FADH_2$ are formed. The remaining energy derived from the breakdown of glucose comes from the reoxidation of NADH and $FADH_2$, with their electrons being transferred through the electron transport chain to (eventually) reduce O_2 to H_2O.

During **oxidative phosphorylation**, the electrons of NADH and $FADH_2$ combine with O_2, and the energy released from the process drives the synthesis of ATP from ADP. The transfer of electrons from NADH to O_2 releases a large amount of free energy: $\Delta G°' = -52.5$ kcal/mol for each pair of electrons transferred. So that this energy can be harvested in usable form, the process takes place gradually by the passage of electrons through a series of carriers, which constitute the **electron transport chain** (Figure 3.14). The components of the electron transport chain are located in the inner

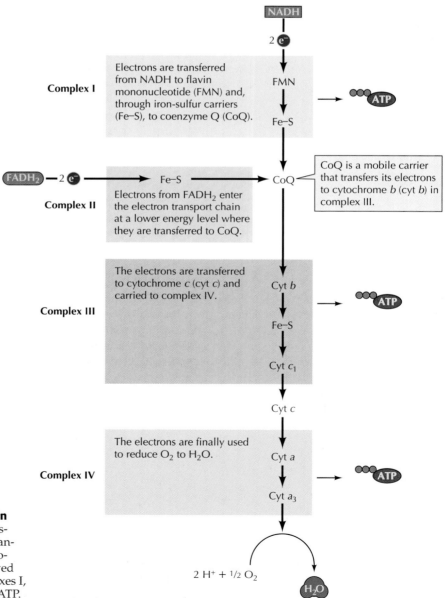

FIGURE 3.14 The electron transport chain Electrons from NADH and $FADH_2$ are transferred to O_2 through a series of carriers organized into four protein complexes in the mitochondrial membrane. The free energy derived from electron transport reactions at complexes I, III, and IV is used to drive the synthesis of ATP.

mitochondrial membrane of eukaryotic cells, and oxidative phosphorylation is considered in more detail when mitochondria are discussed in Chapter 11. In aerobic bacteria, which use a comparable system, components of the electron transport chain are located in the plasma membrane. In either case, the transfer of electrons from NADH to O_2 yields sufficient energy to drive the synthesis of approximately three molecules of ATP. Electrons from $FADH_2$ enter the electron transport chain at a lower energy level, so their transfer to O_2 yields less usable free energy—only two ATP molecules.

It is now possible to calculate the total yield of ATP from the oxidation of glucose. The net gain from glycolysis is two molecules of ATP and two molecules of NADH. The conversion of pyruvate to acetyl CoA and its metabolism via the citric acid cycle yields two additional molecules of ATP, eight of NADH, and two of $FADH_2$. Assuming that three molecules of ATP are derived from the oxidation of each NADH and two from each $FADH_2$, the total yield is 38 molecules of ATP per molecule of glucose. However, this yield is lower in some cells because the two molecules of NADH generated by glycolysis in the cytosol are unable to enter mitochondria directly. Instead, their electrons must be transferred into the mitochondrion via a shuttle system. Depending on the system used, this transfer may result in these electrons entering the electron transport chain at the level of $FADH_2$. In such cases, the two molecules of NADH derived from glycolysis give rise to two rather than three molecules of ATP, reducing the total yield to 36 rather than 38 ATPs per molecule of glucose.

The Derivation of Energy from Other Organic Molecules

Energy in the form of ATP can be derived from the breakdown of other organic molecules, with the pathways involved in glucose degradation again playing a central role. Nucleotides, for example, can be broken down to sugars, which then enter the glycolytic pathway, and amino acids are degraded via the citric acid cycle. The two principal storage forms of energy within cells, polysaccharides and lipids, can also be broken down to produce ATP. Polysaccharides are broken down into free sugars, which are then metabolized as discussed in the previous section. Lipids, however, are an even more efficient energy storage molecule. Because lipids are more reduced than carbohydrates, which consist primarily of hydrocarbon chains, the oxidation of lipids yields substantially more energy per weight of starting material.

Fats (triacylglycerols) are the major storage form of lipids. The first step in their utilization is their breakdown to glycerol and free fatty acids. Each fatty acid is joined to coenzyme A, yielding a fatty acyl-CoA at the cost of one molecule of ATP (Figure 3.15). The fatty acids are then degraded in a stepwise oxidative process, two carbons at a time, yielding acetyl CoA plus a fatty acyl-CoA shorter by one two-carbon unit. Each round of oxidation also yields one molecule of NADH and one of $FADH_2$. The acetyl CoA then enters the citric acid cycle, and degradation of the remainder of the fatty acid continues in the same manner.

The breakdown of a 16-carbon fatty acid thus yields seven molecules of NADH, seven of $FADH_2$, and eight of acetyl CoA. In terms of ATP generation, this yield corresponds to 21 molecules of ATP derived from NADH (3×7), 14 ATPs from $FADH_2$ (2×7), and 96 from acetyl CoA (8×12). Since one ATP was used to start the process, the net gain is 130 ATPs per molecule of a 16-carbon fatty acid. Compare this yield with the net gain of 38 ATPs per molecule of glucose. Since the molecular weight of a saturated 16-carbon fatty acid is 256 and that of glucose is 180, the yield of ATP is approximately

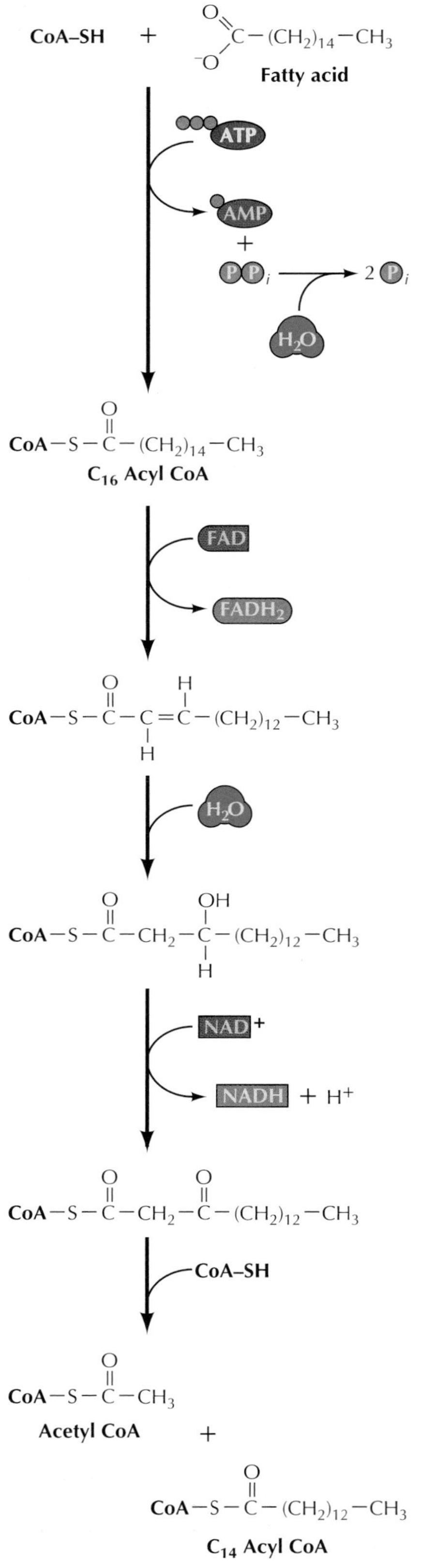

CoA–SH   +

Fatty acid

FIGURE 3.15 Oxidation of fatty acids   The fatty acid (e.g., the 16-carbon saturated fatty acid palmitate) is initially joined to coenzyme A at the cost of one molecule of ATP. Oxidation of the fatty acid then proceeds by stepwise removal of two-carbon units as acetyl CoA, coupled to the formation of one molecule each of NADH and $FADH_2$.

2.5 times greater per gram of the fatty acid—hence, the advantage of lipids over polysaccharides as energy storage molecules.

Photosynthesis

The generation of energy from oxidation of carbohydrates and lipids relies on the degradation of preformed organic compounds. The energy required for the synthesis of these compounds is ultimately derived from sunlight, which is harvested and used by plants and photosynthetic bacteria to drive the synthesis of carbohydrates. By converting the energy of sunlight to a usable form of chemical energy, photosynthesis is the source of virtually all metabolic energy in biological systems.

The overall equation of photosynthesis can be written as follows:

$$6\ CO_2 + 6\ H_2O \xrightarrow{\text{Light}} C_6H_{12}O_6 + 6\ O_2$$

The process is much more complex, however, and takes place in two distinct stages. In the first, called the **light reactions**, energy absorbed from sunlight drives the synthesis of ATP and NADPH (a coenzyme similar to NADH), coupled to the oxidation of H_2O to O_2. The ATP and NADPH generated by the light reactions drive the synthesis of carbohydrates from CO_2 and H_2O in a second set of reactions, called the **dark reactions** because they do not require sunlight. In eukaryotic cells, both the light and dark reactions occur in chloroplasts.

Photosynthetic pigments capture energy from sunlight by absorbing photons. Absorption of light by these pigments causes an electron to move from its normal molecular orbital to one of higher energy, thus converting energy from sunlight into chemical energy. In plants the most abundant photosynthetic pigments are the **chlorophylls** (Figure 3.16), which together absorb visible light of all wavelengths other than green. Additional pigments absorb light of other wavelengths, so essentially the entire spectrum of visible light can be captured and utilized for photosynthesis.

The energy captured by the absorption of light is used to convert H_2O to O_2 (Figure 3.17). The high-energy electrons derived from this process then enter an electron transport chain in which their transfer through a series of carriers is coupled to the synthesis of ATP. In addition, these high energy electrons reduce NADP⁺ to NADPH.

In the dark reactions, the ATP and NADPH produced from the light reactions drive the synthesis of carbohydrates from CO_2 and H_2O. One molecule of CO_2 at a time is added to a cycle of reactions—known as the **Calvin cycle** after its discoverer, Melvin Calvin—that leads to the formation of carbohydrates (Figure 3.18). Overall, the Calvin cycle consumes 18 molecules of ATP and 12 of NADPH for each molecule of glucose synthesized. Two electrons are needed to convert each molecule of NADP⁺ to NADPH, so 24 electrons must pass through the electron transport chain to generate sufficient

FIGURE 3.16   The structure of chlorophyll   Chlorophylls consist of porphyrin ring structures linked to hydrocarbon tails. Chlorophylls *a* and *b* differ by a single functional group in the porphyrin ring.

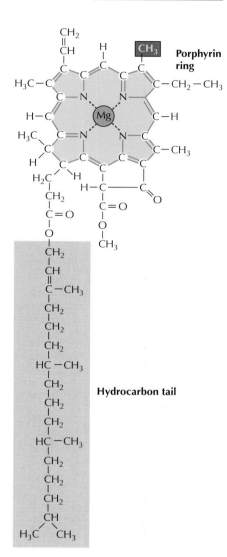

NADPH to synthesize one molecule of glucose. These electrons are obtained by the conversion of 12 molecules of H_2O to six molecules of O_2, consistent with the formation of six molecules of O_2 for each molecule of glucose. It is not clear, however, whether the passage of the same 24 electrons through the electron transport chain is also sufficient to generate the 18 ATPs that are required by the Calvin cycle. Some of these ATP molecules may instead be generated by alternative electron transport chains that use the energy derived from sunlight to synthesize ATP without the synthesis of NADPH (see Chapter 11).

The Biosynthesis of Cell Constituents

The preceding section reviewed the major metabolic reactions by which the cell obtains and stores energy in the form of ATP. This metabolic energy is then used to accomplish various tasks, including the synthesis of macromolecules and other cell constituents. Thus, energy derived from the breakdown of organic molecules (catabolism) is used to drive the synthesis of other required components of the cell. Most catabolic pathways involve the oxidation of organic molecules coupled to the generation of both energy (ATP) and reducing power (NADH). In contrast, biosynthetic (anabolic) pathways generally involve the use of both ATP and reducing power (usually in the form of NADPH) for the production of new organic compounds. One major biosynthetic pathway, the synthesis of carbohydrates from CO_2 and H_2O during the dark reactions of photosynthesis, was discussed in the preceding section. Additional pathways leading to the biosynthesis of

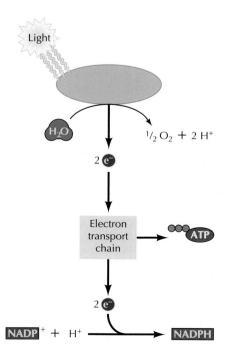

FIGURE 3.17   The light reactions of photosynthesis   Energy from sunlight is used to split H_2O to O_2. The high-energy electrons derived from this process are then transported through a series of carriers and used to convert $NADP^+$ to NADPH. Energy derived from the electron transport reactions also drives the synthesis of ATP. The details of these reactions are discussed in Chapter 11.

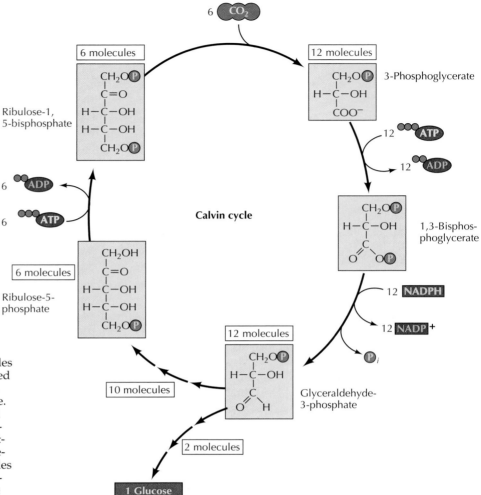

FIGURE 3.18 The Calvin cycle
Shown here is the synthesis of one molecule of glucose from six molecules of CO_2. Each molecule of CO_2 is added to ribulose-1,5-bisphosphate to yield two molecules of 3-phosphoglycerate. These six molecules of CO_2 thus lead to the formation of 12 molecules of 3-phosphoglycerate, which are converted to 12 molecules of glyceraldehyde-3-phosphate at the cost of 12 molecules each of ATP and NADPH. Two molecules of glyceraldehyde-3-phosphate are then used for synthesis of glucose and ten molecules continue in the Calvin cycle to form six molecules of ribulose-5-phosphate. The cycle is then completed by the use of six additional ATP molecules for the synthesis of ribulose-1,5-bisphosphate.

3.5   WEBSITE ANIMATION
The Calvin Cycle In the Calvin cycle of photosynthesis, six molecules of carbon dioxide are used to generate a molecule of glucose.

major cellular constituents (carbohydrates, lipids, proteins, and nucleic acids) are reviewed in the sections that follow.

Carbohydrates

In addition to being obtained directly from food or generated by photosynthesis, glucose can be synthesized from other organic molecules. In animal cells, glucose synthesis (**gluconeogenesis**) usually starts with lactate (produced by anaerobic glycolysis), amino acids (derived from the breakdown of proteins), or glycerol (produced by the breakdown of lipids). Plants (but not animals) are also able to synthesize glucose from fatty acids—a process that is particularly important during the germination of seeds, when energy stored as fats must be converted to carbohydrates to support growth of the plant. In both animal and plant cells, simple sugars are polymerized and stored as polysaccharides.

Gluconeogenesis involves the conversion of pyruvate to glucose—essentially the reverse of glycolysis (Figure 3.19). However, as discussed earlier, the glycolytic conversion of glucose to pyruvate is an energy-yielding pathway, generating two molecules each of ATP and NADH. Although some reactions of glycolysis are readily reversible, others will proceed only in the

FIGURE 3.19 Gluconeogenesis   Glucose is synthesized from two molecules of pyruvate, at the cost of 4 molecules of ATP, 2 molecules of GTP, and 2 molecules of NADH. The energy-requiring steps of gluconeogenesis are indicated by red arrows.

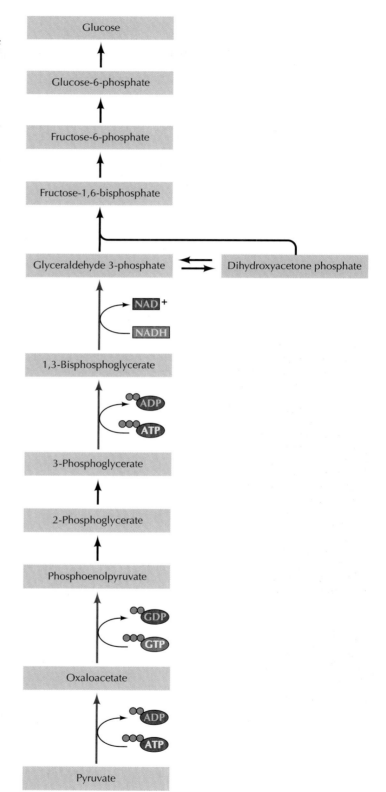

direction of glucose breakdown, because they are associated with a large decrease in free energy. These energetically favorable reactions of glycolysis are bypassed during gluconeogenesis by other reactions (catalyzed by different enzymes) that are coupled to the expenditure of ATP and NADH in order to drive them in the direction of glucose synthesis. Overall, the generation of glucose from two molecules of pyruvate requires four molecules of ATP, two molecules of GTP, and two molecules of NADH. This process is considerably more costly than the simple reversal of glycolysis (which would require two molecules of ATP and two of NADH), illustrating the additional energy required to drive the pathway in the direction of biosynthesis.

In both plant and animal cells, glucose is stored in the form of polysaccharides (starch and glycogen, respectively). The synthesis of polysaccharides, like that of all other macromolecules, is an energy-requiring reaction. As discussed in Chapter 2, the linkage of two sugars by a glycosidic bond can be written as a dehydration reaction, in which H_2O is removed (see Figure 2.3). Such a reaction, however, is energetically unfavorable and therefore unable to proceed in the forward direction. Consequently, the formation of a glycosidic bond must be coupled to an energy-yielding reaction, which is accomplished by the use of nucleotide sugars as intermediates in polysaccharide synthesis (Figure 3.20). Glucose is first phosphorylated in an ATP-driven reaction to glucose-6-phosphate, which is then converted to glucose-1-phosphate. Glucose-1-phosphate reacts with UTP (uridine triphosphate), yielding UDP-glucose plus pyrophosphate, which is hydrolyzed to phosphate with the release of additional free energy. UDP-glucose is an activated intermediate that then donates its glucose residue to a growing polysaccharide chain in an energetically favorable reaction. Thus, chemical energy in the form of ATP and UTP drives the synthesis of polysaccharides from simple sugars.

Lipids

Lipids are important energy storage molecules and the major constituent of cell membranes. They are synthesized from acetyl CoA, which is formed from the breakdown of carbohydrates, in a series of reactions that resemble the reverse of fatty acid oxidation. As with carbohydrate biosynthesis, however, the reactions leading to the synthesis of fatty acids differ from those involved in their degradation and are driven in the biosynthetic direction by being coupled to the expenditure of both energy in the form of ATP and reducing power in the form of NADPH. Fatty acids are synthesized by the stepwise addition of two-carbon units derived from acetyl CoA to a growing chain. The addition of each of these two-carbon units requires the expenditure of one molecule of ATP and two molecules of NADPH.

The major product of fatty acid biosynthesis, which occurs in the cytosol of eukaryotic cells, is the 16-carbon fatty acid palmitate. The principal constituents of cell membranes (phospholipids, sphingomyelin, and glycolipids) are then synthesized from free fatty acids in the endoplasmic reticulum and Golgi apparatus (see Chapter 10).

Proteins

In contrast to carbohydrates and lipids, proteins (as well as nucleic acids) contain nitrogen in addition to carbon, hydrogen, and oxygen. Nitrogen is incorporated into organic compounds from different sources in different organisms (Figure 3.21). Some bacteria can use atmospheric N_2 by a process called **nitrogen fixation** in which N_2 is reduced to NH_3 at the expense of energy in the form of ATP. Although relatively few species of bacteria are capable of nitrogen fixation, most bacteria, fungi, and plants can use nitrate (NO_3^-), which is a common constituent of soil, by reducing it to NH_3 via electrons derived from NADH or NADPH. Finally, all organisms are able to incorporate ammonia (NH_3) into organic compounds.

NH_3 is incorporated into organic molecules primarily during the synthesis of the amino acids glutamate and glutamine, which are derived from the citric acid cycle intermediate α-ketoglutarate. These amino acids then serve as donors of amino groups during the synthesis of the other amino acids, which are also derived from central metabolic pathways, such as glycolysis and the citric acid cycle (Figure 3.22). The raw material for amino acid syn-

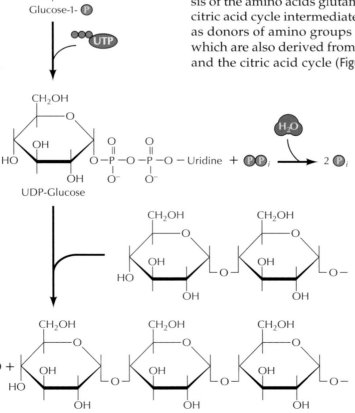

FIGURE 3.20 Synthesis of polysaccharides Glucose is first converted to an activated form, UDP-glucose, at the cost of one molecule each of ATP and UTP. The glucose residue can then be transferred from UDP-glucose to a growing polysaccharide chain in an energetically favorable reaction.

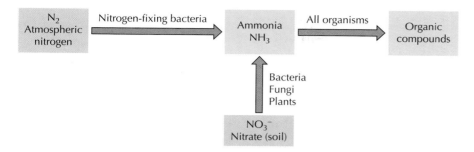

FIGURE 3.21 Assimilation of nitrogen into organic compounds Ammonia is incorporated into organic compounds by all organisms. Some bacteria are capable of converting atmospheric nitrogen to ammonia, and most bacteria, fungi, and plants can utilize nitrate from soil.

thesis is thus obtained from glucose, and the amino acids are synthesized at the cost of both energy (ATP) and reducing power (NADPH). Many bacteria and plants can synthesize all 20 amino acids. Humans and other mammals, however, can synthesize only about half of the required amino acids; the remainder must be obtained from dietary sources (Table 3.2).

The polymerization of amino acids to form proteins also requires energy. Like the synthesis of polysaccharides, the formation of the peptide bond can be considered a dehydration reaction, which must be driven in the

■ A comparison of catalysts: The industrial production of ammonia is carried out by an iron catalyst under high pressure and temperature. The enzyme nitrogenase produces ammonia under physiological conditions with the expenditure of ATP and reducing power in the form of NADH.

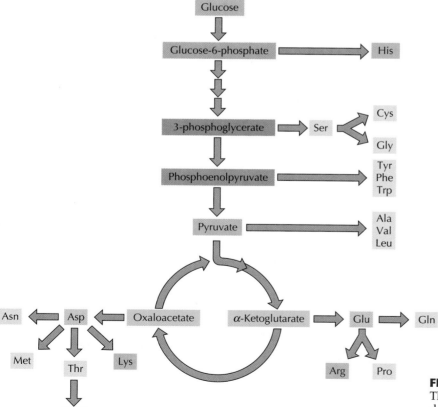

FIGURE 3.22 Biosynthesis of amino acids The carbon skeletons of the amino acids are derived from intermediates in glycolysis and in the citric acid cycle.

MOLECULAR MEDICINE

Phenylketonuria

The Disease

Phenylketonuria, or PKU, is an inborn error of amino acid metabolism with devastating effects. It afflicts approximately one in 10,000 newborn infants and, if untreated, results in severe mental retardation. Fortunately, understanding the nature of the defect responsible for phenylketonuria has allowed early diagnosis and effective treatment.

Molecular and Cellular Basis

Phenylketonuria is caused by a deficiency of the enzyme phenylalanine hydroxylase, which converts phenylalanine to tyrosine. This deficiency causes phenylalanine to accumulate to very high levels and undergo other reactions, such as conversion to phenylpyruvate. Phenylalanine, phenylpyruvate, and other abnormal metabolites accumulate in the blood and are secreted at high levels in the urine (the name of the disease derives from the high levels of phenylpyruvate, a phenylketone found in the urine of affected children). Although the pre-

cise biochemical cause is unknown, mental retardation is a critical consequence of the accumulation of these abnormal phenylalanine metabolites.

Prevention and Treatment

The enzyme deficiency causes no difficulties while the fetus is within the uterus, so children with phenylketonuria are normal at birth. If untreated, however, affected children become permanently and severely retarded within the first year of life. Fortunately, newborns with phenylke-

tonuria can be readily identified by routine screening tests that detect elevated levels of phenylalanine in blood. Mental retardation can be prevented by feeding affected infants a synthetic diet that is low in phenylalanine. This dietary treatment eliminates the accumulation of toxic phenylalanine metabolites and effectively prevents the permanent mental deficiency that would otherwise ensue. Routine screening for phenylketonuria is thus a critical test for all newborn infants.

Abnormal metabolism of phenylalanine in patients with phenylketonuria.

TABLE 3.2 Dietary Requirements for Amino Acids in Humans

Essential	Nonessential
Histidine	Alanine
Isoleucine	Arginine[a]
Leucine	Asparagine
Lysine	Aspartate
Methionine	Cysteine
Phenylalanine	Glutamate
Threonine	Glutamine
Tryptophan	Glycine
Valine	Proline
	Serine
	Tyrosine

The essential amino acids must be obtained from dietary sources; the nonessential amino acids can be synthesized by human cells.

[a]Although arginine is classified as a nonessential amino acid, growing children must obtain additional arginine from their diet.

direction of protein synthesis by being coupled to another source of metabolic energy. In the biosynthesis of polysaccharides, this coupling is accomplished through the conversion of sugars to activated intermediates, such as UDP-glucose. Amino acids are similarly activated before being used for protein synthesis.

A critical difference between the synthesis of proteins and that of polysaccharides is that the amino acids are incorporated into proteins in a unique order, specified by a gene. The order of nucleotides in a gene specifies the amino acid sequence of a protein via translation in which messenger RNA (mRNA) acts as a template for protein synthesis (see Chapter 4). Each amino acid is first attached to a specific transfer RNA (tRNA) molecule in a reaction coupled to ATP hydrolysis (Figure 3.23). The aminoacyl tRNAs then align on the mRNA template bound to ribosomes, and each amino acid is added to the C terminus of a growing peptide chain through a series of reactions that will be discussed in detail in Chapter 8. During the process, two additional molecules of GTP are hydrolyzed, so the incorporation of each amino acid into a protein is coupled to the hydrolysis of one ATP and two GTP molecules.

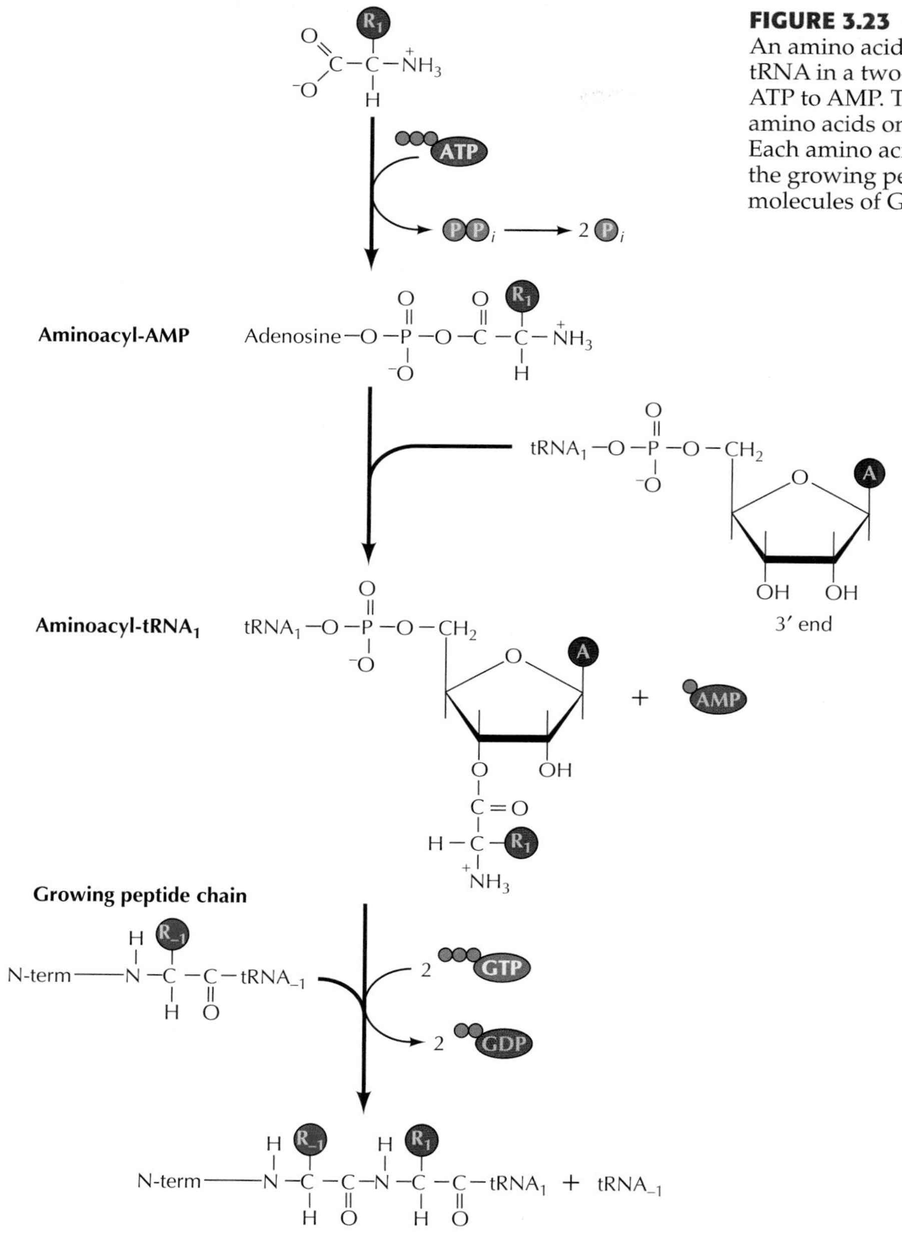

FIGURE 3.23 Formation of the peptide bond
An amino acid is first activated by attachment to its tRNA in a two-step reaction involving the hydrolysis of ATP to AMP. The tRNAs serve as adaptors to align the amino acids on an mRNA template bound to ribosomes. Each amino acid is then transferred to the C terminus of the growing peptide chain at the cost of two additional molecules of GTP.

Aminoacyl-AMP

Aminoacyl-tRNA$_1$

Growing peptide chain

Nucleic Acids

The precursors of nucleic acids, the nucleotides, are composed of phosphorylated five-carbon sugars joined to nucleic acid bases. Nucleotides can be synthesized from carbohydrates and amino acids; they can also be obtained from dietary sources or reused following nucleic acid breakdown. The starting point for nucleotide biosynthesis is the phosphorylated sugar ribose-5-phosphate, which is derived from glucose-6-phosphate. Divergent pathways then lead to the synthesis of purine and pyrimidine ribonucleotides, which are the immediate precursors for RNA synthesis (Figure 3.24). These ribonucleotides are converted to deoxyribonucleotides, which serve as the monomeric building blocks of DNA.

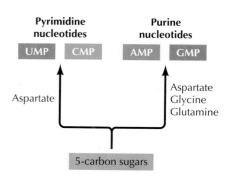

FIGURE 3.24 Biosynthesis of purine and pyrimidine nucleotides
Purine and pyrimidine nucleotides are synthesized from 5-carbon sugars and amino acids.

KEY EXPERIMENT

Antimetabolites and Chemotherapy

Antagonists of Nucleic Acid Derivatives VI. Purines

Gertrude B. Elion, George H. Hitchings, and Henry Vanderwerff

Wellcome Research Laboratories, Tuckahoe, NY

Journal of Biological Chemistry, Volume 192, 1951, pages 505–518

The Context

Gertrude Elion and George Hitchings began a collaboration in 1944 that lasted over 20 years and led to the development of drugs that have proven effective for the treatment of cancer, gout, viruses, and parasitic infections. The principle of their approach to drug development was based on the antimetabolite theory, which initially proposed that some drugs active against bacteria functioned to prevent the utilization of essential nutrients (metabolites) by the bacterial cells. Elion and Hitchings suggested that the growth of rapidly dividing cells, such as cancer cells, might be inhibited by analogs of the nucleic acid bases that interfered with the normal synthesis of DNA. They proceeded to test this hypothesis by synthesizing a large number of compounds that were related to purines and testing their biological effects. These studies led to the finding that 6-mercaptopurine was a potent inhibitor of purine utilization in bacteria, which was rapidly followed by studies showing that it was an effective drug for treatment of acute leukemia in children.

The Experiments

Elion and Hitchings chose to test the potential activities of purine analogs on the growth of the bacterium *Lacto-* *bacillus casei.* In 1948, they found that the compound 2,6-diaminopurine inhibited the growth of *L. casei* by interfering with normal purine metabolism. They extended these findings in their 1951 paper by testing the effects of 100 different purines, bearing substitutions of amino, hydroxyl, methyl, chloro, sulfhydryl and other chemical groups at various positions of the purine ring. Two compounds, 6-mercaptopurine and 6-thioguanine, in which oxygen was substituted by sulfur at the 6-position of guanine and hypoxanthine, were found to be potent inhibitors of bacterial growth (see figure). These compounds were then found to be active inhibitors of the growth of a variety of rodent cancers, leading to the first trial of 6-mercaptopurine as a treatment of childhood leukemia. The results of this trial were a spectacular success, and 6-mercaptopurine was approved for treatment of childhood

---

RNA and DNA are polymers of nucleoside monophosphates. As for other macromolecules, however, direct polymerization of nucleoside monophosphates is energetically unfavorable, and the synthesis of polynucleotides instead uses nucleoside triphosphates as activated precursors (Figure 3.25). A nucleoside 5'-triphosphate is added to the 3' hydroxyl group of a growing polynucleotide chain, with the release and subsequent hydrolysis of pyrophosphate serving to drive the reaction in the direction of polynucleotide synthesis.

COMPANION WEBSITE

Visit the website that accompanies **The Cell** (www.sinauer.com/cooper) for animations, videos, quizzes, problems, and other review material.

FIGURE 3.25 Synthesis of polynucleotides Nucleoside triphosphates ▶ are joined to the 3' end of a growing polynucleotide chain with the release of pyrophosphate.

KEY EXPERIMENT

6-Thioguanine

6-Mercaptopurine

Structures of 6-thioguanine and 6-mercaptopurine.

leukemia by the Food and Drug Administration in 1953, just over two years after its synthesis and initial demonstration of its activity as a purine antimetabolite in bacteria.

The Impact

The success of 6-mercaptopurine in treatment of childhood leukemia provided convincing evidence that inhibitors of nucleic acid metabolism could be effective drugs against cancer. This remains the case today, and 6-mercaptopurine is still one of the drugs being successfully used for leukemia treatment. In addition, a number of other antimetabolites that interfere with nucleic acid metabolism have been found to be useful in cancer therapy.

Gertrude Elion and George Hitchings continued their collaborative work on purine antimetabolites until Hitchings' retirement in 1967. In addition to 6-mercaptopurine, they developed drugs that are used for immunosuppression after tissue transplantation, for treatment of rheumatoid arthritis, for treatment of gout, and for treatment of parasitic infections. Following Hitchings' retirement, Elion focused her research on treatment of viral infec-

Gertrude B. Elion

George H. Hitchings

tions and developed the first drug effective against human viral infections—the guanine analog acyclovir, which is highly active against herpes simplex and other herpes viruses. Subsequent success in development of nucleic acid antimetabolites as antiviral drugs includes the thymine analog AZT, which is widely used as an HIV inhibitor for treatment of AIDS. The pioneering work of Elion and Hitchings has thus opened many new areas of research and had a profound impact on the treatment of a wide range of diseases.

Growing polynucleotide chain
5' end

3' end

+

Nucleoside triphosphate

5' end

3' end

KEY TERMS

enzyme, substrate, product, transition state, activation energy, active site

lock-and-key model, induced fit

prosthetic group, coenzyme, nicotinamide adenine dinucleotide (NAD$^+$)

feedback inhibition, allosteric regulation, phosphorylation

Gibbs free energy (G), adenosine 5'-triphosphate (ATP), high-energy bond

glycolysis, coenzyme A (CoA), citric acid cycle, Krebs cycle, flavin adenine dinucleotide (FADH$_2$), oxidative phosphorylation, electron transport chain

light reactions, dark reactions, photosynthetic pigment, chlorophyll, Calvin cycle

gluconeogenesis

SUMMARY

THE CENTRAL ROLE OF ENZYMES AS BIOLOGICAL CATALYSTS

The Catalytic Activity of Enzymes: Virtually all chemical reactions within cells are catalyzed by enzymes.

Mechanisms of Enzymatic Catalysis: Enzymes increase reaction rates by binding substrates in the proper position, by altering the conformation of substrates to approach the transition state, and by participating directly in chemical reactions.

Coenzymes: Coenzymes function in conjunction with enzymes to carry chemical groups between substrates.

Regulation of Enzyme Activity: The activities of enzymes are regulated to meet the physiological needs of the cell. Enzyme activity can be controlled by the binding of small molecules, by interactions with other proteins, and by covalent modifications.

METABOLIC ENERGY

Free Energy and ATP: ATP serves as a store of free energy, which is used to drive energy-requiring reactions within cells.

The Generation of ATP from Glucose: The breakdown of glucose provides a major source of cellular energy. In aerobic cells, the complete oxidation of glucose yields 36 to 38 molecules of ATP. Most of this ATP is derived from electron transport reactions in which O_2 is reduced to H_2O.

The Derivation of Energy from Other Organic Molecules: ATP can also be derived from the breakdown of organic molecules other than glucose. Because fats are more reduced than carbohydrates, they provide a more efficient form of energy storage.

Photosynthesis: The energy required for the synthesis of organic molecules is ultimately derived from sunlight, which is harvested by plants and photosynthetic bacteria. In the first stage of photosynthesis, energy from sunlight is used to drive the synthesis of ATP and NADPH, coupled to the oxidation of H_2O to O_2. The ATP and NADPH produced by these reactions are then used to synthesize glucose from CO_2 and H_2O.

THE BIOSYNTHESIS OF CELL CONSTITUENTS

Carbohydrates: Glucose can be synthesized from other organic molecules, using energy and reducing power in the forms of ATP and NADH, respectively. Additional ATP is then needed to drive the synthesis of polysaccharides from simple sugars.

Lipids: Lipids are synthesized from acetyl CoA, which is formed from the breakdown of carbohydrates.

SUMMARY

Proteins: The amino acids are synthesized from intermediates in glycolysis and the citric acid cycle. Their polymerization to form proteins requires additional energy in the form of ATP and GTP.

Nucleic Acids: Purine and pyrimidine nucleotides are synthesized from carbohydrates and amino acids. Their polymerization to DNA and RNA is driven by the use of nucleoside triphosphates as activated precursors.

KEY TERMS

nitrogen fixation

Questions

1. The binding pocket of trypsin contains an aspartate residue. How would changing this amino acid to lysine affect the enzyme's activity?

2. What characteristic of histidine allows it to play a role in enzymatic reactions involving the exchange of hydrogen ions?

3. The following is a pathway for the biosynthesis of molecule D catalyzed by the enzymes E1, E2, and E3.

$$A \xrightarrow{E1} B \xrightarrow{E2} C \xrightarrow{E3} D$$

You are studying the reaction catalyzed by E1 in the absence of the other enzymes. You find that the rate of the reaction decreases as you add increasing concentrations of D. What does this tell you about the mechanism of regulation of enzyme E1?

4. Many biochemical reactions are energetically unfavorable under physiological conditions ($\Delta G^{\circ\prime} > 0$). How does the cell carry out these reactions?

5. Consider the reaction:

Fructose-6-phosphate + HPO_4^{2-} ⇌ Fructose-1,6-bisphosphate + H_2O, $\Delta G^{\circ\prime}$ = +4 kcal/mol

Knowing the standard free energy change for ATP hydrolysis (–7.3 kcal/mol), calculate the standard free energy change for the reaction catalyzed by phosphofructokinase.

6. Consider the reaction A ⇌ B + C, in which $\Delta G^{\circ\prime}$ = +3.5 kcal/mol. Calculate ΔG under intracellular conditions, given that the concentration of A is 10^{-2} *M* and the concentrations of B and C are each 10^{-3} *M*. In which direction will the reaction proceed in the cell? For your calculation, R = 1.98×10^{-3} kcal/mol/degree and T = 298 K (25°C). Note, ln (x) = 2.3 $\log_{10}$ (x).

7. How would glycolysis be affected by an increase in the cellular concentration of ATP?

8. Yeast can grow under anaerobic as well as aerobic conditions. For every molecule of glucose consumed, how many molecules of ATP will be generated by yeast grown in anaerobic compared to aerobic conditions?

9. How do anaerobic organisms regenerate NAD^+ from NADH produced during glycolysis?

10. Why are lipids more efficient energy storage molecules than carbohydrates?

11. Differentiate between the light and dark reactions of photosynthesis.

12. Why is gluconeogenesis not a simple reversal of glycolysis?

References and Further Reading

Berg, J. M., J. L. Tymoczko and L. Stryer. 2002. *Biochemistry.* 5th ed. New York: W. H. Freeman.

Mathews, C. K., K. E. van Holde and K. G. Ahern. 2000. *Biochemistry.* 3rd ed. Redwood City, CA: Benjamin Cummings.

Nelson, D. L. and M. M. Cox. 2005. *Lehninger Principles of Biochemistry.* 4th ed. New York: W. H. Freeman.

The Central Role of Enzymes as Biological Catalysts

Fersht, A. 1999. *Structure and Mechanism in Protein Science: A Guide to Enzyme Catalysis and Protein Folding.* New York: W. H. Freeman

Koshland, D. E. 1984. Control of enzyme activity and metabolic pathways. *Trends Biochem. Sci.* 9: 155–159. [R]

Lienhard, G. E. 1973. Enzymatic catalysis and transition-state theory. *Science* 180: 149–154. [R]

Lipscomb, W. N. 1983. Structure and catalysis of enzymes. *Ann. Rev. Biochem.* 52: 17–34. [R]

Monod, J., J.-P. Changeux and F. Jacob. 1963. Allosteric proteins and cellular control systems. *J. Mol. Biol.* 6: 306–329. [P]

Narlikar, G. J. and D. Herschlag. 1997. Mechanistic aspects of enzymatic catalysis: Lessons from comparison of RNA and protein enzymes. *Ann. Rev. Biochem.* 66: 19–59. [R]

Neurath, H. 1984. Evolution of proteolytic enzymes. *Science* 224: 350–357. [R]

Petsko, G. A. and D. Ringe. 2003. *Protein Structure and Function.* Sunderland, MA: Sinauer.

Schramm, V. L. 1998. Enzymatic transition states and transition state analog design. *Ann. Rev. Biochem.* 67: 693–720. [R]

Metabolic Energy

Beinert, H., R. H. Holm and E. Munck. 1997. Iron-sulfur clusters: Nature's modular, multipurpose structures. *Science* 277: 653–659. [R]

Bennett, J. 1979. The protein that harvests sunlight. *Trends Biochem. Sci.* 4: 268–271. [R]

Calvin, M. 1962. The path of carbon in photosynthesis. *Science* 135: 879–889. [R]

Deisenhofer, J. and H. Michel. 1991. Structures of bacterial photosynthetic reaction centers. *Ann. Rev. Cell Biol.* 7: 1–23. [R]

Krebs, H. A. 1970. The history of the tricarboxylic cycle. *Perspect. Biol. Med.* 14: 154–170. [R]

Kuhlbrandt, W., D. N. Wang and Y. Fujiyoshi. 1994. Atomic model of plant light-harvesting complex by electron crystallography. *Nature* 367: 614–621. [P]

Nicholls, D. G. and S. J. Ferguson. 2002. *Bioenergetics*. 3rd ed. London: Academic Press.

Saraste, M. 1999. Oxidative phosphorylation at the fin de siecle. *Science* 283: 1488–1493. [R]

The Biosynthesis of Cell Constituents

Hers, H. G. and L. Hue. 1983. Gluconeogenesis and related aspects of glycolysis. *Ann. Rev. Biochem.* 52: 617–653. [R]

Jones, M. E. 1980. Pyrimidine nucleotide bio synthesis in animals: Genes, enzymes, and regulation of UMP biosynthesis. *Ann. Rev. Biochem.* 49: 253–279. [R]

Kornberg, A. and T. A. Baker. 1991. *DNA Replication*. 2nd ed. New York: W. H. Freeman.

Tolbert, N. E. 1981. Metabolic pathways in peroxisomes and glyoxysomes. *Ann. Rev. Biochem.* 50: 133–157. [R]

Umbarger, H. E. 1978. Amino acid biosynthesis and its regulation. *Ann. Rev. Biochem.* 47: 533–606. [R]

Van den Bosch, H., R. B. H. Schutgens, R. J. A. Wanders and J. M. Tager. 1992. Biochemistry of peroxisomes. *Ann. Rev. Biochem.* 61: 157–197. [R]

Wakil, S. J., J. K. Stoops and V. C. Joshi. 1983. Fatty acid synthesis and its regulation. *Ann. Rev. Biochem.* 52: 537–579. [R]

CHAPTER 4

Fundamentals of Molecular Biology

- **Heredity, Genes, and DNA** 103
- **Expression of Genetic Information** 111
- **Recombinant DNA** 118
- **Detection of Nucleic Acids and Proteins** 129
- **Gene Function in Eukaryotes** 137
- **KEY EXPERIMENT:** The DNA Provirus Hypothesis 116
- **MOLECULAR MEDICINE:** HIV and AIDS 120

CONTEMPORARY MOLECULAR BIOLOGY is concerned principally with understanding the mechanisms responsible for transmission and expression of the genetic information that governs cell structure and function. As reviewed in Chapter 1, all cells share a number of basic properties, and this underlying unity of cell biology is particularly apparent at the molecular level. Such unity has allowed scientists to choose simple organisms (such as bacteria) as models for many fundamental experiments, with the expectation that similar molecular mechanisms are operative in organisms as diverse as *E. coli* and humans. Numerous experiments have established the validity of this assumption, and it is now clear that the molecular biology of cells provides a unifying theme to understanding diverse aspects of cell behavior.

Initial advances in molecular biology were made by taking advantage of the rapid growth and readily manipulable genetics of simple bacteria, such as *E. coli*, and their viruses. The development of recombinant DNA then allowed both the fundamental principles and many of the experimental approaches first developed in prokaryotes to be extended to eukaryotic cells. The application of recombinant DNA technology has had a tremendous impact, initially allowing individual eukaryotic genes to be isolated and characterized in detail and more recently allowing the determination of the complete genome sequences of complex plants and animals, including humans.

Heredity, Genes, and DNA

Perhaps the most fundamental property of all living things is the ability to reproduce. All organisms inherit the genetic information specifying their structure and function from their parents. Likewise, all cells arise from preexisting cells, so the genetic material must be replicated and passed from parent to progeny cell at each cell division. How genetic information is replicated and transmitted from cell to cell and organism to organism thus represents a question that is central to all of biology. Consequently, elucidation of the mechanisms of genetic transmission and identification of the genetic material as DNA were discoveries that formed the foundation of our current understanding of biology at the molecular level.

Genes and Chromosomes

The classical principles of genetics were deduced by Gregor Mendel in 1865, on the basis of the results of breeding experiments with peas. Mendel studied the inheritance of a number of well-defined traits, such as seed color, and was able to deduce general rules for their transmission. In all cases, he could correctly interpret the observed patterns of inheritance by assuming that each trait is determined by a pair of inherited factors, which are now called **genes**. One gene copy (called an **allele**) specifying each trait is inherited from each parent. For example, breeding two strains of peas—one having yellow seeds, and the other green seeds—yields the following results (Figure 4.1). The parental strains each have two identical copies of the gene specifying yellow (Y) or green (y) seeds, respectively. The progeny plants are therefore hybrids, having inherited one gene for yellow seeds (Y) and one for green seeds (y). All these progeny plants (the first filial, or F_1, generation) have yellow seeds, so yellow (Y) is said to be **dominant** and green (y) **recessive**. The **genotype** (genetic composition) of the F_1 peas is thus Yy, and their **phenotype** (physical appearance) is yellow. If one F_1 offspring is bred with another, giving rise to F_2 progeny, the genes for yellow and green seeds segregate in a characteristic manner such that the ratio between F_2 plants with yellow seeds and those with green seeds is 3:1.

Mendel's findings, apparently ahead of their time, were largely ignored until 1900, when Mendel's laws were rediscovered and their importance recognized. Shortly thereafter, the role of **chromosomes** as the carriers of genes was proposed. It was realized that most cells of higher plants and

FIGURE 4.1 Inheritance of dominant and recessive genes

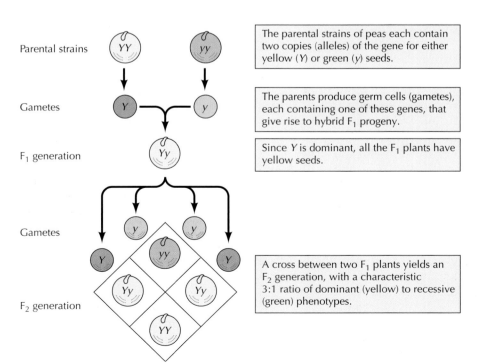

Parental strains

The parental strains of peas each contain two copies (alleles) of the gene for either yellow (Y) or green (y) seeds.

Gametes

The parents produce germ cells (gametes), each containing one of these genes, that give rise to hybrid F_1 progeny.

F_1 generation

Since Y is dominant, all the F_1 plants have yellow seeds.

Gametes

F_2 generation

A cross between two F_1 plants yields an F_2 generation, with a characteristic 3:1 ratio of dominant (yellow) to recessive (green) phenotypes.

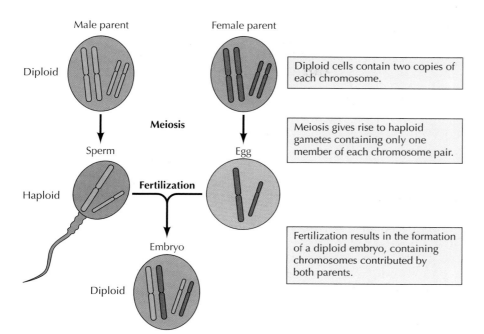

FIGURE 4.2 Chromosomes at meiosis and fertilization
Two chromosome pairs of a hypothetical organism are illustrated.

Male parent

Female parent

Diploid

Diploid cells contain two copies of each chromosome.

Meiosis

Sperm

Egg

Haploid

Meiosis gives rise to haploid gametes containing only one member of each chromosome pair.

Fertilization

Embryo

Diploid

Fertilization results in the formation of a diploid embryo, containing chromosomes contributed by both parents.

animals are **diploid**—containing two copies of each chromosome. Formation of the germ cells (the sperm and egg), however, involves a unique type of cell division (**meiosis**) in which only one member of each chromosome pair is transmitted to each progeny cell (Figure 4.2). Consequently, the sperm and egg are **haploid**, containing only one copy of each chromosome. The union of these two haploid cells at fertilization creates a new diploid organism, now containing one member of each chromosome pair derived from the male and one from the female parent. The behavior of chromosome pairs thus parallels that of genes, leading to the conclusion that genes are carried on chromosomes.

The fundamentals of mutation, genetic linkage, and the relationships between genes and chromosomes were largely established by experiments performed with the fruit fly, *Drosophila melanogaster*. *Drosophila* can be easily maintained in the laboratory, and they reproduce about every two weeks, which is a considerable advantage for genetic experiments. Indeed, these features continue to make *Drosophila* an organism of choice for genetic studies of animals, particularly the genetic analysis of development and differentiation.

In the early 1900s, a number of genetic alterations (**mutations**) were identified in *Drosophila*, usually affecting readily observable characteristics, such as eye color or wing shape. Breeding experiments indicated that some of the genes governing these traits are inherited independently of each other, suggesting that these genes are located on different chromosomes that segregate independently during meiosis (Figure 4.3). Other genes, however, are frequently inherited together as paired characteristics. Such genes are said to be linked to each other by virtue of being located on the same chromosome. The number of groups of linked genes is the same as the number of chromosomes (four in *Drosophila*), supporting the idea that chromosomes are carriers of the genes. By 1915, nearly a hundred genes had been defined and mapped onto the four chromosomes of *Drosophila*, leading to general acceptance of the chromosomal basis of heredity.

(A) Segregation of two hypothetical genes located on different chromosomes (*A/a* = square/round and *B/b* = red/blue)

(B) Linkage of two genes located on the same chromosome

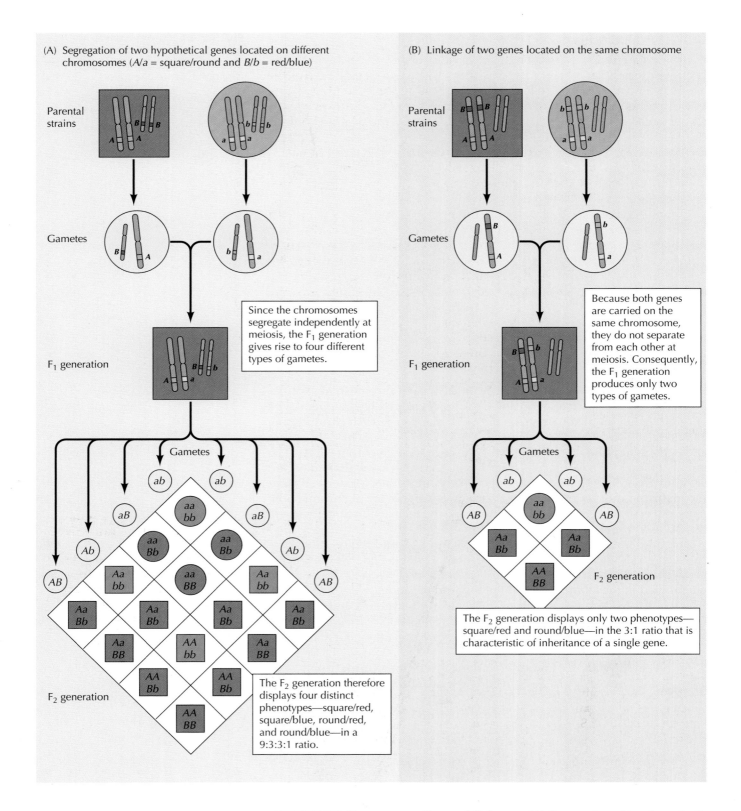

Since the chromosomes segregate independently at meiosis, the F₁ generation gives rise to four different types of gametes.

Because both genes are carried on the same chromosome, they do not separate from each other at meiosis. Consequently, the F₁ generation produces only two types of gametes.

The F₂ generation displays only two phenotypes—square/red and round/blue—in the 3:1 ratio that is characteristic of inheritance of a single gene.

The F₂ generation therefore displays four distinct phenotypes—square/red, square/blue, round/red, and round/blue—in a 9:3:3:1 ratio.

FIGURE 4.3 Gene segregation and linkage (A) Segregation of two hypothetical genes for shape (*A/a* = square/round) and color (*B/b* = red/blue) located on different chromosomes. (B) Linkage of two genes located on the same chromosome.

Genes and Enzymes

Early genetic studies focused on the identification and chromosomal localization of genes that control readily observable characteristics, such as the eye color of *Drosophila*. How these genes lead to the observed phenotypes, however, was unclear. The first insight into the relationship between genes and enzymes came in 1909, when it was realized that the inherited human disease phenylketonuria (see Molecular Medicine in Chapter 3) results from a genetic defect in metabolism of the amino acid phenylalanine. This defect was hypothesized to result from a deficiency in the enzyme needed to catalyze the relevant metabolic reaction, leading to the general suggestion that genes specify the synthesis of enzymes.

Clearer evidence linking genes with the synthesis of enzymes came from experiments of George Beadle and Edward Tatum, performed in 1941 with the fungus *Neurospora crassa*. In the laboratory, *Neurospora* can be grown on minimal or rich media similar to those discussed in Chapter 1 for the growth of *E. coli*. For *Neurospora*, minimal media consist only of salts, glucose, and biotin; rich media are supplemented with amino acids, vitamins, purines, and pyrimidines. Beadle and Tatum isolated mutants of *Neurospora* that grew normally on rich media but could not grow on minimal media. Each mutant was found to require a specific nutritional supplement, such as a particular amino acid, for growth. Furthermore, the requirement for a specific nutritional supplement correlated with the failure of the mutant to synthesize that particular compound. Thus each mutation resulted in a deficiency in a specific metabolic pathway. Since such metabolic pathways were known to be governed by enzymes, the conclusion from these experiments was that each gene specified the structure of a single enzyme—the **one gene–one enzyme hypothesis**. Many enzymes are now known to consist of multiple polypeptides, so the currently accepted statement of this hypothesis is that each gene specifies the structure of a single polypeptide chain.

Identification of DNA as the Genetic Material

Understanding the chromosomal basis of heredity and the relationship between genes and enzymes did not in itself provide a molecular explanation of the gene. Chromosomes contain proteins as well as DNA, and it was initially thought that genes were proteins. The first evidence leading to the identification of DNA as the genetic material came from studies in bacteria. These experiments represent a prototype for current approaches to defining the function of genes by introducing new DNA sequences into cells, as discussed later in this chapter.

The experiments that defined the role of DNA were derived from studies of the bacterium that causes pneumonia (*Pneumococcus*). Virulent strains of *Pneumococcus* are surrounded by a polysaccharide capsule that protects the bacteria from attack by the immune system of the host. Because the capsule gives bacterial colonies a smooth appearance in culture, encapsulated strains are denoted S. Mutant strains that have lost the ability to make a capsule (denoted R) form rough-edged colonies in culture and are no longer lethal when inoculated into mice. In 1928 it was observed that mice inoculated with nonencapsulated (R) bacteria plus heat-killed encapsulated (S) bacteria developed pneumonia and died. Importantly, the bacteria that were then isolated from these mice were of the S type. Subsequent experiments showed that a cell-free extract of S bacteria was similarly capable of converting (or transforming) R bacteria to the S state. Thus a substance in the S extract (called the transforming principle) was responsible for inducing the genetic **transformation** of R to S bacteria.

4.1 WEBSITE ANIMATION

Avery, MacLeod, & McCarty Through a series of experiments in 1944, Oswald Avery, Colin MacLeod, and Maclyn McCarty established that the transforming principle (genetic material) is DNA.

FIGURE 4.4 Transfer of genetic information by DNA DNA is extracted from a pathogenic strain of *Pneumococcus*, which is surrounded by a capsule and forms smooth colonies (S). Addition of the purified S DNA to a culture of nonpathogenic, nonencapsulated bacteria (R for "rough" colonies) results in the formation of S colonies. The purified DNA therefore contains the genetic information responsible for transformation of R to S bacteria.

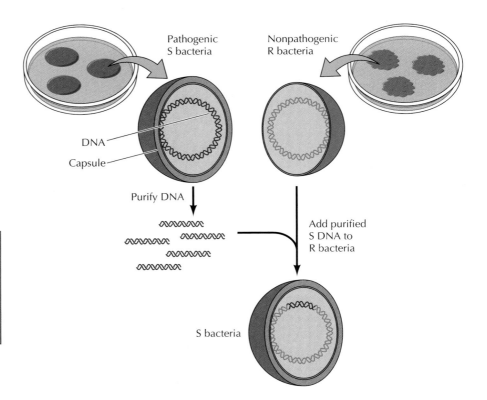

4.2 WEBSITE ANIMATION

Bacterial Transformation
A nonpathogenic strain of *Pneumococcus* can be transformed into a pathogenic strain by taking up DNA fragments from the pathogenic strain and incorporating them into its chromosome.

In 1944 Oswald Avery, Colin MacLeod, and Maclyn McCarty established that the transforming principle was DNA, both by purifying it from bacterial extracts and by demonstrating that the activity of the transforming principle is abolished by enzymatic digestion of DNA but not by digestion of proteins (Figure 4.4). Although these studies did not immediately lead to the acceptance of DNA as the genetic material, they were extended within a few years by experiments with bacterial viruses. In particular, it was shown that, when a bacterial virus infects a cell, the viral DNA rather than the viral protein must enter the cell in order for the virus to replicate. Moreover, the parental viral DNA (but not the protein) is transmitted to progeny virus particles. The concurrence of these results with continuing studies of the activity of DNA in bacterial transformation led to acceptance of the idea that DNA is the genetic material.

The Structure of DNA

Our understanding of the three-dimensional structure of DNA, deduced in 1953 by James Watson and Francis Crick, has been the basis for present-day molecular biology. At the time of Watson and Crick's work, DNA was known to be a polymer composed of four nucleic acid bases—two purines (adenine [A] and guanine [G]) and two pyrimidines (cytosine [C] and thymine [T])—linked to phosphorylated sugars. Given the central role of DNA as the genetic material, elucidation of its three-dimensional structure appeared critical to understanding its function. Watson and Crick's consideration of the problem was heavily influenced by Linus Pauling's description of hydrogen bonding and the α helix, a common element of the secondary structure of proteins (see Chapter 2). Moreover, experimental data on the structure of DNA were available from X-ray crystallography studies by Maurice Wilkins and Rosalind Franklin. Analysis of these data revealed

FIGURE 4.5 The structure of DNA

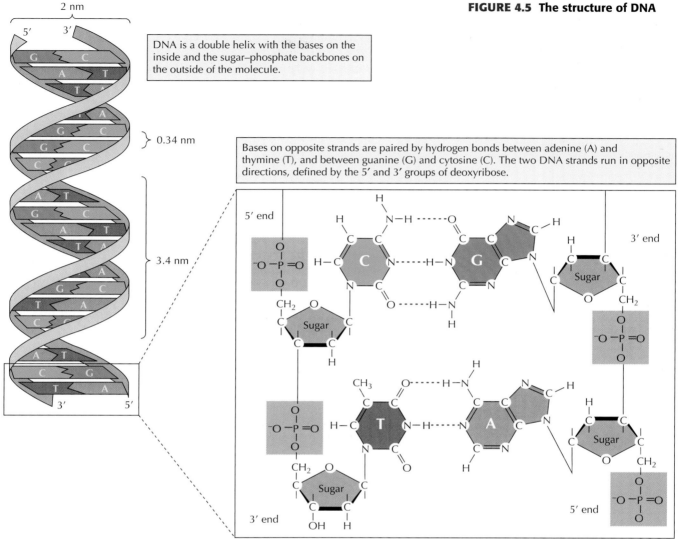

DNA is a double helix with the bases on the inside and the sugar–phosphate backbones on the outside of the molecule.

Bases on opposite strands are paired by hydrogen bonds between adenine (A) and thymine (T), and between guanine (G) and cytosine (C). The two DNA strands run in opposite directions, defined by the 5′ and 3′ groups of deoxyribose.

that the DNA molecule is a helix that turns every 3.4 nm. In addition, the data showed that the distance between adjacent bases is 0.34 nm, so there are ten bases per turn of the helix. An important finding was that the diameter of the helix is approximately 2 nm, suggesting that it is composed of not one but two DNA chains.

From these data, Watson and Crick built their model of DNA (Figure 4.5). The central features of the model are that DNA is a double helix with the sugar–phosphate backbones on the outside of the molecule. The bases are on the inside, oriented such that hydrogen bonds are formed between purines and pyrimidines on opposite chains. The base pairing is very specific: A always pairs with T and G with C. This specificity accounts for the earlier results of Erwin Chargaff, who had analyzed the base composition of various DNAs and found that the amount of adenine was always equal to that of thymine, and the amount of guanine to that of cytosine. Because of this specific base pairing, the two strands of a DNA molecule are complementary: Each strand contains all the information required to specify the sequences of bases on the other.

■ The structure of DNA described by Watson and Crick was a remarkable achievement, but not perfect. In their initial paper, they assumed that both A-T and G-C were paired by two hydrogen bonds, missing the third hydrogen bond in G-C base pairs.

Replication of DNA

The discovery of complementary base pairing between DNA strands immediately suggested a molecular solution to the question of how the genetic material could direct its own replication—a process that is required each time a cell divides. It was proposed that the two strands of a DNA molecule could separate and serve as templates for synthesis of new complementary strands, the sequence of which would be dictated by the specificity of base pairing (Figure 4.6). The process is called **semiconservative replication** because one strand of parental DNA is conserved in each progeny DNA molecule.

Direct support for semiconservative DNA replication was obtained in 1958 as a result of elegant experiments, performed by Matthew Meselson and Frank Stahl, in which DNA was labeled with isotopes that altered its density (Figure 4.7). *E. coli* were first grown in media containing the heavy isotope of nitrogen (^{15}N) in place of the normal light isotope (^{14}N). The DNA of these bacteria consequently contained ^{15}N and was heavier than that of bacteria grown in ^{14}N. Such heavy DNA could be separated from DNA containing ^{14}N by equilibrium centrifugation in a density gradient of CsCl. This ability to separate heavy (^{15}N) DNA from light (^{14}N) DNA enabled the study of DNA synthesis. *E. coli* that had been grown in ^{15}N were transferred to media containing ^{14}N and allowed to replicate one more time. Their DNA was then extracted and analyzed by CsCl density gradient centrifugation. The results of this analysis indicated that all of the heavy DNA had been replaced by newly synthesized DNA with a density intermediate between that of heavy (^{15}N) and that of light (^{14}N) DNA molecules. The implication was that during replication, the two parental strands of heavy DNA separated and served as templates for newly synthesized progeny strands of light DNA, yielding double-stranded molecules of intermediate density. This experiment thus provided direct evidence for semiconservative DNA replication, clearly underscoring the importance of complementary base pairing between strands of the double helix.

The ability of DNA to serve as a template for its own replication was further established with the demonstration that an enzyme purified from *E. coli* (**DNA polymerase**) could catalyze DNA replication *in vitro*. In the presence of DNA to act as a template, DNA polymerase was able to direct the incorporation of nucleotides into a complementary DNA molecule.

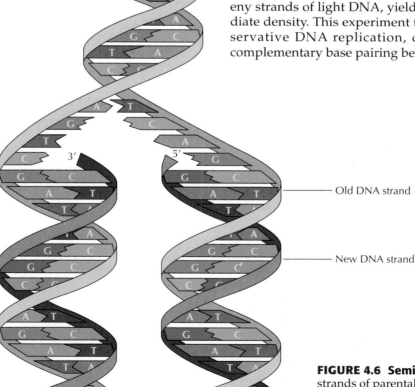

Old DNA strand

New DNA strand

FIGURE 4.6 Semiconservative replication of DNA The two strands of parental DNA separate, and each serves as a template for synthesis of a new daughter strand by complementary base pairing.

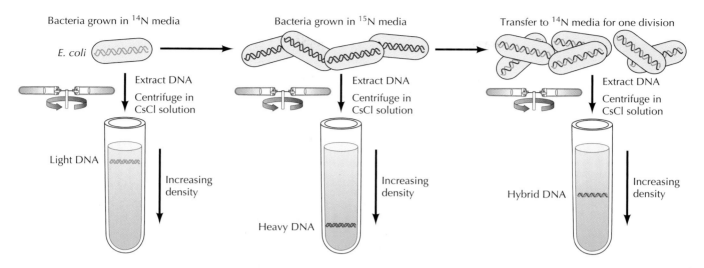

FIGURE 4.7 Experimental demonstration of semiconservative replication
Bacteria grown in medium containing the normal isotope of nitrogen (^{14}N) are transferred into medium containing the heavy isotope (^{15}N) and grown in this medium for several generations. They are then transferred back to medium containing ^{14}N and grown for one additional generation. DNA is extracted from these bacteria and analyzed by equilibrium ultracentrifugation in a CsCl solution. The CsCl sediments to form a density gradient, and the DNA molecules band at a position where their density is equal to that of the CsCl solution. DNA of the bacteria transferred from ^{15}N to ^{14}N medium for a single generation bands at a density intermediate between that of ^{15}N DNA and that of ^{14}N DNA, indicating that it represents a hybrid molecule with one heavy and one light strand.

Expression of Genetic Information

Genes act by determining the structure of proteins, which are responsible for directing cell metabolism through their activity as enzymes. The identification of DNA as the genetic material and the elucidation of its structure revealed that genetic information must be specified by the order of the four bases (A, C, G, and T) that make up the DNA molecule. Proteins, in turn, are polymers of 20 amino acids, the sequence of which determines their structure and function. The first direct link between a genetic mutation and an alteration in the amino acid sequence of a protein was made in 1957, when it was found that patients with the inherited disease sickle-cell anemia had hemoglobin molecules that differed from normal ones by a single amino acid substitution. Deeper understanding of the molecular relationship between DNA and proteins came, however, from a series of experiments that took advantage of *E. coli* and its viruses as genetic models.

Colinearity of Genes and Proteins

The simplest hypothesis to account for the relationship between genes and enzymes was that the order of nucleotides in DNA specified the order of amino acids in a protein. Mutations in a gene would correspond to alterations in the sequence of DNA, which might result from the substitution of one nucleotide for another or from the addition or deletion of nucleotides. These changes in the nucleotide sequence of DNA would then lead to corresponding changes in the amino acid sequence of the protein encoded by the gene in question. This hypothesis predicted that different mutations within a single gene could alter different amino acids in the encoded protein, and

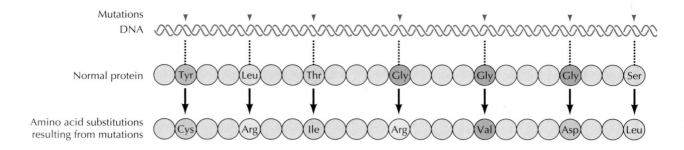

FIGURE 4.8 Colinearity of genes and proteins A series of mutations (arrowheads) were mapped in the *E. coli* gene encoding tryptophan synthetase (top line). The amino acid substitutions resulting from each of the mutations were then determined by sequence analysis of the proteins of mutant bacteria (bottom line). These studies revealed that the order of mutations in DNA was the same as the order of amino acid substitutions in the encoded protein.

that the positions of mutations in a gene should reflect the positions of amino acid alterations in its protein product.

The rapid replication and the simplicity of the genetic system of *E. coli* were of major help in addressing these questions. A variety of mutants of *E. coli* could be isolated, including nutritional mutants that (like the *Neurospora* mutants discussed earlier) require particular amino acids for growth. Importantly, the rapid growth of *E. coli* made feasible the isolation and mapping of multiple mutants in a single gene, leading to the first demonstration of the linear relationship between genes and proteins. In these studies, Charles Yanofsky and his colleagues mapped a series of mutations in the gene that encodes an enzyme required for synthesis of the amino acid tryptophan. Analysis of the enzymes encoded by the mutant genes indicated that the relative positions of the amino acid alterations were the same as those of the corresponding mutations (**Figure 4.8**). Thus the sequence of amino acids in the protein was colinear with that of mutations in the gene, as expected if the order of nucleotides in DNA specifies the order of amino acids in proteins.

The Role of Messenger RNA

Although the sequence of nucleotides in DNA appeared to specify the order of amino acids in proteins, it did not necessarily follow that DNA itself directs protein synthesis. Indeed, this appeared not to be the case, since DNA is located in the nucleus of eukaryotic cells, whereas protein synthesis takes place in the cytoplasm. Some other molecule was therefore needed to convey genetic information from DNA to the sites of protein synthesis (the ribosomes).

RNA appeared a likely candidate for such an intermediate because the similarity of its structure to that of DNA suggested that RNA could be synthesized from a DNA template (**Figure 4.9**). RNA differs from DNA in that it is single-stranded rather than double-stranded, its sugar component is ribose instead of deoxyribose, and it contains the pyrimidine base uracil (U) instead of thymine (T) (see Figure 2.10). However, neither the change in sugar nor the substitution of U for T alters base pairing, so the synthesis of RNA can be readily directed by a DNA template. Moreover, since RNA is located primarily in the cytoplasm, it appeared a logical intermediate to convey information from DNA to the ribosomes. These characteristics of

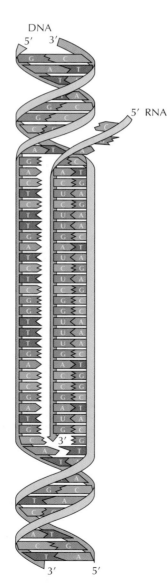

FIGURE 4.9 Synthesis of RNA from DNA The two strands of DNA unwind, and one is used as a template for synthesis of a complementary strand of RNA.

RNA suggested a pathway for the flow of genetic information that is known as the **central dogma** of molecular biology:

$$DNA \rightarrow RNA \rightarrow Protein$$

According to this concept, RNA molecules are synthesized from DNA templates (a process called **transcription**), and proteins are synthesized from RNA templates (a process called **translation**).

Experimental evidence for the RNA intermediates postulated by the central dogma was obtained by Sidney Brenner, Francois Jacob, and Matthew Meselson in studies of *E. coli* infected with the bacteriophage T4. The synthesis of *E. coli* RNA stops following infection by T4, and the only new RNA synthesized in infected bacteria is transcribed from T4 DNA. This T4 RNA becomes associated with bacterial ribosomes, thus conveying the information from DNA to the site of protein synthesis. Because of their role as intermediates in the flow of genetic information, RNA molecules that serve as templates for protein synthesis are called **messenger RNAs (mRNAs)**. They are transcribed by an enzyme (**RNA polymerase**) that catalyzes the synthesis of RNA from a DNA template.

In addition to mRNA, two other types of RNA molecules are important in protein synthesis. **Ribosomal RNA (rRNA)** is a component of ribosomes, and **transfer RNAs (tRNAs)** serve as adaptor molecules that align amino acids along the mRNA template. The structures and functions of these molecules are discussed in the following section and in more detail in Chapters 7 and 8.

The Genetic Code

How is the nucleotide sequence of mRNA translated into the amino acid sequence of a protein? In this step of gene expression, genetic information is transferred between chemically unrelated types of macromolecules—nucleic acids and proteins—raising two new types of problems in understanding the action of genes.

First, since amino acids are structurally unrelated to the nucleic acid bases, direct complementary pairing between mRNA and amino acids during the incorporation of amino acids into proteins seemed impossible. How then could amino acids align on an mRNA template during protein synthesis? This question was solved by the discovery that tRNAs serve as adaptors between amino acids and mRNA during translation (Figure 4.10). Prior to its use in protein synthesis, each amino acid is attached by a specific enzyme to its appropriate tRNA. Base pairing between a recognition sequence on each tRNA and a complementary sequence on the mRNA then directs the attached amino acid to its correct position on the mRNA template.

The second problem in the translation of nucleotide sequence to amino acid sequence was determination of the **genetic code**. How could the information contained in the sequence of four different nucleotides be converted to the sequences of 20 different amino acids in proteins? Because 20 amino acids must be specified by only four nucleotides, at least three nucleotides must be used to encode each amino acid. Used singly, four nucleotides could encode only four amino acids and, used in pairs, four nucleotides could encode only sixteen (4^2) amino acids. Used as triplets, however, four nucleotides could encode 64 (4^3) different amino acids—more than enough to account for the 20 amino acids actually found in proteins.

Direct experimental evidence for the triplet code was obtained by studies of bacteriophage T4 bearing mutations in an extensively studied gene called *rII*. Phages with mutations in this gene form abnormally large

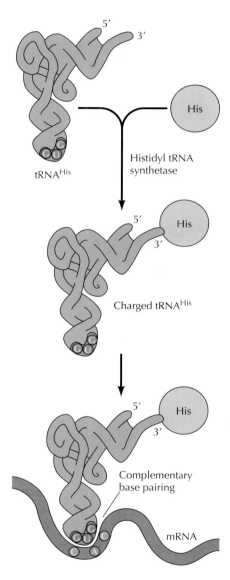

FIGURE 4.10 Function of transfer RNA   Transfer RNA serves as an adaptor during protein synthesis. Each amino acid (e.g., histidine) is attached to the 3′ end of a specific tRNA by an appropriate enzyme (an aminoacyl tRNA synthetase). The charged tRNAs then align on an mRNA template by complementary base pairing.

FIGURE 4.11 Genetic evidence for a triplet code A series of mutations consisting of additions of one, two, or three nucleotides were studied in the *rII* gene of bacteriophage T4. Additions of one or two nucleotides alter the reading frame of the remainder of the gene. Therefore all the subsequent amino acids are abnormal, and an inactive protein is produced, giving rise to mutant phage. Additions of three nucleotides, however, alter only a single amino acid. The reading frame of the remainder of the gene is normal, and an active protein giving rise to wild-type (WT) phage is produced.

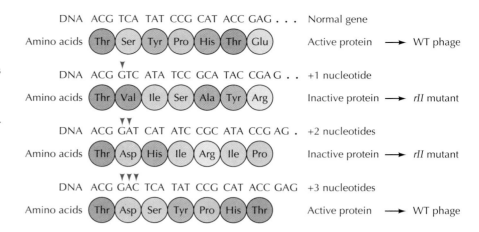

4.4 WEBSITE ANIMATION

DNA Mutations A single base-pair mutation in the coding region of a gene can result in a prematurely terminated polypeptide or in a polypeptide with an incorrect amino acid in its amino acid chain, depending on the particular mutation.

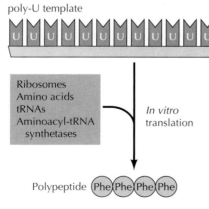

FIGURE 4.12 The triplet UUU encodes phenylalanine *In vitro* translation of a synthetic RNA consisting of repeated uracils (a poly-U template) results in the synthesis of a polypeptide containing only phenylalanine.

plaques, which can be clearly distinguished from those formed by wild-type phages. Hence isolating and mapping a number of *rII* mutants was easy and led to the establishment of a detailed genetic map of this locus. Study of recombinants between *rII* mutants that had arisen by additions or deletions of nucleotides revealed that phages containing additions or deletions of one or two nucleotides always exhibited the mutant phenotype. Phages containing additions or deletions of three nucleotides, however, were frequently wild-type in function (Figure 4.11). These findings suggested that the gene is read in groups of three nucleotides, starting from a fixed point. Additions or deletions of one or two nucleotides would then alter the reading frame of the entire gene, leading to the coding of abnormal amino acids throughout the encoded protein. In contrast, additions or deletions of three nucleotides would lead to the addition or deletion of only a single amino acid; the rest of the amino acid sequence would remain unaltered, frequently yielding an active protein.

Deciphering the genetic code thus became a problem of assigning nucleotide triplets to their corresponding amino acids. This problem was approached using *in vitro* systems that could carry out protein synthesis (*in vitro* **translation**). Cell extracts containing ribosomes, amino acids, tRNAs, and the enzymes responsible for attaching amino acids to the appropriate tRNAs (aminoacyl-tRNA synthetases) were known to catalyze the incorporation of amino acids into proteins. However, such protein synthesis depends on the presence of mRNA bound to the ribosomes, and can be greatly enhanced by the addition of purified mRNA. Since added mRNA directs protein synthesis in such systems, the genetic code could be deciphered by study of the translation of synthetic mRNAs of known base sequence.

The first such experiment, performed by Marshall Nirenberg and Heinrich Matthaei, involved the *in vitro* translation of a synthetic RNA polymer containing only uracil (Figure 4.12). This poly-U template was found to direct the incorporation of only a single amino acid—phenylalanine—into a polypeptide consisting of repeated phenylalanine residues. Therefore the triplet UUU encodes the amino acid phenylalanine. Similar experiments with RNA polymers containing only single nucleotides established that AAA encodes lysine and CCC encodes proline. The remainder of the code was deciphered using RNA polymers containing mixtures of nucleotides, leading to the coding assignment of all 64 possible triplets (called **codons**) (Table 4.1). Of the 64 codons, 61 specify an amino acid; the remaining three

TABLE 4.1 The Genetic Code

First position	Second position				Third position
	U	**C**	**A**	**G**	
U	Phe	Ser	Tyr	Cys	U
	Phe	Ser	Tyr	Cys	C
	Leu	Ser	stop	stop	A
	Leu	Ser	stop	Trp	G
C	Leu	Pro	His	Arg	U
	Leu	Pro	His	Arg	C
	Leu	Pro	Gln	Arg	A
	Leu	Pro	Gln	Arg	G
A	Ile	Thr	Asn	Ser	U
	Ile	Thr	Asn	Ser	C
	Ile	Thr	Lys	Arg	A
	Met	Thr	Lys	Arg	G
G	Val	Ala	Asp	Gly	U
	Val	Ala	Asp	Gly	C
	Val	Ala	Glu	Gly	A
	Val	Ala	Glu	Gly	G

(UAA, UAG, and UGA) are stop codons that signal the termination of protein synthesis. The code is degenerate; that is, many amino acids are specified by more than one codon. With few exceptions (discussed in Chapter 11), all organisms utilize the same genetic code, providing strong support for the conclusion that all present-day cells evolved from a common ancestor.

RNA Viruses and Reverse Transcription

With the elucidation of the genetic code, the fundamental principles of the molecular biology of cells appeared to have been established. According to the central dogma, the genetic material consists of DNA, which is capable of self-replication as well as being transcribed into mRNA, which serves in turn as the template for protein synthesis. However, as noted in Chapter 1, many viruses contain RNA rather than DNA as their genetic material, implying the use of other modes of information transfer.

RNA genomes were first discovered in plant viruses, many of which were found to be composed of only RNA and protein. Direct proof that RNA acts as the genetic material of these viruses was obtained in the 1950s by experiments demonstrating that RNA purified from tobacco mosaic virus could infect new host cells, giving rise to infectious progeny virus. The mode of replication of most viral RNA genomes was subsequently determined by studies of the RNA bacteriophages of *E. coli*. These viruses were found to encode a specific enzyme that could catalyze the synthesis of RNA from an RNA template (RNA-directed RNA synthesis), using the same mechanism of base pairing between complementary strands as is employed during DNA replication or transcription of RNA from DNA.

Although most animal viruses, such as poliovirus or influenza virus, were found to replicate by RNA-directed RNA synthesis, this mechanism did not appear to account for the replication of one family of animal viruses (the RNA tumor viruses), which were of particular interest because of their

KEY EXPERIMENT

The DNA Provirus Hypothesis

Nature of the Provirus of Rous Sarcoma
Howard M. Temin
McArdle Laboratory, University of Wisconsin, Madison, WI
National Cancer Institute Monographs, Volume 17, 1964, pages 557–570

Howard M. Temin

The Context

Rous sarcoma virus (RSV), the first cancer-causing virus to be described, was of considerable interest as an experimental system for studying the molecular biology of cancer. Howard Temin began his research in this area when, as a graduate student in 1958, he developed the first assay for the transformation of normal cells to cancer cells in culture following infection with RSV. The availability of such a quantitative *in vitro* assay provided the tool needed for further studies of both cell transformation and virus replication. As Temin proceeded with these studies, he made a series of unexpected observations indicating that the replication of RSV was fundamentally different from that of other RNA viruses. These experiments led to Temin's proposal of the DNA provirus hypothesis, which stated that the viral RNA was copied into DNA in infected cells—a proposal that ran directly counter to the universally accepted central dogma of molecular biology.

The Experiments

The DNA provirus hypothesis was based on several different types of experimental evidence. First, studies of cell transformation using mutants of RSV indicated that important characteristics of transformed cells were determined by genetic information of the virus. This information was regularly transmitted to daughter cells following cell division, even in the absence of virus replication. Temin therefore proposed that the viral genome was present in infected cells in a stably inherited form, which he called a provirus.

Evidence that the provirus was DNA was then derived from experiments with metabolic inhibitors. First, actinomycin D, which inhibits the synthesis of RNA from a DNA template, was found to inhibit virus production by RSV-infected cells (see figure). Second, inhibitors of DNA synthesis inhibited early stages of cell infection by RSV. Thus, DNA synthesis appeared to be required early in infec-

tion, and DNA-directed RNA synthesis appeared to be needed subsequently for the production of progeny viruses, leading to the proposal that the provirus was a DNA copy of the viral RNA genome. Temin sought further evidence for this proposal by using nucleic acid hybridization to detect viral sequences in infected cell DNA, but the sensitivity of the available techniques was limited and the data were unconvincing.

The Impact

The DNA provirus hypothesis was thus proposed principally on the basis of genetic experiments and the effects of metabolic inhibitors. It was a radical proposal, which contradicted the accepted central dogma of molecular biology. In this setting, Temin's

---

4.5 WEBSITE ANIMATION

HIV Reproduction
As part of a replication cycle within a host cell, a retrovirus uses the enzyme reverse transcriptase to copy its RNA genome into DNA.

ability to cause cancer in infected animals. Although these viruses contain genomic RNA in their viral particles, experiments performed by Howard Temin in the early 1960s indicated that their replication requires DNA synthesis in infected cells, leading to the hypothesis that the RNA tumor viruses (now called **retroviruses**) replicate via synthesis of a DNA intermediate, called a DNA provirus (Figure 4.13). This hypothesis was initially met with widespread disbelief because it involves RNA-directed synthesis of DNA—a reversal of the central dogma. In 1970, however, Temin and David Baltimore independently discovered that the RNA tumor viruses contain a novel enzyme that catalyzes the synthesis of DNA from an RNA template. In addition, clear-cut evidence for the existence of viral DNA sequences in infected cells was obtained. The synthesis of DNA from RNA, now called

KEY EXPERIMENT

hypothesis that RSV replicated by the transfer of information from RNA to DNA not only failed to win the acceptance of the scientific community, but was met with general derision. Nonetheless, Temin persevered through the 1960s, continuing with experiments to test his hypothesis and providing increasingly convincing evidence in its support. These efforts culminated in 1970 with the discovery by Temin and Satoshi Mizutani, and at the same time by David Baltimore, of a viral enzyme, now known as reverse transcriptase, that synthesizes DNA from an RNA template—an unambiguous biochemical demonstration that the central dogma could be reversed.

Temin concluded his 1970 paper with the statement that the results

"constitute strong evidence that the DNA provirus hypothesis is correct and that RNA tumour viruses have a DNA genome when they are in cells and an RNA genome when they are in virions. This result would have strong implications for theories of viral carcinogenesis and, possibly, for theories of information transfer in other biological systems." As Temin predicted, the discovery of RNA-directed DNA synthesis has led to major advances in our understanding of cancer, human retroviruses, and gene rearrangements. Reverse transcriptase has further provided a critical tool for cDNA cloning, thereby impacting virtually all areas of contemporary cell and molecular biology.

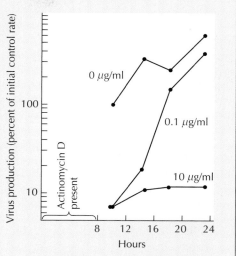

Effect of actinomycin D on RSV replication. RSV-infected cells were cultured with the indicated concentrations of actinomycin D for 8 hours. Actinomycin D was then removed and the amount of virus produced was determined.

reverse transcription, was thus established as a mode of information transfer in biological systems.

Reverse transcription is important not only in the replication of retroviruses but also in at least two other broad aspects of molecular and cellular biology. First, reverse transcription is not restricted to retroviruses; it also occurs in cells and, as discussed in Chapters 5 and 6, is frequently responsible for the transposition of DNA sequences from one chromosomal location to another. Indeed, the sequence of the human genome has revealed that approximately 40% of human genomic DNA is derived from reverse transcription. Second, enzymes that catalyze RNA-directed DNA synthesis (**reverse transcriptases**) can be used experimentally to generate DNA copies of any RNA molecule. The use of reverse transcriptase has thus allowed mRNAs of eukaryotic cells to be studied using the molecular approaches that are currently applied to the manipulation of DNA, as discussed in the following section.

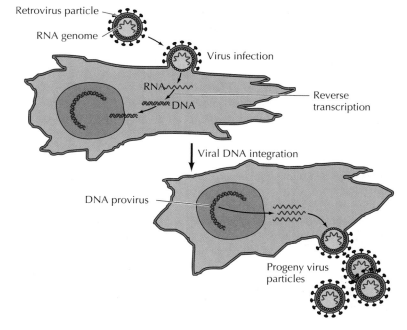

FIGURE 4.13 Reverse transcription and retrovirus replication
Retroviruses contain RNA genomes in their viral particles. When a retrovirus infects a host cell, however, a DNA copy of the viral RNA is synthesized via reverse transcription. This viral DNA is then integrated into chromosomal DNA of the host to form a DNA provirus, which is transcribed to yield progeny virus RNA.

Recombinant DNA

Classical experiments in molecular biology were strikingly successful in developing our fundamental concepts of the nature and expression of genes. Since these studies were based primarily on genetic analysis, their success depended largely on the choice of simple, rapidly replicating organisms (such as bacteria and viruses) as models. It was not clear, however, how these fundamental principles could be extended to provide a molecular understanding of the complexities of eukaryotic cells, since the genomes of most eukaryotes (e.g., the human genome) are up to a thousand times more complex than that of *E. coli*. In the early 1970s, the possibility of studying such genomes at the molecular level seemed daunting. In particular, there appeared to be no way in which individual genes could be isolated and studied.

This obstacle to the progress of molecular biology was overcome by the development of recombinant DNA technology, which provided scientists with the ability to isolate, sequence, and manipulate individual genes derived from any type of cell. The application of recombinant DNA has thus enabled detailed molecular studies of the structure and function of eukaryotic genes and genomes, thereby revolutionizing our understanding of cell biology.

Restriction Endonucleases

The first step in the development of recombinant DNA technology was the characterization of **restriction endonucleases**—enzymes that cleave DNA at specific sequences. These enzymes were identified in bacteria, where they apparently provide a defense against the entry of foreign DNA (e.g., from a virus) into the cell. Bacteria have a variety of restriction endonucleases that cleave DNA at more than a hundred distinct recognition sites, each of which consists of a specific sequence of four to eight base pairs (examples are given in Table 4.2).

Since restriction endonucleases digest DNA at specific sequences, they can be used to cleave a DNA molecule at unique sites. For example, the restriction endonuclease *Eco*RI recognizes the six-base-pair sequence GAATTC. This sequence is present at five sites in DNA of the bacteriophage λ, so *Eco*RI digests λ DNA into six fragments ranging from 3.6 to 21.2 kilo-

---

4.6 WEBSITE ANIMATION

Restriction Endonucleases Restriction endonucleases cleave DNA at specific sequences, leaving staggered or blunt ends on the resulting DNA fragments.

■ **Why isn't the DNA of bacteria digested by their own restriction endonucleases?** Restriction endonucleases do not work alone in bacteria. Instead, they usually work in conjunction with enzymes that modify the bacterial DNA so that it is no longer recognized by the restriction endonuclease. This restriction-modification system ensures that only unmodified foreign DNA (e.g., viral DNA) is digested by the restriction endonucleases expressed by any particular species of bacteria.

TABLE 4.2 Recognition Sites of Representative Restriction Endonucleases

Enzyme[a]	Source	Recognition site[b]
*Bam*HI	*Bacillus amyloliquefaciens* H	GGATCC
*Eco*RI	*Escherichia coli* RY13	GAATTC
*Hae*III	*Haemophilus aegyptius*	GGCC
*Hind*III	*Haemophilus influenzae* Rd	AAGCTT
*Hpa*I	*Haemophilus parainfluenzae*	GTTAAC
*Hpa*II	*Haemophilus parainfluenzae*	CCGG
*Mbo*I	*Moraxella bovis*	GATC
*Not*I	*Nocardia otitidis-caviarum*	GCGGCCGC
*Sfi*I	*Streptomyces fimbriatus*	GGCCNNNNNGGCC
*Taq*I	*Thermus aquaticus*	TCGA

[a] Enzymes are named according to their species of isolation, followed by a number to distinguish different enzymes isolated from the same organism (e.g., *Hpa*I and *Hpa*II).

[b] Recognition sites show the sequence of only one strand of double-stranded DNA. "N" represents any base.

FIGURE 4.14 *Eco*RI digestion and gel electrophoresis of λ DNA
*Eco*RI cleaves λ DNA at five sites (arrows), yielding six DNA fragments. These fragments are then separated by electrophoresis in an agarose gel. The DNA fragments migrate toward the positive electrode, with smaller fragments moving more rapidly through the gel. Following electrophoresis, the DNA is stained with a fluorescent dye and photographed. The sizes of DNA fragments are indicated.

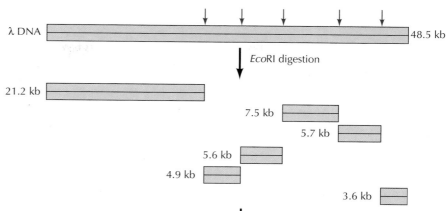

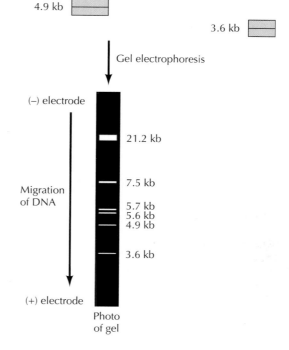

bases long (1 kilobase, or kb = 1000 base pairs) (Figure 4.14). These fragments can be separated according to size by **gel electrophoresis**—a common method in which molecules are separated based on the rates of their migration in an electric field. A gel, usually formed from agarose or polyacrylamide, is placed between two buffer compartments containing electrodes. The sample (e.g., the mixture of DNA fragments to be analyzed) is then pipetted into preformed slots in the gel, and the electric field is turned on. Nucleic acids are negatively charged (because of their phosphate backbone), so they migrate toward the positive electrode. The gel acts like a sieve, selectively retarding the movement of larger molecules. Smaller molecules therefore move through the gel more rapidly, allowing a mixture of nucleic acids to be separated on the basis of size.

In addition to size, the order of restriction fragments can be determined by a variety of methods, yielding (for example) a map of the *Eco*RI sites in λ DNA. The locations of cleavage sites for multiple different restriction endonucleases can be used to generate detailed **restriction maps** of DNA molecules, such as viral genomes (Figure 4.15). In addition, individual DNA fragments produced by restriction endonuclease digestion can be isolated following electrophoresis for further study—including determination of their DNA sequence. The DNAs of many viruses have been characterized by this approach.

Restriction endonuclease digestion alone, however, does not provide sufficient resolution for the analysis of larger DNA molecules, such as cellular genomes. A restriction endonuclease with a six-base-pair recognition site (such as *Eco*RI) cleaves DNA with a statistical frequency of once every 4096

FIGURE 4.15 Restriction maps of λ and adenovirus DNAs The locations of cleavage sites for *Bam*HI, *Eco*RI, and *Hind*III are shown in the DNAs of *E. coli* bacteriophage λ (48.5 kb) and human adenovirus-2 (35.9 kb).

MOLECULAR MEDICINE

HIV and AIDS

The Disease

Acquired immune deficiency syndrome (AIDS) is a new disease, first described in 1981. It has now become a worldwide pandemic, with approximately 70 million people having been infected with HIV and nearly 30 million having died of AIDS. The clinical manifestations of AIDS result principally from failure of the immune system to function normally. In the absence of normal immunity, AIDS patients are sensitive to opportunistic infections by agents (viruses, bacteria, fungi, and protozoans) against which a healthy individual would be resistant. People with AIDS also suffer a high frequency of some types of cancers, particularly lymphomas and Kaposi's sarcoma, although it is the opportunistic infections that are responsible for most deaths.

Molecular and Cellular Basis

AIDS is caused by a retrovirus (human immunodeficiency virus or HIV) that was discovered by the research groups of Robert Gallo and Luc Montagnier in 1983. HIV infects principally a specific type of lymphocyte (the T4 lymphocyte) that is required for a normal immune response. In contrast to many other retroviruses, such as Rous sarcoma virus, HIV does not cause the cells it infects to become cancerous. Instead, HIV eventually kills the cells in which it replicates, ultimately resulting in the depletion of the population of T4 lymphocytes and the failure of the immune system in infected individuals. This failure of the immune system in turn leads to the opportunistic infections and cancers that represent the clinical manifestations of AIDS.

Prevention and Treatment

At present, the only means of preventing AIDS is to avoid HIV infection. HIV is a fragile virus that quickly loses infectivity outside the body, so it cannot be transmitted by casual contact with an infected person. HIV can be transmitted in three ways: through sexual contact, through contaminated blood products, and from mother to child during pregnancy or breast-feeding. Following the isolation of HIV, screening tests were developed to ensure the safety of clotting factors and blood supplies used for transfusions. Prevention of HIV infection by other routes currently depends on individuals minimizing their personal risk of infection by adhering to safe sexual practices and avoiding sources of contaminated blood, such as shared needles used for intravenous drug injection.

Beyond modifying individual behavior to reduce the risk of infection, the identification of HIV as the cause of AIDS opens possibilities for prevention and treatment. A vaccine to prevent HIV infection is being actively pursued, although several features of the biology of HIV pose difficulties to this approach. Alternatively, drugs that inhibit virus replication are now providing effective therapies for HIV-infected individuals. These drugs either act as inhibitors of the HIV reverse transcriptase or of the HIV protease, which is an enzyme required for processing viral proteins. Combinations of such drugs are now prolonging the lives of AIDS patients, although further work is clearly needed to develop drugs that are not only more effective but also less expensive and more practical for use in developing countries.

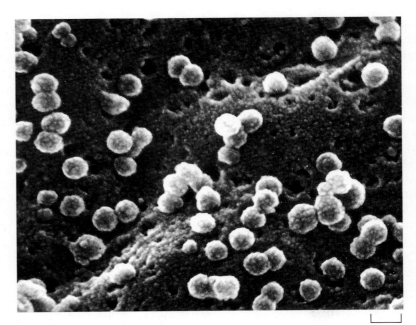

Scanning electron micrograph of HIV budding from T lymphocytes (Cecil Fox/Photo Researchers, Inc.)

0.1 μm

base pairs ($1/4^6$). A molecule the size of λ DNA (48.5 kb) would therefore be expected to yield about ten *Eco*RI fragments, consistent with the results illustrated in Figure 4.14. However, restriction endonuclease digestion of larger genomes yields quite different results. For example, the human genome is approximately 3×10^6 kb long and is therefore expected to yield more than 500,000 *Eco*RI fragments. Such a large number of fragments cannot be separated from one another, so agarose gel electrophoresis of *Eco*RI-digested human DNA yields a continuous smear rather than a discrete pattern of DNA fragments. Because it is impossible to isolate single restriction fragments from such digests, restriction endonuclease digestion alone does not yield a source of homogeneous DNA suitable for further analysis. Quantities of such purified DNA fragments, however, can be obtained through molecular cloning.

Generation of Recombinant DNA Molecules

The basic strategy in **molecular cloning** is to insert a DNA fragment of interest (e.g., a segment of human DNA) into a DNA molecule (called a **vector**) that is capable of independent replication in a host cell. The result is a **recombinant molecule** or **molecular clone**, composed of the DNA insert linked to vector DNA sequences. Large quantities of the inserted DNA can be obtained if the recombinant molecule is allowed to replicate in an appropriate host. For example, fragments of human DNA can be cloned in plasmid vectors (Figure 4.16). **Plasmids** are small circular DNA molecules that can replicate independently—without being associated with chromosomal DNA— in bacteria. Recombinant plasmids carrying human DNA inserts can be introduced into *E. coli*, where they replicate along with the bacteria to yield millions of copies of plasmid DNA. The DNA of these plasmids can then be isolated, yielding large quantities of recombinant molecules containing a single fragment of human DNA. Whereas a typical DNA fragment 1 kb in length would represent less than 1 part in a million of human genomic DNA, it would represent approximately 1 part in 5 after being cloned in a plasmid vector. Moreover, the fragment can be easily isolated from the rest of the vector DNA by restriction endonuclease digestion and gel electrophoresis, allowing a pure fragment of human DNA to be analyzed and further manipulated.

The DNA fragments used to create recombinant molecules are usually generated by digestion with restriction endonucleases. Many of these enzymes cleave their recognition sequences at staggered sites, leaving overhanging or cohesive single-stranded tails that can associate with each other by complementary base pairing (Figure 4.17). The association between such paired complementary ends can be established permanently by treatment with **DNA ligase**, an enzyme that seals breaks in DNA strands (see Chapter 6). Thus two different fragments of DNA (e.g., a human DNA insert and a plasmid DNA vector) prepared by digestion with the same restriction endonuclease can be readily joined to create a recombinant DNA molecule.

The fragments of DNA that can be cloned are not limited to those that terminate in restriction endonuclease cleavage sites. Synthetic DNA "linkers" containing desired restriction endonuclease sites can be added to the

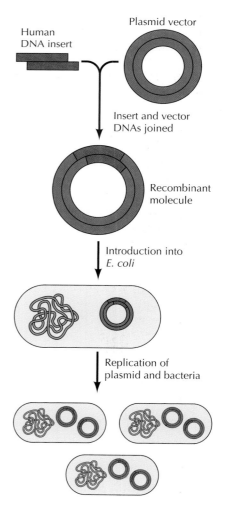

4.7 WEBSITE ANIMATION

Recombinant DNA Molecules The basic strategy in molecular cloning is to insert a DNA fragment of interest into a DNA molecule (called a vector) that is capable of independent replication in a host cell.

Human DNA insert

Plasmid vector

Insert and vector DNAs joined

Recombinant molecule

Introduction into *E. coli*

Replication of plasmid and bacteria

FIGURE 4.16 Generation of a recombinant DNA molecule A fragment of human DNA is inserted into a plasmid DNA vector. The resulting recombinant molecule is then introduced into *E. coli*, where it replicates along with the bacteria to yield a population of bacteria carrying plasmids with the human DNA insert.

FIGURE 4.17 Joining of DNA molecules Vector and insert DNAs are digested with a restriction endonuclease (such as *Eco*RI), which cleaves at staggered sites leaving overhanging single-stranded tails. Vector and insert DNAs can then associate by complementary base pairing, and covalent joining of the DNA strands by DNA ligase yields a recombinant molecule.

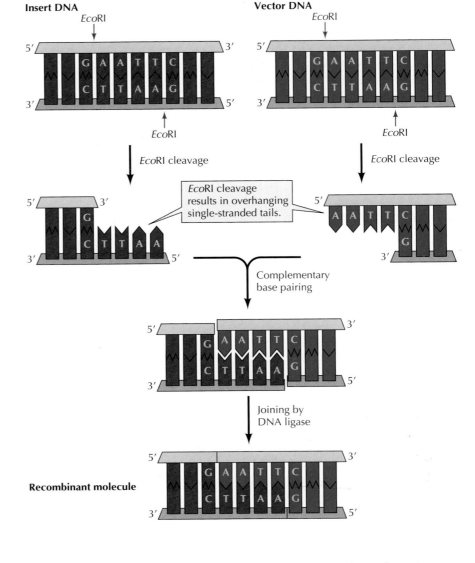

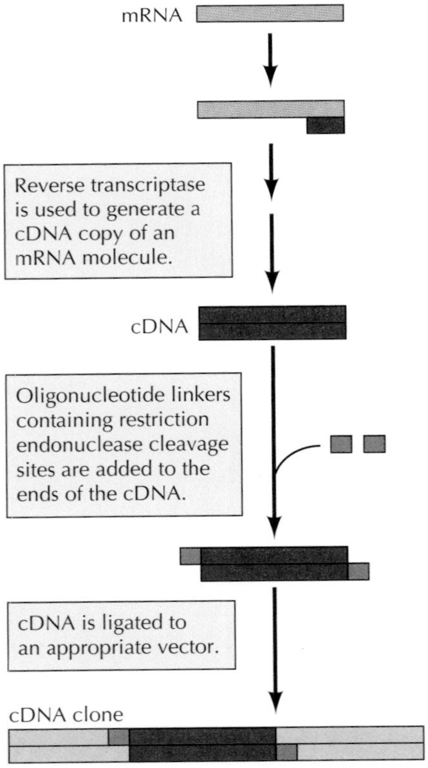

FIGURE 4.18 cDNA cloning

ends of any DNA fragment, allowing virtually any fragment of DNA to be ligated to a vector and isolated as a molecular clone.

Not only DNA, but also RNA sequences can be cloned (Figure 4.18). The first step is to synthesize a DNA copy of the RNA using the enzyme reverse transcriptase. The DNA product (called a **cDNA** because it is *complementary* to the RNA used as a template) can then be ligated to vector DNA as already described. Since eukaryotic genes are usually interrupted by noncoding sequences (introns; see Chapter 5), which are removed from mRNA by splicing, the ability to clone cDNA as well as genomic DNA has been critical for understanding gene structure and function. Moreover, cDNA cloning allows the mRNA corresponding to a single gene to be isolated as a molecular clone.

Vectors for Recombinant DNA

Depending on the size of the insert DNA and the purpose of the experiment, many different types of cloning vectors can be used for the generation of recombinant molecules. The basic vector systems used for the isolation and propagation of cloned DNAs are reviewed here. Other vectors developed for the expression of cloned DNAs and the introduction of recombinant molecules into eukaryotic cells are discussed in subsequent sections.

Plasmids are commonly used for cloning genomic or cDNA inserts of up to a few thousand base pairs. Plasmid vectors usually consist of 2 to 4 kb of DNA, including an **origin of replication**—the DNA sequence that signals the host cell DNA polymerase to replicate the DNA molecule. In addition, plasmid vectors carry genes that confer resistance to antibiotics (e.g., ampicillin resistance), so bacteria carrying the plasmids can be selected. For example, Figure 4.19 illustrates the isolation of human cDNA clones in a plasmid vector. A pool of cDNA fragments are ligated to restriction endonuclease-digested plasmid DNA. The resulting recombinant DNA molecules are then used to transform *E. coli*. Antibiotic-resistant colonies, which contain plasmid DNA, are selected. Since each recombinant plasmid yields a single antibiotic-resistant colony, the bacteria present in any given colony will contain a unique cDNA insert. Plasmid-containing bacteria can then be grown in large quantities and their DNA extracted. The small circular plasmid DNA molecules, of which there are often hundreds of copies per cell, can be separated from the bacterial chromosomal DNA; the result is purified plasmid DNA that is suitable for analysis of the cloned insert.

Bacteriophage λ vectors are also used for the isolation of either genomic or cDNA clones from eukaryotic cells, and will accommodate larger fragments of insert DNA than plasmids. In λ cloning vectors, sequences of the bacteriophage genome that are dispensable for virus replication have been removed and replaced with unique restriction sites for insertion of cloned DNA. These recombinant molecules can be introduced into *E. coli*, where they replicate to yield millions of progeny phages containing a single DNA insert. The DNA of these phages can then be isolated, yielding large quantities of recombinant molecules containing a single fragment of cloned DNA. The DNA inserts can be as large as about 15 kb and still yield a recombinant genome that can be packaged into bacteriophage λ particles.

For many studies involving analysis of genomic DNA, it is desirable to clone larger fragments of DNA than are accommodated by plasmid or λ vectors. There are five major types of vectors that are used for this purpose (Table 4.3). **Cosmid** vectors accommodate inserts of approximately 45 kb. These vectors contain bacteriophage λ sequences that allow efficient packaging of the cloned DNA into phage particles. In addition, cosmids contain origins of replication and the genes for antibiotic resistance that are characteristic of plasmids, so they are able to replicate as plasmids in bacterial cells. Two other types of vectors are derived from bacteriophage P1, rather than from bacteriophage λ. Bacteriophage P1 vectors, which will accommodate DNA fragments of 70–100 kb, contain sequences that allow recombinant molecules to be packaged *in vitro* into P1 phage particles and then to be replicated as plasmids in *E. coli*. **P1 artificial chromosome (PAC)** vectors also contain sequences of bacteriophage P1 but are introduced directly as plasmids into *E. coli* and will accommodate larger inserts of 130–150 kb. **Bacterial artificial chromosome (BAC)** vectors are derived from a naturally occurring plasmid of *E. coli* (called the F factor). The replication origin and other F factor sequences allow BACs to replicate as stable plasmids carrying inserts of 120–300 kb. Even larger fragments of DNA (250–400 kb) can be cloned in **yeast artificial chromosome (YAC)** vectors. These vectors contain yeast origins of replication as well as other sequences (centromeres and telomeres, discussed in Chapter 5) that allow them to replicate as linear chromosome-like molecules in yeast cells.

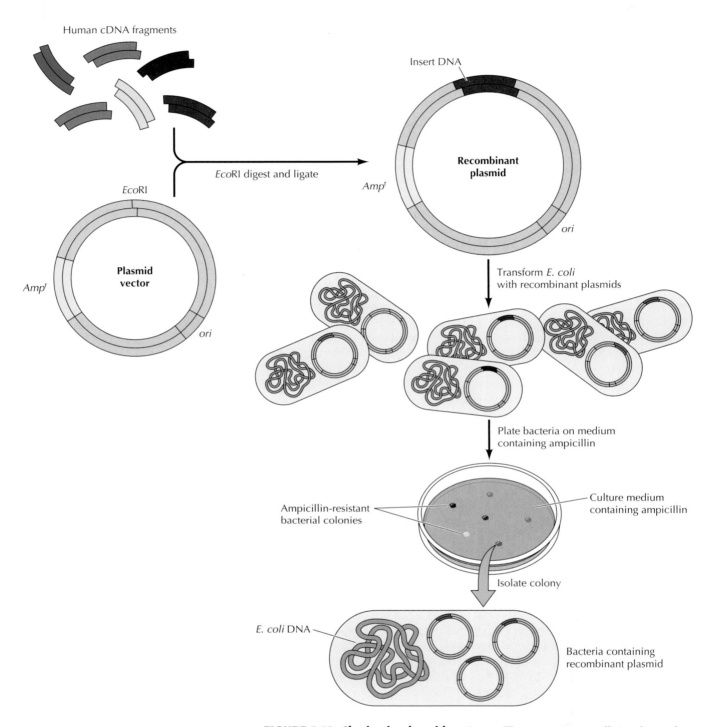

Human cDNA fragments

Insert DNA

Recombinant plasmid

*Amp^r^*

ori

EcoRI

EcoRI digest and ligate

*Amp^r^*

Plasmid vector

*Amp^r^*

ori

Transform *E. coli* with recombinant plasmids

Plate bacteria on medium containing ampicillin

Ampicillin-resistant bacterial colonies

Culture medium containing ampicillin

Isolate colony

E. coli DNA

Bacteria containing recombinant plasmid

FIGURE 4.19 Cloning in plasmid vectors The vector is a small circular molecule that contains an origin of replication (*ori*), a gene conferring resistance to ampicillin (*Amp^r^*), and a restriction site (e.g., *EcoRI*), which can be used to insert foreign DNA. Insert DNA (e.g., human cDNA) is ligated to the vector, and the recombinant plasmids are used to transform *E. coli*. The bacteria are plated on medium containing ampicillin, so that only the bacteria that are ampicillin-resistant because they carry plasmid DNAs are able to form colonies. Individual colonies of plasmid-containing bacteria can then be isolated and grown in large quantities for isolation of recombinant plasmids.

TABLE 4.3 Vectors for Cloning Large Fragments of DNA

Vector	DNA Insert (kb)	Host cell
Cosmids	30–45	*E. coli*
Bacteriophage P1	70–100	*E. coli*
P1 Artificial Chromosomes (PACs)	130–150	*E. coli*
Bacterial Artificial Chromosomes (BACs)	120–300	*E. coli*
Yeast Artificial Chromosomes (YACs)	250–400	Yeast

DNA Sequencing

Molecular cloning allows the isolation of individual fragments of DNA in quantities suitable for detailed characterization, including the determination of nucleotide sequence. Indeed, determination of the nucleotide sequences of many genes has elucidated not only the structure of their protein products but also the properties of DNA sequences that regulate gene expression. Furthermore, the coding sequences of novel genes are frequently related to those of previously studied genes, and the functions of newly isolated genes can often be correctly deduced on the basis of such sequence similarities.

DNA sequencing is usually performed with automated systems that are both rapid and accurate, so determining the sequence of several kilobases of DNA is a straightforward task. Thus it is far easier to clone and sequence DNA than it is to determine the amino acid sequence of a protein. Since the nucleotide sequence of a gene can be readily translated into the amino acid sequence of its encoded protein, the easiest way of determining protein sequence is the sequencing of a cloned gene or cDNA.

The most common method of DNA sequencing is based on premature termination of DNA synthesis resulting from the inclusion of chain-terminating **dideoxynucleotides** (which do not contain the deoxyribose 3′ hydroxyl group) in DNA polymerase reactions (Figure 4.20). DNA synthesis is initiated at a unique site on the cloned DNA from a synthetic primer. The DNA synthesis reaction includes each of the four dideoxynucleotides (A, C, G, and T) in addition to their normal counterparts. Each of the four dideoxynucleotides is labeled with a different fluorescent dye, so their incorporation into DNA can be monitored. Incorporation of a dideoxynucleotide stops further DNA synthesis because no 3′ hydroxyl group is available for addition of the next nucleotide. Thus a series of labeled DNA molecules is generated, each terminating at the base represented by a specific fluorescent dideoxynucleotide. These fragments of DNA are then separated according to size by gel electrophoresis. As the newly synthesized DNA strands are electrophoresed through the gel, they pass through a laser beam that excites the fluorescent labels. The resulting emitted light is then detected by a photomultiplier, and a computer collects and analyzes the data. The size of each fragment is determined by its terminal dideoxynucleotide, marked by a specific color fluorescence, so the DNA sequence can be read from the order of fluorescent-labeled fragments as they migrate through the gel. High throughput automated DNA sequencing of this type has enabled the large-scale analysis required for determination of the sequences of complete genomes, including that of humans.

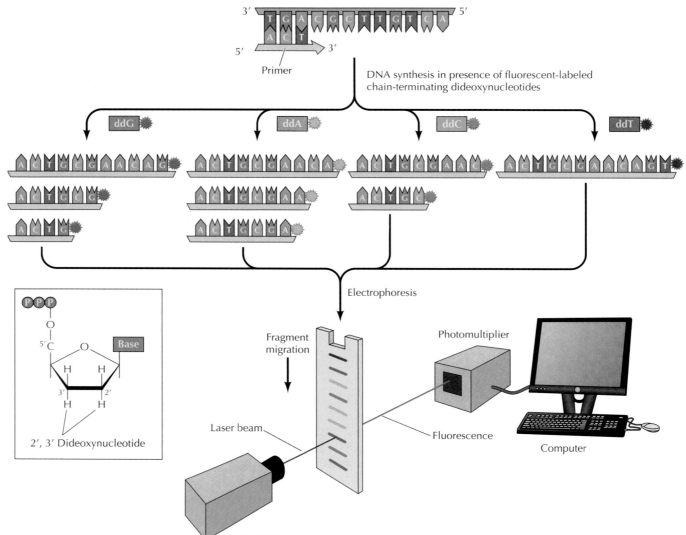

FIGURE 4.20 DNA sequencing   Dideoxynucleotides, which lack OH groups at the 3′ as well as the 2′ position of deoxyribose, are used to terminate DNA synthesis at specific bases. These molecules are incorporated normally into growing DNA strands. Because they lack a 3′ OH, however, the next nucleotide cannot be added, so synthesis of that DNA strand terminates. DNA synthesis is initiated at a specific site with a primer. The reaction contains the four dideoxynucleotides, each labeled with a different fluorescent dye, as well as the four normal deoxynucleotides. When the dideoxynucleotide is incorporated, DNA synthesis stops, so the reaction yields a series of products extending from the primer to the base substituted by a fluorescent dideoxynucleotide. These products are then separated by gel electrophoresis. As the DNA strands migrate through the gel, they pass through a laser beam that excites the fluorescent labels on the dideoxynucleotides. The emitted light is detected by a photomultiplier, which is connected to a computer that collects and analyzes the data to determine the DNA sequence.

Expression of Cloned Genes

In addition to enabling determination of the nucleotide sequences of genes—and hence the amino acid sequences of their protein products—molecular cloning has provided new approaches to obtaining large amounts of proteins for structural and functional characterization. Many proteins of interest are present at only low levels in eukaryotic cells and therefore cannot be purified in significant amounts by conventional bio-

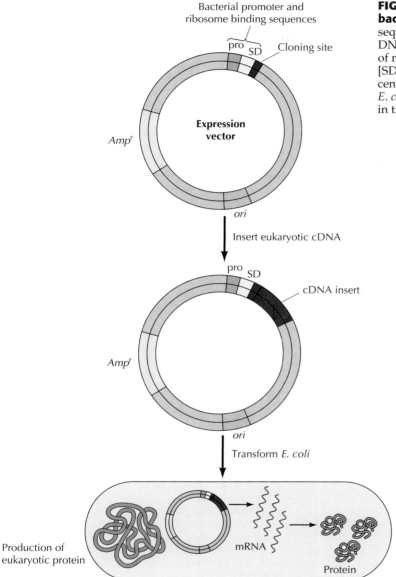

FIGURE 4.21 Expression of cloned genes in bacteria Expression vectors contain promoter sequences (pro) that direct transcription of inserted DNA in bacteria and sequences required for binding of mRNA to bacterial ribosomes (Shine-Dalgarno [SD] sequences). A eukaryotic cDNA inserted adjacent to these sequences can be efficiently expressed in *E. coli*, resulting in production of eukaryotic proteins in transformed bacteria.

chemical techniques. Given a cloned gene, however, this problem can be solved by the engineering of vectors that lead to high levels of gene expression in either bacteria or eukaryotic cells.

To express a eukaryotic gene in *E. coli*, the cDNA of interest is cloned into a plasmid or phage vector (called an **expression vector**) that contains sequences that drive transcription and translation of the inserted gene in bacterial cells (**Figure 4.21**). Inserted genes often can be expressed at levels high enough that the protein encoded by the cloned gene corresponds to as much as 10% of the total bacterial protein. Purifying the protein encoded by the cloned gene in quantities suitable for detailed biochemical or structural studies is then a straightforward matter.

It is frequently useful to express high levels of a cloned gene in eukaryotic cells rather than in bacteria. This mode of expression may be important, for example, to ensure that posttranslational modifications of the protein (such as addition of carbohydrates or lipids) occur normally. Such protein expression in eukaryotic cells can be achieved, as in *E. coli*, by insertion of the cloned gene into a vector (usually derived from a virus) that directs high-level gene expression. One system frequently used for protein expression in eukaryotic cells is infection of insect cells by baculovirus vectors, which direct very high levels of expression of genes inserted in place of a viral structural protein. Alternatively, high levels of protein expression can be achieved using appropriate vectors in mammalian cells.

Expression of cloned genes in yeast is particularly useful because simple methods of yeast genetics can be employed to identify proteins that interact with one another. In this type of analysis, called the **yeast two-hybrid** system, two different cDNAs (for example, from human cells) are joined to two distinct domains of a protein that stimulates expression of a target gene in yeast (Figure 4.22). Yeast are then transformed with the hybrid cDNA clones to test for interactions between the two proteins. If the human proteins do interact with each other, they will bring the two domains of the yeast protein together, resulting in stimulation of target gene expression in the transformed yeast. Expression of the target gene can be easily detected by growth of the yeast in a specific medium or by production of an enzyme that produces a blue yeast colony, so the yeast two-hybrid system provides a straightforward method to test protein-protein interactions. Indeed, high throughput yeast two-hybrid screens have been used to construct large scale interaction maps of thousands of proteins in eukaryotic cells (see Figure 2.33).

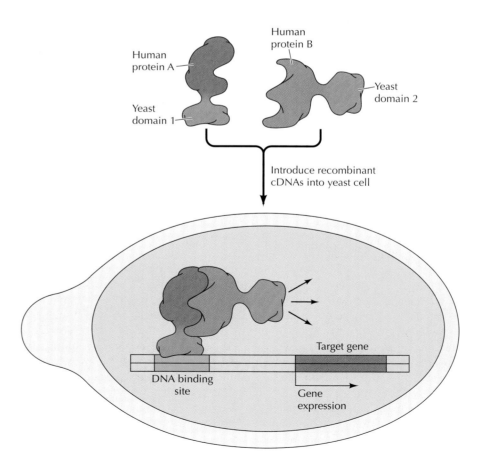

FIGURE 4.22 The yeast two-hybrid system cDNAs of two human proteins are cloned as fusions with two domains (designated 1 and 2) of a yeast protein that stimulates transcription of a target gene. The two recombinant cDNAs are introduced into a yeast cell. If the two human proteins interact with each other, they bring the two domains of the yeast protein together. Domain 1 binds DNA sequences at a site upstream of the target gene, and domain 2 stimulates target gene transcription. The interaction between the two human proteins can thus be detected by expression of the target gene in transformed yeast.

Detection of Nucleic Acids and Proteins

The advent of molecular cloning has enabled the isolation and characterization of individual genes from eukaryotic cells. Understanding the role of genes within cells, however, requires analysis of the intracellular organization and expression of individual genes and their encoded proteins. In this section, the basic procedures used for detection of specific nucleic acids and proteins are discussed. These approaches are important for a wide variety of studies, including the mapping of genes to chromosomes, the analysis of gene expression, and the localization of proteins to subcellular organelles.

Amplification of DNA by the Polymerase Chain Reaction

Molecular cloning allows individual DNA fragments to be propagated in bacteria and isolated in large amounts. An alternative method to isolating large amounts of a single DNA molecule is the **polymerase chain reaction (PCR)**, which was developed by Kary Mullis in 1988. Provided that some sequence of the DNA molecule is known, PCR can achieve a striking amplification of DNA via reactions carried out entirely *in vitro*. Essentially, DNA polymerase is used for repeated replication of a defined segment of DNA. The number of DNA molecules increases exponentially, doubling with each round of replication, so a substantial quantity of DNA can be obtained from a small number of initial template copies. For example, a single DNA molecule amplified through 30 cycles of replication would theoretically yield 2^{30} (approximately 1 billion) progeny molecules. Single DNA molecules can thus be amplified to yield readily detectable quantities of DNA that can be isolated by molecular cloning or further analyzed directly by restriction endonuclease digestion or nucleotide sequencing.

The general procedure for PCR amplification of DNA is illustrated in Figure 4.23. The starting material can be either a cloned DNA fragment or a mixture of DNA molecules—for example, total DNA from human cells. A specific region of DNA can be amplified from such a mixture, provided that the nucleotide sequence surrounding the region is known so that primers can be designed to initiate DNA synthesis at the desired point. Such primers are usually chemically synthesized oligonucleotides containing 15 to 20 bases of DNA. Two primers are used to initiate DNA synthesis in opposite directions from complementary DNA strands. The reaction is started by heating the template DNA to a high temperature (e.g., 95°C) so that the two strands separate. The temperature is then lowered to allow the primers to pair with their complementary sequences on the template strands. DNA polymerase then uses the primers to synthesize a new strand complementary to each template. Thus, in one cycle of amplification, two new DNA molecules are synthesized from one template molecule. The process can be repeated multiple times, with a twofold increase in DNA molecules resulting from each round of replication.

The multiple cycles of heating and cooling involved in PCR are performed by programmable heating blocks called thermocyclers. The DNA polymerases used in these reactions are heat-stable enzymes from bacteria such as *Thermus aquaticus*, which lives in hot springs at temperatures of about 75°C. These polymerases are stable even at the high temperatures used to separate the strands of double-stranded DNA, so PCR amplification can be performed rapidly and automatically. RNA sequences can also be amplified by this method if reverse transcriptase is used to synthesize a cDNA copy prior to PCR amplification.

Starting DNA

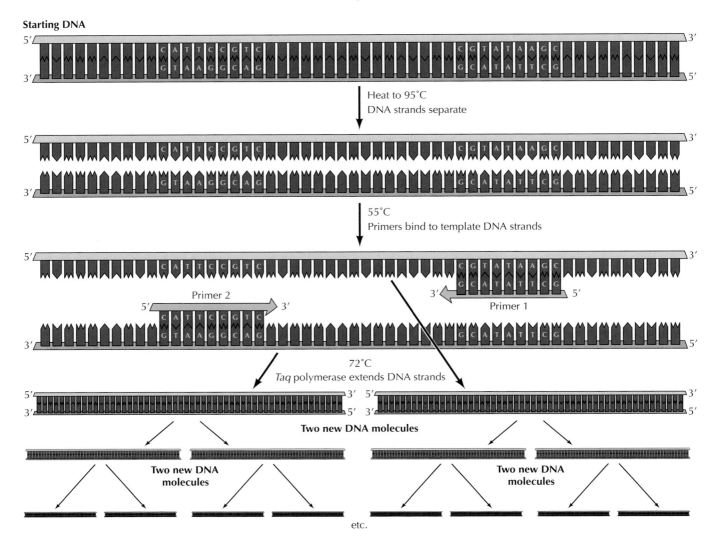

Heat to 95°C
DNA strands separate

55°C
Primers bind to template DNA strands

Primer 2

Primer 1

72°C
Taq polymerase extends DNA strands

Two new DNA molecules

Two new DNA molecules

Two new DNA molecules

etc.

FIGURE 4.23 Amplification of DNA by PCR The region of DNA to be amplified is flanked by two sequences used to prime DNA synthesis. The starting double-stranded DNA is heated to separate the strands and then cooled to allow primers (usually oligonucleotides of 15 to 20 bases) to bind to each strand of DNA. DNA polymerase from *Thermus aquaticus* (*Taq* polymerase) is used to synthesize new DNA strands starting from the primers, resulting in the formation of two new DNA molecules. The process can be repeated for multiple cycles, each resulting in a twofold amplification of DNA.

■ **PCR has become an extremely powerful forensic method. Amplification by PCR allows investigators to obtain a DNA profile from small samples of DNA at a crime scene.**

If enough of the sequence of a gene is known so that primers can be specified, PCR amplification provides an extremely powerful method of detecting small amounts of specific DNA or RNA molecules in a complex mixture of other molecules. The only DNA molecules that will be amplified by PCR are those containing sequences complementary to the primers used in the reaction. Therefore PCR can selectively amplify a specific template from complex mixtures, such as total cell DNA or RNA. This extraordinary sensitivity has made PCR an important method for a variety of applications, including the analysis of gene expression in cells available in only limited quantities.

The DNA segments amplified by PCR can also be directly sequenced or ligated to vectors and propagated as molecular clones. PCR thus allows the amplification and cloning of any segment of DNA for which primers can be designed. Since the complete genome sequences of many organisms are now known, PCR can be used to amplify and clone a vast array of desired DNA fragments, making it a powerful addition to the repertoire of recombinant DNA techniques.

Nucleic Acid Hybridization

The key to detection of specific nucleic acid sequences is base pairing between complementary strands of RNA or DNA. At high temperatures (e.g., 90 to 100°C) the complementary strands of DNA separate (denature), yielding single-stranded molecules. If such denatured DNA strands are then incubated under appropriate conditions (e.g., 65°), they will renature to form double-stranded molecules as dictated by complementary base pairing—a process called **nucleic acid hybridization**. Nucleic acid hybrids can be formed between two strands of DNA, two strands of RNA, or one strand of DNA and one of RNA.

As discussed above, hybridization between the primers and the template DNA provides the specificity to PCR amplification. In addition, a variety of other methods use nucleic acid hybridization as a means for detecting DNA or RNA sequences that are complementary to any isolated nucleic acid, such as a cloned DNA sequence (Figure 4.24). The cloned DNA is labeled with either radioactive nucleotides or with modified nucleotides that can be detected by fluorescence or chemiluminescence. This labeled DNA is then used as a probe for hybridization to complementary DNA or RNA sequences, which are detected by virtue of the radioactivity, fluorescence, or luminescence of the resulting double-stranded hybrids.

Southern blotting (a technique developed by E. M. Southern) is widely used for detection of specific genes in cellular DNA (Figure 4.25). The DNA to be analyzed is digested with a restriction endonuclease, and the digested DNA fragments are separated by gel electrophoresis. The gel is then overlaid with a nitrocellulose filter or nylon membrane to which the DNA fragments are transferred (blotted) to yield a replica of the gel. The filter is then incubated with a labeled probe, which hybridizes to the DNA fragments that contain the complementary sequence, allowing visualization of these specific fragments of cell DNA.

Northern blotting is a variation of the Southern blotting technique (hence its name) that is used for detection of RNA instead of DNA. In this method, total cellular RNAs are extracted and fractionated according to size by gel electrophoresis. As in Southern blotting, the RNAs are transferred to a filter and detected by hybridization with a cloned probe. Northern blotting is frequently used in studies of gene expression—for example, to determine whether specific mRNAs are present in different types of cells.

FIGURE 4.24 Detection of DNA by nucleic acid hybridization   A specific sequence can be detected in total cell DNA by hybridization with a labeled DNA probe, containing radioactive nucleotides or modified nucleotides that can be detected by fluorescence or chemiluminescence. The DNA is denatured by heating to 95°C, yielding single-stranded molecules. The labeled probe is then added and the temperature is lowered to 65°C, allowing complementary DNA strands to renature by pairing with each other. The probe hybridizes to complementary sequences in cell DNA, which can then be detected by incorporation of the labeled probe into double-stranded molecules.

4.9 WEBSITE ANIMATION
Polymerase Chain Reaction The polymerase chain reaction allows the production of millions of copies of a DNA fragment from just one starting DNA molecule.

4.10 WEBSITE ANIMATION
Nucleic Acid Hybridization At high temperatures, the complementary strands of DNA separate, and then when they are cooled, they re-form double-stranded molecules as dictated by complementary base pairing.

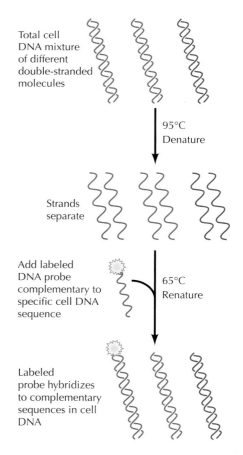

Total cell DNA mixture of different double-stranded molecules

95°C Denature

Strands separate

Add labeled DNA probe complementary to specific cell DNA sequence

65°C Renature

Labeled probe hybridizes to complementary sequences in cell DNA

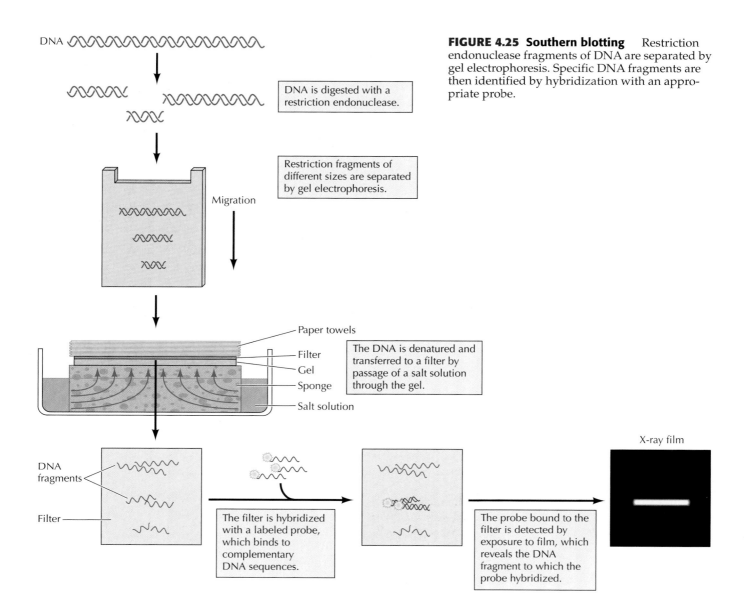

DNA

DNA is digested with a restriction endonuclease.

Restriction fragments of different sizes are separated by gel electrophoresis.

Migration

Paper towels
Filter
Gel
Sponge
Salt solution

The DNA is denatured and transferred to a filter by passage of a salt solution through the gel.

DNA fragments

Filter

The filter is hybridized with a labeled probe, which binds to complementary DNA sequences.

The probe bound to the filter is detected by exposure to film, which reveals the DNA fragment to which the probe hybridized.

X-ray film

FIGURE 4.25 Southern blotting   Restriction endonuclease fragments of DNA are separated by gel electrophoresis. Specific DNA fragments are then identified by hybridization with an appropriate probe.

4.11 WEBSITE ANIMATION

Southern Blotting
DNA fragments are separated by size using gel electrophoresis, and then—as part of the Southern blotting technique—they are incubated with a radioactive DNA probe to identify specific DNA fragments.

Nucleic acid hybridization can also be used to identify molecular clones that contain specific cellular DNA inserts. The first step in isolation of either genomic or cDNA clones is frequently the preparation of **recombinant DNA libraries**—collections of clones that contain all the genomic or mRNA sequences of a particular cell type (Figure 4.26). For example, a genomic library of human DNA might be prepared by cloning random DNA fragments of about 15 kb in a λ vector. Since the human genome is approximately 3 million kb, the complete human genome would be represented in a collection of approximately 500,000 such clones. Any gene for which a probe is available can then be isolated from such a recombinant library. The recombinant phages are plated on *E. coli*, and each phage replicates to produce a plaque on the lawn of bacteria. The plaques are then blotted onto filters in a process similar to the transfer of DNA from a gel to a filter during Southern blotting, and the filters are hybridized with a labeled probe to identify the phage plaques that contain the gene of interest. A variety of

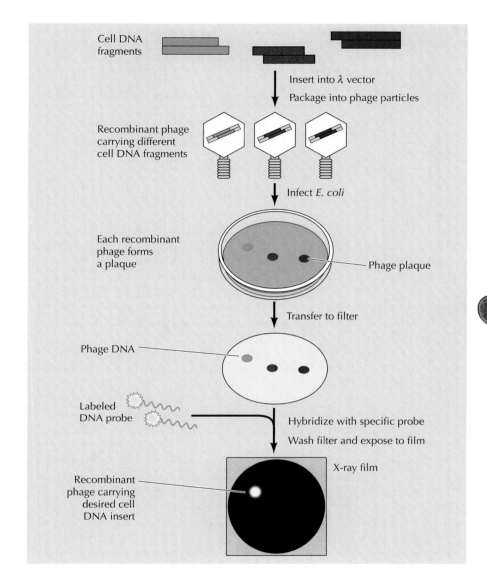

Cell DNA fragments

Insert into λ vector

Package into phage particles

Recombinant phage carrying different cell DNA fragments

Infect *E. coli*

Each recombinant phage forms a plaque

Phage plaque

Transfer to filter

Phage DNA

Labeled DNA probe

Hybridize with specific probe

Wash filter and expose to film

X-ray film

Recombinant phage carrying desired cell DNA insert

FIGURE 4.26 Screening a recombinant library by hybridization
Fragments of cell DNA are cloned in a bacteriophage λ vector and packaged into phage particles, yielding a collection of recombinant phage carrying different cell inserts. The phages are used to infect bacteria, and the culture is overlaid with a filter. Some of the phages in each plaque are transferred to the filter, which is then hybridized with a labeled probe to identify the phage plaque containing the desired gene. The appropriate phage plaque can then be isolated from the original culture plate.

4.12 WEBSITE ANIMATION

Colony Hybridization
Bacteria can house recombinant DNA libraries, and specific recombinant DNA molecules within the library can be identified by the colony hybridization procedure—a procedure in which the DNA is incubated with a specific radioactive DNA probe.

probes can be used for such experiments. For example, a cDNA clone can be used as a probe to isolate the corresponding genomic clone, or a gene cloned from one species (e.g., mouse) can be used to isolate a related gene from a different species (e.g., human). The appropriate plaque can then be isolated from the original plate in order to propagate the recombinant phage that carries the desired cell DNA insert. Similar procedures can be used to screen bacterial colonies carrying plasmid DNA clones, so specific clones can be isolated by hybridization from either phage or plasmid libraries.

Rather than analyzing one gene at a time, as in Southern or Northern blotting, hybridization to **DNA microarrays** allows tens of thousands of genes to be analyzed simultaneously. As the complete sequences of eukaryotic genomes have become available, hybridization to DNA microarrays has enabled researchers to undertake global analyses of sequences present in either cellular DNA or RNA samples. A DNA microarray consists of a glass slide or membrane filter onto which oligonucleotides or fragments of

(A)

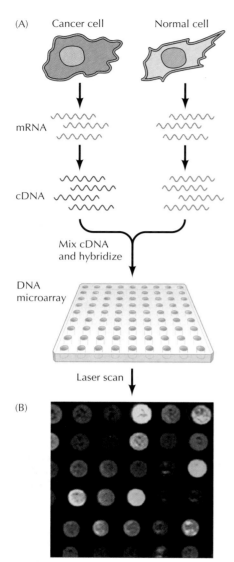

cDNAs are printed by a robotic system in small spots at a high density (**Figure 4.27**). Each spot on the array consists of a single oligonucleotide or cDNA. More than 10,000 unique DNA sequences can be printed onto a typical glass microscope slide, so it is readily possible to produce DNA microarrays containing sequences representing all of the genes in cellular genomes. As illustrated in Figure 4.27, one widespread application of DNA microarrays is in studies of gene expression; for example, a comparison of the genes expressed by two different types of cells. In an experiment of this type, cDNA probes are synthesized from the mRNAs expressed in each of the two cell types (e.g., cancer cells and normal cells). The two cDNAs are labeled with different fluorescent dyes (typically red and green), and a mixture of the cDNAs is hybridized to a DNA microarray in which 10,000 or more human genes are represented as single spots. The array is then analyzed using a high-resolution laser scanner, and the relative extent of transcription of each gene in the cancer cells compared to the normal cells is indicated by the ratio of red to green fluorescence at the appropriate spot on the array.

Nucleic acid hybridization can be used to detect homologous DNA or RNA sequences not only in cell extracts but also in chromosomes or intact cells—a procedure called *in situ* **hybridization** (Figure 4.28). In this case, the hybridization of fluorescent probes to specific cells or subcellular structures is analyzed by microscopic examination. For example, labeled probes can be hybridized to intact chromosomes in order to identify the chromosomal regions that contain a gene of interest. *In situ* hybridization can also be used to detect specific mRNAs in different types of cells within a tissue.

Antibodies as Probes for Proteins

Studies of gene expression and function require the detection not only of DNA and RNA, but also of specific proteins. For these studies, **antibodies** take the place of nucleic acid probes as reagents that can selectively react with unique protein molecules. Antibodies are proteins produced by cells of the immune system (B lymphocytes) that react against molecules (**antigens**) that the host organism recognizes as foreign substances—for example, the protein coat of a virus. The immune systems of vertebrates are capable of producing millions of different antibodies, each of which specifically recognizes a unique antigen, which may be a protein, a carbohydrate, or a nonbiological molecule. An individual lymphocyte produces only a single type

FIGURE 4.27 DNA microarrays (A) An example of comparative analysis of gene expression in cancer cells and normal cells. mRNAs extracted from cancer cells and normal cells are used as templates for synthesis of cDNA probes labeled with different fluorescent dyes (e.g., a red fluorescent label for cancer cell cDNA and green for normal cell cDNA). The two cDNA probes are mixed and hybridized to a DNA microarray containing spots of oligonucleotides corresponding to 10,000 or more distinct human genes. The relative level of expression of each gene in cancer cells compared to normal cells is indicated by the ratio of red to green fluorescence at each position on the microarray. (B) Photograph of a portion of a microarray.

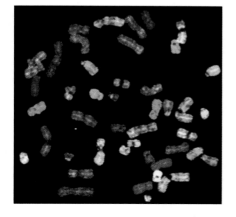

FIGURE 4.28 Fluorescence *in situ* hybridization Hybridization of human chromosomes with chromosome-specific fluorescent probes that label each of the 24 chromosomes a different color. (Courtesy of Thomas Reid and Hesed Padilla-Nash, National Cancer Institute.)

FIGURE 4.29 Western blotting Proteins are separated according to size by SDS-polyacrylamide gel electrophoresis and transferred from the gel to a filter. The filter is incubated with an antibody directed against a protein of interest. The antibody bound to the filter can then be detected by reaction with various reagents, such as a chemiluminescent probe that binds to the antibody.

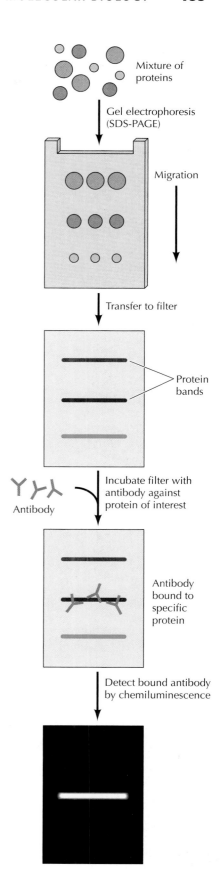

of antibody, but the antibody genes of different lymphocytes vary as a result of programmed gene rearrangements during development of the immune system (see Chapter 6). This variation gives rise to an array of lymphocytes with distinct antibody genes, programmed to respond to different antigens.

Antibodies can be generated by inoculation of an animal with any foreign protein. For example, antibodies against human proteins are frequently raised in rabbits. The sera of such immunized animals contain a mixture of antibodies (produced by different lymphocytes) that react against multiple sites on the immunizing antigen. However, single species of antibodies (**monoclonal antibodies**) can also be produced by the culturing of clonal lines of B lymphocytes from immunized animals (usually mice). Because each lymphocyte is programmed to produce only a single antibody, a clonal line of lymphocytes produces a monoclonal antibody that recognizes only a single antigenic determinant, thereby providing a highly specific immunological reagent.

Although antibodies can be raised against proteins purified from cells, other materials may also be used for immunization. For example, animals may be immunized with intact cells to raise antibodies against unknown proteins expressed by a specific cell type (e.g., a cancer cell). Such antibodies may then be used to identify proteins specifically expressed by the cell type used for immunization. In addition, antibodies are frequently raised against proteins expressed in bacteria as recombinant clones. In this way, molecular cloning allows the production of antibodies against proteins that may be difficult to isolate from eukaryotic cells. Moreover, antibodies can be raised against synthetic peptides that consist of only 10 to 15 amino acids, rather than against intact proteins. Therefore, once the sequence of a gene is known, antibodies against peptides synthesized from part of its predicted protein sequence can be produced. Because antibodies against these synthetic peptides frequently react with the intact protein as well, it is possible to produce antibodies against a protein starting with only the sequence of a gene.

Antibodies can be used in a variety of ways to detect proteins in cell extracts. Two common methods are **immunoblotting** (also called **Western blotting**) and **immunoprecipitation**. Western blotting (Figure 4.29) is another variation of Southern blotting. Proteins in cell extracts are first separated according to size by gel electrophoresis. Because proteins have different shapes and charges, however, this process requires a modification of the methods used for electrophoresis of nucleic acids. Proteins are separated by a method known as **SDS-polyacrylamide gel electrophoresis (SDS-PAGE)** in which they are dissolved in a solution containing the negatively charged detergent sodium dodecyl sulfate (SDS). Each protein binds many detergent molecules, which denature the protein and give the protein an overall negative charge. Under these conditions, all proteins migrate toward the positive electrode—their rates of migration determined (like those of nucleic acids) only by size. Following electrophoresis, the proteins are transferred to a filter, which is then allowed to react with antibodies against the protein of interest. The antibody bound to the filter can be detected by a

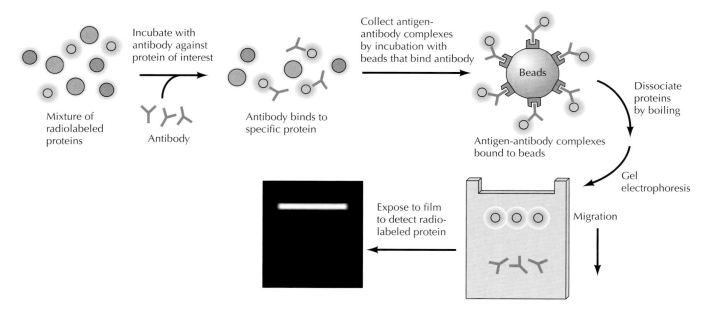

FIGURE 4.30 Immunoprecipitation Radiolabeled proteins are incubated with an antibody, which forms complexes with the protein against which it is directed (the antigen). These antigen-antibody complexes are collected on beads that bind the antibody. The beads are then boiled to dissociate the antigen-antibody complexes, and the recovered proteins are analyzed by SDS-polyacrylamide gel electrophoresis. The radioactive protein that was immunoprecipitated is detected by exposure to film.

variety of methods, such as chemiluminescence, to identify the protein against which the antibody is targeted.

In immunoprecipitation, antibodies are used to isolate the proteins against which they are directed (Figure 4.30). Typically, cells are incubated with radioactive amino acids to label their proteins. Such a radiolabeled cell extract is then incubated with an antibody, which binds to its antigenic target protein. The resulting antigen-antibody complexes are isolated and subjected to electrophoresis, allowing detection of the radioactive antigen.

Immunoprecipitation can also be used to detect protein-protein interactions within cells, by co-immunoprecipitation of two interacting proteins. As discussed in Chapter 2, one approach to identification of protein complexes is to immunoprecipitate a protein from cells under gentle conditions, so that it remains associated with the proteins it normally interacts with within the cell. The immunoprecipitated protein complexes can then be analyzed, for example by gel electrophoresis and mass spectrometry, to identify not only the protein against which the antibody was directed, but also other proteins with which it was associated in the cell extract.

As discussed in Chapter 1, antibodies can be used to visualize proteins in intact cells, as well as in cell lysates. For example, cells can be stained with antibodies labeled with fluorescent dyes, and the subcellular localization of the antigenic proteins visualized by fluorescence microscopy (Figure 4.31). Antibodies can also be labeled with tags that are visible in the electron microscope, such as heavy metals, allowing visualization of antigenic proteins at the ultrastructural level.

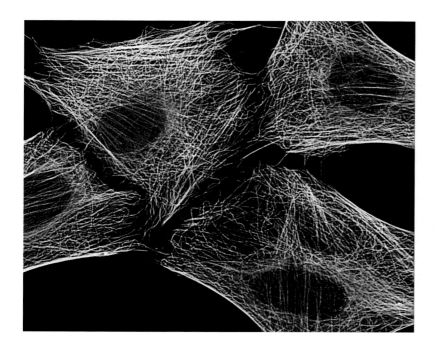

FIGURE 4.31 Immunofluorescence
Human cells in culture were stained with immunofluorescent antibodies against actin (blue) and tubulin (yellow). Nuclei are stained with a red fluorescent dye. (From Dr. Torsten Wittmann/Photo Researchers, Inc.)

Gene Function in Eukaryotes

The recombinant DNA techniques discussed in the preceding sections provide powerful approaches to the isolation and detailed characterization of the genes of eukaryotic cells. Understanding the function of a gene, however, requires analysis of the gene within cells or intact organisms—not simply as a molecular clone in bacteria. In classical genetics, the function of genes has generally been revealed by the altered phenotypes of mutant organisms. The advent of recombinant DNA has added a new dimension to studies of gene function. Namely, it has become possible to investigate the function of a cloned gene directly by reintroducing the cloned DNA into eukaryotic cells. In simpler eukaryotes, such as yeasts, this technique has made possible the isolation of molecular clones corresponding to virtually any mutant gene. In addition, there are several methods by which cloned genes can be introduced into cultured animal and plant cells, as well as intact organisms, for functional analysis. These approaches can be coupled with the ability to introduce mutations in cloned DNA *in vitro*, extending the power of recombinant DNA to allow functional studies of the genes of more complex eukaryotes.

Genetic Analysis in Yeasts

Yeasts are particularly advantageous for studies of eukaryotic molecular biology (see Chapter 1). The genome of *Saccharomyces cerevisiae*, which consists of approximately 1.2×10^7 base pairs, is 200 times smaller than the human genome. Moreover, yeasts can easily be grown in culture, reproducing with a division time of about 2 hours. Thus yeasts offer the same basic advantages—a small genome and rapid reproduction—that are afforded by bacteria.

Mutations in yeasts can be identified as readily as in *E. coli*. For example, yeast mutants that require a particular amino acid or other nutrient for growth can easily be isolated. In addition, yeasts with defects in genes required for fundamental cell processes (in contrast to metabolic defects) can

(A)

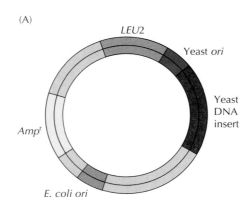

(B)

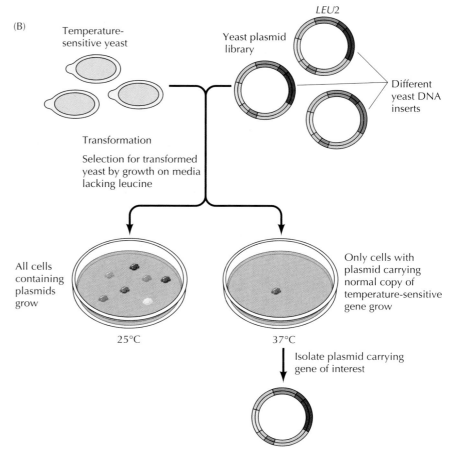

FIGURE 4.32 Cloning of yeast genes
(A) A yeast vector. The vector contains a bacterial origin of replication (*ori*) and an ampicillin resistance gene (*Amp^r*), allowing it to be propagated as a plasmid in *E. coli*. In addition, the vector contains a yeast origin of replication and a marker gene (*LEU2*), allowing the selection of transformed yeast. The *LEU2* gene encodes an enzyme required for synthesis of the amino acid leucine, so transformation of yeast strains lacking this enzyme can be selected for by growth on medium lacking leucine. (B) Isolation of a yeast gene. A gene of interest is identified by a temperature-sensitive mutation, which allows yeast to grow at 25°C but not at 37°C. To isolate a clone of the gene, the temperature-sensitive yeasts are transformed with a plasmid library containing a collection of genes encompassing the entire yeast genome. All yeasts transformed by plasmid DNAs are able to grow on media lacking leucine at 25°C, but only those yeasts transformed by a plasmid carrying a normal copy of the gene of interest are able to grow at 37°C. The desired plasmid can be isolated from transformed yeasts that form colonies at the nonpermissive temperature.

be isolated as **temperature-sensitive mutants**. Such mutants encode proteins that are functional at one temperature (the permissive temperature) but not another (the nonpermissive temperature), whereas normal proteins are functional at both. A yeast with a temperature-sensitive mutation in an essential gene can be identified by its ability to grow only at the permissive temperature. The ability to isolate such temperature-sensitive mutants has allowed the identification of yeast genes controlling many fundamental cell processes, such as RNA synthesis and processing, progression through the cell cycle, and transport of proteins between cellular compartments.

The relatively simple genetics of yeast also enables a gene corresponding to any yeast mutation to be cloned, simply on the basis of its functional activity (Figure 4.32). First, a genomic library of normal yeast DNA is prepared in vectors that replicate as plasmids in yeasts as well as in *E. coli*. The small size of the yeast genome means that a complete library consists of only a few thousand plasmids. A mixture of such plasmids is then used to transform a temperature-sensitive yeast mutant, and transformants that are able to grow at the nonpermissive temperature are selected. Such transformants have acquired a normal copy of the gene of interest on plasmid DNA, which can then be easily isolated from the transformed yeast cells for further characterization.

Yeast genes encoding a wide variety of essential proteins have been identified in this manner. In many cases, such genes isolated from yeasts have also been useful in identifying and cloning related genes from mammalian cells. Thus, the simple genetics of yeast has not only provided an important model for eukaryotic cells but has also led directly to the cloning of related genes from more complex eukaryotes.

Gene Transfer in Plants and Animals

Although the cells of complex eukaryotes are not amenable to the simple genetic manipulations possible in yeasts, gene function can still be assayed by the introduction of cloned DNA into plant and animal cells. Such experiments (generally called **gene transfer**) have proven critical to addressing a wide variety of questions, including studies of the mechanisms that regulate gene expression and protein processing. In addition, as discussed later in the book, gene transfer has enabled the identification and characterization of genes that control animal cell growth and differentiation, including a variety of genes responsible for the abnormal growth of human cancer cells.

The methodology for introduction of DNA into animal cells was initially developed for infectious viral DNAs and is therefore called **transfection** (a word derived from *trans*formation + in*fection*) (Figure 4.33). DNA can be introduced into animal cells in culture by a variety of methods, including direct microinjection into the cell nucleus, coprecipitation of DNA with calcium phosphate to form small particles that are taken up by the cells, incorporation of DNA into lipid vesicles (**liposomes**) that fuse with the plasma membrane, and exposure of cells to a brief electric pulse that transiently opens pores in the plasma membrane (**electroporation**). The DNA taken up by a high fraction of cells is transported to the nucleus, where it can be tran-

FIGURE 4.33 Introduction of DNA into animal cells   A eukaryotic gene of interest is cloned in a plasmid containing a drug resistance marker that can be selected for in cultured animal cells. The plasmid DNA is taken up and expressed by a high fraction of the cells for a few days (transient expression). Stably transformed cells, in which the plasmid DNA becomes integrated into chromosomal DNA, can then be selected for their ability to grow in drug-containing medium.

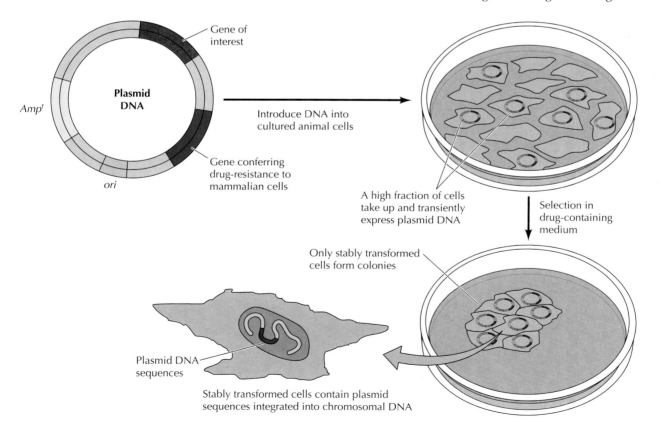

Gene of interest

Plasmid DNA

Amp^r

ori

Gene conferring drug-resistance to mammalian cells

Introduce DNA into cultured animal cells

A high fraction of cells take up and transiently express plasmid DNA

Selection in drug-containing medium

Only stably transformed cells form colonies

Plasmid DNA sequences

Stably transformed cells contain plasmid sequences integrated into chromosomal DNA

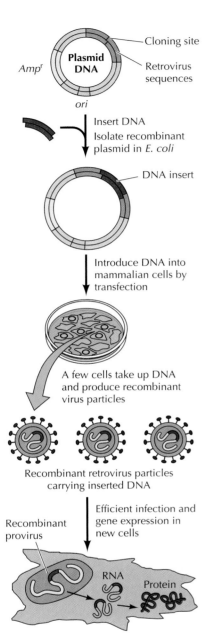

FIGURE 4.34 Retroviral vectors The vector consists of retroviral sequences cloned in a plasmid that can be propagated in *E. coli*. Foreign DNA is inserted into the viral sequences, and recombinant plasmids are isolated in bacteria. Animal cells in culture are then transfected with the recombinant DNA. The DNA is taken up by a small fraction of the cells, which produce recombinant retroviruses carrying the inserted DNA. These recombinant retroviruses can be used to efficiently infect new cells, where the viral genome carrying the inserted gene integrates into chromosomal DNA as a provirus.

scribed for several days—a phenomenon called **transient expression**. In a smaller fraction of cells (usually 1% or less), the foreign DNA becomes stably integrated into the cell genome and is transferred to progeny cells at cell division just as any other cell gene is. These stably transformed cells can be isolated if the transfected DNA contains a selectable marker, such as resistance to a drug that inhibits the growth of normal cells. Thus any cloned gene can be introduced into mammalian cells by being transferred together with a drug resistance marker that can be used to isolate stable transformants. The effects of such cloned genes on cell behavior—for example, cell growth or differentiation—can then be analyzed.

Animal viruses can also be used as vectors for more efficient introduction of cloned DNAs into cells. Retroviruses are particularly useful in this respect, since their life cycle involves the stable integration of viral DNA into the genome of infected cells (Figure 4.34). Consequently, retroviral vectors can be used to efficiently introduce cloned genes into a wide variety of cell types, making them an important vehicle for a broad range of applications.

Cloned genes can also be introduced into the germ line of multicellular organisms, allowing them to be studied in the context of the intact animal rather than in cultured cells. One method used to produce mice that carry such foreign genes (**transgenic mice**) is the direct microinjection of cloned DNA into the pronucleus of a fertilized egg (Figure 4.35). The injected eggs are then transferred to foster mothers and allowed to develop to term. In a fraction of the progeny mice (often about 10%), the foreign DNA will have integrated into the genome of the fertilized egg and is therefore present in all cells of the animal. Since the foreign DNA is present in germ cells as well as in somatic cells, it is transferred by breeding to new progeny mice just as any other cell gene would be.

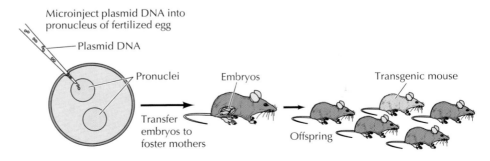

FIGURE 4.35 Production of transgenic mice DNA is microinjected into a pronucleus of a fertilized mouse egg (fertilized eggs contain two pronuclei, one from the egg and one from the sperm). The microinjected eggs are then transferred to foster mothers and allowed to develop. Some of the offspring (transgenic) have incorporated the injected DNA into their genome.

FIGURE 4.36 Introduction of genes into mice via embryonic stem cells Embryonic stem (ES) cells are cultured cells derived from early mouse embryos (blastocysts). DNA can be introduced into these cells in culture, and stably transformed ES cells can be isolated. These transformed ES cells can then be injected into a recipient blastocyst, where they are able to participate in normal development of the embryo. Some of the progeny mice that develop after transfer of injected embryos to foster mothers therefore contain cells derived from transformed ES cells, as well as from the normal cells of the blastocyst. Since these mice are a mixture of two different cell types, they are referred to as chimeric. Offspring carrying the transfected gene can then be produced by the breeding of chimeric mice in which descendants of the transformed ES cells have been incorporated into the germ line.

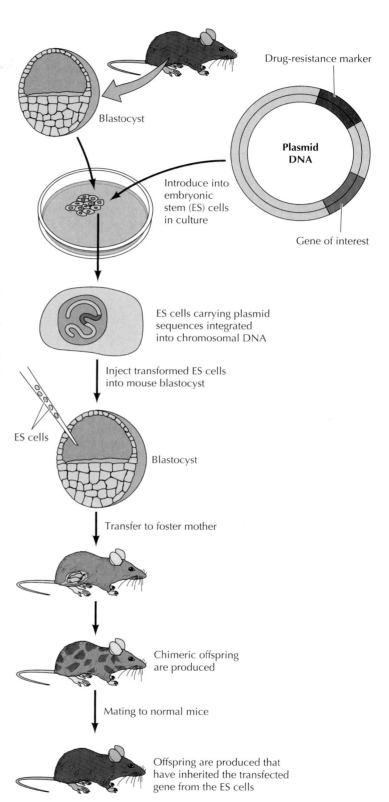

The properties of **embryonic stem (ES) cells** provide an alternative means of introducing cloned genes into mice (**Figure 4.36**). ES cells can be established in culture from early mouse embryos. They can also be reintroduced into early embryos, where they participate normally in development and can give rise to cells in all tissues of the mouse—including germ cells. It is thus possible to introduce cloned DNA into ES cells in culture, select stably transformed cells, and then introduce these cells back into mouse embryos. Such embryos give rise to chimeric offspring in which some cells are derived from the normal embryo cells and some from the transfected ES cells. In some such mice, the transfected ES cells are incorporated into the germ line. Breeding these mice therefore leads to the direct inheritance of the transfected gene by their progeny.

Cloned DNAs can also be introduced into plant cells. One approach is to bombard plant cells with DNA-coated microprojectiles, such as small particles of tungsten. The DNA-coated particles are shot directly into the plant cells; some of the cells are killed, but others survive and become stably transformed. In addition, a plasmid from the bacterium *Agrobacterium tumefaciens* (the **Ti plasmid**) provides a novel vehicle for the introduction of cloned DNAs into many species of plants (**Figure 4.37**). In nature, *Agrobacterium* attaches to the leaves of plants and the Ti plasmid is transferred into plant cells where it becomes incorporated into chromosomal DNA. Vectors developed from the Ti plasmid therefore provide an efficient means of introducing recombinant DNA into sensitive plant cells. Since many plants can be regenerated from single cultured cells (see Chapter 1), transgenic plants can be established directly from cells into which recombinant DNA has been introduced in culture—a much simpler proce-

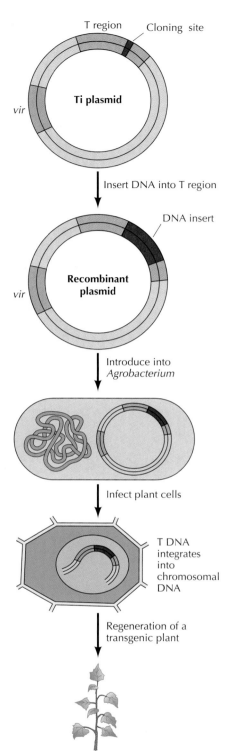

T region

Cloning site

Ti plasmid

vir

Insert DNA into T region

DNA insert

Recombinant plasmid

vir

Introduce into *Agrobacterium*

Infect plant cells

T DNA integrates into chromosomal DNA

Regeneration of a transgenic plant

FIGURE 4.37 Introduction of genes into plant cells via the Ti plasmid The Ti plasmid contains the T region, which is transferred to infected plant cells, and virulence (*vir*) genes, which function in T DNA transfer. In Ti plasmid vectors, foreign DNA is inserted into the T region. The recombinant plasmid is introduced into *Agrobacterium tumefaciens*, which is then used to infect cultured cells. The T region of the plasmid (carrying the inserted DNA) is transferred to the plant cells and integrates into chromosomal DNA. A transgenic plant can then be generated from the transformed cells.

dure than the production of transgenic animals. Indeed, many economically important types of plants, including corn, tomatoes, soybeans and potatoes, are transgenic varieties.

Mutagenesis of Cloned DNAs

In classical genetic studies (e.g., in bacteria or yeasts), mutants are the key to identifying genes and understanding their function by observing the altered phenotype of mutant organisms. In such studies, mutant genes are detected because they result in observable phenotypic changes—for example, temperature-sensitive growth or a specific nutritional requirement. The isolation of genes by recombinant DNA, however, has opened a different approach to mutagenesis. It is now possible to introduce any desired alteration into a cloned gene and to determine the effect of the mutation on gene function. Such procedures have been called **reverse genetics**, since a mutation is introduced into a gene first and its functional consequence is determined second. The ability to introduce specific mutations into cloned DNAs (*in vitro* **mutagenesis**) has proven to be a powerful tool in studying the expression and function of eukaryotic genes.

Cloned genes can be altered by many *in vitro* mutagenesis procedures, which can lead to the introduction of deletions, insertions, or single nucleotide alterations. The most common method of mutagenesis is the use of synthetic oligonucleotides to generate nucleotide changes in a DNA sequence (Figure 4.38). In this procedure a synthetic oligonucleotide bearing the desired mutation is used as a primer for DNA synthesis. Newly synthesized DNA molecules containing the mutation can then be isolated and characterized. For example, specific amino acids of a protein can be altered in order to characterize their role in protein function.

Variations of this approach, combined with the versatility of other methods for manipulating recombinant DNA molecules, can be used to introduce virtually any desired alteration in a cloned gene. The effects of such mutations on gene expression and function can then be determined by introduction of the gene into an appropriate cell type. *In vitro* mutagenesis has thus allowed detailed characterization of the functional roles of both the regulatory and protein-coding sequences of cloned genes.

Introducing Mutations into Cellular Genes

Although the transfer of cloned genes into cells (particularly in combination with *in vitro* mutagenesis) provides a powerful approach to studying gene structure and function, such experiments fall short of defining the role of an unknown gene in a cell or intact organism. The cells used as recipients for transfer of cloned genes usually already have normal copies of the gene in their chromosomal DNAs, and these normal copies continue to perform their roles in the cell. Determining the biological role of a cloned gene therefore requires that the activity of the normal cellular gene copies be eliminated. Several approaches can be used to either inac-

■ The United States is the largest producer of genetically modified plants, with 49.8 million hectares growing genetically modified crops. The first genetically modified plant approved for consumption by the FDA was the Flavr Savr tomato, which had been modified to delay ripening.

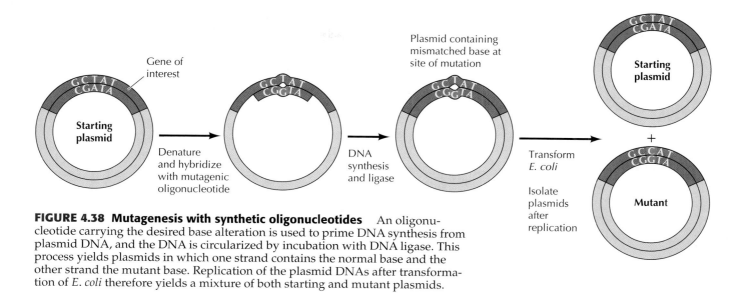

FIGURE 4.38 Mutagenesis with synthetic oligonucleotides   An oligonucleotide carrying the desired base alteration is used to prime DNA synthesis from plasmid DNA, and the DNA is circularized by incubation with DNA ligase. This process yields plasmids in which one strand contains the normal base and the other strand the mutant base. Replication of the plasmid DNAs after transformation of *E. coli* therefore yields a mixture of both starting and mutant plasmids.

tivate the chromosomal copies of a cloned gene or inhibit normal gene function, both in cultured cells and in transgenic mice.

Mutating chromosomal genes is based on the ability of a cloned gene introduced into a cell to undergo **homologous recombination** with its chromosomal copy (Figure 4.39). In homologous recombination, the cloned gene replaces the normal allele, so mutations introduced into the cloned gene *in vitro* become incorporated into the chromosomal copy of the gene. In the simplest case, mutations that inactivate the cloned gene can be introduced in place of the normal gene copy in order to determine its role in cellular processes.

Recombination between transferred DNA and the homologous chromosomal gene occurs frequently in yeast but is a rare event in mammalian cells, so inactivation of mammalian genes by this approach is technically difficult. Possibly because the genomes of mammalian cells are so much larger than that of yeasts, most transfected DNA that integrates into the recipient cell genome does so at random sites by recombination with unrelated sequences. However, various procedures have been developed both to increase the frequency of homologous recombination and to select and isolate the transformed cells in which homologous recombination has occurred, so it is feasible to inactivate any desired gene in mammalian cells by this approach. Importantly, genes can be readily inactivated in mouse embryonic stem cells, which can then be used to generate transgenic mice (Figure 4.40). These mice can be bred to yield progeny containing mutated copies of the targeted gene on both homologous chromosomes, so the effects of inactivation of a gene can be investigated in the context of the intact

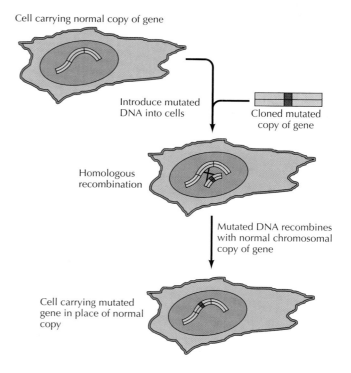

FIGURE 4.39 Gene inactivation by homologous recombination   A mutated copy of the cloned gene is introduced into cultured animal cells. The cloned gene may then replace the normal gene copy by homologous recombination, yielding a cell carrying the desired mutation in its chromosomal DNA.

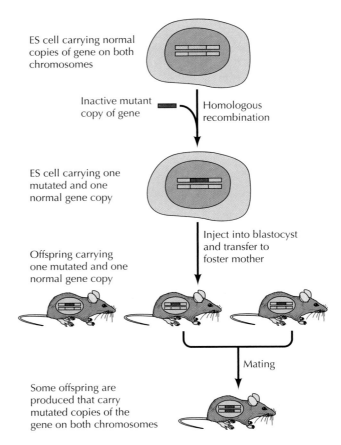

ES cell carrying normal copies of gene on both chromosomes

Inactive mutant copy of gene

Homologous recombination

ES cell carrying one mutated and one normal gene copy

Inject into blastocyst and transfer to foster mother

Offspring carrying one mutated and one normal gene copy

Mating

Some offspring are produced that carry mutated copies of the gene on both chromosomes

FIGURE 4.40 Production of mutant mice by homologous recombination in ES cells Homologous recombination is used to replace a normal gene with an inactive mutant copy in embryonic stem cells (ES cells), resulting in ES cells carrying one normal and one inactive copy of the gene on homologous chromosomes. These ES cells can be transferred to blastocysts and used to generate mice as described in Figure 4.36. These mice, carrying one mutant and one normal gene copy, can then be mated to produce offspring carrying mutant copies of the gene on both chromosomes.

animal. In addition, cells can be cultured from mouse embryos containing the mutated gene copies, so the functions of target genes can also be studied in cell culture. The biological activities of over 2000 mouse genes have been investigated in this way, and such studies have been critically important in revealing the roles of many genes in mouse development.

Homologous recombination has been used to systematically inactivate (or **knockout**) every gene in yeast, so a collection of genome-wide yeast mutants is available for scientists to use to study the function of any desired gene. In mice, approximately 10% of genes have currently been knocked out by homologous recombination. However, a large scale project has been proposed to systematically knockout all mouse genes, potentially yielding a genome-wide collection of mutant mice that would provide a major resource for researchers interested in mammalian development and cell function. Methods have also been developed to conditionally knockout genes in specific mouse tissues, allowing the function of a gene to be studied in a defined cell type (for example, in nerve cells) rather than in all cells of the organism.

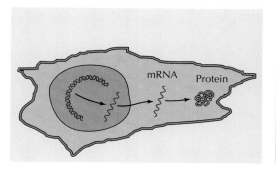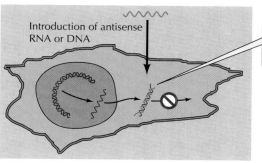

FIGURE 4.41 Inhibition of gene expression by antisense RNA or DNA
Antisense RNA or single-stranded DNA is complementary to the mRNA of a gene
of interest. Antisense nucleic acids therefore form hybrids with their target
mRNAs, blocking the translation of mRNA into protein and leading to mRNA
degradation.

Interfering with Cellular Gene Expression

As an alternative to gene inactivation by homologous recombination, a
variety of approaches can be used to specifically interfere with gene expression or function. One method that has been used to inhibit expression of a
desired target gene is the introduction of **antisense nucleic acids** into cultured cells (Figure 4.41). RNA or single-stranded DNA complementary to the
mRNA of the gene of interest (antisense) hybridizes with the mRNA and
blocks its translation into protein. Moreover, the RNA-DNA hybrids resulting from the introduction of antisense DNA molecules are usually
degraded within the cell. Antisense RNAs can be introduced directly into
cells, or cells can be transfected with vectors that have been engineered to
express antisense RNA. Antisense DNA is usually in the form of short
oligonucleotides (about 20 bases long), which can either be transfected into
cells or, in many cases, taken up by cells directly from the culture medium.

In the last several years, **RNA interference (RNAi)** has emerged as an
extremely effective and widely used method for interfering with gene
expression at the mRNA level. RNA interference was first discovered in *C.
elegans* in 1998, when Andrew Fire, Craig Mello and their colleagues found
that injection of double-stranded RNA inhibited expression of a gene with a
complementary mRNA sequence. This unexpected observation demonstrated an unanticipated role for double-stranded RNAs in gene regulation,
which has subsequently been developed into a powerful experimental tool
for inhibiting expression of target genes. When double-stranded RNAs are
introduced into cells, they are cleaved into short double-stranded molecules
(21–23 nucleotides) by an enzyme called Dicer (Figure 4.42). These short double-stranded molecules, called short interfering RNAs (siRNAs), then associate with a complex of proteins known as the RNA-induced silencing complex (RISC). Within this complex, the two strands of siRNA separate and
the strand complementary to the mRNA (the antisense strand) guides the
complex to the target mRNA by complementary base pairing. The mRNA is
then cleaved by one of the RISC proteins. The RISC-siRNA complex is
released following degradation of the mRNA and can continue to participate in multiple rounds of mRNA cleavage, leading to effective destruction
of the targeted mRNA.

RNAi has been established as a potent method for interfering with gene
expression in *C. elegans, Drosophila, Arabidopsis*, and mammalian cells, and
provides a relatively straightforward approach to investigating the function

■ Both antisense oligonucleotides
and siRNAs are being investigated
as potential therapeutic agents.
The antisense oligonucleotide
fomiversen has been approved by
the FDA for use in the treatment of
eye infections caused by a herpesvirus. An siRNA is currently in clinical trial for the treatment of age
related macular degeneration.

FIGURE 4.42 RNA interference Double-stranded RNA molecules are cleaved by the enzyme Dicer into short interfering RNAs (siRNAs). The siRNAs associate with the RNA-induced silencing complex (RISC) and are unwound. The antisense strand of siRNA then targets RISC to a homologous mRNA, which is cleaved by one of the RISC proteins. The RISC-siRNA complex is released and can participate in further cycles of mRNA degradation.

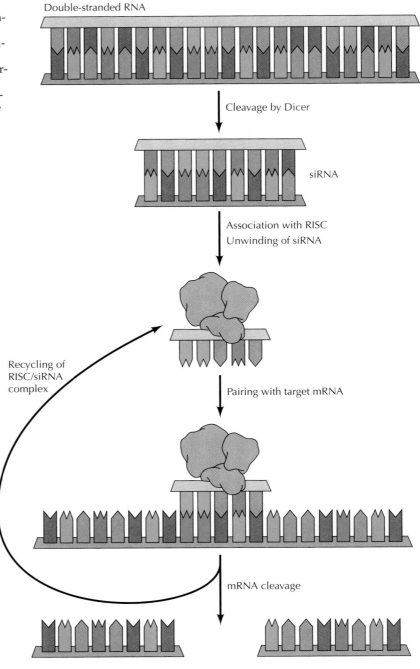

Double-stranded RNA

Cleavage by Dicer

siRNA

Association with RISC
Unwinding of siRNA

Recycling of RISC/siRNA complex

Pairing with target mRNA

mRNA cleavage

of any gene whose sequence is known. In addition, libraries of double-stranded RNAs or siRNAs that cover a large fraction of genes in the genome are being used to screen *C. elegans*, *Drosophila*, and human cells to identify novel genes involved in specific biological functions, such as cell growth or cell survival. Finally, RNAi is not only an experimental tool: as will be discussed in Chapters 7 and 8, RNA interference is also a major regulatory mechanism used by cells to control gene expression at both the transcriptional and translational levels.

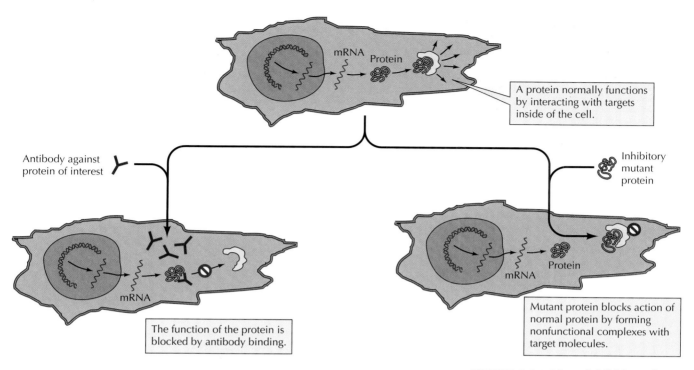

A protein normally functions by interacting with targets inside of the cell.

Antibody against protein of interest

Inhibitory mutant protein

The function of the protein is blocked by antibody binding.

Mutant protein blocks action of normal protein by forming nonfunctional complexes with target molecules.

FIGURE 4.43 Direct inhibition of protein function   Microinjected antibodies can bind to proteins within cells, thereby inhibiting their normal function. In addition, some mutant proteins interfere with the function of normal proteins—for example, by competing with the normal protein for interaction with target molecules.

In addition to inactivating a gene or inducing degradation of an mRNA, it is sometimes possible to interfere directly with the function of proteins within cells (Figure 4.43). One approach is to microinject antibodies that block the activity of the protein against which they are directed. Alternatively, some mutant proteins interfere with the function of their normal counterparts when they are expressed within the same cell—for example, by competing with the normal protein for binding to its target molecule. Cloned DNAs encoding such mutant proteins (called **dominant inhibitory mutants**) can be introduced into cells by gene transfer and used to study the effects of blocking normal gene function.

COMPANION WEBSITE

Visit the website that accompanies **The Cell** (www.sinauer.com/cooper) for animations, videos, quizzes, problems, and other review material.

SUMMARY

HEREDITY, GENES, AND DNA

Genes and Chromosomes: Chromosomes are the carriers of genes.

Genes and Enzymes: A gene specifies the amino acid sequence of a polypeptide chain.

Identification of DNA as the Genetic Material: DNA was identified as the genetic material by bacterial transformation experiments.

KEY TERMS

gene, allele, dominant, recessive, genotype, phenotype, chromosome, diploid, meiosis, haploid, mutation

one gene–one enzyme hypothesis

transformation

KEY TERMS

SUMMARY

The Structure of DNA: DNA is a double helix in which hydrogen bonds form between purines and pyrimidines on opposite strands. Because of specific base pairing—A with T and G with C—the two strands of a DNA molecule are complementary in sequence.

semiconservative replication, DNA polymerase

Replication of DNA: DNA replicates by semiconservative replication in which the two strands separate and each serves as a template for synthesis of a new progeny strand.

EXPRESSION OF GENETIC INFORMATION

Colinearity of Genes and Proteins: The order of nucleotides in DNA specifies the order of amino acids in proteins.

central dogma, transcription, translation, messenger RNA (mRNA), RNA polymerase, ribosomal RNA (rRNA), transfer RNA (tRNA)

The Role of Messenger RNA: Messenger RNA functions as an intermediate to convey information from DNA to the ribosomes, where it serves as a template for protein synthesis.

genetic code, *in vitro* translation, codon

The Genetic Code: Transfer RNAs serve as adaptors between amino acids and mRNA during translation. Each amino acid is specified by a codon consisting of three nucleotides.

retrovirus, reverse transcription, reverse transcriptase

RNA Viruses and Reverse Transcription: DNA can be synthesized from RNA templates, as first discovered in retroviruses.

RECOMBINANT DNA

Restriction Endonucleases: Restriction endonucleases cleave specific DNA sequences, yielding defined fragments of DNA molecules.

restriction endonuclease, gel electrophoresis, restriction map

molecular cloning, vector, recombinant molecule, molecular clone, plasmid, DNA ligase, cDNA

Generation of Recombinant DNA Molecules: Recombinant DNA molecules consist of a DNA fragment of interest ligated to a vector that is able to replicate independently in an appropriate host cell.

origin of replication, cosmid, P1 artificial chromosome (PAC), bacterial artificial chromosome (BAC), yeast artificial chromosome (YAC)

Vectors for Recombinant DNA: A variety of vectors are used to clone different sizes of DNA fragments.

dideoxynucleotide

DNA Sequencing: The nucleotide sequences of cloned DNA fragments can be readily determined.

expression vector, yeast two-hybrid

Expression of Cloned Genes: The proteins encoded by cloned genes can be expressed at high levels in either bacteria or eukaryotic cells.

SUMMARY

DETECTION OF NUCLEIC ACIDS AND PROTEINS

Amplification of DNA by the Polymerase Chain Reaction: PCR allows the amplification and isolation of specific fragments of DNA *in vitro*, providing a sensitive method for detecting small amounts of specific DNA or RNA molecules.

Nucleic Acid Hybridization: Nucleic acid hybridization allows the detection of specific DNA or RNA sequences by base pairing between complementary strands.

Antibodies as Probes for Proteins: Antibodies are used to detect specific proteins in cells or cell extracts.

GENE FUNCTION IN EUKARYOTES

Genetic Analysis in Yeasts: The simple genetics and rapid replication of yeasts facilitate the molecular cloning of a gene corresponding to any yeast mutation.

Gene Transfer in Plants and Animals: Cloned genes can be introduced into complex eukaryotic cells and multicellular organisms for functional analysis.

Mutagenesis of Cloned DNAs: *In vitro* mutagenesis of cloned DNAs is used to study the effect of engineered mutations on gene function.

Introducing Mutations into Cellular Genes: Mutations can be introduced into chromosomal gene copies by homologous recombination with cloned DNA sequences.

Interfering with Cellular Gene Expression: The expression or function of specific genes can be blocked by antisense nucleic acids, RNA interference, or dominant inhibitory mutants.

KEY TERMS

polymerase chain reaction (PCR)

nucleic acid hybridization, Southern blotting, Northern blotting, recombinant DNA library, DNA microarray, *in situ* hybridization

antibody, antigen, monoclonal antibody, immunoblotting, Western blotting, immunoprecipitation, SDS-polyacrylamide gel electrophoresis (SDS-PAGE)

temperature-sensitive mutant

gene transfer, transfection, liposome, electroporation, transient expression, transgenic mice, embryonic stem (ES) cell, Ti plasmid

reverse genetics, *in vitro* mutagenesis

homologous recombination, knockout, RNA interference (RNAi), dominant inhibitory mutant

antisense nucleic acid, RNA interference

Questions

1. How would you determine whether two different genes are linked in *Drosophila*?

2. You grow *E. coli* for several generations in media containing ^{15}N. The cells are then transferred to media containing ^{14}N and grown for two additional generations. What proportion of the DNA isolated from these cells will be heavy, light, or intermediate in density?

3. Addition or deletion of one or two nucleotides in the coding sequence of a gene produces a nonfunctioning protein, whereas addition or deletion of three nucleotides often produces a protein with nearly normal function. Explain.

4. Describe the features that a yeast artificial chromosome must have in order to be used to clone a piece of human DNA cut with *Eco*RI.

5. You are studying an enzyme in which an active-site cysteine residue is encoded by the triplet UGU. How would mutating the third base to a C affect enzyme function? How about mutating it to an A?

6. Digestion of a 4-kb DNA molecule with *Eco*RI yields two fragments of 1 kb and 3 kb each. Digestion of the same molecule with *Hind*III yields fragments of 1.5 kb and 2.5 kb. Finally, digestion with *Eco*RI and *Hind*III in combination yields fragments of 0.5 kb, 1 kb, and 2.5 kb. Draw a restriction map indicating the positions of the *Eco*RI and *Hind*III cleavage sites.

7. Starting with DNA from a single sperm, how many copies of a specific gene sequence will be obtained after 10 cycles of PCR amplification? After 30 cycles?

8. You have cloned a fragment of human genomic DNA in a cosmid vector; approximately how many times would you expect the insert to be cut by the restriction enzyme *Bam*HI?

9. What is the minimum number of BAC clones required to make a genomic library of human DNA?

10. How would you expect actinomycin D to affect the replication of influenza virus?

11. What is the critical feature of a cloning vector that would allow you to isolate stably transfected mammalian cells?

12. Nucleic acids have a net negative charge and can be separated by gel electrophoresis on the basis of their size. In contrast, different proteins have different charges. How then can proteins be separated on the basis of size by electrophoresis?

References and Further Reading

Heredity, Genes, and DNA

Avery, O. T., C. M. MacLeod and M. McCarty. 1944. Studies on the chemical nature of the substance inducing transformation of pneumococcal types. *J. Exp. Med.* 79: 137–158. [P]

Franklin, R. E. and R. G. Gosling. 1953. Molecular configuration in sodium thymonucleate. *Nature* 171: 740–741. [P]

Kornberg, A. 1960. Biologic synthesis of deoxyribonucleic acid. *Science* 131: 1503–1508. [P]

Kresge, N., R. D. Simoni and R. L. Hill. 2005. Launching the age of biochemical genetics, with *Neurospora*: The work of George Wells Beadle. *J. Biol. Chem.* 280: e9–e11. [R]

Lehman, I. R. 2003. Discovery of DNA polymerase. *J. Biol. Chem.* 278: 34733–34738. [R]

Meselson, M. and F. W. Stahl. 1958. The replication of DNA in *Escherichia coli. Proc. Natl. Acad. Sci.* USA 44: 671–682. [P]

Watson, J. D. and F. H. C. Crick. 1953. Genetical implications of the structure of deoxyribonucleic acid. *Nature* 171: 964–967. [P]

Watson, J. D. and F. H. C. Crick. 1953. Molecular structure of nucleic acids: A structure for deoxyribose nucleic acid. *Nature* 171: 737–738. [P]

Wilkins, M. H. F., A. R. Stokes and H. R. Wilson. 1953. Molecular structure of deoxypentose nucleic acids. *Nature* 171: 738–740. [P]

Expression of Genetic Information

Baltimore, D. 1970. RNA-dependent DNA polymerase in virions of RNA tumour viruses. *Nature* 226: 1209–1211. [P]

Brenner, S., F. Jacob and M. Meselson. 1961. An unstable intermediate carrying information from genes to ribosomes for protein synthesis. *Nature* 190: 576–581. [P]

Crick, F. H. C., L. Barnett, S. Brenner and R. J. Watts-Tobin. 1961. General nature of the genetic code for proteins. *Nature* 192: 1227–1232. [P]

Ingram, V. M. 1957. Gene mutations in human hemoglobin: The chemical difference be-

tween normal and sickle cell hemoglobin. *Nature* 180: 326–328. [P]

Nirenberg, M. 2004. Historical review: Deciphering the genetic code—a personal account. *Trends Biochem. Sci.* 29: 46–54. [R]

Nirenberg, M. and P. Leder. 1964. RNA codewords and protein synthesis. *Science* 145: 1399–1407. [P]

Nirenberg, M. W. and J. H. Matthaei. 1961. The dependence of cell-free protein synthesis in *E. coli* upon naturally occurring or synthetic polyribonucleotides. *Proc. Natl. Acad. Sci. USA* 47: 1588–1602. [P]

Temin, H. M. and S. Mizutani. 1970. RNA-dependent DNA polymerase in virions of Rous sarcoma virus. *Nature* 226: 1211–1213. [P]

Yanofsky, C., B. C. Carlton, J. R. Guest, D. R. Helinski and U. Henning. 1964. On the colinearity of gene structure and protein structure. *Proc. Natl. Acad. Sci. USA* 51: 266–272. [P]

Recombinant DNA

Ausubel, F. M., R. Brent, R. E. Kingston, D. D. Moore, J. G. Seidman, J. A. Smith and K. Struhl. eds. 1989. *Current Protocols in Molecular Biology.* New York: Greene Publishing and Wiley Interscience. [R]

Burke, D. T., G. F. Carle and M. V. Olson. 1987. Cloning of large segments of exogenous DNA into yeast by means of artificial chromosome vectors. *Science* 236: 806–812. [P]

Cohen, S. N., A. C. Y. Chang, H. W. Boyer and R. B. Helling. 1973. Construction of biologically functional bacterial plasmids *in vitro. Proc. Natl. Acad. Sci. USA* 70: 3240–3244. [P]

Nathans, D. and H. O. Smith. 1975. Restriction endonucleases in the analysis and restructuring of DNA molecules. *Ann. Rev. Biochem.* 44: 273–293. [R]

Saiki, R. K., D. H. Gelfand, S. Stoffel, S. J. Scharf, R. Higuchi, G. T. Horn, K. B. Mullis and H. A. Erlich. 1988. Primer-directed enzymatic amplification of DNA with a thermostable DNA polymerase. *Science* 239: 487–491. [P]

Sambrook, J. and D. Russell. 2001. *Molecular Cloning: A Laboratory Manual.* 3rd ed. Plainview, N.Y.: Cold Spring Harbor Laboratory Press.

Sanger, F., S. Nicklen and A. R. Coulson. 1977. DNA sequencing with chain-terminating inhibitors. *Proc. Natl. Acad. Sci. USA* 74: 5463–5467. [P]

Detection of Nucleic Acids and Proteins

Ausubel, F. M., R. Brent, R. E. Kingston, D. D. Moore, J. G. Seidman, J. A. Smith and K. Struhl. eds. 1989. *Current Protocols in Molecular Biology.* New York: Greene Publishing and Wiley Interscience.

Brown, P. O. and D. Botstein. 1999. Exploring the new world of the genome with DNA microarrays. *Nature Genetics* 21: 33–37. [R]

Caruthers, M. H. 1985. Gene synthesis machines: DNA chemistry and its uses. *Science* 230: 281–285. [R]

Gerhold, D., T. Rushmore and C. T. Caskey. 1999. DNA chips: Promising toys have become powerful tools. *Trends Biochem. Sci.* 24: 168–173. [R]

Grunstein, M. and D. S. Hogness. 1975. Colony hybridization: A method for the isolation of cloned DNAs that contain a specific gene. *Proc. Natl. Acad. Sci. USA* 72: 3961–3965. [P]

Harlow, E. and D. Lane. 1999. *Using Antibodies: A Laboratory Manual.* Cold Spring Harbor, N.Y.: Cold Spring Harbor Laboratory Press.

Kohler, G. and C. Milstein. 1975. Continuous cultures of fused cells secreting antibody of predefined specificity. *Nature* 256: 495–497. [P]

Sambrook, J., and D. Russell. 2001. *Molecular Cloning: A Laboratory Manual.* 3rd ed. Plainview, N.Y.: Cold Spring Harbor Laboratory Press.

Southern, E. M. 1975. Detection of specific sequences among DNA fragments separated by gel electrophoresis. *J. Mol. Biol.* 98: 503–517. [P]

Gene Function in Eukaryotes

Boutros, M., A. A. Kiger, S. Armknecht, K. Kerr, M. Hild, B. Koch, S. A. Haas, Heidelberg Fly Array Consortium, R. Paro and N. Perrimon. 2004. Genome-wide RNAi analysis of growth and viability in *Drosophila* cells. *Science* 303: 832–835. [P]

Branda, C. S. and S. M. Dymecki. 2004. Talking about a revolution: The impact of site-specific recombinases on genetic analyses in mice. *Devel. Cell* 6: 7–28. [R]

Bronson, S. K. and O. Smithies. 1994. Altering mice by homologous recombination using embryonic stem cells. *J. Biol. Chem.* 269: 27155–27158. [R]

Capecchi, M. R. 1989. Altering the genome by homologous recombination. *Science* 244: 1288–1292. [R]

Carpenter, A. E. and D. M. Sabatini. 2004. Systematic genome-wide screens of gene function. *Nature Rev. Genet.* 5: 11–22. [R]

Downward, J. 2004. Use of RNA interference libraries to investigate oncogenic signaling in mammalian cells. *Oncogene* 23: 8376–8383. [R]

Fire, A., S. Xu, M. K. Montgomery, S. A. Kostas, S. E. Driver and C. C. Mello. 1998. Potent and specific genetic interference by double-stranded RNA in *Caenorhabditis elegans*. *Nature* 391: 806–811. [P]

Gelvin, S. B. 2003. *Agrobacterium*-mediated plant transformation: The biology behind the "gene-jockeying" tool. *Microbiol. Molec. Biol. Rev.* 67: 16–37. [R]

Herskowitz, I. 1987. Functional inactivation of genes by dominant negative mutations. *Nature* 329: 219–222. [R]

Izant, J. G. and H. Weintraub. 1984. Inhibition of thymidine kinase gene expression by antisense RNA: A molecular approach to genetic analysis. *Cell* 36: 1007–1015. [P]

Kuhn, R., F. Schwenk, M. Aguet and K. Rajewsky. 1995. Inducible gene targeting in mice. *Science* 269: 1427–1429. [P]

Mello, C. C. and D. Conte Jr. 2004. Revealing the world of RNA interference. *Nature* 431: 338–342. [R]

Novina, C. D. and P. A. Sharp. 2004. The RNAi revolution. *Nature* 430: 161–164. [R]

Palmiter, R. D. and R. L. Brinster. 1986. Germline transformation of mice. *Ann. Rev. Genet.* 20: 465–499. [R]

Smith, M. 1985. *In vitro* mutagenesis. *Ann. Rev. Genet.* 19: 423–462. [R]

Sontheimer, E. J. 2005. Assembly and function of RNA silencing complexes. *Nature Rev. Mol. Cell Biol.* 6: 127–138. [R]

Struhl, K. 1983. The new yeast genetics. *Nature* 305: 391–397. [R]

The Comprehensive Knockout Mouse Project Consortium. 2004. The knockout mouse project. *Nature Genet.* 36: 921–924. [R]

Tomari, Y. and P. D. Zamore. 2005. Perspective: Machines for RNAi. *Genes Dev.* 19: 517–529. [R]

PART II

The Flow of Genetic Information

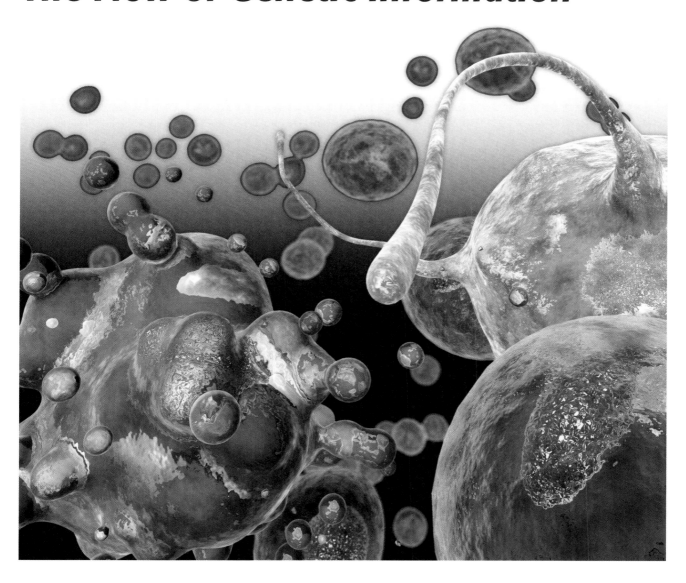

CHAPTER 5 ■ The Organization and Sequences of Cellular Genomes

CHAPTER 6 ■ Replication, Maintenance, and Rearrangements of Genomic DNA

CHAPTER 7 ■ RNA Synthesis and Processing

CHAPTER 8 ■ Protein Synthesis, Processing, and Regulation

CHAPTER 5

The Organization and Sequences of Cellular Genomes

- **The Complexity of Eukaryotic Genomes 155**

- **Chromosomes and Chromatin 166**

- **The Sequences of Complete Genomes 176**

- **Bioinformatics and Systems Biology 192**

- **KEY EXPERIMENT:**
  The Discovery of Introns 158

- **KEY EXPERIMENT:**
  The Human Genome 188

AS THE GENETIC MATERIAL, DNA PROVIDES A BLUEPRINT that directs all cellular activities and specifies the developmental plan of multicellular organisms. An understanding of gene structure and function is therefore fundamental to an appreciation of the molecular biology of cells. The development of gene cloning represented a major step toward this goal, enabling scientists to dissect complex eukaryotic genomes and probe the functions of eukaryotic genes. Continuing advances in recombinant DNA technology have now brought us to the exciting point of determining the sequences of entire genomes, providing a new approach to deciphering the genetic basis of cell behavior.

As reviewed in Chapter 4, the initial applications of recombinant DNA were directed toward the isolation and analysis of individual genes. More recently, large scale sequencing projects have yielded the complete genome sequences of many bacteria, of yeast, and of several species of plants and animals, including humans. The sequences of these complete cellular genomes provide a rich harvest of information, enabling the identification of many hitherto unknown genes and regulatory sequences. The results of these genome sequencing projects can be expected to stimulate many years of future research in molecular and cellular biology, and to have a profound impact on our understanding and treatment of human disease.

The Complexity of Eukaryotic Genomes

The genomes of most eukaryotes are larger and more complex than those of prokaryotes (Figure 5.1). This larger size of eukaryotic genomes is not inherently surprising, since one would expect to find more genes in organisms that are more complex. However, the genome size of many eukaryotes does not appear to be related to genetic complexity. For example, the genomes of salamanders and lilies contain more than ten times the

FIGURE 5.1 Genome size The range of sizes of the genomes of representative groups of organisms is shown on a logarithmic scale.

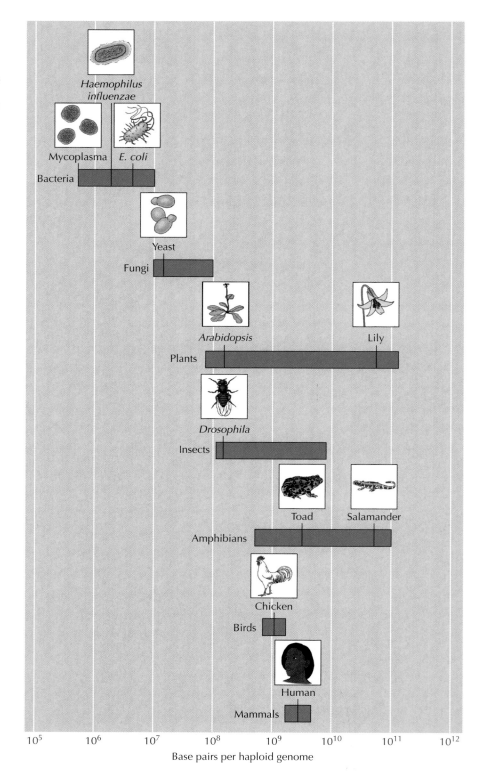

amount of DNA that is in the human genome, yet these organisms are clearly not ten times more complex than humans.

This apparent paradox was resolved by the discovery that the genomes of most eukaryotic cells contain not only functional genes but also large amounts of DNA sequences that do not code for proteins. The difference in the sizes of the salamander and human genomes thus reflects larger

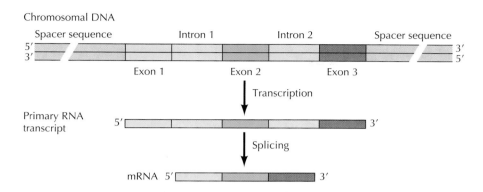

FIGURE 5.2 The structure of eukaryotic genes   Most eukaryotic genes contain segments of coding sequences (exons) interrupted by noncoding sequences (introns). Both exons and introns are transcribed to yield a long primary RNA transcript. The introns are then removed by splicing to form the mature mRNA.

amounts of noncoding DNA, rather than more genes, in the genome of the salamander. The presence of large amounts of noncoding sequences is a general property of the genomes of complex eukaryotes. Thus the thousandfold greater size of the human genome compared to that of *E. coli* is not due solely to a larger number of human genes. The human genome is thought to contain 20,000–25,000 genes—only about 5 times more than *E. coli* has. Much of the complexity of eukaryotic genomes thus results from the abundance of several different types of noncoding sequences, which constitute most of the DNA of higher eukaryotic cells.

Introns and Exons

In molecular terms, a **gene** can be defined as a segment of DNA that is expressed to yield a functional product, which may be either an RNA (e.g., ribosomal and transfer RNAs) or a polypeptide. Some of the noncoding DNA in eukaryotes is accounted for by long DNA sequences that lie between genes (**spacer sequences**). However, large amounts of noncoding DNA are also found within most eukaryotic genes. Such genes have a split structure in which segments of coding sequence (called **exons**) are separated by noncoding sequences (intervening sequences, or **introns**) (Figure 5.2). The entire gene is transcribed to yield a long RNA molecule and the introns are then removed by splicing, so only exons are included in the mRNA. Although most introns have no known function, they account for a substantial fraction of DNA in the genomes of higher eukaryotes.

Introns were first discovered in 1977, independently in the laboratories of Phillip Sharp and Richard Roberts, during studies of the replication of adenovirus in cultured human cells. Adenovirus is a useful model for studies of gene expression, both because the viral genome is only about 3.5×10^4 base pairs long and because adenovirus mRNAs are produced at high levels in infected cells. One approach used to characterize the adenovirus mRNAs was to determine the locations of the corresponding viral genes by examination of RNA-DNA hybrids in the electron microscope. Because RNA-DNA hybrids are distinguishable from single-stranded DNA, the positions of RNA transcripts on a DNA molecule can be determined. Surprisingly, such experiments revealed that adenovirus mRNAs do not hybridize to only a single region of viral DNA (Figure 5.3). Instead, a single mRNA molecule hybridizes to several separated regions of the viral genome. Thus the adenovirus mRNA does not correspond to an uninterrupted transcript of the template DNA; rather the mRNA is assembled from several distinct blocks of sequences that originated from different parts of the viral DNA. This was subsequently shown to occur by **RNA splicing**, which will be discussed in detail in Chapter 7.

KEY EXPERIMENT

The Discovery of Introns

Spliced Segments at the 5' Terminus of Adenovirus 2 Late mRNA
Susan M. Berget, Claire Moore, and Phillip A. Sharp
Massachusetts Institute of Technology, Cambridge, Massachusetts
Proceedings of the National Academy of Sciences USA, Volume 74, 1977,
pages 3171–3175

Phillip Sharp   Richard Roberts

The Context

Prior to molecular cloning, little was known about mRNA synthesis in eukaryotic cells. However, it was clear that this process is more complex in eukaryotes than in bacteria. The synthesis of eukaryotic mRNAs appeared to require not only transcription but also processing reactions that modify the structure of primary transcripts. Most notably, eukaryotic mRNAs appeared to be synthesized as long primary transcripts, found in the nucleus, which were then cleaved to yield much shorter mRNA molecules that were exported to the cytoplasm.

These processing steps were generally assumed to involve the removal of sequences from the 5' and 3' ends of the primary transcripts. In this model, mRNAs embedded within long primary transcripts would be encoded by uninterrupted DNA sequences. This view of eukaryotic mRNA was changed radically by the discovery of splicing, made independently by Berget, Moore, and Sharp, and by Louise Chow, Richard Gelinas, Tom Broker, and Richard Roberts (An amazing sequence arrangement at the 5' ends of adenovirus 2 messenger RNA, 1977. *Cell* 12: 1–8).

The Experiments

Both of the research groups that discovered splicing used adenovirus 2 to investigate mRNA synthesis in human cells. The major advantage of the virus is that it provides a model that is much simpler than the host cell. Viral DNA can be isolated directly from virus particles, and mRNAs encoding the viral structural proteins are present in such high

amounts that they can be purified directly from infected cells. Berget, Moore, and Sharp focused their experiments on an abundant mRNA that encodes a viral structural polypeptide known as the hexon.

To map the hexon mRNA on the viral genome, purified mRNA was hybridized to adenovirus DNA and the hybrid molecules were examined by electron microscopy. As expected, the body of the hexon mRNA formed hybrids with restriction fragments of adenovirus DNA that had previously been shown to contain the hexon gene. Surprisingly, however, sequences at the 5' end of hexon mRNA failed to hybridize to DNA sequences adjacent to those encoding the body of the message, suggesting that the 5' end of the mRNA had arisen from sequences located elsewhere in the viral genome.

This possibility was tested by hybridization of hexon mRNA to a restriction fragment extending

upstream of the hexon gene. The mRNA-DNA hybrids formed in this experiment displayed a complex loop structure (see figure). The body of the mRNA formed a long hybrid region with the previously identified hexon DNA sequences. Strikingly, the 5' end of the hexon mRNA hybridized to three short upstream regions of DNA, which were separated from each other and from the body of the message by large single-stranded DNA loops. The sequences at the 5' end of hexon mRNA thus appeared to be transcribed from three separate regions of the viral genome, which were spliced to the body of the mRNA during the processing of a long primary transcript.

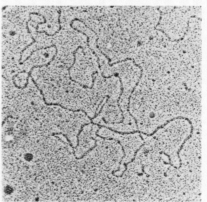
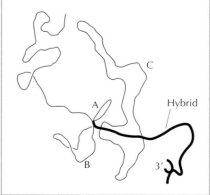

An electron micrograph and tracing of hexon mRNA hybridized to adenovirus DNA. The single-stranded loops designated A, B, and C, correspond to introns.

KEY EXPERIMENT

The Impact

The discovery of splicing in adenovirus mRNA was quickly followed by similar experiments with cellular mRNAs, demonstrating that eukaryotic genes had a previously unexpected structure. Rather than being continuous, their coding sequences were interrupted by introns, which were removed from primary transcripts by splicing. Introns are now known to account for much of the DNA in eukaryotic genomes, and the roles of introns in the evolution and regulation of gene expression continue to be active areas of investigation. The discovery of splicing also stimulated intense interest in the mechanism of this unexpected RNA processing reaction. As discussed in Chapter 7, these studies have not only illuminated new mechanisms of regulating gene expression; they have also revealed novel catalytic activities of RNA and provided critical evidence supporting the hypothesis that early evolution was based on self-replicating RNA molecules. The unexpected structure of adenovirus mRNAs has thus had a major impact on diverse areas of cellular and molecular biology.

Soon after the discovery of introns in adenovirus, similar observations were made on cloned genes of eukaryotic cells. For example, electron microscopic analysis of RNA-DNA hybrids and subsequent nucleotide sequencing of cloned genomic DNAs and cDNAs indicated that the coding region of the mouse β-globin gene (which encodes the β subunit of hemoglobin) is interrupted by two introns that are removed from the mRNA by splicing (Figure 5.4). The intron-exon structure of many eukaryotic genes is quite complicated, and the amount of DNA in the intron sequences is often

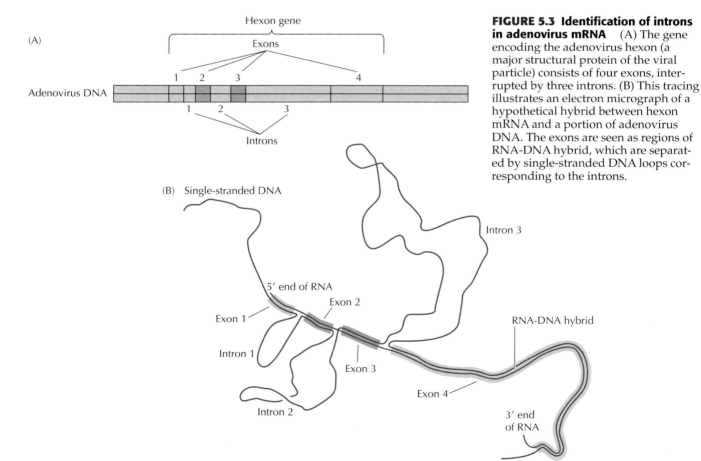

FIGURE 5.3 Identification of introns in adenovirus mRNA   (A) The gene encoding the adenovirus hexon (a major structural protein of the viral particle) consists of four exons, interrupted by three introns. (B) This tracing illustrates an electron micrograph of a hypothetical hybrid between hexon mRNA and a portion of adenovirus DNA. The exons are seen as regions of RNA-DNA hybrid, which are separated by single-stranded DNA loops corresponding to the introns.

FIGURE 5.4 The mouse β-globin gene This gene contains two introns, which divide the coding region among three exons. Exon 1 encodes amino acids 1 to 30, exon 2 encodes amino acids 31 to 104, and exon 3 encodes amino acids 105 to 146. Exons 1 and 3 also contain untranslated regions (UTRs) at the 5′ and 3′ ends of the mRNA, respectively.

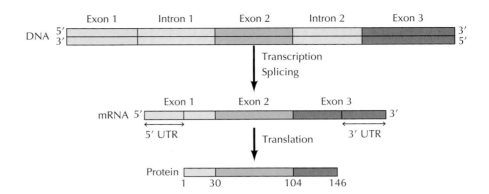

greater than that in the exons. For example, an average human gene contains approximately 9 exons, interrupted by 8 introns and distributed over approximately 30,000 base pairs (30 **kilobases**, or **kb**) of genomic DNA (Table 5.1). The exons generally total only about 2.5 kb, including regions at both the 5′ and 3′ ends of the mRNA that are not translated into protein (5′ and 3′ untranslated regions or UTRs). Introns thus comprise more than 90% of the average human gene.

Introns are present in most genes of complex eukaryotes, although they are not universal. Almost all histone genes, for example, lack introns, so introns are clearly not required for gene function in eukaryotic cells. In addition, introns are not found in most genes of simple eukaryotes, such as yeasts. Conversely, introns *are* present in rare genes of prokaryotes. The presence or absence of introns is therefore not an absolute distinction between prokaryotic and eukaryotic genes, although introns are much more prevalent in higher eukaryotes (both plants and animals), where they account for a substantial amount of total genomic DNA. Many introns are conserved in genes of both plants and animals, indicating that they arose early in evolution, prior to the plant-animal divergence.

Most introns do not specify the synthesis of a cellular product, although a few do encode functional RNAs or proteins. However, introns play important roles in controlling gene expression. For example, the presence of introns allows the exons of a gene to be joined in different combinations, resulting in the synthesis of different proteins from the same gene. This process, called **alternative splicing** (Figure 5.5), occurs frequently in the genes of complex eukaryotes and is thought to be important in extending the functional repertoire of the 20,000–25,000 genes of the human genome.

Introns are also thought to have played an important role in evolution by facilitating recombination between protein-coding regions (exons) of differ-

TABLE 5.1 Characteristics of the Average Human Gene

Number of exons	9
Number of introns	8
Exon Sequence:	
5′ untranslated region	300 base pairs
coding sequence	1400 base pairs
3′ untranslated region	800 base pairs
TOTAL	2500 base pairs
Intron Sequence:	27,000 base pairs

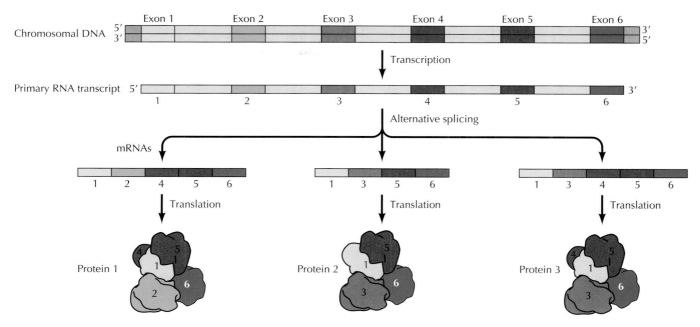

FIGURE 5.5 Alternative splicing   The gene illustrated contains six exons, separated by five introns. Alternative splicing allows these exons to be joined in different combinations, resulting in the formation of three distinct mRNAs and proteins from the single primary transcript.

ent genes—a process known as exon shuffling. Exons frequently encode functionally distinct protein domains, so recombination between introns of different genes would result in new genes containing novel combinations of protein-coding sequences. As predicted by this hypothesis, DNA sequencing studies have demonstrated that some genes are chimeras of exons derived from several other genes, providing direct evidence that new genes can be formed by recombination between intron sequences.

Repetitive DNA Sequences

Introns make a substantial contribution to the large size of higher eukaryotic genomes. In humans, for example, introns account for approximately 20% of the total genomic DNA. However, an even larger portion of complex eukaryotic genomes consists of highly repeated noncoding DNA sequences. These sequences, sometimes present in hundreds of thousands of copies per genome, were first demonstrated by Roy Britten and David Kohne during studies of the rates of reassociation of denatured fragments of cellular DNAs (Figure 5.6). Denatured strands of DNA hybridize to each other (reassociate), re-forming double-stranded molecules (see Figure 4.24). Since DNA reassociation is a bimolecular reaction (two separated strands of denatured DNA must collide with each other in order to hybridize), the rate of reassociation depends on the concentration of DNA strands. When fragments of *E. coli* DNA were denatured and allowed to hybridize with each other, all of the DNA reassociated at the same rate, as expected if each DNA sequence were represented once per genome. However, reassociation of fragments of DNA extracted from mammalian cells showed a very different pattern. Approximately 50% of the DNA fragments reassociated at the rate expected for sequences present once per genome, but the remainder reasso-

FIGURE 5.6 Identification of repetitive sequences by DNA reassociation The kinetics of the reassociation of fragments of *E. coli* and bovine DNAs are illustrated as a function of C_0t, which is the initial concentration of DNA multiplied by the time of incubation. The *E. coli* DNA reassociates at a uniform rate, consistent with each fragment of DNA being represented once in a genome of 4.6×10^6 base pairs. In contrast, the bovine DNA fragments exhibit two distinct steps in their reassociation. About 60% of the DNA fragments (the nonrepeated sequences) reassociate more slowly than *E. coli* DNA, as expected for sequences represented as single copies in the larger bovine genome (3×10^9 base pairs). However, the other 40% of the bovine DNA fragments (the repeated sequences) reassociate more rapidly than *E. coli* DNA, indicating that multiple copies of these sequences are present.

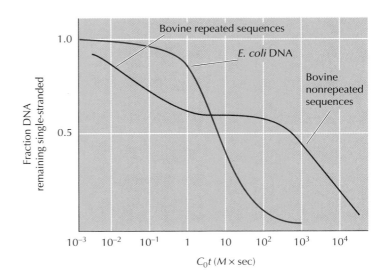

ciated much more rapidly than expected. The interpretation of these results was that some sequences were present in multiple copies and therefore reassociated more rapidly than those sequences that were represented only once per genome. In particular, these experiments indicated that approximately 50% of mammalian DNA consists of highly repetitive sequences, some of which are repeated 10^5 to 10^6 times.

Further analysis, culminating in the sequencing of complete genomes, has identified several types of these highly repeated sequences (Table 5.2). One class (called **simple-sequence repeats**) consists of tandem arrays of up to thousands of copies of short sequences, ranging from 1 to 500 nucleotides. For example, one type of simple-sequence repeat in *Drosophila* consists of tandem repeats of the seven nucleotide unit ACAAACT. Because of their distinct base compositions, many simple-sequence DNAs can be separated from the rest of the genomic DNA by equilibrium centrifugation in CsCl density gradients. The density of DNA is determined by its base composition, with AT-rich sequences being less dense than GC-rich sequences. Therefore an AT-rich simple-sequence DNA bands in CsCl gradients at a lower density than the bulk of *Drosophila* genomic DNA (Figure 5.7). Since such repeat-sequence DNAs band as "satellites" separate from the main band of DNA, they are frequently referred to as **satellite DNAs**. These sequences are repeated millions of times per genome, accounting for about 10% of the DNA of most higher eukaryotes. Simple-sequence DNAs are not transcribed and do not convey functional genetic information.

TABLE 5.2 Repetitive Sequences in the Human Genome

Type of sequence	Number of copies	Fraction of genome
Simple-sequence repeats[a]	>1,000,000	~10%
Retrotransposons		
LINEs	850,000	21%
SINEs	1,500,000	13%
Retrovirus-like elements	450,000	8%
DNA transposons	300,000	3%

[a] The content of simple-sequence repeats is estimated from the fraction of heterochromatin in the human genome.

FIGURE 5.7 Satellite DNA   Equilibrium centrifugation of *Drosophila* DNA in a CsCl gradient separates satellite DNAs (designated I–IV) with buoyant densities (in g/cm³) of 1.672, 1.687, and 1.705 from the main band of genomic DNA (buoyant density 1.701).

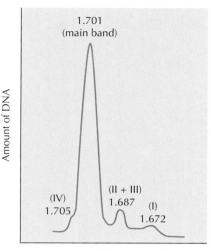

Some, however, play important roles in chromosome structure, as discussed in the next section of this chapter.

Other repetitive DNA sequences are scattered throughout the genome rather than being clustered as tandem repeats. These interspersed repetitive elements are a major contributor to genome size, accounting for approximately 45% of human genomic DNA. The two most prevalent classes of these sequences are called **SINEs** (short interspersed elements) and **LINEs** (long interspersed elements). SINEs are 100–300 base pairs long. About 1.5 million such sequences are dispersed throughout the genome, accounting for approximately 13% of the total human DNA. Although SINES are transcribed into RNA, they do not encode proteins and their function is unknown. The major human LINEs are 4–6 kb long, although many repeated sequences derived from LINEs are shorter, with an average size of about 1 kb. There are approximately 850,000 repeats of LINE sequences in the genome, accounting for about 21% of human DNA. LINEs are transcribed and at least some encode proteins, but like SINEs, they have no known function in cell physiology.

Both SINEs and LINEs are examples of transposable elements, which are capable of moving to different sites in genomic DNA. As discussed in detail in Chapter 6, both SINEs and LINEs are **retrotransposons**, meaning that their transposition is mediated by reverse transcription (**Figure 5.8**). An RNA copy of a SINE or LINE is converted to DNA by reverse transcriptase within the cell, and the new DNA copy is integrated at a new site in the genome. A third class of interspersed repetitive sequences, which closely resemble retroviruses and are called **retrovirus-like elements**, also move within the genome by reverse transcription. Human retrovirus-like elements range from approximately 2–10 kb in length. There are approximately 450,000 retrovirus-like elements in the human genome, accounting for approximately 8% of human DNA. In contrast, the fourth class of inter-

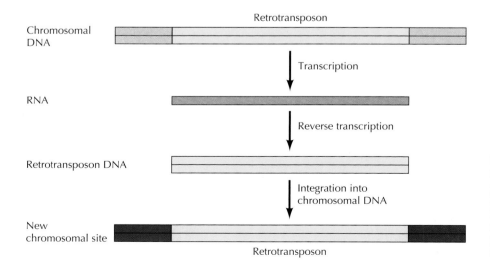

FIGURE 5.8 Movement of retrotransposons   A retrotransposon present at one site in chromosomal DNA is transcribed into RNA, and then converted back into DNA by reverse transcription. The retrotransposon DNA can then integrate into a new chromosomal site.

spersed repetitive elements (**DNA transposons**) moves through the genome by being copied and reinserted as DNA sequences, rather than moving by reverse transcription. In the human genome, there are about 300,000 copies of DNA transposons, ranging from 80–3000 base pairs in length and accounting for approximately 3% of human DNA.

Nearly half of the human genome thus consists of interspersed repetitive elements that have replicated and moved through the genome by either RNA or DNA intermediates. It is noteworthy that the vast majority of these elements transpose via RNA intermediates, so reverse transcription has been responsible for generating more than 40% of the human genome. Some of these sequences may help regulate gene expression, but most interspersed repetitive sequences appear not to make a useful contribution to the cell. Instead, they appear to represent "selfish DNA elements" that have been selected for their own ability to replicate within the genome rather than conferring a selective advantage to their host. In some cases, however, transposable elements have played important evolutionary roles by stimulating gene rearrangements and contributing to the generation of genetic diversity.

Gene Duplication and Pseudogenes

Another factor contributing to the large size of eukaryotic genomes is that many genes are present in multiple copies, some of which are frequently nonfunctional. In some cases, multiple copies of genes are needed to produce RNAs or proteins required in large quantities, such as ribosomal RNAs or histones. In other cases, distinct members of a group of related genes (called a **gene family**) may be transcribed in different tissues or at different stages of development. For example, the α and β subunits of hemoglobin are both encoded by gene families in the human genome, with different members of these families being expressed in embryonic, fetal, and adult tissues (**Figure 5.9**). Members of many gene families (e.g., the globin genes) are clustered within a region of DNA; members of other gene families are dispersed to different chromosomes.

Gene families are thought to have arisen by duplication of an original ancestral gene, with different members of the family then diverging as a consequence of mutations during evolution. Such divergence can lead to the evolution of related proteins that are optimized to function in different tissues or at different stages of development. For example, fetal globins

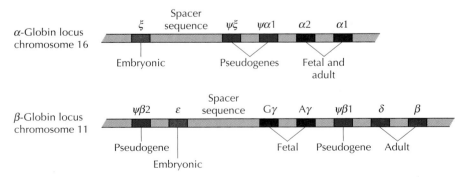

FIGURE 5.9 Globin gene families Members of the human α- and β-globin gene families are clustered on chromosomes 16 and 11, respectively. Each family contains genes that are specifically expressed in embryonic, fetal, and adult tissues, in addition to nonfunctional gene copies (pseudogenes).

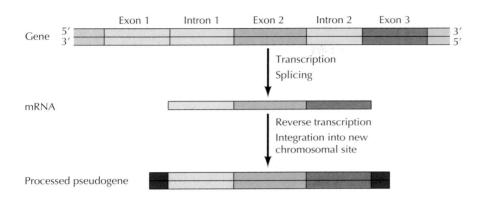

FIGURE 5.10 Formation of a processed pseudogene A gene is transcribed and spliced to yield an mRNA from which the introns have been removed. The mRNA is copied by reverse transcriptase, yielding a cDNA copy lacking introns. Integration into chromosomal DNA results in formation of a processed pseudogene.

have a higher affinity for O_2 than do adult globins—a difference that allows the fetus to obtain O_2 from the maternal circulation.

As might be expected, however, not all mutations enhance gene function. Some gene copies have instead sustained mutations that result in their loss of ability to produce a functional gene product. For example, the human α- and β-globin gene families each contain two genes that have been inactivated by mutations. Such nonfunctional gene copies (called **pseudogenes**) represent evolutionary relics that increase the size of eukaryotic genomes without making a functional genetic contribution. Recent studies have identified more than 20,000 pseudogenes in the human genome. Since this is generally assumed to be an underestimate, it is likely that our genome contains many more pseudogenes than functional genes.

Gene duplications can arise by two distinct mechanisms. The first is duplication of a segment of DNA, which can result in the transfer of a block of DNA sequence to a new location in the genome. Such duplications of DNA segments ranging from 1 kb to more than 50 kb are estimated to account for approximately 5% of the human genome. Alternatively, genes can be duplicated by reverse transcription of an mRNA, followed by integration of the cDNA copy into a new chromosomal site (**Figure 5.10**). This mode of gene duplication, analogous to the transposition of repetitive elements that move via RNA intermediates, results in the formation of gene copies that lack introns and also lack the normal chromosomal sequences that direct transcription of the gene into mRNA. As a result, duplication of a gene by reverse transcription usually yields an inactive gene copy called a **processed pseudogene**. Processed pseudogenes account for about two-thirds of the pseudogenes that have been identified in the human genome.

The Composition of Higher Eukaryotic Genomes

Having discussed several kinds of noncoding DNA that contribute to the genomic complexity of higher eukaryotes, it is of interest to overview the composition of cell genomes. In bacterial genomes, most of the DNA encodes proteins. For example, the genome of *E. coli* is approximately 4.6×10^6 base pairs long and contains about 4000 genes, with nearly 90% of the DNA used as protein-coding sequence. The yeast genome, which consists of 12×10^6 base pairs, is about 2.5 times the size of the genome of *E. coli*, but is still extremely compact. Only 4% of the genes of *Saccharomyces cerevisiae* contain introns, and these usually have only a single small intron near the start of the coding sequence. Approximately 70% of the yeast genome is used as protein-coding sequence, specifying a total of about 6000 proteins.

The relatively simple animal genomes of *C. elegans* and *Drosophila* are about 10 times larger than the yeast genome, but contain only 2–3 times more genes. Instead, these simple animal genomes contain more introns and more repetitive sequence, so that protein-coding sequences correspond to only about 25% of the *C. elegans* genome and about 13% of the genome of *Drosophila*. The genome of the model plant *Arabidopsis* contains a similar number of genes, with approximately 26% of the genome corresponding to protein-coding sequence.

The genomes of higher animals (such as humans) are approximately 20–30 times larger than those of *C. elegans* and *Drosophila*. However, a major surprise from deciphering the human genome sequence was the discovery that the human genome contains only 20,000 to 25,000 genes. It appears that only about 1.2% of the human genome consists of protein-coding sequence. Approximately 20% of the genome consists of introns, and more than 60% is composed of various types of repetitive and duplicated DNA sequences, with the remainder corresponding to pseudogenes, to nonrepetitive spacer sequences between genes, and to exon sequences that are present at the 5' and 3' ends of mRNAs but are not translated into protein. The increased size of the genomes of higher eukaryotes is thus due far more to the presence of large amounts of repetitive sequences and introns than to an increased number of genes.

Chromosomes and Chromatin

Not only are the genomes of most eukaryotes much more complex than those of prokaryotes, but the DNA of eukaryotic cells is also organized differently from that of prokaryotic cells. The genomes of prokaryotes are contained in single chromosomes, which are usually circular DNA molecules. In contrast, the genomes of eukaryotes are composed of multiple chromosomes, each containing a linear molecule of DNA. Although the numbers and sizes of chromosomes vary considerably between different species (Table 5.3), their basic structure is the same in all eukaryotes. The DNA of eukaryotic cells is tightly bound to small basic proteins (histones) that package the DNA in an orderly way in the cell nucleus. This task is substantial, given the DNA content of most eukaryotes. For example, the total extended length of DNA in a human cell is nearly 2 meters, but this DNA must fit into a nucleus with a diameter of only 5 to 10 μm.

Chromatin

The complexes between eukaryotic DNA and proteins are called **chromatin**, which typically contains about twice as much protein as DNA. The major proteins of chromatin are the **histones**—small proteins containing a high proportion of basic amino acids (arginine and lysine) that facilitate binding to the negatively charged DNA molecule. There are five major types of histones—called H1, H2A, H2B, H3, and H4—which are very similar among different species of eukaryotes (Table 5.4). The histones are extremely abundant proteins in eukaryotic cells; together their mass is approximately equal to that of the cell's DNA. In addition, chromatin contains an approximately equal mass of a wide variety of nonhistone chromosomal proteins. There are more than a thousand different types of these proteins, which are involved in a range of activities, including DNA replication and gene expression.

TABLE 5.3 Chromosome Numbers of Eukaryotic Cells

Organism	Genome size (Mb)[a]	Chromosome number[a]
Yeast (*Saccharomyces cerevisiae*)	12	16
Slime mold (*Dictyostelium*)	70	7
Arabidopsis thaliana	125	5
Corn	5000	10
Onion	15,000	8
Lily	50,000	12
Nematode (*Caenorhabditis elegans*)	97	6
Fruit fly (*Drosophila*)	180	4
Toad (*Xenopus laevis*)	3000	18
Lungfish	50,000	17
Chicken	1200	39
Mouse	3000	20
Cow	3000	30
Dog	3000	39
Human	3000	23

[a] Both genome size and chromosome number are for haploid cells.
Mb = millions of base pairs.

The basic structural unit of chromatin, the **nucleosome**, was described by Roger Kornberg in 1974 (Figure 5.11). Two types of experiments led to Kornberg's proposal of the nucleosome model. First, partial digestion of chromatin with micrococcal nuclease (an enzyme that degrades DNA) was found to yield DNA fragments approximately 200 base pairs long. In contrast, a similar digestion of naked DNA (not associated with proteins) yielded a continuous smear of randomly sized fragments. These results suggested that the binding of proteins to DNA in chromatin protects regions of the DNA from nuclease digestion, so that the enzyme can attack DNA only at sites separated by approximately 200 base pairs. Consistent with this notion, electron microscopy revealed that chromatin fibers have a beaded appearance, with the beads spaced at intervals of approximately 200 base pairs. Thus both the nuclease digestion and the electron microscopic studies suggested that chromatin is composed of repeating 200-base-pair units, which were called nucleosomes.

TABLE 5.4 The Major Histone Proteins

Histone[a]	Molecular weight	Number of amino acids	Percentage lysine + arginine
H1	22,500	244	30.8
H2A	13,960	129	20.2
H2B	13,774	125	22.4
H3	15,273	135	22.9
H4	11,236	102	24.5

[a] Data are for rabbit (H1) and bovine histones.

FIGURE 5.11 The organization of chromatin in nucleosomes   (A) The DNA is wrapped around histones in nucleosome core particles and sealed by histone H1. Nonhistone proteins bind to the linker DNA between nucleosome core particles. (B) Gel electrophoresis of DNA fragments obtained by partial digestion of chromatin with micrococcal nuclease. The linker DNA between the nucleosome core particles is preferentially sensitive, so limited digestion of chromatin yields fragments corresponding to multiples of 200 base pairs. (C) An electron micrograph of an extended chromatin fiber, illustrating its beaded appearance. (B, courtesy of Roger Kornberg, Stanford University; C, courtesy of Ada L. Olins and Donald E. Olins, Oak Ridge National Laboratory.)

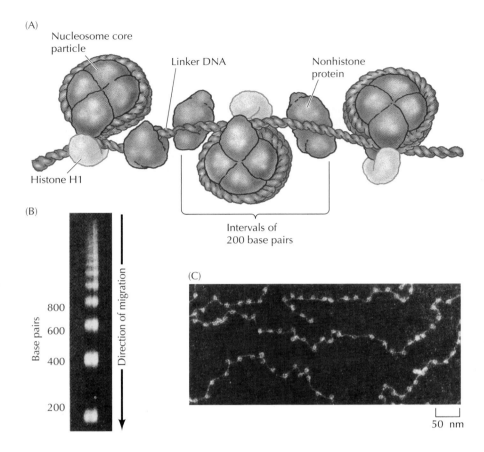

(A)

Nucleosome core particle

Linker DNA

Nonhistone protein

Histone H1

Intervals of 200 base pairs

(B)

Base pairs

Direction of migration

800

600

400

200

(C)

50 nm

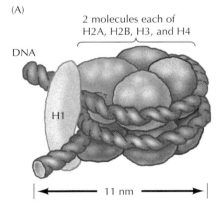

(A)

DNA

2 molecules each of H2A, H2B, H3, and H4

H1

11 nm

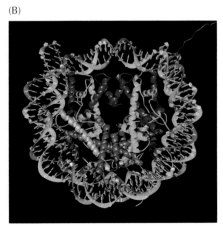

(B)

More extensive digestion of chromatin with micrococcal nuclease was found to yield particles (called **nucleosome core particles**) that correspond to the beads visible by electron microscopy. Detailed analysis of these particles has shown that they contain 147 base pairs of DNA wrapped 1.67 times around a histone core consisting of two molecules each of H2A, H2B, H3, and H4 (the core histones) (Figure 5.12). One molecule of the fifth histone, H1, is bound to the DNA as it enters each nucleosome core particle. This forms a chromatin subunit known as a **chromatosome**, which consists of 166 base pairs of DNA wrapped around the histone core and held in place by H1 (a linker histone).

The packaging of DNA with histones yields a chromatin fiber approximately 10 nm in diameter that is composed of chromatosomes separated by linker DNA segments averaging about 50 base pairs in length (Figure 5.13). In the electron microscope, this 10-nm fiber has the beaded appearance that suggested the nucleosome model. Packaging of DNA into such a 10-nm chromatin fiber shortens its length approximately sixfold. The chromatin

FIGURE 5.12 Structure of a chromatosome   (A) The nucleosome core particle consists of 147 base pairs of DNA wrapped 1.67 turns around a histone octamer consisting of two molecules each of H2A, H2B, H3, and H4. A chromatosome contains two full turns of DNA (166 base pairs) locked in place by one molecule of H1. (B) Model of the nucleosome core particle. The DNA backbones are shown in brown and turquoise. The histones are shown in blue (H3), green (H4), yellow (H2A), and red (H2B). (B, from K. Luger et al., 1997. *Nature* 389: 251.)

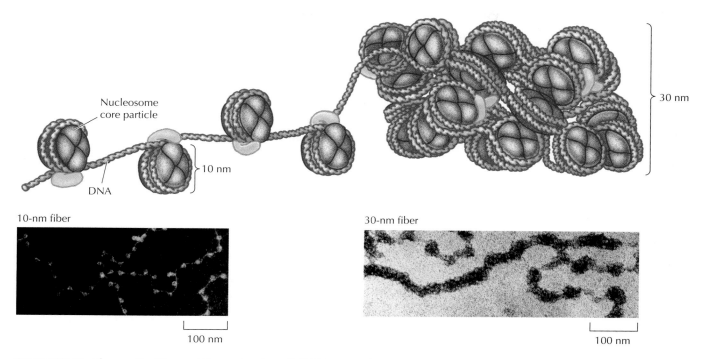

FIGURE 5.13 Chromatin fibers   The packaging of DNA into nucleosomes yields a chromatin fiber approximately 10 nm in diameter. The chromatin is further condensed by coiling into a 30-nm fiber, containing about six nucleosomes per turn. (Photographs courtesy of Ada L. Olins and Donald E. Olins, Oak Ridge National Laboratory.)

can then be further condensed by coiling into 30-nm fibers, resulting in a total condensation of about fiftyfold. Interactions between histone H1 molecules appear to play an important role in this stage of chromatin condensation, which is critical to determining the accessibility of chromosomal DNA for processes such as DNA replication and transcription. Despite its importance, the structure of the 30-nm fiber remained unknown until 2005, when X-ray studies by Timothy Richmond and his colleagues revealed that the fiber is formed by two stacks of nucleosomes, with linker DNA zigzagging back and forth between them. Folding of 30-nm fibers upon themselves can lead to further condensation of chromatin within the cell.

The extent of chromatin condensation varies during the life cycle of the cell and plays an important role in regulating gene expression, as will be discussed in Chapter 7. In interphase (nondividing) cells, most of the chromatin (called **euchromatin**) is relatively decondensed and distributed throughout the nucleus (Figure 5.14). During this period of the cell cycle, genes are transcribed and the DNA is replicated in preparation for cell division. Most of the euchromatin in interphase nuclei appears to be in the form of 30-nm, or somewhat more condensed 60- to 130-nm, chromatin fibers. Genes that are actively transcribed are in a more decondensed state that makes the DNA accessible to the transcription machinery. In contrast to euchromatin, about 10% of interphase chromatin (called **heterochromatin**) is in a very highly condensed state that resembles the chromatin of cells undergoing mitosis. Heterochromatin is transcriptionally inactive and contains highly repeated DNA sequences, such as those present at centromeres and telomeres.

FIGURE 5.14 Interphase chromatin
Electron micrograph of an interphase
nucleus. The euchromatin is distributed throughout the nucleus. The heterochromatin is indicated by arrowheads and the nucleolus by an arrow.
(Courtesy of Ada L. Olins and Donald
E. Olins, Oak Ridge National
Laboratory.)

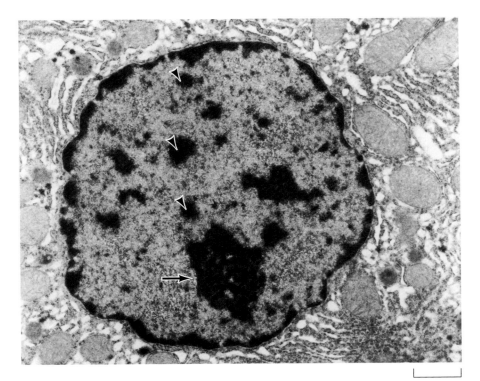

1 µm

As cells enter mitosis, their chromosomes become highly condensed so
that they can be distributed to daughter cells. Loops of 30-nm chromatin
fibers are thought to fold upon themselves to form the compact metaphase
chromosomes of mitotic cells in which the DNA has been condensed nearly
ten thousandfold (Figure 5.15). Such condensed chromatin can no longer be
used as a template for RNA synthesis, so transcription ceases during mitosis.
Electron micrographs indicate that the DNA in metaphase chromosomes is
organized into large loops attached to a protein scaffold (Figure 5.16), but we
currently understand neither the detailed structure of this highly condensed
chromatin nor the mechanism of chromatin condensation.

Metaphase chromosomes are so highly condensed that their morphology
can be studied using the light microscope (Figure 5.17). Several staining techniques yield characteristic patterns of alternating light and dark chromosome bands, which result from the preferential binding of stains or fluores-

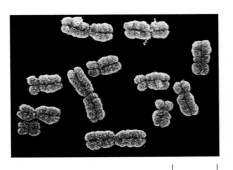

**FIGURE 5.15 Chromatin condensa-
tion during mitosis**   Scanning electron micrograph of metaphase chromosomes. Artificial color has been
added. (Biophoto Associates/Photo
Researchers Inc.)

10 µm

FIGURE 5.16 Structure of metaphase chromosomes An electron micrograph of DNA loops attached to the protein scaffold of metaphase chromosomes that have been depleted of histones. (From J. R. Paulson and U. K. Laemmli, 1977. *Cell* 12: 817.)

Protein scaffold   DNA loops

cent dyes to AT-rich versus GC-rich DNA sequences. These bands are specific for each chromosome and appear to represent distinct chromosome regions. Genes can be localized to specific chromosome bands by *in situ* hybridization, indicating that the packaging of DNA into metaphase chromosomes is a highly ordered and reproducible process.

Centromeres

The **centromere** is a specialized region of the chromosome that plays a critical role in ensuring the correct distribution of duplicated chromosomes to daughter cells during mitosis (Figure 5.18). The cellular DNA is replicated during interphase, resulting in the formation of two copies of each chromosome prior to the beginning of mitosis. As the cell enters mitosis, chromatin condensation leads to the formation of metaphase chromosomes consisting of two identical sister chromatids. These sister chromatids are held together at the centromere, which is seen as a constricted chromosomal region. As mitosis proceeds, microtubules of the mitotic spindle attach to the centromere, and the two sister chromatids separate and move to opposite poles of the spindle. At the end of mitosis, nuclear membranes re-form and the chromosomes decondense, resulting in the formation of daughter nuclei containing one copy of each parental chromosome.

The centromeres thus serve both as the sites of association of sister chromatids and as the attachment sites for microtubules of the mitotic spindle. They consist of specific DNA sequences to which a number of centromere-associated proteins bind, forming a specialized structure called the **kinetochore** (Figure 5.19). The binding of microtubules to kinetochore proteins

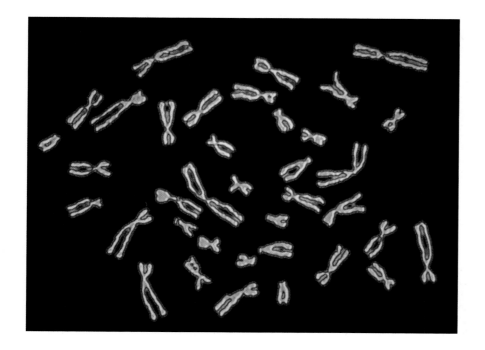

FIGURE 5.17 Human metaphase chromosomes A micrograph of human chromosomes spread from a metaphase cell. (Leonard Lessin/Peter Arnold, Inc.)

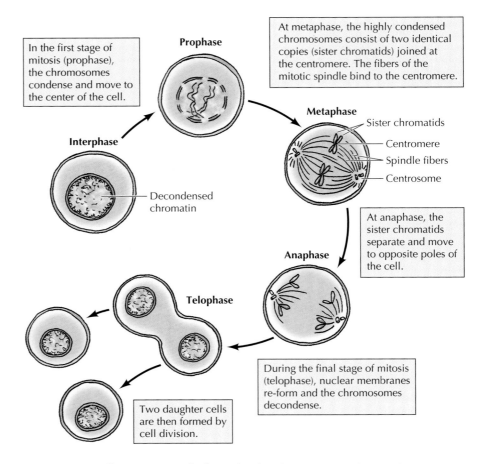

Prophase

In the first stage of mitosis (prophase), the chromosomes condense and move to the center of the cell.

At metaphase, the highly condensed chromosomes consist of two identical copies (sister chromatids) joined at the centromere. The fibers of the mitotic spindle bind to the centromere.

Metaphase

Sister chromatids
Centromere
Spindle fibers
Centrosome

Interphase

Decondensed chromatin

At anaphase, the sister chromatids separate and move to opposite poles of the cell.

Anaphase

Telophase

During the final stage of mitosis (telophase), nuclear membranes re-form and the chromosomes decondense.

Two daughter cells are then formed by cell division.

FIGURE 5.18 Chromosomes during mitosis Since DNA replicates during interphase, the cell contains two identical duplicated copies of each chromosome prior to entering mitosis.

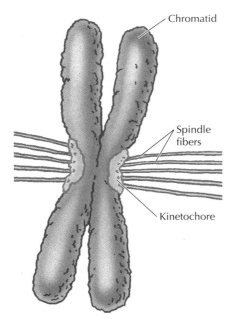

Chromatid

Spindle fibers

Kinetochore

mediates the attachment of chromosomes to the mitotic spindle. Proteins associated with the kinetochore then act as "molecular motors" that drive the movement of chromosomes along the spindle fibers, segregating the chromosomes to daughter nuclei.

Centromeric DNA sequences were initially defined in yeasts, where their function can be assayed by following the segregation of plasmids at mitosis (Figure 5.20). Plasmids that contain functional centromeres segregate like chromosomes and are equally distributed to daughter cells following mitosis. In the absence of a functional centromere, however, the plasmid does not segregate properly, and many daughter cells fail to inherit plasmid DNA. Assays of this type have enabled determination of the sequences required for centromere function. Such experiments first showed that the centromere sequences of the well-studied yeast *Saccharomyces cerevisiae* are contained in approximately 125 base pairs consisting of three sequence ele-

FIGURE 5.19 The centromere of a metaphase chromosome The centromere is the region at which the two sister chromatids remain attached at metaphase. Specific proteins bind to centromeric DNA, forming the kinetochore, which is the site of spindle fiber attachment.

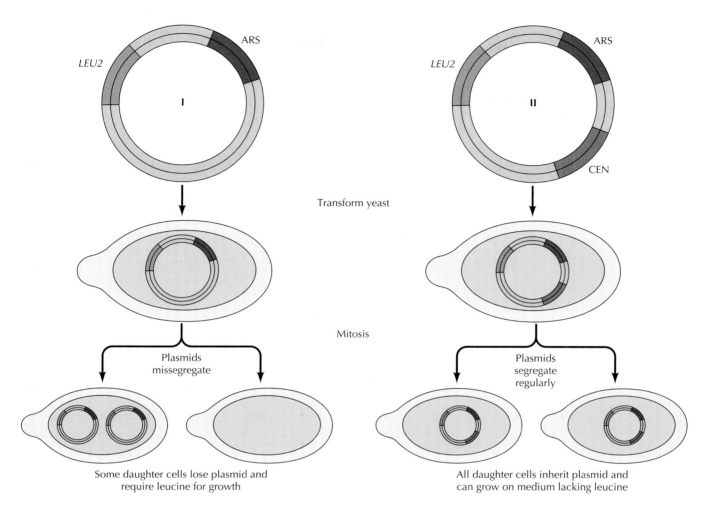

FIGURE 5.20 Assay of a centromere in yeast   Both plasmids shown contain a selectable marker (*LEU2*) and DNA sequences that serve as origins of replication in yeast (ARS, which stands for autonomously replicating sequence). However, plasmid I lacks a centromere and is therefore frequently lost as a result of missegregation during mitosis. In contrast, the presence of a centromere (CEN) in plasmid II ensures its regular transmission to daughter cells.

ments: two short sequences of 8 and 25 base pairs separated by 78 to 86 base pairs of very AT-rich DNA (Figure 5.21A).

The short centromere sequences defined in *S. cerevisiae*, however, do not appear to reflect the situation in other eukaryotes. More recent studies have defined the centromeres of the fission yeast *Schizosaccharomyces pombe* by a similar functional approach. Although *S. cerevisiae* and *S. pombe* are both yeasts, they appear to be as divergent from each other as either is from humans and are quite different in many aspects of their cell biology. These two yeast species thus provide complementary models for simple and easily studied eukaryotic cells. The centromeres of *S. pombe* span 40 to 100 kb of DNA; they are approximately a thousand times larger than those of *S. cerevisiae*. They consist of a central core of 4 to 7 kb of single-copy DNA flanked by repetitive sequences (Figure 5.21B). Not only the central core but also the flanking repeated sequences are required for centromere function, so the

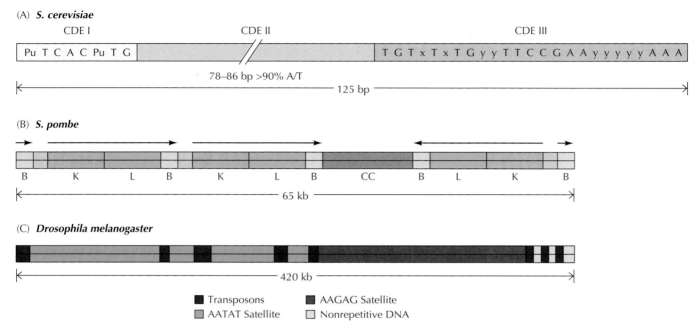

(A) *S. cerevisiae*

CDE I CDE II CDE III

Pu T C A C Pu T G — TGTxTxTGyyTTCCGAAyyyyyAAA

78–86 bp >90% A/T

125 bp

(B) *S. pombe*

B K L B K L B CC B L K B

65 kb

(C) *Drosophila melanogaster*

420 kb

■ Transposons ■ AAGAG Satellite
■ AATAT Satellite □ Nonrepetitive DNA

FIGURE 5.21 Centromeres of *S. cerevisiae*, *S. pombe*, and *Drosophila melanogaster* (A) The *S. cerevisiae* centromere (CEN) sequences consist of two short conserved sequences (CDE I and CDE III) separated by 78 to 86 base pairs (bp) of AT-rich DNA (CDE II). The sequences shown are consensus sequences derived from analysis of the centromere sequences of individual yeast chromosomes. Pu = A or G; x = A or T; y = any base. (B) The arrangement of sequences at the centromere of *S. pombe* chromosome II is illustrated. The centromere consists of a central core (CC) of unique-sequence DNA, flanked by tandem repeats of three repetitive sequence elements (B, K, and L). (C) The *Drosophila* centromere consists of two satellite sequences, transposable elements, and nonrepetitive DNA.

centromeres of *S. pombe* appear to be considerably more complex than those of *S. cerevisiae*.

Studies of a *Drosophila* chromosome provided the first characterization of a centromere in higher eukaryotes (Figure 5.21C). The *Drosophila* centromere spans 420 kb, most of which (more than 85%) consists of two highly repeated satellite DNAs with the sequences AATAT and AAGAG. The remainder of the centromere consists of interspersed transposable elements, which are also found at other sites in the *Drosophila* genome, in addition to a nonrepetitive region of AT-rich DNA. Deletion of the satellite sequences and transposable elements, as well as the nonrepetitive DNA, reduced the activity of the centromere in functional assays. Thus both repetitive and nonrepetitive sequences appear to contribute to kinetochore formation and centromere function. However, there do not appear to be any sequences that are specific to the centromere or that define centromere activity.

Centromeres of other plants and animals are characterized by heterochromatin containing extensive arrays of highly repetitive sequences. In *Arabidopsis*, centromeres consist of 3 million base pairs of an AT-rich 178-base pair satellite DNA. In humans and other primates the primary centromeric sequence is α satellite DNA, which is a 171-base-pair AT-rich sequence arranged in tandem repeats spanning 1–5 million base pairs. The α satellite DNA has been found to bind centromere-associated proteins, and recent experiments have shown that the centromeric α satellite array of the human X chromosome is sufficient to serve as a functional centromere. However, abnormal human chromosomes have also been described with functional centromeres that lack α satellite DNA, so the precise requirements for centromere function in higher eukaryotes remain unclear.

Although specific DNA sequences have not been associated with centromere function, it has been shown that the chromatin at centromeres has a unique structure. In particular, histone H3 is replaced in centromeric chromatin by an H3-like variant histone called centromeric H3 (CenH3). CenH3 is uniformly present at the centromeres of all organisms that have been studied and CenH3-containing nucleosomes are required for assembly of the other kinetochore proteins needed for centromere function. It thus appears that chromatin structure rather than a specific DNA sequence may be the primary determinant of the identity and function of centromeres. However, we still do not understand how centromeric chromatin is specified and stably maintained following cell division, so fundamental questions about the nature of centromeres in higher eukaryotes remain to be answered.

Telomeres

The sequences at the ends of eukaryotic chromosomes, called **telomeres**, play critical roles in chromosome replication and maintenance. Telomeres were initially recognized as distinct structures because broken chromosomes were highly unstable in eukaryotic cells, implying that specific sequences are required at normal chromosomal termini. This was subsequently demonstrated by experiments in which telomeres from the protozoan *Tetrahymena* were added to the ends of linear molecules of yeast plasmid DNA. The addition of these telomeric DNA sequences allowed these plasmids to replicate as linear chromosome-like molecules in yeasts, demonstrating directly that telomeres are required for the replication of linear DNA molecules.

The telomere DNA sequences of a variety of eukaryotes are similar, consisting of repeats of a simple-sequence DNA containing clusters of G residues on one strand (Table 5.5). For example, the sequence of telomere repeats in humans and other mammals is AGGGTT, and the telomere repeat in *Tetrahymena* is GGGGTT. These sequences are repeated hundreds or thousands of times and terminate with a 3′ overhang of single-stranded DNA. The repeated sequences of telomere DNA of some organisms (including humans) form loops at the ends of chromosomes as well as binding a number of proteins that protect the chromosome termini from degradation or from being joined together (Figure 5.22).

TABLE 5.5 Telomeric DNAs

Organism	Telomeric repeat sequence
Yeasts	
Saccharomyces cerevisiae	$G_{1-3}T$
Schizosaccharomyces pombe	$G_{2-5}TTAC$
Protozoans	
Tetrahymena	GGGGTT
Dictyostelium	$G_{1-8}A$
Plant	
Arabidopsis	AGGGTTT
Mammal	
Human	AGGGTT

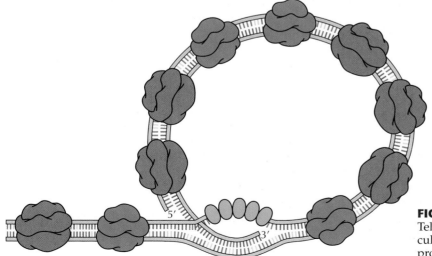

FIGURE 5.22 Structure of a telomere
Telomere DNA loops back on itself to form a circular structure and associates with a number of proteins that protect the ends of chromosomes.

■ Cancer cells have high levels of telomerase, allowing them to maintain the ends of their chromosomes through indefinite divisions. Since normal somatic cells lack telomerase activity and do not divide indefinitely, drugs that inhibit telomerase are being developed as anti-cancer agents.

Telomeres play a critical role in replication of the ends of linear DNA molecules (see Chapter 6). DNA polymerase is able to extend a growing DNA chain but cannot initiate synthesis of a new chain at the terminus of a linear DNA molecule. Consequently, the ends of linear chromosomes cannot be replicated by the normal action of DNA polymerase. This problem has been solved by the evolution of a special enzyme, **telomerase**, which uses reverse transcriptase activity to replicate telomeric DNA sequences. Maintenance of telomeres appears to be an important factor in determining the lifespan and reproductive capacity of cells, so studies of telomeres and telomerase have the promise of providing new insights into conditions such as aging and cancer.

The Sequences of Complete Genomes

Some of the most exciting recent advances in molecular biology have been the results of analyzing the complete nucleotide sequences of both the human genome and the genomes of several model organisms, including *E. coli, Saccharomyces cerevisiae, Caenorhabditis elegans, Drosophila, Arabidopsis,* and the mouse (Table 5.6). The results of whole genome sequencing have taken us beyond the characterization of individual genes to a global view of the organization and gene content of entire genomes. In principle, this approach has the potential of identifying all the genes in an organism, which then become accessible for investigations of their structure and function. Moreover, the availability of complete genome sequences opens the exciting possibility of identifying the sequences that regulate gene expression by genome-wide analysis. While much remains to be learned, the available genome sequences have provided scientists with a unique data-

TABLE 5.6 Representative Sequenced Genomes

Organism	Genome size (Mb)[a]	Number of genes	Protein-coding sequence
Bacteria			
Mycoplasma genitalium	0.58	470	88%
H. influenzae	1.8	1743	89%
E. coli	4.6	4288	88%
Yeasts			
S. cerevisiae	12	6000	70%
S. pombe	12	4800	60%
Invertebrates			
C. elegans	97	19,000	25%
Drosophila	180	13,600	13%
Plants			
Arabidopsis thaliana	125	26,000	25%
Rice	390	37,000	12%
Fish			
Pufferfish	370	20,000–23,000	10%
Birds			
Chicken	1000	20,000–23,000	3%
Mammals			
Human	3200	20,000–25,000	1.2%

[a]Mb = millions of base pairs

base, consisting of the nucleotide sequences of complete sets of genes and their regulatory sequences. Since many of these genes have not been previously identified, determination of their functions will form the basis of many future studies in cell biology.

Prokaryotic Genomes

We now know the complete genome sequences of more than 100 different bacteria, and still more are in the process of being determined. The first complete sequence of a cellular genome, reported in 1995 by a team of researchers led by Craig Venter, was that of the bacterium *Haemophilus influenzae*, a common inhabitant of the human respiratory tract. The genome of *H. influenzae* is approximately 1.8×10^6 base pairs (1.8 **megabases,** or **Mb**), slightly less than half the size of the *E. coli* genome. The complete nucleotide sequence indicated that the *H. influenzae* genome is a circular molecule containing 1,830,137 base pairs of DNA. The sequence was then analyzed to identify the genes encoding rRNAs, tRNAs, and proteins. Potential protein-coding regions were identified by computer analysis of the DNA sequence to detect **open-reading frames**—long stretches of nucleotide sequence that can encode polypeptides because they contain none of the three chain-terminating codons (UAA, UAG, and UGA). Since these chain-terminating codons occur randomly once in every 21 codons (3 chain-terminating codons out of 64 total), open-reading frames that extend for more than a hundred codons usually represent functional genes.

This analysis identified six copies of rRNA genes, 54 different tRNA genes, and 1743 potential protein-coding regions in the *H. influenzae* genome (Figure 5.23). More than a thousand of these could be assigned a biological role (e.g., an enzyme of the citric acid cycle) on the basis of their relationships to known protein sequences, but the others represent genes of

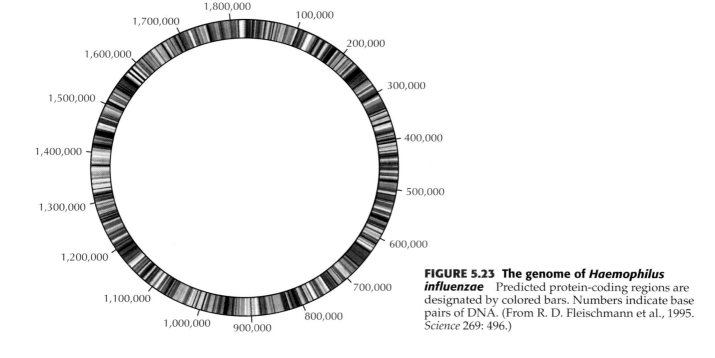

FIGURE 5.23 The genome of *Haemophilus influenzae* Predicted protein-coding regions are designated by colored bars. Numbers indicate base pairs of DNA. (From R. D. Fleischmann et al., 1995. *Science* 269: 496.)

unknown function. The predicted coding sequences have an average size of approximately 900 base pairs, so they cover about 1.6 Mb of DNA, corresponding to nearly 90% of the genome of *H. influenzae*.

The complete sequence of the genome of *Mycoplasma genitalium* is of particular interest because mycoplasmas are the simplest present-day bacteria and contain the smallest genomes of all known cells. The genome of *M. genitalium* is only 580 kb (0.58 Mb) long and may represent the minimal set of genes required to maintain a self-replicating organism. Analysis of its DNA sequence indicates that *M. genitalium* contains only 470 predicted protein-coding sequences, which correspond to approximately 88% of genomic DNA. Many of these sequences were identified as genes encoding proteins involved in DNA replication, transcription, translation, membrane transport, and energy metabolism. However, *M. genitalium* contains many fewer genes for metabolic enzymes than does *H. influenzae*, reflecting its more limited metabolism. For example, many genes known to encode components of biosynthetic pathways are lacking in the genome of *M. genitalium*, consistent with its need to obtain amino acids and nucleotide precursors from a host organism. Interestingly, the *Mycoplasma* genome also includes approximately 150 genes of currently unknown function. Thus, even in the simplest of cells, the biological roles of many genes remain to be determined.

The sequence of the genome of the archaebacterium *Methanococcus jannaschii*, reported in 1996, provided major insights into the evolutionary relationships between the archaebacteria, eubacteria, and eukaryotes. The genome of *M. jannaschii* is 1.7 Mb and contains 1738 predicted protein-coding sequences—similar in size to the genome of *H. influenzae*. However, only about one-third of the protein-coding sequences identified in *M. jannaschii* were related to known genes of either eubacteria or eukaryotes, indicating the distinct genetic composition of the archaebacteria. The genes of *M. jannaschii* encoding proteins involved in energy production and biosynthesis of cell constituents are related to those of eubacteria, suggesting that basic metabolic processes evolved in a common ancestor of both the archaebacteria and the eubacteria. Importantly, however, the *M. jannaschii* genes encoding proteins involved in DNA replication, transcription, and translation are more closely related to those of eukaryotes than to those of eubacteria. Genomic sequencing of this archaebacterium thus indicates that the archaebacteria and eukaryotes are as closely related to each other as either is to the eubacteria (see Figure 1.7).

Although the relative simplicity and facile genetics of *E. coli* have made it a favored organism of molecular biologists, the 4.6-Mb *E. coli* genome was not completely sequenced until 1997. Analysis of the *E. coli* sequence revealed a total of 4288 genes, with protein-coding sequences accounting for approximately 88% of the *E. coli* genome. Of the 4288 genes revealed by sequencing, 1835 had been previously identified and the functions of an additional 821 could be deduced by comparisons to the sequences of characterized genes of other organisms. However, the functions of 1632 *E. coli* genes (nearly 40% of the genome) could not be determined. Thus, even for an organism as thoroughly studied as *E. coli*, genomic sequencing demonstrates that a great deal remains to be learned about prokaryotic cell biology.

The Yeast Genome

As noted already, the simplest eukaryotic genome (1.2×10^7 base pairs of DNA) is found in the yeast *Saccharomyces cerevisiae*. Moreover, yeasts grow rapidly and are subject to simple genetic manipulations. Thus in many

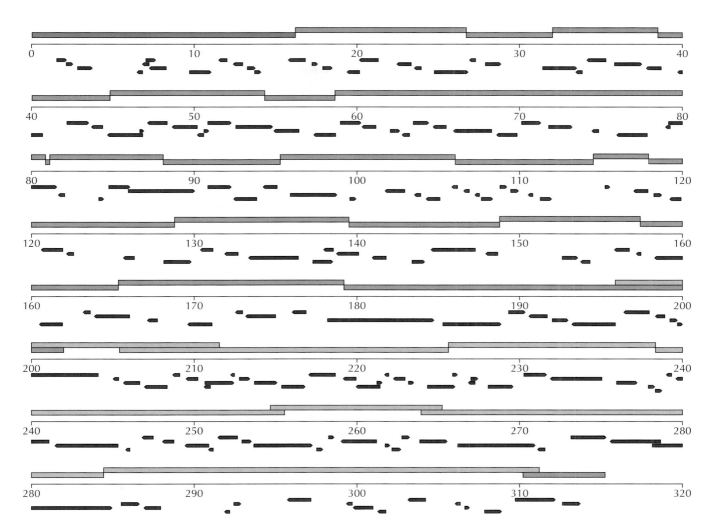

FIGURE 5.24 Yeast chromosome III
The upper blue bars designate the clones used for DNA sequencing. Open-reading frames are indicated by arrows. (From S. G. Oliver et al., 1992. *Nature* 357: 38.)

ways yeasts are model eukaryotic cells that can be studied much more readily than the cells of mammals or other higher eukaryotes. Consequently, the complete sequencing of an entire yeast chromosome in 1992 (Figure 5.24), followed by determination of the sequence of the complete *S. cerevisiae* genome in 1996, were major steps in understanding the molecular biology of eukaryotic cells.

The *S. cerevisiae* genome contains about 6000 genes, including 5885 predicted protein-coding sequences, 140 ribosomal RNA genes, 275 transfer RNA genes, and 40 genes encoding small nuclear RNAs involved in RNA processing (discussed in Chapter 7). Yeasts thus have a high density of protein-coding sequences, similar to bacterial genomes, with protein-coding sequences accounting for approximately 70% of total yeast DNA. Consistent with this, only 4% of yeast genes were found to contain introns. Moreover, those *S. cerevisiae* genes that do contain introns usually have only a single small intron near the beginning of the gene.

Computer analysis was able to assign a predicted function to approximately 3000 of the *S. cerevisiae* protein-coding sequences based on similarities to the sequences of known genes. Based on analysis of these genes, it appears that approximately 11% of yeast proteins function in metabolism,

3% in the production and storage of metabolic energy, 3% in DNA replication, repair, and recombination, 7% in transcription, 6% in translation, and 14% in protein sorting and transport. However, the functions of many of these genes are only known in general terms (such as "transcription factor"), so their precise roles within the cell still need to be determined. Moreover, since half of the proteins encoded by the yeast genome were unrelated to previously described genes, the functions of an additional 3000 unknown proteins remain to be elucidated by genetic and biochemical analyses.

The sequence of the *S. cerevisiae* genome has been more recently followed by the sequence of the genome of the fission yeast, *S. pombe*, as well as the genomes of several other yeast and fungi. As discussed earlier in this chapter, *S. cerevisiae* and *S. pombe* are quite divergent and differ in many aspects of their biology, including the structure of their centromeres (see Figure 5.21). Interestingly, their genomes also display considerable differences. Although both *S. cerevisiae* and *S. pombe* have approximately the same amount of unique sequence DNA (12.5 Mb), *S. pombe* appears to contain only about 4800 genes. Introns are much more prevalent in *S. pombe* than in *S cerevisiae*. Approximately 43% of *S. pombe* genes contain introns and the introns in *S. pombe* are larger than those in *S cerevisiae*, so protein-coding sequence accounts for only about 60% of the *S. pombe* genome. The majority of *S. pombe* genes have homologs in the *S cerevisiae* genome, but approximately 700 genes are unique to *S. pombe*.

Now that yeast genome sequences have been completed, determination of the functions of the many new genes described in both *S. cerevisiae* and *S. pombe* is a major goal. Fortunately, yeasts are particularly amenable to functional analyses of unknown genes because of the facility with which normal chromosomal loci can be inactivated by homologous recombination with cloned sequences (discussed in Chapter 4). Therefore direct functional analysis of yeast genes that were initially identified only on the basis of their nucleotide sequence can be systematically undertaken. Sequencing the yeast genomes has thus opened the door to studying many new areas of the biology of a simple eukaryotic cell. Such studies are expected to reveal the functions of many new genes that are not restricted to yeasts but are common to all eukaryotes, including humans.

The Genomes of *Caenorhabditis elegans* and *Drosophila melanogaster*

The genomes of *C. elegans* and *Drosophila* are relatively simple animal genomes, intermediate in size and complexity between those of yeasts and humans. Distinctive features of each of these organisms make them important models for genome analysis: *C. elegans* is widely used for studies of animal development, and *Drosophila* has been especially well analyzed genetically. The genomes of these organisms, however, are about tenfold larger than those of yeasts, introducing a new order of difficulty in genome mapping and sequencing. Determination of the sequence of *C. elegans* in 1998 therefore represented an important milestone in genome analysis, which extended genome sequencing from unicellular organisms (bacteria and yeast) to a multicellular organism recognized as an important model for animal development.

The initial phases of analysis of the *C. elegans* genome used DNA fragments cloned in cosmids, which accommodate DNA inserts of approximately 30–45 kb (see Table 4.3). This approach, however, was unable to cover the complete genome, which was accomplished by the cloning of much larger pieces of DNA in **yeast artificial chromosome (YAC)** vectors.

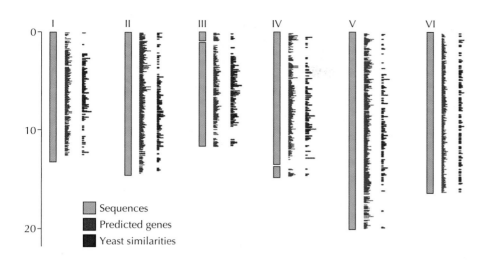

FIGURE 5.25 The *C. elegans* genome The positions of the predicted genes of *C. elegans* on each chromosome are indicated by red bars. Those that are similar to genes of yeast are indicated by purple. (From The *C. elegans* Sequencing Consortium, 1998. *Science* 282: 2012.)

As noted in Chapter 4, the unique feature of YACs is that they contain centromeres and telomeres, allowing them to replicate as linear chromosome-like molecules in yeasts. They can therefore be used to clone DNA fragments the size of yeast chromosomal DNAs, up to thousands of kilobases in length. The large DNA inserts that can be cloned in YACs and other high-capacity vectors are critically important for analysis of complex genomes.

The *C. elegans* genome is 97×10^6 base pairs and contains about 19,000 predicted protein-coding sequences—approximately three times the number of genes in yeast (Figure 5.25). In contrast to the compact genome organization of yeast, genes in *C. elegans* span about 5 kilobases and contain an average of five introns. Protein-coding sequences thus account for only about 25% of the *C. elegans* genome, as compared to 60–70% of *S. pombe* and *S. cerevisiae* and nearly 90% of bacterial genomes.

Approximately 40% of the predicted *C. elegans* proteins displayed significant similarity to known proteins of other organisms. As expected, there are substantially more similarities between the proteins of *C. elegans* and humans than between *C. elegans* and either yeast or bacteria. Proteins that are common between *C. elegans* and yeast may function in the basic cellular processes shared by these organisms, such as metabolism, DNA replication, transcription, translation, and protein sorting. These core biological processes appear to be carried out by a similar number of genes in both organisms, and it is likely that these genes will be shared by all eukaryotic cells. In contrast, the majority of *C. elegans* genes are not found in yeast and may function in the more intricate regulatory activities required for the development of multicellular organisms. Elucidating the functions of these genes is likely to be particularly exciting in terms of understanding animal development. Although adult *C. elegans* contain only 959 somatic cells in the entire body, they have all of the specialized cell types found in more complicated animals. Moreover, the complete pattern of cell divisions leading to *C. elegans* development has been described, including analysis of the connections made by all 302 neurons in the adult animal. Many of the genes involved in *C. elegans* development and differentiation have already been found to be related to genes involved in controlling the proliferation and differentiation of mammalian cells, substantiating the validity of *C. elegans* as a model for more complex animals. With little doubt, many more critical developmental control genes will be uncovered from studies of the *C. elegans* genomic sequence.

FIGURE 5.26 Polytene chromosomes of *Drosophila* A light micrograph of stained salivary gland chromosomes. The four chromosomes (X, 2, 3, and 4) are joined at their centromeres. (Peter J. Bryant/Biological Photo Service.)

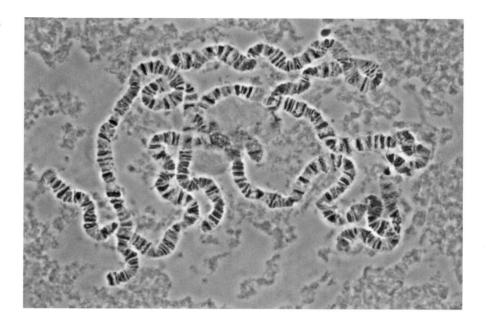

Drosophila is another key model for animal development, which has been particularly well-characterized genetically. The advantages of *Drosophila* for genetic analysis include its relatively simple genome and the fact that it can be easily maintained and bred in the laboratory. In addition, a special tool for genetic analysis in *Drosophila* is provided by the giant **polytene chromosomes** that are found in some tissues, such as the salivary glands of larvae. These chromosomes arise in nondividing cells as a consequence of repeated replication of DNA strands that fail to separate from each other. Thus each polytene chromosome contains hundreds of identical DNA molecules aligned in parallel. Because of their size, these polytene chromosomes are visible in the light microscope, and appropriate staining procedures reveal a distinct banding pattern (Figure 5.26). The banding of polytene chromosomes provides a much greater degree of resolution than that achieved with metaphase chromosomes (e.g., see Figure 5.17). The polytene chromosomes are decondensed interphase chromosomes that contain actively expressed genes. More than 5000 bands are visible, each corresponding to an average length of approximately 20 kb of DNA. In contrast, the bands identified in human metaphase chromosomes contain several megabases of DNA.

The banding pattern of polytene chromosomes thus provides a high-resolution physical map of the *Drosophila* genome. Gene deletions can often be correlated with the loss of a specific chromosomal band, thereby defining the physical location of the gene on the chromosome. In addition, cloned DNAs can be mapped by *in situ* hybridization to polytene chromosomes, often with sufficient resolution to localize cloned genes to specific bands (Figure 5.27). Thus the map positions of cosmid or YAC clones (which span many bands) can readily be determined, providing the base for genomic sequence analysis.

Because of the power of *Drosophila* genetics, the sequencing of the *Drosophila* genome early in 2000 was an important advance in genomic analysis. The genome of *Drosophila* consists of approximately 180×10^6 base pairs, of which about one-third is heterochromatin. The heterochromatin consists principally of simple sequence satellite repeats, in addition to interspersed transposable elements, and was not included in the genomic sequence. The remaining 120×10^6 base pairs of euchromatin was sequenced using a combination of **bacterial artificial chromosome (BAC)** clones, which carry large inserts of DNA (see Table 4.3), and a shotgun approach in which small fragments of DNA were randomly cloned and sequenced in plasmid vectors. The sequences of these small fragments of DNA were then assembled into a large contiguous sequence by identification of overlaps between fragments, and these sequence assemblies were aligned with the BAC clones to yield a complete sequence of the euchromatic portion of the *Drosophila* genome.

The *Drosophila* genome contains approximately 13,600 genes; surprisingly fewer than the number of genes in *C. elegans*, even though *Drosophila* is a more complex organism. However, it is important to note that this difference in gene number does not correspond to a difference in genetic complexity, because many genes are duplicated in both *Drosophila* and *C. elegans*. When these duplications are taken into account, it appears that both *Drosophila* and *C. elegans* contain a similar number of distinct genes, estimated between 10,000 and 15,000. Like *C. elegans*, *Drosophila* genes contain an average of 4 introns, and the total amount of intron sequence is similar to the amount of exon sequence. In total, protein-coding sequence accounts for about 13% of the *Drosophila* genome.

It is especially striking that a complex animal like *Drosophila* has only about twice the number of unique genes found in yeast, which appears to be a much simpler organism. Apparently, the complexity of multicellular organisms is not simply related to a greater number of genes. Part of the increased biological complexity of *Drosophila* and *C. elegans* may arise from the fact that their proteins are generally larger and contain more functional domains than the proteins of yeast. Further studies and functional analysis of the genes that have been uncovered by sequencing the *Drosophila* and *C. elegans* genomes will undoubtedly play a major role in understanding the ways in which these genes act to direct the complex process of animal development.

Plant Genomes

The completion of the genome sequence of *Arabidopsis thaliana* in 2000 extended genome sequencing from animals to plants, and was thus a major event in plant biology. *Arabidopsis thaliana* is a simple flowering plant, which has been widely used as a model for studies of plant molecular biology and development. Its advantages as a model organism for molecular biology and genetics include its relatively small genome of approximately 125×10^6 base pairs, similar in size to the genomes of *C. elegans* and *Drosophila*. Like the *Drosophila* genome, the *Arabidopsis* genome was sequenced principally using BAC vectors to accommodate large DNA inserts.

Surprisingly, analysis of the *Arabidopsis* genome indicated that it contained approximately 26,000 protein-coding genes—significantly more genes than were found in either *C. elegans* or *Drosophila*. However, this unexpectedly large number of genes does not reflect a greater diversity of proteins encoded by the *Arabidopsis* genome. Instead, it appears that the large number of genes in *Arabidopsis* is the result of duplications of large segments of the *Arabidopsis* genome. These duplications involve approximately 60% of the genome, so the number of distinct protein-coding genes in *Arabidopsis* is estimated to be about 15,000—similar to the number of genes in *C. elegans* or *Drosophila*.

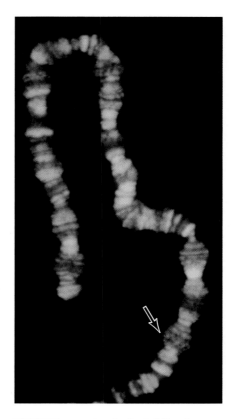

FIGURE 5.27 *In situ* **hybridization to a *Drosophila* polytene chromosome** Hybridization of a YAC clone to a polytene chromosome is illustrated. The region of hybridization is indicated by an arrow. (Courtesy of Daniel L. Hartl, Harvard University.)

FIGURE 5.28 Functions of predicted genes of *Arabidopsis thaliana* The chart illustrates the proportion of *Arabidopsis* in different functional categories. (From The *Arabidopsis* Genome Initiative, 2000. *Nature* 408: 796.)

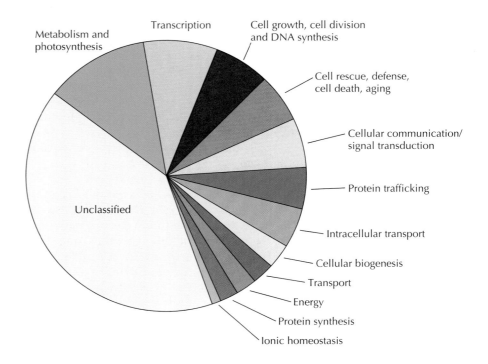

The gene density in *Arabidopsis* is also similar to that of *C. elegans*, with protein-coding sequences accounting for about 25% of the *Arabidopsis* genome. On the average, *Arabidopsis* genes have approximately 4 introns, and the total length of intron sequences is about the same as the total length of exon sequences. Transposable elements account for about 10% of the *Arabidopsis* genome. As in *Drosophila*, transposable element repeats are clustered at the centromeres together with satellite repetitive sequences.

Comparative analysis of the functions of the *Arabidopsis* genes has revealed both interesting similarities and differences between the genes of plants and animals. *Arabidopsis* genes involved in fundamental cellular processes such as DNA replication, repair, transcription, translation, and protein trafficking are similar to those in yeast, *C. elegans,* and *Drosophila*, reflecting the common evolutionary origins of all eukaryotic cells. In contrast, the *Arabidopsis* genes encoding proteins involved in processes such as cell signaling and membrane transport are quite different from those in animals, consistent with the major differences in physiology and development between plants and animals. About one-third of all *Arabidopsis* genes appear unique to plants, as they are not found in yeast or animal genomes. The largest functional group of *Arabidopsis* genes, corresponding to 22% of the genome, encodes proteins involved in metabolism and photosynthesis (Figure 5.28). Another large group of genes (12% of the genome) encodes proteins involved in plant defense. It is also noteworthy that *Arabidopsis* encodes more than 3000 proteins that regulate transcription (accounting for approximately 17% of the genome). This number of gene regulatory proteins (transcription factors) is two or three times more than are found in *Drosophila* and *C. elegans*, respectively. Many of the *Arabidopsis* transcription factors are unique to plants, presumably reflecting distinct features of gene expression in plant development and in the response of plants to the environment.

The sequence of *Arabidopsis* was followed in 2002 by publication of two draft sequences of the rice genome. Rice is of major importance as a cereal crop and is the staple food for more than half the world's population, so sequencing the rice genome has the potential of leading to very significant applications in agriculture and biotechnology. Two groups of researchers reported draft sequences of the genomes of two subspecies of rice: the *indica* subspecies, which is the most widely cultivated subspecies in China and most of the rest of Asia; and the *japonica* subspecies, which is the variety preferred in Japan. These initial draft sequences of the rice genome were followed by a high quality complete sequence of the *japonica* subspecies in 2005.

The rice genome consists of about 390×10^6 base pairs of DNA—about 3 times larger than the genome of *Arabidopsis*. At least 35% of the rice genome consists of transposable elements, in part accounting for its larger size. In addition, rice contain a surprisingly high number of predicted protein-coding genes, estimated at approximately 37,000. Like *Arabidopsis*, rice contains many duplicated genes, which have arisen as a result of duplication of large segments (approximately 60%) of the genome. Nonetheless, the rice genome contains more genes than either *Arabidopsis* or humans, underscoring the fact that gene number does not directly correlate with biological complexity in eukaryotes. Interestingly, approximately 70% of the genes predicted in rice are also found in *Arabidopsis*, and almost 90% of the genes that have been identified in *Arabidopsis* are found in rice. Most of the genes shared between *Arabidopsis* and rice are not found in yeast or animal genomes and therefore appear to be specific for plants.

The Human Genome

For many scientists, the ultimate goal of genome analysis was determination of the complete nucleotide sequence of the human genome: approximately 3×10^9 base pairs of DNA. To understand the magnitude of this undertaking, recall that the human genome is more than ten times larger than that of *Drosophila;* that the smallest human chromosome is several times larger than the entire yeast genome; and that the extended length of DNA that makes up the human genome is about 1 m long. From all of these perspectives, determination of the human genome sequence was a phenomenal undertaking, and its publication in draft form in 2001 was heralded as a scientific achievement of historic magnitude.

The human genome is distributed among 24 chromosomes (22 autosomes and the 2 sex chromosomes), each containing between 45 and 280 Mb of DNA (Figure 5.29). Prior to determination of the genome sequence, several thousand human genes had been identified and mapped to positions on the human chromosomes. One commonly used method to localize genes is *in situ* hybridization of probes labeled with fluorescent dyes to chromosomes—a method generally referred to as **fluorescence *in situ* hybridization,** or **FISH (**Figure 5.30). *In situ* hybridization to metaphase chromosomes allows the mapping of a cloned gene to a locus defined by a chromosome band. Because each band of human metaphase chromosomes contains thousands of kilobases of DNA, *in situ* hybridization to human metaphase chromosomes does not provide the detailed mapping information obtained by hybridization to the polytene chromosomes of *Drosophila*, which allows the localization of genes to interphase chromosome bands containing only 10 to 20 kb of DNA. Higher resolution can be obtained, however, by hybridization to more extended human chromosomes from

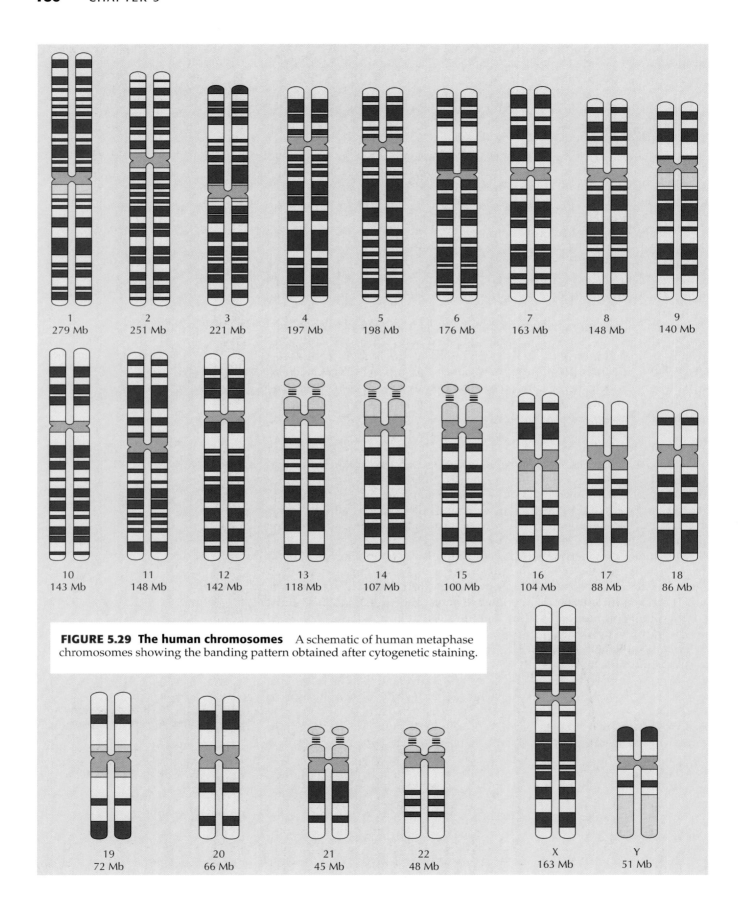

1 279 Mb	2 251 Mb	3 221 Mb	4 197 Mb	5 198 Mb	6 176 Mb	7 163 Mb	8 148 Mb

9
140 Mb

10 143 Mb	11 148 Mb	12 142 Mb	13 118 Mb	14 107 Mb	15 100 Mb	16 104 Mb	17 88 Mb

18
86 Mb

FIGURE 5.29 The human chromosomes A schematic of human metaphase chromosomes showing the banding pattern obtained after cytogenetic staining.

19 72 Mb	20 66 Mb	21 45 Mb	22 48 Mb	X 163 Mb	Y 51 Mb

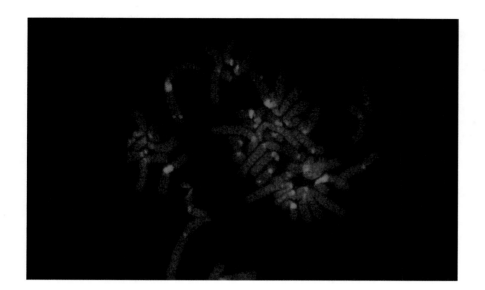

FIGURE 5.30 Fluorescence *in situ* hybridization A fluorescent probe for the gene encoding lamin B receptor is hybridized to stained human metaphase chromosomes (blue). Single gene hybridization signals are detected as red fluorescence. (Courtesy of K. L. Wydner and J. B. Lawrence, University of Massachusetts Medical Center.)

prometaphase or interphase cells, allowing the use of fluorescence *in situ* hybridization to map cloned genes to regions of about 100 kb. In addition to FISH, genetic linkage analysis and the physical mapping of cloned genomic and cDNA sequences were used to establish physical and genetic maps of the human genome, which provided a background for genomic sequencing.

The draft sequences of the human genome published in 2001 were produced by two independent teams of researchers, who used different approaches. One research team, The International Human Genome Sequencing Consortium, used BAC clones that had been mapped to sites on the human chromosomes as the substrates for sequencing. The other team, led by Craig Venter of Celera Genomics, used a shotgun approach in which small fragments were cloned and sequenced, and overlaps between the sequences of these fragments were then used to assemble the sequence of the genome. Both of these sequences were initially incomplete drafts in which approximately 90% of the euchromatin portion of the genome had been sequenced and assembled. Continuing efforts have closed the gaps and improved the accuracy of the draft sequences, leading to publication of a high-quality human genome sequence in 2004.

The sequenced euchromatin portion of the genome encompasses approximately 2.9×10^6 kb of DNA (Figure 5.31). The total size of the genome is approximately 3.2×10^6 kb, with the remaining 10% of the genome (0.3×10^6

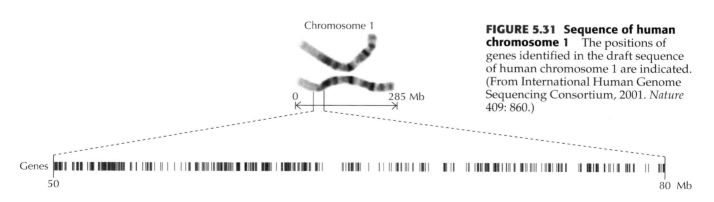

Chromosome 1

0   285 Mb

Genes

50   80 Mb

FIGURE 5.31 Sequence of human chromosome 1 The positions of genes identified in the draft sequence of human chromosome 1 are indicated. (From International Human Genome Sequencing Consortium, 2001. *Nature* 409: 860.)

KEY EXPERIMENT

The Human Genome

Initial Sequencing and Analysis of the Human Genome
International Human Genome Sequencing Consortium
Nature, Volume 409, 2001, pages 860–921

The Sequence of the Human Genome
J. Craig Venter and 273 others
Science, Volume 291, 2001, pages 1304–1351

The Context

The idea of sequencing the entire human genome was first conceived in the mid-1980s. It was initially met with broad skepticism among biologists, most of whom felt it was simply not a feasible undertaking. At the time, the largest genome that had been completely sequenced was that of Epstein-Barr virus, which totaled approximately 180,000 base pairs of DNA. From this perspective, sequencing the human genome, which was almost 20,000 times larger, seemed inconceivable to many. However, the idea of such a massive project in biology captivated the imagination of others, including Charles DeLisi who was then head of the Office of Health and Environmental Research at the Department of Energy. In 1986 DeLisi succeeded in launching the Human Genome Initiative as a project within the Department of Energy.

The project gained broader support in 1988 when it was endorsed by a committee of the National Research Council. This committee recommended a broader effort, including sequencing the genomes of several model organisms and the parallel development of detailed genetic and physical maps of the human chromosomes. This effort was centered at the National Institutes of Health, initially under the direction of James Watson (codiscoverer of the structure of DNA), and then under the leadership of Frances Collins.

The first complete genome to be sequenced was that of the bacterium *Haemophilus influenzae*, reported by Craig Venter and colleagues in 1995. Venter had been part of the genome sequencing effort at the National Institutes of Health but had left to head a nonprofit company, The Institute for Genomic Research, in 1991. In the meantime, considerable progress had been made in mapping the human genome, and the initial sequence of *H. influenzae* was followed by the sequences of other bacteria, yeast, and *C. elegans* in 1998.

In 1998 Venter formed a new company, Celera Genomics, and announced plans to use advanced sequencing technologies to obtain the entire human genome sequence in 3 years. Collins and other leaders of the publicly funded genome project responded by accelerating their efforts, resulting in a race that eventually led to the publication of two draft sequences of the human genome in February, 2001.

The Experiments

The two groups of scientists used different approaches to obtain the human genome sequence. The publicly funded team, The International Human Genome Sequencing Consortium, headed by Eric Lander, sequenced DNA fragments derived from BAC clones that had been previously mapped to human chromosomes, similar to the approach used to determine the sequence of the yeast and *C. elegans* genomes (see figure). In contrast, the Celera Genomics team used a whole-genome shotgun sequencing approach that Venter and colleagues had first used to sequence the genome of *H. influenzae*. In this approach, DNA fragments were sequenced at random, and overlaps between fragments were then used to reassemble a complete genome sequence. Both sequences covered only the euchromatin portion of the human genome—approximately 2900

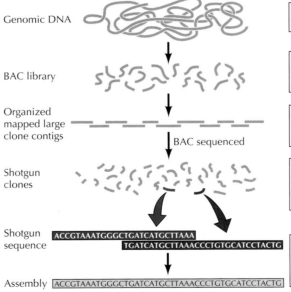

Strategy for genome sequencing using BAC clones that had been organized into overlapping clusters (contigs) and mapped to human chromosomes.

Mb of DNA—with the heterochromatin repeat-rich portion of the genome (approximately 300 Mb) remaining unsequenced.

Both of these initially published versions were draft, rather than completed, sequences. Subsequent efforts completed the sequence, leading to publication of a highly accurate sequence of the human genome in 2004.

The Impact

Several important conclusions immediately emerged from the human genome sequences. First, the number of human genes was surprisingly small and appears to be between 20,000 and 25,000 in the completed sequence. Interestingly, however, alternative splicing appears to be common in the human genome, so

many genes may encode more than 1 protein. Introns account for about 20% of the human genome and repetitive sequences for about 60%. It is noteworthy that over 40% of human DNA is composed of sequences derived by reverse transcription, emphasizing the importance of this mode of information transfer in shaping our genome.

Beyond these immediate conclusions, the sequence of the human genome, together with the genome sequences of other organisms, will provide a new basis for biology and medicine in the years to come. The impact of the genome sequence will be felt in discovering new genes and their functions, understanding gene regulation, elucidating the basis of human diseases, and developing new strategies for prevention and treatment

Eric Lander

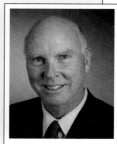

Craig Venter

based on the genetic makeup of individuals. Knowledge of the human genome may ultimately contribute to meeting what Venter and colleagues refer to as "The real challenge of human biology…to explain how our minds have come to organize thoughts sufficiently well to investigate our own existence."

---

kb) corresponding to highly repetitive sequences in heterochromatin. As discussed earlier in this chapter, interspersed repetitive sequences, the majority of which are transposable elements that have moved throughout the genome by reverse transcription of RNA intermediates, account for approximately 45% of the human euchromatin sequence. Another 5% of the genome consists of duplicated segments of DNA, so about 60% of the human genome consists of repetitive DNA sequences.

A major surprise from the genome sequence is the unexpectedly low number of human genes. The human genome consists of only 20,000–25,000 genes, which is not much larger than the number of genes in simpler animals like *C. elegans* and *Drosophila*. In fact, humans have fewer genes than rice, emphasizing one of the major conclusions that has emerged from the results of genome sequencing: the biological complexity of an organism is not simply a function of the number of genes in its genome. On the other hand, there appears to be a significant amount of alternative splicing in human genes, allowing a single gene to specify more than one protein (see Figure 5.5). Although the extent of alternative splicing in humans is not yet clear, it may substantially expand the number of proteins that can be encoded by the human genome.

Human genes are spread over much larger distances and contain more intron sequence than genes in *Drosophila* or *C. elegans*. The average protein-coding sequence in human genes is approximately 1400 base pairs, similar to that in *Drosophila* and *C. elegans*. However, the average human gene spans about 30 kb of DNA, with more than 90% of the gene corresponding to introns. Approximately 20% of the genome thus consists of introns, with only about 1.2% of the human genome corresponding to protein-coding sequences.

> ■ For many years, scientists generally accepted an estimate of approximately 100,000 genes in the human genome. On publication of the draft genome sequence in 2001, the number was drastically reduced to between 30,000 and 40,000. Current estimates, based on the high quality sequence published in 2004 and using improved computational tools to identify genes, reduce the number of human genes still further, to approximately 20,000 to 25,000.

Over 40% of the predicted human proteins are related to proteins in other sequenced organisms, including *Drosophila* and *C. elegans*. Many of these conserved proteins function in basic cellular processes, such as metabolism, DNA replication and repair, transcription, translation, and protein trafficking. Most of the proteins that are unique to humans are made up of protein domains that are also found in other organisms, but these domains are arranged in novel combinations to yield distinct proteins in humans. Compared to *Drosophila* and *C. elegans*, the human genome contains expanded numbers of genes involved in functions related to the greater complexity of vertebrates, such as the immune response, the nervous system, and blood clotting, as well as increased numbers of genes involved in development, cell signaling, and the regulation of transcription.

The Genomes of Other Vertebrates

In addition to the human genome, a large and growing number of vertebrate genomes have been sequenced in the last few years, including the genomes of fish, chickens, and other mammals (Figure 5.32). These sequences provide interesting comparisons to that of the human genome and are expected to facilitate the identification of a variety of different types of functional sequences, including regulatory elements that control gene expression.

The genome of the pufferfish *Fugu rubripes* was chosen for sequencing because it is unusually compact for a vertebrate genome. Consisting of only 3.7×10^8 base pairs of DNA, the pufferfish genome is only about one-eighth the size of the human genome. Although the pufferfish and human genomes contain a similar number of genes, the pufferfish has far less repetitive sequence and smaller introns. In particular, repetitive sequences account for only about 15% of the pufferfish genome (corresponding to approximately 50 million base pairs of DNA) as compared to about 60% of the human genome (approximately 2 billion base pairs). Because of this reduced amount of repetitive sequence, genes are more closely packed in the pufferfish and occupy about one-third of its genome. Pufferfish and human genes contain similar numbers of introns, but introns are shorter in the pufferfish, so that protein coding sequence corresponds to approximately one-third of the average gene or about 10% of the pufferfish genome (as compared to 1.2% of the human genome). The pufferfish thus provides a compact model of a vertebrate genome in which genes and critical regulatory sequences are highly concentrated, facilitating efforts to focus continuing studies on these functional genomic elements.

The chicken is intermediate between the pufferfish and mammals, both in evolutionary divergence and in the size of its genome. Consisting of approximately 10^9 base pairs, the chicken genome is about one-third the size of the human genome. However, it is estimated to contain 20,000 to 23,000 genes, similar to the gene content of humans. The smaller size of the chicken genome is largely the result of a substantial reduction in the amount of repetitive sequences and pseudogenes compared to mammalian genomes.

The mammalian genomes that have been sequenced, in addition to the human genome, include the genomes of the mouse, rat, dog, and chimpanzee. These genomes are all similar in size to the human genome and contain similar numbers of genes. However, each offers particular advantages for further understanding gene regulation and function. As discussed in earlier chapters, the mouse is the key model system for experimental studies of mammalian genetics and development, so the availability of the

> ■ **Pufferfish contain a very powerful neurotoxin, called tetrodotoxin, in some of their tissues. In Japan, pufferfish are considered a delicacy and prepared by specially trained chefs in licensed restaurants.**

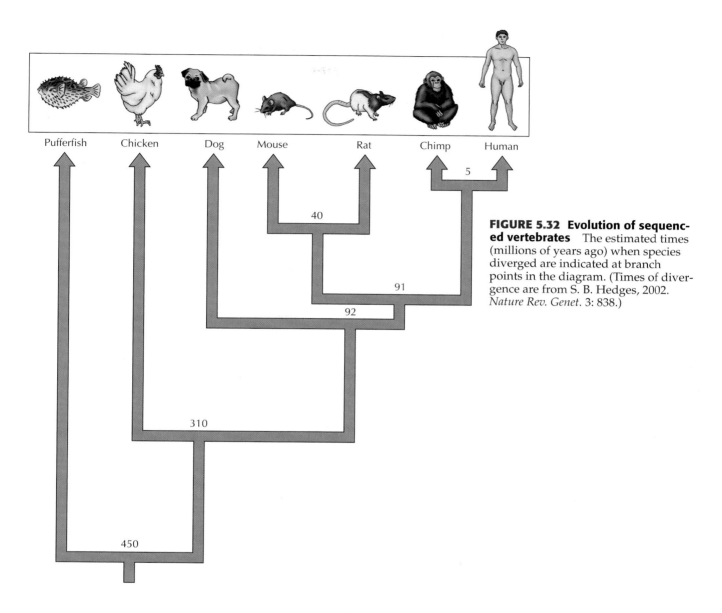

FIGURE 5.32 Evolution of sequenced vertebrates The estimated times (millions of years ago) when species diverged are indicated at branch points in the diagram. (Times of divergence are from S. B. Hedges, 2002. *Nature Rev. Genet.* 3: 838.)

mouse genome sequence provides an essential database for research in these areas. Likewise, the rat is an important model for human physiology and medicine, and these studies will be facilitated by the availability of the rat genome sequence. Mice, rats, and humans have 90% of their genes in common, providing a clear genetic foundation for the use of the mouse and rat as models for human development and disease.

The many distinct breeds of pet dogs make the sequence of the dog genome particularly important in understanding the genetic basis of morphology, behavior, and a variety of complex diseases that afflict both dogs and humans. There are approximately 300 breeds of dogs, which differ in their physical and behavioral characteristics as well as in their susceptibility to a variety of diseases, including several types of cancer, blindness, deafness, and metabolic disorders. Susceptibility to particular diseases is a highly specific property of different breeds, greatly facilitating identification of the responsible genes. Since many of these diseases are common to both dogs and humans, genetic studies in dogs can be expected to impact human health as well as veterinary medicine. An interesting example is provided by studies of sleep disorders in which the gene responsible for a rare inherited form of narcolepsy was identified in Doberman pinschers. Subsequent studies implicated related defects in human narcolepsy and possibly other sleep disorders. Similar types of genetic analysis are underway to understand the genetic basis of other complex diseases, such as hip dysplasia and rheumatoid arthritis, that are common in some breeds of dogs, and the results of these studies will undoubtedly benefit both dogs and humans. In the future, we can also expect genetic analysis of behavior in dogs. Since many canine behaviors, such as separation anxiety, are also common in humans, psychologists may have much to learn from the species that has been our closest companion for thousands of years.

The sequence of the genome of the chimpanzee, our nearest evolutionary relative, is expected to help pinpoint the unique features of our genome that distinguish humans from other primates. Interestingly however, comparison of the chimpanzee and human genome sequences does not suggest an easy answer to the question of what makes us human. The nucleotide sequences of the chimpanzee and human genomes are nearly 99% identical. The difference between the sequences of these closely related species (approximately 1 nucleotide in 100) is about ten times greater than the difference between the genomes of individual humans (approximately 1 nucleotide in 1000). Perhaps surprisingly, the sequence differences between humans and chimpanzees are not restricted to noncoding sequences. Instead, they frequently alter the coding sequences of genes, leading to changes in the amino acid sequences of most of the proteins encoded by chimpanzees and humans. Although many of these amino acid changes may not affect protein function, it appears that there are changes in the structure as well as in the expression of thousands of genes between chimpanzees and humans, so identifying those differences that are key to the origin of humans will not be a simple task.

Bioinformatics and Systems Biology

The human genome sequence, together with the sequences of other genomes, provides a wealth of information that forms a new framework for studies of cell and molecular biology and opens new possibilities in medical practice. In addition, the genome sequencing projects have raised new questions and substantially changed the way in which many problems in biology are being approached. Traditionally, molecular biologists have studied one or a few genes or proteins at a time. This has been changed by the genome sequencing projects, which introduced new large-scale experimental approaches in which vast amounts of data were generated. Handling the enormous amounts of data generated by whole genome sequencing required sophisticated computational analysis and spawned the new field of **bioinformatics**, which lies at the interface between biology and computer science and is focused on developing the computational methods needed to analyze and extract useful biological information from the sequence of billions of bases of DNA. The development of such computational methods has also led to other types of large-scale biological experimentation, including simultaneous analysis of the expression of thousands of mRNAs or proteins and the development of high-throughput methods to determine gene function using RNA interference. These large-scale experimental approaches form the basis of the new field of **systems biology**, which seeks a quantitative understanding of the integrated dynamic behavior of complex biological systems and processes. Systems biology thus combines large-scale biological experimentation with quantitative analysis and the development of testable models for complex biological processes. The global analysis of cell proteins (proteomics), discussed in Chapter 2, is one example of these new large-scale experimental/computational approaches. Some of the additional research areas that are amenable to large-scale experimentation, bioinformatics, and systems biology are discussed below.

Systematic Screens of Gene Function

The identification of all of the genes in an organism opens the possibility for a large-scale systematic analysis of gene function. One approach is to systematically inactivate (or knockout) each gene in the genome by homologous recombination with an inactive mutant allele (see Figure 4.39). As noted in Chapter 4, this has been done in yeast to produce a collection of yeast strains with mutations in all known genes, which can then be analyzed to determine which genes are involved in any biological property of interest. Alternatively, large-scale screens based on RNA interference (RNAi) are being used to systematically dissect gene function in a variety of organisms, including *Drosophila, C. elegans,* and mammalian cells in culture.

In RNAi screens, double-stranded RNAs are used to induce degradation of the homologous mRNAs in cells (see Figure 4.42). With the availability of complete genome sequences, libraries of double-stranded RNAs can be designed and used in genome-wide screens to identify all of the genes involved in any biological process that can be assayed in a high-throughput manner. For example, genome-wide RNAi analysis has been used to identify genes required for the growth and viability of *Drosophila* cells in culture (Figure 5.33). Individual double-stranded RNAs from the genome-wide library are tested in microwells in a high-throughput format to identify those that interfere with the growth of cultured cells, thereby characterizing the entire set of genes in the *Drosophila* genome that are required for cell growth or survival. Similar RNAi screens have been used to identify genes involved in a variety of biological processes, including cell signaling path-

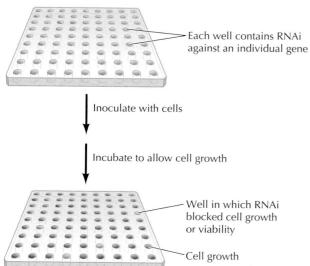

Each well contains RNAi against an individual gene

Inoculate with cells

Incubate to allow cell growth

Well in which RNAi blocked cell growth or viability

Cell growth

FIGURE 5.33 Genome-wide RNAi screen for cell growth and viability   Each microwell contains RNAi corresponding to an individual gene in the *Drosophila* genome. *Drosophila* tissue culture cells are added to each well and incubated to allow cell growth. Those wells in which cells fail to grow identify genes required for cell growth or viability.

ways, protein degradation, and transmission at synapses in the nervous system.

Regulation of Gene Expression

Genome sequences can in principle reveal not only the protein-coding sequences of genes, but also the regulatory elements that control gene expression. As discussed in subsequent chapters, regulation of gene expression is critical to many aspects of cell function, including the development of complex multicellular organisms. Understanding the mechanisms that control gene expression is therefore a central undertaking in contemporary cell and molecular biology, and it is expected that the availability of genome sequences will contribute substantially to this task. Unfortunately, it is far more difficult to identify gene regulatory sequences than it is to identify protein-coding sequences. Most regulatory elements are short sequences of DNA, typically spanning only about 10 base pairs. Consequently, sequences resembling regulatory elements occur frequently by chance in genomic DNA, so physiologically significant elements can not be identified from DNA sequence alone. The identification of functional regulatory elements and elucidation of the signaling networks that control gene expression therefore represent major challenges in bioinformatics and systems biology.

The availability of genomic sequences has enabled scientists to undertake global studies of gene expression in which the expression levels of all genes in a cell can be assayed simultaneously. These experiments employ DNA microarrays in which each gene is represented by an oligonucleotide corresponding to a small dot on a slide (see Figure 4.27). Hybridization of fluorescent-labeled cDNA copies of cellular mRNAs to such a microarray allows simultaneous determination of the mRNA levels of all cellular genes. This approach has been particularly valuable in revealing global changes in gene regulation associated with discrete cell behaviors, such as cell differentiation or the response of cells to a particular hormone or growth factor. Since genes that are coordinately regulated within a cell may be controlled by similar mechanisms, analyzing changes in the expression of multiple genes can help to pinpoint shared regulatory elements.

COMPANION WEBSITE

Visit the website that accompanies **The Cell** (www.sinauer.com/cooper) for animations, videos, quizzes, problems, and other review material.

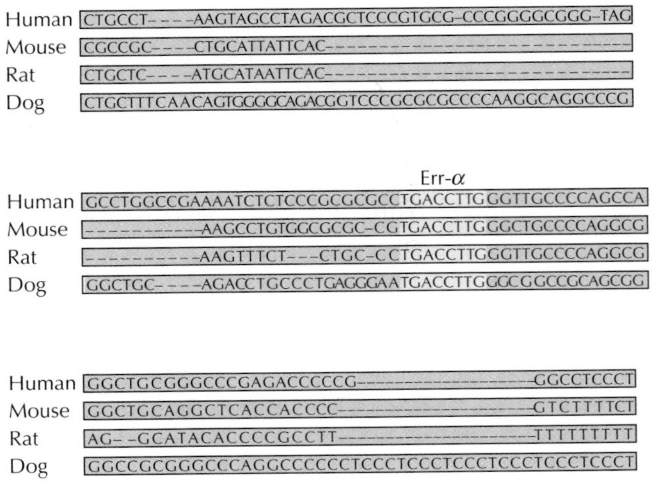

FIGURE 5.34 Conservation of functional gene regulatory elements Human, mouse, rat, and dog sequences near the transcription start site of a gene contain a functional regulatory element that binds the transcriptional regulatory protein Err-α. These sequences (highlighted in yellow) are conserved in all four genomes, whereas the surrounding sequences are not. (From X. Xie et al., 2005. *Nature* 434: 338.)

A variety of computational approaches are also being used to characterize functional regulatory elements. One approach is comparative analysis of the genome sequences of related organisms. This is based on the assumption that functionally important sequences are conserved in evolution, whereas nonfunctional segments of DNA diverge more rapidly. Computational analysis based on this approach has recently identified gene regulatory sequences that are conserved between the mouse, rat, dog, and human genomes (Figure 5.34). In addition, functional regulatory elements often occur in clusters, reflecting the fact that genes are generally regulated by the interactions of multiple transcription factors (see Chapter 7). Computer algorithms designed to detect clusters of transcription factor binding sites in genomic DNA have also proven useful in identifying sequences that regulate gene expression.

The combination of large-scale experimental methods and computational analysis has been successful in providing at least an initial indication of the transcriptional regulatory elements that govern expression of genes in yeast. However, extending these approaches to the far more complicated genomes of humans and other mammals remains a major challenge for future research.

Variation among Individuals and Genomic Medicine

Comparisons of genome sequences of related species is helpful in understanding the basis of differences between species, as well as in identifying genes and regulatory sequences that have been conserved in evolution. A different type of information can be gained by comparing the genome sequences of different individuals. Variations between individual genomes underlie differences in physical and mental characteristics, including susceptibility to many diseases. One of the major applications of the human genome sequence will be helping to uncover new genes involved in many of the diseases that afflict mankind, including cancer, heart disease, and degenerative diseases of the nervous system such as Parkinson's and Alzheimer's disease. In addition, understanding our unique genetic makeup as individuals is expected to lead to the development of new tailor-made strategies for disease prevention and treatment.

The genomes of two unrelated people differ in about one of every thousand bases. Most of this variation is in the form of single base changes, known as single nucleotide polymorphisms (SNPs), which are found at about 10 million positions in the genome. Over a million commonly occurring SNPs have been mapped in the human genome. These SNPs are distributed relatively uniformly in genomic DNA (Figure 5.35), and it is noteworthy that more than 90% of protein-coding genes contain at least one SNP. It is likely that these SNPs are responsible for most genetic differences in individual characteristics, so a substantial effort is being directed towards using SNPs to map the genes responsible for inherited differences in disease susceptibility. Analysis of these variations among individuals will not only allow specific genes to be associated with susceptibility to different diseases but will also enable physicians to tailor strategies for disease prevention and treatment to match the genetic makeup of individual patients. Comparisons between the genomes of different individuals may also help to elucidate the contribution of our genes to other unique characteristics, such as athletic ability or intelligence, and to better understand the interactions between genes and environment that lead to complex human behaviors.

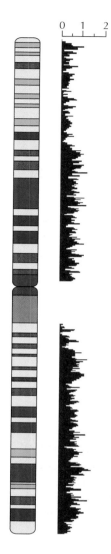

FIGURE 5.35 Single nucleotide polymorphisms (SNPs) in human chromosome 1   The distribution of SNPs (frequency per kilobase) is indicated. (From D. A. Hinds et al., 2005. *Science* 307: 1072.)

■ Current efforts aim to develop technologies that would be capable of sequencing the genome of individuals at very low cost. Such affordable 'personal genome projects' might provide better health care options tailored to the needs of individual patients.

KEY TERMS

gene, spacer sequence, exon, intron, RNA splicing, kilobase (kb), alternative splicing

simple-sequence repeat, satellite DNA, SINE, LINE, retrotransposon, DNA transposon, retrovirus-like element

gene family, pseudogene, processed pseudogene

chromatin, histone, nucleosome, nucleosome core particle, chromatosome, euchromatin, heterochromatin

centromere, kinetochore

telomere, telomerase

megabase (Mb), open-reading frame

SUMMARY

THE COMPLEXITY OF EUKARYOTIC GENOMES

Introns and Exons: Most eukaryotic genes have a split structure in which segments of coding sequence (exons) are interrupted by noncoding sequences (introns). In complex eukaryotes, introns account for more than ten times as much DNA as exons.

Repetitive DNA Sequences: Over 50% of mammalian DNA consists of highly repetitive DNA sequences, some of which are present in 10^5 to 10^6 copies per genome. These sequences include simple-sequence repeats as well as repetitive elements that have moved throughout the genome by either RNA or DNA intermediates.

Gene Duplications and Pseudogenes: Many eukaryotic genes are present in multiple copies, called gene families, which have arisen by duplication of ancestral genes. Some members of gene families function in different tissues or at different stages of development. Other members of gene families (pseudogenes) have been inactivated by mutations and no longer represent functional genes. Gene duplications can occur either by duplication of a segment of DNA or by reverse transcription of an mRNA, giving rise to a processed pseudogene. Approximately 5% of the human genome consists of duplicated DNA segments. In addition, there are more than 10,000 processed pseudogenes in the human genome.

The Composition of Higher Eukaryotic Genomes: Only a small fraction of the genome in complex eukaryotes corresponds to protein-coding sequences. The human genome is estimated to contain 20,000–25,000 genes, with protein-coding sequence corresponding to only about 1.2% of the DNA. Approximately 20% of the human genome consists of introns, and more than 60% is composed of repetitive and duplicated DNA sequences.

CHROMOSOMES AND CHROMATIN

Chromatin: The DNA of eukaryotic cells is wrapped around histones to form nucleosomes. Chromatin can be further compacted by the folding of nucleosomes into higher-order structures, including the highly condensed metaphase chromosomes of cells undergoing mitosis.

Centromeres: Centromeres are specialized regions of eukaryotic chromosomes that serve as the sites of association between sister chromatids and the sites of spindle fiber attachment during mitosis.

Telomeres: Telomeres are specialized sequences required to maintain the ends of eukaryotic chromosomes.

THE SEQUENCES OF COMPLETE GENOMES

Prokaryotic Genomes: The genomes of more than 100 different bacteria, including *E. coli*, have been completely sequenced. The *E. coli* genome contains 4288 genes, with protein-coding sequences accounting for nearly 90% of the DNA.

SUMMARY

The Yeast Genome: The first eukaryotic genome to be sequenced was that of the yeast *S. cerevisiae*. The *S. cerevisiae* genome contains about 6000 genes, and protein-coding sequences account for approximately 70% of the genome. The genome of the fission yeast *S. pombe* contains fewer genes (about 5000) and more introns than *S. cerevisiae*, with protein-coding sequence corresponding to about 60% of the *S. pombe* genome.

The Genomes of Caenorhabditis elegans *and* Drosophila melanogaster: The genome of *C. elegans* was the first sequenced genome of a multicellular organism. The *C. elegans* genome contains about 19,000 protein-coding sequences, which account for only about 25% of the genome. The genome of *Drosophila* contains approximately 14,000 genes, with protein-coding sequences accounting for about 13% of the genome. Although *Drosophila* contains fewer genes than *C. elegans,* many genes in both species are duplicated, and it appears that both species contain 10,000–15,000 unique genes. Some of these genes are shared between *Drosophila, C. elegans,* and yeast—these genes may encode proteins with common functions in all eukaryotic cells. However, the majority of *Drosophila* and *C. elegans* genes are not found in yeast and are likely to function in the regulation and development of multicellular animals.

Plant Genomes: The genome of the small flowering plant *Arabidopsis thaliana* contains approximately 26,000 genes—surprisingly more genes than were found in either *Drosophila* or *C. elegans*. However, many of these genes are the result of duplications of large segments of the *Arabidopsis* genome, so the number of unique genes in *Arabidopsis* is about 15,000. Many of these genes are unique to plants, including genes involved in plant physiology, development, and defense. The sequence of the rice genome is of particular agricultural interest because rice is the staple food for more than half the world's population. The draft sequence of the rice genome is estimated to contain approximately 37,000 genes, many of which are duplicated and may have arisen by duplication of large genome segments.

The Human Genome: The human genome appears to contain 20,000–25,000 genes—not much more than the number of genes found in simpler animals like *Drosophila* and *C. elegans*. Over 40% of the predicted human proteins are related to proteins found in other sequenced organisms, including *Drosophila* and *C. elegans*. In addition, the human genome contains expanded numbers of genes involved in the nervous system, the immune system, blood clotting, development, cell signaling, and the regulation of gene expression.

The Genomes of Other Vertebrates: The genomes of fish, chickens, mice, rats, dogs, and chimpanzees provide important comparisons to the human genome. All of these vertebrates contain similar numbers of genes but in some cases differ substantially in their content of repetitive sequences.

KEY TERMS

yeast artificial chromosome (YAC), polytene chromosome, bacterial artificial chromosome (BAC)

fluorescence *in situ* hybridization (FISH)

KEY TERMS

bioinformatics, systems biology

SUMMARY

BIOINFORMATICS AND SYSTEMS BIOLOGY

Systematic Screens of Gene Function: The genome sequencing projects have introduced large-scale experimental and computational approaches to research in cell and molecular biology. Genome-wide screens using RNA interference can systematically identify all of the genes in an organism that are involved in any biological process that can be assayed in a high-throughput format.

Regulation of Gene Expression: The identification of gene regulatory sequences and elucidation of the signaling networks that control gene expression are major challenges in bioinformatics and systems biology. These problems are being approached by genome-wide studies of gene expression combined with the development of computational approaches to identify functional regulatory elements.

Variation among Individuals and Genomic Medicine: Variations in our genomes are responsible for the characteristics of individual people, including susceptibility to many diseases. Analysis of these variations will allow the identification of genes responsible for disease susceptibility and enable the development of new strategies for disease prevention and treatment that match the genetic makeup of different individuals.

Questions

1. Many eukaryotic organisms have genome sizes that are much larger than their complexity would seem to require. Explain this paradox.

2. How were introns discovered during studies of adenovirus mRNAs?

3. How do intron sequences in the human genome increase the diversity of proteins expressed from the limited number of 20,000–25,000 genes?

4. How can simple-sequence repetitive DNA be separated from the bulk of the nuclear DNA?

5. Yeast (*S. cerevisiae*) centromeres form a kinetochore that attaches to a single microtubule, whereas multiple microtubules are attached to the kinetochores of most animal cells. How does the structure of *S. cerevisiae* centromeres reflect this difference?

6. When circular plasmids are provided with a centromere sequence and inserted into yeast cells, they reproduce and segregate normally each cell division. However, if a linear chromosome is generated by cutting the plasmid at a single site with a restriction endonuclease, the plasmid genes are quickly lost from the yeast. Explain. What additional experiment could you perform to test your hypothesized explanation?

7. What is the average distance between genes in the human genome?

8. Approximately how many molecules of histone H1 are bound to yeast genomic DNA?

9. What is the average length of an intron in a human gene?

10. You have made a library in a plasmid vector containing complete human cDNAs. What is the expected average size of an insert?

11. How was the approach used by Celera Genomics to sequence the human genome different from that used by the International Human Genome Sequencing Consortium?

12. Why is it more difficult to identify regulatory sequences than it is to identify protein coding sequences? What are the different approaches used to identify functional regulatory sequences?

13. What is a SNP? What results are expected from the study of SNPs?

References and Further Reading

The Complexity of Eukaryotic Genomes

Berget, S. M., C. Moore and P. A. Sharp. 1977. Spliced segments at the 5' terminus of adenovirus 2 late mRNA. *Proc. Natl. Acad. Sci. USA* 74: 3171–3175. [P]

Breathnach, R., J. L. Mandel and P. Chambon. 1977. Ovalbumin gene is split in chicken DNA. *Nature* 270: 314–319. [P]

Britten, R. J. and D. E. Kohne. 1968. Repeated sequences in DNA. *Science* 161: 529–540. [P]

Chow, L. T., R. E. Gelinas, T. R. Broker and R. J. Roberts. 1977. An amazing sequence arrangement at the 5' ends of adenovirus 2 messenger RNA. *Cell* 12: 1–8. [P]

Fritsch, E. F., R. M. Lawn and T. Maniatis. 1980. Molecular cloning and characterization of the human β-like globin gene cluster. *Cell* 19: 959–972. [P]

International Human Genome Sequencing Consortium. 2001. Initial sequencing and analysis of the human genome. *Nature* 409: 860–921. [P]

International Human Genome Sequencing Consortium. 2004. Finishing the euchromatic sequence of the human genome. *Nature* 431: 931–945. [P]

Kazazian, H. H., Jr. 2004. Mobile elements: Drivers of genome evolution. *Science* 303: 1626–1632. [R]

Little, P. F. R. 1982. Globin pseudogenes. *Cell* 28: 683–684. [R]

Maniatis, T. and B. Tasic. 2002. Alternative pre-mRNA splicing and proteome expansion in metazoans. *Nature* 418: 236–243. [R]

Roy, S. W. and W. Gilbert. 2005. Complex early genes. *Proc. Natl. Acad. Sci. USA* 102: 1986–1991. [P]

Stoltzfus, A., D. F. Spencer, M. Zuker, J. M. Logsdon, Jr. and W. F. Doolittle. 1994. Testing the exon theory of genes: The evidence from protein structure. *Science* 265: 202–207. [R]

Tilghman, S. M., P. J. Curtis, D. C. Tiemeier, P. Leder and C. Weissmann. 1978. The intervening sequence of a mouse β-globin gene is transcribed within the 15S β-globin mRNA precursor. *Proc. Natl. Acad. Sci. USA* 75: 1309–1313. [P]

Venter, J. C. and 273 others. 2001. The sequence of the human genome. *Science* 291: 1304–1351. [P]

Zhang, Z. and M. Gerstein. 2004. Large-scale analysis of pseudogenes in the human genome. *Curr. Opin. Genet. Dev.* 14: 328–335. [R]

Chromosomes and Chromatin

Blackburn, E. H. 2001. Switching and signaling at the telomere. *Cell* 106: 661–673. [R]

Blackburn, E. H. 2005. Telomeres and telomerase: Their mechanisms of action and the effects of altering their functions. *FEBS Letters* 579: 859–862. [R]

Blasco, M. A. 2005. Telomeres and human disease: Ageing, cancer and beyond. *Nature Rev. Genet.* 6: 611–622. [R]

Carbon, J. 1984. Yeast centromeres: Structure and function. *Cell* 37: 351–353. [R]

Clarke, L. 1990. Centromeres of budding and fission yeasts. *Trends Genet.* 6: 150–154. [R]

Dorigo, B., T. Schalch, A. Kulangara, S. Duda, R. R. Schroeder and T. J. Richmond. 2004. Nucleosome arrays reveal the two-start organization of the chromatin fiber. *Science* 306: 1571–1573. [P]

Felsenfeld, G. and M. Groudine. 2003. Controlling the double helix. *Nature* 421: 448–453. [R]

Ferreira, M. G., K. M. Miller and J. P. Cooper. 2004. Indecent exposure: When telomeres become uncapped. *Mol. Cell* 13: 7–18. [R]

Greider, C. W. 1999. Telomeres do D-loop-T-loop. *Cell* 97: 419–422. [R]

Henikoff, S., and Y. Dalal. 2005. Centromeric chromatin: What makes it unique? *Curr. Opin. Genet. Dev.* 15: 177–184. [R]

Kornberg, R. D. 1974. Chromatin structure: A repeating unit of histones and DNA. *Science* 184: 868–871. [P]

Koshland, D. and A. Strunnikov. 1996. Mitotic chromosome condensation. *Ann. Rev. Cell Biol.* 12: 305–333. [R]

Luger, K., A. W. Mader, R. K. Richmond, D. F. Sargent and T. J. Richmond. 1997. Crystal structure of the nucleosome core particle at 2.8 Å resolution. *Nature* 389: 251–260. [P]

Pardue, M. L. and J. G. Gall. 1970. Chromosomal localization of mouse satellite DNA. *Science* 168: 1356–1358. [P]

Paulson, J. R. and U. K. Laemmli. 1977. The structure of histone-depleted metaphase chromosomes. *Cell* 12: 817–828. [P]

Richmond, T. J., J. T. Finch, B. Rushton, D. Rhodes and A. Klug. 1984. Structure of the nucleosome core particle at 7 Å resolution. *Nature* 311: 532–537. [P]

Schalch, T., S. Duda, D. F. Sargent and T. J. Richmond. 2005. X-ray structure of a tetranucleosome and its implications for the chromatin fibre. *Nature* 436: 138–141. [P]

Schueler, M. G., A. W. Higgins, M. Katharine Rudd, K. Gustashaw and H. F. Willard. 2001. Genomic and genetic definition of a functional human centromere. *Science* 294: 109–115. [P]

Sun, X., J. Wahlstrom and G. Karpen. 1997. Molecular structure of a functional *Drosophila* centromere. *Cell* 91: 1007–1019. [P]

Szostak, J. W. and E. H. Blackburn. 1982. Cloning yeast telomeres on linear plasmid vectors. *Cell* 29: 245–255. [P]

Zakian, V. A. 1995. Telomeres: Beginning to understand the end. *Science* 270: 1601–1607. [R]

The Sequences of Complete Genomes

Adams, M. D. and 194 others. 2000. The genome sequence of *Drosophila melanogaster*. *Science* 287: 2185–2195. [P]

Aparicio, S. and 40 others. 2002. Whole-genome shotgun assembly and analysis of the genome of *Fugu rubripes*. *Science* 297: 1301–1310. [P]

Baltimore, D. 2001. Our genome unveiled. *Nature* 409: 814–816. [R]

Blattner, F. R., G. Plunkett III, C. A. Bloch, N. T. Perna, V. Burland, M. Riley, J. Collado-Vides, J. D. Glasner, C. K. Rode, G. F. Mayhew, J. Gregor, N. W. Davis, H. A. Kirkpatrick, M. A. Goeden, D. J. Rose, B. Mau and Y. Shao. 1997. The complete genome sequence of *Escherichia coli* K–12. *Science* 277: 1453–1462. [P]

Bult, C. J. and 39 others. 1996. Complete genome sequence of the methanogenic Archaeon, *Methanococcus jannaschii*. *Science* 273: 1058–1073. [P]

Chakravarti, A. 2001. Single nucleotide polymorphisms: To a future of genetic medicine. *Nature* 409: 822–823. [R]

Chervitz, S. A., L. Aravind, G. Sherlock, C. A. Ball, E. V. Koonin, S. S. Dwight, M. A. Harris, K. Dolinski, S. Mohr, T. Smith, S. Weng, J. M. Cherry and D. Botstein. 1998. Comparison of the complete protein sets of worm and yeast: Orthology and divergence. *Science* 282: 2022–2028. [R]

Ellegren, H. 2005. The dog has its day. *Nature* 438: 745–746. [R]

Fleischmann, R. D., and 39 others. 1995. Whole-genome random sequencing and assembly of *Haemophilus influenzae* Rd. *Science* 269: 496–512. [P]

Fraser, C. M. and 28 others. 1995. The minimal gene complement of *Mycoplasma genitalium*. *Science* 270: 397–403. [P]

Goff, S. A. and 54 others. 2002. A draft sequence of the rice genome (*Oryza sativa* L. ssp. *Japonica*). *Science* 296: 92–100. [P]

Goffeau, A. and 15 others. 1996. Life with 6000 genes. *Science* 274: 546–567. [P]

Goldstein, D. B. and G. L. Cavalleri. 2005. Understanding human diversity. *Nature* 437: 1241–1242. [R]

International Chicken Genome Sequencing Consotium. 2004. Sequence and comparative analysis of the chicken genome provide unique perspectives on vertebrate evolution. *Nature* 432: 695–715. [P]

International Human Genome Sequencing Consortium. 2001. Initial sequencing and analysis of the human genome. *Nature* 409: 860–921. [P]

International Human Genome Sequencing Consortium. 2004. Finishing the euchromatic sequence of the human genome. *Nature* 431: 931–945. [P]

International Rice Genome Sequencing Project. 2005. The map-based sequence of the rice genome. *Nature* 436: 793–800. [P]

Kirkness, E. F., V. Bafna, A. L. Halpern, S. Levy, K. Remington, D. B. Rusch, A. L. Delcher, M. Pop, W. Wang, C. M. Fraser and J. C. Venter. 2003. The dog genome: Survey sequencing and comparative analysis. *Science* 301: 1898–1903. [P]

Lindblad-Toh, K. and 46 others. 2005. Genome sequence, comparative analysis and haplotype structure of the domestic dog. *Nature* 438: 803–819. [P]

Martienssen, R. and W. R. McCombie. 2001. The first plant genome. *Cell* 105: 571–574. [R]

Mouse Genome Sequencing Consortium, 2002. Initial sequence and comparative analysis of the mouse genome. *Nature* 420: 520–562. [P]

Oliver, S. G. and 146 others. 1992. The complete DNA sequence of yeast chromosome III. *Nature* 357: 38–46. [P]

Peltonen, L. and V. A. McKusick. 2001. Dissecting human disease in the postgenomic era. *Science* 291: 1224–1229. [R]

Rat Genome Sequencing Project Consortium. 2004. Genome sequence of the Brown Norway rat yields insights into mammalian evolution. *Nature* 428: 493–521. [P]

Rubin, G. M. 2001. The draft sequences: Comparing species. *Nature* 409: 820–821. [R]

Rubin, G. M. and 54 others. 2000. Comparative genomics of the eukaryotes. *Science* 287: 2204–2215. [R]

Sutter, N. B. and E. A Ostrander. 2004. Dog star rising: The canine genetic system. *Nature Rev. Genet.* 5: 900–910. [R]

The *Arabidopsis* Genome Initiative. 2000. Analysis of the genome sequence of the flowering plant *Arabidopsis thaliana*. *Nature* 408: 796–815. [P]

The *C. elegans* Sequencing Consortium. 1998. Genome sequence of the nematode *C. elegans*: A platform for investigating biology. *Science* 282: 2012–2018. [P]

The Chimpanzee Sequencing and Analysis Consortium. 2005. Initial sequence of the chimpanzee genome and comparison with the human genome. *Nature* 437: 69–87. [P]

The International Chimpanzee Chromosome 22 Consortium. 2004. DNA sequence and comparative analysis of chimpanzee chromosome 22. *Nature* 429: 382–388. [P]

Venter, J. C. and 273 others. 2001. The sequence of the human genome. *Science* 291: 1304–1351. [P]

Walbot, V. 2000. A green chapter in the book of life. *Nature* 408: 794–795. [R]

Wood, V. and 132 others. 2002. The genome sequence of *Schizosaccharomyces pombe*. *Nature* 415: 871–880. [P]

Yu, J. and 99 others. 2002. A draft sequence of the rice genome (*Oryza sativa* L. ssp. *Indica*). *Science* 296: 79–92. [P]

Bioinformatics and Systems Biology

Bell, J. 2004. Predicting disease using genomics. *Nature* 429: 453–456. [R]

Ehrenberg, M., J. Elf, E. Aurell, R. Sandberg and J. Tegner. 2003. Systems biology is taking off. *Genome Res.* 13: 2377–2380. [R]

Friedman, A. and N. Perrimon. 2004. Genome-wide high-throughput screens in functional genomics. *Curr. Opin. Genet. Dev.* 14: 470–476. [R]

Ge, H., A. J. M. Walhout and M. Vidal. 2003. Integrating 'omic' information: A bridge between genomics and systems biology. *Trends Genet.* 19: 551–559. [R]

Guttmacher, A. E. and F. S. Collins. 2002. Genomic medicine—a primer. *N. Engl. J. Med.* 347: 1512–1520. [R]

Harbison, C. T. and 19 others. 2004. Transcriptional regulatory code of a eukaryotic genome. *Nature* 431: 99–104. [P]

Hinds, D. A., L. L. Stuve, G. B. Nilsen, E. Halperin, E. Eskin, D. G. Ballinger, K. A. Frazer and D. R. Cox. 2005. Whole-genome patterns of common DNA variation in three human populations. *Science* 307: 1072–1079. [P]

Kirschner, M. W. 2005. The meaning of systems biology. *Cell* 121: 503–504. [R]

Kitano, H. 2002. Systems biology: A brief overview. *Science* 295: 1662–1664. [R]

Sieburth, D., Q. Ch'ng, M. Dybbs, M. Tavazoie, S. Kennedy, D. Wang, D. Dupuy, J.-F. Rual, D. E. Hill, M. Vidal, G. Ruvkun and J. M. Kaplan. 2005. Systematic analysis of genes required for synapse structure and function. *Nature* 436: 510–517. [P]

Vavouri, T. and G. Elgar. 2005. Prediction of *cis*-regulatory elements using binding site matrices—the successes, the failures and the reasons for both. *Curr. Opin. Genet. Dev.* 15: 395–402. [R]

Wasserman, W. W. and A. Sandelin. 2004. Applied bioinformatics for the identification of regulatory elements. *Nature Rev. Genet.* 5: 276–287. [R]

Xie, X., J. Lu, E. J. Kulbokas, T. R. Golub, V. Mootha, K. Lindblad-Toh, E. S. Lander and M. Kellis. 2005. Systematic discovery of regulatory motifs in human promoters and 3' UTRs by comparison of several mammals. *Nature* 434: 338–345. [P]

Replication, Maintenance, and Rearrangements of Genomic DNA

- **DNA Replication 202**

- **DNA Repair 216**

- **Recombination between Homologous DNA Sequences 227**

- **DNA Rearrangements 233**

- **KEY EXPERIMENT:**
  Rearrangement of Immunoglobulin Genes 240

- **MOLECULAR MEDICINE:**
  Colon Cancer and DNA Repair 224

THE FUNDAMENTAL BIOLOGICAL PROCESS OF REPRODUCTION requires the faithful transmission of genetic information from parent to off-spring. Thus, the accurate replication of genomic DNA is essential to the lives of all cells and organisms. Each time a cell divides, its entire genome must be duplicated, and complex enzymatic machinery is required to copy the large DNA molecules that make up both prokaryotic and eukaryotic chromosomes. In addition, cells have evolved mechanisms to correct mistakes that sometimes occur during DNA replication and to repair DNA damage that can result from the action of environmental agents, such as radiation. Abnormalities in these processes result in a failure of accurate replication and maintenance of genomic DNA—a failure that can have disastrous consequences, such as the development of cancer.

Despite the importance of accurate DNA replication and maintenance, cell genomes are far from static. In order for species to evolve, mutations and gene rearrangements are needed to maintain genetic variation between individuals. Recombination between homologous chromosomes during meiosis plays an important role in this process by allowing parental genes to be rearranged into new combinations in the next generation. Rearrangements of DNA sequences within the genome are also thought to contribute to evolution by creating novel combinations of genetic information. In addition, some DNA rearrangements are programmed to regulate gene expression during the differentiation and development of individual cells and organisms. In humans, a prominent example is the rearrangement of antibody genes during development of the immune system. A careful balance between maintenance and variation of genetic information is thus critical to the development of individual organisms as well as to evolution of the species.

DNA Replication

As discussed in Chapter 4, DNA replication is a semiconservative process in which each parental strand serves as a template for the synthesis of a new complementary daughter strand. The central enzyme involved is DNA polymerase, which catalyzes the joining of deoxyribonucleoside 5'-triphosphates (dNTPs) to form the growing DNA chain. However, DNA replication is much more complex than a single enzymatic reaction. Other proteins are involved, and proofreading mechanisms are required to ensure that the accuracy of replication is compatible with the low frequency of errors that is needed for cell reproduction. Additional proteins and specific DNA sequences are also needed both to initiate replication and to copy the ends of eukaryotic chromosomes.

DNA Polymerases

DNA polymerase was first identified in lysates of *E. coli* by Arthur Kornberg in 1956. The ability of this enzyme to accurately copy a DNA template provided a biochemical basis for the mode of DNA replication that was initially proposed by Watson and Crick, so its isolation represented a landmark discovery in molecular biology. Ironically, this first DNA polymerase to be identified (now called DNA polymerase I) is not the major enzyme responsible for *E. coli* DNA replication. Instead, polymerase I is principally involved in repair of damaged DNA, and it is now clear that both prokaryotic and eukaryotic cells contain multiple different DNA polymerases that play distinct roles in DNA replication and repair. In prokaryotic cells, DNA polymerase III is the major polymerase responsible for DNA replication. Eukaryotic cells contain 3 DNA polymerases (α, δ, and ϵ) that function in replication of nuclear DNA. A distinct DNA polymerase (γ) is localized to mitochondria and is responsible for replication of mitochondrial DNA.

All known DNA polymerases share two fundamental properties that have critical implications for DNA replication (Figure 6.1). First, all polymerases synthesize DNA only in the 5' to 3' direction, adding a dNTP to the 3' hydroxyl group of a growing chain. Second, DNA polymerases can add a new deoxyribonucleotide only to a preformed primer strand that is hydrogen-bonded to the template; they are not able to initiate DNA synthesis *de novo* by catalyzing the polymerization of free dNTPs. In this respect, DNA polymerases differ from RNA polymerases, which can initiate the synthesis of a new strand of RNA in the absence of a primer. As discussed later in this chapter, these properties of DNA polymerases appear critical for maintaining the high fidelity of DNA replication that is required for cell reproduction.

The Replication Fork

DNA molecules in the process of replication were first analyzed by John Cairns in experiments in which *E. coli* were grown in the presence of radioactive thymidine, which allowed subsequent visualization of newly replicated DNA by autoradiography (Figure 6.2). In some cases, complete circular molecules in the process of replicating could be observed. These DNA molecules contained two **replication forks**, representing the regions of active DNA synthesis. At each fork the parental strands of DNA separated and two new daughter strands were synthesized.

The synthesis of new DNA strands complementary to both strands of the parental molecule posed an important problem to understanding the biochemistry of DNA replication. Since the two strands of double-helical DNA run in opposite (antiparallel) directions, continuous synthesis of two new

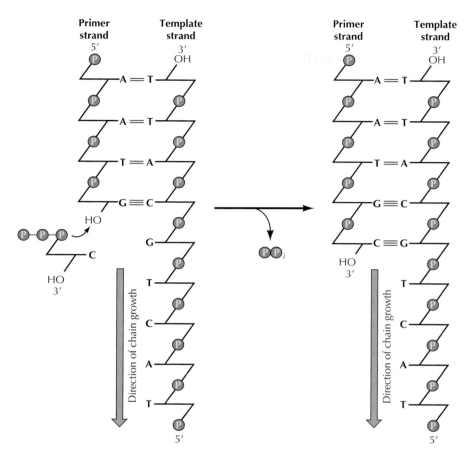

FIGURE 6.1 The reaction catalyzed by DNA polymerase All known DNA polymerases add a deoxyribonucleoside 5′-triphosphate to the 3′ hydroxyl group of a growing DNA chain (the primer strand).

strands at the replication fork would require that one strand be synthesized in the 5′ to 3′ direction while the other is synthesized in the opposite (3′ to 5′) direction. But DNA polymerase catalyzes the polymerization of dNTPs only in the 5′ to 3′ direction. How, then, can the other progeny strand of DNA be synthesized?

This enigma was resolved by experiments showing that only one strand of DNA is synthesized in a continuous manner in the direction of overall DNA replication; the other is formed from short (1–3 kb), discontinuous pieces of DNA that are synthesized backward with respect to the direction of movement of the replication fork (Figure 6.3). These small pieces of newly synthesized DNA (called **Okazaki fragments** after their discoverer, the Japanese biochemist Reiji Okazaki) are joined by the action of **DNA ligase**, forming an intact new DNA strand. The continuously synthesized strand is called the **leading strand**, since its elongation in the direction of replication fork movement exposes the template used for the synthesis of Okazaki fragments (the **lagging strand**).

Although the discovery of discontinuous synthesis of the lagging strand provided a mechanism for the elongation of both strands of DNA at the replication fork, it raised another question: Since DNA polymerase requires a primer and cannot initiate synthesis *de novo*, how is the synthesis of

FIGURE 6.2 Replication of *E. coli* DNA (A) An autoradiograph showing bacteria that were grown in [³H]thymidine for two generations to label the DNA, which was then extracted and visualized by exposure to photographic film. (B) This schematic illustrates the two replication forks shown in (A). (From J. Cairns, 1963. *Cold Spring Harbor Symp. Quant. Biol.* 28: 43.)

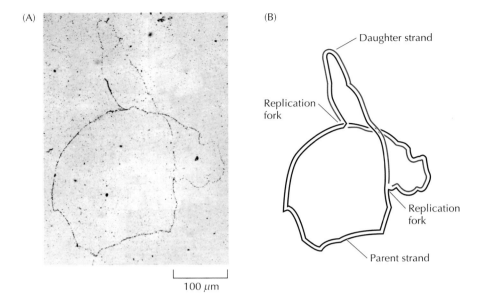

Okazaki fragments initiated? The answer is that short fragments of RNA serve as primers for DNA replication (**Figure 6.4**). In contrast to DNA synthesis, the synthesis of RNA can initiate *de novo*, and an enzyme called **primase** synthesizes short fragments of RNA (e.g., three to ten nucleotides long) complementary to the lagging strand template at the replication fork. Okazaki fragments are then synthesized via extension of these RNA primers by DNA polymerase. An important consequence of such RNA priming is that newly synthesized Okazaki fragments contain an RNA-DNA joint, the discovery of which provided critical evidence for the role of RNA primers in DNA replication.

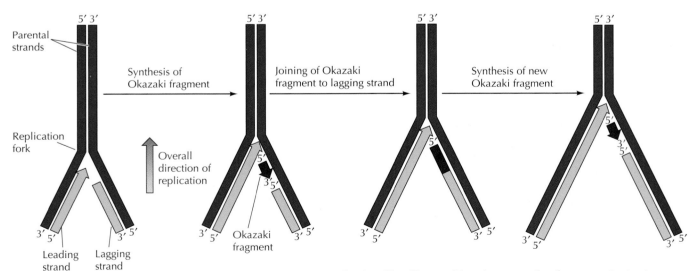

FIGURE 6.3 Synthesis of leading and lagging strands of DNA The leading strand is synthesized continuously in the direction of replication fork movement. The lagging strand is synthesized in small pieces (Okazaki fragments) backward from the overall direction of replication. The Okazaki fragments are then joined by the action of DNA ligase.

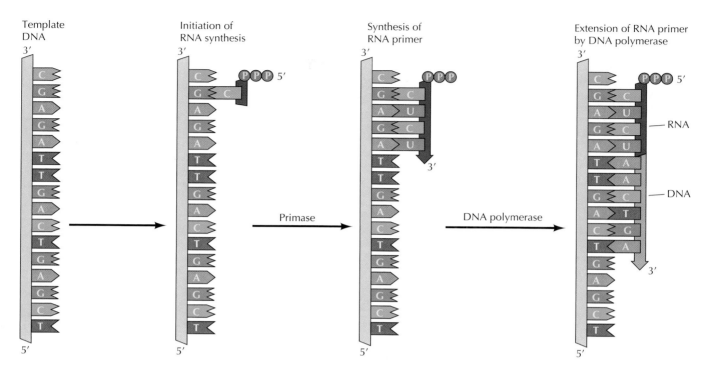

FIGURE 6.4 Initiation of Okazaki fragments with RNA primers
Short fragments of RNA serve as primers that can be extended by DNA polymerase.

To form a continuous lagging strand of DNA, the RNA primers must eventually be removed from the Okazaki fragments and replaced with DNA (Figure 6.5). In prokaryotes, RNA primers are removed by the action of polymerase I. In addition to its DNA polymerase activity, polymerase I acts as an **exonuclease** that can hydrolyze DNA (or RNA) in either the 3′ to 5′ or 5′ to 3′ direction. The action of polymerase I as a 5′ to 3′ exonuclease removes ribonucleotides from the 5′ ends of Okazaki fragments, allowing them to be replaced with deoxyribonucleotides to yield fragments consisting entirely of DNA. In eukaryotic cells, RNA primers are removed by the combined action of **RNase H**, an enzyme that degrades the RNA strand of RNA-DNA hybrids, and 5′ to 3′ exonucleases. The resulting gaps are then filled by polymerase δ and the DNA fragments are joined by DNA ligase, yielding an intact lagging strand.

As noted earlier, different DNA polymerases play distinct roles at the replication fork in both prokaryotic and eukaryotic cells (Figure 6.6). In *E. coli*, polymerase III is the major replicative polymerase, functioning in the synthesis both of the leading strand of DNA and of Okazaki fragments by the extension of RNA primers. In eukaryotic cells, three distinct DNA polymerases (α, δ, and ε) are involved in the replication of nuclear DNA. The roles of these DNA polymerases have been studied in two types of experiments. First, replication of the DNA of some animal viruses, such as SV40, can be studied in cell-free extracts, allowing direct biochemical analysis of the activities of different DNA polymerases as well as other proteins involved in DNA replication. Second, polymerases α, δ, and ε are found in yeasts as well as in mammalian cells, enabling the use of the powerful approaches of yeast genetics (see Chapter 4) to test their biological roles. Biochemical analysis has established that polymerase α is found in a complex with primase, and functions in conjunction with primase to synthesize short RNA-DNA fragments during lagging strand synthesis. Polymerase δ

FIGURE 6.5 Removal of RNA primers and joining of Okazaki fragments
RNA primers are removed and DNA polymerase fills the gaps between Okazaki fragments with DNA. The resultant DNA fragments can then be joined by DNA ligase.

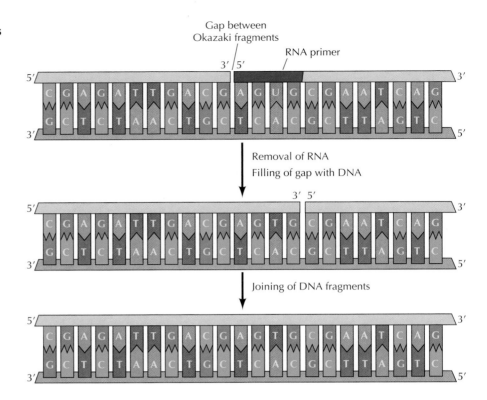

can then synthesize both the leading and lagging strands, acting to extend the RNA-DNA primers initially synthesized by the polymerase α-primase complex. Polymerase ε is an essential gene in yeast, but it is not required for SV40 DNA replication *in vitro*, so its role remains to be understood. However, its activities seem to be similar to those of polymerase δ, so it probably functions like polymerase δ in the replication process.

Not only polymerases and primase but also a number of other proteins act at the replication fork. These additional proteins have been identified both by the analysis of *E. coli* and yeast mutants defective in DNA replication and by the purification of the mammalian proteins required for *in vitro* replication of SV40 DNA. One class of proteins required for replication

FIGURE 6.6 Roles of DNA polymerases in *E. coli* and mammalian cells   The leading strand is synthesized by polymerase III (pol III) in *E. coli* and by polymerases δ and ε (pol δ/ε) in mammalian cells. In *E. coli*, lagging strand synthesis is initiated by primase, and RNA primers are extended by polymerase III. In mammalian cells, lagging strand synthesis is initiated by a complex of primase with polymerase α (pol α). The short RNA-DNA fragments synthesized by this complex are then extended by polymerases δ and ε.

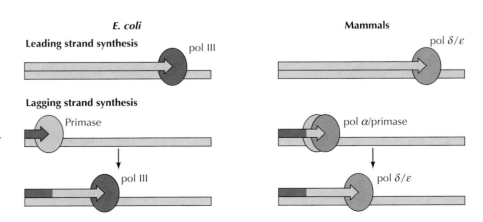

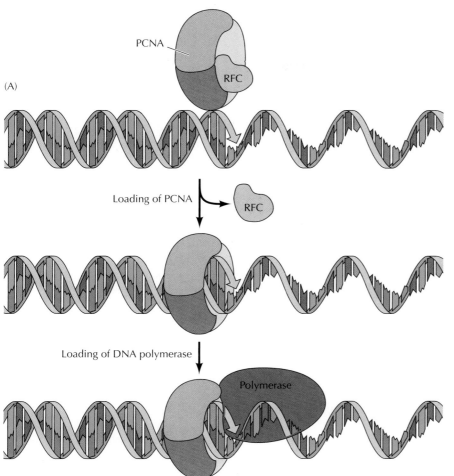

(A)

PCNA

RFC

Loading of PCNA

RFC

Loading of DNA polymerase

Polymerase

FIGURE 6.7 Polymerase accessory proteins (A) A complex of the clamp-loading and sliding-clamp proteins (RFC and PCNA in mammalian cells, respectively) binds DNA at the junction between primer and template. RFC is then released, loading PCNA onto the DNA. DNA polymerase then binds to PCNA. (B) Model of PCNA bound to DNA. (B, from T. S. Krishna, X. P. Kong, S. Gary, P. M. Burgers and J. Kuriyan, 1994. *Cell* 79: 1233.)

(B)

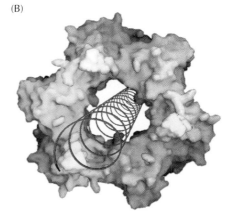

binds to DNA polymerases, increasing the activity of the polymerases and causing them to remain bound to the template DNA so that they continue synthesis of a new DNA strand. Both *E. coli* polymerase III and eukaryotic polymerases δ and ε are associated with two types of accessory proteins (sliding-clamp proteins and clamp-loading proteins) that load the polymerase onto the primer and maintain its stable association with the template (Figure 6.7). The clamp-loading proteins (called replication factor C [RFC] in eukaryotes) and the sliding-clamp proteins (proliferating cell nuclear antigen [PCNA] in eukaryotes) form a complex that specifically recognizes and binds DNA at the junction between the primer and template. The clamp loader then releases the sliding clamp protein, which forms a ring around the template DNA. The sliding clamp protein then loads the DNA polymerase onto DNA at the primer-template junction. The ring formed by the sliding clamp maintains the association of the polymerase with its template as replication proceeds, allowing the uninterrupted synthesis of many thousands of nucleotides of DNA.

Other proteins unwind the template DNA and stabilize single-stranded regions (Figure 6.8). **Helicases** are enzymes that catalyze the unwinding of parental DNA, coupled to the hydrolysis of ATP, ahead of the replication fork. **Single-stranded DNA-binding proteins** (e.g., eukaryotic replication protein A [RPA]) then stabilize the unwound template DNA, keeping it in an extended single-stranded state so that it can be copied by the polymerase.

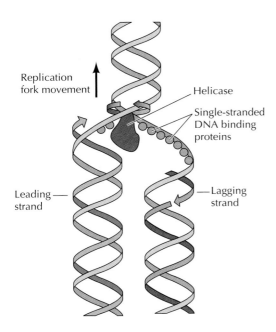

FIGURE 6.8 Action of helicases and single-stranded DNA-binding proteins   Helicases unwind the two strands of parental DNA ahead of the replication fork. The unwound DNA strands are then stabilized by single-stranded DNA-binding proteins so that they can serve as templates for new DNA synthesis.

As the strands of parental DNA unwind, the DNA ahead of the replication fork is forced to rotate. Unchecked, this rotation would cause circular DNA molecules (such as SV40 DNA or the *E. coli* chromosome) to become twisted around themselves, eventually blocking replication (Figure 6.9). This problem is solved by **topoisomerases,** enzymes that catalyze the reversible breakage and rejoining of DNA strands. There are two types of these enzymes: Type I topoisomerases break just one strand of DNA; type II topoisomerases introduce simultaneous breaks in both strands. The breaks introduced by type I and type II topoisomerases serve as "swivels" that allow the two strands of template DNA to rotate freely around each other so that replication can proceed without twisting the DNA ahead of the fork (see Figure 6.9). Although eukaryotic chromosomes are composed of linear rather than circular DNA molecules, their replication also requires topoisomerases; otherwise, the complete chromosomes would have to rotate continually during DNA synthesis.

Type II topoisomerase is needed not only to unwind DNA but also to unravel newly replicated circular DNA molecules that become intertwined with each other. In eukaryotic cells, topoisomerase II appears to be involved in mitotic chromosome condensation. In addition, studies of yeast mutants, as well as experiments in *Drosophila* and mammalian cells, indicate that topoisomerase II is required for the separation of daughter chromatids at mitosis, suggesting that it functions to untangle newly replicated loops of DNA in the chromosomes of eukaryotes.

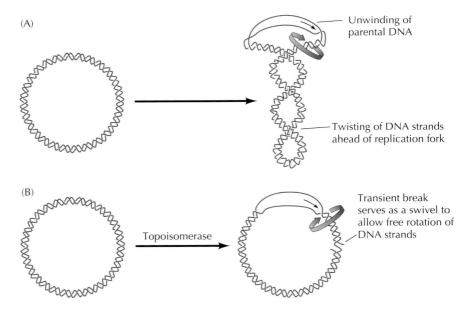

FIGURE 6.9 Action of topoisomerases during DNA replication   (A) As the two strands of template DNA unwind, the DNA ahead of the replication fork is forced to rotate in the opposite direction, causing circular molecules to become twisted around themselves. (B) This problem is solved by topoisomerases, which catalyze the reversible breakage and rejoining of DNA strands. The transient breaks introduced by these enzymes serve as swivels that allow the two strands of DNA to rotate freely around each other.

> ■ An inhibitor of topoisomerase II, a drug called etoposide, is used as a chemotherapeutic agent for the treatment of several types of cancer.

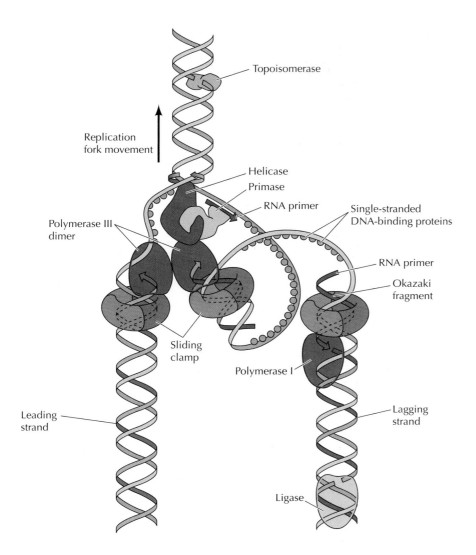

FIGURE 6.10 Model of the *E. coli* replication fork Helicase, primase, and two molecules of DNA polymerases III carry out coordinated synthesis of both the leading and lagging strands of DNA. The lagging strand template is folded so that the polymerase responsible for lagging strand synthesis moves in the same direction as overall movement of the fork. Topoisomerase acts as a swivel ahead of the fork. Behind the fork, RNA primers are removed by DNA polymerase I and Okazaki fragments are joined by DNA ligase.

6.1 WEBSITE ANIMATION

DNA Replication Fork
Helicase, primase, and two molecules of DNA polymerase III carry out the coordinated synthesis of both the leading and lagging strands of DNA at a replication fork.

The enzymes involved in DNA replication act in a coordinated manner to synthesize both leading and lagging strands of DNA simultaneously at the replication fork (Figure 6.10). This task is accomplished by the formation of dimers of the replicative DNA polymerases, each with its appropriate accessory proteins. One molecule of polymerase then acts in synthesis of the leading strand while the other acts in synthesis of the lagging strand. The lagging strand template is thought to form a loop at the replication fork so that the polymerase subunit engaged in lagging strand synthesis moves in the same overall direction as the other subunit, which is synthesizing the leading strand.

In eukaryotic cells, nucleosomes are also disrupted during DNA replication. The histones bound to the parental chromatin are divided between the two daughter strands of DNA, and new histones are then added to reassemble nucleosomes by additional proteins (chromatin assembly factors) that travel along with the replication fork.

The Fidelity of Replication

The accuracy of DNA replication is critical to cell reproduction, and estimates of mutation rates for a variety of genes indicate that the frequency of errors during replication corresponds to only one incorrect base per 10^8 to 10^9 nucleotides incorporated. This error frequency is much lower than would be predicted simply on the basis of complementary base pairing. In particular, the free energy differences resulting from the changes in hydrogen bonding between correctly matched and mismatched bases are only large enough to favor the formation of correctly matched base pairs by about a hundredfold. Consequently, base selection determined simply by hydrogen bonding between complementary bases would result in an error frequency corresponding to the incorporation of one incorrect base in every 100–1000 nucleotides of newly synthesized DNA. The much higher degree of fidelity actually achieved results largely from the activities of DNA polymerase.

One mechanism by which DNA polymerase increases the fidelity of replication is by helping to select the correct base for insertion into newly synthesized DNA. The polymerase does not simply catalyze incorporation of whatever nucleotide is hydrogen-bonded to the template strand. Instead, it actively discriminates against incorporation of a mismatched base by adapting to the conformation of a correct base pair. In particular, recent structural studies of several DNA polymerases indicate that the binding of correctly matched dNTPs induces conformational changes in DNA polymerase that lead to the incorporation of the nucleotide into DNA. This ability of DNA polymerase to select matched nucleotides for incorporation appears to increase the accuracy of replication about a thousandfold, reducing the expected error frequency from greater than 10^{-3} to between 10^{-5} and 10^{-6}.

The other major mechanism responsible for the accuracy of DNA replication is the **proofreading** activity of DNA polymerase. The replicative DNA polymerases (eukaryotic polymerases δ and ε and *E. coli* polymerase III) have an exonuclease activity that can hydrolyze DNA in the 3′ to 5′ direction. This 3′ to 5′ exonuclease operates in the reverse direction of DNA synthesis, and participates in proofreading newly synthesized DNA (**Figure 6.11**). Proofreading is effective because DNA polymerase requires a primer and is not able to initiate synthesis *de novo*. Primers that are hydrogen-bonded to the template are preferentially used, so when an incorrect base is incorporated, it is likely to be removed by the 3′ to 5′ exonuclease activity rather than being used to continue synthesis. The 3′ to 5′ exonucleases of the replicative DNA polymerases selectively excise mismatched bases that have been incorporated at the end of a growing DNA chain, thereby increasing the accuracy of replication by a hundred- to a thousandfold.

The importance of proofreading may explain the fact that DNA polymerases require primers and catalyze the growth of DNA strands only in the 5′ to 3′ direction. When DNA is synthesized in the 5′ to 3′ direction, the energy required for polymerization is derived from hydrolysis of the 5′ triphosphate group of a free dNTP as it is added to the 3′ hydroxyl group of the growing chain (see Figure 6.1). If DNA were to be extended in the 3′ to 5′ direction, the energy of polymerization would instead have to be derived from hydrolysis of the 5′ triphosphate group of the terminal nucleotide already incorporated into DNA. This arrangement would eliminate the possibility of proofreading, because removal of a mismatched terminal nucleotide would also remove the 5′ triphosphate group needed as an energy source for further chain elongation. Thus, although the ability of DNA polymerase to extend a primer only in the 5′ to 3′ direction appears to make replication a complicated process, it is necessary for ensuring accurate duplication of the genetic material.

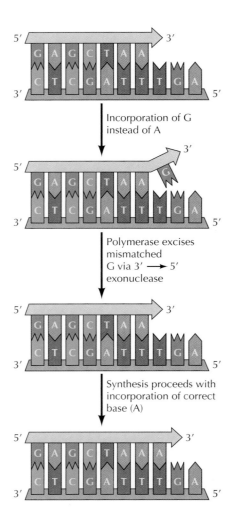

FIGURE 6.11 Proofreading by DNA polymerase   G is incorporated in place of A as a result of mispairing with T on the template strand. Because it is mispaired, the 3′ terminal G is not hydrogen-bonded to the template strand. This mismatch at the 3′ terminus of the growing chain is recognized and excised by the 3′ to 5′ exonuclease activity of DNA polymerase, which requires a primer hydrogen-bonded to the template strand in order to continue synthesis. Following excision of the mismatched G, DNA synthesis can proceed with incorporation of the correct nucleotide (A).

Incorporation of G instead of A

Polymerase excises mismatched G via 3′ → 5′ exonuclease

Synthesis proceeds with incorporation of correct base (A)

Combined with the ability to discriminate against the insertion of mismatched bases, the proofreading activity of DNA polymerases is sufficient to reduce the error frequency of replication to about one mismatched base per 10^8 to 10^9. Additional mechanisms (discussed in the section "DNA Repair") act to remove mismatched bases that have been incorporated into newly synthesized DNA, and further ensure correct replication of the genetic information.

Origins and the Initiation of Replication

The replication of both prokaryotic and eukaryotic DNAs starts at sites called **origins of replication**, which serve as binding sites for proteins that initiate the replication process. The first origin to be defined was that of *E. coli*, in which genetic analysis indicated that replication always begins at a unique site on the bacterial chromosome. The *E. coli* origin has since been studied in detail and found to consist of 245 base pairs of DNA, elements of which serve as binding sites for proteins required to initiate DNA replication (Figure 6.12). The key step is the binding of an initiator protein to specific DNA sequences within the origin. The initiator protein begins to unwind the origin DNA and recruits the other proteins involved in DNA synthesis. Helicase and single-stranded DNA-binding proteins then act to continue unwinding and exposing the template DNA, and primase initiates the synthesis of leading strands. Two replication forks are formed and move in opposite directions along the circular *E. coli* chromosome.

The origins of replication of several animal viruses, such as SV40, have been studied as models for the initiation of DNA synthesis in eukaryotes. SV40 has a single origin of replication (consisting of 64 base pairs) that functions both in infected cells and in cell-free systems. Replication is initiated by a virus-encoded protein (called T antigen) that binds to the origin and also acts as a helicase. A single-stranded DNA-binding protein is required to stabilize the unwound template, and the DNA polymerase α-primase complex then initiates DNA synthesis.

Although single origins are sufficient to direct the replication of bacterial and viral genomes, multiple origins are needed to replicate the much larger genomes of eukaryotic cells within a reasonable period of time. For exam-

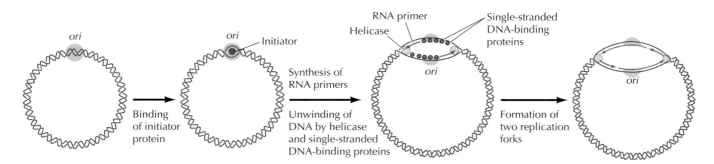

FIGURE 6.12 Origin of replication in *E. coli* Replication initiates at a unique site on the *E. coli* chromosome, designated the origin (*ori*). The first event is the binding of an initiator protein to *ori* DNA, which leads to partial unwinding of the template. The DNA continues to unwind by the actions of helicase and single-stranded DNA-binding proteins, and RNA primers are synthesized by primase. The two replication forks formed at the origin then move in opposite directions along the circular DNA molecule.

FIGURE 6.13 Replication origins in eukaryotic chromosomes
Replication initiates at multiple origins (*ori*), each of which produces two replication forks.

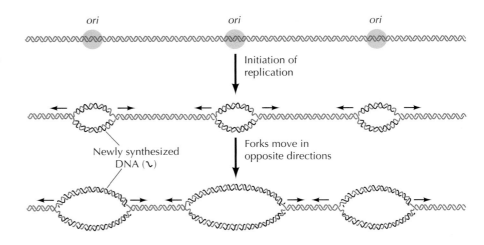

ple, the entire genome of *E. coli* (4×10^6 base pairs) is replicated from a single origin in approximately 30 minutes. If mammalian genomes (3×10^9 base pairs) were replicated from a single origin at the same rate, DNA replication would require about 3 weeks (30,000 minutes). The problem is further exacerbated by the fact that the rate of DNA replication in mammalian cells is actually about tenfold lower than in *E. coli*, probably as a result of the packaging of eukaryotic DNA in chromatin. Nonetheless, the genomes of mammalian cells are typically replicated within a few hours, necessitating the use of thousands of replication origins.

The presence of multiple replication origins in eukaryotic cells was first demonstrated by the exposure of cultured mammalian cells to radioactive thymidine for different time intervals, followed by autoradiography to detect newly synthesized DNA. The results of such studies indicated that DNA synthesis is initiated at multiple sites, from which it then proceeds in both directions along the chromosome (Figure 6.13). The replication origins in mammalian cells are spaced at intervals of approximately 50 to 300 kb; thus, the human genome has about 30,000 origins of replication. The genomes of simpler eukaryotes also have multiple origins; for example, replication in yeasts initiates at origins separated by intervals of approximately 40 kb.

The origins of replication of eukaryotic chromosomes were first studied in the yeast *S. cerevisiae*, in which they were identified as sequences that can support the replication of plasmids in transformed cells (Figure 6.14). This has provided a functional assay for these sequences, and several such elements (called **autonomously replicating sequences**, or **ARSs**) have been isolated. Their role as origins of replication has been verified by direct biochemical analysis, not only in plasmids but also in yeast chromosomal DNA.

Functional ARS elements span about 100 base pairs, including an 11-base-pair core sequence common to many different ARSs (Figure 6.15). This core sequence is essential for ARS function and has been found to be the binding site of a six-subunit protein complex (called the **origin recognition complex**, or **ORC**) that is required for initiation of DNA replication at *S. cerevisiae* origins. The ORC complex functions to recruit other proteins (including the MCM DNA helicase) to the origin, leading to the initiation of replication. The mechanism of initiation of DNA replication in *S. cerevisiae* thus appears similar to that in bacteria and eukaryotic viruses; that is, an initiator protein specifically binds to specific DNA sequences that define an origin of replication.

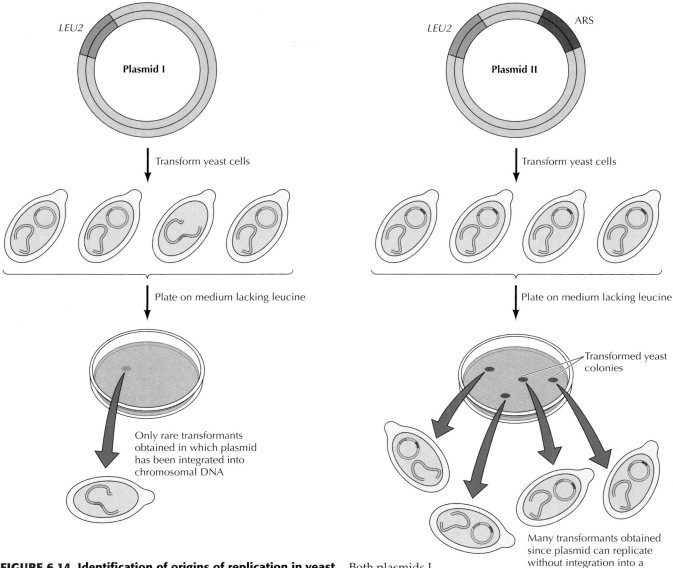

FIGURE 6.14 Identification of origins of replication in yeast Both plasmids I and II contain a selectable marker gene (*LEU2*) that allows transformed cells to grow on medium lacking leucine. Only plasmid II, however, contains an origin of replication (ARS). The transformation of yeasts with plasmid I yields only rare transformants in which the plasmid has integrated into chromosomal DNA. Plasmid II, however, is able to replicate without integration into a yeast chromosome (autonomous replication), so many more transformants result from its introduction into yeast cells.

Subsequent studies have shown that the role of ORC proteins as initiators of replication is conserved in all eukaryotes, from yeasts to mammals. However, replication origins in other eukaryotes are much less well defined than the ARS elements of *S. cerevisiae*. In the fission yeast *S. pombe*, origin sequences are spread over 0.5 to 1 kb of DNA. The *S. pombe* origins lack the clearly defined ORC binding site of the *S. cerevisiae* ARS elements, but they contain repeats of AT-rich sequences that appear to serve as binding sites for the *S. pombe* ORC complex. In *Drosophila* and mammalian cells, however,

FIGURE 6.15 A yeast ARS element
An *S. cerevisiae* ARS element contains an 11-base-pair ARS consensus sequence (ACS) and three additional elements (B1, B2, and B3) that contribute to origin function. The origin recognition complex (ORC) binds to the ACS and B1. ORC then recruits additional proteins, including the MCM DNA helicase, to the origin.

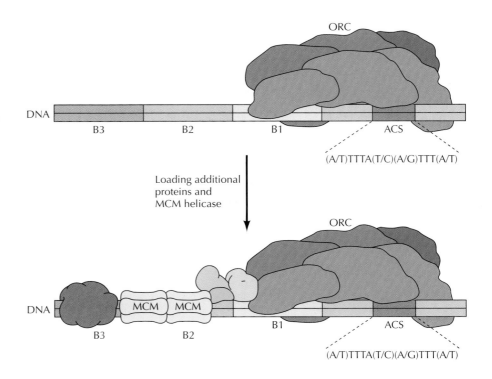

Telomeres and Telomerase: Maintaining the Ends of Chromosomes

Because DNA polymerases extend primers only in the 5′ to 3′ direction, they are unable to copy the extreme 5′ ends of linear DNA molecules. Consequently, special mechanisms are required to replicate the terminal sequences of the linear chromosomes of eukaryotic cells. These sequences (**telomeres**) consist of tandem repeats of simple-sequence DNA (see Chapter 5). They are maintained by the action of a unique enzyme called **telomerase**, which is able to catalyze the synthesis of telomeres in the absence of a DNA template.

Telomerase is a **reverse transcriptase**, one of a class of DNA polymerases, first discovered in retroviruses (see Chapter 4), that synthesize DNA from an RNA template. Importantly, telomerase carries its own template RNA, which is complementary to the telomere repeat sequences, as part of the enzyme complex. The use of this RNA as a template allows telomerase to generate multiple copies of the telomeric repeat sequences, thereby maintaining telomeres in the absence of a conventional DNA template to direct their synthesis.

The mechanism of telomerase action was initially elucidated in 1985 by Carol Greider and Elizabeth Blackburn during studies of the protozoan *Tetrahymena* (**Figure 6.16**). The *Tetrahymena* telomerase is complexed to a 159-nucleotide-long RNA that includes the sequence 3′–AACCCCAAC–5′. This sequence is complementary to the *Tetrahymena* telomeric repeat (5′–TTGGGG–3′) and serves as the template for the synthesis of telomeric

The ORC proteins do not appear to recognize specific DNA sequences. Instead, the sites of ORC binding and initiation of DNA replication in higher eukaryotes may be determined by aspects of chromatin structure that remain to be elucidated by future research.

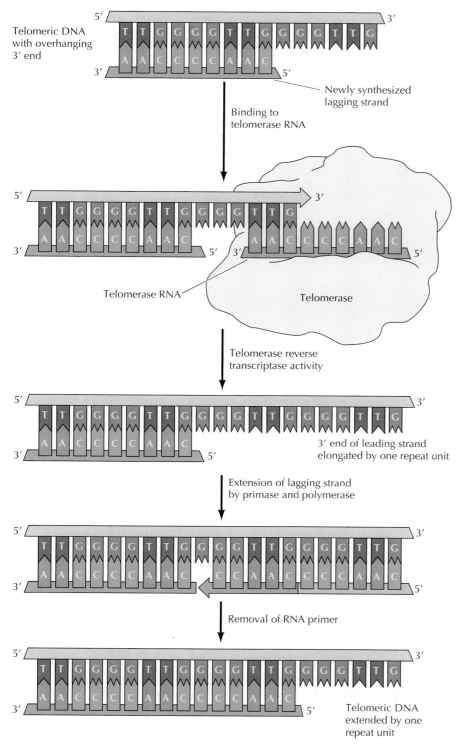

FIGURE 6.16 Action of telomerase   Telomeric DNA is a simple repeat sequence with an overhanging 3′ end on the newly synthesized leading strand. Telomerase carries its own RNA molecule, which is complementary to telomeric DNA, as part of the enzyme complex. The overhanging end of telomeric DNA binds to the telomerase RNA, which then serves as a template for extension of the leading strand by one repeat unit. The lagging strand of telomeric DNA can then be elongated by conventional RNA priming and DNA polymerase activity.

DNA. The use of this RNA as a template allows telomerase to extend the 3' end of chromosomal DNA by one repeat unit beyond its original length. The complementary strand can then be synthesized by the polymerase α-primase complex using conventional RNA priming. Removal of the RNA primer leaves an overhanging 3' end of chromosomal DNA, which can form loops at the ends of eukaryotic chromosomes (see Figure 5.22).

Telomerase has been identified in a variety of eukaryotes, and genes encoding telomerase RNAs have been cloned from *Tetrahymena*, yeasts, mice, and humans. In each case, the telomerase RNA contains sequences complementary to the telomeric repeat sequence of that organism (see Table 5.5). Moreover, the introduction of mutant telomerase RNA genes into yeasts has been shown to result in corresponding alterations of the chromosomal telomeric repeat sequences, directly demonstrating the function of telomerase in maintaining the termini of eukaryotic chromosomes.

Defects in telomerase and the normal maintenance of telomeres are associated with several human diseases. The activity of telomerase is regulated in dividing cells to maintain telomeres at their normal length. In human cells, for example, telomeres span approximately 10 kb with a 3' single-strand overhang of 150–200 bases. Although this length is maintained in germ cells, most somatic cells do not have sufficiently high levels of telomerase to maintain the length of their telomeres for an indefinite number of cell divisions. Consequently, telomeres gradually shorten as cells age, and this shortening eventually leads to cell death or senescence. Interestingly, several premature aging syndromes are characterized by an abnormally high rate of telomere loss, and some of these diseases have been found to be caused directly by mutations in telomerase. Conversely, cancer cells express abnormally high levels of telomerase, allowing them to continue dividing indefinitely.

DNA Repair

DNA, like any other molecule, can undergo a variety of chemical reactions. Because DNA uniquely serves as a permanent copy of the cell genome, however, changes in its structure are of much greater consequence than are alterations in other cell components, such as RNAs or proteins. Mutations can result from the incorporation of incorrect bases during DNA replication. In addition, various chemical changes occur in DNA either spontaneously (Figure 6.17) or as a result of exposure to chemicals or radiation (Figure 6.18). Such damage to DNA can block replication or transcription, and can result in a high frequency of mutations—consequences that are unacceptable from the standpoint of cell reproduction. To maintain the integrity of their genomes, cells have therefore had to evolve mechanisms to repair damaged DNA. These mechanisms of DNA repair can be divided into two general classes: (1) direct reversal of the chemical reaction responsible for DNA damage, and (2) removal of the damaged bases followed by their replacement with newly synthesized DNA. Where DNA repair fails, additional mechanisms have evolved to enable cells to cope with the damage.

Direct Reversal of DNA Damage

Most damage to DNA is repaired by removal of the damaged bases followed by resynthesis of the excised region. Some lesions in DNA, however, can be repaired by direct reversal of the damage, which may be a more efficient way of dealing with specific types of DNA damage that occur fre-

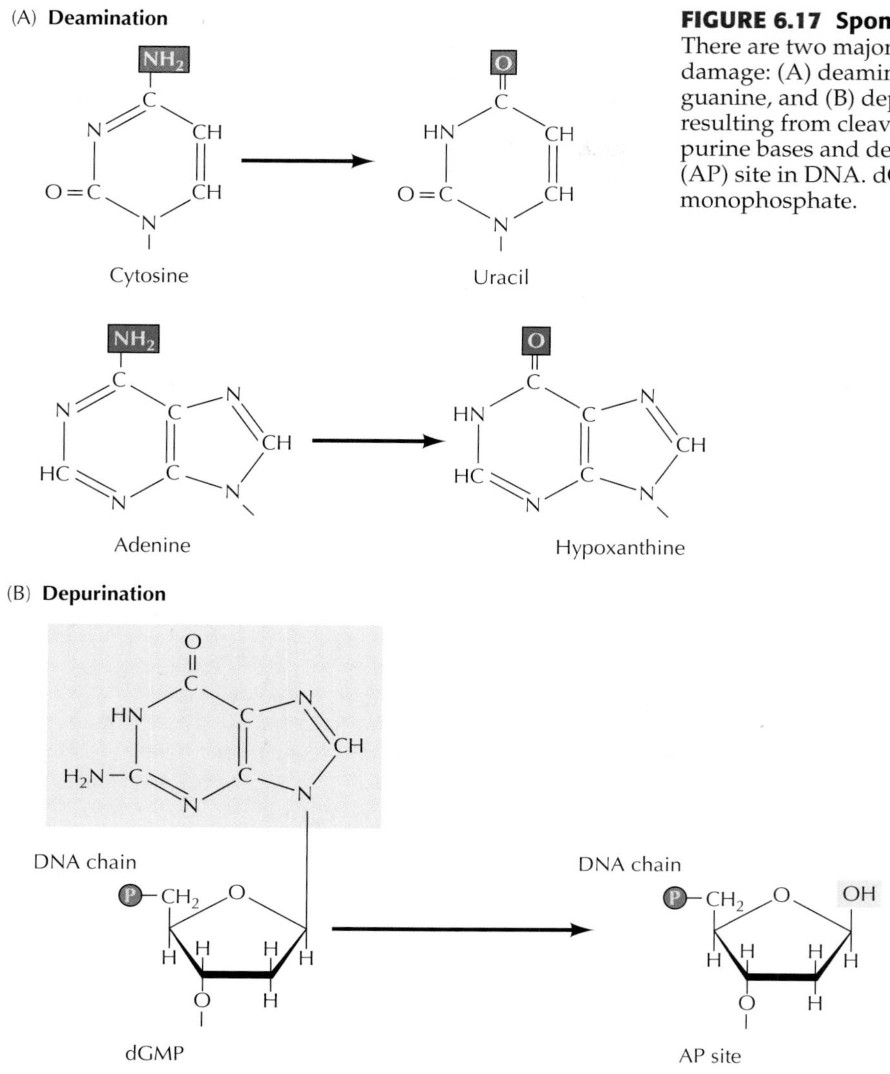

(A) **Deamination**

Cytosine → Uracil

Adenine → Hypoxanthine

(B) **Depurination**

dGMP → AP site

FIGURE 6.17 Spontaneous damage to DNA
There are two major forms of spontaneous DNA damage: (A) deamination of adenine, cytosine, and guanine, and (B) depurination (loss of purine bases) resulting from cleavage of the bond between the purine bases and deoxyribose, leaving an apurinic (AP) site in DNA. dGMP = deoxyguanosine monophosphate.

quently. Only a few types of DNA damage are repaired in this way, particularly pyrimidine dimers resulting from exposure to ultraviolet (UV) light and alkylated guanine residues that have been modified by the addition of methyl or ethyl groups at the O^6 position of the purine ring.

UV light is one of the major sources of damage to DNA and is also the most thoroughly studied form of DNA damage in terms of repair mechanisms. Its importance is illustrated by the fact that exposure to solar UV irradiation is the cause of almost all skin cancer in humans. The major type of damage induced by UV light is the formation of **pyrimidine dimers,** in which adjacent pyrimidines on the same strand of DNA are joined by the formation of a cyclobutane ring resulting from saturation of the double bonds between carbons 5 and 6 (see Figure 6.18A). The formation of such dimers distorts the structure of the DNA chain and blocks transcription or replication past the site of damage, so their repair is closely correlated with the ability of cells to survive UV irradiation. One mechanism of repairing UV-induced pyrimidine dimers is direct reversal of the dimerization reac-

(A) **Exposure to UV light**

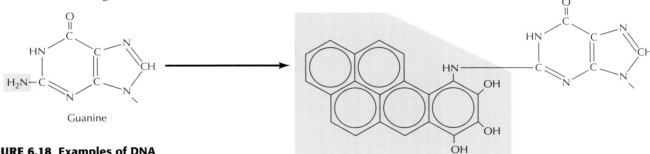

Adjacent thymines in DNA

Thymine dimer

Cyclobutane ring

(B) **Alkylation**

Guanine

O^6-methylguanine

(C) **Reaction with carcinogen**

Guanine

Bulky group addition

FIGURE 6.18 Examples of DNA damage induced by radiation and chemicals (A) UV light induces the formation of pyrimidine dimers in which two adjacent pyrimidines (e.g., thymines) are joined by a cyclobutane ring structure. (B) Alkylation is the addition of methyl or ethyl groups to various positions on the DNA bases. In this example, alkylation of the O^6 position of guanine results in formation of O^6-methylguanine. (C) Many carcinogens (e.g., benzo[*a*]pyrene) react with DNA bases, resulting in the addition of large bulky chemical groups to the DNA molecule.

tion. The process is called **photoreactivation** because energy derived from visible light is utilized to break the cyclobutane ring structure (Figure 6.19). The original pyrimidine bases remain in DNA, now restored to their normal state. As might be expected from the fact that solar UV irradiation is a major source of DNA damage for diverse cell types, the repair of pyrimidine dimers by photoreactivation is common to a variety of prokaryotic and eukaryotic cells, including *E. coli*, yeasts, and some species of plants and animals. Curiously, however, photoreactivation is not universal; many species (including humans) lack this mechanism of DNA repair.

Another form of direct repair deals with damage resulting from the reaction between alkylating agents and DNA. Alkylating agents are reactive compounds that can transfer methyl or ethyl groups to a DNA base, thereby chemically modifying the base (see Figure 6.18B). A particularly important type of damage is methylation of the O^6 position of guanine, because the product, O^6-methylguanine, forms complementary base pairs with thymine instead of cytosine. This lesion can be repaired by an enzyme (called O^6-methylguanine methyltransferase) that transfers the methyl group from O^6-methylguanine to a cysteine residue in its active site (Figure 6.20). The potentially mutagenic chemical modification is thus removed, and the original guanine is restored. Enzymes that catalyze this direct repair reaction are widespread in both prokaryotes and eukaryotes, including humans.

Excision Repair

Although direct repair is an efficient way of dealing with particular types of DNA damage, excision repair is a more general means of repairing a wide variety of chemical alterations to DNA. Consequently, the various types of excision repair are the most important DNA repair mechanisms in both prokaryotic and eukaryotic cells. In excision repair, the damaged DNA is recognized and removed, either as free bases or as nucleotides. The resulting gap is then filled in by synthesis of a new DNA strand, using the undamaged complementary strand as a template. Three types of excision repair—base-excision, nucleotide-excision, and mismatch repair—enable cells to cope with a variety of different kinds of DNA damage.

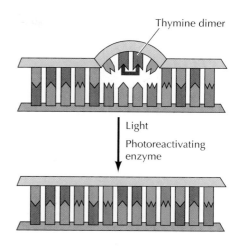

FIGURE 6.19 Direct repair of thymine dimers UV-induced thymine dimers can be repaired by photoreactivation, in which energy from visible light is used to split the bonds forming the cyclobutane ring.

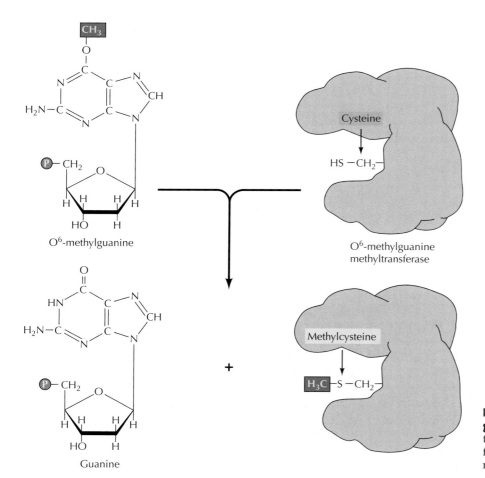

FIGURE 6.20 Repair of O^6-methylguanine O^6-methylguanine methyltransferase transfers the methyl group from O^6-methylguanine to a cysteine residue in the enzyme's active site.

DNA containing U formed by
deamination of C

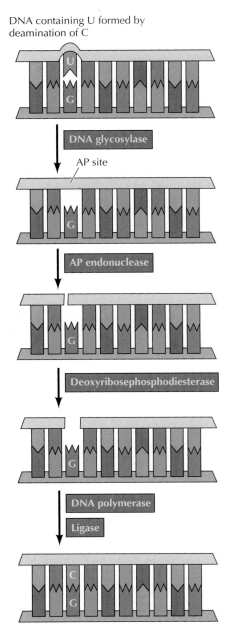

DNA glycosylase

AP site

AP endonuclease

Deoxyribosephosphodiesterase

DNA polymerase

Ligase

FIGURE 6.21 Base-excision repair In this example, uracil (U) has been formed by deamination of cytosine (C) and is therefore opposite a guanine (G) in the complementary strand of DNA. The bond between uracil and the deoxyribose is cleaved by a DNA glycosylase, leaving a sugar with no base attached in the DNA (an AP site). This site is recognized by AP endonuclease, which cleaves the DNA chain. The remaining deoxyribose is removed by deoxyribosephosphodiesterase. The resulting gap is then filled by DNA polymerase and sealed by ligase, leading to incorporation of the correct base (C) opposite the G.

The repair of uracil-containing DNA is a good example of **base-excision repair**, in which single damaged bases are recognized and removed from the DNA molecule (Figure 6.21). Uracil can arise in DNA by two mechanisms: (1) Uracil (as dUTP [deoxyuridine triphosphate]) is occasionally incorporated in place of thymine during DNA synthesis, and (2) uracil can be formed in DNA by the deamination of cytosine (see Figure 6.17A). The second mechanism is of much greater biological significance because it alters the normal pattern of complementary base pairing and thus represents a mutagenic event. The excision of uracil in DNA is catalyzed by **DNA glycosylase**, an enzyme that cleaves the bond linking the base (uracil) to the deoxyribose of the DNA backbone. This reaction yields free uracil and an apyrimidinic site—a sugar with no base attached. DNA glycosylases also recognize and remove other abnormal bases, including hypoxanthine formed by the deamination of adenine, pyrimidine dimers, alkylated purines other than O^6-alkylguanine, and bases damaged by oxidation or ionizing radiation.

The result of DNA glycosylase action is the formation of an apyrimidinic or apurinic site (generally called an AP site) in DNA. Similar AP sites are formed as a result of the spontaneous loss of purine bases (see Figure 6.17B), which occurs at a significant rate under normal cellular conditions. For example, each cell in the human body is estimated to lose several thousand purine bases daily. These sites are repaired by **AP endonuclease**, which cleaves adjacent to the AP site (see Figure 6.21). The remaining deoxyribose moiety is then removed, and the resulting single-base gap is filled by DNA polymerase and ligase.

Whereas DNA glycosylases recognize only specific forms of damaged bases, other excision repair systems recognize a wide variety of damaged bases that distort the DNA molecule, including UV-induced pyrimidine dimers and bulky groups added to DNA bases as a result of the reaction of many carcinogens with DNA (see Figure 6.18C). This widespread form of DNA repair is known as **nucleotide-excision repair**, because the damaged bases (e.g., a thymine dimer) are removed as part of an oligonucleotide containing the lesion (Figure 6.22).

In *E. coli*, nucleotide-excision repair is catalyzed by the products of three genes (*uvrA*, *uvrB*, and *uvrC*) that were identified because mutations at these loci result in extreme sensitivity to UV light. The protein UvrA recognizes damaged DNA and recruits UvrB and UvrC to the site of the lesion. UvrB and UvrC then cleave on the 3' and 5' sides of the damaged site, respectively, thus excising an oligonucleotide consisting of 12 or 13 bases. The UvrABC complex is frequently called an **excinuclease**, a name that reflects its ability to directly *exci*se an oligonucleotide. The action of a helicase is then required to remove the damage-containing oligonucleotide from the double-stranded DNA molecule, and the resulting gap is filled by DNA polymerase and sealed by ligase.

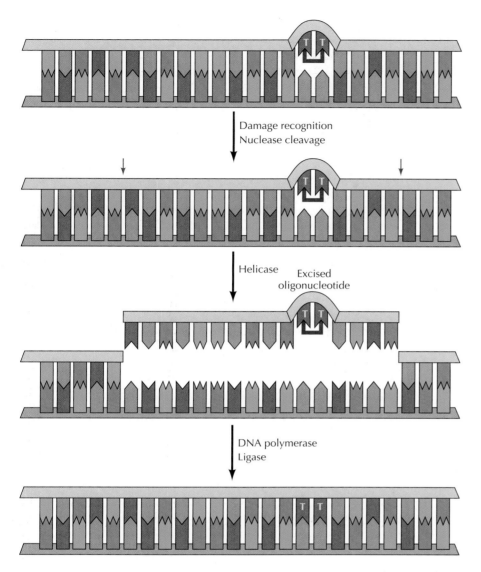

FIGURE 6.22 Nucleotide-excision repair of thymine dimers Damaged DNA is recognized and then cleaved on both sides of a thymine dimer by 3' and 5' nucleases. Unwinding by a helicase results in excision of an oligonucleotide containing the damaged bases. The resulting gap is then filled by DNA polymerase and sealed by ligase.

Nucleotide-excision repair systems have also been studied extensively in eukaryotes, including yeasts, rodents, and humans. In yeasts, as in *E. coli*, several genes involved in DNA repair (called *RAD* genes for *rad*iation sensitivity) have been identified by the isolation of mutants with increased sensitivity to UV light. In humans, DNA repair genes have been identified largely by studies of individuals suffering from inherited diseases resulting from deficiencies in the ability to repair DNA damage. The most extensively studied of these diseases is xeroderma pigmentosum (XP), a rare genetic disorder that affects approximately one in 250,000 people. Individuals with this disease are extremely sensitive to UV light and develop multiple skin cancers on the regions of their bodies that are exposed to sunlight. In 1968, James Cleaver made the key discovery that cultured cells from XP patients were deficient in the ability to carry out nucleotide-excision repair. This observation not only provided the first link between DNA repair and cancer, but also suggested the use of XP cells as an experimental system to identify human DNA repair genes. The identification of human DNA repair

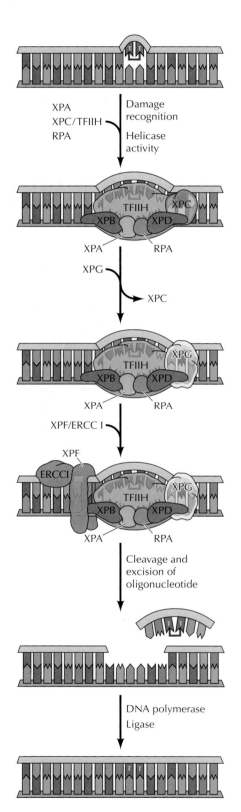

genes has been accomplished by studies not only of XP cells, but also of two other human diseases resulting from DNA repair defects (Cockayne's syndrome and trichothiodystrophy) and of UV-sensitive mutants of rodent cell lines. The availability of mammalian cells with defects in DNA repair has allowed the cloning of repair genes based on the ability of wild-type alleles to restore normal UV sensitivity to mutant cells in gene transfer assays, thereby opening the door to experimental analysis of nucleotide-excision repair in mammalian systems.

Molecular cloning has identified seven different repair genes (designated *XPA* through *XPG*) that are mutated in cases of xeroderma pigmentosum, as well as genes that are mutated in Cockayne's syndrome, trichothiodystrophy, and UV-sensitive mutants of rodent cells. The proteins encoded by these mammalian DNA repair genes are closely related to proteins encoded by yeast *RAD* genes, indicating that nucleotide-excision repair is highly conserved throughout eukaryotes. With cloned yeast and mammalian repair genes available, it has been possible to purify their encoded proteins and develop *in vitro* systems to study their roles in the repair process (Figure 6.23). The initial step in excision repair in mammalian cells involves recognition of disrupted base pairing by the cooperative binding of XPA, XPC, and the single-stranded DNA binding protein replication protein A (RPA) discussed earlier in this chapter (see Figure 6.8). XPC is associated with a multisubunit transcription factor called TFIIH, which is required to initiate transcription of eukaryotic genes (see Chapter 7). Two of the subunits of TFIIH are the XPB and XPD proteins, which act as helicases to unwind approximately 25 base pairs of DNA around the site of damage. The XPG protein is then recruited to the complex, displacing XPC, followed by recruitment of XPF as a heterodimer with ERCC1 (a repair protein identified in UV-sensitive rodent cells). XPF/ERCC1 and XPG are endonucleases, which cleave DNA on the 5' and 3' sides of the damaged site, respectively. This cleavage excises an oligonucleotide consisting of approximately 30 bases. The resulting gap is then filled by DNA polymerase δ or ε (in association with RFC and PCNA) and sealed by ligase.

Although damaged DNA can be recognized throughout the genome, an alternative form of nucleotide-excision repair, called **transcription-coupled repair**, is specifically dedicated to repairing damage within actively transcribed genes. A connection between transcription and repair was first suggested by experiments showing that transcribed strands of DNA are repaired more rapidly than nontranscribed strands in both *E. coli* and mammalian cells. Since DNA damage blocks transcription, this transcription-repair coupling is thought to be advantageous by allowing the cell to preferentially repair damage to genes that are actively expressed. In both *E. coli* and mammalian cells, the mechanism of transcription-repair coupling involves recognition of RNA polymerase stalled at a lesion in the DNA strand being transcribed. In *E. coli*, the stalled RNA polymerase is recognized by a protein called transcription-repair coupling factor, which dis-

FIGURE 6.23 Nucleotide-excision repair in mammalian cells   DNA damage (e.g., a thymine dimer) is recognized by XPA, RPA, and XPC, which is complexed to the transcription factor TFIIH, containing the XPB and XPD helicases. Following unwinding of the DNA by XPB and XPD, XPG is recruited to the complex, replacing XPC. The XPF/ERCC1 complex is then recruited and the DNA is then cleaved by the XPF/ERCC1 and XPG endonucleases, excising the damaged oligonucleotide. The resulting gap is filled by DNA polymerase and sealed by ligase.

places RNA polymerase and recruits the UvrABC excinuclease to the site of damage. In mammalian cells, transcription-coupled repair involves recognition of stalled RNA polymerase by two proteins (CSA and CSB) that are encoded by genes responsible for Cockayne's syndrome. In contrast to patients with xeroderma pigmentosum, patients with Cockayne's syndrome are specifically defective in transcription-coupled repair, consistent with the role of CSA and CSB as transcription-repair coupling factors.

A third excision repair system recognizes mismatched bases that are incorporated during DNA replication. Many such mismatched bases are removed by the proofreading activity of DNA polymerase. The ones that are missed are subject to later correction by the **mismatch repair** system, which scans newly replicated DNA. If a mismatch is found, the enzymes of this repair system are able to identify and excise the mismatched base specifically from the newly replicated DNA strand, allowing the error to be corrected and the original sequence restored.

In *E. coli*, the ability of the mismatch repair system to distinguish between parental DNA and newly synthesized DNA is based on the fact that DNA of this bacterium is modified by the methylation of adenine residues within the sequence GATC to form 6-methyladenine (Figure 6.24). Since methylation occurs after replication, newly synthesized DNA strands are not methylated and thus can be specifically recognized by the mismatch repair enzymes. Mismatch repair is initiated by the protein MutS, which recognizes the mismatch and forms a complex with two other proteins called MutL and MutH. The MutH endonuclease then cleaves the unmethylated DNA strand at a GATC sequence. MutL and MutS then act together with an exonuclease and a helicase to excise the DNA between the strand break and the mismatch, with the resulting gap being filled by DNA polymerase and ligase.

Eukaryotes have a similar mismatch repair system, although the mechanism by which eukaryotic cells identify newly replicated DNA differs from that used by *E. coli*. Eukaryotic cells do not have a homolog of MutH and the strand-specificity of mismatch repair is not determined by DNA methylation. Instead, the presence of single-strand breaks in newly replicated DNA appears to specify the strand to be repaired. Eukaryotic homologs of MutS and MutL bind to the mismatched base, as in *E. coli* and direct excision of the DNA between a strand break and the mismatch. The importance of this repair system is dramatically illustrated by the fact that mutations in the human homologs of *MutS* and *MutL* are responsible for a common type of inherited colon cancer (hereditary nonpolyposis colorectal cancer, or HNPCC). HNPCC is one of the most common inherited diseases; it affects as many as one in 200 people and is responsible for about 15% of all colorectal cancers in this country. The relationship between HNPCC and defects in mismatch repair was discovered in 1993, when two groups of researchers cloned the human homolog of *MutS* and found that mutations in this gene were responsible for about half of all HNPCC cases. Subsequent studies have shown that most of the remaining cases of HNPCC are caused by mutations in one of three human genes that are homologs of *MutL*. Defects in these genes appear to result in a high frequency of mutations in other cell genes, with a correspondingly high likelihood that some of these mutations will eventually lead to the development of cancer.

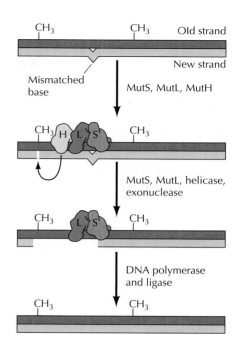

FIGURE 6.24 Mismatch repair in *E. coli* The mismatch repair system detects and excises mismatched bases in newly replicated DNA, which is distinguished from the parental strand because it has not yet been methylated. MutS binds to the mismatched base, followed by MutL. The binding of MutL activates MutH, which cleaves the unmodified strand opposite a site of methylation. MutS and MutL, together with a helicase and an exonuclease, then excise the portion of the unmodified strand that contains the mismatch. The gap is then filled by DNA polymerase and sealed by ligase.

Translesion DNA Synthesis

The direct reversal and excision repair systems act to correct DNA damage before replication, so that replicative DNA synthesis can proceed using an undamaged DNA strand as a template. Should these systems fail, however, the cell has alternative mechanisms for dealing with damaged DNA at the replication fork. Pyrimidine dimers and many other types of lesions cannot be copied by the normal action of DNA polymerases, so replication is

MOLECULAR MEDICINE

Colon Cancer and DNA Repair

The Disease

Cancers of the colon and rectum (colorectal cancers) are some of the most common types of cancer in Western countries, accounting for about 140,000 cancer cases per year in the United States (approximately 10% of the total cancer incidence). Most colon cancers (like other types of cancer) are not inherited diseases; that is, they are not transmitted directly from parent to offspring. However, two inherited forms of colon cancer have been described. In both of these syndromes, the inheritance of a cancer susceptibility gene results in a very high likelihood of cancer development. One inherited form of colon cancer (familial adenomatous polyposis) is extremely rare, accounting for less than 1% of total colon cancer incidence. The second inherited form of colon cancer (hereditary nonpolyposis colorectal cancer, or HNPCC) is much more common and accounts for up to 15% of all colon cancer cases. Indeed, HNPCC is one of the most common inherited diseases, affecting as many as one in 200 people. Although colon cancers are the most common manifestation of this disease, affected individuals also suffer an increased incidence of other types of cancer, including cancers of the ovaries and endometrium.

Molecular and Cellular Basis

Like other cancers, colorectal cancer results from mutations in genes that regulate cell proliferation, leading to the uncontrolled growth of cancer cells. In most cases these mutations occur sporadically in somatic cells. In

hereditary cancers, however, inherited germ-line mutations predispose the individual to cancer development.

A striking advance was made in 1993 with the discovery that a gene responsible for approximately 50% of HNPCC cases encodes an enzyme involved in mismatch repair of DNA; this gene is a human homolog of the *E. coli MutS* gene. Subsequent studies have shown that three other genes, responsible for most remaining cases of HNPCC, are homologs of *MutL* and thus are also involved in the mismatch repair pathway. Defects in these genes appear to result in a high frequency of mutations in other cell genes, with a correspondingly high likelihood that some of these mutations will eventually lead to the development of cancer by affecting genes that regulate cell proliferation.

Prevention and Treatment

As with other inherited diseases, identification of the genes responsible for HNPCC allows individuals at risk for this inherited cancer to be identified by genetic testing. Moreover, prenatal genetic diagnosis may be of great importance to carriers of HNPCC mutations who are planning a family. However, the potential benefits of detecting these mutations are not limited to preventing the transmission of mutant genes to the next generation; their detection may also help prevent the development of cancer in affected individuals.

In terms of disease prevention, a key characteristic of colon cancer is that it develops gradually over several years. Early diagnosis of the disease substantially improves the chances for

patient survival. The initial stage of colon cancer development is the outgrowth of small benign polyps, which eventually become malignant and invade the surrounding connective tissue. Prior to the development of malignancy, however, polyps can be easily removed surgically, effectively preventing the outgrowth of a malignant tumor. Polyps and early stages of colon cancer can be detected by examination of the colon with a thin lighted tube (colonoscopy), so frequent colonoscopy of HNPCC patients may allow polyps to be removed before cancer develops. In addition, several drugs are being tested as potential inhibitors of colon cancer development, and these drugs may be of significant benefit to HNPCC patients. By allowing the timely application of such preventive measures, the identification of mutations responsible for HNPCC may make a significant contribution to disease prevention.

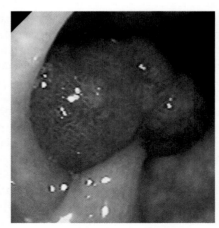

A colon polyp visualized by colonoscopy. (David M. Martin, M.D./SPL/Photo Researches, Inc.)

FIGURE 6.25 Translesion DNA synthesis   Normal replication is blocked by a thymine dimer, but specialized DNA polymerases such as polymerase V (pol V) recognize and continue DNA synthesis across the lesion. Replication can then be resumed by the normal replicative DNA polymerase, and the thymine dimer subsequently removed by nucleotide-excision repair.

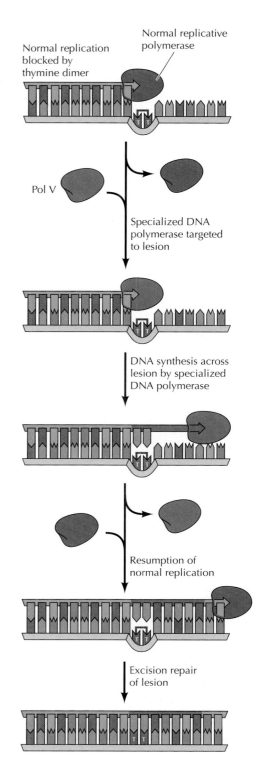

blocked at the sites of such damage. However, cells also possess several specialized DNA polymerases that are capable of replicating across a site of DNA damage. The replication of damaged DNA by these specialized polymerases, called **translesion DNA synthesis**, provides a mechanism by which the cell can bypass DNA damage at the replication fork, which can then be corrected after replication is complete.

The first specialized DNA polymerase responsible for translesion DNA synthesis was discovered in *E. coli* in 1999. This enzyme, called polymerase V, is induced in response to extensive UV irradiation and can synthesize a new DNA strand across from a thymine dimer (**Figure 6.25**). Two other specialized *E. coli* DNA polymerases, polymerases II and IV, are similarly induced by DNA damage and function in translesion synthesis. Eukaryotic cells also contain multiple specialized DNA polymerases, with ten such enzymes having been identified to date in humans. All of these specialized DNA polymerases exhibit low fidelity when copying undamaged DNA, with error rates ranging from 100 to 10,000 times higher than the error rates of the normal replicative DNA polymerases (e.g., polymerase III in *E. coli* or polymerases δ and ε in eukaryotes). In addition, the specialized polymerases lack the 3'–> 5' proofreading activity that is characteristic of normal replicative DNA polymerases (see Figure 6.11).

Importantly, however, the specialized polymerases involved in translesion synthesis have evolved to insert the correct base opposite specific lesions in damaged DNA, and are therefore able to accurately synthesize a new strand using some forms of damaged DNA as a template. For example, *E. coli* polymerase V specifically recognizes thymine dimers and correctly inserts AA on the opposite strand. On the other hand, polymerase V makes a high frequency of errors when it synthesizes a new DNA strand opposite other forms of DNA damage. Thus, these enzymes are able to specifically insert correct bases opposite some forms of DNA damage, although they are "error-prone" in inserting bases opposite other forms of damaged DNA or in the synthesis of DNA from a normal undamaged template.

Recombinational Repair

Another means of DNA repair, **recombinational repair**, relies on replacement of the damaged DNA by recombination with an undamaged molecule. This mechanism is frequently used to repair damage encountered during DNA replication, where the presence of thymine dimers or other lesions that cannot be copied by the normal replicative DNA polymerases block the progress of a replication fork. Recombinational repair depends on the fact that one strand of the parental DNA was undamaged and therefore was copied during replication to yield a normal daughter molecule, which can then be used to repair the damaged strand.

The molecular mechanisms of recombinational repair are not entirely understood and may vary between different types of cells, but an illus-

FIGURE 6.26 Recombinational repair The presence of a thymine dimer blocks replication, but DNA polymerase can bypass the lesion and reinitiate replication at a new site downstream of the dimer. The result is a gap opposite the dimer in the newly synthesized DNA strand. This gap is filled by recombination with the undamaged parental strand. Although this leaves a gap in the previously intact parental strand, the gap can be filled by the actions of polymerase and ligase, using the intact daughter strand as a template. Two intact DNA molecules are thus formed, and the remaining thymine dimer eventually can be removed by excision repair.

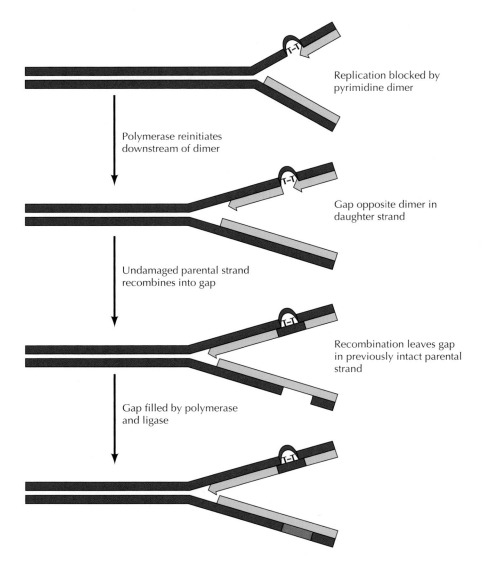

Replication blocked by pyrimidine dimer

Polymerase reinitiates downstream of dimer

Gap opposite dimer in daughter strand

Undamaged parental strand recombines into gap

Recombination leaves gap in previously intact parental strand

Gap filled by polymerase and ligase

trative model is presented in Figure 6.26. In this example, normal replication is blocked by the presence of a thymine dimer in one strand of DNA. Downstream of the damaged site, however, replication can be initiated again by the synthesis of an Okazaki fragment and can proceed along the damaged template strand. The result is a daughter strand that has a gap opposite the site of damage to the parental strand. The undamaged parental strand, which has been replicated to yield a normal daughter molecule, can then be used to fill the gap opposite the site of damage by recombination between homologous DNA sequences (see the next section). Because the resulting gap in the previously intact parental strand is opposite an undamaged strand, it can be filled in by DNA polymerase. Although the other parent molecule still retains the original damage (e.g., a thymine dimer), the damage now lies opposite a normal strand and can be dealt with later by excision repair.

Recombinational repair also provides a major mechanism for repair of double strand breaks, which can be introduced into DNA by ionizing radiation (such as X-rays) and some chemicals (Figure 6.27). Since this type of damage affects both strands of DNA, it is particularly difficult to repair.

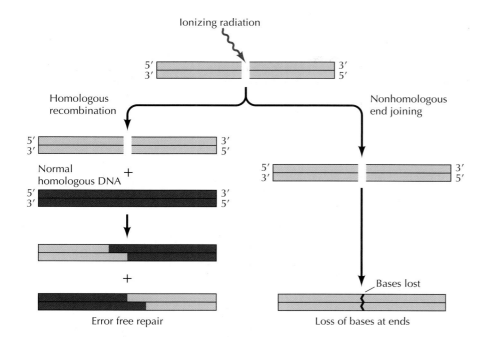

FIGURE 6.27 Repair of double strand breaks   Ionizing radiation and some chemicals induce double strand breaks in DNA. These breaks can be repaired by homologous recombination with a normal chromosome, leading to restoration of the original DNA sequence. Alternatively, the ends of the broken molecule can be rejoined by nonhomologous end joining, with the frequent loss of bases around the site of damage.

Recombination with homologous DNA sequences on an undamaged chromosome (discussed in detail in the next section of this chapter) provides a mechanism for repairing such damage and restoring the normal DNA sequence. Alternatively, double strand breaks can be repaired simply by rejoining the broken ends of a single DNA molecule, but this leads to a high frequency of errors resulting from deletion of bases around the site of damage. It is noteworthy that the genes responsible for inherited breast cancer (*BRCA1* and *BRCA2*) encode proteins that are involved in the repair of double strand breaks by homologous recombination, suggesting that defects in this type of DNA repair can lead to the development of one of the most common cancers in women.

■ **DNA repair champion:** The bacterium *Deinococcus radiodurans* was isolated in 1956 from cans of meat that were thought to have been sterilized by irradiation. The bacteria survived because of their extreme radiation resistance. *D. radiodurans* is over a hundredfold more resistant to ionizing radiation than *E. coli.*

Recombination between Homologous DNA Sequences

Accurate DNA replication and repair of DNA damage are essential to maintaining genetic information and ensuring its accurate transmission from parent to offspring. As discussed in the previous section, recombination is an important mechanism for repairing damaged DNA. In addition, recombination is key to the generation of genetic diversity, which is critical from the standpoint of evolution. Genetic differences between individuals provide the essential starting material of natural selection, which allows species to evolve and adapt to changing environmental conditions. Recombination plays a central role in this process by allowing genes to be reassorted into different combinations. For example, genetic recombination results in the exchange of genes between paired homologous chromosomes during meiosis. In addition, recombination is involved in rearrangements of specific DNA sequences that alter the expression and function of some genes during development and differentiation. Thus, recombination plays important roles in the lives of individual cells and organisms, as well as contributing to the genetic diversity of the species.

This section discusses the molecular mechanism of **homologous recombination**, which involves the exchange of information between DNA molecules that share sequence homology over hundreds of bases. Examples include recombination between paired chromosomes during meiosis, as well as homologous recombination during DNA repair. Since this type of recombination involves the exchange of genetic information between two homologous DNA molecules, it does not alter the overall arrangement of the genes on a chromosome. Other types of recombination, however, do not require extensive sequence homology and therefore can occur between unrelated DNAs. Recombination events of this type lead to gene rearrangements, which are discussed later in the chapter.

Models of Homologous Recombination

Recombination results from the breakage and rejoining of two parental DNA molecules, leading to reassortment of the genetic information of the two parental chromosomes. During homologous recombination, this takes place without any other alteration in the genetic information. Thus, a critical question is: How can two parental DNA molecules be broken at precisely the same point, so that they can rejoin without mutations resulting from the gain or loss of nucleotides at the break point? During recombination between homologous DNA molecules, this alignment is provided, not surprisingly, by base pairing between complementary DNA strands (Figure 6.28). Overlapping single strands are exchanged between homologous DNA molecules, leading to the formation of a heteroduplex region, in which the two strands of the recombinant double helix are derived from different parents. If the heteroduplex region contains a genetic difference, the result is a single progeny DNA molecule that contains two genetic markers. In some cases, mispaired bases in a heteroduplex may be recognized and corrected by mismatch repair systems, as discussed in preceding sections of this chapter. Genetic evidence for the formation and repair of such heteroduplex regions, obtained in studies of recombination in fungi as well as in bacteria, led to the development of a molecular model for recombination in 1964. This model, known as the **Holliday model** (after Robin Holliday), has continued to provide the basis for current thinking about recombination mechanisms, although it has been modified as new data have been obtained.

The original version of the Holliday model proposed that recombination is initiated by the introduction of nicks at the same position on the two parental DNA molecules (Figure 6.29). The nicked DNA strands partially unwind, and each invades the other molecule by pairing with the complementary unbroken strand. Ligation of the broken strands then produces a crossed-strand intermediate, known as a **Holliday junction**, which is the central intermediate in recombination. The direct demonstration of Holliday junctions by electron microscopy has provided clear support for this model of recombination (Figure 6.30).

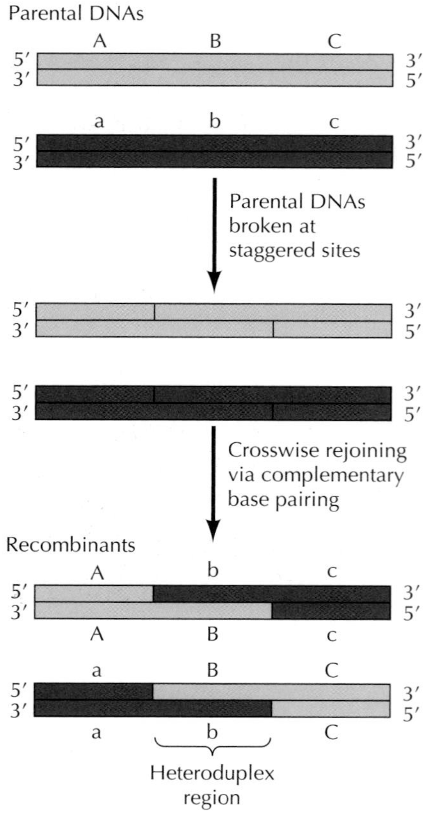

FIGURE 6.28 Homologous recombination by complementary base pairing Parental DNAs are broken at staggered sites, and overlapping single-stranded regions are exchanged via base pairing with homologous sequences. The result is a heteroduplex region in which the two DNA strands are derived from different parental molecules.

FIGURE 6.29 The Holliday model for homologous recombination Single-strand nicks are introduced at the same position on both parental molecules. The nicked strands then exchange by complementary base pairing, and ligation produces a crossed-strand intermediate called a Holliday junction.

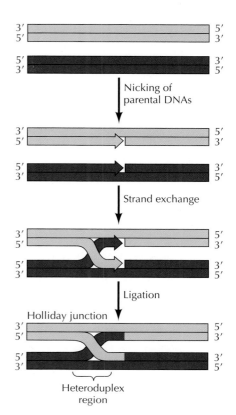

Once a Holliday junction is formed, it can be resolved by cutting and rejoining of the crossed strands to yield recombinant molecules (**Figure 6.31**). This can occur in two different ways, depending on the orientation of the Holliday junction, which can readily form two different isomers. In the isomer resulting from the initial strand exchange, the crossed strands are those that were nicked at the start of the recombination process. However, simple rotation of this structure yields a different isomer in which the unbroken parental strands are crossed. Resolution of these different isomers has distinct genetic consequences. In the first case, the progeny molecules have heteroduplex regions but are nonrecombinant for DNA that flanks these regions. However, if isomerization occurs, cutting and rejoining of the crossed strands results in progeny molecules that are recombinant for DNA that flanks the heteroduplex regions. The structure of the Holliday junction thus provides the possibility of generating both recombinant and nonrecombinant heteroduplexes, consistent with the genetic data upon which the Holliday model was based.

As initially proposed, the Holliday model failed to explain how recombination was initiated by simultaneously nicking both parental molecules at the same position. It now appears that recombination is generally initiated at double strand breaks, both during DNA repair and during recombination between homologous chromosomes during meiosis (**Figure 6.32**). Both strands of DNA at the double strand break are first resected by nucleases that digest DNA in the 5′ to 3′ direction, yielding single-stranded ends. These single strands then invade the other parental molecule by homologous base pairing. The gaps are then filled by repair synthesis and the strands are joined by ligation to yield a molecule with a double Holliday junction, which can be resolved to yield either recombinant or nonrecombinant heteroduplex molecules as already described.

Enzymes Involved in Homologous Recombination

Most of the enzymes currently known to be involved in recombination were first identified by analysis of recombination-defective mutants of *E. coli*. Such genetic analysis has established that recombination requires specific enzymes, in addition to proteins (such as DNA polymerase, ligase, and single-stranded DNA-binding proteins) that function in multiple aspects of DNA metabolism. The identification of genes required for efficient recombination in *E. coli* has been followed by considerable progress in eukaryotic cells, leading to the isolation and characterization of the proteins that catalyze the formation and resolution of Holliday junctions.

6.2 WEBSITE ANIMATION

Homologous Recombination In the Holliday model for recombination between two homologous double-stranded DNA molecules, the process begins by nicking one strand of each double-stranded molecule, followed by strand exchange and joining of the nicked strands on the opposite molecules.

FIGURE 6.30 Identification of Holliday junctions by electron microscopy
Electron micrograph of a Holliday junction that was detected during recombination of plasmid DNAs in *E. coli*. An interpretive drawing of the structure is shown below. The molecule illustrates a Holliday junction in the open configuration resulting from rotation of the crossed-strand intermediate (see Figure 6.31). (Courtesy of Huntington Potter, University of South Florida, and David Dressler, University of Oxford.)

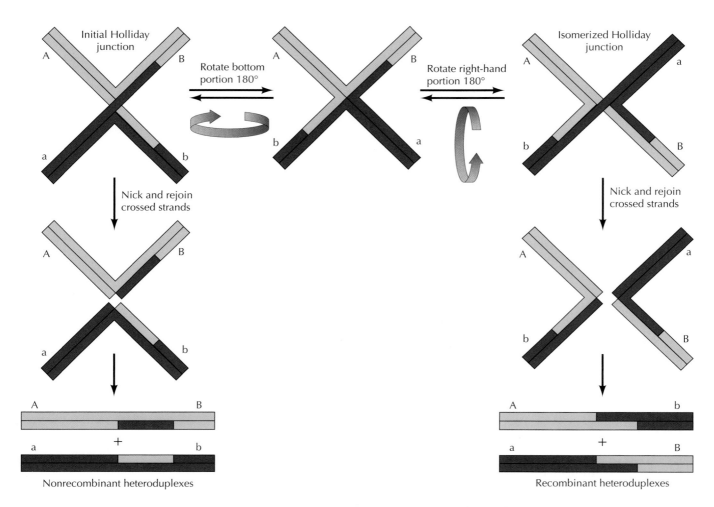

FIGURE 6.31 Isomerization and resolution of Holliday junctions Holliday junctions are resolved by cutting and rejoining of the crossed strands. If the Holliday junction formed by the initial strand exchange is resolved, the resulting progeny are heteroduplexes but are not recombinant for genetic markers outside of the heteroduplex region. Two rotations of the crossed-strand molecule, however, produce an isomer in which the unbroken parental strands, rather than the initially nicked strands, are crossed. Cutting and rejoining of the crossed strands of this isomer yield progeny that are recombinant heteroduplexes.

The key protein involved in the central steps of homologous recombination in *E. coli* is **RecA.** The RecA protein promotes the exchange of strands between homologous DNAs that causes heteroduplexes to form (Figure 6.33). The action of RecA can be considered in three stages. First, the RecA protein binds to single-stranded DNA, coating the DNA to form a protein-DNA filament. Because RecA has 3 distinct DNA binding sites, the RecA protein bound to single-stranded DNA is able to bind a second, double-stranded DNA molecule, forming a complex between the two DNAs. This nonspecific RecA-mediated association is followed by specific base pairing between the single-stranded DNA and its complement. The RecA protein then catalyzes strand exchange, with the single strand originally coated with RecA displacing its homologous strand to form a heteroduplex.

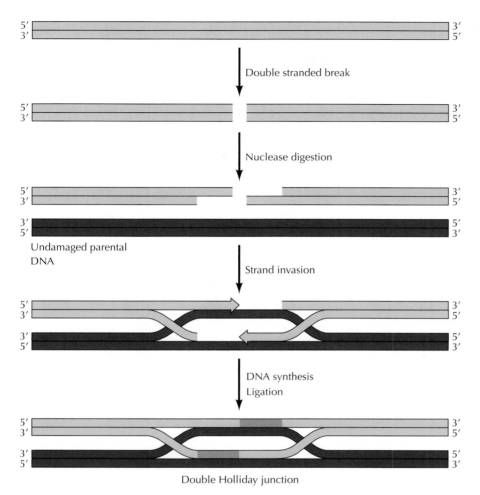

FIGURE 6.32 Initiation of recombination by double strand breaks   Both strands of DNA at the double strand break are digested by nucleases in the 5′ to 3′ direction. The single strands then invade the other parental molecule by homologous base pairing. The gaps are then filled by repair synthesis and sealed by ligation, yielding a double Holliday junction.

In eukaryotic cells, a closely-related protein, **Rad51**, functions similarly to RecA. Rad51 was first identified in yeast, as a protein required for genetic recombination as well as for the repair of double strand breaks. Rad51 binds to single-stranded DNA to form protein-DNA filaments similar to those formed by RecA (Figure 6.34) and is also able to catalyze strand exchange reactions *in vitro*. Proteins related to Rad51 have been identified in complex eukaryotes, including humans, indicating that the RecA/Rad51 family of proteins plays key roles in homologous recombination in both prokaryotic and eukaryotic cells.

In *E. coli*, Holliday junctions are resolved by a complex of three proteins, RuvA, RuvB, and RuvC (Figure 6.35). RuvA recognizes the Holliday junction and recruits RuvB. RuvA and RuvB then act as a motor to drive migration of the site at which the DNA strands are crossed, thereby varying the extent of the heteroduplex region and the position at which the crossed strands will be cut and rejoined. RuvC then resolves Holliday junctions by cleaving the

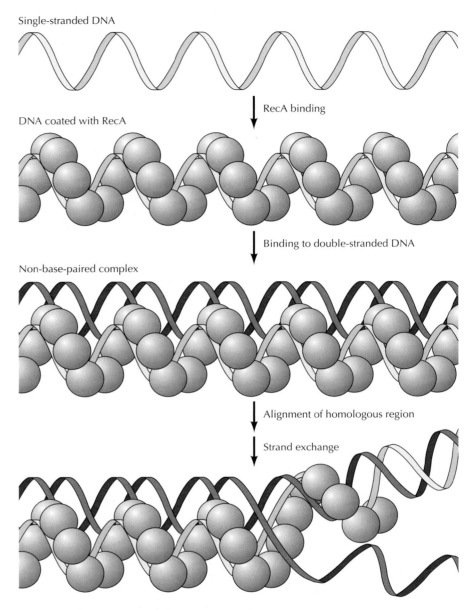

Single-stranded DNA

RecA binding

DNA coated with RecA

Binding to double-stranded DNA

Non-base-paired complex

Alignment of homologous region

Strand exchange

FIGURE 6.33 Function of the RecA protein   RecA initially binds to single-stranded DNA to form a protein-DNA filament. The RecA protein that coats the single-stranded DNA then binds to a second, double-stranded DNA molecule to form a non-base-paired complex. Complementary base pairing and strand exchange follow, forming a heteroduplex region.

crossed DNA strands. Rejoining of the cleaved strands by ligation completes the process, yielding two recombinant molecules. Eukaryotic cells do not have homologs of the *E. coli* RuvA, RuvB, and RuvC proteins. Instead, the resolution of Holliday junctions in eukaryotic cells appears to be mediated by other proteins, which remain to be fully characterized. Recently, an endonuclease called Mus81 has been identified as a possible Holliday junction resolvase in yeast and mammalian cells, but the function of Mus81 and other potential resolvase activities still remain to be fully understood.

(A) Rec A

(B) Rad 51

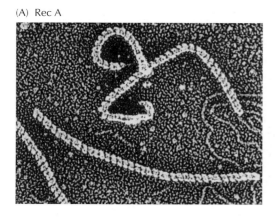
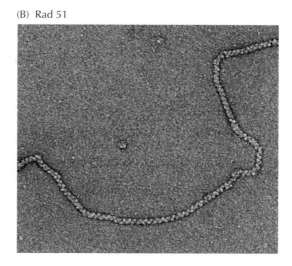

FIGURE 6.34 RecA and Rad51 protein-DNA filaments   Electron micrographs of filaments formed by the binding of *E. coli* RecA and human Rad51 proteins to DNA. (From S. C. West 2003. *Nature Rev. Mol. Cell Biol.* 4: 1. Courtesy of A. Stasiak and S. West.)

DNA Rearrangements

Homologous recombination results in the reassortment of genes between chromosome pairs without altering the arrangement of genes within the genome. In contrast, other types of recombinational events lead to rearrangements of genomic DNA. Some of these DNA rearrangements are important in controlling gene expression in specific cell types; others may play an evolutionary role by contributing to genetic diversity.

The discovery that genes can move to different chromosomal locations came from Barbara McClintock's studies of corn in the 1940s. Purely on the basis of genetic analysis, McClintock described novel genetic elements that could move to different locations in the genome and alter the expression of adjacent genes. Nearly three decades elapsed, however, before the physical basis of McClintock's work was elucidated by the discovery of transposable elements in bacteria and the notion of movable genetic elements became widely accepted by scientists. Several types of DNA rearrangements, including the transposition of elements initially described by McClintock, are now recognized in both prokaryotic and eukaryotic cells. Moreover, we now know that transposable elements constitute a large fraction of the genomes of plants and animals, including nearly half of the human genome.

Site-Specific Recombination

In contrast to general homologous recombination, which occurs at any extensive region of sequence homology, **site-specific recombination** occurs between specific DNA sequences, which are usually homologous over only a short stretch of DNA. The principal interaction in this process is mediated by proteins that recognize the specific DNA target sequences rather than by complementary base pairing. Site-specific recombination thus leads to programmed DNA rearrangements that can play important roles in development and the regulation of gene expression.

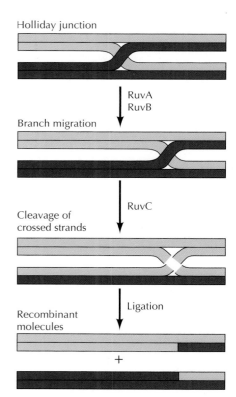

FIGURE 6.35 Branch migration and resolution of Holliday junctions RuvA recognizes the Holliday junction and recruits RuvB, which catalyzes the movement of the crossed-strand site (branch migration). RuvC resolves the Holliday junctions by cleaving the crossed strands, which are then joined by ligase.

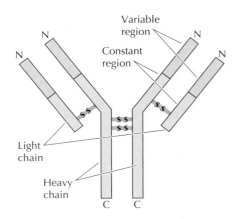

FIGURE 6.36 Structure of an immunoglobulin Immunoglobulins are composed of two heavy chains and two light chains, joined by disulfide bonds. Both the heavy and the light chains consist of variable and constant regions.

6.3 WEBSITE ANIMATION

Light-Chain Genes
During B cell development, site-specific recombination joins regions of the immunoglobulin light-chain genes, leading to the production of unique immunoglobulin light chains.

6.4 WEBSITE ANIMATION

Heavy-Chain Genes
During B cell development, site-specific recombination joins regions of the immunoglobulin heavy-chain genes, leading to the production of unique immunoglobulin heavy chains.

Development of the vertebrate immune system, which recognizes foreign substances (**antigens**) and provides protection against infectious agents, is a prime example of site-specific recombination in higher eukaryotes. There are two major classes of immune responses, which are mediated by B and T lymphocytes. B lymphocytes secrete antibodies (**immunoglobulins**) that react with soluble antigens; T lymphocytes express cell surface proteins (called **T cell receptors**) that react with antigens expressed on the surfaces of other cells. The key feature of both immunoglobulins and T cell receptors is their enormous diversity, which enables different antibody or T cell receptor molecules to recognize a vast array of foreign antigens. For example, each individual is capable of producing more than 10^{11} different antibody molecules, which is far in excess of the total number of genes in mammalian genomes (20,000–25,000). Rather than being encoded in germline DNA, these diverse antibodies (and T cell receptors) are encoded by unique lymphocyte genes that are formed during development of the immune system as a result of site-specific recombination between distinct segments of immunoglobulin and T cell receptor genes.

The role of site-specific recombination in the formation of immunoglobulin genes was first demonstrated by Susumu Tonegawa in 1976. Immunoglobulins consist of pairs of identical heavy and light polypeptide chains (Figure 6.36). Both the heavy and light chains are composed of C-terminal constant regions and N-terminal variable regions. The variable regions, which have different amino acid sequences in different immunoglobulin molecules, are responsible for antigen binding, and it is the diversity of variable region amino acid sequences that allows different individual antibodies to recognize unique antigens. Although every individual is capable of producing a vast spectrum of different antibodies, each B lymphocyte produces only a single type of antibody. Tonegawa's key discovery was that each antibody is encoded by unique genes formed by site-specific recombination during B lymphocyte development. These gene rearrangements create different immunoglobulin genes in different individual B lymphocytes, so the population of approximately 10^{12} B lymphocytes in the human body includes cells capable of producing antibodies against a diverse array of foreign antigens.

The genes that encode immunoglobulin light chains consist of three regions: a V region that encodes the 95 to 96 N-terminal amino acids of the polypeptide variable region; a joining (J) region that encodes the 12 to 14 C-terminal amino acids of the polypeptide variable region; and a C region that encodes the polypeptide constant region (Figure 6.37). The major class of light-chain genes in the mouse is formed from combinations of approximately 250 V regions and 4 J regions with a single C region. Site-specific recombination during lymphocyte development leads to a gene rearrangement in which a single V region recombines with a single J region to generate a functional light-chain gene. Different V and J regions are rearranged in different B lymphocytes, so the possible combinations of 250 V regions with 4 J regions can generate approximately 1000 (4 × 250) unique light chains.

The heavy-chain genes include a fourth region (known as the diversity, or D, region), which encodes amino acids lying between V and J (Figure 6.38). Assembly of a functional heavy-chain gene requires two recombination events: A D region first recombines with a J region, and a V region then recombines with the rearranged DJ segment. In the mouse, there are about 500 heavy-chain V regions, 12 D regions, and 4 J regions, so the total number of heavy chains that can be generated by the recombination events is 24,000 (500 × 12 × 4).

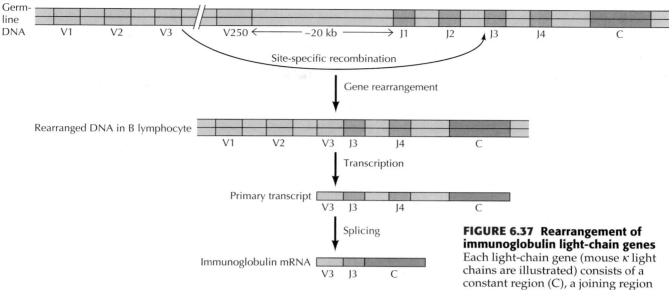

FIGURE 6.37 Rearrangement of immunoglobulin light-chain genes
Each light-chain gene (mouse κ light chains are illustrated) consists of a constant region (C), a joining region (J), and a variable region (V). There are approximately 250 different V regions, which are separated from J and C regions by about 20 kb in germ-line DNA. During the development of B lymphocytes, site-specific recombination joins one of the V regions to one of the four J regions. This rearrangement activates transcription, resulting in the formation of a primary transcript containing the rearranged VJ region together with the remaining J regions and C region. The remaining unused J regions and the introns between J and C regions are then removed by splicing, yielding a functional mRNA.

Combinations between the 1000 different light chains and 24,000 different heavy chains formed by site-specific recombination can generate approximately 2×10^7 different immunoglobulin molecules. This diversity is further increased because the joining of immunoglobulin gene segments often involves the loss or gain of one to several nucleotides, as discussed below. The mutations resulting from these deletions and insertions increase the diversity of immunoglobulin variable regions approximately a hundredfold, corresponding to the formation of about 10^5 different light chains and 2×10^6 heavy chains, which can then combine to form more than 10^{11} distinct antibodies.

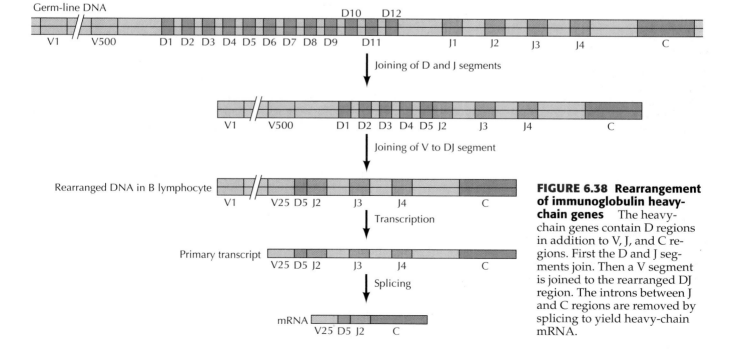

FIGURE 6.38 Rearrangement of immunoglobulin heavy-chain genes The heavy-chain genes contain D regions in addition to V, J, and C regions. First the D and J segments join. Then a V segment is joined to the rearranged DJ region. The introns between J and C regions are removed by splicing to yield heavy-chain mRNA.

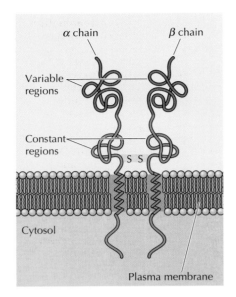

FIGURE 6.39 Structure of a T cell receptor T cell receptors consist of two polypeptide chains (α and β) that span the plasma membrane and are joined by disulfide bonds. Both the α and β chains are composed of variable and constant regions.

T cell receptors similarly consist of two chains (called α and β), each of which contains variable and constant regions (Figure 6.39). The genes encoding these polypeptides are generated by recombination between V and J segments (the α chain) or between V, D, and J segments (the β chain), analogous to the formation of immunoglobulin genes. Site-specific recombination between these distinct segments of DNA, in combination with mutations introduced during recombination, generates a degree of diversity in T cell receptors that is similar to that in immunoglobulins.

V(D)J recombination is mediated by a complex of two proteins, called RAG1 and RAG2, which are specifically expressed in lymphocytes. The RAG proteins recognize recombination signal (RS) sequences adjacent to the coding sequences of each gene segment, and initiate DNA rearrangements by introducing a double strand break between the RS sequences and the coding sequences (Figure 6.40). The coding ends of the gene segments are then joined to yield a rearranged immunoglobulin or T cell receptor gene. Since these double strand breaks are joined by a nonhomologous end joining process (see Figure 6.27), the joining reaction is accompanied by the frequent loss of nucleotides. In addition, lymphocytes contain a specialized enzyme (terminal deoxynucleotide transferase) that adds nucleotides at random to the ends of DNA molecules, so mutations corresponding to both the loss and the gain of nucleotides are introduced during the joining reaction. As noted above, these mutations contribute substantially to the diversity of immunoglobulins and T cell receptors.

Still further antibody diversity is generated after the formation of rearranged immunoglobulin genes by two processes that occur only in B lymphocytes: class switch recombination and somatic hypermutation. **Class switch recombination** results in the association of rearranged V(D)J regions with different heavy chain constant regions, leading to the production of antibodies with distinct functional roles in the immune response. Mammals produce four different classes of immunoglobulins—IgM, IgG, IgE, and IgA—with heavy chains encoded by a variable region joined to the Cμ, Cγ, Cϵ, and Cα constant regions, respectively. In addition, there are several subclasses of IgG, which are encoded by different Cγ regions. The different classes of immunoglobulins are specialized to remove antigens in different ways. IgM activates complement (a group of serum proteins that destroy invading cells or viruses), so IgM antibodies are an effective first line of defense against bacterial or viral infections. IgG antibodies, the most abundant immunoglobulins in serum, not only activate complement but also bind receptors on phagocytic cells. In addition, IgG antibodies can cross the placenta from the maternal circulation, providing immune protection to the fetus. IgA antibodies are secreted into a variety of bodily fluids, such as nasal mucus and saliva, where they can bind and eliminate invading bacteria or viruses to prevent infection. IgA antibodies are also secreted into the milk of nursing mothers, so they provide immune protection to newborns. IgE antibodies are effective in protecting against parasitic infections, and are also the class of antibodies responsible for allergies.

The initial V(D)J rearrangement produces a variable region joined to Cμ, resulting in the production of IgM antibodies. Class switch recombination then transfers a rearranged variable region to a new downstream constant region, with deletion of the intervening DNA (Figure 6.41). Recombination occurs between highly repetitive sequences in switch (S) regions that are located immediately upstream of each C region. The switch regions are 2–10 kb in length and recombination can take place anywhere within these regions, so class switching is more properly referred to as a region specific

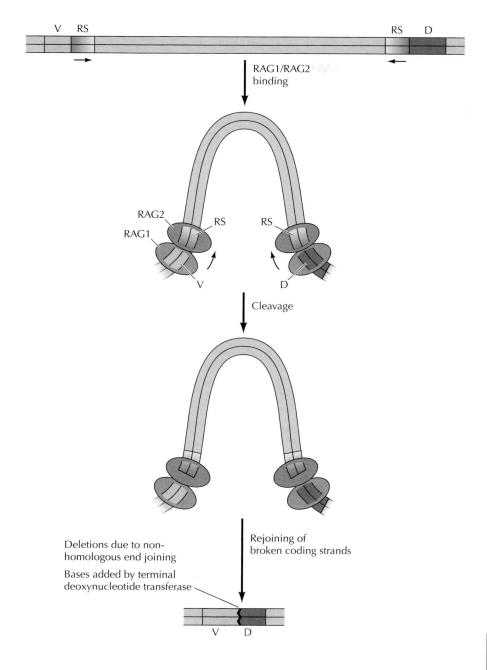

FIGURE 6.40 V(D)J recombination
The coding segments of immunoglobulin and T cell receptor genes (e.g., a V and D segment) are flanked by short recombination signal (RS) sequences, which are in opposite orientations at the 5′ and 3′ ends of the coding sequences. The RS sequences are recognized by a complex of the lymphocyte-specific recombination proteins RAG1 and RAG2, which cleave the DNA between the coding sequence and the RS sequence. The broken coding strands are then rejoined to yield a rearranged gene segment. Mutations result from the loss of bases at the ends during nonhomologous end joining, as well as from the addition of bases by terminal deoxynucleotide transferase.

rather than a site specific recombination event. Because the switch regions are within introns, the precise site at which class switch recombination takes place does not affect the immunoglobulin coding sequence.

Somatic hypermutation increases the diversity of immunoglobulins by producing multiple mutations within rearranged variable regions of both heavy and light chains. These mutations, principally single base substitutions, occur with frequencies as high as 10^{-3}, approximately a million times higher than normal rates of spontaneous mutation. They lead to the production of immunoglobulins with a substantially increased affinity for antigen, and therefore are an important contributor to an effective immune response.

■ Mutations in the genes for RAG1 and RAG2 can lead to severe combined immunodeficiency (SCID). These patients are born without a functional immune system and develop lethal infections if left untreated. Treatments include living in a germ free environment (in plastic "bubbles"), transplantation with stem cells giving rise to a functional immune system, and gene therapy.

FIGURE 6.41 Class switch recombination Class switching takes place by recombination between repetitive switch (S) regions upstream of a series of constant (C) regions in the heavy chain locus (the mouse locus is shown). In the example shown, a V(D)J region is transferred from $C\mu$ to $C\gamma1$ by recombination between the $S\mu$ and $S\gamma1$ switch regions. The intervening DNA is excised as a circular molecule.

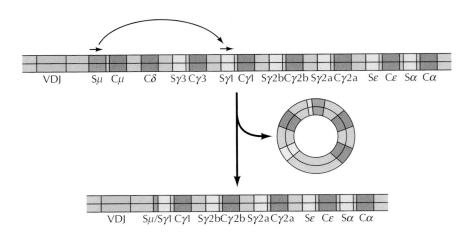

Class switch recombination and somatic hypermutation are novel types of programmed genetic alterations, and the molecular mechanisms involved are very active areas of investigation. A key player in both processes is an enzyme called **activation-induced deaminase (AID)**, which was discovered by Tasuku Honjo and his collaborators in 1999. AID is expressed only in B lymphocytes, and it is required for both class switch recombination and somatic hypermutation. AID catalyzes the deamination of cytosine in DNA to form uracil (Figure 6.42). Its action results in the conversion of C→U in the variable regions and switch regions of immunoglobulin genes, leading to class switch recombination and somatic hypermutation, respectively. The mechanisms by which C→U mutations stimulate these processes are not completely understood, but an important step appears to be the removal of U by base-excision repair (see Figure 6.21). This leads to the formation of single-strand breaks, and it is thought that the formation of multiple breaks in the switch regions (which have a high content of GC base pairs) results in class switch recombination. In the variable regions, somatic hypermutation is thought to result from a high frequency of errors during repair of the C→U mutations, possibly resulting from repair by specialized error-prone DNA polymerases (see Figure 6.25). Although the details of these processes remain to be elucidated, it is clear that AID is an extremely interesting enzyme with the novel role of introducing mutations into DNA at a specific stage of development.

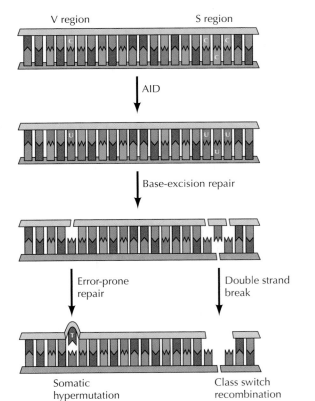

FIGURE 6.42 Model for the role of activation-induced deaminase (AID) in somatic hypermutation and class switch recombination AID deaminates C to U in DNA. Removal of U by base-excision repair (see Figure 6.21) leaves a single-strand gap in the DNA. In the variable regions, errors during repair may lead to somatic hypermutation, possibly resulting from the action of a specialized error-prone DNA polymerase (see Figure 6.25). In the switch regions, the presence of multiple sites of base excision may result in double strand breaks that stimulate class switch recombination.

Transposition via DNA Intermediates

Site-specific recombination occurs between two specific sequences that contain at least a small core of homology. In contrast, transposition involves the movement of sequences throughout the genome and has no requirement for sequence homology. Elements that move by transposition, such as those first described by McClintock, are called **transposable elements,** or **transposons**. They are divided into two general classes, depending on whether they transpose via DNA intermediates or via RNA intermediates. The first class of transposable elements is discussed here; transposition via RNA intermediates is considered in the next section.

The first transposons that were characterized in detail are those of bacteria, which move via DNA intermediates (**Figure 6.43**). The simplest of these elements are the insertion sequences, ranging in size from about 800 to 2000 nucleotides. Insertion sequences encode a gene for the enzyme involved in transposition (transposase) flanked by short inverted repeats, which are the sites at which transposase acts.

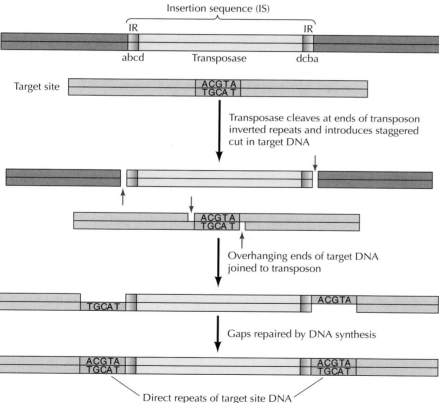

FIGURE 6.43 Bacterial transposons   Insertion sequences (IS) range from 800 to 2000 nucleotides and contain a gene for transposase flanked by inverted repeats (IR) of about 20 nucleotides. Transposase cleaves at both ends of the transposon and introduces a staggered cut in the target DNA. The overhanging ends of target DNA are then joined to the transposon, and gaps resulting from the staggered cuts at the target site are repaired. The result is the formation of short direct repeats of target-site DNA (5 to 10 nucleotides long) flanking the integrated transposon.

KEY EXPERIMENT

Rearrangement of Immunoglobulin Genes

Evidence for Somatic Rearrangement of Immunoglobulin Genes Coding for Variable and Constant Regions

Nobumichi Hozumi and Susumu Tonegawa

Basel Institute for Immunology, Basel, Switzerland

Proceedings of the National Academy of Sciences, USA, Volume 73, 1976, pages 3628–3632

The Context

The ability of the vertebrate immune system to recognize a seemingly infinite variety of foreign molecules implies that lymphocytes can produce a correspondingly vast array of antibodies. Since this antibody diversity is key to immune recognition, understanding the mechanism by which an apparently unlimited number of distinct immunoglobulins are encoded in genomic DNA is a central issue in immunology.

Prior to the experiments of Hozumi and Tonegawa, protein sequencing of multiple immunoglobulins had demonstrated that both heavy and light chains consist of distinct variable and constant regions. Genetic studies further indicated that mice inherit only single copies of the constant-region genes. These observations first led to the proposal that immunoglobulins are encoded by multiple variable-region genes that can associate with a single constant-region gene. The discovery of immunoglobulin gene rearrangements by Hozumi and Tonegawa provided the first direct experimental support for this hypothesis and laid the groundwork for understanding the molecular basis of antibody diversity.

The Experiments

Hozumi and Tonegawa tested the possibility that the genes encoding immunoglobulin variable and constant regions were joined at the DNA level during lymphocyte development. Their experimental approach was to use restriction endonuclease

digestion to compare the organization of variable-region and constant-region sequences in DNAs extracted from mouse embryos and from cells of a mouse plasmacytoma (a B lymphocyte tumor that produces a single species of immunoglobulin).

Embryo and plasmacytoma DNAs were digested with the restriction endonuclease *Bam*HI, and DNA fragments of different sizes were separated by electrophoresis in an agarose gel.

The gel was then cut into slices, and DNA extracted from each slice was hybridized with radiolabeled probes that had been prepared from immunoglobulin mRNA isolated from the plasmacytoma cells. Two probes were used, corresponding either to the complete immunoglobulin mRNA or to the 3' half of the mRNA, consisting only of constant-region sequences.

The critical result was that completely different patterns of variable-region and constant-region sequences were detected in embryo versus plasmacytoma DNAs (see figure). In embryo DNA, the complete probe hybridized to two *Bam*HI fragments of approximately 8.6 and 5.6 kb, respectively. Only the 8.6-kb fragment hybridized to the 3' probe, suggesting that the 8.6-kb fragment contained

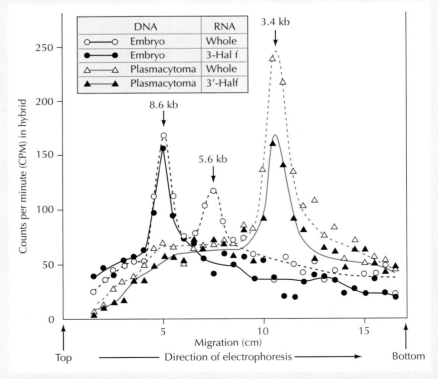

Gel electrophoresis of embryo and plasmacytoma DNAs digested with BamHI and hybridized to probes corresponding to either the whole or the 3' half of the plasmacytoma mRNA. Data are presented as the radioactivity detected in hybrid molecules with DNA from each gel slice.

KEY EXPERIMENT

constant-region sequences and the 5.6-kb fragment contained variable-region sequences. In striking contrast, both probes hybridized to only a single 3.4-kb fragment in plasmacytoma DNA. The interpretation of these results was that the variable- and constant-region sequences were separated in embryo DNA but rearranged to form a single immunoglobulin gene during lymphocyte development.

The Impact

The initial results of Hozumi and Tonegawa, based on the relatively indirect approach of restriction endonuclease mapping, were confirmed and extended by the molecular cloning and sequencing of immunoglobulin genes. Such studies have now unambiguously established that these genes are generated by site-specific recombination between distinct segments of DNA in B lymphocytes. In T lymphocytes, similar DNA rearrangements are responsible for formation of the genes encoding T cell receptors. Thus, site-specific recombination and programmed gene rearrangements are central to the development of the immune system.

Further studies have shown that the variable regions of immunoglobulins and T cell receptors are generated by rearrangements of two or three distinct segments of DNA. The ability of these segments to recombine, together with a high frequency of mutations introduced at the recombination sites, is largely responsible for immunoglobulin and T cell receptor diversity. The discovery of immunoglobulin gene rearrangements thus provided the basis for understanding how the immune system can recognize and respond to a virtually unlimited range of foreign substances.

Susumu Tonegawa

Insertion sequences move from one chromosomal site to another without replicating their DNA. Transposase introduces a staggered break in the target DNA and cleaves at the ends of the transposon inverted-repeat sequences. Although transposase acts specifically at the transposon inverted-repeats, it is usually less specific with respect to the sequence of the target DNA, so it catalyzes the movement of transposons throughout the genome. Following the cleavage of transposon and target site DNAs, transposase joins the overhanging ends of the target DNA to the transposable element. The resulting gap in the target-site DNA is repaired by DNA synthesis, followed by ligation to the other strand of the transposon. The result of this process is a short direct repeat of the target-site DNA on both sides of the transposable element—a hallmark of transposon integration.

This transposition mechanism causes the transposon to move from one chromosomal site to another. Other types of transposons move by a more complex mechanism, in which the transposon is replicated in concert with its integration into a new target site. This mechanism results in the integration of one copy of the transposon into a new position in the genome, while another copy remains at its original location.

Transposons that move via DNA intermediates are present in eukaryotes as well as in bacteria. For example, the human genome contains approximately 300,000 DNA transposons, which account for about 3% of human DNA. The original transposable elements described by McClintock in corn move by a nonreplicative mechanism, as do most transposable elements in other plants and animals. Like bacterial transposons, these elements move to many different target sites throughout the genome. The movement of these transposons to nonspecific sites in the genome is not likely to be useful to the cells in which it occurs, but has undoubtedly played a major role in evolution by promoting DNA rearrangements.

In yeasts and protozoans, however, transposition by a replicative mechanism is responsible for programmed DNA rearrangements that regulate gene expression. In these cases, transposition is initiated by the action of a site-specific nuclease that cleaves a specific target site, at which a copy of the transposable element is then inserted. Transposable elements are thus capable not only of moving to nonspecific sites throughout the genome, but also of participating in specific gene rearrangements that result in programmed changes in gene expression.

Transposition via RNA Intermediates

Most transposons in eukaryotic cells are **retrotransposons**, which move via reverse transcription of RNA intermediates. In humans, there are almost 3 million copies of retrotransposons, accounting for more than 40% of the genome (see Table 5.2). The mechanism of transposition of these elements is similar to the replication of retroviruses, which have provided the prototype system for studying this class of movable DNA sequences.

Retroviruses contain RNA genomes in their virus particles but replicate via the synthesis of a DNA provirus, which is integrated into the chromosomal DNA of infected cells (see Figure 4.13). A DNA copy of the viral RNA is synthesized by the viral enzyme **reverse transcriptase**. The mechanism by which this occurs results in the synthesis of a DNA molecule that contains direct repeats of several hundred nucleotides at both ends (Figure 6.44). These repeated sequences, called **long terminal repeats**, or **LTRs**, arise from duplication of the sites on viral RNA at which primers bind to initiate DNA synthesis. The LTR sequences thus play central roles in reverse transcription, in addition to being involved in the integration and subsequent transcription of proviral DNA.

Like all DNA polymerases, reverse transcriptase requires a primer, which in the case of retroviruses, is a tRNA molecule bound at a specific site (the primer-binding site) close to the 5′ terminus of the viral RNA (Figure 6.45). Since DNA synthesis proceeds in the 5′ to 3′ direction, only a short piece of DNA is synthesized before reverse transcriptase reaches the end of its template. Continuation of DNA synthesis then depends on the ability of reverse transcriptase to "jump" to the 3′ end of the template RNA molecule. This is accomplished via an RNase H activity of reverse transcriptase, which degrades the RNA strand of DNA-RNA hybrids. As a result, the newly synthesized DNA is converted to a single-stranded molecule, which can hybridize to a short repeated sequence present at both the 5′ and the 3′ ends of the viral RNA. DNA synthesis can then continue, yielding a single-stranded DNA complementary to viral RNA. The viral RNA is then degraded, and synthesis of the opposite strand of DNA is initiated by a fragment of viral RNA that acts as a primer, at a site near the 3′ end of the template DNA strand. Again the result is a short piece of DNA, which includes the primer-binding site copied from the tRNA used as the initial primer for reverse transcription. The primer-binding sequence of the tRNA

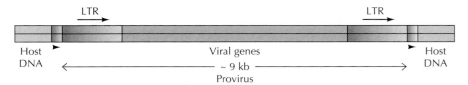

FIGURE 6.44 The organization of retroviral DNA The integrated proviral DNA is flanked by long terminal repeats (LTRs), which are direct repeats of several hundred nucleotides. Viral genes, including genes for reverse transcriptase, integrase, and structural proteins of the virus particle, are located between the LTRs. The integrated provirus is flanked by short direct repeats of host DNA.

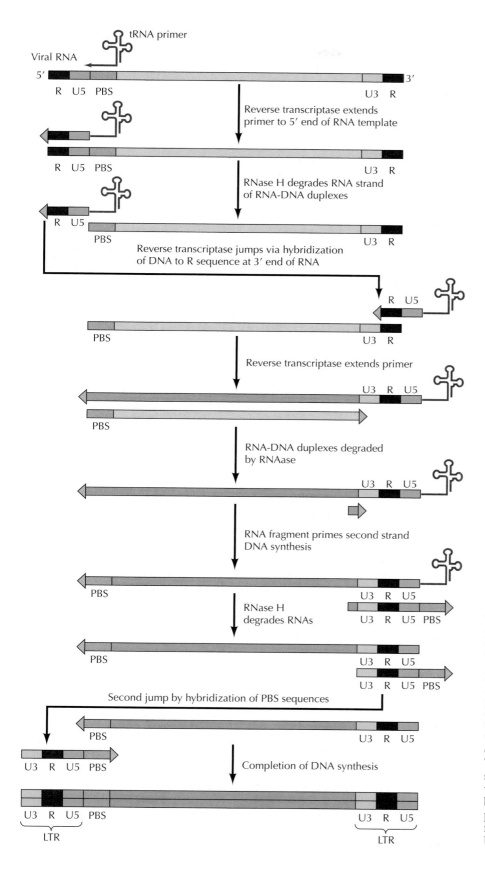

FIGURE 6.45 Generation of LTRs during reverse transcription LTRs consist of three sequence elements: a short repeat sequence (R) of about 20 nucleotides that is present at both ends of the viral RNA; a sequence unique to the 5' end of viral RNA (U5); and a sequence unique to the 3' end of viral RNA (U3). Repeats of these sequences are generated during DNA synthesis as reverse transcriptase jumps twice between the ends of its template. Synthesis is initiated using a tRNA primer bound to a primer-binding site (PBS) adjacent to U5 at the 5' end of the viral RNA. The polymerase copies U5 and R, and the RNA strand of the RNA-DNA hybrid is then degraded by RNase H. The polymerase then jumps to the 3' end of the viral RNA in order to synthesize a complete DNA strand complementary to the RNA template. The polymerase jumps again during synthesis of the second strand of DNA, which is also initiated by a primer bound close to the 5' end of its template. The result of these jumps is the formation of LTRs that contain U3-R-U5 sequences.

is then degraded by RNase H, leaving an overhanging DNA strand that again "jumps" to pair with its complementary sequence at the other end of the template. DNA synthesis can then continue once more, finally yielding a linear DNA with LTRs at both ends.

The linear viral DNA integrates into the host cell chromosome by a process that resembles the integration of DNA transposable elements. Integration is catalyzed by a viral integrase and occurs at many different target sequences in cellular DNA. The integrase cleaves two bases from the ends of viral DNA and introduces a staggered cut at the target site in cellular DNA. The overhanging ends of cellular DNA are then joined to the termini of viral DNA, and the gap is filled by DNA synthesis. The integrated provirus is therefore flanked by a direct repeat of cell sequences, similar to the repeats that flank DNA transposons.

The viral life cycle continues with transcription of the integrated provirus, which yields viral genomic RNA as well as mRNAs that direct the synthesis of viral proteins (including reverse transcriptase and integrase). The genomic RNA is then packaged into viral particles, which are released from the host cell. These progeny viruses can infect a new cell, initiating another round of DNA synthesis and integration. The net effect can be viewed as the movement of the provirus from one chromosomal site to another, via the synthesis and reverse transcription of an RNA intermediate.

Other retrotransposons differ from retroviruses in that they are not packaged into infectious particles and therefore cannot spread from one cell to another. However, these retrotransposons can move to new chromosomal sites within the same cell, via mechanisms fundamentally similar to those involved in retrovirus replication.

Some retrotransposons (called retrovirus-like elements or LTR retrotransposons) are structurally similar to retroviruses (Figure 6.46). Retrotransposons of this type account for about 8% of the human genome. They have LTR sequences at both ends; they encode reverse transcriptase and integrase; and they transpose (like retroviruses) via transcription into RNA, synthesis of a new DNA copy by reverse transcriptase, and integration into cellular DNA.

The non-LTR retrotransposons differ from retroviruses in that they do not contain LTR sequences, although they do encode their own reverse transcriptase and an endonuclease involved in transposition. In mammals, the major class of these retrotransposons consists of the highly repetitive long interspersed elements (**LINEs**), which are repeated approximately 850,000 times in the genome and account for about 21% of genomic DNA (see Chapter 5). A full-length LINE element is about 6 kb long, although most members of the family are truncated at their 5′ end (Figure 6.47). At

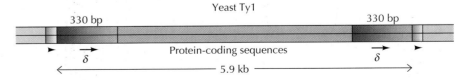

FIGURE 6.46 Structure of a LTR retrotransposon The yeast Ty1 transposable element displays the same organization as a retrovirus. Protein-coding sequences, including genes for reverse transcriptase and integrase, are flanked by LTRs (called δ elements) of about 330 base pairs (bp). The integrated transposon is flanked by short direct repeats of target-site DNA.

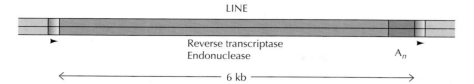

FIGURE 6.47 Structure of human LINEs LINEs lack LTRs, but they do encode reverse transcriptase and an endonuclease. They have tracts of A-rich sequences (designated A_n) at their 3′ ends, which are thought to arise from reverse transcription of poly-A tails added to the 3′ end of mRNAs. Like other transposable elements, LINEs are flanked by short direct repeats of target-site DNA.

their 3′ end, LINEs have tracts of A-rich sequences thought to be derived by reverse transcription of the poly-A tails that are added to mRNAs following transcription (see Chapter 7). Like other transposable elements, LINEs are flanked by short direct repeats of the target-site DNA, indicating that integration involves staggered cuts and repair synthesis.

Since LINEs do not contain LTR sequences, the mechanism of their reverse transcription and subsequent integration into chromosomal DNA must differ from that of retroviruses and LTR-containing retrotransposons. In particular, reverse transcription is primed by a broken end of chromosomal DNA at the integration target site, resulting from cleavage of the target site DNA by the endonuclease encoded by the retrotransposon (Figure 6.48). Reverse transcription then initiates within the poly-A tract at the 3′ end of the transposon RNA and continues along the molecule. The opposite strand

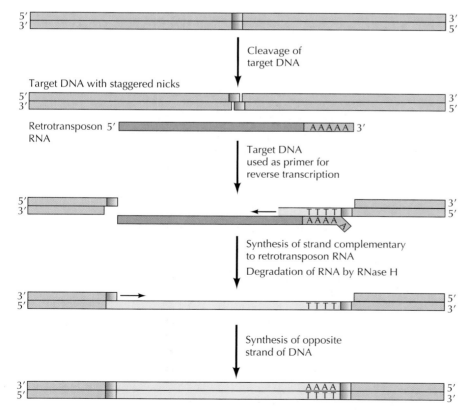

FIGURE 6.48 Model for reverse transcription and integration of LINEs Target site DNA is cleaved by an endonuclease encoded by the retrotransposon. Reverse transcription, primed by a broken end of the target DNA, initiates within the poly-A tail at the 3′ end of retrotransposon RNA. Synthesis of the opposite strand of retrotransposon DNA is similarly primed by the other strand of DNA at the target site.

of DNA is synthesized using the other broken end of the target-site DNA as a primer, resulting in simultaneous synthesis and integration of the retro-transposon DNA.

Other sequence elements, which do not encode their own reverse transcriptase, also transpose via RNA intermediates. These elements include the highly repetitive short interspersed elements (**SINEs**), of which there are approximately 1,500,000 copies in mammalian genomes (see Chapter 5). The major family of these elements in humans consists of the *Alu* sequences, which are about 300 bases long. These sequences have A-rich tracts at their 3′ end and are flanked by short duplications of target-site DNA sequences, a structure similar to that of non-LTR retrotransposons (e.g., LINEs). SINEs arose by reverse transcription of small RNAs, including tRNAs and small cytoplasmic RNAs involved in protein transport. Since SINEs no longer encode functional RNA products, they represent pseudogenes that arose via RNA-mediated transposition. Pseudogenes of many protein-coding genes (called **processed pseudogenes**) have similarly arisen by reverse transcription of mRNAs (Figure 6.49). Such processed pseudogenes are readily recognized not only because they terminate in an A-rich tract but also because the introns present in the corresponding normal gene have been removed during mRNA processing. The transposition of SINEs and of other processed pseudogenes is thought to proceed similarly to the transposition of LINEs. However, since these elements do not include genes for reverse transcriptase or a nuclease, their transposition presumably involves the action of reverse transcriptases and nucleases that are encoded elsewhere in the genome—probably by other retrotransposons, such as LINEs.

Although the highly repetitive SINEs and LINEs account for a significant fraction of genomic DNA, their transpositions to random sites in the genome are not likely to be useful for the cell in which they are located.

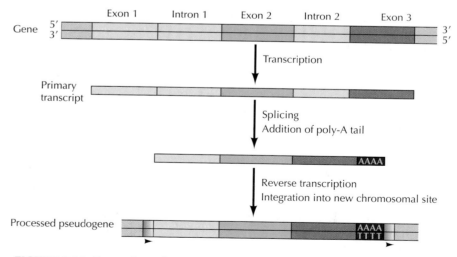

FIGURE 6.49 Formation of a processed pseudogene The gene illustrated contains three exons, separated by two introns. The introns are removed from the primary transcript by splicing, and a poly-A tail is added to the 3′ end of the mRNA. Reverse transcription and integration then yield a processed pseudogene, which does not contain introns and has an A-rich tract at its 3′ end. The processed pseudogene is flanked by short direct repeats of target-site DNA that were generated during its integration.

These transposons induce mutations when they integrate at a new target site, and like mutations induced by other agents, most mutations resulting from transposon integration are expected to be harmful to the cell. For example, mutations resulting from the transposition of both LINEs and SINEs have been associated with some cases of hemophilia, muscular dystrophy, breast cancer, and colon cancer. On the other hand, some mutations resulting from the movement of transposable elements may be beneficial, contributing in a positive way to evolution of the species. For example, some retrotransposons in mammalian genomes have been found to contain regulatory sequences that control the expression of adjacent genes.

In addition to their role as mutagens, retrotransposons have played a major role in shaping the genome by stimulating DNA rearrangements. For example, rearrangements of chromosomal DNA can result from recombination between LINEs integrated at different sites in the genome. Moreover, sequences of cellular DNA adjacent to LINEs are frequently carried along during the process of transposition. Consequently, the transposition of LINEs can result in the movement of cellular DNA sequences to new genomic sites. Since LINEs can integrate into active genes, the associated transposition of cellular DNA sequences can lead to the formation of new combinations of regulatory and/or coding sequences and contribute directly to the evolution of new genes.

The vast majority of the transposable elements in the human genome are inactive, with only about 100 copies of LINEs still retaining the protein-coding sequences required for their transposition. All of the human DNA transposons and most retrotransposons thus represent evolutionary relics rather than currently functional elements. However, this is not the case in other species, including *Arabidopsis, C. elegans, Drosophila,* and mice, which have a much higher level of ongoing transposon activity. In the mouse, for example, LTR retrotransposons, LINEs, and SINEs are all active. As a consequence, it is estimated that about 10% of all mutations in mice are the result of transposons, compared to only about 1 in 1000 mutations in humans. There is thus a dramatic and intriguing difference in transposon activity between mice and humans, the explanation for which remains to be determined.

Gene Amplification

The DNA rearrangements that have been discussed so far alter the position of a DNA sequence within the genome. **Gene amplification** may be viewed as a different type of alteration in genome structure; it increases the number of copies of a gene within a cell. Gene amplification results from repeated rounds of DNA replication, yielding multiple copies of a particular region (Figure 6.50). The amplified DNA sequences can be found either as free extrachromosomal molecules or as tandem arrays of sequences within a chromosome. In either case, the result is increased expression of the amplified gene, simply because more copies of the gene are available to be transcribed.

In some cases, gene amplification is responsible for developmentally programmed increases in gene expression. The prototypical example is amplification of the ribosomal RNA genes in amphibian oocytes (eggs). Eggs are extremely large cells, with correspondingly high requirements for protein synthesis. Amphibian oocytes in particular are about a million times larger in volume than typical somatic cells and must support large amounts of protein synthesis during early development. This requires increased synthesis of ribosomal RNAs, which is accomplished in part by amplification of the ribosomal RNA genes. As discussed in Chapter 5, there are already several hundred copies of ribosomal RNA genes per genome, so that

FIGURE 6.50 DNA amplification
Repeated rounds of DNA replication yield multiple copies of a particular chromosomal region.

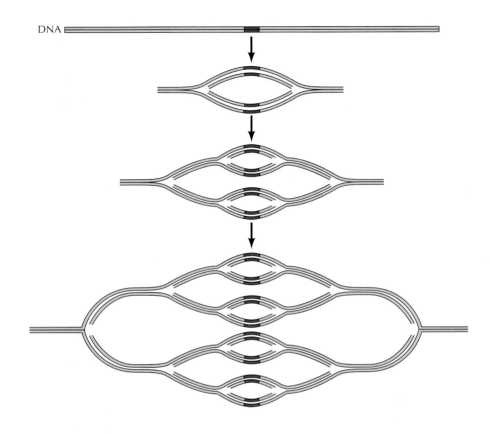

DNA

enough ribosomal RNA can be produced to meet the needs of somatic cells. In amphibian eggs, these genes are amplified an additional two thousand-fold, to approximately 1 million copies per oocyte. Another example of programmed gene amplification occurs in *Drosophila*, where the genes that encode eggshell proteins (chorion genes) are amplified in ovarian cells to support the requirement for large amounts of these proteins. Like other programmed gene rearrangements, however, gene amplification is a relatively infrequent event that occurs in highly specialized cell types; it is not a common mechanism of gene regulation.

Gene amplification also occurs as an abnormal event in cancer cells, where it results in the increased expression of genes that contribute to uncontrolled cell growth. Such gene amplification was first recognized in cancer cells that had become resistant to methotrexate, a drug commonly used in cancer chemotherapy. Methotrexate inhibits the enzyme dihydrofolate reductase, which is involved in the synthesis of dTTP and is therefore required for DNA synthesis. Resistance to methotrexate frequently develops by amplification of the dihydrofolate reductase gene, leading to increased production of the enzyme and consequently the loss of effective inhibition by methotrexate. In addition, gene amplification in cancer cells frequently results in the increased expression of genes that drive cell proliferation (oncogenes) and thereby directly contributes to tumor development (see Chapter 18). For example, amplification of the oncogene *erb*B-2 is frequently involved in human breast cancers. Thus, as with other types of DNA rearrangements, gene amplification can have either beneficial or deleterious consequences for the cell or organism in which it occurs.

SUMMARY	KEY TERMS

DNA REPLICATION

DNA Polymerases: Different DNA polymerases play distinct roles in DNA replication and repair in both prokaryotic and eukaryotic cells. All known DNA polymerases synthesize DNA only in the 5'to 3' direction by the addition of dNTPs to a preformed primer strand of DNA.

DNA polymerase

The Replication Fork: Parental strands of DNA separate and serve as templates for the synthesis of two new strands at the replication fork. One new DNA strand (the leading strand) is synthesized in a continuous manner; the other strand (the lagging strand) is formed by the joining of small fragments of DNA that are synthesized backward with respect to the overall direction of replication. DNA polymerases and various other proteins act in a coordinated manner to synthesize both leading and lagging strands of DNA.

replication fork, Okazaki fragment, DNA ligase, leading strand, lagging strand, primase, exonuclease, RNase H, helicase, single-stranded DNA-binding protein, topoisomerase

The Fidelity of Replication: DNA polymerases increase the accuracy of replication both by selecting the correct base for insertion and by proofreading newly synthesized DNA to eliminate mismatched bases.

proofreading

Origins and the Initiation of Replication: DNA replication starts at specific origins of replication, which contain binding sites for proteins that initiate the process. In higher eukaryotes, origins may be defined by chromatin structure rather than DNA sequence.

origin of replication, autonomously replicating sequence (ARS), origin recognition complex (ORC)

Telomeres and Telomerase: Maintaining the Ends of Chromosomes: Telomeric repeat sequences at the ends of chromosomes are maintained by the action of a reverse transcriptase (telomerase) that carries its own template RNA.

telomere, telomerase, reverse transcriptase

DNA REPAIR

Direct Reversal of DNA Damage: A few types of common DNA lesions, such as pyrimidine dimers and alkylated guanine residues, are repaired by direct reversal of the damage.

pyrimidine dimer, photoreactivation

Excision Repair: Most types of DNA damage are repaired by excision of the damaged DNA. The resulting gap is filled by newly synthesized DNA, using the undamaged complementary strand as a template. In base-excision repair, specific types of single damaged bases are removed from the DNA molecule. In contrast, nucleotide excision repair systems recognize a wide variety of lesions that distort the structure of DNA and remove the damaged bases as part of an oligonucleotide. A third excision repair system specifically removes mismatched bases from newly synthesized DNA strands.

base-excision repair, DNA glycosylase, AP endonuclease, nucleotide-excision repair, excinuclease, transcription-coupled repair, mismatch repair

Translesion DNA Synthesis: Specialized DNA polymerases are capable of replicating DNA across from a site of DNA damage, although the action of these polymerases may result in a high frequency of incorporation of incorrect bases.

translesion DNA synthesis

Recombinational Repair: Damaged DNA can be replaced by recombination with an undamaged molecule. This mechanism plays an important role in repairing damage encountered during DNA replication as well as in the repair of double strand breaks.

recombinational repair

KEY TERMS

homologous recombination, Holliday model, Holliday junction

RecA, Rad51

site-specific recombination, antigen, immunoglobulin, T cell receptor, class switch recombination, somatic hypermutation, activation-induced deaminase (AID)

transposable element, transposon

retrotransposon, retrovirus, reverse transcriptase, long terminal repeat (LTR), LINE, SINE, processed pseudogene

gene amplification

SUMMARY

RECOMBINATION BETWEEN HOMOLOGOUS DNA SEQUENCES

Models of Homologous Recombination: Recombination involves the breaking and rejoining of parental DNA molecules. Alignment between homologous DNA molecules is provided by complementary base pairing. Nicked strands of parental DNA invade the other parental molecule, yielding a crossed-strand intermediate known as a Holliday junction. Recombinant molecules are then formed by cleavage and rejoining of the crossed strands.

Enzymes Involved in Homologous Recombination: The central enzyme of homologous recombination is RecA (Rad51 in eukaryotes), which catalyzes the exchange of strands between homologous DNAs. Other enzymes nick and unwind parental DNAs and resolve Holliday junctions.

DNA REARRANGEMENTS

Site-Specific Recombination: Site-specific recombination takes place between specific DNA sequences that are recognized by proteins that mediate the process. In vertebrates, site-specific recombination plays a critical role in generating immunoglobulin and T cell receptor genes during development of the immune system. Additional diversity is provided to immunoglobulin genes by somatic hypermutation and class switch recombination.

Transposition via DNA Intermediates: Most DNA transposons move throughout the genome with no requirement for specific DNA sequences at their sites of insertion. In yeasts and protozoans, however, the transposition of some DNA sequences to specific target sites results in programmed DNA rearrangements that regulate gene expression.

Transposition via RNA Intermediates: Most transposons in eukaryotic cells move by reverse transcription of RNA intermediates, similar to the replication of retroviruses. These retrotransposons include the highly repeated LINE and SINE sequences of mammalian genomes.

Gene Amplification: Gene amplification results from repeated replication of a chromosomal region. In some cases, gene amplification provides a mechanism for increasing gene expression during development. Gene amplification also frequently occurs in cancer cells, where it can result in the elevated expression of genes that contribute to uncontrolled cell proliferation.

Questions

1. You have isolated a temperature sensitive strain of *E. coli* with a mutation in DNA polymerase I. What defects, if any, would you observe in bacteria carrying this mutation at high temperature?

2. Compare and contrast the actions of topoisomerases I and II.

3. What is the approximate number of Okazaki fragments synthesized during replication of the yeast genome?

4. Why is the synthesis of Okazaki fragments initiated by primase instead of a DNA polymerase?

5. What is the function of 3' to 5' exonuclease activity in DNA polymerases?

What would be the consequence of mutating this activity of DNA polymerase III on the fidelity of replication of *E. coli* DNA?

6. How would you test a sequence of DNA in a yeast cell to determine whether it contains an origin of replication?

7. Yeast cells require telomerase to completely replicate their genome, whereas *E. coli* does not need this special enzyme. Why?

8. What mechanism does a cell use to repair double strand breaks in its DNA? How does this differ from the repair of single strand breaks?

9. Patients with xeroderma pigmentosum suffer an extremely high incidence of skin cancer but have not been found to have correspondingly high incidences of cancers of internal organs (e.g., colon cancer). What might this suggest about the kinds of DNA damage responsible for most internal cancers?

10. Many of the drugs in clinical use for the treatment of AIDS are inhibitors of the HIV reverse transcriptase. What other cellular processes might be affected by these inhibitors?

11. The mismatch repair system in humans includes homologs of *E. coli Mut S* and *Mut L* but not of *Mut H*. Why?

12. What phenotype would you predict for a mutant mouse lacking one of the genes required for site-specific recombination in lymphocytes?

References and Further Reading

DNA Replication

Annunziato, A. T. 2005. Split decision: What happens to nucleosomes during DNA replication? *J. Biol. Chem.* 280: 12065–12068. [R]

Bell, S. P. and A. Dutta. 2002. DNA replication in eukaryotic cells. *Ann. Rev. Biochem.* 71: 333–374. [R]

Benkovic, S. J., A. M. Valentine and F. Salinas. 2001. Replisome-mediated DNA replication. *Ann. Rev. Biochem.* 70: 181–208. [R]

Blackburn, E. H. 2005. Telomeres and telomerase: Their mechanisms of action and the effects of altering their functions. *FEBS Lett.* 579: 859–862. [R]

Blasco, M. A. 2005. Telomeres and human disease: Ageing, cancer and beyond. *Nature Rev. Genet.* 6: 611–622. [R]

Bowman, G. D., E. R. Goedken, S. L. Kazmirski, M. O'Donnell and J. Kuriyan. 2005. DNA polymerase clamp loaders and DNA recognition. *FEBS Lett.* 579: 863–867. [R]

Champoux, J. J. 2001. DNA topoisomerases: Structure, function, and mechanism. *Ann. Rev. Biochem.* 70: 369–413. [R]

Cvetic, C. and J. C. Walter. 2005. Eukaryotic origins of DNA replication: Could you please be more specific? *Sem. Cell Dev. Biol.* 16: 343–353. [R]

Frick, D. N. and C. C. Richardson. 2001. DNA primases. *Ann. Rev. Biochem.* 70: 39–80. [R]

Gilbert, D. M. 2004. In search of the holy replicator. *Nature Rev. Mol. Cell Biol.* 5: 1–8. [R]

Huberman, J. A. and A. D. Riggs. 1968. On the mechanism of DNA replication in mammalian chromosomes. *J. Mol. Biol.* 32: 327–341. [P]

Hubscher, U., G. Maga and S. Spadari. 2002. Eukaryotic DNA polymerases. *Ann. Rev. Biochem.* 71: 133–163. [R]

Kornberg, A., I. R. Lehman, M. J. Bessman and E. S. Simms. 1956. Enzymic synthesis of deoxyribonucleic acid. *Biochim. Biophys. Acta* 21: 197–198. [P]

Kunkel, T. A. 2004. DNA replication fidelity. *J. Biol. Chem.* 279: 16895–16898. [R]

McEachern, M. J., A. Krauskopf and E. H. Blackburn. 2000. Telomeres and their control. *Ann. Rev. Genet.* 34: 331–358. [R]

Ogawa, T. and T. Okazaki. 1980. Discontinuous DNA replication. *Ann. Rev. Biochem.* 49: 421–457. [R]

Stillman, B. 2004. Origin recognition and the chromosome cycle. *FEBS Lett.* 579: 877–884. [R]

Stinthcomb, D. T., K. Struhl and R. W. Davis. 1979. Isolation and characterization of a yeast chromosomal replicator. *Nature* 282: 39–43. [P]

Takeda, D. Y. and A. Dutta. 2005. DNA replication and progression through S phase. *Oncogene* 24: 2827–2843. [R]

Waga, S. and B. Stillman. 1998. The DNA replication fork in eukaryotic cells. *Ann. Rev. Biochem.* 67: 721–751. [R]

DNA Repair

Cleaver, J. E. 1968. Defective repair replication of DNA in xeroderma pigmentosum. *Nature* 218: 652–656. [P]

Cox, M. M. 2001. Recombinational DNA repair of damaged replication forks in *Escherichia coli*: Questions. *Ann. Rev. Genet.* 35: 53–82. [R]

Daley, J. M., P. L. Palmbos, D. Wu and T. E. Wilson. 2005. Nonhomologous end joining in yeast. *Ann. Rev. Genet.* 39: 431–451. [R]

Fishel, R., M. K. Lescoe, M. R. S. Rao, N. G. Copeland, N. A. Jenkins, J. Garber, M. Kane and R. Kolodner. 1993. The human mutator gene homolog *MSH2* and its association with hereditary nonpolyposis colon cancer. *Cell* 75: 1027–1038. [P]

Friedberg, E. C., A. R. Lehmann and R. P. P. Fuchs. 2005. Trading places: How do DNA polymerases switch during translesion DNA synthesis? *Mol. Cell* 18: 499–505. [R]

Friedberg, E. C., R. Wagner and M. Radman. 2002. Specialized DNA polymerases, cellular survival, and the genesis of mutations. *Science* 296: 1627–1630. [R]

Friedberg, E. C., G. C. Walker and W. Siede. 1995. *DNA Repair and Mutagenesis.* Washington, D.C.: ASM Press.

Goodman, M. F. 2002. Error-prone repair DNA polymerases in prokaryotes and eukaryotes. *Ann. Rev. Biochem.* 71: 17–50. [R]

Hoeijmakers, J. H. J. 2001. Genome maintenance mechanisms for preventing cancer. *Nature* 411: 366–374. [R]

Khanna, K. K. and S. P. Jackson. 2001. DNA double strand breaks: Signaling, repair and the cancer connection. *Nature Genetics* 27: 247–254. [R]

Kunkel, T. A. and D. A. Erie. 2005. DNA mismatch repair. *Ann. Rev. Biochem.* 74: 681–710. [R]

Leach, F. S. and 34 others. 1993. Mutations of a *mutS* homolog in hereditary nonpolyposis colorectal cancer. *Cell* 75: 1215–1225. [P]

Pellegrini, L. and A. Venkitaraman. 2004. Emerging functions of BRCA2 in DNA recombination. *Trends Biochem. Sci.* 29: 310–316. [R]

Sancar, A., L. A. Lindsey-Boltz, K. Unsal-Kacmaz and S. Linn. 2004. Molecular mechanisms of mammalian DNA repair and the DNA damage checkpoints. *Ann. Rev. Biochem.* 73: 39–85. [R]

Recombination between Homologous DNA Sequences

Cox, M. M. 2003. The bacterial RecA protein as a motor protein. *Ann. Rev. Microbiol.* 57: 551–577. [R]

DasGupta, C., A. M. Wu, R. Kahn, R. P. Cunningham and C. M. Radding. 1981. Concerted strand exchange and formation of Holliday structures by *E. coli* RecA protein. *Cell* 25: 507–516. [P]

Holliday, R. 1964. A mechanism for gene conversion in fungi. *Genet. Res.* 5: 282–304. [P]

Hollingsworth, N. M. and S. J. Brill. 2004. The Mus81 solution to resolution: Generating meiotic crossovers without Holliday junctions. *Genes Dev.* 18: 117–125. [R]

Kowalczykowski, S. C. and A. K. Eggleston. 1994. Homologous pairing and DNA strand-exchange proteins. *Ann. Rev. Biochem.* 63: 991–1043. [R]

Krogh, B. O. and L. S. Symington. 2004. Recombination proteins in yeast. *Ann. Rev. Genet.* 38: 233–271. [R]

Liu, Y. and S. C. West. 2004. Happy Hollidays: 40th anniversary of the Holliday junction. *Nature Rev. Mol. Cell Biol.* 5: 937–946. [R]

Potter, H. and D. Dressler. 1976. On the mechanism of genetic recombination: Electron microscopic observation of recombination intermediates. *Proc. Natl. Acad. Sci. USA* 73: 3000–3004. [P]

Stahl, F. 1996. Meiotic recombination in yeast: Coronation of the double-strand-break repair model. *Cell* 87: 965–968. [R]

Szostak, J. W., T. L. Orr-Weaver, R. J. Rothstein and F. W. Stahl. 1983. The double-strand-break repair model for recombination. *Cell* 33: 25–35. [P]

Taylor, A. F. 1992. Movement and resolution of Holliday junctions by enzymes from *E. coli*. *Cell* 69: 1063–1065. [R]

West, S. C. 2003. Molecular views of recombination proteins and their control. *Nature Rev. Mol. Cell Biol.* 4: 1–11. [R]

DNA Rearrangements

Boeke, J. D., D. J. Garfinkel, C. A. Styles and G. R. Fink. 1985. Ty elements transpose through an RNA intermediate. *Cell* 40: 491–500. [P]

Chaudhuri, J. and F. W. Alt. 2004. Class-switch recombination: Interplay of transcription, DNA deamination and DNA repair. *Nature Rev. Immunol.* 4: 541–552. [R]

Claycomb, J. M. and T. L. Orr-Weaver. 2005. Developmental gene amplification: Insights into DNA replication and gene expression. *Trends Genet.* 21: 149–162. [R]

Fedoroff, N. and D. Botstein. 1992. *The Dynamic Genome: Barbara McClintock's Ideas in the Century of Genetics.* Plainview, N.Y.: Cold Spring Harbor Laboratory Press.

Gilboa, E., S. W. Mitra, S. Goff and D. Baltimore. 1979. A detailed model of reverse transcription and tests of crucial aspects. *Cell* 18: 93–100. [P]

Haren, L., B. Ton-Hoang and M. Chandler. 1999. Integrating DNA: Transposases and retroviral integrases. *Ann. Rev. Microbiol.* 53: 245–281. [R]

Honjo, T., H. Nagaoka, R. Shinkura and M. Muramatsu. 2005. AID to overcome the limitations of genomic information. *Nature Immunol.* 6: 655–661. [R]

Hozumi, N. and S. Tonegawa. 1976. Evidence for somatic rearrangement of immunoglobulin genes coding for variable and constant regions. *Proc. Natl. Acad. Sci. USA* 73: 3628–3632. [P]

Jung, D. and F. W. Alt. 2004. Unraveling V(D)J recombination: Insights into gene regulation. *Cell* 116: 299–311. [R]

Kazazian, H. H. Jr. 2004. Mobile elements: Drivers of genome evolution. *Science* 303: 1626–1632. [R]

Li, Z., C. J. Woo, M. D. Iglesias-Ussel, D. Ronai and M. D. Scharff. 2004. The generation of antibody diversity through somatic hypermutation and class switch recombination. *Genes Dev.* 18: 1–11. [R]

Maizels, N. 2005. Immunoglobulin gene diversification. *Ann. Rev. Genet.* 39: 23–46. [R]

McClintock, B. 1956. Controlling elements and the gene. *Cold Spring Harbor Symp. Quant. Biol.* 21: 197–216. [P]

Tonegawa, S. 1983. Somatic generation of antibody diversity. *Nature* 302: 575–581. [R]

Tower, J. 2004. Developmental gene amplification and origin regulation. *Ann. Rev. Genet.* 38: 273–304. [R]

CHAPTER 7

RNA Synthesis and Processing

■ **Transcription in Prokaryotes 254**

■ **Eukaryotic RNA Polymerases and General Transcription Factors 262**

■ **Regulation of Transcription in Eukaryotes 268**

■ **RNA Processing and Turnover 287**

■ **KEY EXPERIMENT:** Isolation of a Eukaryotic Transcription Factor 276

■ **KEY EXPERIMENT:** The Discovery of snRNPs 294

CHAPTERS 5 AND 6 DISCUSSED THE ORGANIZATION and maintenance of genomic DNA, which can be viewed as the set of genetic instructions governing all cellular activities. These instructions are implemented via the synthesis of RNAs and proteins. Importantly, the behavior of a cell is determined not only by what genes it inherits but also by which of those genes are expressed at any given time. Regulation of gene expression allows cells to adapt to changes in their environments and is responsible for the distinct activities of the multiple differentiated cell types that make up complex plants and animals. Muscle cells and liver cells, for example, contain the same genes; the functions of these cells are determined not by differences in their genomes, but by regulated patterns of gene expression that govern development and differentiation.

The first step in expression of a gene, the transcription of DNA into RNA, is the initial level at which gene expression is regulated in both prokaryotic and eukaryotic cells. RNAs in eukaryotic cells are then modified in various ways—for example, introns are removed by splicing—to convert the primary transcript into its functional form. Different types of RNA play distinct roles in cells: Messenger RNAs (mRNAs) serve as templates for protein synthesis; ribosomal RNAs (rRNAs) and transfer RNAs (tRNAs) function in mRNA translation. Still other small RNAs function in gene regulation, mRNA splicing, rRNA processing, and protein sorting in eukaryotes. In fact, some of the most exciting advances in recent years have pertained to the roles of noncoding RNAs (microRNAs) as regulators of both transcription and translation in eukaryotic cells. Transcription and RNA processing are discussed in this chapter. The final step in gene expression, the translation of mRNA to protein, is the subject of Chapter 8.

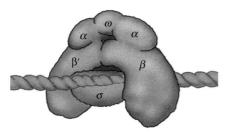

FIGURE 7.1 *E. coli* **RNA polymerase**
The complete enzyme consists of six
subunits: two α, one β, one β′, one ω
and one σ. The σ subunit is relatively
weakly bound and can be dissociated
from the other five subunits, which
constitute the core polymerase.

Transcription in Prokaryotes

As in most areas of molecular biology, studies of *E. coli* have provided the
model for subsequent investigations of transcription in eukaryotic cells. As
reviewed in Chapter 4, mRNA was discovered first in *E. coli*. *E. coli* was also
the first organism from which RNA polymerase was purified and studied.
The basic mechanisms by which transcription is regulated were likewise elucidated by pioneering experiments in *E. coli* in which regulated gene expression allows the cell to respond to variations in the environment, such as
changes in the availability of nutrients. An understanding of transcription in
E. coli has thus provided the foundation for studies of the far more complex
mechanisms that regulate gene expression in eukaryotic cells.

RNA Polymerase and Transcription

The principal enzyme responsible for RNA synthesis is **RNA polymerase**,
which catalyzes the polymerization of ribonucleoside 5′-triphosphates
(NTPs) as directed by a DNA template. The synthesis of RNA is similar to
that of DNA, and like DNA polymerase, RNA polymerase catalyzes the
growth of RNA chains always in the 5′ to 3′ direction. Unlike DNA polymerase, however, RNA polymerase does not require a preformed primer to
initiate the synthesis of RNA. Instead, transcription initiates *de novo* at specific sites at the beginning of genes. The initiation process is particularly
important because this is a major step at which transcription is regulated.

RNA polymerase, like DNA polymerase, is a complex enzyme made up
of multiple polypeptide chains. The intact bacterial enzyme consists of five
different types of subunits, called α, β, β′, ω and σ (Figure 7.1). The σ subunit
is relatively weakly bound and can be separated from the other subunits,
yielding a core polymerase consisting of two α, one β, one β′ and one ω subunits. The core polymerase is fully capable of catalyzing the polymerization
of NTPs into RNA, indicating that σ is not required for the basic catalytic
activity of the enzyme. However, the core polymerase does not bind specifically to the DNA sequences that signal the normal initiation of transcription; therefore the α subunit is required to identify the correct sites for transcription initiation. The selection of these sites is a critical element of
transcription because synthesis of a functional RNA must start at the beginning of a gene.

The DNA sequence to which RNA polymerase binds to initiate transcription of a gene is called the **promoter**. The DNA sequences involved in promoter function were first identified by comparisons of the nucleotide
sequences of a series of different genes isolated from *E. coli*. These comparisons revealed that the region upstream of the transcription initiation site
contains two sets of sequences that are similar in a variety of genes. These
common sequences encompass six nucleotides each and are located approximately 10 and 35 base pairs upstream of the transcription start site (**Figure
7.2**). They are called the –10 and –35 elements, denoting their position relative to the transcription initiation site, which is defined as the +1 position.
The sequences at the –10 and –35 positions in different promoters are not

**FIGURE 7.2 Sequences of *E. coli*
promoters** *E. coli* promoters are
characterized by two sets of sequences
located 10 and 35 base pairs upstream
of the transcription start site (+1). The
consensus sequences shown correspond to the bases most frequently
found in different promoters.

identical, but they are all similar enough to establish consensus sequences—the bases most frequently found at each position.

Several types of experimental evidence support the functional importance of the −10 and −35 promoter elements. First, genes with promoters that differ from the consensus sequences are transcribed less efficiently than genes whose promoters match the consensus sequences more closely. Second, mutations introduced in either the −35 or −10 consensus sequences have strong effects on promoter function. Third, the sites at which RNA polymerase binds to promoters have been directly identified by **footprinting** experiments, which are widely used to determine the sites at which proteins bind to DNA (**Figure 7.3**). In experiments of this type, a DNA frag-

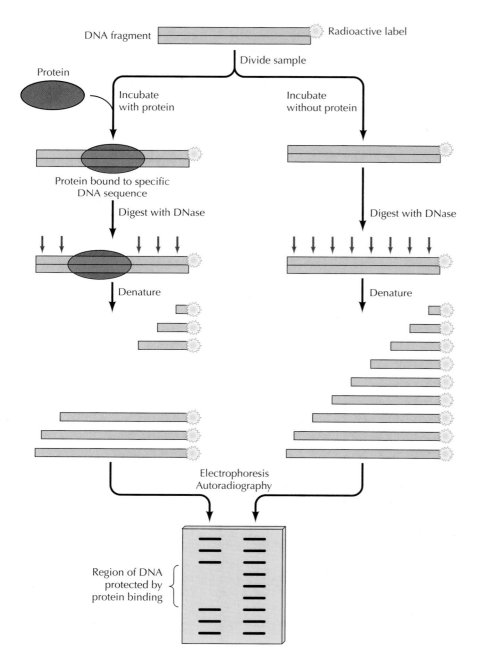

FIGURE 7.3 DNA footprinting A sample containing fragments of DNA radiolabeled at one end is divided in two, and one half of the sample is incubated with a protein that binds to a specific DNA sequence within the fragment. Both samples are then digested with DNase, under conditions such that the DNase introduces an average of one cut per molecule. The region of DNA bound to the protein is protected from DNase digestion. The DNA-protein complexes are then denatured, and the sizes of the radiolabeled DNA fragments produced by DNase digestion are analyzed by electrophoresis. Fragments of DNA resulting from DNase cleavage within the region protected by protein binding are missing from the sample of DNA that was incubated with protein.

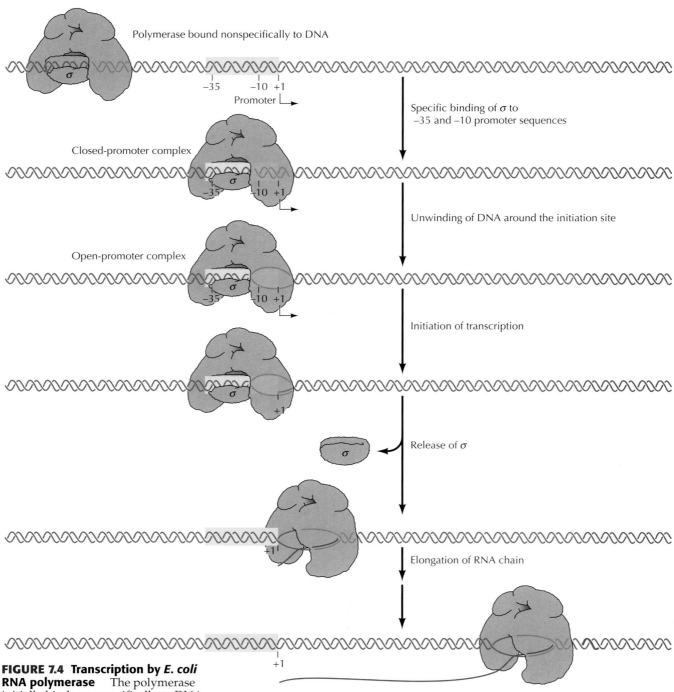

FIGURE 7.4 Transcription by *E. coli* RNA polymerase The polymerase initially binds nonspecifically to DNA and migrates along the molecule until the σ subunit binds to the −35 and −10 promoter elements, forming a closed-promoter complex. The polymerase then unwinds DNA around the initiation site, and transcription is initiated by the polymerization of free NTPs. The σ subunit then dissociates from the core polymerase, which migrates along the DNA and elongates the growing RNA chain.

ment is radiolabeled at one end. The labeled DNA is incubated with the protein of interest (e.g., RNA polymerase) and then subjected to partial digestion with DNase. The principle of the method is that the regions of DNA to which the protein binds are protected from DNase digestion. These regions can therefore be identified by comparison of the digestion products of the protein-bound DNA with those resulting from identical DNase treatment of a parallel sample of DNA that was not incubated with protein. Variations of this basic method, which employ chemical reagents to modify and cleave DNA at particular nucleotides, can be used to identify the specific DNA bases that are in contact with protein. Such footprinting analysis has shown that RNA polymerase generally binds to promoters over approximately a 60-base-pair region, extending from −40 to +20 (i.e., from 40 nucleotides upstream to 20 nucleotides downstream of the transcription start site). The σ subunit binds specifically to sequences in both the −35 and −10 promoter regions, substantiating the importance of these sequences in promoter function. In addition, some *E. coli* promoters have a third sequence, located upstream of the −35 region, that serves as a specific binding site for the RNA polymerase α subunit.

In the absence of σ, RNA polymerase binds nonspecifically to DNA with low affinity. The role of σ is to direct the polymerase to promoters by binding specifically to both the −35 and −10 sequences, leading to the initiation of transcription at the beginning of a gene (Figure 7.4). The initial binding between the polymerase and a promoter is referred to as a closed-promoter complex because the DNA is not unwound. The polymerase then unwinds 12–14 bases of DNA, from about −12 to +2, to form an open-promoter complex in which single-stranded DNA is available as a template for transcription. Transcription is initiated by the joining of two free NTPs. After addition of about the first 10 nucleotides, σ is released from the polymerase, which then leaves the promoter and moves along the template DNA to continue elongation of the growing RNA chain.

During elongation, the polymerase remains associated with its template while it continues synthesis of mRNAs. As it travels, the polymerase unwinds the template DNA ahead of it and rewinds the DNA behind it, maintaining an unwound region of about 15 base pairs in the region of transcription. Within this unwound portion of DNA, 8–9 bases of the growing RNA chain are bound to the complementary template DNA strand. High resolution structural analysis of bacterial RNA polymerase indicates that the β and β′ subunits form a crab claw-like structure that grips the DNA template (Figure 7.5). An internal channel between the β and β′ subunits accommodates approximately 20 base pairs of DNA and contains the polymerase active site.

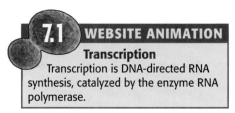

7.1 WEBSITE ANIMATION

Transcription
Transcription is DNA-directed RNA synthesis, catalyzed by the enzyme RNA polymerase.

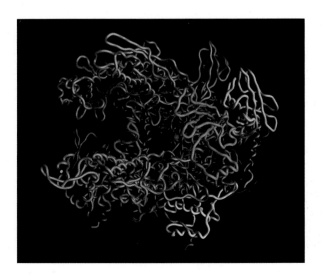

FIGURE 7.5 Structure of bacterial RNA polymerase The α subunits of the polymerase are colored dark green and light green, β blue, β′ pink, and ω yellow. (Courtesy of Seth Darst, Rockefeller University.)

FIGURE 7.6 Transcription termination The termination of transcription is signaled by a GC-rich inverted repeat followed by seven A residues. The inverted repeat forms a stable stem-loop structure in the RNA, causing the RNA to dissociate from the DNA template.

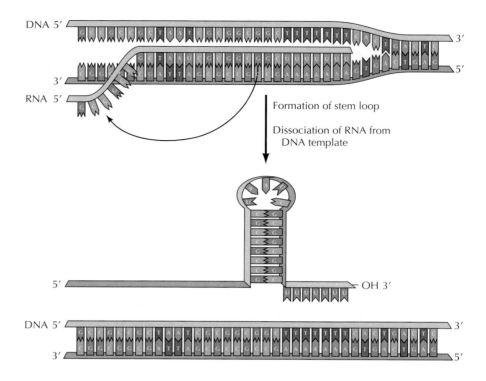

RNA synthesis continues until the polymerase encounters a termination signal, at which point transcription stops, the RNA is released from the polymerase, and the enzyme dissociates from its DNA template. There are two alternative mechanisms for termination of transcription in *E. coli*. The simplest and most common type of termination signal consists of a symmetrical inverted repeat of a GC-rich sequence followed by approximately seven A residues (Figure 7.6). Transcription of the GC-rich inverted repeat results in the formation of a segment of RNA that can form a stable stem-loop structure by complementary base pairing. The formation of such a self-complementary structure in the RNA disrupts its association with the DNA template and terminates transcription. Because hydrogen bonding between A and U is weaker than that between G and C, the presence of A residues downstream of the inverted repeat sequences is thought to facilitate the dissociation of the RNA from its template. Alternatively, the transcription of some genes is terminated by a specific termination protein (called Rho), which binds extended segments (greater than 60 nucleotides) of single-stranded RNA. Since mRNAs in bacteria become associated with ribosomes and are translated while they are being transcribed, such extended regions of single-stranded RNA are exposed only at the end of an mRNA.

Repressors and Negative Control of Transcription

Transcription can be regulated at the stages of both initiation and elongation, but most transcriptional regulation in bacteria operates at the level of initiation. The pioneering studies of gene regulation in *E. coli* were carried out by François Jacob and Jacques Monod in the 1950s. These investigators and their colleagues analyzed the expression of enzymes involved in the metabolism of lactose, which can be used as a source of carbon and energy via cleavage to glucose and galactose (Figure 7.7). The enzyme that catalyzes the cleavage of lactose (β-galactosidase) and other enzymes involved in lactose metabolism are expressed only when lactose is available for use by the bacteria. Otherwise, the cell is able to economize by not investing energy in the synthesis of unnecessary RNAs and proteins. Thus lactose induces the synthesis of enzymes involved in its own metabolism. In addition to requiring β-galactosidase, lactose metabolism involves the products of two other closely linked genes: lactose permease, which transports lactose into the cell, and a transacetylase, which is thought to inactivate toxic thiogalactosides that are transported into the cell along with lactose by the permease. On the basis of purely genetic experiments, Jacob and Monod deduced the mechanism by which the expression of these genes was regulated, thereby formulating a model that remains fundamental to our understanding of transcriptional regulation.

The starting point in this analysis was the isolation of mutants that were defective in regulation of the genes involved in lactose utilization. These mutants were of two types: constitutive mutants, which expressed all three genes even when lactose was not available, and noninducible mutants, which failed to express the genes even in the presence of lactose. Genetic mapping localized these regulatory mutants to two distinct loci, called o and i, with o located immediately upstream of the structural gene for β-galactosidase. Mutations affecting o resulted in constitutive expression; mutants of i were either constitutive or noninducible.

The function of these regulatory genes was probed by experiments in which two strains of bacteria were mated, resulting in diploid cells containing genes derived from both parents (Figure 7.8). Analysis of gene expression in such diploid bacteria provided critical insights by defining which alleles of these regulatory genes are dominant and which are recessive. For example, when bacteria containing a normal i gene (i^+) were mated with bacteria carrying an i gene mutation resulting in constitutive expression (an i^- mutation), the resulting diploid bacteria displayed normal inducibility; therefore

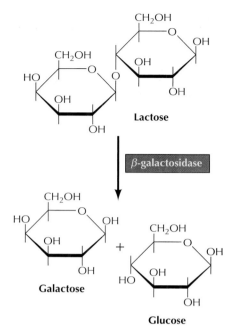

FIGURE 7.7 Metabolism of lactose β-galactosidase catalyzes the hydrolysis of lactose to glucose and galactose.

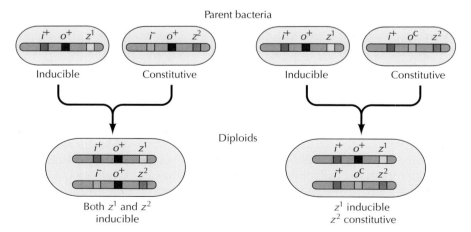

FIGURE 7.8 Regulation of β-galactosidase in diploid _E. coli_ The mating of two bacterial strains results in diploid cells that contain genes from both parents. In these examples, it is assumed that the genes encoding β-galactosidase (the z genes) can be distinguished on the basis of structural gene mutations, designated z^1 and z^2. In an i^+/i^- diploid (left), both structural genes are inducible; therefore i^+ is dominant over i^- and affects expression of z genes on both chromosomes. In contrast, in an o^c/o^+ diploid (right), the z gene linked to o^c is constitutively expressed, whereas that linked to o^+ is inducible. Therefore o affects expression of only the adjacent z gene on the same chromosome.

FIGURE 7.9 Negative control of the *lac* operon The *i* gene encodes a repressor which, in the absence of lactose (top), binds to the operator (*o*) and interferes with the binding of RNA polymerase to the promoter, blocking transcription of the three structural genes (*z*, β-galactosidase; *y*, permease; and *a*, transacetylase). Lactose induces expression of the operon by binding to the repressor (bottom), which prevents the repressor from binding to the operator. P = promoter; Pol = polymerase.

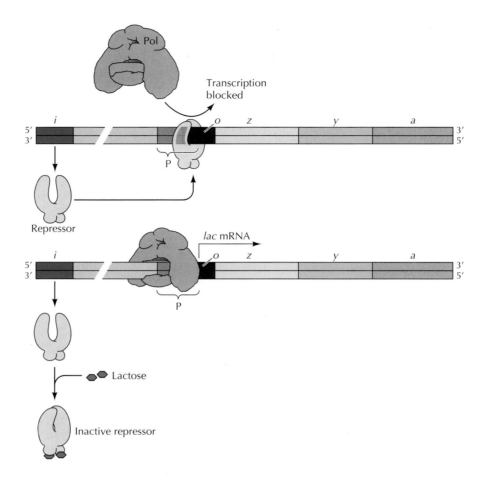

the normal i^+ gene was dominant over the i^- mutant. In contrast, matings between normal bacteria and bacteria with an o^c mutation (constitutive expression) yielded diploids with the constitutive expression phenotype, indicating that o^c is dominant over o^+. Additional experiments in which mutations in *o* and *i* were combined with different mutations in the structural genes showed that *o* affects the expression of only the genes to which it is physically linked, whereas *i* affects the expression of genes on both chromosome copies in diploid bacteria. Thus, in an o^c/o^+ cell, only the structural genes that are linked to o^c are constitutively expressed. In contrast, in an i^+/i^+ cell, structural genes on both chromosomes are regulated normally. These results led to the conclusion that *o* represents a region of DNA that controls the transcription of adjacent genes, whereas the *i* gene encodes a regulatory factor (e.g., a protein) that can diffuse throughout the cell and control genes on both chromosomes.

The model of gene regulation developed on the basis of these experiments is illustrated in Figure 7.9. The genes encoding β-galactosidase, permease, and transacetylase are expressed as a single unit, called an **operon**. Transcription of the operon is controlled by *o* (the **operator**), which is adjacent to the transcription initiation site. The *i* gene encodes a protein that regulates transcription by binding to the operator. Since i^- mutants (which result in constitutive gene expression) are recessive, it was concluded that these mutants failed to make a functional gene product. This result implies that the normal *i* gene product is a **repressor**, which blocks transcription when bound to *o*. The addition of lactose leads to induction of the operon because lactose binds to the repressor, thereby preventing it from binding to the operator DNA. In noninducible *i* mutants (which are dominant over i^+), the repressor fails to bind lactose, so expression of the operon cannot be induced.

The model neatly fits the results of the genetic experiments from which it was derived. In i^- cells, the repressor is not made, so the *lac* operon is constitutively expressed. Diploid i^+/i^- cells are normally inducible, since the functional repressor is encoded by the i^+ allele. Finally, in o^c mutants, a functional operator has been lost and the repressor cannot be bound. Consequently, o^c mutants are dominant but affect the expression only of linked structural genes.

Confirmation of this basic model has since come from a variety of experiments, including Walter Gilbert's isolation, in the 1960s, of the *lac* repressor and analysis of its binding to operator DNA. Molecular analysis has defined the operator as approximately 20 base pairs of DNA, starting a few bases before the transcription initiation site. Footprinting analysis has identified this region as the site to which the repressor binds, blocking transcription by interfering with the binding of RNA polymerase to the promoter. As predicted, lactose binds to the repressor, which then no longer binds to operator DNA. Also as predicted, o^c mutations alter sequences within the operator, thereby preventing repressor binding and resulting in constitutive gene expression.

The central principle of gene regulation exemplified by the lactose operon is that control of transcription is mediated by the interaction of regulatory proteins with specific DNA sequences. This general mode of regulation is broadly applicable to both prokaryotic and eukaryotic cells. Regulatory sequences like the operator are called ***cis*-acting control elements**, because they affect the expression of only linked genes on the same DNA molecule. On the other hand, proteins like the repressor are called ***trans*-acting factors** because they can affect the expression of genes located on other chromosomes within the cell. The *lac* operon is an example of negative control because binding of the repressor blocks transcription. This, however, is not always the case; many *trans*-acting factors are activators rather than inhibitors of transcription.

Positive Control of Transcription

The best-studied example of positive control in *E. coli* is the effect of glucose on the expression of genes that encode enzymes involved in the breakdown (catabolism) of other sugars (including lactose) that provide alternative sources of carbon and energy. Glucose is preferentially utilized; so as long as glucose is available, enzymes involved in catabolism of alternative energy sources are not expressed. For example, if *E. coli* are grown in medium containing both glucose and lactose, the *lac* operon is not induced and only glucose is used by the bacteria. Thus glucose represses the *lac* operon even in the presence of the normal inducer (lactose).

Glucose repression (generally called catabolite repression) is now known to be mediated by a positive control system, which is coupled to levels of cyclic AMP (cAMP) (Figure 7.10). In bacteria, the enzyme adenylyl cyclase, which converts ATP to cAMP, is regulated such that levels of cAMP increase when glucose levels drop. cAMP then binds to a transcriptional regulatory protein called catabolite activator protein (CAP). The binding of cAMP stimulates the binding of CAP to its target DNA sequences, which in the *lac* operon are located approximately 60 bases upstream of the transcription start

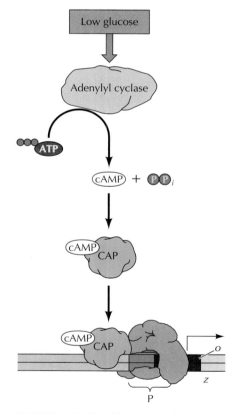

FIGURE 7.10 Positive control of the *lac* operon by glucose Low levels of glucose activate adenylyl cyclase, which converts ATP to cyclic AMP (cAMP). Cyclic AMP then binds to the catabolite activator protein (CAP) and stimulates its binding to regulatory sequences of various operons concerned with the metabolism of alternative sugars, such as lactose. CAP interacts with the α subunit of RNA polymerase to facilitate the binding of polymerase to the promoter.

TABLE 7.1 Classes of Genes Transcribed by Eukaryotic RNA Polymerases

Type of RNA synthesized	RNA polymerase
Nuclear genes	
mRNA	II
tRNA	III
rRNA	
5.8S, 18S, 28S	I
5S	III
snRNA and scRNA	II and III[a]
Mitochondrial genes	Mitochrondrial[b]
Chloroplast genes	Chloroplast[b]

[a]Some small nuclear (sn) and small cytoplasmic (sc) RNAs are transcribed by polymerase II and others by polymerase III.

[b]The mitochondrial and chloroplast RNA polymerases are similar to bacterial enzymes.

site. CAP then interacts with the α subunit of RNA polymerase, facilitating the binding of polymerase to the promoter and activating transcription.

Eukaryotic RNA Polymerases and General Transcription Factors

Although transcription proceeds by the same fundamental mechanisms in all cells, it is considerably more complex in eukaryotic cells than in bacteria. This is reflected in three distinct differences between the prokaryotic and eukaryotic systems. First, whereas all genes are transcribed by a single core RNA polymerase in bacteria, eukaryotic cells contain multiple different RNA polymerases that transcribe distinct classes of genes. Second, eukaryotic RNA polymerases need to interact with a variety of additional proteins to specifically initiate and regulate transcription. Finally, transcription in eukaryotes takes place on chromatin rather than on free DNA, and regulation of chromatin structure is an important factor in the transcriptional activity of eukaryotic genes. This increased complexity of eukaryotic transcription presumably facilitates the sophisticated regulation of gene expression needed to direct the activities of the many different cell types of multicellular organisms.

Eukaryotic RNA Polymerases

Eukaryotic cells contain three distinct nuclear RNA polymerases that transcribe different classes of genes (Table 7.1). Protein-coding genes are transcribed by RNA polymerase II to yield mRNAs; ribosomal RNAs (rRNAs) and transfer RNAs (tRNAs) are transcribed by RNA polymerases I and III. RNA polymerase I is specifically devoted to transcription of the three largest species of rRNAs, which are designated 28S, 18S, and 5.8S according to their rates of sedimentation during velocity centrifugation. RNA polymerase III transcribes the genes for tRNAs and for the smallest species of ribosomal RNA (5S rRNA). Some of the small RNAs involved in splicing and protein transport (snRNAs and scRNAs) are also transcribed by RNA polymerase III, while others are polymerase II transcripts. RNA polymerase II also transcribes microRNAs, which are important regulators of both transcription and translation in eukaryotic cells. In addition, separate RNA polymerases (which are similar to bacterial RNA polymerases) are found in chloroplasts and mitochondria, where they specifically transcribe the DNAs of those organelles.

All three of the nuclear RNA polymerases are complex enzymes, consisting of 12 to 17 different subunits each. Although they recognize different promoters and transcribe distinct classes of genes, they share several features in common with each other as well as with bacterial RNA polymerase. In particular, all three eukaryotic RNA polymerases contain nine conserved subunits, five of which are related to the α, β, β', and ω subunits of bacterial RNA polymerase. The structure of yeast RNA polymerase II is strikingly similar to that of the bacterial enzyme (Figure 7.11), suggesting that all RNA polymerases utilize fundamentally conserved mechanisms to transcribe DNA.

General Transcription Factors and Initiation of Transcription by RNA Polymerase II

Because RNA polymerase II is responsible for the synthesis of mRNA from protein-coding genes, it has been the focus of most studies of transcription

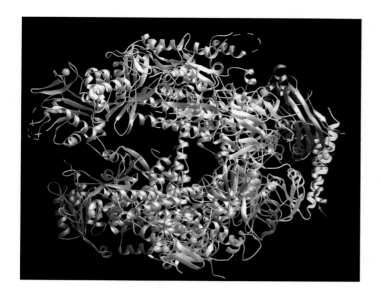

FIGURE 7.11 Structure of yeast RNA polymerase II
Individual subunits are distinguished by colors. (From P. D. Kramer et al., 2001. *Science* 292: 1863.)

in eukaryotes. Early attempts at studying this enzyme indicated that its activity is different from that of prokaryotic RNA polymerase. The accurate transcription of bacterial genes that can be accomplished *in vitro* simply by the addition of purified RNA polymerase to DNA containing a promoter is not possible in eukaryotic systems. The basis of this difference was elucidated in 1979, when Robert Roeder and his colleagues discovered that RNA polymerase II is able to initiate transcription only if additional proteins are added to the reaction. Thus transcription in the eukaryotic system appeared to require distinct initiation factors that (in contrast to bacterial σ factors) were not associated with the polymerase.

Biochemical fractionation of nuclear extracts subsequently led to the identification of specific proteins (called **transcription factors**) that are required for RNA polymerase II to initiate transcription. **General transcription factors** are involved in transcription from all polymerase II promoters and therefore constitute part of the basic transcription machinery. Additional gene-specific transcription factors (discussed later in the chapter) bind to DNA sequences that control the expression of individual genes and are thus responsible for regulating gene expression. It is estimated that about 10% of the genes in the human genome encode transcription factors, emphasizing the importance of these proteins.

Five general transcription factors are minimally required for initiation of transcription by RNA polymerase II in reconstituted *in vitro* systems (Figure 7.12). The promoters of many genes transcribed by polymerase II contain a sequence similar to TATAA located 25 to 30 nucleotides upstream of the transcription start site. This sequence (called the **TATA box**) resembles the −10 sequence element of bacterial promoters, and the results of introducing mutations into TATAA sequences have demonstrated their role in the initiation of transcription. The promoters of many genes transcribed by RNA polymerase II also contain a second important sequence element (an initiator, or Inr, sequence) that spans the transcription start site. Although many RNA polymerase II promoters contain both of these elements, some contain only a TATA box and others contain only an Inr element, with no TATA box. Many promoters that lack a TATA box but contain an Inr element also contain an additional downstream promoter element (DPE), located approxi-

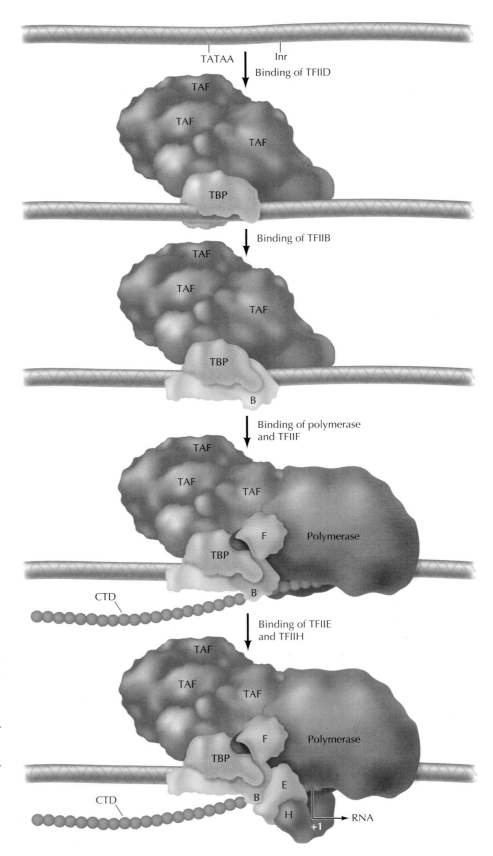

FIGURE 7.12 Formation of a polymerase II transcription initiation complex   Many polymerase II promoters have a TATA box (consensus sequence TATAA) 25 to 30 nucleotides upstream of the transcription start site and an initiator (Inr) element that spans the transcription start site. Formation of a transcription complex is initiated by the binding of transcription factor TFIID. One subunit of TFIID, the TATA-binding protein or TBP, binds to the TATA box; other subunits (TBP-associated factors or TAFs) may bind to the Inr. TFIIB(B) then binds to TBP, followed by binding of the polymerase in association with TFIIF(F). Finally, TFIIE(E) and TFIIH(H) associate with the complex.

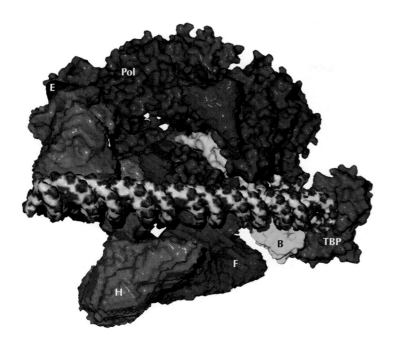

FIGURE 7.13 Model of the polymerase II transcription initiation complex A molecular model of RNA polymerase II (grey), TBP (green), TFIIB (yellow), TFIIE (purple), TFIIF (blue), and TFIIH (beige) assembled on a promoter. (From D. A. Bushnell et al., 2004. *Science* 303: 983.)

mately 30 base pairs downstream of the transcription start site, which functions cooperatively with the Inr sequence.

The first step in formation of a transcription complex is the binding of a general transcription factor called TFIID to the promoter (*TF* indicates *t*ranscription *f*actor; *II* indicates polymerase *II*). TFIID is itself composed of multiple subunits, including the **TATA-binding protein** (**TBP**) and at least 14 other polypeptides, called **TBP-associated factors** (**TAFs**). TBP binds specifically to the TATA box while other subunits of TFIID (TAFs) appear to bind to the Inr and DPE sequences. The binding of TFIID is followed by recruitment of a second general transcription factor (TFIIB), which binds to TBP as well as to DNA sequences that are present upstream of the TATA box in some promoters. TFIIB in turn serves as a bridge to RNA polymerase II, which binds to the TBP-TFIIB complex in association with a third factor, TFIIF.

Following recruitment of RNA polymerase II to the promoter, the binding of two additional factors (TFIIE and TFIIH) completes formation of the initiation complex (Figure 7.13). TFIIH is a multisubunit factor that appears to play at least two important roles. First, two subunits of TFIIH are helicases, which unwind DNA around the initiation site. (These subunits of TFIIH are the XPB and XPD proteins which are also required for nucleotide excision repair, as discussed in Chapter 6.) Another subunit of TFIIH is a protein kinase that phosphorylates repeated sequences present in the C-terminal domain of the largest subunit of RNA polymerase II. The polymerase II C-terminal domain (or CTD) consists of tandem repeats (27 repeats in yeast and 52 in humans) of seven amino acids with the consensus sequence Tyr-Ser-Pro-Thr-Ser-Pro-Ser. Phosphorylation of these amino acids releases the polymerase from its association with the preinitiation complex and leads to the recruitment of other proteins that allow the polymerase to initiate transcription and begin synthesis of a growing mRNA chain.

Although the sequential recruitment of five general transcription factors and RNA polymerase II described here represents the minimal system required for transcription *in vitro*, additional factors are needed to stimulate

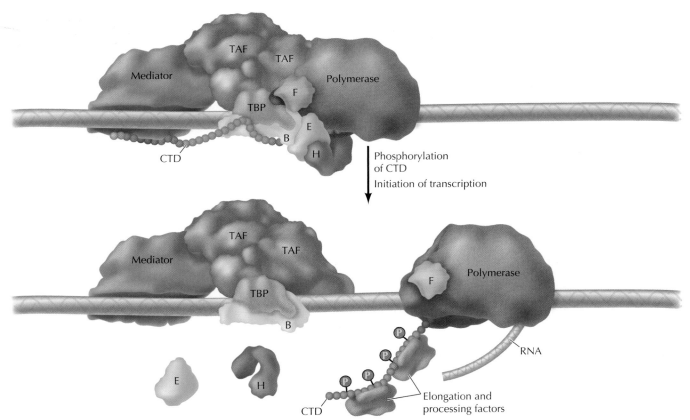

FIGURE 7.14 RNA polymerase II/ Mediator complexes RNA polymerase II is associated with Mediator proteins, as well as with the general transcription factors, at the promoter. The Mediator complex binds to the nonphosphorylated CTD of polymerase II and is released following phosphorylation of the CTD when transcription initiates. The phosphorylated CTD then binds elongation and processing factors that facilitate mRNA synthesis and processing.

transcription within the cell. These factors include a large protein complex, called **Mediator,** which consists of more than 20 distinct subunits. The Mediator complex not only stimulates basal transcription, but also plays a key role in linking the general transcription factors to the gene-specific transcription factors that regulate gene expression. The Mediator proteins are associated with the C-terminal domain of polymerase II, and are released from the polymerase following assembly of the preinitiation complex and phosphorylation of the polymerase C-terminal domain (Figure 7.14). The phosphorylated CTD then binds other proteins that facilitate transcriptional elongation and function in mRNA processing, as discussed later in this chapter.

Transcription by RNA Polymerases I and III

As previously discussed, distinct RNA polymerases are responsible for the transcription of genes encoding ribosomal RNAs, transfer RNAs, and some small noncoding RNAs in eukaryotic cells. All three RNA polymerases, however, require additional transcription factors to associate with appropriate promoter sequences. Furthermore, although the three different polymerases in eukaryotic cells recognize distinct types of promoters, a common transcription factor—the TATA-binding protein (TBP)—appears to be required for initiation of transcription by all three enzymes.

RNA polymerase I is devoted solely to the transcription of ribosomal RNA genes, which are present in tandem repeats. Transcription of these genes yields a large 45S pre-rRNA, which is then processed to yield the 28S, 18S, and 5.8S rRNAs (Figure 7.15). The promoters of ribosomal RNA genes

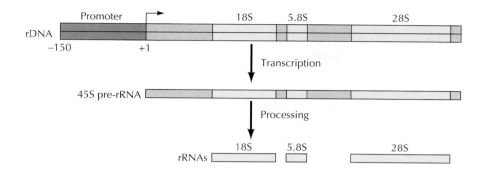

FIGURE 7.15 The ribosomal RNA gene Ribosomal DNA (rDNA) is transcribed to yield a large RNA molecule (45S pre-rRNA), which is then cleaved into 28S, 18S, and 5.8S rRNAs.

span about 150 base pairs just upstream of the transcription initiation site. These promoter sequences are recognized by two transcription factors, UBF (upstream binding factor) and SL1 (selectivity factor 1), which bind cooperatively to the promoter and then recruit polymerase I to form an initiation complex (Figure 7.16). The SL1 transcription factor is composed of four protein subunits, one of which is TBP. Since the promoters for ribosomal RNA genes do not contain a TATA box, TBP does not bind to specific promoter sequences. Instead, the association of TBP with ribosomal RNA genes is mediated by the binding of other proteins in the SL1 complex to the promoter, a situation similar to the association of TBP with the Inr sequences of polymerase II genes that lack TATA boxes.

The genes for tRNAs, 5S rRNA, and some of the small RNAs involved in splicing and protein transport are transcribed by polymerase III. These genes are transcribed from three distinct classes of promoters, two of which lie within, rather than upstream of, the transcribed sequence (Figure 7.17). The most thoroughly studied of the genes transcribed by polymerase III are the 5S rRNA genes of *Xenopus*. TFIIIA (which is the first transcription factor to have been purified) initiates assembly of a transcription complex by binding to specific DNA sequences in the 5S rRNA promoter. This binding is followed by the sequential binding of TFIIIC, TFIIIB, and the polymerase. The promoters for the tRNA genes differ from the 5S rRNA promoter in that they do not contain the DNA sequence recognized by TFIIIA. Instead, TFIIIC binds directly to the promoter of tRNA genes, serving to recruit TFIIIB and polymerase to form a transcription complex. Promoters of the third class of genes transcribed by polymerase III, including genes encoding some of the small nuclear RNAs involved in splicing, are located upstream of the transcription start site. These promoters contain a TATA box (like promoters for polymerase II genes) as well as a binding site for another factor, called SNAP. SNAP and TFIIIB bind cooperatively to these promoters, with

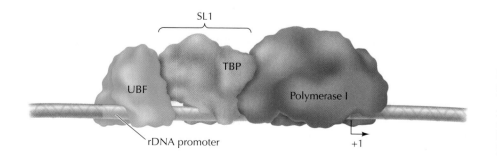

FIGURE 7.16 Initiation of rDNA transcription Two transcription factors, UBF and SL1, bind cooperatively to the rDNA promoters and recruit RNA polymerase I to form an initiation complex. One subunit of SL1 is the TATA-binding protein (TBP).

FIGURE 7.17 Transcription of RNA polymerase III genes Genes transcribed by polymerase III are expressed from three types of promoters. The promoters of 5S rRNA and tRNA genes are downstream of the transcription initiation site. Transcription of the 5S rRNA gene is initiated by the binding of TFIIIA, followed by the binding of TFIIIC, TFIIIB, and the polymerase. The tRNA promoters do not contain a binding site for TFIIIA, and TFIIIA is not required for their transcription. Instead, TFIIIC initiates the transcription of tRNA genes by binding to promoter sequences, followed by the association of TFIIIB and polymerase. The promoter of the U6 snRNA gene is upstream of the transcription start site and contains a TATA box, which is recognized by the TATA-binding protein (TBP) subunit of TFIIIB, in cooperation with another factor called SNAP.

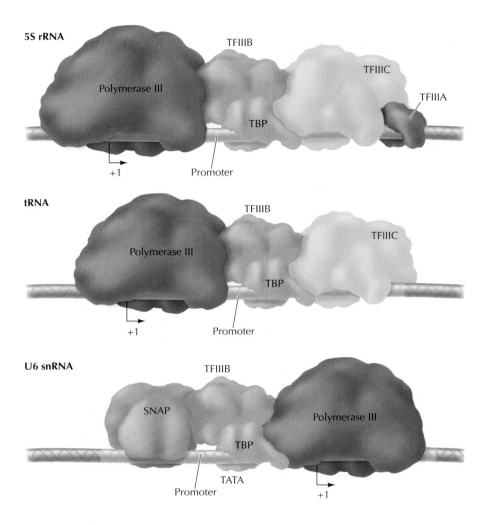

TFIIIB binding directly to the TATA box. This is mediated by the TATA-binding protein, TBP, which is one of the subunits of TFIIIB. As in the case of the promoters of other RNA polymerase III genes, TFIIIB then recruits the polymerase to the transcription complex.

Regulation of Transcription in Eukaryotes

The control of gene expression is far more complex in eukaryotes than in bacteria, although some of the same basic principles apply. The expression of eukaryotic genes is controlled primarily at the level of initiation of transcription, although in many cases transcription is also regulated during elongation. As in bacteria, transcription in eukaryotic cells is controlled by proteins that bind to specific regulatory sequences and modulate the activity of RNA polymerase. An important difference between transcriptional regulation in prokaryotes and eukaryotes, however, results from the packaging of eukaryotic DNA into chromatin, which limits its availability as a template for transcription. As a result, modifications of chromatin structure play key roles in the control of transcription in eukaryotic cells. A particularly exciting area of current research is based on the discovery that noncod-

FIGURE 7.18 Identification of eukaryotic regulatory sequences   The regulatory sequence of a cloned eukaryotic gene is ligated to a reporter gene that encodes an easily detectable enzyme. The resulting plasmid is then introduced into cultured recipient cells by transfection. An active regulatory sequence directs transcription of the reporter gene, expression of which is then detected in the transfected cells.

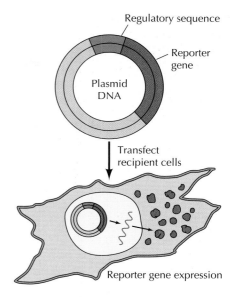

ing RNAs, as well as proteins, regulate transcription in eukaryotic cells via modifications in chromatin structure.

cis-Acting Regulatory Sequences: Promoters and Enhancers

As already discussed, transcription in bacteria is regulated by the binding of proteins to *cis*-acting sequences (e.g., the *lac* operator) that control the transcription of adjacent genes. Similar *cis*-acting sequences regulate the expression of eukaryotic genes. These sequences have been identified in mammalian cells largely by the use of gene transfer assays to study the activity of suspected regulatory regions of cloned genes (Figure 7.18). The eukaryotic regulatory sequences are usually ligated to a reporter gene that encodes an easily detectable enzyme. The expression of the reporter gene following its transfer into cultured cells then provides a sensitive assay for the ability of the cloned regulatory sequences to direct transcription. Biologically active regulatory regions can thus be identified, and *in vitro* mutagenesis can be used to determine the roles of specific sequences within the region.

Genes transcribed by RNA polymerase II have core promoter elements, including the TATA box and the Inr sequence, that serve as specific binding sites for general transcription factors. Other *cis*-acting sequences serve as binding sites for a wide variety of regulatory factors that control the expression of individual genes. These *cis*-acting regulatory sequences are frequently, though not always, located upstream of the TATA box. For example, two regulatory sequences that are found in many eukaryotic genes were identified by studies of the promoter of the herpes simplex virus gene that encodes thymidine kinase (Figure 7.19). Both of these sequences are located within 100 base pairs upstream of the TATA box: Their consensus sequences are CCAAT and GGGCGG (called a GC box). Specific proteins that bind to these sequences and stimulate transcription have since been identified.

In contrast to the relatively simple organization of CCAAT and GC boxes in the herpes thymidine kinase promoter, many genes in mammalian cells are controlled by regulatory sequences located farther away (sometimes more than 10 kilobases) from the transcription start site. These sequences,

FIGURE 7.19 A eukaryotic promoter   The promoter of the thymidine kinase gene of herpes simplex virus contains three sequence elements upstream of the TATA box that are required for efficient transcription: a CCAAT box and two GC boxes (consensus sequence GGGCGG).

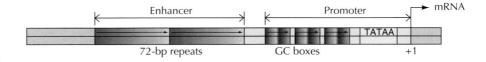

FIGURE 7.20 The SV40 enhancer
The SV40 promoter for early gene expression contains a TATA box and six GC boxes arranged in three sets of repeated sequences. In addition, efficient transcription requires an upstream enhancer consisting of two 72-base-pair (bp) repeats.

called **enhancers**, were first identified during studies of the promoter of another virus, SV40 (Figure 7.20). In addition to a TATA box and a set of six GC boxes, two 72-base-pair repeats located farther upstream are required for efficient transcription from this promoter. These sequences were found to stimulate transcription from other promoters as well as from that of SV40, and, surprisingly, their activity depended on neither their distance nor their orientation with respect to the transcription initiation site (Figure 7.21). They could stimulate transcription when placed either upstream or downstream of the promoter, in either a forward or backward orientation.

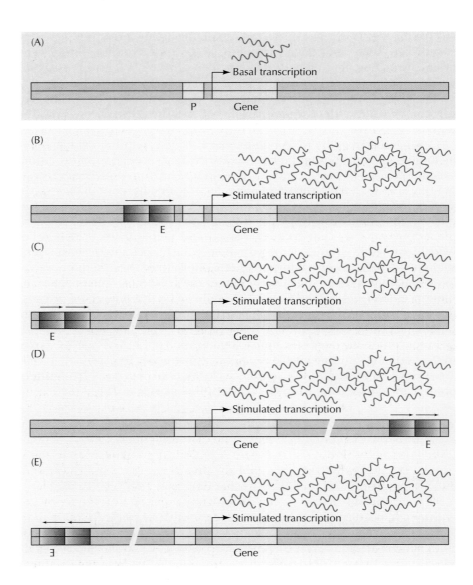

FIGURE 7.21 Action of enhancers
Without an enhancer, the gene is transcribed at a low basal level (A). Addition of an enhancer, E—for example, the SV40 72-base-pair repeats—stimulates transcription. The enhancer is active not only when placed just upstream of the promoter (B), but also when inserted up to several kilobases either upstream or downstream from the transcription start site (C and D). In addition, enhancers are active in either the forward or backward orientation (E).

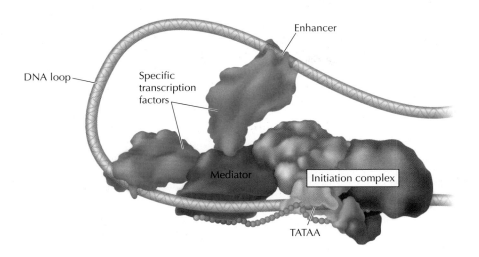

FIGURE 7.22 DNA looping Transcription factors bound at distant enhancers are able to interact with the RNA polymerase II/Mediator complex at the promoter because the intervening DNA can form loops. There is therefore no fundamental difference between the action of transcription factors bound to DNA just upstream of the gene in the promoter and to distant enhancers.

The ability of enhancers to function even when separated by long distances from transcription initiation sites at first suggested that they work by mechanisms different from those of promoters. However, this has turned out not to be the case: Enhancers, like promoters, function by binding transcription factors that then regulate RNA polymerase. This is possible because of DNA looping, which allows a transcription factor bound to a distant enhancer to interact with proteins associated with the RNA polymerase/Mediator complex at the promoter (Figure 7.22). Transcription factors bound to distant enhancers can thus work by the same mechanisms as those bound adjacent to promoters, so there is no fundamental difference between the actions of enhancers and those of *cis*-acting regulatory sequences adjacent to transcription start sites. Interestingly, although enhancers were first identified in mammalian cells, they have subsequently been found in bacteria—an unusual instance in which studies of eukaryotes served as a model for the simpler prokaryotic systems.

The binding of specific transcriptional regulatory proteins to enhancers is responsible for the control of gene expression during development and differentiation, as well as during the response of cells to hormones and growth factors. An example of a thoroughly studied mammalian enhancer controls the transcription of immunoglobulin genes in B lymphocytes. Gene transfer experiments have established that the immunoglobulin enhancer is active in lymphocytes, but not in other types of cells. Thus this regulatory sequence is at least partly responsible for tissue-specific expression of the immunoglobulin genes in the appropriate differentiated cell type.

An important aspect of enhancers is that they usually contain multiple functional sequence elements that bind different transcriptional regulatory proteins. These proteins work together to regulate gene expression. The immunoglobulin heavy-chain enhancer, for example, spans approximately 200 base pairs and contains at least nine distinct sequence elements that serve as protein-binding sites (Figure 7.23). Mutation of any one of these sequences reduces but does not abolish enhancer activity, indicating that the functions of individual proteins that bind to the enhancer are at least partly redundant. Many of the individual sequence elements of the immunoglobulin enhancer by themselves stimulate transcription in non-lymphoid cells. The restricted activity of the intact enhancer in B lymphocytes therefore does not result from the tissue-specific function of each of its components. Instead, tissue-specific expression results from the combina-

■ Enhancers can function over very long distances, sometimes even from different chromosomes. This process, termed transvection, has been best studied in *Drosophila* and involves *trans*-acting enhancers from one gene regulating the expression of its homolog on a separate chromosome.

FIGURE 7.23 The immunoglobulin enhancer The immunoglobulin heavy-chain enhancer spans about 200 bases and contains nine functional sequence elements (E, μE1–5, π, μB, and OCT), which together stimulate transcription in B lymphocytes.

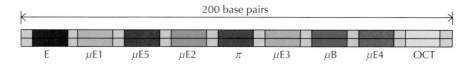

tion of the individual sequence elements that make up the complete enhancer. These elements include some *cis*-acting regulatory sequences that bind transcriptional activators that are expressed specifically in B lymphocytes, as well as other regulatory sequences that bind repressors in nonlymphoid cells. Thus the immunoglobulin enhancer contains negative regulatory elements that inhibit transcription in inappropriate cell types, as well as positive regulatory elements that activate transcription in B lymphocytes. The overall activity of the enhancer is greater than the sum of its parts, reflecting the combined action of the proteins associated with each of its individual sequence elements.

Although DNA looping allows enhancers to act at a considerable distance from promoters, the activity of any given enhancer is specific for the promoter of its appropriate target gene. This specificity is maintained in part by **insulators** or **barrier elements**, which divide chromosomes into independent domains and prevent enhancers from acting on promoters located in an adjacent domain. Insulators also prevent the chromatin structure of one domain from spreading to its neighbors, thereby maintaining independently regulated regions of the genome. It is thought that insulators function by organizing independent domains of chromatin within the nucleus, but their mechanism of action remains to be elucidated by further research.

Transcription Factor Binding Sites

The binding sites of transcriptional regulatory proteins in promoter or enhancer sequences have commonly been identified by two types of experiments. The first, footprinting, was described earlier in connection with the binding of RNA polymerase to prokaryotic promoters (see Figure 7.3). The second approach is the **electrophoretic-mobility shift assay** in which a radiolabeled DNA fragment is incubated with a protein preparation and then subjected to electrophoresis through a nondenaturing gel (Figure 7.24). Protein binding is detected as a decrease in the electrophoretic mobility of the DNA fragment, since its migration through the gel is slowed by the bound protein. The combined use of footprinting and electrophoretic-mobility shift assays has led to the correlation of protein-binding sites with the regulatory elements of enhancers and promoters, indicating that these sequences generally constitute the recognition sites of specific DNA-binding proteins.

The binding sites of most transcription factors consist of short DNA sequences, typically spanning 6–10 base pairs. In most cases, these binding sites are degenerate, meaning that the transcription factor will bind not only to the consensus sequence but also to sequences that differ from the consensus at one or more positions. It is thus common to represent transcription factor binding sites as pictograms, representing the frequency of each base at all positions of known binding sites for a given factor (Figure 7.25). Because of their short degenerate nature, sequences matching transcription factor binding sites occur frequently in genomic DNA, so physiologically significant regulatory sequences can not be identified from DNA sequence alone.

■ A major hurdle for gene therapy is that introduced genes are often aberrantly regulated or inactivated because of the nearby chromatin structure. The addition of insulator elements is a potential solution to this problem.

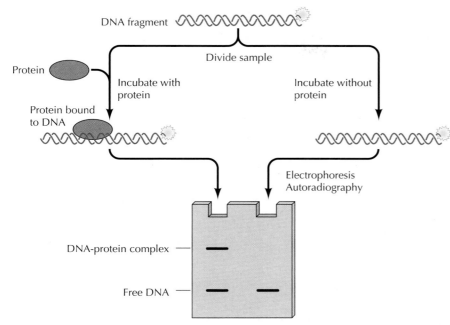

FIGURE 7.24 Electrophoretic-mobility shift assay A sample containing radiolabeled fragments of DNA is divided in two, and one half of the sample is incubated with a protein that binds to a specific DNA sequence. Samples are then analyzed by electrophoresis in a nondenaturing gel so that the protein remains bound to DNA. Protein binding is detected by the slower migration of DNA-protein complexes compared to that of free DNA. Only a fraction of the DNA in the sample is actually bound to protein, so both DNA-protein complexes and free DNA are detected following incubation of the DNA with protein.

As discussed in Chapter 5, identifying functional regulatory sequences in genomic DNA thus remains one of the prime challenges in bioinformatics and systems biology.

An important experimental approach for determining the regions of DNA that bind a transcription factor within the cell has been provided by **chromatin immunoprecipitation** (Figure 7.26). Cells are first treated with formaldehyde, which cross links proteins to DNA. As a result, transcription factors are covalently linked to the DNA sequences to which they were bound within the living cell. Chromatin is then extracted and sheared to fragments of about 500 base pairs. Fragments of DNA linked to a transcription factor of interest can then be isolated by immunoprecipitation with an antibody against the transcription factor (see Figure 4.30). The formaldehyde crosslinks are then reversed, and the immunoprecipitated DNA isolated and analyzed to determine the sites to which the specific transcription factor was bound within the cell. In yeast, chromatin immunoprecipitation has been combined with computational analysis of transcription factor binding sites to produce a global map of transcription factor binding sites throughout the genome.

Transcriptional Regulatory Proteins

A variety of transcriptional regulatory proteins have been isolated based on their binding to specific DNA sequences. One of the proto-

FIGURE 7.25 Representative transcription factor binding sites The binding sites of three mammalian transcription factors (AP1, Myc, and SRF) are shown as pictograms in which the frequency of each nucleotide is represented by the height of the corresponding letter at each position of the binding site.

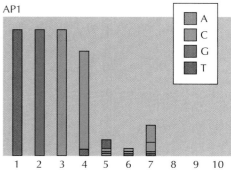

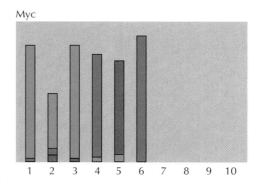

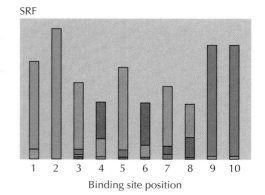

Binding site position

FIGURE 7.26 Chromatin immuno-precipitation Cells are treated with formaldehyde to crosslink DNA and proteins and then sonicated to produce fragments of chromatin. The chromatin fragments are incubated with an antibody against a specific transcription factor, and chromatin fragments bound to the antibody are collected as described in Figure 4.30. The crosslinks are then reversed, and DNA is purified to yield DNA fragments containing binding sites for the specific transcription factor of interest.

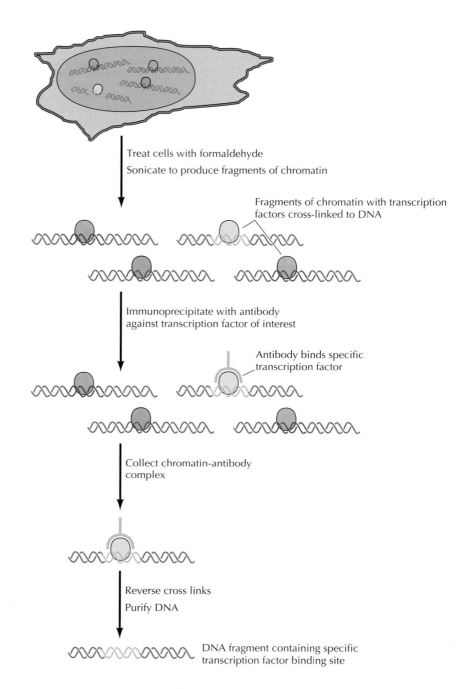

Treat cells with formaldehyde
Sonicate to produce fragments of chromatin

Fragments of chromatin with transcription factors cross-linked to DNA

Immunoprecipitate with antibody against transcription factor of interest

Antibody binds specific transcription factor

Collect chromatin-antibody complex

Reverse cross links
Purify DNA

DNA fragment containing specific transcription factor binding site

types of eukaryotic transcription factors was initially identified by Robert Tjian and his colleagues during studies of the transcription of SV40 DNA. This factor (called Sp1, for specificity protein 1) was found to stimulate transcription via binding to GC boxes in the SV40 promoter. Importantly, the specific binding of Sp1 to the GC box not only established the action of Sp1 as a sequence-specific transcription factor but also suggested a general approach to the purification of transcription factors. The isolation of these proteins initially presented a formidable challenge because they are present in very small quantities (e.g., only 0.001% of total cell protein) that are difficult to purify by conventional biochemical techniques. This problem was

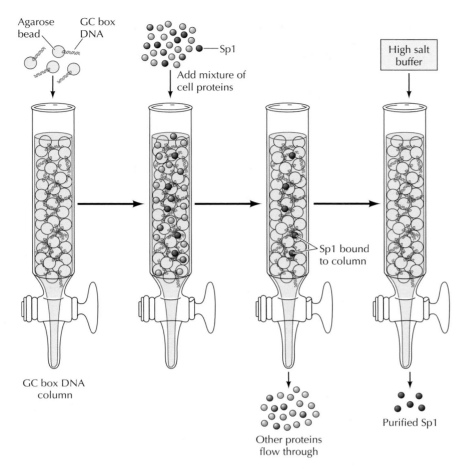

FIGURE 7.27 Purification of Sp1 by DNA-affinity chromatography   A double-stranded oligonucleotide containing repeated GC box sequences is bound to agarose beads, which are poured into a column. A mixture of cell proteins containing Sp1 is then applied to the column; because Sp1 specifically binds to the GC box oligonucleotide, it is retained on the column while other proteins flow through. Washing the column with high salt buffer then dissociates Sp1 from the GC box DNA, yielding purified Sp1.

overcome in the purification of Sp1 by **DNA-affinity chromatography** (Figure 7.27). Multiple copies of oligonucleotides corresponding to the GC box sequence were bound to a solid support, and cell extracts were passed through the oligonucleotide column. Because Sp1 bound to the GC box with high affinity, it was specifically retained on the column while other proteins were not. Highly purified Sp1 could thus be obtained and used for further studies.

The general method of DNA-affinity chromatography, first optimized for the purification of Sp1, has been used successfully to isolate a wide variety of sequence-specific DNA-binding proteins from eukaryotic cells. Genes encoding other transcription factors have been isolated by screening cDNA expression libraries to identify recombinant proteins that bind to specific DNA sequences. The cloning and sequencing of transcription factor cDNAs has led to the accumulation of a great deal of information on the structure and function of these critical regulatory proteins.

KEY EXPERIMENT

Isolation of a Eukaryotic Transcription Factor

Affinity Purification of Sequence-Specific DNA-Binding Proteins

James T. Kadonaga and Robert Tjian
University of California, Berkeley
Proceedings of the National Academy of Sciences, USA, 1986, Volume 83, pages 5889–5893

Robert Tjian

The Context

Starting with studies of the *lac* operon by Jacob and Monod, it became clear that transcription is regulated by proteins that bind to specific DNA sequences. One of the prototype systems for studies of gene expression in eukaryotic cells was the monkey virus SV40 in which several regulatory DNA sequences were identified in the early 1980s. In 1983 William Dynan and Robert Tjian first demonstrated that one of these sequence elements (the GC box) is the specific binding site of a protein detectable in nuclear extracts of human cells. This protein (called Sp1, for specificity protein 1) not only binds to the GC box sequence; it also stimulates transcription *in vitro*, demonstrating that it is a sequence-specific transcriptional activator.

To study the mechanism of Sp1 action, it then became necessary to obtain the transcription factor in pure form and eventually to clone the Sp1 gene. The isolation of pure Sp1 thus became a high priority, but it also posed a daunting technical challenge.

Sp1 and other transcription factors appeared to represent only about 0.001% of total cell protein, so they could not readily be purified by conventional biochemical techniques.

James Kadonaga and Robert Tjian solved this problem by developing a method of DNA-affinity chromatography that led to the purification not only of Sp1 but also of many other eukaryotic transcription factors, thereby opening the door to molecular analysis of transcriptional regulation in eukaryotic cells.

The Experiments

The DNA-affinity chromatography method developed by Kadonaga and Tjian exploited the specific high-affinity binding of Sp1 to the GC box sequence, GGGCGG. Synthetic oligonucleotides containing multiple copies of this sequence were coupled to solid beads, and a crude nuclear extract was passed through a column consisting of beads linked to GC box DNA. The beads were then washed to remove proteins that had failed to bind specifically to the oligonucleotides. Finally, the beads were washed with a high salt buffer (0.5 M KCl), which disrupted the binding of Sp1 to DNA, thereby releasing Sp1 from the column.

Gel electrophoresis demonstrated that the crude nuclear extract ini-

tially applied to the column was a complex mixture of proteins (see figure). In contrast, approximately 90% of the protein recovered after two cycles of DNA-affinity chromatography corresponded to only two polypeptides, which were identified as Sp1 by DNA binding and by their activity in *in vitro* transcription assays. Thus Sp1 had been successfully purified by DNA-affinity chromatography.

The Impact

In their 1986 paper, Kadonaga and Tjian stated that the DNA-affinity chromatography technique "should be generally applicable for the purification of other sequence-specific DNA binding proteins." This prediction has been amply verified; many eukaryotic transcription factors have been purified by this method. The genes that encode still other transcription factors have been isolated by an alternative approach (developed independently in 1988 in the laboratories of Phillip Sharp and Steven McKnight) in which cDNA expression libraries are screened with oligonucleotide probes to detect recombinant proteins that bind specifically to the desired DNA sequences. The ability to isolate sequence-specific DNA-binding proteins by these methods has led to detailed characterization of the structure and function of a wide variety of transcriptional regulatory proteins, providing the basis for our current understanding of gene expression in eukaryotic cells.

205.0

116.0

97.4

66.0

45.0

29.0

1 2 3

Purification of Sp1. Gel electrophoresis of proteins initially present in the crude nuclear extract (lane 1) and of proteins obtained after either one or two sequential cycles of DNA-affinity chromatography (lanes 2 and 3, respectively). The sizes of marker proteins (in kilodaltons) are indicated to the left of the gel, and the Sp1 polypeptides are indicated by arrows.

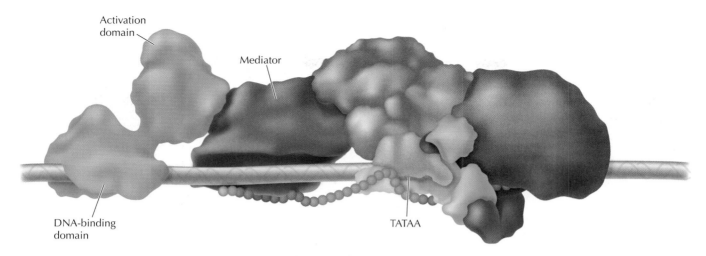

Activation domain

Mediator

DNA-binding domain

TATAA

FIGURE 7.28 Structure of transcriptional activators Transcriptional activators consist of two independent domains. The DNA-binding domain recognizes a specific DNA sequence, and the activation domain interacts with Mediator or other components of the transcriptional machinery.

Structure and Function of Transcriptional Activators

Because transcription factors are central to the regulation of gene expression, understanding the mechanisms of their action is a major area of ongoing research in cell and molecular biology. The most thoroughly studied of these proteins are **transcriptional activators**, which, like Sp1, bind to regulatory DNA sequences and stimulate transcription. In general, these factors consist of two independent domains: One region of the protein specifically binds DNA; the other region stimulates transcription by interacting with other proteins, including Mediator or other components of the transcriptional machinery (Figure 7.28). The basic function of the DNA-binding domain is to anchor the transcription factor to the proper site on DNA; the activation domain then independently stimulates transcription through protein-protein interactions.

Many different transcription factors have now been identified in eukaryotic cells, with about 2500 encoded in the human genome. They contain many distinct types of DNA-binding domains, some of which are illustrated in Figure 7.29. The most common is the **zinc finger domain**, which contains repeats of cysteine and histidine residues that bind zinc ions and fold into looped structures ("fingers") that bind DNA. These domains were initially identified in the polymerase III transcription factor TFIIIA but are also common among transcription factors that regulate polymerase II promoters, including Sp1. Other examples of transcription factors that contain zinc finger domains are the **steroid hormone receptors**, which regulate gene transcription in response to hormones such as estrogen and testosterone.

The **helix-turn-helix** motif was first recognized in prokaryotic DNA-binding proteins, including the *E. coli* catabolite activator protein (CAP). In these proteins, one helix makes most of the contacts with DNA, while the other helices lie across the complex to stabilize the interaction. In eukaryotic cells, helix-turn-helix proteins include the **homeodomain** proteins, which play critical roles in the regulation of gene expression during embryonic development. The genes encoding these proteins were first discovered as developmental mutants in *Drosophila*. Some of the earliest recognized *Drosophila* mutants (termed homeotic mutants in 1894) resulted in the development of flies in which one body part was transformed into another. For example, in the homeotic mutant called *Antennapedia*, legs rather than

■ Artificial transcription factors with zinc-finger domains designed to bind specific sequences within the genome have been developed. By ligating different effector domains to the DNA binding domain, the target gene can be either activated or repressed. This technology can be applied to gene therapy and the development of transgenic plants and animals.

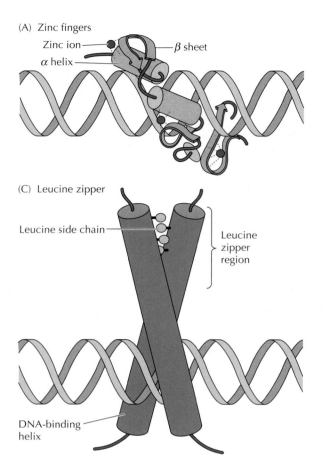

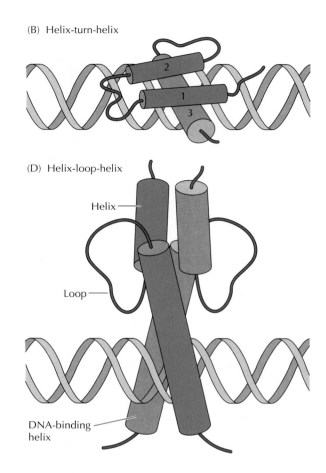

FIGURE 7.29 Examples of DNA-binding domains (A) Zinc finger domains consist of loops in which an α helix and a β sheet coordinately bind a zinc ion. (B) Helix-turn-helix domains consist of three (or in some cases four) helical regions. One helix (helix 3) makes most of the contacts with DNA, while helices 1 and 2 lie on top and stabilize the interaction. (C) The DNA-binding domains of leucine zipper proteins are formed from two distinct polypeptide chains. Interactions between the hydrophobic side chains of leucine residues exposed on one side of a helical region (the leucine zipper) are responsible for dimerization. Immediately following the leucine zipper is a DNA-binding helix, which is rich in basic amino acids. (D) Helix-loop-helix domains are similar to leucine zippers, except that the dimerization domains of these proteins each consist of two helical regions separated by a loop.

antennae grow out of the head of the fly (**Figure 7.30**). Genetic analysis of these mutants, pioneered by Ed Lewis in the 1940s, has shown that *Drosophila* contains nine homeotic genes, each of which specifies the identity of a different body segment. Molecular cloning and analysis of these genes then indicated that they contain conserved sequences of 180 base pairs (called **homeoboxes**) that encode DNA-binding domains (homeodomains) of transcription factors. A wide variety of similar homeodomain proteins have since been identified in fungi, plants, and other animals, including humans.

Two other families of DNA-binding proteins, **leucine zipper** and **helix-loop-helix** proteins, contain DNA-binding domains formed by dimerization of two polypeptide chains. The leucine zipper contains four or five leucine residues spaced at intervals of seven amino acids, resulting in their hydrophobic side chains being exposed at one side of a helical region. This region serves as the dimerization domain for the two protein subunits, which are held together by hydrophobic interactions between the leucine side chains. Immediately following the leucine zipper is a region rich in positively charged amino acids (lysine and arginine) that binds DNA. The helix-loop-helix proteins are similar in structure, except that their dimerization domains are each formed by two helical regions separated by a loop. An important feature of both leucine zipper and helix-loop-helix transcription factors is that different members of each family can dimerize with one

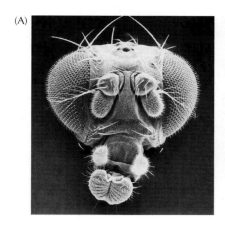 

FIGURE 7.30 The Antennapedia mutation *Antennapedia* mutant flies have legs growing out of their heads in place of antennae. (A) Head of a normal fly. (B) Head of an *Antennapedia* mutant. (Courtesy of F. Rudolf Turner, Indiana University.)

another. Thus the combination of distinct protein subunits can form an expanded array of factors that can differ both in DNA sequence recognition and in transcription-stimulating activities. Both leucine zipper and helix-loop-helix proteins play important roles in regulating tissue-specific and inducible gene expression, and the formation of dimers between different members of these families is a critical aspect of the control of their function.

The activation domains of transcription factors are not as well characterized as their DNA-binding domains. Some, called acidic activation domains, are rich in negatively charged residues (aspartate and glutamate); others are rich in proline or glutamine residues. The activation domains of eukaryotic transcription factors stimulate transcription by two distinct mechanisms (Figure 7.31). First, they interact with Mediator proteins and general transcription factors, such as TFIIB or TFIID, to recruit RNA polymerase and facilitate the assembly of a transcription complex on the promoter, similar to transcriptional activators in bacteria (see Figure 7.10). In addition, eukaryotic transcription factors interact with a variety of **coactivators** that stimulate transcription by modifying chromatin structure, as dis-

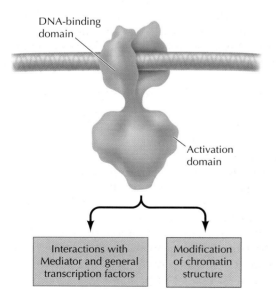

FIGURE 7.31 Action of transcriptional activators Eukaryotic activators stimulate transcription by two mechanisms: 1) they interact with Mediator proteins and general transcription factors to facilitate the assembly of a transcription complex and stimulate transcription, and 2) they interact with coactivators that facilitate transcription by modifying chromatin structure.

cussed later in this chapter. It is also important to note that activators not only regulate the initiation of transcription: Elongation and RNA processing can also be regulated, both by direct modulation of the activity of RNA polymerase and by effects on chromatin structure.

Eukaryotic Repressors

Gene expression in eukaryotic cells is regulated by repressors as well as by transcriptional activators. Like their prokaryotic counterparts, eukaryotic repressors bind to specific DNA sequences and inhibit transcription. In some cases, eukaryotic repressors simply interfere with the binding of other transcription factors to DNA (Figure 7.32A). For example, the binding of a repressor near the transcription start site can block the interaction of RNA polymerase or general transcription factors with the promoter, which is similar to the action of repressors in bacteria. Other repressors compete with activators for binding to specific regulatory sequences. Some such repressors contain the same DNA-binding domain as the activator but lack its activation domain. As a result, their binding to a promoter or enhancer blocks the binding of the activator, thereby inhibiting transcription.

In contrast to repressors that simply interfere with activator binding, many repressors (called active repressors) contain specific functional domains that inhibit transcription via protein-protein interactions (Figure 7.32B). The first such active repressor was described in 1990 during studies

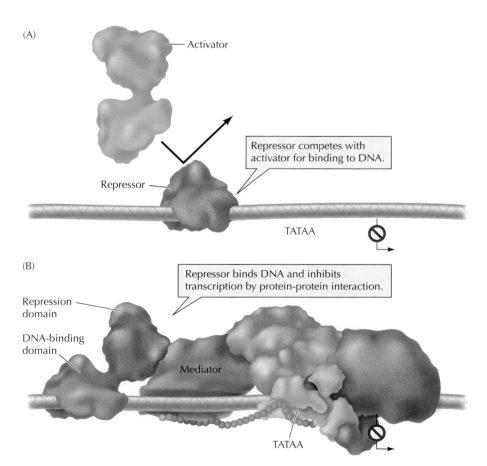

(A)

Activator

Repressor competes with activator for binding to DNA.

Repressor

TATAA

(B)

Repressor binds DNA and inhibits transcription by protein-protein interaction.

Repression domain

DNA-binding domain

Mediator

TATAA

FIGURE 7.32 Action of eukaryotic repressors   (A) Some repressors block the binding of activators to regulatory sequences. (B) Other repressors have active repression domains that inhibit transcription by interactions with Mediator proteins or general transcription factors, as well as with corepressors that act to modify chromatin structure.

of a gene called *Krüppel*, which is involved in embryonic development in *Drosophila*. Molecular analysis of the Krüppel protein demonstrated that it contains a discrete repression domain, linked to a zinc finger DNA-binding domain, which inhibits transcription via protein-protein interactions.

Many active repressors have since been found to play key roles in the regulation of transcription in animal cells, in many cases serving as critical regulators of cell growth and differentiation. As with transcriptional activators, several distinct types of repression domains have been identified. For example, the repression domain of Krüppel is rich in alanine residues, whereas other repression domains are rich in proline or acidic residues. The functional targets of repressors are also diverse: Repressors can inhibit transcription by interacting with specific activator proteins, with Mediator proteins or general transcription factors, and with **corepressors** that act by modifying chromatin structure.

The regulation of transcription by repressors as well as by activators considerably extends the range of mechanisms that control the expression of eukaryotic genes. One important role of repressors may be to inhibit the expression of tissue-specific genes in inappropriate cell types. For example, as noted earlier, a repressor-binding site in the immunoglobulin enhancer is thought to contribute to its tissue-specific expression by suppressing transcription in nonlymphoid cell types. Other repressors play key roles in the control of cell proliferation and differentiation in response to hormones and growth factors.

Relationship of Chromatin Structure to Transcription

As noted in the preceding discussion, both activators and repressors regulate transcription in eukaryotes not only by interacting with Mediator and other components of the transcriptional machinery but also by inducing changes in the structure of chromatin. Rather than being present within the nucleus as naked DNA, the DNA of all eukaryotic cells is tightly bound to histones. The basic structural unit of chromatin is the nucleosome, which consists of 147 base pairs of DNA wrapped around two molecules each of histones H2A, H2B, H3, and H4, with one molecule of histone H1 bound to the DNA as it enters the nucleosome core particle (see Figure 5.12). The chromatin is then further condensed by being coiled into higher-order structures organized into large loops of DNA. This packaging of eukaryotic DNA in chromatin clearly has important consequences in terms of its availability as a template for transcription, so chromatin structure is a critical aspect of gene expression in eukaryotic cells.

Actively transcribed genes are found in relatively decondensed chromatin, probably corresponding to the 30-nm chromatin fibers discussed in Chapter 5 (see Figure 5.13). For example, microscopic visualization of the polytene chromosomes of *Drosophila* indicates that regions of the genome that are actively engaged in RNA synthesis correspond to decondensed chromosome regions (Figure 7.33). Nonetheless, actively transcribed genes remain bound to histones and packaged in nucleosomes, so transcription factors and RNA polymerase are still faced with the problem of interacting with chromatin rather than with naked DNA. The tight winding of DNA around the nucleosome core particle is a major obstacle to transcription, affecting both the ability of transcription factors to bind DNA and the ability of RNA polymerase to transcribe through a chromatin template.

Several modifications are characteristic of transcriptionally active chromatin, including modifications of histones, rearrangements of nucleosomes, and the association of two nonhistone chromosomal proteins, called

FIGURE 7.33 Decondensed chromosome regions in *Drosophila*
A light micrograph showing decondensed regions of polytene chromosomes (arrows), which are active in RNA synthesis. (Courtesy of Joseph Gall, Carnegie Institute.)

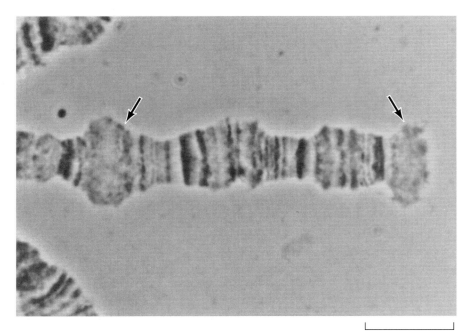

10 µm

HMGN proteins, with the nucleosomes of actively transcribed genes. The binding sites of the HMGN proteins on nucleosomes overlap the binding site of histone H1, and it appears that HMGN proteins stimulate transcription by affecting modifications of the core histones and altering the interaction of histone H1 with nucleosomes to maintain a decondensed chromatin structure.

Histone acetylation has been correlated with transcriptionally active chromatin in a wide variety of cell types (**Figure 7.34**). The core histones (H2A, H2B, H3 and H4) have two domains: a histone fold domain, which is involved in interactions with other histones and in wrapping DNA around the nucleosome core particle, and an amino-terminal tail domain, which extends outside of the nucleosome. The amino-terminal tail domains are rich in lysine and can be modified by acetylation at specific lysine residues.

Studies from two groups of researchers in 1996 provided direct links between histone acetylation and transcriptional regulation by demonstrating that transcriptional activators and repressors are associated with histone acetyltransferases and deacetylases, respectively. This association was first revealed by cloning a gene encoding a histone acetyltransferase from *Tetrahymena*. Unexpectedly, the sequence of this histone acetyltransferase was closely related to a previously known yeast transcriptional coactivator called Gcn5p. Further experiments revealed that Gcn5p has histone acetyltransferase activity, suggesting that transcriptional activation results directly from histone acetylation. These results have been extended by demonstrations that histone acetyltransferases are also associated with a number of mammalian transcriptional coactivators, as well as with the general transcription factor TFIID. Conversely, many transcriptional corepressors in both yeast and mammalian cells function as histone deacetylases, which remove the acetyl groups from histone tails. Histone acetylation is thus targeted directly by both transcriptional activators and repressors, indicating that it plays a key role in regulation of eukaryotic gene expression.

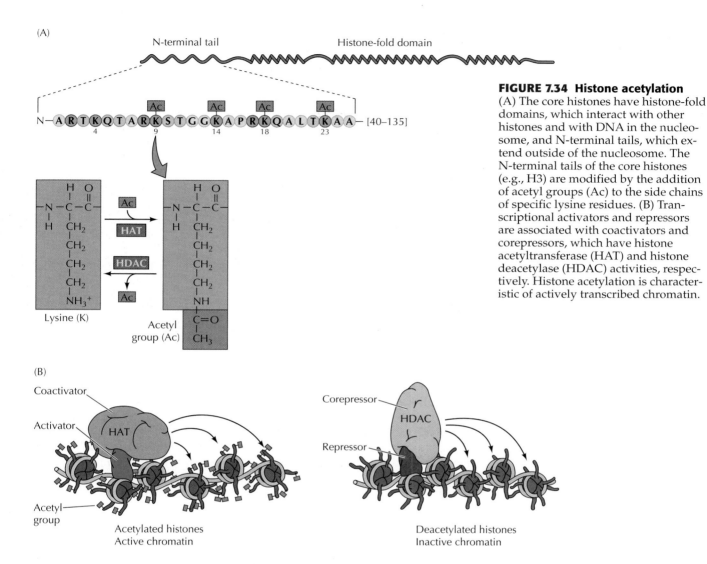

FIGURE 7.34 Histone acetylation
(A) The core histones have histone-fold domains, which interact with other histones and with DNA in the nucleosome, and N-terminal tails, which extend outside of the nucleosome. The N-terminal tails of the core histones (e.g., H3) are modified by the addition of acetyl groups (Ac) to the side chains of specific lysine residues. (B) Transcriptional activators and repressors are associated with coactivators and corepressors, which have histone acetyltransferase (HAT) and histone deacetylase (HDAC) activities, respectively. Histone acetylation is characteristic of actively transcribed chromatin.

Histones are modified not only by acetylation, but also by phosphorylation of serine residues, methylation of lysine and arginine residues, and addition of ubiquitin (a small peptide discussed in Chapter 8) to lysine residues. Like acetylation, these modifications occur at specific amino acid residues in the histone tails and are associated with changes in transcriptional activity (Figure 7.35). It appears that specific histone modifications affect gene expression by providing binding sites for other transcriptional regulatory proteins. According to this hypothesis, combinations of histone modifications constitute a "**histone code**" that regulates gene expression by recruiting other regulatory proteins to the chromatin template. For example, transcriptionally active chromatin is associated with several specific modifications of histone H3, including methylation of lysine-4, phosphorylation of serine-10, acetylation of lysines 9, 14, 18, and 23, and methylation of arginines 17 and 26. In contrast, methylation of lysines 9, 27, and 36 is associated with repression and chromatin condensation. Enzymes that catalyze the methylation of H3 lysine-9 are recruited to target genes by corepressors. The methylated H3 lysine-9 residues have further been shown to

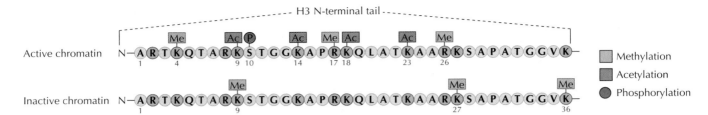

FIGURE 7.35 Patterns of histone modification Transcriptional activity of chromatin is affected by methylation and phosphorylation of specific amino acid residues in histone tails, as well as by their acetylation. Distinct patterns of histone modification are characteristic of transcriptionally active and inactive chromatin.

serve as binding sites for proteins that induce chromatin condensation, directly linking this histone modification to transcriptional repression and the formation of heterochromatin. It is notable that these modifications of histone tails may also regulate one another, leading to the establishment of distinct patterns of histone modification that correlate with stable modifications in transcriptional activity.

In contrast to the enzymes that regulate chromatin structure by modifying histones, **nucleosome remodeling factors** are protein complexes that alter the arrangement or structure of nucleosomes, without removing or covalently modifying the histones (Figure 7.36). One mechanism by which nucleosome remodeling factors act is to catalyze the sliding of histone octamers along the DNA molecule, thereby repositioning nucleosomes to change the accessibility of specific DNA sequences to transcription factors. Alternatively, nucleosome remodeling factors may act by inducing changes in the conformation of nucleosomes, again affecting the ability of specific DNA sequences to interact with transcriptional regulatory proteins. Like histone modifying enzymes, nucleosome remodeling factors can be recruited to DNA in association with either transcriptional activators or repressors, and can alter the arrangement of nucleosomes to either stimulate or inhibit transcription.

The recruitment of histone modifying enzymes and nucleosome remodeling factors by transcriptional activators stimulates the initiation of transcription by altering the chromatin structure of enhancer and promoter regions. However, following the initiation of transcription, RNA polymerase is still faced with the problem of transcriptional elongation through a chromatin template. This is facilitated by **elongation factors** that become associated with the phosphorylated C-terminal domain of RNA polymerase II when transcription is initiated (see Figure 7.14). These elongation factors include histone modifying enzymes (acetyltransferases and methyltransferases), as well as proteins that transiently disrupt the structure of nucleosomes during transcription. Transcriptional elongation is thus a complex process, which involves multiple proteins in addition to RNA polymerase. Although it has not been studied as thoroughly as transcription initiation, the regulation of transcriptional elongation presents an additional level at which gene expression can be controlled in eukaryotic cells.

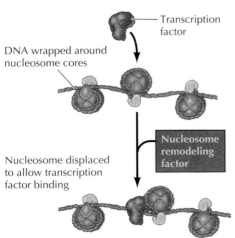

FIGURE 7.36 Nucleosome remodeling factors Nucleosome remodeling factors alter the arrangement or structure of nucleosomes. For example, a nucleosome remodeling factor can facilitate the binding of transcription factors to chromatin by repositioning nucleosomes on the DNA.

Regulation of Transcription by Noncoding RNAs

A series of recent advances indicate that gene expression can be regulated not only by the transcriptional regulatory proteins discussed so far but also by noncoding regulatory RNA molecules. One mode of action of noncoding regulatory RNAs is to inhibit translation by RNA interference—a phenomenon in which short double-stranded RNAs induce degradation of a homologous mRNA (see Figure 4.42). In addition, noncoding RNAs can repress transcription by inducing histone modifications that lead to chromatin condensation and the formation of heterochromatin. **MicroRNAs (miRNAs)** are naturally occurring short noncoding RNAs that function as normal regulators of gene expression. Hundreds of genes encode miRNAs in both plants and animals, so it appears that gene regulation by these noncoding RNAs is a widespread phenomenon, even though the functions of most miRNAs have yet to be determined.

miRNAs are transcribed as precursors containing inverted repeats that form stem-loop structures (Figure 7.37). These precursors are then cleaved to yield mature miRNAs, which are short double-stranded RNAs of approximately 20–25 nucleotides. In RNA interference, miRNAs associate with the RNA-induced silencing complex (RISC), within which the two strands of miRNA separate and target homologous mRNAs for cleavage (see Figure 4.42). In transcriptional repression, the miRNAs associate with a different protein complex called the RITS (RNA-induced transcriptional silencing) complex. The separated miRNA strands then guide the RITS complex to the homologous gene, most likely by base pairing with the mRNA transcript in association with RNA polymerase II. RITS then represses transcription by recruiting a histone methyltransferase that methylates histone H3 lysine-9, leading to the formation of heterochromatin.

The phenomenon of **X chromosome inactivation** provides another example of the role of a noncoding RNA in regulating gene expression in mammals. In many animals, including humans, females have two X chromosomes, and males have one X and one Y chromosome. The X chromosome contains hundreds of genes that are not present on the much smaller Y chromosome (see Figure 5.29). Thus females have twice as many copies of most X chromosome genes as males have. Despite this difference, female and male cells contain equal amounts of the proteins encoded by the majority of X chromosome genes. This results from a dosage compensation mechanism in which most of the genes on one of the two X chromosomes in female cells are inactivated by being converted to heterochromatin early in development. Consequently, only one copy of most genes located on the X chromosome is available for transcription in either female or male cells.

Although the mechanism of X chromosome inactivation is not yet fully understood, the key element appears to be a noncoding RNA transcribed from a regulatory gene, called *Xist*, on the inactive X chromosome. *Xist* RNA remains localized to the inactive X, binding to and coating this chro-

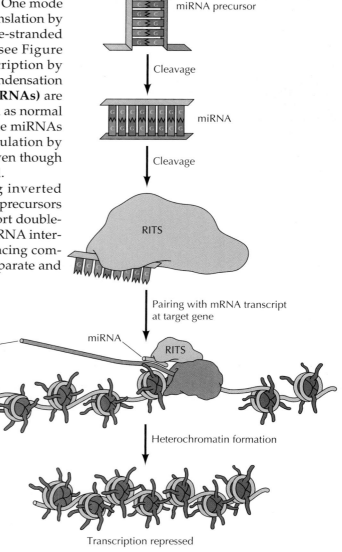

FIGURE 7.37 Regulation of transcription by miRNAs miRNAs are transcribed as precursors containing stem-loop structures, which are cleaved to yield short double-stranded miRNAs. Transcriptional repression is mediated by association of miRNAs with the RITS complex. The miRNA then guides RITS to the homologous target gene, most likely by base pairing with nascent mRNA transcripts associated with RNA polymerase II. RITS then recruits a histone methyltransferase that methylates lysine-9 of histone H3, leading to formation of heterochromatin and repression of transcription.

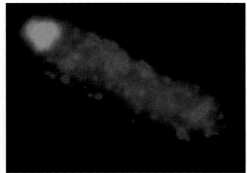

FIGURE 7.38 X chromosome inactivation The inactive X chromosome (blue) is coated by *Xist* RNA (red). (From B. Panning and R. Jaenisch, 1998. *Cell* 93: 305.)

mosome (Figure 7.38). This leads to the recruitment of a protein complex that induces methylation of histone H3 lysine-27 and lysine-9, leading to chromatin condensation and conversion of most of the inactive X to heterochromatin.

DNA Methylation

The methylation of DNA is another general mechanism that controls transcription in eukaryotes. Cytosine residues in DNA of fungi, plants, and animals can be modified by the addition of methyl groups at the 5-carbon position (Figure 7.39). DNA is methylated specifically at the cytosines (C) that precede guanines (G) in the DNA chain (CpG dinucleotides), and this methylation is correlated with transcriptional repression. Methylation commonly occurs within transposable elements, and it appears that methylation plays a key role in suppressing the movement of transposons throughout the genome. In addition, DNA methylation is associated with transcriptional repression of some genes, in concert with alterations in chromatin structure. In plants, miRNAs direct DNA methylation as well as chromatin modifications of repressed genes, although it is not clear whether this also occurs in animals. However, genes on the inactive X chromosome in mammals become methylated following transcriptional repression by *Xist* RNA, so both DNA methylation as well as histone modification appear to play an important role in X inactivation.

One important regulatory role of DNA methylation has been established in the phenomenon known as **genomic imprinting**, which controls the expression of some genes involved in the development of mammalian embryos. In most cases, both the paternal and maternal alleles of a gene are expressed in diploid cells. However, there are some imprinted genes (about 70 have been described in mice and humans) whose expression depends on whether they are inherited from the mother or from the father. In some cases, only the paternal allele of an imprinted gene is expressed, and the maternal allele is transcriptionally inactive. For other imprinted genes, the maternal allele is expressed and the paternal allele is inactive.

DNA methylation appears to play a key role in distinguishing between the paternal and maternal alleles of imprinted genes. A good example is the gene *H19*, which is transcribed only from the maternal copy (Figure 7.40). The *H19* gene is specifically methylated during the development of male, but not female, germ cells. The union of sperm and egg at fertilization therefore yields an embryo containing a methylated paternal allele and an unmethylated maternal allele of the gene. These differences in methylation are maintained following DNA replication by an enzyme that specifically methylates CpG sequences of a daughter strand that is hydrogen-bonded to a methylated parental strand (Figure 7.41). The paternal *H19* allele therefore remains methylated, and transcriptionally inactive, in embryonic cells and somatic tissues. However, the paternal *H19* allele becomes demethylated in the germ line, allowing a new pattern of methylation to be established for transmittal to the next generation.

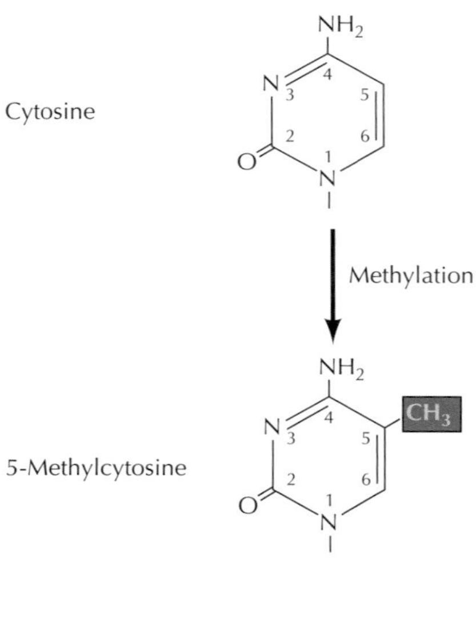

Cytosine

Methylation

5-Methylcytosine

FIGURE 7.39 DNA methylation A methyl group is added to the 5-carbon position of cytosine residues in DNA.

FIGURE 7.40 Genomic imprinting The *H19* gene is specifically methylated during development of male germ cells. Therefore sperm contain a methylated *H19* allele and eggs contain an unmethylated allele. Following fertilization, the methylated paternal allele remains transcriptionally inactive, and only the unmethylated maternal allele is expressed in the embryo.

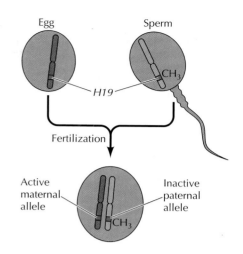

RNA Processing and Turnover

Although transcription is the first and most highly regulated step in gene expression, it is usually only the beginning of the series of events required to produce a functional RNA. Most newly synthesized RNAs must be modified in various ways to be converted to their functional forms. Bacterial mRNAs are an exception; they are used immediately as templates for protein synthesis while still being transcribed. However, the primary transcripts of both rRNAs and tRNAs must undergo a series of processing steps in prokaryotic as well as eukaryotic cells. Primary transcripts of eukaryotic mRNAs similarly undergo extensive modifications, including the removal of introns by splicing, before they are transported from the nucleus to the cytoplasm to serve as templates for protein synthesis. Regulation of these processing steps provides an additional level of control of gene expression, as does regulation of the rates at which different mRNAs are subsequently degraded within the cell.

Processing of Ribosomal and Transfer RNAs

The basic processing of ribosomal and transfer RNAs in prokaryotic and eukaryotic cells is similar, as might be expected given the fundamental roles of these RNAs in protein synthesis. As discussed previously, eukaryotes

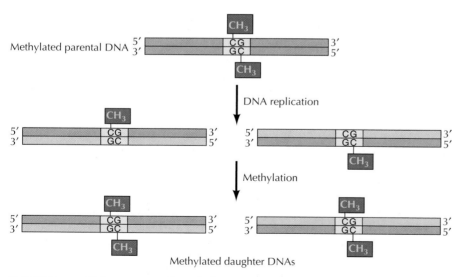

FIGURE 7.41 Maintenance of methylation patterns In parental DNA, both strands are methylated at complementary CpG sequences. Following replication, only the parental strand of each daughter molecule is methylated. The newly synthesized daughter strands are then methylated by an enzyme that specifically recognizes CpG sequences opposite a methylation site.

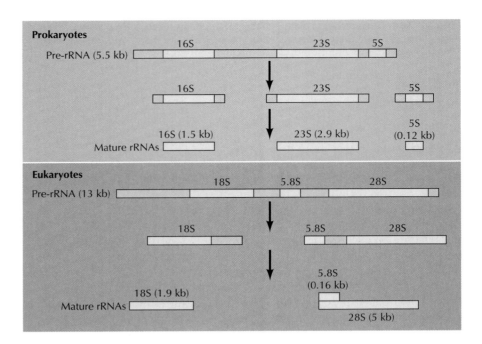

FIGURE 7.42 Processing of ribosomal RNAs Prokaryotic cells contain three rRNAs (16S, 23S, and 5S), which are formed by cleavage of a pre-rRNA transcript. Eukaryotic cells (e.g., human cells) contain four rRNAs. One of these (5S rRNA) is transcribed from a separate gene; the other three (18S, 28S, and 5.8S) are derived from a common pre-rRNA. Following cleavage, the 5.8S rRNA (which is unique to eukaryotes) becomes hydrogen-bonded to 28S rRNA.

have four species of ribosomal RNAs (see Table 7.1), three of which (the 28S, 18S, and 5.8S rRNAs) are derived by cleavage of a single long precursor transcript, called a **pre-rRNA** (Figure 7.42). Prokaryotes have three ribosomal RNAs (23S, 16S, and 5S), which are equivalent to the 28S, 18S, and 5S rRNAs of eukaryotic cells and are also formed by the processing of a single pre-rRNA transcript. The only rRNA that is not processed extensively is the 5S rRNA in eukaryotes, which is transcribed from a separate gene.

Prokaryotic and eukaryotic pre-rRNAs are processed in several steps. Initial cleavages of bacterial pre-rRNA yield separate precursors for the three individual rRNAs; these are then further processed by secondary cleavages to the final products. In eukaryotic cells, pre-rRNA is first cleaved at a site adjacent to the 5.8S rRNA on its 5′ side, yielding two separate precursors that contain the 18S and the 28S + 5.8S rRNAs, respectively. Further cleavages then convert these to their final products, with the 5.8S rRNA becoming hydrogen-bonded to the 28S molecule. In addition to these cleavages, rRNA processing involves the addition of methyl groups to the bases and sugar moieties of specific nucleotides and the conversion of some uridines to pseudouridines. Processing of rRNA takes place within the nucleolus of eukaryotic cells, and will be discussed in detail in Chapter 9.

Like rRNAs, tRNAs in both bacteria and eukaryotes are synthesized as longer precursor molecules (**pre-tRNAs**), some of which contain several individual tRNA sequences (Figure 7.43). In bacteria, some tRNAs are included in the pre-rRNA transcripts. The processing of the 5′ end of pre-tRNAs involves cleavage by an enzyme called **RNase P**, which is of special interest because it is a prototypical model of a reaction catalyzed by an RNA enzyme. RNase P consists of RNA and protein molecules, both of which are required for maximal activity. In 1983 Sidney Altman and his colleagues demonstrated that the isolated RNA component of RNase P is itself capable of catalyzing pre-tRNA cleavage. These experiments established that RNase P is a **ribozyme**—an enzyme in which RNA rather than protein is responsible for catalytic activity.

The 3′ end of tRNAs is generated by the action of a conventional protein RNase, but the processing of this end of the tRNA molecule also involves an unusual activity: the addition of a CCA terminus. All tRNAs have the sequence CCA at their 3′ ends. This sequence is the site of amino acid attachment, so it is required for tRNA function during protein synthesis. The CCA terminus is encoded in the DNA of some tRNA genes, but in others it is not, instead being added as an RNA processing step by an enzyme that recognizes and adds CCA to the 3′ end of all tRNAs that lack this sequence.

Another unusual aspect of tRNA processing is the extensive modification of bases in tRNA molecules. Approximately 10% of the bases in tRNAs are altered to yield a variety of modified nucleotides at specific positions in tRNA molecules (see Figure 7.43). The functions of most of these modified bases are unknown, but some play important roles in protein synthesis by altering the base-pairing properties of the tRNA molecule (see Chapter 8).

Some pre-tRNAs, as well as pre-rRNAs in a few organisms, contain introns that are removed by splicing. In contrast to other splicing reactions,

FIGURE 7.43 Processing of transfer RNAs (A) Transfer RNAs are derived from pre-tRNAs, some of which contain several individual tRNA molecules. Cleavage at the 5′ end of the tRNA is catalyzed by the RNase P ribozyme; cleavage at the 3′ end is catalyzed by a conventional protein RNase. A CCA terminus is then added to the 3′ end of many tRNAs in a posttranscriptional processing step. Finally, some bases are modified at characteristic positions in the tRNA molecule. In this example, these modified nucleosides include dihydrouridine (DHU), methylguanosine (mG), inosine (I), ribothymidine (T), and pseudouridine (ψ). (B) Structure of modified bases. Ribothymidine, dihydrouridine, and pseudouridine are formed by modification of uridines in tRNA. Inosine and methylguanosine are formed by the modification of guanosines.

(B) Modified bases

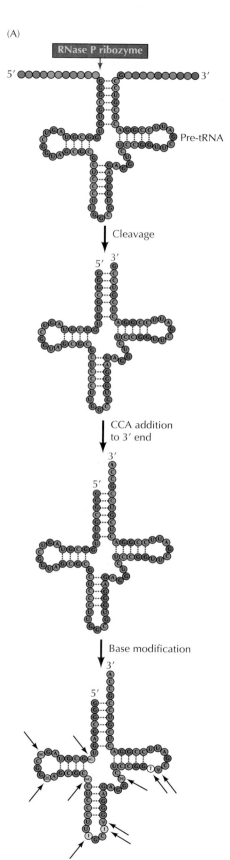

Dihydrouridine (DHU)

Ribothymidine (T)

Pseudouridine (ψ)

Inosine (I)

^2N-Methylguanosine (mG)

which (as discussed in the next section) involve the activities of catalytic RNAs, tRNA splicing is mediated by conventional protein enzymes. An endonuclease cleaves the pre-tRNA at the splice sites to excise the intron, followed by joining of the exons to form a mature tRNA molecule.

Processing of mRNA in Eukaryotes

In contrast to the processing of ribosomal and transfer RNAs, the processing of messenger RNAs represents a major difference between prokaryotic and eukaryotic cells. In bacteria, ribosomes have immediate access to mRNA and translation begins on the nascent mRNA chain while transcription is still in progress. In eukaryotes, mRNA synthesized in the nucleus must first be transported to the cytoplasm before it can be used as a template for protein synthesis. Moreover, the initial products of transcription in eukaryotic cells (**pre-mRNAs**) are extensively modified before export from the nucleus. The processing of mRNA includes modification of both ends of the initial transcript, as well as the removal of introns from its middle (**Figure 7.44**). Rather than occurring as independent events following synthesis of a pre-mRNA, these processing reactions are coupled to transcription so that mRNA synthesis and processing are closely coordinated steps in gene expression. The C-terminal domain (CTD) of RNA polymerase II plays a key role in coordinating these processes by serving as a binding site for the enzyme complexes involved in mRNA processing. The association of these processing enzymes with the CTD of polymerase II accounts for their speci-

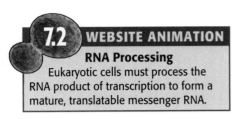

7.2 WEBSITE ANIMATION

RNA Processing
Eukaryotic cells must process the RNA product of transcription to form a mature, translatable messenger RNA.

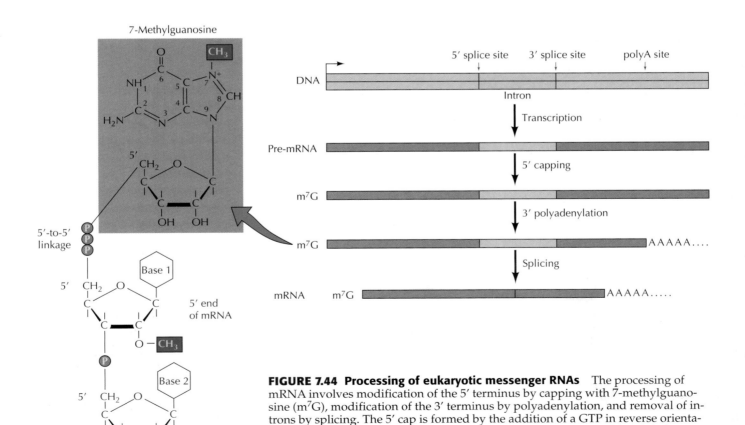

FIGURE 7.44 Processing of eukaryotic messenger RNAs The processing of mRNA involves modification of the 5′ terminus by capping with 7-methylguanosine (m^{7}G), modification of the 3′ terminus by polyadenylation, and removal of introns by splicing. The 5′ cap is formed by the addition of a GTP in reverse orientation to the 5′ end of the mRNA, forming a 5′-to-5′ linkage. The added G is then methylated at the N-7 position, and methyl groups are added to the riboses of the first one or two nucleotides in the mRNA.

ficity in processing mRNAs; polymerases I and III lack a CTD, so their transcripts are not processed by the same enzyme complexes.

The first step in mRNA processing is the modification of the 5′ end of the transcript by the addition of a structure called a **7-methylguanosine cap**. The enzymes responsible for capping are recruited to the phosphorylated CTD following initiation of transcription, and the cap is added after transcription of the first 20–30 nucleotides of the RNA. Capping is initiated by the addition of a GTP in reverse orientation to the 5′ terminal nucleotide of the RNA. Then methyl groups are added to this G residue and to the ribose moieties of one or two 5′ nucleotides of the RNA chain. The 5′ cap stabilizes the RNA, as well as aligning eukaryotic mRNAs on the ribosome during translation (see Chapter 8).

The 3′ end of most eukaryotic mRNAs is defined not by termination of transcription but by cleavage of the primary transcript and addition of a **poly-A tail**—a processing reaction called **polyadenylation** (Figure 7.45). The signals for polyadenylation include a highly conserved hexanucleotide (AAUAAA in mammalian cells), which is located 10 to 30 nucleotides upstream of the site of polyadenylation, and a G-U rich downstream sequence element. In addition, some genes have a U-rich sequence element upstream of the AAUAAA. These sequences are recognized by a complex of proteins, including an endonuclease that cleaves the RNA chain and a separate poly-A polymerase that adds a poly-A tail of about 200 nucleotides to the transcript. These processing enzymes are associated with the phosphorylated CTD of RNA polymerase II, and may travel with the polymerase all the way from the transcription initiation site. Cleavage and polyadenylation is followed by degradation of the RNA that has been synthesized downstream of the site of poly-A addition, resulting in the termination of transcription.

Almost all mRNAs in eukaryotes are polyadenylated, and poly-A tails have been shown to regulate both translation and mRNA stability. In addition, polyadenylation plays an important regulatory role in early development, where changes in the length of poly-A tails control mRNA translation. For example, many mRNAs are stored in unfertilized eggs in an untranslated form with short poly-A tails (usually 30 to 50 nucleotides long). Fertilization stimulates the lengthening of the poly-A tails of these stored mRNAs, which in turn activates their translation and the synthesis of proteins required for early embryonic development.

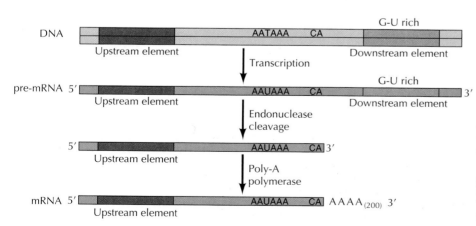

FIGURE 7.45 Formation of the 3′ ends of eukaryotic mRNAs Polyadenylation signals in mammalian cells consist of the hexanucleotide AAUAAA in addition to upstream and downstream (G-U rich) elements. An endonuclease cleaves the pre-mRNA 10 to 30 nucleotides downstream of the AAUAAA, usually at a CA sequence. Poly-A polymerase then adds a poly-A tail consisting of about 200 adenines (A) to the 3′ end of the RNA.

The most striking modification of pre-mRNAs is the removal of introns by splicing. As discussed in Chapter 5, the coding sequences of most eukaryotic genes are interrupted by noncoding sequences (introns) that are precisely excised from the mature mRNA. In mammals, most genes contain multiple introns, which typically account for about ten times more pre-mRNA sequences than do the exons. The unexpected discovery of introns in 1977 generated an active research effort directed toward understanding the mechanism of splicing, which had to be highly specific to yield functional mRNAs. Further studies of splicing have not only illuminated new mechanisms of gene regulation; they have also revealed novel catalytic activities of RNA molecules.

Splicing Mechanisms

The key to understanding pre-mRNA splicing was the development of *in vitro* systems that efficiently carried out the splicing reaction (Figure 7.46). Pre-mRNAs were synthesized *in vitro* by the cloning of structural genes (with their introns) adjacent to promoters for bacteriophage RNA polymerases, which could readily be isolated in large quantities. Transcription

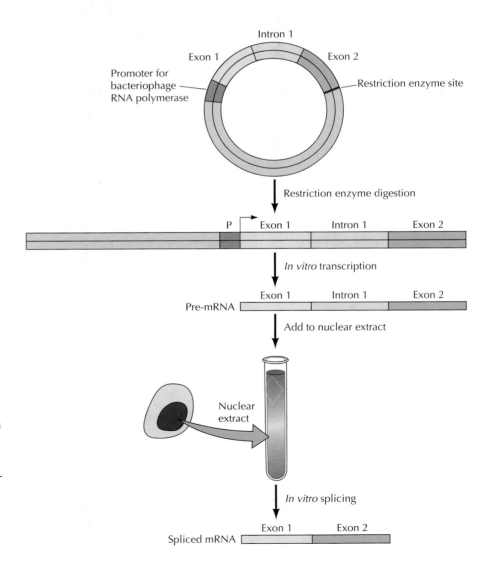

FIGURE 7.46 *In vitro* splicing A gene containing an intron is cloned downstream of a promoter (P) recognized by a bacteriophage RNA polymerase. The plasmid is digested with a restriction enzyme that cleaves at the 3′ end of the inserted gene to yield a linear DNA molecule. This DNA is then transcribed *in vitro* with the bacteriophage polymerase to produce pre-mRNA. Splicing reactions can then be studied *in vitro* by addition of this pre-mRNA to nuclear extracts of mammalian cells.

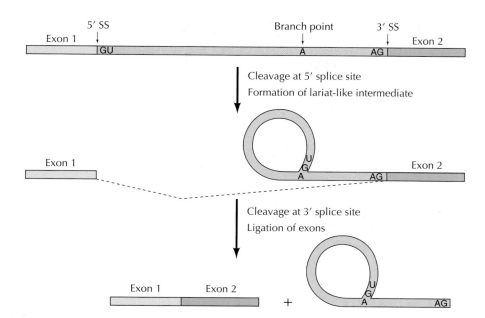

FIGURE 7.47 Splicing of pre-mRNA
The splicing reaction proceeds in two steps. The first step involves cleavage at the 5′ splice site (SS) and joining of the 5′ end of the intron to an A within the intron (the branch point). This reaction yields a lariat-like intermediate in which the intron forms a loop. The second step is cleavage at the 3′ splice site and simultaneous ligation of the exons, resulting in excision of the intron as a lariat-like structure.

of these plasmids could then be used to prepare large amounts of pre-mRNAs that, when added to nuclear extracts of mammalian cells, were found to be correctly spliced. As with transcription, the use of such *in vitro* systems has allowed splicing to be analyzed in much greater detail than would have been possible in intact cells.

Analysis of the reaction products and intermediates formed *in vitro* revealed that pre-mRNA splicing proceeds in two steps (Figure 7.47). First, the pre-mRNA is cleaved at the 5′ splice site, and the 5′ end of the intron is joined to an adenine nucleotide within the intron (near its 3′ end). In this step, an unusual bond forms between the 5′ end of the intron and the 2′ hydroxyl group of the adenine nucleotide. The resulting intermediate is a lariat-like structure in which the intron forms a loop. The second step in splicing then proceeds with simultaneous cleavage at the 3′ splice site and ligation of the two exons. The intron is thus excised as a lariat-like structure, which is then linearized and degraded within the nucleus of intact cells.

These reactions define three critical sequence elements of pre-mRNAs: sequences at the 5′ splice site, sequences at the 3′ splice site, and sequences within the intron at the branch point (the point at which the 5′ end of the intron becomes ligated to form the lariat-like structure) (see Figure 7.47). Pre-mRNAs contain similar consensus sequences at each of these positions, allowing the splicing apparatus to recognize pre-mRNAs and carry out the cleavage and ligation reactions involved in the splicing process.

Biochemical analysis of nuclear extracts has revealed that splicing takes place in large complexes, called **spliceosomes**, composed of proteins and RNAs. The RNA components of the spliceosome are five types of **small nuclear RNAs (snRNAs)** called U1, U2, U4, U5, and U6. These snRNAs, which range in size from approximately 50 to nearly 200 nucleotides, are complexed with six to ten protein molecules to form **small nuclear ribonucleoprotein particles (snRNPs)**, which play central roles in the splicing process. The U1, U2, and U5 snRNPs each contain a single snRNA molecule, whereas U4 and U6 snRNAs are complexed to each other in a single snRNP.

KEY EXPERIMENT

The Discovery of snRNPs

Antibodies to small nuclear RNAs complexed with proteins are produced by patients with systemic lupus erythematosus

Michael R. Lerner and Joan A. Steitz
Yale University, New Haven, Connecticut

Proceedings of the National Academy of Sciences, USA, 1979, Volume 76, pages 5495–5499

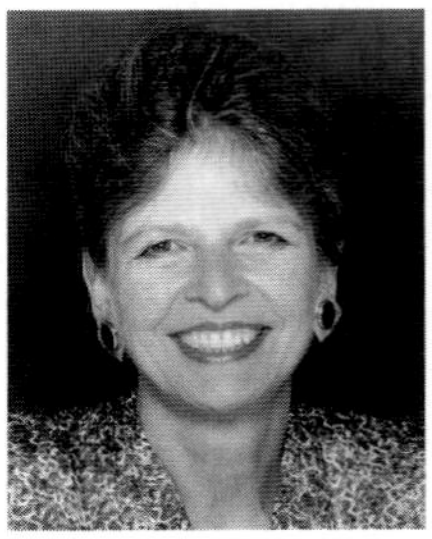

Joan Steitz

The Context

The discovery of introns in 1977 implied that a totally unanticipated processing reaction was required to produce mRNA in eukaryotic cells. Introns had to be precisely excised from pre-mRNA, followed by the joining of exons to yield a mature mRNA molecule. Given the unexpected nature of pre-mRNA splicing, understanding the mechanism of the splicing reaction captivated the attention of many molecular biologists. One of the major steps in elucidating this mechanism was the discovery of snRNPs and their involvement in pre-mRNA splicing.

Small nuclear RNAs were first identified in eukaryotic cells in the late 1960s. However, the function of snRNAs remained unknown. In this 1979 paper, Michael Lerner and Joan Steitz demonstrated that the most abundant snRNAs were present as RNA-protein complexes called snRNPs. In addition, they provided the first suggestion that these RNA-protein complexes might function in pre-mRNA splicing. This identification of snRNPs led to a variety of experiments that confirmed their roles and elucidated the mechanism by which pre-mRNA splicing takes place.

The Experiments

The identification of snRNPs was based on the use of antisera from patients with systemic lupus erythematosus, an autoimmune disease in which patients produce antibodies against their own normal cell constituents. Many of the antibodies produced by systemic lupus erythematosus patients are directed against components of the nucleus, including DNA, RNA, and histones. The discovery of snRNPs arose from studies in which Lerner and Steitz sought to characterize two antigens, called ribonucleoprotein (RNP) and Sm, which were recognized by antibodies from systemic lupus erythematosus patients. Indirect data suggested that RNP consisted of both protein and RNA, as its name implies, but neither RNP nor Sm had been characterized at the molecular level.

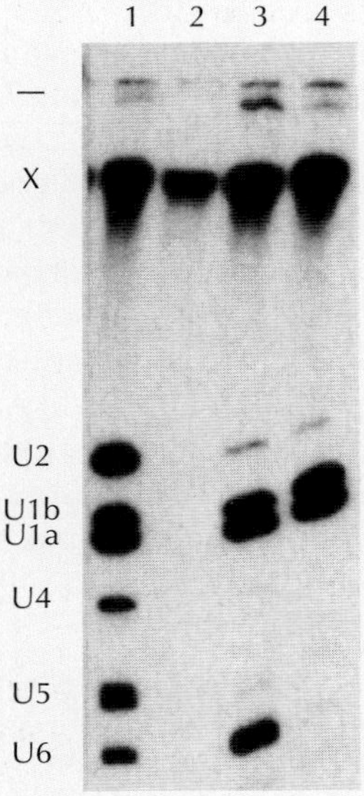

Immunoprecipitation of snRNAs with antisera from systemic lupus erythematosus patients. Lane 1, anti-Sm; lane 2, normal control serum; lane 3, antiserum recognizing primarily the RNP antigen; lane 4, anti-RNP. Note that a nonspecific RNA designated X is present in all immunoprecipitates, including the control.

To identify possible RNA components of the RNP and Sm antigens, nuclear RNAs of mouse cells were radiolabeled with ^{32}P and immunoprecipitated with antisera from different systemic lupus erythematosus patients (see Figure 4.30). Six specific species of snRNAs were found to be selectively immunoprecipitated by antisera from different patients but not by serum from a normal control patient (see figure). Anti-Sm serum immunoprecipitated all six of these snRNAs, which were designated U1a, U1b, U2, U4, U5, and U6. Anti-RNP serum immunoprecipitated only U1a and U1b, and serum from a third patient (which had been characterized as mostly anti-RNP) immunoprecipitated U1a, U1b, and U6. The immunoprecipitated snRNAs were further characterized by sequence analysis, which demonstrated that U1a, U1b, and U2 were identical to the most abundant snRNAs previously reported in mammalian nuclei, with U1a and U1b representing sequence variants of a single species of U1 snRNA present in human cells. In contrast, the U4, U5, and U6 snRNAs were newly identified by Lerner and Steitz in these experiments.

Importantly, the immunoprecipitation of these snRNAs demonstrated that they were components of RNA-protein complexes. The anti-Sm serum, which immunoprecipitated all six of the snRNAs, had previously been shown to be directed against a protein antigen. Similarly, protein was known to be required for antigen recognition by anti-RNP serum. Moreover, Lerner and Steitz showed that none of the snRNAs could be immunoprecipitated if protein was first removed by extraction of the RNAs with phenol. Further analysis of cells in which proteins had been radiolabeled with ^{35}S-methionine identified seven prominent nuclear proteins that were immunoprecipitated along with the snRNAs by anti-Sm and anti-RNP sera. These data therefore indicated that each of the six snRNAs was present in an snRNP complex with specific nuclear proteins.

The Impact

The finding that snRNAs were components of snRNPs that were recognized by specific antisera opened a new approach to studying snRNA function. Lerner and Steitz noted that a "most intriguing" possible role for snRNAs might be in pre-mRNA splicing, and pointed out that sequences near the 5' terminus of U1 snRNA were complementary to splice sites.

Steitz and her colleagues then proceeded with a series of experiments that established the critical involvement of snRNPs in splicing. These studies included more extensive sequence analysis that demonstrated the complementarity of conserved 5' sequences of U1 snRNA to the consen-

sus sequences of 5' splice sites, suggesting that U1 functioned in 5' splice site recognition. In addition, antisera against snRNPs were used to demonstrate that U1 was required for pre-mRNA splicing both in isolated nuclei and in *in vitro* splicing extracts. Further studies have gone on to show that the snRNAs themselves play critical roles not only in the identification of splice sites, but also as catalysts of the splicing reaction. The initial discovery that snRNAs were components of snRNPs that could be recognized by specific antisera thus opened the door to understanding the mechanism of pre-mRNA splicing.

The first step in spliceosome assembly is the binding of U1 snRNP to the 5' splice site of pre-mRNA (Figure 7.48). This recognition of 5' splice sites involves base pairing between the 5' splice site consensus sequence and a complementary sequence at the 5' end of U1 snRNA (Figure 7.49). U2 snRNP then binds to the branch point, by similar complementary base pairing between U2 snRNA and branch point sequences. A preformed complex consisting of U4/U6 and U5 snRNPs is then incorporated into the spliceosome, with U5 binding to sequences upstream of the 5' splice site. The splicing reaction is then accompanied by rearrangements of the snRNAs. Prior to the first reaction step (formation of the lariat-like intermediate, see Figure 7.47), U6 dissociates from U4 and displaces U1 at the 5' splice site. U5 then binds to sequences at the 3' splice site, followed by excision of the intron and ligation of the exons.

Not only do the snRNAs recognize consensus sequences at the branch points and splice sites of pre-mRNAs, but they also catalyze the splicing reaction directly. The catalytic role of RNAs in splicing was demonstrated by the discovery that some RNAs are capable of **self-splicing**; that is, they can catalyze the removal of their own introns in the absence of other protein or RNA factors. Self-splicing was first described by Tom Cech and his colleagues during studies of the 28S rRNA of the protozoan *Tetrahymena*. This RNA contains an intron of approximately 400 bases that is precisely removed following incubation of the pre-rRNA in the absence of added proteins. Further studies have revealed that splicing is catalyzed by the intron, which acts as a ribozyme to direct its own excision from the pre-rRNA molecule. The discovery of self-splicing of *Tetrahymena* rRNA, together with the studies of RNase P already discussed, provided the first demonstrations of the catalytic activity of RNA.

Additional studies have revealed self-splicing RNAs in mitochondria, chloroplasts, and bacteria. These self-splicing RNAs are divided into two

FIGURE 7.48 Assembly of the spliceosome The first step in spliceosome assembly is the binding of U1 snRNP to the 5′ splice site (SS), followed by the binding of U2 snRNP to the branch point. A preformed complex consisting of U4/U6 and U5 snRNPs then enters the spliceosome, with U5 binding to sequences upstream of the 5′ splice site. U6 then dissociates from U4 and displaces U1, leading to formation of the lariat-like intermediate. U5 then binds to the 3′ splice site, followed by excision of the intron and ligation of the exons.

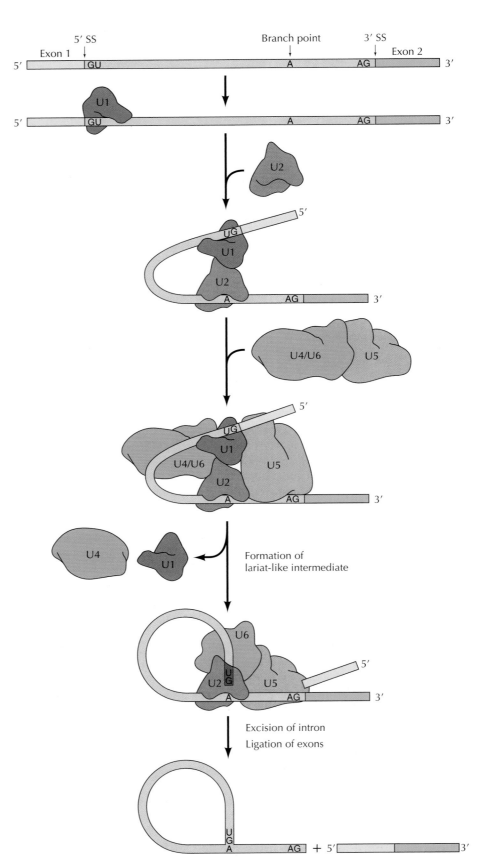

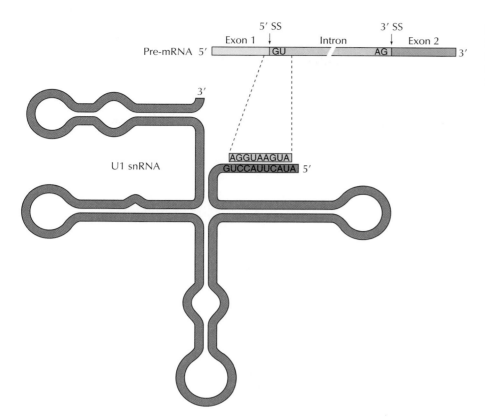

FIGURE 7.49 Binding of U1 snRNA to the 5′ splice site   The 5′ terminus of U1 snRNA binds to consensus sequences at 5′ splice sites by complementary base pairing.

classes on the basis of their reaction mechanisms (**Figure 7.50**). The first step in splicing for group I introns (e.g., *Tetrahymena* pre-rRNA) is cleavage at the 5′ splice site mediated by a guanosine cofactor. The 3′ end of the free exon then reacts with the 3′ splice site to excise the intron as a linear RNA. In contrast, the self-splicing reactions of group II introns (e.g., some mitochondrial pre-mRNAs) closely resemble those characteristic of nuclear pre-mRNA splicing in which cleavage of the 5′ splice site results from attack by an adenosine nucleotide in the intron. As with pre-mRNA splicing, the result is a lariat-like intermediate, which is then excised.

The similarity between spliceosome-mediated pre-mRNA splicing and self-splicing of group II introns strongly suggested that the active catalytic components of the spliceosome were RNAs rather than proteins. In particular, these similarities suggested that pre-mRNA splicing was catalyzed by the snRNAs of the spliceosome. Continuing studies of pre-mRNA splicing have provided clear support for this view, including the demonstration that U2 and U6 snRNAs, in the absence of proteins, can catalyze the first step in pre-mRNA splicing. Pre-mRNA splicing is thus considered to be an RNA-based reaction, catalyzed by spliceosome snRNAs acting analogously to group II self-splicing introns. Within the cell, protein components of the snRNPs are also required, however, and participate in both assembly of the spliceosome and the splicing reaction.

A number of protein splicing factors that are not snRNP components also play critical roles in spliceosome assembly, particularly in identification of the correct splice sites in pre-mRNAs. Mammalian pre-mRNAs typically contain multiple short exons (an average of 150 nucleotides in humans) sep-

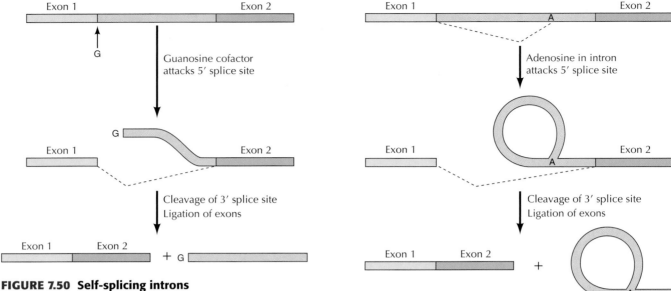

FIGURE 7.50 Self-splicing introns
Group I and group II self-splicing introns are distinguished by their reaction mechanisms. In group I introns, the first step in splicing is cleavage of the 5' splice site by reaction with a guanosine cofactor. The result is a linear intermediate with a G added to the 5' end of the intron. In group II introns (as in pre-mRNA splicing), the first step is cleavage of the 5' splice site by reaction with an A within the intron, forming a lariat-like intermediate. In both cases, the second step is simultaneous cleavage of the 3' splice site and ligation of the exons.

arated by much larger introns (average of 3500 nucleotides). Introns frequently contain many sequences that resemble splice sites, so the splicing machinery must be able to identify the appropriate 5' and 3' splice sites at intron/exon boundaries to produce a functional mRNA. Splicing factors serve to direct spliceosomes to the correct splice sites by binding to specific RNA sequences and then recruiting U1 and U2 snRNPs to the appropriate sites on pre-mRNA by protein-protein interactions. For example, the SR splicing factors bind to specific sequences within exons and act to recruit U1 snRNP to the 5' splice site (Figure 7.51). SR proteins also interact with another splicing factor (U2AF), which binds to pyrimidine-rich sequences at 3' splice sites and recruits U2 snRNP to the branchpoint. In addition to recruiting the components of the spliceosome to pre-mRNA, splicing factors couple splicing to transcription by associating with the phosphorylated CTD of RNA polymerase II. This anchoring of the splicing machinery to RNA polymerase is thought to be important in ensuring that exons are joined in the correct order as the pre-mRNA is synthesized.

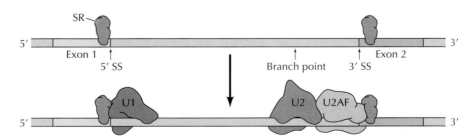

FIGURE 7.51 Role of splicing factors in spliceosome assembly SR splicing factors bind to specific sequences within exons. The SR proteins recruit U1 snRNP to the 5' splice site and an additional splicing factor (U2AF) to the 3' splice site. U2AF then recruits U2 snRNP to the branchpoint.

Alternative Splicing

The central role of splicing in the processing of pre-mRNA opens the possibility of regulation of gene expression by control of the splicing machinery. Since most pre-mRNAs contain multiple introns, different mRNAs can be produced from the same gene by different combinations of 5' and 3' splice sites. The possibility of joining exons in varied combinations provides a novel means of controlling gene expression by generating multiple mRNAs (and therefore multiple proteins) from the same pre-mRNA. This process, called **alternative splicing**, occurs frequently in genes of complex eukaryotes. For example, it is estimated that about 50% of human genes produce transcripts that are alternatively spliced, considerably increasing the diversity of proteins that can be encoded by the estimated 20,000–25,000 genes in mammalian genomes. Because patterns of alternative splicing can vary in different tissues and in response to extracellular signals, alternative splicing provides an important mechanism for tissue-specific and developmental regulation of gene expression.

One well-studied example of tissue-specific alternative splicing is provided by sex determination in *Drosophila*, where alternative splicing of the same pre-mRNA determines whether a fly is male or female (Figure 7.52). Alternative splicing of the pre-mRNA of a gene called *transformer* is controlled by a protein (SXL) that is only expressed in female flies. The *transformer* pre-mRNA has three exons, but a different second exon is incorporated into the mRNA as a result of using alternate 3' splice sites in the two different sexes. In males, exon 1 is joined to the most upstream of these 3' splice sites, which is selected by the binding of the U2AF splicing factor. In females, the SXL protein binds to this 3' splice site, blocking the binding of U2AF. Consequently, the upstream 3' splice site is skipped in females, and exon 1 is instead joined to an alternate 3' splice site that is further downstream. The exon 2 sequences included in the male *transformer* mRNA contain a translation termination codon, so no protein is produced. This termination codon is not included in the female mRNA, so female flies express a functional *transformer* protein, which acts a key regulator of sex determination.

The alternative splicing of *transformer* illustrates the action of a repressor (the SXL protein) that functions by blocking the binding of a splicing factor (U2AF), and a large group of proteins similarly regulate alternative splicing by binding to silencer sequences in pre-mRNAs. In other cases, alternative

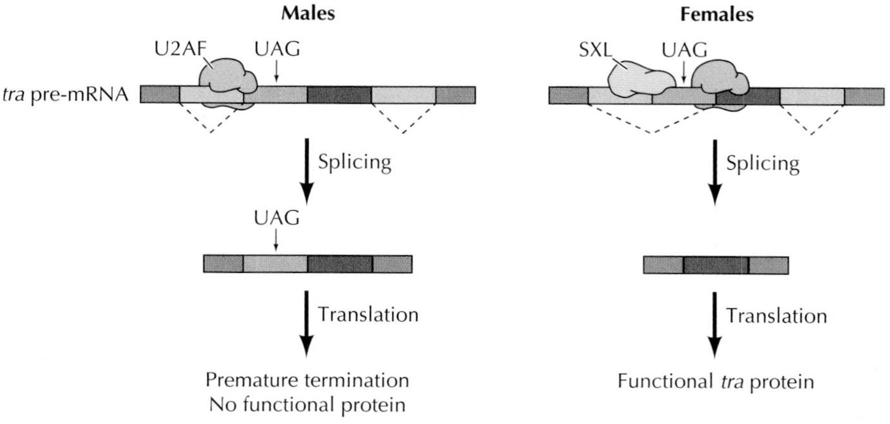

■ It is estimated that aberrant splicing may cause as many as 15% of all inherited diseases. These include diseases like β-thalassemia and several types of cancer.

FIGURE 7.52 Alternative splicing in *Drosophila* sex determination
Alternative splicing of *transformer (tra)* mRNA is regulated by the SXL protein, which is only expressed in female flies. In males, the first exon of *tra* mRNA is joined to a 3' splice site that yields a second exon containing a translation termination codon (UAG), so no *tra* protein is expressed. In females, the binding of SXL protein blocks the binding of U2AF to this 3' splice site, resulting in the use of an alternative site further downstream in exon 2. This alternative 3' splice site is downstream of the translation termination codon, so the mRNA expressed in females directs the synthesis of functional *tra* protein.

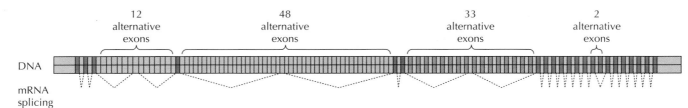

FIGURE 7.53 Alternative splicing of *Dscam* The *Dscam* gene contains four sets of alternative exons: 12 for exon 4, 48 for exon 6, 33 for exon 9, and 2 for exon 17. Any single exon from each of these sets can be incorporated into the mature mRNA, so alternative splicing can produce a total of 38,016 different mRNAs ($12 \times 48 \times 33 \times 2 = 38,016$).

splicing is controlled by activators that recruit splicing factors to splice sites that would otherwise not be recognized. The best-studied splicing activators are members of the SR protein family (see Figure 7.51), which bind to specific splicing enhancer sequences.

Multiple mechanisms can thus regulate alternative splicing, and variations in alternative splicing make a major contribution to the diversity of proteins expressed during development and differentiation. One of the most striking examples is the alternative splicing of a *Drosophila* cell surface protein (called Dscam) that is involved in specifying connections between neurons. The *Dscam* gene contains four sets of alternative exons, with a single exon from each set being incorporated into the spliced mRNA (Figure 7.53). These exons can be joined in any combination, so alternative splicing can potentially yield 38,016 different mRNAs and proteins from this single gene: more than twice the total number of genes in the *Drosophila* genome. It is thought that the functional diversity provided by these alternatively spliced forms of Dscam makes an important contribution to specifying the correct connections between neurons during development of the fly brain.

RNA Editing

RNA editing refers to RNA processing events (other than splicing) that alter the protein-coding sequences of some mRNAs. This unexpected form of RNA processing was first discovered in mitochondrial mRNAs of trypanosomes in which U residues are added and deleted at multiple sites along the pre-mRNA in order to generate the mRNA. More recently, editing has also been described in mitochondrial mRNAs of other organisms, chloroplast mRNAs of higher plants, and nuclear mRNAs of some mammalian genes.

Editing in mammalian nuclear mRNAs, as well as in mitochondrial and chloroplast RNAs of higher plants, involves single base changes as a result of base modification reactions, similar to those involved in tRNA processing. In mammalian cells, RNA editing reactions include the deamination of cytosine to uridine and of adenosine to inosine. One of the best-studied examples is editing of the mRNA for apolipoprotein B, which transports lipids in the blood. In this case, tissue-specific RNA editing results in two different forms of apolipoprotein B (Figure 7.54). In humans, Apo-B100 (4536 amino acids) is synthesized in the liver by translation of the unedited mRNA. However, a shorter protein (Apo-B48, 2152 amino acids) is synthesized in the intestine as a result of translation of an edited mRNA in which a C has been changed to a U by deamination. This alteration changes the

■ The mammalian ear contains hair cells that are tuned to respond to sounds of different frequency. The tuning of hair cells is thought to be mediated in part by the alternative splicing of a gene encoding a channel protein.

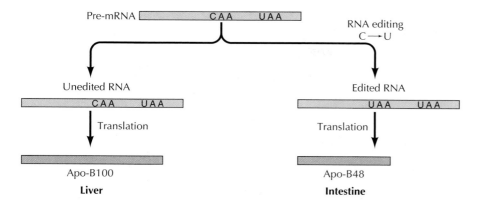

FIGURE 7.54 Editing of apolipoprotein B mRNA In human liver, unedited mRNA is translated to yield a 4536-amino-acid protein called Apo-B100. In human intestine, however, the mRNA is edited by a base modification that changes a specific C to a U. This modification changes the codon for glutamine (CAA) to a termination codon (UAA), resulting in synthesis of a shorter protein (Apo-B48, consisting of only 2152 amino acids).

codon for glutamine (CAA) in the unedited mRNA to a translation termination codon (UAA) in the edited mRNA, resulting in synthesis of the shorter Apo-B protein. Tissue-specific editing of Apo-B mRNA thus results in the expression of structurally and functionally different proteins in liver and intestine. The full-length Apo-B100 produced by the liver transports lipids in the circulation; Apo-B48 functions in the absorption of dietary lipids by the intestine.

RNA editing by the deamination of adenosine to inosine is the most common form of nuclear RNA editing in mammals. This form of editing plays an important role in the nervous system, where A-to-I editing results in single amino acid changes in ion channels and receptors on the surface of neurons. For example, the mRNAs encoding receptors for the neurotransmitter serotonin can be edited at up to five sites, potentially yielding 24 different versions of the receptor with different signaling activities. The importance of A-to-I editing in the nervous system is further demonstrated by the finding that *C. elegans, Drosophila*, and mouse mutants lacking the editing enzyme suffer from a variety of neurological defects.

RNA Degradation

The processing steps discussed in the previous section result in the formation of mature mRNAs, which are then transported to the cytoplasm and function to direct protein synthesis. However, most of the sequences transcribed into pre-mRNA are instead degraded within the nucleus. Over 90% of pre-mRNA sequences are introns, which are degraded within the nucleus following their excision by splicing. This is carried out by an enzyme that recognizes the unique 2′-5′ bond formed at the branchpoint, as well as by enzymes that recognize either the 5′ or 3′ ends of RNA molecules and catalyze degradation of the RNA in either direction. The 5′ and 3′ ends of processed mRNAs are protected from this degradation machinery by capping and polyadenylation, respectively, while the unprotected ends of introns are recognized and degraded.

In addition to degrading introns, cells possess a quality-control system (called **nonsense-mediated mRNA decay**) that leads to the degradation of mRNAs that lack complete open-reading frames. This eliminates defective mRNA molecules and prevents the synthesis of abnormal truncated proteins. In yeast, nonsense-mediated mRNA decay takes place in the cytoplasm and is triggered when a premature termination codon is encountered by a ribosome during protein synthesis. In mammals, however, at least some nonsense-mediated mRNA decay may take place while mRNAs are associated with the nucleus. However, it is not clear whether this occurs during nuclear export or within the nucleus, which would require translation of nuclear mRNAs.

What may be considered the final aspect of the processing of an RNA molecule is its eventual degradation in the cytoplasm. Since the intracellular level of any RNA is determined by a balance between synthesis and degradation, the rate at which individual RNAs are degraded is another level at which gene expression can be controlled. Both ribosomal and transfer RNAs are very stable, and this stability largely accounts for the high levels of these RNAs (greater than 90% of all RNA) in both prokaryotic and eukaryotic cells. In contrast, bacterial mRNAs are rapidly degraded, usually having half-lives of only 2 to 3 minutes. This rapid turnover of bacterial mRNAs allows the cell to respond quickly to alterations in its environment, such as changes in the availability of nutrients required for growth. In eukaryotic cells, however, different mRNAs are degraded at different rates, providing an additional parameter to the regulation of eukaryotic gene expression.

The cytoplasmic degradation of most eukaryotic mRNAs is initiated by shortening of their poly-A tails. Then follows removal of the 5′ cap and degradation of the RNA by nucleases acting from both ends. The half-lives of mRNAs in mammalian cells vary from less than 30 minutes to approximately 20 hours. The unstable mRNAs frequently code for regulatory proteins, such as transcription factors, whose levels within the cell vary rapidly in response to environmental stimuli. These mRNAs often contain specific AU-rich sequences near their 3′ ends that appear to signal rapid degradation by promoting deadenylation. In contrast, mRNAs encoding structural proteins or central metabolic enzymes generally have long half-lives.

The degradation of some mRNAs can also be triggered by internal cleavages. In many cases, cleavage is induced by siRNAs or miRNAs acting through the RISC complex (see Figure 4.42). Alternatively, the cleavage of some mRNAs is mediated by proteins and can be regulated in response to extracellular signals. A good example is provided by the mRNA that encodes transferrin receptor—a cell surface protein involved in the uptake of iron by mammalian cells. The amount of transferrin receptor within cells is regulated by the availability of iron, largely as a result of modulation of the stability of its mRNA (Figure 7.55). In the presence of adequate amounts of iron, transferrin receptor mRNA is rapidly degraded as a result of specific nuclease cleavage at a sequence near its 3′ end. If an adequate supply of iron is not available, however, the mRNA is stabilized, resulting in increased synthesis of transferrin receptor and more iron uptake by the cell. This regulation is mediated by an iron regulatory protein (IRP) that binds to specific sequences (called the iron response element, or IRE) near the 3′ end of transferrin receptor mRNA and protects the mRNA from cleavage. The activity of the IRP is in turn controlled by the levels of iron within the cell: If

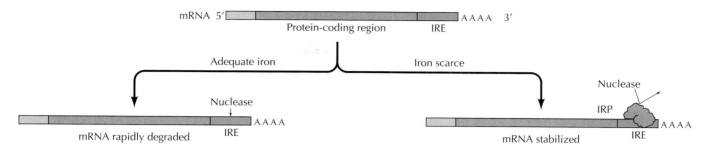

FIGURE 7.55 Regulation of transferrin receptor mRNA stability   The levels of transferrin receptor mRNA are regulated by the availability of iron. If the supply of iron is adequate, the mRNA is rapidly degraded as a result of nuclease cleavage near the 3' end. If iron is scarce, however, a regulatory protein (called the iron regulatory protein, or IRP) binds to a sequence near the 3' end of the mRNA (the iron response element, or IRE), protecting the mRNA from nuclease cleavage.

iron is scarce, the IRP binds to the IRE and protects transferrin receptor mRNA from degradation. Similar changes in the stability of other mRNAs are involved in the regulation of gene expression by certain hormones. Thus, although transcription remains the primary level at which gene expression is regulated, variations in the rate of mRNA degradation also play an important role in controlling steady-state levels of mRNAs within the cell.

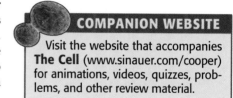

COMPANION WEBSITE

Visit the website that accompanies **The Cell** (www.sinauer.com/cooper) for animations, videos, quizzes, problems, and other review material.

SUMMARY

KEY TERMS

TRANSCRIPTION IN PROKARYOTES

RNA Polymerase and Transcription: *E. coli* RNA polymerase consists of α, β, β', ω, and σ subunits. Transcription is initiated by the binding of σ to promoter sequences. After synthesis of about the first ten nucleotides of RNA, the core polymerase dissociates from σ and travels along the template DNA as it elongates the RNA chain. Transcription then continues until the polymerase encounters a termination signal.

RNA polymerase, promoter, footprinting

Repressors and Negative Control of Transcription: The prototype model for gene regulation in bacteria is the *lac* operon, which is regulated by the binding of a repressor to specific DNA sequences overlapping the promoter.

operon, operator, repressor, *cis*-acting control element, *trans*-acting factor

Positive Control of Transcription: Some bacterial genes are regulated by transcriptional activators rather than repressors.

EUKARYOTIC RNA POLYMERASES AND GENERAL TRANSCRIPTION FACTORS

Eukaryotic RNA Polymerases: Eukaryotic cells contain three distinct nuclear RNA polymerases that transcribe genes encoding mRNAs (polymerase II), rRNAs (polymerases I and III), and tRNAs (polymerase III).

KEY TERMS

transcription factor, general transcription factor, TATA box, TATA-binding protein (TBP), TBP-associated factor (TAF), Mediator

enhancer, insulator, barrier element

electrophoretic-mobility shift assay, chromatin immunoprecipitation

DNA-affinity chromatography

transcriptional activator, zinc finger domain, steroid hormone receptor, helix-turn-helix, homeodomain, homeobox, leucine zipper, helix-loop-helix, coactivator

corepressor

HMGN proteins, histone acetylation, histone code, nucleosome remodeling factor, elongation factor

SUMMARY

General Transcription Factors and Initiation of Transcription by RNA Polymerase II: Eukaryotic RNA polymerases do not bind directly to promoter sequences; they require additional proteins (general transcription factors) to initiate transcription. The promoter sequences of many polymerase II genes are recognized by the TATA-binding protein, which recruits additional transcription factors and RNA polymerase to the promoter.

Transcription by RNA Polymerases I and III: RNA polymerases I and III also require additional transcription factors to bind to the promoters of rRNA, tRNA, and some snRNA genes.

REGULATION OF TRANSCRIPTION IN EUKARYOTES

cis-*Acting Regulatory Sequences: Promoters and Enhancers:* Transcription of eukaryotic genes is controlled by proteins that bind to regulatory sequences, which can be located up to several kilobases away from the transcription start site. Enhancers typically contain binding sites for multiple proteins that work together to regulate gene expression.

Transcription Factor Binding sites: Eukaryotic transcription factors bind to short DNA sequences, usually 6–10 base pairs, in promoters or enhancers.

Transcriptional Regulatory Proteins: Many eukaryotic transcription factors have been isolated on the basis of their binding to specific DNA sequences.

Structure and Function of Transcriptional Activators: Transcriptional activators are modular proteins, consisting of distinct DNA-binding and activation domains. DNA-binding domains mediate association with specific regulatory sequences; activation domains stimulate transcription by interacting with Mediator proteins and general transcription factors, as well as with coactivators that modify chromatin structure.

Eukaryotic Repressors: Gene expression in eukaryotic cells is regulated by repressors as well as by activators. Some repressors interfere with the binding of activators or general transcription factors to DNA. Other repressors contain discrete repression domains that inhibit transcription by interacting with Mediator proteins, general transcription factors, transcriptional activators, or corepressors that affect chromatin structure.

Relationship of Chromatin Structure to Transcription: The packaging of DNA in nucleosomes presents an impediment to transcription in eukaryotic cells. Modification of histones by acetylation is tightly linked to transcriptional regulation and enzymes that catalyze histone acetylation are associated with transcriptional activators, whereas histone deacetylases are associated with repressors. Histones are also modified by phosphorylation and methylation, and specific modifications of histones affect gene expression by serving as binding sites for other regulatory proteins. In addition, nucleosome remodeling factors facilitate the binding of transcription factors to DNA by altering the arrangement or structures of nucleosomes.

SUMMARY

Regulation of Transcription by Noncoding RNAs: Transcription can be regulated by noncoding RNAs, as well as by regulatory proteins. MicroRNAs repress transcription of homologous genes by associating with a protein complex (RITS) that induces histone modifications resulting in formation of heterochromatin. X chromosome inactivation in mammals is also mediated by a noncoding RNA.

DNA Methylation: Methylation of cytosine residues can inhibit the transcription of eukaryotic genes and is important in silencing transposable elements. Regulation of gene expression by methylation also plays an important role in genomic imprinting, which controls the transcription of some genes involved in mammalian development.

RNA PROCESSING AND TURNOVER

Processing of Ribosomal and Transfer RNAs: Ribosomal and transfer RNAs are derived by cleavage of long primary transcripts in both prokaryotic and eukaryotic cells. Methyl groups are added to rRNAs, and various bases are modified in tRNAs.

Processing of mRNA in Eukaryotes: Eukaryotic pre-mRNAs are modified by the addition of 7-methylguanosine caps and 3′ poly-A tails, in addition to the removal of introns by splicing.

Splicing Mechanisms: Splicing of nuclear pre-mRNAs takes place in large complexes, called spliceosomes, composed of proteins and small nuclear RNAs (snRNAs). The snRNAs recognize sequences at the splice sites of pre-mRNAs and catalyze the splicing reaction. Some mitochondrial, chloroplast, and bacterial RNAs undergo self-splicing in which the splicing reaction is catalyzed by intron sequences.

Alternative Splicing: Exons can be joined in various combinations as a result of alternative splicing, which provides an important mechanism for tissue-specific control of gene expression in complex eukaryotes.

RNA Editing: Some mRNAs are modified by processing events that alter their protein-coding sequences. Editing of mitochondrial mRNAs in some protozoans involves the addition and deletion of U residues at multiple sites in the molecule. Other forms of RNA editing in plant and mammalian cells involve the modification of specific bases.

RNA Degradation: Introns are degraded within the nucleus, and abnormal mRNAs lacking complete open-reading frames are eliminated by nonsense-mediated mRNA decay. Functional mRNAs in eukaryotic cells are degraded at different rates, providing an additional mechanism for control of gene expression. In some cases, rates of mRNA degradation are regulated by extracellular signals.

KEY TERMS

microRNA (miRNA), X chromosome inactivation

genomic imprinting

pre-rRNA, pre-tRNA, RNase P, ribozyme

pre-mRNA, 7-methylguanosine cap, poly-A tail, polyadenylation

spliceosome, small nuclear RNA (snRNA), small nuclear-ribonucleoprotein particle (snRNP), self-splicing

alternative splicing

RNA editing

nonsense-mediated mRNA decay

Questions

1. How does footprinting identify a protein-binding site on DNA?

2. What is the role of sigma (σ) factors in bacterial RNA synthesis?

3. What is the major mechanism for termination of *E. coli* mRNAs?

4. You are working with two strains of *E. coli*. One contains a wild-type β-galactosidase gene and an i^- mutation; the other contains a temperature-sensitive β-galactosidase gene and an o^c mutation. After mating these strains, you assay for the production of β-galactosidase by the diploid cells at both permissive and nonpermissive temperatures in the absence of lactose. What do you expect to find?

5. You are comparing the requirements for *in vitro* transcription of two RNA polymerase II genes, one containing a TATA box and the other containing only an Inr sequence. Does transcription from these promoters require TBP or TFIID?

6. How do enhancers differ from promoters as *cis*-acting regulatory sequences in eukaryotes?

7. You are studying the enhancer of a gene that normally is expressed only in neurons. Constructs in which this enhancer is linked to a reporter gene are expressed in neuronal cells but not in fibroblasts. However, if you mutate a specific sequence element within the enhancer, you find expression in both fibroblasts and neuronal cells. What type of regulatory protein do you expect binds the sequence element?

8. What are the functions of insulators?

9. Explain the mechanism of X chromosome inactivation in human females.

10. What property of Sp1 was exploited for its purification by DNA affinity chromatography? How would you determine that the purified protein is indeed Sp1?

11. You have developed an *in vitro* splicing reaction in which unspliced premRNA is processed to the mature mRNA. What results would you expect if you added anti-Sm antiserum to the reaction?

12. How can noncoding RNAs regulate the concentration of a specific mRNA within a cell?

13. What is the function of splicing factors that are not components of snRNPs?

14. How are two structurally and functionally different forms of apolipoprotein B synthesized in human liver and intestine?

15. What is nonsense-mediated mRNA decay? What is its significance in the cell?

References and Further Reading

Transcription in Prokaryotes

Borukhov, S., J. Lee and O. Laptenko. 2005. Bacterial transcription elongation factors: New insights into molecular mechanism of action. *Molec. Microbiol.* 55: 1315–1324. [R]

Borukhov, S. and E. Nudler. 2003. RNA polymerase holoenzyme: Structure, function and biological implications. *Curr. Opin. Microbiol.* 6: 93–100. [R]

Gilbert, W. and B. Muller-Hill. 1966. Isolation of the *lac* repressor. *Proc. Natl. Acad. Sci. USA* 56: 1891–1899. [P]

Greive, S. J. and P. H. von Hippel. 2005. Thinking quantitatively about transcriptional regulation. *Nature Rev. Mol. Cell Biol.* 6: 221–232. [R]

Jacob, F. and J. Monod. 1961. Genetic and regulatory mechanisms in the synthesis of proteins. *J. Mol. Biol.* 3: 318–356. [P]

Lawson, C. L., D. Swigon, K. S. Murakami, S. A. Darst, H. M. Berman and R. H. Ebright. 2004. Catabolite activator protein: DNA binding and transcription activation. *Curr. Opin. Struc. Biol.* 14: 10–20. [R]

Murakami, K. S. and S. A. Darst. 2003. Bacterial RNA polymerases: The wholo story. *Curr. Opin. Struc. Biol.* 13: 31–39. [R]

Murakami, K. S., S. Masuda and S. A. Darst. 2002. Structural basis of transcription initiation: RNA polymerase holoenzyme at 4 Å resolution. *Science* 296: 1280–1284. [P]

Ptashne, M. and A. Gann. 2002. *Genes and Signals*. Cold Spring Harbor, NY: Cold Spring Harbor Laboratory Press.

Richardson, J. P. 2003. Loading Rho to terminate transcription. *Cell* 114: 157–159. [R]

Uptain, S. M., C. M. Kane and M. J. Chamberlin. 1997. Basic mechanisms of transcript elongation and its regulation. *Ann. Rev. Biochem.* 66: 117–172. [R]

Vassylyev, D. G., S. Sekine, O. Laptenko, J. Lee, M. N. Vassylyeva, S. Borukhov and S. Yokoyama. 2002. Crystal structure of a bacterial RNA polymerase holoenzyme at 2.6 Å resolution. *Nature* 417: 712–719. [P]

Zhang, G., E. A. Campbell, E. A., L. Minakhin, C. Richter, K. Severinov and S. A. Darst. 1999. Crystal structure of *Thermus aquaticus* core RNA polymerase at 3.3 Å resolution. *Cell* 98: 811–824. [P]

Eukaryotic RNA Polymerases and General Transcription Factors

Boeger, H., D. A. Bushnell, R. Davis, J. Griesenbeck, Y. Lorch, J. S. Strattan, K. D. Westover and R. D. Kornberg. 2005. Structural basis of eukaryotic gene transcription. *FEBS Lett.* 579: 899–903. [R]

Bushnell, D. A., K. D. Westover, R. E. Davis and R. D. Kornberg. 2004. Structural basis of transcription: An RNA polymerase II-TFIIB cocrystal at 4.5 Angstroms. *Science* 303: 983–988. [P]

Butler, J. E. F. and J. T. Kadonaga. 2002. The RNA polymerase II core promoter: A key component in the regulation of gene expression. *Genes Dev.* 16: 2583–2592. [R]

Conaway, J. W., L. Florens, S. Sato, C. Tomomori-Sato, T. J. Parmely, T. Yao, S. K. Swanson, C. A. S. Banks, M. P. Washburn and R. C. Conaway. 2005. The mammalian Mediator complex. *FEBS Lett.* 579: 904–908. [R]

Conaway, J. W., A. Shilatifard, A. Dvir and R. C. Conaway. 2000. Control of elongation by RNA polymerase II. *Trends Biochem. Sci.* 25: 375–380. [R]

Cramer, P., D. A. Bushnell and R. D. Kornberg. 2001. Structural basis of transcription: RNA polymerase II at 2.8 Å resolution. *Science* 292: 1863–1876. [P]

Ebright, R. H. 2000. RNA polymerase: Structural similarities between bacterial RNA polymerase and eukaryotic RNA polymerase II. *J. Mol. Biol.* 304: 687–698. [R]

Kim, Y.-J. and J. T. Lis. 2005. Interactions between subunits of *Drosophila* Mediator and activator proteins. *Trends Biochem. Sci.* 30: 245–249. [R]

Kornberg, R. D. 2005. Mediator and the mechanism of transcriptional activation. *Trends Biochem. Sci.* 30: 235–239. [R]

Lee, T. I. and R. A. Young. 2000. Transcription of eukaryotic protein-coding genes. *Ann. Rev. Genet.* 34: 77–137. [R]

Malik, S. and R. G. Roeder. 2005. Dynamic regulation of pol II transcription by the mammalian Mediator complex. *Trends Biochem. Sci.* 30: 256–263. [R]

Matsui, T., J. Segall, P. A. Weil and R. G. Roeder. 1980. Multiple factors are required for accurate initiation of transcription by purified RNA polymerase II. *J. Biol. Chem.* 255: 11992–11996. [P]

Orphanides, G. and D. Reinberg. 2002. A unified theory of gene expression. *Cell* 108: 439–451. [R]

Reese, J. C. 2003. Basal transcription factors. *Curr. Opin. Genet. Dev.* 13: 114–118. [R]

Russell, J. and J. C. B. M. Zomerdijk. 2005. RNA polymerase I-directed rDNA transcription, life and works. *Trends Biochem. Sci.* 30: 87–96. [R]

Schramm, L. and N. Hernandez. 2002. Recruitment of RNA polymerase III to its target promoters. *Genes Dev.* 16: 2593–2620. [R]

Weil, P. A., D. S. Luse, J. Segall and R. G. Roeder. 1979. Selective and accurate transcription at the Ad2 major late promoter in a soluble system dependent on purified RNA polymerase II and DNA. *Cell* 18: 469–484. [P]

Woychik, N. A. and M. Hampsey. 2002. The RNA polymerase II machinery: Structure illuminates function. *Cell* 108: 453–463. [R]

Regulation of Transcription in Eukaryotes

Ambros, V. 2004. The functions of animal microRNAs. *Nature* 431: 350–355. [R]

Arndt, K. M. and C. M. Kane. 2003. Running with RNA polymerase: Eukaryotic transcript elongation. *Trends Genet.* 19: 543–550. [R]

Bartel, D. P. 2004. MicroRNAs: Genomics, biogenesis, mechanism, and function. *Cell* 116: 281–297. [R]

Bayne, E. H. and R. C. Allshire. 2005. RNA-directed transcriptional gene silencing in mammals. *Trends Genet.* 21: 370–373. [R]

Belosserkovskaya, R. and D. Reinberg. 2004. Facts about FACT and transcript elongation through chromatin. *Curr. Opin. Genet. Dev.* 14: 139–146. [R]

Bird, A. 2002. DNA methylation patterns and epigenetic memory. *Genes Dev.* 16: 6–21. [R]

Brasset, E. and C. Vaury. 2005. Insulators are fundamental components of the eukaryotic genomes. *Heredity* 94: 571–576. [R]

Brent, R. and M. Ptashne. 1985. A eukaryotic transcriptional activator bearing the DNA specificity of a prokaryotic repressor. *Cell* 43: 729–736. [P]

Brownell, J. E., J. Zhou, T. Ranalli, R. Kobayashi, D. G. Edmondson, S. Y. Roth and C. D. Allis. 1996. *Tetrahymena* histone acetyltransferase A: A homolog to yeast Gcn5p linking histone acetylation to gene activation. *Cell* 84: 843–851. [P]

Buratowski, S. and D. Moazed. 2005. Expression and silencing coupled. *Nature* 435: 1174–1175. [R]

Bustin, M. 2001. Chromatin unfolding and activation by HMGN chromosomal proteins. *Trends Biochem. Sci.* 26: 431–437. [R]

Capelson, M. and V. G. Corces. 2004. Boundary elements and nuclear organization. *Biol. Cell* 96: 617–629. [R]

Cohen, D. E. and J. T. Lee. 2002. X-chromosome inactivation and the search for chromosome-wide silencers. *Curr. Opin. Genet. Dev.* 12: 219–224. [R]

Conaway, J. W., A. Shilatifard, A. Dvir and R. C. Conaway. 2000. Control of elongation by RNA polymerase II. *Trends Biochem. Sci.* 25: 375–380. [R]

Courey, A. J. and S. Jia. 2001. Transcriptional repression: The long and the short of it. *Genes Dev.* 15: 2786–2796. [R]

Dynan, W. S. and R. Tjian. 1983. The promoter-specific transcription factor Sp1 binds to upstream sequences in the SV40 early promoter. *Cell* 35: 79–87. [P]

Ekwall, K. 2004. The RITS complex—A direct link between small RNA and heterochromatin. *Mol. Cell* 13: 304–305. [R]

Fischle, W., Y. Wang and C. D. Allis. 2003. Binary switches and modification cassettes in histone biology and beyond. *Nature* 425: 475–479. [R]

Freitag, M. and E. U. Selker. 2005. Controlling DNA methylation: Many roads to one modification. *Curr. Opin. Genet. Dev.* 15: 191–199. [R]

Goll, M. G. and T. H. Bestor. 2005. Eukaryotic cytosine methyltransferases. *Ann. Rev. Biochem.* 74: 481–514. [R]

Green, M. R. 2005. Eukaryotic transcription: Right on target. *Mol. Cell* 18: 399–402. [R]

Grewal, S. I. S. and D. Moazed. 2003. Heterochromatin and epigenetic control of gene expression. *Science* 301: 798–802. [R]

Hajkova, P. and M. A. Surani. 2004. Programming the X chromosome. *Science* 303: 633–634. [R]

Hanna-Rose, W. and U. Hansen. 1996. Active repression mechanisms of eukaryotic transcription repressors. *Trends Genet.* 12: 229–234. [R]

Harbison, C. T. and 19 others. 2004. Transcriptional regulatory code of a eukaryotic genome. *Nature* 431: 99–104. [P]

Horn, P. J. and C. L. Peterson. 2002. Chromatin higher order folding: Wrapping up transcription. *Science* 297: 1824–1827. [R]

Jenuwein, T. and C. D. Allis. 2001. Translating the histone code. *Science* 293: 1074–1080. [R]

Kadonaga, J. T. 2004. Regulation of RNA polymerase II transcription by sequence-specific DNA binding factors. *Cell* 116: 247–257. [R]

Kadonaga, J. T. and R. Tjian. 1986. Affinity purification of sequence-specific DNA binding proteins. *Proc. Natl. Acad. Sci. USA* 83: 5889–5893. [P]

Khorasanizadeh, S. 2004. The nucleosome: From genomic organization to genomic regulation. *Cell* 116: 259–272. [R]

Lippman, Z. and R. Martienssen. 2004. The role of RNA interference in heterochromatic silencing. *Nature* 431: 364–370. [R]

Lonard, D. M. and B. W. O'Malley. 2005. Expanding functional diversity of the coactivators. *Trends Biochem. Sci.* 30: 126–132. [R]

Meister, G. and T. Tuschl. 2004. Mechanisms of gene silencing by double-stranded RNA. *Nature* 431: 343–349. [R]

Morey, C. and P. Avner. 2004. Employment opportunities for non-coding RNAs. *FEBS Lett.* 567: 27–34. [R]

Nowak, S. J. and V. G. Corces. 2004. Phosphorylation of histone H3: A balancing act between chromosome condensation and transcriptional activation. *Trends Genet.* 20: 214–220. [R]

Panning, B. and R. Jaenisch. 1998. RNA and the epigenetic regulation of X chromosome inactivation. *Cell* 93: 305–308. [R]

Peterson, C. L. 2000. ATP-dependent chromatin remodeling: Going mobile. *FEBS Lett.* 476: 68–72. [R]

Richards, E. J. and S. C. R. Elgin. 2002. Epigenetic codes for heterochromatin formation and silencing: Rounding up the usual suspects. *Cell* 108: 489–500. [R]

Roeder, R. G. 2005. Transcriptional regulation and the role of diverse coactivators in animal cells. *FEBS Lett.* 579: 909–915. [R]

Sims, R. J. III, R. Belotserkovskaya and D. Reinberg. 2004. Elongation by RNA polymerase II: The short and long of it. *Genes Dev.* 18: 2437–2468. [R]

Staudt, L. M. and M. J. Lenardo. 1991. Immunoglobulin gene transcription. *Ann. Rev. Immunol.* 9: 373–398. [R]

Taunton, J., C. A. Hassig and S. L. Schreiber. 1996. A mammalian histone deacetylase related to the yeast transcriptional regulator Rpd3p. *Science* 272: 408–411. [P]

Teixeira da Rocha, S. and A. C. Ferguson-Smith. 2004. Genomic imprinting. *Curr. Biol.* 14: R646–649. [R]

Turner, B. M. 2002. Cellular memory and the histone code. *Cell* 111: 285–291. [R]

RNA Processing and Turnover

Abelson, J., C. R. Trotta and H. Li. 1998. tRNA splicing. *J. Biol. Chem.* 273:12685–12688. [R]

Black, D. L. 2003. Mechanisms of alternative pre-messenger RNA splicing. *Ann. Rev. Biochem.* 72: 291–336. [R]

Blanc, V. and N. O. Davidson. 2003. C-to-U RNA editing: Mechanisms leading to genetic diversity. *J. Biol. Chem.* 278: 1395–1398. [R]

Dodson, R. E. and D. J. Shapiro. 2002. Regulation of pathways of mRNA destabilization and stabilization. *Prog. Nucl. Acid Res.* 72: 129–164. [R]

Guerrier-Takada, C., K. Gardiner, T. Marsh, N. Pace and S. Altman. 1983. The RNA moiety of ribonuclease P is the catalytic subunit of the enzyme. *Cell* 35: 849–857. [P]

Haugen, P., D. M. Simon and D. Bhattacharya. 2005. The natural history of group I introns. *Trends Genet.* 21: 111–119. [R]

Hentze, M. W., M. U. Muckenthaler and N. C. Andrews. 2004. Balancing acts: Molecular control of mammalian iron metabolism. *Cell* 117: 285–297. [R]

Hopper, A. K. and E. M. Phizicky. 2003. tRNA transfers to the limelight. *Genes Dev.* 17: 162–180. [R]

Kruger, K., P. J. Grabowski, A. Zaug, A. J. Sands, D. E. Gottschling and T. R. Cech. 1982. Self-splicing RNA: Autoexcision and autocyclization of the ribosomal RNA intervening sequence of *Tetrahymena*. *Cell* 31: 147–157. [P]

Maas, S., A. Rich and K. Nishikura. 2003. A-to-I RNA editing: Recent news and residual mysteries. *J. Biol. Chem.* 278: 1391–1394. [R]

Maniatis, T. and R. Reed. 2002. An extensive network of coupling among gene expression machines. *Nature* 416: 499–506. [R]

Maquat, L. E. 2004. Nonsense-mediated mRNA decay: Splicing, translation and mRNP dynamics. *Nature Rev. Mol. Cell Biol.* 5: 89–99. [R]

Matlin, A. J., F. Clark and C. W. J. Smith. 2005. Understanding alternative splicing: Towards a cellular code. *Nature Rev. Mol. Cell Biol.* 6: 386–398. [R]

Moore, M. J. 2002. Nuclear RNA turnover. *Cell* 108: 431–434. [R]

Padgett, R. A., M. M. Konarska, P. J. Grabowski, S. F. Hardy and P. A. Sharp. 1984. Lariat RNAs as intermediates and products in the splicing of messenger RNA precursors. *Science* 225: 898–903. [P]

Padgett, R. A., S. M. Mount, J. A. Steitz and P. A. Sharp. 1983. Splicing of messenger RNA precursors is inhibited by antisera to small nuclear ribonucleoprotein. *Cell* 35: 101–107. [P]

Parker, R. and H. Song. 2004. The enzymes and control of eukaryotic mRNA turnover. *Nature Struc. Mol. Biol.* 11: 121–127. [R]

Proudfoot, N. J. 2004. New perspectives on connecting messenger RNA 3′ end formation to transcription. *Curr. Opin. Cell Biol.* 16: 272–278. [R]

Schmicker, D. and J. G. Flanagan. 2004. Generation of recognition diversity in the nervous system. *Neuron* 44: 219–222. [R]

Seeburg, P. H. and J. Hartner. 2003. Regulation of ion channel/neurotransmitter receptor function by RNA editing. *Curr. Opin. Neurobiol.* 13: 279–283. [R]

Shin, C. and J. L. Manley. 2004. Cell signaling and the control of pre-mRNA splicing. *Nature Rev. Mol. Cell Biol.* 5: 727–738. [R]

Smith, C. W. J. and J. Valcarcel. 2000. Alternative pre-mRNA splicing: The logic of combinatorial control. *Trends Biochem. Sci.* 25: 381–388. [R]

Stetefeld, J. and M. A. Ruegg. 2005. Structural and functional diversity generated by alternative mRNA splicing. *Trends Biochem. Sci.* 30: 515–521. [R]

Stuart, K. D., A. Schnaufer, N. L. Ernst and A. K. Panigrahi. 2005. Complex management: RNA editing in trypanosomes. *Trends Biochem. Sci.* 30: 97–105. [R]

Valadkhan, S. and J. L. Manley. 2001. Splicing-related catalysis by protein-free snRNAs. *Nature* 413: 701–7078. [P]

Villa, T., J. A. Pleiss and C. Guthrie. 2002. Spliceosomal snRNAs: Mg^{2+}-dependent chemistry at the catalytic core? *Cell* 109: 149–152. [R]

Weiner, A. M. 2004. tRNA maturation: RNA polymerization without a nucleic acid template. *Curr. Biol.* 14: R883–R885. [R]

CHAPTER 8

Protein Synthesis, Processing, and Regulation

- **Translation of mRNA** *309*

- **Protein Folding and Processing** *329*

- **Regulation of Protein Function** *339*

- **Protein Degradation** *344*

- **KEY EXPERIMENT:**
  Catalytic Role of Ribosomal RNA *316*

- **MOLECULAR MEDICINE:**
  Antibiotic Resistance and the
  Ribosome *320*

TRANSCRIPTION AND RNA PROCESSING ARE FOLLOWED BY TRANSLATION, the synthesis of proteins as directed by mRNA templates. Proteins are the active players in most cell processes, implementing the myriad tasks that are directed by the information encoded in genomic DNA. Protein synthesis is thus the final stage of gene expression. However, the translation of mRNA is only the first step in the formation of a functional protein. The polypeptide chain must then fold into the appropriate three-dimensional conformation and, frequently, undergo various processing steps before being converted to its active form. These processing steps, particularly in eukaryotes, are intimately related to the sorting and transport of different proteins to their appropriate destinations within the cell.

Although the expression of most genes is regulated primarily at the level of transcription (see Chapter 7), gene expression can also be controlled at the level of translation, and this control is an important element of gene regulation in both prokaryotic and eukaryotic cells. Of even broader significance, however, are the mechanisms that control the activities of proteins within cells. Once synthesized, most proteins can be regulated in response to extracellular signals by either covalent modifications or by association with other molecules. In addition, the levels of proteins within cells can be controlled by differential rates of protein degradation. These multiple controls of both the amounts and activities of intracellular proteins ultimately regulate all aspects of cell behavior.

Translation of mRNA

Proteins are synthesized from mRNA templates by a process that has been highly conserved throughout evolution (reviewed in Chapter 4). All mRNAs are read in the 5′ to 3′ direction, and polypeptide chains are synthesized from the amino to the carboxy terminus. Each amino acid is specified by three bases (a codon) in the mRNA, according to a nearly universal genetic code. The basic mechanics of protein synthesis are also the same in all cells: Translation is carried out on ribosomes, with tRNAs serving as adaptors between the mRNA template and the amino acids being incorporated into protein. Protein synthesis thus involves interactions between three types of RNA molecules (mRNA templates, tRNAs, and rRNAs), as well as various proteins that are required for translation.

Transfer RNAs

During translation, each of the 20 amino acids must be aligned with its corresponding codon on the mRNA template. All cells contain a variety of **tRNAs** that serve as adaptors for this process. As might be expected, given their common function in protein synthesis, different tRNAs share similar overall structures. However, they also possess unique identifying sequences that allow the correct amino acid to be attached and aligned with the appropriate codon in mRNA.

Transfer RNAs are approximately 70 to 80 nucleotides long and have characteristic cloverleaf structures that result from complementary base pairing between different regions of the molecule (Figure 8.1). X-ray crystallography studies have further shown that all tRNAs fold into similar compact L shapes, which are required for the tRNAs to fit onto ribosomes during the translation process. The adaptor function of the tRNAs involves two separated regions of the molecule. All tRNAs have the sequence CCA at their 3' terminus, and amino acids are covalently attached to the ribose of the terminal adenosine. The mRNA template is then recognized by the **anticodon** loop, located at the other end of the folded tRNA, which binds to the appropriate codon by complementary base pairing.

The incorporation of the correctly encoded amino acids into proteins depends on the attachment of each amino acid to an appropriate tRNA, as well as on the specificity of codon-anticodon base pairing. The attachment of amino acids to specific tRNAs is mediated by a group of enzymes called **aminoacyl tRNA synthetases**, which were discovered by Paul Zamecnik and Mahlon Hoagland in 1957. Each of these 20 enzymes recognizes a sin-

FIGURE 8.1 Structure of tRNAs The structure of yeast phenylalanyl tRNA is illustrated in open "cloverleaf" form (A) to show complementary base pairing. Modified bases are indicated as mG, methylguanosine; mC, methylcytosine; DHU, dihydrouridine; T, ribothymidine; Y, a modified purine (usually adenosine); and ψ, pseudouridine. The folded form of the molecule is shown in (B) and a space-filling model in (C). (C, courtesy of Dan Richardson.)

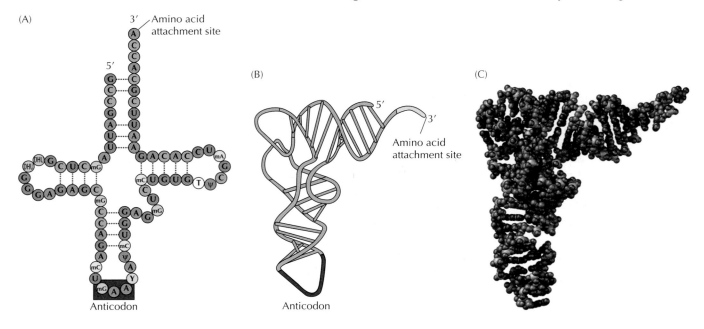

FIGURE 8.2 Attachment of amino acids to tRNAs   In the first reaction step, the amino acid is joined to AMP, forming an aminoacyl AMP intermediate. In the second step, the amino acid is transferred to the 3' CCA terminus of the acceptor tRNA and AMP is released. Both steps of the reaction are catalyzed by aminoacyl tRNA synthetases.

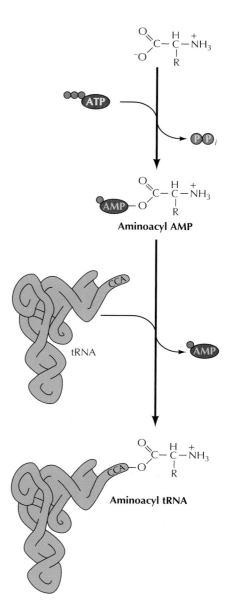

gle amino acid, as well as the correct tRNA (or tRNAs) to which that amino acid should be attached. The reaction proceeds in two steps (Figure 8.2). First, the amino acid is activated by reaction with ATP to form an aminoacyl AMP synthetase intermediate. The activated amino acid is then joined to the 3' terminus of the tRNA. The aminoacyl tRNA synthetases must be highly selective enzymes that recognize both individual amino acids and specific base sequences that identify the correct acceptor tRNAs. In some cases, the high fidelity of amino acid recognition results in part from a proofreading function by which incorrect aminoacyl AMPs are hydrolyzed rather than being joined to tRNA during the second step of the reaction. Recognition of the correct tRNA by the aminoacyl tRNA synthetase is also highly selective; the synthetase recognizes specific nucleotide sequences (in most cases including the anticodon) that uniquely identify each species of tRNA.

After being attached to tRNA, an amino acid is aligned on the mRNA template by complementary base pairing between the mRNA codon and the anticodon of the tRNA. Codon-anticodon base pairing is somewhat less stringent than the standard A-U and G-C base pairing discussed in preceding chapters. The significance of this unusual base pairing in codon-anticodon recognition relates to the redundancy of the genetic code. Of the 64 possible codons, 3 are stop codons that signal the termination of translation; the other 61 encode amino acids (see Table 4.1). Thus most of the amino acids are specified by more than one codon. In part, this redundancy results from the attachment of many amino acids to more than one species of tRNA. *E. coli*, for example, contain about 40 different tRNAs that serve as acceptors for the 20 different amino acids. In addition, some tRNAs are able to recognize more than one codon in mRNA, as a result of nonstandard base pairing (called wobble) between the tRNA anticodon and the third position of some complementary codons (Figure 8.3). Relaxed base pairing at this position results partly from the formation of G-U base pairs and partly from the modification of guanosine to inosine in the anticodons of several tRNAs during processing (see Figure 7.43). Inosine can base-pair with either C, U, or A in the third position, so its inclusion in the anticodon allows a single tRNA to recognize three different codons in mRNA templates.

The Ribosome

Ribosomes are the sites of protein synthesis in both prokaryotic and eukaryotic cells. First characterized as particles detected by ultracentrifugation of cell lysates, ribosomes are usually designated according to their rates of sedimentation: 70S for bacterial ribosomes and 80S for the somewhat larger ribosomes of eukaryotic cells. Both prokaryotic and eukaryotic ribosomes are composed of two distinct subunits, each containing characteristic proteins and **rRNAs**. The fact that cells typically contain many ribosomes reflects the central importance of protein synthesis in cell metabolism. *E. coli*, for example, contain about 20,000 ribosomes, which account for approximately 25% of the dry weight of the cell, and rapidly growing mammalian cells contain about 10 million ribosomes.

FIGURE 8.3 Nonstandard codon-anticodon base pairing Base pairing at the third codon position is relaxed, allowing G to pair with U, and inosine (I) in the anticodon to pair with U, C, or A. Two examples of abnormal base pairing, allowing phenylalanyl (Phe) tRNA to recognize either UUC or UUU codons and alanyl (Ala) tRNA to recognize GCU, GCC, or GCA, are illustrated.

Phenylalanyl tRNA pairing

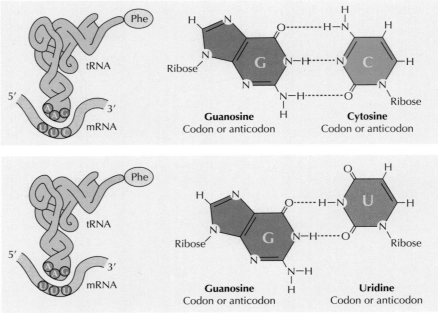

Guanosine
Codon or anticodon

Cytosine
Codon or anticodon

Guanosine
Codon or anticodon

Uridine
Codon or anticodon

Alanyl tRNA pairing

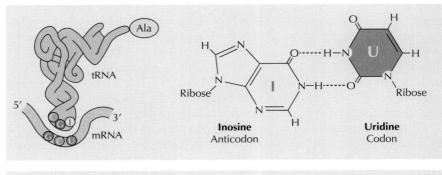

Inosine
Anticodon

Uridine
Codon

Inosine
Anticodon

Cytosine
Codon

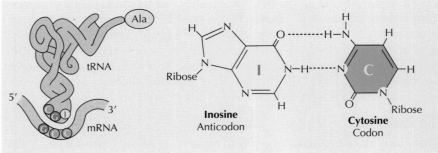

Inosine
Anticodon

Adenine
Codon

The general structures of prokaryotic and eukaryotic ribosomes are similar, although they differ in some details (Figure 8.4). The small subunit (designated 30S) of *E. coli* ribosomes consists of the 16S rRNA and 21 proteins; the large subunit (50S) is composed of the 23S and 5S rRNAs and 34 proteins. Each ribosome contains one copy of the rRNAs and one copy of each of the ribosomal proteins, with one exception: One protein of the 50S subunit is present in four copies. The subunits of eukaryotic ribosomes are larger and contain more proteins than their prokaryotic counterparts. The small subunit (40S) of eukaryotic ribosomes is composed of the 18S rRNA and approximately 30 proteins; the large subunit (60S) contains the 28S, 5.8S, and 5S rRNAs and about 45 proteins. Because of their large size and complexity, high-resolution structural analysis of ribosomes by X-ray crystallography was not accomplished until 2000, when structures of both the 50S and 30S subunits were first reported. As discussed below, understand-

(A)

Prokaryotic 70S ribosome

50S — 23S and 5S rRNAs (34 proteins)

30S — 16S rRNA (21 proteins)

Eukaryotic 80S ribosome

60S — 28S, 5.8S, and 5S rRNAs (~45 proteins)

40S — 18S rRNA (~30 proteins)

FIGURE 8.4 Ribosome structure
(A) Components of prokaryotic and eukaryotic ribosomes. Intact prokaryotic and eukaryotic ribosomes are designated 70S and 80S, respectively, on the basis of their sedimentation rates in ultracentrifugation. They consist of large and small subunits, which contain both ribosomal proteins and rRNAs. (B–C) High resolution X-ray crystal structures of 30S (B) and 50S (C) ribosomal subunits. (B, from B. T. Wimberly, D. E. Brodersen, W. M. Clemons, R. J. Morgan-Warren, A. P. Carter, C. Vonrhein, T. Hartrsch and V. Ramakrishnan, 2000. *Nature* 407: 327. C, from N. Ban, P. Nissen, J. Hansen, P. B. Moore and T. A. Steitz, 2000. *Science* 289: 905.)

(B)

(C)

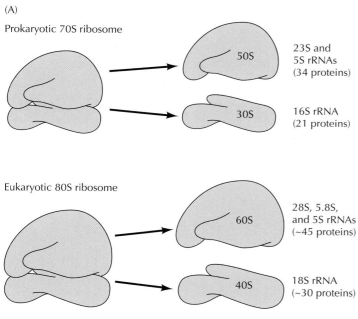

FIGURE 8.5 Structure of 16S rRNA
Complementary base pairing results in the formation of a distinct secondary structure. (From M. M. Yusupov, G. Z. Yusupova, A. Baucom, K. Lieberman, T. N. H. Earnest, J. H. D. Cate and H. F. Noller. 2001. *Science* 292: 883.)

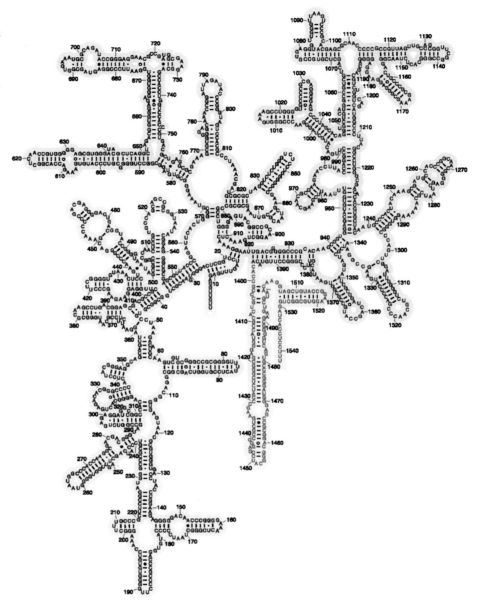

■ Although the sequence of ribosomal RNAs is highly conserved, substitutions have occurred throughout the course of evolution. The comparison of rRNA sequences is a powerful tool in determining evolutionary relationships between different species.

ing the structure of ribosomes at the atomic level has had a major impact on our understanding of ribosome function.

A noteworthy feature of ribosomes is that they can be formed *in vitro* by self-assembly of their RNA and protein constituents. As first described in 1968 by Masayasu Nomura, purified ribosomal proteins and rRNAs can be mixed together and, under appropriate conditions, will re-form a functional ribosome. Although ribosome assembly *in vivo* (particularly in eukaryotic cells) is considerably more complicated, the ability of ribosomes to self-assemble *in vitro* has provided an important experimental tool, allowing analysis of the roles of individual proteins and rRNAs.

Like tRNAs, rRNAs form characteristic secondary structures by complementary base pairing (Figure 8.5). In association with ribosomal proteins, the rRNAs fold further, into distinct three-dimensional structures. Initially, rRNAs were thought to play a structural role, providing a scaffold upon which ribosomal proteins assemble. However, with the discovery of the catalytic activity of other RNA molecules (e.g., RNase P and the self-splicing introns discussed in Chapter 7), the possible catalytic role of rRNA became widely considered. Consistent with this hypothesis, rRNAs were found to be absolutely required for the *in vitro* assembly of functional ribosomes. On the other hand, the omission of many ribosomal proteins resulted in a decrease, but not a complete loss, of ribosomal activity.

Direct evidence for the catalytic activity of rRNA first came from experiments of Harry Noller and his colleagues in 1992. These investigators demonstrated that the large ribosomal subunit is able to catalyze the formation of peptide bonds (the peptidyl transferase reaction) even after approximately 90% of the ribosomal proteins have been removed by standard protein extraction procedures. In contrast, treatment with RNase completely abolishes peptide bond formation, providing strong support for the hypothesis that the formation of a peptide bond is an RNA-catalyzed reaction. However, some of the ribosomal proteins could not be removed under conditions that left the ribosomal RNA intact, so the role of ribosomal proteins as catalysts of peptide bond formation could not be definitively ruled out.

Unambiguous evidence that protein synthesis is catalyzed by rRNA came from the first high-resolution structural analysis of the 50S ribosomal subunit, which was reported by Peter Moore, Thomas Seitz, and their colleagues in 2000 (Figure 8.6). This atomic-level view of ribosome structure revealed that ribosomal proteins were strikingly absent from the site at which the peptidyl transferase reaction occurred, making it evident that rRNA was responsible for catalyzing peptide bond formation. The large

FIGURE 8.6 Structure of the 50S ribosomal subunit A high-resolution model of the 50S ribosomal subunit with three tRNA molecules bound to the A, P, and E sites of the ribosome (see Figure 8.12). Ribosomal proteins are shown in pink and the rRNA in blue. (From P. Nissen, J. Hansen, N. Ban, P. B. Moore and T. A. Steitz. 2000. *Science* 289: 920.)

KEY EXPERIMENT

Catalytic Role of Ribosomal RNA

Unusual Resistance of Peptidyl Transferase to Protein Extraction Procedures

Harry F. Noller, Vernita Hoffarth, and Ludwika Zimniak
University of California at Santa Cruz
Science, Volume 256, 1992, pages 1416–1419

Harry F. Noller

The Context

The role of ribosomes in protein synthesis was elucidated in the 1960s. During this period, ribosomes were characterized as particles consisting of both proteins and RNAs, and the reconstitution of functional ribosomes from purified components was accomplished. At that time peptide bond formation (the peptidyl transferase reaction) was generally assumed to be catalyzed by ribosomal proteins, with the rRNAs delegated to a supporting role in ribosome structure. As early as the 1970s, however, evidence began to suggest that the rRNAs might play a more direct role in protein synthesis. For example, many ribosomal proteins were found to be nonessential for ribosome function. Conversely, the sequences of some parts of rRNA were shown to be extremely well conserved in evolution, suggesting a critical functional role for these portions of the rRNA molecule.

In the early 1980s, the catalytic activity of RNA molecules was established by Tom Cech's studies of the *Tetrahymena* ribozyme and Sidney Altman's studies of RNase P. These discoveries of RNA catalysis provided a precedent for the hypothesis that rRNA is directly involved in catalyzing peptide bond formation. Compelling evidence in support of such a role for rRNA was provided in this 1992 paper by Harry Noller and his colleagues.

The Experiments

To study the catalytic activity of rRNA, Noller and colleagues used a simplified model reaction to assay

peptidyl transferase activity. This reaction measures the transfer of radioactively labeled N-formylmethionine from a fragment of tRNA to the amino group of puromycin, an antibiotic that resembles an aminoacyl tRNA and can form peptide bonds with a growing peptide chain. The advantage of this peptidyl transferase model reaction is that it can be carried out by isolated 50S ribosomal subunits; small ribosomal subunits, other protein factors, and mRNA are not required.

The investigators then tested the role of rRNA by assaying the peptidyl transferase activity of 50S subunits from which the ribosomal proteins had been removed by standard protein extraction procedures. An important aspect of the experiments was the use of ribosomes from the bacterium *Thermus aquaticus*. Because these bacteria live at high temperatures, the structure of their rRNA was thought to be more stable than that of *E. coli* rRNA. The critical result was that the peptidyl transferase activity of *T. aquaticus* ribosomes was completely resistant to a vigorous extraction procedure involving treatment with detergents, proteases, and phenol (see figure). Most strikingly, full peptidyl transferase activity was retained even after repeated extractions that removed 90% of the ribosomal protein. In contrast, peptidyl transferase of either intact or extracted ribosomes was extremely sensitive to even a brief treatment with RNase. Although these experiments could not exclude a possible role for the remaining ribosomal proteins, they provided strong support for the direct participation of 23S rRNA in the peptidyl transferase reaction.

The Impact

The results of Noller's experiments were definitively confirmed by the high-resolution structural analysis of the 50S ribosomal subunit reported by Peter Moore, Thomas Steitz and their colleagues in 2000. The ribosomal site of the peptidyl transferase reaction contained only rRNA, with no ribosomal proteins in the vicinity. These results remove any doubts as to the catalytic role of RNA in this reaction and demonstrate that the fundamental reaction of protein synthesis is cat-

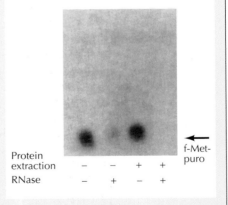

| Protein extraction | − | − | + | + |
| RNase | − | + | − | + |

f-Met-puro

The peptidyl transferase reaction is assayed by the formation of radioactive N-formyl-methionine-puromycin (f-Met-puro), which is detected by electrophoresis and autoradiography. Ribosomes from *T. aquaticus* were subjected to protein extraction or treatment with RNase, as indicated.

alyzed by ribosomal RNA. These findings not only have a striking impact on our understanding of ribosome function but also profoundly extend the previously described catalytic activities of RNA molecules and provide substantial new support for the hypothesis that an early period of evolution was an RNA world populated by self-replicating RNA molecules. This hypothesis had previously been based on the ability of RNA molecules to catalyze the reactions required for their own replication. The discovery that RNA could catalyze reactions involved in protein synthesis provided a clear link between the RNA world and the flow of genetic information in present-day cells, with rRNA continuing to function in the key reaction of peptide bond formation.

ribosomal subunit thus functions as a ribozyme, with the fundamental reaction of protein synthesis being catalyzed by ribosomal RNA. Rather than being the primary catalytic constituents of ribosomes, ribosomal proteins are now thought to play a largely structural role.

The direct involvement of rRNA in the peptidyl transferase reaction has important evolutionary implications. RNAs are thought to have been the first self-replicating macromolecules (see Chapter 1). This notion is strongly supported by the fact that ribozymes, such as RNase P and self-splicing introns, can catalyze reactions that involve RNA substrates. The role of rRNA in the formation of peptide bonds extends the catalytic activities of RNA beyond self-replication to direct involvement in protein synthesis. Additional studies indicate that the *Tetrahymena* rRNA ribozyme can catalyze the attachment of amino acids to RNA, lending credence to the possibility that the original aminoacyl tRNA synthetases were RNAs rather than proteins. The ability of RNA molecules to catalyze the reactions required for protein synthesis as well as for self-replication may provide an important link for understanding the early evolution of cells.

The Organization of mRNAs and the Initiation of Translation

Although the mechanisms of protein synthesis in prokaryotic and eukaryotic cells are similar, there are also differences, particularly in the signals that determine the positions at which synthesis of a polypeptide chain is initiated on an mRNA template (Figure 8.7). Translation does not simply begin at the 5′ end of the mRNA; it starts at specific initiation sites. The 5′ terminal portions of both prokaryotic and eukaryotic mRNAs are therefore noncoding sequences, referred to as **5′ untranslated regions**. Eukaryotic mRNAs usually encode only a single polypeptide chain, but many prokaryotic mRNAs encode multiple polypeptides that are synthesized independently from distinct initiation sites. For example, the *E. coli lac* operon consists of three genes that are translated from the same mRNA (see Figure 7.9). Messenger RNAs that encode multiple polypeptides are called **polycistronic**, whereas **monocistronic** mRNAs encode a single polypeptide chain. Finally, both prokaryotic and eukaryotic mRNAs end in noncoding **3′ untranslated regions**.

In both prokaryotic and eukaryotic cells, translation always initiates with the amino acid methionine, usually encoded by AUG. Alternative initiation codons, such as GUG, are used occasionally, but when they occur at the beginning of a polypeptide chain, these codons direct the incorporation of methionine rather than the amino acid they normally encode (GUG normally encodes valine). In most bacteria, protein synthesis is initiated with a

FIGURE 8.7 Prokaryotic and eukaryotic mRNAs Both prokaryotic and eukaryotic mRNAs contain untranslated regions (UTRs) at their 5′ and 3′ ends. Eukaryotic mRNAs also contain 5′ 7-methylguanosine (m⁷G) caps and 3′ poly-A tails. Prokaryotic mRNAs are frequently polycistronic: They encode multiple proteins, each of which is translated from an independent start site. Eukaryotic mRNAs are usually monocistronic, encoding only a single protein.

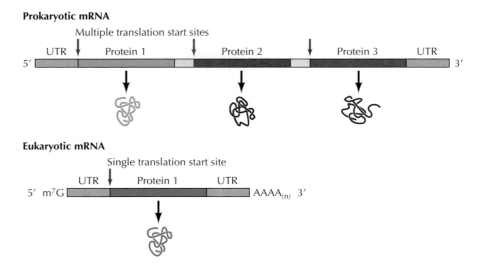

modified methionine residue (*N*-formylmethionine), whereas unmodified methionines initiate protein synthesis in eukaryotes (except in mitochondria and chloroplasts, whose ribosomes resemble those of bacteria).

The signals that identify initiation codons are different in prokaryotic and eukaryotic cells, consistent with the distinct functions of polycistronic and monocistronic mRNAs (Figure 8.8). Initiation codons in bacterial mRNAs are preceded by a specific sequence (called a **Shine-Dalgarno sequence**, after its discoverers, John Shine and Lynn Dalgarno) that aligns the mRNA on the ribosome for translation by base-pairing with a complementary sequence near the 3′ terminus of 16S rRNA. This base-pairing interaction enables bacterial ribosomes to initiate translation not only at the 5′ end of an mRNA but also at the internal initiation sites of polycistronic messages. In contrast, ribosomes recognize most eukaryotic mRNAs by binding to the 7-methylguanosine cap at their 5′ terminus (see Figure 7.44). The ribosomes then scan downstream of the 5′ cap until they encounter the

FIGURE 8.8 Signals for translation initiation Initiation sites in prokaryotic mRNAs are characterized by a Shine-Dalgarno sequence that precedes the AUG initiation codon. Base pairing between the Shine-Dalgarno sequence and a complementary sequence near the 3′ terminus of 16S rRNA aligns the mRNA on the ribosome. In contrast, eukaryotic mRNAs are bound to the 40S ribosomal subunit by their 5′ 7-methylguanosine caps. The ribosome then scans along the mRNA until it encounters an AUG initiation codon.

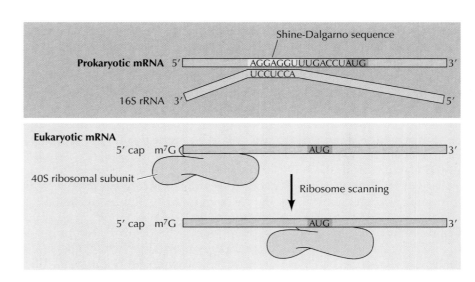

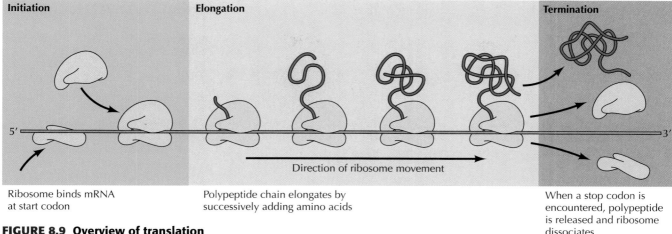

Initiation	Elongation	Termination

Direction of ribosome movement

Ribosome binds mRNA
at start codon

Polypeptide chain elongates by
successively adding amino acids

When a stop codon is
encountered, polypeptide
is released and ribosome
dissociates

FIGURE 8.9 Overview of translation

initiation codon (usually AUG). Sequences that surround AUGs affect the efficiency of initiation, so in some cases the first AUG in the mRNA is bypassed and translation initiates at an AUG farther downstream. It should also be noted that several viral and cellular mRNAs have internal ribosome entry sites at which translation can initiate by ribosome binding to an internal position on the mRNA. Neither the biological function nor the molecular mechanism by which translation initiates at these internal sites is understood, so the role of internal ribosomal entry sites in eukaryotic mRNAs remains an active area of investigation.

The Process of Translation

Translation is generally divided into three stages: initiation, elongation, and termination (Figure 8.9). In both prokaryotes and eukaryotes the first step of the initiation stage is the binding of a specific initiator methionyl tRNA and the mRNA to the small ribosomal subunit. The large ribosomal subunit then joins the complex, forming a functional ribosome on which elongation of the polypeptide chain proceeds. A number of specific nonribosomal proteins are also required for the various stages of the translation process (Table 8.1).

The first translation step in bacteria is the binding of three **initiation factors** (IF1, IF2, and IF3) to the 30S ribosomal subunit (Figure 8.10). The mRNA and initiator N-formylmethionyl tRNA then join the complex, with IF2 (which is bound to GTP) specifically recognizing the initiator tRNA. IF3 is then released, allowing a 50S ribosomal subunit to associate with the complex. This association triggers the hydrolysis of GTP bound to IF2, which

8.1 **WEBSITE ANIMATION**
Translation Translation is
RNA-directed polypeptide synthesis.

TABLE 8.1 Translation Factors

Role	Prokaryotes	Eukaryotes
Initiation	IF1, IF2, IF3	eIF1, eIF1A, eIF2, eIF2B, eIF3, eIF4A, eIF4B, eIF4E, eIF4G, eIF4H, eIF5, eIF5B
Elongation	EF-Tu, EF-Ts, EF-G	eEF1α, eEF1$\beta\gamma$, eEF2
Termination	RF1, RF2, RF3	eRF1, eRF3

MOLECULAR MEDICINE

Antibiotic Resistance and the Ribosome

The Disease

Bacteria are responsible for a wide variety of potentially lethal infectious diseases, including tuberculosis, bacterial pneumonia, childhood meningitis, infections of wounds and burns, syphilis, and gonorrhea. Prior to the 1940s, physicians had no effective treatments for these bacterial infections. At that time, the first antibiotics became available for clinical use. Previously untreatable infections became curable, and a significant increase in average life span has been attributed to the introduction of antibiotics into clinical practice. More than a hundred different antibiotics are now in use, providing effective treatment for bacterial infections. However, bacteria are increasingly becoming resistant to the antibiotics in current use, so the development of new drugs is a continuing challenge.

Molecular and Cellular Basis

To be effective clinically, an antibiotic must kill or inhibit the growth of bacteria without being toxic to humans. Thus most clinically useful antibiotics are directed against targets that are present in bacteria but not in human cells. Penicillin, for example, inhibits synthesis of the bacterial cell wall (see Chapter 14). Many other commonly used antibiotics, however, inhibit protein synthesis by binding to the ribosome. Some of these antibiotics—including streptomycin, tetracycline,

chloramphenicol, and erythromycin—are specific for prokaryotic ribosomes and therefore are effective agents for the treatment of bacterial infections. Other antibiotic inhibitors of protein synthesis act against both prokaryotes and eukaryotes (e.g., puromycin) or against eukaryotes only (e.g., cycloheximide). Although these antibiotics are obviously not useful clinically, they have provided important experimental tools for studies of protein synthesis in both prokaryotic and eukaryotic cells.

Prevention and Treatment

The use of antibiotics has had a major impact on modern medicine by enabling physicians to cure otherwise life-threatening bacterial infections. Unfortunately, however, mutations can lead to the development of antibiotic-resistant strains of bacteria. Many strains of bacteria are now resistant to one or more antibiotics, so physicians sometimes have to try several different antibiotics before they find one that is effective. Bacterial resistance to antibiotics is increasing rapidly, particularly in hospitals. It is estimated that about two million patients a year acquire bacterial infections in hospitals, and more than half of these bacteria are resistant to at least one antibiotic. Most seriously, strains of bacteria that are resistant to multiple antibiotics have emerged, and a few strains of bacteria, frequently found in hospitals, are now resistant to all

but one or a few known antibiotics. The emergence of these strains with multiple resistance raises the specter of untreatable infections caused by the spread of antibiotic-resistant bacteria—a scenario that would herald a return to the pre-antibiotic era of infectious diseases. The continuing development of new antibiotics thus represents an important concern in drug development. Our increasing understanding of the structures of several antibiotics bound to bacterial ribosomes may provide a rational approach to this problem and allow the targeted development of new antibiotics to replace those against which bacterial resistance has emerged.

References

Leeb, M. 2004. A shot in the arm. *Nature* 431: 892–893.

Steitz, T. A. 2005. On the structural basis of peptide-bond formation and antibiotic resistance from atomic structures of the large ribosomal subunit. *FEBS Lett.* 579: 955–958.

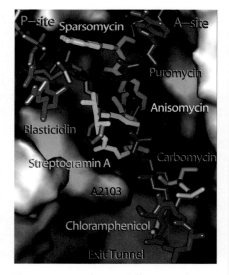

The structures of seven different antibiotics bound to a bacterial ribosome. (From T. A. Steitz. 2005. *FEBS Lett.* 579: 955.)

Antibiotic Inhibitors of Protein Synthesis

Antibiotic	Target cells	Effect
Streptomycin	Prokaryotic	Inhibits initiation and causes misreading
Tetracycline	Prokaryotic	Inhibits binding of aminoacyl tRNAs
Chloramphenicol	Prokaryotic	Inhibits peptidyl transferase activity
Erythromycin	Prokaryotic	Inhibits translocation
Puromycin	Prokaryotic and eukaryotic	Causes premature chain termination
Cycloheximide	Eukaryotic	Inhibits peptidyl transferase activity

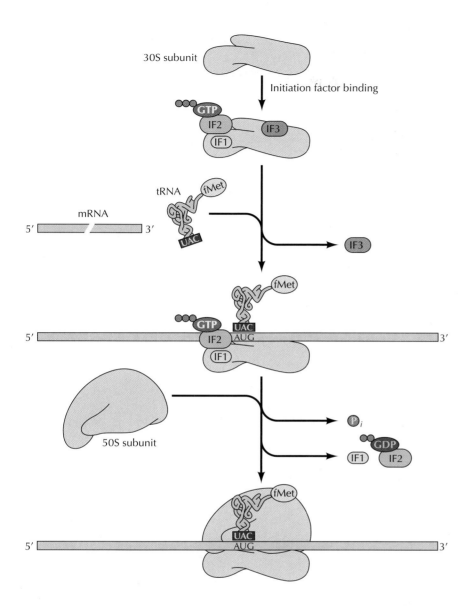

FIGURE 8.10 Initiation of translation in bacteria Three initiation factors (IF1, IF2, and IF3) first bind to the 30S ribosomal subunit. This step is followed by binding of the mRNA and the initiator *N*-formylmethionyl (fMet) tRNA, which is recognized by IF2 bound to GTP. IF3 is then released, and a 50S subunit binds to the complex, triggering the hydrolysis of bound GTP, followed by the release of IF1 and IF2 bound to GDP.

leads to the release of IF1 and IF2 (bound to GDP). The result is the formation of a 70S initiation complex (with mRNA and initiator tRNA bound to the ribosome) that is ready to begin peptide bond formation during the elongation stage of translation.

Initiation in eukaryotes is more complicated and requires at least twelve proteins (each consisting of multiple polypeptide chains), which are designated eIFs (*e*ukaryotic *i*nitiation *f*actors; see Table 8.1). The factors eIF1A and eIF3 bind to the 40S ribosomal subunit, and eIF2 (in a complex with GTP) binds to the initiator methionyl tRNA (Figure 8.11). A preinitiation complex is then formed by association of the 40S subunit, the initiator tRNA, and eIF1. The mRNA is recognized and brought to the ribosome by the eIF4 group of factors. The 5′ cap of the mRNA is recognized by eIF4E, which forms a complex with eIF4A and eIF4G. eIF4G also binds to poly-A binding protein (PABP), which is associated with the poly-A tail at the 3′ end of the mRNA. Eukaryotic initiation factors thus recognize both the 5′ and 3′ ends of mRNAs, accounting for the stimulatory effect of polyadeny-

FIGURE 8.11 Initiation of translation in eukaryotic cells Initiation factors eIF1A and eIF3 bind to the 40S ribosomal subunit. The initiator methionyl tRNA is bound by eIF2 (complexed to GTP), and forms a complex with the 40S subunit and eIF1. The mRNA is brought to the 40S subunit by eIF4E (which binds to the 5′ cap), eIF4G (which binds to both eIF4E at the 5′ cap and PABP at the 3′ poly-A tail), eIF4A, and eIF4B. The ribosome then scans down the mRNA to identify the first AUG initiation codon. Scanning requires energy and is accompanied by ATP hydrolysis. When the initiating AUG is identified, eIF5 triggers the hydrolysis of GTP bound to eIF2, followed by the release of eIF2 (complexed to GDP) and other initiation factors. The 60S ribosomal subunit then joins the 40S complex, facilitated by eIF5B.

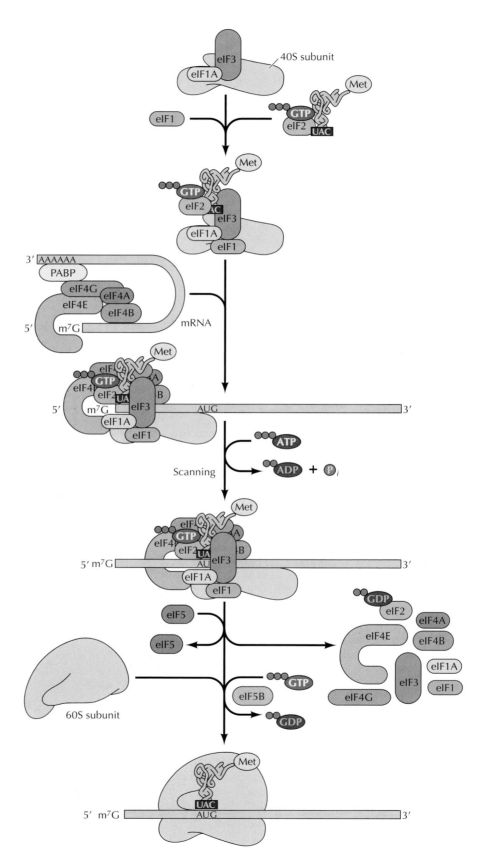

FIGURE 8.12 Elongation stage of translation The ribosome has three tRNA-binding sites, designated P (peptidyl), A (aminoacyl), and E (exit). The initiating methionyl tRNA is positioned in the P site, leaving an empty A site. The second aminoacyl tRNA (e.g., alanyl tRNA) is then brought to the A site by eEF1α (complexed with GTP). Following GTP hydrolysis, eEF1α (complexed with GDP) leaves the ribosome, with alanyl tRNA inserted into the A site. A peptide bond is then formed, resulting in the transfer of methionine to the aminoacyl tRNA at the A site. The ribosome then moves three nucleotides along the mRNA. This movement translocates the peptidyl (Met-Ala) tRNA to the P site and the uncharged tRNA to the E site, leaving an empty A site ready for addition of the next amino acid. Translocation is mediated by eEF2, coupled to GTP hydrolysis. The process, illustrated here for eukaryotic cells, is very similar in prokaryotes. (Table 8.1 gives the names of the prokaryotic elongation factors.)

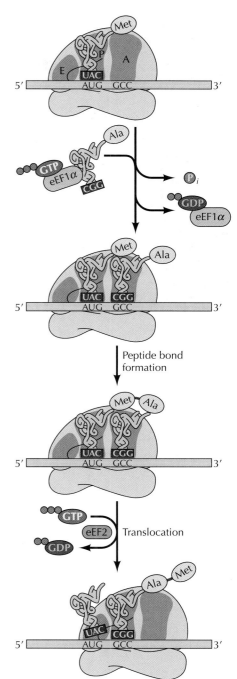

lation on translation. The initiation factors eIF4E and eIF4G, in association with eIF4A and eIF4B, then bring the mRNA to the 40S ribosomal subunit, with eIF4G interacting with eIF3. The 40S ribosomal subunit, in association with the bound methionyl tRNA and eIFs, then scans the mRNA to identify the AUG initiation codon. When the AUG codon is reached, eIF5 triggers the hydrolysis of GTP bound to eIF2. Initiation factors (including eIF2 bound to GDP) are then released, and a 60S subunit binds to the 40S subunit to form the 80S initiation complex of eukaryotic cells.

After the initiation complex has formed, translation proceeds by elongation of the polypeptide chain. The mechanism of elongation in prokaryotic and eukaryotic cells is very similar (Figure 8.12). The ribosome has three sites for tRNA binding, designated the P (peptidyl), A (aminoacyl), and E (exit) sites. The initiator methionyl tRNA is bound at the P site. The first step in elongation is the binding of the next aminoacyl tRNA to the A site by pairing with the second codon of the mRNA. The aminoacyl tRNA is escorted to the ribosome by an **elongation factor** (EF-Tu in prokaryotes, eEF1α in eukaryotes), which is complexed to GTP. The selection of the correct aminoacyl tRNA for incorporation into the growing polypeptide chain is the critical step that determines the accuracy of protein synthesis. Although this selection is based on base pairing between the codon on mRNA and the anticodon on tRNA, base pairing alone is not sufficient to account for the accuracy of protein synthesis, which has an error rate of less than 10^{-3}. This accuracy is provided by a "decoding center" in the small ribosomal subunit, which recognizes correct codon-anticodon base pairs and discriminates against mismatches. Insertion of a correct aminoacyl tRNA into the A site triggers a conformational change that induces the hydrolysis of GTP bound to eEF1α and release of the elongation factor bound to GDP. A notable result of the recent structural studies of the ribosome is that recognition of correct codon-anticodon pairing in the decoding center, like peptidyl transferase activity, is principally based on the activity of ribosomal RNA rather than proteins.

Once eEF1α has left the ribosome, a peptide bond can be formed between the initiator methionyl tRNA at the P site and the second aminoacyl tRNA at the A site. This reaction is catalyzed by the large ribosomal subunit, with the rRNA playing a critical role (as already discussed). The result is the transfer of methionine to the aminoacyl tRNA at the A site of the ribosome, forming a peptidyl tRNA at this position and leaving the uncharged initiator tRNA at the P site. The next step in elongation is translocation, which requires another elongation factor (EF-G in prokaryotes, eEF2 in

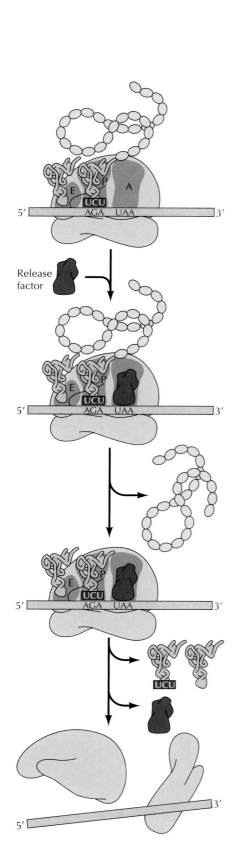

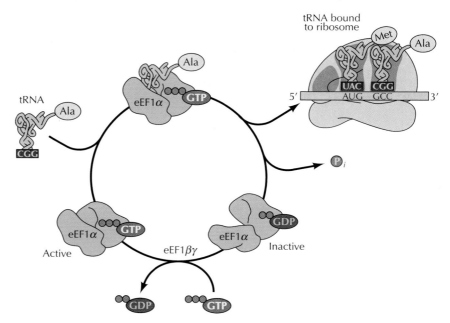

FIGURE 8.13 Regeneration of eEF1α/GTP eEF1α complexed to GTP escorts the aminoacyl tRNA to the ribosome. The bound GTP is hydrolyzed as the correct tRNA is inserted, so eEF1α complexed to GDP is released. The eEF1α/GDP complex is inactive and unable to bind another tRNA. In order for translation to continue, the active eEF1α/GTP complex must be regenerated by another factor, eEF1βγ, which stimulates the exchange of the bound GDP for free GTP.

eukaryotes) and is again coupled to GTP hydrolysis. During translocation, the ribosome moves three nucleotides along the mRNA, positioning the next codon in an empty A site. This step translocates the peptidyl tRNA from the A site to the P site, and the uncharged tRNA from the P site to the E site. The ribosome is then left with a peptidyl tRNA bound at the P site, and an empty A site. The binding of a new aminoacyl tRNA to the A site then induces the release of the uncharged tRNA from the E site, leaving the ribosome ready for insertion of the next amino acid in the growing polypeptide chain.

As elongation continues, the eEF1α (or EF-Tu) that is released from the ribosome bound to GDP must be reconverted to its GTP form (Figure 8.13). This conversion requires a third elongation factor, eEF1βγ (EF-Ts in prokaryotes), which binds to the eEF1α/GDP complex and promotes the exchange of bound GDP for GTP. This exchange results in the regeneration of eEF1α /GTP, which is now ready to escort a new aminoacyl tRNA to the A site of the ribosome, beginning a new cycle of elongation. The regulation of eEF1α by GTP binding and hydrolysis illustrates a common means of the regulation of protein activities. As will be discussed in later chapters, similar mechanisms control the activities of a wide variety of proteins involved in the regulation of cell growth and differentiation, as well as in protein transport and secretion.

FIGURE 8.14 Termination of translation A termination codon (e.g., UAA) at the A site is recognized by a release factor rather than by a tRNA. The result is the release of the completed polypeptide chain, followed by the dissociation of tRNA and mRNA from the ribosome.

Elongation of the polypeptide chain continues until a stop codon (UAA, UAG, or UGA) is translocated into the A site of the ribosome. Cells do not contain tRNAs with anticodons complementary to these termination signals; instead, they have **release factors** that recognize the signals and terminate protein synthesis (Figure 8.14). Prokaryotic cells contain two release factors that recognize termination codons: RF1 recognizes UAA or UAG, and RF2 recognizes UAA or UGA (see Table 8.1). In eukaryotic cells a single release factor (eRF1) recognizes all three termination codons. Both prokaryotic and eukaryotic cells also contain release factors (RF3 and eRF3, respectively) that do not recognize specific termination codons but act together with RF1 (or eRF1) and RF2. The release factors bind to a termination codon at the A site and stimulate hydrolysis of the bond between the tRNA and the polypeptide chain at the P site, resulting in release of the completed polypeptide from the ribosome. The tRNA is then released, and the ribosomal subunits and the mRNA template dissociate.

Messenger RNAs can be translated simultaneously by several ribosomes in both prokaryotic and eukaryotic cells. Once one ribosome has moved away from the initiation site, another can bind to the mRNA and begin synthesis of a new polypeptide chain. Thus mRNAs are usually translated by a series of ribosomes, spaced at intervals of about 100 to 200 nucleotides (Figure 8.15). The group of ribosomes bound to an mRNA molecule is called a polyribosome, or **polysome**. Each ribosome within the group functions independently to synthesize a separate polypeptide chain.

Regulation of Translation

Although transcription is the initial level at which gene expression is controlled, the translation of mRNAs is also regulated in both prokaryotic and eukaryotic cells. The translation of particular mRNAs can be regulated by translational repressor proteins, noncoding microRNAs, and controlled polyadenylation. In addition, the global translational activity of cells is modulated in response to cell stress, nutrient availability, and growth factor stimulation.

One mechanism of translational regulation is the binding of repressor proteins (which block translation) to specific mRNA sequences. A well understood example of this mechanism in eukaryotic cells is regulation of the synthesis of ferritin, a protein that stores iron within the cell. The trans-

■ In some cases, selenocysteine, which has been termed the 21st amino acid, is inserted at stop codons. Selenocysteine is a selenium-containing amino acid that is found in proteins from diverse organisms, including humans. Another modified amino acid, pyrrolysine, can also be inserted at stop codons. Pyrrolysine is a modified lysine found in proteins from some archaeabacteria and is important in the production of methane.

(A)

(B)

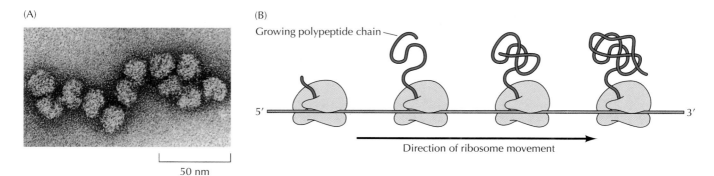

FIGURE 8.15 Polysomes   Messenger RNAs are translated by a series of multiple ribosomes (a polysome). (A) Electron micrograph of a eukaryotic polysome. (B) Schematic of a generalized polysome. Note that the ribosomes closer to the 3' end of the mRNA have longer polypeptide chains. (A, from M. Boublik et al., 1990. *The Ribosome*, p. 117. Courtesy of American Society for Microbiology.)

FIGURE 8.16 Translational regulation of ferritin The mRNA contains an iron response element (IRE) near its 5′ cap. In the presence of adequate supplies of iron, translation of the mRNA proceeds normally. If iron is scarce, however, a protein (called the iron regulatory protein, or IRP) binds to the IRE, blocking translation of the mRNA.

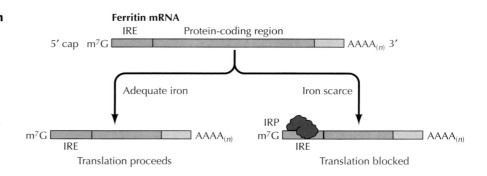

lation of ferritin mRNA is regulated by the supply of iron: More ferritin is synthesized if iron is abundant (**Figure 8.16**). This regulation is mediated by a protein, which (in the absence of iron) binds to a sequence (the iron response element, or IRE) in the 5′ untranslated region of ferritin mRNA, blocking its translation. In the presence of iron, the repressor no longer binds to the IRE and ferritin translation is able to proceed.

It is interesting to note that the regulation of translation of ferritin mRNA by iron is similar to the regulation of transferrin receptor mRNA stability, which was discussed in the previous chapter (see Figure 7.55). Namely, the stability of transferrin receptor mRNA is regulated by protein binding to an IRE in its 3′ untranslated region. The same protein binds to the IREs of both ferritin and transferrin receptor mRNAs. However, the consequences of protein binding to the two IREs are quite different. Protein bound to the transferrin receptor IRE protects the mRNA from degradation rather than inhibiting its translation. These distinct effects presumably result from the different locations of the IRE in the two mRNAs. To function as a repressor-binding site, the IRE must be located within 70 nucleotides of the 5′ cap of ferritin mRNA, suggesting that protein binding to the IRE blocks translation by interfering with cap recognition and binding of the 40S ribosomal subunit. Rather than inhibiting translation, protein binding to the same sequence in the 3′ untranslated region of transferrin receptor mRNA protects the mRNA from nuclease degradation. Binding of the same regulatory protein to different sites on mRNA molecules can thus have distinct effects on gene expression, in one case inhibiting translation and in the other case stabilizing the mRNA to increase protein synthesis.

Translation can also be regulated by proteins that bind to specific sequences in the 3′ untranslated regions of some mRNAs. In some cases, these translational repressors function by interacting with the initiation factor eIF4E to inhibit the initiation of translation (**Figure 8.17**). Proteins that bind to the 3′ untranslated regions of mRNAs are also responsible for localizing mRNAs to specific regions of cells, allowing proteins to be produced in specific subcellular locations. Localization of mRNAs is an important part of translational regulation in a variety of cell types, including eggs, embryos, nerve cells, and moving fibroblasts. For example, localization of mRNAs to specific regions of eggs or embryos plays an important role in development by allowing the encoded proteins to be synthesized at the appropriate sites in the developing embryo (**Figure 8.18**). The localization of mRNAs is coupled to regulation of their translation, so that their encoded proteins are synthesized when the mRNA becomes properly localized at the appropriate developmental stage.

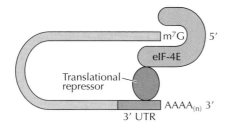

FIGURE 8.17 Translational repressor binding to 3′ untranslated sequences Translational repressors can bind to regulatory sequences in the 3′ untranslated region (UTR) and inhibit translation by binding to the initiation factor eIF4E, bound to the 5′ cap. This interferes with translation by blocking the binding of eIF4E to eIF4G in a normal initiation complex (see Figure 8.11).

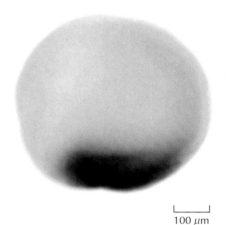

FIGURE 8.18 Localization of mRNA in *Xenopus* oocytes *In situ* hybridization illustrating the localization of Xlerk mRNA to the vegetal cortex of *Xenopus* oocytes. (Courtesy of James Deshler, Boston University.)

100 μm

Translational regulation is particularly important during early development. As discussed in Chapter 7, a variety of mRNAs are stored in oocytes in an untranslated form; the translation of these stored mRNAs is activated at fertilization or later stages of development. One mechanism of such translational regulation is the controlled polyadenylation of oocyte mRNAs. Many untranslated mRNAs are stored in oocytes with short poly-A tails (approximately 30–50 nucleotides). These stored mRNAs are subsequently recruited for translation at the appropriate stage of development by the lengthening of their poly-A tails to several hundred nucleotides. This allows the binding of poly-A binding protein (PABP), which stimulates translation by interacting with eIF4G (see Figure 8.11).

As discussed in Chapter 7, gene regulation by noncoding RNAs is a recently discovered phenomenon, the full significance of which is just beginning to be appreciated by cell biologists. **MicroRNAs (miRNAs)** are short (approximately 20–25 bases) double-stranded RNAs that can regulate gene expression at both the transcriptional and translational levels. Transcriptional regulation is mediated by the association of miRNAs with a protein complex called RITS, which represses transcription by chromatin modification of target genes (see Figure 7.37). Translational regulation is mediated by association of miRNAs with a distinct protein complex (called RISC), which targets homologous mRNAs (Figure 8.19). In some cases, the miRNAs direct cleavage of the targeted mRNA by RISC, as discussed in chapter 4 with respect to RNA interference. In other cases, however, the miRNA/RISC complex represses translation without inducing cleavage of the mRNA. It appears that cleavage is generally induced by miRNAs that pair perfectly with their mRNA targets, whereas translational repression in the absence of cleavage results from miRNAs that are mismatched with their targets. The mechanism of translational repression is not yet understood and remains an area of active investigation. It is noteworthy that sequence analysis of human mRNAs has predicted that approximately one-third of human genes are targets for regulation by miRNAs, implying that translational regulation by miRNAs is an extremely widespread mechanism of gene regulation.

Another mechanism of translational regulation in eukaryotic cells, resulting in global effects on overall translational activity rather than on the translation of specific mRNAs, involves modulation of the activity of initiation factors, particularly eIF2 and eIF4E. As already discussed, eIF2 (complexed

■ Specific mRNAs are localized at synapses between neurons. The regulated translation of these localized mRNAs appears to be required for the formation and maintenance of synapses and may play a role in the process of learning.

FIGURE 8.19 Regulation of translation by miRNAs miRNAs associate with the RISC complex in which the two strands of the miRNA are unwound. The miRNA then targets RISC to a homologous mRNA, leading either to mRNA cleavage or repression of translation. Cleavage generally results from miRNAs that pair perfectly with their targets, whereas translational repression results from miRNAs that are mismatched with their targets.

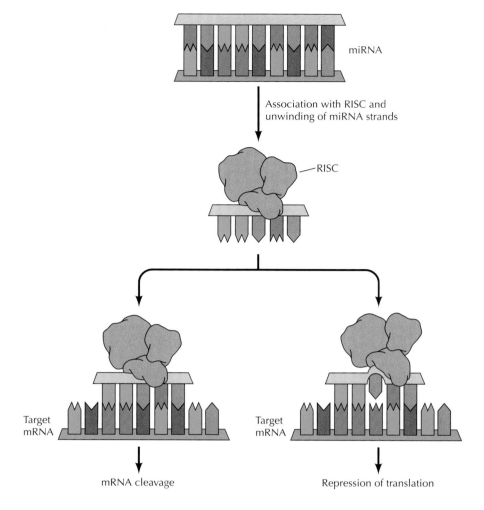

with GTP) binds to the initiator methionyl tRNA, bringing it to the ribosome. The subsequent release of eIF2 is accompanied by GTP hydrolysis, leaving eIF2 as an inactive GDP complex. To participate in another cycle of initiation, the eIF2/GTP complex must be regenerated by the exchange of bound GDP for GTP (Figure 8.20). This exchange is mediated by another factor, eIF2B. The control of eIF2 activity by GTP binding and hydrolysis is thus similar to that of eEF1α (see Figure 8.13). However, the regulation of eIF2 provides a critical control point in a variety of eukaryotic cells. In particular, both eIF2 and eIF2B can be phosphorylated by regulatory protein kinases. These phosphorylations inhibit the exchange of bound GDP for GTP, thereby inhibiting initiation of translation. For example, if mammalian cells are subjected to stress or starved of growth factors, protein kinases that phosphorylate eIF2 and eIF2B become activated, inhibiting further protein synthesis.

Regulation of the activity of eIF4E, which binds to the 5' cap of mRNAs, is another critical point at which growth factors act to control protein synthesis. For example, growth factors that stimulate protein synthesis in mammalian cells activate protein kinases that phosphorylate regulatory proteins (called eIF4E binding proteins or 4E-BPs) that bind to eIF4E. In the absence of the appropriate growth factors, the nonphosphorylated 4E-BPs

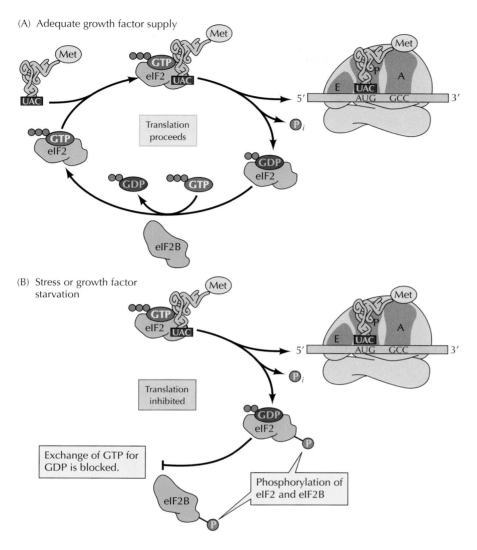

(A) Adequate growth factor supply

Translation proceeds

(B) Stress or growth factor starvation

Translation inhibited

Exchange of GTP for GDP is blocked.

Phosphorylation of eIF2 and eIF2B

FIGURE 8.20 Regulation of translation by phosphorylation of eIF2 and eIF2B The active form of eIF2 (complexed with GTP) escorts initiator methionyl tRNA to the ribosome (see Figure 8.11). The eIF2 is then released from the ribosome in an inactive GDP-bound form. In order to continue translation, eIF2 must be reactivated by eIF2B, which stimulates the exchange of GTP for the bound GDP. Translation can be inhibited (for example, if cells are stressed or starved of growth factors) by regulatory protein kinases that phosphorylate either eIF2 or eIF2B. These phosphorylations block the exchange of GTP for GDP, so eIF2/GTP cannot be regenerated.

bind to eIF4E and inhibit translation by interfering with the interaction of eIF4E with eIF4G (see Figure 8.11). When growth factors are present in adequate supply, phosphorylation of the 4E-BPs prevents their interaction with eIF4E, leading to increased rates of translation initiation.

Protein Folding and Processing

Translation completes the flow of genetic information within the cell. The sequence of nucleotides in DNA has now been converted to the sequence of amino acids in a polypeptide chain. The synthesis of a polypeptide, however, is not equivalent to the production of a functional protein. To be useful, polypeptides must fold into distinct three-dimensional conformations, and in many cases multiple polypeptide chains must assemble into a functional complex. In addition, many proteins undergo further modifications, including cleavage and the covalent attachment of carbohydrates and lipids that are critical for the function and correct localization of proteins within the cell.

Chaperones and Protein Folding

The three-dimensional conformations of proteins result from interactions between the side chains of their constituent amino acids, as reviewed in Chapter 2. The classic principle of protein folding is that all the information required for a protein to adopt the correct three-dimensional conformation is provided by its amino acid sequence. This was initially established by Christian Anfinsen's experiments demonstrating that denatured RNase can spontaneously refold *in vitro* to its active conformation (see Figure 2.17). Protein folding thus appeared to be a self-assembly process that did not require additional cellular factors. More recent studies, however, have shown that this is not an adequate description of protein folding within the cell. The proper folding of proteins within cells is mediated by the activities of other proteins.

Proteins that facilitate the folding of other proteins are called molecular **chaperones**. The term "chaperone" was first used by Ron Laskey and his colleagues to describe a protein (nucleoplasmin) that is required for the assembly of nucleosomes from histones and DNA. Nucleoplasmin binds to histones and mediates their assembly into nucleosomes, but nucleoplasmin itself is not incorporated into the final nucleosome structure. Chaperones thus act as catalysts that facilitate assembly without being part of the assembled complex. Subsequent studies have extended the concept to include proteins that mediate a variety of other assembly processes, particularly protein folding.

It is important to note that chaperones do not convey additional information required for the folding of polypeptides into their correct three-dimensional conformations; the folded conformation of a protein is determined solely by its amino acid sequence. Rather, chaperones catalyze protein folding by assisting the self-assembly process. They appear to function by binding to and stabilizing unfolded or partially folded polypeptides that are intermediates along the pathway leading to the final correctly folded state. In the absence of chaperones, unfolded or partially folded polypeptide chains would be unstable within the cell, frequently folding incorrectly or aggregating into insoluble complexes. The binding of chaperones stabilizes these unfolded polypeptides, thereby preventing incorrect folding or aggregation and allowing the polypeptide chain to fold into its correct conformation.

A good example is provided by chaperones that bind to nascent polypeptide chains that are still being translated on ribosomes, thereby preventing incorrect folding or aggregation of the amino-terminal portion of the polypeptide before synthesis of the chain is finished (Figure 8.21). Proteins fold into domains of approximately 100–300 amino acids, so it is necessary to protect the nascent chain from aberrant folding or aggregation with other proteins until synthesis of the entire domain is complete and the protein can fold into its correct conformation. Chaperone binding stabilizes the amino-terminal portion in an unfolded conformation until the rest of the polypeptide chain is synthesized and the completed protein can fold correctly. Chaperones also stabilize unfolded polypeptide chains during their transport into subcellular organelles—for example, during the transfer of proteins into mitochondria from the cytosol (Figure 8.22). Proteins are transported across the mitochondrial membrane in partially unfolded conformations that are stabilized by chaperones in the cytosol. Chaperones within the mitochondrion then facilitate transfer of the polypeptide chain across the membrane and its subsequent folding within the organelle. In addition, chaperones are involved in the assembly of proteins that consist of

FIGURE 8.21 Action of chaperones during translation Chaperones bind to the amino (N) terminal portion of the nascent polypeptide chain, stabilizing it in an unfolded configuration until synthesis of the polypeptide is completed. The completed protein is then released from the ribosome and is able to fold into its correct three-dimensional conformation.

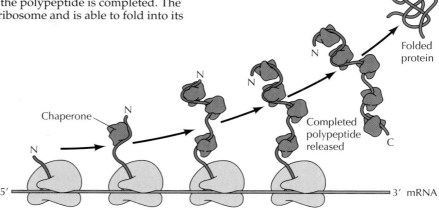

multiple polypeptide chains and in the assembly of macromolecular structures (e.g., nucleoplasmin).

Many of the proteins now known to function as chaperones were initially identified as heat-shock proteins, a group of proteins expressed in cells that have been subjected to elevated temperatures. The heat-shock proteins (abbreviated Hsp) are thought to stabilize and facilitate the refolding of proteins that have been partially denatured as a result of exposure to elevated temperatures. Two families of chaperone proteins, the Hsp70 chaperones and the chaperonins, act in a general pathway of protein folding in both prokaryotic and eukaryotic cells (Figure 8.23). Members of the Hsp70 and chaperonin families are found in the cytosol and in subcellular

◼ **Protein misfolding can have dire consequences for the cell.** Some misfolded proteins can aggregate and form insoluble fibers, called amyloid fibers, that accumulate in extracellular spaces and within cells. Amyloid fibers are hallmarks of several human neurodegenerative diseases, such as Alzheimer's and Parkinson's disease.

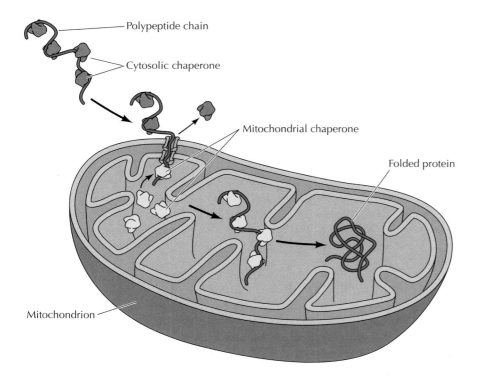

FIGURE 8.22 Action of chaperones during protein transport A partially unfolded polypeptide is transported from the cytosol to a mitochondrion. Cytosolic chaperones stabilize the unfolded configuration. Mitochondrial chaperones facilitate transport and subsequent folding of the polypeptide chain within the organelle.

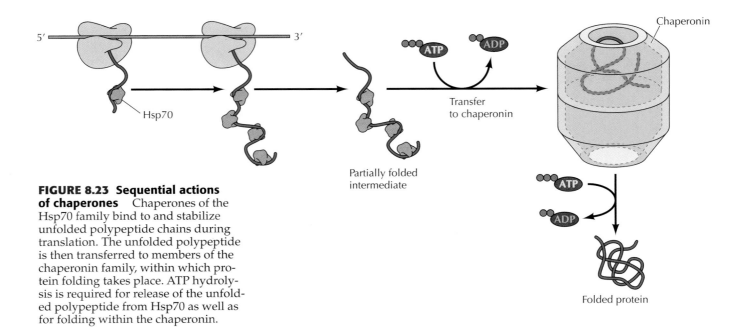

FIGURE 8.23 Sequential actions of chaperones Chaperones of the Hsp70 family bind to and stabilize unfolded polypeptide chains during translation. The unfolded polypeptide is then transferred to members of the chaperonin family, within which protein folding takes place. ATP hydrolysis is required for release of the unfolded polypeptide from Hsp70 as well as for folding within the chaperonin.

organelles (endoplasmic reticulum, mitochondria, and chloroplasts) of eukaryotic cells, as well as in bacteria. Members of the Hsp70 family stabilize unfolded polypeptide chains during translation (see, for example, Figure 8.21) as well as during the transport of polypeptides into a variety of subcellular compartments, such as mitochondria and the endoplasmic reticulum. These proteins bind to short hydrophobic segments (approximately seven amino acid residues) of unfolded polypeptides, maintaining the polypeptide chain in an unfolded configuration and preventing aggregation.

The unfolded polypeptide chain is then transferred from an Hsp70 chaperone to a chaperonin, within which protein folding takes place, yielding a protein correctly folded into its functional three-dimensional conformation. The chaperonins consist of multiple protein subunits arranged in two stacked rings to form a double-chambered structure. Unfolded polypeptide chains are shielded from the cytosol within the chamber of the chaperonin. In this isolated environment, protein folding can proceed while aggregation of unfolded segments of the polypeptide chain with other unfolded polypeptides is prevented.

Both bacteria and eukaryotic cells also contain additional families of chaperones, and the number of chaperones is considerably larger in eukaryotes. For example, an alternative pathway for the folding of some proteins in the cytosol and endoplasmic reticulum of eukaryotic cells involves the sequential actions of Hsp70 and Hsp90 family members. The majority of substrates for folding by Hsp90 are proteins that are involved in cell signaling, including receptors for steroid hormones and a variety of protein kinases.

Enzymes that Catalyze Protein Folding

In addition to chaperones that facilitate protein folding by binding to and stabilizing partially folded intermediates, cells contain at least two types of enzymes that act as chaperones by catalyzing protein folding. The formation of disulfide bonds between cysteine residues is important in stabilizing

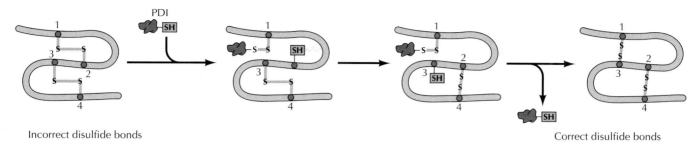

Incorrect disulfide bonds   Correct disulfide bonds

FIGURE 8.24 The action of protein disulfide isomerase   Protein disulfide isomerase (PDI) catalyzes the breakage and rejoining of disulfide bonds, resulting in exchanges between paired disulfides in a polypeptide chain. The enzyme forms a disulfide bond with a cysteine residue of the polypeptide and then exchanges its paired disulfide with another cysteine residue. In this example, PDI catalyzes the conversion of two incorrect disulfide bonds (1-2 and 3-4) to the correct pairing (1-3 and 2-4).

the folded structures of many proteins (see Figure 2.16). **Protein disulfide isomerase (PDI)**, which was discovered by Christian Anfinsen in 1963, catalyzes disulfide bond formation. For proteins that contain multiple cysteine residues, PDI plays an important role by promoting rapid exchanges between paired disulfides, thereby allowing the protein to attain the pattern of disulfide bonds that is compatible with its stably folded conformation (Figure 8.24). Disulfide bonds are generally restricted to secreted proteins and some membrane proteins because the cytosol contains reducing agents that maintain cysteine residues in their reduced (—SH) form, thereby preventing the formation of disulfide (S—S) linkages. In eukaryotic cells, disulfide bonds form in the endoplasmic reticulum in which an oxidizing environment is maintained. PDI is a critical chaperone and catalyst of protein folding in the endoplasmic reticulum and is one of the most abundant proteins in that organelle.

The second enzyme that plays a role in protein folding catalyzes the isomerization of peptide bonds that involve proline residues (Figure 8.25). Proline is an unusual amino acid in that the equilibrium between the *cis* and *trans* conformations of peptide bonds that precede proline residues is only slightly in favor of the *trans* form. In contrast, peptide bonds between other amino acids are almost always in the *trans* form. Isomerization between the *cis* and *trans* configurations of prolyl-peptide bonds, which could otherwise represent a rate-limiting step in protein folding, is catalyzed by the enzyme **peptidyl prolyl isomerase**. This enzyme is widely distributed in both prokaryotic and eukaryotic cells and plays an important role in the folding of some proteins.

Protein Cleavage

Cleavage of the polypeptide chain (**proteolysis**) is an important step in the maturation of many proteins. A simple example is removal of the initiator methionine from the amino terminus of many polypeptides, which occurs soon after the amino terminus of the growing polypeptide chain emerges from the ribosome. Additional chemical groups, such as acetyl groups or fatty acid chains (discussed shortly), are then frequently added to the amino-terminal residues.

Proteolytic modifications of the amino terminus also play a part in the translocation of many proteins across membranes, including secreted pro-

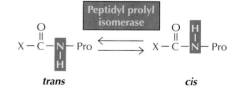

FIGURE 8.25 The action of peptidyl prolyl isomerase   Peptidyl prolyl isomerase catalyzes the isomerization of peptide bonds that involve proline between the *cis* and *trans* conformations.

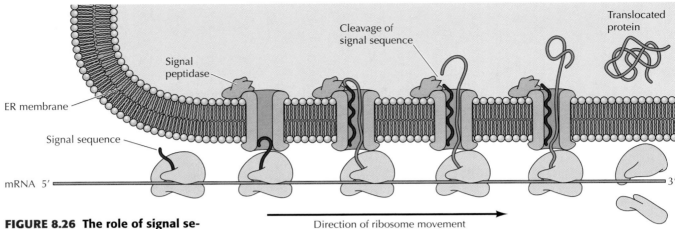

ER membrane

Signal sequence

mRNA 5′

3′

Direction of ribosome movement

FIGURE 8.26 The role of sequences in membrane translocation
Signal sequences target the translocation of polypeptide chains across the plasma membrane of bacteria or into the endoplasmic reticulum of eukaryotic cells (shown here). The signal sequence, a stretch of hydrophobic amino acids at the amino terminus of the polypeptide chain, inserts into a membrane channel as it emerges from the ribosome. The rest of the polypeptide is then translocated through the channel and the signal sequence is cleaved by the action of signal peptidase, releasing the mature translocated protein.

teins in both bacteria and eukaryotes as well as proteins destined for incorporation into the plasma membrane, lysosomes, mitochondria, and chloroplasts of eukaryotic cells. These proteins are targeted for transport to their destinations by amino-terminal sequences that are removed by proteolytic cleavage as the protein crosses the membrane. For example, amino-terminal **signal sequences**, usually about 20 amino acids long, target many secreted proteins to the plasma membrane of bacteria or to the endoplasmic reticulum of eukaryotic cells while translation is still in progress (Figure 8.26). The signal sequence, which consists predominantly of hydrophobic amino acids, is inserted into a membrane channel as it emerges from the ribosome. The remainder of the polypeptide chain passes through the channel membrane as translation proceeds. The signal sequence is then cleaved by a specific membrane protease (**signal peptidase**), and the mature protein is released. In eukaryotic cells, the translocation of growing polypeptide chains into the endoplasmic reticulum is the first step in targeting proteins for secretion, incorporation into the plasma membrane, or incorporation into lysosomes. The mechanisms that direct the transport of proteins to these destinations, as well as the role of other targeting sequences in directing the import of proteins into mitochondria and chloroplasts, will be discussed in detail in Chapters 10 and 11.

In other important instances of proteolytic processing, active enzymes or hormones form via cleavage of larger precursors. Insulin, which is synthesized as a longer precursor polypeptide, is a good example. Insulin forms by two cleavages. The initial precursor (preproinsulin) contains an amino-terminal signal sequence that targets the polypeptide chain to the endoplasmic reticulum (Figure 8.27). Removal of the signal sequence during transfer to the endoplasmic reticulum yields a second precursor, called proinsulin. This precursor is then converted to insulin (which consists of two chains held together by disulfide bonds) by proteolytic removal of an internal peptide. Other proteins activated by similar cleavage processes include digestive enzymes, proteins involved in blood clotting, and a cascade of proteases that regulate programmed cell death in animals.

It is interesting to note that the proteins of many animal viruses are derived from the cleavage of larger precursors. One particularly important example of the role of proteolysis in virus replication is provided by HIV. In the replication of HIV, a virus-encoded protease cleaves precursor polypep-

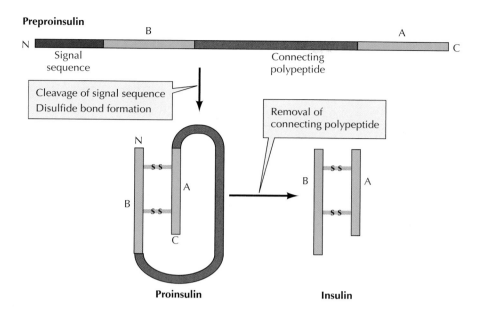

Preproinsulin

FIGURE 8.27 Proteolytic processing of insulin The mature insulin molecule consists of two polypeptide chains (A and B) joined by disulfide bonds. It is synthesized as a precursor polypeptide (preproinsulin) containing an amino-terminal signal sequence that is cleaved during transfer of the growing polypeptide chain to the endoplasmic reticulum. This cleavage yields a second precursor (proinsulin), which is converted to insulin by further proteolysis, removing the internal connecting polypeptide.

tides to form the viral structural proteins. Because of its central role in virus replication, the HIV protease (in addition to reverse transcriptase) is an important target for the development of drugs used for treating AIDS. Indeed, such protease inhibitors are now among the most effective agents available for combating this disease.

Glycosylation

Many proteins, particularly in eukaryotic cells, are modified by the addition of carbohydrates, a process called **glycosylation**. The proteins to which carbohydrate chains have been added (called **glycoproteins**) are usually secreted or localized to the cell surface, although many nuclear and cytosolic proteins are also glycosylated. The carbohydrate moieties of glycoproteins play important roles in protein folding in the endoplasmic reticulum, in the targeting of proteins for delivery to the appropriate intracellular compartments, and as recognition sites in cell-cell interactions.

Glycoproteins are classified as either *N*-linked or *O*-linked, depending on the site of attachment of the carbohydrate side chain (**Figure 8.28**). In *N*-linked glycoproteins, the carbohydrate is attached to the nitrogen atom in the side chain of asparagine. In *O*-linked glycoproteins, the oxygen atom in the side chain of serine or threonine is the site of carbohydrate attachment. The sugars directly attached to these positions are usually either *N*-acetylglucosamine or *N*-acetylgalactosamine, respectively.

Most glycoproteins in eukaryotic cells are destined either for secretion or for incorporation into the plasma membrane. These proteins are usually transferred into the endoplasmic reticulum while their translation is still in progress. Glycosylation is also initiated in the endoplasmic reticulum before

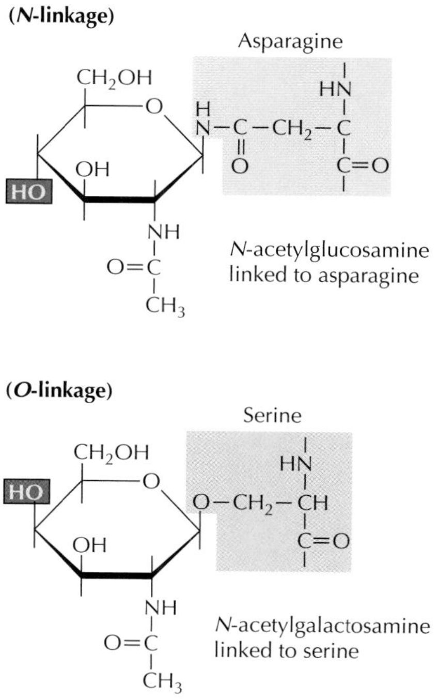

FIGURE 8.28 Linkage of carbohydrate side chains to glycoproteins The carbohydrate chains of *N*-linked glycoproteins are attached to asparagine; those of *O*-linked glycoproteins are attached to either serine (shown) or threonine. The sugars joined to the amino acids are usually either *N*-acetylglucosamine (*N*-linked) or *N*-acetylgalactosamine (*O*-linked).

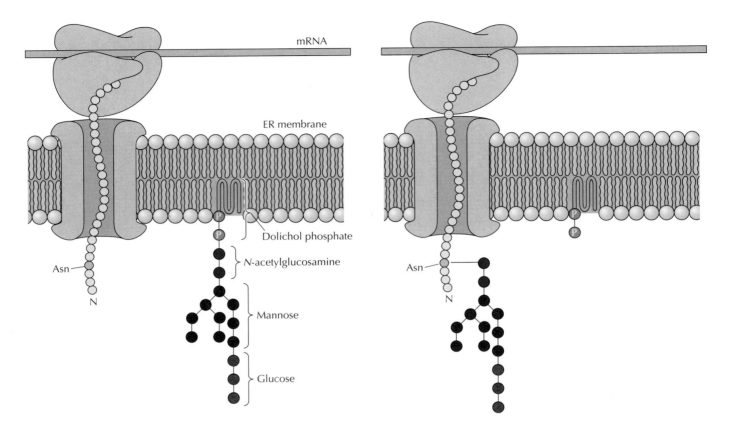

FIGURE 8.29 Synthesis of N-linked glycoproteins The first step in glycosylation is the addition of an oligosaccharide consisting of 14 sugar residues to a growing polypeptide chain in the endoplasmic reticulum (ER). The oligosaccharide (consisting of two N-acetylglucosamine, nine mannose, and three glucose residues) is assembled on a lipid carrier (dolichol phosphate) in the ER membrane. It is then transferred as a unit to an acceptor asparagine residue of the polypeptide.

translation is complete. The first step is the transfer of a common oligosaccharide consisting of 14 sugar residues (two N-acetylglucosamine, three glucose, and nine mannose) to an asparagine residue of the growing polypeptide chain (Figure 8.29). The oligosaccharide is assembled within the endoplasmic reticulum on a lipid carrier (**dolichol phosphate**). It is then transferred as an intact unit to an acceptor asparagine (Asn) residue within the sequence Asn-X-Ser or Asn-X-Thr (where X is any amino acid other than proline).

In further processing, the common N-linked oligosaccharide is modified (Figure 8.30). Three glucose residues are removed while the glycoprotein is in the endoplasmic reticulum. As discussed in Chapter 10, these modifications play a critical role in protein folding. The oligosaccharide is then further modified in the Golgi apparatus, to which glycoproteins are transferred from the endoplasmic reticulum. These modifications (which will be discussed in Chapter 10) include both the removal and addition of carbohydrate residues as the glycoprotein is transported through the compartments of the Golgi. The N-linked oligosaccharides of different glycoproteins are processed to different extents, depending on both the enzymes present in different cells and on the accessibility of the oligosaccharide to the enzymes

FIGURE 8.30 Processing of *N*-linked oligosaccharides
Various oligosaccharides form from further modifications of the common 14-sugar unit initially added in the endoplasmic reticulum. Three glucose residues are removed in the endoplasmic reticulum. The glycoprotein is then transferred to the Golgi apparatus in which mannose residues are removed and other sugars are added. The structure shown is a representative example.

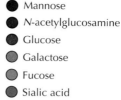

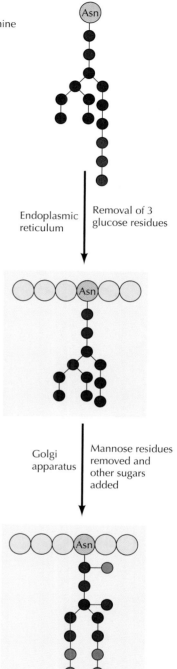

that catalyze its modification. Glycoproteins with inaccessible oligosaccharides do not have new sugars added to them in the Golgi. In contrast, glycoproteins with accessible oligosaccharides are processed extensively, resulting in the formation of a variety of complex oligosaccharides.

O-linked oligosaccharides are also added within the Golgi apparatus. In contrast to the *N*-linked oligosaccharides, *O*-linked oligosaccharides are formed by the addition of one sugar at a time and usually consist of only a few residues (Figure 8.31). Many cytoplasmic and nuclear proteins, including a variety of transcription factors, are also modified by the addition of single *O*-linked *N*-acetylglucosamine residues, catalyzed by a different enzyme system. Glycosylation of these cytoplasmic and nuclear proteins is thought to play a role in regulating their activities, although the consequences of *O*-glycosylation remain to be fully understood.

Attachment of Lipids

Some proteins in eukaryotic cells are modified by the attachment of lipids to the polypeptide chain. Such modifications frequently target and anchor these proteins to the plasma membrane, with which the hydrophobic lipid is able to interact (see Figure 2.25). Three general types of lipid additions—*N*-myristoylation, prenylation, and palmitoylation—are common in eukaryotic proteins associated with the cytosolic face of the plasma membrane. A fourth type of modification, the addition of glycolipids, plays an important role in anchoring some cell surface proteins to the extracellular face of the plasma membrane.

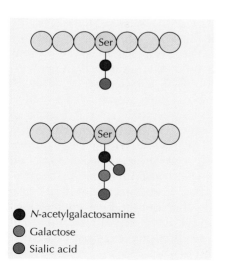

FIGURE 8.31 Examples of *O*-linked oligosaccharides *O*-linked oligosaccharides usually consist of only a few carbohydrate residues, which are added one sugar at a time.

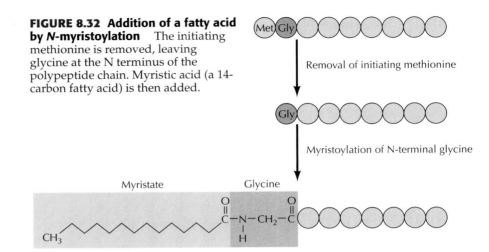

FIGURE 8.32 Addition of a fatty acid by *N*-myristoylation The initiating methionine is removed, leaving glycine at the N terminus of the polypeptide chain. Myristic acid (a 14-carbon fatty acid) is then added.

In some proteins, a fatty acid is attached to the amino terminus of the growing polypeptide chain during translation. In this process, called ***N*-myristoylation**, myristic acid (a 14-carbon fatty acid) is attached to an N-terminal glycine residue (Figure 8.32). The glycine is usually the second amino acid incorporated into the polypeptide chain; the initiator methionine is removed by proteolysis before fatty acid addition. Many proteins that are modified by *N*-myristoylation are associated with the inner face of the plasma membrane, and the role of the fatty acid in this association has been clearly demonstrated by analysis of mutant proteins in which the N-terminal glycine is changed to an alanine. This substitution prevents myristoylation and blocks the function of the mutant proteins by inhibiting their membrane association.

Lipids can also be attached to the side chains of cysteine, serine, and threonine residues. One important example of this type of modification is **prenylation** in which specific types of lipids (prenyl groups) are attached to the sulfur atoms in the side chains of cysteine residues located near the C terminus of the polypeptide chain (Figure 8.33). Many plasma membrane-associated proteins involved in the control of cell growth and differentiation are modified in this way, including the Ras oncogene proteins, which are responsible for the uncontrolled growth of many human cancers (see Chapter 18). Prenylation of these proteins proceeds by three steps. First, the prenyl group is added to a cysteine located three amino acids from the carboxy terminus of the polypeptide chain. The prenyl groups added in this reaction are either farnesyl (15 carbons, as shown in Figure 8.33) or geranylgeranyl (20 carbons). The amino acids following the cysteine residue are then removed, leaving cysteine at the carboxy terminus. Finally, a methyl group is added to the carboxyl group of the C-terminal cysteine residue.

The biological significance of prenylation is indicated by the fact that mutations of the critical cysteine block the membrane association and function of Ras proteins. Because farnesylation is a relatively rare modification of cellular proteins, interest in this reaction has been stimulated by the possibility that inhibitors of the key enzyme (farnesyl transferase) might prove useful as drugs for the treatment of cancers that involve Ras proteins. Such inhibitors of farnesylation have been found to interfere with the growth of cancer cells in experimental models and are undergoing evaluation of their efficacy against human tumors in clinical trials.

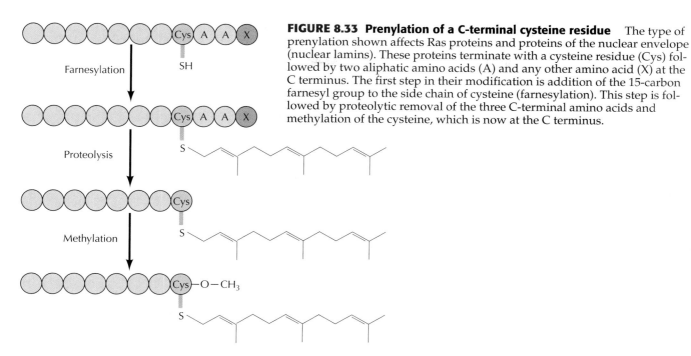

FIGURE 8.33 Prenylation of a C-terminal cysteine residue The type of prenylation shown affects Ras proteins and proteins of the nuclear envelope (nuclear lamins). These proteins terminate with a cysteine residue (Cys) followed by two aliphatic amino acids (A) and any other amino acid (X) at the C terminus. The first step in their modification is addition of the 15-carbon farnesyl group to the side chain of cysteine (farnesylation). This step is followed by proteolytic removal of the three C-terminal amino acids and methylation of the cysteine, which is now at the C terminus.

In the third type of fatty acid modification, **palmitoylation**, palmitic acid (a 16-carbon fatty acid) is added to sulfur atoms of the side chains of internal cysteine residues (Figure 8.34). Like *N*-myristoylation and prenylation, palmitoylation plays an important role in the association of some proteins with the cytosolic face of the plasma membrane.

Finally, lipids linked to oligosaccharides (**glycolipids**) are added to the C-terminal carboxyl groups of some proteins, where they serve as anchors that attach the proteins to the external face of the plasma membrane. Because the glycolipids attached to these proteins contain phosphatidylinositol, they are usually called **glycosylphosphatidylinositol**, or **GPI anchors** (Figure 8.35). The oligosaccharide portions of GPI anchors are attached to the terminal carboxyl group of polypeptide chains. The inositol head group of phosphatidylinositol is in turn attached to the oligosaccharide, so the carbohydrate serves as a bridge between the protein and the fatty acid chains of the phospholipid. The GPI anchors are synthesized and added to proteins as a preassembled unit within the endoplasmic reticulum. Their addition is accompanied by cleavage of a peptide consisting of about 20 amino acids from the C terminus of the polypeptide chain. The modified protein is then transported to the cell surface, where the fatty acid chains of the GPI anchor mediate its attachment to the plasma membrane.

Regulation of Protein Function

A critical function of proteins is their activity as enzymes, which are needed to catalyze almost all biological reactions. Regulation of enzyme activity thus plays a key role in governing cell behavior. This is accomplished in part at the level of gene expression, which determines the amount of any enzyme (protein) synthesized by the cell. A further level of control is then obtained by regulation of protein function, which allows the cell to regulate

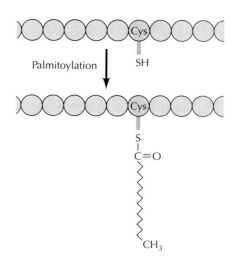

FIGURE 8.34 Palmitoylation Palmitate (a 16-carbon fatty acid) is added to the side chain of an internal cysteine residue.

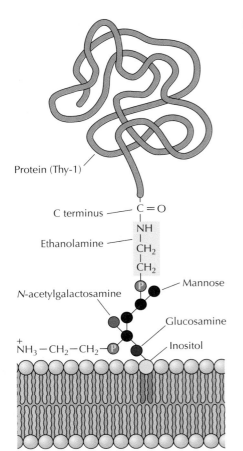

Protein (Thy-1)

C terminus — C=O

Ethanolamine — NH / CH₂ / CH₂

N-acetylgalactosamine

Mannose

Glucosamine

$^+NH_3-CH_2-CH_2-$ P

Inositol

FIGURE 8.35 Structure of a GPI anchor The GPI anchor, attached to the C terminus, anchors the protein in the plasma membrane. The anchor is joined to the C-terminal amino acid by an ethanolamine, which is linked to an oligosaccharide that consists of mannose, N-acetylgalactosamine, and glucosamine residues. The oligosaccharide is in turn joined to the inositol head group of phosphatidylinositol. The two fatty acid chains of the lipid are embedded in the plasma membrane. The GPI anchor shown here is that of a rat protein, Thy-1.

not only the amounts but also the activities of its protein constituents. Regulation of the activities of some of the proteins involved in transcription and translation has already been discussed in this and the preceding chapter, and many further examples of regulated protein function in the control of cell behavior will be evident throughout the remainder of this book. This section discusses three general mechanisms by which the activities of cellular proteins are controlled.

Regulation by Small Molecules

Most enzymes are controlled by changes in their conformation, which in turn alter catalytic activity. In many cases such conformational changes result from the binding of small molecules, such as amino acids or nucleotides, that regulate enzyme activity. This type of regulation commonly is responsible for controlling metabolic pathways by feedback inhibition. For example, the end products of many biosynthetic pathways (e.g., amino acids) inhibit the enzymes that catalyze the first step in their synthesis, thus ensuring an adequate supply of the product while preventing the synthesis of excess amounts (Figure 8.36). Feedback inhibition is an example of **allosteric regulation** in which a regulatory molecule binds to a site on an enzyme that is distinct from the catalytic site (*allo* = "other"; *steric* = "site"). The binding of such a regulatory molecule alters the conformation of the protein, thereby changing the shape of the catalytic site and affecting catalytic activity (see Figure 3.8). Many transcription factors (discussed in Chapter 7) are also regulated by the binding of small molecules. For

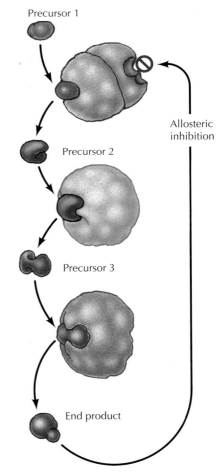

Precursor 1

Precursor 2

Precursor 3

End product

Allosteric inhibition

FIGURE 8.36 Feedback inhibition The end product of a biochemical pathway acts as an allosteric inhibitor of the enzyme that catalyzes the first step in its synthesis.

FIGURE 8.37 Conformational differences between active and inactive Ras proteins   The Ras proteins alternate between active GTP-bound and inactive GDP-bound forms. The major effect of GTP binding versus GDP binding is alteration of the conformation of two regions of the molecule, designated the switch I and switch II regions. The backbone of the GTP complex is shown here in white; the backbone of the GDP complex of switch I and switch II are in blue and yellow, respectively. The guanine nucleotide is in red and Mg^{2+} is the yellow sphere. (Courtesy of Sung-Hou Kim, University of California, Berkeley.)

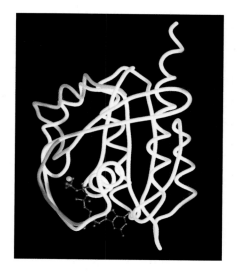

example, the binding of lactose to the *E. coli lac* repressor induces a conformational change that prevents the repressor from binding DNA (see Figure 7.9). In eukaryotic cells, steroid hormones similarly control gene expression by binding to transcriptional regulatory proteins.

The regulation of translation factors such as $eEF1\alpha$ by GTP binding (see Figure 8.13) illustrates another common mechanism by which the activities of intracellular proteins are controlled. In this case, the GTP-bound form of the protein is its active conformation, while the GDP-bound form is inactive. Many cellular proteins are similarly regulated by GTP or GDP binding. These proteins include the Ras oncogene proteins, which have been studied intensively because of their roles in the control of cell proliferation and in human cancers. X-ray crystallography of these proteins has been particularly interesting, revealing subtle but functionally very important conformational differences between the inactive GDP-bound and active GTP-bound forms (Figure 8.37). This small difference in protein conformation determines whether Ras (in the active GTP-bound form) can interact with its target molecule, which signals the cell to divide. The importance of such subtle differences in protein conformation is dramatically illustrated by the fact that mutations in *ras* genes contribute to the development of about 20% of human cancers. Such mutations alter the structure of the Ras proteins so that they are locked in the active GTP-bound conformation and continually signal cell division, thereby driving the uncontrolled growth of cancer cells. In contrast, normal Ras proteins alternate between the GTP- and GDP-bound conformations, such that they are active only following stimulation by the hormones and growth factors that normally control cell proliferation in multicellular organisms.

Protein Phosphorylation

The examples discussed in the previous section involve noncovalent associations of proteins with small-molecule inhibitors or activators. Since no covalent bonds form, the binding of these regulatory molecules to the protein is readily reversible, allowing the cell to respond rapidly to environmental changes. The activity of many proteins, however, is also regulated by covalent modifications. One example of this type of regulation is the activation of some enzymes by proteolytic cleavage of inactive precursors. As noted previously in this chapter, digestive enzymes and proteins involved in blood clotting are regulated by this mechanism. Since proteolysis is irreversible, however, it provides a means of controlling enzyme activation rather than of turning proteins on and off in response to changes in the environment. In contrast, other covalent modifications—particularly phosphorylation—are readily reversible within the cell and function, as allosteric regulation does, to reversibly activate or inhibit a wide variety of cellular proteins in response to environmental signals.

FIGURE 8.38 Protein kinases and phosphatases Protein kinases catalyze the transfer of a phosphate group from ATP to the side chains of serine and threonine (protein-serine/threonine kinases) or tyrosine (protein-tyrosine kinases) residues. Protein phosphatases catalyze the removal of phosphate groups from the same amino acids by hydrolysis.

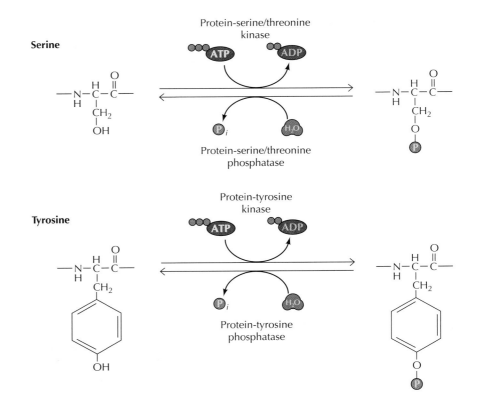

Protein phosphorylation is catalyzed by **protein kinases**, most of which transfer phosphate groups from ATP to the hydroxyl groups of the side chains of serine, threonine, or tyrosine residues (Figure 8.38). The protein kinases are one of the largest protein families in eukaryotes, accounting for approximately 2% of eukaryotic genes. Most protein kinases phosphorylate either serine and threonine or tyrosine residues: These enzymes are called **protein-serine/threonine kinases** or **protein-tyrosine kinases**, respectively. Protein phosphorylation is reversed by **protein phosphatases**, which catalyze the hydrolysis of phosphorylated amino acid residues. Like protein kinases, most protein phosphatases are specific either for serine and threonine or for tyrosine residues, although some protein phosphatases recognize all three phosphoamino acids.

The combined action of protein kinases and protein phosphatases mediates the reversible phosphorylation of many cellular proteins. Frequently, protein kinases function as components of signal transduction pathways in which one kinase activates a second kinase, which may act on yet another kinase. The sequential action of a series of protein kinases can transmit a signal received at the cell surface to target proteins within the cell, resulting in changes in cell behavior in response to environmental stimuli.

The prototype of the action of protein kinases came from studies of glycogen metabolism by Ed Fischer and Ed Krebs in 1955. In muscle cells the hormone epinephrine (adrenaline) signals the breakdown of glycogen to glucose-1-phosphate, providing an available source of energy for increased muscular activity. Glycogen breakdown is catalyzed by the enzyme glycogen phosphorylase, which is regulated by a protein kinase (Figure 8.39). Epinephrine binds to a cell surface receptor that triggers the conversion of ATP to cyclic AMP (cAMP), which then binds to and activates

FIGURE 8.39 Regulation of glycogen breakdown by protein phosphorylation
The binding of epinephrine (adrenaline) to its cell surface receptor triggers the production of cyclic AMP (cAMP), which activates cAMP-dependent protein kinase. cAMP-dependent protein kinase phosphorylates and activates phosphorylase kinase, which in turn phosphorylates and activates glycogen phosphorylase. Glycogen phosphorylase then catalyzes the breakdown of glycogen to glucose-1-phosphate.

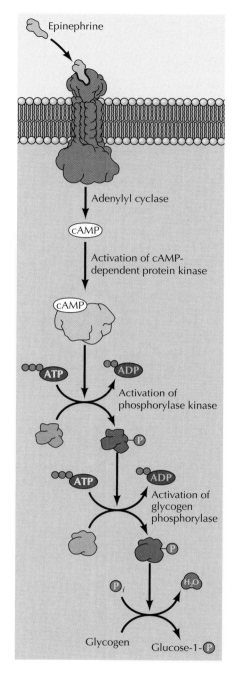

a protein kinase, called cAMP-dependent protein kinase. This kinase phosphorylates and activates a second protein kinase, called phosphorylase kinase. Phosphorylase kinase in turn phosphorylates and activates glycogen phosphorylase, leading to glucose production. The activating phosphorylations of both phosphorylase kinase and glycogen phosphorylase can be reversed by specific phosphatases, so removal of the initial stimulus (epinephrine) inhibits further glycogen breakdown.

The signaling pathway that leads to activation of glycogen phosphorylase is initiated by the binding of small molecules at the cell surface—epinephrine binding to its receptor and cAMP binding to cAMP-dependent protein kinase. The signal is then transmitted to its intracellular target by the sequential action of protein kinases. Similar signaling pathways, in which protein kinases and phosphatases play central roles, are involved in regulating almost all aspects of the behavior of eukaryotic cells (see Chapters 15 and 16). Aberrations in these pathways, frequently involving abnormalities of protein kinases, are also responsible for many diseases associated with improper regulation of cell growth and differentiation, particularly the development of cancer.

Although phosphorylation is the most common and best studied type of covalent modification that regulates protein activity, several other types of protein modifications also play important roles. These include methylation and acetylation of lysine residues (discussed in Chapter 7), as well as **nitrosylation**—the addition of NO groups to the side chains of cysteine residues (Figure 8.40). As noted earlier in this chapter, *O*-linked glycosylation of nuclear and cytosolic proteins may also play a regulatory role. In addition, some proteins are regulated by the covalent attachment of small polypeptides, such as ubiquitin and SUMO, discussed in the next section of this chapter.

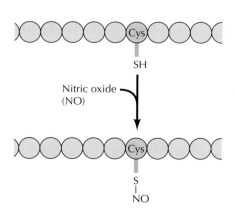

FIGURE 8.40 Nitrosylation
Ntiric oxide (NO) can react with the side chain of cysteine residues.

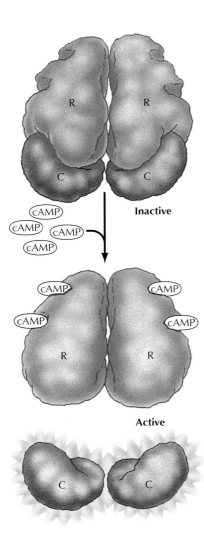

FIGURE 8.41 Regulation of cAMP-dependent protein kinase In the inactive state, the enzyme consists of two regulatory (R) and two catalytic (C) subunits. Cyclic AMP binds to the regulatory subunits, inducing a conformational change that leads to their dissociation from the catalytic subunits. The free catalytic subunits are enzymatically active.

Protein-Protein Interactions

Many proteins consist of multiple subunits, each of which is an independent polypeptide chain. In some proteins the subunits are identical; other proteins are composed of two or more distinct polypeptides. In either case, interactions between the polypeptide chains are important in regulation of protein activity. The importance of these interactions is evident in many allosteric enzymes in which the binding of a regulatory molecule alters protein conformation by changing the interactions between subunits.

Many other enzymes are similarly regulated by protein-protein interactions. A good example is cAMP-dependent protein kinase, which is composed of two regulatory and two catalytic subunits (Figure 8.41). In this state, the enzyme is inactive; the regulatory subunits inhibit the enzymatic activity of the catalytic subunits. The enzyme is activated by cAMP, which binds to the regulatory subunits and induces a conformational change leading to dissociation of the complex; the free catalytic subunits are then enzymatically active protein kinases. Cyclic AMP thus acts as an allosteric regulator by altering protein-protein interactions. As discussed in later chapters, similar protein-protein interactions, which can themselves be regulated by the binding of small molecules and phosphorylation, play critical roles in the control of many different aspects of cell behavior.

Protein Degradation

The levels of proteins within cells are determined not only by rates of synthesis but also by rates of degradation. The half-lives of proteins within cells vary widely, from minutes to several days, and differential rates of protein degradation are an important aspect of cell regulation. Many rapidly degraded proteins function as regulatory molecules, such as transcription factors. The rapid turnover of these proteins is necessary to allow their levels to change quickly in response to external stimuli. Other proteins are rapidly degraded in response to specific signals, providing another mechanism for the regulation of intracellular enzyme activity. In addition, faulty or damaged proteins are recognized and rapidly degraded within cells, thereby eliminating the consequences of mistakes made during protein synthesis. In eukaryotic cells, two major pathways—the ubiquitin-proteasome pathway and lysosomal proteolysis—mediate protein degradation.

The Ubiquitin-Proteasome Pathway

The major pathway of selective protein degradation in eukaryotic cells uses **ubiquitin** as a marker that targets cytosolic and nuclear proteins for rapid proteolysis (Figure 8.42). Ubiquitin is a 76-amino-acid polypeptide that is highly conserved in all eukaryotes (yeasts, animals, and plants). Proteins are marked for degradation by the attachment of ubiquitin to the amino group of the side chain of a lysine residue. Additional ubiquitins are then added to form a multiubiquitin chain. Such polyubiquinated proteins are

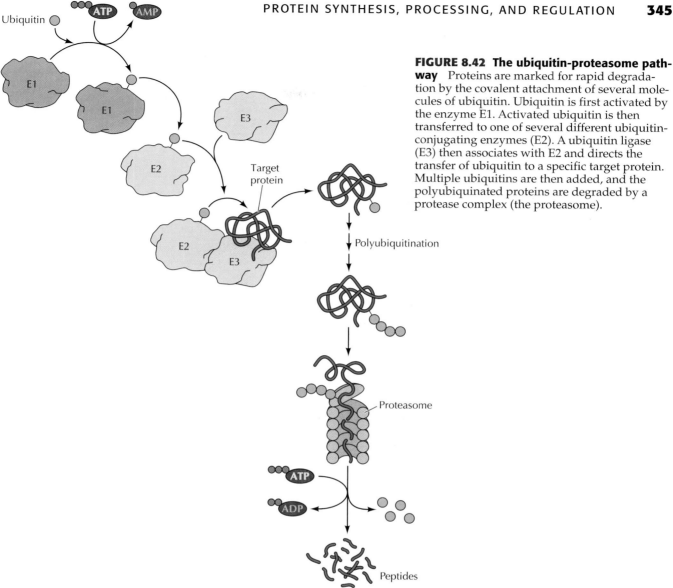

FIGURE 8.42 The ubiquitin-proteasome pathway Proteins are marked for rapid degradation by the covalent attachment of several molecules of ubiquitin. Ubiquitin is first activated by the enzyme E1. Activated ubiquitin is then transferred to one of several different ubiquitin-conjugating enzymes (E2). A ubiquitin ligase (E3) then associates with E2 and directs the transfer of ubiquitin to a specific target protein. Multiple ubiquitins are then added, and the polyubiquinated proteins are degraded by a protease complex (the proteasome).

recognized and degraded by a large, multisubunit protease complex, called the **proteasome**. Ubiquitin is released in the process, so it can be reused in another cycle. Both the attachment of ubiquitin and the degradation of marked proteins require energy in the form of ATP.

Since the attachment of ubiquitin marks proteins for rapid degradation, the stability of many proteins is determined by whether they become ubiquitinated. Ubiquitination is a multistep process. First, ubiquitin is activated by being attached to a ubiquitin-activating enzyme, E1. The ubiquitin is then transferred to a second enzyme, called ubiquitin-conjugating enzyme (E2). The ubiquitin is then transferred to the target protein in association with a third enzyme, called ubiquitin ligase or E3, which is responsible for the selective recognition of appropriate substrate proteins. Most cells contain a single E1 but have several E2s and a large number of E3 enzymes. Different E3s recognize different substrate proteins, and the specificity of these enzymes is what selectively targets cellular proteins for degradation by the ubiquitin-proteasome pathway.

A number of proteins that control fundamental cellular processes, such as gene expression and cell proliferation, are targets for regulated ubiquiti-

FIGURE 8.43 Cyclin degradation during the cell cycle The progression of eukaryotic cells through the division cycle is controlled in part by the synthesis and degradation of cyclin B, which is a regulatory subunit of the Cdk1 protein kinase. Synthesis of cyclin B during interphase leads to the formation of an active cyclin B–Cdk1 complex, which induces entry into mitosis. Rapid degradation of cyclin B then leads to inactivation of the Cdk1 kinase, allowing the cell to exit mitosis and return to interphase of the next cell cycle.

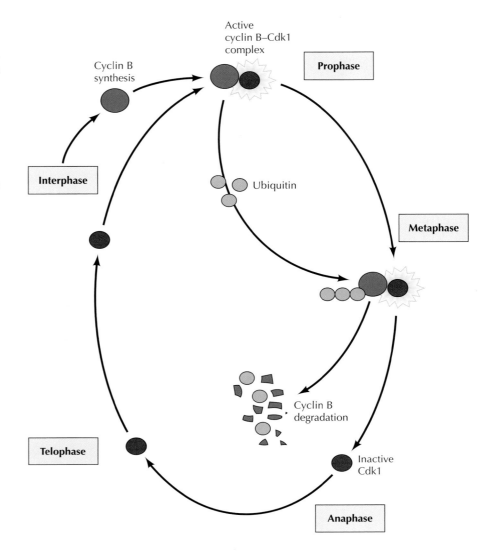

nation and proteolysis. An interesting example of such controlled degradation is provided by proteins (known as cyclins) that regulate progression through the division cycle of eukaryotic cells (Figure 8.43). The entry of all eukaryotic cells into mitosis is controlled in part by cyclin B, which is a regulatory subunit of a protein kinase called Cdk1 (see Chapter 16). The association of cyclin B with Cdk1 is required for activation of the Cdk1 kinase, which initiates the events of mitosis (including chromosome condensation and nuclear envelope breakdown) by phosphorylating various cellular proteins. Cdk1 also activates a ubiquitin ligase that targets cyclin B for degradation toward the end of mitosis. This degradation of cyclin B inactivates Cdk1, allowing the cell to exit mitosis and progress to interphase of the next cell cycle. The ubiquitination of cyclin B is a highly selective reaction, targeted by a 9-amino-acid cyclin B sequence called the destruction box. Mutations of this sequence prevent cyclin B proteolysis and lead to the arrest of dividing cells in mitosis, demonstrating the importance of regulated protein degradation in controlling the fundamental process of cell division.

Although ubiquitination usually targets proteins for degradation, the addition of ubiquitin to some proteins serves other functions. For example, ubiquitination of some proteins serves as a marker for endocytosis, and ubiquitination of histones may constitute an element of the "histone code" discussed in Chapter 7. In addition, proteins can be modified by the attachment of other ubiquitin-related polypeptides such as SUMO (small *u*biquitin-related *mo*difier). SUMO does not target proteins for degradation but instead serves as a marker for protein localization and a regulator of protein activity. Many of the proteins modified by SUMO are transcription factors and other nuclear proteins with roles in maintenance of chromatin structure and DNA repair.

Lysosomal Proteolysis

The other major pathway of protein degradation in eukaryotic cells involves the uptake of proteins by **lysosomes**. Lysosomes are membrane-enclosed organelles that contain an array of digestive enzymes, including several proteases (see Chapter 10). They have several roles in cell metabolism, including the digestion of extracellular proteins taken up by endocytosis as well as the turnover of cytoplasmic organelles and cytosolic proteins.

The containment of proteases and other digestive enzymes within lysosomes prevents uncontrolled degradation of the contents of the cell. Therefore, in order to be degraded by lysosomal proteolysis, cellular proteins must first be taken up by lysosomes. The principal pathway for this uptake of cellular proteins, **autophagy**, involves the formation of vesicles (autophagosomes) in which small areas of cytoplasm or cytoplasmic organelles are enclosed in membranes derived from the endoplasmic reticulum (Figure 8.44). These vesicles then fuse with lysosomes, and the degradative lysosomal enzymes digest their contents. The uptake of pro-

■ The ubiquitin-proteasome pathway is responsible for the degradation of several important regulatory proteins, including proteins that control cell proliferation and cell survival. Since the growth of cancer cells depends on the destruction of these regulatory proteins, the proteasome has emerged as a target for anti-cancer drugs. Bortezomib was the first proteasome inhibitor approved for the treatment of a human cancer, multiple myeloma.

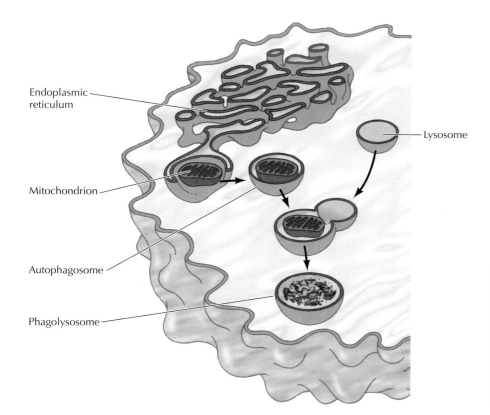

Endoplasmic reticulum

Lysosome

Mitochondrion

Autophagosome

Phagolysosome

FIGURE 8.44 Autophagy
Lysosomes contain various digestive enzymes, including proteases. Lysosomes take up cellular proteins by fusion with autophagosomes, which are formed by the enclosure of areas of cytoplasm or organelles (e.g., a mitochondrion) in fragments of the endoplasmic reticulum. This fusion yields a phagolysosome, which digests the contents of the autophagosome.

teins into autophagosomes appears to be nonselective, so it results in the eventual slow degradation of long-lived cytoplasmic proteins.

Autophagy is regulated both in response to the availability of nutrients and during development of multicellular organisms. Autophagy is generally activated under conditions of nutrient starvation, allowing cells to degrade nonessential proteins and organelles so that their components can be reutilized. In addition, autophagy plays an important role in many developmental processes, such as insect metamorphosis, which involve extensive tissue remodeling and degradation of cellular components. Recent studies have also linked defects in autophagy to several human diseases, including neurodegenerative diseases and cancer.

KEY TERMS

tRNA, anticodon, aminoacyl tRNA synthetase

ribosome, rRNA

5' untranslated region, polycistronic, monocistronic, 3' untranslated region, Shine-Dalgarno sequence

initiation factor, elongation factor, release factor, polysome,

microRNA (miRNA)

SUMMARY

TRANSLATION OF mRNA

Transfer RNAs: Transfer RNAs serve as adaptors that align amino acids on the mRNA template. Aminoacyl tRNA synthetases attach amino acids to the appropriate tRNAs, which then bind to mRNA codons by complementary base pairing.

The Ribosome: Ribosomes consist of two subunits, which are composed of proteins and ribosomal RNAs. The 23S rRNA is the primary catalyst of peptide bond formation.

The Organization of mRNAs and the Initiation of Translation: Translation of both prokaryotic and eukaryotic mRNAs initiates with a methionine residue. In bacteria, initiation codons are preceded by a sequence that aligns the mRNA on the ribosome by base pairing with 16S rRNA. In eukaryotes, most initiation codons are identified by scanning from the 5' end of the mRNA, which is recognized by its 7-methylguanosine cap.

The Process of Translation: Translation is initiated by the binding of methionyl tRNA and mRNA to the small ribosomal subunit. The large ribosomal subunit then joins the complex, and the polypeptide chain elongates until the ribosome reaches a termination codon in the mRNA. A variety of nonribosomal factors are required for initiation, elongation, and termination of translation in both prokaryotic and eukaryotic cells.

Regulation of Translation: Translation of specific mRNAs can be regulated by the binding of repressor proteins and by proteins that localize mRNAs to specific regions of cells. Controlled polyadenylation of mRNA is also an important mechanism for the regulation of translation during early development. In addition, the translation of many mRNAs may be controlled by noncoding microRNAs that either repress translation or target homologous mRNAs for degradation. Finally, the general translational activity of cells can be regulated by modification of initiation factors.

SUMMARY	KEY TERMS

PROTEIN FOLDING AND PROCESSING

Chaperones and Protein Folding: Molecular chaperones facilitate protein folding by binding to and stabilizing unfolded or partially folded polypeptide chains.

chaperone

Enzymes that Catalyze Protein Folding: At least two types of enzymes, protein disulfide isomerase and peptidyl prolyl isomerase, catalyze protein folding.

protein disulfide isomerase (PDI), peptidyl prolyl isomerase

Protein Cleavage: Proteolysis is an important step in the processing of many proteins. For example, secreted proteins and proteins incorporated into most eukaryotic organelles are targeted to their destinations by amino-terminal sequences that are removed by proteolytic cleavage as the polypeptide chain crosses the membrane.

proteolysis, signal sequence, signal peptidase

Glycosylation: Many eukaryotic proteins, particularly secreted and plasma membrane proteins, are modified by the addition of carbohydrates in the endoplasmic reticulum and Golgi apparatus.

glycosylation, glycoprotein, dolichol phosphate

Attachment of Lipids: Covalently attached lipids frequently target and anchor proteins to the plasma membrane.

N-myristoylation, prenylation, pamitoylation, glycolipid, glycosylphosphatidylinositol (GPI) anchor

REGULATION OF PROTEIN FUNCTION

Regulation by Small Molecules: Many proteins are regulated by the binding of small molecules, such as amino acids and nucleotides, which induce changes in protein conformation and activity.

allosteric regulation

Protein Phosphorylation: Reversible phosphorylation, which controls the activities of a wide variety of cellular proteins, results from the action of protein kinases and phosphatases. Other modifications, such as nitrosylation, also regulate the activities of some proteins.

protein kinase, protein-serine/threonine kinase, protein-tyrosine kinase, protein phosphatase, nitrosylation

Protein-Protein Interactions: Interactions between polypeptide chains are important in the regulation of allosteric enzymes and other cellular proteins.

PROTEIN DEGRADATION

The Ubiquitin-Proteasome Pathway: The major pathway of selective protein degradation in eukaryotic cells uses ubiquitin as a marker that targets proteins for rapid proteolysis by the proteasome.

ubiquitin, proteasome

Lysosomal Proteolysis: Lysosomal proteases degrade extracellular proteins taken up by endocytosis and are responsible for the degradation of cytoplasmic organelles and long-lived cytosolic proteins. Autophagy is activated as a response to cell starvation.

lysosome, autophagy

Questions

1. *E. coli* contain 64 different codons in their mRNAs, 61 of which code for amino acids. How can they synthesize proteins when they have only 40 different tRNAs?

2. You wish to express a cloned eukaryotic cDNA in bacteria. What type of sequence must you add in order for the mRNA to be translated on prokaryotic ribosomes?

3. Discuss the evidence that ribosomal RNA is the major catalytic component of the ribosome.

4. What effect would an inhibitor of polyadenylation have on protein synthesis in fertilized eggs?

5. What are chaperones? Why is it beneficial for the synthesis of heat-shock proteins to be induced by exposure of cells to elevated temperatures?

6. You are interested in studying a protein expressed on the surface of liver cells. How could treatment of these cells with a phospholipase (an enzyme that cleaves phospholipids) enable you to determine whether your protein is a transmembrane protein or one that is attached to the cell surface by a GPI anchor?

7. What was the first evidence that ubiquitination and degradation of specific proteins by proteasomes requires a specific target sequence on the protein?

8. Does ubiquitination of a protein always signal its destruction by the proteasome?

9. How do miRNAs regulate the translation of specific mRNAs?

10. What is the function of 3′ UTRs in mRNAs?

11. Why did Noller and colleagues use ribosomes of *T. aquaticus* for their studies?

12. Why is regulated proteolytic cleavage important for the activity of certain proteins?

13. How does the ribosome ensure that the correct aminoacyl tRNA is inserted opposite a codon?

14. You are studying the pathway responsible for the secretion of ribonuclease (RNase) in cultured pancreatic cells, by assaying the activity of secreted RNase in the culture medium. How would the expression of siRNA targeted against protein disulfide isomerase (PDI) affect the amount of active RNase you detect in your experiments?

References and Further Reading

Translation of mRNA

Ban, N., P. Nissen, J. Hansen, P. B. Moore and T. A. Steitz. 2000. The complete atomic structure of the large ribosomal subunit at 2.4 Å resolution. *Science* 289: 905–920. [P]

Bartel, D. P. 2004. MicroRNAs: Genomics, biogenesis, mechanism, and function. *Cell* 116: 281–297. [R]

Crick, F. H. C. 1966. Codon-anticodon pairing: The wobble hypothesis. *J. Mol. Biol.* 19: 548–555. [P]

Holcik, M. and N. Sonenberg. 2005. Translational control in stress and apoptosis. *Nature Rev. Cel Mol. Biol.* 6: 318–327. [R]

Hopper, A. K. and E. M. Phizicky. 2003. tRNA transfers to the limelight. *Genes Dev.* 17:162–180. [R]

Ibba, M. and D. Soll. 2004. Aminoacyl-tRNAs: Setting the limits of the genetic code. *Genes Dev.* 18: 731–738. [R]

Kisselev, L. L. and R. H. Buckingham. 2000. Translational termination comes of age. *Trends Biochem. Sci.* 25: 561–566. [R]

Klausner, R. D., T. A. Rouault and J. B. Harford. 1993. Regulating the fate of mRNA: The control of cellular iron metabolism. *Cell* 72: 19–28. [R]

Komar, A. A. and M. Hatzoglou. 2005. Internal ribosome entry sites in cellular mRNAs: Mystery of their existence. *J. Biol. Chem.* 280: 23425–23428. [R]

Lewis, B. P., C. B. Burge and D. P. Bartel. 2005. Conserved seed pairing, often flanked by adenosines, indicates that thousands of human genes are microRNA targets. *Cell* 120: 15–20. [P]

Meister, G. and T. Tuschl. 2004. Mechanisms of gene silencing by double-stranded RNA. *Nature* 431: 343–349. [R]

Moore, P. B. and T. A. Steitz. 2002. The involvement of RNA in ribosome function. *Nature* 418: 229–235. [R]

Nissen, P., J. Hansen, N. Ban, P. B. Moore and T. A. Steitz. 2000. The structural basis of ribosome activity in peptide bond synthesis. *Science* 289: 920–930. [P]

Noller, H. F. 2005. RNA structure: Reading the ribosome. *Science* 309: 1508–1514. [R]

Noller, H. F., V. Hoffarth and L. Zimniak. 1992. Unusual resistance of peptidyl transferase to protein extraction procedures. *Science* 256: 1416–1419. [P]

Nomura, M. 1997. Reflections on the days of ribosome reconstitution research. *Trends Biochem. Sci.* 22: 275–279. [R]

Novina, C. D. and P. A. Sharp. 2004. The RNAi revolution. *Nature* 430: 161–164. [R]

Ramakrishnan, V. 2002. Ribosome structure and the mechanism of translation. *Cell* 108: 557–572. [R]

Richter, J. D. and N. Sonenberg. 2005. Regulation of cap-dependent translation by eIF4E inhibitory proteins. *Nature* 433: 477–480. [R]

Saks, M. E., J. R. Sampson and J. N. Abelson. 1994. The transfer RNA identity problem: A search for rules. *Science* 263: 191–197. [R]

St. Johnston, D. 2005. Moving messages: The intracellular localization of mRNAs. *Nature Rev. Mol. Cell Biol.* 6: 363–375. [R]

Steitz, T. A. 2005. On the structural basis of peptide-bond formation and antibiotic resistance from atomic structures of the large ribosomal subunit. *FEBS Lett.* 579: 955–958. [R]

Wimberly, B. T., D. E. Brodersen, W. M. Clemons, R. J. Morgan-Warren, A. P. Carter, C. Vonrhein, T. Hartrsch and V. Ramakrishnan. 2000. Structure of the 30S ribosomal subunit. *Nature* 407: 327–339. [P]

Yusupov, M. M., G. Z. Yusupova, A. Baucom, K. Lieberman, T. N. Earnest, J. H. Cate and H. F. Noller. 2001. Crystal structure of the ribosome at 5.5 Å resolution. *Science* 292: 883–896. [P]

Protein Folding and Processing

Farazi, T. A., G. Waksman and J. I. Gordon. 2001. The biology and enzymology of protein N-myristoylation. *Ann. Rev. Biochem.* 276: 39501–39504. [R]

Hebert, D. N., S. C. Garman and M. Molinari. 2005. The glycan code of the endoplasmic reticulum: Asparagine-linked carbohydrates as protein maturation and quality-control tags. *Trends Cell Biol.* 15: 364–370. [R]

Helenius, A. and M. Aebi. 2004. Roles of N-linked glycans in the endoplasmic reticulum. *Ann. Rev. Biochem.* 73: 1019–1049. [R]

Iyer, S. P. N. and G. W. Hart. 2003. Dynamic nuclear and cytoplasmic glycosylation: Enzymes of O-GlcNAc cycling. *Biochem.* 42: 2493–2499. [R]

Paetzel, M., A. Karla, N. C. Strynadka and R. E. Dalbey. 2002. Signal peptidases. *Chem. Rev.* 102: 4549–4580. [R]

Roth, J. 2002. Protein *N*-glycosylation along the secretory pathway: Relationship to organelle topography and function, protein quality control, and cell interactions. *Chem. Rev.* 102: 285–303. [R]

Schiene, C. and G. Fischer. 2000. Enzymes that catalyse the restructuring of proteins. *Curr. Opin. Struc. Biol.* 10: 40–45. [R]

Spiro, R. G. 2002. Protein glycosylation: Nature, distribution, enzymatic formation, and disease implications of glycopeptide bonds. *Glycobiol.* 12: 43R–56R. [R]

Udenfriend, S. and K. Kodukula. 1995. How glycosylphosphatidylinositol-anchored membrane proteins are made. *Ann. Rev. Biochem.* 64: 563–591. [R]

Wells, L., K. Vosseller and G. W. Hart. 2001. Glycosylation of nucleocytoplasmic proteins: Signal transduction and *O*-GlcNAc. *Science* 291: 2376–2378. [R]

Wilkinson, B. and H. F. Gilbert. 2004. Protein disulfide isomerase. *Biochim. Biophys. Acta* 1699: 35–44. [R]

Winter-Vann, A. M. and P. J. Casey. 2005. Post-prenylation-processing enzymes as new targets in oncogenesis. *Nature Rev. Cancer* 5: 405–412. [R]

Young, J. C., V. R. Agashe, K. Siegers and F. U. Hartl. 2004. Pathways of chaperone-mediated protein folding in the cytosol. *Nature Rev. Mol. Cell Biol.* 5: 781–791. [R]

Zhang, F. L. and P. J. Casey. 1996. Protein prenylation: Molecular mechanisms and functional consequences. *Ann. Rev. Biochem.* 65: 241–269. [R]

Regulation of Protein Function

Alonso, A., J. Sasin, N. Bottini, I. Friedberg, I. Friedberg, A. Osterman, A. Godzik, T. Hunter, J. Dixon and T. Mustelin. 2004. Protein tyrosine phosphatases in the human genome. *Cell* 117: 699–711. [R]

Barford, D. 1996. Molecular mechanisms of the protein serine/threonine phosphatases. *Trends Biochem. Sci.* 21: 407–412. [R]

Fauman, E. B. and M. A. Saper. 1996. Structure and function of the protein tyrosine phosphatases. *Trends Biochem. Sci.* 21: 413–417. [R]

Fischer, E. H. and E. G. Krebs. 1989. Commentary on "The phosphorylase β to α converting enzyme of rabbit skeletal muscle." *Biochim. Biophys. Acta* 1000: 297–301. [R]

Hanks, S. K., A. M. Quinn and T. Hunter. 1988. The protein kinase family: Conserved features and deduced phylogeny of the catalytic domains. *Science* 241: 42–52. [R]

Hess, D. T., A. Matsumoto, S.-O. Kim, H. E. Marshall and J. S. Stamler. 2005. Protein *S*-nitrosylation: Purview and parameters. *Nature Rev. Mol. Cell Biol.* 6: 150–166. [R]

Hunter, T. 1995. Protein kinases and phosphatases: The yin and yang of protein phosphorylation and signaling. *Cell* 80: 225–236. [R]

Huse, M. and J. Kuriyan. 2002. The conformational plasticity of protein kinases. *Cell* 109: 275–282. [R]

Manning, G., G. D. Plowman, T. Hunter and S. Sudarsanam. 2002. Evolution of protein kinase signaling from yeast to man. *Trends Biochem. Sci.* 27: 514–520. [R]

Marianayagam, N. J., M. Sunde and J. M. Matthews. 2004. The power of two: Protein dimerization in biology. *Trends Biochem. Sci.* 29: 618–625. [R]

Milburn, M. V., L. Tong, A. M. DeVos, A. Brunger, Z. Yamaizumi, S. Nishimura and S.-H. Kim. 1990. Molecular switch for signal transduction: Structural differences between active and inactive forms of protooncogenic *ras* proteins. *Science* 247: 939–945. [P]

Monod, J., J.-P. Changeux and F. Jacob. 1963. Allosteric proteins and cellular control systems. *J. Mol. Biol.* 6: 306–329. [P]

Taylor, S. S., D. R. Knighton, J. Zheng, L. F. R. Eyck and J. M. Sowadski. 1992. Structural framework for the protein kinase family. *Ann. Rev. Cell Biol.* 8: 429–462. [R]

Vetter, I. R. and A. Wittinghofer. 2001. The guanine nucleotide-binding switch in three dimensions. *Science* 294: 1299–1304. [R]

Protein Degradation

Cuervo, A. M. 2004. Autophagy: In sickness and health. *Trends Cell Biol.* 14: 70–77. [R]

Gil, G. 2004. SUMO and ubiquitin in the nucleus: Different functions, similar mechanisms? *Genes Dev.* 18: 2046–2059. [R]

Glotzer, M., A. W. Murray and M. W. Kirschner. 1991. Cyclin is degraded by the ubiquitin pathway. *Nature* 349: 132–138. [P]

Hershko, A. and A. Ciechanover. 1998. The ubiquitin system. *Ann. Rev. Biochem.* 67: 425–479. [R]

Hilgarth, R. S., L. A. Murphy, H. S. Skaggs, D. C. Wilkerson, H. Xing and K. D. Sarge. 2004. Regulation and function of SUMO modification. *J. Biol. Chem.* 279: 53889–53902. [R]

Laney, J. D. and M. Hochstrasser. 1999. Substrate targeting in the ubiquitin system. *Cell* 97: 427–430. [R]

Levine, B. and D. J. Klionsky. 2004. Development by self-digestion: Molecular mechanisms and biological functions of autophagy, *Dev. Cell* 6: 463–477. [R]

Pickart, C. M. and R. E. Cohen. 2004. Proteasomes and their kin: Proteases in the machine age. *Nature Rev. Mol. Cell Biol.* 5: 177–187. [R]

Reed, S. I. 2003. Ratchets and clocks: The cell cycle, ubiquitylation and protein turnover. *Nature Rev. Mol. Cell Biol.* 4: 855–864. [R]

Welchman, R. L., C. Gordon and R. J. Mayer. 2005. Ubiquitin and ubiquitin-like proteins as multifunctional signals. *Nature Rev. Mol. Cell Biol* 6: 599–609. [R]

PART III

Cell Structure and Function

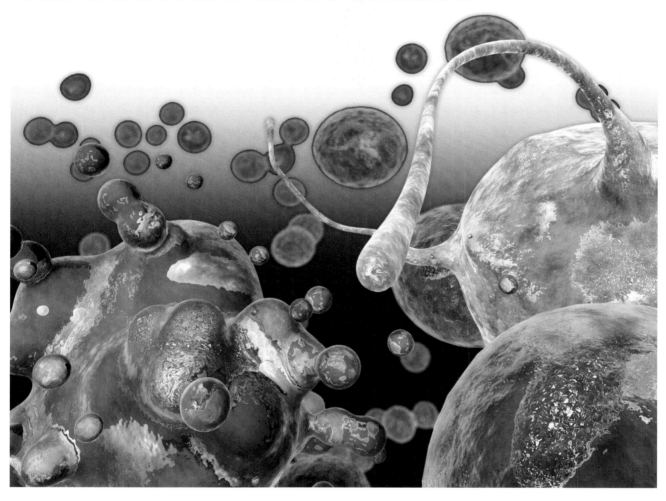

CHAPTER **9** ■ **The Nucleus**

CHAPTER **10** ■ **Protein Sorting and Transport**

CHAPTER **11** ■ **Bioenergetics and Metabolism**

CHAPTER **12** ■ **The Cytoskeleton and Cell Movement**

CHAPTER **13** ■ **The Plasma Membrane**

CHAPTER **14** ■ **Cell Walls, the Extracellular Matrix, and Cell Interactions**

CHAPTER 9

The Nucleus

■ **The Nuclear Envelope and Traffic between the Nucleus and the Cytoplasm 355**

■ **Internal Organization of the Nucleus 371**

■ **The Nucleolus and rRNA Processing 375**

■ **KEY EXPERIMENT:** Identification of Nuclear Localization Signals 364

■ **MOLECULAR MEDICINE:** Nuclear Lamina Diseases 359

THE PRESENCE OF A NUCLEUS IS THE PRINCIPAL FEATURE that distinguishes eukaryotic from prokaryotic cells. By housing the cell's genome, the nucleus serves both as the repository of genetic information and as the cell's control center. DNA replication, transcription, and RNA processing all take place within the nucleus, with only the final stage of gene expression (translation) localized to the cytoplasm.

By separating the genome from the cytoplasm, the nuclear envelope allows gene expression to be regulated by mechanisms that are unique to eukaryotes. Whereas prokaryotic mRNAs are translated while their transcription is still in process, eukaryotic mRNAs undergo several forms of posttranscriptional processing before being transported from the nucleus to the cytoplasm. The presence of a nucleus thus allows gene expression to be regulated by posttranscriptional mechanisms, such as alternative splicing. By limiting the access of selected proteins to the genetic material, the nuclear envelope also provides novel opportunities for the control of gene expression at the level of transcription. For example, the expression of some eukaryotic genes is controlled by the regulated transport of transcription factors from the cytoplasm to the nucleus —a form of transcriptional regulation unavailable to prokaryotes. Separation of the genome from the site of mRNA translation thus plays a central role in eukaryotic gene expression.

The Nuclear Envelope and Traffic between the Nucleus and the Cytoplasm

The nuclear envelope separates the contents of the nucleus from the cytoplasm and provides the structural framework of the nucleus. The two envelope membranes, acting as barriers that prevent the free passage of molecules between the nucleus and the cytoplasm, maintain the nucleus as a distinct biochemical compartment. The only channels through the

(A)

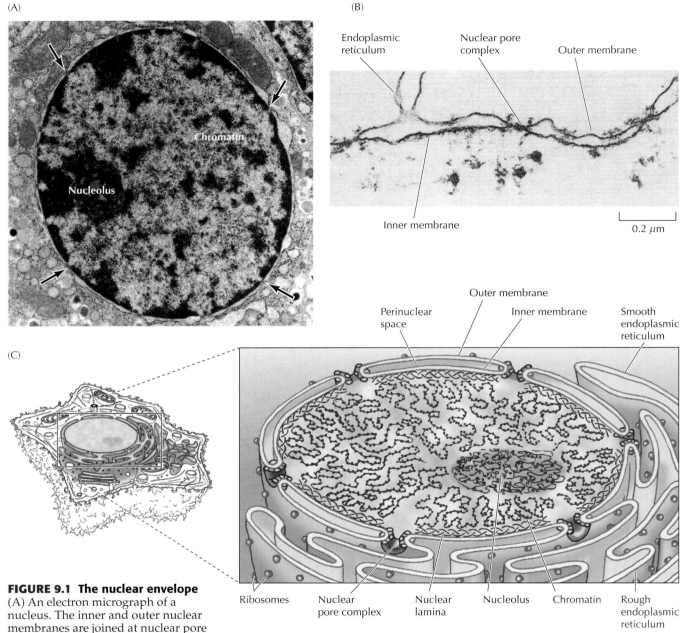

(B)

(C)

FIGURE 9.1 The nuclear envelope
(A) An electron micrograph of a nucleus. The inner and outer nuclear membranes are joined at nuclear pore complexes (arrows). (B) An electron micrograph illustrating the continuity of the outer nuclear membrane with the endoplasmic reticulum. (C) Schematic of the nuclear envelope. The inner nuclear membrane is lined by the nuclear lamina, which serves as an attachment site for chromatin. (A, David M. Phillips/Photo Researchers, Inc.; B, courtesy of Dr. Werner W. Franke, German Cancer Research Center, Heidelberg.)

nuclear envelope are provided by the nuclear pore complexes, which allow the regulated exchange of molecules between the nucleus and the cytoplasm. The selective traffic of proteins and RNAs through the nuclear pore complexes not only establishes the internal composition of the nucleus but also plays a critical role in regulating eukaryotic gene expression.

Structure of the Nuclear Envelope

The **nuclear envelope** has a complex structure consisting of two nuclear membranes, an underlying nuclear lamina, and nuclear pore complexes (Figure 9.1). The nucleus is surrounded by a system of two concentric membranes, called the inner and outer **nuclear membranes**. The outer nuclear

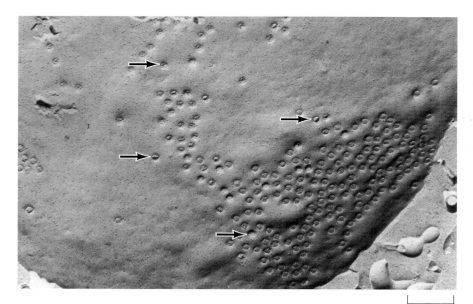

0.5 μm

membrane is continuous with the endoplasmic reticulum, so the space between the inner and outer nuclear membranes is directly connected with the lumen of the endoplasmic reticulum. In addition, the outer nuclear membrane is functionally similar to the membranes of the endoplasmic reticulum (see Chapter 10) and has ribosomes bound to its cytoplasmic surface. In contrast, the inner nuclear membrane carries proteins that are specific to the nucleus, such as those that bind the nuclear lamina (discussed below).

The critical function of the nuclear membranes is to act as a barrier that separates the contents of the nucleus from the cytoplasm. Like other cell membranes, each nuclear membrane is a phospholipid bilayer permeable only to small nonpolar molecules (see Figure 2.27). Other molecules are unable to diffuse through the bilayer. The inner and outer nuclear membranes are joined at nuclear pore complexes—the sole channels through which small polar molecules and macromolecules pass through the nuclear envelope (Figure 9.2). As discussed in the next section, the nuclear pore complex is a complicated structure that is responsible for the selective traffic of proteins and RNAs between the nucleus and the cytoplasm.

Underlying the inner nuclear membrane is the **nuclear lamina**, a fibrous meshwork that provides structural support to the nucleus (Figure 9.3). The nuclear lamina is composed of 60- to 80-kilodalton (kd) fibrous proteins called **lamins**. Lamins are a class of intermediate filament proteins; the other classes are found in the cytoskeleton (see Chapter 12). Mammalian cells have three lamin genes, designated A, B, and C, which code for at least seven distinct proteins. Like other intermediate filament proteins, the lamins associate with each other to form higher order structures (Figure 9.4), although the extent and polarity of this association is thought to differ from that of other intermediate filaments. The first stage of this association is the interaction of two lamins to form a dimer in which the α-helical regions of two polypeptide chains are wound around each other in a structure called a coiled coil. These lamin dimers then associate with each other to form the nuclear lamina.

■ Erythrocytes (or red blood cells) in mammals are devoid of nuclei. As the erythrocytes develop from precursor cells, the nucleus is extruded from the cell.

FIGURE 9.3 Electron micrograph of the nuclear lamina The lamina is a meshwork of filaments underlying the inner nuclear membrane and extending into the interior of the nucleus. (From U. Aebi, J. Cohn, L. Buhle and L. Gerace, 1986. *Nature* 323: 560.)

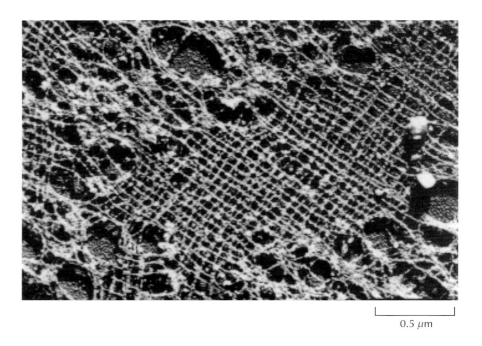

0.5 μm

The association of lamins with the inner nuclear membrane is facilitated by the posttranslational addition of lipid—in particular, prenylation of C-terminal cysteine residues (see Figure 8.33). In addition, the lamins bind to specific inner nuclear membrane proteins such as emerin and the lamin B receptor, mediating their attachment to the nuclear envelope and localizing

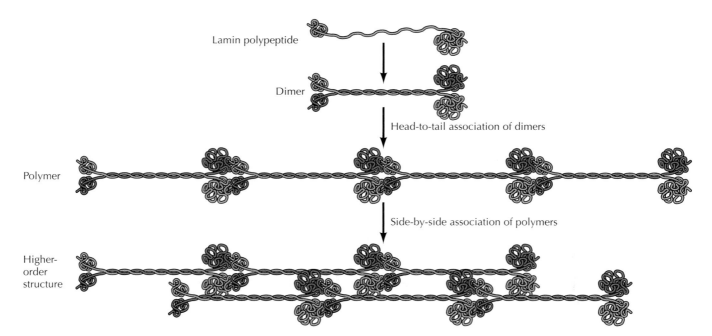

FIGURE 9.4 Model of lamin assembly The lamin polypeptides form dimers in which the central α-helical regions of two polypeptide chains are wound around each other. Further assembly may involve the head-to-tail association of dimers to form linear polymers and the side-by-side association of polymers to form higher order structures.

MOLECULAR MEDICINE

Nuclear Lamina Diseases

The Diseases

In 1966 Emery and Dreifuss described a new muscular dystrophy linked to the X-chromosome. Early in the disease the elbows, neck, and heels of affected individuals become stiff, and often there is a conduction block in the heart. These symptoms are seen by age 10 and include "toe-walking" because of stiff Achilles' tendons in the heels and difficulty bending the elbows. Heart problems develop by age 20 and may require a pacemaker. There is a gradual wasting and weakness of the shoulder and upper arm muscles and the calf muscles of the legs, but this occurs slowly and is often not a problem until late in life.

Nearly 30 years later researchers showed that mutations in a novel transmembrane protein were responsible for this X-linked Emery–Dreifuss muscular dystrophy. They named the protein emerin, after Alan Emery. Soon several groups found that emerin was a protein localized to the inner nuclear membrane and absent in patients with the X-linked Emery–Dreifuss muscular dystrophy. This was unexpected; mutations in a nuclear envelope protein expressed in all cells apparently caused a tissue-specific disease. While all cells in the body were missing the protein, the pathology occurred only

in muscle. Subsequent investigators found that the same dystrophy could also be inherited in a non-sex-linked manner. Families with this non-sex-linked Emery–Dreifuss muscular dystrophy had mutations in *LMNA*, the single gene encoding nuclear lamins A and C. So mutations in two genes—one coding for an inner nuclear membrane protein and one coding for a major nuclear lamin—caused clinically identical muscular dystrophy.

More surprising was that parallel investigations on different diseases, Dunnigan-type partial lipodystrophy, Charcot–Marie–Tooth disorder type 2B1, and a disease that causes premature ageing—Hutchinson-Gilford progeria syndrome—traced them to different mutations in the *LMNA* gene. Previously, physicians classified these as distinct diseases based on their clinical features and modes of inheritance. Recent work shows that mutations in another protein in the inner nuclear membrane, the lamin B receptor, are the basis for Pelger-Huët anomaly.

Molecular and Cellular Basis

Most biologists thought that mutations in lamins would cause generalized defects in nuclear architecture and serious problems in rapidly dividing cells. However, only minor aberrations of nuclear structure occur in the patients. Thus the puzzle is

how mutations in nuclear lamins or lamin-binding proteins cause different tissue-specific diseases. The answer is not yet known but there are two major hypotheses. The first is the "gene expression" hypothesis. This posits that the correct interaction of the two lamin proteins, A and C, with the nuclear envelope is essential for normal tissue-specific expression of certain genes. Transcriptionally inactive genes are located preferentially at the nuclear periphery, whereas expressed genes are concentrated in the center of the nucleus with a cell-type specificity. Thus the basis of these diseases would be a change in gene expression caused by defective protein interactions.

In the "mechanical stress" hypothesis the mutations in the nuclear lamins-emerin complex are thought to weaken the structural integrity of an integrated cytoskeletal network. In all cells the lamina, inner nuclear membrane, and nuclear pore complexes are tightly connected. This hypothesis, which works best for the muscular dystrophies, suggests that through filaments attached to the nuclear pore complex, the lamina could be connected indirectly with the muscle cell cytoskeleton.

(Continued on next page)

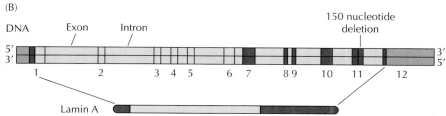

(A) A child with Hutchinson-Gilford progeria. (B) Diagram of the intron-exon structure of the *LMNA* gene and lamin A protein, with the globular domains indicated in red and the rod domains in yellow. In the mutant gene shown, the gene has a 150-bp deletion (black) in exon 11. (A, courtesy of Maggie Bartlett, NHGRI.)

MOLECULAR MEDICINE

Prevention and Treatment

The discovery that mutations in commonly expressed proteins of the nuclear lamina complex cause different inherited tissue-specific diseases has come as a surprise and has changed the way scientists look at nuclear lamins. Further research is needed to learn whether the bases of the pathologies in each of these diseases is mechanical regulation or gene expression. However, the known molecular nature of the diseases greatly simplifies their diagnosis and makes eventual treatment more likely.

The recent development of a mouse model where the *LMNA* gene is knocked out represents a first step. As the embryos develop they show symptoms of Emery–Dreifuss muscular dystrophy. Finally, researchers are now aware that several slow-developing congenital diseases might be new members of the nuclear "laminopathies."

References

De Sandre-Giovannoli, A., M. Chaouch, S. Kozlov, J. M. Vallat, M. Tazir, N. Kassouri, P. Szepetowski, T. Hammadouche, A. Vandenberghe, C. L. Stewart, D. Grid and N. Levy. 2002. Homozygous defects in LMNA, encoding lamin A/C nuclear-envelope proteins, cause autosomal recessive axonal neuropathy in human (Charcot-Marie-Tooth disorder type 2) and mouse. *Am. J. Hum. Genet.* 70: 726–736.

Gruenbaum, Y., A. Margalit, R. D. Goldman, D. K. Shumaker and K. L. Wilson. 2005. The nuclear lamina comes of age. *Nat. Rev. Mol. Cell Biol.* 6: 21–31.

and organizing them within the nucleus (Figure 9.5). The nuclear lamina also binds to chromatin through histones H2A and H2B as well as other chromatin proteins. While it binds DNA directly, it is not clear if this interaction is important in the cell. Lamins also extend in a loose meshwork throughout the interior of the nucleus. Many nuclear proteins that function in DNA synthesis, transcription, or chromatin modification are known to bind lamins, although the significance of these interactions is only beginning to be understood.

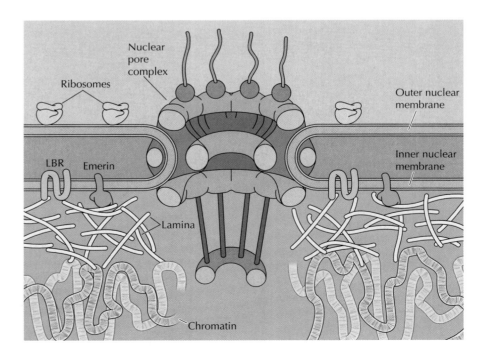

FIGURE 9.5 The nuclear lamina
The inner nuclear membrane contains several integral proteins, such as emerin and the lamin B receptor (LBR) that interact with nuclear lamins. The lamins also interact with chromatin.

The Nuclear Pore Complex

The **nuclear pore complexes** are the only channels through which small polar molecules, ions, and macromolecules (proteins and RNAs) can travel between the nucleus and the cytoplasm. The nuclear pore complex is an extremely large structure with a diameter of about 120 nm and an estimated molecular mass of approximately 125 million daltons—about 30 times the size of a ribosome. In vertebrates, the nuclear pore complex is composed of 30–50 different pore proteins (called nucleoporins), most of which are present in multiple copies. By controlling the traffic of molecules between the nucleus and the cytoplasm, the nuclear pore complex plays a fundamental role in the physiology of all eukaryotic cells. RNAs synthesized in the nucleus must be efficiently exported to the cytoplasm where they function in protein synthesis. Conversely, proteins required for nuclear functions (e.g., transcription factors) must be transported to the nucleus from their sites of synthesis in the cytoplasm. In addition, many proteins shuttle continuously between the nucleus and the cytoplasm.

Depending on their size, molecules can travel through the nuclear pore complex by one of two different mechanisms (Figure 9.6). Small molecules and some proteins with molecular mass less than approximately 20–40 kd pass freely through the pore in either direction: cytoplasm to nucleus or nucleus to cytoplasm. These molecules diffuse passively through open aqueous channels, estimated to have diameters of approximately 9 nm, in the nuclear pore complex. Most proteins and RNAs, however, are unable to pass through these open channels. Instead, these macromolecules pass through the nuclear pore complex by an active process in which appropriate proteins and RNAs are recognized and selectively transported in a specific direction (nucleus to cytoplasm or cytoplasm to nucleus).

Visualization of nuclear pore complexes by electron microscopy reveals a structure with eightfold symmetry organized around a large central channel (Figure 9.7), which is the route through which proteins and RNAs cross the nuclear envelope. Detailed structural studies, including computer-

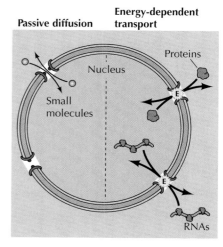

FIGURE 9.6 Molecular traffic through nuclear pore complexes Small molecules are able to pass rapidly through open channels in the nuclear pore complex by passive diffusion. In contrast, macromolecules (proteins and RNAs) are transported by a selective, energy-dependent mechanism.

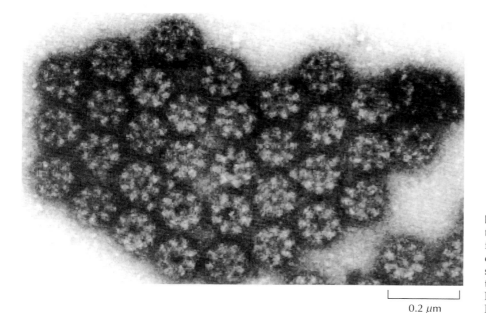

0.2 μm

FIGURE 9.7 Electron micrograph of nuclear pore complexes In this face-on view, isolated nuclear pore complexes appear to consist of eight structural subunits surrounding a central channel. (Courtesy of Dr. Ron Milligan, The Scripps Research Institute.)

FIGURE 9.8 Model of the nuclear pore complex The complex consists of an assembly of eight spokes attached to rings on the cytoplasmic and nuclear sides of the nuclear envelope. The spoke-ring assembly surrounds a central channel. Cytoplasmic filaments extend from the cytoplasmic ring, and filaments forming the nuclear basket extend from the nuclear ring.

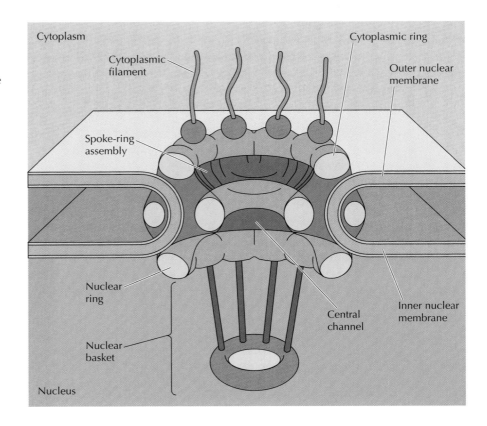

based image analysis, have led to the development of three-dimensional models of the nuclear pore complex (Figure 9.8). These studies show that the nuclear pore complex consists of an assembly of eight spokes arranged around a central channel. The spokes are connected to rings at the nuclear and cytoplasmic surfaces, and the spoke-ring assembly is anchored within the nuclear envelope at sites of fusion between the inner and outer nuclear membranes. Protein filaments extend from both the cytoplasmic and nuclear rings, forming a distinct basketlike structure on the nuclear side.

Selective Transport of Proteins to and from the Nucleus

Several million macromolecules selectively pass between the nucleus and the cytoplasm every minute. The basis for selective traffic across the nuclear envelope is best understood for proteins imported from the cytoplasm to the nucleus. Such proteins are responsible for all aspects of genome structure and function; they include histones, DNA polymerases, RNA polymerases, transcription factors, splicing factors, and many others. These proteins are targeted to the nucleus by specific amino acid sequences called **nuclear localization signals** that are recognized by transport receptors and direct the transport of the proteins through the nuclear pore complex.

The first nuclear localization signal to be mapped in detail was characterized by Alan Smith and colleagues in 1984. These investigators studied simian virus 40 (SV40) T antigen, a virus-encoded protein that initiates viral DNA replication in infected cells (see Chapter 6). As expected for a replication protein, T antigen is normally localized to the nucleus. The signal responsible for its nuclear localization was first identified by the finding that mutation of a single lysine residue prevents nuclear import, resulting instead in the accumulation of T antigen in the cytoplasm. Subsequent studies defined the T antigen nuclear localization signal as the seven-amino-acid sequence Pro-Lys-Lys-Lys-Arg-Lys-Val. Not only was this sequence necessary for the nuclear transport of T antigen but its addition to other, normally cytoplasmic, proteins was sufficient to direct their accumulation in the nucleus.

Nuclear localization signals have since been identified in many other proteins. Some of these sequences, like that of T antigen, are short stretches rich in basic amino acid residues (lysine and arginine). Often, however, the amino acids that form the nuclear localization signal are close together but not immediately adjacent to each other. For example, the nuclear localization signal of nucleoplasmin (a protein involved in chromatin assembly) consists of two parts: a Lys-Arg pair followed by four lysines located ten amino acids farther downstream (Figure 9.9). Both the Lys-Arg and Lys-Lys-Lys-Lys sequences are required for nuclear targeting, but the ten amino acids between these sequences can be mutated without affecting nuclear localization. Because this nuclear localization sequence is composed of two separated elements, it is called bipartite. Similar bipartite motifs appear to function as the localization signals of many nuclear proteins; thus they may be more common than the simpler nuclear localization signal of T antigen. While many nuclear localization signals consist of these basic amino acid residues—often termed the basic or "classical" nuclear localization signal— the amino acid sequences and structures of other nuclear localization signals vary considerably. Some are far apart in the amino acid sequence and depend on normal folding of the protein for their activity.

Nuclear localization signals are recognized by proteins that function as **nuclear transport receptors**. Most are members of the **karyopherin** protein family and function either as **importins**, which transport macromolecules to the nucleus from the cytoplasm, or **exportins**, which transport macromolecules from the nucleus to the cytoplasm (Table 9.1). Some importins (Kapβ1) act with an adapter karyopherin (Kapα) in a heterodimer to import proteins containing the basic nuclear localization signal, and some karyopherins act as importins for one protein and exportins for another.

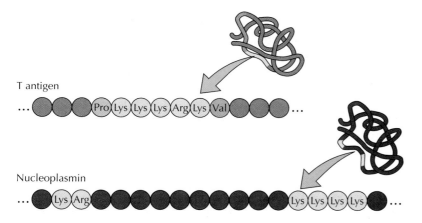

FIGURE 9.9 Nuclear localization signals   The T antigen nuclear localization signal is a single stretch of amino acids. In contrast, the nuclear localization signal of nucleoplasmin is bipartite, consisting of a Lys-Arg sequence, followed by a Lys-Lys-Lys-Lys sequence located ten amino acids farther downstream.

KEY EXPERIMENT

Identification of Nuclear Localization Signals

A Short Amino Acid Sequence Able to Specify Nuclear Location

Daniel Kalderon, Bruce L. Roberts, William D. Richardson, and Alan E. Smith

National Institute for Medical Research, Mill Hill, London

Cell, Volume 39, 1984, pages 499–509

Alan Smith

The Context

Maintaining the nucleus as a distinct biochemical compartment requires a mechanism by which proteins are segregated between the nucleus and the cytoplasm. Studies in the 1970s established that small molecules diffuse rapidly across the nuclear envelope but that most proteins are unable to do so. It therefore appeared likely that nuclear proteins are specifically recognized and selectively imported to the nucleus from their sites of synthesis on cytoplasmic ribosomes.

Earlier experiments of Günter Blobel and his colleagues had established that proteins are targeted to the endoplasmic reticulum by signal sequences consisting of short stretches of amino acids (see Chapter 10). In this 1984

paper, Alan Smith and his colleagues extended this principle to the targeting of nuclear proteins by identifying a short amino acid sequence that serves as a nuclear localization signal.

The Experiments

The viral protein SV40 T antigen was used as a model for studies of nuclear localization in animal cells. T antigen is a 94-kd protein that is required for SV40 DNA replication and is normally localized to the nucleus of SV40-infected cells. Previous studies in both Alan Smith's laboratory and in the laboratory of Janet Butel (Lanford and Butel, 1984, *Cell* 37: 801–813) had shown that mutation of Lys-128 to either Thr or Asn prevented the normal nuclear accumulation of T

antigen in both rodent and monkey cells. Rather than being transported to the nucleus these mutant T antigens remained in the cytoplasm, suggesting that Lys-128 was part of a nuclear localization signal. Kalderon and colleagues tested this hypothesis using two distinct experimental approaches.

First, they determined the effects of different deletions on the subcellular localization of T antigen. Mutant T antigens bearing deletions that eliminated amino acids either between residues 1 and 126 or between residue

TABLE 9.1 Karyopherins with Known Substrates

Karyopherin	Substrates
Import	
Kapα/Kapβ1 dimer	Proteins with a basic amino acid nuclear localization signal (e.g., nucleoplasmin)
Snurportin/Kapβ1	snRNPs (U1, U2, U4, U5)
Kapβ1 alone	Cdk/cyclin complexes
Kapβ2 (transportin)	mRNA binding proteins, ribosomal proteins
Importin7/Kapβ1 dimer	Histone H1, ribosomal proteins
Export	
Crm1	Proteins with a leucine-rich nuclear export signal, snurportin
CAS	Kapα
Exportin-t	tRNAs
Exportin4	Elongation factor 5A

KEY EXPERIMENT

136 and the C terminus were found to accumulate normally in the nucleus. In contrast, a mutant with a deletion of amino acids 127 to 132 remained in the cytoplasm. Thus the amino acid sequence extending from residue 127 to 132 appeared to be responsible for nuclear localization of T antigen.

To determine whether this amino acid sequence was able to target other proteins to the nucleus the investigators constructed chimeras in which the T antigen amino acid sequence was fused to proteins that were normally cytoplasmic. These experiments established that the addition of T antigen amino acids 126 to 132 to either β-galactosidase or pyruvate kinase is sufficient to specify the nuclear accumulation of these otherwise cytoplasmic proteins (see figure). This short amino acid sequence of SV40 T antigen thus functions as a nuclear localization signal, which is both necessary and sufficient to target proteins for nuclear import.

The Impact

As Kalderon and colleagues suggested in their 1984 paper, the nuclear localization signal of SV40 T antigen has proved to "represent a prototype of similar sequences in other nuclear proteins." By targeting proteins for nuclear import, these signals are key to establishing the biochemical identity of the nucleus and maintaining the fundamental division of eukaryotic cells into nuclear and cytoplasmic compartments. Nuclear localization signals are now known to be recognized by cytoplasmic receptors that transport their substrate proteins through the nuclear pore complex. The identification of nuclear localization signals was thus a key advance in understanding nuclear protein import.

(A)

(B)

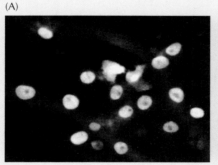 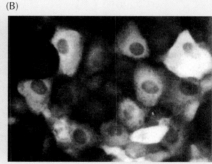

Cells were microinjected with plasmid DNAs encoding chimeric proteins in which SV40 amino acids were fused to pyruvate kinase. Cellular localization of the fusion proteins was then determined by immunofluorescence microscopy. (A) The fusion protein contains an intact SV40 nuclear localization signal (amino acids 126 to 132). (B) The nuclear localization signal has been inactivated by deletion of amino acids 131 and 132.

Movement of macromolecules through the nuclear pore is regulated by a protein called **Ran**. This is one of several types of small GTP-binding proteins whose conformation and activity are regulated by GTP binding and hydrolysis. Ras (see Figure 8.37), several of the translation factors involved in protein synthesis (see Figure 8.13), Arf and Rab (discussed in Chapter 10), and Rac, Rho and Cdc42 (discussed in Chapter 15) are other examples of small GTP-binding proteins. For Ran, enzymes that stimulate GTP hydrolysis to GDP are localized to the cytoplasmic side of the nuclear envelope, whereas enzymes that stimulate the exchange of GDP for GTP are localized to the nuclear side (Figure 9.10). Consequently, there is an unequal distribution of Ran/GTP across the nuclear pore, with a high concentration of Ran/GTP in the nucleus. This high concentration of Ran/GTP in the nucleus determines the directionality of nuclear transport of cargo proteins.

Ran regulates movement through the nuclear pore by controlling the activity of the nuclear transport receptors. Protein import through the nuclear pore complex begins when a specific importin binds to the nuclear

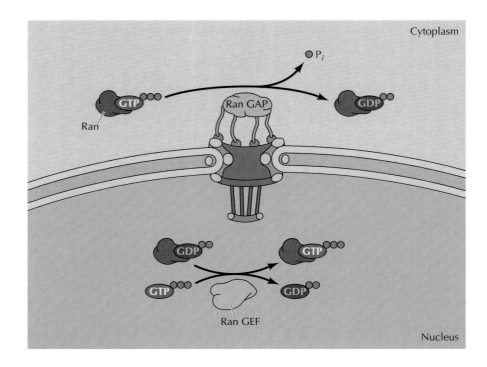

FIGURE 9.10 Distribution of Ran/GTP across the nuclear envelope   An unequal distribution of Ran/GTP across the nuclear envelope is maintained by the localization of Ran GTPase-activating protein (Ran GAP) in the cytoplasm and Ran guanine nucleotide exchange factor (Ran GEF) in the nucleus. In the cytoplasm, Ran GAP (which is bound to the cytoplasmic filaments of the nuclear pore complex) stimulates the hydrolysis of GTP bound to Ran, leading to the conversion of Ran/GTP to Ran/GDP. In the nucleus, Ran GEF stimulates the exchange of GDP bound to Ran for GTP, leading to the conversion of Ran/GDP to Ran/GTP. Consequently, a high concentration of Ran/GTP is maintained within the nucleus.

■ Many viruses must gain entry to the nucleus in order to replicate. Following infection of a cell, retroviruses, such as HIV, reverse transcribe their genomic RNA to synthesize a DNA provirus in the cytoplasm. HIV has evolved special mechanisms for the transport of proviral DNA to the nucleus where it can be transcribed.

localization signal of a cargo protein in the cytoplasm (Figure 9.11). This importin/cargo complex then binds to proteins in the cytoplasmic filaments of the nuclear pore complex, and transport proceeds by sequential binding to specific nuclear pore proteins located further and further toward the nuclear side of the pore complex. At the nuclear side of the pore complex the importin/cargo complex is disrupted by the binding of Ran/GTP. This causes a change in the conformation of the importin, which displaces the cargo protein and releases it into the nucleus.

The importin-Ran/GTP complex is then exported back through the nuclear pore complex. In the cytoplasm the GTP is hydrolyzed to GDP. This releases the importin so that it can bind to a new cargo protein in the cytoplasm and participate in another round of transport. The Ran/GDP formed in the cytoplasm is then transported back to the nucleus by its own import receptor (a protein called NTF2), where Ran/GTP is regenerated.

Some proteins remain within the nucleus following their import from the cytoplasm, but many others shuttle back and forth between the nucleus and the cytoplasm. Some of these proteins act as carriers in the transport of other molecules, such as RNAs; others coordinate nuclear and cytoplasmic functions (e.g., by regulating the activities of transcription factors). Proteins

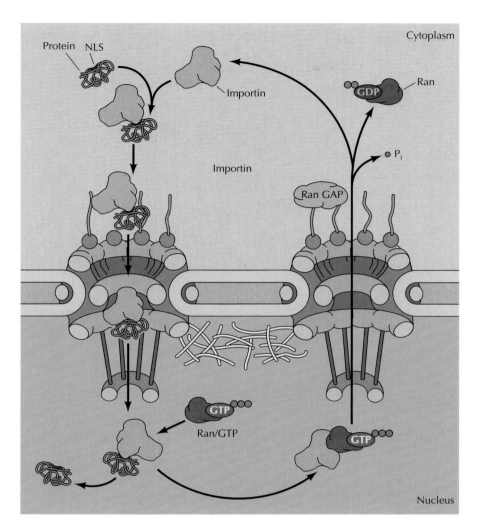

FIGURE 9.11 Protein import through the nuclear pore complex
Transport of a protein through the nuclear pore complex begins when its nuclear localization sequence (NLS) is recognized by an importin nuclear transport receptor. The cargo (the protein with the nuclear localization sequence)/importin complex binds to specific nuclear pore proteins in the cytoplasmic filaments. By sequential binding to more interior nuclear pore proteins, the complex is translocated through the nuclear pore. At the nuclear side of the pore, the cargo/importin complex is disrupted by the binding of Ran/GTP to the importin. The change in conformation of the importin displaces the cargo protein and releases it into the nucleus. The importin-Ran/GTP complex is re-exported through the nuclear pore and the GTPase-activating protein (Ran GAP) in the cytoplasm hydrolyzes the GTP on Ran to GDP, releasing the importin.

are targeted for export from the nucleus by specific amino acid sequences, called **nuclear export signals**. Like nuclear localization signals, nuclear export signals are recognized by receptors within the nucleus—exportins that direct protein transport through the nuclear pore complex to the cytoplasm. Many exportins are also members of the karyopherin protein family (see Table 9.1). Like importins these exportins bind to Ran, which is required for nuclear export as well as for nuclear import (Figure 9.12). However, Ran/GTP promotes the formation of stable complexes between exportins and their cargo proteins, whereas it dissociates the complexes between importins and their cargos. This effect of Ran/GTP binding on exportins dictates the movement of proteins containing nuclear export signals from the nucleus to the cytoplasm. Thus exportins form stable complexes with their cargo proteins in association with Ran/GTP within the nucleus. Following transport to the cytosolic side of the nuclear envelope, GTP hydrolysis and release of Ran/GDP leads to dissociation of the cargo protein, which is released into the cytoplasm.

FIGURE 9.12 Nuclear export
Complexes between cargo proteins bearing nuclear export signals (NES), exportins, and Ran/GTP form in the nucleus. Following transport through the nuclear pore complex, Ran GAP stimulates the hydrolysis of bound GTP, leading to formation of Ran/GDP and release of the cargo protein and exportin into the cytoplasm. Exportin is then transported back to the nucleus.

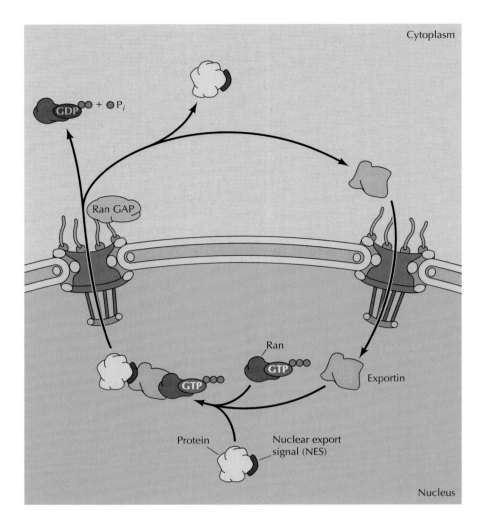

Regulation of Nuclear Protein Import

The transport of proteins to the nucleus is yet another level at which the activities of nuclear proteins can be controlled. Transcription factors, for example, are functional only when they are present in the nucleus, so regulation of their import to, and export from, the nucleus is a novel means of controlling gene expression. As will be discussed in Chapter 15, the regulated nuclear import of both transcription factors and protein kinases plays an important role in controlling the behavior of cells in response to changes in the environment, because it provides a mechanism by which signals received at the cell surface can be transmitted to the nucleus. The importance of regulated nuclear import is demonstrated by the finding that changes in nuclear transport receptor affinity of only two nuclear pore complex proteins apparently contributed to the evolutionary split between *Drosophila melanogaster* and *Drosophila simulans*.

In one mechanism of regulation, transcription factors (or other proteins) associate with cytoplasmic proteins that mask their nuclear localization signals; because their signals are no longer recognizable, these proteins remain in the cytoplasm. A good example is provided by the transcription factor NF-κB, which is activated in response to a variety of extracellular signals in

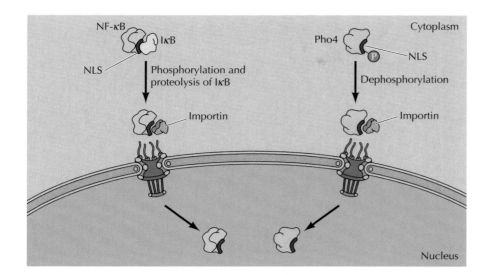

FIGURE 9.13 Regulation of nuclear import of transcription factors The transcription factor NF-κB is maintained as an inactive complex with IκB, which masks its nuclear localization sequence (NLS) in the cytoplasm. In response to appropriate extracellular signals, IκB is phosphorylated and degraded by proteolysis, allowing the import of NF-κB to the nucleus. The yeast transcription factor Pho4 is maintained in the cytoplasm by phosphorylation in the vicinity of its nuclear localization sequence. Regulated dephosphorylation exposes the NLS and allows Pho4 to be transported to the nucleus.

mammalian cells (Figure 9.13). In unstimulated cells, NF-κB is found as an inactive complex with an inhibitory protein (IκB) in the cytoplasm. Binding to IκB masks the NF-κB nuclear localization signal, thus preventing NF-κB from being transported to the nucleus. In stimulated cells, IκB is phosphorylated and degraded by ubiquitin-mediated proteolysis, allowing NF-κB to enter the nucleus and activate transcription of its target genes.

The nuclear import of other transcription factors is regulated directly by their phosphorylation rather than by association with inhibitory proteins (see Figure 9.13). For example, the yeast transcription factor Pho4 is phosphorylated at a serine residue adjacent to its nuclear localization signal. Phosphorylation at this site inhibits Pho4 by interfering with its nuclear import. Under appropriate conditions, regulated dephosphorylation of this site activates Pho4 by permitting its translocation to the nucleus.

Transport of RNAs

Whereas many proteins are selectively transported from the cytoplasm to the nucleus, most RNAs are exported from the nucleus to the cytoplasm. Since proteins are synthesized in the cytoplasm, the export of mRNAs, rRNAs, and tRNAs is a critical step in gene expression in eukaryotic cells. Like protein import, the export of all RNAs through the nuclear pore complex is an active, energy-dependent process requiring the transport receptors to interact with the nuclear pore complex. Karyopherin importins and exportins (see Table 9.1) transport most tRNAs, rRNAs, and small nuclear RNAs in a Ran/GTP–dependent manner. However, mRNAs are exported by a complex of two proteins (the "mRNA exporter"), one of which is related to the Ran/GDP transporter, NTF2. This transport of mRNAs appears to be independent of Ran.

RNAs are transported across the nuclear envelope as ribonucleoprotein complexes (RNPs) (Figure 9.14). Ribosomal RNAs are first associated with both ribosomal proteins and specific RNA processing proteins in the nucleolus, and nascent 60S and 40S ribosomal subunits are then transported to the cytoplasm (see Figure 9.30). Their export from the nucleus is mediated by nuclear export signals present on proteins within the subunit complex. Pre-mRNAs and mRNAs are associated with a set of at least 20 proteins

(A) (B) (C) (D)

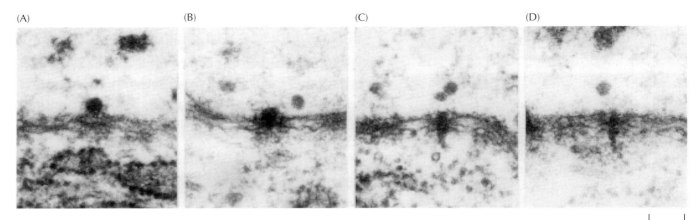

0.1 μm

FIGURE 9.14 Transport of a ribonucleoprotein complex Insect salivary gland cells produce large ribonucleoprotein complexes (RNPs), which contain 35 to 40 kb of RNA and have a total mass of approximately 30 million daltons. This series of electron micrographs shows the attachment of such an RNP to a nuclear pore complex (A) and the unfolding of the RNA during its translocation to the cytoplasm (B–D). (From H. Mehlin et al., 1992. *Cell* 69: 605.)

throughout their processing in the nucleus and eventual transport to the cytoplasm, which is mediated by the mRNA exporter complex after its recruitment to the processed mRNA. tRNAs are exported from the nucleus by exportin-t, which binds directly to the tRNAs.

In contrast to mRNAs, tRNAs, and rRNAs, which function in the cytoplasm, many small RNAs (snRNAs and snoRNAs) function within the nucleus as components of the RNA processing machinery. Perhaps surprisingly, snRNAs are initially transported from the nucleus to the cytoplasm, where they associate with proteins to form functional snRNPs and then return to the nucleus (Figure 9.15). Transport receptor proteins that bind to the 5′ caps of snRNAs appear to be involved in the export of the snRNAs to the cytoplasm. In contrast, sequences present on the snRNP proteins are responsible for the transport of snRNPs from the cytoplasm to the nucleus.

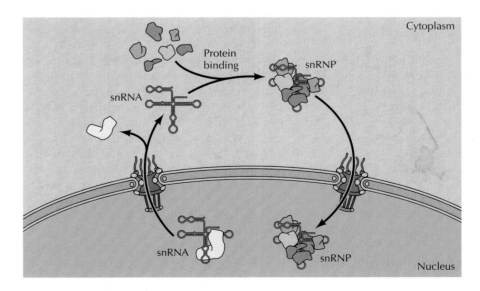

FIGURE 9.15 Transport of snRNAs between nucleus and cytoplasm Small nuclear RNAs (snRNAs) are initially exported from the nucleus to the cytoplasm, where they associate with proteins to form snRNPs. The assembled snRNPs are then transported back to the nucleus.

Internal Organization of the Nucleus

The nucleus is more than a container in which chromatin, RNAs, and nuclear proteins move freely in aqueous solution. Instead, the nucleus has an internal structure that organizes the genetic material and localizes nuclear functions. A loosely organized matrix of nuclear lamins extends from the nuclear lamina into the interior of the nucleus. These lamins serve as sites of chromatin attachment and organize other proteins into functional nuclear bodies. Chromatin within the nucleus is organized into large loops of DNA, and specific regions of these loops are bound to the lamin matrix by lamin-binding proteins in the chromatin. Many other nuclear proteins form lamin-dependent complexes, and these complexes form nuclear bodies that have roles in DNA repair, chromatin organization, gene regulation, and signal transduction. It is thought that this role of the nuclear lamina and lamin proteins in localizing DNA repair and gene transcription is the basis for the variety of lamin-related genetic diseases.

Chromosomes and Higher-Order Chromatin Structure

Chromatin becomes highly condensed during mitosis to form the compact metaphase chromosomes that are distributed to daughter nuclei (see Figure 5.15). During interphase, some of the chromatin (**heterochromatin**) remains highly condensed and is transcriptionally inactive; the remainder of the chromatin (**euchromatin**) is decondensed and distributed throughout the nucleus (Figure 9.16). Cells contain two types of heterochromatin: Constitutive heterochromatin, which contains DNA sequences that are generally not transcribed, such as the satellite sequences present at centromeres; and facultative heterochromatin, which contains sequences that are not transcribed in the cell being examined but are transcribed in other cell types. Conse-

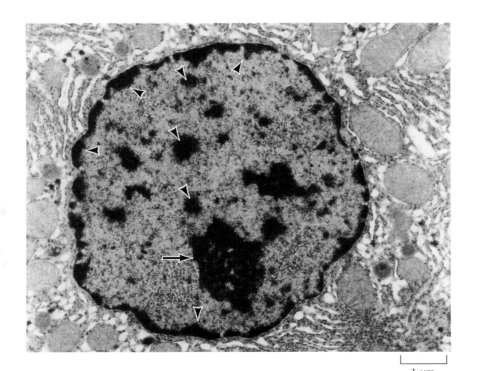

1 μm

FIGURE 9.16 Heterochromatin in interphase nuclei   The euchromatin is distributed throughout the nucleus. The heterochromatin is indicated by arrowheads and the nucleolus by an arrow. (Courtesy of Ada L. Olins and Donald E. Olins, Oak Ridge National Laboratory.)

FIGURE 9.17 Chromosome organization Reproduction of hand-drawn sketches of chromosomes in salamander cells. (A) Complete chromosomes. (B) Telomeres only (located at the nuclear membrane). (From C. Rabl, 1885. *Morphologisches Jahrbuch* 10: 214.)

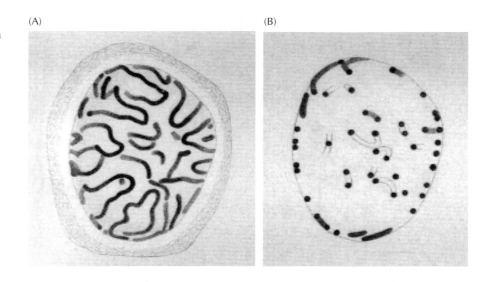

> ■ Ciliated protozoa contain two types of nuclei: a polyploid macronucleus that contains transcriptionally active genes, and one or more diploid transcriptionally inactive micronuclei that participate in sexual reproduction.

quently, the amount of facultative heterochromatin varies depending on the transcriptional activity of the cell.

Although interphase chromatin appears to be uniformly distributed, the chromosomes are actually arranged in an organized fashion and divided into discrete functional domains that play an important role in regulating gene expression. The nonrandom distribution of chromatin within the interphase nucleus was first suggested in 1885 by Carl Rabl who proposed that each chromosome occupies a distinct territory, with centromeres and telomeres attached to opposite sides of the nuclear envelope (Figure 9.17). This basic model of chromosome organization was confirmed nearly a hundred years later (in 1984) by detailed studies of polytene chromosomes in *Drosophila* salivary glands. Rather than randomly winding around one another, each chromosome was found to occupy a discrete region of the nucleus (Figure 9.18). The chromosomes are closely associated with the nuclear envelope at many sites.

Individual chromosomes also occupy distinct territories within the nuclei of mammalian cells (Figure 9.19). Actively transcribed genes are

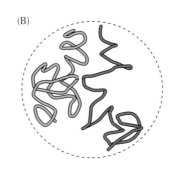

FIGURE 9.18 Organization of *Drosophila* chromosomes (A) A model of the nucleus, showing the five chromosome arms in different colors. The positions of telomeres and centromeres are indicated. (B) The two arms of chromosome 3 are shown to illustrate the topological separation between chromosomes. (From D. Mathog et al., 1984. *Nature* 308: 414.)

FIGURE 9.19 Organization of chromosomes in the mammalian nucleus (A) Probes to repeated sequences on chromosome 4 were hybridized to a human cell. The two copies of chromosome 4, identified by yellow fluorescence, occupy distinct territories in the nucleus. (B) A model of chromosome organization. The chromosomes occupy discrete territories, separated by intrachromosomal domains in which RNA processing and transport are thought to occur. (A, courtesy of Thomas Cremer, Ludwig Maximilians University, from A. I. Lamond and W. C. Earnshaw, 1999. *Science* 280: 547.)

(A)

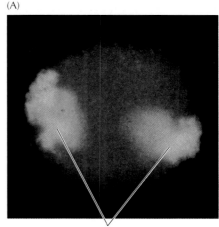

Copies of chromosome 4

(B)

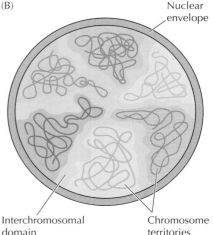

Nuclear envelope

Interchromosomal domain

Chromosome territories

localized to the periphery of these territories, adjacent to channels separating the chromosomes. Newly transcribed RNAs are thought to be released into these channels between chromosomes where RNA processing takes place. Much of the heterochromatin is localized to the periphery of the nucleus because proteins associated with heterochromatin bind to the matrix of the nuclear lamina. Since different cell types express different genes their facultative heterochromatin is different, and varying regions of the chromosomes interact with the nuclear lamina in different cells and tissues. Some cells have their centromeres and telomeres clustered at opposite poles while others have their chromosomes arranged radially. While the locations of the chromosomes within the nucleus are not random, they differ across different tissues, different organisms, and during cell differentiation.

Like the DNA in metaphase chromosomes (see Figure 5.16), the chromatin in interphase nuclei is organized into looped domains containing approximately 50 to 100 kb of DNA. A good example of this looped-domain organization is provided by the highly transcribed chromosomes of amphibian oocytes in which actively transcribed regions of DNA can be visualized as large loops of extended chromatin (Figure 9.20). These chromatin domains appear to represent discrete functional units, which independently regulate gene expression (see Chapter 7).

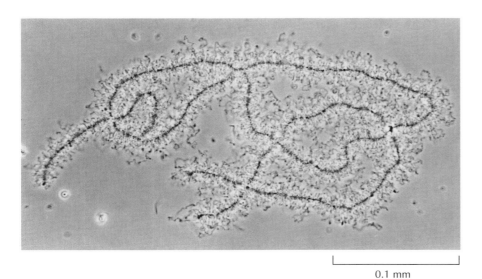

0.1 mm

FIGURE 9.20 Looped chromatin domains Light micrograph of a chromosome of amphibian oocytes showing decondensed loops of actively transcribed chromatin extending from an axis of highly condensed nontranscribed chromatin. (Courtesy of Joseph Gall, Carnegie Institute.)

FIGURE 9.21 Clustered sites of DNA replication Newly replicated DNA was labeled by a brief exposure of cells to bromodeoxyuridine, which is incorporated into DNA in place of thymidine. This substitution allows detection of newly synthesized DNA by immunofluorescence following staining with an antibody against bromodeoxyuridine. Note that the newly replicated DNA is present in discrete clusters distributed throughout the nucleus. The two panels show the distribution early and late in DNA synthesis, respectively. (From B. K. Kennedy et al. 2000. *Genes Dev.* 14: 2855)

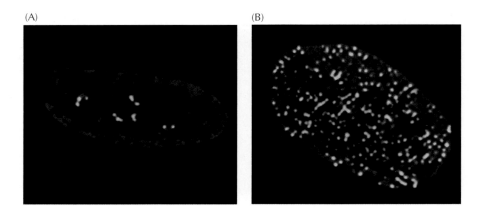

FIGURE 9.22 Localization of splicing components Staining with immunofluorescent antibodies indicates that splicing factors are concentrated in discrete bodies within the nucleus. (Courtesy of David L. Spector, Cold Spring Harbor Laboratory.)

Sub-Compartments within the Nucleus

The internal organization of the nucleus is further demonstrated by the localization of most nuclear processes to distinct regions of the nucleus. Many important enzymes and other proteins of the nucleus are localized to discrete subnuclear bodies that have a low-density, sponge-like structure that allows macromolecules from the rest of the nucleus to move in and out. Targeting or retention signals for some of these structures have been identified but not yet characterized. The nature and function of these nuclear substructures are not yet clear, and understanding the organization of specific biochemical processes within the nucleus is an incompletely explored area of cell biology.

The nuclei of mammalian cells contain clustered sites of DNA replication within which the replication of multiple DNA molecules takes place. These discrete sites of DNA replication have been defined by experiments in which newly synthesized DNA was visualized within cell nuclei by labeling cells with bromodeoxyuridine, an analog of thymidine that is incorporated into DNA and then detected by staining with fluorescent antibodies (Figure 9.21). At the beginning of DNA synthesis, the newly replicated DNA was detected in approximately 20 discrete clusters distributed around the nucleolus. These perinucleolar sites were associated with nuclear lamins. Later in DNA synthesis the process spread to several hundred sites that were distributed over the nucleus. Since approximately 4000 origins of replication can be active in a diploid mammalian cell at any given time, each of these clustered sites of DNA replication must contain multiple replication forks. Thus DNA replication appears to take place in large structures that contain multiple replication complexes organized into distinct functional bodies, which have been called replication factories.

While actively transcribed genes appear to be distributed throughout the nucleus, components of the mRNA splicing machinery are concentrated in discrete nuclear bodies termed nuclear speckles. Immunofluorescent staining with antibodies against snRNPs and splicing factors showed that, rather than being distributed uniformly throughout the nucleus, these components of the RNA splicing apparatus are concentrated in these 20 to 50 discrete structures (Figure 9.22). It is thought that the speckles are storage sites of splicing components, which are then recruited to actively transcribed genes where pre-mRNA processing occurs.

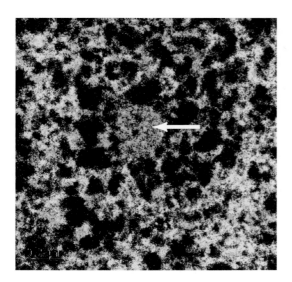

FIGURE 9.23 A PML body   The PML body (arrow) is surrounded by chromatin. (From G. Dellaire, R. Nisman and D. P. Bazett-Jones. 2004. *Met. Enzymol.* 375: 456; courtesy of D. Bazett-Jones.)

In addition to speckles, nuclei contain several other types of distinct structures. Besides nucleoli (discussed below), these include PML bodies and Cajal bodies. PML bodies (typically 5–20 within a nucleus) were first identified as distinct sites of localization of a transcriptional regulatory protein involved in acute *promyelocytic leukemia*. PML bodies are known to interact with chromatin (Figure 9.23) and are the sites of accumulation of transcription factors and chromatin-modifying proteins (such as histone deacetylases), which may be targeted to PML bodies by the small polypeptide SUMO (discussed in Chapter 8). However, the function of PML bodies remains largely unknown. Cajal bodies contain the characteristic protein coilin (Figure 9.24) and are enriched in small RNPs. They are believed to function as sites of RNP assembly, especially in the final steps of snRNP processing.

The Nucleolus and rRNA Processing

The most prominent nuclear body is the **nucleolus** (see Figure 9.1), which is the site of rRNA transcription and processing as well as aspects of ribosome assembly. As discussed in the preceding chapter, cells require large num-

(A)

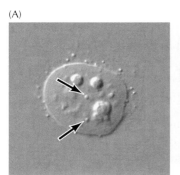

(B)

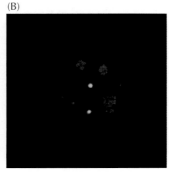

FIGURE 9.24 Cajal bodies in the nucleus   (A) Differential interference contrast microscope image of the nucleus of a HeLa cell. The arrows indicate the two Cajal bodies. (B) Immunofluorescent staining of the same nucleus with antibodies to the proteins Coilin (green) and Fibrillarin (red). Fibrillarin is present in both the dense fibrillar zones of the nucleoli and in the Cajal bodies. Coilin is detectable only in the Cajal bodies. (From J. G. Gall, 2000. *Ann. Rev. Cell Dev. Biol.* 16: 273.)

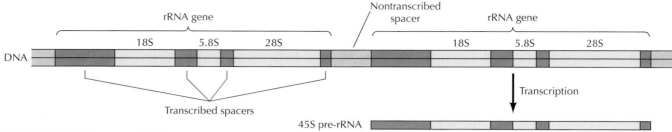

FIGURE 9.25 Ribosomal RNA genes
Each rRNA gene is a single transcription unit containing the 18S, 5.8S, and 28S rRNAs as well as transcribed spacer sequences. The rRNA genes are organized in tandem arrays, separated by nontranscribed spacer DNA.

bers of ribosomes at specific times to meet their needs for protein synthesis. Actively growing mammalian cells, for example, contain 5 million to 10 million ribosomes that must be synthesized each time the cell divides. The nucleolus is a ribosome production factory, designed to fulfill the need for regulated and efficient production of rRNAs and assembly of the ribosomal subunits. Recent evidence suggests that nucleoli also have a more general role in RNA modification and that several types of RNA move in and out of the nucleolus at specific stages during their processing.

Ribosomal RNA Genes and the Organization of the Nucleolus

The nucleolus, which is not surrounded by a membrane, is associated with the chromosomal regions that contain the genes for the 5.8S, 18S, and 28S rRNAs. Ribosomes of higher eukaryotes contain four types of RNA designated the 5S, 5.8S, 18S, and 28S rRNAs (see Figure 8.4). The 5.8S, 18S, and 28S rRNAs are transcribed as a single unit within the nucleolus by RNA polymerase I, yielding a 45S ribosomal precursor RNA (Figure 9.25). The 45S pre-rRNA is processed to the 18S rRNA of the 40S (small) ribosomal subunit and to the 5.8S and 28S rRNAs of the 60S (large) ribosomal subunit. Transcription of the 5S rRNA, which is also found in the 60S ribosomal subunit, takes place outside the nucleolus in higher eukaryotes and is catalyzed by RNA polymerase III.

To meet the need for transcription of large numbers of rRNA molecules, all cells contain multiple copies of the rRNA genes. The human genome, for example, contains about 200 copies of the gene that encodes the 5.8S, 18S, and 28S rRNAs and approximately 2000 copies of the gene that encodes 5S rRNA. The genes for 5.8S, 18S, and 28S rRNAs are clustered in tandem arrays on five different human chromosomes (chromosomes 13, 14, 15, 21, and 22); the 5S rRNA genes are present in a single tandem array on chromosome 1.

The importance of ribosome production is particularly evident in oocytes in which the rRNA genes are amplified to support the synthesis of the large numbers of ribosomes required for early embryonic development. In *Xenopus* oocytes, the rRNA genes are amplified approximately two thousand-fold, resulting in about one million copies per cell. These amplified rRNA genes are distributed to thousands of nucleoli (Figure 9.26), which support the accumulation of nearly 10^{12} ribosomes per oocyte.

Morphologically, nucleoli consist of three distinguishable regions: the fibrillar center, dense fibrillar component, and granular component (Figure 9.27).

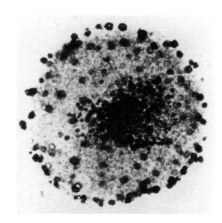

FIGURE 9.26 Nucleoli in amphibian oocytes The amplified rRNA genes of *Xenopus* oocytes are clustered in multiple nucleoli (darkly stained spots). (From D. D. Brown and I. B. Dawid, 1969. *Science* 160: 272.)

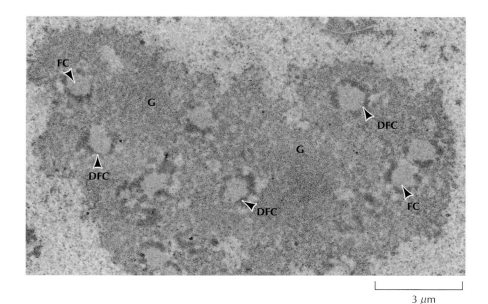

FIGURE 9.27 Structure of the nucleolus An electron micrograph illustrating the fibrillar center (FC), dense fibrillar component (DFC), and granular component (G) of a nucleolus. (Courtesy of David L. Spector, Cold Spring Harbor Laboratory.)

3 μm

These different regions are thought to represent the sites of progressive stages of rRNA transcription, processing, and ribosome assembly. Modification of other small RNAs, such as that of the signal recognition particle (see Chapter 10), occurs elsewhere in the nucleolus.

Following each cell division, nucleoli become associated with the chromosomal regions that contain the 5.8S, 18S, and 28S rRNA genes, which are therefore called **nucleolar organizing regions**. The formation of nucleoli requires the transcription of 45S pre-rRNA, which appears to lead to the fusion of small prenucleolar bodies that contain processing factors and other components of the nucleolus. In most cells, the initially separate nucleoli then fuse to form a single nucleolus. The size of the nucleolus depends on the metabolic activity of the cell, with large nucleoli found in cells that are actively engaged in protein synthesis. This variation is due primarily to differences in the size of the granular component, reflecting the levels of ribosome assembly.

Transcription and Processing of rRNA

Each nucleolar organizing region contains a cluster of tandemly repeated rRNA genes separated from each other by nontranscribed spacer DNA. These genes are very actively transcribed by RNA polymerase I, allowing their transcription to be readily visualized by electron microscopy (Figure 9.28). In such electron micrographs, each of the tandemly arrayed rRNA

FIGURE 9.28 Transcription of rRNA genes An electron micrograph of nucleolar chromatin showing three rRNA genes separated by nontranscribed spacer DNA. Each rRNA gene is surrounded by an array of growing RNA chains, resulting in a Christmas tree appearance. (Courtesy of O. L. Miller, Jr.)

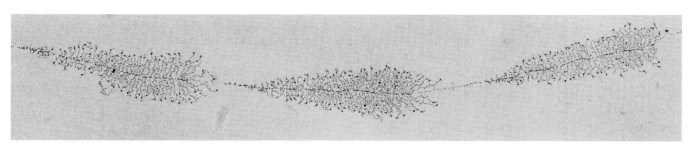

FIGURE 9.29 Processing of pre-rRNA The higher eukaryote 45S pre-rRNA transcript contains external transcribed spacers (ETS) at both ends and internal transcribed spacers (ITS) between the sequences of 18S, 5.8S, and 28S rRNAs. The pre-rRNA is processed via a series of cleavages to yield the mature rRNA species.

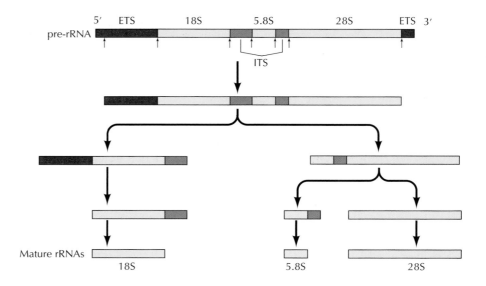

genes is surrounded by densely packed growing RNA chains forming a structure that looks like a Christmas tree. The high density of growing RNA chains reflects that of RNA polymerase molecules, which are present at a maximal density of approximately one polymerase per hundred base pairs of template DNA.

In higher eukaryotes the primary transcript of the rRNA genes is the large 45S pre-rRNA, which contains the 18S, 5.8S, and 28S rRNAs as well as transcribed spacer regions (Figure 9.29). External transcribed spacers are present at both the 5′ and 3′ ends of the pre-rRNAs, and two internal transcribed spacers lie between the 18S, 5.8S, and 28S rRNA sequences. Early steps in the processing of this transcript are cleavages within the external transcribed spacer near the 5′ end of the pre-rRNA and removal of the external transcribed spacer at the 3′ end of the molecule. Additional cleavages then result in formation of the mature rRNAs. Processing follows a similar pattern in all eukaryotes, although there may be differences in the order or the number of cleavages.

In addition to cleavage, the processing of pre-rRNA involves a substantial amount of base modification resulting both from the addition of methyl groups to specific bases and ribose residues and from the conversion of uridine to pseudouridine (see Figure 7.43). In animal cells, pre-rRNA processing involves the methylation of approximately a hundred ribose residues and ten bases, in addition to the formation of about a hundred pseudouridines. Most of these modifications occur during or shortly after synthesis of the pre-rRNA, although a few take place at later stages of pre-rRNA processing.

The processing of pre-rRNA requires the action of both proteins and RNAs that are localized to the nucleolus. The involvement of small nuclear RNAs (snRNAs) in pre-mRNA splicing was discussed in Chapter 7. Nucleoli contain more than 300 proteins and a large number (about 200) of **small nucleolar RNAs (snoRNAs)** that function in pre-rRNA processing. Like the spliceosomal snRNAs, the snoRNAs are complexed with proteins, forming snoRNPs. Individual snoRNPs consist of single snoRNAs associated with eight to ten proteins. The snoRNPs then assemble on the pre-rRNA to form processing complexes in a manner analogous to the formation of spliceosomes on pre-mRNA.

FIGURE 9.30 Role of snoRNAs in base modification of pre-rRNA   The snoRNAs contain short sequences complementary to rRNA. Base pairing between snoRNAs and pre-rRNA targets the enzymes that catalyze base modification (e.g., methylation) to the appropriate sites on pre-rRNA.

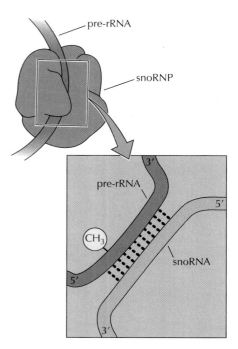

Some snoRNAs are responsible for the cleavages of pre-rRNA into 18S, 5.8S, and 28S products. For example, the most abundant nucleolar snoRNA is U3, which is present in about 200,000 copies per cell and is required for cleavage of pre-rRNA within the 5′ external transcribed spacer sequences. Similarly, U8 snoRNA is required for cleavage of pre-rRNA to 5.8S and 28S rRNAs, and U22 snoRNA is required for cleavage of pre-rRNA to 18S rRNA.

Most snoRNAs, however, function in rRNA synthesis as guide RNAs to direct the specific base modifications of pre-rRNA, including the methylation of specific ribose residues and the formation of pseudouridines (Figure 9.30). Most of the snoRNAs contain short sequences of approximately 15 nucleotides that are complementary to 18S or 28S rRNA. Importantly, these regions of complementarity include the sites of base modification in the rRNA. By base pairing with specific regions of the pre-rRNA, the snoRNAs act as guide RNAs that target the enzymes responsible for ribose methylation or pseudouridylation to the correct site on the pre-rRNA molecule. Other RNAs besides rRNA require modified bases, and it is the localization of snoRNPs in the nucleolus that is thought to be the basis for its more general role in RNA modification. One example is the signal recognition particle RNA (see Chapter 10).

Ribosome Assembly

The formation of ribosomes involves the assembly of the ribosomal precursor RNA with both ribosomal proteins and 5S rRNA (Figure 9.31). The genes that encode ribosomal proteins are transcribed outside the nucleolus by RNA polymerase II yielding mRNAs that are translated on cytoplasmic ribosomes. The ribosomal proteins are then transported from the cytoplasm to the nucleolus where they are assembled with rRNAs to form preribosomal particles. Although the genes for 5S rRNA are also transcribed outside the nucleolus—in this case by RNA polymerase III—5S rRNAs similarly are assembled into preribosomal particles within the nucleolus.

The association of ribosomal proteins with rRNA begins while the pre-rRNA is still being synthesized, and more than half of the ribosomal proteins are complexed with the pre-rRNA before its cleavage. The remaining ribosomal proteins and the 5S rRNA are incorporated into preribosomal particles as cleavage of the pre-rRNA proceeds. Early in ribosome assembly, the processing of the two nascent ribosomal subunits separates. Processing of the smaller subunit, which contains only the 18S rRNA, is simpler and involves only four endonuclease cleavages. In higher eukaryotes this is completed within the nucleus but in yeast the final cleavage to the mature 18S rRNA actually occurs after export of the 40S subunit to the cytosol. Processing of the larger subunit, which contains the 28S, 5.8S and 5S rRNAs, involves extensive nuclease cleavages and is completed within the nucleolus. Consequently, most of the preribosomal particles in the nucleolus represent precursors to the large (60S) subunit. The final stages of ribosome maturation follow the export of preribosomal particles to the cytoplasm, forming the active 40S and 60S subunits of eukaryotic ribosomes.

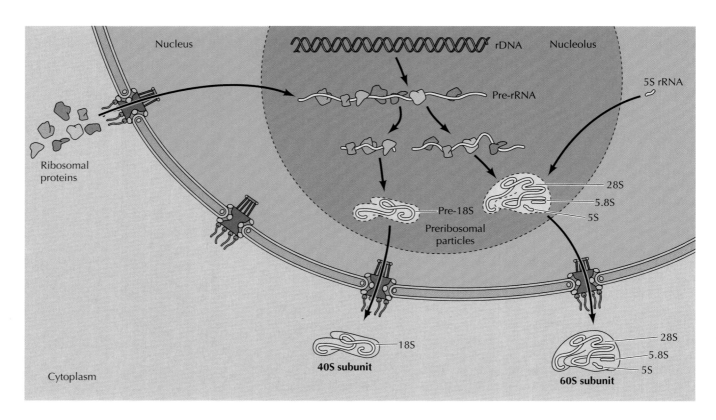

FIGURE 9.31 Ribosome assembly Ribosomal proteins are imported to the nucleolus from the cytoplasm and begin to assemble on pre-rRNA prior to its cleavage. As the pre-rRNA is processed, additional ribosomal proteins and the 5S rRNA (which is synthesized elsewhere in the nucleus) assemble to form preribosomal particles. The final steps of maturation follow the export of preribosomal particles to the cytoplasm, yielding the 40S and 60S ribosomal subunits.

COMPANION WEBSITE

Visit the website that accompanies **The Cell** (www.sinauer.com/cooper) for animations, videos, quizzes, problems, and other review material.

KEY TERMS

nuclear envelope, nuclear membrane, nuclear lamina, lamin

nuclear pore complex

SUMMARY

THE NUCLEAR ENVELOPE AND TRAFFIC BETWEEN THE NUCLEUS AND THE CYTOPLASM

Structure of the Nuclear Envelope: The nuclear envelope separates the contents of the nucleus from the cytoplasm, maintaining the nucleus as a distinct biochemical compartment that houses the genetic material and serves as the site of transcription and RNA processing in eukaryotic cells. The nuclear envelope consists of the inner and outer nuclear membranes (which are joined at nuclear pore complexes) and an underlying nuclear lamina.

The Nuclear Pore Complex: Nuclear pore complexes are large structures that provide the only routes through which molecules can travel between the nucleus and the cytoplasm. Small molecules can diffuse freely through open channels in the nuclear pore complex. Macromolecules are selectively transported in an energy-dependent process.

SUMMARY

KEY TERMS

Selective Transport of Proteins to and from the Nucleus: Proteins destined for import to the nucleus contain nuclear localization signals that are recognized by receptors that direct transport through the nuclear pore complex. Proteins that shuttle back and forth between the nucleus and the cytoplasm contain nuclear export signals that target them for transport from the nucleus to the cytoplasm. In most cases, the small GTP-binding protein Ran is required for translocation through the nuclear pore complex and determines the directionality of transport.

nuclear localization signal, nuclear transport receptor, karyopherin, importin, exportin, Ran, nuclear export signal

Regulation of Nuclear Protein Import: The activity of some proteins, such as transcription factors, is controlled by regulation of both their import to, and export from, the nucleus.

Transport of RNAs: RNAs are transported through the nuclear pore complex as ribonucleoprotein complexes. Messenger RNAs, ribosomal RNAs, and transfer RNAs are exported from the nucleus to function in protein synthesis. Several classes of small nuclear RNAs are initially transported from the nucleus to the cytoplasm where they associate with proteins to form RNPs; they then return to the nucleus.

INTERNAL ORGANIZATION OF THE NUCLEUS

Chromosomes and Higher-Order Chromatin Structure: The interphase nucleus contains transcriptionally inactive, highly condensed heterochromatin as well as decondensed euchromatin. Interphase chromosomes are organized within the nucleus and divided into large looped domains that function as independent units.

heterochromatin, euchromatin

Sub-Compartments within the Nucleus: Some nuclear processes, such as DNA replication and mRNA metabolism, may be localized to discrete subnuclear structures.

THE NUCLEOLUS AND rRNA PROCESSING

Ribosomal RNA Genes and the Organization of the Nucleolus: The nucleolus is associated with the genes for ribosomal RNAs. It is the site of rRNA transcription, rRNA processing, ribosome assembly, and the modification of several small RNAs.

nucleolus, nucleolar organizing region

Transcription and Processing of rRNA: The primary transcript of the rRNA genes is 45S pre-rRNA, which is processed to yield 18S, 5.8S, and 28S rRNAs. Processing of pre-rRNA and other small RNAs is mediated by small nucleolar RNAs (snoRNAs).

small nucleolar RNA (snoRNA)

Ribosome Assembly: Ribosomal subunits are assembled within the nucleolus from rRNAs and ribosomal proteins.

Questions

1. By separating transcription from translation, the nuclear envelope allows eukaryotes to regulate gene expression by processes that are not found in prokaryotes. What are these regulatory mechanisms that are unique to eukaryotes?

2. What roles do lamins play in nuclear structure and function?

3. You inject a frog egg with two globular proteins, one 15 kd and the other 100 kd—both of which lack nuclear localization signals. Will either protein enter the nucleus?

4. What determines the directionality of nuclear import?

5. Describe how the activity of a transcription factor can be regulated by nuclear import.

6. You are studying a transcription factor that is regulated by phosphorylation of serine residues that inactivate its nuclear localization signal. How would mutating these serines to alanines affect subcellular localization of the transcription factor and expression of its target gene?

7. How would mutational inactivation of the nuclear export signal of a protein that normally shuttles back and forth between the nucleus and cytoplasm affect its subcellular distribution?

8. DNA replication appears to occur in specific sites or replication factories. How would you locate these sites in mammalian cells in culture?

9. How did Kalderon and colleagues demonstrate that the T antigen amino acid sequence 126 to 132 was sufficient for nuclear accumulation of the protein?

10. What is the significance of nuclear speckles?

11. What is the function of snoRNAs?

12. How would RNAi against human exportin-t affect human fibroblasts in culture?

References and Further Reading

The Nuclear Envelope and Traffic between the Nucleus and the Cytoplasm

Ben-Efraim, I. and L. Gerace. 2001. Gradient of increasing affinity of importin beta for nucleoporins along the pathway of nuclear import. *J. Cell Biol.* 152: 411–417. [P]

Beck, M., F. Forster, M. Ecke, J. M. Plitzko, F. Melchior, G. Gerisch, W. Baumeister and O. Medalia. 2004. Nuclear pore complex structure and dynamics revealed by cryoelectron tomography. *Science* 306: 1387–1390. [P]

Chook, Y. M. and G. Blobel. 1999. Structure of the nuclear transport complex karyopherin-2-Ran.GppNHp. *Nature* 399: 230–237. [P]

Chook, Y. M. and G. Blobel. 2001. Karyopherins and nuclear import. *Curr. Opin. Struct. Biol.* 11: 703–715. [R]

Cronshaw, J. M., A. N. Krutchinsky, W. Zhang, B. T. Chait and M. J. Matunis. 2002. Proteomic analysis of the mammalian nuclear pore complex. *J. Cell Biol.* 158: 915–927. [P]

Erkmann, J. A. and U. Kutay. 2004. Nuclear export of mRNA: From the site of transcription to the cytoplasm. *Exp. Cell Res.* 296: 12–20. [R]

Goldman, R. D., Y. Gruenbaum, R. D. Moir, D. K. Shumaker and T. P. Spann. 2002. Nuclear lamins: Building blocks of nuclear architecture. *Genes Dev.* 16: 533–547. [R]

Gruenbaum, Y., A. Margalit, R. D. Goldman, D. K. Shumaker and K. L. Wilson. 2005. The nuclear lamina comes of age. *Nat. Rev. Mol. Cell Biol.* 6: 21–31. [R]

Izaurralde, E. 2004. Directing mRNA export. *Nat. Struct. Mol. Biol.* 11: 210–212. [R]

Mosammaparast, N. and L. F. Pemberton. 2004. Karyopherins: From nuclear-transport mediators to nuclear-function regulators. *Trends Cell Biol.* 14: 547–556. [R]

Ohno, M., M. Fornerod and I. W. Mattaj. 1998. Nucleocytoplasmic transport: The last 200 nanometers. *Cell* 92: 327–336. [R]

Polesello, C. and F. Payre 2004. Small is beautiful: What flies tell us about ERM protein function in development. *Trends Cell Biol.* 14: 294–302. [R]

Presgraves, D. C., L. Balagopalan, S. M. Abmayr and H. A. Orr. 2003. Adaptive evolution drives divergence of a hybrid inviability gene between two species of *Drosophila. Nature* 423: 715–719. [P]

Quimby, B. B. and M. Dasso. 2003. The small GTPase Ran: Interpreting the signs. *Curr. Opin. Cell Biol.* 15: 338–344. [R]

Ryan, K. J. and S. R. Wente. 2000. The nuclear pore complex: A protein machine bridging the nucleus and cytoplasm. *Curr. Opin. Cell Biol.* 2: 361–371. [R]

Vetter, I. R., A. Arndt, U. Kutay, D. Gorlich and A. Wittinghofer. 1999. Structural view of the Ran-importin β interaction at 2.3 Å resolution. *Cell* 67: 635–646. [P]

Weis, K. 2003. Regulating access to the genome: Nucleocytoplasmic transport throughout the cell cycle. *Cell* 112: 441–451. [R]

Internal Organization of the Nucleus

Gall, J. G. 2000. Cajal bodies: The first 100 years. *Ann. Rev. Cell Dev. Biol.* 16: 273–300. [R]

Grosshans, H., K. Deinert, E. Hurt and G. Simos. 2001. Biogenesis of the signal recognition particle (SRP) involves import of SRP proteins into the nucleolus, assembly with the SRP-RNA, and Xpo1p-mediated export. *J. Cell Biol.* 153: 745–762. [P]

Johnson, A. W., E. Lund and J. Dahlberg. 2002. Nuclear export of ribosomal subunits. *Trends Biochem. Sci.* 27: 580–585. [R]

Kennedy, B. K., D. A. Barbie, M. Classon, N. Dyson and E. Harlow. 2000. Nuclear organization of DNA replication in primary mammalian cells. *Genes Dev.* 14: 2855–2868. [P]

Lamond, A. I. and W. C. Earnshaw. 1998. Structure and function in the nucleus. *Science* 280: 547–553. [R]

Misteli, T. 2004. Spatial positioning: A new dimension in genome function. *Cell* 119: 153–156. [R]

Misteli, T., J. F. Caceres and D. L. Spector. 1997. The dynamics of a pre-mRNA splicing factor in living cells. *Nature* 387: 523–527. [P]

Shumaker, D. K., E. R. Kuczmarski and R. D. Goldman. 2003. The nucleoskeleton: Lamins and actin are major players in essential nuclear functions. *Curr. Opin. Cell Biol.* 15: 358–366. [R]

Spector, D. L. 2003. The dynamics of chromosome organization and gene regulation. *Annu. Rev. Biochem.* 72: 573–608. [R]

Taddei, A., F. Hediger, F. R. Neumann and S. M. Gasser. 2004. The function of nuclear architecture: A genetic approach. *Ann. Rev. Genet.* 38: 305–345. [R]

Takahashi, Y., V. Lallemand-Breitenbach, J. Zhu and T. H. de Thé. 2004. PML nuclear bodies and apoptosis. *Oncogene* 23: 2819–2824. [R]

The Nucleolus and rRNA Processing

Fatica, A. and D. Tollervey. 2002. Making ribosomes. *Curr. Opin. Cell Biol.* 14: 313–318. [R]

Granneman, S. and S. J. Baserga. 2004. Ribosome biogenesis: Of knobs and RNA processing. *Exp. Cell Res.* 296: 43–50. [R]

Kiss, T. 2002. Small nucleolar RNAs: An abundant group of noncoding RNAs with diverse cellular functions. *Cell* 109: 145–148. [R]

Miller, O. L., Jr. and B. Beatty. 1969. Visualization of nucleolar genes. *Science* 164: 955–957. [P]

Nazar, R. N. 2004. Ribosomal RNA processing and ribosome biogenesis in eukaryotes. *IUBMB Life* 56: 457–465. [R]

Olson, M. O., K. Hingorani and A. Szebeni. 2002. Conventional and nonconventional roles of the nucleolus. *Int. Rev. Cytol.* 219: 199–266. [R]

Thiry, M. and D. L. Lafontaine. 2005. Birth of a nucleolus: The evolution of nucleolar compartments. *Trends Cell Biol.* 15: 194–199. [R]

CHAPTER 10

Protein Sorting and Transport
The Endoplasmic Reticulum, Golgi Apparatus, and Lysosomes

■ **The Endoplasmic Reticulum 386**

■ **The Golgi Apparatus 408**

■ **The Mechanism of Vesicular Transport 417**

■ **Lysosomes 424**

■ **KEY EXPERIMENT:** The Signal Hypothesis 390

■ **MOLECULAR MEDICINE:** Gaucher Disease 426

IN ADDITION TO THE PRESENCE OF A NUCLEUS, eukaryotic cells are distinguished from prokaryotic cells by the presence of membrane-enclosed organelles within their cytoplasm. These organelles provide discrete compartments in which specific cellular activities take place, and the resulting subdivision of the cytoplasm allows eukaryotic cells to function efficiently in spite of their large size—at least a thousand times the volume of bacteria.

Because of the complex internal organization of eukaryotic cells, the sorting and targeting of proteins to their appropriate destinations are considerable tasks. The first step of protein sorting takes place while translation is still in progress. Many proteins destined for the endoplasmic reticulum, the Golgi apparatus, lysosomes, the plasma membrane, and secretion from the cell are synthesized on ribosomes that are bound to the membrane of the endoplasmic reticulum. As translation proceeds, the polypeptide chains are transported into the endoplasmic reticulum where protein folding and processing take place. From the endoplasmic reticulum, proteins are transported in vesicles to the Golgi apparatus where they are further processed and sorted for transport to endosomes, lysosomes, the plasma membrane, or secretion from the cell. Some of these organelles also participate in the sorting and transport of proteins being taken up from outside the cell (see Chapter 13). The endoplasmic reticulum, Golgi apparatus, endosomes, and lysosomes are thus distinguished from other cytoplasmic organelles by their common involvement in protein processing and connection by vesicular transport.

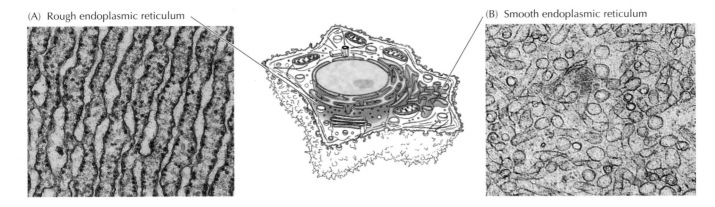

(A) Rough endoplasmic reticulum

(B) Smooth endoplasmic reticulum

FIGURE 10.1 The endoplasmic reticulum (ER) (A) Electron micrograph of rough ER in rat liver cells. Ribosomes are attached to the cytosolic face of the ER membrane. (B) Electron micrograph of smooth ER in Leydig cells of the testis, which are active in steroid hormone synthesis. (A, Richard Rodewald, University of Virginia/Biological Photo Service; B, Don Fawcett/Photo Researchers, Inc.)

■ The endoplasmic reticulum also plays a key role in signal transduction by acting as a major repository of intracellular calcium. The release of calcium from the ER in response to appropriate signals alters the activity of key cytosolic proteins and plays a very important role in muscle contraction (see Chapters 12 and 15).

The Endoplasmic Reticulum

The **endoplasmic reticulum (ER)** is a network of membrane enclosed tubules and sacs (cisternae) that extends from the nuclear membrane throughout the cytoplasm (Figure 10.1). The entire endoplasmic reticulum is enclosed by a continuous membrane and is the largest organelle of most eukaryotic cells. Its membrane may account for about half of all cell membranes, and the space enclosed by the ER (the lumen, or cisternal space) may represent about 10% of the total cell volume. As discussed below, there are three contiguous membrane domains within the ER that perform different functions within the cell. The **rough ER**, which is covered by ribosomes on its outer surface, and the **transitional ER**, where vesicles exit to the Golgi apparatus, both function in protein processing. The **smooth ER** is not associated with ribosomes and is involved in lipid, rather than protein, metabolism.

The Endoplasmic Reticulum and Protein Secretion

The role of the endoplasmic reticulum in protein processing and sorting was first demonstrated by George Palade and his colleagues in the 1960s (Figure 10.2). These investigators studied the fate of newly synthesized proteins in specialized cells of the pancreas (pancreatic acinar cells) that secrete digestive enzymes into the small intestine. Because most proteins synthesized by these cells are secreted, Palade and coworkers were able to study the pathway taken by secreted proteins simply by labeling newly synthesized proteins with radioactive amino acids. The location of the radiolabeled proteins within the cell was then determined by autoradiography, revealing the cellular sites involved in the events leading to protein secretion. After a brief exposure of pancreatic acinar cells to radioactive amino acids, newly synthesized proteins were detected in the rough ER, which was therefore identified as the site of synthesis of proteins destined for secretion. If the cells were then incubated for a short time in media containing nonradioactive amino acids (a process known as a "chase"), the radiolabeled proteins were detected in the Golgi apparatus. Following longer chase periods, the radiolabeled proteins traveled from the Golgi apparatus to the cell surface in **secretory vesicles**, which then fused with the plasma membrane to release their contents outside of the cell.

These experiments defined a pathway taken by secreted proteins—the **secretory pathway**: rough ER → Golgi → secretory vesicles → cell exterior. Further studies extended these results and demonstrated that this pathway

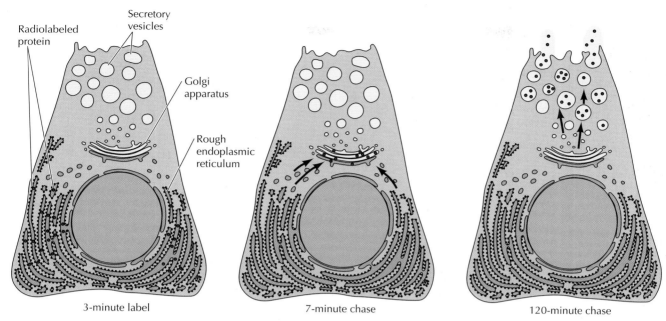

Radiolabeled protein

Secretory vesicles

Golgi apparatus

Rough endoplasmic reticulum

3-minute label

7-minute chase

120-minute chase

FIGURE 10.2 The secretory pathway Pancreatic acinar cells, which secrete most of their newly synthesized proteins into the digestive tract, were labeled with radioactive amino acids to study the intracellular pathway taken by secreted proteins. After a short incubation with radioactive amino acids (3-minute label), autoradiography revealed that newly synthesized proteins were localized to the rough ER. Following further incubation with nonradioactive amino acids (a "chase"), proteins were found to move from the ER to the Golgi apparatus and then, within secretory vesicles, to the plasma membrane and cell exterior.

is not restricted to proteins destined for secretion from the cell. Portions of it are shared by proteins destined for other compartments. Plasma membrane and lysosomal proteins also travel from the rough ER to the Golgi and then to their final destinations. Still other proteins travel through the initial steps of the secretory pathway but are then retained and function within either the ER or the Golgi apparatus.

The entrance of proteins into the ER thus represents a major branch point for the traffic of proteins within eukaryotic cells. Proteins destined for secretion or incorporation into the ER, Golgi apparatus, lysosomes, or plasma membrane are initially targeted to the ER. In mammalian cells most proteins are transferred into the ER while they are being translated on membrane bound ribosomes (Figure 10.3). In contrast, proteins destined to remain in the cytosol or to be incorporated into the nucleus, mitochondria, chloroplasts, or peroxisomes are synthesized on free ribosomes and released into the cytosol when their translation is complete.

Targeting Proteins to the Endoplasmic Reticulum

Proteins can be translocated into the ER either during their synthesis on membrane-bound ribosomes (cotranslational translocation) or after their translation has been completed on free ribosomes in the cytosol (posttranslational translocation). In mammalian cells, most proteins enter the ER cotranslationally, whereas both cotranslational and posttranslational pathways are used in yeast. The first step in the cotranslational pathway is the association of ribosomes with the ER. Ribosomes are targeted for binding to

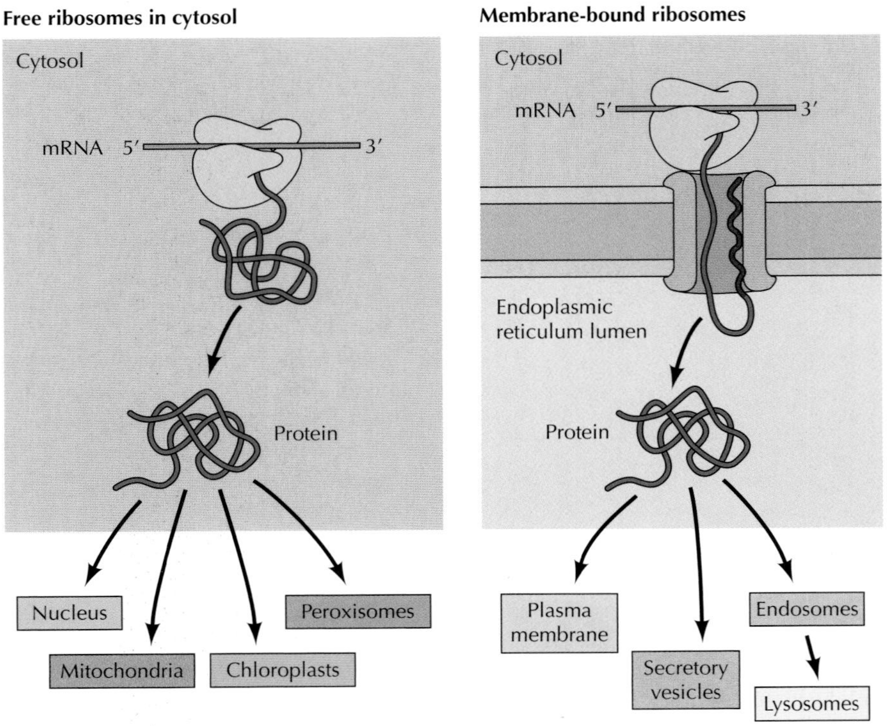

FIGURE 10.3 Overview of protein sorting In higher eukaryotic cells, the initial sorting of proteins to the ER takes place while translation is in progress. Proteins synthesized on free ribosomes either remain in the cytosol or are transported to the nucleus, mitochondria, chloroplasts, or peroxisomes. In contrast, proteins synthesized on membrane-bound ribosomes are translocated into the ER while their translation is in progress. They may be either retained within the ER or transported to the Golgi apparatus and, from there, to lysosomes, the plasma membrane, or the cell exterior via secretory vesicles.

Free ribosomes in cytosol

Cytosol

mRNA 5' 3'

Protein

Nucleus | Peroxisomes

Mitochondria | Chloroplasts

Membrane-bound ribosomes

Cytosol

mRNA 5' 3'

Endoplasmic reticulum lumen

Protein

Plasma membrane | Endosomes

Secretory vesicles | Lysosomes

FIGURE 10.4 Isolation of rough ER When cells are disrupted, the ER fragments into small vesicles called microsomes. The microsomes derived from the rough ER (rough microsomes) are lined with ribosomes on their outer surface. Because ribosomes contain large amounts of RNA, the rough microsomes are denser than smooth microsomes and can be isolated by equilibrium density-gradient centrifugation.

the ER membrane by the amino-acid sequence of the polypeptide chain being synthesized, rather than by intrinsic properties of the ribosome itself. Free and membrane bound ribosomes are functionally indistinguishable, and all protein synthesis initiates on ribosomes that are free in the cytosol. Ribosomes engaged in the synthesis of proteins that are destined for secretion are then targeted to the endoplasmic reticulum by a **signal sequence** at the amino terminus of the growing polypeptide chain. These signal sequences are short stretches of hydrophobic amino acids that are usually cleaved from the polypeptide chain during its transfer into the ER lumen.

The general role of signal sequences in targeting proteins to their appropriate locations within the cell was first elucidated by studies of the import of secretory proteins into the ER. These experiments used *in vitro* preparations of rough ER, which were isolated from cell extracts by density-gradient centrifugation (Figure 10.4). When cells are disrupted and the nuclei cen-

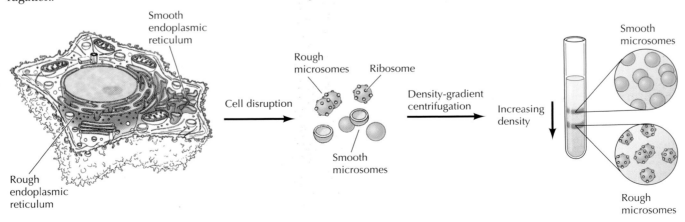

Smooth endoplasmic reticulum

Rough endoplasmic reticulum

Cell disruption

Rough microsomes | Ribosome

Smooth microsomes

Density-gradient centrifugation

Increasing density

Smooth microsomes

Rough microsomes

Translation on free ribosomes

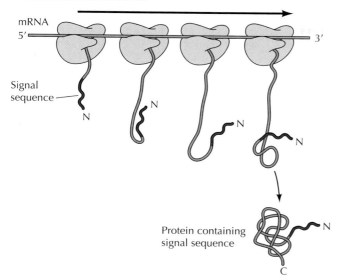

Translation with microsomes present

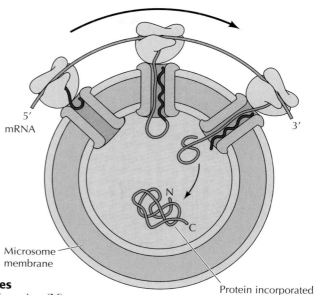

FIGURE 10.5 Incorporation of secretory proteins into microsomes
Secretory proteins are targeted to the ER by a signal sequence at their amino (N) terminus, which is removed during incorporation of the growing polypeptide chain into the ER. This was demonstrated by experiments showing that translation of secretory protein mRNAs on free ribosomes yielded proteins that retained their signal sequences and were therefore slightly larger than the normal secreted proteins. However, when microsomes were added to the system, the growing polypeptide chains were incorporated into the microsomes and the signal sequences were removed by proteolytic cleavage.

trifuged out, the ER breaks up into small vesicles called **microsomes**. Because the vesicles derived from the rough ER are covered with ribosomes, they can be separated from similar vesicles derived from the smooth ER or from other membranes (e.g., the plasma membrane). In particular, the large amount of RNA within ribosomes increases the density of the membrane vesicles to which they are attached, allowing purification of vesicles derived from the rough ER (rough microsomes) by equilibrium centrifugation in density gradients.

David Sabatini and Günter Blobel first proposed in 1971 that the signal for ribosome attachment to the ER might be an amino acid sequence near the amino terminus of the growing polypeptide chain. This hypothesis was supported by the results of *in vitro* translation of mRNAs encoding secreted proteins, such as immunoglobulins (Figure 10.5). If an mRNA encoding a secreted protein was translated on free ribosomes *in vitro*, it was found that the protein produced was slightly larger than the normal secreted protein. If microsomes were added to the system, however, the *in vitro* translated protein was incorporated into the microsomes and cleaved to the correct size. These experiments led to a more detailed formulation of the signal hypothesis, which proposed that an amino terminal leader sequence targets the polypeptide chain to the microsomes and is then cleaved by a microsomal protease. Many subsequent findings have substantiated this model, including recombinant DNA experiments demonstrating that addition of a signal sequence to a normally nonsecreted protein is sufficient to direct the incorporation of the recombinant protein into the rough ER.

KEY EXPERIMENT

The Signal Hypothesis

Transfer of Proteins across Membranes. I. Presence of Proteolytically Processed and Unprocessed Nascent Immunoglobulin Light Chains on Membrane-Bound Ribosomes of Murine Myeloma

Günter Blobel and Bernhard Dobberstein
Rockefeller University, New York
Journal of Cell Biology, 1975, Volume 67, pages 835–851

Günter Blobel

The Context

How are specific polypeptide chains transferred across the appropriate membranes? Studies in the 1950s and 1960s indicated that secreted proteins are synthesized on membrane-bound ribosomes and transferred across the membrane during their synthesis. However, this did not explain why ribosomes—engaged in the synthesis of secreted proteins—attach to membranes, while ribosomes synthesizing cytosolic proteins do not. A hypothesis to explain this difference was first suggested by Günter Blobel and David Sabatini in 1971. At that time, they proposed that (1) mRNAs to be translated on membrane-bound ribosomes contain a unique set of codons just 3′ of the initiation site, (2) translation of these codons yields a unique sequence at the amino terminus of the growing polypeptide chain (the signal sequence), and (3) the signal sequence triggers attachment of the ribosome to the membrane. In 1975 Blobel and Dobberstein reported a series of experiments that provided critical support for this notion. In addition, they proposed "a somewhat more detailed version of this hypothesis, henceforth referred to as the signal hypothesis."

The Experiments

Myelomas are cancers of B lymphocytes that actively secrete immunoglobulins, so they provide a good model for studies of secreted proteins. Previous studies in Cesar Milstein's laboratory had shown that the proteins produced by *in vitro* translation of immunoglobulin light-chain mRNA contain about 20 amino acids at their amino terminus that are not present in the secreted light chains. This result led to the suggestion that these amino acids direct binding of the ribosome to the membrane. To test this idea, Blobel and Dobberstein investigated the synthesis of light chains by membrane-bound ribosomes from myeloma cells.

As expected from earlier work, *in vitro* translation of light-chain mRNA on free ribosomes yielded a protein that was larger than the secreted light chain (see figure). In contrast, *in vitro* translation of mRNA associated with membrane-bound ribosomes from myeloma cells yielded a protein that was the same size as the normally secreted light chain. Moreover, the light chains synthesized by ribosomes that remained bound to microsomes were resistant to digestion by added proteases, indicating that the light chains had been transferred into the microsomes.

These results indicated that an amino-terminal signal sequence is removed by a microsomal protease as growing polypeptide chains are transferred across the membrane. The results were interpreted in terms of a more detailed version of the signal

The mechanism by which secretory proteins are targeted to the ER during their translation (the cotranslational pathway) is now well understood. The signal sequences span about 20 amino acids, including a stretch of hydrophobic residues usually located at the amino terminus of the polypeptide chain (Figure 10.6). As they emerge from the ribosome, signal

FIGURE 10.6 The signal sequence of growth hormone Most signal sequences contain a stretch of hydrophobic amino acids preceded by basic residues (e.g., arginine).

Cleavage site of signal peptidase
↓

Met–Ala–Thr–Gly–Ser–Arg–Thr–Ser–Leu–Leu–Leu–Ala–Phe–Gly–Leu–Leu–Cys–Leu–Pro–Trp–Leu–Gln–Glu–Gly–Ser–Ala–Phe–Pro–Thr

KEY EXPERIMENT

hypothesis. As stated by Blobel and Dobberstein, "the essential feature of the signal hypothesis is the occurrence of a unique sequence of codons, located immediately to the right of the initiation codon, which is present only in those mRNAs whose translation products are to be transferred across a membrane."

In vitro translation of immunoglobulin light-chain mRNA on free ribosomes (lane 1) yields a product that migrates slower than secreted light chains (lane S) in gel electrophoresis. In contrast, light chains synthesized by *in vitro* translation on membrane-bound ribosomes (lane 2) are the same size as secreted light chains. In addition, the products of *in vitro* translation on membrane-bound ribosomes were unaffected by subsequent digestion with proteases (lane 3), indicating that they were protected from the proteases by insertion into microsomes.

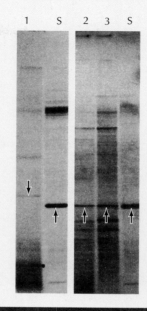

The Impact

The selective transfer of proteins across membranes is critical to the maintenance of the membrane-enclosed organelles of eukaryotic cells. To maintain the identity of these organelles, proteins must be translocated specifically across the appropriate membranes. The signal hypothesis provided the conceptual basis for understanding this phenomenon. Not only has this basic model been firmly substantiated for the transfer of secreted proteins into the endoplasmic reticulum, but it also has provided the framework for understanding the targeting of proteins to all the other membrane-enclosed compartments of the cell, thereby impacting virtually all areas of cell biology.

sequences are recognized and bound by a **signal recognition particle (SRP)** consisting of six polypeptides and a small cytoplasmic RNA (**srpRNA**) (Figure 10.7). The SRP binds the ribosome as well as the signal sequence, inhibiting further translation and targeting the entire complex (the SRP, ribosome, and growing polypeptide chain) to the rough ER by binding to the **SRP receptor** on the ER membrane (Figure 10.8). Recent structural studies have shown that both the SRP proteins and the srpRNA are involved in interac-

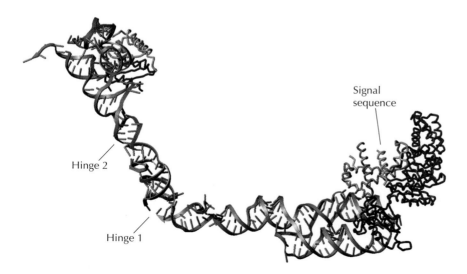

FIGURE 10.7 Structure of the SRP   The SRP proteins (blue) are associated with the SRP RNA (orange), which has two flexible hinge regions. The signal sequence on the nascent protein (green) binds to a pocket in the SRP proteins. (After Halic et al., 2004. *Nature* 427: 808.)

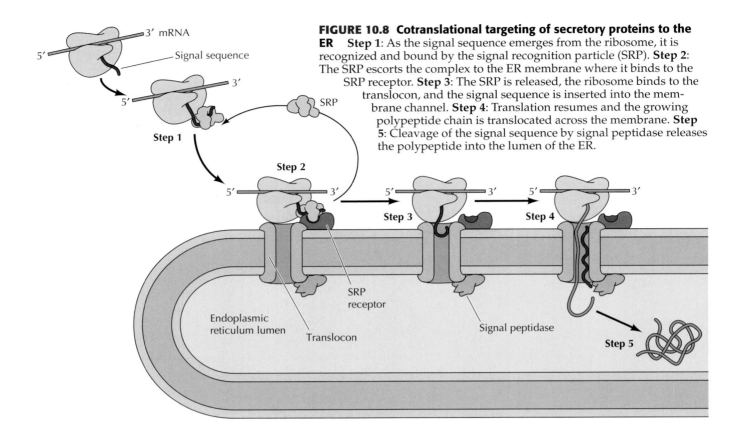

FIGURE 10.8 Cotranslational targeting of secretory proteins to the ER **Step 1**: As the signal sequence emerges from the ribosome, it is recognized and bound by the signal recognition particle (SRP). **Step 2**: The SRP escorts the complex to the ER membrane where it binds to the SRP receptor. **Step 3**: The SRP is released, the ribosome binds to the translocon, and the signal sequence is inserted into the membrane channel. **Step 4**: Translation resumes and the growing polypeptide chain is translocated across the membrane. **Step 5**: Cleavage of the signal sequence by signal peptidase releases the polypeptide into the lumen of the ER.

tion with the ribosome, with srpRNA binding to both proteins and rRNA of the large ribosomal subunit. Binding to the receptor releases the SRP from both the ribosome and the signal sequence of the growing polypeptide chain. The ribosome then binds to a protein translocation complex in the ER membrane, and the signal sequence is inserted into a membrane channel or **translocon**. This entire process is coordinated by GTP binding to both the SRP and the SRP receptor, with hydrolysis of GTP to GDP leading to dissociation of SRP from both the receptor and the ribosome-mRNA complex.

In both yeast and mammalian cells, the translocons through the ER membrane are complexes of three transmembrane proteins called the Sec61 proteins. The yeast and mammalian translocon proteins are closely related to the plasma membrane proteins that translocate secreted polypeptides in bacteria, demonstrating a striking conservation of the protein secretion machinery in prokaryotic and eukaryotic cells. Transfer of the ribosome-mRNA complex from the SRP to the translocon opens the gate on the translocon and allows translation to resume, and the growing polypeptide chain is transferred directly into the translocon channel and across the ER membrane as translation proceeds. Thus the process of protein synthesis directly drives the transfer of growing polypeptide chains through the translocon and into the ER. As translocation proceeds, the signal sequence is cleaved by **signal peptidase** and the polypeptide is released into the lumen of the ER.

Many proteins in yeast, as well as a few proteins in mammalian cells, are targeted to the ER after their translation is complete (posttranslational translocation) rather than being transferred into the ER during synthesis on

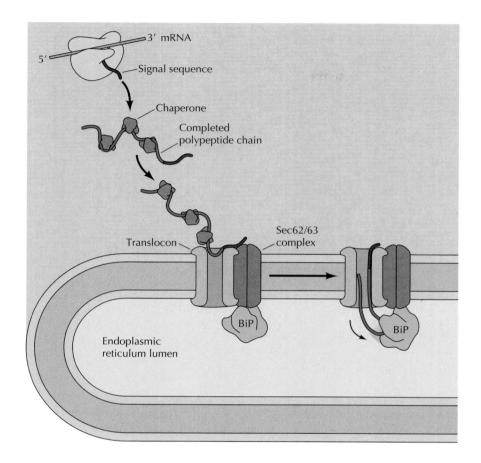

FIGURE 10.9 Posttranslational translocation of proteins into the ER Proteins destined for posttranslational import to the ER are synthesized on free ribosomes and maintained in an unfolded conformation by cytosolic chaperones. Their signal sequences are recognized by the Sec62/63 complex, which is associated with the translocon in the ER membrane. The Sec63 protein is also associated with a chaperone protein (BiP), which acts as a molecular ratchet to drive protein translocation into the ER.

membrane bound ribosomes. These proteins are synthesized on free cytosolic ribosomes, and their posttranslational incorporation into the ER does not require SRP. Instead, their signal sequences are recognized by distinct receptor proteins (the Sec62/63 complex) associated with the translocon in the ER membrane (Figure 10.9). Cytosolic Hsp70 chaperones are required to maintain the polypeptide chains in an unfolded conformation so they can enter the translocon, and another Hsp70 chaperone within the ER (called BiP) is required to pull the polypeptide chain through the channel and into the ER. BiP acts as a ratchet to drive the posttranslational translocation of proteins into the ER, whereas the cotranslational translocation of growing polypeptide chains is driven directly by the process of protein synthesis.

Insertion of Proteins into the ER Membrane

Proteins destined for secretion from the cell or residence within the lumen of the ER, Golgi apparatus, endosomes, or lysosomes are translocated across the ER membrane and released into the lumen of the ER as already described. However, proteins destined for incorporation into the plasma membrane or the membranes of these compartments are initially inserted into the ER membrane instead of being released into the lumen. From the ER membrane, they proceed to their final destination along the same pathway as that of secretory proteins: ER → Golgi → plasma membrane or endosomes → lysosomes. However, these proteins are transported along this pathway as membrane components rather than as soluble proteins.

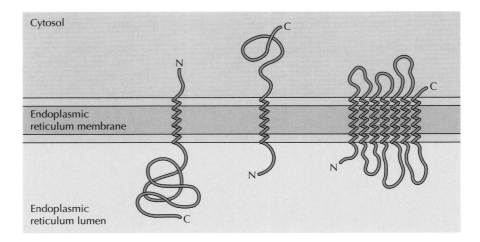

Cytosol

Endoplasmic
reticulum membrane

Endoplasmic
reticulum lumen

FIGURE 10.10 Orientations of membrane proteins   Integral membrane proteins span the membrane via α-helical regions of 20 to 25 hydrophobic amino acids, which can be inserted in a variety of orientations. The two proteins at left and center each span the membrane once, but they differ in whether the amino (N) or carboxy (C) terminus is on the cytosolic side. On the right is an example of a protein that has multiple membrane-spanning regions.

Integral membrane proteins are embedded in the membrane by hydrophobic sequences that span the phospholipid bilayer (see Figure 2.25). The membrane spanning portions of these proteins are usually α helical regions consisting of 20 to 25 hydrophobic amino acids. The formation of an α helix maximizes hydrogen bonding between the peptide bonds, and the hydrophobic amino acid side chains interact with the fatty acid tails of the phospholipids in the bilayer. However, different integral membrane proteins vary in how they are inserted (Figure 10.10). For example, whereas some integral membrane proteins span the membrane only once, others have multiple membrane spanning regions. In addition, some proteins are oriented in the membrane with their amino terminus on the cytosolic side; others have their carboxy terminus exposed to the cytosol. These orientations of proteins inserted into the ER, Golgi, lysosomal, and plasma membranes are generally established as the growing polypeptide chains are translocated into the ER. The lumen of the ER is topologically equivalent to the exterior of the cell, so the domains of plasma membrane proteins that are exposed on the cell surface correspond to the regions of polypeptide chains that are translocated into the ER lumen (Figure 10.11).

The most straightforward mode of insertion into the ER membrane results in the synthesis of transmembrane proteins oriented with their carboxy termini exposed to the cytosol (Figure 10.12). These proteins have a normal amino terminal signal sequence, which is cleaved by signal peptidase during translocation of the polypeptide chain across the ER membrane through the translocon. They are then anchored in the membrane by a second membrane spanning α helix in the middle of the protein. This transmembrane sequence, called a stop-transfer sequence, signals a change in the translocon channel. Further translocation of the polypeptide chain across the ER membrane is thus blocked, so the carboxy terminal portion of

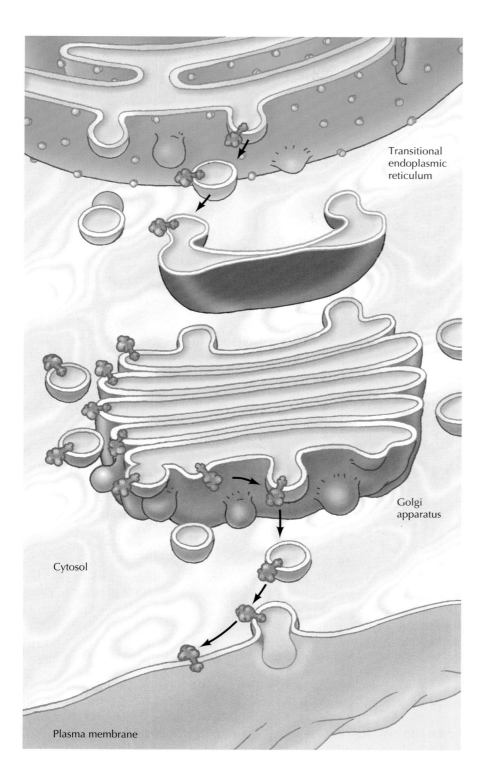

FIGURE 10.11 Topology of the secretory pathway   The lumens of the endoplasmic reticulum and Golgi apparatus are topologically equivalent to the exterior of the cell. Consequently, those portions of polypeptide chains that are translocated into the ER are exposed on the cell surface following transport to the plasma membrane.

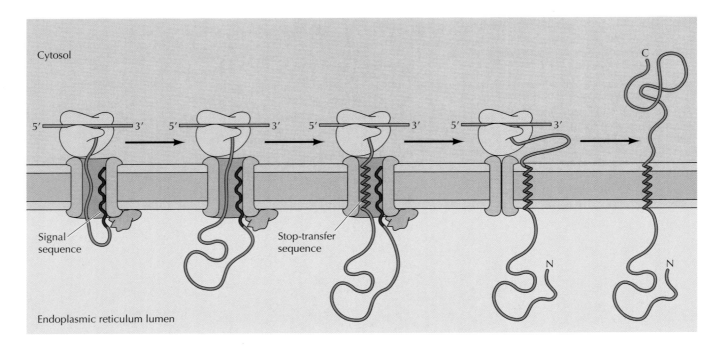

Cytosol

Signal
sequence

Stop-transfer
sequence

Endoplasmic reticulum lumen

FIGURE 10.12 Insertion of a membrane protein with a cleavable signal sequence and a single stop-transfer sequence The signal sequence is cleaved as the polypeptide chain crosses the membrane, so the amino terminus of the polypeptide chain is exposed in the ER lumen. However, translocation of the polypeptide chain across the membrane is halted by a transmembrane stop-transfer sequence that closes the translocon and exits the channel laterally to anchor the protein in the ER membrane. Continued translation results in a membrane-spanning protein with its carboxy terminus on the cytosolic side.

the growing polypeptide chain remains in the cytosol. As the subunits of the translocon separate, the transmembrane domain of the protein enters the lipid bilayer. The insertion of these transmembrane proteins into the membrane thus involves the sequential action of two distinct elements: a cleavable amino terminal signal sequence that initiates translocation across the membrane, and a transmembrane stop-transfer sequence that anchors the protein in the membrane.

Proteins can also be anchored in the ER membrane by internal signal sequences that are not cleaved by signal peptidase (Figure 10.13). These internal signal sequences are recognized by the SRP and brought to the ER membrane as already discussed. Because they are not cleaved by signal peptidase, however, these signal sequences act as transmembrane α helices that exit the translocon and anchor proteins in the ER membrane. Importantly, internal signal sequences can be oriented so as to direct the translocation of either the amino or carboxy terminus of the polypeptide chain across the membrane. Therefore, depending on the orientation of the signal sequence, proteins inserted into the membrane by this mechanism can have either their amino or carboxy terminus exposed to the cytosol.

Proteins that span the membrane multiple times are thought to be inserted as a result of an alternating series of internal signal sequences and transmembrane stop-transfer sequences. For example, an internal signal

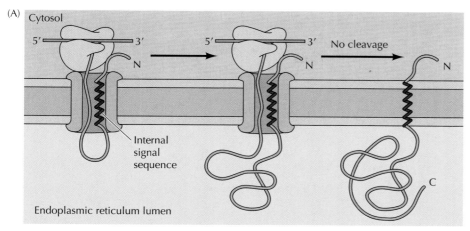

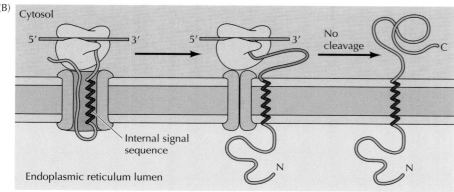

FIGURE 10.13 Insertion of membrane proteins with internal noncleavable signal sequences   Internal noncleavable signal sequences can lead to the insertion of polypeptide chains in either orientation in the ER membrane. (A) The signal sequence directs insertion of the polypeptide such that its amino terminus is exposed on the cytosolic side. The remainder of the polypeptide chain is translocated into the ER as translation proceeds. The signal sequence is not cleaved, so it acts as a membrane-spanning sequence that anchors the protein in the membrane with its carboxy terminus in the lumen of the ER. (B) Other internal signal sequences are oriented to direct the transfer of the amino-terminal portion of the polypeptide across the membrane. Continued translation results in a protein that spans the ER membrane with its amino terminus in the lumen and its carboxy terminus in the cytosol. Note that this orientation is the same as that resulting from insertion of a protein that contains a cleavable signal sequence followed by a stop-transfer sequence (see Figure 10.12).

sequence can result in membrane insertion of a polypeptide chain with its amino terminus on the cytosolic side (Figure 10.14). If a stop-transfer sequence is then encountered, the polypeptide will form a loop in the ER lumen, and protein synthesis will continue on the cytosolic side of the membrane. If a second signal sequence is encountered, the growing polypeptide chain will again be inserted into the ER, forming another looped domain on the cytosolic side of the membrane. This can be followed by yet another stop-transfer sequence and so forth, so that an alternating series of signal and stop-transfer sequences can result in the insertion of proteins that span the membrane multiple times, with looped domains exposed on both the lumenal and cytosolic sides.

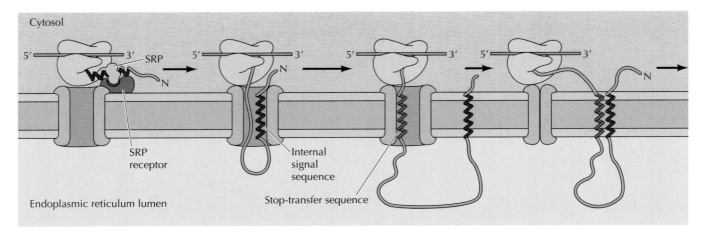

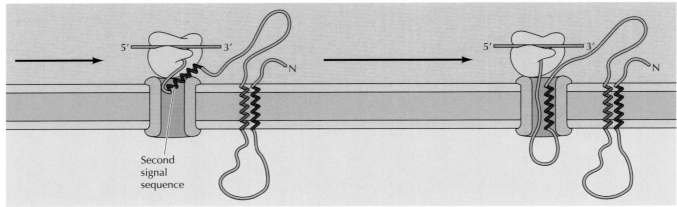

FIGURE 10.14 Insertion of a protein that spans the membrane multiple times In this example, an internal signal sequence results in insertion of the polypeptide chain with its amino terminus on the cytosolic side of the membrane. A stop-transfer sequence then signals closure of the translocation channel causing the polypeptide chain to form a loop within the lumen of the ER; translation continues in the cytosol. A second internal sequence reopens the channel, triggering reinsertion of the polypeptide chain into the ER membrane and forming a loop in the cytosol. The process can be repeated many times, resulting in the insertion of proteins with multiple membrane-spanning regions.

As discussed below, most transmembrane proteins destined for other compartments in the secretory pathway are delivered to them in transport vesicles. However, the inner nuclear membrane is continuous with the ER and it is thought that proteins destined for the inner nuclear membrane (such as emerin or LBR; see Chapter 9) diffuse in the plane of the membrane and then are retained in the inner nuclear membrane by interactions with nuclear components, such as lamins or chromatin. In addition, recent studies suggest that inner nuclear membrane proteins contain specific transmembrane sequences that alter their interaction with the translocon and signal their transport to the inner nuclear membrane.

Protein Folding and Processing in the ER

The folding of polypeptide chains into their correct three-dimensional conformations, the assembly of polypeptides into multisubunit proteins, and the covalent modifications involved in protein processing were discussed in Chapter 8. For proteins that enter the secretory pathway, many of these events occur either during translocation across the ER membrane or within the ER lumen. One such processing event is the proteolytic cleavage of the signal peptide as the polypeptide chain is translocated across the ER membrane. The ER is also the site of protein folding, assembly of multisubunit proteins, disulfide bond formation, the initial stages of glycosylation, and the addition of glycolipid anchors to some plasma membrane proteins. In

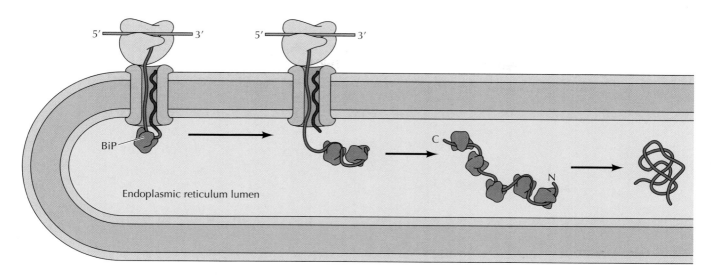

FIGURE 10.15 Protein folding in the ER   The molecular chaperone BiP binds to polypeptide chains as they cross the ER membrane and facilitates protein folding and assembly within the ER.

fact, the primary role of lumenal ER proteins is to assist the folding and assembly of newly translocated polypeptides.

As already discussed, proteins are translocated across the ER membrane as unfolded polypeptide chains while their translation is still in progress. These polypeptides, therefore, fold into their three-dimensional conformations within the ER, assisted by molecular chaperones that facilitate the folding of polypeptide chains (see Chapter 8). The Hsp70 chaperone, BiP, is thought to bind to the unfolded polypeptide chain as it crosses the membrane and then mediate protein folding and the assembly of multisubunit proteins within the ER (Figure 10.15). Correctly assembled proteins are released from BiP (and other chaperones) and are available for transport to the Golgi apparatus. Abnormally folded or improperly assembled proteins are targets for degradation, as will be discussed later.

The formation of disulfide bonds between the side chains of cysteine residues is an important aspect of protein folding and assembly within the ER. These bonds do not form in the cytosol, which is characterized by a reducing environment that maintains cysteine residues in their reduced (—SH) state. In the ER, however, an oxidizing environment promotes disulfide (S—S) bond formation, and disulfide bonds formed in the ER play important roles in the structure of secreted and cell surface proteins. Disulfide bond formation is facilitated by the enzyme **protein disulfide isomerase** (see Figure 8.24), which is located in the ER lumen.

Proteins are also glycosylated on specific asparagine residues (*N*-linked glycosylation) within the ER while their translation is still in process (Figure 10.16). As discussed in Chapter 8 (see Figures 8.29 and 8.30), oligosaccharide units consisting of 14 sugar residues are added to acceptor asparagine residues of growing polypeptide chains as they are translocated into the ER. The oligosaccharide is synthesized on a lipid (dolichol) carrier anchored in the ER membrane. It is then transferred as a unit to acceptor asparagine residues in the consensus sequence Asn-X-Ser/Thr by a membrane-bound enzyme called oligosaccharyl transferase. Three glucose residues are

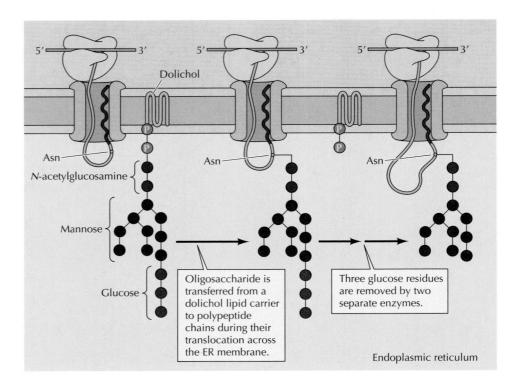

Dolichol

Asn

N-acetylglucosamine

Mannose

Glucose

Oligosaccharide is transferred from a dolichol lipid carrier to polypeptide chains during their translocation across the ER membrane.

Three glucose residues are removed by two separate enzymes.

Endoplasmic reticulum

FIGURE 10.16 Protein glycosylation in the ER

removed while the protein is still within the ER, and the protein is modified further after being transported to the Golgi apparatus (discussed later in this chapter).

Some proteins are attached to the plasma membrane by glycolipids rather than by membrane-spanning regions of the polypeptide chain. Because these membrane-anchoring glycolipids contain phosphatidylinositol, they are called **glycosylphosphatidylinositol (GPI) anchors**, the structure of which was illustrated in Figure 8.35. The GPI anchors are assembled in the ER membrane. They are then added immediately after completion of protein synthesis to the carboxy terminus of some proteins that are retained in the membrane by a C-terminal hydrophobic sequence (Figure 10.17). The C-terminal sequence of the protein is cleaved and exchanged for the GPI anchor, so these proteins remain attached to the membrane only by their associated glycolipid. Like transmembrane proteins, they are transported to the cell surface as membrane components via the secretory pathway. Their orientation within the ER dictates that GPI-anchored proteins are exposed on the outside of the cell, with the GPI anchor mediating their attachment to the plasma membrane.

Quality Control in the ER

Many proteins synthesized in the ER are rapidly degraded, primarily because they fail to fold correctly; others reside in the ER for several hours while they are properly folded. Thus an important role of the ER is to identify misfolded proteins, mark them, and divert them to a degradation pathway. Because they assist proteins in correct folding, chaperones and protein processing enzymes in the ER lumen often act as sensors of misfolded pro-

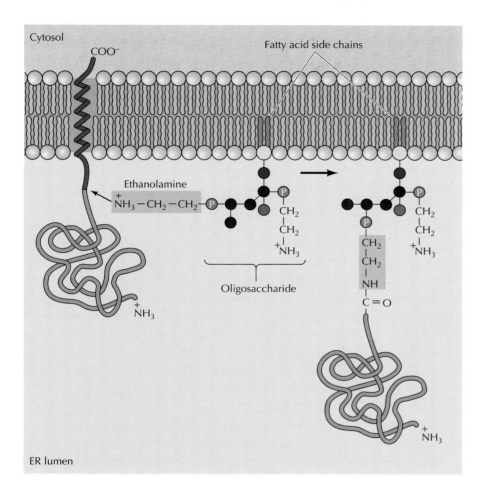

FIGURE 10.17 Addition of GPI anchors   Glycosylphosphatidylinositol (GPI) anchors contain two fatty acid chains, an oligosaccharide portion consisting of inositol and other sugars, and ethanolamine (see Figure 8.35 for a more detailed structure). The GPI anchors are assembled in the ER and added to polypeptides anchored in the membrane by a carboxy-terminal membrane-spanning region. The membrane-spanning region is cleaved, and the new carboxy terminus is joined to the NH_2 group of ethanolamine immediately after translation is completed, leaving the protein attached to the membrane by the GPI anchor.

teins. The process of ER protein quality control is complex and involves at least four chaperones, protein disulfide isomerase, and many supporting proteins. While the process is not fully understood, much is known about the roles of the glycoprotein chaperones, calnexin and calreticulin (Figure 10.18). These proteins bind sugar residues on the partially folded glycoproteins before translocation is complete and assist the glycoprotein in folding correctly. If the glycoprotein fails to fold after multiple cycles, the chaperone complex will send it into a degradation pathway that involves retro-translocation of the protein back through the translocon channel. In the cytosol it is marked by ubiquitination and degraded in the proteasome as described in Chapter 8.

   In addition to acting as a chaperone, BiP plays a pivotal role as a sensor of the general state of protein folding within the cell. If an excess of

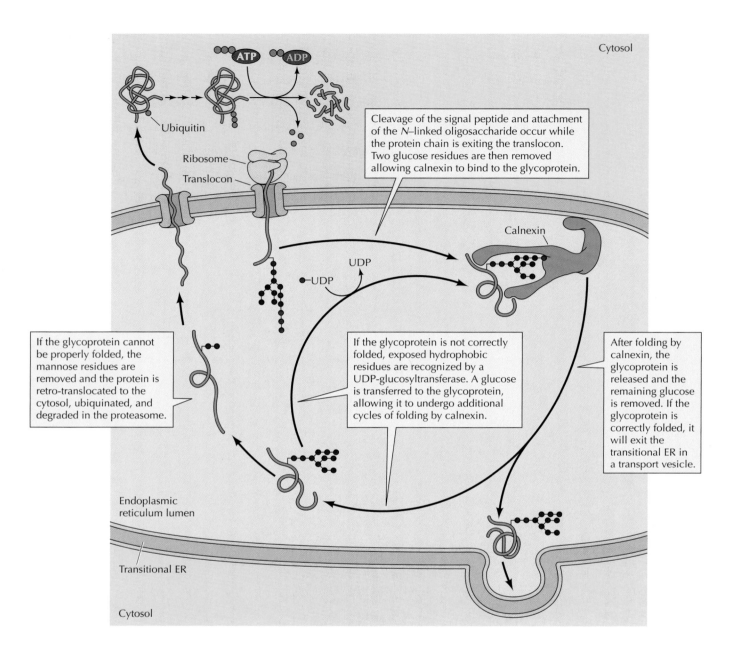

Cytosol

Ubiquitin

Ribosome

Translocon

Cleavage of the signal peptide and attachment of the *N*–linked oligosaccharide occur while the protein chain is exiting the translocon. Two glucose residues are then removed allowing calnexin to bind to the glycoprotein.

Calnexin

UDP

UDP

If the glycoprotein cannot be properly folded, the mannose residues are removed and the protein is retro-translocated to the cytosol, ubiquinated, and degraded in the proteasome.

If the glycoprotein is not correctly folded, exposed hydrophobic residues are recognized by a UDP-glucosyltransferase. A glucose is transferred to the glycoprotein, allowing it to undergo additional cycles of folding by calnexin.

After folding by calnexin, the glycoprotein is released and the remaining glucose is removed. If the glycoprotein is correctly folded, it will exit the transitional ER in a transport vesicle.

Endoplasmic reticulum lumen

Transitional ER

Cytosol

FIGURE 10.18 Glycoprotein folding by calnexin

unfolded proteins accumulates, as may result from a variety of types of cellular stress, signaling via BiP initiates a process known as the **unfolded protein response** (Figure 10.19). The levels of BiP in the ER lumen are normally sufficient not only to function in protein import and folding but also to bind signaling molecules, keeping them in an inactive state. However, if an excess of unfolded proteins accumulates, they compete for the available BiP. This releases the molecules that signal the unfolded protein response, which includes general inhibition of protein synthesis, increased expression of chaperones (such as calnexin, calreticulin, protein disulfide isomerase, and BiP itself), and an increase in the activity of the proteasome to enhance degradation of misfolded proteins (see Figure 10.18).

(A) Normal protein folding

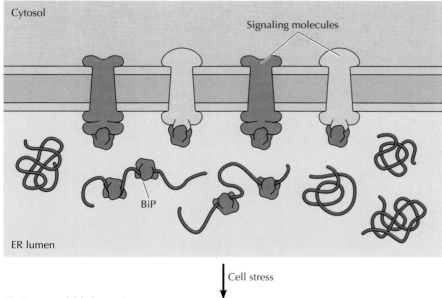

(B) Excess unfolded proteins

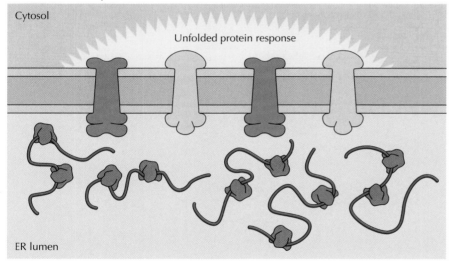

FIGURE 10.19 Unfolded protein response The chaperone protein, BiP, participates in the folding of proteins in the ER lumen. (A) In an unstressed cell there is sufficient BiP available to both fold newly synthesized proteins and keep several types of ER membrane signal molecules inactive. (B) Cellular stress, such as heat, chemical insult, or viral infection interferes with protein folding, so unfolded proteins accumulate in the ER. BiP has a higher affinity for unfolded proteins than for the ER membrane signal molecules, so the latter are released to become active and initiate the unfolded protein response.

The Smooth ER and Lipid Synthesis

In addition to its activities in the processing of secreted and membrane proteins, the ER is the major site at which membrane lipids are synthesized in eukaryotic cells. Because they are extremely hydrophobic, membrane lipids are synthesized in association with already existing cellular membranes rather than in the aqueous environment of the cytosol. Although some lipids are synthesized in association with other membranes, most are synthesized in the ER. They are then transported from the ER to their ultimate destinations either in vesicles or by carrier proteins, as will be discussed later in this chapter and in Chapter 13.

The membranes of eukaryotic cells are composed of three main types of lipids: phospholipids, glycolipids, and cholesterol. Most of the phospho-

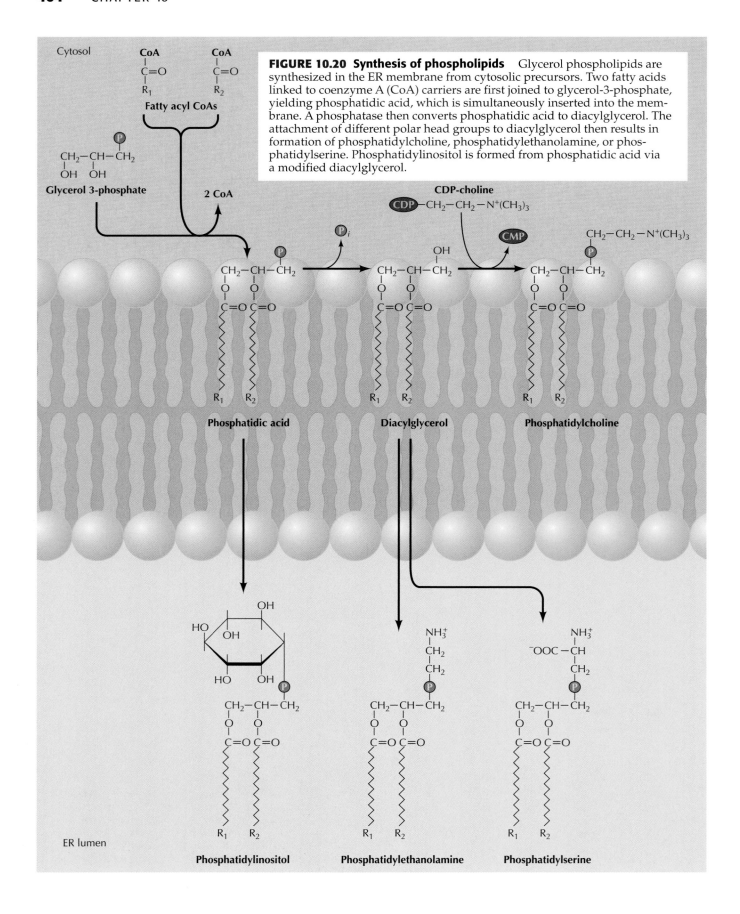

FIGURE 10.20 Synthesis of phospholipids Glycerol phospholipids are synthesized in the ER membrane from cytosolic precursors. Two fatty acids linked to coenzyme A (CoA) carriers are first joined to glycerol-3-phosphate, yielding phosphatidic acid, which is simultaneously inserted into the membrane. A phosphatase then converts phosphatidic acid to diacylglycerol. The attachment of different polar head groups to diacylglycerol then results in formation of phosphatidylcholine, phosphatidylethanolamine, or phosphatidylserine. Phosphatidylinositol is formed from phosphatidic acid via a modified diacylglycerol.

lipids, which are the basic structural components of the membrane, are derived from glycerol. They are synthesized on the cytosolic side of the ER membrane from water-soluble cytosolic precursors (Figure 10.20). Fatty acids are first transferred from coenzyme A carriers to glycerol-3-phosphate by membrane-bound enzymes, and the resulting phospholipid (phosphatidic acid) is inserted into the membrane. Enzymes on the cytosolic face of the ER membrane then either modify phosphatidic acid or directly catalyze the addition of different polar head groups, resulting in formation of phosphatidylcholine, phosphatidylserine, phosphatidylethanolamine, or phosphatidylinositol.

The synthesis of these phospholipids on the cytosolic side of the ER membrane allows the hydrophobic fatty acid chains to remain buried in the membrane while membrane-bound enzymes catalyze their reactions with water-soluble precursors (e.g., CDP-choline) in the cytosol. Because of this topography, however, new phospholipids are added only to the cytosolic half of the ER membrane (Figure 10.21). To maintain a stable membrane some of these newly synthesized phospholipids must therefore be transferred to the other (lumenal) half of the ER bilayer. This transfer does not occur spon-

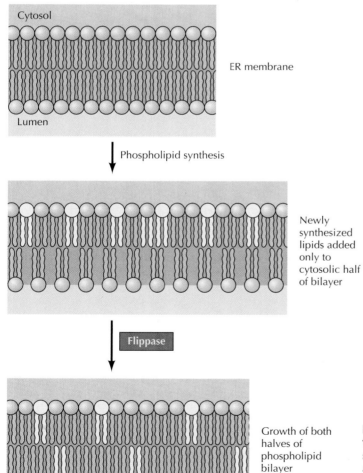

Cytosol

ER membrane

Lumen

Phospholipid synthesis

Newly synthesized lipids added only to cytosolic half of bilayer

Flippase

Growth of both halves of phospholipid bilayer

FIGURE 10.21 Translocation of phospholipids across the ER membrane Because phospholipids are synthesized on the cytosolic side of the ER membrane, they are added only to the cytosolic half of the bilayer. They are then translocated across the membrane by phospholipid flippases, resulting in even growth of both halves of the phospholipid bilayer.

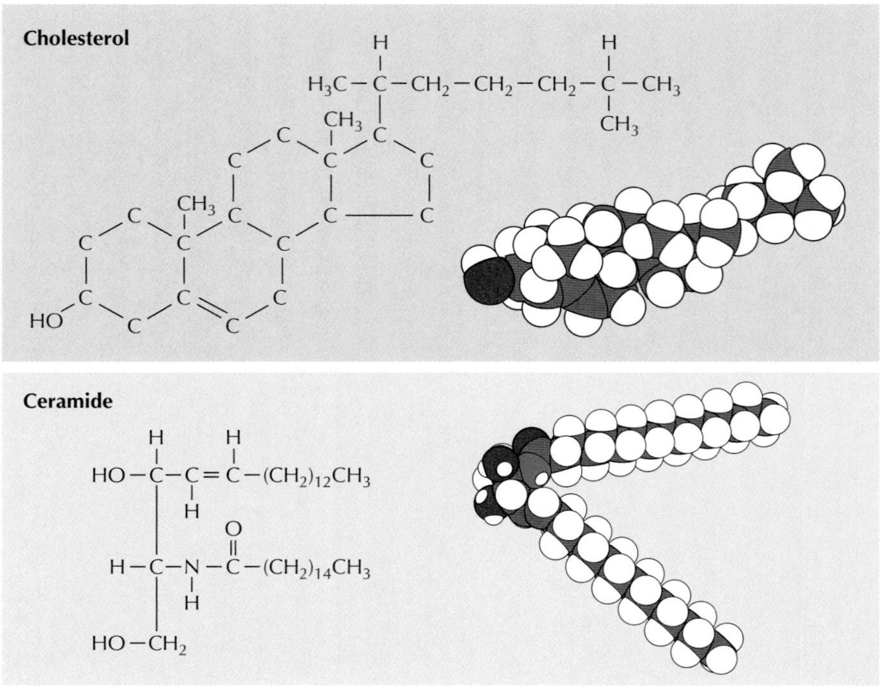

FIGURE 10.22 Structure of cholesterol and ceramide The hydrogens attached to the ring carbons of cholesterol are not shown.

taneously because it requires the passage of a polar head group through the membrane. Instead, membrane proteins called **flippases** catalyze the rapid translocation of phospholipids across the ER membrane resulting in even growth of both halves of the bilayer. At least some flippases are phospholipid-specific and ATP-dependent.

In addition to its role in synthesis of the glycerol phospholipids, the ER also serves as the major site of synthesis of two other membrane lipids: cholesterol and ceramide (Figure 10.22). As discussed later, ceramide is converted to either glycolipids or sphingomyelin (the only membrane phospholipid not derived from glycerol) in the Golgi apparatus. The ER is thus responsible for synthesis of either the final products or the precursors of all the major lipids of eukaryotic membranes. Cholesterol and sphingomyelin are important components of lipid rafts as discussed in Chapter 13.

Smooth ER is abundant in cell types that are particularly active in lipid metabolism. For example, steroid hormones are synthesized (from cholesterol) in the ER, so large amounts of smooth ER are found in steroid-producing cells, such as those in the testis and ovary. In addition, smooth ER is abundant in the liver where it contains enzymes that metabolize various lipid-soluble compounds. These detoxifying enzymes inactivate a number of potentially harmful drugs (e.g., phenobarbital) by converting them to water-soluble compounds that can be eliminated from the body in the urine. The smooth ER is thus involved in multiple aspects of the metabolism of lipids and lipid-soluble compounds.

Export of Proteins and Lipids from the ER

Both proteins and phospholipids travel along the secretory pathway in transport vesicles, which bud from the membrane of one organelle and then fuse with the membrane of another. Thus molecules are exported from the

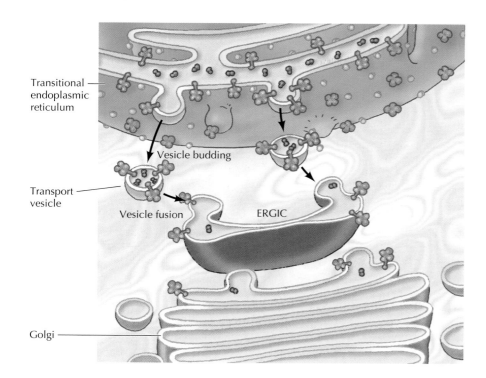

Transitional endoplasmic reticulum

Vesicle budding

Transport vesicle

Vesicle fusion

ERGIC

Golgi

FIGURE 10.23 Vesicular transport from the ER to the Golgi Proteins and lipids are carried from the ER to the Golgi in transport vesicles that bud from the membrane of the transitional ER, fuse to form the vesicles and tubules of the ER-Golgi intermediate compartment (ERGIC), and are then carried to the Golgi. Lumenal ER proteins are taken up by the vesicles and released into the lumen of the Golgi. Membrane proteins maintain the same orientation in the Golgi as in the ER.

ER in vesicles that bud from the transitional ER and carry their cargo through the ER-Golgi intermediate compartment and then to the Golgi apparatus (Figure 10.23). Subsequent steps in the secretory pathway involve vesicular transport between different compartments of the Golgi and from the Golgi to lysosomes or to the plasma membrane. In most cases, proteins within the lumen of one organelle are packaged into the budding transport vesicle and then released into the lumen of the recipient organelle following vesicle fusion. Membrane proteins and lipids are transported similarly, and it is significant that their topological orientation is maintained as they travel from one membrane-enclosed organelle to another. For example, the domains of a protein exposed on the cytosolic side of the ER membrane will also be exposed on the cytosolic side of the Golgi and plasma membranes, whereas protein domains exposed on the lumenal side of the ER membrane will be exposed on the lumenal side of the Golgi and on the exterior of the cell (see Figure 10.11).

Most proteins that enter the transitional ER move through the ER-Golgi intermediate compartment and on to the Golgi. These proteins are marked by sequences that signal either their export from or retention within the ER (Figure 10.24). Many transmembrane proteins possess di-acidic or di-hydrophobic amino acid sequences in their cytosolic domains that function as ER export signals. Both GPI-anchored proteins (which are marked for export by their GPI-anchors) and lumenal secretory proteins appear to be recognized and sequestered by these transmembrane receptor proteins. Very few ER export signals have been detected on lumenal secretory proteins and their recognition may depend on the shape of the correctly folded protein. It is also possible that there is a default pathway where otherwise unmarked proteins in the ER lumen move to the Golgi and beyond.

FIGURE 10.24 ER export signals
Three types of secretory proteins are recruited into budding vesicles by cytosolic adaptor proteins. Transmembrane proteins are recognized by di-acidic (e.g., Asp-Asp or Glu-Glu) or di-hydrophobic (e.g., Met-Met) signal sequences in their cytosolic segments. Some of these are also receptors for lumenal and GPI-anchored membrane secretory proteins. The transmembrane receptors recognize the GPI-anchors and signal sequences or signal patches characteristic of the folded lumenal proteins.

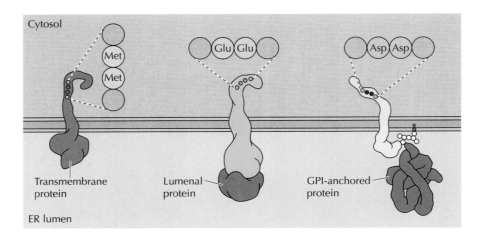

If proteins that function within the ER (including BiP, signal peptidase, protein disulfide isomerase, and other enzymes discussed earlier) are allowed to proceed along the secretory pathway, they will be lost to the cell. Thus many such proteins have a targeting sequence Lys-Asp-Glu-Leu (KDEL, in the single-letter code) at their carboxy terminus that directs their retrieval back to the ER. If this sequence is deleted from a protein that normally functions in the ER (e.g., BiP or protein disulfide isomerase), the mutated protein is instead transported to the Golgi and secreted from the cell. Conversely, addition of the KDEL sequence to the carboxy terminus of proteins that are normally secreted blocks their secretion. Some ER transmembrane proteins are similarly marked by short C-terminal sequences that contain two lysine residues (KKXX sequences).

Interestingly, the KDEL and KKXX signals do not prevent ER proteins from being packaged into vesicles and carried to the Golgi. Instead, these signals cause these ER resident proteins to be selectively retrieved from the ER-Golgi intermediate compartment or the Golgi complex and returned to the ER via a recycling pathway (Figure 10.25). Proteins bearing the KDEL and KKXX sequences bind to specific recycling receptors in the membranes of these compartments and are then selectively transported back to the ER. The KDEL and KKXX sequences are the best-characterized retention/retrieval signals but there may be others. Other proteins are retrieved because they specifically bind to KDEL-bearing proteins such as BiP. Thus continued movement along the secretory pathway or retrieval back from the Golgi to the ER is the first branch point encountered by proteins being sorted to their correct destinations in the secretory pathway. Similar branch points arise at each subsequent stage of transport, such as retention in the Golgi versus export to lysosomes or the plasma membrane. In each case, specific localization signals target proteins to their correct intracellular destinations.

The Golgi Apparatus

The **Golgi apparatus**, or **Golgi complex**, functions as a factory in which proteins received from the ER are further processed and sorted for transport to their eventual destinations: endosomes, lysosomes, the plasma

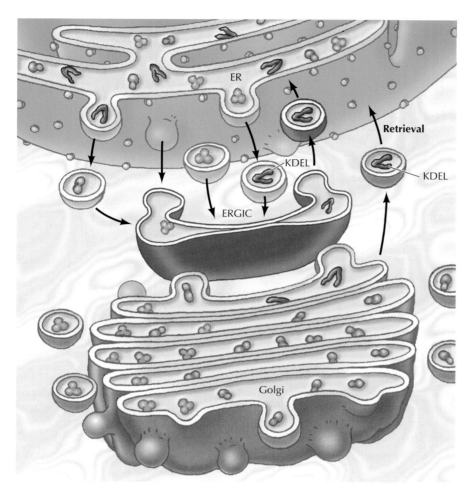

FIGURE 10.25 Retrieval of resident ER proteins Proteins destined to remain in the lumen of the ER are marked by the sequence Lys-Asp-Glu-Leu (KDEL) at their carboxy terminus. These proteins are exported from the ER to the Golgi, but they are recognized by a receptor in the ERGIC or the Golgi apparatus and selectively returned to the ER.

membrane, or secretion. In addition, as noted earlier, most glycolipids and sphingomyelin are synthesized within the Golgi. In plant cells, the Golgi apparatus further serves as the site at which the complex polysaccharides of the cell wall are synthesized. The Golgi apparatus is thus involved in processing the broad range of cellular constituents that travel along the secretory pathway.

Organization of the Golgi

In most cells, the Golgi is composed of flattened membrane-enclosed sacs (cisternae) and associated vesicles (Figure 10.26). A striking feature of the Golgi apparatus is its distinct polarity in both structure and function. Proteins from the ER enter at its *cis* face (entry face), which is convex and usually oriented toward the nucleus. They are then transported through the Golgi and exit from its concave *trans* face (exit face). The importance of this dynamic process for the structure of the Golgi is shown by the disappearance of the Golgi as an organized structure if vesicle transport from the ER is blocked. As they pass through the Golgi, proteins are modified and sorted for transport to their eventual destinations within the cell.

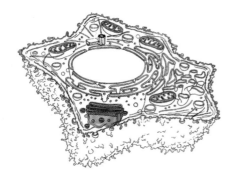

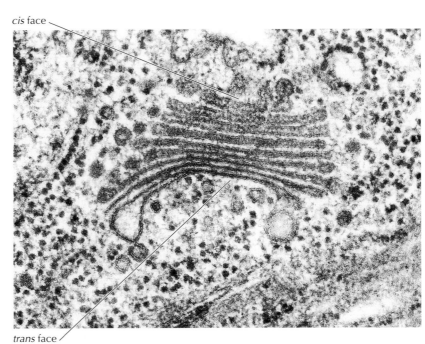

cis face

trans face

FIGURE 10.26 Electron micrograph of a Golgi apparatus The Golgi apparatus consists of a stack of flattened cisternae and associated vesicles. Proteins and lipids from the ER enter the Golgi apparatus at its *cis* face and exit near its *trans* face. (Courtesy of Dr. L. Andrew Staehelin, University of Colorado at Boulder.)

Distinct processing and sorting events appear to take place in an ordered sequence within different regions of the Golgi complex, so the Golgi is usually considered to consist of multiple discrete compartments. Although the number of such compartments has not been established, the Golgi is most commonly viewed as consisting of four functionally distinct regions: the **cis Golgi network**, the **Golgi stack** (which is divided into the *medial* and *trans* subcompartments), and the **trans Golgi network** (Figure 10.27). Proteins from the ER are transported to the ER-Golgi intermediate compartment and then enter the Golgi apparatus at the *cis* Golgi network. They then progress to the *medial* and *trans* compartments of the Golgi stack within which most metabolic activities of the Golgi apparatus take place. The modified proteins, lipids, and polysaccharides then move to the *trans* Golgi network, which acts as a sorting and distribution center, directing molecular traffic to endosomes, lysosomes, the plasma membrane, or the cell exterior.

Although the Golgi apparatus was first described over 100 years ago, the mechanism by which proteins move through the Golgi apparatus has still not been established and is an area of controversy among cell biologists. One possibility is that transport vesicles carry proteins between the cisternae of the Golgi compartments. However, there is considerable experimental support for proteins being carried through compartments of the Golgi within the Golgi cisternae, which gradually mature and progressively move through the Golgi in the *cis* to *trans* direction. It is likely that transport occurs by both of these processes. Recent work suggests that the dynamic structure of the Golgi is maintained by interactions of proteins in the cisternae membranes with the cytoskeleton.

Protein Glycosylation within the Golgi

Protein processing within the Golgi involves the modification and synthesis of the carbohydrate portions of glycoproteins. One of the major aspects of this processing is the modification of the *N*-linked oligosaccharides that

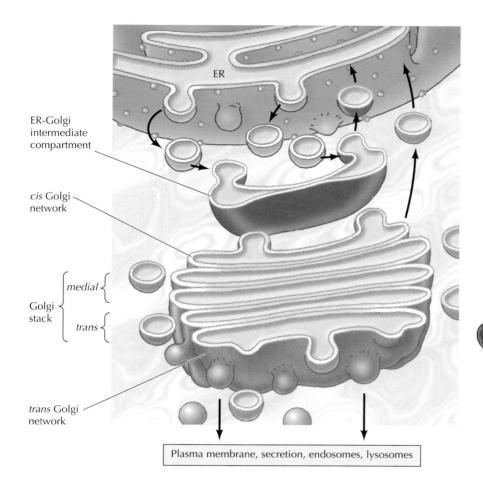

ER

ER-Golgi intermediate compartment

cis Golgi network

Golgi stack

medial

trans

trans Golgi network

Plasma membrane, secretion, endosomes, lysosomes

FIGURE 10.27 Regions of the Golgi apparatus Vesicles from the ER fuse to form the ER-Golgi intermediate compartment, and proteins from the ER are then transported to the *cis* Golgi network. Resident ER proteins are returned from the ER-Golgi intermediate compartment and the *cis* Golgi network via the recycling pathway. The *medial* and *trans* compartments of the Golgi stack correspond to the cisternae in the middle of the Golgi complex and are the sites of most protein modifications. Proteins are then carried to the *trans* Golgi network where they are sorted for transport to the plasma membrane, secretion, endosomes, or lysosomes.

10.1 WEBSITE ANIMATION

Organization of the Golgi The Golgi apparatus is composed of flattened membrane-enclosed sacs that receive proteins from the ER, process them, and sort them to their eventual destinations.

were added to proteins in the ER. As discussed earlier in this chapter, proteins are modified within the ER by the addition of an oligosaccharide consisting of 14 sugar residues (see Figure 10.16). Three glucose residues are removed while the polypeptides are still in the ER. Following transport to the Golgi apparatus, the *N*-linked oligosaccharides of these glycoproteins are subject to extensive further modifications.

N-linked oligosaccharides are processed within the Golgi apparatus in an ordered sequence of reactions (Figure 10.28). In most cases, the first modification of proteins destined for secretion or for the plasma membrane is the removal of four mannose residues. This is followed by the sequential addition of an *N*-acetylglucosamine, the removal of two more mannoses, and the addition of a fucose and two more *N*-acetylglucosamines. Finally, three galactose and three sialic acid residues are added. As noted in Chapter 8, different glycoproteins are modified to different extents during their passage through the Golgi, depending on both the structure of the protein and on the amount of processing enzymes that are present within the Golgi complexes of different types of cells. Consequently, proteins can emerge from the Golgi with a variety of different *N*-linked oligosaccharides. The enzymes that carry out the addition of sugar residues, **glycosyltransferases**, and those that remove them, **glycosidases**, are well-characterized, but the basis of their localization to specific cisternae of the Golgi is not known.

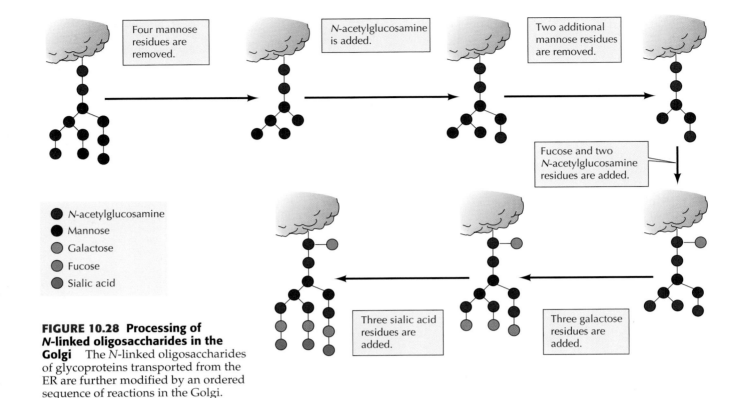

FIGURE 10.28 Processing of N-linked oligosaccharides in the Golgi The N-linked oligosaccharides of glycoproteins transported from the ER are further modified by an ordered sequence of reactions in the Golgi.

The processing of the N-linked oligosaccharide of lysosomal proteins differs from that of secreted and plasma membrane proteins. Rather than the initial removal of mannose residues, proteins destined for incorporation into lysosomes are modified by mannose phosphorylation. In the first step of this reaction, N-acetylglucosamine phosphates are added to specific mannose residues, probably while the protein is still in the *cis* Golgi network (Figure 10.29). This is followed by removal of the N-acetylglucosamine group, leaving **mannose-6-phosphate** residues on the N-linked oligosaccharide. Because of this modification these residues are not removed during further processing. Instead, these phosphorylated mannose residues are specifically recognized by a mannose-6-phosphate receptor in the *trans* Golgi network, which directs the transport of these proteins to endosomes and on to lysosomes.

The phosphorylation of mannose residues is thus a critical step in sorting lysosomal proteins to their correct intracellular destinations. The specificity of this process resides in the enzyme that catalyzes the first step in the reaction sequence—the selective addition of N-acetylglucosamine phosphates to lysosomal proteins. This enzyme recognizes a structural determinant that is present on lysosomal proteins but not on proteins destined for the plasma membrane or secretion. This recognition determinant is not a simple sequence of amino acids; rather, it is formed in the folded protein by the juxtaposition of amino acid sequences from different regions of the polypeptide chain. In contrast to the signal sequences that direct protein translocation to the ER, the recognition determinant that leads to mannose phosphorylation, and thus ultimately targets proteins to lysosomes, depends on the three-dimensional conformation of the folded protein. Such determinants are called **signal patches**, in contrast to the linear targeting signals discussed earlier in this chapter.

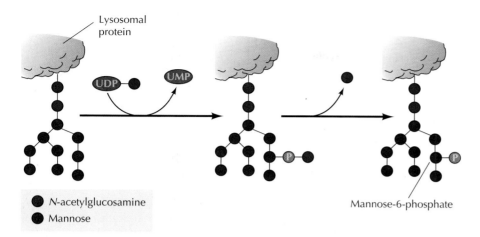

- ● N-acetylglucosamine
- ● Mannose

Mannose-6-phosphate

FIGURE 10.29 Targeting of lysosomal proteins by phosphorylation of mannose residues Proteins destined for incorporation into lysosomes are specifically recognized and modified by the addition of phosphate groups to the number 6 position of mannose residues. In the first step of the reaction, N-acetylglucosamine phosphates are transferred to mannose residues from UDPN-acetylglucosamine. The N-acetylglucosamine groups are then removed, leaving mannose-6-phosphates.

Proteins can also be modified by the addition of carbohydrates to the side chains of acceptor serine and threonine residues within specific sequences of amino acids (O-linked glycosylation) (see Figure 8.31). These modifications take place in the Golgi apparatus by the sequential addition of single sugar residues. The serine or threonine is usually linked directly to N-acetylgalactosamine to which other sugars can then be added. In some cases these sugars are further modified by the addition of sulfate groups.

Lipid and Polysaccharide Metabolism in the Golgi

In addition to its activities in processing and sorting glycoproteins, the Golgi apparatus functions in lipid metabolism—in particular, in the synthesis of glycolipids and sphingomyelin. As discussed earlier, the glycerol phospholipids, cholesterol, and ceramide are synthesized in the ER. Sphingomyelin and glycolipids are then synthesized from ceramide in the Golgi apparatus (Figure 10.30). Sphingomyelin (the only nonglycerol phospholipid in cell membranes) is synthesized by the transfer of a phosphorylcholine group from phosphatidylcholine to ceramide. Alternatively, the addition of carbohydrates to ceramide can yield a variety of different glycolipids.

Sphingomyelin is synthesized on the lumenal surface of the Golgi, but glucose is added to ceramide on the cytosolic side. Glucosylceramide then apparently

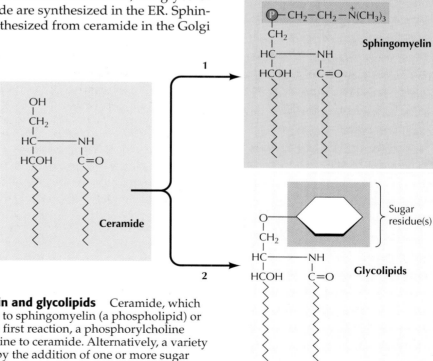

FIGURE 10.30 Synthesis of sphingomyelin and glycolipids Ceramide, which is synthesized in the ER, is converted either to sphingomyelin (a phospholipid) or to glycolipids in the Golgi apparatus. In the first reaction, a phosphorylcholine group is transferred from phosphatidylcholine to ceramide. Alternatively, a variety of different glycolipids can be synthesized by the addition of one or more sugar residues (e.g., glucose).

flips, however, and additional carbohydrates are added on the lumenal side of the membrane. Glycolipids are not able to translocate across the Golgi membrane, so they are found only in the lumenal half of the Golgi bilayer as is most sphingomyelin. Following vesicular transport they are correspondingly localized to the exterior half of the plasma membrane, with their polar head groups exposed on the cell surface. As will be discussed in Chapter 13, the oligosaccharide portions of glycolipids are important surface markers in cell-cell recognition.

In plant cells, the Golgi apparatus has the additional task of serving as the site where complex polysaccharides of the cell wall are synthesized. As discussed further in Chapter 14, the plant cell wall is composed of three major types of polysaccharides. Cellulose, the predominant constituent, is a simple linear polymer of glucose residues. It is synthesized at the cell surface by enzymes in the plasma membrane. The other cell wall polysaccharides (hemicelluloses and pectins), however, are complex branched chain molecules that are synthesized in the Golgi apparatus and then transported in vesicles to the cell surface. The synthesis of these cell wall polysaccharides is a major cellular function, and as much as 80% of the metabolic activity of the Golgi apparatus in plant cells may be devoted to polysaccharide synthesis.

Protein Sorting and Export from the Golgi Apparatus

Proteins as well as lipids and polysaccharides are transported from the Golgi apparatus to their final destinations through the secretory pathway. This involves the sorting of proteins into different kinds of transport vesicles, which bud from the *trans* Golgi network and deliver their contents to the appropriate cellular locations (Figure 10.31). Some proteins are carried from the Golgi to the plasma membrane by a constitutive secretory pathway, which accounts for the incorporation of new proteins and lipids into the plasma membrane as well as for the continuous secretion of proteins from the cell. Other proteins are transported to the cell surface by a distinct pathway of regulated secretion or are specifically targeted to other intracellular destinations, such as lysosomes in animal cells or vacuoles in yeast.

Proteins that function within the Golgi apparatus must be retained within that organelle rather than being transported along the secretory pathway. In contrast to the ER, all of the proteins known to be retained within the Golgi complex are associated with the Golgi membrane rather than being soluble proteins within the lumen. The signals responsible for retention of some proteins within the Golgi have been localized to their transmembrane domains, which prevent the protein from being packaged in the transport vesicles that leave the *trans* Golgi network. In addition, like the KKXX sequences of resident ER membrane proteins, signals in the cytoplasmic tails of some Golgi proteins mediate the retrieval of these proteins from subsequent compartments along the secretory pathway.

The constitutive secretory pathway, which operates in all cells, leads to continual unregulated protein secretion. However, some cells also possess a distinct regulated secretory pathway in which specific proteins are secreted in response to environmental signals. Examples of regulated secretion include the release of hormones from endocrine cells, the release of neurotransmitters from neurons, and the release of digestive enzymes from the pancreatic acinar cells discussed at the beginning of this chapter (see Figure 10.2). Proteins are sorted into the regulated secretory pathway in the *trans* Golgi network where they are packaged into specialized secretory vesicles. These immature secretory vesicles, which are larger than transport vesicles,

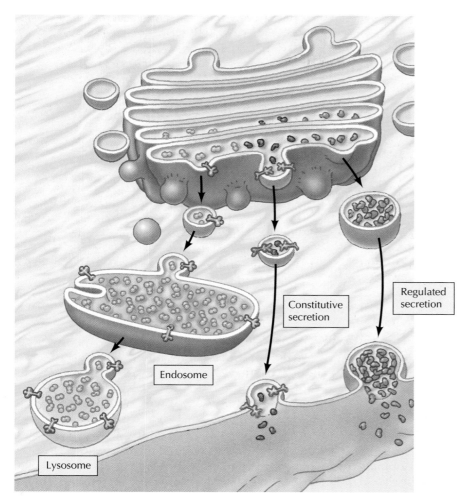

FIGURE 10.31 Transport from the Golgi apparatus Proteins are sorted in the *trans* Golgi network and transported in vesicles to their final destinations. In the absence of specific targeting signals, proteins are carried to the plasma membrane by constitutive secretion. Alternatively, proteins can be diverted from the constitutive secretion pathway and targeted to other destinations, such as endosomes and lysosomes or regulated secretion from the cells.

Regulated
secretion

Constitutive
secretion

Endosome

Lysosome

further process their protein contents and often fuse with each other. They also store their contents until specific signals direct their fusion with the plasma membrane. For example, the digestive enzymes produced by pancreatic acinar cells are stored in mature secretory vesicles until the presence of food in the stomach and small intestine triggers their secretion. The sorting of proteins into the regulated secretory pathway appears to involve the recognition of signal patches shared by multiple proteins that enter this pathway. These proteins selectively aggregate in cisternae of the *trans* Golgi network and are then released by budding as immature secretory vesicles. Proteins destined for very different locations, such as the plasma membrane versus lysosomes, are known to aggregate in different *trans* Golgi cisternae.

A further complication in the transport of proteins to the plasma membrane arises in many epithelial cells, which are polarized when they are organized into tissues. The plasma membrane of such cells is divided into two separate regions, the **apical domain** and the **basolateral domain**, that contain specific proteins related to their particular functions. For example, the apical membrane of intestinal epithelial cells faces the lumen of the intestine and is specialized for the efficient absorption of nutrients; the

FIGURE 10.32 Transport to the plasma membrane of polarized cells
The plasma membranes of polarized epithelial cells are divided into apical and basolateral domains. In this example (intestinal epithelium), the apical surface of the cell faces the lumen of the intestine, the lateral surfaces are in contact with neighboring cells, and the basal surface rests on a sheet of extracellular matrix (the basal lamina). The apical membrane is characterized by the presence of microvilli, which facilitate the absorption of nutrients by increasing surface area. Specific proteins are targeted to the apical or basolateral membranes either in the *trans* Golgi network or in an endosome. Tight junctions between neighboring cells maintain the identity of the apical and basolateral membranes by preventing the diffusion of proteins between these domains.

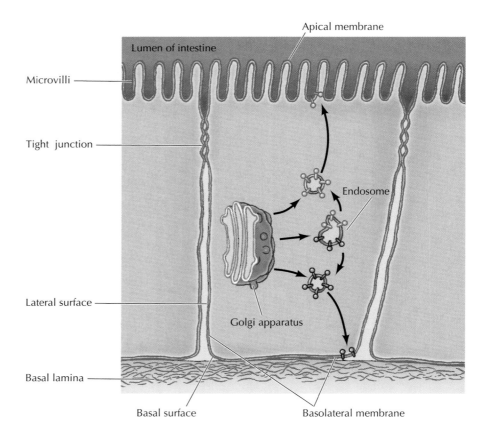

remainder of the cell is covered by the basolateral membrane (Figure 10.32). Distinct domains of the plasma membrane are present not only in epithelial cells but also in other cell types. Thus the constitutive secretory pathway must selectively transport proteins from the *trans* Golgi network to these distinct domains of the plasma membrane. This is accomplished by the selective packaging of proteins into at least two types of constitutive secretory vesicles targeted specifically for either the apical or basolateral plasma membrane domains of the cell. This may occur in the *trans* Golgi network or in an endosome.

The best-characterized pathway of protein sorting in the Golgi is the selective transport of proteins to lysosomes. As already discussed, lumenal lysosomal proteins are marked by mannose-6-phosphates that are formed by modification of their *N*-linked oligosaccharides shortly after entry into the Golgi apparatus. A specific receptor in the membrane of the *trans* Golgi network then recognizes these mannose-6-phosphate residues. The resulting complexes of receptor plus lysosomal enzyme are packaged into transport vesicles destined for lysosomes. Lysosomal membrane proteins are targeted by sequences in their cytoplasmic tails rather than by mannose-6-phosphates.

In yeasts and plant cells, both of which lack lysosomes, proteins are transported from the Golgi apparatus to an additional destination: the **vacuole** (Figure 10.33). Vacuoles assume the functions of lysosomes in these cells as well as perform a variety of other tasks, such as the storage of nutrients and the maintenance of turgor pressure and osmotic balance. In contrast to lysosomal targeting, proteins are directed to vacuoles by short peptide sequences instead of carbohydrate markers.

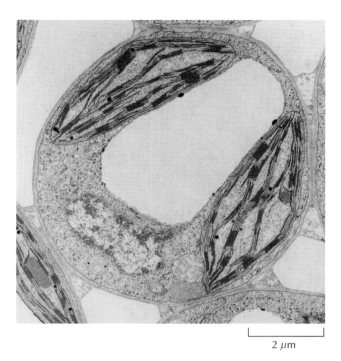

FIGURE 10.33 A plant cell vacuole   The large central vacuole functions as a lysosome in addition to storing nutrients and maintaining osmotic balance. (E. H. Newcombe/Biological Photo Service.)

The Mechanism of Vesicular Transport

As is evident from the preceding sections of this chapter, transport vesicles play a central role in the traffic of molecules between different membrane-enclosed compartments of the secretory pathway. As discussed in Chapter 13, vesicles are similarly involved in the transport of materials taken up at the cell surface. Vesicular transport is thus a major cellular activity, responsible for molecular traffic between a variety of specific membrane-enclosed compartments. The selectivity of such transport is therefore key to maintaining the functional organization of the cell. For example, lysosomal enzymes must be transported specifically from the Golgi apparatus to lysosomes—not to the plasma membrane or to the ER. Some of the signals that target proteins to specific organelles, such as lysosomes, were discussed earlier in this chapter. These proteins are transported within vesicles, so the specificity of transport is based on the selective packaging of the intended cargo into vesicles that recognize and fuse only with the appropriate target membrane. Because of the central importance of vesicular transport to the organization of eukaryotic cells, understanding the molecular mechanisms that control vesicle packaging, budding, and fusion is a major area of research in cell biology.

Experimental Approaches to Understanding Vesicular Transport

Progress toward elucidating the molecular mechanisms of vesicular transport has been made by three distinct experimental approaches: (1) isolation of yeast mutants that are defective in protein transport and sorting; (2) reconstitution of vesicular transport in cell-free systems; and (3) biochemical analysis of synaptic vesicles, which are responsible for the regulated secretion of neurotransmitters by neurons. Each of these experimental systems has distinct advantages for understanding particular aspects of the transport process. Most important, however, is the fact that the results from all three of these avenues of investigation have converged, indicating that similar molecular mechanisms regulate secretion in cells as different as yeasts and mammalian neurons.

As in other areas of cell biology, yeasts have proven to be advantageous in studying the secretory pathway because they are readily amenable to genetic analysis. In particular, Randy Schekman and his colleagues have pioneered the isolation of yeast mutants defective in vesicular transport. These include mutants that are defective at various stages of protein secretion (*sec* mutants), mutants that are unable to transport proteins to the vacuole, and mutants that are unable to retain resident ER proteins. The isolation of such mutants in yeasts led directly to the molecular cloning and analysis of the corresponding genes, thereby identifying a number of proteins involved in various steps of the secretory pathway. For example, the role of Sec61 as a major component of the protein translocation channel in the endoplasmic reticulum was discussed earlier in this chapter.

Biochemical studies of vesicular transport using reconstituted systems have complemented these genetic studies and have enabled the direct isolation of transport proteins from mammalian cells. The first cell-free transport system was developed by James Rothman and colleagues, who analyzed protein transport between compartments of the Golgi apparatus. Similar reconstituted systems have been developed to analyze transport between other compartments, including transport from the ER to the Golgi and transport from the Golgi to secretory vesicles, vacuoles, and the plasma membrane. The development of these *in vitro* systems has enabled biochemical studies of the transport process and functional analysis of proteins identified by mutations in yeasts, as well as direct isolation of some of the proteins involved in vesicle budding and fusion.

Insights into the molecular mechanisms of vesicular transport have come from studies of synaptic transmission in neurons, which represents a highly specialized form of regulated secretion. A synapse is the junction of a neuron with another cell, which may be either another neuron or an effector, such as a muscle cell. Information is transmitted across the synapse by chemical neurotransmitters, such as acetylcholine, which are stored within the neuron in **synaptic vesicles**. Stimulation of the transmitting neuron triggers the fusion of these synaptic vesicles with the plasma membrane causing neurotransmitters to be released into the synapse and stimulating the postsynaptic neuron or effector cell. Synaptic vesicles are extremely abundant in the brain, allowing them to be purified in large amounts for biochemical analysis. Some of the proteins isolated from synaptic vesicles are closely related to proteins that have been shown to play critical roles in vesicular transport by yeast genetics and reconstitution experiments, so biochemical analysis of these proteins has provided important insights into the molecular mechanisms of vesicle fusion.

Recent studies using GFP fusion proteins have allowed transport vesicles carrying specific proteins to be visualized by immunofluorescence as they move through the secretory pathway. In these experiments, cells are transfected with cDNA constructs encoding secretory proteins tagged with green fluorescent protein (GFP) (see Figure 1.27). The progress of the GFP-labeled proteins through the secretory pathway can then be followed in living cells, allowing characterization of many aspects of the dynamics and molecular interactions involved in vesicular transport.

Cargo Selection, Coat Proteins, and Vesicle Budding

Most transport vesicles that carry secretory proteins from the ER to the Golgi and from the Golgi to other targets are coated with cytosolic coat proteins and thus are called coated vesicles. Initially, the secretory proteins are sorted from proteins targeted for other destinations and from proteins that need to remain behind (see Figure 10.24). The coats assemble as the secretory protein-containing vesicle buds off the donor membrane and are generally removed from the vesicle in the cytosol before it reaches its target (Figure 10.34). Some of the remaining proteins allow the vesicles to travel along microtubules to their targets by interacting with specific tubulin-based molecular motors as described in Chapter 12. At the target membrane, the vesicles dock and fuse with the membrane, emptying their lumenal cargo and inserting their membrane proteins into the target membrane.

The formation of coated vesicles is regulated by small GTP-binding proteins related to Ras and Ran. Two families of GTP-binding proteins play roles in transport vesicle budding: ADP-ribosylation factors (ARFs 1–3 & Sar1) and a large family of Rab proteins. These regulate adaptor proteins that interact directly with a vesicle coat protein. The binding of GTP-binding proteins and adaptor proteins establishes a "platform" on the membrane for a specific process, such as assembly and budding of a transport vesicle directed from the transitional ER to the Golgi or from the *trans* Golgi network to endosomes and lysosomes. Individual proteins in the complex (coat proteins, adaptor proteins, and GTP-binding proteins) may participate in assembly of transport vesicles directed elsewhere, or in vesicle fusion (described later in the chapter), but each protein complex is apparently unique to a particular budding, transport, or fusion pathway.

Three kinds of coated vesicles have been characterized: two types of **COP-coated vesicles, COPI** and **COPII** (COP indicates coat protein), and

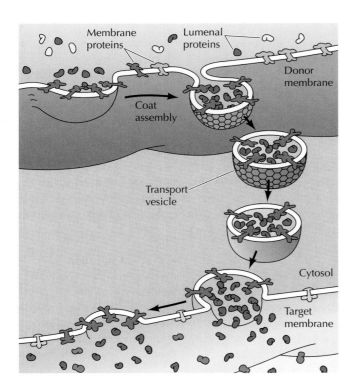

FIGURE 10.34 Formation and fusion of a transport vesicle   Membrane and lumenal secretory proteins are collected into selected regions of a donor membrane where the formation of a cytosolic coat results in the budding off of a transport vesicle. During transport the coat is disassembled, and the transport vesicle docks and fuses with the target membrane.

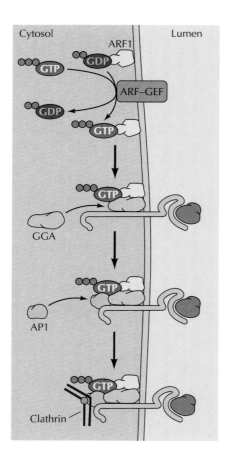

Cytosol   Lumen

ARF1

GTP   GDP

ARF–GEF

GDP

GTP

GTP

GGA

GTP

AP1

GTP

Clathrin

FIGURE 10.35 Initiation of a clathrin-coated vesicle by ARF1   The small GTP-binding protein, ARF1, can initiate the formation of a clathrin-coated vesicle on the *trans* Golgi membrane. After being delivered to the membrane, ARF/GDP is activated to ARF/GTP by a specific ARF-guanine nucleotide exchange factor (ARF-GEF). ARF/GTP recruits a GGA adaptor protein to the membrane and this protein recruits a transmembrane receptor with its bound lumenal cargo by interacting with the cytosolic tail of the receptor. GGA then recruits a second adaptor protein, AP1, which serves as a binding site for assembly of a clathrin coat.

clathrin-coated vesicles. COPII-coated vesicles carry secretory proteins from the ER to the ER-Golgi intermediate compartment or Golgi apparatus, budding from the transitional ER and carrying their cargo forward along the secretory pathway. COPI-coated vesicles bud from the ER-Golgi intermediate compartment, or the Golgi apparatus. COPI is the coat protein on vesicles moving from one Golgi cisternae to another during secretion and on the retrieval vesicles that return resident ER proteins marked by the KDEL or KKXX retrieval signals back to the ER from the ER-Golgi intermediate compartment or the *cis* Golgi network.

Clathrin-coated vesicles are the best understood and are responsible for the uptake of extracellular molecules from the plasma membrane by endocytosis (see Chapter 13) as well as the transport of molecules from the *trans* Golgi network to endosomes, lysosomes, or the plasma membrane. The formation of clathrin-coated vesicles on the *trans* Golgi network requires **clathrin**, the GTP-binding protein, ARF1, and at least two types of adaptor proteins (Figure 10.35). First, ARF/GDP binds to proteins in the Golgi membrane. An ARF-guanine nucleotide exchange factor in the membrane then stimulates the exchange of the GDP for GTP. ARF/GTP initiates the budding process by recruiting adaptor proteins, which then serve as binding sites for both transmembrane receptors and for clathrin. Clathrin actually plays a structural role in vesicle budding by assembling into a basketlike lattice structure that distorts the membrane and initiates the bud. This is illustrated for vesicles targeted to the lysosome in Figure 10.36. During transport, the GTP bound to ARF1 is hydrolyzed to GDP and the ARF/GDP is released from the membrane to recycle. Loss of ARF1 and the action of uncoating enzymes weakens the cooperative binding of the clathrin coat complex, allowing chaperone proteins in the cytoplasm to dissociate most of the coat from the vesicle membrane (see Figure 10.34).

While COPI- and COPII-coated vesicles have limited targets, clathrin-coated vesicles exit the *trans* Golgi for different destinations: endosomes, lysosomes, or different plasma membrane domains. Since these targets require specific cargoes, different adaptor proteins play a role in the assembly of vesicles for different destinations.

Vesicle Fusion

The fusion of a transport vesicle with its target involves two types of events. First, the transport vesicle must recognize the correct target membrane; for example, a vesicle carrying lysosomal enzymes has to deliver its cargo only to lysosomes. Second, the vesicle and target membranes must fuse, delivering the contents of the vesicle to the target organelle. Research over the last several years has led to the development of several models of vesicle fusion in which specific recognition between a vesicle and its target (tethering) is

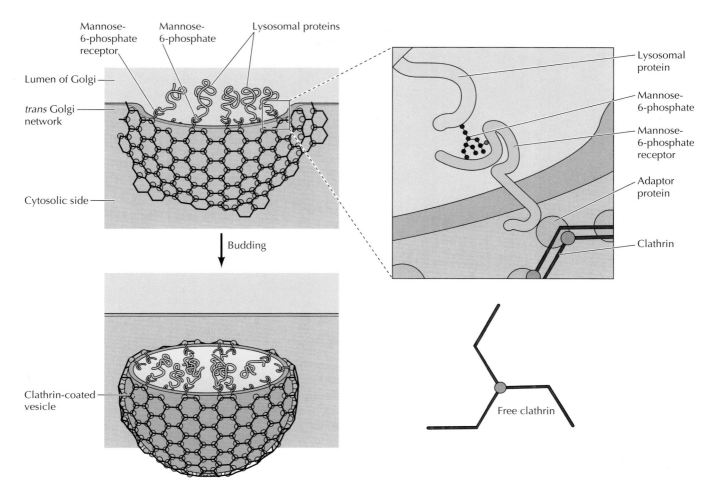

FIGURE 10.36 Incorporation of lysosomal proteins into clathrin-coated vesicles Proteins targeted for lysosomes are marked by mannose-6-phosphates, which bind to mannose-6-phosphate receptors in the *trans* Golgi network. The mannose-6-phosphate receptors span the Golgi membrane and serve as binding sites for cytosolic adaptor proteins, which in turn bind clathrin. Clathrins consist of three protein chains that associate with each other to form a basketlike lattice that distorts the membrane and drives vesicle budding.

mediated by interactions between proteins on the vesicle and the target membranes, followed by fusion between the phospholipid bilayers.

Proteins involved in vesicle fusion were initially identified in James Rothman's laboratory by biochemical analysis of reconstituted vesicular transport systems from mammalian cells. Analysis of the proteins involved in vesicle fusion in these systems led Rothman and his colleagues to propose a general model, called the **SNARE hypothesis**, in which vesicle fusion is mediated by interactions between specific pairs of transmembrane proteins, called SNAREs, on the vesicle and target membranes (v-SNAREs and t-SNAREs, respectively). According to the hypothesis, the formation of complexes between v-SNAREs on the vesicle and t-SNAREs on the target membranes leads to membrane fusion. This hypothesis was supported by the identification of SNAREs that were present on synaptic vesicles and by the finding of yeast secretion mutants that appeared to encode SNAREs

TABLE 10.1 Rab GTP-Binding Proteins and Their Sites of Action

Transport step	Rab proteins involved
Exocytosis	
Transitional ER to Golgi	Rab1, Rab1b, Rab2
Golgi back to ER	Rab6, Rab6b
Intra-Golgi	Rab1, Rab6, Rab6b
trans Golgi network to plasma membrane	Rab11a, Rab11b
Endocytosis	
Plasma membrane to early endosome	Rab5a, Rab5b, Rab5c
Early endosome to plasma membrane	Rab4, Rab15, Rab18
Early endosome to late endosome	Rab7
Special roles	
Exocytosis of secretory granules	Rab8b
Late endosome to *trans* Golgi network	Rab9, Rab11a, Rab11b
trans Golgi network to basolateral membrane	Rab8a
trans Golgi network to apical membrane	Rab21

Examples of the more than 60 mammalian Rab proteins whose locations and presumptive functions are known.

> ■ Griscelli syndrome is a rare disease caused by mutations in the gene encoding Rab27a. Rab27a appears to play a key role in the transport of pigment-containing vesicles (melanosomes) to the skin and hair and in the exocytosis of vesicles in T lymphocytes. Patients with Griscelli syndrome exhibit partial albinism (lack of pigment) and are immunodeficient.

required for a variety of vesicle transport events. For example, transport from the ER to the Golgi in yeast requires specific SNAREs that are located on both the vesicle and target membranes.

Recent research has confirmed that SNAREs are required for vesicle fusion with a target membrane and that SNARE-SNARE pairing provides the energy to bring the two bilayers sufficiently close to destabilize them and result in fusion. However, the docking, tethering, and fusion of transport vesicles to specific target membranes appears to be mediated by a sequentially assembled protein complex much like that which led to transport vesicle budding. Members of the **Rab** family of small GTP-binding proteins play key roles in this docking of transport vesicles. Rab proteins, like the ARF family, participate in many of the vesicle budding and fusion reactions during vesicular transport. More than 60 different Rab proteins have been identified and shown to function in specific vesicle transport processes (Table 10.1). They function in many steps of vesicle trafficking, including interacting with SNAREs to regulate and facilitate the formation of SNARE/SNARE complexes.

Individual Rab proteins or combinations of Rab proteins mark different organelles and transport vesicles, so their localization on the correct membrane is key to establishing the specificity of vesicular transport (Figure 10.37). The Rab proteins are carried through the cytosol in their GDP-bound form by GDP-dissociation inhibitors (GDIs). At a membrane, they are

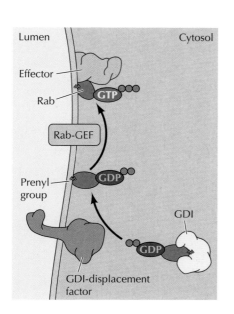

FIGURE 10.37 Delivery of Rab to a membrane The small GTP-binding protein Rab is modified by the addition of a prenyl group, which allows it to insert into a membrane (see Figure 8.33). Rab is carried in the cytosol bound to a GDP–dissociation inhibitor (GDI) keeping it in the Rab/GDP state. At a membrane, a non-specific GDI-displacement factor can remove the Rab/GDP from the GDI and insert it into the membrane. If a specific Rab-guanine nucleotide exchange factor (Rab-GEF) is present, the GDP on Rab will be exchanged for GTP and the active Rab/GTP can interact with effector proteins. If the correct Rab-guanine nucleotide exchange factor is not present, the Rab/GDP will be removed by a GDI and carried to another membrane.

FIGURE 10.38 Vesicle fusion Vesicle fusion is initiated by Rab/GTP. Specific Rab proteins on the vesicle and target membranes bind effector proteins to tether the vesicle to the target membrane. Tethering allows the v-SNAREs and t-SNAREs to interact. The coiled-coil domains of the SNAREs zip together, providing energy to bring the vesicle and target membranes into close proximity. This close proximity of the membranes destabilizes the lipid bilayers, and the vesicle and target membranes fuse. The changes in protein-protein interactions recruit NSF and SNAPs to the SNARE complex, and they disassemble the complex using energy from the hydrolysis of ATP.

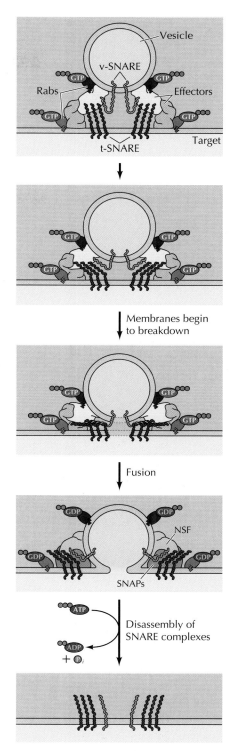

removed from GDIs by GDI-displacement factors. Specific guanine nucleotide exchange factors then convert Rab/GDP to the active Rab/GTP state. Individual guanine nucleotide exchange factors are localized to specific membranes and act on specific members of the Rab family, so they are, in part, responsible for formation of active Rab/GTP complexes at the correct membrane sites. In the absence of the appropriate exchange factor, Rab proteins remain as Rab/GDP and are removed from the membrane. Thus full activation of a Rab protein requires that it both bind GTP and be associated with a membrane.

To initiate transport vesicle fusion, Rab/GTP on the transport vesicle interacts with effector proteins and v-SNAREs to assemble a pre-fusion complex (Figure 10.38). A different Rab protein on the target membrane similarly organizes other effector proteins and t-SNAREs. When the transport vesicle encounters this target membrane, the effector proteins link the membranes by protein-protein interactions. This tethering of the vesicle to the target membrane stimulates Rab/GTP hydrolysis and allows the v-SNAREs to contact the t-SNAREs. All SNARE proteins have a long central coiled-coil domain like that found in nuclear lamins (see Figure 9.4). As in the lamins, this domain binds strongly to other coiled-coil domains and, in effect, zips the SNAREs together, bringing the two membranes into nearly direct contact. The simplest hypothesis is that this creates instability in the lipid bilayers and they fuse. Following membrane fusion, the NSF/SNAP complex disassembles the SNARE complex, allowing the SNAREs to be reused for subsequent rounds of vesicle transport. As the energy of SNARE-SNARE interaction drives the fusion of the membranes, energy from hydrolysis of ATP is required to separate the SNAREs.

Membrane fusion is a general process that occurs whenever a transport vesicle fuses with a target membrane. However, specific types of fusion may involve specialized sites on the plasma membrane. One of these is exocytosis, the fusion of a transport vesicle with the plasma membrane, resulting in secretion of the vesicle contents. Many types of exocytosis occur at specific protein complexes, called **exocysts,** on the plasma membrane. This eight protein complex was first discovered to be required for secretion in the yeast *Saccharomyces cerevisiae,* but it also plays an important role in secretion in polarized mammalian cells. The structure of exocysts is not well understood but their assembly appears to require sequential interactions among eight exocyst proteins localized on both the transport vesicle and the target membrane site (Figure 10.39). Interaction of these proteins results in efficient targeting of the transport vesicle to a specific location on the plasma membrane. Several small GTP-binding proteins are also associated with exocysts; some like the Rab proteins are involved in vesicle docking and fusion but others may play a role in localizing exocysts to apical or basolateral membranes or to axons or dendrites.

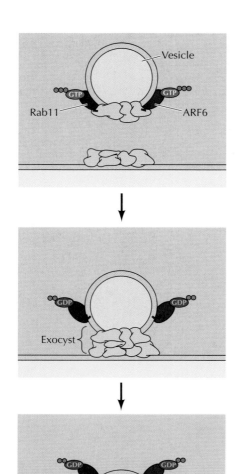

FIGURE 10.39 Exocyst assembly and vesicle targeting Exocysts are complexes of eight different proteins formed during exocytosis from proteins present on both transport vesicles and specific regions of the plasma membrane. Tethering and docking at exocysts results in normal SNARE-mediated vesicle fusion. Small GTP-binding proteins including Rab11 and ARF6 regulate assembly of the exocyst complex on the transport vesicle and coordinate its movement to the target site.

Lysosomes

Lysosomes are membrane-enclosed organelles that contain an array of enzymes capable of breaking down all types of biological polymers—proteins, nucleic acids, carbohydrates, and lipids. Lysosomes function as the digestive system of the cell, serving both to degrade material taken up from outside the cell and to digest obsolete components of the cell itself. In their simplest form, lysosomes are visualized as dense spherical vacuoles, but they can display considerable variation in size and shape as a result of differences in the materials that have been taken up for digestion (Figure 10.40). Lysosomes thus represent morphologically diverse organelles defined by the common function of degrading intracellular material.

Lysosomal Acid Hydrolases

Lysosomes contain about 50 different degradative enzymes that can hydrolyze proteins, DNA, RNA, polysaccharides, and lipids. Mutations in the genes that encode these enzymes are responsible for more than 30 different human genetic diseases, which are called **lysosomal storage diseases** because undegraded material accumulates within the lysosomes of affected individuals. Most of these diseases result from deficiencies in single lysosomal enzymes. For example, Gaucher disease (the most common of these disorders) results from a mutation in the gene that encodes a lysosomal enzyme required for the breakdown of glycolipids. An intriguing exception is I-cell disease, which is caused by a deficiency in the enzyme that catalyzes the first step in the tagging of lysosomal enzymes with mannose-6-phosphate in the Golgi apparatus (see Figure 10.29). The result is a general failure of lysosomal enzymes to be incorporated into lysosomes.

Most lysosomal enzymes are acid hydrolases, which are active at the acidic pH (about 5) that is maintained within lysosomes but not at the neutral pH (about 7.2) characteristic of the rest of the cytoplasm (Figure 10.41). The requirement of these lysosomal hydrolases for acidic pH provides double protection against uncontrolled digestion of the contents of the cytosol; even if the lysosomal membrane were to break down, the released acid hydrolases would be inactive at the neutral pH of the cytosol. To maintain their acidic internal pH, lysosomes must actively concentrate H^+ ions (protons). This is accomplished by a proton pump in the lysosomal membrane, which actively transports protons into the lysosome from the cytosol. This pumping requires expenditure of energy in the form of ATP hydrolysis in order to maintain approximately a hundredfold higher H^+ concentration inside the lysosome.

Endocytosis and Lysosome Formation

One of the major functions of lysosomes is the digestion of material taken up from outside the cell by **endocytosis**, which is discussed in detail in Chapter 13. However, the role of lysosomes in the digestion of material taken up by endocytosis relates not only to the function of lysosomes but

■ The antimalarial drug chloroquine is uncharged and able to cross membranes at physiological pH. At acidic pH, chloroquine is positively charged and accumulates within the parasite's digestive vacuoles (analogous to lysosomes). The high concentration of chloroquine within the digestive vacuoles is thought to be responsible for its activity against the parasite.

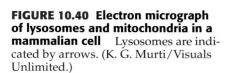

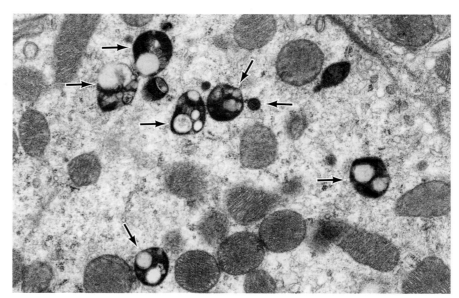

0.5 μm

FIGURE 10.40 Electron micrograph of lysosomes and mitochondria in a mammalian cell Lysosomes are indicated by arrows. (K. G. Murti/Visuals Unlimited.)

also to their formation. In particular, lysosomes are formed when transport vesicles from the *trans* Golgi network fuse with endosomes, which contain molecules taken up by endocytosis at the plasma membrane.

The formation of **endosomes** and lysosomes thus represents an intersection between the secretory pathway through which lysosomal proteins are processed, and the endocytic pathway through which extracellular molecules are taken up at the cell surface (Figure 10.42). Material from outside the cell is taken up in clathrin-coated endocytic vesicles, which bud from the plasma membrane and then fuse with early endosomes. Membrane components are then recycled to the plasma membrane (discussed in detail in Chapter 13) and the early endosomes gradually mature into late endosomes, which are the precursors to lysosomes. One of the important changes during endosome maturation is the lowering of the internal pH to about 5.5, which plays a key role in the delivery of lysosomal acid hydrolases from the *trans* Golgi network.

As discussed earlier, acid hydrolases are targeted to lysosomes by mannose-6-phosphate residues, which are recognized by mannose-6-phosphate receptors in the *trans* Golgi network and packaged into clathrin-coated vesicles. Following removal of the clathrin coat, these transport vesicles fuse with late endosomes, and the acidic internal pH causes the hydrolases to dissociate from the mannose-6-phosphate receptor (see Figure 10.42). The hydrolases are thus released into the lumen of the endosome, while the receptors remain in the membrane and are eventually recycled to the Golgi. Late endosomes then mature into lysosomes as they acquire a full complement of acid hydrolases, which digest the molecules originally taken up by endocytosis.

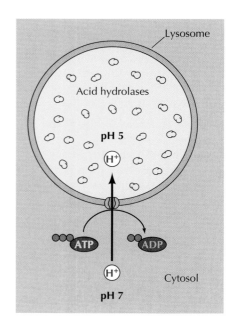

FIGURE 10.41 Organization of the lysosome Lysosomes contain a variety of acid hydrolases that are active at the acidic pH maintained within the lysosome but not at the neutral pH of the cytosol. The acidic internal pH of lysosomes results from the action of a proton pump in the lysosomal membrane, which imports protons from the cytosol coupled to ATP hydrolysis.

MOLECULAR MEDICINE

Gaucher Disease

The Disease

Gaucher disease is the most common of the lysosomal storage diseases, which are caused by a failure of lysosomes to degrade substances that they normally break down. The resulting accumulation of nondegraded compounds leads to an increase in the size and number of lysosomes within the cell, eventually resulting in cellular malfunction and pathological consequences to affected organs. Gaucher disease is found primarily in the Jewish population, where it has a frequency of about 1 in 2500 individuals. There are three types of Gaucher disease, which differ in severity and nervous system involvement. In the most common form of the disease (type I), the nervous system is not involved; the disease is manifest as spleen and liver enlargement and development of bone lesions. Many patients with this form of the disease have no serious symptoms, and their life span is unaffected. The more severe forms of the disease (types II and III) are much rarer and found in both Jewish and non-Jewish populations. The most devastating is type II disease in which extensive neurological involvement is evident in infancy, and patients die early in life. Type III disease, intermediate in severity between types I and II, is characterized by the onset of neurological symptoms (including dementia and spasticity) by about age ten.

Molecular and Cellular Basis

Gaucher disease is caused by a deficiency of the lysosomal enzyme glucocerebrosidase, which catalyzes the hydrolysis of glucocerebroside to glucose and ceramide (see figure). This enzyme deficiency was demonstrated in 1965, and the responsible gene was cloned in 1985. Since then more than 30 different mutations responsible for Gaucher disease have been identified. Interestingly, the severity of the disease can be largely predicted from the

nature of these mutations. For example, patients with a mutation leading to the relatively conservative amino acid substitution of serine for asparagine have type I disease; patients with a mutation leading to substitution of proline for leucine have more severe enzyme deficiencies and develop either type II or III disease.

Except for the very rare type II and III forms of the disease, the only cells affected in Gaucher disease are macrophages. Because their function is to eliminate aged and damaged cells by phagocytosis, macrophages continually ingest large amounts of lipids, which are normally degraded in lysosomes. Deficiencies of glucocerebrosidase are therefore particularly evident in macrophages of both the spleen and the liver, consistent with these organ abnormalities being the primary manifestation in most cases of Gaucher disease.

Prevention and Treatment

Gaucher disease is a prime example of a disease that can be treated by enzyme replacement therapy in which exogenous administration of an enzyme is used to correct an enzyme defect. This approach to treatment of lysosomal storage diseases was suggested by Christian de Duve in the 1960s, based on the idea that exogenously administered enzymes might be taken up by endocytosis and trans-

ported to lysosomes. In type I Gaucher disease, this approach is particularly attractive because the single target cell is the macrophage. In the 1970s it was discovered that macrophages express cell surface receptors that bind mannose residues on extracellular glycoproteins and then internalize these proteins by endocytosis. This finding suggested that exogenously administered glucocerebrosidase could be specifically targeted to macrophages by modifications that would expose mannose residues. Enzyme prepared from human placenta was appropriately modified, and clinical studies have clearly demonstrated its effectiveness in the treatment of Gaucher disease.

Unfortunately, enzyme replacement therapy for Gaucher disease is expensive. The expense of this treatment puts it far beyond the resources of individual patients and has raised a number of societal issues concerning the cost of expensive drugs for the treatment of rare disorders.

References

Futerman, A. H. and G. van Meer. 2004. The cell biology of lysosomal storage disorders. *Nat. Rev. Mol. Cell Biol.* 5: 554–565.

Jmoudiak, M. and A. H. Futerman. 2005. Gaucher disease: Pathological mechanisms and modern management. *Br. J. Haematol.* 129: 178–188.

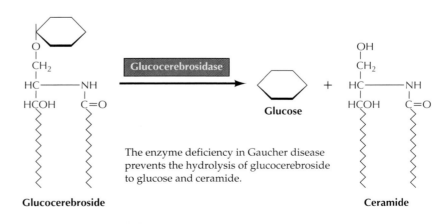

The enzyme deficiency in Gaucher disease prevents the hydrolysis of glucocerebroside to glucose and ceramide.

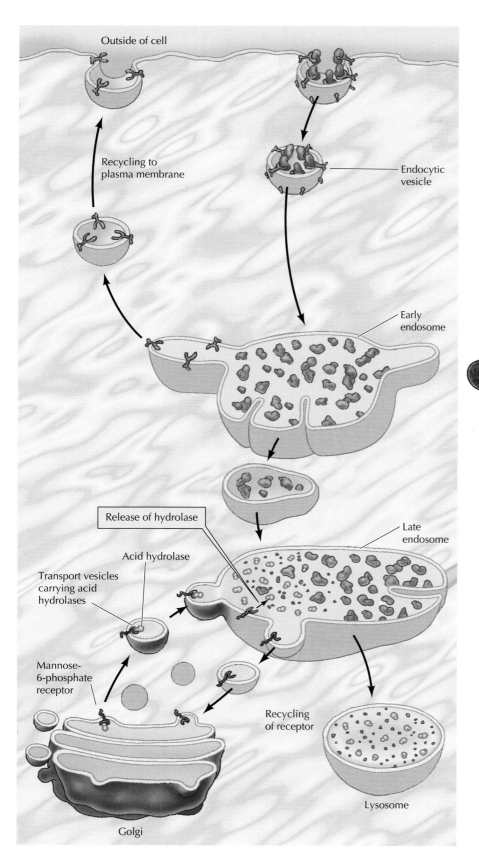

Outside of cell

Recycling to
plasma membrane

Endocytic
vesicle

Early
endosome

Release of hydrolase

Acid hydrolase

Late
endosome

Transport vesicles
carrying acid
hydrolases

Mannose-
6-phosphate
receptor

Recycling
of receptor

Golgi

Lysosome

FIGURE 10.42 Endocytosis and lysosome formation Molecules are taken up from outside the cell in endocytic vesicles, which fuse with early endosomes. Membrane components are recycled as the early endosomes mature into late endosomes. Transport vesicles carrying acid hydrolases from the Golgi apparatus then fuse with late endosomes, which mature into lysosomes as they acquire a full complement of lysosomal enzymes. The acid hydrolases dissociate from the mannose-6-phosphate receptors when the transport vesicles fuse with late endosomes, and the receptors are recycled to the Golgi apparatus.

13.3 WEBSITE ANIMATION

Endocytosis In endocytosis, the plasma membrane pinches off small vesicles (pinocytosis) or large vesicles (phagocytosis) to take in extracellular materials.

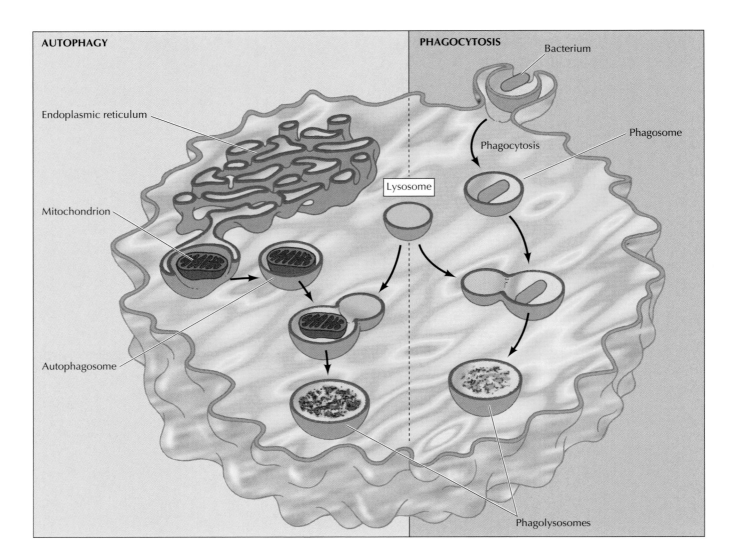

FIGURE 10.43 Lysosomes in phago-cytosis and autophagy In phagocy-tosis, large particles (such as bacteria) are taken up into phagocytic vacuoles or phagosomes. In autophagy, internal organelles (such as mitochondria) are enclosed by membrane fragments from the ER, forming autophagosomes. Both phagosomes and autophagosomes fuse with lysosomes to form large phago-lysosomes in which their contents are digested.

Phagocytosis and Autophagy

In addition to degrading molecules taken up by endocytosis, lysosomes digest material derived from two other routes: phagocytosis and autophagy (Figure 10.43). In **phagocytosis**, specialized cells, such as macrophages, take up and degrade large particles, including bacteria, cell debris, and aged cells that need to be eliminated from the body. Such large particles are taken up in phagocytic vacuoles (**phagosomes**), which then fuse with lysosomes, resulting in digestion of their contents. The lysosomes formed in this way (**phagolysosomes**) can be quite large and heterogeneous since their size and shape is determined by the content of material that is being digested.

Lysosomes are also responsible for **autophagy**, the gradual turnover of the cell's own components (discussed in Chapter 8). In contrast to phagocy-tosis, autophagy is a function of all cells and is critical at certain stages of embryonic development when it allows rapid reallocation of the cell's resources. The first step of autophagy appears to be the enclosure of an organelle (e.g., a mitochondrion) in membrane derived from the ER. The resulting vesicle (an **autophagosome**) then fuses with a lysosome, and its contents are digested (see Figure 10.43).

SUMMARY	KEY TERMS

THE ENDOPLASMIC RETICULUM

The Endoplasmic Reticulum and Protein Secretion: The endoplasmic reticulum is the first branch point in protein sorting. In mammalian cells, proteins destined for secretion, lysosomes, or the plasma membrane are translated on membrane-bound ribosomes and transferred into the rough ER as their translation proceeds.

endoplasmic reticulum (ER), rough ER, transitional ER, smooth ER, secretory vesicle, secretory pathway

Targeting Proteins to the Endoplasmic Reticulum: Proteins can be targeted to the ER either while their translation is still in progress or following completion of translation in the cytosol. In mammalian cells, most proteins are translocated into the ER while they are being translated on membrane-bound ribosomes. Ribosomes engaged in the synthesis of secreted proteins are targeted to the endoplasmic reticulum by signal sequences at the amino terminus of the polypeptide chain. Growing polypeptide chains are then translocated into the ER through protein channels and released into the ER lumen by cleavage of the signal sequence.

signal sequence, microsome, signal recognition particle (SRP), srpRNA, SRP receptor, translocon, signal peptidase

Insertion of Proteins into the ER Membrane: Integral membrane proteins of the plasma membrane or the membranes of the ER, Golgi apparatus, and lysosomes are initially inserted into the membrane of the ER. Rather than being translocated into the ER lumen, these proteins are anchored by membrane-spanning α helices that stop the transfer of the growing polypeptide chain across the membrane.

Protein Folding and Processing in the ER: Polypeptide chains are folded into their correct three-dimensional conformations within the ER. The ER is also the site of *N*-linked glycosylation and addition of GPI anchors.

protein disulfide isomerase, glycosylphosphatidylinositol (GPI) anchor

Quality Control in the ER: Many secretory proteins are not folded correctly the first time. Chaperones detect incorrectly folded proteins and recycle them through the folding pathway. Those proteins that cannot be correctly folded are diverted from the secretory pathway and marked for degradation.

unfolded protein response

The Smooth ER and Lipid Synthesis: The ER is the major site of lipid synthesis in eukaryotic cells, and the smooth ER is abundant in cells that are active in lipid metabolism and detoxification of lipidsoluble drugs.

flippase

Export of Proteins and Lipids from the ER: Proteins and lipids are transported in vesicles from the ER to the Golgi apparatus. Resident ER proteins are marked by sequences that signal their return from the Golgi to the ER by a recycling pathway. Other targeting sequences mediate the selective packaging of exported proteins into vesicles that transport them to the Golgi.

THE GOLGI APPARATUS

Organization of the Golgi: The Golgi apparatus functions in protein processing and sorting as well as in the synthesis of lipids and polysaccharides. Proteins are transported from the endoplasmic reticulum to the *cis* Golgi network. From there they are transported to the Golgi stack, which represents the site of most metabolic activities of the Golgi apparatus. Modified proteins are transported from the Golgi stack to the *trans* Golgi network, where they are sorted and packaged in vesicles for transport to endosomes, lysosomes, the plasma membrane, or the exterior of the cell.

Golgi apparatus, Golgi complex, *cis* Golgi network, Golgi stack, *trans* Golgi network

KEY TERMS

mannose-6-phosphate, glycosyltransferase, glycosidase, signal patch

apical domain, basolateral domain, vacuole

synaptic vesicle

COP-coated vesicle, COPI, COPII, clathrin-coated vesicle, clathrin

SNARE hypothesis, Rab, exocyst

lysosome, lysosomal storage disease

endocytosis, endosome

phagocytosis, phagosome, phagolysosome, autophagy, autophagosome

SUMMARY

Protein Glycosylation within the Golgi: The *N*-linked oligosaccharides added to proteins in the ER are modified within the Golgi. Those proteins destined for lysosomes are specifically phosphorylated on mannose residues, and mannose-6-phosphate serves as a targeting signal that directs their transport to lysosomes from the *trans* Golgi network. *O*-linked glycosylation also takes place within the Golgi.

Lipid and Polysaccharide Metabolism in the Golgi: The Golgi apparatus is the site of synthesis of glycolipids, sphingomyelin, and the complex polysaccharides of plant cell walls.

Protein Sorting and Export from the Golgi Apparatus: Proteins are sorted in the *trans* Golgi network for packaging into transport vesicles targeted for secretion, the plasma membrane, lysosomes, or yeast and plant vacuoles. In polarized cells, proteins are specifically targeted to the apical and basolateral domains of the plasma membrane.

THE MECHANISM OF VESICULAR TRANSPORT

Experimental Approaches to Understanding Vesicular Transport: The mechanism of vesicular transport has been elucidated through studies of yeast mutants, reconstituted cell-free systems, and synaptic vesicles.

Cargo Selection, Coat Proteins, and Vesicle Budding: The cytoplasmic surfaces of vesicles are coated with proteins that drive vesicle budding and select the specific molecules to be transported.

Vesicle Fusion: Vesicle binding to the correct target membrane is mediated by interactions between pairs of transmembrane proteins, which leads to membrane fusion. Some types of fusion with the plasma membrane (exocytosis) occur at specific multiprotein complexes called exocysts.

LYSOSOMES

Lysosomal Acid Hydrolases: Lysosomes contain an array of acid hydrolases that degrade proteins, nucleic acids, polysaccharides, and lipids. These enzymes function specifically at the acidic pH maintained within lysosomes.

Endocytosis and Lysosome Formation: Extracellular molecules taken up by endocytosis are transported to endosomes, which mature to lysosomes as lysosomal acid hydrolases are delivered from the Golgi.

Phagocytosis and Autophagy: Lysosomes are responsible for the degradation of large particles taken up by phagocytosis and for the gradual digestion of the cell's own components by autophagy.

Questions

1. What was the original experimental evidence for the secretory pathway from rough ER → Golgi apparatus → secretory vesicles → secreted protein?

2. How did *in vitro* translation of mRNAs provide evidence for the existence of a signal sequence that targets secretory proteins to the rough endoplasmic reticulum?

3. Compare and contrast cotranslational and posttranslational translocation of polypeptide chains into the endoplasmic reticulum.

4. Sec61 is a critical component of the protein channel through the ER membrane. In Sec61 mutant yeast, what is the fate of proteins that are normally localized to the Golgi apparatus?

5. Why are the carbohydrate groups of glycoproteins always exposed on the surface of the cell?

6. What would be the effect of mutating the KDEL sequence of a resident ER protein like BiP? Would this effect be similar or different from that of mutating the KDEL receptor protein?

7. How is a lysosomal protein targeted to a lysosome? What effect would the addition of a lysosome-targeting signal patch have on the subcellular localization of a protein that is normally cytosolic? How would it affect localization of a protein that is normally secreted?

8. What is the predicted fate of lysosomal acid hydrolases in I-cell disease in which cells are deficient in the enzyme required for formation of mannose-6-phosphate residues?

9. What processes result in glycolipids and sphingomyelin being found in the outer—but not the inner—half of the plasma membrane bilayer?

10. What experimental evidence demonstrated vesicular transport of proteins between Golgi cisternae?

11. A patient comes to your clinic with an accumulation of glucocerebrosides in macrophage lysosomes. What is your diagnosis, and what therapy would you suggest if price is not a limiting factor?

12. Lysosomes contain powerful hydrolytic enzymes, which are transported there from the site of their synthesis in the ER via the Golgi apparatus. Why don't these enzymes damage the constituents of these organelles?

13. What is the source of energy for fusion between target and vesicle membranes?

14. Why does activation of Rab proteins require association with a membrane?

References and Further Reading

The Endoplasmic Reticulum

Bickford, L. C., E. Mossessova, and J. Goldberg. 2004. A structural view of the COPII vesicle coat. *Curr. Opin. Struct. Biol.* 14: 147–153. [R]

Blobel, G. and B. Dobberstein. 1975. Transfer of proteins across the membrane. I. Presence of proteolytically processed and unprocessed nascent immunoglobulin light chains on membrane-bound ribosomes of murine myeloma. *J. Cell Biol.* 67: 835–851. [P]

Ellgaard, L. and A. Helenius. 2003. Quality control in the endoplasmic reticulum. *Nat. Rev. Mol. Cell Biol.* 4: 181–191. [R]

Glick, B. S. 2001. ER export: More than one way out. *Curr. Biol.* 11: R361–R363. [R]

Halic, M., T. Becker, M. R. Pool, C. M. Spahn, R. A. Grassucci, J. Frank and R. Beckmann. 2004. Structure of the signal recognition particle interacting with the elongation-arrested ribosome. *Nature* 427: 808–814. [P]

Halic, M. and R. Beckmann. 2005. The signal recognition particle and its interactions during protein targeting. *Curr. Opin. Struct. Biol.* 15: 116–125. [R]

Haucke, V. 2003. Vesicle budding: A coat for the COPs. *Trends Cell Biol.* 13: 59–60. [R]

Kent, C. 1995. Eukaryotic phospholipid biosynthesis. *Ann. Rev. Biochem.* 64: 315–343. [R]

Kleizen, B. and I. Braakman. 2004. Protein folding and quality control in the endoplasmic reticulum. *Curr. Opin. Cell Biol.* 16: 343–349. [R]

Kol, M. A., A. I. de Kroon, J. A. Killian and B. de Kruijff. 2004. Transbilayer movement of phospholipids in biogenic membranes. *Biochemistry* 43: 2673–2681. [R]

Lee, M. C., E. A. Miller, J. Goldberg, L. Orci and R. Schekman. 2004. Bi-directional protein transport between the ER and Golgi. *Ann. Rev. Cell Dev. Biol.* 20: 87–123. [R]

Lippincott-Schwartz, J., T. H. Roberts and K. Hirschberg. 2000. Secretory protein trafficking and organelle dynamics in living cells. *Ann. Rev. Cell Dev. Biol.* 16: 557–589. [R]

Mackinnon, R. 2005. Structural biology. Membrane protein insertion and stability. *Science* 307: 1425–1426. [R]

Menon, A. K. 1995. Flippases. *Trends Cell Biol.* 5: 355–360. [R]

Powers, T. and P. Walter. 1997. A ribosome at the end of the tunnel. *Science* 278: 2072–2073. [R]

Presley, J. F., N. B. Cole, T. A. Schroer, K. Hirschberg, K. J. Zaal and J. Lippincott-Schwartz. 1997. ER-to-Golgi transport visualized in living cells. *Nature* 389: 81–85. [P]

Rutkowski, D. T. and R. J. Kaufman. 2004. A trip to the ER: Coping with stress. *Trends Cell Biol.* 14: 20–28. [R]

Schroder, M. and R. J. Kaufman. 2005. The mammalian unfolded protein response. *Ann. Rev. Biochem.* 74: 739–789. [R]

Udenfriend, S. and K. Kodukula. 1995. How glycosylphosphatidylinositol-anchored membrane proteins are made. *Ann. Rev. Biochem.* 64: 563–591. [R]

Watanabe, R. and H. Riezman. 2004. Differential ER exit in yeast and mammalian cells. *Curr. Opin. Cell Biol.* 16: 350–355. [R]

The Golgi Apparatus

Altan-Bonnet, N., R. Sougrat and J. Lippincott-Schwartz. 2004. Molecular basis for Golgi maintenance and biogenesis. *Curr. Opin. Cell Biol.* 16: 364–372. [R]

Baranski, T. J., P. L. Faust and S. Kornfeld. 1990. Generation of a lysosomal enzyme targeting signal in the secretory protein pepsinogen. *Cell* 63: 281–291. [P]

Barr, F. A. 2002. The Golgi apparatus: Going round in circles? *Trends Cell Biol.* 12: 101–104. [R]

Brodsky, F. M., C. Y. Chen, C. Knuehl, M. C. Towler and D. E. Wakeham. 2001. Biological basket weaving: Formation and function of clathrin-coated vesicles. *Ann. Rev. Cell Dev. Biol.* 17: 517–568. [R]

Chrispeels, M. J. and N. V. Raikhel. 1992. Short peptide domains target proteins to plant vacuoles. *Cell* 68: 613–616. [R]

Conibear, E. and T. H. Stevens. 1995. Vacuolar biogenesis in yeast: Sorting out the sorting proteins. *Cell* 83: 513–516. [R]

de Graffenried, C. L. and C. R. Bertozzi. 2004. The roles of enzyme localisation and complex formation in glycan assembly within the Golgi apparatus. *Curr. Opin. Cell Biol.* 16: 356–363. [R]

Folsch, H., H. Ohno, J. S. Bonifacino and I. Mellman. 1999. A novel clathrin adaptor complex mediates basolateral targeting in polarized epithelial cells. *Cell* 99: 189–196. [P]

Fries, E., and J. E. Rothman. 1980. Transport of vesicular stomatitis virus glycoprotein in a cell-free extract. *Proc. Natl. Acad. Sci. U.S.A.* 77: 3870–3874. [P]

Glick, B. S. 2002. Can the Golgi form *de novo*? *Nat. Rev. Mol. Cell Biol.* 3: 615–619. [R]

Kirchhausen, T. 2002. Clathrin adaptors really adapt. *Cell* 109: 413–416. [R]

Kornfeld, R. and S. Kornfeld. 1985. Assembly of asparagine-linked oligosaccharides. *Ann. Rev. Biochem.* 54: 631–664. [R]

Pearse, B. M. 1975. Coated vesicles from pig brain: Purification and biochemical characterization. *J. Mol. Biol.* 97: 93–98. [P]

Tooze, S. A., G. J. Martens and W. B. Huttner. 2001. Secretory granule biogenesis: Rafting to the SNARE. *Trends Cell Biol.* 11: 116–122. [R]

The Mechanism of Vesicular Transport

Balch, W. E., W. G. Dunphy, W. A. Braeli and J. E. Rothman. 1984. Reconstitution of the transport of protein between successive compartments of the Golgi measured by the coupled incorporation of *N*-acetylglucosamine. *Cell* 39: 405–416. [P]

Bernards, A. and J. Settleman. 2004. GAP control: Regulating the regulators of small GTPases. *Trends Cell Biol.* 14: 377–385. [R]

Bonifacino, J. S., and B. S. Glick, 2004. The mechanisms of vesicle budding and fusion. *Cell* 116: 153–166. [R]

Camonis, J. H. and M. A. White. 2005. Ral GTPases: Corrupting the exocyst in cancer cells. *Trends Cell Biol.* 15: 327–332. [R]

Chen, Y. A. and R. H. Scheller. 2001. SNARE-mediated membrane fusion. *Nat. Rev. Mol. Cell Biol.* 2: 98–106. [R]

Gruenberg, J. 2001. The endocytic pathway: A mosaic of domains. *Nat. Rev. Mol. Cell Biol.* 2: 721–730. [R]

Hsu, S. C., D. TerBush, M. Abraham and W. Guo. 2004. The exocyst complex in polarized exocytosis. *Int. Rev. Cytol.* 233: 243–265. [R]

Jahn, R., T. Lang and T. C. Sudhof. 2003. Membrane fusion. *Cell* 112: 519–533. [R]

Maxfield, F. R. and T. E. McGraw. 2004. Endocytic recycling. *Nat. Rev. Mol. Cell Biol.* 5: 121–132. [R]

Mayor, S. and H. Riezman. 2004. Sorting GPI-anchored proteins. *Nat. Rev. Mol. Cell Biol.* 5: 110–120. [R]

McMahon, H. T. and I. G. Mills. 2004. COP and clathrin-coated vesicle budding: Different pathways, common approaches. *Curr. Opin. Cell Biol.* 16: 379–391. [R]

Novick, P., C. Field and R. Schekman. 1980. Identification of 23 complementation groups required for post-translational events in the yeast secretory pathway. *Cell* 21: 205–215. [P]

Owen, D. J., B. M. Collins and P. R. Evans. 2004. Adaptors for clathrin coats: Structure and function. *Ann. Rev. Cell Dev. Biol.* 20: 153–191. [R]

Palmer, K. J. and D. J. Stephens. 2004. Biogenesis of ER-to-Golgi transport carriers: Complex roles of COPII in ER export. *Trends Cell Biol.* 14: 57–61. [R]

Pelham, H. R. 2001. SNAREs and the specificity of membrane fusion. *Trends Cell Biol.* 11: 99–101. [R]

Pfeffer, S. 2003. Membrane domains in the secretory and endocytic pathways. *Cell* 112: 507–517. [R]

Rizo, J. and T. C. Sudhof. 2002. Snares and Munc18 in synaptic vesicle fusion. *Nat. Rev. Neurosci.* 3: 641–653. [R]

Robinson, M. S. 2004. Adaptable adaptors for coated vesicles. *Trends Cell Biol.* 14: 167–174. [R]

Segev, N. 2001. Ypt and Rab GTPases: Insight into functions through novel interactions. *Curr. Opin. Cell Biol.* 13: 500–511. [R]

Söllner, T., S. W. Whiteheart, M. Brunner, H. Erdjument-Bromage, S. Geromanos, P. Tempst and J. E. Rothman. 1993. SNAP receptors implicated in vesicle targeting and fusion. *Nature* 362: 318–324. [P]

Sorensen, J. B. 2005. SNARE complexes prepare for membrane fusion. *Trends Neurosci.* 28: 453–455. [R]

Sorkin, A. 2004. Cargo recognition during clathrin-mediated endocytosis: A team effort. *Curr. Opin. Cell Biol.* 16: 392–399. [R]

Lysosomes

Cuervo, A. M. 2004. Autophagy: In sickness and in health. *Trends Cell Biol.* 14: 70–77. [R]

Forgac, M. 1999. Structure and properties of the vacuolar (H^+)-ATPases. *J. Biol. Chem.* 274: 12951–12954. [R]

Fukuda, M. 1991. Lysosomal membrane glycoproteins. *J. Biol. Chem.* 266: 21327–21330. [R]

Ghosh, P., N. M. Dahms and S. Kornfeld. 2003. Mannose 6-phosphate receptors: New twists in the tale. *Nat. Rev. Mol. Cell Biol.* 4: 202–213. [R]

Neufeld, E. F. 1991. Lysosomal storage diseases. *Ann. Rev. Biochem.* 60: 257–280. [R]

CHAPTER 11

Bioenergetics and Metabolism
Mitochondria, Chloroplasts, and Peroxisomes

■ *Mitochondria 434*

■ *The Mechanism of Oxidative Phosphorylation 443*

■ *Chloroplasts and Other Plastids 451*

■ *Photosynthesis 458*

■ *Peroxisomes 462*

■ *KEY EXPERIMENT:*
The Chemiosmotic Theory 448

■ *MOLECULAR MEDICINE:*
Diseases of Mitochondria: Leber's Hereditary Optic Neuropathy 438

IN ADDITION TO BEING INVOLVED IN PROTEIN SORTING AND TRANS-PORT, cytoplasmic organelles provide specialized compartments in which a variety of metabolic activities take place. The generation of metabolic energy is a major activity of all cells, and two cytoplasmic organelles are specifically devoted to energy metabolism and the production of ATP. Mitochondria are responsible for generating most of the useful energy derived from the breakdown of lipids and carbohydrates, and chloroplasts use energy captured from sunlight to generate both ATP and the reducing power needed to synthesize carbohydrates from CO_2 and H_2O. The third organelle discussed in this chapter, the peroxisome, contains enzymes involved in a variety of different metabolic pathways, including the breakdown of fatty acids and the metabolism of a byproduct of photosynthesis.

Mitochondria, chloroplasts, and peroxisomes differ from the organelles discussed in the preceding chapter not only in their functions but also in their mechanism of assembly. Rather than being synthesized on membrane-bound ribosomes and translocated into the endoplasmic reticulum, most proteins destined for peroxisomes, mitochondria, and chloroplasts are synthesized on free ribosomes in the cytosol and imported into their target organelles as completed polypeptide chains. Mitochondria and chloroplasts also contain their own genomes, which include some genes that are transcribed and translated within the organelle. Protein sorting to the cytoplasmic organelles discussed in this chapter is thus distinct from the pathways of vesicular transport that connect the endoplasmic reticulum, Golgi apparatus, lysosomes, and plasma membrane.

Mitochondria

Mitochondria play a critical role in the generation of metabolic energy in eukaryotic cells. As reviewed in Chapter 3, they are responsible for most of the useful energy derived from the breakdown of carbohydrates and fatty acids, which is converted to ATP by the process of oxidative phosphorylation. Most mitochondrial proteins are translated on free cytosolic ribosomes and imported into the organelle by specific targeting signals. In addition, mitochondria are unique among the cytoplasmic organelles already discussed in that they contain their own DNA, which encodes tRNAs, rRNAs, and some mitochondrial proteins. The assembly of mitochondria thus involves proteins encoded by their own genomes and translated within the organelle, as well as proteins encoded by the nuclear genome and imported from the cytosol.

Organization and Function of Mitochondria

Mitochondria are surrounded by a double-membrane system, consisting of inner and outer mitochondrial membranes separated by an intermembrane space (Figure 11.1). The inner membrane forms numerous folds (**cristae**), which extend into the interior (or **matrix**) of the organelle. Each of these components plays distinct functional roles, with the matrix and inner membrane representing the major working compartments of mitochondria.

The matrix contains the mitochondrial genetic system as well as the enzymes responsible for the central reactions of oxidative metabolism (Figure 11.2). As discussed in Chapter 3, the oxidative breakdown of glucose and fatty acids is the principal source of metabolic energy in animal cells. The initial stages of glucose metabolism (glycolysis) occur in the cytosol, where glucose is converted to pyruvate (see Figure 3.11). Pyruvate is then transported into the mitochondria, where its complete oxidation to CO_2 yields the bulk of usable energy (ATP) obtained from glucose metabolism. This involves the initial oxidation of pyruvate to acetyl CoA, which is then broken down to CO_2 via the citric acid cycle (see Figures 3.12 and 3.13). The oxidation of fatty acids also yields acetyl CoA (see Figure 3.15), which is similarly metabolized by the citric acid cycle in mitochondria. The enzymes of the citric acid cycle

FIGURE 11.1 Structure of a mitochondrion Mitochondria are bounded by a double-membrane system, consisting of inner and outer membranes. Folds of the inner membrane (cristae) extend into the matrix. (Micrograph by K. R. Porter/Photo Researchers, Inc.)

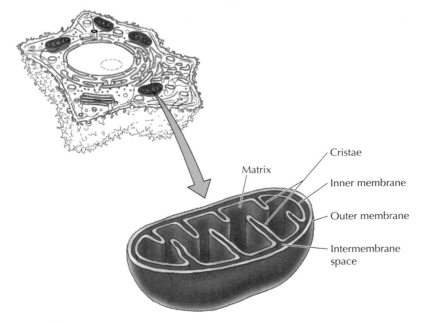

Matrix

Cristae

Inner membrane

Outer membrane

Intermembrane space

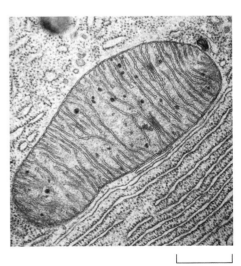

0.5 µm

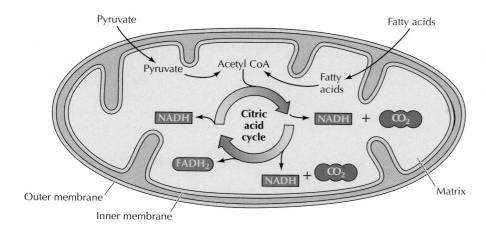

FIGURE 11.2 Metabolism in the matrix of mitochondria Pyruvate and fatty acids are imported from the cytosol and converted to acetyl CoA in the mitochondrial matrix. Acetyl CoA is then oxidized to CO_2 via the citric acid cycle, the central pathway of oxidative metabolism.

(located in the matrix of mitochondria) thus are central players in the oxidative breakdown of both carbohydrates and fatty acids.

The oxidation of acetyl CoA to CO_2 is coupled to the reduction of NAD^+ and FAD to NADH and $FADH_2$, respectively. Most of the energy derived from oxidative metabolism is then produced by the process of oxidative phosphorylation (discussed in detail in the next section), which takes place in the inner mitochondrial membrane. The high-energy electrons from NADH and $FADH_2$ are transferred through a series of carriers in the membrane to molecular oxygen. The energy derived from these electron transfer reactions is converted to potential energy stored in a proton gradient across the membrane, which is then used to drive ATP synthesis. The inner mitochondrial membrane thus represents the principal site of ATP generation, and this critical role is reflected in its structure. First, its surface area is substantially increased by its folding into cristae. In addition, the inner mitochondrial membrane contains an unusually high percentage (greater than 70%) of proteins, which are involved in oxidative phosphorylation as well as in the transport of metabolites (e.g., pyruvate and fatty acids) between the cytosol and mitochondria. Otherwise, the inner membrane is impermeable to most ions and small molecules—a property critical to maintaining the proton gradient that drives oxidative phosphorylation.

In contrast to the inner membrane, the outer mitochondrial membrane is highly permeable to small molecules. This is because it contains proteins called **porins**, which form channels that allow the free diffusion of molecules smaller than about 1000 daltons. The composition of the intermembrane space is therefore similar to the cytosol with respect to ions and small molecules. Consequently, the inner mitochondrial membrane is the functional barrier to the passage of small molecules between the cytosol and the matrix, and it maintains the proton gradient that drives oxidative phosphorylation.

The Genetic System of Mitochondria

Mitochondria contain their own genetic system, which is separate and distinct from the nuclear genome of the cell. As reviewed in Chapter 1, mitochondria are thought to have evolved from bacteria that developed a symbiotic relationship in which they lived within larger cells (**endosymbiosis**). The genomes of living organisms that are most similar to the mitochondrial genome are those of free living α-proteobacteria, which code for 6700–8300 proteins. An intracellular parasite like the α-proteobacterium *Rickettsia*

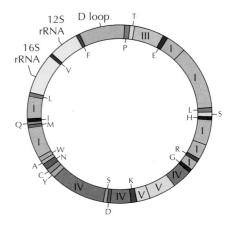

FIGURE 11.3 The human mitochondrial genome   The genome contains 13 protein-coding sequences, which are designated as components of respiratory complexes I, III, IV, or V. In addition, the genome contains genes for 12S and 16S rRNAs and for 22 tRNAs, which are designated by the one-letter code for the corresponding amino acid. The region of the genome designated "D loop" contains an origin of DNA replication and transcriptional promoter sequences.

prowazekii has a smaller genome, about 830 protein-coding genes. Like mitochondria, *Rickettsia prowazekii* is able to reproduce only within eukaryotic cells, but unlike mitochondria it still transcribes and translates most of its own genes.

Mitochondrial genomes are usually circular DNA molecules like those of bacteria, which are present in multiple copies per organelle. They vary considerably in size between different species. The genomes of human and most other animal mitochondria are only about 16 kb, but substantially larger mitochondrial genomes are found in yeasts (approximately 80 kb) and plants (more than 200 kb). However, these larger mitochondrial genomes are composed predominantly of noncoding sequences and do not appear to contain significantly more genetic information. For example, the largest sequenced mitochondrial genome is that of the plant *Arabidopsis thaliana*. Although *Arabidopsis* mitochondrial DNA is approximately 367 kb, it encodes only 31 proteins: just more than twice the number encoded by human mitochondrial DNA. The largest number of mitochondrial genes has been found in mitochondrial DNA of the protozoan *Reclinomonas americana*, which is 69 kb and contains 67 genes. The smallest mitochondrial genome is that of the protist *Plasmodium falciparum*, which is 6 kb and codes for only 3 proteins. In contrast, the genomes of the free-living α-proteobacteria are 7–10 Mb. Most present-day mitochondrial genomes encode only a small number of proteins that are essential components of the oxidative phosphorylation system. In addition, mitochondrial genomes encode all of the ribosomal RNAs and most of the transfer RNAs needed for translation of these protein-coding sequences within mitochondria. Other mitochondrial proteins are encoded by nuclear genes, which are thought to have been transferred to the nucleus from the ancestral mitochondrial genome.

The human mitochondrial genome encodes 13 proteins involved in electron transport and oxidative phosphorylation (Figure 11.3). In addition, human mitochondrial DNA encodes 16S and 12S rRNAs and 22 tRNAs, which are required for translation of the proteins encoded by the organelle genome. The two rRNAs are the only RNA components of animal and yeast mitochondrial ribosomes, in contrast to the three rRNAs of bacterial ribosomes (23S, 16S, and 5S). Plant mitochondrial DNAs, however, also encode a third rRNA: 5S. The mitochondria of plants and protozoans also differ in importing and utilizing tRNAs encoded by the nuclear as well as the mitochondrial genome, whereas in animal mitochondria, all the tRNAs are encoded by the organelle genome.

The small number of tRNAs encoded by the mitochondrial genome highlights an important feature of the mitochondrial genetic system—the use of a slightly different genetic code, which is distinct from the "universal" genetic code used by both prokaryotic and eukaryotic cells (Table 11.1). As discussed in Chapter 4, there are 64 possible triplet codons, of which 61 encode the 20 different amino acids incorporated into proteins (see Table 4.1). Many tRNAs in both prokaryotic and eukaryotic cells are able to recog-

■ Since almost all mitochondria are inherited from the mother, it is possible to trace the human maternal lineage back to our most recent common female ancestor: Mitochondrial Eve.

nize more than a single codon in mRNA because of "wobble," which allows some mispairing between the tRNA anticodon and the third position of certain complementary codons (see Figure 8.3). However, at least 30 different tRNAs are required to translate the universal code according to the wobble rules. Yet human mitochondrial DNA encodes only 22 tRNA species, and these are the only tRNAs used for translation of mitochondrial mRNAs. This is accomplished by an extreme form of wobble in which U in the anticodon of the tRNA can pair with any of the four bases in the third codon position of mRNA, allowing four codons to be recognized by a single tRNA. In addition, some codons specify different amino acids in mitochondria than in the universal code.

Like the DNA of nuclear genomes, mitochondrial DNA can be altered by mutations, which are frequently deleterious to the organelle. Since almost all the mitochondria of fertilized eggs are contributed by the oocyte rather than by the sperm, germ-line mutations in mitochondrial DNA are transmitted to the next generation by the mother. Such mutations have been associated with a number of diseases. Mutations in one mitochondrial tRNA gene are associated with metabolic syndrome, the human condition associated with obesity and diabetes. In addition, Leber's hereditary optic neuropathy, a disease that leads to blindness, can be caused by mutations in mitochondrial genes that encode components of the electron transport chain. In addition, the progressive accumulation of mutations in mitochondrial DNA during the lifetime of individuals has been suggested to contribute to the process of aging.

In contrast to the mitochondrial genome, the proteins present in mitochondria are much less-well understood. Mammalian mitochondria are thought to contain between 1000–2000 different proteins, representing ~5% of the proteins encoded by mammalian genomes. However, nearly half of all mitochondrial proteins remain unidentified. Protein analysis has been difficult because mitochondria from different tissues contain different proteins. For example, only half the proteins in human mitochondria are found in all tissues. Some of these differences can be traced to tissue-specific functions of the mitochondria, such as complex steroid synthesis in adrenal cells or heme biosynthesis in bone marrow; others likely play roles in poorly understood processes such as control of mitochondrial shape or number, and mitochondrial inheritance.

Protein Import and Mitochondrial Assembly

In contrast to the RNA components of the mitochondrial translation apparatus (rRNAs and tRNAs), most mitochondrial genomes do not encode the proteins required for DNA replication, transcription, or translation. Instead, the genes that encode proteins required for the replication and expression of mitochondrial DNA are contained in the nucleus. In addition, the nucleus contains the genes that encode most of the mitochondrial proteins required for oxidative phosphorylation and all of the enzymes involved in mitochondrial metabolism (e.g., enzymes of the citric acid cycle). Some of these genes were transferred to the nucleus from the original prokaryotic ancestor of mitochondria.

Approximately 1000 proteins encoded by nuclear genes (more than 95% of mitochondrial proteins) are synthesized on free cytosolic ribosomes and imported into mitochondria as completed polypeptide chains. Because of the double-membrane structure of mitochondria, the import of proteins is considerably more complicated than the transfer of a polypeptide across a single phospholipid bilayer. Proteins targeted to the matrix have to cross

TABLE 11.1 Differences between the Universal and Mitochondrial Genetic Codes

Codon	Universal code	Human mitochondrial code
UGA	Stop	Trp
AGA	Arg	Stop
AGG	Arg	Stop
AUA	Ile	Met

Other codons vary from the universal code in yeast and plant mitochondria.

Diseases of Mitochondria: Leber's Hereditary Optic Neuropathy

The Disease

Leber's hereditary optic neuropathy (LHON) is a rare inherited disease that results in blindness because of degeneration of the optic nerve. Vision loss usually occurs between the ages of 15 and 35, and is generally the only manifestation of the disease. Not all individuals who inherit the genetic defects responsible for LHON develop the disease, and females are affected less frequently than males. This propensity to affect males might suggest that LHON is an X-linked disease. This is not the case, however, because males never transmit LHON to their offspring. Instead, the inheritance of LHON is entirely by maternal transmission. This characteristic is consistent with cytoplasmic rather than nuclear inheritance of LHON, since the cytoplasm of fertilized eggs is derived almost entirely from the oocyte.

Molecular and Cellular Basis

In 1988 Douglas Wallace and his colleagues identified a mutation in the mitochondrial DNA of LHON patients. This mutation (at base pair 11778) affects one of the subunits of complex I of the electron transport chain (NADH dehydrogenase), resulting in the substitution of a histidine for an arginine. The 11778 mutation accounts for approximately half of all cases of LHON. Three other mutations of mitochondrial DNA have also been identified as primary causes of LHON. Two of these mutations affect other subunits of complex I, while the third affects cytochrome *b*, which is a component of complex III (see figure). Together, these four mutations account for more than 80% of LHON cases. A fifth mutation (at base pair 14459), affecting a complex I subunit, can cause either LHON or muscular disorders.

The mutations causing LHON reduce the capacity of mitochondria to carry out oxidative phosphorylation

and generate ATP. This has the greatest effect on those tissues that are most dependent on oxidative phosphorylation, so defects in components of mitochondria can lead to clinical manifestations in specific organs, rather than to systemic disease. The central nervous system (including the brain and optic nerve) is most highly dependent on oxidative metabolism, consistent with blindness being the primary clinical manifestation resulting from the mitochondrial DNA mutations responsible for LHON.

As already noted, inheritance of LHON mutations does not always

lead to development of the disease; only about 10% of females and 50% of males possessing a mutation suffer vision loss. One factor that may contribute to this low incidence of disease among carriers of LHON mutations is that each cell contains thousands of copies of mitochondrial DNA, which can be present in mixtures of mutant and normal mitochondria. These mitochondria are randomly distributed to daughter cells at cell division, so the population of mitochondria can change as cells divide, leading to the formation of cells containing either greater or lesser proportions of mutant

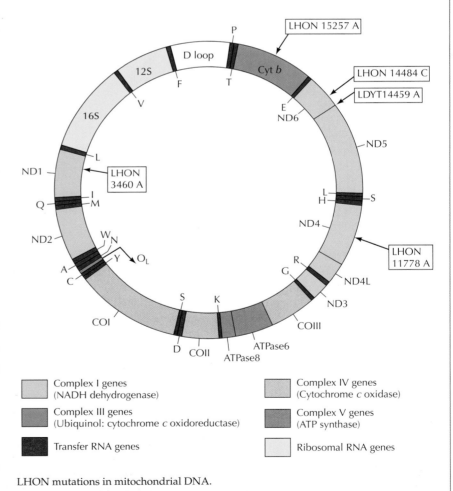

LHON mutations in mitochondrial DNA.

organelles. Importantly, however, many individuals who bear predominantly mutant mitochondrial DNAs still fail to develop the disease. Thus additional genetic or environmental factors, which have yet to be identified, appear to play a significant role in the development of LHON.

Prevention and Treatment

The identification of mitochondrial DNA mutations responsible for LHON allows molecular diagnosis of the disease, which can be important in establishing a definitive diagnosis of patients without a family history. However, the detection of mutations in mitochondrial DNA is of little value for screening members of affected families or for family planning. This contrasts to the utility of detecting inherited mutations of nuclear genes,

where molecular analysis can determine whether a family member or embryo has inherited a mutant or wild-type allele. In LHON, however, mutant mitochondria are present in large numbers and are maternally transmitted to all offspring. As noted above, not all such offspring develop the disease, but this cannot be predicted by genetic analysis.

The finding that LHON is caused by mutations of mitochondrial DNA suggests the potential of new therapies. One approach is metabolic therapy intended to enhance oxidative phosphorylation by administration of substrates or cofactors in the electron transport pathway, such as succinate or coenzyme Q. Another possibility that has been considered for treatment of LHON is gene therapy designed to

relocate a normal gene allele to the nucleus. An appropriate targeting signal would be added to direct the gene product to mitochondria, where it could substitute for the defective mitochondrial-encoded protein.

References

Brown, M. D., D. S. Voljavec, M. T. Lott, I. MacDonald and D. C. Wallace. 1992. Leber's hereditary optic neuropathy: A model for mitochondrial neurodegenerative diseases. *FASEB J.* 6: 2791–2799.

Howell, N., J. L. Elson, P. F. Chinnery and D. M. Turnbull. 2005. mtDNA mutations and common neurodegenerative disorders. *Trends Genet.* 21: 583–586.

Riordan-Eva, P. and A. E. Harding. 1995. Leber's hereditary optic neuropathy: The clinical relevance of different mitochondrial DNA mutations. *J. Med. Genet.* 32: 81–87.

both the outer and inner mitochondrial membranes, while other proteins need to be sorted to distinct compartments within the organelle (e.g., the intermembrane space).

The import of proteins to the matrix is the best-understood aspect of mitochondrial protein sorting (Figure 11.4). Most proteins are targeted to mitochondria by amino-terminal sequences of 20 to 35 amino acids (called **presequences**) that are removed by proteolytic cleavage following their import into the organelle. The presequences of mitochondrial proteins, first characterized by Gottfried Schatz, contain multiple positively charged amino acid residues, usually in an amphipathic α helix. The first step in protein import is the binding of these presequences to receptors on the surface of mitochondria. These receptors are part of a protein complex that directs translocation across the outer membrane (the *t*ranslocase of the *o*uter *m*embrane or **Tom complex**). Individual Tom proteins are designated according to their molecular weights, so the receptors are called Tom20, Tom22, and Tom5. From these receptors, proteins are transferred to the Tom40 pore protein and translocated across the outer membrane. The proteins are then transferred to a second protein complex in the inner membrane (one of two different *t*ranslocases of the *i*nner *m*embrane or **Tim complexes**). Proteins with presequences cross the inner membrane through the Tim23 complex. Continuing protein translocation requires the electrochemical potential established across the inner mitochondrial membrane during electron transport. As discussed in the next section of this chapter, the transfer of high-energy electrons from NADH and FADH$_2$ to molecular oxygen is coupled to the transfer of protons from the mitochondrial matrix to the

FIGURE 11.4 Import of mitochondrial matrix proteins Proteins are targeted to the Tom complex in the mitochondrial outer membrane by amino-terminal presequences containing positively charged amino acids. The presequence first binds to Tom20 and is transferred to Tom5 and then to the import pore, Tom40. Following passage through the outer membrane, the presequence binds the intermembrane domain of Tom22 and is passed to the Tim23 complex in the inner membrane. In the matrix, an Hsp70 chaperone associated with Tim44 acts as a ratchet, using ATP hydrolysis to translocate the protein across the inner membrane. Most proteins destined for the mitochondrial matrix have their presequences removed by the mitochondrial matrix processing peptidase (MPP) and associate with soluble Hsp70 proteins that assist in their folding.

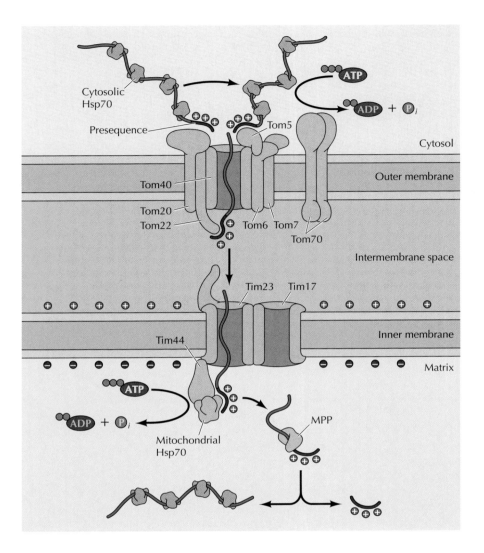

intermembrane space. Since protons are charged particles, this transfer establishes an electric potential across the inner membrane, with the matrix being negative. During protein import, this electric potential drives translocation of the positively charged presequence.

To be translocated across the mitochondrial membrane, proteins must be at least partially unfolded. Consequently, protein import into mitochondria requires molecular chaperones in addition to the membrane proteins involved in translocation (see Figure 11.4). On the cytosolic side, members of the Hsp70 family of chaperones both maintain proteins in a partially unfolded state and present them to the translocase so that they can be inserted into the mitochondrial membrane. As they cross the inner membrane, the unfolded polypeptide chains are bound by another Hsp70 chaperone, which is associated with the Tim23 complex and acts as a ratchet that uses repeated ATP hydrolysis to drive protein import (Figure 11.5). In most cases, the presequence is then cleaved by the **matrix processing peptidase (MPP)** and the polypeptide chain is bound by other matrix Hsp70 chaperones that facilitate its folding. Some polypeptides are then transferred to a

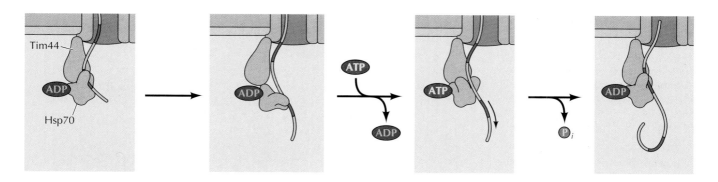

FIGURE 11.5 Binding cycle of an Hsp70 chaperone Hsp70 chaperones provide the force for translocating proteins across many membranes. To do this they exploit their ability to reversibly bind short hydrophobic sequences in polypeptides and, when bound to another structure such as Tim44, can act as a ratchet. In the ADP-bound state, the binding pocket for the substrate protein is closed and the hydrophobic sequence in the polypeptide is tightly bound. The displacement of ADP by ATP opens the pocket, allowing Hsp70 to release the polypeptide. Recognition of a second hydrophobic sequence results in hydrolysis of ATP by Hsp70, which then closes the pocket and tightly binds the polypeptide.

chaperone of the Hsp60 family (a chaperonin, see Figure 8.23), within which additional protein folding takes place. These interactions of polypeptide chains with molecular chaperones depend on ATP, so protein import requires ATP both outside and inside the mitochondria as well as the electric potential across the inner membrane.

As noted above, some mitochondrial proteins are targeted to the outer membrane, inner membrane, or intermembrane space rather than to the matrix, so additional mechanisms are needed to direct these proteins to the correct submitochondrial compartment. Many of the proteins in the inner membrane are small molecule transporters, which are multi-pass transmembrane proteins that exchange nucleotides and ions between the mitochondria and the cytosol. These proteins do not contain presequences but instead have multiple internal mitochondrial import signals. Consequently, they are not recognized by Tom20. Instead, these inner membrane proteins, in association with an Hsp90 chaperone, are recognized by a distinct receptor on the mitochondrial outer membrane (Tom70), and then translocated across the membrane through the Tom40 channel (Figure 11.6). In the intermembrane space, the proteins are recognized by mobile components of a distinct Tim complex, that of Tim22. These small Tim proteins (called "tiny Tim" proteins) serve as both chaperones and shuttle proteins that escort proteins to Tim22. The proteins are then partially translocated through Tim22, before internal stop-transfer signals cause them to exit the Tim22 pore laterally and insert into the inner membrane.

Some proteins destined for the outer or inner membranes or the intermembrane space have both a presequence and internal signal sequences. Since they contain presequences, these proteins are recognized by the Tom20 receptor on the outer membrane and translocated through the Tom40 channel (Figure 11.7). Some outer membrane proteins exit the Tom40 channel laterally, although the porin proteins and Tom40 itself pass through the Tom complex and interact with a second protein complex to insert into

FIGURE 11.6 Import of small molecule transport proteins into the mitochondrial inner membrane
These multiple-pass transmembrane proteins have internal signal sequences, rather than N-terminal presequences. The internal signal sequences in association with Hsp90 chaperones interact with the Tom70 receptors, from which the transmembrane protein is transferred to the Tom40 channel. In the intermembrane space, the protein is bound by small mobile Tim proteins, the "tiny Tim" proteins that guide it to the Tim22 complex in the inner membrane. The tiny Tim proteins transfer the protein to Tim54 and then to the Tim22 import pore. Internal stop-transfer sequences halt translocation, and the protein is transferred laterally into the mitochondrial inner membrane.

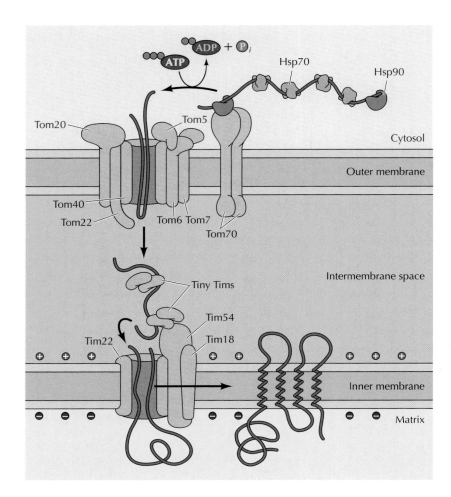

the outer membrane from the intermembrane space. Other proteins with complex signal sequences pass through the outer membrane but remain in the intermembrane space instead of entering Tim23. Other proteins destined for the intermembrane space, as well as some inner membrane proteins, are first transported across the inner membrane into the mitochondrial matrix through the Tim23 complex. They are then targeted for further transport by a second sorting signal that is uncovered by the removal of the presequence in the matrix. This second sorting signal targets them to another translocase, Oxa1, where they are either passed into the intermembrane space or arrested in transit by internal stop-transfer signals and inserted into the inner membrane. Oxa1 is also the translocase for those intermembrane and inner membrane proteins that are encoded by the mitochondrial genome and are synthesized on mitochondrial ribosomes in the matrix.

Not only the proteins but also the phospholipids of mitochondrial membranes are imported from the cytosol. In animal cells, phosphatidylcholine and phosphatidylethanolamine are synthesized in the ER and carried to mitochondria by **phospholipid transfer proteins**, which extract single phospholipid molecules from the membrane of the ER. The lipid can then be transported through the aqueous environment of the cytosol, buried in a hydrophobic binding site of the protein, and released when the complex

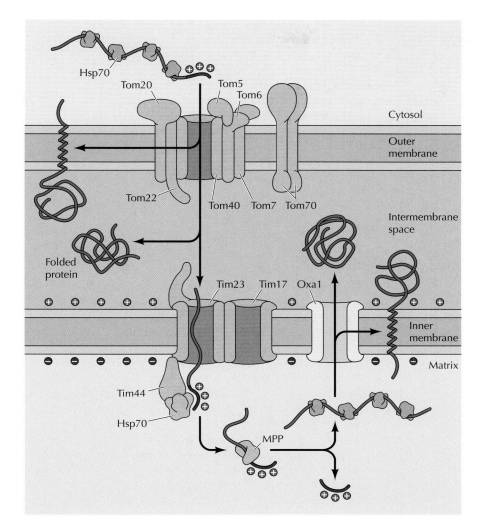

FIGURE 11.7 Sorting of proteins containing presequences to different mitochondrial compartments Mitochondrial proteins with N-terminal presequences can be imported to the outer membrane, inner membrane, or intermembrane space. The presequences of these proteins are recognized by the Tom20 receptor and transferred to Tom40. Proteins destined for the outer membrane halt translocation in the Tom40 complex and pass laterally into the membrane. Some proteins destined for the intermembrane space are translocated through Tom40 but remain in the intermembrane space rather than interacting with the Tim23 complex. Other proteins are transferred through Tim23 into the mitochondrial matrix. Removal of the presequence within the matrix then exposes a second sorting signal that targets these proteins back into either the inner membrane or the intermembrane space through the Oxa1 translocation pore.

reaches a new membrane, such as that of mitochondria. The mitochondria then synthesize phosphatidylserine from phosphatidylethanolamine, in addition to catalyzing the synthesis of the unusual phospholipid **cardiolipin**, which contains four fatty acid chains (Figure 11.8).

The Mechanism of Oxidative Phosphorylation

Most of the usable energy obtained from the breakdown of carbohydrates or fats is derived by oxidative phosphorylation, which takes place within mitochondria. For example, the breakdown of glucose by glycolysis and the citric acid cycle yields a total of four molecules of ATP, ten molecules of NADH, and two molecules of $FADH_2$ (see Chapter 3). Electrons from NADH and $FADH_2$ are then transferred to molecular oxygen, which is coupled to the formation of an additional 32 to 34 ATP molecules by oxidative phosphorylation. Electron transport and oxidative phosphorylation are critical activities of protein complexes in the inner mitochondrial membrane which ultimately serve as the major sources of cellular energy.

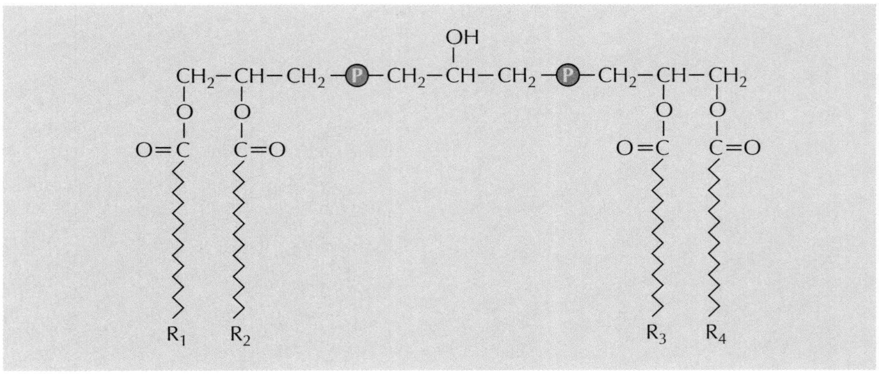

FIGURE 11.8 Structure of cardiolipin
Cardiolipin is an unusual "double" phospholipid, containing four fatty acid chains, that is found primarily in the inner mitochondrial membrane.

| ■ Rotenone, an inhibitor of electron transfer from complex I to coenzyme Q, is used as a broad-spectrum insecticide. |

The Electron Transport Chain

During **oxidative phosphorylation**, electrons derived from NADH and $FADH_2$ combine with O_2, and the energy released from these oxidation/reduction reactions is used to drive the synthesis of ATP from ADP. The transfer of electrons from NADH to O_2 is a very energy-yielding reaction, with $\Delta G^{o'} = -52.5$ kcal/mol for each pair of electrons transferred. To be harvested in usable form this energy must be produced gradually by the passage of electrons through a series of carriers, which constitute the **electron transport chain**. These carriers are organized into four complexes in the inner mitochondrial membrane. A fifth protein complex then serves to couple the energy-yielding reactions of electron transport to ATP synthesis.

Electrons from NADH enter the electron transport chain in complex I, which consists of nearly 40 polypeptide chains (Figure 11.9). These electrons are initially transferred from NADH to flavin mononucleotide and then, through an iron-sulfur carrier, to coenzyme Q—an energy-yielding process $\Delta G^{o'} = -16.6$ kcal/mol. **Coenzyme Q** (also called **ubiquinone**) is a small, lipid-soluble molecule that carries electrons from complex I through the membrane to complex III, which consists of about ten polypeptides. In complex III, electrons are transferred from cytochrome b to cytochrome c—an energy-yielding reaction with $\Delta G^{o'} = -10.1$ kcal/mol. **Cytochrome c**, a peripheral membrane protein bound to the outer face of the inner membrane then carries electrons to complex IV (**cytochrome oxidase**), where they are finally transferred to O_2 ($\Delta G^{o'} = -25.8$ kcal/mol).

A distinct protein complex (complex II), which consists of four polypeptides, receives electrons from the citric acid cycle intermediate, succinate (Figure 11.10). These electrons are transferred to $FADH_2$, rather than to NADH, and then to coenzyme Q. From coenzyme Q, electrons are transferred to complex III and then to complex IV as already described. In contrast to the transfer of electrons from NADH to coenzyme Q at complex I, the transfer of electrons from $FADH_2$ to coenzyme Q is not associated with a significant decrease in free energy and, therefore, is not coupled to ATP synthesis. Consequently, the passage of electrons derived from $FADH_2$ through the electron transport chain yields free energy only at complexes III and IV.

The free energy derived from the passage of electrons through complexes I, III, and IV is harvested by being coupled to the synthesis of ATP. Importantly, the mechanism by which the energy derived from these electron transport reactions is coupled to ATP synthesis, is fundamentally different from the synthesis of ATP during glycolysis or the citric acid cycle. In

FIGURE 11.9 Transport of electrons from NADH

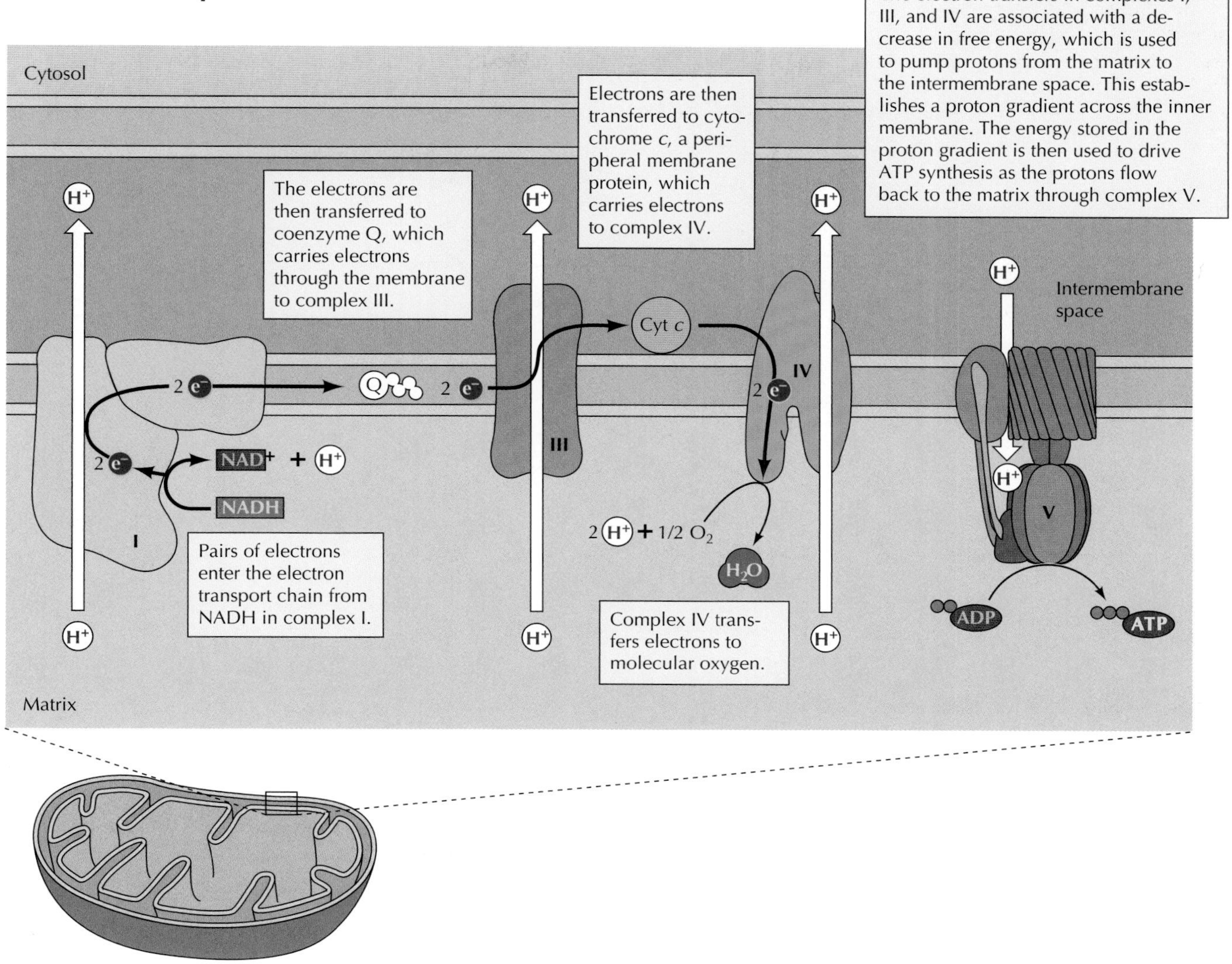

Cytosol

The electron transfers in complexes I, III, and IV are associated with a decrease in free energy, which is used to pump protons from the matrix to the intermembrane space. This establishes a proton gradient across the inner membrane. The energy stored in the proton gradient is then used to drive ATP synthesis as the protons flow back to the matrix through complex V.

The electrons are then transferred to coenzyme Q, which carries electrons through the membrane to complex III.

Electrons are then transferred to cytochrome c, a peripheral membrane protein, which carries electrons to complex IV.

Intermembrane space

Pairs of electrons enter the electron transport chain from NADH in complex I.

Complex IV transfers electrons to molecular oxygen.

Matrix

the latter cases, a high-energy phosphate is transferred directly to ADP from the other substrate of an energy-yielding reaction. For example, in the final reaction of glycolysis, the high-energy phosphate of phosphoenolpyruvate is transferred to ADP, yielding pyruvate plus ATP (see Figure 3.11). Such direct transfer of high-energy phosphate groups does not occur during electron transport. Instead, the energy derived from electron transport is coupled to the generation of a proton gradient across the inner mitochondrial membrane. The potential energy stored in this gradient is then harvested by a fifth protein complex, which couples the energetically favorable flow of protons back across the membrane to the synthesis of ATP.

■ In addition to its role in electron transport, cytochrome c is a key regulator of programmed cell death in mammalian cells (discussed in Chapter 16).

Chemiosmotic Coupling

The mechanism of coupling electron transport to ATP generation, **chemiosmotic coupling**, is a striking example of the relationship between structure and function in cell biology. The hypothesis of chemiosmotic coupling was first proposed in 1961 by Peter Mitchell, who suggested that ATP is gener-

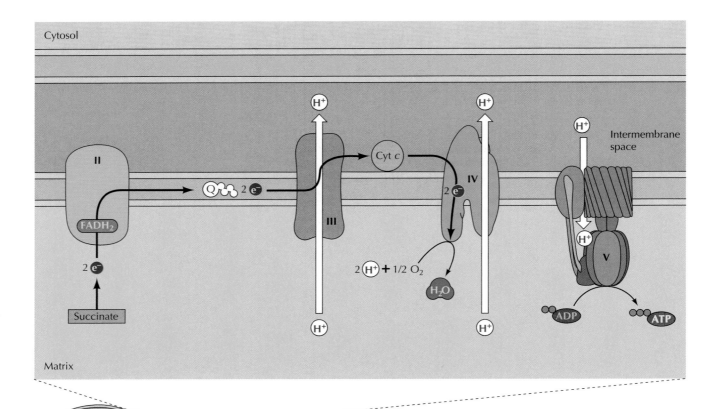

FIGURE 11.10 Transport of electrons from FADH$_2$ Electrons from succinate enter the electron transport chain via FADH$_2$ in complex II. They are then transferred to coenzyme Q and carried through the rest of the electron transport chain as described in Figure 11.9. The transfer of electrons from FADH$_2$ to coenzyme Q is not associated with a significant decrease in free energy, so protons are not pumped across the membrane at complex II.

ated by the use of energy stored in the form of proton gradients across biological membranes, rather than by direct chemical transfer of high-energy groups. Biochemists were initially highly skeptical of this proposal, and the chemiosmotic hypothesis took more than a decade to win general acceptance in the scientific community. Overwhelming evidence eventually accumulated in its favor, however, and chemiosmotic coupling is now recognized as a general mechanism of ATP generation, operating not only in mitochondria but also in chloroplasts and in bacteria, where ATP is generated via a proton gradient across the plasma membrane.

Electron transport through complexes I, III, and IV is coupled to the transport of protons out of the interior of the mitochondrion (see Figure 11.9). Thus the energy-yielding reactions of electron transport are coupled to the transfer of protons from the matrix to the intermembrane space, which establishes a proton gradient across the inner membrane. Complexes I and IV appear to act as proton pumps that transfer protons across the membrane as a result of conformational changes induced by electron transport. In complex III, protons are carried across the membrane by coenzyme Q, which accepts protons from the matrix at complexes I or II and releases them into the intermembrane space at complex III. Complexes I and III each transfer four protons across the membrane per pair of electrons. In complex

IV, two protons per pair of electrons are pumped across the membrane and another two protons per pair of electrons are combined with O_2 to form H_2O within the matrix. Thus the equivalent of four protons per pair of electrons is transported out of the mitochondrial matrix at each of these three complexes. This transfer of protons from the matrix to the intermembrane space plays the critical role of converting the energy derived from the oxidation/reduction reactions of electron transport to the potential energy stored in a proton gradient.

Because protons are electrically charged particles, the potential energy stored in the proton gradient is electric as well as chemical in nature. The electric component corresponds to the voltage difference across the inner mitochondrial membrane, with the matrix of the mitochondrion being negative and the intermembrane space being positive. The corresponding free energy is given by the equation

$$\Delta G = -F\Delta V$$

where F is the Faraday constant and ΔV is the membrane potential. The additional free energy corresponding to the difference in proton concentration across the membrane is given by the equation

$$\Delta G = RT \ln \frac{[H^+]_i}{[H^+]_o}$$

where $[H^+]_i$ and $[H^+]_o$ refer, respectively, to the proton concentrations inside and outside the mitochondria.

In metabolically active cells, protons are typically pumped out of the matrix such that the proton gradient across the inner membrane corresponds to about one pH unit, or a tenfold lower concentration of protons within mitochondria (Figure 11.11). The pH of the mitochondrial matrix is

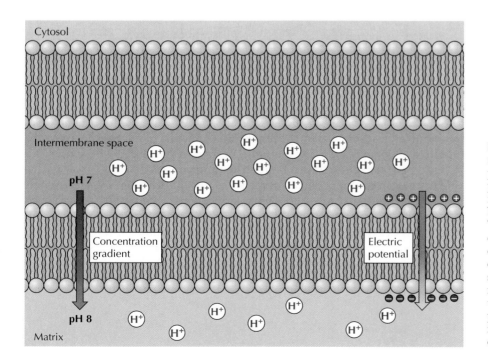

FIGURE 11.11 The electrochemical nature of the proton gradient Since protons are positively charged, the proton gradient established across the inner mitochondrial membrane has both chemical and electric components. The chemical component is the proton concentration, or pH, gradient, which corresponds to about a tenfold higher concentration of protons on the cytosolic side of the inner mitochondrial membrane (a difference of one pH unit). In addition, there is an electric potential across the membrane, resulting from the net increase in positive charge on the cytosolic side.

KEY EXPERIMENT

The Chemiosmotic Theory

Coupling of Phosphorylation to Electron and Hydrogen Transfer by a Chemiosmotic Type of Mechanism

Peter Mitchell

University of Edinburgh, Edinburgh, Scotland

Nature, 1961, Volume 191, pages 144–148

Peter Mitchell

The Context

By the 1950s it had been clearly established that oxidative phosphorylation involved the stepwise transfer of electrons through a series of carriers to molecular oxygen. But how the energy derived from these electron transfer reactions was converted to ATP remained a mystery. The natural assumption was that ADP was converted to ATP by direct transfer of high-energy phosphate groups from some other intermediate, as was known to occur during glycolysis. Thus it was postulated that high-energy intermediates were produced as a result of electron transfer reactions and that these intermediates then drove ATP synthesis by phosphate group transfer.

The search for these postulated high-energy intermediates became a central goal of research during the 1950s and 1960s. But despite many false leads, no such intermediates were to be found. Moreover, several features of oxidative phosphorylation were difficult to reconcile with the orthodox hypothesis that ATP synthesis was driven by simple phosphate group transfer. In particular, phosphorylation was closely associated with membranes and was inhibited by a variety of compounds that disrupted membrane structure. These considerations led Peter Mitchell to propose a fundamentally different mechanism of energy coupling in which ATP synthesis was driven by an electrochemical gradient across a membrane rather

than by the elusive high-energy intermediates sought by other investigators.

The Experiments

The fundamental proposal of the chemiosmotic hypothesis was that the "intermediate" that coupled electron transport to ATP synthesis was a proton electrochemical gradient across the membrane. Mitchell postulated that such a gradient was produced by electron transport and that the flow of protons back across the membrane in the energetically favorable direction

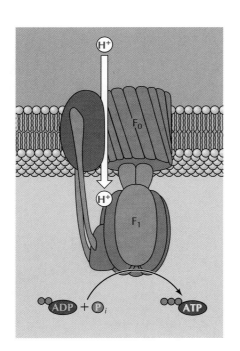

therefore about 8, compared to the neutral pH (approximately 7) of the cytosol and intermembrane space. This gradient also generates an electric potential of approximately 0.14 V across the membrane, with the matrix being negative. Both the pH gradient and the electric potential drive protons back into the matrix from the cytosol, so they combine to form an **electrochemical gradient** across the inner mitochondrial membrane, corresponding to a ΔG of approximately -5 kcal/mol per proton.

Because the phospholipid bilayer is impermeable to ions, protons are able to cross the membrane only through a protein channel. This restriction allows the energy in the electrochemical gradient to be harnessed and converted to ATP as a result of the action of the fifth complex involved in oxidative phosphorylation, complex V, or **ATP synthase** (see Figure 11.9).

FIGURE 11.12 Structure of ATP synthase The mitochondrial ATP synthase (complex V) consists of two multisubunit components, F_0 and F_1, which are linked by a slender stalk. F_0 spans the lipid bilayer, forming a channel through which protons can cross the membrane. F_1 harvests the free energy derived from proton movement down the electrochemical gradient by catalyzing the synthesis of ATP.

KEY EXPERIMENT

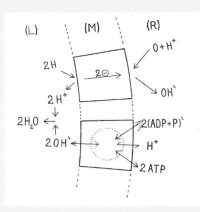

Mitchell's representation of chemiosmotic coupling between an electron transport system (top) and an ATP-generating system (bottom) in a membrane (M) enclosing aqueous phase L within aqueous phase R.

was then coupled to ATP synthesis (see figure).

The hypothesis of chemiosmotic coupling clearly explained the lack of success in identifying a chemical high-energy intermediate, as well as the fact that intact membranes were required to synthesize ATP. Yet it was a radical concept that went against the biochemical dogma of the time. In a concluding paragraph of this 1961 paper, Mitchell took a philosophical view of his revolutionary proposal:

In the exact sciences, cause and effect are no more than events linked in sequence. Biochemists now generally accept the idea that metabolism is the cause of membrane transport. The underlying thesis of the hypothesis put forward here is that if the processes that we call metabolism and transport represent events in a sequence, not only can metabolism be the cause of transport, but also transport can be the cause of metabolism.

The Impact

Mitchell's hypothesis was greeted with skepticism and remained the subject of acrimonious debate for more than a decade. However, the wealth of supporting evidence obtained by Mitchell and his colleagues, as well as by other investigators, eventually led to general acceptance of the chemiosmotic hypothesis—which became known instead as the chemiosmotic theory. It is now accepted not only as the basis for the generation of ATP during oxidative phosphorylation and photosynthesis in bacteria, mitochondria, and chloroplasts but also for the energy-requiring transport of a variety of molecules across cell membranes.

Mitchell's work was recognized with a Nobel Prize in 1978. The lecture he delivered on that occasion began as follows:

Although I had hoped that the chemiosmotic rationale of vectorial metabolism and biological energy transfer might one day come to be generally accepted, it would have been presumptuous of me to expect it to happen. Was it not Max Planck who remarked that a new scientific idea does not triumph by convincing its opponents, but rather because its opponents eventually die? The fact that what began as the chemiosmotic hypothesis has now been acclaimed as the chemiosmotic theory … has therefore both astonished and delighted me, particularly because those who were formerly my most capable opponents are still in the prime of their scientific lives.

ATP synthase is organized into two structurally distinct components, F_0 and F_1, which are linked by a slender stalk (**Figure 11.12**). The F_0 portion is an electrically driven motor that spans the inner membrane and provides a channel through which protons are able to flow back from the intermembrane space to the matrix. The energetically favorable return of protons to the matrix is coupled to ATP synthesis by the F_1 subunit, which catalyzes the synthesis of ATP from ADP and phosphate ions (P_i). Detailed structural studies have established the mechanism of ATP synthase action, which involves mechanical coupling between the F_0 and F_1 subunits. In particular, the flow of protons through F_0 drives the rotation of a part of F_1, which acts as a rotary motor to drive ATP synthesis.

It appears that the flow of four protons back across the membrane through F_0 is required to drive the synthesis of one molecule of ATP by F_1, consistent with the proton transfers at complexes I, III, and IV each contributing sufficient free energy to the proton gradient to drive the synthesis of one ATP molecule. The oxidation of one molecule of NADH thus leads to the synthesis of three molecules of ATP, whereas the oxidation of $FADH_2$, which enters the electron transport chain at complex II, yields only two ATP molecules.

■ **Another example of an electrochemical rotary motor is the bacterial flagellum.**

FIGURE 11.13 Transport of metabolites across the mitochondrial inner membrane · The transport of small molecules across the inner membrane of mitochondria is mediated by membrane-spanning proteins and driven by the electrochemical gradient. For example, ATP is exported from mitochondria to the cytosol by a transporter that exchanges it for ADP. The voltage component of the electrochemical gradient drives this exchange: ATP carries a greater negative charge (–4) than ADP (–3), so ATP is exported from the mitochondrial matrix to the cytosol while ADP is imported into mitochondria. In contrast, the transport of phosphate (P_i) and pyruvate is coupled to an exchange for hydroxyl ions (OH^-); in this case, the pH component of the electrochemical gradient drives the export of hydroxyl ions, coupled to the transport of P_i and pyruvate into mitochondria.

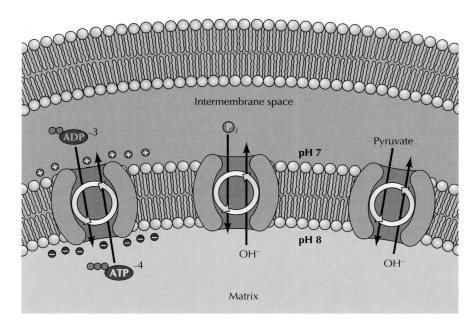

■ **All newborn mammals (and certain adult mammals) contain a specialized tissue called brown fat. The mitochondria in cells of brown fat contain an uncoupling protein called thermogenin that utilizes the proton gradient to generate heat. Brown fat is very important for thermoregulation in neonates and in hibernating mammals.**

Transport of Metabolites across the Inner Membrane

In addition to driving the synthesis of ATP, the potential energy stored in the electrochemical gradient drives the transport of small molecules into and out of mitochondria. For example, the ATP synthesized within mitochondria has to be exported to the cytosol, while ADP and P_i need to be imported from the cytosol for ATP synthesis to continue. The electrochemical gradient generated by proton pumping provides energy required for the transport of these molecules and other metabolites that need to be concentrated within mitochondria (Figure 11.13).

The transport of ATP and ADP across the inner membrane is mediated by an integral membrane protein, the adenine nucleotide translocator, which transports one molecule of ADP into the mitochondrion in exchange for one molecule of ATP transferred from the mitochondrion to the cytosol. Because ATP carries more negative charge than ADP (–4 compared to –3), this exchange is driven by the voltage component of the electrochemical gradient. Since the proton gradient establishes a positive charge on the cytosolic side of the membrane, the export of ATP in exchange for ADP is energetically favorable.

The synthesis of ATP within the mitochondrion requires phosphate ions (P_i) as well as ADP, so P_i must also be imported from the cytosol. This is mediated by another membrane transport protein, which imports phosphate ($H_2PO_4^-$) and exports hydroxyl ions (OH^-). This exchange is electrically neutral because both phosphate and hydroxyl ions have a charge of –1. However, the exchange is driven by the proton concentration gradient; the higher pH within mitochondria corresponds to a higher concentration of hydroxyl ions, favoring their translocation to the cytosolic side of the membrane.

Energy from the electrochemical gradient is similarly used to drive the transport of other metabolites into mitochondria. For example, the import of pyruvate from the cytosol (where it is produced by glycolysis) is mediated by a transport protein that exchanges pyruvate for hydroxyl ions. Other intermediates of the citric acid cycle are able to shuttle between mitochondria and the cytosol by similar exchange mechanisms.

Chloroplasts and Other Plastids

Chloroplasts, the organelles responsible for photosynthesis, are in many respects similar to mitochondria. Both chloroplasts and mitochondria function to generate metabolic energy, both have evolved by endosymbiosis, both contain their own genetic systems, and both replicate by division. However, chloroplasts are larger and more complex than mitochondria, and they perform several critical tasks in addition to the generation of ATP. Most importantly, chloroplasts are responsible for the photosynthetic conversion of CO_2 to carbohydrates. In addition, chloroplasts synthesize amino acids, fatty acids, and the lipid components of their own membranes. The reduction of nitrite (NO_2^-) to ammonia (NH_3), an essential step in the incorporation of nitrogen into organic compounds, also occurs in chloroplasts. Moreover, chloroplasts are only one of several types of related organelles (plastids) that play a variety of roles in plant cells.

The Structure and Function of Chloroplasts

Plant chloroplasts are large organelles (5 to 10 μm long) that, like mitochondria, are bounded by a double membrane called the chloroplast envelope (Figure 11.14). In addition to the inner and outer membranes of the envelope, chloroplasts have a third internal membrane system, called the thylakoid membrane. The **thylakoid membrane** forms a network of flattened discs called thylakoids, which are frequently arranged in stacks called grana. Because of this three-membrane structure, the internal organization of chloroplasts is more complex than that of mitochondria. In particular, their three membranes divide chloroplasts into three distinct internal compartments: (1) the intermembrane space between the two membranes of the chloroplast envelope; (2) the **stroma**, which lies inside the envelope but outside the thylakoid membrane; and (3) the thylakoid lumen.

Despite this greater complexity, the membranes of chloroplasts have clear functional similarities with those of mitochondria—as expected, given the role of both organelles in the chemiosmotic generation of ATP. The outer membrane of the chloroplast envelope, like that of mitochondria, contains porins and is therefore freely permeable to small molecules. In contrast, the inner membrane is impermeable to ions and metabolites, which are therefore able to enter chloroplasts only via specific membrane transporters. These properties of the inner and outer membranes of the chloroplast envelope are similar to the inner and outer membranes of mitochondria: In both cases the inner membrane restricts the passage of molecules between the cytosol and the interior of the organelle. The chloroplast stroma is also equivalent in function to the mitochondrial matrix: It contains the chloroplast genetic system and a variety of metabolic enzymes, including those responsible for the critical conversion of CO_2 to carbohydrates during photosynthesis.

The major difference between chloroplasts and mitochondria, in terms of both structure and function, is the thylakoid membrane. This membrane is of central importance in chloroplasts, where it fills the role of the inner mitochondrial membrane in electron transport and the chemiosmotic gen-

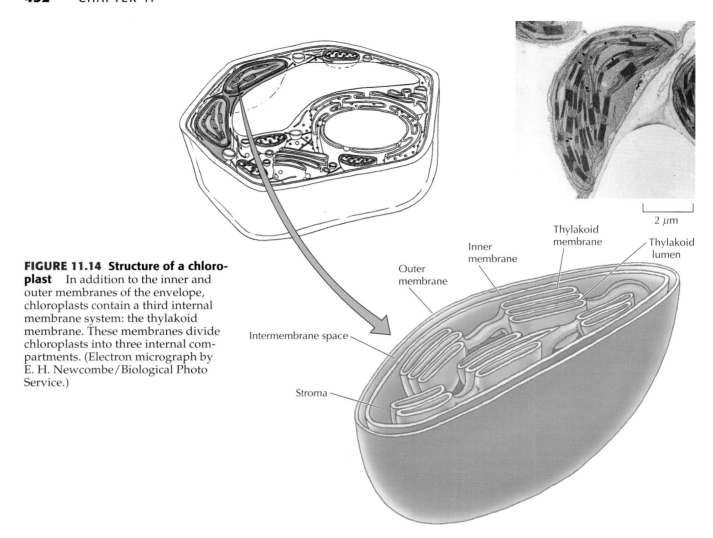

FIGURE 11.14 Structure of a chloroplast In addition to the inner and outer membranes of the envelope, chloroplasts contain a third internal membrane system: the thylakoid membrane. These membranes divide chloroplasts into three internal compartments. (Electron micrograph by E. H. Newcombe/Biological Photo Service.)

eration of ATP (Figure 11.15). The inner membrane of the chloroplast envelope (which is not folded into cristae) does not function in either electron transport or photosynthesis. Instead, the chloroplast electron transport system is located in the thylakoid membrane, and protons are pumped across this membrane from the stroma to the thylakoid lumen. The resulting electrochemical gradient then drives ATP synthesis as protons cross back into the stroma. In terms of its role in generation of metabolic energy, the thylakoid membrane of chloroplasts is thus equivalent to the inner membrane of mitochondria.

The Chloroplast Genome

Like mitochondria, chloroplasts contain their own genetic system, reflecting their evolutionary origins from photosynthetic bacteria. The 6–9 Mb genomes of present day free living photosynthetic cyanobacteria code for between 5400 and 7200 proteins. Like those of mitochondria, the genomes of chloroplasts consist of circular DNA molecules present in multiple copies per organelle. However, chloroplast genomes are larger and more complex than those of mitochondria, ranging from 120 to 160 kb and containing approximately 150 genes.

The chloroplast genomes of several plants have been completely sequenced, leading to the identification of many of the genes contained in

Mitochondrion

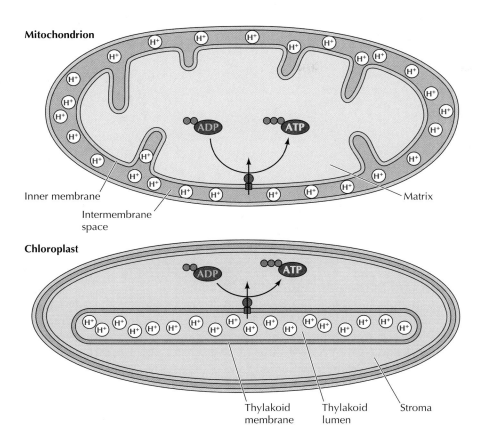

FIGURE 11.15 Chemiosmotic generation of ATP in chloroplasts and mitochondria In mitochondria, electron transport generates a proton gradient across the inner membrane, which is then used to drive ATP synthesis in the matrix. In chloroplasts, the proton gradient is generated across the thylakoid membrane and used to drive ATP synthesis in the stroma.

the organelle DNAs. These chloroplast genes encode both RNAs and proteins involved in gene expression, as well as a variety of proteins that function in photosynthesis (Table 11.2). Both the ribosomal and transfer RNAs used for translation of chloroplast mRNAs are encoded by the organelle genome. These include four rRNAs (23S, 16S, 5S, and 4.5S) and 30 tRNA species. In contrast to the smaller number of tRNAs encoded by the mitochondrial genome, the chloroplast tRNAs are sufficient to translate all the mRNA codons according to the universal genetic code. In addition to these

TABLE 11.2 Genes Encoded by Chloroplast DNA

Function	Number of genes
Genes for the genetic apparatus	
rRNAs (23S, 16S, 5S, 4.5S)	4
tRNAs	30
Ribosomal proteins	21
RNA polymerase subunits	4
Genes for photosynthesis	
Photosystem I	5
Photosystem II	12
Cytochrome *bf* complex	4
ATP synthase	6
Ribulose bisphosphate carboxylase	1

Sequence analysis indicates that chloroplast genomes contain about 30 genes in addition to those listed here. Some of these encode proteins involved in respiration, but most remain to be identified.

RNA components of the translation system, the chloroplast genome encodes about 20 ribosomal proteins, which represent approximately a third of the proteins of chloroplast ribosomes. Some subunits of RNA polymerase are also encoded by chloroplasts, although additional RNA polymerase subunits and other factors needed for chloroplast gene expression are encoded in the nucleus.

The chloroplast genome also encodes approximately 30 proteins that are involved in photosynthesis, including components of photosystems I and II, of the cytochrome *bf* complex, and of ATP synthase. In addition, one of the subunits of ribulose bisphosphate carboxylase (rubisco) is encoded by chloroplast DNA. Rubisco is the critical enzyme that catalyzes the addition of CO_2 to ribulose-1,5-bisphosphate during the Calvin cycle (see Figure 3.18). Not only is it the major protein component of the chloroplast stroma, but it is also thought to be the single most abundant protein on Earth, so it is noteworthy that one of its subunits is encoded by the chloroplast genome.

Import and Sorting of Chloroplast Proteins

Although chloroplasts encode more of their own proteins than mitochondria, about 3500 or 95% of chloroplast proteins are still encoded by nuclear genes. As with mitochondria, these proteins are synthesized on cytosolic ribosomes and then imported into chloroplasts as completed polypeptide chains. They must then be sorted to their appropriate location within chloroplasts—an even more complicated task than protein sorting in mitochondria, since chloroplasts contain three separate membranes that divide them into three distinct internal compartments.

Proteins are targeted for import into chloroplasts by N-terminal sequences of 30 to 100 amino acids, called **transit peptides**, which direct protein translocation across the two membranes of the chloroplast envelope and are then removed by proteolytic cleavage (Figure 11.16). A **guidance complex** initially recognizes the transit peptides and directs them to the translocase of the chloroplast outer member (the **Toc complex**), where they bind to the Toc159 and Toc34 receptors. In contrast to the presequences for mitochondrial import, transit peptides are not positively charged and the chloroplast inner membrane does not have a strong electric potential. Protein import into chloroplasts requires Hsp70 molecules to keep the preprotein in an unfolded state. In addition, Hsp70 molecules are attached to the Toc complex where they drive protein import by hydrolysis of ATP (see Figure 11.5). At least one Toc protein, Toc 34, binds GTP, and hydrolysis of GTP may provide an additional source of energy for translocation.

After preproteins are transported through the Toc complex, they are transferred to the translocase of the inner membrane (the **Tic complex**) and transported across the inner membrane into the stroma. Like the Toc, Tom, and Tim complexes, Tic is a multiprotein complex with one or more protein channels. However, little is known about the specific proteins, perhaps because there is more than one type of Tic complex. A chaperone of the Hsp100 family (another family of chaperones in addition to those that were discussed in Chapter 8) is associated with the stromal side of the Tic complex. This chloroplast Hsp100 acts to draw the preprotein through the inner membrane. In the stroma, the transit peptide is cleaved by a **stromal processing peptidase (SPP)**, and the protein associates with stromal Hsp70 chaperones. As in the mitochondrial matrix, some proteins that remain in the stroma complete their folding within an Hsp60 chaperonin.

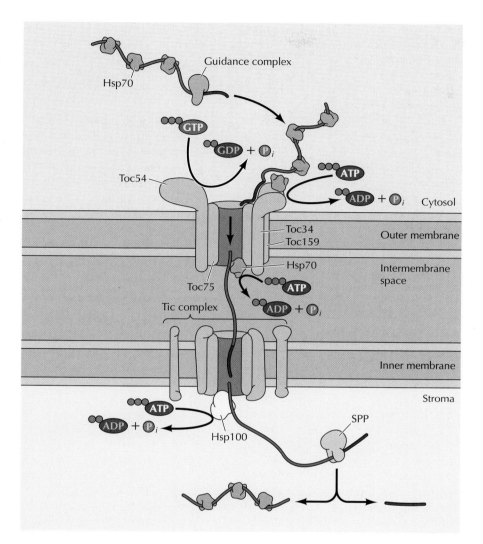

FIGURE 11.16 Import of proteins into the chloroplast stroma
Proteins with N-terminal transit peptides are recognized by a guidance complex that targets them to the Toc complex in the chloroplast outer membrane. The transit peptide binds Toc159 and Toc34, which are associated with Hsp70, before being passed to the Toc75 import pore. Passage through the outer membrane also requires ATP hydrolysis by Hsp70 in the intermembrane space, and possibly the hydrolysis of GTP by Toc34. Once through the chloroplast outer membrane, the transit peptide is passed to the Tic complex in the inner membrane. The preprotein is drawn through the Tic complex by the action of an Hsp100. In the stroma, the transit peptide is removed by the chloroplast stromal processing peptidase (SPP) and the protein interacts with Hsp70.

Little is known about how proteins are targeted to the outer or inner membranes of the chloroplast. Proteins to be incorporated into the thylakoid membrane or lumen are transported to their destination in two steps. They are first imported into the stroma, as already described, and then most are targeted for translocation across the thylakoid membrane by a second signal sequence, which is exposed following cleavage of the transit peptide. Proteins are translocated into the thylakoid lumen by at least three different pathways (Figure 11.17). In the Sec pathway, the thylakoid signal sequence is recognized by the SecA protein and translocated in an ATP-dependent manner through the Sec translocon. The second pathway (the TAT pathway) uses a twin-arginine signal sequence and depends on the proton gradient across the thylakoid membrane to translocate fully folded proteins. A third pathway (the SRP pathway) is used for thylakoid membrane proteins, which are recognized by a stromal signal recognition particle (SRP). In addition, some proteins may insert directly into the thylakoid membrane.

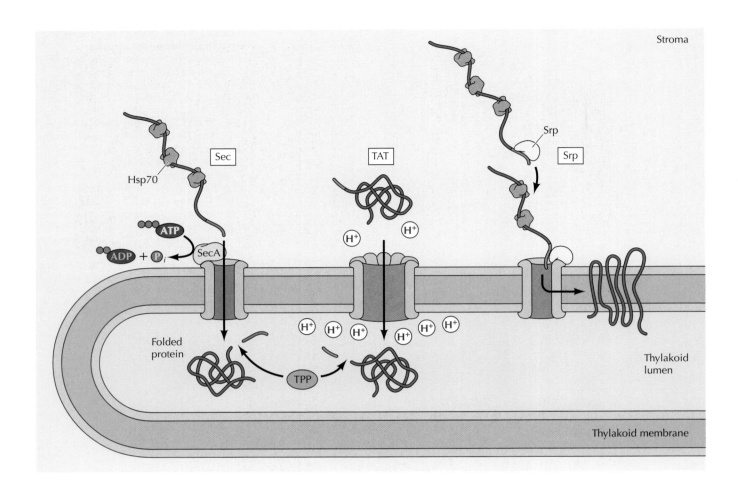

FIGURE 11.17 Import of proteins into the thylakoid lumen or membrane
Three characterized pathways transfer proteins from the stroma to the thylakoid lumen or membrane. In the Sec pathway, the SecA protein recognizes a thylakoid signal sequence and targets the protein to the Sec translocon, using energy derived from ATP hydrolysis to transfer the protein to the lumen. In the lumen, the thylakoid signal sequence is cleaved by a thylakoid processing protease (TPP). In the twin-arginine-translocation (TAT) pathway, a cleavable thylakoid transfer sequence containing two arginines near the amino terminus targets fully folded proteins directly to a novel translocon, which inserts them into the thylakoid lumen. Both assembly of the translocon and translocation of the protein are dependent on the proton gradient across the thylakoid membrane. In the Srp pathway, transmembrane proteins are recognized by the chloroplast signal recognition particle (Srp) and inserted into the thylakoid membrane.

Other Plastids

Chloroplasts are only one, albeit the most prominent, member of a larger family of plant organelles called **plastids**. All plastids contain the same genome as chloroplasts, but they differ in both structure and function. Chloroplasts are specialized for photosynthesis and are unique in that they contain the internal thylakoid membrane system. Other plastids, which are involved in different aspects of plant cell metabolism (synthesis of amino acids, fatty acids and lipids, plant hormones, nucleotides, vitamins, and sec-

(A)

(B)

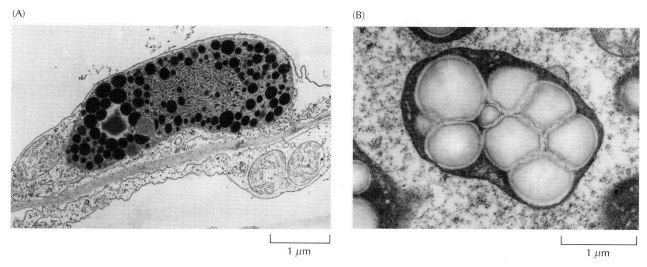

1 μm

1 μm

FIGURE 11.18 Electron micrographs of chromoplasts and amyloplasts
(A) Chromoplasts contain lipid droplets in which carotenoids are stored.
(B) Amyloplasts contain large starch granules. (A, Biophoto Associates/Photo Researchers, Inc.; B, Dr. Jeremy Burgess/Photo Researchers, Inc.)

ondary metabolites), are bounded by the two membranes of the plastid envelope but lack both the thylakoid membranes and other components of the photosynthetic apparatus.

The different types of plastids are frequently classified according to the kinds of pigments they contain. Chloroplasts are so named because they contain chlorophyll. **Chromoplasts** (Figure 11.18A) lack chlorophyll but contain carotenoids; they are responsible for the yellow, orange, and red colors of some flowers and fruits, although their precise function in cell metabolism is not clear. **Leucoplasts** are nonpigmented plastids, which store a variety of energy sources in nonphotosynthetic tissues. **Amyloplasts** (Figure 11.18B) and **elaioplasts** are examples of leucoplasts that store starch and lipids, respectively.

All plastids, including chloroplasts, develop from **proplastids**, small (0.5 to 1 μm in diameter) undifferentiated organelles present in the rapidly dividing cells of plant roots and shoots. Proplastids then develop into the various types of mature plastids according to the needs of differentiated cells. In addition, mature plastids are able to change from one type to another. Chromoplasts develop from chloroplasts, for example, during the ripening of fruit (e.g., tomatoes). During this process, chlorophyll and the thylakoid membranes break down, while new types of carotenoids are synthesized.

An interesting feature of plastids is that their development is controlled both by environmental signals and by intrinsic programs of cell differentiation. In the photosynthetic cells of leaves, for example, proplastids develop into chloroplasts (Figure 11.19). During this process, the thylakoid membrane is formed by vesicles budding from the inner membrane of the plastid envelope and the various components of the photosynthetic apparatus are synthesized and assembled. However, chloroplasts develop only in the presence of light. If plants are kept in the dark, the development of proplastids in leaves is arrested at an intermediate stage (called **etioplasts**) in

FIGURE 11.19 Development of chloroplasts Chloroplasts develop from proplastids in the photosynthetic cells of leaves. Proplastids contain only the inner and outer envelope membranes; the thylakoid membrane is formed by vesicle budding from the inner membrane during chloroplast development. If the plant is kept in the dark, chloroplast development is arrested at an intermediate stage (etioplasts). Etioplasts lack chlorophyll and contain semicrystalline arrays of membrane tubules. In the presence of light, they continue their development to chloroplasts.

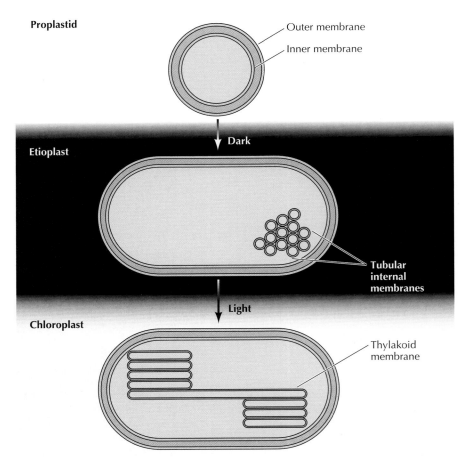

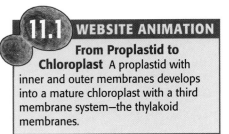

WEBSITE ANIMATION

From Proplastid to Chloroplast A proplastid with inner and outer membranes develops into a mature chloroplast with a third membrane system—the thylakoid membranes.

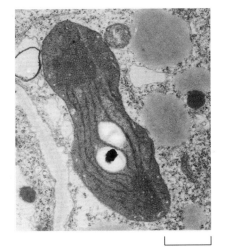

FIGURE 11.20 Electron micrograph of an etioplast (John N. A. Lott/Biological Photo Service.)

which a semicrystalline array of tubular internal membranes has formed but chlorophyll has not been synthesized (Figure 11.20). If dark-grown plants are then exposed to light, the etioplasts continue their development to chloroplasts. It is noteworthy that this dual control of plastid development involves the coordinated expression of genes within both the plastid and nuclear genomes. The mechanisms responsible for such coordinated gene expression are largely unknown, and their elucidation represents a challenging problem in plant molecular biology.

Photosynthesis

During photosynthesis, energy from sunlight is harvested and used to drive the synthesis of glucose from CO_2 and H_2O. By converting the energy of sunlight to a usable form of potential chemical energy, photosynthesis is the ultimate source of metabolic energy for all biological systems. Photosynthesis takes place in two distinct stages. In the light reactions, energy from sunlight drives the synthesis of ATP and NADPH, coupled to the formation of O_2 from H_2O. In the dark reactions (so named because they do not require sunlight) the ATP and NADPH produced by the light reactions drive glucose synthesis. In eukaryotic cells, both the light and dark reactions of photosynthesis occur within chloroplasts—the light reactions in the thylakoid

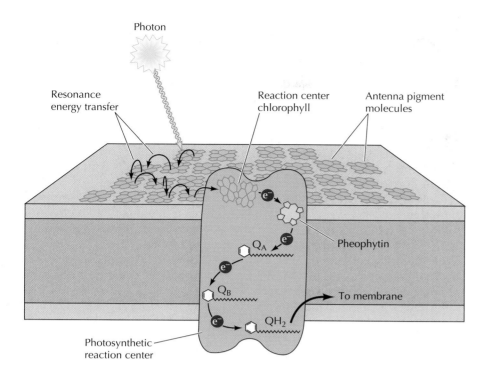

FIGURE 11.21 Organization of a photocenter Each photocenter consists of hundreds of antenna pigment molecules, which absorb photons and transfer energy to a reaction center chlorophyll. The reaction center chlorophyll then transfers its excited electron to an acceptor in the electron transport chain. The reaction center illustrated is that of photosystem II in which electrons are transferred from the reaction center chlorophyll to pheophytin and then to quinones (Q_A, Q_B, and QH_2).

membrane and the dark reactions within the stroma. This section discusses the light reactions of photosynthesis, which are related to oxidative phosphorylation in mitochondria. The dark reactions were discussed in detail in Chapter 3.

Electron Flow through Photosystems I and II

Sunlight is absorbed by photosynthetic pigments, the most abundant of which in plants are the **chlorophylls**. Absorption of light excites an electron to a higher energy state, thus converting the energy of sunlight to potential chemical energy. The photosynthetic pigments are organized into **photocenters** in the thylakoid membrane, each of which contains hundreds of pigment molecules (Figure 11.21). The many pigment molecules in each photocenter act as antennae to absorb light and transfer the energy of their excited electrons to a chlorophyll molecule that serves as a reaction center. The reaction center chlorophyll then transfers its high-energy electron to an acceptor molecule in an electron transport chain. High-energy electrons are then transferred through a series of membrane carriers, coupled to the synthesis of ATP and NADPH.

The proteins involved in the light reactions of photosynthesis are organized into multiprotein complexes in photosynthetic membranes (Figure 11.22). The earliest characterized photosynthetic reaction center was that of the bacterium *Rhodopseudomonas viridis*, the structure of which was determined by Johann Deisenhofer, Hartmut Michel, Robert Huber, and their colleagues in 1985 (Figure 11.23). The reaction center consists of three trans-

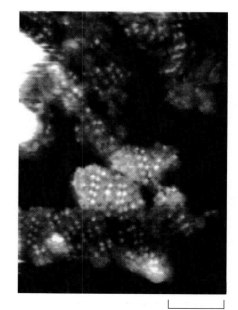

FIGURE 11.22 Photosynthetic complexes Atomic force micrograph of photosynthetic membrane fragments from the purple bacterium *Rhodobacter sphaeroides* showing the high density of photosynthetic complexes. (From S. Bahatyrova et al., 2004. *Nature* 430: 1059.)

1 nm

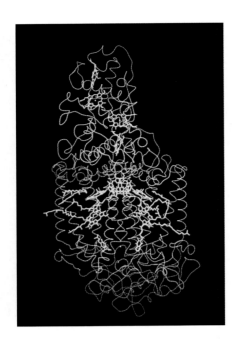

FIGURE 11.23 Structure of a photosynthetic reaction center The reaction center of *R. viridis* consists of three transmembrane proteins (purple, blue, and beige) and a *c*-type cytochrome (green). Chlorophylls and other prosthetic groups are colored yellow. (Courtesy of Johann Deisenhofer, University of Texas Medical Center and The Nobel Foundation, 1989.)

■ A variety of compounds that bind photosystem II and inhibit electron transfer are used as herbicides for the control of weeds.

membrane polypeptides, bound to a *c*-type cytochrome on the exterior side of the membrane. Energy from sunlight is captured by a pair of chlorophyll molecules known as the special pair. Electrons are then transferred from the special pair to another pair of chlorophylls and from there to other prosthetic groups (pheophytins and quinones). From there the electrons are transferred to a cytochrome *bc* complex in which electron transport is coupled to the generation of a proton gradient. The electrons are then transferred to the reaction center cytochrome and finally returned to the chlorophyll special pair. The reaction center thus converts the energy of sunlight to high-energy electrons, the potential energy of which is converted to a proton gradient by the cytochrome *bc* complex.

The proteins involved in the light reactions of photosynthesis in plants are organized into five multiprotein complexes in the thylakoid membrane (Figure 11.24). Two of these complexes are photosystems (**photosystems I and II**) in which light is absorbed and transferred to reaction center chlorophylls. High-energy electrons are then transferred through a series of carriers in both photosystems and in a third protein complex, the **cytochrome *bf* complex**. As in mitochondria, these electron transfers are coupled to the transfer of protons into the thylakoid lumen, thereby establishing a proton gradient across the thylakoid membrane. The energy stored in this proton gradient is then harvested by a fourth protein complex in the thylakoid membrane, ATP synthase, which (like the mitochondrial enzyme) couples proton flow back across the membrane to the synthesis of ATP.

One important difference between electron transport in chloroplasts and that in mitochondria is that the energy derived from sunlight during photosynthesis not only is converted to ATP but also is used to generate the NADPH required for subsequent conversion of CO_2 to carbohydrates. This is accomplished by the use of two different photosystems in the light reactions of photosynthesis, one to generate ATP and the other to generate NADPH. Electrons are transferred sequentially between the two photosystems, with photosystem I acting to generate NADPH and photosystem II acting to generate ATP.

The pathway of electron flow starts at photosystem II, which is homologous to the photosynthetic reaction center of *R. viridis*, already described. However, at photosystem II the energy derived from absorption of photons is used to split water molecules to molecular oxygen and protons (see Figure 11.24). This reaction takes place within the thylakoid lumen, so the release of protons from H_2O establishes a proton gradient across the thylakoid membrane. The high-energy electrons derived from this process are transferred through a series of carriers to plastoquinone, a lipid-soluble carrier similar to coenzyme Q (ubiquinone) of mitochondria. Plastoquinone carries electrons from photosystem II to the cytochrome *bf* complex, within which electrons are transferred to plastocyanin and additional protons are pumped into the thylakoid lumen. Electron transport through photosystem II is thus coupled to establishment of a proton gradient, which drives the chemiosmotic synthesis of ATP.

From photosystem II, electrons are carried by plastocyanin (a peripheral membrane protein) to photosystem I, where the absorption of additional photons again generates high-energy electrons. Photosystem I, however, does not act as a proton pump; instead, it uses these high-energy electrons to reduce $NADP^+$ to NADPH. The reaction center chlorophyll of photosystem I transfers its excited electrons through a series of carriers to ferrodoxin, a small protein on the stromal side of the thylakoid membrane, which then complexes with the enzyme **NADP reductase** and transfers electrons from

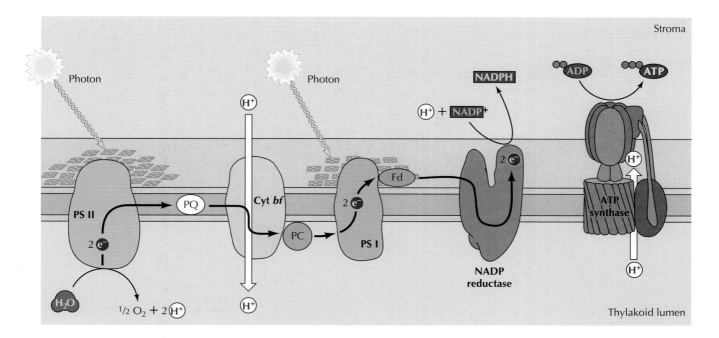

FIGURE 11.24 Electron transport and ATP synthesis during photosynthesis
Four protein complexes in the thylakoid membrane function in electron transport and the synthesis of ATP and NADPH. Photons are absorbed by complexes of pigment molecules associated with photosystems I and II (PS I and PS II). At photosystem II, energy derived from photon absorption is used to split a water molecule within the thylakoid lumen. Electrons are then carried by plastoquinone (PQ) to the cytochrome *bf* complex, where they are transferred to a lower energy state and protons are pumped into the thylakoid lumen. Electrons are then transferred to photosystem I by plastocyanin (PC). At photosystem I, energy derived from light absorption again generates high-energy electrons, which are transferred to ferrodoxin (Fd) and used to reduce $NADP^+$ to NADPH in the stroma. ATP synthase then uses the energy stored in the proton gradient to convert ADP to ATP.

11.2 WEBSITE ANIMATION

The Light Reactions
During the light reactions of photosynthesis, energy absorbed from sunlight drives the synthesis of ATP and NADPH, coupled to the oxidation of H_2O to O_2.

ferrodoxin to $NADP^+$, generating NADPH. The passage of electrons through photosystems I and II thus generates both ATP and NADPH, which are used by the Calvin cycle enzymes in the chloroplast stroma to convert CO_2 to carbohydrates (see Figure 3.18).

Cyclic Electron Flow

A second electron transport pathway, called **cyclic electron flow**, produces ATP without the synthesis of NADPH, thereby supplying additional ATP for other metabolic processes. In cyclic electron flow, light energy harvested at photosystem I is used for ATP synthesis rather than NADPH synthesis (Figure 11.25). Instead of being transferred to $NADP^+$, high-energy electrons from photosystem I are transferred to the cytochrome *bf* complex. Electron transfer through the cytochrome *bf* complex is then coupled, as in photosystem II, to the establishment of a proton gradient across the thylakoid membrane. Plastocyanin then returns these electrons to photosystem I in a lower energy state, completing a cycle of electron transport in which light energy harvested at photosystem I is used to pump protons at the cytochrome *bf* complex. Electron transfer from photosystem I can thus generate either ATP or NADPH, depending on the metabolic needs of the cell.

FIGURE 11.25 The pathway of cyclic electron flow   Light energy absorbed at photosystem I (PS I) is used for ATP synthesis rather than NADPH synthesis. High-energy electrons generated by photon absorption are transferred to the cytochrome *bf* complex rather than to NADP⁺. At the cytochrome *bf* complex, electrons are transferred to a lower energy state and protons are pumped into the thylakoid lumen. The electrons are then returned to photosystem I by plastocyanin (PC).

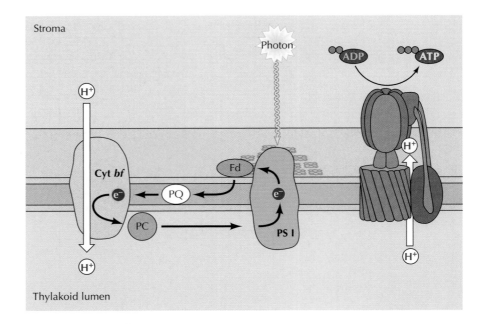

ATP Synthesis

The ATP synthase of the thylakoid membrane is similar to the mitochondrial enzyme. However, the energy stored in the proton gradient across the thylakoid membrane, in contrast to the inner mitochondrial membrane, is almost entirely chemical in nature. This is because the thylakoid membrane, although impermeable to protons, differs from the inner mitochondrial membrane in being permeable to other ions, particularly Mg^{2+} and Cl^-. The free passage of these ions neutralizes the voltage component of the proton gradient, so the energy derived from photosynthesis is conserved mainly as the difference in proton concentration (pH) across the thylakoid membrane. However, because the thylakoid lumen is a closed compartment, this difference in proton concentration can be quite large, corresponding to a differential of more than three pH units between the stroma and the thylakoid lumen. Because of the magnitude of this pH differential, the total free energy stored across the thylakoid membrane is similar to that stored across the inner mitochondrial membrane.

For each pair of electrons transported, two protons are transferred across the thylakoid membrane at photosystem II and two to four protons at the cytochrome *bf* complex. Since four protons are needed to drive the synthesis of one molecule of ATP, passage of each pair of electrons through photosystems I and II by noncyclic electron flow yields between 1 and 1.5 ATP molecules. Cyclic electron flow has a lower yield, corresponding to between 0.5 and 1 ATP molecules per pair of electrons.

Peroxisomes

Peroxisomes are small, single membrane-enclosed organelles (Figure 11.26) that contain enzymes involved in a variety of metabolic reactions, including several aspects of energy metabolism. Most human cells contain about 500 peroxisomes. Peroxisomes do not have their own genomes and all their pro-

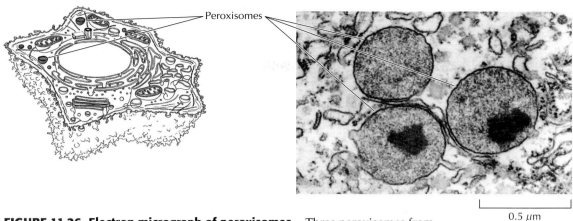

FIGURE 11.26 Electron micrograph of peroxisomes   Three peroxisomes from rat liver are shown. Two contain dense regions, which are paracrystalline arrays of the enzyme urate oxidase. (Don Fawcett/Photo Researchers, Inc.)

teins, **peroxins** (Pex1, Pex2, etc.), are synthesized from the nuclear genome. Most peroxins are synthesized on free ribosomes and then imported into peroxisomes as completed polypeptide chains. Like mitochondria and chloroplasts, peroxisomes can replicate by division. However, unlike those organelles, peroxisomes can also be rapidly regenerated even if entirely lost to the cell. While many mitochondrial and plastid proteins resemble those of prokaryotes, reflecting their endosymbiotic origin, the peroxins resemble typical eukaryotic proteins.

Functions of Peroxisomes

Peroxisomes contain at least 50 different enzymes, which are involved in a variety of biochemical pathways in different types of cells. Peroxisomes originally were defined as organelles that carry out oxidation reactions leading to the production of hydrogen peroxide. Because hydrogen peroxide is harmful to the cell, peroxisomes also contain the enzyme **catalase**, which decomposes hydrogen peroxide either by converting it to water or by using it to oxidize another organic compound. A variety of substrates are broken down by such oxidative reactions in peroxisomes, including uric acid, amino acids, purines, methanol, and fatty acids. The oxidation of fatty acids (Figure 11.27) is a particularly important example, since it provides a major source of metabolic energy. In animal cells, fatty acids are oxidized in

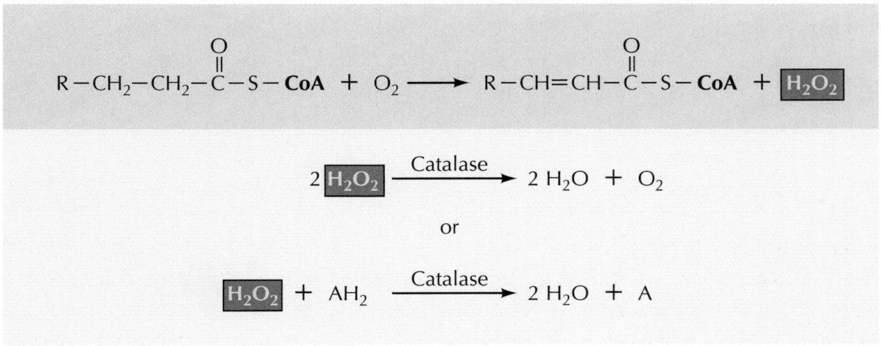

$$R-CH_2-CH_2-\overset{\overset{\textstyle O}{\|}}{C}-S-\textbf{CoA} + O_2 \longrightarrow R-CH=CH-\overset{\overset{\textstyle O}{\|}}{C}-S-\textbf{CoA} + \boxed{H_2O_2}$$

$$2\,\boxed{H_2O_2} \xrightarrow{\text{Catalase}} 2\,H_2O + O_2$$

or

$$\boxed{H_2O_2} + AH_2 \xrightarrow{\text{Catalase}} 2\,H_2O + A$$

FIGURE 11.27 Fatty acid oxidation in peroxisomes   The oxidation of a fatty acid is accompanied by the production of hydrogen peroxide (H_2O_2) from oxygen. The hydrogen peroxide is decomposed by catalase, either by conversion to water or by oxidation of another organic compound (designated AH_2).

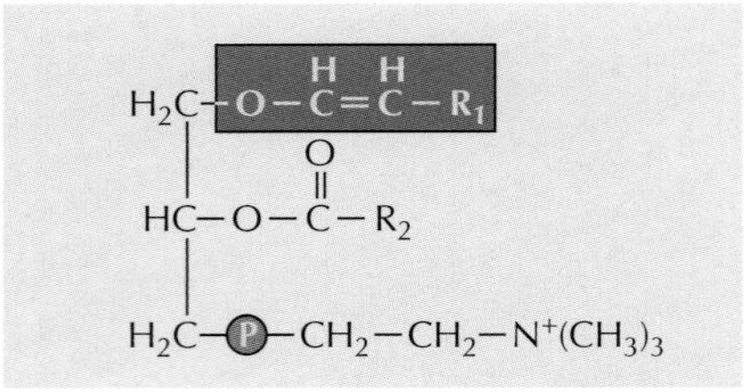

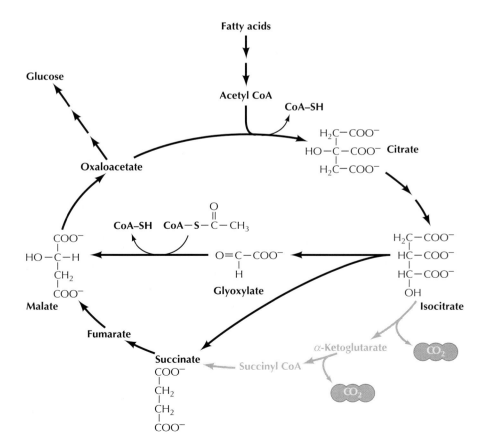

FIGURE 11.28 Structure of a plasmalogen The plasmalogen shown is analogous to phosphatidylcholine. However, one of the fatty acid chains is joined to glycerol by an ether, rather than an ester, bond.

both peroxisomes and mitochondria, but in yeasts and plants, fatty acid oxidation is restricted to peroxisomes.

In addition to providing a compartment for oxidation reactions, peroxisomes are involved in biosynthesis of lipids and the amino acid, lysine. In animal cells, cholesterol and dolichol are synthesized in peroxisomes as well as in the ER. In the liver, peroxisomes are also involved in the synthesis of bile acids, which are derived from cholesterol. In addition, peroxisomes contain enzymes required for the synthesis of **plasmalogens**—a family of phospholipids in which one of the hydrocarbon chains is joined to glycerol by an ether bond rather than an ester bond (Figure 11.28). Plasmalogens are important membrane components in some tissues, particularly heart and brain, although they are absent in others. Peroxisomes carry out different biochemical reactions in different tissues. However, it is currently unknown whether there are subpopulations of peroxisomes that specialize in one or a limited number of processes within a cell.

Peroxisomes play two particularly important roles in plants. First, peroxisomes in seeds are responsible for the conversion of stored fatty acids to carbohydrates, which is critical to providing energy and raw materials for growth of the germinating plant. This occurs via a series of reactions termed the **glyoxylate cycle**, which is a variant of the citric acid cycle (Figure 11.29). The peroxisomes in which this takes place are sometimes called **glyoxysomes**.

Second, peroxisomes in leaves are involved in **photorespiration**, which serves to metabolize a side product formed during photosynthesis (Figure 11.30). CO_2 is converted to carbohydrates during photosynthesis via a series

FIGURE 11.29 The glyoxylate cycle Plants are capable of synthesizing carbohydrates from fatty acids via the glyoxylate cycle, which is a variant of the citric acid cycle (see Figure 3.13). As in the citric acid cycle, acetyl CoA combines with oxaloacetate to form citrate, which is converted to isocitrate. However, instead of being degraded to CO_2 and α-ketoglutarate (gray arrows), isocitrate is converted to succinate and glyoxylate. Glyoxylate then reacts with another molecule of acetyl CoA to yield malate, which is converted to oxaloacetate and used for glucose synthesis.

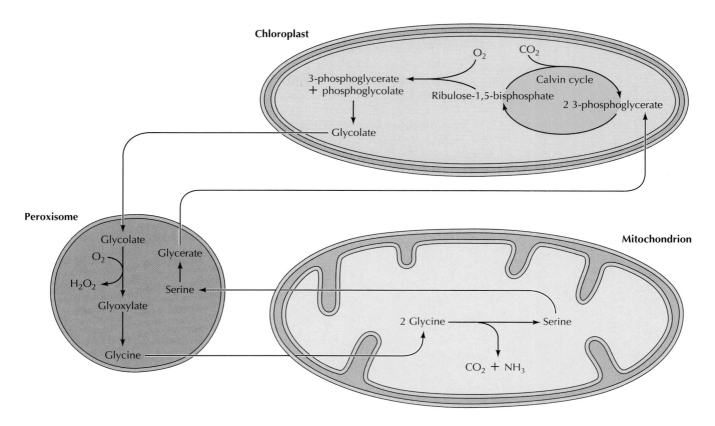

FIGURE 11.30 Role of peroxisomes in photorespiration During photosynthesis CO_2 is converted to carbohydrates by the Calvin cycle, which initiates with the addition of CO_2 to the five-carbon sugar ribulose-1,5-bisphosphate. However, the enzyme involved sometimes catalyzes the addition of O_2 instead, resulting in production of the two-carbon compound phosphoglycolate. Phosphoglycolate is converted to glycolate, which is then transferred to peroxisomes, where it is oxidized and converted to glycine. Glycine is then transferred to mitochondria and converted to serine. The serine is returned to peroxisomes and converted to glycerate, which is transferred back to chloroplasts.

of reactions called the Calvin cycle (see Figure 3.18). The first step is the addition of CO_2 to the five-carbon sugar ribulose-1,5-bisphosphate, yielding two molecules of 3-phosphoglycerate (three carbons each). However, the enzyme involved (ribulose bisphosphate carboxylase or rubisco) sometimes catalyzes the addition of O_2 instead of CO_2, producing one molecule of 3-phosphoglycerate and one molecule of phosphoglycolate (two carbons). This is a side reaction, and phosphoglycolate is not a useful metabolite. It is first converted to glycolate and then transferred to peroxisomes, where it is oxidized and converted to glycine. Glycine is then transferred to mitochondria, where two molecules of glycine are converted to one molecule of serine, with the loss of CO_2 and NH_3. The serine is then returned to peroxisomes, where it is converted to glycerate. Finally, the glycerate is transferred back to chloroplasts, where it reenters the Calvin cycle. Photorespiration does not appear to be beneficial for the plant, since it is essentially the reverse of photosynthesis—O_2 is consumed and CO_2 is released without any gain of ATP. However, the occasional utilization of O_2 in place of CO_2 appears to be an inherent property of rubisco, so photorespiration is a general accompaniment of photosynthesis. Peroxisomes thus play an important role by allowing most of the carbon in glycolate to be recovered and utilized.

Peroxisome Assembly

Although most peroxins are synthesized on free cytosolic ribosomes and then imported to peroxisomes, recent experiments indicate that peroxisome assembly begins on the rough ER, where two peroxins, Pex3 and Pex19, initially localize (Figure 11.31). Pex3 is an integral transmembrane protein while Pex19 is a farnesylated protein found largely in the cytosol. Pex3 recruits Pex19 to the ER membrane, where their interaction causes Pex3/Pex19-containing vesicles to bud off the ER. These vesicles may then fuse either with pre-existing peroxisomes or with one another to form new peroxisomes.

Pex3, Pex19, and other peroxisomal membrane proteins then act as receptors for import of the other peroxins, which are translated on free cytosolic ribosomes and then transported into peroxisomes as completed and folded polypeptides. They are targeted to the interior of peroxisomes by at least two pathways, which are conserved from yeasts to humans. Most peroxins are targeted by the simple amino acid sequence Ser-Lys-Leu

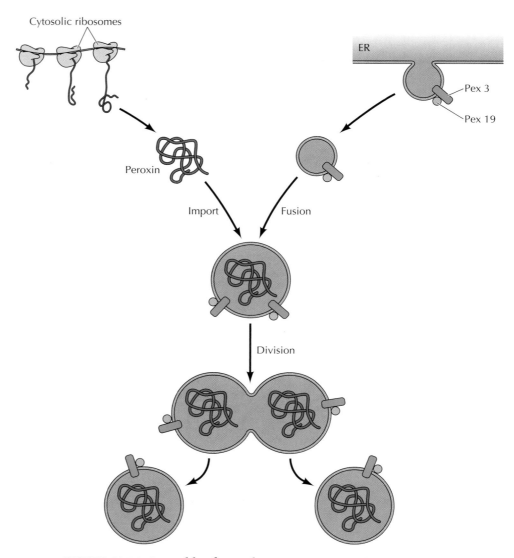

FIGURE 11.31 Assembly of peroxisomes Initiation of peroxisome assembly begins in the rough ER when the transmembrane protein peroxin 3 (Pex3) recruits the soluble farnesylated protein peroxin 19 (Pex19) and initiates budding of a nascent peroxisome. These nascent peroxisomes fuse either with each other or with existing peroxisomes. Additional peroxins are synthesized on free cytosolic ribosomes and imported as completed polypeptide chains to form functional peroxisomes, which grow larger and divide.

at their carboxy terminus (peroxisome targeting signal 1, or PTS1). A small number of proteins are targeted by a sequence of nine amino acids at their amino terminus (peroxisome targeting signal 2, or PTS2). PTS1 and PTS2 are recognized by distinct cytosolic receptors and then passed through a poorly understood channel in the peroxisomal membrane into the matrix. The receptors are subsequently retrieved from the peroxisome and recycled. Unlike the translocation of polypeptide chains across the membranes of the endoplasmic reticulum, mitochondria, and chloroplasts, the targeting signals are usually not cleaved during the import of proteins into peroxisomes and the mechanism of translocation is not known.

Protein import, together with the continuing addition of lipids from the rough ER, results in peroxisome growth, and new peroxisomes can be formed by division of old ones. In addition, peroxisomes undergo a complex maturation process that involves the import of different classes of proteins from the cytosol at different times. As a result, the enzyme content, and therefore the metabolic activities, of peroxisomes may change as they mature.

Interestingly, some components of peroxisome import pathways have been identified not only as mutants of yeasts but also as mutations associated with serious human diseases involving disorders of peroxisomes. In some such diseases, only a single peroxisomal enzyme is deficient. However, in other diseases resulting from defects in peroxisome function, multiple peroxisomal enzymes fail to be imported to peroxisomes and instead are localized in the cytosol. The latter group of diseases results from deficiencies in the PTS1 or PTS2 pathways responsible for peroxisomal protein import. The prototypical example is Zellweger syndrome, which is lethal within the first ten years of life. Zellweger syndrome can result from mutations in at least ten different genes affecting peroxisomal protein import, one of which has been identified as the gene encoding the receptor for the peroxisome targeting signal PTS1.

COMPANION WEBSITE

Visit the website that accompanies **The Cell** (www.sinauer.com/cooper) for animations, videos, quizzes, problems, and other review material.

SUMMARY

MITOCHONDRIA

Organization and Function of Mitochondria: Mitochondria, which play a critical role in the generation of metabolic energy, are surrounded by a double-membrane system. The matrix contains the enzymes of the citric acid cycle; the inner membrane contains protein complexes involved in electron transport and oxidative phosphorylation. In contrast to the inner membrane, the outer membrane is freely permeable to small molecules.

The Genetic System of Mitochondria: Mitochondria contain their own genomes, which encode rRNAs, tRNAs, and some of the proteins that are involved in oxidative phosphorylation.

Protein Import and Mitochondrial Assembly: Most mitochondrial proteins are encoded by the nuclear genome. These proteins are translated on free ribosomes and imported into mitochondria as completed polypeptide chains. Positively charged presequences target proteins for import to the mitochondrial matrix. Phospholipids are carried to mitochondria from the endoplasmic reticulum by phospholipid transfer proteins.

KEY TERMS

mitochondria, crista, matrix, porin

endosymbiosis

presequence, Tom complex, Tim complex, matrix processing peptidase (MPP), phospholipid transfer protein, cardiolipin

KEY TERMS

oxidative phosphorylation, electron transport chain, coenzyme Q, ubiquinone, cytochrome *c*, cytochrome oxidase

chemiosmotic coupling, electrochemical gradient, ATP synthase

chloroplast, thylakoid membrane, stroma

transit peptide, guidance complex, Toc complex, Tic complex, stromal processing peptidase (SPP)

plastid, chromoplast, leucoplast, amyloplast, elaioplast, proplastid, etioplast

chlorophyll, photocenter, photosystem I, photosystem II, cytochrome *bf* complex, NADP reductase

SUMMARY

THE MECHANISM OF OXIDATIVE PHOSPHORYLATION

The Electron Transport Chain: Most of the energy derived from oxidative metabolism comes from the transfer of electrons from NADH and $FADH_2$ to O_2. In order to harvest this energy in usable form, electrons are transferred through a series of carriers organized into four protein complexes in the inner mitochondrial membrane.

Chemiosmotic Coupling: The energy-yielding reactions of electron transport are coupled to the generation of a proton gradient across the inner mitochondrial membrane. The potential energy stored in this gradient is harvested by a fifth protein complex, ATP synthase, which couples ATP synthesis to the energetically favorable return of protons to the mitochondrial matrix.

Transport of Metabolites across the Inner Membrane: In addition to driving ATP synthesis, potential energy stored in the proton gradient drives the transport of ATP, ADP, and other metabolites into and out of mitochondria.

CHLOROPLASTS AND OTHER PLASTIDS

The Structure and Function of Chloroplasts: Chloroplasts are large organelles that function in photosynthesis and a variety of other metabolic activities. Like mitochondria, chloroplasts are bounded by a double-membrane envelope. In addition, chloroplasts have an internal thylakoid membrane, which is the site of electron transport and the chemiosmotic generation of ATP.

The Chloroplast Genome: Chloroplast genomes contain more than 100 genes, including genes encoding rRNAs, tRNAs, some ribosomal proteins, and some proteins involved in photosynthesis.

Import and Sorting of Chloroplast Proteins: Most chloroplast proteins are synthesized on free ribosomes in the cytosol and targeted for import to chloroplasts by amino-terminal transit peptides. Most proteins incorporated into the thylakoid lumen are first imported into the chloroplast stroma and then targeted for transport across the thylakoid membrane by several different pathways.

Other Plastids: The chloroplast is only one member of a family of related organelles, all of which contain the same genome. Other plastids serve to store energy sources, such as starch and lipids, and function in other aspects of plant metabolism.

PHOTOSYNTHESIS

Electron Flow through Photosystems I and II: During photosynthesis, energy from sunlight is harvested and converted to usable forms of potential chemical energy. Absorption of light by chlorophylls excites electrons to a higher energy state. These high-energy electrons are then transferred through a series of carriers organized into two photosystems and the cytochrome *bf* complex in the thylakoid membrane. The sequential flow of electrons through both photosystems is coupled to the syn-

SUMMARY

KEY TERMS

thesis of ATP at photosystem II and the reduction of NADP$^+$ to NADPH at photosystem I. Both ATP and NADPH are then used in the synthesis of carbohydrates from CO_2, which takes place in the chloroplast stroma.

Cyclic Electron Flow: The alternative pathway of cyclic electron flow allows light energy harvested at photosystem I to be converted to ATP, rather than NADPH.

cyclic electron flow

ATP Synthesis: The chemiosmotic synthesis of ATP is driven by a proton gradient across the thylakoid membrane.

PEROXISOMES

Functions of Peroxisomes: Peroxisomes are small organelles, bounded by a single membrane, that contain enzymes involved in a variety of metabolic reactions, including fatty acid oxidation, the glyoxylate cycle, and photorespiration.

peroxisome, peroxin, catalase, plasmalogen, glyoxylate cycle, glyoxysome, photorespiration

Peroxisome Assembly: Peroxisome assembly begins on the ER with the formation of specific vesicles. However, most peroxisomal proteins are synthesized on free ribosomes in the cytosol and imported to peroxisomes as complete polypeptide chains. At least two types of signals target proteins to the interior of peroxisomes, but the mechanism of protein import is not well understood.

Questions

1. What two properties of the mitochondrial inner membrane allow it to have unusually high metabolic activity?

2. According to standard "wobble" rules, protein synthesis requires a minimum of 30 different tRNAs. How do human mitochondria manage to translate mRNAs into proteins with only 22 tRNA species?

3. Assume that the electric potential across the inner mitochondrial membrane is dissipated, so the electrochemical gradient is composed solely of a proton concentration gradient corresponding to one pH unit. Calculate the free energy stored in this gradient. For your calculation, use $R = 1.98 \times 10^{-3}$ kcal/mol/deg, $T = 298°K$ (25°C), and note that $\ln(x) = 2.3 \log_{10}(x)$.

4. What roles do molecular chaperones play in mitochondrial protein import?

5. Cytochrome *b2* is synthesized in the cytosol and has a second, hydrophobic signal sequence behind the usual positively charged mitochondrial presequence. Suggest a pathway that will get it to its final location in the mitochondrial intermembrane space.

6. What are the roles of coenzyme Q and cytochome *c* in the electron transport chain?

7. ATP synthase consists of two multisubunit complexes, F_0 and F_1. Where is each one located in mitochondria and in chloroplasts, and what is the function of each?

8. Why do the transit peptides of chloroplast proteins, in contrast to the presequences of mitochondrial proteins, not need to be positively charged?

9. How many high-energy electrons are required to drive the synthesis of one molecule of glucose during photosynthesis, coupled to the formation of six molecules of O_2? How many molecules of ATP and NADPH are generated by passage of these electrons through photosystems I and II?

10. What fraction of the carbon atoms converted to glycolate during photorespiration is salvaged by peroxisomes?

11. How are proteins targeted to peroxisomes?

12. Why is the energy stored across the thylakoid membrane almost entirely chemical?

References and Further Reading

Mitochondria

Chen, X. J. and R. A. Butow. 2005. The organization and inheritance of the mitochondrial genome. *Nat. Rev. Genet.* 6: 815–825. [R]

Dietrich, A., J. H. Weil and L. Marechal-Drouard. 1992. Nuclear-encoded transfer RNAs in plant mitochondria. *Ann. Rev. Cell Biol.* 8: 115–131. [R]

Gabriel, K., S. K. Buchanan and T. Lithgow. 2001. The alpha and the beta: Protein translocation across mitochondrial and plastid outer membranes. *Trends Biochem. Sci.* 26: 36–40. [R]

Howell, N., J. L. Elson, P. F. Chinnery and D. M. Turnbull. 2005. mtDNA mutations and common neurodegenerative disorders. *Trends Genet.* 21: 583–586. [R]

Jensen, R. E. and C. D. Dunn. 2002. Protein import into and across the mitochondrial inner membrane: Role of the TIM23 and TIM22 translocons. *Biochim. Biophys. Acta* 1592: 25–34. [R]

Jensen, R. E., C. D. Dunn, M. J. Youngman and H. Sesaki. 2004. Mitochondrial building blocks. *Trends Cell Biol.* 14: 215–218. [R]

Millar, A. H., J. L. Heazlewood, B. K. Kristensen, H. P. Braun and I. M. Moller. 2005. The plant mitochondrial proteome. *Trends Plant Sci.* 10: 36–43. [R]

Neupert, W. and M. Brunner. 2002. The protein import motor of mitochondria. *Nat. Rev. Mol. Cell Biol.* 3: 555–565. [R]

Paschen, S. A., T. Waizenegger, T. Stan, M. Preuss, M. Cyrklaff, K. Hell, D. Rapaport and W. Neupert. 2003. Evolutionary conservation of biogenesis of beta-barrel membrane proteins. *Nature* 426: 862–866. [P]

Peeters, N. and I. Small. 2001. Dual targeting to mitochondria and chloroplasts. *Biochim. Biophys. Acta* 1541: 54–63. [R]

Pfanner, N. and A. Chacinska. 2002. The mitochondrial import machinery: Preprotein-conducting channels with binding sites for presequences. *Biochim. Biophys. Acta* 1592: 15–24. [R]

Pfanner, N. and A. Geissler. 2001. Versatility of the mitochondrial protein import machinery. *Nat. Rev. Mol. Cell Biol.* 2: 339–349. [R]

Pfanner, N. and N. Wiedemann. 2002. Mitochondrial protein import: Two membranes, three translocases. *Curr. Opin. Cell Biol.* 14: 400–411. [R]

Rapaport, D. 2002. Biogenesis of the mitochondrial TOM complex. *Trends Biochem. Sci.* 27: 191–197. [R]

Schnell, D. J. and D. N. Hebert. 2003. Protein translocons: Multifunctional mediators of protein translocation across membranes. *Cell* 112: 491–505.]R]

Stuart, R. 2002. Insertion of proteins into the inner membrane of mitochondria: The role of the Oxa1 complex. *Biochim. Biophys. Acta* 1592: 79–87. [R]

Wiedemann, N., V. Kozjak, A. Chacinska, B. Schonfisch, S. Rospert, M. T. Ryan, N. Pfanner and C. Meisinger. 2003. Machinery for protein sorting and assembly in the mitochondrial outer membrane. *Nature* 424: 565–571. [P]

Wilson, F. H., A. Hariri, A. Farhi, H. Zhao, K. F. Petersen, H. R. Toka, C. Nelson-Williams, K. M. Raja, M. Kashgarian, G. I. Shulman, S. J. Scheinman and R. P. Lifton. 2004. A cluster of metabolic defects caused by mutation in a mitochondrial tRNA. *Science* 306: 1190–1194. [P]

Yi, L. and R. E. Dalbey. 2005. Oxa1/Alb3/YidC system for insertion of membrane proteins in mitochondria, chloroplasts and bacteria (review). *Mol. Membr. Biol.* 22: 101–111. [R]

Young, J. C., Hoogenraad, N. J. and F. U. Hartl. 2003. Molecular chaperones Hsp90 and Hsp70 deliver preproteins to the mitochondrial import receptor Tom70. *Cell* 112: 41–50. [P]

The Mechanism of Oxidative Phosphorylation

Junge, W. and N. Nelson. 2005. Structural biology. Nature's rotary electromotors. *Science* 308: 642–644. [R]

Kinosita, K., Jr., K. Adachi and H. Itoh. 2004. Rotation of F1-ATPase: How an ATP-driven molecular machine may work. *Ann. Rev. Biophys. Biomol. Struct.* 33: 245–268. [R]

Michel, H. 1998. The mechanism of proton pumping by cytochrome *c* oxidase. *Proc. Natl. Acad. Sci. USA* 95: 12819–12824. [P]

Mitchell, P. 1979. Keilin's respiratory chain concept and its chemiosmotic consequences. *Science* 206: 1148–1159. [R]

Nicholls, D. G. and S. J. Ferguson. 2002. *Bioenergetics*, 3rd ed. London: Academic Press.

Racker, E. 1980. From Pasteur to Mitchell: A hundred years of bioenergetics. *Fed. Proc.* 39: 210–215. [R]

Reichert, A. S. and W. Neupert. 2004. Mitochondriomics or what makes us breathe. *Trends Genet.* 20: 555–562. [R]

Saraste, M. 1999. Oxidative phosphorylation at the *fin de siecle*. *Science* 283: 1488–1493. [R]

Tielens, A. G., C. Rotte, J. J. Van Hellemond and W. Martin. 2002. Mitochondria as we don't know them. *Trends Biochem. Sci.* 27: 564–572. [R]

Vogel, R., L. Nijtmans, L., Ugalde, C., L. van den Heuvel and J. Smeitink. 2004. Complex I assembly: A puzzling problem. *Curr. Opin. Neurol.* 17: 179–186. [R]

Yoshida M., E. Muneyuki and T. Hisabori. 2001. ATP synthase—A marvelous rotary engine of the cell. *Nat. Rev. Mol. Cell Biol.* 2: 669–677. [R]

Chloroplasts and Other Plastids

Jackson-Constan, D., M. Akita and K. Keegstra. 2001. Molecular chaperones involved in chloroplast protein import. *Biochim. Biophys. Acta* 1541: 102–113. [R]

Jarvis, P. and C. Robinson. 2004. Mechanisms of protein import and routing in chloroplasts. *Curr. Biol.* 14: R1064–R1077. [R]

Jarvis, P. and J. Soll. 2002. Toc, Tic, and chloroplast protein import. *Biochim. Biophys. Acta* 1590: 177–189. [R]

Kessler, F. and D. J. Schnell. 2004. Chloroplast protein import: Solve the GTPase riddle for entry. *Trends Cell Biol.* 14: 334–338. [R]

Rochaix, J.-D. 1992. Post-transcriptional steps in the expression of chloroplast genes. *Ann. Rev. Cell Biol.* 8: 1–28. [R]

Schleiff, E. and J. Soll. 2005. Membrane protein insertion: Mixing eukaryotic and prokaryotic concepts. *EMBO Rep.* 6: 1023–1027. [R]

Subramanian, A. R. 1993. Molecular genetics of chloroplast ribosomal proteins. *Trends Biochem. Sci.* 18: 177–180. [R]

van Dooren, G. G., S. D. Schwartzbach, T. Osafune and G. I. McFadden. 2001. Translocation of proteins across the multiple membranes of complex plastids. *Biochim. Biophys. Acta* 1541: 34–53. [R]

Photosynthesis

Allen, J. F. and J. Forsberg. 2001. Molecular recognition in thylakoid structure and function. *Trends Plant Sci.* 6: 317–326. [R]

Arnon, D. I. 1984. The discovery of photosynthetic phosphorylation. *Trends Biochem. Sci.* 9: 258–262. [R]

Bahatyrova, S., R. N. Frese, C. A. Siebert, J. D. Olsen, K. O. Van der Werf, G. R. Van, R. A. Niederman, P. A. Bullough, C. Otto and C. N. Hunter. 2004. The native architecture of a photosynthetic membrane. *Nature* 430: 1058–1062. [P]

Barber, J. and B. Andersson. 1994. Revealing the blueprint of photosynthesis. *Nature* 370: 31–34. [R]

Deisenhofer, J., O. Epp, K. Miki, R. Huber and H. Michel. 1985. Structure of the protein subunits in the photosynthetic reaction centre of *Rhodopseudomonas viridis*. *Nature* 318: 618–624. [P]

Leister, D. 2003. Chloroplast research in the genomic age. *Trends Genet.* 19: 47–56. [R]

Minagawa, J. and Y. Takahashi. 2004. Structure, function and assembly of photosystem II and its light-harvesting proteins. *Photosynth. Res.* 82: 241–263. [R]

Nicholls, D. G. and S. J. Ferguson. 2002. *Bioenergetics,* 3rd ed. San Diego, CA: Academic Press.

Rhee, K.-H., E. P. Morris, J. Barber and W. Kuhlbrandt. 1998. Three-dimensional structure of the plant photosystem II reaction centre at 8 Å resolution. *Nature* 396: 283–286. [P]

Peroxisomes

Gould, S. J. and C. S. Collins. 2002. Peroxisomal-protein import: Is it really that complex? *Nat. Rev. Mol. Cell Biol.* 3: 382–389. [R]

Hoepfner, D., D. Schildknegt, L. Braakman, P. Philippsen and H. F. Tabak. 2005. Contribution of the endoplasmic reticulum to peroxisome formation. *Cell* 122: 85–95. [P]

Purdue, P. E. and P. B. Lazarow. 2001. Peroxisome biogenesis. *Ann. Rev. Cell Dev. Biol.* 17: 701–752. [R]

Schekman, R. 2005. Peroxisomes: Another branch of the secretory pathway? *Cell* 122: 1–2. [R]

Titorenko, V. I. and R. A. Rachubinski. 2001. The life cycle of the peroxisome. *Nat. Rev. Mol. Cell Biol.* 2: 357–368. [R]

Tolbert, N. E. 1981. Metabolic pathways in peroxisomes and glyoxysomes. *Ann. Rev. Biochem.* 50: 133–157. [R]

Van den Bosch, H., R. B. H. Schutgens, R. J. A. Wanders and J. M. Tager. 1992. Biochemistry of peroxisomes. *Ann. Rev. Biochem.* 61: 157–197. [R]

Wickner, W. and R. Schekman. 2005. Protein translocation across biological membranes. *Science* 310: 1452–1456. [R]

The Cytoskeleton and Cell Movement

■ *Structure and Organization of Actin Filaments 473*

■ *Actin, Myosin, and Cell Movement 486*

■ *Intermediate Filaments 497*

■ *Microtubules 505*

■ *Microtubule Motors and Movement 511*

■ *KEY EXPERIMENT:*
Expression of Mutant Keratin Causes Abnormal Skin Development 502

■ *KEY EXPERIMENT:*
The Isolation of Kinesin 514

THE MEMBRANE-ENCLOSED ORGANELLES discussed in the preceding chapters constitute one level of the organizational substructure of eukaryotic cells. A further level of organization is provided by the cytoskeleton, which consists of a network of protein filaments extending throughout the cytoplasm of all eukaryotic cells. The cytoskeleton provides a structural framework for the cell, serving as a scaffold that determines cell shape, the positions of organelles, and the general organization of the cytoplasm. In addition to playing this structural role, the cytoskeleton is responsible for cell movements. These include not only the movements of entire cells, but also the internal transport of organelles and other structures (such as mitotic chromosomes) through the cytoplasm. Importantly, the cytoskeleton is much less rigid and permanent than its name implies. Rather, it is a dynamic structure that is continually reorganized as cells move and change shape—for example, during cell division.

The cytoskeleton is composed of three principal types of protein filaments: actin filaments, intermediate filaments, and microtubules, which are held together and linked to subcellular organelles and the plasma membrane by a variety of accessory proteins. This chapter discusses the structure and organization of each of these three major components of the cytoskeleton as well as their roles in cell motility, organelle transport, cell division, and other types of cell movements.

Structure and Organization of Actin Filaments

The major cytoskeletal protein of most cells is **actin**, which polymerizes to form actin filaments—thin, flexible fibers approximately 7 nm in diameter and up to several micrometers in length (Figure 12.1). Within the cell, actin filaments (also called **microfilaments**) are organized into higher-order structures, forming bundles or three-dimensional networks with the prop-

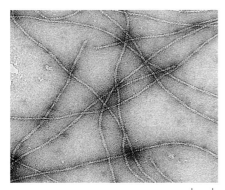

|__|
50 nm

FIGURE 12.1 Actin filaments
Electron micrograph of actin filaments. (Courtesy of Roger Craig, University of Massachusetts Medical Center.)

12.1 WEBSITE ANIMATION

Assembly of an Actin Filament Actin monomers polymerize to form actin filaments, a process that is reversible and is regulated within the cell by actin-binding proteins.

erties of semisolid gels. The assembly and disassembly of actin filaments, their cross-linking into bundles and networks, and their association with other cell structures (such as the plasma membrane) are regulated by a variety of actin-binding proteins, which are critical components of the actin cytoskeleton. Actin filaments are particularly abundant beneath the plasma membrane where they form a network that provides mechanical support, determines cell shape, and allows movement of the cell surface, thereby enabling cells to migrate, engulf particles, and divide.

Assembly and Disassembly of Actin Filaments

Actin was first isolated from muscle cells, in which it constitutes approximately 20% of total cell protein, in 1942. Although actin was initially thought to be uniquely involved in muscle contraction, it is now known to be an extremely abundant protein (typically 5 to 10% of total protein) in all types of eukaryotic cells. Yeasts have only a single actin gene, but higher eukaryotes have several distinct types of actin, which are encoded by different members of the actin gene family. Mammals, for example, have six distinct actin genes: Four are expressed in different types of muscle and two are expressed in nonmuscle cells. All of the actins, however, are very similar in amino acid sequence and have been highly conserved throughout the evolution of eukaryotes. Yeast actin, for example, is 90% identical in amino acid sequence to the actins of mammalian cells. The prokaryotic ancestor of actin is a protein called MreB, which gives rod-shaped bacteria their structure.

The three-dimensional structures of both individual actin molecules and actin filaments were determined in 1990 by Kenneth Holmes, Wolfgang Kabsch, and their colleagues. Individual actin molecules are globular proteins of 375 amino acids (43 kd). Each actin monomer (**globular [G] actin**) has tight binding sites that mediate head-to-tail interactions with two other actin monomers, so actin monomers polymerize to form filaments (**filamentous [F] actin**) (Figure 12.2). Each monomer is rotated by 166° in the filaments, which therefore have the appearance of a double-stranded helix. Because all the actin monomers are oriented in the same direction, actin filaments have a distinct polarity and their ends (called barbed or plus ends, and pointed or minus ends) are distinguishable from one another. This polarity of actin filaments is important both in their assembly and in establishing a specific direction of myosin movement relative to actin, as discussed later in the chapter.

While actin monomers and filaments are tightly controlled within cells, much of their behavior is a property of the actin monomers and filaments themselves. In solutions of low ionic strength *in vitro*, actin filaments depolymerize to monomers. Actin then polymerizes spontaneously if the ionic strength is increased to physiological levels. The first step in actin polymerization (called nucleation) is the formation of a small aggregate consisting of three actin monomers. Actin filaments are then able to grow by the reversible addition of monomers to both ends, but one end (the barbed end) elongates five to ten times faster than the pointed end. The actin monomers also bind ATP, which is hydrolyzed to ADP following filament assembly. Although ATP is not required for polymerization, actin monomers to which ATP is bound polymerize more readily than those to which ADP is bound. As discussed below, ATP binding and hydrolysis play a key role in regulating the assembly and dynamic behavior of actin filaments.

Because actin polymerization is reversible, filaments can depolymerize by the dissociation of actin subunits, allowing actin filaments to be broken down when necessary (Figure 12.3). *In vitro*, an equilibrium exists between

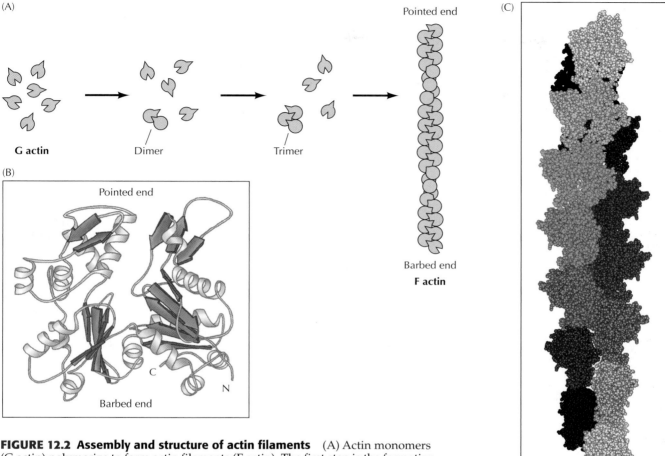

FIGURE 12.2 Assembly and structure of actin filaments (A) Actin monomers (G actin) polymerize to form actin filaments (F actin). The first step is the formation of dimers and trimers, which then grow by the addition of monomers to both ends. (B) Structure of an actin monomer. (C) Space-filling model of F actin. Fourteen actin monomers are represented in different colors. (C, based on data from Chen et al., 2002. *J. Struct. Biol.* 138: 92.)

actin monomers and filaments. The rate at which actin monomers are incorporated into filaments is proportional to their concentration, so there is a critical concentration of actin monomers at which the rate of their polymerization into filaments equals the rate of dissociation. At this critical concentration, monomers and filaments are in apparent equilibrium.

As noted earlier, the two ends of an actin filament grow at different rates, with monomers being added to the fast-growing barbed end five to ten times faster than to the slow-growing pointed end. Because ATP-actin dissociates less readily than ADP-actin, this results in a difference in the criti-

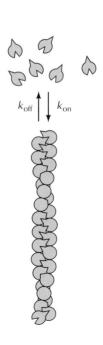

FIGURE 12.3 Reversible polymerization of actin monomers Actin polymerization is a reversible process in which monomers both associate with and dissociate from the ends of actin filaments. The rate of subunit dissociation (k_{off}) is independent of monomer concentration, while the rate of subunit association is proportional to the concentration of free monomers and given by $C \times k_{on}$ (C = concentration of free monomers). An apparent equilibrium is reached at the critical concentration of monomers (C_c), where $k_{off} = C_c \times k_{on}$.

FIGURE 12.4 Treadmilling and the role of ATP in microfilament polymerization The pointed ends of actin filaments grow less rapidly than the barbed ends. This difference in growth rate is reflected in a difference in the critical concentration for addition of monomers to the two ends of the filament. Actin bound to ATP associates with the rapidly growing barbed ends, and the ATP bound to actin is then hydrolyzed to ADP. Because ADP-actin dissociates from filaments more readily than ATP-actin, the critical concentration of actin monomers is higher for addition to the pointed end than to the barbed end of actin filaments. Treadmilling takes place at monomer concentrations intermediate between the critical concentrations for the barbed and pointed ends. Under these conditions, there is a net dissociation of monomers (bound to ADP) from the pointed end, balanced by the addition of monomers (bound to ATP) to the barbed end.

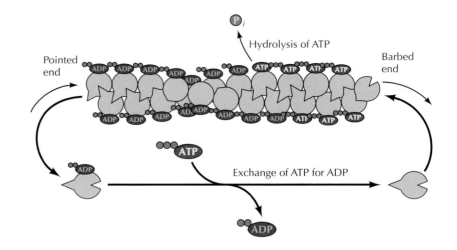

■ Phalloidin is derived from the deathcap mushroom (*Amanita phalloides*). This mushroom also produces α-amanitin, a specific inhibitor of RNA polymerase II. The toxins produced by the deathcap and related mushrooms primarily attack the liver and kidney and can cause death if left untreated.

---

cal concentration of monomers needed for polymerization at the two ends. This difference can result in the phenomenon known as **treadmilling**, which illustrates the dynamic behavior of actin filaments (Figure 12.4). For the *in vitro* system to be at an overall steady state, the concentration of free actin monomers must be intermediate between the critical concentrations required for polymerization at the barbed and pointed ends of the filaments. Under these conditions, there is a net loss of monomers from the pointed end, which is balanced by a net addition to the barbed end. Treadmilling requires ATP, with ATP-actin polymerizing at the barbed end of filaments while ADP-actin dissociates from the pointed end. As discussed later in this chapter, treadmilling is important in the formation of cell processes and in cell movement.

It is noteworthy that several drugs useful in cell biology act by binding to actin and affecting its polymerization. For example, the **cytochalasins** bind to the barbed ends of actin filaments and block their elongation. This results in changes in cell shape as well as inhibition of some types of cell movements (e.g., cell division following mitosis), indicating that actin polymerization is required for these processes. Another drug, **phalloidin**, binds tightly to actin filaments and prevents their dissociation into individual actin molecules. Phalloidin labeled with a fluorescent dye is frequently used to visualize actin filaments by fluorescence microscopy.

Within the cell, the concentration of actin filaments and monomers is far from equilibrium. In parts of the cell, the turnover of actin filaments can be 100 times faster than it is *in vitro* while in other parts of the cell, filaments can be stabilized against treadmilling. The assembly and disassembly of actin filaments within the cell is regulated by a diverse group of **actin-binding proteins** (Table 12.1). These proteins can regulate the formation and stability of the actin cytoskeleton at several different levels (Figure 12.5). Some of these proteins bind to actin filaments, either stabilizing them or cross-linking them to one another. Other actin-binding proteins stabilize filaments by capping their ends and preventing the dissociation of actin monomers. Conversely, some proteins act to disassemble actin filaments either by severing them or stimulating their depolymerization. Finally, some actin-binding proteins bind monomeric actin and control its assembly into filaments by regulating the exchange of ATP for ADP.

The initial and rate-limiting step in the formation of actin filaments is nucleation, which requires monomers to interact correctly. Two types of

TABLE 12.1 Actin-binding Proteins

Cellular Role	Representative Proteins
Filament initiation and polymerization	Arp2/3, formin
Filament stabilization	Nebulin, tropomyosin
Filament cross-linking	α-actinin, filamin, fimbrin, villin
End-capping	CapZ, tropomodulin
Filament severing/depolymerization	ADF/cofilin, gelsolin, thymosin
Monomer binding	Profilin, twinfilin
Actin filament linkage to other proteins	α-catenin, dystrophin, spectrin, talin, vinculin

proteins, **formin** and the **Arp2/3 complex** (*a*ctin-*r*elated *p*rotein), determine where filaments are formed within the cell by nucleating actin filaments. Formins are a family of large (140–200 kd) barbed-end tracking proteins that both nucleate the initial actin monomers and then move along the growing filament, adding new monomers to the barbed end (**Figure 12.6**). It is thought that formins nucleate long unbranched actin filaments (see Figure 12.1) that make up stress fibers and the thin filaments of muscle cells (discussed later in this chapter). Many of these filaments are relatively stable due to filament-stabilizing proteins, such as members of the **tropomyosin** family. Tropomyosins are 30–36 kd fibrous proteins that bind lengthwise along the groove of actin filaments.

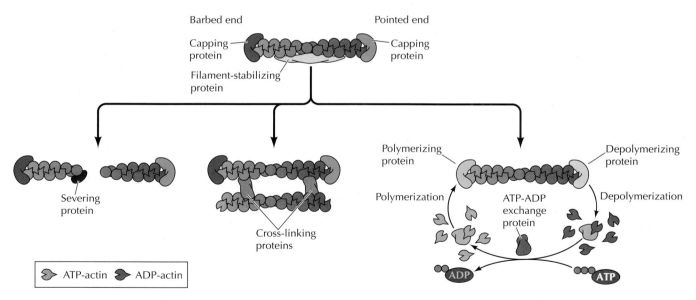

FIGURE 12.5 Regulation of actin filament organization by actin-binding proteins   Actin-binding proteins play several roles in the dynamic behavior of actin filaments. Actin filaments can be stabilized by filament-stabilizing proteins that bind along their length. Both the barbed and pointed ends can also be capped and the filaments themselves can be cross-linked. Intact filaments can also be split by filament-severing proteins. The equilibrium between actin monomers and filaments can be regulated by filament-depolymerizing proteins, filament-polymerizing proteins, or proteins that modulate the exchange of ATP for ADP on actin monomers.

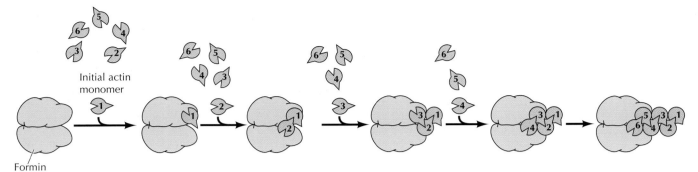

FIGURE 12.6 Initiation of actin filaments by formin The rate-limiting step of actin filament formation is nucleation, which requires the correct alignment of the first three actin monomers to allow subsequent polymerization. Within the cell, nucleation is facilitated by the actin-binding protein, formin. Each subunit of a formin dimer binds an actin monomer. The actin monomers are held in the correct configuration to allow binding of the third monomer, followed by rapid polymerization during which formin continues to track the barbed end.

In contrast, at the leading edge of cell processes or moving cells, actin filaments both actively treadmill and branch extensively. The densely packed and highly branched actin filaments in these areas are nucleated by the Arp2/3 complex, which binds actin/ATP near the barbed ends of microfilaments (Figure 12.7). The Arp2/3 complex contains seven proteins, two of which are similar to actin. The complex by itself has very low activity but is activated by several proteins that bind and activate it. Once activated, the Arp2/3 complex binds to the side of an existing actin filament near the barbed end and forms a new branch.

Another type of actin binding protein remodels or modifies existing filaments. One family of proteins responsible for actin filament remodeling

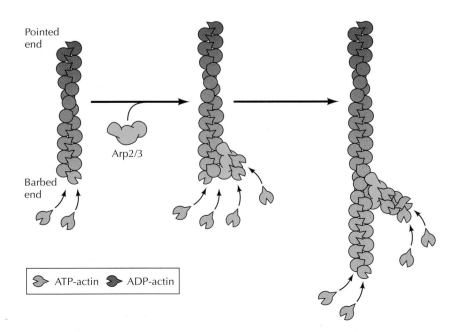

FIGURE 12.7 Initiation of actin filament branches The Arp2/3 complex binds to actin filaments near their barbed ends and initiates the formation of branches.

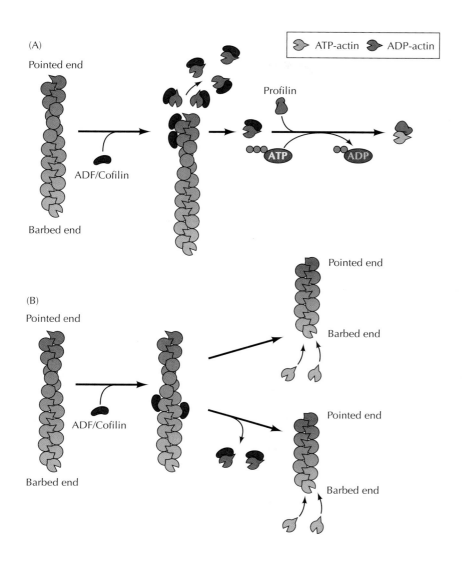

(A)

Pointed end

ADF/Cofilin

Barbed end

ATP-actin   ADP-actin

Profilin

ATP   ADP

(B)

Pointed end

ADF/Cofilin

Barbed end

Pointed end

Barbed end

Pointed end

Barbed end

FIGURE 12.8 Effects of ADF/cofilin and profilin on actin filaments
ADF/cofilin has two different activities. (A) It is an *a*ctin *d*epolymerization *f*actor, binding to actin filaments and increasing the rate of dissociation of actin monomers from the pointed end. ADF/cofilin remains bound to the actin/ADP monomers, preventing their reassembly into filaments. Profilin can stimulate the exchange of bound ADP for ATP, resulting in the formation of actin/ATP monomers that can be repolymerized into filaments. (B) ADF/cofilin can also bind to and sever actin filaments, creating new plus ends.

within the cell is the **ADF/cofilin** (*a*ctin *d*epolymerizing *f*actor) family (Figure 12.8). These proteins bind to actin filaments and enhance the rate of dissociation of actin/ADP monomers from the pointed end. ADF/cofilin can also sever actin filaments. ADF/cofilin preferentially binds to ADP-actin, so it remains bound to actin monomers following filament disassembly and sequesters them in the ADP-bound form, preventing their reincorporation into filaments. However, another actin-binding protein, **profilin**, can reverse this effect of cofilin and stimulate the incorporation of actin monomers into filaments. Profilin acts by stimulating the exchange of bound ADP for ATP, resulting in the formation of actin/ATP monomers, which dissociate from cofilin and are then available for assembly into filaments.

As might be expected, the activities of these proteins are controlled by a variety of cell signaling mechanisms (discussed in Chapter 15), allowing actin polymerization to be appropriately regulated in response to environmental signals. ADF/cofilin, profilin, formin, and the Arp2/3 complex (as well as other actin-binding proteins) can thus act together to promote the rapid turnover of actin filaments and remodeling of the actin cytoskeleton that is required for a variety of cell movements and changes in cell shape, as discussed later in this chapter. This is a major undertaking, and in some cell types, microfilament assembly and disassembly are responsible for half the hydrolysis and turnover of ATP in the cell.

Several of the actin-related proteins (Arp4–8) and actin itself occur in the nucleus. The actin-related proteins are involved in chromatin remodeling in both plants and animals and may participate in the assembly of the nucleus after cell division. The role of nuclear actin remains controversial.

Organization of Actin Filaments

Individual actin filaments are assembled into two general types of structures called **actin bundles** and **actin networks**, which play different roles in the cell (Figure 12.9). In bundles, the actin filaments are cross-linked into closely packed parallel arrays. In networks, the actin filaments are cross-linked in orthogonal arrays that form three-dimensional meshworks with the properties of semisolid gels. As mentioned previously, the formation of these structures is governed by a variety of actin-binding proteins that cross-link actin filaments in distinct patterns (see Table 12.1).

All of the actin-binding proteins involved in cross-linking contain at least two domains that bind actin, allowing them to bind and cross-link two different actin filaments. The nature of the association between these filaments is then determined by the size and shape of the cross-linking proteins (see Figure 12.9). The proteins that cross-link actin filaments into bundles (called **actin-bundling proteins**) usually are small rigid proteins that force the filaments to align closely with one another. In contrast, the proteins that organize actin filaments into networks tend to be large flexible proteins that can cross-link perpendicular filaments. Actin cross-linking proteins are modular proteins consisting of related structural units. In particular, the actin-binding domains of many of these proteins are similar in structure. They are separated by regulatory domains and spacer sequences that vary in length and flexibility, and it is these differences in the spacer sequences that are responsible for the distinct cross-linking properties of different actin-binding proteins.

There are at least two structurally and functionally distinct types of actin bundles involving different actin-bundling proteins (Figure 12.10). The first type of bundle, containing closely spaced actin filaments aligned in parallel, supports projections of the plasma membrane, such as microvilli (see Figures 12.18 and 12.19). In these bundles, all the filaments have the same polarity, with their barbed ends adjacent to the plasma membrane. An example of a bundling protein involved in the formation of these structures is **fimbrin**, which was first isolated from intestinal microvilli and later found in surface projections of a wide variety of cell types. Fimbrin is a 68

FIGURE 12.9 Actin bundles and networks (A) Electron micrograph of actin bundles (arrowheads) projecting from the actin network (arrows) underlying the plasma membrane of a macrophage. The bundles support cell surface projections called filopodia (see Figure 12.20). (B) Schematic organization of bundles and networks. Actin filaments in bundles are cross-linked into parallel arrays by small proteins that align the filaments closely with one another. In contrast, networks are formed by large flexible proteins that cross-link orthogonal filaments. (A, courtesy of John H. Hartwig, Brigham & Women's Hospital.)

(A)

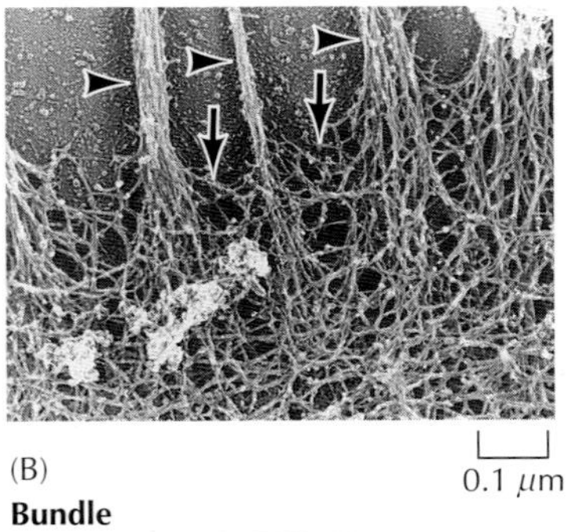

(B)

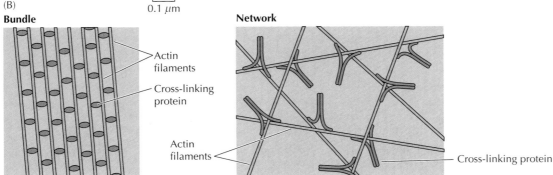

0.1 μm

Bundle

Actin filaments
Cross-linking protein

Network

Actin filaments
Cross-linking protein

FIGURE 12.10 Actin-bundling proteins   Actin filaments are associated into two types of bundles by different actin-bundling proteins. Fimbrin has two adjacent actin-binding domains (ABD) and cross-links actin filaments into closely packed parallel bundles in which the filaments are approximately 14 nm apart. In contrast, the two separated actin-binding domains of α-actinin dimers cross-link filaments into more loosely spaced contractile bundles in which the filaments are separated by 40 nm. Both fimbrin and α-actinin contain two related Ca^{2+}-binding domains, and α-actinin contains four repeated α-helical spacer domains.

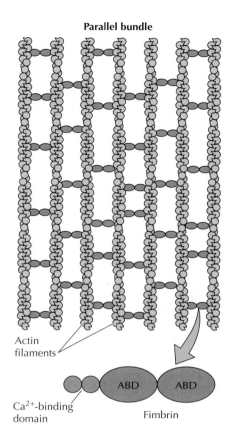

kd protein containing two adjacent actin-binding domains. It binds to actin filaments as a monomer, holding two parallel filaments close together.

The second type of actin bundle is composed of filaments that are more widely spaced, allowing the bundle to contract. The more open structure of these bundles (which are called **contractile bundles**) reflects the properties of the cross-linking protein, **α-actinin**. In contrast to fimbrin, α-actinin binds to actin as a dimer, each subunit of which is a 102 kd protein containing a single actin-binding site. Filaments cross-linked by α-actinin are consequently separated by a greater distance than those cross-linked by fimbrin (40 nm apart instead of 14 nm). The increased spacing between filaments allows the motor protein myosin to interact with the actin filaments in these bundles, which (as discussed later) enables the bundles to contract.

The actin filaments in networks are held together by large actin-binding proteins, such as **filamin** (Figure 12.11). Filamin binds actin as a dimer of two 280 kd subunits. The actin-binding domains and dimerization domains are at opposite ends of each subunit, so the filamin dimer is a flexible V-shaped molecule with actin-binding domains at the ends of each arm. As a result, filamin forms cross-links between orthogonal actin filaments, creating a loose three-dimensional meshwork. As discussed in the next section, such networks of actin filaments underlie the plasma membrane and support the surface of the cell.

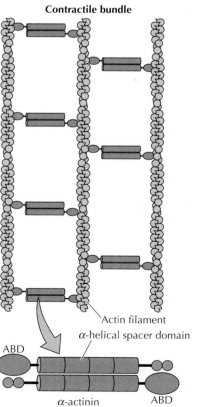

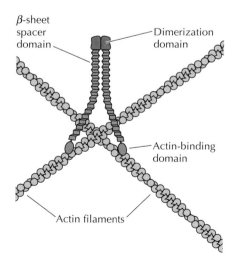

FIGURE 12.11 Actin networks and filamin   Filamin is a dimer of two large (280 kd) subunits, forming a flexible V-shaped molecule that cross-links actin filaments into orthogonal networks. The carboxy-terminal dimerization domain is separated from the amino-terminal actin-binding domain by repeated β-sheet spacer domains.

FIGURE 12.12 Morphology of red blood cells Scanning electron micrograph of red blood cells illustrating their biconcave shape. (Omikron/Photo Researchers, Inc.)

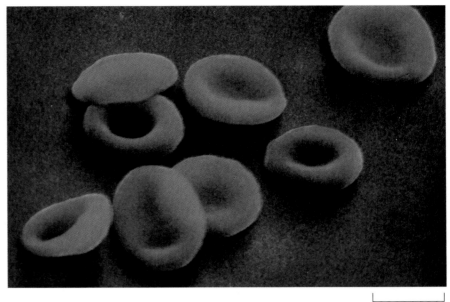

5 μm

Association of Actin Filaments with the Plasma Membrane

Actin filaments are highly concentrated at the periphery of the cell where they form a three-dimensional network beneath the plasma membrane (see Figure 12.9). This network of actin filaments and associated actin-binding proteins (called the **cell cortex**) determines cell shape and is involved in a variety of cell surface activities, including movement. The association of the actin cytoskeleton with the plasma membrane is thus central to cell structure and function.

Red blood cells (erythrocytes) have proven particularly useful for studies of both the plasma membrane (discussed in the next chapter) and the cortical cytoskeleton. The principal advantage of red blood cells for these studies is that they contain no nucleus or internal organelles, so their plasma membrane and associated proteins can be easily isolated without contamination by the various internal membranes that are abundant in other cell types. In addition, human erythrocytes lack other cytoskeletal components (microtubules and intermediate filaments), so the cortical actin cytoskeleton is the principal determinant of their distinctive shape as biconcave discs (Figure 12.12).

The major protein that provides the structural basis for the cortical cytoskeleton in erythrocytes is the actin-binding protein **spectrin** (Figure 12.13). Spectrin is a member of the large calponin family of actin-binding proteins, which includes α-actinin, filamin, and fimbrin. Erythrocyte spec-

FIGURE 12.13 Structure of spectrin Spectrin is a tetramer consisting of two α and two β chains. Each β chain has a single actin-binding domain (ABD) at its amino terminus. Both α and β chains contain multiple repeats of α-helical spacer domains, which separate the two actin-binding domains of the tetramer. The α chain has two Ca^{2+} binding domains at its carboxy terminus.

Spectrin tetramer

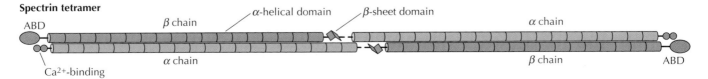

ABD β chain α-helical domain β-sheet domain α chain

α chain β chain ABD

Ca^{2+}-binding domain

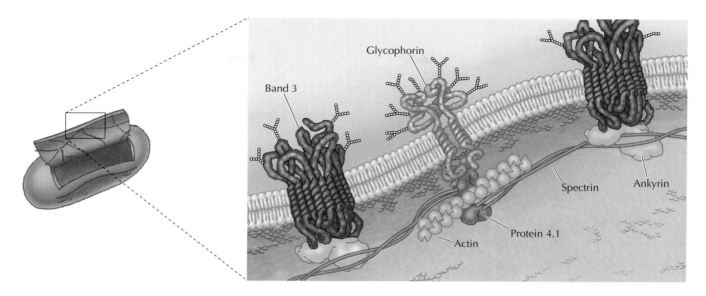

FIGURE 12.14 Association of the erythrocyte cortical cytoskeleton with the plasma membrane The plasma membrane is associated with a network of spectrin tetramers cross-linked by short actin filaments in association with protein 4.1. The spectrin-actin network is linked to the membrane by ankyrin, which binds to both spectrin and the abundant transmembrane protein, band 3. An additional link is provided by the binding of protein 4.1 to glycophorin.

trin is a tetramer consisting of two distinct polypeptide chains called α and β, with molecular weights of 240 and 220 kd, respectively. The β chain has a single actin-binding domain at its amino terminus. The α and β chains associate laterally to form dimers, which then join head-to-head to form tetramers with two actin-binding domains separated by approximately 200 nm. The ends of the spectrin tetramers then associate with short actin filaments, resulting in the spectrin-actin network that forms the cortical cytoskeleton of red blood cells (Figure 12.14). The major link between the spectrin-actin network and the plasma membrane is provided by a protein called **ankyrin**, which binds both to spectrin and to the cytoplasmic domain of an abundant transmembrane protein called band 3. An additional link between the spectrin-actin network and the plasma membrane is provided by protein 4.1, which binds to spectrin-actin junctions as well as recognizing the cytoplasmic domain of glycophorin (another abundant transmembrane protein).

Other types of cells contain linkages between the cortical cytoskeleton and the plasma membrane that are similar to those observed in red blood cells. Proteins related to spectrin (nonerythroid spectrin is also called **fodrin**), ankyrin, and protein 4.1 are expressed in a wide range of cell types where they fulfill functions analogous to those described for erythrocytes. For example, a family of proteins related to protein 4.1 (**ERM proteins**) link actin filaments to the plasma membranes of many different kinds of cells, and another member of the calponin family, filamin (see Figure 12.11), constitutes a major link between actin filaments and the plasma membrane of blood platelets. An additional member of the calponin family, **dystrophin**, is of particular interest because it is the product of the gene responsible for two types of muscular dystrophy (Duchenne's and Becker's). These X-linked inherited diseases result in progressive degeneration of skeletal muscle, and patients with the more severe form of the disease (Duchenne's muscular dystrophy) usually die in their teens or early twenties. Molecular cloning of the gene responsible for this disorder revealed that it encodes a large protein (427 kd) that is either absent or abnormal in patients with Duchenne's or Becker's muscular dystrophy, respectively. The sequence of dystrophin fur-

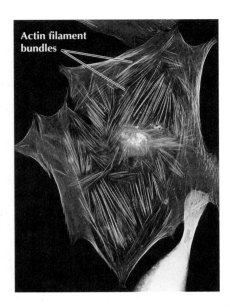

FIGURE 12.15 Stress fibers and focal adhesions Fluorescence microscopy of a human fibroblast in which actin filaments have been stained with a fluorescent dye. Stress fibers are revealed as bundles of actin filaments anchored at sites of cell attachment to the culture dish surface (focal adhesions). (Don Fawcett/Photo Researchers, Inc.)

ther indicated that it has a single actin-binding domain at its amino terminus and a membrane-binding domain at its carboxy terminus. Like spectrin, dystrophin links actin filaments to transmembrane proteins of the muscle cell plasma membrane. These transmembrane proteins in turn link the cytoskeleton to the extracellular matrix, which plays an important role in maintaining cell stability during muscle contraction.

In contrast to the uniform surface of red blood cells, most cells have specialized regions of the plasma membrane that form contacts with adjacent cells, the extracellular matrix (discussed in Chapter 14), or other substrata (such as the surface of a culture dish). These regions can also serve as attachment sites for bundles of actin filaments that anchor the cytoskeleton to areas of cell contact. These attachments of actin filaments are particularly evident in fibroblasts maintained in tissue culture (Figure 12.15). Such cultured fibroblasts secrete extracellular matrix proteins that stick to the surface of the culture dish. The fibroblasts then attach to this extracellular matrix on the culture dish via the binding of transmembrane proteins (called **integrins**). The sites of attachment are discrete regions (called **focal adhesions**) that also serve as attachment sites for large bundles of actin filaments called **stress fibers**.

Stress fibers are contractile bundles of actin filaments, cross-linked by α-actinin and stabilized by tropomyosin, which anchor the cell and exert tension against the substratum. They are attached to the plasma membrane at focal adhesions via interactions with integrin. These complex associations are mediated by several other proteins, including **talin** and **vinculin** (Figure 12.16). For example, both talin and α-actinin bind to the cytoplasmic domains of integrins. Talin also binds to vinculin and both proteins also bind actin. Other proteins found at focal adhesions also participate in the attachment of actin filaments, and a combination of these interactions is responsible for the linkage of actin filaments to the plasma membrane.

The actin cytoskeleton is similarly anchored to regions of cell-cell contact called **adherens junctions** (Figure 12.17). In sheets of epithelial cells, these junctions form a continuous beltlike structure (called an **adhesion belt**) around each cell in which an underlying contractile bundle of actin filaments is linked to the plasma membrane. Contact between cells at adherens

FIGURE 12.16 Attachment of stress fibers to the plasma membrane at focal adhesions Focal adhesions are mediated by the binding of integrins to proteins of the extracellular matrix. Stress fibers (bundles of actin filaments cross-linked by α-actinin) are then bound to the cytoplasmic domain of integrins by complex associations involving a number of proteins. Two possible associations are illustrated: 1) talin binds to both integrin and vinculin, and both talin and vinculin bind to actin, and 2) integrin binds to α-actinin. A number of other proteins (not shown) are also present at focal adhesions and may be involved in anchoring stress fibers to the plasma membrane.

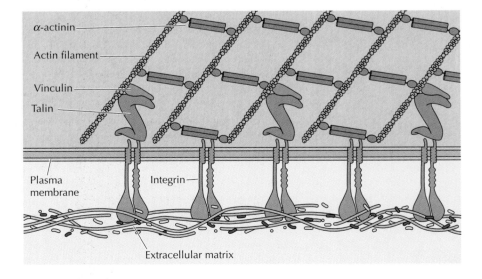

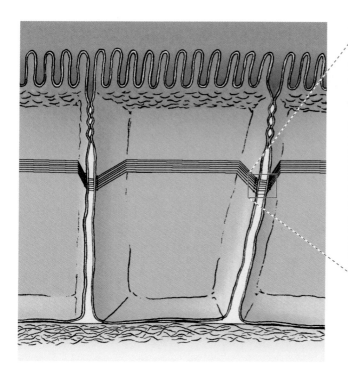

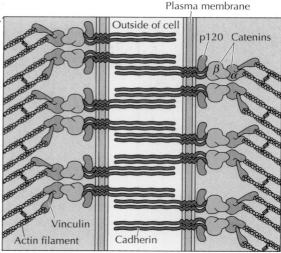

FIGURE 12.17 Attachment of actin filaments to adherens junctions
Cell-cell contacts at adherens junctions are mediated by cadherins, which serve as sites for attachment of actin filaments. In sheets of epithelial cells, these junctions form a continuous belt of actin filaments around each cell. The transmembrane cadherins bind β-catenin at their cytoplasmic domains. β-catenin interacts with α-catenin, which binds both actin filaments and vinculin. p120 regulates the stability of the junction.

junctions is mediated by transmembrane proteins called **cadherins**, which are discussed further in Chapter 14. The cadherins form a complex with cytoplasmic proteins called **catenins**, which associate with actin filaments.

Protrusions of the Cell Surface

The surfaces of most cells have a variety of protrusions or extensions that are involved in cell movement, phagocytosis, or specialized functions, such as absorption of nutrients. Most of these cell surface extensions are based on actin filaments, which are organized into either relatively permanent or rapidly rearranging bundles or networks.

The best-characterized of these actin-based cell surface protrusions are **microvilli**, fingerlike extensions of the plasma membrane that are particularly abundant on the surfaces of cells involved in absorption, such as the epithelial cells lining the intestine (Figure 12.18). The microvilli of these cells form a layer on the apical surface (called a **brush border**) that consists of approximately a thousand microvilli per cell and increases the exposed surface area available for absorption by ten to twentyfold. In addition to their role in absorption, specialized forms of microvilli, the **stereocilia** of auditory hair cells, are responsible for hearing by detecting sound vibrations.

Their abundance and ease of isolation have facilitated detailed structural analysis of intestinal microvilli, which contain closely packed parallel bundles of 20 to 30 actin filaments (Figure 12.19). The filaments in these bundles are cross-linked in part by fimbrin, an actin-bundling protein (discussed earlier) that is present in surface projections of a variety of cell types. However, the major actin-bundling protein in intestinal microvilli is **villin**, a 95 kd protein present in microvilli of only a few specialized types of cells, such as those lining the intestine and kidney tubules. Along their length, the actin bundles of microvilli are attached to the plasma membrane by lateral arms consisting of the calcium-binding protein calmodulin in association

■ In addition to its structural role at adherens junctions, β-catenin is a critical component of the Wnt signaling pathway where it functions as a transcriptional activator (see Chapter 15).

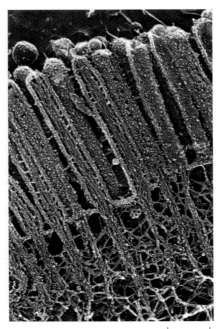

0.25 μm

FIGURE 12.18 Electron micrograph of microvilli   The microvilli of intestinal epithelial cells are fingerlike projections of the plasma membrane. They are supported by actin bundles anchored in a dense region of the cortex called the terminal web. (Courtesy of Nobutaka Hirokawa.)

with myosin I, which may be involved in movement of the plasma membrane along the actin bundle of the microvillus. At their base, the actin bundles are anchored in a spectrin-rich region of the actin cortex called the terminal web, which cross-links and stabilizes the microvilli.

In contrast to microvilli many surface protrusions are transient structures that form in response to environmental stimuli. Several types of these structures extend from the leading edge of a moving cell and are involved in cell locomotion (Figure 12.20). **Pseudopodia** are extensions of moderate width, based on actin filaments cross-linked into a three-dimensional network, that are responsible for phagocytosis and for the movement of amoebas across a surface. **Lamellipodia** are broad, sheetlike extensions at the leading edge of fibroblasts, which similarly contain a network of actin filaments. Many cells also extend **microspikes** or **filopodia**, thin projections of the plasma membrane supported by actin bundles. The formation and retraction of these structures is based on the regulated assembly and disassembly of actin filaments, as discussed later in this chapter.

Actin, Myosin, and Cell Movement

Actin filaments, often in association with myosin, are responsible for many types of cell movements. **Myosin** is the prototype of a **molecular motor**—a protein that converts chemical energy in the form of ATP to mechanical energy, thus generating force and movement. The most striking variety of such movement is muscle contraction, which has provided the model for understanding actin-myosin interactions and the motor activity of myosin molecules. However, interactions of actin and myosin are responsible not only for muscle contraction but also for a variety of

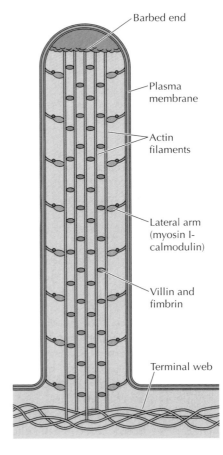

Barbed end

Plasma membrane

Actin filaments

Lateral arm (myosin I-calmodulin)

Villin and fimbrin

Terminal web

FIGURE 12.19 Organization of microvilli   The core actin filaments of microvilli are cross-linked into closely packed bundles by fimbrin and villin. They are attached to the plasma membrane along their length by lateral arms, consisting of myosin I and calmodulin. The barbed ends of the actin filaments are embedded in a cap of unidentified proteins at the tip of the microvillus.

(A)
(B)
(C)

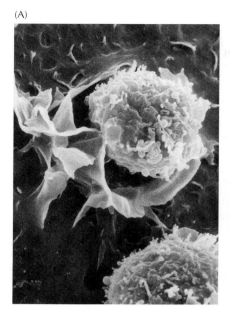
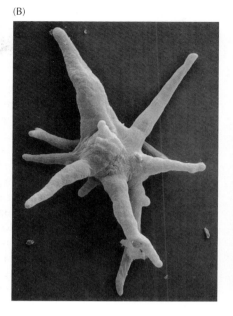
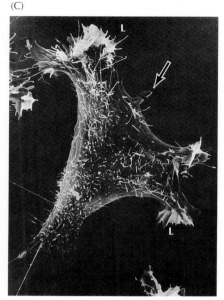

FIGURE 12.20 Examples of cell surface projections involved in phagocytosis and movement (A) Scanning electron micrograph showing pseudopodia of a macrophage engulfing a tumor cell during phagocytosis. (B) An amoeba with several extended pseudopodia. (C) A tissue culture cell illustrating lamellipodia (L) and filopodia (arrow). (A, K. Wassermann/Visuals Unlimited; B, Stanley Flegler/Visuals Unlimited; C, Don Fawcett/Photo Researchers, Inc.)

movements of nonmuscle cells, including cell division, so these interactions play a central role in cell biology.

Muscle Contraction

Muscle cells are highly specialized for a single task—contraction—and it is this specialization in structure and function that has made muscle the prototype for studying movement at the cellular and molecular levels. There are three distinct types of muscle cells in vertebrates: skeletal muscle, which is responsible for all voluntary movements; cardiac muscle, which pumps blood from the heart; and smooth muscle, which is responsible for involuntary movements of organs such as the stomach, intestine, uterus, and blood vessels. In both skeletal and cardiac muscle, the contractile elements of the cytoskeleton are present in highly organized arrays that give rise to characteristic patterns of cross-striations. It is the characterization of these structures in skeletal muscle that has led to our current understanding of muscle contraction and other actin-based cell movements at the molecular level.

Skeletal muscles are bundles of **muscle fibers**, which are single large cells (approximately 50 μm in diameter and up to several centimeters in length) formed by the fusion of many individual cells during development (Figure 12.21). Most of the cytoplasm consists of **myofibrils**, which are cylindrical bundles of two types of filaments: thick filaments of myosin (about 15 nm in diameter) and thin filaments of actin (about 7 nm in diameter). Each myofibril is organized as a chain of contractile units called **sarcomeres**, which are responsible for the striated appearance of skeletal and cardiac muscle.

■ Molecules that regulate the activity of smooth muscle cells are important drugs. For example, albuterol, a drug that relaxes smooth muscles, is used in the treatment of asthma and to prevent premature labor in pregnant women. In contrast, oxytocin, a hormone that stimulates smooth muscle contraction, is administered to induce labor.

FIGURE 12.21 Structure of muscle cells   Muscles are composed of bundles of single large cells (called muscle fibers) that form by cell fusion and contain multiple nuclei. Each muscle fiber contains many myofibrils, which are bundles of actin and myosin filaments organized into a chain of repeating units called sarcomeres.

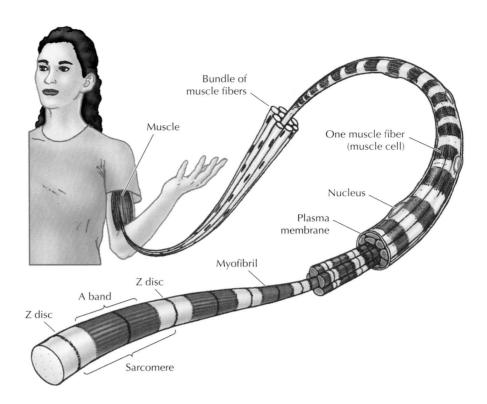

The sarcomeres (which are approximately 2.3 μm long) consist of several distinct regions discernible by electron microscopy, which provided critical insights into the mechanism of muscle contraction (Figure 12.22). The ends of each sarcomere are defined by the Z disc. Within each sarcomere, dark bands (called A bands because they are *a*nisotropic when viewed with

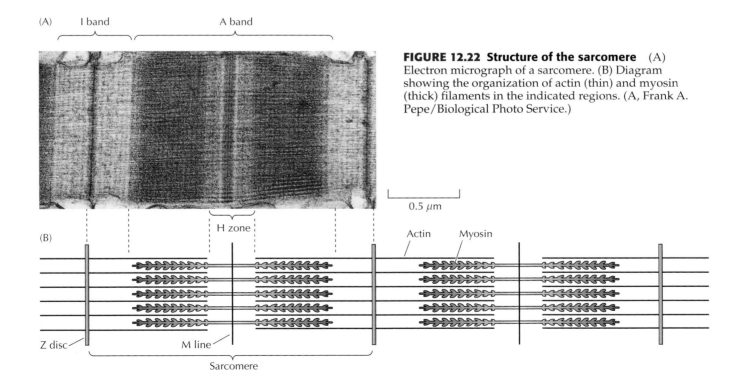

FIGURE 12.22 Structure of the sarcomere   (A) Electron micrograph of a sarcomere. (B) Diagram showing the organization of actin (thin) and myosin (thick) filaments in the indicated regions. (A, Frank A. Pepe/Biological Photo Service.)

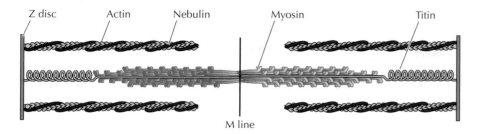

FIGURE 12.23 Titin and nebulin
Molecules of titin extend from the Z disc to the M line and act as springs to keep myosin filaments centered in the sarcomere. Molecules of nebulin extend from the Z disc and are thought to determine the length of associated actin filaments.

polarized light) alternate with light bands (called I bands for *i*sotropic). These bands correspond to the presence or absence of myosin filaments. The I bands contain only thin (actin) filaments, whereas the A bands contain thick (myosin) filaments. The myosin and actin filaments overlap in peripheral regions of the A band, whereas a middle region (called the H zone) contains only myosin. The actin filaments are attached at their barbed ends to the Z disc, which includes the cross-linking protein α-actinin. The myosin filaments are anchored at the M line in the middle of the sarcomere.

Two additional proteins (**titin** and **nebulin**) also contribute to sarcomere structure and stability (Figure 12.23). Titin is an extremely large protein (3000 kd), and single titin molecules extend from the M line to the Z disc. These long molecules of titin are thought to act like springs that keep the myosin filaments centered in the sarcomere and maintain the resting tension that allows a muscle to snap back if overextended. Nebulin filaments are associated with actin and are thought to regulate the assembly of actin filaments by acting as rulers that determine their length.

The basis for understanding muscle contraction is the **sliding filament model**, first proposed in 1954 both by Andrew Huxley and Ralph Niedergerke and by Hugh Huxley and Jean Hanson (Figure 12.24). During muscle contraction each sarcomere shortens, bringing the Z discs closer together. There is no change in the width of the A band, but both the I bands and the H zone almost completely disappear. These changes are explained by the

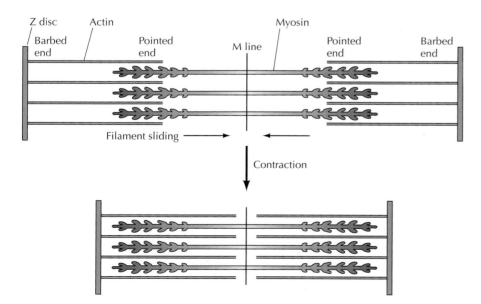

FIGURE 12.24 Sliding-filament model of muscle contraction The actin filaments slide past the myosin filaments toward the middle of the sarcomere. The result is shortening of the sarcomere without any change in filament length.

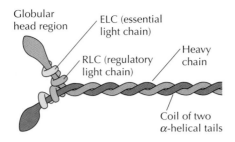

Globular
head region

ELC (essential
light chain)

RLC (regulatory
light chain)

Heavy
chain

Coil of two
α-helical tails

FIGURE 12.25 Myosin II The myosin II molecule consists of two heavy chains and two pairs of light chains (called the essential and regulatory light chains). The heavy chains have globular head regions and long α-helical tails, which coil around each other to form dimers.

actin and myosin filaments sliding past one another so that the actin filaments move into the A band and H zone. Muscle contraction thus results from an interaction between the actin and myosin filaments that generates their movement relative to one another. The molecular basis for this interaction is the binding of myosin to actin filaments, allowing myosin to function as a motor that drives filament sliding.

The type of myosin present in muscle (**myosin II**) is a very large protein (about 500 kd) consisting of two identical heavy chains (about 200 kd each) and two pairs of light chains (about 20 kd each) (Figure 12.25). Each heavy chain consists of a globular head region and a long α-helical tail. The α-helical tails of two heavy chains twist around each other in a coiled-coil structure to form a dimer, and two light chains associate with the neck of each head region to form the complete myosin II molecule.

The thick filaments of muscle consist of several hundred myosin molecules associated in a parallel staggered array by interactions between their tails (Figure 12.26). The globular heads of myosin bind actin, forming cross-bridges between the thick and thin filaments. It is important to note that the orientation of myosin molecules in the thick filaments reverses at the M line of the sarcomere. The polarity of actin filaments (which are attached to Z discs at their barbed ends) similarly reverses at the M line, so the relative orientation of myosin and actin filaments is the same on both halves of the sarcomere. As discussed later, the motor activity of myosin moves its head groups along the actin filament in the direction of the barbed end. This movement slides the actin filaments from both sides of the sarcomere toward the M line, shortening the sarcomere and resulting in muscle contraction.

In addition to binding actin the myosin heads bind and hydrolyze ATP, which provides the energy to drive filament sliding. This translation of chemical energy to movement is mediated by changes in the shape of myosin resulting from ATP binding. The generally accepted model (the swinging-cross-bridge model) is that ATP hydrolysis drives repeated cycles of interaction between myosin heads and actin. During each cycle, conformational changes in myosin result in the movement of myosin heads along actin filaments.

Although the molecular mechanisms are still not fully understood, a plausible working model for myosin function has been derived both from

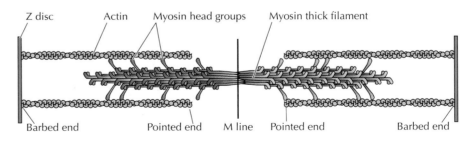

Z disc   Actin   Myosin head groups   Myosin thick filament

Barbed end   Pointed end   M line   Pointed end   Barbed end

FIGURE 12.26 Organization of myosin thick filaments Thick filaments are formed by the association of several hundred myosin II molecules in a staggered array. The globular heads of myosin bind actin, forming cross-bridges between the myosin and actin filaments. The orientation of both actin and myosin filaments reverses at the M line, so their relative polarity is the same on both sides of the sarcomere.

FIGURE 12.27 Model for myosin action The binding of ATP dissociates myosin from actin. ATP hydrolysis then induces a conformational change that displaces the myosin head group. This is followed by binding of the myosin head to a new position on the actin filament and release of ADP and P_i. The return of the myosin head to its original conformation drives actin filament sliding.

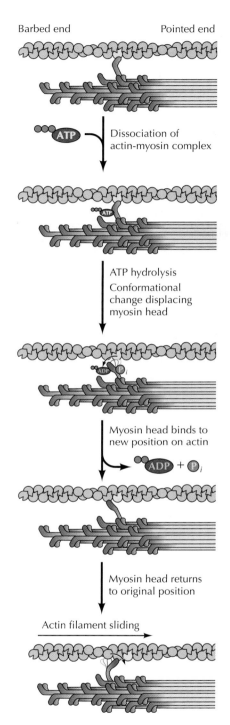

Barbed end   Pointed end

Dissociation of actin-myosin complex

ATP hydrolysis Conformational change displacing myosin head

Myosin head binds to new position on actin

ADP + P$_i$

Myosin head returns to original position

Actin filament sliding

in vitro studies of myosin movement along actin filaments (a system developed by James Spudich and Michael Sheetz) and from determination of the three-dimensional structure of myosin by Ivan Rayment and his colleagues (Figure 12.27). The cycle starts with myosin (in the absence of ATP) tightly bound to actin. ATP binding dissociates the myosin-actin complex and the hydrolysis of ATP then induces a conformational change in myosin. This change affects the neck region of myosin that binds the light chains (see Figure 12.25), which acts as a lever arm to displace the myosin head by about 5 nm. The products of hydrolysis (ADP and P_i) remain bound to the myosin head, which is said to be in the "cocked" position. The myosin head then rebinds at a new position on the actin filament, resulting in the release of ADP and P_i and triggering the "power stroke" in which the myosin head returns to its initial conformation, thereby sliding the actin filament toward the M line of the sarcomere.

The contraction of skeletal muscle is triggered by nerve impulses, which stimulate the release of Ca^{2+} from the **sarcoplasmic reticulum**—a specialized network of internal membranes (similar to the endoplasmic reticulum) that stores high concentrations of Ca^{2+} ions. The release of Ca^{2+} from the sarcoplasmic reticulum increases the concentration of Ca^{2+} in the cytosol from approximately 10^{-7} to 10^{-5} M. The increased Ca^{2+} concentration signals muscle contraction via the action of two actin filament binding proteins: tropomyosin and **troponin** (Figure 12.28). In striated muscle each tropomyosin molecule is bound to troponin, which is a complex of three polypeptides: troponin C (Ca^{2+}-binding), troponin I (inhibitory), and troponin T (tropomyosin-binding). When the concentration of Ca^{2+} is low, the complex of the troponins with tropomyosin blocks the interactions of almost all actins with the myosin head groups, so the muscle does not contract. At high concentrations, Ca^{2+} binding to troponin C shifts the position of the complex, allowing access of the myosin head groups to an increasing number of actins and allowing contraction to proceed.

Contractile Assemblies of Actin and Myosin in Nonmuscle Cells

Contractile assemblies of actin and myosin resembling small-scale versions of muscle fibers are present also in nonmuscle cells. As in muscle, the actin filaments in these contractile assemblies are interdigitated with bipolar filaments of myosin II, consisting of 15 to 20 myosin II molecules, which produce contraction by sliding the actin filaments relative to one another (Figure 12.29). The actin filaments in contractile bundles in nonmuscle cells are also associated with tropomyosin, which facilitates their interaction with myosin II.

Two examples of contractile assemblies in nonmuscle cells, stress fibers, and adhesion belts, were discussed earlier with respect to attachment of the actin cytoskeleton to regions of cell-substratum and cell-cell contacts (see Figures 12.16 and 12.17). The contraction of stress fibers produces tension across the cell, allowing the cell to pull on a substratum (e.g., the extracellular matrix) to which it is anchored. The contraction of adhesion belts alters

FIGURE 12.28 Association of tropomyosin and troponins with actin filaments (A) Tropomyosin binds lengthwise along actin filaments, and in striated muscle, is associated with a complex of three troponins: troponin I (TnI), troponin C (TnC), and troponin T (TnT). In the absence of Ca^{2+}, the tropomyosin-troponin complex blocks the binding of myosin to actin. Binding of Ca^{2+} to TnC shifts the complex, relieving this inhibition and allowing contraction to proceed. (B) Cross-sectional view.

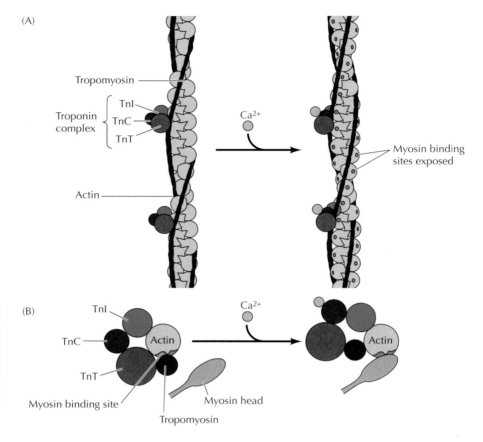

12.2 WEBSITE ANIMATION

A Thin Filament The thin filaments of skeletal muscle consist of actin filaments decorated with tropomyosin-troponin complexes covering the myosin binding sites.

the shape of epithelial cell sheets: a process that is particularly important during embryonic development when sheets of epithelial cells fold into structures such as tubes.

The most dramatic example of actin-myosin contraction in nonmuscle cells, however, is provided by **cytokinesis**—the division of a cell into two cells following mitosis (Figure 12.30). Toward the end of mitosis in yeast and animal cells, a **contractile ring** consisting of actin filaments and myosin II assembles just underneath the plasma membrane. Its contraction pulls the plasma membrane progressively inward, constricting the center of the cell and pinching it in two. Interestingly, the thickness of the contractile ring remains constant as it contracts, implying that actin filaments disassemble as contraction proceeds. The ring then disperses completely following cell

FIGURE 12.29 Contractile assemblies in nonmuscle cells Bipolar filaments of myosin II produce contraction by sliding actin filaments in opposite directions.

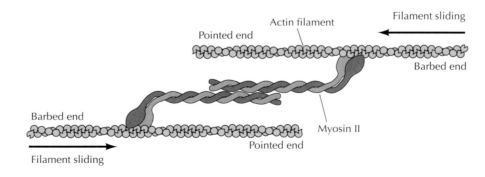

FIGURE 12.30 Cytokinesis Following completion of mitosis (nuclear division), a contractile ring consisting of actin filaments and myosin II divides the cell in two.

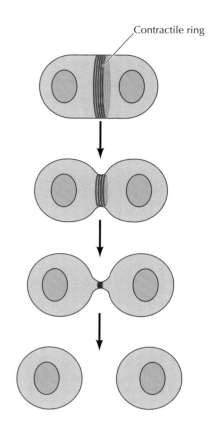

division. This mechanism appears to be evolutionarily ancient, as the bacterial MreB protein also functions in cell division.

The regulation of actin-myosin contraction in striated muscle, discussed earlier, is mediated by the binding of Ca^{2+} to troponin. In nonmuscle cells and in smooth muscle, however, contraction is regulated primarily by phosphorylation of one of the myosin light chains called the regulatory light chain (Figure 12.31). Phosphorylation of the regulatory light chain in these cells has at least two effects: It promotes the assembly of myosin into filaments, and it increases myosin catalytic activity enabling contraction to proceed. The enzyme that catalyzes this phosphorylation, called **myosin light-chain kinase**, is itself regulated by association with the Ca^{2+}-binding protein **calmodulin**. Increases in cytosolic Ca^{2+} promote the binding of calmodulin to the kinase resulting in phosphorylation of the myosin regulatory light chain. Increases in cytosolic Ca^{2+} are thus responsible, albeit indirectly, for activating myosin in smooth muscle and nonmuscle cells, as well as in striated muscle.

Nonmuscle Myosins

In addition to myosin II (the two-headed myosins found in muscle cells), several other types of myosin are found in nonmuscle cells. These myosins do not have tails able to form coiled-coils, so they do not form filaments and are not involved in contraction. However, they are important in a variety of cell movements, such as the transport of membrane vesicles and organelles along actin filaments, phagocytosis, and extension of pseudopods in amoebae (see Figure 12.20).

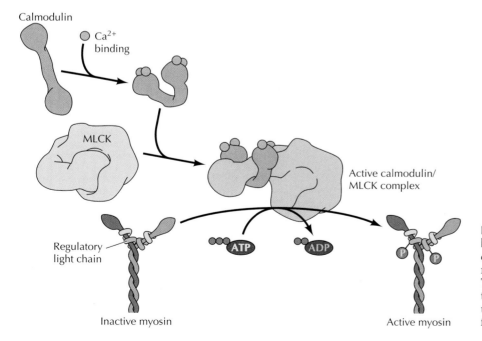

FIGURE 12.31 Regulation of myosin by phosphorylation Ca^{2+} binds to calmodulin, which in turn binds to myosin light-chain kinase (MLCK). The active calmodulin-MLCK complex then phosphorylates the myosin II regulatory light chain converting myosin from an inactive to an active state.

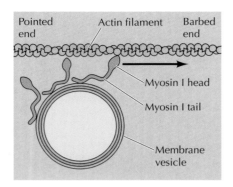

FIGURE 12.32 Myosin I Myosin I contains a head group similar to myosin II, but it has a comparatively short tail and does not form dimers or filaments. Although it cannot induce contraction, myosin I can move along actin filaments (toward the barbed end), carrying a variety of cargoes (such as membrane vesicles) attached to its tail.

One family of nonmuscle myosins are members of the **myosin I** family (Figure 12.32). The myosin I proteins contain a globular head group that acts as a molecular motor, like that of myosin II, but myosin I proteins are much smaller molecules (about 110 kd in mammalian cells) that lack the long tail of myosin II. Rather than forming dimers, their tails bind to other structures, such as membrane vesicles or organelles. The movement of myosin I along an actin filament can then transport its attached cargo. One function of myosin I, discussed earlier, is to form the lateral arms that link actin bundles to the plasma membrane of intestinal microvilli (see Figure 12.19). In these structures, the motor activity of myosin I may move the plasma membrane along the actin bundles toward the tip of the microvillus. Additional functions of myosin I may include the transport of vesicles and organelles along actin filaments and the movement of the plasma membrane during phagocytosis and pseudopod extension.

In addition to myosins I and II at least 12 other classes of nonmuscle myosins (III through XIV) have been identified. Some of these nonmuscle myosins are one-headed like myosin I, whereas others like myosin V (Figure 12.33) are two-headed like myosin II. The functions of most of these nonmuscle myosins remain to be determined, but some have been shown to play important roles in cargo transport and organelle movement (myosins V and VI) and in sensory functions such as vision (myosin III) and hearing (myosins VI and VII). Myosin VI is apparently unique among myosins in that it moves toward the pointed ends of actin filaments. Finally, some of

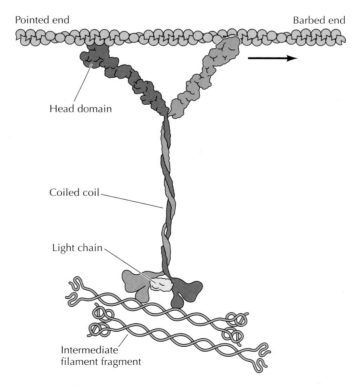

FIGURE 12.33 Myosin V Myosin V is a two-headed myosin like myosin II. It transports organelles and other cargo (for example, intermediate filaments) toward the barbed ends of actin filaments. The model shown is based on X-ray crystallography data (R. D. Vale, 2003. *Cell* 112: 467).

FIGURE 12.34 Cell migration   The movement of cells across a surface can be viewed as three stages of coordinated movements: (1) extension of the leading edge, (2) attachment of the leading edge to the substratum, and (3) retraction of the rear of the cell into the cell body.

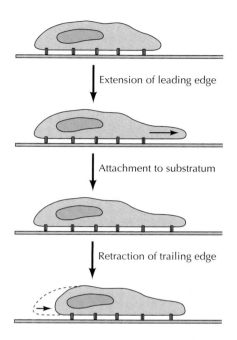

Extension of leading edge

Attachment to substratum

Retraction of trailing edge

the other myosins do not move cargo but participate in actin filament reorganization or anchor actin filaments to the plasma membrane.

Formation of Protrusions and Cell Movement

The movement of cells across a surface represents a basic form of cell locomotion employed by a wide variety of different kinds of cells. Examples include the crawling of amoebas, the migration of embryonic cells during development, the invasion of tissues by white blood cells to fight infection, the migration of cells involved in wound healing, and the spread of cancer cells during the metastasis of malignant tumors. Similar types of movement are also responsible for phagocytosis and for the extension of nerve cell processes during development of the nervous system. All of these movements are based on local specializations and extensions of the plasma membrane driven by the dynamic properties of the actin cytoskeleton.

Cell movement or the extension of long cellular processes involves a coordinated cycle of movements, which can be viewed in several stages (Figure 12.34). First, cells must develop an initial polarity via specialization of the plasma membrane or the cell cortex. Second, protrusions such as pseudopodia, lamellipodia, or filopodia (see Figure 12.20) must be extended to establish a leading edge of the cell. These extensions must then attach to the substratum across which the cell is moving. Finally, during cell migration the trailing edge of the cell must dissociate from the substratum and retract into the cell body. A variety of experiments indicate that extension of the leading edge involves the branching and polymerization of actin filaments. For example, inhibition of actin polymerization (e.g., by treatment with cytochalasin) blocks the formation of cell surface protrusions. As described below, the process that underlies the extension of cell protrusions is the force of actin filament polymerization against the cell membrane at the leading edge.

In most cases, cells move in response to signals from other cells or the environment. For example, in wound healing, cells at the edge of a cut extend lamellipodia and move across the extracellular matrix or the underlying cells to cover the wound. The signals that stimulate cell movement activate receptors in a small area of the cell membrane, leading to the recruitment of membrane proteins and the formation of specialized lipids in that area. These proteins and lipids in turn recruit actin binding proteins, including the Arp2/3 complex, its activator the **WASP/Scar complex,** and barbed-end tracking proteins that connect the growing actin filaments to the plasma membrane (Figure 12.35). WASP/Scar activates the Arp2/3 complex, which then initiates actin filament branches near the barbed ends, thereby increasing the number of growing barbed ends that are able to push against the cell membrane. At high local concentrations of ATP-actin, the growth of the barbed ends of actin filaments is energetically favorable and can generate considerable force. While a single filament would not exert enough force to extend the cell membrane, many filaments can easily do so. As the barbed ends of the actin filaments at the leading edge branch and grow, the pointed ends of the filaments are actively disassembled by

■ Certain pathogenic bacteria subvert the normal dynamics of the actin cytoskeleton to move through the host cell cytoplasm and infect neighboring cells.

FIGURE 12.35 Actin filament branching at the leading edge The Arp2/3 complex, WASP/Scar proteins, and barbed end-tracking proteins are recruited to a small area of the plasma membrane at the leading edge of a cell. WASP/Scar activates the Arp2/3 complex to initiate actin filament branches near the barbed ends, which are connected to the plasma membrane by barbed end-tracking proteins. At the pointed ends of the filaments, monomers of ADP-actin are removed by ADF/cofilin. The ADP-actin monomers are carried to the growing barbed ends by twinfilin and reactivated through ADP/ATP exchange by profilin.

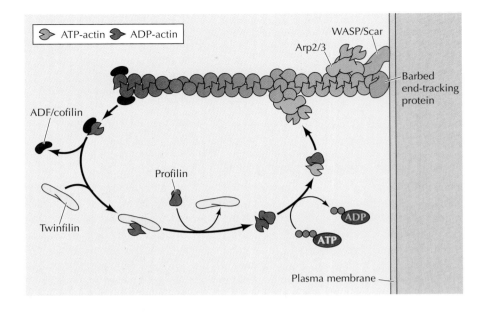

ADF/cofilin. The ADP-actin monomers are reactivated through ADP/ATP exchange by profilin and carried to the growing barbed ends by **twinfilin**. As the new microfilaments extend into the growing cell process, reorganized microtubules and the new microfilaments provide pathways for the delivery of lipid vesicles and proteins needed for continued extension. In neurons, myosin V is required to provide the new membrane components for extension of filopodia. The regulation of this entire process involves small GTP-binding proteins of the Rho family, which are discussed in Chapter 15.

Cell attachment to the surface requires rebuilding cell-substratum or cell-cell adhesions. For slow-moving cells, such as epithelial cells or fibroblasts, attachment involves the formation of focal adhesions (see Figure 12.16). Among the cargo proteins delivered to the growing cell protrusion by actin filaments and microtubules are actin-bundling proteins (which are needed for the formation of actin bundles and stress fibers immediately behind the leading edge) as well as focal adhesion proteins, such as talin and vinculin. Intermediate filament proteins are also transported toward the leading edge where they are used for the reorganization of the intermediate filament meshwork.

Reconstruction of focal adhesions occurs in two steps: the appearance of small focal complexes containing a few microfilaments attached to integrins and the growth of those focal complexes into mature focal contacts (illustrated in Figure 12.16). Vinculin and talin are activated by contact with the cell membrane lipids. They activate integrins to bind to the extracellular matrix as well as connecting the integrins to actin filaments. The appearance of mature focal contacts also requires the development of tension between the cell and the substratum, which is generated by myosin motors acting on actin bundles or stress fibers. Cells moving more rapidly, such as amoebas or white blood cells, form more diffuse contacts with the substratum, the molecular composition of which is not known.

The final stage of cell migration—retraction of the trailing edge—involves the action of small GTP-binding proteins of the ARF family (see Chapter 10), which break down existing focal adhesions at the trailing edge of the cell. The myosin II-mediated sliding of microfilaments in actin bundles and stress fibers connected to the new focal adhesions formed at the leading edge then generates the force necessary to pull the trailing edge of the cell forward.

Intermediate Filaments

Intermediate filaments have diameters between 8 and 11 nm, which is intermediate between the diameters of the two other principal elements of the cytoskeleton, actin filaments (about 7 nm) and microtubules (about 25 nm). In contrast to actin filaments and microtubules the intermediate filaments are not directly involved in cell movements. Instead, they appear to play basically a structural role by providing mechanical strength to cells and tissues.

Intermediate Filament Proteins

Whereas actin filaments and microtubules are polymers of single types of proteins (actin and tubulin, respectively), intermediate filaments are composed of a variety of proteins that are expressed in different types of cells. More than 65 different intermediate filament proteins have been identified and classified into six groups based on similarities between their amino acid sequences (Table 12.2). Types I and II consist of two groups of **keratins**, each consisting of about 15 different proteins, which are expressed in epithelial cells. Each type of epithelial cell synthesizes at least one type I (acidic) and one type II (neutral/basic) keratin, which copolymerize to form filaments. Some type I and type II keratins (called hard keratins) are used for production of structures such as hair, nails, and horns. The other type I and type II keratins (soft keratins) are abundant in the cytoplasm of epithelial cells, with different keratins being expressed in various differentiated cell types.

TABLE 12.2 Intermediate Filament Proteins

Type	Protein	Size (kd)	Site of expression
I	Acidic keratins (~15 proteins)	40–60	Epithelial cells
II	Neutral or basic keratins (~15 proteins)	50–70	Epithelial cells
III	Vimentin	54	Fibroblasts, white blood cells, and other cell types
	Desmin	53	Muscle cells
	Glial fibrillary acidic protein	51	Glial cells
	Peripherin	57	Peripheral neurons
IV	Neurofilament proteins		
	NF-L	67	Neurons
	NF-M	150	Neurons
	NF-H	200	Neurons
	α-internexin	66	Neurons
V	Nuclear lamins	60–75	Nuclear lamina of all cell types
VI	Nestin	200	Stem cells, especially of the central nervous system

Nestins are sometimes classified as type IV rather than type VI intermediate filaments.

The type III intermediate filament proteins include **vimentin**, which is found in several different kinds of cells, including fibroblasts, smooth muscle cells, and white blood cells. Unlike actin filaments, vimentin forms a network extending out from the nucleus toward the cell periphery. Another type III protein, **desmin**, is specifically expressed in muscle cells where it connects the Z discs of individual contractile elements. A third type III intermediate filament protein is specifically expressed in glial cells, and a fourth is expressed in neurons of the peripheral nervous system.

The type IV intermediate filament proteins include the three **neurofilament (NF) proteins** (designated NF-L, NF-M, and NF-H for light, medium, and heavy, respectively). These proteins form the major intermediate filaments of many types of mature neurons. They are particularly abundant in the axons of motor neurons and are thought to play a critical role in supporting these long, thin processes, which can extend more than a meter in length. Another type IV protein (α-internexin) is expressed at an earlier stage of neuron development prior to expression of the neurofilament proteins. The type V intermediate filament proteins are the nuclear lamins, which are found in most eukaryotic cells. Rather than being part of the cytoskeleton, the nuclear lamins are components of the nucleus where they assemble to form an orthogonal meshwork underlying the nuclear membrane and extending more diffusely into the nucleus (see Figure 9.5). Nestins (the type VI intermediate filaments) are expressed during embryonic development in several types of stem cells. They differ from other intermediate filaments in that they only polymerize if other intermediate filaments are present in the cell; nestins are sometimes classified as another type IV rather a type VI intermediate filament.

Despite considerable diversity in size and amino acid sequence, the various intermediate filament proteins share a common structural organization (Figure 12.36). All of the intermediate filament proteins have a central α-helical rod domain of approximately 310 amino acids (350 amino acids in the nuclear lamins). This central rod domain is flanked by amino- and carboxy-terminal domains, which vary among the different intermediate filament proteins in size, sequence, and secondary structure. As discussed next, the α-helical rod domain plays a central role in filament assembly, while the variable head and tail domains presumably determine the specific functions of the different intermediate filament proteins.

Assembly of Intermediate Filaments

The first stage of filament assembly is the formation of dimers in which the central rod domains of two polypeptide chains are wound around each other in a coiled-coil structure similar to that formed by myosin II heavy chains (Figure 12.37). The dimers of cytoskeletal intermediate filaments then

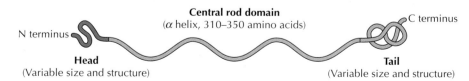

N terminus   **Head** (Variable size and structure)   **Central rod domain** (α helix, 310–350 amino acids)   **Tail** (Variable size and structure)   C terminus

FIGURE 12.36 Structure of intermediate filament proteins Intermediate filament proteins contain a central α-helical rod domain of approximately 310 amino acids (350 amino acids in the nuclear lamins). The N-terminal head and C-terminal tail domains vary in size and shape.

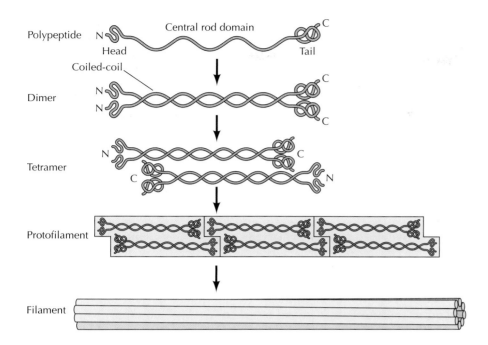

FIGURE 12.37 Assembly of intermediate filaments The central rod domains of two polypeptides wind around each other in a coiled-coil structure to form dimers. Dimers then associate in a staggered antiparallel fashion to form tetramers. Tetramers associate end-to-end to form protofilaments and laterally to form filaments. Each filament contains approximately eight protofilaments wound around each other in a ropelike structure.

associate in a staggered antiparallel fashion to form tetramers, which can assemble end-to-end to form protofilaments. A common step is the interaction of approximately eight protofilaments wound around each other in a ropelike structure. Because they are assembled from antiparallel tetramers, both ends of intermediate filaments are equivalent. Consequently, in contrast to actin filaments and microtubules, intermediate filaments are apolar—they do not have distinct plus and minus ends.

Filament assembly requires interactions between specific types of intermediate filament proteins. For example, keratin filaments are always assembled from heterodimers containing one type I and one type II polypeptide. In contrast, the type III proteins can assemble into filaments containing only a single polypeptide (e.g., vimentin) or consisting of two different type III proteins (e.g., vimentin plus desmin). The type III proteins do not, however, form copolymers with the keratins. Among the type IV proteins, α-internexin can assemble into filaments by itself, whereas the three neurofilament proteins copolymerize to form heteropolymers.

Intermediate filaments are generally more stable than actin filaments or microtubules and do not exhibit the dynamic behavior associated with these other elements of the cytoskeleton (e.g., the treadmilling of actin filaments illustrated in Figure 12.4). However, intermediate filament proteins are frequently modified by phosphorylation, which can regulate their assembly and disassembly within the cell. One example is phosphorylation of the nuclear lamins (see Chapter 16), which results in disassembly of the nuclear lamina and breakdown of the nuclear envelope during mitosis. Cytoplasmic intermediate filaments, such as vimentin, are also phosphorylated, which can lead to their disassembly and reorganization in dividing or migrating cells.

Intracellular Organization of Intermediate Filaments

Intermediate filaments form an elaborate network in the cytoplasm of most cells, extending from a ring surrounding the nucleus to the plasma mem-

FIGURE 12.38 Intracellular organization of keratin filaments Micrograph of epithelial cells stained with fluorescent antibodies to keratin (green). Nuclei are stained blue. The keratin filaments extend from a ring surrounding the nucleus to the plasma membrane. (Nancy Kedersha/Immunogen/Photo Researchers, Inc.)

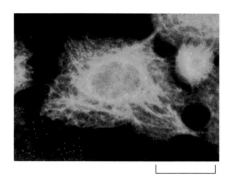

10 μm

brane (Figure 12.38). Both keratin and vimentin filaments attach to the nuclear envelope, apparently serving to position and anchor the nucleus within the cell. In addition, intermediate filaments can associate not only with the plasma membrane but also with the other elements of the cytoskeleton, actin filaments and microtubules. Intermediate filaments thus provide a scaffold that integrates the components of the cytoskeleton and organizes the internal structure of the cell.

The keratin filaments of epithelial cells are tightly anchored to the plasma membrane at two areas of specialized cell contacts: **desmosomes** and **hemidesmosomes** (Figure 12.39). Desmosomes are junctions between adjacent cells at which cell-cell contacts are mediated by transmembrane proteins related to the cadherins. On their cytoplasmic side, desmosomes are associated with a characteristic dense plaque of intracellular proteins to which keratin filaments are attached. These attachments are mediated by desmoplakin, a member of a family of proteins called **plakins** that bind intermediate filaments and link them to other cellular structures. Hemidesmosomes are morphologically similar junctions between epithelial cells and underlying connective tissue at which keratin filaments are linked by different members of the plakin family (e.g., plectin) to integrins. Desmosomes and hemidesmosomes thus anchor intermediate filaments to regions of cell-cell and cell-substratum contact, respectively, similar to the attachment of the actin cytoskeleton to the plasma membrane at adherens junctions and focal adhesions. It is important to note that the keratin filaments anchored to both sides of desmosomes serve as a mechanical link between adjacent cells in an epithelial layer, thereby providing mechanical stability to the entire tissue.

In addition to linking intermediate filaments to cell junctions, some plakins link intermediate filaments to other elements of the cytoskeleton. Plectin, for example, binds actin filaments and microtubules in addition to intermediate filaments, so it can provide bridges between these cytoskeletal components (Figure 12.40). These bridges to intermediate filaments are thought to brace and stabilize actin filaments and microtubules, thereby increasing the mechanical stability of the cell.

Two types of intermediate filaments—desmin and the neurofilaments—play specialized roles in muscle and nerve cells, respectively. Desmin connects the individual actin-myosin assemblies of muscle cells both to one another and to the plasma membrane, thereby linking the actions of individual contractile elements. Neurofilaments are the major intermediate filaments in most mature neurons. They are particularly abundant in the long axons of motor neurons where they appear to be anchored to actin filaments and microtubules by neuronal members of the plakin family. Neuro-

(A)

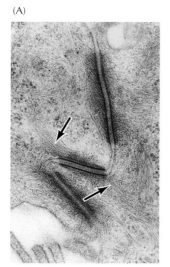

(B) **Desmosome**

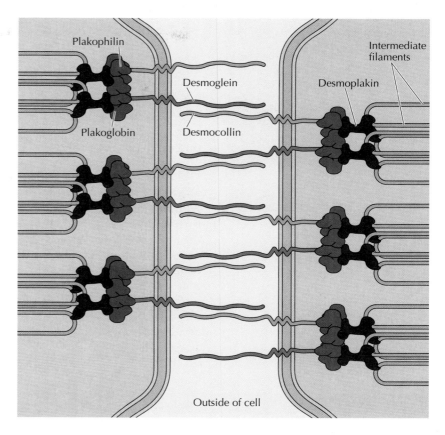

(C) **Hemidesmosome**

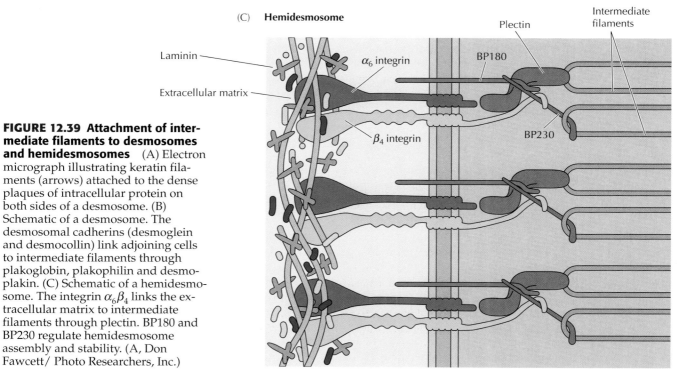

FIGURE 12.39 Attachment of intermediate filaments to desmosomes and hemidesmosomes (A) Electron micrograph illustrating keratin filaments (arrows) attached to the dense plaques of intracellular protein on both sides of a desmosome. (B) Schematic of a desmosome. The desmosomal cadherins (desmoglein and desmocollin) link adjoining cells to intermediate filaments through plakoglobin, plakophilin and desmoplakin. (C) Schematic of a hemidesmosome. The integrin $\alpha_6\beta_4$ links the extracellular matrix to intermediate filaments through plectin. BP180 and BP230 regulate hemidesmosome assembly and stability. (A, Don Fawcett/ Photo Researchers, Inc.)

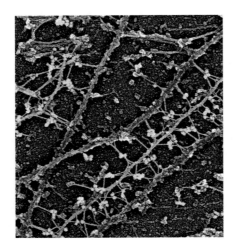

FIGURE 12.40 Electron micrograph of plectin bridges between intermediate filaments and microtubules Micrograph of a fibroblast stained with antibody against plectin. The micrograph has been artificially colored to show plectin (green), antibodies against plectin (yellow), intermediate filaments (blue), and microtubules (red). (Courtesy of Tatyana Svitkina and Gary Borisy, University of Wisconsin, Madison.)

filaments are thought to play an important role in providing mechanical support and stabilizing other elements of the cytoskeleton in these long, thin extensions of nerve cells.

Functions of Keratins and Neurofilaments: Diseases of the Skin and Nervous System

Although intermediate filaments have long been thought to provide structural support to the cell, direct evidence for their function has only recently been obtained. Some cells in culture make no intermediate filament proteins, indicating that these proteins are not required for the growth of cells *in vitro*. Similarly, injection of cultured cells with antibody against vimentin disrupts intermediate filament networks without affecting cell growth or movement. Therefore, it has been thought that intermediate filaments are most needed to strengthen the cytoskeleton of cells in the tissues of multicellular organisms where they are subjected to a variety of mechanical stresses that do not affect cells in the isolated environment of a culture dish.

KEY EXPERIMENT

Expression of Mutant Keratin Causes Abnormal Skin Development

Mutant Keratin Expression in Transgenic Mice Causes Marked Abnormalities Resembling a Human Genetic Skin Disease

Robert Vassar, Pierre A. Coulombe, Linda Degenstein, Kathryn Albers and Elaine Fuchs

University of Chicago

Cell, 1991, Volume 64, pages 365–380

Elaine Fuchs

The Context

By 1991 intermediate filaments in epithelial cells were well known and the developmental appearance of different forms of type I and type II keratins in the skin was being studied. What remained unknown for any intermediate filaments was their function. While all vertebrate cells contain intermediate filaments, cell lines that lack them survive in culture and continue to proliferate. Thus whatever function intermediate filaments had, they were important not for cells in culture but perhaps for cells within the tissues of multicellular organisms.

Fuchs and coworkers knew that during early development the epithelium of the skin expresses keratin 5 (type I) and keratin 14 (type II). Because keratins polymerize as heterodimers, it was thought that expression of an abnormal protein might interfere with the formation of normal intermediate filaments. Fuchs and her colleagues initially tested this possibility in cultured skin cells and demonstrated that expression of a truncated keratin 14 interfered with keratin filament formation. This suggested that similar expression of the mutant keratin in transgenic mice might cause a defect in the intermediate filament network in the skin cells of an embryo. If this occurred, it would provide a test of the role of intermediate filaments in an intact tissue.

KEY EXPERIMENT

In the experiments described here, Fuchs and her colleagues demonstrated that expression of a mutant keratin in transgenic mice not only disrupted the intermediate filament network of skin cells but also led to severe defects in the organization and tissue stability of the skin. These experiments thus provided the first demonstration of a physiological role for intermediate filaments.

The Experiments

For their earlier experiments in cultured cells, Fuchs and colleagues had constructed a mutant keratin 14 gene in which a truncated protein missing 30% of the central α-helical domain and all of the carboxy-terminal tail was expressed from the normal keratin promoter. To investigate the role of keratin 14 in early mouse development, they introduced the plasmid encoding this mutant keratin 14 into fertilized mouse eggs, which were transferred to foster mothers and allowed to develop into offspring. All of the offspring were analyzed for keratin 14, and some were found to be transgenic and express the mutant keratin 14 protein.

Most of the transgenic animals died within 24 hours of birth. Those that survived longer showed severe skin abnormalities, including blisters due to epidermal cell lysis following mild mechanical trauma, such as rubbing of the skin. Analysis of stained tissue sections of the skin from transgenic animals demonstrated severe disorganization of the epidermis in the most affected animals (see figure) and patches of disorganized tissue in others. Patchy expression is characteristic of a mosaic animal where some tissue develops from normal embryo

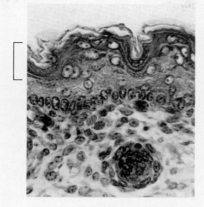

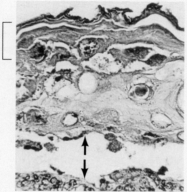

Skin from a normal and a transgenic mouse. (Top) Skin from a normal mouse showing the highly organized outer layers (brackets) with no intervening spaces between the intact underlying tissue. (Bottom) Skin from a transgenic mouse showing severe disruption of the outer layers, which contain spaces due to abrasion by mechanical trauma and are separated from the underlying tissue (arrows).

cells and some from cells carrying the transgene. By analyzing for the mutant protein, they found that the areas of disorganized tissue correlated with expression of the mutant keratin

14. In addition, there was a clear correlation between the amount of disorganized epidermis and susceptibility to skin damage and death during the trauma of birth.

Fuchs and her colleagues further noted that the pattern of tissue disorganization in the transgenic mice resembled that seen in a group of human skin diseases called epidermolysis bullosa simplex. Thus they compared sections of the transgenic mouse tissue with sections obtained from the skin of a human patient and found very similar patterns of tissue disruption. From this, Fuchs and coworkers concluded that defects in keratins or intermediate filament-related proteins might be a cause of human genetic diseases of the skin.

The Impact

The skin abnormalities of these transgenic mice provided the first direct support for the presumed role of keratins in providing mechanical strength to epithelial cells in tissues. It is now known that the intermediate filament cytoskeleton is critical for the structure of tissues such as skin, intestine, heart, and skeletal muscle that are subject to mechanical stress. In contrast, single-celled eukaryotes such as yeast and many small invertebrates survive perfectly well without intermediate filaments, often modifying the actin or tubulin cytoskeleton to serve a structural role. The results of Fuchs and her colleagues also suggested a basis for several human diseases. In fact, there are now more than 17 different keratins known to be defective in human disease; these include keratin 5 and keratin 14, both of which were initially studied in the transgenic mice.

Experimental evidence for such an *in vivo* role of intermediate filaments was first provided in 1991 by studies in the laboratory of Elaine Fuchs. These investigators used transgenic mice to investigate the *in vivo* effects of expressing a keratin deletion mutant encoding a truncated polypeptide that disrupted the formation of normal keratin filaments (Figure 12.41). This

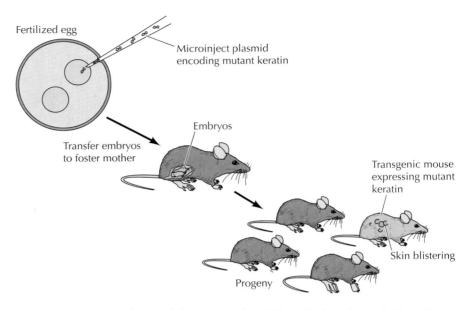

Fertilized egg

Microinject plasmid encoding mutant keratin

Transfer embryos to foster mother

Embryos

Transgenic mouse expressing mutant keratin

Skin blistering

Progeny

FIGURE 12.41 Experimental demonstration of keratin function   A plasmid encoding a mutant keratin that interferes with the normal assembly of keratin filaments was microinjected into one pronucleus of a fertilized egg. Microinjected embryos were then transferred to a foster mother, and some of the offspring were found to have incorporated the mutant keratin gene into their genome. Expression of the mutant gene in these transgenic mice disrupted the keratin cytoskeleton of cells of the epidermis, resulting in severe skin blistering due to cell lysis following mild mechanical stress.

■ **The physical properties of intermediate filaments are well suited to their role in providing structural support. They are not normally very rigid but harden and resist breakage when subjected to high stress.**

mutant keratin gene was introduced into transgenic mice where it was expressed in basal cells of the epidermis and disrupted formation of a normal keratin cytoskeleton. This resulted in the development of severe skin abnormalities, including blisters due to epidermal cell lysis following mild mechanical trauma, such as rubbing of the skin. The skin abnormalities of these transgenic mice thus provided direct support for the presumed role of keratins in providing mechanical strength to epithelial cells in tissues. Subsequently, the same result has been shown in mice in which the gene for the same keratin was inactivated by homologous recombination.

These experiments also pointed to the molecular basis of a human genetic disease, epidermolysis bullosa simplex (EBS). Like the transgenic mice expressing mutant keratin genes, patients with this disease develop skin blisters resulting from cell lysis after minor trauma. This similarity prompted studies of the keratin genes in EBS patients, leading to the demonstration that EBS is caused by keratin gene mutations that interfere with the normal assembly of keratin filaments. Thus both experimental studies in transgenic mice and molecular analysis of a human genetic disease have demonstrated the role of keratins in allowing skin cells to withstand mechanical stress. Continuing studies have shown that mutations in other keratins are responsible for several other inherited skin diseases, which are similarly characterized by abnormal fragility of epidermal cells.

Other studies in transgenic mice have implicated abnormalities of neurofilaments in diseases of motor neurons, particularly amyotrophic lateral sclerosis (ALS). ALS, known as Lou Gehrig's disease (and the disease afflicting the renowned physicist Stephen Hawking), results from progressive loss of motor neurons, which in turn leads to muscle atrophy, paralysis, and eventual death. ALS and other types of motor neuron disease are characterized by the accumulation and abnormal assembly of neurofilaments, suggesting that neurofilament abnormalities might contribute to these pathologies. Consistent with this possibility, overexpression of NF-L or NF-H in transgenic mice has been found to result in the development of a condition similar to ALS. Although the mechanism involved remains to be understood, these experiments clearly suggest the involvement of neurofilaments in the pathogenesis of motor neuron disease.

Microtubules

Microtubules, the third principal component of the cytoskeleton, are rigid hollow rods approximately 25 nm in diameter. Like actin filaments, microtubules are dynamic structures that undergo continual assembly and disassembly within the cell. They function both to determine cell shape and in a variety of cell movements, including some forms of cell locomotion, the intracellular transport of organelles, and the separation of chromosomes during mitosis.

Structure and Dynamic Organization of Microtubules

In contrast to intermediate filaments, which are composed of a variety of different fibrous proteins, microtubules are composed of a single type of globular protein called **tubulin**. The building blocks of microtubules are tubulin dimers consisting of two closely related 55 kd polypeptides: α-tubulin and β-tubulin. Like actin, both α- and β-tubulin are encoded by small families of related genes. In addition, a third type of tubulin (γ-tubulin) is concentrated in the centrosome where it plays a critical role in initiating microtubule assembly (discussed shortly). The evolutionary ancestor of all plant and animal tubulins appears to be a protein similar to the prokaryote protein, FtsZ.

Like actin, tubulin polymerization can be studied *in vitro*. Tubulin dimers polymerize to form microtubules, which generally consist of 13 linear protofilaments assembled around a hollow core (**Figure 12.42**). The protofilaments, which are composed of head-to-tail arrays of tubulin dimers, are arranged in parallel. Consequently, microtubules (like actin filaments) are polar structures with two distinct ends: a fast-growing plus end and a slow-growing minus end. This polarity is an important consideration in determining the direction of movement along microtubules, just as the polarity of actin filaments defines the direction of myosin movement.

Tubulin dimers can depolymerize as well as polymerize, and microtubules can undergo rapid cycles of assembly and disassembly. Both α- and β-tubulin bind GTP, which functions analogously to the ATP bound to actin to regulate polymerization. In particular, the GTP bound to β-tubulin (though not that bound to α-tubulin) is hydrolyzed to GDP during or shortly after polymerization. This GTP hydrolysis weakens the binding affinity of tubulin for adjacent molecules, thereby favoring depolymeriza-

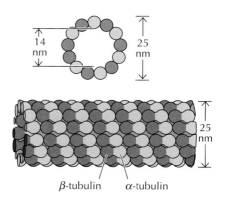

FIGURE 12.42 Structure of microtubules Dimers of α- and β-tubulin polymerize to form microtubules, which are composed of 13 protofilaments assembled around a hollow core.

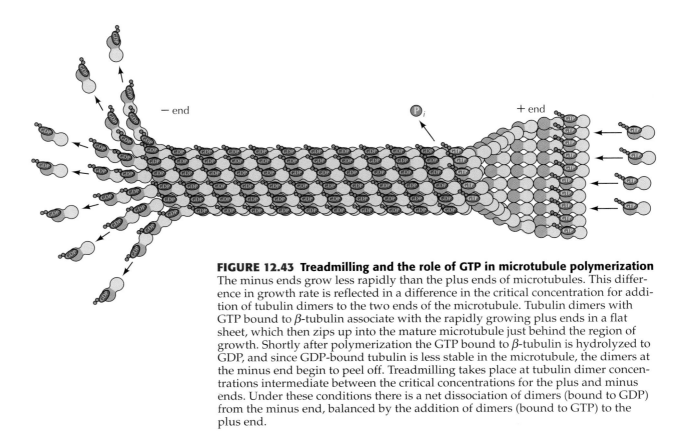

FIGURE 12.43 Treadmilling and the role of GTP in microtubule polymerization
The minus ends grow less rapidly than the plus ends of microtubules. This difference in growth rate is reflected in a difference in the critical concentration for addition of tubulin dimers to the two ends of the microtubule. Tubulin dimers with GTP bound to β-tubulin associate with the rapidly growing plus ends in a flat sheet, which then zips up into the mature microtubule just behind the region of growth. Shortly after polymerization the GTP bound to β-tubulin is hydrolyzed to GDP, and since GDP-bound tubulin is less stable in the microtubule, the dimers at the minus end begin to peel off. Treadmilling takes place at tubulin dimer concentrations intermediate between the critical concentrations for the plus and minus ends. Under these conditions there is a net dissociation of dimers (bound to GDP) from the minus end, balanced by the addition of dimers (bound to GTP) to the plus end.

tion and resulting in the dynamic behavior of microtubules. Like actin filaments, microtubules undergo treadmilling (Figure 12.43), a dynamic behavior in which tubulin molecules bound to GDP are continually lost from the minus end and replaced by the addition of tubulin molecules bound to GTP to the plus end of the same microtubule. In microtubules, rapid GTP hydrolysis also results in the behavior known as **dynamic instability** in which individual microtubules alternate between cycles of growth and shrinkage (Figure 12.44). Whether a microtubule grows or shrinks is determined in part by the rate of tubulin addition relative to the rate of GTP hydrolysis. As long as new GTP-bound tubulin molecules are added more rapidly than GTP is hydrolyzed, the microtubule retains a GTP cap at its plus end and microtubule growth continues. However, if the rate of polymerization slows, the GTP bound to tubulin at the plus end of the microtubule will be hydrolyzed to GDP. If this occurs, the GDP-bound tubulin will dissociate, resulting in rapid depolymerization and shrinkage of the microtubule.

Dynamic instability, described by Tim Mitchison and Marc Kirschner in 1984, allows for the continual and rapid turnover of many microtubules within the cell, some of which have half-lives of only several minutes. As discussed later, this rapid turnover of microtubules is particularly critical for the remodeling of the cytoskeleton that occurs during mitosis. Because of the central role of microtubules in mitosis, drugs that affect microtubule assembly are useful not only as experimental tools in cell biology but also in the treatment of cancer. **Colchicine** and **colcemid** are examples of com-

High concentration of tubulin bound to GTP

Low concentration of tubulin bound to GTP

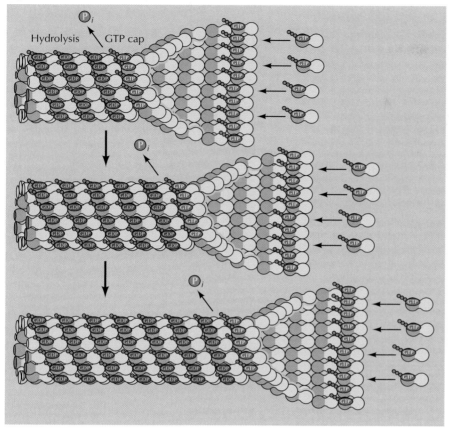

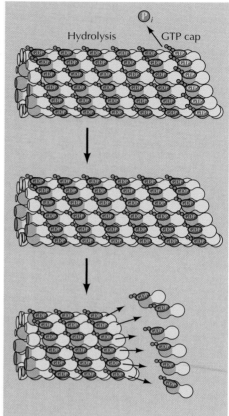

FIGURE 12.44 Dynamic instability of microtubules Dynamic instability results from the hydrolysis of GTP bound to β-tubulin during or shortly after polymerization, which reduces its binding affinity for adjacent molecules. Growth of microtubules continues as long as there is a high concentration of tubulin bound to GTP. New GTP-bound tubulin molecules are then added more rapidly than GTP is hydrolyzed, so a GTP cap is retained at the growing end. However, if GTP is hydrolyzed more rapidly than new subunits are added, the presence of GDP-bound tubulin at the plus end of the microtubule leads to disassembly and shrinkage.

12.3 WEBSITE ANIMATION
Microtubule Assembly
Tubulin dimers can polymerize as well as depolymerize to assemble or break down microtubules.

■ Taxol was originally isolated from the bark of Pacific yew trees. The Pacific yew is one of the slowest growing trees and several trees were needed to extract enough taxol for the treatment of one patient, limiting the availability of taxol. Fortunately, taxol and its derivatives are now produced economically via a semisynthetic process wherein a compound similar to taxol is extracted from related trees and modified chemically.

monly used experimental drugs that bind tubulin and inhibit microtubule polymerization, which in turn blocks mitosis. Two related drugs (**vincristine** and **vinblastine**) are used in cancer chemotherapy because they selectively inhibit rapidly dividing cells. Another useful drug, **taxol**, stabilizes microtubules rather than inhibiting their assembly. Such stabilization also blocks cell division, and taxol is used as an anticancer agent as well as an experimental tool.

Assembly of Microtubules

In animal cells, most microtubules extend outward from the **centrosome** (first described by Theodor Boveri in 1888), which is located adjacent to the nucleus near the center of interphase (nondividing) cells (Figure 12.45). During mitosis, microtubules similarly extend outward from duplicated centrosomes to form the mitotic spindle, which is responsible for the separation

FIGURE 12.45 Intracellular organization of microtubules The minus ends of microtubules are anchored in the centrosome. In interphase cells, the centrosome is located near the nucleus and microtubules extend outward to the cell periphery. During mitosis, duplicated centrosomes separate and microtubules reorganize to form the mitotic spindle.

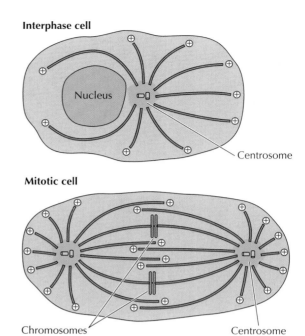

and distribution of chromosomes to daughter cells. The centrosome thus plays a key role in determining the intracellular organization of microtubules. Plant cells do not have an organized centrosome; instead, microtubules in most plants extend outward from the nucleus.

The centrosome is now known to function as a microtubule-organizing center in which the minus ends of the microtubules are anchored. It serves as the initiation site for the assembly of microtubules, which then grow outward from the centrosome toward the periphery of the cell. This can be clearly visualized in cells that have been treated with colcemid to disassemble their microtubules (Figure 12.46). When the drug is removed, the cells

FIGURE 12.46 Growth of microtubules from the centrosome
Microtubules in mouse fibroblasts are visualized by immunofluorescence microscopy using an antibody against tubulin. (A) The distribution of microtubules in a normal interphase cell. (B) This cell was treated with colcemid for one hour to disassemble microtubules. The drug was then removed and the cell allowed to recover for 30 minutes, allowing the visualization of new microtubules growing out of the centrosome. (From M. Osborn and K. Weber, 1976. *Proc. Natl. Acad. Sci. U.S.A.* 73: 867.)

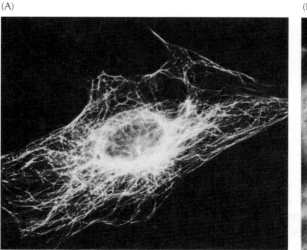

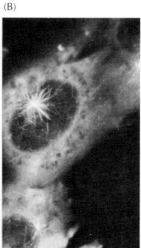

10 μm

(A)

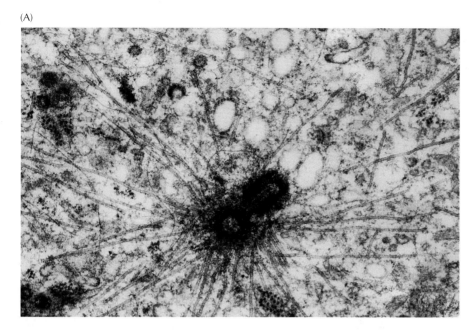

(B)

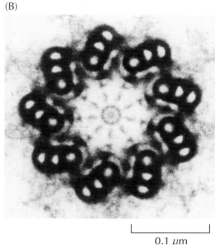

0.1 μm

FIGURE 12.47 Structure of centrosomes   (A) Electron micrograph of a centrosome showing microtubules radiating from the pericentriolar material that surrounds a pair of centrioles. (B) Transverse section of a centriole illustrating its nine triplets of microtubules. (A, © Cytographics; B, Don Fawcett, Photo Researchers, Inc.)

recover and new microtubules can be seen growing outward from the centrosome. Thus the initiation of microtubule growth at the centrosome establishes the polarity of microtubules within the cell. Note that microtubules grow by the addition of tubulin to their plus ends, which extend outward from the centrosome toward the cell periphery, so the role of the centrosome is to *initiate* microtubule growth. The key protein in the centrosome is γ-tubulin, a minor species of tubulin first identified in fungi, which nucleates assembly of microtubules. γ-tubulin is associated with eight or more other proteins in a ring-shaped structure called the **γ-tubulin ring complex**. This is thought to act as a seed for rapid microtubule growth, bypassing the rate-limiting nucleation step. Once assembled, microtubules can be released from the microtubule organizing center to organize elsewhere in the cell. This is especially evident in polarized epithelial cells or nerve cells as well as in plant root cells during growth of root hairs. Cells can initiate microtubules in the absence of a centrosome because γ-tubulin is also found in the cytosol, and in plant cells γ-tubulin is concentrated at the periphery of the nucleus.

The centrosomes of most animal cells contain a pair of **centrioles**, oriented perpendicular to each other and surrounded by amorphous **pericentriolar material** (Figure 12.47). The centrioles are cylindrical structures based on nine triplets of microtubules, similar to the basal bodies of cilia and flagella (discussed later in the chapter). Centrioles are necessary to form basal bodies, cilia, and flagella, and play a poorly understood role in coordinating the cell cycle. However, they are not necessary for the microtubule-organizing functions of the centrosome and are not found in plant cells, many unicellular eukaryotes, and most meiotic animal cells (such as mouse eggs). It is the pericentriolar material, not the centrioles, that initiates microtubule assembly. However, removal of the centrioles from animal cells results in dispersion of the centrosome contents and a decline in microtubule turnover.

FIGURE 12.48 Structure of a centriole Centrioles are highly polar structures with a cartwheel protein structure at one end and several types of processes extending from the other end. These processes (called satellites and appendages) are thought to interact with specific proteins in the centrosome matrix. Other centrosome matrix proteins such as γ-tubulin are associated with the lumen of the centriole. The triplet microtubules contain highly modified α- and β-tubulins and the unique δ-tubulin. Centrin fibers extend out from the triplet microtubules and connect to the other centriole. (Modified from W. F. Marshall, 2001. *Curr. Biol.* 11: R487.)

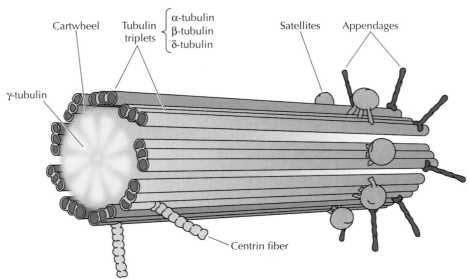

Centrioles are complex structures with a clear polarity, a cartwheel-like protein structure at one end, and numerous extensions into the centrosome (Figure 12.48). The extensions are thought to help organize the centrosome matrix. Centrioles also contain several unique proteins, including δ-tubulin, which is part of the characteristic microtubule triplet. During interphase the two centrioles are connected by one or more fibers that contain centrin (a Ca^{2+}-binding protein related to calmodulin).

Organization of Microtubules within Cells

Because of their inherent dynamic instability, most microtubules would be frequently disassembled within the cell. This dynamic behavior can, however, be modified by the interactions of microtubules with **microtubule-associated proteins** (**MAPs**). As with microfilaments, some of these proteins stabilize microtubules by capping their ends, while others act to disassemble microtubules, either by severing them or by increasing the rate of tubulin depolymerization at their ends. Several MAPs are plus-end tracking proteins, which bind to tubulin/GTP and act to track growing microtubules toward specific cellular locations. Such interactions allow the cell to stabilize microtubules in particular locations and provide an important mechanism for determining cell shape and polarity.

Many different MAPs have been identified and they vary depending on the type of cell. Among these are MAP-1, MAP-2, and tau (isolated from neuronal cells), and MAP-4, which is present in all non-neuronal vertebrate cell types. The tau protein has been extensively studied because it is the main component of the characteristic lesions found in the brains of Alzheimer's patients. The activity of many MAPs is regulated by phosphorylation, allowing the cell to control microtubule stability.

A good example of the role of stable microtubules in determining cell polarity is provided by nerve cells, which consist of two distinct types of processes (axons and dendrites) extending from a cell body (Figure 12.49). Both axons and dendrites are supported by stable microtubules, together with the neurofilaments discussed in the preceding section of this chapter. However, the microtubules in axons and dendrites are organized differently and associated with distinct MAPs. In axons, the microtubules are all ori-

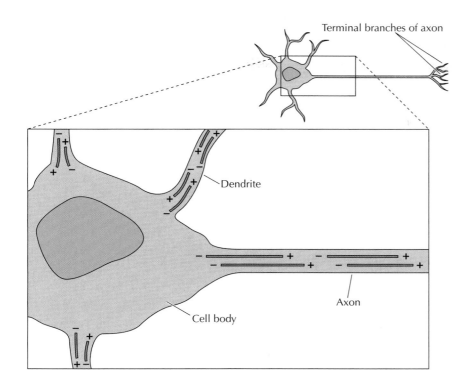

FIGURE 12.49 Organization of microtubules in nerve cells Two distinct types of processes extend from the cell body of nerve cells (neurons). Dendrites are short processes that receive stimuli from other nerve cells. The single long axon then carries impulses from the cell body to other cells, which may be either other neurons or an effector cell, such as a muscle. Stable microtubules in both axons and dendrites terminate in the cytoplasm rather than being anchored in the centrosome. In dendrites, microtubules are oriented in both directions, with their plus ends pointing both toward and away from the cell body. In contrast, all of the axon microtubules are oriented with their plus ends pointing toward the tip of the axon.

ented with their plus ends away from the cell body, similar to the general orientation of microtubules in other cell types. The minus ends of most of the microtubules in axons, however, are not anchored in the centrosome; instead, both the plus and minus ends of these microtubules terminate in the cytoplasm of the axon. In dendrites, the microtubules are oriented in both directions; some plus ends point toward the cell body and some point toward the cell periphery. These distinct microtubule arrangements are paralleled by differences in MAPs: Axons contain tau proteins, but no MAP-2, whereas dendrites contain MAP-2, but no tau proteins, and it appears that these differences in MAP-2 and tau distribution are responsible for the distinct organization of stable microtubules in axons and dendrites.

Microtubule Motors and Movement

Microtubules are responsible for a variety of cell movements, including the intracellular transport and positioning of membrane vesicles and organelles, the separation of chromosomes at mitosis, and the beating of cilia and flagella. As discussed for actin filaments earlier in this chapter, movement along microtubules is based on the action of motor proteins that utilize energy derived from ATP hydrolysis to produce force and movement. Members of two large families of motor proteins—the **kinesins** and the **dyneins**—are responsible for powering the variety of movements in which microtubules participate.

Identification of Microtubule Motor Proteins

Kinesin and dynein, the prototypes of microtubule motor proteins, move along microtubules in opposite directions—most kinesins toward the plus end and dyneins toward the minus end (Figure 12.50). The first of these

FIGURE 12.50 Microtubule motor proteins Kinesin and dynein move in opposite directions along microtubules, toward the plus and minus ends, respectively. Kinesin consists of two heavy chains (wound around each other in a coiled-coil structure) and two light chains. The globular head domains of the heavy chains bind microtubules and are the motor domains of the molecule. Dynein consists of two or three heavy chains (two are shown here) in association with multiple light and intermediate chains. The globular head domains of the heavy chains are the motor domains. The models shown are based on X-ray crystallography data (After R. D. Vale, 2003. *Cell* 112: 467).

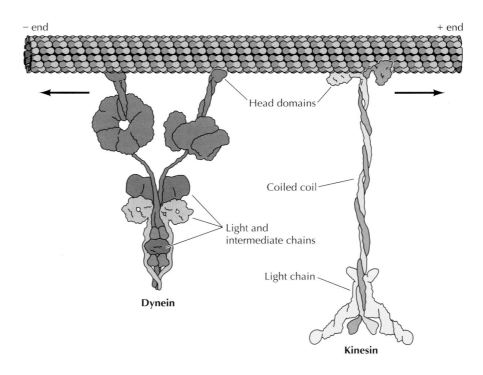

12.4 WEBSITE ANIMATION

Kinesin Kinesin is a motor protein that moves vesicles and organelles toward the plus ends of microtubules.

microtubule motor proteins to be identified was dynein, which was isolated by Ian Gibbons in 1965. The purification of this form of dynein (called **axonemal dynein**) was facilitated because it is a highly abundant protein in cilia, just as the abundance of myosin facilitated its isolation from muscle cells. The identification of other microtubule-based motors, however, was more problematic because the proteins responsible for processes such as chromosome movement and organelle transport are present at comparatively low concentrations in the cytoplasm. Isolation of these proteins therefore depended on the development of new experimental methods to detect the activity of molecular motors in cell-free systems.

The development of *in vitro* assays for cytoplasmic motor proteins was based on the use of **video-enhanced microscopy** (developed by Robert Allen and Shinya Inoué in the early 1980s) to study the movement of membrane vesicles and organelles along microtubules in squid axons. In this method, a video camera is used to increase the contrast of images obtained with the light microscope, substantially improving the detection of small objects and allowing the movement of organelles to be followed in living cells. Using this approach, Allen, Scott Brady, and Ray Lasek demonstrated that organelle movements also took place in a cell-free system in which the plasma membrane had been removed and a cytoplasmic extract had been spread on a glass slide. These observations led to the development of an *in vitro* reconstructed system, which provided an assay capable of detecting cellular proteins responsible for organelle movement. In 1985 Brady, as well as Ronald Vale, Thomas Reese, and Michael Sheetz, capitalized on these developments to identify kinesin (now know as "conventional" kinesin, or kinesin I) as a novel microtubule motor protein present in both squid axons and bovine brain.

Further studies demonstrated that kinesin I translocates along microtubules in only a single direction—toward the plus end. Because the plus ends of microtubules in axons are all oriented away from the cell body (see

Figure 12.49), the movement of kinesin I in this direction transports vesicles and organelles away from the cell body toward the tip of the axon. Within intact axons, however, vesicles and organelles also had been observed to move back toward the cell body, implying that a different motor protein might be responsible for movement along microtubules in the opposite direction—toward the minus end. Consistent with this prediction, further experiments showed that a protein previously identified as the microtubule-associated protein MAP-1C was in fact a motor protein that moved along microtubules in the minus end direction. Subsequent analysis demonstrated that MAP-1C is related to the dynein isolated from cilia (axonemal dynein), so MAP-1C is now referred to as **cytoplasmic dynein**.

Kinesin I is a molecule of approximately 380 kd consisting of two heavy chains (120 kd each) and two light chains (64 kd each) (see Figure 12.50). The heavy chains have long α-helical regions that wind around each other in a coiled-coil structure. The amino-terminal globular head domains of the heavy chains are the motor domains of the molecule. They bind to both microtubules and ATP, the hydrolysis of which provides the energy required for movement. Although the motor domain of kinesin (approximately 340 amino acids) is much smaller than that of myosin (about 850 amino acids), X-ray crystallography indicates that kinesin and myosin (see Figure 12.33) evolved from a common ancestor and are structurally similar. The tail portion of the kinesin molecule consists of the light chains in association with the carboxy-terminal domains of the heavy chains. This portion of kinesin is responsible for binding to other cell components (such as membrane vesicles and organelles) that are transported along microtubules by the action of kinesin motors.

Cytoplasmic dynein is an extremely large molecule (up to 2000 kd), which consists of two or three heavy chains (each about 500 kd) complexed with a variable number of light and intermediate chains, which range from 14 to 120 kd (see Figure 12.50). As in kinesin, the heavy chains form globular ATP-binding motor domains that are responsible for movement along microtubules. The basal portion of the molecule, including the light and intermediate chains, is thought to bind to other subcellular structures, such as organelles and vesicles. In many situations cytoplasmic dynein acts along with a protein called **dynactin** to move cargoes over long distances along the microtubules.

Like the myosins, both kinesin and dynein define families of related motor proteins. Following the initial isolation of kinesin in 1985, a variety of kinesin-related proteins have been identified. Eighteen different kinesins are encoded in the genome of *C. elegans*, and there are forty-five kinesins known in humans. Most of these, like kinesin I, move along microtubules in the plus end direction (see Figure 12.50). Other kinesins move in the opposite direction toward the minus end, and three kinesins do not move along microtubules at all but use their tubulin-bind properties to anchor microtubules to other structures (discussed later). There is a direct correlation between where in the molecule the motor domain is located and the direction of movement along the microtubule; plus end-directed kinesins have N-terminal motor domains, minus end-directed kinesins have C-terminal motor domains, and those that do not move along microtubules have motor domains in the middle of the heavy chain (middle motor kinesins). Different members of the kinesin family vary in the sequences of their carboxy-terminal tails and are responsible for the movements of different types of "cargo" along microtubules. There are at least two types of cytoplasmic dyneins and several types of axonemal dyneins. While all members of the

KEY EXPERIMENT

The Isolation of Kinesin

Identification of a Novel Force-Generating Protein, Kinesin, Involved in Microtubule-Based Motility

Ronald D. Vale, Thomas S. Reese, and Michael P. Sheetz

National Institute of Neurological and Communicative Disorders and Stroke, Marine Biological Laboratory, Woods Hole, MA; University of Connecticut Health Center, Farmington, CT; Stanford University School of Medicine, Stanford, CA

Cell, Volume 42, 1985, pages 39–50

R. D. Vale

T. S. Reese

M. P. Sheetz

The Context

The transport and positioning of cytoplasmic organelles is key to the organization of eukaryotic cells, so understanding the mechanisms responsible for vesicle and organelle transport is a fundamental question in cell biology. In 1982 Robert Allen, Scott Brady, Ray Lasek, and their colleagues used video-enhanced microscopy to visualize the movement of organelles along cytoplasmic filaments in squid giant axons, both *in vivo* and in a cell-free system. These filaments were then identified as microtubules by electron microscopy, but the motor proteins responsible for organelle movement were unknown. The only microtubule motor identified at the time was axonemal dynein, which was present only in cilia and flagella. In 1985 Ronald Vale, Thomas Reese, and Michael Sheetz described the isolation of a novel motor protein, kinesin, which was responsible for the movement of organelles along micro-

tubules. Similar experiments were reported at the same time by Scott Brady (*Nature*, 1985, 317: 73–75).

The Experiments

Two experimental strategies were key to the isolation of kinesin. The first, based on the work of Allen and his colleagues, was the use of an *in vitro* system in which motor protein activity could be detected. Vale, Reese, and Sheetz used a system in which proteins in the cytoplasm of axons were found to power the movement of microtubules across the surface of glass coverslips. This cell-free system provided a sensitive and rapid functional assay for the activity of kinesin as a molecular motor.

The second important approach used in the isolation of kinesin was to take advantage of its binding to microtubules. Lasek and Brady (*Nature*, 1985, 316: 645–647) had made the key observation that *in vitro* movements of microtubules require ATP and are

inhibited by an ATP analog (adenylyl imidodiphosphate, AMP-PNP) that cannot be hydrolyzed and therefore does not provide a usable source of energy. In the presence of AMP-PNP, organelles remained attached to microtubules, suggesting that the motor protein responsible for organelle movement might also remain bound to microtubules under these conditions.

dynein family move toward the minus ends of microtubules, different cytoplasmic dyneins may transport different cargoes.

Cargo Transport and Intracellular Organization

One of the major roles of microtubules is to transport macromolecules, membrane vesicles, and organelles through the cytoplasm of eukaryotic cells. As already discussed, such cytoplasmic organelle transport is particularly evident in nerve cell axons, which may extend more than a meter in length. Ribosomes are present only in the cell body and dendrites, so proteins, membrane vesicles, and organelles (e.g., mitochondria) must be transported from the cell body to the axon. Via video-enhanced microscopy, the

KEY EXPERIMENT

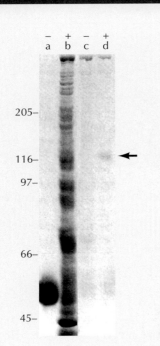

Binding of a motor protein to microtubules in the presence of AMP-PNP. Protein samples were analyzed by electrophoresis through a polyacrylamide gel, stained, and photographed. Molecular weights of marker proteins are indicated in kilodaltons at the left. Lane a represents purified microtubules in which only tubulin (55 kd) is detected. Lane b is a soluble extract of squid axon cytoplasm containing many different polypeptides. This soluble extract was incubated with microtubules either without (lane c) or with (lane d) AMP-PNP. Microtubules were then recovered and incubated with ATP to release bound proteins, which were subjected to electrophoresis. Note that a 110 kd polypeptide was specifically bound to microtubules in the presence of AMP-PNP and then released by ATP. The proteins bound in the presence of AMP-PNP and released by ATP were also found to induce microtubule movement (indicated by "+" above the gel).

On the basis of these findings, Vale and colleagues incubated microtubules with cytoplasmic proteins from squid axon in the presence of AMP-PNP. The microtubules were then recovered and incubated with ATP to release proteins that were specifically bound in the presence of the nonhydrolyzable ATP analog. This experiment identified a 110 kd polypeptide that was bound to microtubules in the presence of AMP-PNP and released by subsequent incubation with ATP (see figure). Furthermore,

the proteins bound to microtubules and then released by ATP were also shown to support the movement of microtubules *in vitro*. Binding to microtubules in the presence of AMP-PNP thus provided an efficient approach to isolation of the motor protein, which was shown by further biochemical studies to contain the 110 kd polypeptide complexed to polypeptides of 60 to 70 kd. By a similar approach, a related protein was also purified from bovine brain. The authors concluded that these proteins "represent a novel class of motility proteins that are structurally as well as enzymatically distinct from dynein, and we propose to call these translocators kinesin (from the Greek *kinein*, to move)."

The Impact
The movement of vesicles and organelles along microtubules is fundamental to the organization of

eukaryotic cells, so the motor proteins responsible for these movements play critical roles in cell biology. Subsequent experiments using *in vitro* assays similar to those described here established that kinesin moves along microtubules in the plus-end direction, whereas cytoplasmic dynein is responsible for the transport of vesicles and organelles toward microtubule minus ends. Moreover, a large family of kinesin-related proteins has since been identified. In addition to their roles in vesicle and organelle transport, members of these families of motor proteins are responsible for the separation and distribution of chromosomes during mitosis. The identification of kinesin thus opened the door to understanding a variety of microtubule-based movements that are critical to the structure and function of eukaryotic cells.

transport of membrane vesicles and organelles in both directions can be visualized along axon microtubules where kinesin and dynein carry their cargoes to and from the tips of the axons, respectively. For example, secretory vesicles containing neurotransmitters are carried from the Golgi apparatus to the terminal branches of the axon by kinesin. In the reverse direction, cytoplasmic dynein transports endocytic vesicles from the axon back to the cell body.

Microtubules similarly transport macromolecules, membrane vesicles, and organelles in other types of cells. Because microtubules are usually oriented with their minus end anchored in the centrosome and their plus end extending toward the cell periphery, different members of the kinesin and

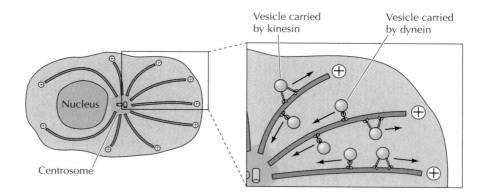

FIGURE 12.51 Transport of vesicles along microtubules Kinesin and other plus end-directed members of the kinesin family transport vesicles and organelles in the direction of microtubule plus ends, which extend toward the cell periphery. In contrast, dynein and minus end-directed members of the kinesin family carry their cargo in the direction of microtubule minus ends, which are anchored in the center of the cell.

dynein families are thought to transport cargo in opposite directions through the cytoplasm (Figure 12.51). Kinesin I and other plus end-directed members of the kinesin family carry their cargo toward the cell periphery, whereas cytoplasmic dyneins and minus end-directed members of the kinesin family transport materials toward the center of the cell. Cargo selection can be very specific. Kinesin II, a plus end-directed motor, transports selected mRNAs toward the cell cortex in *Xenopus* oocytes, a process that appears to be general to all vertebrates. Kinesin I similarly transports actin mRNA in fibroblasts, and dynein moves specific mRNAs toward one side of the *Drosophila* embryo. In addition to transporting membrane vesicles in the endocytic and secretory pathways, microtubules and associated motor proteins position membrane-enclosed organelles (such as the endoplasmic reticulum, Golgi apparatus, lysosomes, and mitochondria) within the cell. For example, the endoplasmic reticulum extends to the periphery of the cell in association with microtubules (Figure 12.52). Drugs that depolymerize microtubules cause the endoplasmic reticulum to retract toward the cell center, indicating that association with microtubules is required to maintain the endoplasmic reticulum in its extended state. This positioning of the endoplasmic reticulum appears to involve the action of kinesin I (or possibly multiple members of the kinesin family), which pulls the endoplasmic reticulum along microtubules in the plus-end direction, toward the cell periphery. Similarly, kinesin appears to play a key role in the positioning of lysosomes away from the center of the cell, and three different members of the kinesin family have been implicated in the movements of mitochondria.

Conversely, cytoplasmic dynein is thought to play a role in positioning the Golgi apparatus. The Golgi apparatus is located in the center of the cell near the centrosome. If microtubules are disrupted, either by a drug or when the cell enters mitosis, the Golgi breaks up into small vesicles that disperse throughout the cytoplasm. When the microtubules re-form, the Golgi apparatus also reassembles, with the Golgi vesicles apparently being transported to the center of the cell (toward the minus end of the microtubules) by cytoplasmic dynein. Movement along microtubules is thus responsible

(A)

(B)

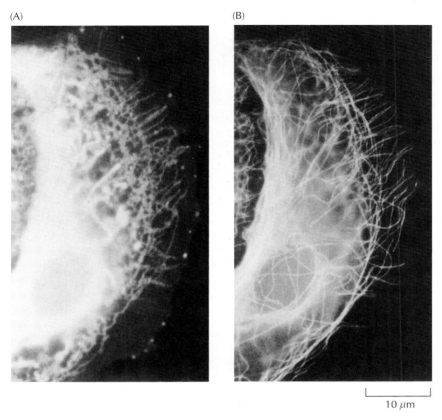

10 μm

FIGURE 12.52 Association of the endoplasmic reticulum with microtubules
Fluorescence microscopy of the endoplasmic reticulum (A) and microtubules (B) in an epithelial cell. The endoplasmic reticulum is stained with a fluorescent dye, and microtubules are stained with an antibody against tubulin. Note the close correlation between the endoplasmic reticulum and microtubules at the periphery of the cell. (From M. Terasaki, L. B. Chen and K. Fujiwara, 1986. *J. Cell Biol.* 103: 1557.)

not only for vesicle transport but also for establishing the positions of membrane-enclosed organelles within the cytoplasm of eukaryotic cells.

Cilia and Flagella

Cilia and **flagella** are microtubule-based projections of the plasma membrane that are responsible for movement of a variety of eukaryotic cells. Many bacteria also have flagella, but these prokaryotic flagella are quite different from those of eukaryotes. Bacterial flagella (which are not discussed further here) are protein filaments projecting from the cell surface, rather than projections of the plasma membrane supported by microtubules.

Eukaryotic cilia and flagella are very similar structures, each with a diameter of approximately 0.25 μm (Figure 12.53). Many cells are covered by numerous cilia, which are about 10 μm in length. Cilia beat in a coordinated back-and-forth motion, which either moves the cell through fluid or moves fluid over the surface of the cell. For example, the cilia of some protozoans (such as *Paramecium*) are responsible both for cell motility and for sweeping food organisms over the cell surface and into the oral cavity. In animals, an important function of cilia is to move fluid or mucus over the surface of epithelial cell sheets. A good example is provided by the ciliated cells lining

(A)

(C)

(B)

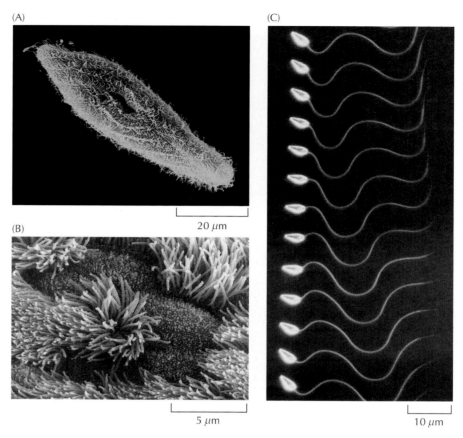

20 μm

5 μm

10 μm

FIGURE 12.53 Examples of cilia and flagella   (A) Scanning electron micrograph showing numerous cilia covering the surface of *Paramecium*. (B) Scanning electron micrograph of ciliated epithelial cells lining the surface of a trachea. (C) Multiple-flash photograph (500 flashes per second) showing the wavelike movement of a sea urchin sperm flagellum. (A, Karl Aufderheide/Visuals Unlimited; B, Fred E. Hossler/Visuals Unlimited; and C, C. J. Brokaw, California Institute of Technology.)

the respiratory tract, which clear mucus and dust from the respiratory passages. Flagella differ from cilia in their length (they can be as long as 200 μm) and in their wavelike pattern of beating. Cells usually have only one or two flagella, which are responsible for the locomotion of a variety of protozoans and of sperm.

The fundamental structure of both cilia and flagella is the **axoneme**, which is composed of microtubules and their associated proteins (Figure 12.54). The microtubules are arranged in a characteristic "9 + 2" pattern in which a central pair of microtubules is surrounded by nine outer microtubule doublets. The two fused microtubules of each outer doublet are distinct: One (called the A tubule) is a complete microtubule consisting of 13 protofilaments; the other (the B tubule) is incomplete, containing only 10 or 11 protofilaments fused to the A tubule. The outer microtubule doublets are connected to the central pair by radial spokes and to each other by links of a protein called **nexin**. In addition, two arms of dynein are attached to each A tubule, and it is the motor activity of these axonemal dyneins that drives the beating of cilia and flagella.

(A)

(B)

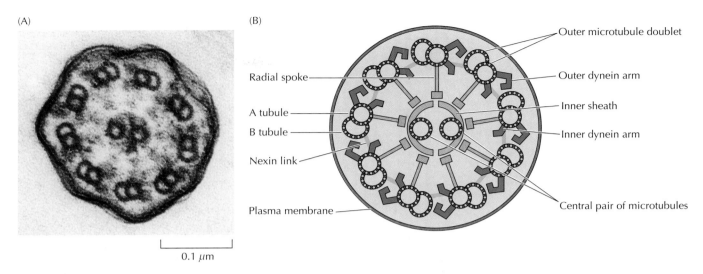

Outer microtubule doublet

Radial spoke

Outer dynein arm

A tubule

Inner sheath

B tubule

Inner dynein arm

Nexin link

Plasma membrane

Central pair of microtubules

0.1 μm

FIGURE 12.54 Structure of the axoneme of cilia and flagella (A) Computer-enhanced electron micrograph of a cross-section of the axoneme of a rat sperm flagellum. (B) Schematic cross-section of an axoneme. The nine outer doublets consist of one complete (A) microtubule and one incomplete (B) microtubule, containing only 10 or 11 protofilaments. The outer doublets are joined to each other by nexin links and to the central pair of microtubules by radial spokes. Each outer microtubule doublet is associated with inner and outer dynein arms. (A, K. G. Murti/Visuals Unlimited.)

The minus ends of the microtubules of cilia and flagella are anchored in a **basal body**, which is similar in structure to a centriole and contains nine triplets of microtubules (Figure 12.55). Centrioles were discussed earlier as components of the centrosome in which their function is complex and poorly understood. Basal bodies, however, play a clear role in organization of the axoneme microtubules. Namely, each of the outer microtubule doublets of the axoneme is initiated by extension of two of the microtubules present in the triplets of the basal body. Basal bodies thus serve to initiate

FIGURE 12.55 Electron micrographs of basal bodies (A) A longitudinal view of cilia anchored in basal bodies. (B) A cross-section of basal bodies. Each basal body consists of nine triplets of microtubules. (A, Conly L. Reider/Biological Photo Service; B, W. L. Dentler, Biological Photo Service.)

(A)

(B)

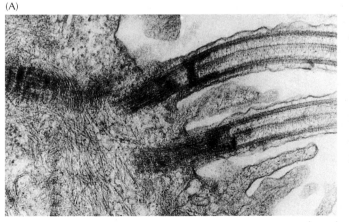

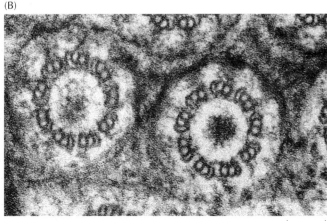

0.1 μm

0.1 μm

**FIGURE 12.56 Movement of micro-
tubules in cilia and flagella** The
bases of dynein arms are attached to A
tubules, and the motor head groups
interact with the B tubules of adjacent
doublets. Movement of the dynein
head groups in the minus end direction
(toward the base of the cilium) then
causes the A tubule of one doublet to
slide toward the base of the adjacent
B tubule. Because both microtubule
doublets are connected by nexin links,
this sliding movement forces them to
bend.

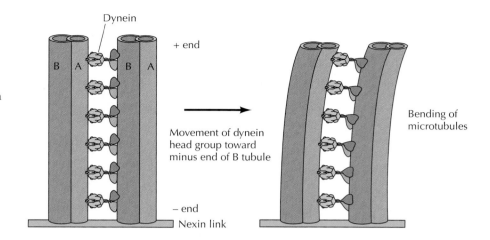

the growth of axonemal microtubules as well as anchoring cilia and flagella
to the surface of the cell.

The movements of cilia and flagella result from the sliding of outer
microtubule doublets relative to one another, powered by the motor activity of the axonemal dyneins (Figure 12.56). The dynein bases bind to the A
tubules while the dynein head groups bind to the B tubules of adjacent doublets. Movement of the dynein head groups in the minus end direction then
causes the A tubule of one doublet to slide toward the basal end of the adjacent B tubule. Because the microtubule doublets in an axoneme are connected by nexin links, the sliding of one doublet along another causes them
to bend, forming the basis of the beating movements of cilia and flagella. It
is apparent, however, that the activities of dynein molecules in different
regions of the axoneme must be carefully regulated to produce the coordinated beating of cilia and the wavelike oscillations of flagella—a process
about which little is currently understood.

Reorganization of Microtubules during Mitosis

As noted earlier, microtubules completely reorganize during mitosis, providing a dramatic example of the importance of their dynamic nature. The
microtubule array present in interphase cells disassembles and the free
tubulin subunits are reassembled to form the **mitotic spindle**, which is
responsible for the separation of daughter chromosomes (Figure 12.57). This
restructuring of the microtubule cytoskeleton is directed by duplication of
the centrosome to form two separate microtubule-organizing centers at
opposite poles of the mitotic spindle.

The centrioles and other components of the centrosome are duplicated in
interphase cells, but they remain together on one side of the nucleus until
the beginning of mitosis (Figure 12.58). The two centrosomes then separate
and move to opposite sides of the nucleus, forming the two poles of the
mitotic spindle. As the cell enters mitosis, the dynamics of microtubule

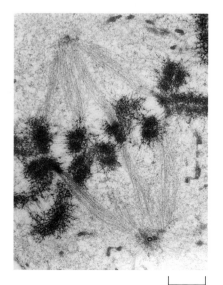

2 μm

FIGURE 12.57 Electron micrograph of the mitotic spindle The spindle microtubules are attached to condensed chromosomes at metaphase. (From C. L. Rieder
and S. S. Bowser, 1985. *J. Histochem. Cytochem.* 33: 165/Biological Photo Service.)

FIGURE 12.58 Formation of the mitotic spindle The centrioles and centrosomes duplicate during interphase. During prophase of mitosis the duplicated centrosomes separate and move to opposite sides of the nucleus. The nuclear envelope then disassembles, and microtubules reorganize to form the mitotic spindle. Kinetochore microtubules are attached to the kinetochores of condensed chromosomes and chromosomal microtubules are attached to their ends. Polar microtubules overlap with each other in the center of the cell, and astral microtubules extend outward to the cell periphery. At metaphase, the condensed chromosomes are aligned at the center of the spindle.

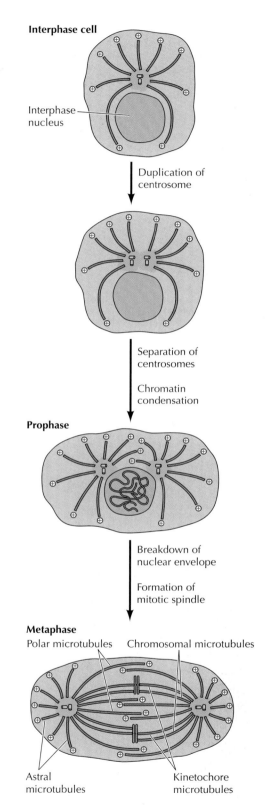

assembly and disassembly also change dramatically. First, the rate of microtubule disassembly increases about tenfold, resulting in overall depolymerization and shrinkage of microtubules. At the same time, the number of microtubules emanating from the centrosome increases by five- to tenfold. In combination, these changes result in disassembly of the interphase microtubules and the outgrowth of large numbers of short microtubules from the centrosomes.

As first proposed by Marc Kirschner and Tim Mitchison in 1986, formation of the mitotic spindle involves the selective stabilization of some of the microtubules radiating from the centrosomes. These microtubules are of four types, three of which make up the mitotic spindle. **Kinetochore microtubules** attach to the condensed chromosomes of mitotic cells at their centromeres, which are associated with specific proteins to form the kinetochore (see Figure 5.19). Attachment to the kinetochore stabilizes these microtubules, which as discussed below play a critical role in separation of the mitotic chromosomes. Also emanating from the centrosomes are the **chromosomal microtubules,** which connect to the ends of the chromosomes via chromokinesin. The third type of microtubules found in the mitotic spindle (**polar microtubules**) is not attached to chromosomes. Instead, the polar microtubules are stabilized by overlapping with each other in the center of the cell. **Astral microtubules** extend outward from the centrosomes to the cell periphery and have freely exposed plus ends. As discussed later, both the polar and astral microtubules contribute to chromosome movement by pushing the spindle poles apart.

As mitosis proceeds, the condensed chromosomes first align on the metaphase plate and then separate, with the two chromatids of each chromosome being pulled to opposite poles of the spindle. Chromosome movement is mediated by motor proteins associated with the spindle microtubules, as will be discussed shortly. In the final stage of mitosis, nuclear envelopes re-form, the chromosomes decondense, and cytokinesis takes place. Each daughter cell then contains one centrosome, which nucleates the formation of a new network of interphase microtubules.

Chromosome Movement

After the two centrosomes move to opposite sides of the cell at the beginning of mitosis, the duplicated chromosomes attach to kinetochore and chromosomal microtubules and align on the metaphase plate, equidistant from the two spindle poles. This alignment of chromosomes is mediated by rapid growth of the kinetochore microtubules and capture of the kinetochores by plus-end tracking proteins. In addi-

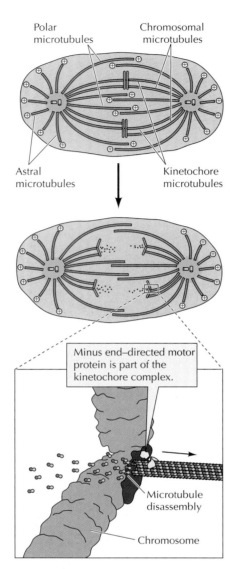

Polar
microtubules

Chromosomal
microtubules

Astral
microtubules

Kinetochore
microtubules

Minus end–directed motor
protein is part of the
kinetochore complex.

Microtubule
disassembly

Chromosome

FIGURE 12.59 Anaphase A chromosome movement Chromosomes move toward the spindle poles along the kinetochore microtubules. Chromosome movement is driven by minus end–directed motor proteins associated with the kinetochore. The action of these motor proteins is coupled to disassembly and shortening of both the kinetochore and chromosomal microtubules.

tion, the chromosome ends are pushed toward the metaphase plate by chromokinesin moving along the chromosomal microtubules. Once all of the chromosomes have aligned on the metaphase plate, the links between the sister chromatids are severed and anaphase begins. During anaphase of mitosis, the sister chromatids separate and move to opposite poles of the spindle. Chromosome movement proceeds by two distinct mechanisms, referred to as **anaphase A** and **anaphase B**, which involve different types of spindle microtubules.

Anaphase A consists of the movement of chromosomes toward the spindle poles along the kinetochore microtubules, which shorten as chromosome movement proceeds (Figure 12.59). Movement of chromosomes along the spindle microtubules in the minus end direction (toward the centrosomes) is driven by kinetochore-associated motor proteins. Cytoplasmic

FIGURE 12.60 Spindle pole separation in anaphase B The separation of spindle poles results from two types of movement. First, overlapping polar microtubules slide past each other to push the spindle poles apart, probably as a result of the action of plus end-directed motor proteins. Second, the spindle poles are pulled apart by the astral microtubules. The driving force appears to be a minus end-directed motor anchored to the cell cortex.

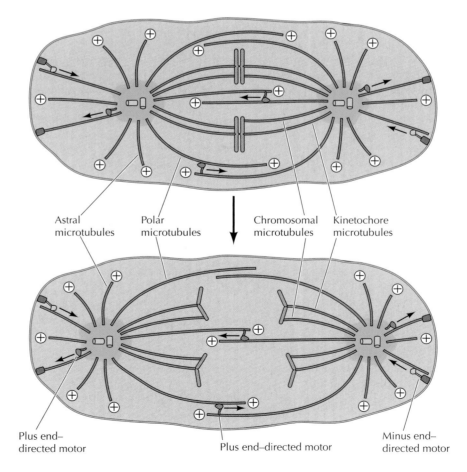

Astral
microtubules

Polar
microtubules

Chromosomal
microtubules

Kinetochore
microtubules

Plus end–
directed motor

Plus end–directed motor

Minus end–
directed motor

dynein is associated with kinetochores and may play a role in poleward chromosome movement, as may minus end-directed members of the kinesin family. The action of these kinetochore motor proteins is coupled to disassembly and shortening of both the kinetochore and chromosomal microtubules, which is mediated by middle motor kinesins that act as microtubule-depolymerizing enzymes.

Anaphase B refers to the separation of the spindle poles themselves (Figure 12.60). Spindle-pole separation is accompanied by elongation of the polar microtubules and is similar to the initial separation of duplicated centrosomes to form the spindle poles at the beginning of mitosis (see Figure 12.58). During anaphase B the overlapping polar microtubules slide against one another, pushing the spindle poles apart. This type of movement has been found to result from the action of several plus end–directed kinesins, which cross-link polar microtubules and move them toward the plus end of their overlapping microtubule away from the opposite spindle pole. In addition, the spindle poles are pulled apart by the astral microtubules. This type of movement involves the action of cytoplasmic dynein anchored to the cell cortex. The movement of this anchored dynein along astral microtubules in the minus end direction pulls the spindle poles apart toward the periphery of the cell. There is a simultaneous depolymerization of the astral microtubules by middle motor kinesins, leading to separation of the spindle poles and their movement to the periphery of the cell, prior to the formation of two daughter cells at the end of mitosis.

COMPANION WEBSITE

Visit the website that accompanies **The Cell** (www.sinauer.com/cooper) for animations, videos, quizzes, problems, and other review material.

SUMMARY

STRUCTURE AND ORGANIZATION OF ACTIN FILAMENTS

Assembly and Disassembly of Actin Filaments: Actin filaments are formed by the head-to-tail polymerization of actin monomers into a helix. A variety of actin-binding proteins regulate the assembly and disassembly of actin filaments within the cell.

Organization of Actin Filaments: Actin filaments are cross-linked by actin-binding proteins to form bundles or three-dimensional networks.

Association of Actin Filaments with the Plasma Membrane: A network of actin filaments and other cytoskeletal proteins underlies the plasma membrane and determines cell shape. Actin bundles also attach to the plasma membrane and anchor the cell at regions of cell-cell and cell-substratum contact.

Protrusions of the Cell Surface: Actin filaments support permanent protrusions of the cell surface, such as microvilli, as well as transient extensions that are responsible for phagocytosis and cell locomotion.

KEY TERMS

actin, microfilament, globular [G] actin, filamentous [F] actin, treadmilling, cytochalasin, phalloidin, actin-binding protein, formin, Arp2/3 complex, tropomyosin, ADF/cofilin, profilin

actin bundle, actin network, actin-bundling protein, fimbrin, contractile bundle, α-actinin, filamin

cell cortex, spectrin, ankyrin, fodrin, ERM proteins, dystrophin, integrin, focal adhesion, stress fiber, talin, vinculin, adherens junction, adhesion belt, cadherin, catenin

microvillus, brush border, stereocilia, villin, pseudopodium, lamellipodium, microspike, filopodium

KEY TERMS

myosin, molecular motor, muscle fiber, myofibril, sarcomere, titin, nebulin, sliding filament model, myosin II, sarcoplasmic reticulum, troponin

cytokinesis, contractile ring, myosin light-chain kinase, calmodulin

myosin I

WASP/Scar complex, twinfilin

intermediate filament, keratin, vimentin, desmin, neurofilament (NF) protein

desmosome, hemidesmosome, plakin

SUMMARY

ACTIN, MYOSIN, AND CELL MOVEMENT

Muscle Contraction: Studies of muscle established the role of myosin as a motor protein that uses the energy derived from ATP hydrolysis to generate force and movement. Muscle contraction results from the sliding of actin and myosin filaments past each other. ATP hydrolysis drives repeated cycles of interaction between myosin and actin during which conformational changes result in movement of the myosin head group along actin filaments.

Contractile Assemblies of Actin and Myosin in Nonmuscle Cells: Assemblies of actin and myosin II are responsible for a variety of movements of nonmuscle cells, including cytokinesis.

Nonmuscle Myosins: Other types of myosin that do not function in contraction serve to transport membrane vesicles and organelles along actin filaments.

Formation of Protrusions and Cell Movement: Extension of cell protrusions is mediated by the growth of multiple actin filament branches at the leading edge of the cell. Cell movement is a complex process in which adhesions form at the ends of the new cell protrusions, the cell body is brought forward by the action of myosin II along stress fibers, and the trailing edge retracts into the cell body.

INTERMEDIATE FILAMENTS

Intermediate Filament Proteins: Intermediate filaments are polymers of more than 50 different proteins that are expressed in various types of cells. They are not involved in cell movement but provide mechanical support to cells and tissues.

Assembly of Intermediate Filaments: Intermediate filaments are formed from dimers of two polypeptide chains wound around each other in a coiled-coil structure. The dimers then associate to form tetramers, which assemble into protofilaments. Intermediate filaments are formed from protofilaments wound around one another in a ropelike structure.

Intracellular Organization of Intermediate Filaments: Intermediate filaments form a network extending from a ring surrounding the nucleus, to the plasma membrane of most cell types. In epithelial cells, intermediate filaments are anchored to the plasma membrane at regions of specialized cell contacts (desmosomes and hemidesmosomes). Intermediate filaments also play specialized roles in muscle and nerve cells.

Functions of Keratins and Neurofilaments: Diseases of the Skin and Nervous System: The importance of intermediate filaments in providing mechanical strength to cells in tissues has been demonstrated by the introduction of mutant keratin genes into transgenic mice. Similar keratin gene mutations are responsible for some human skin diseases, and abnormalities of neurofilaments have been implicated in the development of motor neuron disease.

SUMMARY

MICROTUBULES

Structure and Dynamic Organization of Microtubules: Microtubules are formed by the reversible polymerization of tubulin. They display dynamic instability and undergo continual cycles of assembly and disassembly as a result of GTP hydrolysis following tubulin polymerization.

Assembly of Microtubules: The microtubules in most cells extend outward from a microtubule-organizing center, or centrosome, located near the center of the cell. In animal cells, the centrosome usually contains a pair of centrioles surrounded by pericentriolar material. The growth of microtubules is initiated in the pericentriolar material, which then serves to anchor their minus ends.

Organization of Microtubules within Cells: Selective stabilization of microtubules by microtubule-associated proteins can determine cell shape and polarity, such as the extension of nerve cell axons and dendrites.

MICROTUBULE MOTORS AND MOVEMENT

Identification of Microtubule Motor Proteins: Two families of motor proteins, the kinesins and the dyneins, are responsible for movement along microtubules. Kinesin and most kinesin-related proteins move in the plus-end direction, whereas the dyneins and some members of the kinesin family move toward microtubule minus ends.

Cargo Transport and Intracellular Organization: Movement along microtubules transports macromolecules, membrane vesicles, and organelles through the cytoplasm, as well as positioning cytoplasmic organelles within the cell.

Cilia and Flagella: Cilia and flagella are microtubule-based extensions of the plasma membrane. Their movements result from the sliding of microtubules driven by the action of dynein motors.

Reorganization of Microtubules during Mitosis: Microtubules reorganize at the beginning of mitosis to form the mitotic spindle, which is responsible for chromosome separation.

Chromosome Movement: The duplicated chromosomes align on the metaphase plate. During anaphase of mitosis, daughter chromosomes separate and move to opposite poles of the mitotic spindle. Chromosome separation results from several types of movements in which different classes of spindle microtubules and motor proteins participate.

KEY TERMS

microtubule, tubulin, dynamic instability, colchicine, colcemid, vincristine, vinblastine, taxol

centrosome, γ-tubulin ring complex, centriole, pericentriolar material

microtubule-associated protein (MAP)

kinesin, dynein, axonemal dynein, video-enhanced microscopy, cytoplasmic dynein, dynactin

cilium, flagellum, axoneme, nexin, basal body

mitotic spindle, kinetochore microtubule, chromosomal microtubule, polar microtubule, astral microtubule

anaphase A, anaphase B

Questions

1. Why do actin filaments have a distinct polarity? Why is the polarity of actin filaments important in muscle contraction?

2. What is treadmilling, and at what concentration of monomers does it occur?

3. How would cytochalasin and phalloidin affect treadmilling actin filaments?

4. How do ADF/cofilin, profilin, and the Arp2/3 complex regulate actin filament assembly and turnover?

5. Which bands or zones of a muscle sarcomere change length during contraction? Why doesn't the A band change length?

6. How does Ca^{2+} regulate the contraction of smooth muscle cells?

7. How would expression of siRNAs targeted against vimentin affect the growth of fibroblasts in culture?

8. Why are intermediate filaments apolar, even though they are assembled from monomers that have distinct ends?

9. What key observation helped Vale and colleagues devise a strategy for the isolation of kinesin?

10. How would the removal of nexin affect the beating of cilia?

11. You are studying the transport of secretory vesicles containing insulin along microtubules in cultured pancreatic cells. How would treatment with colcemid affect the transport of these vesicles?

12. What is the cellular function of γ-tubulin?

References and Further Reading

General Reference

Bray, D. 2001. *Cell Movements*, 2nd ed., New York: Garland Publishing.

Schliwa, M. and G. Woehlke. 2003. Molecular motors. *Nature* 422: 759?765. [R]

Structure and Organization of Actin Filaments

Bamburg, J. R. and O. P. Wiggan. 2002. ADF/cofilin and actin dynamics in disease. *Trends Cell Biol.* 12: 598–605. [R]

Bennett, V. and D. M. Gilligan. 1993. The spectrin-based membrane skeleton and microscale organization of the plasma membrane. *Ann. Rev. Cell Biol.* 9: 27–66. [R]

Blessing, C. A., G. T. Ugrinova and H. V. Goodson. 2004. Actin and ARPs: Action in the nucleus. *Trends Cell Biol.* 14: 435–442. [R]

Campbell, K. P. 1995. Three muscular dystrophies: Loss of cytoskeleton-extracellular matrix linkage. *Cell* 80: 675–679. [R]

Dominguez, R. 2004. Actin-binding proteins—a unifying hypothesis. *Trends Biochem. Sci.* 29: 572–578. [R]

Dos Remedios, C. G., D. Chhabra, M. Kekic, I. V. Dedova, M. Tsubakihara, D. A. Berry and N. J. Nosworthy. 2003. Actin binding proteins: Regulation of cytoskeletal microfilaments. *Physiol Rev.* 83: 433–473. [R]

Cooper, J. A., M. A. Wear and A. M. Weaver. 2001. Arp2/3 complex: Advances on the inner workings of a molecular machine. *Cell* 107: 703–705. [R]

Gunning, P. W., G. Schevzov, A. J. Kee and E. C. Hardeman. 2005. Tropomyosin isoforms: Divining rods for actin cytoskeleton function. *Trends Cell Biol.* 15: 333–341. [R]

Holmes, K. C., D. Popp, W. Gebhard and W. Kabsch. 1990. Atomic model of the actin filament. *Nature* 347: 44–49. [P]

Kabsch, W., H. G. Mannherz, D. Suck, E. F. Pai and K. C. Holmes. 1990. Atomic structure of the actin: DNase I complex. *Nature* 347: 37–44. [P]

Kandasamy, M. K., R. B. Deal, E. C. McKinney and R. B. Meagher. 2004. Plant actin-related proteins. *Trends Plant Sci.* 9: 196–202. [R]

Paavilainen, V. O., E. Bertling, S. Falck and P. Lappalainen. 2004. Regulation of cytoskeletal dynamics by actin-monomer-binding proteins. *Trends Cell Biol.* 14: 386–394. [R]

Revenu, C., R. Athman, S. Robine and D. Louvard. 2004. The co-workers of actin filaments: From cell structures to signals. *Nat. Rev. Mol. Cell Biol.* 5: 635–646. [R]

Wear, M. A. and J. A. Cooper. 2004. Capping protein: New insights into mechanism and regulation. *Trends Biochem. Sci.* 29: 418–428. [R]

Winder, S. J. 2003. Structural insights into actin-binding, branching and bundling proteins. *Curr. Opin. Cell Biol.* 15: 14–22. [R]

Witke, W. 2004. The role of profilin complexes in cell motility and other cellular processes. *Trends Cell Biol.* 14: 461–469. [R]

Yap, A. S., W. M. Brieher and B. M. Gumbiner. 1997. Molecular and functional analysis of cadherin-based adherens junctions. *Ann. Rev. Cell Dev. Biol.* 13: 119–146. [R]

Zigmond, S. H. 2004. Formin-induced nucleation of actin filaments. *Curr. Opin. Cell Biol.* 16: 99–105. [R]

Actin, Myosin, and Cell Movement

Bailly, M. 2003. Connecting cell adhesion to the actin polymerization machinery: Vinculin as the missing link? *Trends Cell Biol.* 13: 163–165. [R]

Bakolitsa, C., D. M. Cohen, L. A. Bankston, A. A. Bobkov, G. W. Cadwell, L. Jennings, D. R. Critchley, S. W. Craig and R. C. Liddington. 2004. Structural basis for vinculin activation at sites of cell adhesion. *Nature* 430: 583–586. [P]

Berg, J. S., B. C. Powell and R. E. Cheney. 2001. A millennial myosin census. *Mol. Biol. Cell* 12: 780–794. [R]

Buss, F., G. Spudich and J. Kendrick-Jones. 2004. MYOSIN VI: Cellular functions and motor properties. *Ann. Rev. Cell Dev. Biol.* 20: 649–676. [R]

Carragher, N. O. and M. C. Frame. 2004. Focal adhesion and actin dynamics: A place where kinases and proteases meet to promote invasion. *Trends Cell Biol.* 14: 241–249. [R]

Chou, Y. H., B. T. Helfand and R. D. Goldman. 2001. New horizons in cytoskeletal dynamics: Transport of intermediate filaments along microtubule tracks. *Curr. Opin. Cell Biol.* 13: 106–109. [R]

Dawe, H. R., L. S. Minamide, J. R. Bamburg and L. P. Cramer. 2003. ADF/cofilin controls cell polarity during fibroblast migration. *Curr. Biol.* 13: 252–257. [P]

De Mali, K. A. 2004. Vinculin—A dynamic regulator of cell adhesion. *Trends Biochem. Sci.* 29: 565–567. [R]

Erickson, H. P. 1997. Stretching single protein molecules: Titin is a weird spring. *Science* 276: 1090–1092. [R]

Finer, J. T., R. M. Simmons and J. A. Spudich. 1994. Single myosin molecule mechanics: Piconewton forces and nanometre steps. *Nature* 368: 113–119. [P]

Frank, D. J., T. Noguchi and K. G. Miller. 2004. Myosin VI: A structural role in actin organization important for protein and organelle localization and trafficking. *Curr. Opin. Cell Biol.* 16: 189–194. [R]

Geeves, M. A. and K. C. Holmes. 1999. Structural mechanism of muscle contraction. *Ann. Rev. Biochem.* 68: 687–728. [R]

Higuchi, H. and S. A. Endow. 2002. Directionality and processivity of molecular motors. *Curr. Opin. Cell Biol.* 14: 50–57. [R]

Huxley, H. E. 1969. The mechanism of muscle contraction. *Science* 164: 1356–1366. [R]

Huxley, H. E. and J. Hanson. 1954. Changes in the cross-striations of muscle contraction and their structural interpretation. *Nature* 173: 973–976. [P]

Huxley, A. F. and R. Niedergerke. 1954. Interference microscopy of living muscle fibres. *Nature* 173: 971–973. [P]

Janmey, P. A. and U. Lindberg. 2004. Cytoskeletal regulation: Rich in lipids. *Nat. Rev. Mol. Cell Biol.* 5: 658–666. [R]

Manes, S. and A. Martinez. 2004. Cholesterol domains regulate the actin cytoskeleton at the leading edge of moving cells. *Trends Cell Biol.* 14: 275–278. [R]

Nayal, A., D. J. Webb and A. F. Horwitz. 2004. Talin: An emerging focal point of adhesion dynamics. *Curr. Opin. Cell Biol.* 16: 94–98. [R]

Paavilainen, V. O., E. Bertling, S. Falck and P. Lappalainen. 2004. Regulation of cytoskeletal dynamics by actin-monomer-binding proteins. *Trends Cell Biol.* 14: 386–394. [R]

Pantaloni, D., C. Le Clainche and M. F. Carlier. 2001. Mechanism of actin-based motility. *Science* 292: 1502–1506. [R]

Raftopoulou, M. and A. Hall. 2004. Cell migration: Rho GTPases lead the way. *Dev. Biol.* 265: 23–32. [R]

Rayment, I., W. R. Rypniewski, K. Schmidt-Base, R. Smith, D. R. Tomchick, M. M. Benning, D. A. Winkelmann, G. Wesenberg and H. M. Holden. 1993. Three-dimensional structure of myosin subfragment-1: A molecular motor. *Science* 261: 50–58. [P]

Rayment, I., H. M. Holden, M. Whittaker, C. B. Yohn, M. Lorenz, K. C. Kolmes and R. A. Milligan. 1993. Structure of the actin-myosin complex and its implications for muscle contraction. *Science* 261: 58–65. [P]

Revenu, C., R. Athman, S. Robine and D. Louvard. 2004. The co-workers of actin filaments: From cell structures to signals. *Nat. Rev. Mol. Cell Biol.* 5: 635–646. [R]

Schafer, D. A. 2004. Cell biology: Barbed ends rule. *Nature* 430, 734–735. [R]

Small, J. V., T. Stradal, E. Vignal and K. Rottner. 2002. The lamellipodium: Where motility begins. *Trends Cell Biol.* 12: 112–120. [R]

Stradal, T. E., K. Rottner, A. Disanza, S. Confalonieri, M. Innocenti and G. Scita. 2004. Regulation of actin dynamics by WASP and WAVE family proteins. *Trends Cell Biol.* 14: 303–311. [R]

Tan, J. L., S. Ravid and J. A. Spudich. 1992. Control of nonmuscle myosins by phosphorylation. *Ann. Rev. Biochem.* 61: 721– 759. [R]

Trinick, J. and L. Tskhovrebova. 1997. Titin: A molecular control freak. *Trends Cell Biol.* 9: 377–380. [R]

Turner, C. E. and M. C. Brown. 2001. Cell motility: ARNO and ARF6 at the cutting edge. *Curr. Biol.* 11: R875–R877. [R]

Wear, M. A. and J. A. Cooper. 2004. Capping protein: New insights into mechanism and regulation. *Trends Biochem. Sci.* 29: 418–428. [R]

Witke, W. 2004. The role of profilin complexes in cell motility and other cellular processes. *Trends Cell Biol.* 14: 461–469. [R]

Intermediate Filaments

Bonifas, J. M., A. L. Rothman and E. H. Epstein, Jr. 1991. Epidermolysis bullosa simplex: Evidence in two families for keratin gene abnormalities. *Science* 254: 1202–1205. [P]

Brown, R. H., Jr. 1995. Amyotrophic lateral sclerosis: Recent insights from genetics and transgenic mice. *Cell* 80: 687–692. [R]

Coulombe, P. A., M. E. Hutton, A. Letai, A. Hebert, A. S. Paller and E. Fuchs. 1991. Point mutations in human keratin 14 genes of epidermolysis bullosa simplex patients: Genetic and functional analyses. *Cell* 66: 1301–1311. [P]

Coulombe, P. A. and M. B. Omary. 2002. 'Hard' and 'soft' principles defining the structure, function and regulation of keratin intermediate filaments. *Curr. Opin. Cell Biol.* 14: 110–122. [R]

Fuchs, E. and D. W. Cleveland. 1998. A structural scaffolding of intermediate filaments in health and disease. *Science* 279: 514–519. [R]

Fuchs, E. and S. Raghavan. 2002. Getting under the skin of epidermal morphogenesis. *Nat. Rev. Genet.* 3: 199–209. [R]

Fuchs, E. and Y. Yang. 1999. Crossroads on cytoskeletal highways. *Cell* 98: 547–550. [R]

Herrmann, H. and U. Aebi. 2004. Intermediate filaments: Molecular structure, assembly mechanism, and integration into functionally distinct intracellular scaffolds. *Ann. Rev. Biochem.* 73: 749–789. [R]

Schwarz, M. A., K. Owaribe, J. Kartenbeck and W. W. Franke. 1990. Desmosomes and hemidesmosomes: Constitutive molecular components. *Ann. Rev. Cell Biol.* 6: 461–491. [R]

Vassar, R., P. A. Coulombe, L. Degenstein, K. Albers and E. Fuchs. 1991. Mutant keratin expression in transgenic mice causes marked abnormalities resembling a human genetic skin disease. *Cell* 64: 365–380. [P]

Microtubules

Bornens, M. 2002. Centrosome composition and microtubule anchoring mechanisms. *Curr. Opin. Cell Biol.* 14: 25–34. [R]

Dogterom, M., J. W. Kerssemakers, G. Romet-Lemonne and M. E. Janson. 2005. Force generation by dynamic microtubules. *Curr. Opin. Cell Biol.* 17: 67–74. [R]

Hays, T. and M. Li. 2001. Kinesin transport: Driving kinesin in the neuron. *Curr. Biol.* 11: R136–R139. [R]

Job, D., O. Valiron and B. Oakley. 2003. Microtubule nucleation. *Curr. Opin. Cell Biol.* 15: 111–117. [R]

Karsenti, E. and I. Vernos. 2001. The mitotic spindle: A self-made machine. *Science* 294: 543–547. [R]

Mandelkow, E. and E. M. Mandelkow. 2002. Kinesin motors and disease. *Trends Cell Biol.* 12: 585–591. [R]

Marshall, W. F. 2001. Centrioles take center stage. *Curr. Biol.* 11: R487–R496. [R]

Mitchison, T. and M. Kirschner. 1984. Dynamic instability of microtubule growth. *Nature* 312: 237–242. [P]

Mitchison, T. and M. Kirschner. 1984. Microtubule assembly nucleated by isolated centrosomes. *Nature* 312: 232–237. [P]

Nogales, E., M. Whittaker, R. A. Milligan and K. H. Downing. 1999. High-resolution model of the microtubule. *Cell* 96: 79–88. [P]

Oakley, B. R., C. E. Oakley, Y. Yoon and M. K. Jung. 1990. γ-tubulin is a component of the spindle pole body that is essential for microtubule function in *Aspergillus nidulans*. *Cell* 61: 1289–1301. [P]

Osborn, M. and K. Weber. 1976. Cytoplasmic microtubules in tissue culture cells appear to grow from an organizing structure towards the plasma membrane. *Proc. Natl. Acad. Sci. USA* 73: 867–871. [P]

Zheng, Y., M. L. Wong, B. Alberts and T. Mitchison. 1995. Nucleation of microtubule assembly by a γ-tubulin-containing ring complex. *Nature* 378: 578–583. [P]

Microtubule Motors and Movement

Asbury, C. L. 2005. Kinesin: World's tiniest biped. *Curr. Opin. Cell Biol.* 17: 89–97. [R]

Betley, J. N., B. Heinrich, I. Vernos, C. Sardet, F. Prodon and J. O. Deshler. 2004. Kinesin II mediates Vg1 mRNA transport in *Xenopus* oocytes. *Curr. Biol.* 14: 219–224. [P]

Brady, S. T. 1985. A novel brain ATPase with properties expected for the fast axonal motor. *Nature* 317: 73–75. [P]

Brady, S. T., R. J. Lasek and R. D. Allen. 1982. Fast axonal transport in extruded axoplasm from squid giant axon. *Science* 218: 1129–1131. [P]

Desai, A., S. Verma, T. J. Mitchison and C. E. Walczak. 1999. Kin I kinesins are microtubule-destabilizing enzymes. *Cell* 96: 69–78. [P]

Gibbons, I. R. 1981. Cilia and flagella of eukaryotes. *J. Cell Biol.* 91: 107s–124s. [R]

Gibbons, I. R. and A. Rowe. 1965. Dynein: A protein with adenosine triphosphatase activity from cilia. *Science* 149: 424–426. [P]

Holzbaur, E. L. 2004. Motor neurons rely on motor proteins. *Trends Cell Biol.* 14: 233–240. [R]

Koonce, M. P. and M. Samso. 2004. Of rings and levers: The dynein motor comes of age. *Trends Cell Biol.* 14: 612–619. [R]

Lasek, R. J. and S. T. Brady. 1985. Attachment of transported vesicles to microtubules in axoplasm is facilitated by AMP-PNP. *Nature* 316: 645–647. [P]

Moore, A. and L. Wordeman. 2004. The mechanism, function and regulation of depolymerizing kinesins during mitosis. *Trends Cell Biol.* 14: 537–546. [R]

Oiwa, K. and H. Sakakibara. 2005. Recent progress in dynein structure and mechanism. *Curr. Opin. Cell Biol.* 17: 98–103. [R]

Rieder, C. L. and E. D. Salmon. 1998. The vertebrate cell kinetochore and its roles during mitosis. *Trends Cell Biol.* 8: 310–318. [R]

Roegiers, F. 2003. Insights into mRNA transport in neurons. *Proc. Natl. Acad. Sci. U. S. A.* 100: 1465–1466. [R]

Salmon, E. D. 1995. VE-DIC light microscopy and the discovery of kinesin. *Trends Cell Biol.* 5: 154–157. [R]

Scholey, J. M., I. Brust-Mascher and A. Mogilner. 2003. Cell division. *Nature* 422: 746–752. [R]

Vale, R. D. 2003. The molecular motor toolbox for intracellular transport. *Cell* 112: 467–480. [R]

Vale, R. D., T. S. Reese and M. P. Sheetz. 1985. Identification of a novel force-generating protein, kinesin, involved in microtubule-based motility. *Cell* 42: 39–50. [P]

Vallee, R. B. and M. P. Sheetz. 1996. Targeting of motor proteins. *Science* 271: 1539–1544. [R]

Vaughan, K. T. 2004. Surfing, regulating and capturing: Are all microtubule-tip-tracking proteins created equal? *Trends Cell Biol.* 14: 491–496. [R]

Yildiz, A. and P. R. Selvin. 2005. Kinesin: Walking, crawling or sliding along? *Trends Cell Biol.* 15: 112–120. [R]

CHAPTER 13

The Plasma Membrane

■ **Structure of the Plasma Membrane 529**

■ **Transport of Small Molecules 540**

■ **Endocytosis 556**

■ **KEY EXPERIMENT:** The LDL Receptor 559

■ **MOLECULAR MEDICINE:** Cystic Fibrosis 554

ALL CELLS—BOTH PROKARYOTIC AND EUKARYOTIC—are surrounded by a plasma membrane, which defines the boundary of the cell and separates its internal contents from the environment. By serving as a selective barrier to the passage of molecules, the plasma membrane determines the composition of the cytoplasm. This ultimately defines the very identity of the cell, so the plasma membrane is one of the most fundamental structures of cellular evolution. Indeed, as discussed in Chapter 1, the first cell is thought to have arisen by the enclosure of self-replicating RNA in a membrane of phospholipids.

The plasma membranes of present-day cells are composed of both lipids and proteins. The basic structure of the plasma membrane is the phospholipid bilayer, which is impermeable to most water-soluble molecules. The passage of ions and most biological molecules across the plasma membrane is therefore mediated by proteins, which are responsible for the selective traffic of molecules into and out of the cell. Other proteins of the plasma membrane control the interactions between cells of multicellular organisms and serve as sensors through which the cell receives signals from its environment. The plasma membrane thus plays a dual role: It both isolates the cytoplasm and mediates interactions between the cell and its environment.

Structure of the Plasma Membrane

Like all other cellular membranes, the plasma membrane consists of both lipids and proteins. The fundamental structure of the membrane is the phospholipid bilayer, which forms a stable barrier between two aqueous compartments. In the case of the plasma membrane, these compartments are the inside and the outside of the cell. Proteins embedded within the phospholipid bilayer carry out the specific functions of the plasma membrane, including selective transport of molecules and cell-cell recognition.

The Phospholipid Bilayer

The plasma membrane is the most thoroughly studied of all cell membranes, and it is largely through investigations of the plasma membrane that our current concepts of membrane structure have evolved. The plasma membranes of mammalian red blood cells (erythrocytes) have been particularly useful as a model for studies of membrane structure. Mammalian red blood cells do not contain nuclei or internal membranes, so they represent a source from which pure plasma membranes can be easily isolated for biochemical analysis. Indeed, studies of the red blood cell plasma membrane provided the first evidence that biological membranes consist of lipid bilayers. In 1925, two Dutch scientists (Edwin Gorter and F. Grendel) extracted the membrane lipids from a known number of red blood cells corresponding to a known surface area of plasma membrane. They then determined the surface area occupied by a monolayer of the extracted lipid spread out at an air-water interface. The surface area of the lipid monolayer turned out to be twice that occupied by the erythrocyte plasma membranes, leading to the conclusion that the membranes consisted of lipid bilayers rather than monolayers.

The bilayer structure of the erythrocyte plasma membrane is clearly evident in high-magnification electron micrographs (Figure 13.1). The plasma membrane appears as two dense lines separated by an intervening space—a morphology frequently referred to as a "railroad track" appearance. This image results from the binding of the electron-dense heavy metals used as stains in transmission electron microscopy (see Chapter 1) to the polar head groups of the phospholipids, which therefore appear as dark lines. These dense lines are separated by the lightly stained interior portion of the membrane, which contains the hydrophobic fatty acid chains.

As discussed in Chapter 2, the plasma membranes of animal cells contain four major phospholipids (**phosphatidylcholine**, **phosphatidylethanolamine**, **phosphatidylserine**, and **sphingomyelin**), which together account for more than half of the lipid in most membranes. These phospholipids are asymmetrically distributed between the two halves of the membrane bilayer (Figure 13.2). The outer leaflet of the plasma membrane consists mainly of phosphatidylcholine and sphingomyelin, whereas phosphatidylethanolamine and phosphatidylserine are the predominant phos-

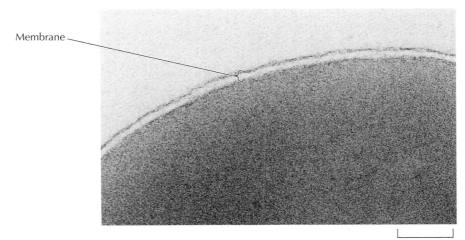

Membrane

FIGURE 13.1 Bilayer structure of the plasma membrane Electron micrograph of a human red blood cell. Note the "railroad track" appearance of the plasma membrane. (Courtesy of J. David Robertson, Duke University Medical Center.)

20 nm

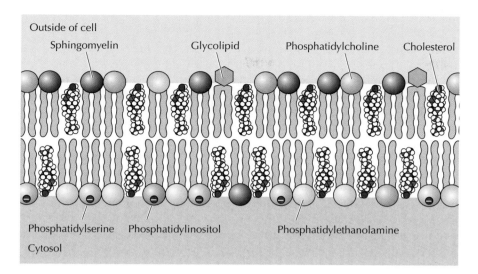

Outside of cell

Sphingomyelin   Glycolipid   Phosphatidylcholine   Cholesterol

Phosphatidylserine   Phosphatidylinositol   Phosphatidylethanolamine

Cytosol

FIGURE 13.2 Lipid components of the plasma membrane The outer leaflet consists predominantly of phosphatidylcholine, sphingomyelin, and glycolipids, whereas the inner leaflet contains phosphatidylethanolamine, phosphatidylserine, and phosphatidylinositol. Cholesterol is distributed in both leaflets. The net negative charge of the head groups of phosphatidylserine and phosphatidylinositol is indicated. The structures of phospholipids, glycolipids, and cholesterol are shown in Figures 2.7, 2.8, and 2.9, respectively.

pholipids of the inner leaflet. The head groups of both phosphatidylserine and phosphatidylinositol are negatively charged, so their predominance in the inner leaflet results in a net negative charge on the cytosolic face of the plasma membrane. A fifth phospholipid, **phosphatidylinositol**, is also localized to the inner half of the plasma membrane. Although phosphatidylinositol is a quantitatively minor membrane component, it plays an important role in cell signaling, as discussed in Chapter 15.

In addition to the phospholipids, the plasma membranes of animal cells contain **glycolipids** and **cholesterol**. The glycolipids are found exclusively in the outer leaflet of the plasma membrane, with their carbohydrate portions exposed on the cell surface. They are relatively minor membrane components, constituting only about 2% of the lipids of most plasma membranes. Cholesterol, on the other hand, is a major membrane constituent of animal cells, being present in about the same molar amounts as the phospholipids.

Two general features of phospholipid bilayers are critical to membrane function. First, the structure of phospholipids is responsible for the basic function of membranes as barriers between two aqueous compartments. Because the interior of the phospholipid bilayer is occupied by hydrophobic fatty acid chains, the membrane is impermeable to water-soluble molecules, including ions and most biological molecules. Second, bilayers of the naturally occurring phospholipids are viscous fluids, not solids. The fatty acids of most natural phospholipids have one or more double bonds, which introduce kinks into the hydrocarbon chains and make them difficult to pack together. The long hydrocarbon chains of the fatty acids therefore move freely in the interior of the membrane, so the membrane itself is soft and flexible. In addition, both phospholipids and proteins are free to diffuse laterally within the membrane—a property that is critical for many membrane functions.

Because of its rigid ring structure, cholesterol plays a distinct role in membrane structure. Cholesterol will not form a membrane by itself but inserts into a bilayer of phospholipids with its polar hydroxyl group close to the phospholipid head groups (see Figure 13.2). Depending on the temperature, cholesterol has distinct effects on membrane fluidity. At high temperatures, cholesterol interferes with the movement of the phospholipid fatty acid chains, making the outer part of the membrane less fluid and reducing its permeability to small molecules. At low temperatures, however, cholesterol has the opposite effect: By interfering with interactions between fatty acid chains, cholesterol prevents membranes from freezing and maintains membrane fluidity. Although cholesterol is not present in bacteria, it is an essential component of animal cell plasma membranes. Plant cells also lack cholesterol, but they contain related compounds (sterols) that fulfill a similar function.

Rather than diffusing freely in the plasma membrane, cholesterol and the sphingolipids (sphingomyelin and glycolipids) form discrete membrane domains known as **lipid rafts**. These clusters of sphingolipids and cholesterol move laterally within the plasma membrane and associate with specific membrane proteins. Although the functions of lipid rafts remain to be fully understood, they play important roles in processes such as cell movement and the uptake of extracellular molecules by endocytosis (as discussed later in this chapter) as well as in cell signaling (discussed in Chapter 15).

Membrane Proteins

While lipids are the fundamental structural elements of membranes, proteins are responsible for carrying out specific membrane functions. Most plasma membranes consist of approximately 50% lipid and 50% protein by weight, with the carbohydrate portions of glycolipids and glycoproteins constituting 5 to 10% of the membrane mass. Since proteins are much larger than lipids, this percentage corresponds to about one protein molecule per every 50 to 100 molecules of lipid. In 1972 Jonathan Singer and Garth Nicolson proposed the **fluid mosaic model** of membrane structure, which is now generally accepted as the basic paradigm for the organization of all biological membranes. In this model, membranes are viewed as two-dimensional fluids in which proteins are inserted into lipid bilayers (**Figure 13.3**).

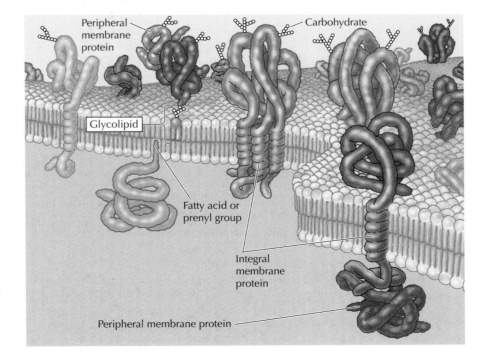

FIGURE 13.3 Fluid mosaic model of the plasma membrane Integral membrane proteins are inserted into the lipid bilayer, whereas peripheral proteins are bound to the membrane indirectly by protein-protein interactions. Most integral membrane proteins are transmembrane proteins with portions exposed on both sides of the lipid bilayer. The extracellular portions of these proteins are usually glycosylated, as are the peripheral membrane proteins bound to the external face of the membrane.

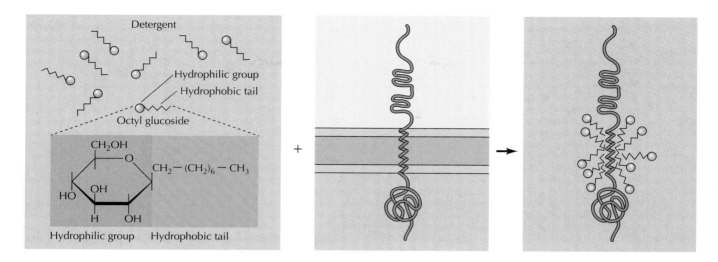

Detergent

Hydrophilic group
Hydrophobic tail

Octyl glucoside

CH₂OH

CH₂ — (CH₂)₆ — CH₃

HO   OH

H   OH

Hydrophilic group   Hydrophobic tail

+

FIGURE 13.4 Solubilization of integral membrane proteins by detergents Detergents (e.g., octyl glucoside) are amphipathic molecules containing hydrophilic head groups and hydrophobic tails. The hydrophobic tails bind to the hydrophobic regions of integral membrane proteins, forming detergent-protein complexes that are soluble in aqueous solution.

Singer and Nicolson distinguished two classes of membrane-associated proteins, which they called **peripheral** and **integral membrane proteins**. Peripheral membrane proteins were operationally defined as proteins that dissociate from the membrane following treatments with polar reagents, such as solutions of extreme pH or high salt concentration that do not disrupt the phospholipid bilayer. Once dissociated from the membrane, peripheral membrane proteins are soluble in aqueous buffers. These proteins are not inserted into the hydrophobic interior of the lipid bilayer. Instead, they are indirectly associated with membranes through protein-protein interactions. These interactions frequently involve ionic bonds, which are disrupted by extreme pH or high salt.

In contrast to the peripheral membrane proteins, integral membrane proteins can be released only by treatments that disrupt the phospholipid bilayer. Portions of these integral membrane proteins are inserted into the lipid bilayer, so they can be dissociated only by reagents that disrupt hydrophobic interactions. The most commonly used reagents for solubilization of integral membrane proteins are detergents, which are small amphipathic molecules containing both hydrophobic and hydrophilic groups (Figure 13.4). The hydrophobic portions of detergents displace the membrane lipids and bind to the hydrophobic portions of integral membrane proteins. Because the other end of the detergent molecule is hydrophilic, the detergent-protein complexes are soluble in aqueous solutions.

Many integral proteins are **transmembrane proteins**, which span the lipid bilayer with portions exposed on both sides of the membrane. These proteins can be visualized in electron micrographs of plasma membranes prepared by the freeze-fracture technique (see Figure 1.36). In these specimens, the membrane is split and separated into its two leaflets. Transmembrane proteins are then apparent as particles on the internal faces of the membrane (Figure 13.5).

The membrane-spanning portions of transmembrane proteins are usually α helices of 20 to 25 hydrophobic amino acids that are inserted into the membrane of the endoplasmic reticulum during synthesis of the polypeptide chain (see Figures 10.12–10.14). These proteins are then transported in membrane vesicles from the endoplasmic reticulum to the Golgi apparatus and from there to the plasma membrane. Carbohydrate groups are added

FIGURE 13.5 Freeze-fracture electron micrograph of human red blood cell membranes The particles in the membrane are transmembrane proteins. (Harold H. Edwards/Visuals Unlimited.)

0.2 μm

to the polypeptide chains in both the endoplasmic reticulum and Golgi apparatus, so most transmembrane proteins of the plasma membrane are glycoproteins with their oligosaccharides exposed on the surface of the cell.

Studies of red blood cells have provided good examples of both peripheral and integral proteins associated with the plasma membrane. The membranes of human erythrocytes contain about a dozen major proteins, which were originally identified by gel electrophoresis of membrane preparations. Most of these are peripheral membrane proteins that have been identified as components of the cortical cytoskeleton, which underlies the plasma membrane and determines cell shape (see Chapter 12). For example, the most abundant peripheral membrane protein of red blood cells is spectrin, which is the major cytoskeletal protein of erythrocytes. Other peripheral membrane proteins of red blood cells include actin, ankyrin, and band 4.1. Ankyrin serves as the principal link between the plasma membrane and the cytoskeleton by binding to both spectrin and the integral membrane protein band 3 (see Figure 12.14). An additional link between the membrane and the cytoskeleton is provided by band 4.1, which binds to the junctions of spectrin and actin, as well as to glycophorin (the other major integral membrane protein of erythrocytes).

The two major integral membrane proteins of red blood cells—glycophorin and band 3—provide well-studied examples of transmembrane protein structure (Figure 13.6). Glycophorin is a small glycoprotein of 131 amino acids, with a molecular weight of about 30,000, half of which is protein and half carbohydrate. Glycophorin crosses the membrane with a single membrane-spanning α helix of 23 amino acids, with its glycosylated amino-terminal portion exposed on the cell surface. Although glycophorin was one of the first transmembrane proteins to be characterized, its precise function remains unknown. In contrast, the function of the other major transmembrane protein of red blood cells is well understood. This protein, originally known as band 3, is the anion transporter responsible for the passage of bicarbonate (HCO_3^-) and chloride (Cl^-) ions across the red blood cell membrane. The band 3 polypeptide chain is 929 amino acids and is thought to have 14 membrane-spanning α-helical regions. Within the membrane, dimers of band 3 form globular structures containing internal channels through which ions are able to travel across the lipid bilayer.

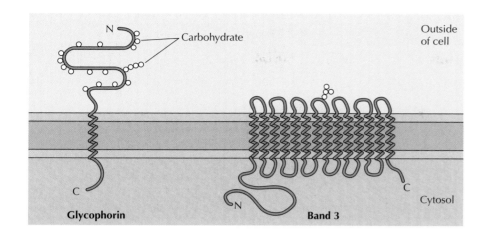

FIGURE 13.6 Integral membrane proteins of red blood cells Glycophorin (131 amino acids) contains a single transmembrane α helix. It is heavily glyocosylated, with oligosaccharides attached to 16 sites on the extracellular portion of the polypeptide chain. Band 3 (929 amino acids) has multiple transmembrane α helices and is thought to cross the membrane 14 times.

Because of their amphipathic character, transmembrane proteins have proved difficult to crystallize, as required for three-dimensional structural analysis by X-ray diffraction. The first transmembrane protein to be analyzed by X-ray crystallography was the photosynthetic reaction center of the bacterium *Rhodopseudomonas viridis* whose structure was reported in 1985 (Figure 13.7). The reaction center contains three transmembrane proteins designated L, M, and H (light, medium, and heavy) according to their apparent sizes indicated by gel electrophoresis. The L and M subunits each have five membrane-spanning α helices. The H subunit has only a single transmembrane α helix, with the bulk of the polypeptide chain on the cytosolic side of the membrane. The fourth subunit of the reaction center is a cytochrome, which is a peripheral membrane protein bound to the complex by protein-protein interactions.

Although most transmembrane proteins span the membrane by α-helical regions, this is not always the case. A well-characterized exception is provided by the **porins**—a class of proteins that form channels in the outer membranes of some bacteria. Many bacteria, including *E. coli*, have a dual membrane system in which the plasma membrane (or inner membrane) is surrounded by the cell wall and a distinct outer membrane (Figure 13.8). In contrast to the plasma membrane, the outer membrane is highly permeable to ions and small polar molecules (in the case of *E. coli*, with molecular weights up to 600). This permeability results from the porins, which form aqueous channels through the lipid bilayer. As discussed in Chapter 11, proteins related to the bacterial porins are also found in the outer membranes of mitochondria and chloroplasts.

Structural analysis shows that the porins do not contain hydrophobic β-helical regions. Instead, they cross the membrane as β barrels in which 8-22 β sheets fold up into a barrel-like structure enclosing an aqueous pore. The side chains of polar amino acids line the pore, whereas side chains of hydrophobic amino acids interact with the interior of the membrane. Some

FIGURE 13.7 A bacterial photosynthetic reaction center The reaction center consists of three transmembrane proteins designated L (red), M (yellow), and H (green). The L and M subunits each have five transmembrane α helices, whereas the H subunit has only one. The fourth subunit of the reaction center is a cytochrome (white), which is a peripheral membrane protein.

Cytosol

FIGURE 13.8 Bacterial outer membranes The plasma membrane of some bacteria is surrounded by a cell wall and a distinct outer membrane. The outer membrane contains porins, which form aqueous channels allowing the free passage of ions and small molecules. Porins cross the membrane as β-barrels.

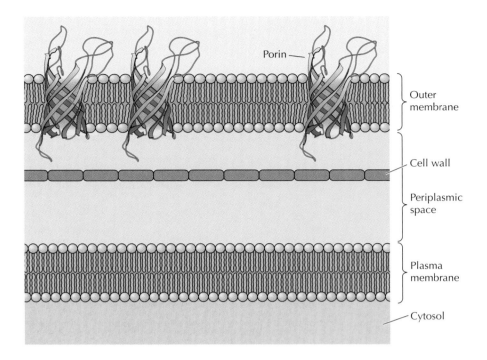

Porin

Outer membrane

Cell wall

Periplasmic space

Plasma membrane

Cytosol

porins exist in the membrane as monomers while others associate to form stable multimers, containing multiple channels through which polar molecules can diffuse across the membrane.

In contrast to transmembrane proteins, a variety of proteins (many of which behave as integral membrane proteins) are anchored in the plasma membrane by covalently attached lipids or glycolipids (Figure 13.9). Members of one class of these proteins are inserted into the outer leaflet of the plasma membrane by **glycosylphosphatidylinositol (GPI) anchors**. GPI anchors are added to certain proteins that have been transferred into the endoplasmic reticulum and are anchored in the membrane by a C-terminal transmembrane region (see Figure 10.17). The transmembrane region is cleaved as the GPI anchor is added, so these proteins remain attached to the membrane only by the glycolipid. Since the polypeptide chains of GPI-anchored proteins are transferred into the endoplasmic reticulum, they are glycosylated and exposed on the surface of the cell following transport to the plasma membrane.

Other proteins are anchored in the inner leaflet of the plasma membrane by covalently attached lipids. Rather than being processed through the secretory pathway, these proteins are synthesized on free cytosolic ribosomes and then modified by the addition of lipids. These modifications include the addition of myristic acid (a 14-carbon fatty acid) to the amino terminus of the polypeptide chain, the addition of prenyl groups (15 or 20 carbons) to the side chains of carboxy-terminal cysteine residues, and the addition of palmitic acid (16 carbons) to the side chains of cysteine residues (see Figures 8.32–8.34). In some cases, these proteins (many of which behave as peripheral membrane proteins) are targeted to the plasma membrane by positively charged regions of the polypeptide chain as well as by the attached lipids. These positively charged protein domains may interact with the negatively charged head groups of phosphatidylserine on the

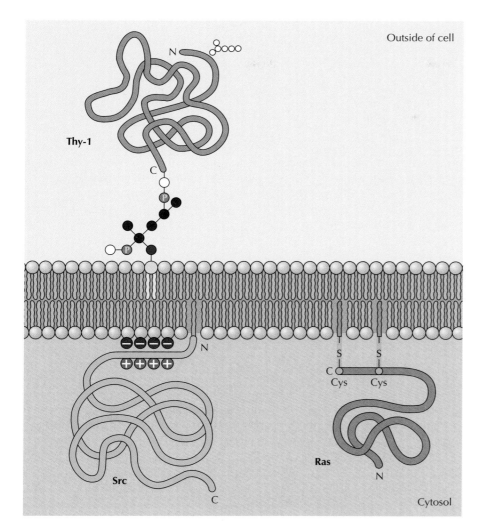

Outside of cell

Thy-1

Src

Ras

N

Cytosol

FIGURE 13.9 Examples of proteins anchored in the plasma membrane by lipids and glycolipids   Some proteins (e.g., the lymphocyte protein Thy-1) are anchored in the outer leaflet of the plasma membrane by GPI anchors added to their C terminus in the endoplasmic reticulum. These proteins are glycosylated and exposed on the cell surface. Other proteins are anchored in the inner leaflet of the plasma membrane following their translation on free cytosolic ribosomes. The Ras protein illustrated is anchored by a prenyl group attached to the side chain of a C-terminal cysteine and by a palmitoyl group attached to a cysteine located five amino acids upstream. The Src protein is anchored by a myristoyl group attached to its N terminus. A positively charged region of Src also plays a role in membrane association, perhaps by interacting with the negatively charged head groups of phosphatidylserine. The structures of these lipid and glycolipid groups are illustrated in Figures 8.32 through 8.34.

cytosolic face of the plasma membrane. It is noteworthy that many of the proteins anchored in the inner leaflet of the plasma membrane (including the Src and Ras proteins illustrated in Figure 13.9) play important roles in the transmission of signals from cell surface receptors to intracellular targets, as discussed in Chapter 15.

Mobility of Membrane Proteins

Membrane proteins and phospholipids are unable to move back and forth between the inner and outer leaflets of the membrane at an appreciable rate. However, because they are inserted into a fluid lipid bilayer, both proteins and lipids are able to diffuse laterally through the membrane. This lateral movement was first shown directly in an experiment reported by Larry Frye and Michael Edidin in 1970, which provided support for the fluid mosaic model. Frye and Edidin fused human and mouse cells in culture to produce human-mouse cell hybrids (Figure 13.10). They then analyzed the distribution of proteins in the membranes of these hybrid cells using antibodies that specifically recognize proteins of human and mouse origin. These antibodies were labeled with different fluorescent dyes, so the

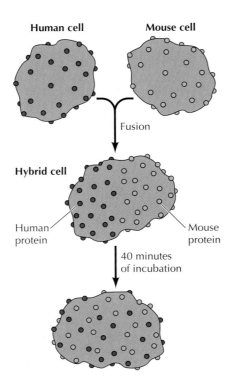

Human cell   Mouse cell

Fusion

Hybrid cell

Human protein   Mouse protein

40 minutes of incubation

FIGURE 13.10 Mobility of membrane proteins Human and mouse cells were fused to produce hybrid cells. The distribution of cell surface proteins was then analyzed using anti-human and anti-mouse antibodies labeled with different fluorescent dyes (red and green, respectively). The human and mouse proteins were detected in different halves of the hybrid cells immediately after fusion but had intermingled over the cell surface following 40 minutes of incubation.

human and mouse proteins could be distinguished by fluorescence microscopy. Immediately after fusion, human and mouse proteins were localized to different halves of the hybrid cells. However, after a brief period of incubation at 37°C the human and mouse proteins were completely intermixed over the cell surface, indicating that they moved freely through the plasma membrane.

However, not all proteins are able to diffuse freely through the membrane. In some cases, the mobility of membrane proteins is restricted by their association with the cytoskeleton. For example, a fraction of band 3 in the red blood cell membrane is immobilized as a result of its association with ankyrin and spectrin. In other cases, the mobility of membrane proteins may be restricted by their associations with other membrane proteins, with proteins on the surface of adjacent cells, or with the extracellular matrix.

In contrast to blood cells, epithelial cells are polarized when they are organized into tissues, with different parts of the cell responsible for performing distinct functions. Consequently, the plasma membranes of many epithelial cells are divided into distinct **apical** and **basolateral domains** that differ in function and protein composition (Figure 13.11). For example, epithelial cells of the small intestine function to absorb nutrients from the

FIGURE 13.11 A polarized intestinal epithelial cell The apical surface of the cell contains microvilli and is specialized for absorption of nutrients from the intestinal lumen. The basolateral surface is specialized for the transfer of absorbed nutrients to the underlying connective tissue, which contains blood capillaries. Tight junctions separate the apical and basolateral domains of the plasma membrane. Membrane proteins are free to diffuse within each domain but are not able to cross from one domain to the other.

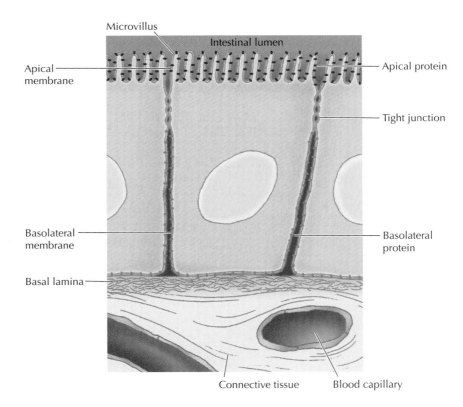

Microvillus

Intestinal lumen

Apical membrane

Apical protein

Tight junction

Basolateral membrane

Basolateral protein

Basal lamina

Connective tissue   Blood capillary

digestive tract. The apical surface of these cells, which faces the intestinal lumen, is therefore covered by microvilli, which increase its surface area and facilitate nutrient absorption. The basolateral surface, which faces underlying connective tissue and the blood supply, is specialized to mediate the transfer of absorbed nutrients into the circulation. In order to maintain these distinct functions, the mobility of plasma membrane proteins must be restricted to the appropriate domains of the cell surface. At least part of the mechanism by which this occurs involves the formation of tight junctions (which are discussed in Chapter 14) between adjacent cells of the epithelium. These junctions not only seal the space between cells but also serve as barriers to the movement of membrane lipids and proteins. As a result, proteins are able to diffuse within either the apical or basolateral domains of the plasma membrane but are not able to cross from one domain to the other.

Lipid composition can also perturb the free diffusion of membrane proteins. The melting temperatures of the sphingolipids (sphingomyelin and glycolipids) differ from those of the phospholipids derived from glycerol. Sphingomyelin and glycolipids tend to cluster in small semisolid patches termed lipid rafts, which are also enriched in cholesterol because of the packing affinity of cholesterol and the sphingolipids (Figure 13.12). Rafts are also enriched in GPI-anchored proteins, and several proteins found in the bulk phospholipid bilayer, such as GTP-binding proteins and several protein kinases, move in and out of rafts. The transient presence of these proteins in rafts allows the clustering necessary for processes such as endocytosis, extension of cell processes, and receptor-mediated signaling.

FIGURE 13.12 Structure of lipid rafts
Lipid rafts are organized by the interactions of sphingomyelin, glycolipids, and cholesterol. GPI-anchored proteins are preferentially found in lipid rafts, and several other types of membrane proteins are present transiently in rafts to mediate cell signaling or endocytosis.

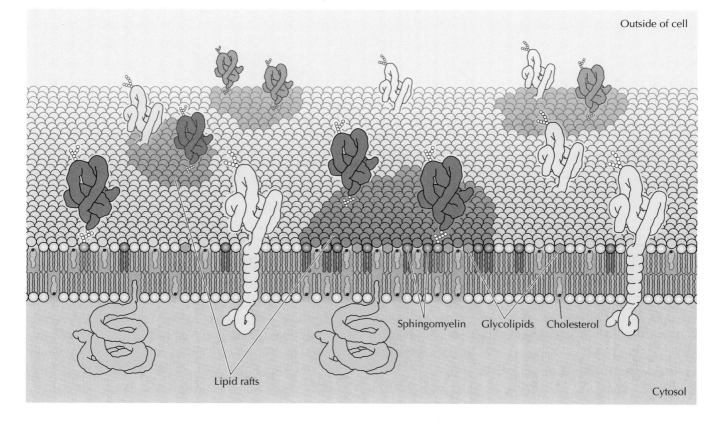

Outside of cell

Sphingomyelin Glycolipids Cholesterol

Lipid rafts

Cytosol

FIGURE 13.13 The glycocalyx   An electron micrograph of intestinal epithelium illustrating the glycocalyx (arrows). (Don Fawcett/Visuals Unlimited.)

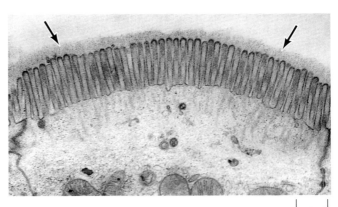

1 μm

Lipid rafts can also be stabilized by interactions with the cytoskeleton through peripheral membrane proteins. One such peripheral protein is caveolin, which is associated with a subclass of lipid rafts called caveolae. As discussed later in this chapter, caveolae function in endocytosis.

The Glycocalyx

As already discussed, the extracellular portions of plasma membrane proteins are generally glycosylated. Likewise, the carbohydrate portions of glycolipids are exposed on the outer face of the plasma membrane. Consequently, the surface of the cell is covered by a carbohydrate coat known as the **glycocalyx**, which is formed by the oligosaccharides of glycolipids and transmembrane glycoproteins (Figure 13.13).

Part of the role of the glycocalyx is to protect the cell surface. In addition, the oligosaccharides of the glycocalyx serve as markers for a variety of cell-cell interactions. A well-studied example of these interactions is the adhesion of white blood cells (leukocytes) to the endothelial cells that line blood vessels—a process that allows the leukocytes to leave the circulatory system and mediate the inflammatory response in injured tissues. The initial step in adhesion between leukocytes and endothelial cells is mediated by a family of transmembrane proteins called **selectins**, which recognize specific carbohydrates on the cell surface (Figure 13.14). Two members of the selectin family (E-selectin and P-selectin) expressed by endothelial cells and platelets bind to specific oligosaccharides expressed on the surface of leukocytes. A different selectin (L-selectin) is expressed by leukocytes and recognizes an oligosaccharide on the surface of endothelial cells. The oligosaccharides exposed on the cell surface thus provide a set of markers that help identify the distinct cell types of multicellular organisms.

Transport of Small Molecules

The internal composition of the cell is maintained because the plasma membrane is selectively permeable to small molecules. Most biological molecules are unable to diffuse through the phospholipid bilayer, so the plasma membrane forms a barrier that blocks the free exchange of molecules between the cytoplasm and the external environment of the cell. Specific transport proteins (carrier proteins and channel proteins) then mediate the selective passage of small molecules across the membrane, allowing the cell to control the composition of its cytoplasm.

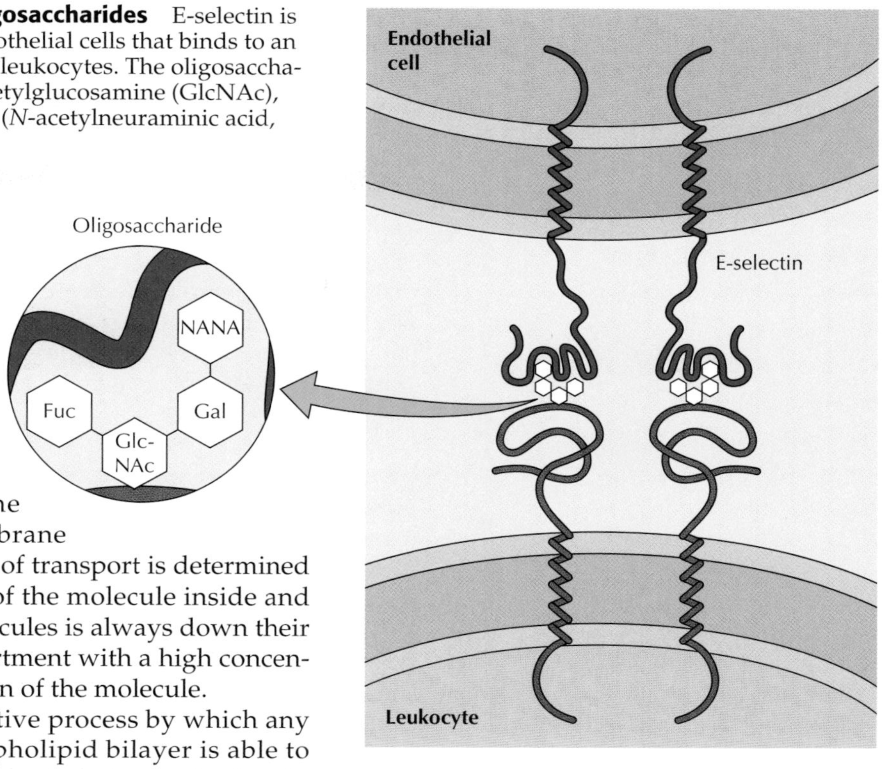

FIGURE 13.14 Binding of selectins to oligosaccharides   E-selectin is a transmembrane protein expressed by endothelial cells that binds to an oligosaccharide expressed on the surface of leukocytes. The oligosaccharide recognized by E-selectin contains *N*-acetylglucosamine (GlcNAc), fucose (Fuc), galactose (Gal), and sialic acid (*N*-acetylneuraminic acid, NANA).

Passive Diffusion

The simplest mechanism by which molecules can cross the plasma membrane is **passive diffusion**. During passive diffusion, a molecule simply dissolves in the phospholipid bilayer, diffuses across it, and then dissolves in the aqueous solution at the other side of the membrane. No membrane proteins are involved and the direction of transport is determined simply by the relative concentrations of the molecule inside and outside of the cell. The net flow of molecules is always down their concentration gradient—from a compartment with a high concentration to one with a lower concentration of the molecule.

Passive diffusion is thus a nonselective process by which any molecule able to dissolve in the phospholipid bilayer is able to cross the plasma membrane and equilibrate between the inside and outside of the cell. Importantly, only small, relatively hydrophobic molecules are able to diffuse across a phospholipid bilayer at significant rates (Figure 13.15). Thus gases (such as O_2 and CO_2), hydrophobic molecules (such as benzene), and small polar but uncharged molecules (such as H_2O and ethanol) are able to passively diffuse across the plasma membrane. Other biological molecules, however, are unable to dissolve in the hydrophobic interior of the phospholipid bilayer. Consequently, larger

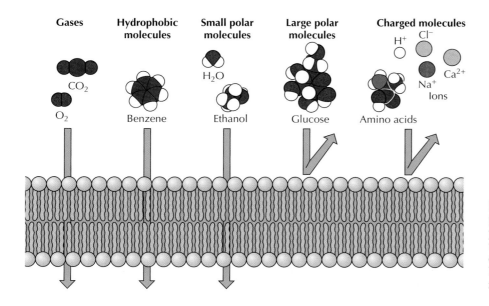

FIGURE 13.15 Permeability of phospholipid bilayers   Gases, hydrophobic molecules, and small polar uncharged molecules can passively diffuse through phospholipid bilayers. Larger polar molecules and charged molecules cannot.

uncharged polar molecules such as glucose are unable to cross the plasma membrane by passive diffusion, as are charged molecules of any size (including small ions such as H^+, Na^+, K^+, and Cl^-). The passage of these molecules across the membrane instead requires the activity of specific transport and channel proteins, which therefore control the traffic of most biological molecules into and out of the cell.

Facilitated Diffusion and Carrier Proteins

Facilitated diffusion, like passive diffusion, involves the movement of molecules in the direction determined by their relative concentrations inside and outside of the cell. No external source of energy is provided, so molecules travel across the membrane in the direction determined by their concentration gradients and, in the case of charged molecules, by the electric potential across the membrane. However, facilitated diffusion differs from passive diffusion in that the transported molecules do not dissolve in the phospholipid bilayer. Instead, their passage is mediated by proteins that enable the transported molecules to cross the membrane without directly interacting with its hydrophobic interior. Facilitated diffusion therefore allows polar and charged molecules, such as carbohydrates, amino acids, nucleosides, and ions, to cross the plasma membrane.

Two classes of proteins that mediate facilitated diffusion have generally been distinguished: carrier proteins and channel proteins. **Carrier proteins** bind specific molecules to be transported on one side of the membrane. They then undergo conformational changes that allow the molecule to pass through the membrane and be released on the other side. In contrast, **channel proteins** form open pores through the membrane, allowing the free diffusion of any molecule of the appropriate size and charge.

Carrier proteins are responsible for the facilitated diffusion of sugars, amino acids, and nucleosides across the plasma membranes of most cells. The uptake of glucose, which serves as a primary source of metabolic energy, is one of the most important transport functions of the plasma membrane, and the glucose transporter provides a well-studied example of a carrier protein. The glucose transporter was initially identified as a 55-kd protein in human red blood cells in which it represents approximately 5% of total membrane protein. Subsequent isolation and sequence analysis of a cDNA clone revealed that the glucose transporter has 12 α-helical transmembrane segments—a structure typical of many carrier proteins (Figure 13.16). These transmembrane α-helices contain predominantly hydrophobic amino acids, but several also contain polar amino acid residues that are thought to form the glucose-binding site in the interior of the protein.

As with many membrane proteins, the three-dimensional structure of the glucose transporter is not known, so the molecular mechanism of glucose transport remains an open question. However, kinetic studies indicate that the glucose transporter functions by alternating between two conformational states (Figure 13.17). In the first conformation, a glucose-binding site faces the outside of the cell. The binding of glucose to this exterior site induces a conformational change in the transporter such that the glucose-binding site now faces the interior of the cell. Glucose can then be released into the cytosol, followed by the return of the transporter to its original conformation.

Most cells, including erythrocytes, are exposed to extracellular glucose concentrations that are higher than those inside the cell, so facilitated diffusion results in the net inward transport of glucose. Once glucose is taken up by these cells it is rapidly metabolized, so intracellular glucose concentra-

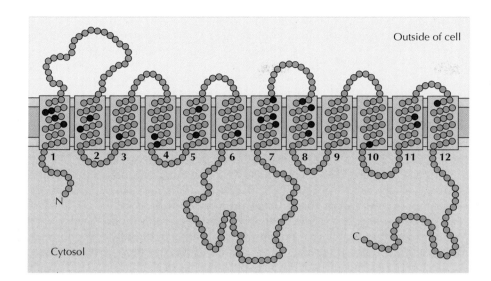

FIGURE 13.16 Structure of the glucose transporter The glucose transporter has 12 transmembrane α helices. Polar amino acid residues located within the phospholipid bilayer are indicated as dark purple circles. (Adapted from G. I. Bell, C. F. Burant, J. Takeda and G. W. Gould, 1993. *J. Biol. Chem.* 268: 19161.)

tions remain low and glucose continues to be transported into the cell from the extracellular fluids. Because the conformational changes of the glucose transporter are reversible, however, glucose can be transported in the opposite direction simply by reversing the steps in Figure 13.17. Such reverse flow occurs, for example, in liver cells in which glucose is synthesized and released into the circulation.

Ion Channels

In contrast to carrier proteins, channel proteins form open pores in the membrane, allowing small molecules of the appropriate size and charge to pass freely through the lipid bilayer. One group of channel proteins discussed earlier is the porins, which permit the free passage of ions and small

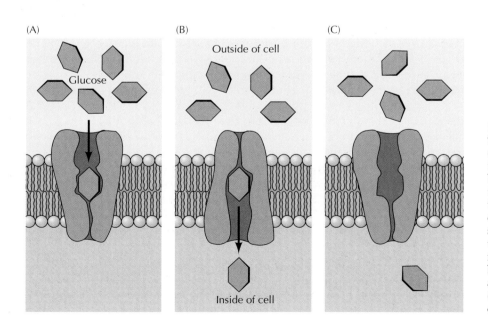

FIGURE 13.17 Model for the facilitated diffusion of glucose The glucose transporter alternates between two conformations in which a glucose-binding site is alternately exposed on the outside and the inside of the cell. In the first conformation shown (A), glucose binds to a site exposed on the outside of the plasma membrane. The transporter then undergoes a conformational change such that the glucose-binding site faces the inside of the cell and glucose is released into the cytosol (B). The transporter then returns to its original conformation (C).

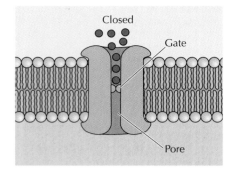

Closed
Gate
Pore

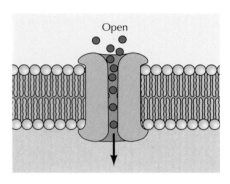

Open

FIGURE 13.18 Model of an ion channel In the closed conformation, the flow of ions is blocked by a gate. Opening of the gate allows ions to flow rapidly through the channel. The channel contains a narrow pore that restricts passage to ions of the appropriate size and charge.

polar molecules through the outer membranes of bacteria (see Figure 13.8). Gap junctions, which are discussed in Chapter 14, contain channel proteins that permit the passage of molecules between connected cells. The plasma membranes of many cells also contain water channel proteins (aquaporins) through which water molecules are able to cross the membrane much more rapidly than they can diffuse through the phospholipid bilayer. The best-characterized channel proteins, however, are the **ion channels**, which mediate the passage of ions across plasma membranes. Although ion channels are present in the membranes of all cells, they have been especially well studied in nerve and muscle, where their regulated opening and closing is responsible for the transmission of electric signals.

Three properties of ion channels are central to their function (Figure 13.18). First, transport through channels is extremely rapid. More than a million ions per second flow through open channels—a flow rate approximately a thousand times greater than the rate of transport by carrier proteins. Second, ion channels are highly selective because narrow pores in the channel restrict passage to ions of the appropriate size and charge. Thus specific channel proteins allow the passage of Na^+, K^+, Ca^{2+}, and Cl^- across the membrane. Third, most ion channels are not permanently open. Instead, the opening of ion channels is regulated by "gates" that transiently open in response to specific stimuli. Some channels (called **ligand-gated channels**) open in response to the binding of neurotransmitters or other signaling molecules; others (**voltage-gated channels**) open in response to changes in electric potential across the plasma membrane.

The fundamental role of ion channels in the transmission of electric impulses was elucidated through a series of elegant experiments reported by Alan Hodgkin and Andrew Huxley in 1952. These investigators used the giant nerve cells of the squid as a model. The axons of these giant neurons have a diameter of about 1 mm, making it possible to insert electrodes and measure the changes in membrane potential that take place during the transmission of nerve impulses. Using this approach, Hodgkin and Huxley demonstrated that these changes in membrane potential result from the regulated opening and closing of Na^+ and K^+ channels in the plasma membrane. It subsequently became possible to study the activity of individual ion channels using the **patch clamp technique** developed by Erwin Neher and Bert Sakmann in 1976 (Figure 13.19). In this method, a micropipette with a tip diameter of about 1 μm is used to isolate a small patch of membrane, allowing the flow of ions through a single channel to be analyzed and greatly increasing the precision with which the activities of ion channels can be studied.

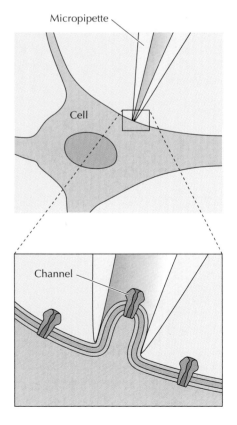

Micropipette

Cell

Channel

FIGURE 13.19 The patch clamp technique A small patch of membrane is isolated in the tip of a micropipette. Stimuli can then be applied from within the pipette, allowing the behavior of the trapped channel to be measured. (Adapted from E. Neher and B. Sakmann, 1992. *Sci. Am.* 266(3): 44.)

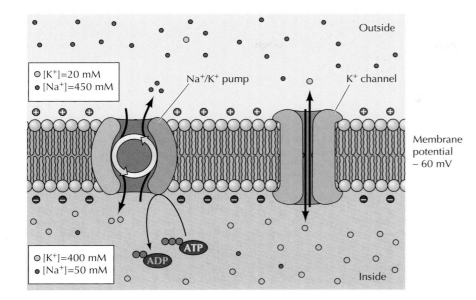

FIGURE 13.20 Ion gradients and resting membrane potential of the giant squid axon Only the concentrations of Na^+ and K^+ are shown because these are the ions that function in the transmission of nerve impulses. Na^+ is pumped out of the cell while K^+ is pumped in, so the concentration of Na^+ is higher outside than inside of the axon, whereas the concentration of K^+ is higher inside than out. The resting membrane is more permeable to K^+ than to Na^+ or other ions because it contains open K^+ channels. The flow of K^+ through these channels makes the major contribution to the resting membrane potential of –60 mV, which is therefore close to the K^+ equilibrium potential.

The flow of ions through membrane channels is dependent on the establishment of ion gradients across the plasma membrane. All cells, including nerve and muscle, contain ion pumps (discussed in the next section) that use energy derived from ATP hydrolysis to actively transport ions across the plasma membrane. As a result, the ionic composition of the cytoplasm is substantially different from that of extracellular fluids (Table 13.1). For example, Na^+ is actively pumped out of cells while K^+ is pumped in. Therefore, in the squid axon the concentration of Na^+ is about 10 times higher in extracellular fluids than inside the cell, whereas the concentration of K^+ is approximately 20 times higher in the cytosol than in the surrounding medium.

Because ions are electrically charged, their transport results in the establishment of an electric gradient across the plasma membrane. In resting squid axons there is an electric potential of about 60 mV across the plasma membrane, with the inside of the cell negative with respect to the outside (Figure 13.20). This electric potential arises both from ion pumps and from the flow of ions through channels that are open in the resting cell plasma membrane. The plasma membrane of resting squid axons contains open K^+ channels, so it is more permeable to K^+ than to Na^+ or other ions. Consequently, the flow of K^+ makes the largest contribution to the resting membrane potential.

As discussed in Chapter 11, the flow of ions across a membrane is driven by both the concentration and voltage components of an electrochemical gradient. For example, the twentyfold higher concentration of K^+ inside the squid axon as compared to the extracellular fluid drives the flow of K^+ out of the cell. However, because K^+ is positively charged this efflux of K^+ from the cell generates an electric potential across the membrane, with the inside of the cell becoming negatively charged. This membrane potential opposes the continuing flow of K^+ out of the cell, and the system approaches the equilibrium state in which the membrane potential balances the K^+ concentration gradient.

■ The squid giant axon is approximately 100 times larger in diameter than typical mammalian axons. Its large diameter allows high-speed transduction of nerve impulses, which is vital for the rapid escape response of squids from predators.

TABLE 13.1 Extracellular and Intracellular Ion Concentrations

Ion	Concentration (mM)	
	Intracellular	Extracellular
Squid axon		
K^+	400	20
Na^+	50	450
Cl^-	40–150	560
Ca^{2+}	0.0001	10
Mammalian cell		
K^+	140	5
Na^+	5–15	145
Cl^-	4	110
Ca^{2+}	0.0001	2.5–5

(A)

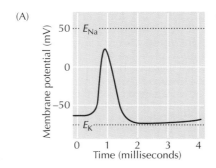

FIGURE 13.21 Membrane potential and ion channels during an action potential (A) Changes in membrane potential at one point on a squid giant axon following a stimulus. E_{Na} and E_K are the equilibrium potentials for Na$^+$ and K$^+$, respectively. (B) The membrane potential first increases as voltage-gated Na$^+$ channels open. The membrane potential then falls below its resting value as the Na$^+$ channels are inactivated and voltage-gated K$^+$ channels open. The voltage-gated K$^+$ channels are then inactivated, and the membrane potential returns to its resting value.

(B)

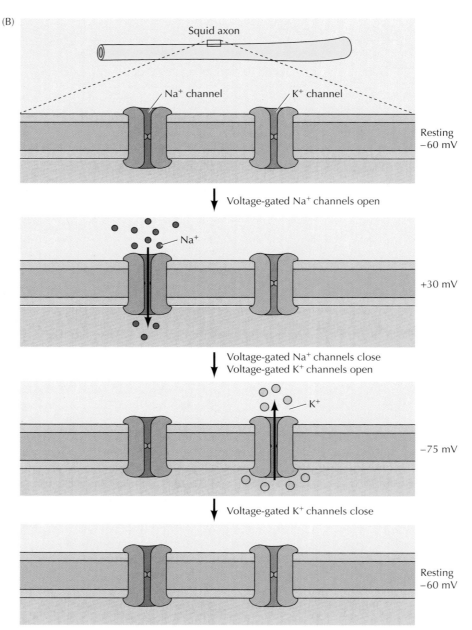

Quantitatively, the relationship between ion concentration and membrane potential is given by the **Nernst equation**:

$$V = \frac{RT}{zF} \ln \frac{C_o}{C_i}$$

where V is the equilibrium potential in volts, R is the gas constant, T is the absolute temperature, z is the charge of the ion, F is Faraday's constant, and C_o and C_i are the concentrations of the ion outside and inside of the cell, respectively. An equilibrium potential exists separately for each ion, and the membrane potential is determined by the flow of all the ions that cross the plasma membrane. However, because resting squid axons are more permeable to K⁺ than to Na⁺ or other ions (including Cl⁻), the resting membrane potential (–60 mV) is close to the equilibrium potential determined by the intracellular and extracellular K^+ concentrations (–75 mV).

As nerve impulses (**action potentials**) travel along axons, the membrane depolarizes (Figure 13.21). The membrane potential changes from –60 mV to approximately +30 mV in less than a millisecond, after which it becomes negative again and returns to its resting value. These changes result from the rapid sequential opening and closing of voltage-gated Na^+ and K^+ channels. Relatively small initial changes in membrane potential (from –60 mV to about –40 mV) lead to the rapid but transient opening of Na^+ channels. This allows Na^+ to flow into the cell, driven by both its concentration gradient and the membrane potential. The sudden entry of Na^+ leads to a large change in membrane potential, which increases to nearly +30 mV, approaching the Na^+ equilibrium potential of approximately +50 mV. At this time, the Na^+ channels are inactivated and voltage-gated K^+ channels open, substantially increasing the permeability of the membrane to K^+. K^+ then flows rapidly out of the cell, driven by both the membrane potential and the K^+ concentration gradient, leading to a rapid decrease in membrane potential to about –75 mV (the K^+ equilibrium potential). This change in membrane potential inactivates the voltage-gated K^+ channels and the membrane potential returns to its resting level of –60 mV, determined by the flow of K^+ and other ions through the channels that remain open in unstimulated cells.

Depolarization of adjacent regions of the plasma membrane allows action potentials to travel down the length of nerve cell axons as electric signals, resulting in the rapid transmission of nerve impulses over long distances. For example, the axons of human motor neurons can be more than a meter long. The arrival of action potentials at the terminus of most neurons then signals the release of neurotransmitters, such as acetylcholine, which carry signals between cells at a synapse (Figure 13.22). Neurotransmitters released from presynaptic cells bind to receptors on the membranes of postsynaptic cells, where they act to open ligand-gated ion channels. One of the best-characterized of these channels is the nicotinic acetylcholine receptor of muscle

■ Decreased concentrations of the neurotransmitter serotonin at neuronal synapses are thought to play a key role in clinical depression. Drugs like Prozac, which are used for the treatment of depression, are selective serotonin re-uptake inhibitors (SSRIs). As the name suggests, SSRIs selectively block transporters responsible for the re-uptake of serotonin by the presynaptic neuron, leading to increased concentrations of serotonin at the synapse.

13.1 WEBSITE ANIMATION

A Chemical Synapse
The arrival of a nerve impulse at the terminus of a neuron triggers synaptic vesicles to fuse with the plasma membrane, thereby releasing neurotransmitter.

FIGURE 13.22 Signaling by neurotransmitter release at a synapse
The arrival of a nerve impulse at the terminus of the neuron signals the fusion of synaptic vesicles with the plasma membrane, resulting in the release of neurotransmitter from the presynaptic cell into the synaptic cleft. The neurotransmitter binds to receptors and opens ligand-gated ion channels in the target cell plasma membrane.

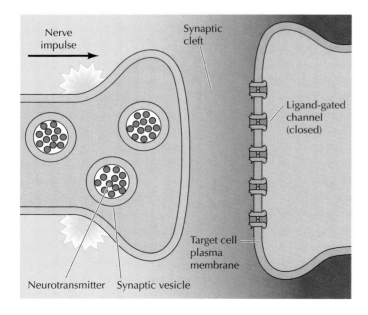

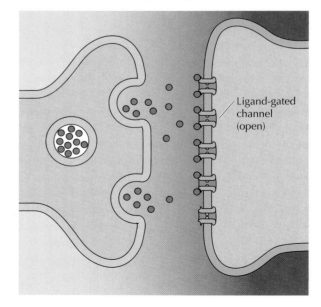

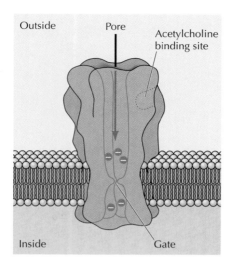

Outside Pore Acetylcholine binding site

Inside Gate

FIGURE 13.23 Model of the nicotinic acetylcholine receptor The receptor consists of five subunits arranged around a central pore. The binding of acetylcholine to a site in the extracellular region of the receptor induces allosteric changes that open the channel gate. The channel is lined by negatively charged amino acids that prevent the flow of negatively charged ions. (Adapted from N. Unwin, 1993. *Cell* 72/*Neuron* 10 (Suppl.): 31.)

cells. Binding of acetylcholine opens a channel that is permeable to both Na^+ and K^+. This permits the rapid influx of Na^+, which depolarizes the muscle cell membrane and triggers an action potential. The action potential then results in the opening of voltage-gated Ca^{2+} channels, leading to the increase in intracellular Ca^{2+} that signals contraction (see Figure 12.28).

This acetylcholine receptor, initially isolated from the electric organ of *Torpedo* rays in the 1970s, is the prototype of ligand-gated channels. The receptor consists of five subunits arranged as a cylinder in the membrane (Figure 13.23). In its closed state, the channel pore is thought to be blocked by the side chains of hydrophobic amino acids. The binding of acetylcholine induces a conformational change in the receptor such that these hydrophobic side chains shift out of the channel, opening a pore that allows the passage of positively charged ions, including Na^+ and K^+. However, the channel remains impermeable to negatively charged ions, such as Cl^-, because it is lined by negatively charged amino acids.

A greater degree of ion selectivity is displayed by the voltage-gated Na^+ and K^+ channels. Na^+ channels are more than ten times more permeable to Na^+ than to K^+, whereas K^+ channels are more than a thousand times more permeable to K^+ than to Na^+. The selectivity of the Na^+ channel can be explained, at least in part, on the basis of a narrow pore that acts as a size filter. The ionic radius of Na^+ (0.95 Å) is smaller than that of K^+ (1.33 Å), and it is thought that the Na^+ channel pore is narrow enough to interfere with the passage of K^+ or larger ions (Figure 13.24).

■ Nicotinic acetylcholine receptors are so named because they respond to nicotine. Nicotine binds to the receptor and keeps the channel in its open state. Molecules that block the activity of nicotinic acetylcholine receptors are potent neurotoxins; these include several snake venom toxins and curare.

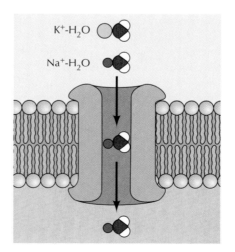
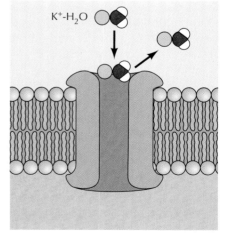

K^+-H_2O
Na^+-H_2O

K^+-H_2O

FIGURE 13.24 Ion selectivity of Na^+ channels A narrow pore permits the passage of Na^+ bound to a single water molecule but interferes with the passage of K^+ or larger ions.

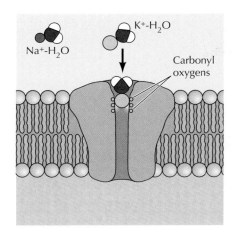 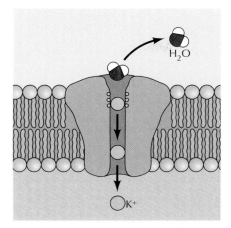 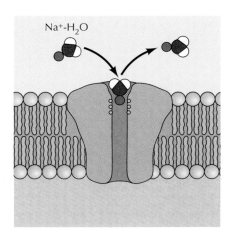

FIGURE 13.25 Selectivity of K⁺ channels The K⁺ channel contains a narrow selectivity filter lined with carbonyl oxygen (C=O) atoms. The pore is just wide enough to allow the passage of dehydrated K⁺ from which all associated water molecules have been displaced as a result of interactions between K⁺ and these carbonyl oxygens. Na⁺ is too small to interact with the carbonyl oxygens of the selectivity filter, so it remains bound to water in a complex that is too large to pass through the channel pore.

K⁺ channels also have narrow pores, which prevent the passage of larger ions. However, since Na⁺ has a smaller ionic radius, this does not account for the selective permeability of these channels to K⁺. Selectivity of the K⁺ channel is based on a different mechanism, which was elucidated with the determination of the three-dimensional structure of a K⁺ channel by X-ray crystallography in 1998 (Figure 13.25). The channel pore contains a narrow selectivity filter that is lined with carbonyl oxygen (C=O) atoms from the polypeptide backbone. When a K⁺ ion enters the selectivity filter, interactions with these carbonyl oxygens displace the water molecules to which K⁺ is bound, allowing dehydrated K⁺ to pass through the pore. In contrast, a dehydrated Na⁺ is too small to interact with these carbonyl oxygens in the selectivity filter, which is held rigidly open. Consequently, Na⁺ remains bound to water molecules in a hydrated complex that is too large to pass through the channel.

Voltage-gated Na⁺, K⁺, and Ca²⁺ channels all belong to a large family of related proteins (Figure 13.26). For example, the genome sequence of *C. elegans* has revealed nearly 200 genes encoding ion channels, which presumably are needed to play diverse roles in cell signaling. K⁺ channels consist of four identical subunits, each containing several transmembrane α helices. Na⁺ and Ca²⁺ channels consist of a single polypeptide chain, but each polypeptide contains four repeated domains that correspond to the K⁺ channel subunits. Voltage gating is mediated by α helix number 4, which contains multiple positively charged amino acids and is thought to move in response to changes in the membrane potential. However, the mechanism by which the movement of these charges opens and closes the channel remains controversial.

A wide variety of ion channels (including Ca²⁺ and Cl⁻ channels) respond to different neurotransmitters or open and close with different kinetics following membrane depolarization. The concerted actions of these multiple channels are responsible for the complexities of signaling in the nervous system. Moreover, as discussed in the next chapter, the roles of ion channels are not restricted to the electrically excitable cells of nerve and muscle; they also play critical roles in signaling in other cell types. The regulated opening and closing of ion channels thus provides cells with a sensitive and versatile mechanism for responding to a variety of environmental stimuli.

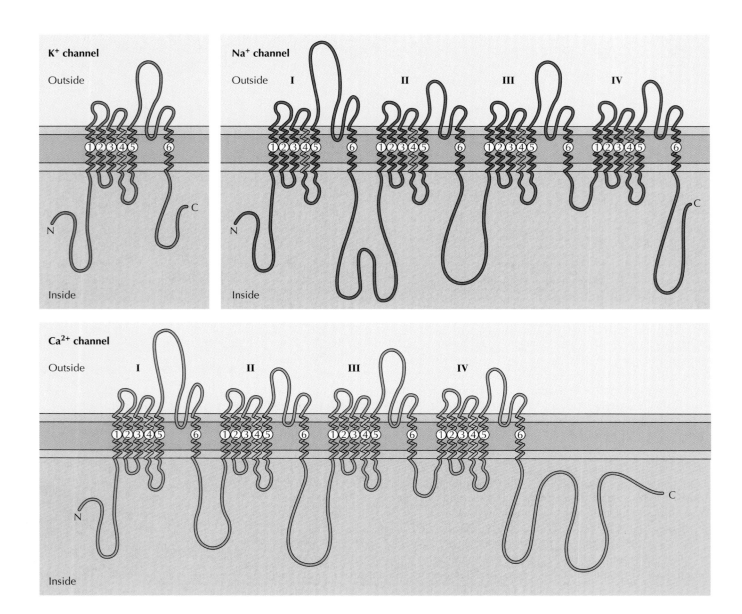

FIGURE 13.26 Structures of voltage-gated cation channels The K⁺, Na⁺, and Ca²⁺ channels belong to a family of related proteins. The K⁺ channel is formed from the association of four identical subunits, one of which is shown. The Na⁺ channel consists of a single polypeptide chain containing four repeated domains, each of which is similar to one K⁺ channel subunit. The Ca²⁺ channel is similar to the Na⁺ channel. Each subunit or domain contains six α helices. The α helix designated 4 contains multiple positively charged amino acids and acts as the voltage sensor that mediates channel opening in response to changes in membrane potential.

Active Transport Driven by ATP Hydrolysis

The net flow of molecules by facilitated diffusion, through either carrier proteins or channel proteins, is always energetically downhill in the direction determined by electrochemical gradients across the membrane. In many cases, however, the cell must transport molecules against their concentration gradients. In **active transport**, energy provided by another coupled reaction (such as the hydrolysis of ATP) is used to drive the uphill transport of molecules in the energetically unfavorable direction.

The **ion pumps** responsible for maintaining gradients of ions across the plasma membrane provide important examples of active transport driven directly by ATP hydrolysis. As discussed earlier (see Table 13.1), the concentration of Na⁺ is approximately ten times higher outside than inside of cells, whereas the concentration of K⁺ is higher inside than out. These ion gradients are maintained by the **Na⁺-K⁺ pump** (also called the **Na⁺-K⁺ ATPase**)

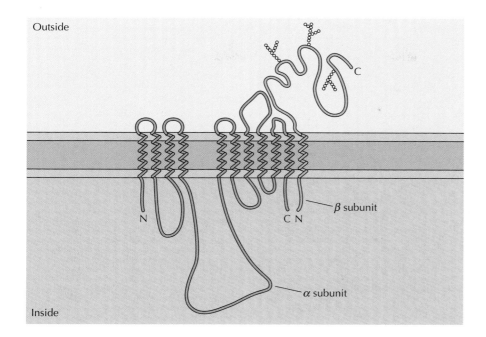

Outside

C

β subunit

N   C N

α subunit

Inside

FIGURE 13.27 Structure of the Na⁺-K⁺ pump The Na^+-K^+ pump is a heterodimer with a 10 transmembrane domain α subunit and a single transmembrane domain β subunit. The β subunit is heavily glycosylated.

(Figure 13.27), which uses energy derived from ATP hydrolysis to transport Na^+ and K^+ against their electrochemical gradients. This process is a result of ATP-driven conformational changes in the pump (Figure 13.28). First, Na^+ ions bind to high-affinity sites inside the cell. This binding stimulates the hydrolysis of ATP and phosphorylation of the pump, inducing a conformational change that exposes the Na^+-binding sites to the outside of the cell and reduces their affinity for Na^+. Consequently, the bound Na^+ is released into the extracellular fluids. At the same time, high-affinity K^+-binding sites are exposed on the cell surface. The binding of extracellular K^+ to these sites then stimulates hydrolysis of the phosphate group bound to the pump, which induces a second conformational change, exposing the K^+-binding sites to the cytosol and lowering their binding affinity so that K^+ is released inside the cell. The pump has three binding sites for Na^+ and two for K^+, so each cycle transports three Na^+ and two K^+ ions across the plasma membrane at the expense of one molecule of ATP.

The importance of the Na^+-K^+ pump is indicated by the fact that it is estimated to consume nearly 25% of the ATP utilized by many animal cells. One critical role of the Na^+ and K^+ gradients established by the pump is the propagation of electric signals in nerve and muscle. As will be discussed shortly, the Na^+ gradient established by the pump is also utilized to drive the active transport of a variety of other molecules. Yet another important role of the Na^+-K^+ pump in most animal cells is to maintain osmotic balance and cell volume. The cytoplasm contains a high concentration of organic molecules, including macromolecules, amino acids, sugars, and nucleotides. In the absence of a counterbalance, this would drive the inward flow of water by osmosis, which if unchecked would result in swelling and eventual bursting of the cell. The required counterbalance is provided by the ion gradients established by the Na^+-K^+ pump (Figure 13.29). In particular, the pump establishes a higher concentration of Na^+ outside than inside the cell. As already discussed, the flow of K^+ through open channels further establishes an electric potential across the plasma

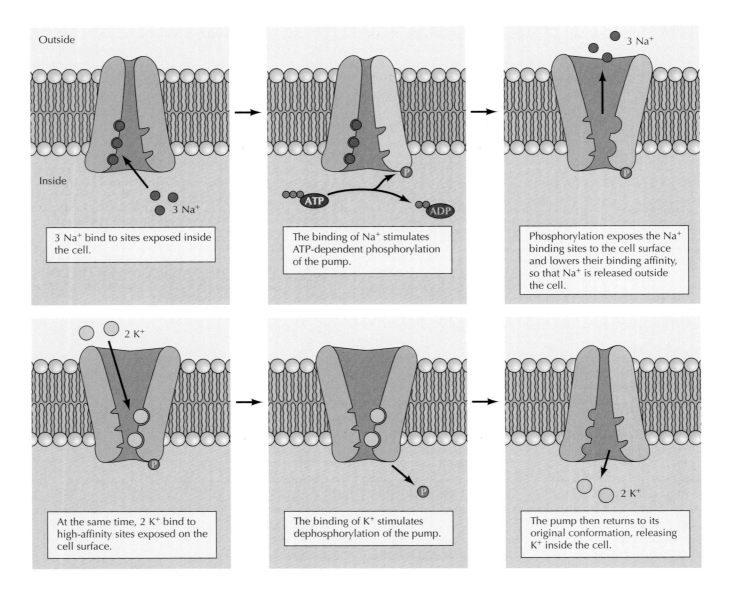

FIGURE 13.28 Model for operation of the Na⁺-K⁺ pump

Outside

Inside

3 Na⁺ bind to sites exposed inside the cell.

The binding of Na⁺ stimulates ATP-dependent phosphorylation of the pump.

ATP

ADP

Phosphorylation exposes the Na⁺ binding sites to the cell surface and lowers their binding affinity, so that Na⁺ is released outside the cell.

3 Na⁺

2 K⁺

At the same time, 2 K⁺ bind to high-affinity sites exposed on the cell surface.

The binding of K⁺ stimulates dephosphorylation of the pump.

The pump then returns to its original conformation, releasing K⁺ inside the cell.

2 K⁺

13.2 WEBSITE ANIMATION

The Sodium-Potassium Pump The sodium and potassium ion gradients across the plasma membrane are maintained by the Na⁺-K⁺ pump, which hydrolyzes ATP to fuel the transport of these ions against their electrochemical gradients.

membrane. This membrane potential in turn drives Cl^- out of the cell, so the concentration of Cl^- (like that of Na^+) is about ten times higher in extracellular fluids than in the cytoplasm. These differences in ion concentrations balance the high concentrations of organic molecules inside cells, equalizing the osmotic pressure and preventing the net influx of water.

The active transport of Ca^{2+} across the plasma membrane is driven by a Ca^{2+} pump that is structurally related to the Na^+-K^+ pump and is similarly powered by ATP hydrolysis. The Ca^{2+} pump transports Ca^{2+} out of the cell, so intracellular Ca^{2+} concentrations are extremely low: approximately 0.1 μM in comparison to extracellular concentrations of about 1 mM. This low intracellular concentration of Ca^{2+} makes the cell sensitive to small increases in intracellular Ca^{2+} levels. Such transient and highly localized increases in intracellular Ca^{2+} play important roles in cell signaling as noted already with respect to muscle contraction (see Figure 12.28) and discussed further in Chapter 15.

Similar ion pumps in the plasma membranes of bacteria, yeasts, and plant cells are responsible for the active transport of H^+ out of the cell. In addition, H^+ is actively pumped out of cells lining the stomach, resulting in the acidity of gastric fluids. Structurally distinct pumps are responsible for the active transport of H^+ into lysosomes and endosomes (see Figure 10. 41). Yet a third type of H^+ pump is exemplified by the ATP synthases of mitochondria and chloroplasts: In these cases the pumps can be viewed as operating in reverse, with the movement of ions down the electrochemical gradient being used to drive ATP synthesis.

The largest family of membrane transporters consists of the **ABC transporters**, so called because they are characterized by highly conserved ATP-binding domains or *ATP-binding cassettes* (Figure 13.30). More than a hundred family members have been identified in both prokaryotic and eukaryotic cells and all use energy derived from ATP hydrolysis to transport molecules in one direction. In bacteria, most ABC transporters bring a wide range of nutrients, including ions, sugars, and amino acids into the cell, while in eukaryotic cells they transport toxic substances out of the cell. The first eukaryotic ABC transporter was discovered as the product of a gene (called the multidrug resistance, or *mdr*, gene) that makes cancer cells resistant to a variety of drugs used in chemotherapy. Two MDR transporters have now been identified. They are normally expressed in a variety of cells, where they function to remove potentially toxic foreign compounds. For example, expression of an MDR transporter in capillary endothelial cells of the brain appears to play an important role in protecting the brain from toxic chemicals. Unfortunately, the MDR transporters are frequently expressed at high levels in cancer cells where they recognize a variety of drugs and pump them out of cells, thereby making the cancer cells resistant to a broad spectrum of chemotherapeutic agents and posing a major obstacle to successful cancer treatment.

Another medically important member of the ABC transporter family is the gene responsible for cystic fibrosis. Although it is a member of the ABC family, the product of this gene (called the cystic fibrosis transmembrane conductance regulator, or CFTR) functions as a Cl^- channel in epithelial cells, and defective Cl^- transport is characteristic of the disease. The CFTR Cl^- channel is also unusual in that it appears to require both ATP hydrolysis

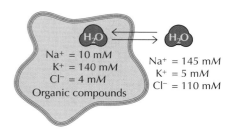

FIGURE 13.29 Ion gradients across the plasma membrane of a typical mammalian cell   The concentrations of Na^+ and Cl^- are higher outside than inside the cell, whereas the concentration of K^+ is higher inside than out. The low concentrations of Na^+ and Cl^- balance the high intracellular concentration of organic compounds, equalizing the osmotic pressure and preventing the net influx of water.

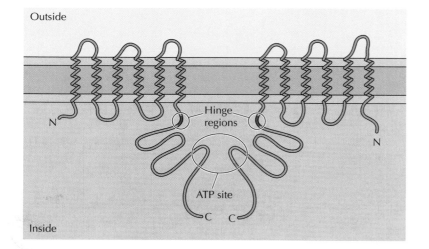

FIGURE 13.30 Structure of an ABC transporter   Many ABC transporters are dimers, with each polypeptide consisting of six transmembrane domains connected via a hinge region to an ATP-binding cassette.

MOLECULAR MEDICINE

Cystic Fibrosis

The Disease

Cystic fibrosis is a recessive genetic disease affecting children and young adults. It is the most common lethal inherited disease of Caucasians, with approximately one in 2500 newborns affected, although it is rare in other races. The characteristic dysfunction in cystic fibrosis is the production of abnormally thick sticky mucus by several types of epithelial cells, including the cells lining the respiratory and gastrointestinal tracts. The primary clinical manifestation is respiratory disease resulting from obstruction of the pulmonary airways by thick plugs of mucus, followed by the development of recurrent bacterial infections. In most patients, the pancreas is also involved because the pancreatic ducts are blocked by mucus. Sweat glands also function abnormally, and the presence of excessive salt in sweat is diagnostic of cystic fibrosis.

Current management of the disease includes physical therapy to promote bronchial drainage, antibiotic administration, and pancreatic enzyme replacement. Although such treatment has extended the survival of affected individuals to about 30 years of age, cystic fibrosis is ultimately fatal, with lung disease being responsible for 95% of mortality.

Molecular and Cellular Basis

The hallmark of cystic fibrosis is defective Cl⁻ transport in affected epithelia, including sweat ducts and the cells lining the respiratory tract. In 1984 it was demonstrated that Cl⁻ channels fail to function normally in epithelial cells from cystic fibrosis patients. The molecular basis of the disease was then elucidated in 1989 with the isolation of the cystic fibrosis gene as a molecular clone. The sequence of the gene revealed that it encodes a protein (called CFTR for *cystic fibrosis transmembrane con-*

ductance regulator) belonging to the ABC transporter family. A variety of subsequent studies then demonstrated that CFTR functions as a Cl⁻ channel and that the inherited mutations responsible for cystic fibrosis result directly in defective Cl⁻ transport. More than 70% of these are a single point mutation at phenylalanine 508 that disrupts folding or assembly of the protein.

Prevention and Treatment

As with other inherited diseases, isolation of the cystic fibrosis gene opens the possibility of genetic screening to identify individuals carrying mutant alleles. In some populations, the frequency of heterozygote carriers of mutant genes is as high as one in 25 individuals, suggesting the possibility of general population screening to identify couples at risk and provide genetic counseling. In addition, understanding the function of CFTR as a Cl⁻ channel has suggested new approaches to treatment. One possibility is the use of drugs that stimulate the opening of other Cl⁻ channels in affected epithelia. Alternatively, gene therapy provides the potential of replacing normal CFTR genes in the respiratory epithelium of cystic fibrosis patients.

The potential of gene therapy has been supported by experiments demonstrating that introduction of a normal CFTR gene into cultured cells of cystic fibrosis patients is sufficient to restore Cl⁻ channel function. The possible application of gene therapy to cystic fibrosis is also enhanced by the accessibility of the epithelial cells lining the airway to aerosol delivery systems. Studies with experimental animals have demonstrated that viral vectors can transmit CFTR cDNA to the respiratory epithelium, and the first human trial of gene therapy for cystic fibrosis was initiated in 1993. Trials to date have demonstrated that the CFTR can be safely delivered to

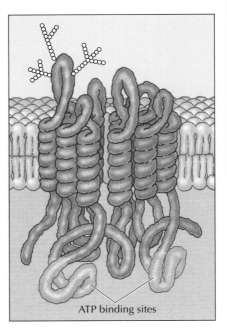

Model of the cystic fibrosis transmembrane conductance regulator (CFTR).

and expressed in the bronchial epithelial cells of cystic fibrosis patients. So far, however, the efficiency of gene transfer has been low and expression of the transferred CFTR cDNA has persisted for less than a month. Thus the principle of successful gene transfer has been established, but significant obstacles in developing an effective gene therapy protocol remain to be overcome.

References

Collins, F. S. 1992. Cystic fibrosis: Molecular biology and therapeutic implications. *Science* 256: 774–779.

Driskell, R. R. and J. F. Engelhardt. 2003. Current status of gene therapy for inherited lung disease. *Ann. Rev. Physiol.* 65: 585–612.

Riordan, J. R. 1993. The cystic fibrosis transmembrane conductance regulator. *Ann. Rev. Physiol.* 55: 609–630.

and cAMP-dependent phosphorylation in order to open. Most cases of cystic fibrosis are the result of a single point mutation in the Cl^- conductance channel that interferes with both protein folding and Cl^- transport.

Active Transport Driven by Ion Gradients

The ion pumps and ABC transporters discussed in the previous section utilize energy derived directly from ATP hydrolysis to transport molecules against their electrochemical gradients. Other molecules are transported against their concentration gradients using energy derived not from ATP hydrolysis but from the coupled transport of a second molecule in the energetically favorable direction. The Na^+ gradient established by the Na^+-K^+ pump provides a source of energy that is frequently used to power the active transport of sugars, amino acids, and ions in mammalian cells. The H^+ gradients established by the H^+ pumps of bacteria, yeast, and plant cells play similar roles.

The epithelial cells lining the intestine provide a good example of active transport driven by the Na^+ gradient. These cells use active-transport systems in the apical domains of their plasma membranes to take up dietary sugars and amino acids from the lumen of the intestine. The uptake of glucose, for example, is carried out by a transporter that coordinately transports two Na^+ ions and one glucose molecule into the cell (Figure 13.31). The flow of Na^+ down its electrochemical gradient provides the energy required to take up dietary glucose and to accumulate high intracellular glucose concentrations. Glucose is then released into the underlying connective tissue (which contains blood capillaries) at the basolateral surface of the intestinal epithelium, where it is transported down its concentration gradient by facilitated diffusion (Figure 13.32). The uptake of glucose from the intestinal lumen and its release into the circulation thus provides a good example of the polarized function of epithelial cells, which results from the specific localization of active transport and facilitated diffusion carriers to the apical and basolateral domains of the plasma membrane, respectively.

Intestinal lumen

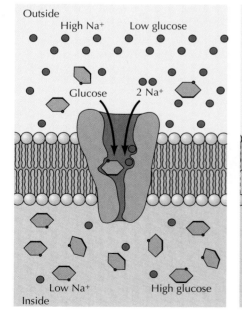 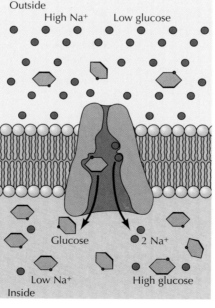

FIGURE 13.31 Active transport of glucose Active transport driven by the Na^+ gradient is responsible for the uptake of glucose from the intestinal lumen. The transporter coordinately binds and transports one glucose molecule and two Na^+ ions into the cell. The transport of Na^+ in the energetically favorable direction drives the uptake of glucose against its concentration gradient.

FIGURE 13.32 Glucose transport by intestinal epithelial cells A transporter in the apical domain of the plasma membrane is responsible for the active uptake of glucose (by cotransport with Na^+) from the intestinal lumen. As a result, dietary glucose is absorbed and concentrated within intestinal epithelial cells. Glucose is then transferred from these cells to the underlying connective tissue and blood supply by facilitated diffusion, mediated by a transporter in the basolateral domain of the plasma membrane. The system is driven by the Na^+-K^+ pump, which is also found in the basolateral domain. Note that the uptake of glucose from the digestive tract and its transfer to the circulation is dependent on the restricted localization of glucose transporters mediating active transport and facilitated diffusion to the apical and basolateral domains of the plasma membrane, respectively.

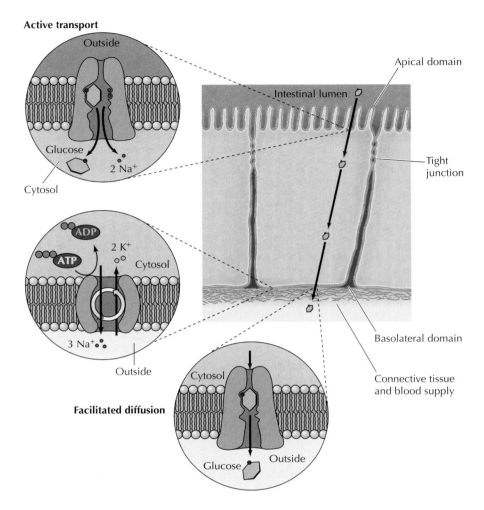

The coordinated uptake of glucose and Na^+ is an example of **symport**, the transport of two molecules in the same direction. In contrast, the facilitated diffusion of glucose is an example of **uniport**, the transport of only a single molecule. Active transport can also take place by **antiport** in which two molecules are transported in opposite directions (Figure 13.33). For example, Ca^{2+} is exported from cells not only by the Ca^{2+} pump but also by a Na^+-Ca^{2+} antiporter that transports Na^+ into the cell and Ca^{2+} out. Another example is provided by the Na^+-H^+ exchange protein, which functions in the regulation of intracellular pH. The Na^+-H^+ antiporter couples the transport of Na^+ into the cell with the export of H^+, thereby removing excess H^+ produced by metabolic reactions and preventing acidification of the cytoplasm.

Endocytosis

The carrier and channel proteins discussed in the preceding section transport small molecules through the phospholipid bilayer. Eukaryotic cells are also able to take up macromolecules and particles from the surrounding medium by a distinct process called **endocytosis**. In endocytosis the mate-

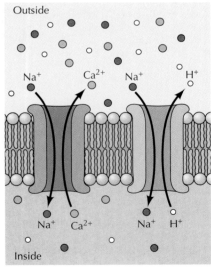

FIGURE 13.33 Examples of antiport Ca^{2+} and H^+ are exported from cells by antiporters, which couple their export to the energetically favorable import of Na^+.

FIGURE 13.34 Phagocytosis   Binding of a bacterium to the cell surface stimulates the extension of a pseudopodium, which eventually engulfs the bacterium. Fusion of the pseudopodium membranes then results in formation of a large intracellular vesicle (a phagosome). The phagosome fuses with lysosomes to form a phagolysosome within which the ingested bacterium is digested.

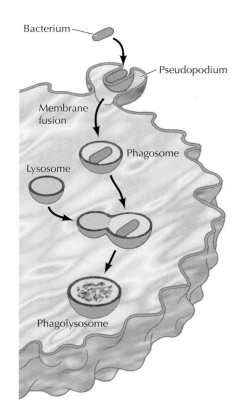

rial to be internalized is surrounded by an area of plasma membrane, which then buds off inside the cell to form a vesicle containing the ingested material. The term "endocytosis" was coined by Christian deDuve in 1963 to include both the ingestion of large particles (such as bacteria) and the uptake of fluids or macromolecules in small vesicles. The former of these activities is known as **phagocytosis** (cell eating) and occurs largely in specialized types of cells. **Pinocytosis** (cell drinking) is a property of all eukaryotic cells and takes place by several different mechanisms.

Phagocytosis

During phagocytosis cells engulf large particles such as bacteria, cell debris, or even intact cells (Figure 13.34). Binding of the particle to receptors on the surface of the phagocytic cell triggers the extension of pseudopodia—an actin-based movement of the cell surface. The pseudopodia eventually surround the particle and their membranes fuse to form a large intracellular vesicle (>0.25 μm in diameter) called a **phagosome**. The phagosomes then fuse with lysosomes, producing **phagolysosomes** in which the ingested material is digested by the action of lysosomal acid hydrolases (see Chapter 10). During maturation of the phagolysosome, some of the internalized membrane proteins are recycled to the plasma membrane as discussed in the next section for receptor-mediated endocytosis.

The ingestion of large particles by phagocytosis plays distinct roles in different kinds of cells (Figure 13.35). Many amoebas use phagocytosis to capture food particles, such as bacteria or other protozoans. In multicellular animals, the major roles of phagocytosis are to provide a defense against

13.3 WEBSITE ANIMATION

Endocytosis In endocytosis, the plasma membrane pinches off small vesicles (pinocytosis) or large vesicles (phagocytosis) to take in extracellular materials.

(A)

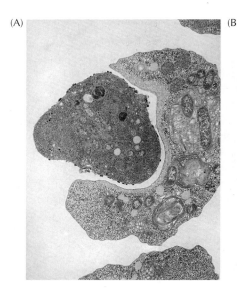

(B)

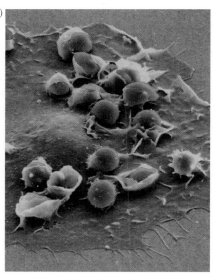

FIGURE 13.35 Examples of phagocytic cells   (A) An amoeba engulfing another protist. (B) Macrophages ingesting red blood cells. False color has been added to the micrograph. (A, R. N. Band and H. S. Pankratz/Biological Photo Service; B, courtesy of Joel Swanson.)

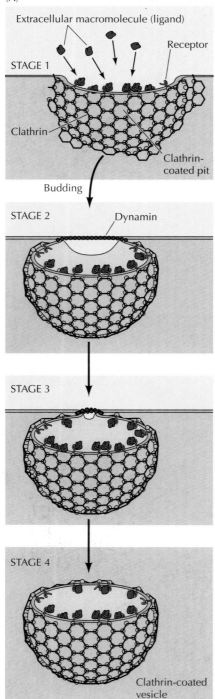

(A)

Extracellular macromolecule (ligand)

STAGE 1

Receptor

Clathrin

Clathrin-coated pit

Budding

STAGE 2

Dynamin

STAGE 3

STAGE 4

Clathrin-coated vesicle

13.4 WEBSITE ANIMATION

Clathrin-Coated Pits and Vesicles Receptor molecules in the plasma membrane become bound to specific macromolecules from outside the cell, coalesce into clathrin-coated pits, and then pinch off in the form of clathrin-coated vesicles.

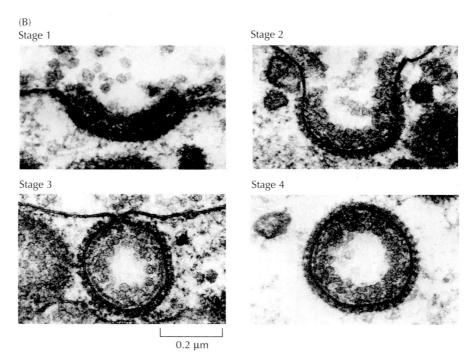

(B)

Stage 1 Stage 2

Stage 3 Stage 4

0.2 μm

FIGURE 13.36 Clathrin-coated vesicle formation (A) Extracellular macromolecules (ligands) bind to cell surface receptors that are concentrated in clathrin-coated pits. With the assistance of the GTP-binding protein dynamin, these pits bud from the plasma membrane to form intracellular clathrin-coated vesicles. (B) Electron micrographs showing four stages in the formation of a clathrin-coated vesicle from a clathrin-coated pit. (B, M. M. Perry, 1979. *J. Cell Science* 34: 266.)

invading microorganisms and to eliminate aged or damaged cells from the body. In mammals, phagocytosis is the function of primarily two types of white blood cells—macrophages and neutrophils—which are frequently referred to as "professional phagocytes." Both macrophages and neutrophils play critical roles in the body's defense systems by eliminating microorganisms from infected tissues. In addition, macrophages eliminate aged or dead cells from tissues throughout the body. A striking example of the scope of this activity is provided by the macrophages of the human spleen and liver, which are responsible for the disposal of more than 10^{11} aged blood cells on a daily basis.

Receptor-Mediated Endocytosis

The best-characterized form of pinocytosis is **receptor-mediated endocytosis**, which provides a mechanism for the selective uptake of specific macromolecules (Figure 13.36). The macromolecules to be internalized first bind to specific cell surface receptors. Most of these receptors are concentrated in specialized regions of the plasma membrane called **clathrin-coated pits**. With assistance from the membrane-associated GTP-binding protein, **dynamin**, these pits bud from the membrane to form small **clathrin-coated vesicles** containing the receptors and their bound macromolecules (**ligands**). The clathrin-coated vesicles then fuse with early endosomes in which their contents are sorted for transport to lysosomes or recycling to the plasma membrane.

FIGURE 13.37 Structure of LDL   Each particle of LDL contains approximately 1500 molecules of cholesteryl esters in an oily core. The core is surrounded by a coat containing 500 molecules of cholesterol, 800 molecules of phospholipid, and one molecule of apoprotein B100.

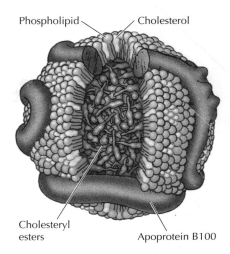

Phospholipid   Cholesterol

Cholesteryl esters

Apoprotein B100

The uptake of cholesterol by mammalian cells has provided a key model for understanding receptor-mediated endocytosis at the molecular level. Cholesterol is transported through the bloodstream in the form of lipoprotein particles, the most common of which is called **low-density lipoprotein**, or **LDL** (Figure 13.37). Studies in the laboratories of Michael Brown and Joseph Goldstein demonstrated that the uptake of LDL by mammalian cells requires the binding of LDL to a specific cell surface receptor that is concentrated in clathrin-coated pits and internalized by endocytosis. As discussed in the next section, the receptor is then recycled to the plasma membrane while LDL is transported to lysosomes where cholesterol is released for use by the cell.

KEY EXPERIMENT

The LDL Receptor

Familial Hypercholesterolemia: Defective Binding of Lipoproteins to Cultured Fibroblasts Associated with Impaired Regulation of 3-Hydroxy-3-Methylglutaryl Coenzyme A Reductase Activity

Michael S. Brown and Joseph L. Goldstein

University of Texas Southwestern Medical School, Dallas

Proceedings of the National Academy of Science USA, 1974, Volume 71, pages 788–792.

Michael S. Brown

Joseph L. Goldstein

The Context

Familial hypercholesterolemia (FH) is a genetic disease in which patients have greatly elevated levels of serum cholesterol and suffer from heart attacks early in life. Michael Brown and Joseph Goldstein began their efforts to understand this disease in 1972 with the idea that cholesterol overproduction results from a defect in the control mechanisms that normally regulate cholesterol biosynthesis. Consistent with this hypothesis, they found that addition of LDL to the culture medium of normal human fibroblasts inhibits the activity of 3-hydroxy-3-methylglutaryl coenzyme A reductase (HMG-CoA reductase), the rate-limiting enzyme in the cho-

lesterol biosynthetic pathway. In contrast, HMG-CoA reductase activity is unaffected by addition of LDL to the cells of FH patients, resulting in overproduction of cholesterol by FH cells.

Perhaps surprisingly, subsequent experiments indicated that this abnormality in HMG-CoA reductase regulation is not the result of a mutation in the HMG-CoA reductase gene. Instead, the abnormal regulation of HMG-CoA reductase appeared to be due to the inability of FH cells to extract cholesterol from LDL. In 1974 Brown and Goldstein demonstrated that the lesion in FH cells is a defect in LDL binding to a receptor on the cell surface. This identification of the LDL receptor led to a series of ground-

breaking experiments in which Brown, Goldstein, and their colleagues delineated the pathway of receptor-mediated endocytosis.

The Experiments

In their 1974 paper, Brown and Goldstein reported the results of experi-

Continued on next page

KEY EXPERIMENT

ments in which they investigated the binding of radiolabeled LDL to fibroblasts from either normal individuals or FH patients. Small amounts of radioactive LDL were added to the culture media, and the amount of radioactivity bound to the cells was determined after varying times of incubation (see figure). Increasing amounts of radioactive LDL bound to normal cells as a function of incubation time. Importantly, addition of excess unlabeled LDL reduced the binding of radioactive LDL, indicating that binding was due to a specific interaction of LDL with a limited number of sites on the cell surface. The specificity of the interaction was further demonstrated by the failure of excess amounts of other lipoproteins to interfere with LDL binding.

In contrast to these results with normal fibroblasts, the cells of FH patients failed to specifically bind radioactive LDL. It therefore appeared that normal fibroblasts possessed a specific LDL receptor that was absent or defective in FH cells. Brown and Goldstein concluded that the defect in LDL binding observed in FH cells "may represent the primary genetic lesion in this disorder," accounting for the inability of LDL to inhibit HMG-CoA reductase and the resultant overproduction of cholesterol. Additional experiments showed that the LDL bound to normal fibroblasts is associated with the mem-

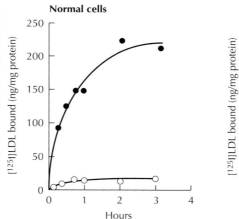
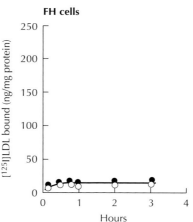

Time course of the binding of radioactive LDL to normal and FH cells. Cells were incubated with radioactive [^{125}I]LDL in the presence (open symbols) or absence (closed symbols) of excess unlabeled LDL. Cells were then harvested, and the amount of radioactive LDL bound was determined. Data are presented as nanograms of LDL bound per milligram of cell protein.

brane fraction of the cell, suggesting that the LDL receptor is a cell surface protein.

The Impact
Following this identification of the LDL receptor, Brown and Goldstein demonstrated that LDL bound to the cell surface is rapidly internalized and degraded to free cholesterol in lysosomes. In collaboration with Richard Anderson, they further established that the LDL receptor is internalized by endocytosis from coated pits. In

addition, their early studies demonstrated that the LDL receptor is recycled to the plasma membrane after dissociation of its ligand inside the cell. Experiments that had been initiated with the goal of understanding the regulation of cholesterol biosynthesis thus led to the elucidation of a major pathway by which eukaryotic cells internalize specific macromolecules—a striking example of the way in which science and scientists can proceed in unanticipated but exciting new directions.

■ **Statins are an important class of drugs, used in the treatment of hypercholesterolemia. They exert their effect by inhibiting the enzyme HMG-CoA reductase and blocking cholesterol biosynthesis.**

The key insights into this process came from studies of patients with the inherited disease known as familial hypercholesterolemia. Patients with this disease have very high levels of serum cholesterol and suffer heart attacks early in life. Brown and Goldstein found that cells of these patients are unable to internalize LDL from extracellular fluids, resulting in the accumulation of high levels of cholesterol in the circulation. Further experiments demonstrated that cells of normal individuals possess a receptor for LDL, which is concentrated in clathrin-coated pits, and that familial hypercholesterolemia results from inherited mutations in the LDL receptor. These mutations are of two types. Cells from most patients with familial hypercholesterolemia simply fail to bind LDL, demonstrating that a specific cell surface receptor was required for LDL uptake. In addition, a few patients

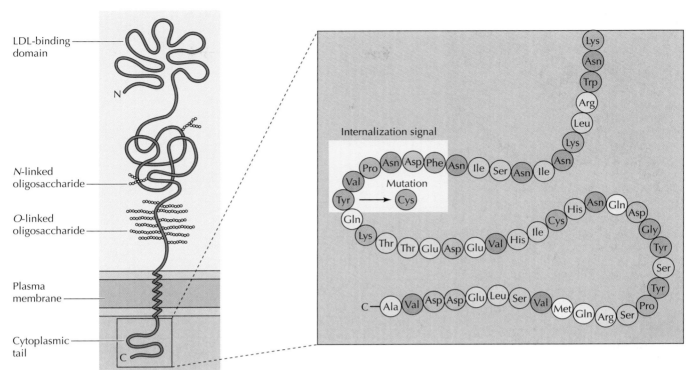

FIGURE 13.38 The LDL receptor
The LDL receptor includes 700 extra-cellular amino acids, a transmembrane α helix of 22 amino acids, and a cytoplasmic tail of 50 amino acids. The N-terminal 292 amino acids constitute the LDL-binding domain. Six amino acids within the cytoplasmic tail define the internalization signal, first recognized because the mutation of Tyr to Cys in a case of familial hypercholesterolemia prevented concentration of the receptor in coated pits.

were identified whose cells bound LDL but were unable to internalize it. The LDL receptors of these patients failed to concentrate in coated pits, providing direct evidence for the central role of coated pits in receptor-mediated endocytosis.

The mutations that prevent the LDL receptor from concentrating in coated pits lie within the cytoplasmic tail of the receptor and can be as subtle as the change of a single tyrosine to cysteine (Figure 13.38). Further studies have defined the internalization signal of the LDL receptor as a sequence of six amino acids, including the essential tyrosine. Similar internalization signals, frequently including tyrosine residues, are found in the cytoplasmic tails of other receptors taken up via clathrin-coated pits. These internalization signals bind to adaptor proteins, which in turn bind clathrin on the cytosolic side of the membrane, similar to the way in which clathrin-coated vesicles form during the transport of lysosomal hydrolases from the *trans* Golgi network (see Figure 10.35). Clathrin assembles into a basketlike structure that distorts the membrane, forming invaginated pits (Figure 13.39). The GTP-binding protein dynamin assembles into rings around the necks of these invaginated pits, eventually leading to the release of coated vesicles inside the cell.

Receptor-mediated endocytosis is a major activity of the plasma membranes of eukaryotic cells. More than 20 different receptors have been shown to be selectively internalized by this pathway. Extracellular fluids are also incorporated into coated vesicles as they bud from the plasma membrane, so receptor-mediated endocytosis results in the nonselective uptake of extracellular fluids and their contents (**fluid phase endocytosis**) in addition to the internalization of specific macromolecules. Clathrin-coated pits typically occupy 1 to 2% of the surface area of the plasma mem-

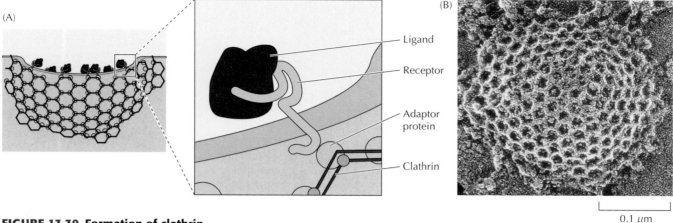

FIGURE 13.39 Formation of clathrin-coated pits (A) Adaptor proteins bind both to clathrin and to the internalization signals present in the cytoplasmic tails of receptors. (B) Electron micrograph of a clathrin-coated pit showing the basketlike structure of the clathrin network. (B, courtesy of John E. Heuser, Washington University School of Medicine.)

brane and are estimated to have a lifetime of 1 to 2 minutes. From these figures, one can calculate that receptor-mediated endocytosis results in the internalization of an area of cell surface equivalent to the entire plasma membrane approximately every 2 hours.

Cells also possess clathrin-independent pathways of endocytosis. One pathway of clathrin-independent endocytosis involves the uptake of molecules in **caveolae**, small invaginations of the plasma membrane (50 to 80 nm in diameter) that are organized by **caveolin** (Figure 13.40). Caveolins are a family of proteins that interact with cholesterol in lipid rafts, insert into the cell membrane, and interact with one another to form the structure of the caveolae. Caveolae are relatively stable structures and the regulation of their internalization is not well-understood. They carry out receptor-mediated endocytosis via specific transmembrane receptors, but the caveolae lipids and caveolin itself also serve as "receptors" for the uptake of specific molecules, including high-density lipoprotein (HDL). Recent studies have demonstrated that there are additional endocytosis pathways that are independent of both caveolin and clathrin. In addition, large vesicles (0.15 to 5.0 μm in diameter) can mediate the uptake of fluids in a process known as **macropinocytosis**. Thus while clathrin-dependent endocytosis clearly pro-

FIGURE 13.40 Caveolae Electron micrographs of caveolae. (A, courtesy of John E. Heuser, Washington University School of Medicine; B, courtesy of R. G. W. Anderson, University of Texas Southwestern Medical School, Dallas.)

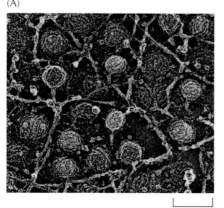

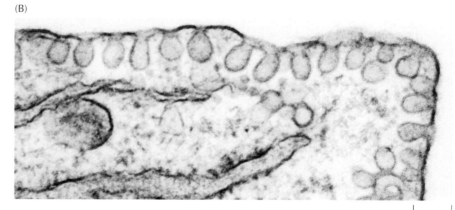

vides a major pathway for the uptake of specific macromolecules, cells also use several clathrin-independent mechanisms.

Protein Trafficking in Endocytosis

Following their internalization, clathrin-coated vesicles rapidly shed their coats and fuse with early **endosomes,** which are vesicles with tubular extensions located at the periphery of the cell. The fusion of endocytic vesicles with endosomes is mediated by Rab GTP-binding proteins, their effectors, and complementary pairs of transmembrane proteins of the vesicle and target membranes (SNARE proteins) (see Figure 10.38). The early endosomes serve as a sorting compartment (they are sometimes called sorting endosomes) from which molecules taken up by endocytosis are either recycled to the plasma membrane or transported to lysosomes for degradation. In addition, the early endosomes of polarized cells can transfer endocytosed proteins between different domains of the plasma membrane—for example, between the apical and basolateral domains of epithelial cells.

An important feature of early endosomes is that they maintain an acidic internal pH (about 6.0 to 6.2) as the result of the action of a membrane H^+ pump. This acidic pH leads to the dissociation of many ligands from their receptors within the early endosome compartment. Following this uncoupling, the receptors and their ligands can be transported to different intracellular destinations. A classic example is provided by LDL, which dissociates from its receptor within early endosomes (Figure 13.41). The receptor is then returned to the plasma membrane via transport vesicles that bud from the tubular extensions of endosomes. In contrast, LDL is transported (along with other soluble contents of the endosome) to lysosomes, where its degradation releases cholesterol.

Recycling to the plasma membrane is the major fate of membrane proteins taken up by receptor-mediated endocytosis, with many receptors (like the LDL receptor) being returned to the plasma membrane following dissociation of their bound ligands in early endosomes. The recycling of these receptors results in the continuous internalization of their ligands. Each LDL receptor, for example, makes a round-trip from the plasma membrane to endosomes and back approximately every 10 minutes. The importance of the recycling pathway is further emphasized by the magnitude of membrane traffic resulting from endocytosis. As already noted, approximately 50% of the plasma membrane is internalized by receptor-mediated endocytosis

FIGURE 13.41 Sorting in early endosomes   LDL bound to its receptor is internalized in clathrin-coated vesicles, which shed their coats and fuse with early endosomes. At the acidic pH of early endosomes, LDL dissociates from its receptor and the endocytosed materials are sorted for degradation in lysosomes or recycling to the plasma membrane. LDL is transported from early to late endosomes in large carrier vesicles that move along microtubules. Transport vesicles carrying lysosomal hydrolases from the Golgi apparatus then fuse with late endosomes, which mature to lysosomes where LDL is degraded and cholesterol is released. In contrast, the LDL receptor is recycled from early endosomes to the plasma membrane.

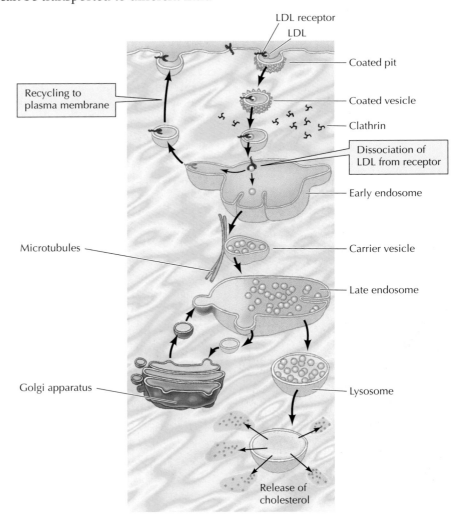

Certain bacterial protein toxins, like cholera toxin, bind to a glyco-lipid found in lipid rafts and are taken up by endocytosis. Once inside endocytic vesicles the toxin commandeers the secretory pathway to gain entry into the cytoplasm, where it exerts its toxic effect.

every hour and must therefore be replaced at an equivalent rate. Most of this replacement is the result of receptor recycling; only about 5% of the cell surface is newly synthesized per hour.

Ligands and membrane proteins destined for degradation in lysosomes are transported from early endosomes to late endosomes, which are located near the nucleus (see Figure 13.41). Transport from early to late endosomes is mediated by the movement of large endocytic carrier vesicles along microtubules. The late endosomes are more acidic than early endosomes (pH about 5.5 to 6.0) and, as discussed in Chapter 10, are able to fuse with transport vesicles carrying lysosomal hydrolases from the Golgi apparatus. Late endosomes then mature into lysosomes as they acquire a full complement of lysosomal enzymes and become still more acidic (pH about 5). Within lysosomes the endocytosed materials are degraded by the action of acid hydrolases.

Although many receptors (like the LDL receptor) are recycled to the plasma membrane, others follow different fates. Some are transported to lysosomes and degraded along with their ligands. For example, the cell surface receptors for several growth factors (discussed in Chapter 15) are internalized following growth factor binding and eventually degraded in lysosomes. The effect of this process is to remove the receptor-ligand complexes from the plasma membrane, thereby terminating the response of the cell to growth factor stimulation—a phenomenon known as **receptor down-regulation**.

A specialized kind of recycling from endosomes plays an important role in the transmission of nerve impulses across synapses (Figure 13.42). As discussed earlier in this chapter, the arrival of an action potential at the terminus of most neurons signals the fusion of synaptic vesicles with the plasma membrane, releasing the neurotransmitters that carry the signal to postsynaptic cells. The empty synaptic vesicles are then recovered from the

FIGURE 13.42 Recycling of synaptic vesicles

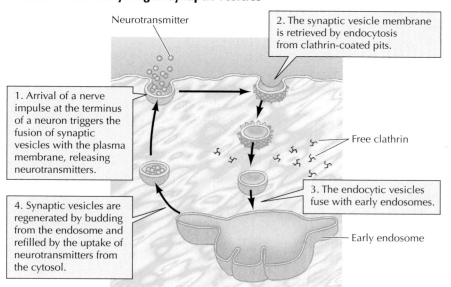

Neurotransmitter

1. Arrival of a nerve impulse at the terminus of a neuron triggers the fusion of synaptic vesicles with the plasma membrane, releasing neurotransmitters.

2. The synaptic vesicle membrane is retrieved by endocytosis from clathrin-coated pits.

Free clathrin

3. The endocytic vesicles fuse with early endosomes.

4. Synaptic vesicles are regenerated by budding from the endosome and refilled by the uptake of neurotransmitters from the cytosol.

Early endosome

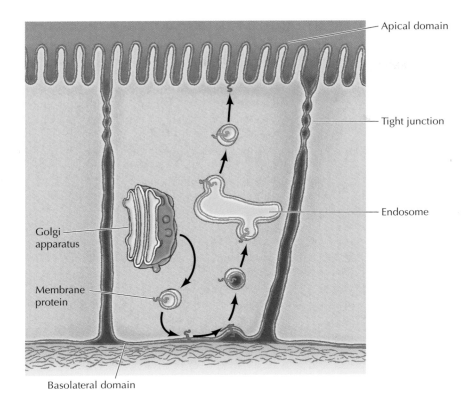

Apical domain

Tight junction

Endosome

Golgi apparatus

Membrane protein

Basolateral domain

FIGURE 13.43 Protein sorting by transcytosis A protein destined for the apical domain of the plasma membrane is first transported from the Golgi apparatus to the basolateral domain. It is then endocytosed and selectively transported to the apical domain from early endosomes.

plasma membrane in clathrin-coated vesicles, which fuse with early endosomes. The synaptic vesicles are then regenerated directly by budding from endosomes. They accumulate a new supply of neurotransmitter and recycle to the plasma membrane, ready for the next cycle of synaptic transmission.

In polarized cells (e.g., epithelial cells) internalized receptors can also be transferred across the cell to the opposite domain of the plasma membrane—a process called **transcytosis**. For example, a receptor endocytosed from the basolateral domain of the plasma membrane can be sorted in early endosomes for transport to the apical membrane. In some cells, this is an important mechanism for sorting membrane proteins (Figure 13.43). Rather than being sorted for delivery to the apical or basolateral domains in the *trans* Golgi network (see Figure 10.31), proteins are initially delivered to the basolateral membrane. Proteins targeted for the apical membrane are then transferred to that site by transcytosis. In addition, transcytosis provides a mechanism for the transfer of extracellular macromolecules across epithelial cell sheets. For example, endothelial cells and many kinds of epithelial cells transport antibodies from the blood to a variety of secreted fluids, such as milk. The antibodies bind to receptors on the basolateral surface and are then transcytosed along with their receptors to the apical surface. The receptors are then cleaved, releasing the antibodies into extracellular secretions.

COMPANION WEBSITE

Visit the website that accompanies **The Cell** (www.sinauer.com/cooper) for animations, videos, quizzes, problems, and other review material.

KEY TERMS	**SUMMARY**

STRUCTURE OF THE PLASMA MEMBRANE

phosphatidylcholine, phosphatidylethanolamine, phosphatidylserine, sphingo-myelin, phosphatidylinositol, glycolipid, cholesterol, lipid raft

The Phospholipid Bilayer: The fundamental structure of the plasma membrane is a phospholipid bilayer, which also contains glycolipids and cholesterol.

fluid mosaic model, peripheral membrane protein, integral membrane protein, transmembrane protein, porin, glycosylphosphatidylinositol (GPI) anchor

Membrane Proteins: Associated proteins are responsible for carrying out specific membrane functions. Membranes are viewed as fluid mosaics in which proteins are inserted into phospholipid bilayers.

apical domain, basolateral domain

Mobility of Membrane Proteins: Proteins are free to diffuse laterally through the phospholipid bilayer. However, the mobility of some proteins is restricted by their associations with other proteins or specific lipids. In addition, tight junctions prevent proteins from moving between distinct plasma membrane domains of epithelial cells.

glycocalyx, selectin

The Glycocalyx: The cell surface is covered by a carbohydrate coat called the glycocalyx. Cell surface carbohydrates serve as markers for cell-cell recognition.

TRANSPORT OF SMALL MOLECULES

passive diffusion

Passive Diffusion: Small hydrophobic molecules are able to cross the plasma membrane by diffusing through the phospholipid bilayer.

facilitated diffusion, carrier protein, channel protein

Facilitated Diffusion and Carrier Proteins: The passage of most biological molecules is mediated by carrier or channel proteins that allow polar and charged molecules to cross the plasma membrane without interacting with its hydrophobic interior.

ion channel, ligand-gated channel, voltage-gated channel, patch clamp technique, Nernst equation, action potential

Ion Channels: Ion channels mediate the rapid passage of selected ions across the plasma membrane. They are particularly well characterized in nerve and muscle cells, where they are responsible for the transmission of electric signals.

active transport, ion pump, Na^+-K^+ pump, Na^+-K^+ ATPase, ABC transporter

Active Transport Driven by ATP Hydrolysis: Energy derived from ATP hydrolysis can drive the transport of molecules against their electrochemical gradients.

symport, uniport, antiport

Active Transport Driven by Ion Gradients: Ion gradients are frequently used as a source of energy to drive the active transport of other molecules.

ENDOCYTOSIS

endocytosis, phagocytosis, pinocytosis, phagosome, phagolysosome

Phagocytosis: Cells ingest large particles, such as bacteria and cell debris, by phagocytosis.

SUMMARY

Receptor-Mediated Endocytosis: The best-characterized form of endocytosis is receptor-mediated endocytosis, which provides a mechanism for the selective uptake of specific macromolecules.

Protein Trafficking in Endocytosis: Molecules taken up by endocytosis are transported to endosomes, where they are sorted for recycling to the plasma membrane or degradation in lysosomes.

KEY TERMS

receptor-mediated endocytosis, clathrin-coated pit, dynamin, clathrin-coated vesicle, ligand, low-density lipoprotein (LDL), fluid phase endocytosis, caveolae, caveolin, macropinocytosis

endosome, receptor down-regulation, transcytosis

Questions

1. How does cholesterol extend the functional temperature range of a lipid bilayer?

2. How are peripheral membrane proteins distinguished from integral membrane proteins?

3. How did the cell fusion experiments of Frye and Edidin support the fluid mosaic model of membrane structure? What results would they have obtained if they had incubated fused cells at 2° C?

4. What are lipid rafts and what are the cellular processes they are involved in?

5. What are the main functions of the glycocalyx?

6. The concentration of K^+ is about 20 times higher inside squid axons than in extracellular fluids, generating an equilibrium membrane potential of –75 mV. What would be the expected equilibrium membrane potential if the K^+ concentration were only 10 times higher inside than outside the cell? Why does the actual resting membrane potential (–60 mV) differ from the K^+ equilibrium potential of –75 mV? (Given: R = 1.98×10^{-3} kcal / mol / deg, T = 298 K, ln(x) = 2.3 $\log_{10}(x)$, F = 23 kcal / V / mol)

7. Curare binds to nicotinic acetylcholine receptors and prevents them from opening. How would it affect the contraction of muscles?

8. How does the selectivity filter of K^+ channels differentiate between K^+ and Na^+ ions?

9. How can glucose be transported against its concentration gradient without the direct expenditure of ATP in intestinal epithelial cells?

10. How does the *mdr* gene confer drug resistance upon cancer cells?

11. How did Brown and Goldstein determine that LDL binds to a limited number of specific binding sites on the surface of normal cells?

12. What have studies on cells from children with familial hypercholesterolemia told us about the mechanisms of receptor-mediated endocytosis?

References and Further Reading

Structure of the Plasma Membrane

Branton, D., C. M. Cohen and J. Tyler. 1981. Interaction of cytoskeletal proteins on the human erythrocyte membrane. *Cell* 24: 24–32. [P]

Diesenhofer, J., O. Epp, K. Miki, R. Huber and H. Michel. 1985. The structure of the protein subunits in the photosynthetic reaction centre of *Rhodopseudomonas viridis* at 3-Å resolution. *Nature* 318: 618–624. [P]

Edidin, M. 2001. Shrinking patches and slippery rafts: Scales of domains in the plasma membrane. *Trends Cell Biol.* 11: 492–496. [R]

Engelman, D. M. 2005. Membranes are more mosaic than fluid. *Nature* 438: 578–580. [R]

Englund, P. T. 1993. The structure and biosynthesis of glycosyl phosphatidylinositol anchors. *Ann. Rev. Biochem.* 62: 121–138. [R]

Ikonen, E. 2001. Roles of lipid rafts in membrane transport. *Curr. Opin. Cell Biol.* 13: 470–477. [R]

Jay, D. G. 1996. Role of band 3 in homeostasis and cell shape. *Cell* 86: 853–854. [R]

Lasky, L. A. 1995. Selectin-carbohydrate interactions and the initiation of the inflammatory response. *Ann. Rev. Biochem.* 64: 113–139. [R]

Lee, A. 2001. Membrane structure. *Curr. Biol.* 11: R811–R814. [R]

London, E. 2002. Insights into lipid raft structure and formation from experiments in model membranes. *Curr. Opin. Struct. Biol.* 12: 480–486. [R]

Montell, C., L. Birnbaumer and V. Flockerzi. 2002. The TRP channels, a remarkably functional family. *Cell* 108: 595–598. [R]

Mukherjee, S. and F. R. Maxfield. 2004. Membrane domains. *Ann. Rev. Cell Dev. Biol.* 20: 839–866. [R]

Rees, D. C., H. Komiya, T. O. Yeates, J. P. Allen and G. Feher. 1989. The bacterial photosynthetic reaction center as a model for membrane proteins. *Ann. Rev. Biochem.* 58: 607–633. [R]

Singer, S. J. 1990. The structure and insertion of integral proteins in membranes. *Ann. Rev. Cell Biol.* 6: 247–296. [R]

Singer, S. J. and G. L. Nicolson. 1972. The fluid mosaic model of the structure of cell membranes. *Science* 175: 720–731. [P]

Tamm, L. K., A. Arora and J. H. Kleinschmidt. 2001. Structure and assembly of beta-barrel membrane proteins. *J. Biol. Chem.* 276: 32399–32402. [R] van Meer, G. and H. Sprong. 2004. Membrane lipids and vesicular traffic. *Curr. Opin. Cell Biol.* 16: 373–378. [R]

Yu, J., D. A. Fischman and T. L. Steck. 1973. Selective solubilization of proteins and phospholipids from red blood cell membranes by nonionic detergents. *J. Supramol. Struct.* 1: 233–248. [P]

Zhang, F. L. and P. J. Casey. 1996. Protein prenylation: Molecular mechanisms and functional consequences. *Ann. Rev. Biochem.* 65: 241–269. [R]

Transport of Small Molecules

Bell, G. I., C. F. Burant, J. Takeda and G. W. Gould. 1993. Structure and function of mammalian facilitative sugar transporters. *J. Biol. Chem.* 268: 19161–19164. [R]

Borgnia, M., S. Nielsen, A. Engel and P. Agre. 1999. Cellular and molecular biology of the aquaporin water channels. *Ann. Rev. Biochem.* 68: 425–458. [R]

Borst, P. and A. H. Schinkel. 1997. Genetic dissection of the function of mammalian P-glycoproteins. *Trends Genet.* 13: 217–222. [R]

Clapham, D. E. 1999. Unlocking family secrets: K^+ channel transmembrane domains. *Cell* 97: 547–550. [R]

Cole, S. P. and R. G. Deeley. 1998. Multidrug resistance mediated by the ATP-binding cassette transporter protein MRP. *Bioessays* 20: 931–940. [R]

Cooper, E. C. and L. Y. Jan. 1999. Ion channel genes and human neurological disease: Recent progress, prospects, and challenges. *Proc. Natl. Acad. Sci. USA* 96: 4759–4766. [R]

Riordan, J. R. 1993. The cystic fibrosis transmembrane conductance regulator. *Ann. Rev. Physiol.* 55: 609–630. [R]

Garcia, M. L. 2004. Ion channels: Gate expectations. *Nature* 430: 153–155. [R]

Gottesman, M. M. and I. M. Pastan. 1993. Biochemistry of multidrug resistance mediated by the multidrug transporter. *Ann. Rev. Biochem.* 62: 385–427. [R]

Hille, B. 2001. *Ionic Channels of Excitable Membranes.* 3rd ed. Sunderland, MA: Sinauer Associates.

Hodgkin, A. L. and A. F. Huxley. 1952. A quantitative description of membrane current and its application to conduction and excitation in nerve. *J. Physiol.* 117: 500–544. [P]

Jiang, Y., V. Ruta, J. Chen, A. Lee and R. MacKinnon. 2003. The principle of gating charge movement in a voltage-dependent K^+ channel. *Nature* 423: 42–48. [P]

Jongsma, H. J. and R. Wilders. 2001. Channelopathies: Kir2.1 mutations jeopardize many cell functions. *Curr. Biol.* 11: R747–R750. [R]

Kaplan, J. H. 2002. Biochemistry of Na, K-ATPase. *Ann. Rev. Biochem.* 71: 511–535. [R]

Lester, H. A., M. I. Dibas, D. S. Dahan, J. F. Leite and D.A Dougherty. 2004. Cys-loop receptors: New twists and turns. *Trends Neurosci.* 27: 329–336. [R]

Locher, K. P. 2004. Structure and mechanism of ABC transporters. *Curr. Opin. Struct. Biol.* 14: 426–431. [R]

MacLennan, D. H., W. J. Rice and N. M. Green. 1997. The mechanism of Ca^{2+} transport by sarco(endo)plasmic reticulum Ca^{2+}-ATPases. *J. Biol. Chem.* 272: 28815–28818. [R]

Neher, E. and B. Sakmann. 1992. The patch clamp technique. *Sci. Am.* 266(3): 44–51. [R]

Sakmann, B. 1992. Elementary steps in synaptic transmission revealed by currents through single ion channels. *Science* 256: 503–512. [R]

Sansom, M. S. and R. J. Law. 2001. Membrane proteins: Aquaporins—Channels without ions. *Curr. Biol.* 11: R71–R73. [R]

Sigworth, F. J. 2003. Structural biology: Life's transistors. *Nature* 423: 21–22. [R]

Swartz, K. J. 2004. Towards a structural view of gating in potassium channels. *Nat. Rev. Neurosci.* 5: 905–916. [R]

Toyoshima, C. and G. Inesi. 2004. Structural basis of ion pumping by Ca^{2+}-ATPase of the sarcoplasmic reticulum. *Ann. Rev. Biochem.* 73: 269–292. [R]

Unwin, N. 1993. Neurotransmitter action: Opening of ligand-gated ion channels. *Cell* 72/*Neuron* 10 (Suppl.): 31–41. [R]

Welsh, M. J. and A. E. Smith. 1993. Molecular mechanisms of CFTR chloride channel dysfunction in cystic fibrosis. *Cell* 73: 1251–1254. [R]

Endocytosis

Brown, M. S. and J. L. Goldstein. 1986. A receptor-mediated pathway for cholesterol homeostasis. *Science* 232: 34–47. [R]

Conner, S. D. and S. L. Schmid. 2003. Regulated portals of entry into the cell. *Nature* 422: 37–44. [R]

Felberbaum-Corti, M., F. G. Van Der Goot and J. Gruenberg. 2003. Sliding doors: Clathrin-coated pits or caveolae? *Nat. Cell Biol.* 5: 382–384. [R]

Gruenberg, J. 2001. The endocytic pathway: A mosaic of domains. *Nat. Rev. Mol. Cell Biol.* 2: 721–730. [R]

Gruenberg, J. and H. Stenmark. 2004. The biogenesis of multivesicular endosomes. *Nat. Rev. Mol. Cell Biol.* 5: 317–323. [R]

Liu, P., M. Rudick and R. G. Anderson. 2002. Multiple functions of caveolin-1. *J. Biol. Chem.* 277: 41295–41298. [R]

Mostov, K., T. Su and M. ter Beest. 2003. Polarized epithelial membrane traffic: Conservation and plasticity. *Nat. Cell Biol.* 5: 287–293. [R]

Nelson, W. J. 1992. Regulation of cell surface polarity from bacteria to mammals. *Science* 258: 948–955. [R]

Orth, J. D. and M.A. McNiven. 2003. Dynamin at the actin-membrane interface. *Curr. Opin. Cell Biol.* 1: 31–39. [R]

Parton, R. G. 2003. Caveolae—From ultrastructure to molecular mechanisms. *Nat. Rev. Mol. Cell Biol.* 4: 162–167. [R]

Pelkmans, L., T. Burli, M. Zerial and A. Helenius. 2004. Caveolin-stabilized membrane domains as multifunctional transport and sorting devices in endocytic membrane traffic. *Cell* 118: 767–780. [R]

Smith, C. J. and B. M. Pearse. 1999. Clathrin: Anatomy of a coat protein. *Trends Cell Biol.* 9: 335–338. [R]

Sudhof, T. C. 1995. The synaptic vesicle cycle: A cascade of protein-protein interactions. *Nature* 375: 645–653. [R]

Takei, K. and V. Haucke. 2001. Clathrin-mediated endocytosis: Membrane factors pull the trigger. *Trends Cell Biol.* 11: 385–391. [R]

Van Deurs, B., K. Roepstorff, A. M. Hommelgaard and K. Sandvig. 2003. Caveolae: Anchored, multifunctional platforms in the lipid ocean. *Trends Cell Biol.* 13: 92–100. [R]

CHAPTER 14

Cell Walls, the Extracellular Matrix, and Cell Interactions

■ *Cell Walls* 569

■ *The Extracellular Matrix and Cell-Matrix Interactions* 575

■ *Cell-Cell Interactions* 584

■ *KEY EXPERIMENT:* The Characterization of Integrins 582

■ *MOLECULAR MEDICINE:* Gap Junction Diseases 591

ALTHOUGH CELL BOUNDARIES ARE DEFINED BY THE PLASMA MEMBRANE, many cells are surrounded by an insoluble array of secreted macromolecules. Cells of bacteria, fungi, algae, and higher plants are surrounded by rigid cell walls, which are an integral part of the cell. Although animal cells are not surrounded by cell walls, most of the cells in animal tissues are embedded in an extracellular matrix consisting of secreted proteins and polysaccharides. The extracellular matrix not only provides structural support to cells and tissues, but also plays important roles in regulating cell behavior. Interactions of cells with the extracellular matrix anchor the cytoskeleton and regulate cell shape and movement. Likewise, direct interactions between cells are key to the organization of cells in the tissues of both plants and animals, as well as providing channels through which cells can communicate with their neighbors.

Cell Walls

The rigid cell walls that surround bacteria and many types of eukaryotic cells (fungi, algae, and higher plants) determine cell shape and prevent cells from swelling and bursting as a result of osmotic pressure. Despite their common functions, the cell walls of bacteria and eukaryotes are structurally very different. Bacterial cell walls consist of polysaccharides cross-linked by short peptides, which form a covalent shell around the entire cell. In contrast, the cell walls of eukaryotes are composed principally of polysaccharides embedded in a gel-like matrix. Rather than being fixed structures, plant cell walls can be modified both during development of the plant and in response to signals from the environment, so the cell walls of plants play critical roles in determining the organization of plant tissues and the structure of entire plants.

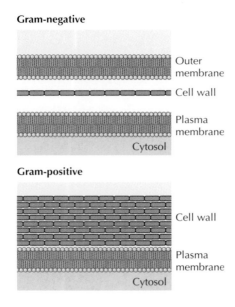

Gram-negative

Outer membrane

Cell wall

Plasma membrane

Cytosol

Gram-positive

Cell wall

Plasma membrane

Cytosol

■ The outer membrane of Gram-negative bacteria contains a lipopolysaccharide (LPS) known as endotoxin. The release of LPS into the bloodstream leads to fever, lowered blood pressure, and inflammation.

FIGURE 14.1 Bacterial cell walls The plasma membrane of Gram-negative bacteria is surrounded by a thin cell wall beneath the outer membrane. Gram-positive bacteria lack outer membranes and have thick cell walls.

Bacterial Cell Walls

The rigid cell walls of bacteria determine the characteristic shapes of different kinds of bacterial cells. For example, some bacteria (such as *E. coli*) are rod-shaped, whereas others are spherical (e.g., *Pneumococcus* and *Staphylococcus*) or spiral-shaped (e.g., the spirochete *Treponema pallidum*, which causes syphilis). In addition, the structure of their cell walls divides bacteria into two broad classes that can be distinguished by a staining procedure known as the Gram stain, developed by Christian Gram in 1884 (Figure 14.1). As described in Chapter 13, Gram-negative bacteria (such as *E. coli*) have a dual membrane system in which the plasma membrane is surrounded by a permeable outer membrane (see Figure 13.8). These bacteria have thin cell walls located between their inner and outer membranes. In contrast, Gram-positive bacteria (such as the common human pathogen *Staphylococcus aureus*) have only a single plasma membrane, which is surrounded by a much thicker cell wall.

Despite these structural differences, the principal component of the cell walls of both Gram-positive and Gram-negative bacteria is a **peptidoglycan** (Figure 14.2), which consists of linear polysaccharide chains cross-linked by short peptides. Because of this cross-linked structure, the peptidoglycan forms a strong covalent shell around the entire bacterial cell. Interestingly, the unique structure of their cell walls also makes bacteria vulnerable to some antibiotics. Penicillin, for example, inhibits the enzyme responsible for forming cross-links between different strands of the peptidoglycan, thereby interfering with cell wall synthesis and blocking bacterial growth.

Eukaryotic Cell Walls

In contrast to bacteria, the cell walls of eukaryotes (including fungi, algae, and higher plants) are composed principally of polysaccharides (Figure 14.3). The basic structural polysaccharide of fungal cell walls is **chitin,** which also forms the shells of crabs and the exoskeletons of insects and other arthropods. Chitin is a linear polymer of *N*-acetylglucosamine residues. The cell walls of most algae and higher plants are composed principally of **cellulose,** which is the most abundant polymer on earth. Cellulose is a linear polymer of glucose residues, often containing more than 10,000 glucose monomers. The sugar residues in both chitin and cellulose are joined by β (1→4) linkages, which allow the polysaccharides to form long straight chains. Following transport across the plasma membrane to the extracellular space, single polysaccharide chains associate in parallel with one another to form microfibrils. Cellulose microfibrils can extend for many micrometers in length.

Within the plant cell wall, cellulose microfibrils are embedded in a matrix consisting of proteins and two other types of polysaccharides: **hemicelluloses** and **pectins** (Figure 14.4). Hemicelluloses are highly branched polysaccharides that are hydrogen-bonded to the surface of cellulose microfibrils (Figure 14.5). This stabilizes the cellulose microfibrils into a tough fiber, which is responsible for the mechanical strength of plant cell walls. The cellulose microfibrils are cross-linked by pectins, which are

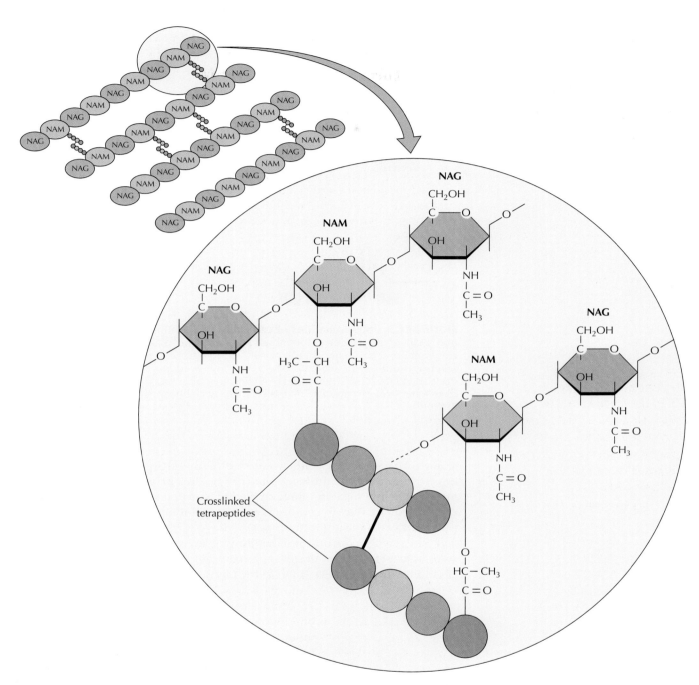

FIGURE 14.2 The peptidoglycan of *E. coli* Polysaccharide chains consist of alternating *N*-acetylglucosamine (NAG) and *N*-acetylmuramic acid (NAM) residues joined by β (1→4) glycosidic bonds. Parallel chains are cross-linked by tetrapeptides attached to the NAM residues. The amino acids forming the tetrapeptides vary in different species of bacteria.

branched polysaccharides containing a large number of negatively charged galacturonic acid residues. Because of these multiple negative charges, pectins bind positively charged ions (such as Ca^{2+}) and trap water molecules to form gels. An illustration of their gel-forming properties is seen in jams and jellies that are produced by the addition of pectins to fruit juice. In the cell wall, the pectins form a gel-like network that is interlocked with the cross-linked cellulose microfibrils. In addition, cell walls contain a variety of glycoproteins that are incorporated into the matrix and are thought to provide further structural support.

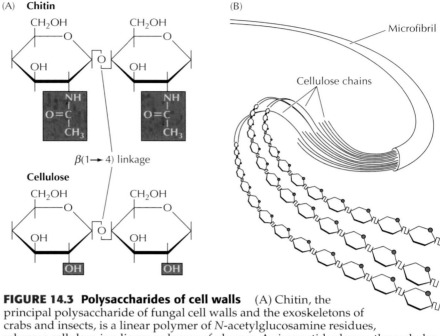

FIGURE 14.3 Polysaccharides of cell walls (A) Chitin, the principal polysaccharide of fungal cell walls and the exoskeletons of crabs and insects, is a linear polymer of *N*-acetylglucosamine residues, whereas cellulose is a linear polymer of glucose. As in peptidoglycan, the carbohydrate monomers are joined by β (1→4) linkages, allowing the polysaccharides to form long, straight chains. (B) Parallel chains of cellulose associate to form microfibrils.

Both the structure and function of cell walls change as plant cells develop. The walls of growing plant cells (called **primary cell walls**) are relatively thin and flexible, allowing the cell to expand in size. Once cells have ceased growth, they frequently lay down **secondary cell walls** between the plasma membrane and the primary cell wall (Figure 14.6). Such secondary cell walls, which are both thicker and more rigid than primary walls, are particularly important in cell types responsible for conducting water and providing mechanical strength to the plant.

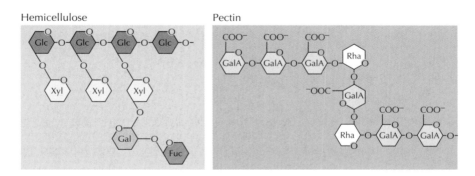

FIGURE 14.4 Structures of hemicellulose and pectin A representative hemicellulose (xyloglucan) consists of a backbone of glucose (Glc) residues with side chains of xylose (Xyl), galactose (Gal), and fucose (Fuc). The backbone of rhamnogalacturonan (a representative pectin) contains galacturonic acid (GalA) and rhamnose (Rha) residues to which numerous side chains are also attached

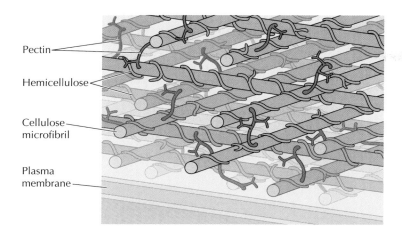

Pectin

Hemicellulose

Cellulose
microfibril

Plasma
membrane

FIGURE 14.5 Model of a plant cell wall
Cellulose is organized into microfibrils that are oriented in layers. Hemicelluloses (green) are tightly associated with the surface of cellulose microfibrils, which are cross-linked by pectins (red).

Primary and secondary cell walls differ in composition as well as in thickness. Primary cell walls contain approximately equal amounts of cellulose, hemicelluloses, and pectins. In contrast, the more rigid secondary walls generally lack pectin and contain 50 to 80% cellulose. Many secondary walls are further strengthened by **lignin**, a complex polymer of phenolic residues that is responsible for much of the strength and density of wood. The orientation of cellulose microfibrils also differs in primary and secondary cell walls. The cellulose fibers of primary walls appear to be randomly arranged, whereas those of secondary walls are highly ordered (see Figure 14.6). Secondary walls are frequently laid down in layers in which the cellulose fibers differ in orientation, forming a laminated structure that greatly increases cell wall strength.

One of the critical functions of plant cell walls is to prevent cell swelling as a result of osmotic pressure. In contrast to animal cells, plant cells do not maintain an osmotic balance between their cytosol and extracellular fluids. Consequently, osmotic pressure continually drives the flow of water into the cell. This water influx is tolerated by plant cells because their rigid cell walls prevent swelling and bursting. Instead, an internal hydrostatic pressure (called **turgor pressure**) builds up within the cell, eventually equalizing the osmotic pressure and preventing the further influx of water.

Turgor pressure is responsible for much of the rigidity of plant tissues as is readily apparent from examination of a dehydrated, wilted plant. In addition, turgor pressure provides the basis for a form of cell growth that is unique to plants. In particular, plant cells frequently expand by taking up water without synthesizing new cytoplasmic components (Figure 14.7). Cell expansion by this mechanism is signaled by plant hormones (**auxins**) that activate proteins called expansins. The expansins act to weaken a region of the cell wall, allowing turgor pressure to drive the expansion of the cell in

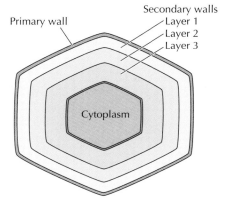

Primary wall

Secondary walls
Layer 1
Layer 2
Layer 3

Cytoplasm

Primary wall

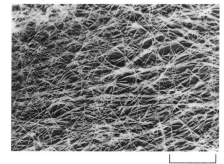

1 μm

Secondary wall

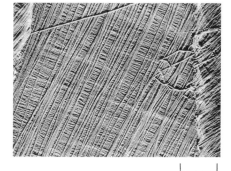

0.2 μm

FIGURE 14.6 Primary and secondary cell walls   Secondary cell walls are laid down between the primary cell wall and the plasma membrane. Secondary walls frequently consist of three layers, which differ in the orientation of their cellulose microfibrils. Electron micrographs show cellulose microfibrils in primary and secondary cell walls. (Primary wall, courtesy of F. C. Steward; secondary wall, Biophoto Associates/Photo Researchers, Inc.)

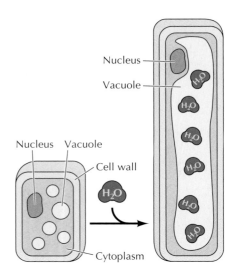

FIGURE 14.7 Expansion of plant cells Turgor pressure drives the expansion of plant cells by the uptake of water, which is accumulated in a large central vacuole.

that direction. As this occurs, the water that flows into the cell accumulates within a large central vacuole, so the cell expands without increasing the volume of its cytosol. Such expansion can result in a ten to one hundredfold increase in the size of plant cells during development.

As cells expand, new components of the cell wall are deposited outside the plasma membrane. While matrix components, including hemicelluloses and pectins, are synthesized in the Golgi apparatus and secreted, cellulose is synthesized by the plasma membrane enzyme complex **cellulose synthase** (Figure 14.8). Cellulose synthase is a transmembrane enzyme that synthesizes cellulose from UDP-glucose in the cytosol. The growing cellulose chain remains bound to the enzyme as it is synthesized and is translocated across the plasma membrane to the exterior of the cell through a pore created by multiple enzyme subunits. A similar mechanism is used to synthesize chitin and hyaluronan, a component of the extracellular matrix discussed later in this chapter.

In expanding cells, the newly synthesized cellulose microfibrils are deposited at right angles to the direction of cell elongation—an orientation that is thought to play an important role in determining the direction of further cell expansion (see Figure 14.8). Interestingly, the cellulose microfibrils in elongating cell walls are laid down in parallel to cortical microtubules underlying the plasma membrane. These microtubules appear to define the orientation of newly synthesized cellulose microfibrils, possibly by determining the direction of movement of the cellulose synthase complexes in the membrane. The cortical microtubules thus define the direction of cell wall growth, which in turn determines the direction of cell expansion and ultimately the shape of the entire plant.

FIGURE 14.8 Cellulose synthesis during cell elongation Cellulose synthases are transmembrane enzymes that synthesize cellulose from UDP-glucose. UDP-glucose binds to the catalytic domain of cellulose synthase in the cytosol and the growing cellulose chain is translocated to the outside of the cell through a pore created by two subunits of the enzyme in the plasma membrane. Complexes of 8–10 cellulose synthase dimers track microtubules beneath the plasma membrane such that new cellulose microfibrils are laid down at right angles to the direction of cell elongation.

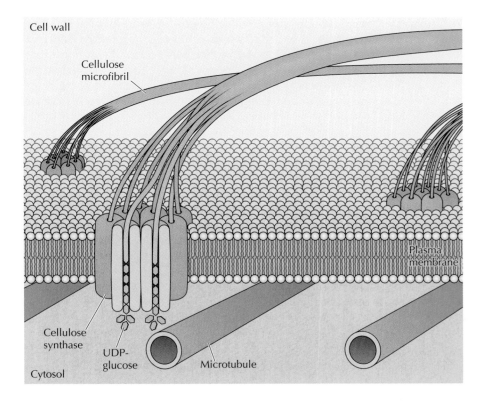

FIGURE 14.9 Examples of extracellular matrix   Sheets of epithelial cells rest on a thin layer of extracellular matrix called a basal lamina. Beneath the basal lamina is loose connective tissue, which consists largely of extracellular matrix secreted by fibroblasts. The extracellular matrix contains fibrous structural proteins embedded in a gel-like polysaccharide ground substance.

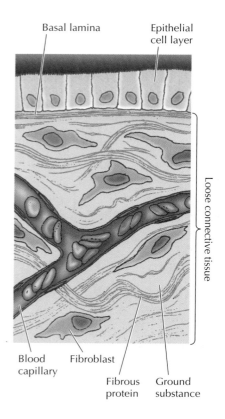

Basal lamina

Epithelial cell layer

Loose connective tissue

Blood capillary

Fibroblast

Fibrous protein

Ground substance

Cell walls from different plant tissues, such as leaves, stems, roots and flowers, consist primarily of cellulose but differ in their matrix components and perhaps in the organization of their cellulose fibrils. There are at least ten different cellulose synthase enzymes and each plasma membrane enzyme complex contains two or more different forms, perhaps accounting for the varying organization of cell walls in different tissues.

The Extracellular Matrix and Cell-Matrix Interactions

Although animal cells are not surrounded by cell walls, most animal cells in tissues are embedded in an **extracellular matrix** that fills the spaces between cells and binds cells and tissues together. There are several types of extracellular matrices, which consist of a variety of secreted proteins and polysaccharides. One type of extracellular matrix is exemplified by the thin, sheetlike **basal laminae,** previously called basement membranes, upon which layers of epithelial cells rest (Figure 14.9). In addition to supporting sheets of epithelial cells, basal laminae surround muscle cells, adipose cells, and peripheral nerves. Extracellular matrix, however, is most abundant in connective tissues. For example, the loose connective tissue beneath epithelial cell layers consists predominantly of an extracellular matrix in which fibroblasts are distributed. Other types of connective tissue, such as bone, tendon, and cartilage, similarly consist largely of extracellular matrix, which is principally responsible for their structure and function. Several of the proteins and polysaccharides found in the extracellular matrix are also found closely associated with the plasma membrane. These can be seen on the plasma membrane of endothelial cells or the apical membrane of intestinal epithelial cells as a polysaccharide rich coat termed the glycocalyx (see Figure 13.13).

Matrix Structural Proteins

Extracellular matrices are composed of tough fibrous proteins embedded in a gel-like polysaccharide ground substance—a design basically similar to that of plant cell walls. In addition to fibrous structural proteins and polysaccharides, the extracellular matrix contains adhesion proteins that link components of the matrix both to one another and to attached cells. The differences between various types of extracellular matrices result from both quantitative variations in the types or amounts of these different constituents and from modifications in their organization. For example, tendons contain a high proportion of fibrous proteins, whereas cartilage contains a high concentration of polysaccharides that form a firm compression-resistant gel. In bone, the extracellular matrix is hardened by deposition of calcium phosphate crystals. The sheetlike structure of basal laminae results from a matrix composition that differs from that found in connective tissues.

The major structural protein of the extracellular matrix is **collagen,** which is the single most abundant protein in animal tissues. The collagens

■ Cancer cells secrete proteases that digest proteins of the extracellular matrix, allowing the cancer cells to invade surrounding tissue and metastasize to other parts of the body.

(A) **Collagen triple helix**

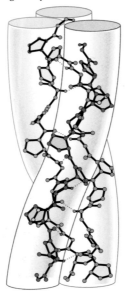

FIGURE 14.10 Structure of collagen (A) Three polypeptide chains coil around one another in a characteristic triple helix structure. (B) The amino acid sequence of a collagen triple helix domain consists of Gly-X-Y repeats in which X is frequently proline and Y is frequently hydroxyproline (Hyp).

(B) **Amino acid sequence**

Gly Pro Ser Gly Pro Arg Gly Pro Hyp Gly Pro Hyp Gly Ala Hyp Gly Pro Gln Gly Phe Gln Gly Pro Hyp

Pro   Pro

Gly   Gly

Hyp   Hyp

■ **Prolyl hydroxylase, the enzyme that catalyzes the modification of proline to hydroxyproline residues in collagen, requires vitamin C for its activity. Vitamin C deficiency causes scurvy, a disorder characterized by skin lesions and blood vessel hemorrhages as a result of weakened connective tissue.**

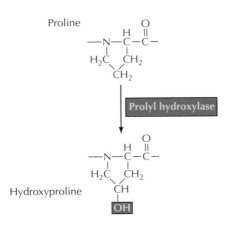

FIGURE 14.11 Formation of hydroxyproline Prolyl hydroxylase converts proline residues in collagen to hydroxyproline.

are a large family of proteins containing at least 27 different members. They are characterized by the formation of triple helices in which three polypeptide chains are wound tightly around one another in a ropelike structure (Figure 14.10). The different collagen polypeptides can assemble into 42 different trimers. The triple helix domains of the collagens consist of repeats of the amino acid sequence Gly-X-Y. A glycine (the smallest amino acid, with a side chain consisting only of a hydrogen) is required in every third position in order for the polypeptide chains to pack together close enough to form the collagen triple helix. Proline is frequently found in the X position and hydroxyproline in the Y position; because of their ring structure these amino acids stabilize the helical conformations of the polypeptide chains.

The unusual amino acid hydroxyproline is formed within the endoplasmic reticulum by modification of proline residues that have already been incorporated into collagen polypeptide chains (Figure 14.11). Lysine residues in collagen are also frequently converted to hydroxylysines. The hydroxyl groups of these modified amino acids are thought to stabilize the collagen triple helix by forming hydrogen bonds between polypeptide chains. These amino acids are rarely found in other proteins although hydroxyproline is also common in some of the glycoproteins of plant cell walls.

The most abundant type of collagen (type I collagen) is one of the fibril-forming collagens that are the basic structural components of connective tissues (Table 14.1). The polypeptide chains of these collagens consist of approximately a thousand amino acids or 330 Gly-X-Y repeats. After being secreted from the cell these collagens assemble into **collagen fibrils** in which the triple helical molecules are associated in regular staggered arrays (Figure 14.12). These fibrils do not form within the cell because the fibril-

FIGURE 14.12 Collagen fibrils   (A) Collagen molecules assemble in a regular staggered array to form fibrils. The molecules overlap by one-fourth of their length, and there is a short gap between the N terminus of one molecule and the C terminus of the next. The assembly is strengthened by covalent cross-links between side chains of lysine or hydroxylysine residues, primarily at the ends of the molecules. (B) Electron micrograph of collagen fibrils. The staggered arrangement of collagen molecules and the gaps between them are responsible for the characteristic cross-striations in the fibrils. (B, J. Gross, F. O. Sahmitt, and D. Fawcett/Visuals Unlimited.)

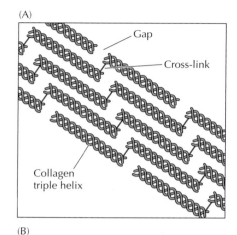

(A)

Gap

Cross-link

Collagen triple helix

(B)

1 μm

forming collagens are synthesized as soluble precursors (**procollagens**) that contain nonhelical segments at both ends of the polypeptide chain. Procollagen is cleaved to collagen after its secretion, so the assembly of collagen into fibrils takes place only outside the cell. The association of collagen molecules in fibrils is further strengthened by the formation of covalent cross-links between the side chains of lysine and hydroxylysine residues. Frequently, the fibrils further associate with one another to form collagen fibers, which can be several micrometers in diameter.

Several other types of collagen do not form fibrils but play distinct roles in various kinds of extracellular matrices (see Table 14.1). In addition to the fibril-forming collagens, connective tissues contain fibril-associated collagens, which bind to the surface of collagen fibrils and link them both to one another and to other matrix components. Basal laminae form from a different type of collagen (type IV collagen), which is a network-forming collagen (Figure 14.13). The Gly-X-Y repeats of these collagens are frequently inter-

TABLE 14.1 Representative Members of the Collagen Family

Collagen class	Types	Tissue distribution
Fibril-forming	I	Most connective tissues
	II	Cartilage and vitreous humor
	III	Extensible connective tissues (e.g., skin and lung)
	V	Tissues containing collagen I
	XI	Cartilage
	XXIV	Bone and cornea
	XXVII	Eye, ear, and lung
Fibril-associated	IX	Cartilage
	XII	Tissues containing collagen I
	XIV	Tissues containing collagen I
	XVI	Many tissues
	XIX	Many tissues
	XX	Cornea
	XXI	Many tissues
	XXII	Cell junctions
	XXVI	Testis and ovary
Network-forming	IV	Basal laminae
	VIII	Many tissues
	X	Cartilage
Anchoring fibrils	VII	Attachments of basal laminae to underlying connective tissue
Transmembrane	XVII	Skin hemidesmosomes
	XXV	Nerve cells

(A)

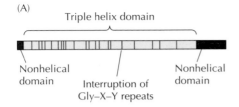

Triple helix domain

Nonhelical domain

Interruption of Gly–X–Y repeats

Nonhelical domain

FIGURE 14.13 Type IV collagen (A) The Gly-X-Y repeat structure of type IV collagen (yellow) is interrupted by multiple nonhelical sequences (bars). (B) Electron micrograph of a type IV collagen network. (B, P. D. Yurchenco and J. C. Schittny, 1990. *FASEB J.* 4: 1577.)

(B)

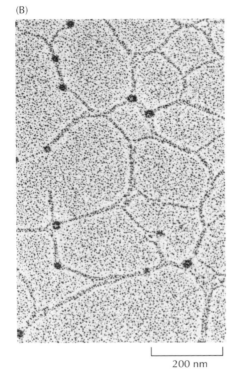

200 nm

rupted by short nonhelical sequences. Because of these interruptions, the network-forming collagens are more flexible than the fibril-forming collagens. Consequently, they assemble into two-dimensional cross-linked networks instead of fibrils. Yet another type of collagen forms anchoring fibrils, which link some basal laminae to underlying connective tissues. Other types of collagen are transmembrane proteins that participate in cell-matrix interactions.

Connective tissues also contain **elastic fibers**, which are particularly abundant in organs that regularly stretch and then return to their original shape. The lungs, for example, stretch each time a breath is inhaled and return to their original shape with each exhalation. Elastic fibers are composed principally of a protein called **elastin**, which is cross-linked into a network by covalent bonds formed between the side chains of lysine residues (similar to those found in collagen). This network of cross-linked elastin chains behaves like a rubber band, stretching under tension and then snapping back when the tension is released.

Matrix Polysaccharides

The fibrous structural proteins of the extracellular matrix are embedded in gels formed from polysaccharides called **glycosaminoglycans (GAGs)**, which consist of repeating units of disaccharides (Figure 14.14). One sugar of the disaccharide is either N-acetylglucosamine or N-acetylgalactosamine and the second is usually acidic (either glucuronic acid or iduronic acid). With the exception of hyaluronan, these sugars are modified by the addition of sulfate groups. Consequently, GAGs are highly negatively charged.

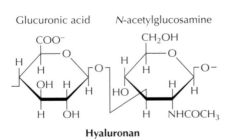

Glucuronic acid   N-acetylglucosamine

Hyaluronan

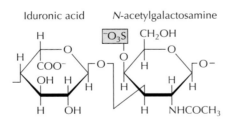

Iduronic acid   N-acetylgalactosamine

Dermatan sulfate

FIGURE 14.14 Major types of glycosaminoglycans Glycosaminoglycans consist of repeating disaccharide units. With the exception of hyaluronan, the sugars frequently contain sulfate.

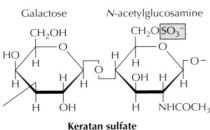

Glucuronic acid   N-acetylgalactosamine

Chondroitin sulfate

Galactose   N-acetylglucosamine

Keratan sulfate

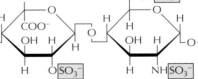

Iduronic acid   N-acetylglucosamine

Heparan sulfate

FIGURE 14.15 Complexes of aggrecan and hyaluronan Aggrecan is a large proteoglycan consisting of more than 100 chondroitin sulfate chains joined to a core protein. Multiple aggrecan molecules bind to long chains of hyaluronan forming large complexes in the extracellular matrix of cartilage. This association is stabilized by link proteins.

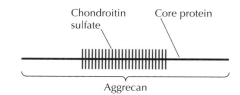

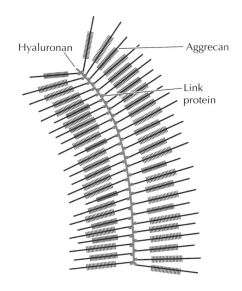

Like the pectins of plant cell walls, they bind positively charged ions and trap water molecules to form hydrated gels, thereby providing mechanical support to the extracellular matrix. The common sulfated GAGs are chondroitin sulfate, dermatan sulfate, heparan sulfate, and keratan sulfate.

Hyaluronan is the only GAG that occurs as a single long polysaccharide chain. Like cellulose and chitin, it is synthesized at the plasma membrane by a transmembrane hyaluronan synthase. All of the other GAGs are linked to proteins to form **proteoglycans**, which can consist of up to 95% polysaccharide by weight. Proteoglycans can contain as few as one or as many as more than a hundred GAG chains attached to serine residues of a core protein. A variety of core proteins (ranging from 10 to >500 kd) have been identified, so the proteoglycans are a diverse group of macromolecules. In addition to being components of the extracellular matrix, some proteoglycans (syndecans and glypicans) are cell surface proteins that function along with integrins in cell-cell and cell-matrix adhesion.

A number of proteoglycans interact with hyaluronan to form large complexes in the extracellular matrix. A well-characterized example is aggrecan, the major proteoglycan of cartilage (Figure 14.15). More than a hundred chains of chondroitin sulfate are attached to a core protein of about 250 kd, forming a proteoglycan of about 3000 kd. Multiple aggrecan molecules then associate with chains of hyaluronan, forming large aggregates (>100,000 kd) that become trapped in the collagen network. Proteoglycans also interact with both collagen and other matrix proteins to form gel-like networks in which the fibrous structural proteins of the extracellular matrix are embedded. For example, perlecan (the major heparan sulfate proteoglycan of basal laminae) binds to both type IV collagen and to the adhesion protein laminin, which is discussed shortly.

Matrix Adhesion Proteins

Matrix adhesion proteins, the final class of extracellular matrix constituents, are responsible for linking the components of the matrix to one another and to the surfaces of cells. They interact with collagen and proteoglycans to specify matrix organization and are the major binding sites for integrins.

The prototype of these molecules is **fibronectin**, which is the principal adhesion protein of connective tissues. Fibronectin is a dimeric glycoprotein consisting of two polypeptide chains, each containing nearly 2500 amino acids (Figure 14.16). Within the extracellular matrix, fibronectin is often cross-linked into fibrils. Fibronectin has binding sites for both collagen and GAGs so it cross-links these matrix components. A distinct site on the fibronectin molecule is recognized by cell surface receptors (such as integrins) and is

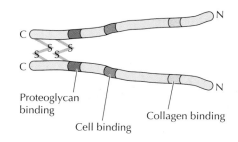

FIGURE 14.16 Structure of fibronectin Fibronectin is a dimer of similar polypeptide chains joined by disulfide bonds near the C terminus. Sites for binding to proteoglycans, cells, and collagen are indicated. The molecule also contains additional binding sites that are not shown.

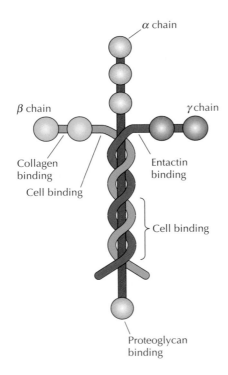

α chain

β chain

γ chain

Collagen binding

Entactin binding

Cell binding

Cell binding

Proteoglycan binding

FIGURE 14.17 Structure of laminins   Laminins consist of three polypeptide chains designated α, β, and γ. Some of the binding sites for entactin, type IV collagen, proteoglycans, and cell surface receptors are indicated.

thus responsible for the attachment of cells to the extracellular matrix. Fibronectin proteins vary greatly from tissue to tissue but all are derived by alternative splicing of the mRNA of a single gene.

Basal laminae contain distinct adhesion proteins of the **laminin** family (Figure 14.17). Laminins are heterotrimers of α, β, and γ subunits which are the products of five α genes, four β genes, and three γ genes. Like type IV collagen, laminins can self-assemble into meshlike polymers. Such laminin networks are the major structural components of the basal laminae synthesized in very early embryos, which do not contain collagen. The laminins also have binding sites for cell surface receptors such as integrins, type IV collagen, and the heparan sulfate proteoglycan, perlecan. In addition, laminins are tightly associated with another adhesion protein, called **entactin,** which also binds to type IV collagen. As a result of these multiple interactions, laminin, entactin, type IV collagen, and perlecan form cross-linked networks in the basal lamina.

Cell-Matrix Interactions

The major cell surface receptors responsible for the attachment of cells to the extracellular matrix are the **integrins**. The integrins are a family of transmembrane proteins consisting of two subunits, designated α and β (Figure 14.18). More than 24 different integrins formed from combinations of 18 known α subunits and 8 known β subunits have been identified. The integrins bind to short amino acid sequences present in multiple components of the extracellular matrix, including collagen, fibronectin, and laminin. The first such integrin-binding site to be characterized was the sequence Arg-Gly-Asp in fibronectin, which is recognized by several members of the integrin family. Other integrins, however, bind to distinct peptide sequences, including recognition sequences in collagens and laminin. Transmembrane proteoglycans on the surface of a variety of cells also bind to components of the extracellular matrix and modulate cell-matrix interactions.

In addition to attaching cells to the extracellular matrix the integrins serve as anchors for the cytoskeleton (Figure 14.19). The resulting linkage of the cytoskeleton to the extracellular matrix is responsible for the stability of cell-matrix junctions. Distinct interactions between integrins and the cytoskeleton are found at two types of cell-matrix junctions: **focal adhesions** and **hemidesmosomes**, both of which were discussed in Chapter 12. Focal adhesions attach a variety of cells, including fibroblasts, to the extracellular matrix. At focal adhesions, the cytoplasmic domains of the β subunits of integrins anchor the actin cytoskeleton by associating with bundles of actin filaments through actin-binding proteins such as α-actinin, talin, and vinculin. Hemidesmosomes mediate epithelial cell attachments at which a specific integrin (designated $\alpha_6\beta_4$) interacts through plectin and BP230 with intermediate filaments instead of with actin. This is mediated

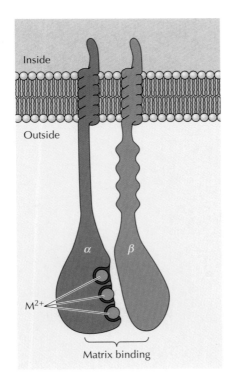

Inside

Outside

α

β

M^{2+}

Matrix binding

FIGURE 14.18 Structure of integrins   The integrins are heterodimers of two transmembrane subunits, designated α and β. The α subunit binds divalent cations, designated M^{2+}. The matrix-binding region is composed of portions of both subunits.

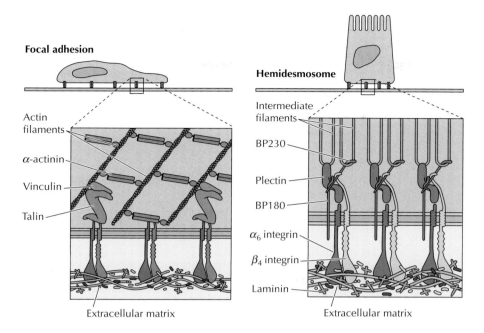

Focal adhesion

Actin filaments

α-actinin

Vinculin

Talin

Extracellular matrix

Hemidesmosome

Intermediate filaments

BP230

Plectin

BP180

α_6 integrin

β_4 integrin

Laminin

Extracellular matrix

FIGURE 14.19 Cell-matrix junctions mediated by integrins Integrins mediate two types of stable junctions in which the cytoskeleton is linked to the extracellular matrix. In focal adhesions, bundles of actin filaments are anchored to the β subunits of most integrins via associations with a number of other proteins, including α-actinin, talin, and vinculin. In hemidesmosomes, $\alpha_6\beta_4$ integrin links the basal lamina to intermediate filaments via plectin and BP230. BP180 functions in hemidesmosome assembly and stability.

by the long cytoplasmic tail of the β_4 integrin subunit. The $\alpha_6\beta_4$ integrin binds to laminin, so hemidesmosomes anchor epithelial cells to the basal lamina. An additional protein, BP180, is important in hemidesmosome assembly and stability. Plectin and BP230 are members of the **plakin** family, which is a family of proteins involved in forming links to intermediate filaments at both hemidesmosomes and desmosomes (discussed in the next section of this chapter). BP180 shares sequence homology with the transmembrane collagens.

Cell-matrix interactions, like the cell-cell interactions discussed in the next section, develop by stepwise recruitment of specific junctional molecules to the cell membrane. Focal adhesions develop from a small cluster of integrins, termed **focal complexes**, by the sequential recruitment of talin, vinculin, and α-actinin. These form the initial connections to the actin cytoskeleton and recruit formin, which initiates actin bundle formation (see Figure 12.6), and myosin II, which leads to the development of tension at the point of adhesion. Tension allows the formation of a larger area of cell-matrix adhesion and the recruitment of signaling molecules (discussed in Chapter 15) to the cell junctions. Focal adhesions can be very stable interactions involved in tissue structure, or they can turn over rapidly as cells move (see Figure 12.34). During cell migration, the formation of new focal complexes at the leading edge of the cell results in the loss of tension at the old focal adhesions, leading to the inactivation of integrin binding to the extracellular matrix.

Recent progress in analyzing the structure of integrins has permitted an understanding of the regulated changes in integrin activity that underlie the rapid assembly and disassembly of focal adhesions during cell movement. The ability of integrins to reversibly bind matrix components is dependent on their ability to change conformation between active and inactive states (Figure 14.20). In the inactive state integrins are unable to bind to the matrix because the head groups containing the ligand-binding site are held close to the cell surface. Signals from the cytosol, via talin or vinculin binding, change the conformation of the cytosolic and extracellular

FIGURE 14.20 Integrin activation In the inactive state, integrin head groups are held close to the cell surface. Signals from the cytosol activate integrins, extending the head groups and enabling them to bind to the extracellular matrix. This leads to recruitment of additional integrins and formation of a focal adhesion. In migrating cells, the reverse set of conformational changes can dissociate integrins from the matrix.

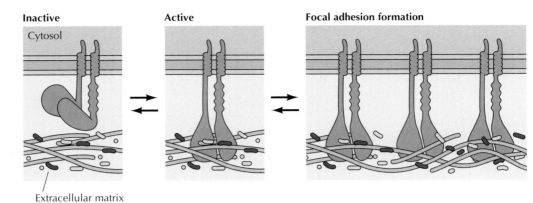

Inactive

Cytosol

Active

Focal adhesion formation

Extracellular matrix

domains of the integrins, extending the head groups into the matrix and allowing binding of the ligand. This in turn conveys a signal to the cytosol that allows the cell to respond to integrin binding and recruits additional integrins to the site of adhesion, leading to focal adhesion formation. Subsequent signaling from inside the cell can cause the reverse set of conformational changes and dissociate the integrin from its ligand, leading to disassembly of focal adhesions.

KEY EXPERIMENT

The Characterization of Integrin

Structure of Integrin, a Glycoprotein Involved in the Transmembrane Linkage between Fibronectin and Actin

John W. Tamkun, Douglas W. DeSimone, Deborah Fonda, Ramila S. Patel, Clayton Buck, Alan F. Horwitz and Richard O. Hynes

Massachusetts Institute of Technology, Cambridge, Massachusetts (JWT, DWD, DF, RSP and ROH), The Wistar Institute, Philadelphia, Pennsylvania (CB) and The University of Pennsylvania School of Medicine (AFH)

Cell, Volume 46, 1986, pages 271–282

Richard O. Hynes

The Context

The molecular basis of cell adhesion to the extracellular matrix has been of interest to cell biologists since it was observed that adhesion to the matrix was reduced in cancer cells, potentially related to their abnormal growth and ability to metastasize or spread throughout the body. In the late 1970s and early 1980s work from several laboratories, including that of Richard Hynes, established that there was a physical link between the stress fibers of the actin cytoskeleton and fibronectin in the extracellular matrix. Several of the cytoskeletal proteins

involved in this linkage, including vinculin and talin, had been identified, but the critical transmembrane proteins that linked cells to the extracellular matrix remained unknown.

Candidates for transmembrane proteins involved in cell adhesion to the extracellular matrix had been identified using antibodies prepared by several groups of scientists, including Alan Horwitz and Clayton Buck. These antibodies recognized a complex of 140 kd glycoproteins that appeared to be transmembrane proteins. Immunofluorescence and immunoelectron microscopy localized these glycopro-

teins to points of cell adhesion to the matrix. Additional studies showed that the 140 kd glycoproteins bound to fibronectin and were involved in cell adhesion. It thus appeared that these 140 kd glycoproteins were likely candidates for the transmembrane proteins responsible for cell-matrix adhesion. In the experiments described by Tamkun et al. these antibodies were used to isolate a molecular clone encoding one of these glycoproteins, thereby providing the first molecular characterization of integrin.

KEY EXPERIMENT

(A)

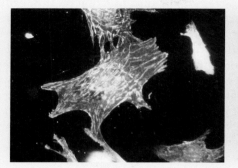

(B)

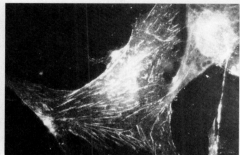

Immunofluorescence of chick embryo fibroblasts stained with the original antiserum to the 140 kd glycoprotein complex (A) or with antibodies purified against the protein expressed from the cDNA clones (B).

The Experiments

To isolate a molecular clone encoding one of the 140 kd glycoproteins Tamkun et al. prepared a cDNA library from mRNA of chick embryo fibroblasts. This cDNA library was prepared in a bacteriophage λ expression vector that directed high level transcription and translation of the eukaryotic cDNA inserts in *E. coli* (see Figure 4.21). The cDNA library contained approximately 100,000 independent cDNA inserts, sufficient to represent clones of all the mRNAs expressed in the chick embryo fibroblasts from which it was prepared. The library was then screened with antibodies against the 140 kd glycoproteins to identify those recombinant phages carrying the cDNAs of interest. Plaques produced by individual recombinant phages were transferred to nitrocellulose filters as would be done to screen a recombinant library by nucleic acid hybridization (see Figure 4.26). However, to screen the expression library, the filters were probed with an antibody to identify clones expressing the desired protein. By such antibody screening, Tamkun et al. successfully identified several recombinant clones expressing proteins that were recognized by antibodies against the 140 kd glycoproteins.

Their next challenge was to determine whether the cloned cDNAs actually encoded one or more of the 140 kd glycoproteins. To do this they used proteins expressed from the clones to purify antibodies that recognized the cloned proteins. In addition, they injected rabbits with the cloned proteins to raise new antibodies specifically directed against the proteins encoded by the cloned cDNAs. The resulting antibodies recognized one of the proteins of the 140 kd complex from chick embryo fibroblasts in immunoblot assays, establishing the relationship of the cloned cDNAs to this protein. In addition, antibodies purified against the cloned proteins stained cells similarly to the original antibody in immunofluorescence assays, yielding a staining pattern corresponding to the sites of stress fiber attachment to the extracellular matrix (see figure). The results of both immunoblotting and immunofluorescence thus indicated that the cloned cDNAs encoded one of the proteins of the 140 kd glycoprotein complex.

The cDNA was then sequenced and found to encode a protein consisting of 803 amino acids. The protein included an amino terminal signal sequence and an apparent transmembrane α helix of 23 hydrophobic amino acids near the carboxy terminus. The sequence predicted a short cytosolic domain and a large extracellular domain containing multiple glycosylation sites consistent with the expected structure of a transmembrane glycoprotein.

The Impact

Hynes and his colleagues concluded that they had cloned a cDNA encoding a transmembrane protein that functioned to link fibronectin in the extracellular matrix to the cytoskeleton. They named the protein integrin "to denote its role as an integral membrane complex involved in the transmembrane association between the extracellular matrix and the cytoskeleton."

The initial cloning of integrin led to our current understanding of the molecular basis of stable cell junctions. At focal adhesions, integrins link the extracellular matrix to actin filaments. In addition, integrins mediate attachment of epithelial cells to the extracellular matrix at hemidesmosomes where they link the extracellular matrix to intermediate filaments. Thus, as suggested by Hynes and his colleagues, integrins play a general role in cell-matrix adhesion. Subsequent studies have shown that integrins also play an important role as signaling complexes in cells, relaying signals from outside of the cell to control multiple aspects of cell movement, proliferation, and survival (discussed in Chapter 15). The characterization of integrins thus opened the door to understanding not only the nature of cell attachment to the matrix but also to elucidating novel signaling mechanisms that regulate cell behavior.

Cell-Cell Interactions

Direct interactions between cells, as well as between cells and the extracellular matrix, are critical to the development and function of multicellular organisms. Some cell-cell interactions are transient, such as the interactions between cells of the immune system and the interactions that direct white blood cells to sites of tissue inflammation. In other cases, stable cell-cell junctions play a key role in the organization of cells in tissues. For example, several different types of stable cell-cell junctions are critical to the maintenance and function of epithelial cell sheets. In addition to mediating cell adhesion, specialized types of junctions provide mechanisms for rapid communication between cells. Plant cells also associate with their neighbors not only by interactions between their cell walls but also by specialized junctions between their plasma membranes.

Adhesion Junctions

Cell-cell adhesion is a selective process such that cells adhere only to other cells of specific types. This selectivity was first demonstrated in classical studies of embryo development, which showed that cells from one tissue (e.g., liver) specifically adhere to cells of the same tissue rather than to cells of a different tissue (e.g., brain). Such selective cell-cell adhesion is mediated by transmembrane proteins called **cell adhesion molecules**, which can be divided into four major groups: the **selectins**, the **integrins**, the **immunoglobulin (Ig) superfamily** (so named because they contain structural domains similar to immunoglobulins), and the **cadherins** (Table 14.2). Cell adhesion mediated by the selectins, integrins, and most cadherins requires Ca^{2+}, Mg^{2+} or Mn^{2+}, so many adhesive interactions between cells are divalent cation-dependent.

The selectins mediate transient interactions between leukocytes and endothelial cells or blood platelets. There are three members of the selectin family: L-selectin, which is expressed on leukocytes; E-selectin, which is expressed on endothelial cells; and P-selectin, which is expressed on platelets. As discussed in Chapter 13, the selectins recognize cell surface carbohydrates (see Figure 13.14). One of their critical roles is to initiate the interactions between leukocytes and endothelial cells during the migration of leukocytes from the circulation to sites of tissue inflammation (Figure 14.21). The selectins mediate the initial adhesion of leukocytes to endothelial cells. This is followed by the formation of more stable adhesions in which integrins on the surface of leukocytes bind to intercellular adhesion molecules (ICAMs), which are members of the Ig superfamily expressed on the surface of endothelial cells. The firmly attached leukocytes are then able to

TABLE 14.2 Cell Adhesion Molecules

Family	Ligands recognized	Stable cell junctions
Selectins	Carbohydrates	No
Integrins	Extracellular matrix	Focal adhesions and hemidesmosomes
	Members of Ig superfamily	No
Ig superfamily	Integrins	No
	Other Ig superfamily proteins	No
Cadherins	Other cadherins	Adherens junctions and desmosomes

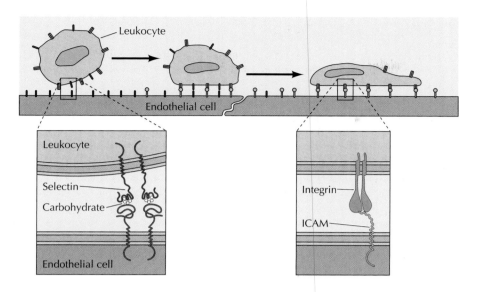

FIGURE 14.21 Adhesion between leukocytes and endothelial cells
Leukocytes leave the circulation at sites of tissue inflammation by interacting with the endothelial cells of capillary walls. The first step in this interaction is the binding of leukocyte selectins to carbohydrate ligands on the endothelial cell surface. This step is followed by more stable interactions between leukocyte integrins and members of the Ig superfamily (ICAMs) on endothelial cells.

penetrate the walls of capillaries and enter the underlying tissue by migrating between endothelial cells.

The binding of ICAMs to integrins is an example of a **heterophilic interaction** in which an adhesion molecule on the surface of one cell (e.g., an ICAM) recognizes a different molecule on the surface of another cell (e.g., an integrin). Other members of the Ig superfamily mediate **homophilic interactions** in which an adhesion molecule on the surface of one cell binds to the same molecule on the surface of another cell. Such homophilic binding can lead to selective adhesion between cells of the same type. For example, nerve cell adhesion molecules (N-CAMs) are members of the Ig superfamily expressed on nerve cells, and homophilic binding between N-CAMs contributes to the formation of selective associations between nerve cells during development. There are more than 100 members of the Ig superfamily, which mediate a variety of cell-cell interactions.

The fourth class of cell adhesion molecules is the cadherins. Cadherins are involved in selective adhesion between embryonic cells, formation of specific synapses in the nervous system, and are the proteins primarily responsible for the maintenance of stable junctions between cells in tissues. Cadherins are a large family of proteins (about 80 members) that share a highly conserved extracellular domain that mediates largely homophilic interactions. For example, E-cadherin is expressed on epithelial cells so homophilic interactions between E-cadherins lead to the selective adhesion of epithelial cells to one another. It is noteworthy that loss of E-cadherin can contribute to the development of cancers arising from epithelial tissues, illustrating the importance of cell-cell interactions in controlling cell behavior. Different members of the cadherin family, such as N-cadherin (neural cadherin) and P-cadherin (placental cadherin), mediate selective adhesion of other cell types. There are several subfamilies of cadherins (classic cadherins, desmosomal cadherins, fat-like cadherins, and seven transmembrane domain cadherins) which differ primarily in their transmembrane and cytosolic domains.

The cell-cell interactions mediated by the selectins, integrins, and most members of the Ig superfamily are transient adhesions in which the cytoskeletons of adjacent cells are not linked to one another. Stable adhesion junctions involving the cytoskeletons of adjacent cells instead are based largely on cadherins. As discussed in Chapter 12, these cell-cell junctions are of two types: **adherens junctions** and **desmosomes.** At these junctions, classic and desmosomal cadherins are linked to actin bundles and intermediate filaments, respectively, by the interaction of their cytosolic tails with β-catenin or desmoplakin. The role of the cadherins in linking the cytoskeletons of adjacent cells is thus analogous to that of the integrins in forming stable junctions between cells and the extracellular matrix.

The basic structural unit of an adherens junction includes β-catenin and α-catenin in addition to the classic transmembrane cadherins (Figure 14.22). β-catenin binds directly to the cytosolic tail of the cadherins. α-catenin binds to β-catenin as well as to actin filaments and actin filament-binding proteins, such as vinculin. Thus α- and β-catenin directly link the actin cytoskeleton of one cell, through the transmembrane cadherins, to the actin cytoskeleton of an adjacent cell. Associated with the cadherins is the β-catenin related protein p120, which is not part of the structural link but is critical to maintain the stability of the adherens junction. β-catenin and p120

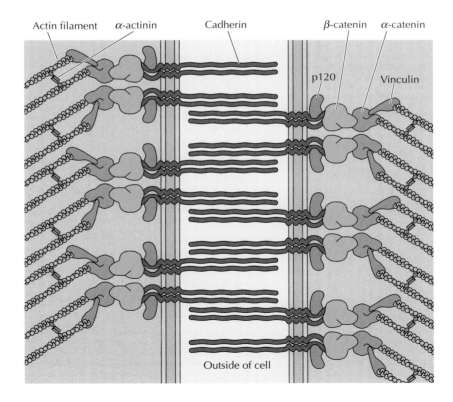

FIGURE 14.22 Adherens junctions In adherens junctions, cadherins link to actin filaments. β-catenin (an armadillo family protein) binds to the cytosolic tails of the cadherins. β-catenin also binds α-catenin, which binds both actin filaments and vinculin. This forms a direct link between the transmembrane cadherins and the actin cytoskeleton. A second armadillo family protein, p120, also binds the cytosolic tails of cadherins and regulates the stability of the junction.

Actin filament α-actinin   Cadherin β-catenin α-catenin

p120   Vinculin

Outside of cell

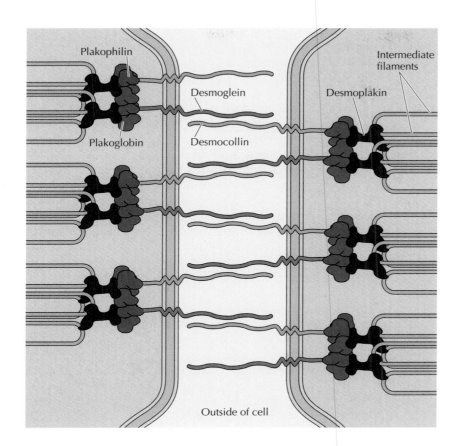

FIGURE 14.23 Desmosomes
In desmosomes, the desmosomal cadherins (desmoglein and desmo-collin) link to intermediate filaments. The armadillo family proteins plako-globin and plakophilin bind to the cytosolic tails of the cadherins. They also bind desmoplakin, which binds to intermediate filaments.

are members of the **armadillo protein family,** named after the *Drosophila β*-catenin protein. Despite its name, *α*-catenin is not part of the armadillo family but is more closely related to vinculin.

In addition to the cadherins, a second type of cell adhesion molecule, **nectin,** is also present at adherens junctions. There are four known nectins and a related family of nectin-like proteins, which are members of the Ig superfamily. Like cadherins, nectins interact in a homophilic manner, but nectins can also bind heterophilically to different nectins in the membrane of the apposing cell. Nectins are also involved in links to the actin cytoskeleton, but their major role appears to be in junction formation. In particular, it appears that cell adhesions are initially mediated by interactions between nectins, which then recruit cadherins to the adhesion sites. Cadherins then recruit actin-binding proteins, such as vinculin, to begin assembly of a mature adherens junction.

Desmosomes directly link the intermediate filament cytoskeletons of adjacent cells (Figure 14.23). The transmembrane cadherins, **desmoglein** and **desmocollin,** bind by heterophilic interactions across the junction. The armadillo family proteins plakoglobin and plakophilin bind to the cytosolic tails of the cadherins and provide a direct link to the intermediate filament-binding protein, desmoplakin. Desmoplakin is a member of the plakin family and is related to plectin, which functions analogously at hemidesmo-somes (see Figure 14.19). The strength of desmosomal links between cells is a property of both the intermediate filaments to which they are attached and the multiple interactions between plakoglobin, plakophilin, and desmoplakin that link the intermediate filaments to the cadherins.

Tight Junctions

Tight junctions are critically important to the function of epithelial cell sheets as barriers between fluid compartments. For example, the intestinal epithelium separates the lumen of the intestine from the underlying connective tissue, which contains blood capillaries. Tight junctions play two roles in allowing epithelia to fulfill such barrier functions. First, tight junctions form seals that prevent the free passage of molecules (including ions) between the cells of epithelial sheets. Second, tight junctions separate the apical and basolateral domains of the plasma membrane by preventing the free diffusion of lipids and membrane proteins between them. Consequently, specialized transport systems in the apical and basolateral domains are able to control the traffic of molecules between distinct extracellular compartments, such as the transport of glucose between the intestinal lumen and the blood supply (see Figure 13.32). While tight junctions are very effective seals of the extracellular space, they provide minimal adhesive strength between the apposing cells, so they are usually associated with adherens junctions and desmosomes in a **junctional complex** (Figure 14.24).

Tight junctions are the closest known contacts between adjacent cells. They were originally described as sites of apparent fusion between the outer leaflets of the plasma membranes, although it is now clear that the

■ The epithelial cell layers constitute significant barriers against invading microorganisms. Pathogenic bacteria have evolved strategies that disrupt junctional complexes, allowing them to penetrate between the cells of epithelial sheets.

(A)

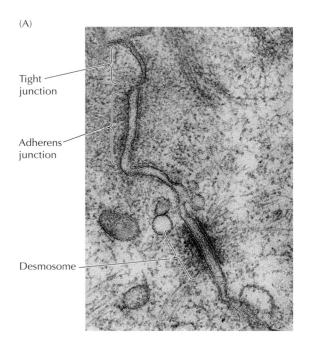

Tight junction

Adherens junction

Desmosome

(B)

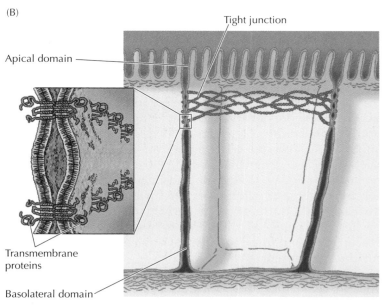

Tight junction

Apical domain

Transmembrane proteins

Basolateral domain

FIGURE 14.24 A junctional complex (A) Electron micrograph of epithelial cells joined by a junctional complex, including a tight junction, an adherens junction, and a desmosome. (B) Tight junctions are formed by interactions between strands of transmembrane proteins on adjacent cells. (A, Don Fawcett/Photo Researchers, Inc.)

FIGURE 14.25 Tight junction proteins   There are three major transmembrane proteins in a tight junction: occludin, claudin, and the junctional adhesion molecule (JAM). JAM has two Ig domains and interacts with a JAM on the apposing cell via the most N-terminal of these domains. Occludin and claudin interact with similar molecules on the apposing cell. All three transmembrane proteins interact with zonula occludens proteins that link to actin filaments.

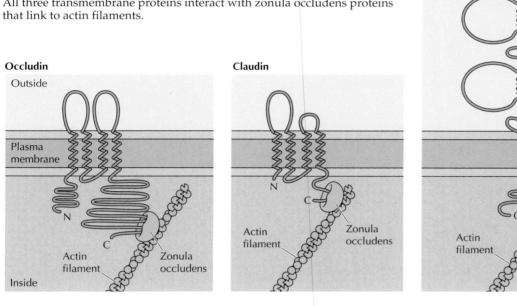

membranes do not fuse. Instead, tight junctions are formed by a network of protein strands that continues around the entire circumference of the cell (see Figure 14.24B). Each strand in these networks is composed of transmembrane proteins of the claudin, occludin, and junctional adhesion molecule (JAM) families (Figure 14.25). All three of these proteins bind to similar proteins on adjacent cells, thereby sealing the space between their plasma membranes. The cytosolic tails of claudins, occludins, and JAMs are also associated with proteins of the zonula occludens family, which link the tight junction complex to the actin cytoskeleton and hold the tight junction in place on the plasma membrane. Nectins may also be present in tight junctions but their major role seems to be in recruiting claudins to initiate tight junction formation, analogous to their role in formation of adherens junctions.

Gap Junctions

The activities of individual cells in multicellular organisms need to be closely coordinated. This can be accomplished by signaling molecules that are released from one cell and act on another, as discussed in Chapter 15. However, within an individual tissue, such as the liver, cells are often linked by **gap junctions,** which provide direct connections between the cytoplasms of adjacent cells. Gap junctions are open channels through the plasma membrane, allowing ions and small molecules (less than approximately a thousand daltons) to diffuse freely between neighboring cells, but preventing the passage of proteins and nucleic acids. Consequently, gap junctions couple both the metabolic activities and the electric responses of the cells they connect. Most cells in animal tissues—including epithelial cells, endothelial cells, and the cells of cardiac and smooth muscle—

FIGURE 14.26 Gap junctions (A) Electron micrograph of a gap junction (arrows) between two liver cells. (B) Gap junctions consist of assemblies of six connexins, which form open channels through the plasma membranes of adjacent cells. (A, Don Fawcett and R. Wood/Photo Researchers, Inc.)

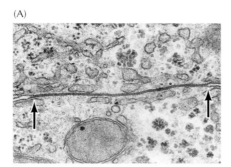

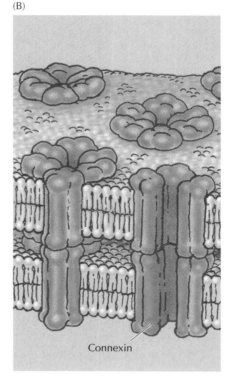

Connexin

communicate by gap junctions. In electrically excitable cells, such as heart muscle cells, the direct passage of ions through gap junctions couples and synchronizes the contractions of neighboring cells. Gap junctions also allow the passage of some intracellular signaling molecules, such as cAMP and Ca^{2+}, between adjacent cells, potentially coordinating the responses of cells in tissues.

Gap junctions are constructed of transmembrane proteins of the **connexin** family, which consists of at least 21 different human proteins. Six connexins assemble to form a cylinder with an open aqueous pore in its center (Figure 14.26). Such an assembly of connexins, a **connexon**, in the plasma membrane of one cell then aligns with a connexon of an adjacent cell, forming an open channel between the two cytoplasms. The plasma membranes of the two cells are separated by a gap corresponding to the space occupied by the connexin extracellular domains—hence the term "gap junction," which was coined by electron microscopists. Many cells express more than one member of the connexin family, and combinations of different connexin proteins may give rise to gap junctions with varying properties.

Specialized assemblies of gap junctions occur on specific nerve cells in all eukaryotes and form an **electrical synapse.** The individual connexons within the electrical synapse can be opened or closed in response to several types of signals but, when open, allow the rapid passage of ions between the two nerve cells. The importance of gap junctions—especially in the nervous system—is illustrated by the number of human diseases associated with connexon mutations.

MOLECULAR MEDICINE

Gap Junction Diseases

The Diseases

Several unrelated human diseases have been found to result from mutations in genes that encode the connexin proteins of gap junctions. The first of these diseases to be described is the X-linked form of Charcot-Marie-Tooth disease (CMT), which was mapped to mutations in the gene encoding connexin 32 in 1993. CMT is an inherited disease leading to progressive degeneration of peripheral nerves, with slow loss of muscle control and eventual muscle degeneration. CMT can also be caused by mutations in several different genes encoding myelin proteins, defects which lead directly to degeneration of myelinated nerves. In addition, one form of CMT has been mapped to a mutation in the human gene for nuclear lamin A (see Chapter 9 Molecular Medicine, Nuclear Lamina Diseases).

The finding that CMT could be caused by mutations in a connexin implied that gap junctions played a critical role in myelinated nerves. Moreover, CMT was only the first of several inherited diseases that now have been traced back to mutations in genes encoding human connexins. The most common pathological consequences of connexin mutations are cataracts, skin disorders, and deafness.

Molecular and Cellular Basis

The human genome contains 21 genes encoding different connexins (Cx genes), which are divided into three subfamilies (α, β, and γ). Mutations in 8 of these genes have been identified as causes of human disease (see table). For example, mutations in Cx32/β1 are associated with both Charcot-Marie-Tooth disease and deafness. The connexin proteins are widely expressed; Cx32/β1 is expressed in peripheral nerve, liver, and brain tissues, and Cx43α1 (which

Gap Junction Mutations in Human Disease

Disease	Connexin protein
Charcot-Marie-Tooth disease	Cx32/β1
Deafness	Cx26/β2, Cx30/β6, Cx31/β3, Cx32/β1, Cx43/α1
Skin diseases	
Erythrokeratoderma variabilis	Cx31/β3, Cx30.3/β4
Vohwinkel's syndrome	Cx26/β2
Cataracts	Cx46/α3, Cx50/α8

Cx proteins are named either by molecular weight or as a member of one subfamily.

is also associated with deafness) is expressed in over 35 different human tissues. Thus it is somewhat surprising that there are fewer than ten different types of human diseases known to be caused by connexin mutations. Since most tissues express several different connexin proteins, the simple explanation for the lack of a defect in many tissues resulting from the loss of any single connexin is that other connexins compensate for the one that is missing. The question then is why any mutation in a single connexin gene causes a human disease.

The initial discovery of Cx32/β1 mutations in CMT was based on eight different mutations in various families: six single base changes that resulted in nonconservative amino acid substitutions, one frameshift that yielded a shorter protein, and one mutation in the promoter region of the gene. These eight mutations in Cx32/β1, as well as 262 others that have been subsequently discovered, cause clinically identical CMT, suggesting that myelinated nerve might be a tissue particularly sensitive to gap junction defects. Three other tissues also appear sensitive to connexin mutations: the lens of the eye, the sensory epithelium of the inner ear, and the skin. The sensitivity of the lens is easiest to understand, because the lens fiber cells lose most of their organelles during development and fill up with crystalline protein to allow the passage of light. They must

obtain nutrients and ions from the lens epithelial cells through gap junctions, and loss of this input leads to cataract formation. The sensitivity of the sensory epithelium of the inner ear is related to the need for the epithelial cells to rapidly exchange K^+ via gap junctional communication. Less clear is the basis for overgrowth of the outer layers of the skin, but it is thought that gap junctions play a role in the balance between proliferation and differentiation, and gap junctional mutations perturb the balance.

Each of these tissues expresses additional connexins, and their failure to compensate for loss of the mutant connexin appears to be based on two phenomena. Although gap junctions can be formed from combinations of different connexins, not all connexins are able to cooperate to form functional connexons. Thus the connexins in sensitive tissues may not be able to compensate effectively for one that is mutated. In addition, gap junction connexons are assembled not at the cell surface but in the Golgi apparatus or earlier in the secretory process. Thus a single mutant connexin that cannot be properly processed and exported might act as a dominant negative and interfere with the processing of the normal connexins expressed in that tissue. This has recently been reported for a mutant of Cx46/α3 that causes congenital cataracts: the mutant

(Continued on next page)

MOLECULAR MEDICINE

protein fails to exit the ERGIC or Golgi complex and traps normal connexins along with it. It thus appears that the relationship of particular disorders to connexin mutations may result from the gap junctional requirements of a given tissue, as well as the nature of interactions between the connexins that are expressed in that tissue.

Prevention and Treatment

The identification of several human diseases caused by mutations in connexin genes illustrates the importance of gap junctions to normal tissue function. Like other junctions critical to tissue structure or function, gap junctions would seem to be obvious

locations for hereditary human diseases, yet such diseases are uncommon. Recent discoveries on the processing and assembly of gap junction connexons have started to explain the bases of these rare diseases. While it is thought that all the human connexin genes have been described, the tissue distribution of the proteins—especially during embryonic development—remains poorly understood. New knowledge about the distribution and interactions of the connexin proteins will provide a basis for understanding each of the pathologies resulting from mutations in connexin genes. Only then can possible compensatory therapies—for exam-

ple, regulating cell proliferation in the skin or modulating the exchange of K⁺ in the inner ear—be developed.

References

Bergoffen, J., S. S. Scherer, S. Wang, M. O. Scott, L. J. Bone, D. L. Paul, K. Chen, M. W. Lensch, P.F. Chance and K.H. Fischbeck. 1993. Connexin mutations in X-linked Charcot-Marie-Tooth disease. *Science* 262: 2039–2042.

Minogue, P. J., X. Liu, L. Ebihara, E. C. Beyer and V. M. Berthoud. 2005. An aberrant sequence in a connexin 46 mutant underlies congenital cataracts. *J. Biol. Chem.* 280: 40788–40795.

Wei, C. J., X. Xu and C. W. Lo, 2004. Connexins and cell signaling in development and disease. *Ann. Rev. Cell Dev. Biol.* 20: 811–838.

Plasmodesmata

Adhesion between plant cells is mediated by their cell walls rather than by transmembrane proteins. In particular, a specialized pectin-rich region of the cell wall called the **middle lamella** acts as a glue to hold adjacent cells together. Because of the rigidity of plant cell walls, stable associations between plant cells do not require the formation of cytoskeletal links such as those provided by the desmosomes and adherens junctions of animal cells. However, adjacent plant cells communicate with each other through cytoplasmic connections called **plasmodesmata** (singular, plasmodesma) (Figure 14.27). Although distinct in structure, plasmodesmata function analogously to gap junctions as a means of direct communication between adjacent cells in tissues.

Plasmodesmata form from incomplete separation of daughter cells following plant cell mitosis. At each plasmodesma, the plasma membrane of one cell is continuous with that of its neighbor, creating a channel between the two cytosols (Figure 14.28). An extension of the smooth endoplasmic reticulum passes through the pore, leaving a ring of surrounding cytoplasm through which ions and small molecules are able to pass freely between the cells. Plasmodesmata are dynamic structures that can open or close in response to appropriate stimuli, permitting the regulated passage of macromolecules between adjacent cells. In addition, there is evidence that proteins and lipids can be targeted to plasmodesmata in response to specific signals. Plasmodesmata may thus play a key role in plant development by controlling the trafficking of regulatory molecules, such as transcription factors or RNAs, between cells.

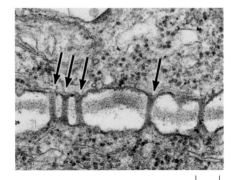

0.1 μm

FIGURE 14.27 Plasmodesmata Electron micrograph of plasmodesmata (arrows). (E. H. Newcomb, University of Wisconsin/Biological Photo Service.)

(A)

(B)

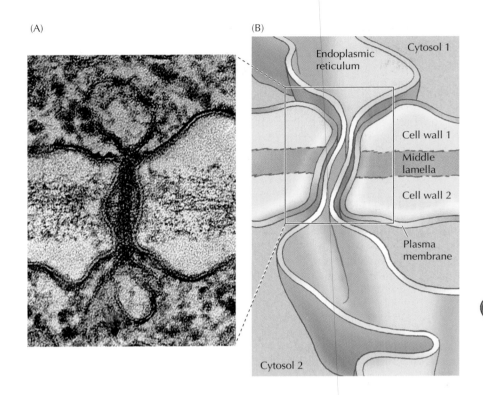

Endoplasmic reticulum

Cytosol 1

Cell wall 1

Middle lamella

Cell wall 2

Plasma membrane

Cytosol 2

FIGURE 14.28 Structure of a plasmodesma In a plasmodesma, the plasma membranes of neighboring cells are continuous, forming cytoplasmic channels through the adjacent cell walls. An extension of the endoplasmic reticulum usually passes through the channel. (A, From Tilney, L., T. J. Cooke, P. S. Connelly and M. S. Tilney. 1991. *J. Cell Biol.* 112: 739–748.)

COMPANION WEBSITE

Visit the website that accompanies **The Cell** (www.sinauer.com/cooper) for animations, videos, quizzes, problems, and other review material.

SUMMARY

KEY TERMS

CELL WALLS

Bacterial Cell Walls: The principal component of bacterial cell walls is a peptidoglycan consisting of polysaccharide chains cross-linked by short peptides.

Eukaryotic Cell Walls: The cell walls of fungi, algae, and higher plants are composed of fibrous polysaccharides (e.g., cellulose) embedded in a gel-like matrix of polysaccharides and proteins. Their rigid cell walls allow plant cells to expand rapidly by the uptake of water.

peptidoglycan

chitin, cellulose, hemicellulose, pectin, primary cell wall, secondary cell wall, lignin, turgor pressure, auxin, cellulose synthase

THE EXTRACELLULAR MATRIX AND CELL-MATRIX INTERACTIONS

Matrix Structural Proteins: The major structural proteins of the extracellular matrix are members of the large collagen protein family. Collagens form the fibrils that characterize the extracellular matrix of connective tissues, as well as forming networks in basal laminae.

Matrix Polysaccharides: Polysaccharides in the form of glycosaminoglycans and proteoglycans make up the bulk of the extracellular matrix. They bind to and modify the collagen fibrils and interact with all other matrix molecules.

extracellular matrix, basal lamina, collagen, collagen fibril, procollagen, elastic fiber, elastin

glycosaminoglycan (GAG), proteoglycan

KEY TERMS

fibronectin, laminin, entactin

integrin, focal adhesion, hemidesmosome, plakin, focal complex

cell adhesion molecule, selectin, integrin, immunoglobulin (Ig) superfamily, cadherin, heterophilic interaction, homophilic interaction, armadillo protein family, nectin, desmoglein, desmocollin

tight junction, junctional complex

gap junction, connexin, connexon, electrical synapse

middle lamella, plasmodesma

SUMMARY

Matrix Adhesion Proteins: Matrix adhesion proteins link the components of the extracellular matrix to one another and are the major binding sites for integrins, which mediate most cell-matrix adhesions.

Cell-Matrix Interactions: Integrins are the major cell surface receptors that attach cells to the extracellular matrix. At focal adhesions and hemidesmosomes, integrins provide stable links between the extracellular matrix and the actin and intermediate filament cytoskeletons, respectively.

CELL-CELL INTERACTIONS

Adhesion Junctions: Selective cell-cell interactions are mediated by four major groups of cell adhesion proteins: selectins, integrins, immunoglobulin (Ig) superfamily members, and cadherins. The cadherins link the cytoskeletons of adjacent cells at stable cell-cell junctions.

Tight Junctions: Tight junctions prevent the free passage of molecules between epithelial cells and separate the apical and basolateral domains of epithelial cell plasma membranes.

Gap Junctions: Gap junctions are open channels connecting the cytosols of adjacent animal cells. Electrical synapses are gap junctions that mediate signaling between cells of the nervous system.

Plasmodesmata: Adjacent plant cells are linked by cytoplasmic connections called plasmodesmata.

Questions

1. How do the cell walls and adjacent membranes differ between Gram-positive and Gram-negative bacteria?

2. An important function of the Na^+-K^+ pump in animal cells is the maintenance of osmotic equilibrium. Why is this unnecessary for plant cells?

3. How do hemicelluloses impart structural strength to plant cell walls?

4. What is the importance of selectively targeting different glucose transporters to the apical and basolateral domains of the plasma membrane of intestinal epithelial cells? What is the role of tight junctions in this process?

5. How would an inhibitor of the enzyme lysyl hydroxylase (the enzyme responsible for hydroxylating lysine residues in collagen) affect the stability of collagen synthesized by a cell?

6. Why do fibril-forming collagens not assemble into triple helices inside cells?

7. What property of GAGs allows them to form hydrated gels?

8. You are studying a transporter found only in the apical plasma membrane of epithelial cells. You treat the epithelial cells with a synthetic peptide that is similar to the extracellular domain of a claudin family protein. The transporter is now found in both the apical and basolateral domains of the plasma membrane. What is the most likely mechanism by which the peptide disturbs the localization of the transporter?

9. You have overexpressed a dominant negative E-cadherin lacking an extracellular domain in epithelial cells. How will this mutant affect cell-cell adhesion?

10. A mutation in an epithelial cell leads to the expression of $\alpha_6\beta_4$ integrin with a deleted cytoplasmic domain. How will this mutation affect adhesion of the epithelial cell to the basal lamina?

11. What is an electrical synapse? What function does it serve?

12. How are gap junctions and plasmodesmata similar? Are they likely to be analogous or homologous structures in animals and plants?

References and Further Reading

Cell Walls

Cosgrove, D. J. 2001. Plant cell walls: Wall-associated kinases and cell expansion. *Curr. Biol.* 11: R558–R559. [R]

Kohorn, B. D. 2001. WAKs: Cell wall associated kinases. *Curr. Opin. Cell Biol.* 13: 529–533. [R]

Merzendorfer, H. 2005. Insect chitin synthases: A review. *J. Comp Physiol [B]* 1–15. [R]

Perrin, R. M. 2001. Cellulose: How many cellulose synthases to make a plant? *Curr. Biol.* 11: R213–R216. [R]

Reiter, W. D. 2002. Biosynthesis and properties of the plant cell wall. *Curr. Opin. Plant Biol.* 5: 536–542. [R]

Somerville, C., S. Bauer, G. Brininstool, M. Facette, T. Hamann, J. Milne, E. Osborne, A. Paredez, S. Persson, Raab, T. S. Vorwerk and H. Youngs. 2004. Toward a systems approach to understanding plant cell walls. *Science* 306: 2206–2211. [R]

Stals, H. and D. Inze. 2001. When plant cells decide to divide. *Trends Plant Sci.* 6: 359–364. [R]

The Extracellular Matrix and Cell-Matrix Interactions

ffrench-Constant, C. and H. Colognato. 2004. Integrins: Versatile integrators of extracellular signals. *Trends Cell Biol.* 14: 678–686. [R]

Ginsberg, M. H., A. Partridge and S. J. Shattil. 2005. Integrin regulation. *Curr. Opin. Cell Biol.* 17: 509–516. [R]

Handel, T. M., Z. Johnson, S. E. Crown, E. K. Lau and A. E. Proudfoot. 2005. Regulation of protein function by glycosaminoglycans—As exemplified by chemokines. *Ann. Rev. Biochem.* 74: 385–410. [R]

Lecuit, T. 2005. Adhesion remodeling underlying tissue morphogenesis. *Trends Cell Biol.* 15: 34–42. [R]

Lin, X. 2004. Functions of heparan sulfate proteoglycans in cell signaling during development. *Development* 131: 6009–6021. [R]

Mao, Y. and J. E. Schwarzbauer. 2005. Fibronectin fibrillogenesis, a cell-mediated matrix assembly process. *Matrix Biol.* 24: 389–399 [R]

Miner, J. H. and P. D. Yurchenco. 2004. Laminin functions in tissue morphogenesis. *Ann. Rev. Cell Dev. Biol.* 20: 255–284. [R]

Mott, J. D. and Z. Werb. 2004. Regulation of matrix biology by matrix metalloproteinases. *Curr. Opin. Cell Biol.* 16: 558–564. [R]

Mould, A. P. and M. J. Humphries. 2004. Regulation of integrin function through conformational complexity: Not simply a knee-jerk reaction? *Curr. Opin. Cell Biol.* 16: 544–551. [R]

Myllyharju, J. and K. I. Kivirikko. 2004. Collagens, modifying enzymes and their mutations in humans, flies and worms. *Trends Genet.* 20: 33–43. [R]

Ponta, H., L. Sherman and P. A. Herrlich. 2003. CD44: From adhesion molecules to signalling regulators. *Nat. Rev. Mol. Cell Biol.* 4: 33–45. [R]

Sakisaka, T. and Y. Takai. 2004. Biology and pathology of nectins and nectin-like molecules. *Curr. Opin. Cell Biol.* 16: 513–521. [R]

Tamkun, J. W., D. W. DeSimone, D. Fonda, R. S. Patel, C. Buck, A. F. Horwitz, and R. O. Hynes. 1986. Structure of integrin, a glycoprotein involved in the transmembrane linkage between fibronectin and actin. *Cell* 46: 271–282. [P]

Toole, B. P. 2004. Hyaluronan: From extracellular glue to pericellular cue. *Nat. Rev. Cancer* 4: 528–539. [R]

Wight, T. N. 2002. Versican: A versatile extracellular matrix proteoglycan in cell biology. *Curr. Opin. Cell Biol.* 14: 617–623. [R]

Woods, A. and J. R. Couchman. 2001. Syndecan-4 and focal adhesion function. *Curr. Opin. Cell Biol.* 13: 578–583. [R]

Wozniak, M. A., K. Modzelewska, L. Kwong and P. J. Keely. 2004. Focal adhesion regulation of cell behavior. *Biochim. Biophys. Acta* 1692: 103–119. [R]

Xiao, T., J. Takagi, B. S. Coller, J. H. Wang and T. A. Springer. 2004. Structural basis for allostery in integrins and binding to fibrinogen-mimetic therapeutics. *Nature* 432: 59–67. [P]

Yurchenco, P. D. and W. G. Wadsworth. 2004. Assembly and tissue functions of early embryonic laminins and netrins. *Curr. Opin. Cell Biol.* 16: 572–579. [R]

Cell-Cell Interactions

Bamji, S. X. 2005. Cadherins: Actin with the cytoskeleton to form synapses. *Neuron* 47: 175–178. [R]

Bergoffen, J., S. S. Scherer, S. Wang, M. O. Scott, L. J. Bone, D. L. Paul, K. Chen, M. W. Lensch, P. F. Chance and K. H. Fischbeck. 1993. Connexin mutations in X-linked Charcot-Marie-Tooth disease. *Science* 262: 2039–2042. [P]

Cilia, M. L. and D. Jackson. 2004. Plasmodesmata form and function. *Curr. Opin. Cell Biol.* 16: 500–506. [R]

Garrod, D. R., A. J. Merritt and Z. Nie. 2002. Desmosomal cadherins. *Curr. Opin. Cell Biol.* 14: 537–545. [R]

Gonzalez-Mariscal, L., A. Betanzos, P. Nava and B. E. Jaramillo. 2003. Tight junction proteins. *Prog. Biophys. Mol. Biol.* 81: 1–44. [R]

Heinlein, M. 2002. Plasmodesmata: Dynamic regulation and role in macromolecular cell-to-cell signaling. *Curr. Opin. Plant Biol.* 5: 543–552. [R]

Herve, J. C., N. Bourmeyster and D. Sarrouilhe. 2004. Diversity in protein-protein interactions of connexins: Emerging roles. *Biochim. Biophys. Acta* 1662: 22–41. [R]

Hynes, R. O. 2002. Integrins: Bidirectional, allosteric signaling machines. *Cell* 110: 673–687. [R]

Knust, E. and O. Bossinger. 2002. Composition and formation of intercellular junctions in epithelial cells. *Science* 298: 1955–1959. [R]

Kobielak, A. and E. Fuchs. 2004. Alpha-catenin: At the junction of intercellular adhesion and actin dynamics. *Nat. Rev. Mol. Cell Biol.* 5: 614–625. [R]

Miranti, C. K. and J. S. Brugge. 2002. Sensing the environment: A historical perspective on integrin signal transduction. *Nat. Cell Biol.* 4: E83–E90. [R]

Miyoshi, J. and Y. Takai. 2005. Molecular perspective on tight-junction assembly and epithelial polarity. *Adv. Drug Deliv. Rev.* 57: 815–855. [R]

Rela, L. and L. Szczupak. 2004. Gap junctions: Their importance for the dynamics of neural circuits. *Mol. Neurobiol.* 30: 341–357. [R]

Richard, G. 2005. Connexin disorders of the skin. *Clin. Dermatol.* 23: 23–32. [R]

Schneeberger, E. E. and R. D. Lynch. 2004. The tight junction: A multifunctional complex. *Am. J. Physiol Cell Physiol* 286: C1213–C1228. [R]

Springer, T. A. 1994. Traffic signals for lymphocyte recirculation and leukocyte emigration: The multistep paradigm. *Cell* 76: 301–314. [R]

Tsukita, S. and M. Furuse. 2002. Claudin-based barrier in simple and stratified cellular sheets. *Curr. Opin. Cell Biol.* 14: 531–536. [R]

van Steensel, M. A. 2004. Gap junction diseases of the skin. *Am. J. Med. Genet. C. Semin. Med. Genet.* 131C: 12–19. [R]

Wei, C. J., X. Xu and C. W. Lo. 2004. Connexins and cell signaling in development and disease. *Ann. Rev. Cell Dev. Biol.* 20: 811–838. [R]

Yagi, T. and M. Takeichi. 2000. Cadherin superfamily genes: Functions, genomic organization, and neurologic diversity. *Genes Dev.* 14: 1169–1180. [R]

Yap, A. S. and E. M. Kovacs. 2003. Direct cadherin-activated cell signaling: A view from the plasma membrane. *J. Cell Biol.* 160: 11–16. [R]

Yin, T. and K. J. Green. 2004. Regulation of desmosome assembly and adhesion. *Semin. Cell Dev. Biol.* 15: 665–677. [R]

PART IV

Cell Regulation

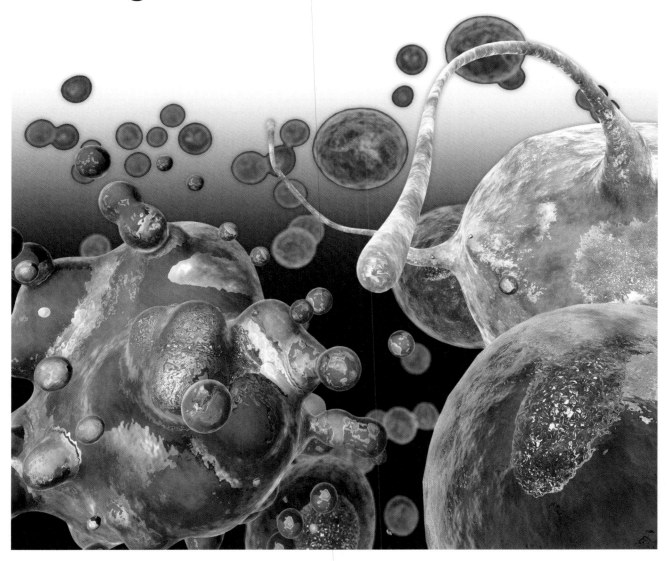

CHAPTER 15 ■ Cell Signaling

CHAPTER 16 ■ The Cell Cycle

CHAPTER 17 ■ Cell Death and Cell Renewal

CHAPTER 18 ■ Cancer

CHAPTER 15

Cell Signaling

- ■ **Signaling Molecules and Their Receptors** 600

- ■ **Functions of Cell Surface Receptors** 609

- ■ **Pathways of Intracellular Signal Transduction** 617

- ■ **Signal Transduction and the Cytoskeleton** 637

- ■ **Signaling Networks** 640

- ■ **KEY EXPERIMENT:**
  The Src Protein-Tyrosine Kinase 612

- ■ **MOLECULAR MEDICINE:**
  Cancer: Signal Transductlion and the *ras* Oncogenes 629

ALL CELLS RECEIVE AND RESPOND TO SIGNALS from their environment. Even the simplest bacteria sense and swim toward high concentrations of nutrients, such as glucose or amino acids. Many unicellular eukaryotes also respond to signaling molecules secreted by other cells, allowing cell-cell communication. Mating between yeast cells, for example, is signaled by peptides that are secreted by one cell and bind to receptors on the surface of another. It is in multicellular organisms, however, that cell-cell communication reaches its highest level of sophistication. Whereas the cells of prokaryotes and unicellular eukaryotes are largely autonomous, the behavior of each individual cell in multicellular plants and animals must be carefully regulated to meet the needs of the organism as a whole. This is accomplished by a variety of signaling molecules that are secreted or expressed on the surface of one cell and bind to receptors expressed by other cells, thereby integrating and coordinating the functions of the many individual cells that make up organisms as complex as human beings.

The binding of most signaling molecules to their receptors initiates a series of intracellular reactions that regulate virtually all aspects of cell behavior, including metabolism, movement, proliferation, survival, and differentiation. Understanding the molecular components of these pathways and how they are regulated has thus become a major area of research in contemporary cell biology. Interest in this area is further heightened by the fact that many cancers arise as a result of a breakdown in the signaling pathways that control normal cell proliferation and survival. In fact, many of our current insights into cell signaling mechanisms have come from the study of cancer cells—a striking example of the fruitful interplay between medicine and basic research in cell and molecular biology.

Direct Cell-Cell Signaling

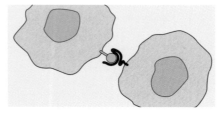

Signaling by Secreted Molecules

(A) Endocrine signaling

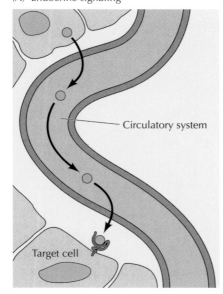

(B) Paracrine signaling

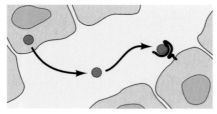

(C) Autocrine signaling

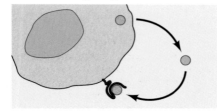

Signaling Molecules and Their Receptors

Many different kinds of molecules transmit information between the cells of multicellular organisms. Although all these molecules act as ligands that bind to receptors expressed by their target cells, there is considerable variation in the structure and function of the different types of molecules that serve as signal transmitters. Structurally, the signaling molecules used by plants and animals range in complexity from simple gases to proteins. Some of these molecules carry signals over long distances, whereas others act locally to convey information between neighboring cells. In addition, signaling molecules differ in their mode of action on their target cells. Some signaling molecules are able to cross the plasma membrane and bind to intracellular receptors in the cytoplasm or nucleus, whereas most bind to receptors expressed on the target cell surface. The sections that follow discuss the major types of signaling molecules and the receptors with which they interact. Subsequent discussion in this chapter focuses on the mechanisms by which cell surface receptors then function to regulate cell behavior.

Modes of Cell-Cell Signaling

Cell signaling can result either from the direct interaction of a cell with its neighbor or from the action of secreted signaling molecules (Figure 15.1). Signaling by direct cell-cell (or cell-matrix) interactions plays a critical role in regulating the behavior of cells in animal tissues. For example, the integrins and cadherins (which were discussed in the previous chapter) function not only as cell adhesion molecules but also as signaling molecules that regulate cell proliferation and survival in response to cell-cell and cell-matrix contacts. In addition, cells express a variety of cell surface receptors that interact with signaling molecules on the surface of neighboring cells. Signaling via such direct cell-cell interactions plays a critical role in regulating the many interactions between different types of cells that take place during embryonic development as well as in the maintenance of adult tissues.

The multiple varieties of signaling by secreted molecules are frequently divided into three general categories based on the distance over which signals are transmitted. In **endocrine signaling**, the signaling molecules (**hormones**) are secreted by specialized endocrine cells and carried through the circulation to act on target cells at distant body sites. A classic example is provided by the steroid hormone estrogen, which is produced by the ovary and stimulates development and maintenance of the female reproductive system and secondary sex characteristics. In animals, more than 50 different hormones are produced by endocrine glands, including the pituitary, thyroid, parathyroid, pancreas, adrenal glands, and gonads.

In contrast to hormones, some signaling molecules act locally to affect the behavior of nearby cells. In **paracrine signaling**, a molecule released by one cell acts on neighboring target cells. An example is provided by the action of neurotransmitters in carrying signals between nerve cells at a synapse. Finally, some cells respond to signaling molecules that they themselves produce. One important example of such **autocrine signaling** is the

FIGURE 15.1 Modes of cell-cell signaling Cell signaling can take place either through direct cell-cell contacts or through the action of secreted signaling molecules. (A) In endocrine signaling, hormones are carried through the circulatory system to act on distant target cells. (B) In paracrine signaling, a molecule released from one cell acts locally to affect nearby target cells. (C) In autocrine signaling, a cell produces a signaling molecule to which it also responds.

response of cells of the vertebrate immune system to foreign antigens. Certain types of T lymphocytes respond to antigenic stimulation by synthesizing a growth factor that drives their own proliferation, thereby increasing the number of responsive T lymphocytes and amplifying the immune response. It is also noteworthy that abnormal autocrine signaling frequently contributes to the uncontrolled growth of cancer cells (see Chapter 18). In this situation, a cancer cell produces a growth factor to which it also responds, thereby continuously driving its own unregulated proliferation.

Steroid Hormones and the Nuclear Receptor Superfamily

As already noted, all signaling molecules act by binding to receptors expressed by their target cells. In many cases these receptors are expressed on the target cell surface, but some receptors are intracellular proteins located in the cytosol or in the nucleus. These intracellular receptors respond to small hydrophobic signaling molecules that are able to diffuse across the plasma membrane. The **steroid hormones** are the classic examples of this group of signaling molecules, which also includes thyroid hormone, vitamin D_3, and retinoic acid (Figure 15.2).

The steroid hormones (including testosterone, estrogen, progesterone, the corticosteroids, and ecdysone) are all synthesized from cholesterol. **Testosterone**, **estrogen**, and **progesterone** are the sex steroids, which are produced by the gonads. The **corticosteroids** are produced by the adrenal gland. They include the **glucocorticoids**, which act on a variety of cells to stimulate production of glucose, and the **mineralocorticoids**, which act on

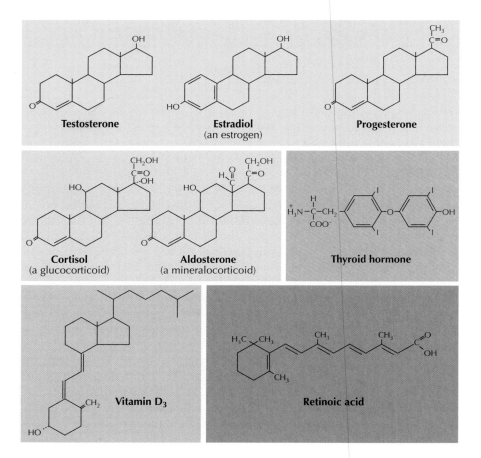

FIGURE 15.2 Structure of steroid hormones, thyroid hormone, vitamin D_3, and retinoic acid   The steroids include the sex hormones (testosterone, estrogen, and progesterone), glucocorticoids, and mineralocorticoids.

FIGURE 15.3 Estrogen action
Estrogen diffuses across the plasma membrane and binds to its receptor in the nucleus. In the absence of hormone, estrogen receptor is bound to Hsp90. Estrogen binding displaces the receptor from Hsp90 and allows the formation of receptor dimers, which bind DNA, associate with coactivators with histone acetyltransferase (HAT) activity, and stimulate transcription of their target genes.

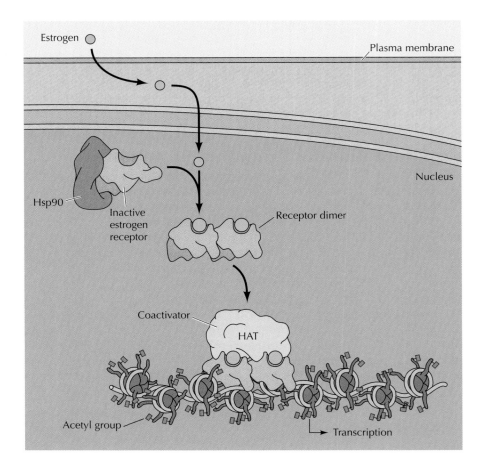

the kidney to regulate salt and water balance. **Ecdysone** is an insect hormone that plays a key role in development by triggering the metamorphosis of larvae to adults. The **brassinosteroids** are plant-specific steroid hormones that control a number of developmental processes, including cell growth and differentiation.

Although thyroid hormone, vitamin D_3, and retinoic acid are structurally and functionally distinct from the steroids, they share a common mechanism of action in their target cells. **Thyroid hormone** is synthesized from tyrosine in the thyroid gland; it plays important roles in development and regulation of metabolism. **Vitamin D_3** regulates Ca^{2+} metabolism and bone growth. **Retinoic acid** and related compounds (**retinoids**) synthesized from vitamin A play important roles in vertebrate development.

Because of their hydrophobic character the steroid hormones, thyroid hormone, vitamin D_3, and retinoic acid are all able to enter cells by diffusing across the plasma membrane. Once inside the cell, they bind to intracellular receptors that are expressed by the hormonally responsive target cells. These receptors, which are members of a family of proteins known as the **nuclear receptor superfamily**, are transcription factors that contain related domains for ligand binding, DNA binding, and transcriptional activation. Ligand binding regulates their function as activators or repressors of their target genes, so the steroid hormones and related molecules directly regulate gene expression.

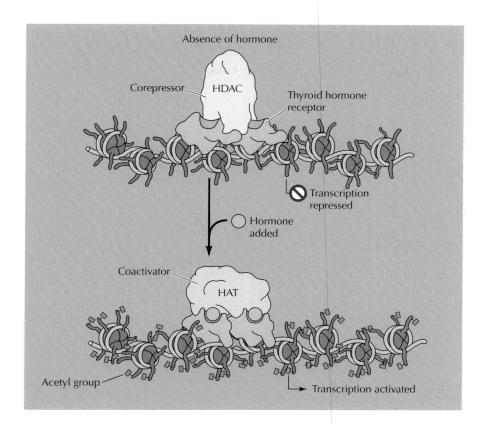

FIGURE 15.4 Gene regulation by the thyroid hormone receptor Thyroid hormone receptor binds DNA in either the presence or absence of hormone. However, hormone binding changes the function of the receptor from a repressor to an activator of target gene transcription. In the absence of hormone, the receptor associates with corepressors with histone deacetylase (HDAC) activity. In the presence of hormone, the receptor associates with coactivators with histone acetyltransferase (HAT) activity.

Ligand binding has distinct effects on different receptors. Some members of the steroid receptor superfamily are unable to bind to DNA in the absence of hormone. Estrogen receptor, for example, is bound to Hsp90 chaperones in the absence of hormone (Figure 15.3). The binding of estrogen induces a conformational change in the receptor, displacing Hsp90 and leading to the formation of receptor dimers that bind to regulatory DNA sequences and activate transcription of target genes. In other cases, the receptor binds DNA in either the presence or absence of hormone, but hormone binding alters the activity of the receptor as a transcriptional regulatory molecule. For example, in the absence of hormone, thyroid hormone receptor is associated with a corepressor complex and represses transcription of its target genes (Figure 15.4). Hormone binding induces a conformational change that results in the interaction of the receptor with coactivators rather than corepressors, leading to transcriptional activation of thyroid hormone-inducible genes.

Nitric Oxide and Carbon Monoxide

The simple gas nitric oxide (NO) is a major paracrine signaling molecule in the nervous, immune, and circulatory systems. Like the steroid hormones, NO is able to diffuse directly across the plasma membrane of its target cells. The molecular basis of NO action, however, is distinct from that of steroid action; rather than binding to a receptor that regulates transcription, NO alters the activity of intracellular target enzymes.

Nitric oxide is synthesized from the amino acid arginine by the enzyme nitric oxide synthase (Figure 15.5). Once synthesized, NO diffuses out of the

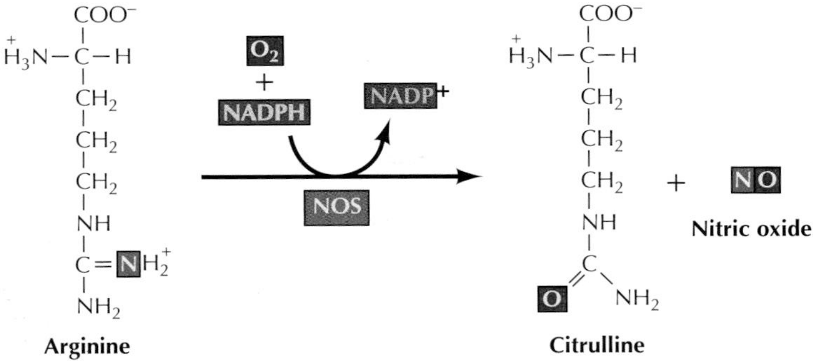

Arginine   Citrulline

Acetylcholine

$CH_3 - C - O - CH_2 - CH_2 - N^+ - (CH_3)_3$

Glycine

$H_3\overset{+}{N} - CH_2 - C - O^-$

Glutamate

$H_3\overset{+}{N} - CH - CH_2 - CH_2 - C - O^-$
         $|$
         $C = O$
         $|$
         O^-

Dopamine

$CH_2 - CH_2 - NH_3^+$

Norepinephrine

$CH - CH_2 - NH_3^+$
$|$
OH

Epinephrine

$CH - CH_2 - NH_2^+ - CH_3$
$|$
OH

Serotonin

$CH_2 - CH_2 - NH_3^+$

Histamine

$HC = C - CH_2 - CH_2 - NH_3^+$

γ-Aminobutyric acid (GABA)

$H_3\overset{+}{N} - CH_2 - CH_2 - CH_2 - C - O^-$

cell and can act locally to affect nearby cells. Its action is restricted to such local effects because NO is extremely unstable, with a half-life of only a few seconds. The major intracellular target of NO is guanylyl cyclase. NO binds to a heme group at the active site of this enzyme, stimulating synthesis of the second messenger cyclic GMP (discussed later in this chapter). In addition, NO may directly modify some target proteins by nitrosylation of cysteine residues (see Figure 8.40). A well-characterized example of NO action is signaling the dilation of blood vessels. The first step in this process is the release of neurotransmitters, such as acetylcholine, from the termini of nerve cells in the blood vessel wall. These neurotransmitters act on endothelial cells to stimulate NO synthesis. NO then diffuses to neighboring smooth muscle cells where it activates guanylyl cyclase resulting in synthesis of cyclic GMP, which induces muscle cell relaxation and blood vessel dilation. For example, NO is responsible for signaling the dilation of blood vessels that leads to penile erection. It is also interesting to note that the medical use of nitroglycerin in treatment of heart disease is based on its conversion to NO, which dilates coronary blood vessels and increases blood flow to the heart.

Another simple gas, carbon monoxide (CO), also functions as a signaling molecule in the nervous system. CO is closely related to NO and appears to act similarly as a neurotransmitter and mediator of blood vessel dilation. The synthesis of CO in brain cells, like that of NO, is stimulated by neurotransmitters. In addition, CO can stimulate guanylate cyclase, which may also represent the major physiological target of CO signaling.

Neurotransmitters

The **neurotransmitters** carry signals between neurons or from neurons to other types of target cells (such as muscle cells). They are a diverse group of small hydrophilic molecules including acetylcholine, dopamine, epinephrine (adrenaline), serotonin, histamine, glutamate, glycine, and γ-aminobutyric acid (GABA) (Figure 15.6). The release of neurotransmitters is signaled by the arrival of an action potential at the terminus of a neuron (see Figure 13.22). The neurotransmitters then diffuse across the synaptic cleft and bind to receptors on the target cell surface. Note that some neurotransmitters can also act as hormones—for example, epinephrine functions both as a neurotransmitter and as a hormone produced by the adrenal gland to signal glycogen breakdown in muscle cells.

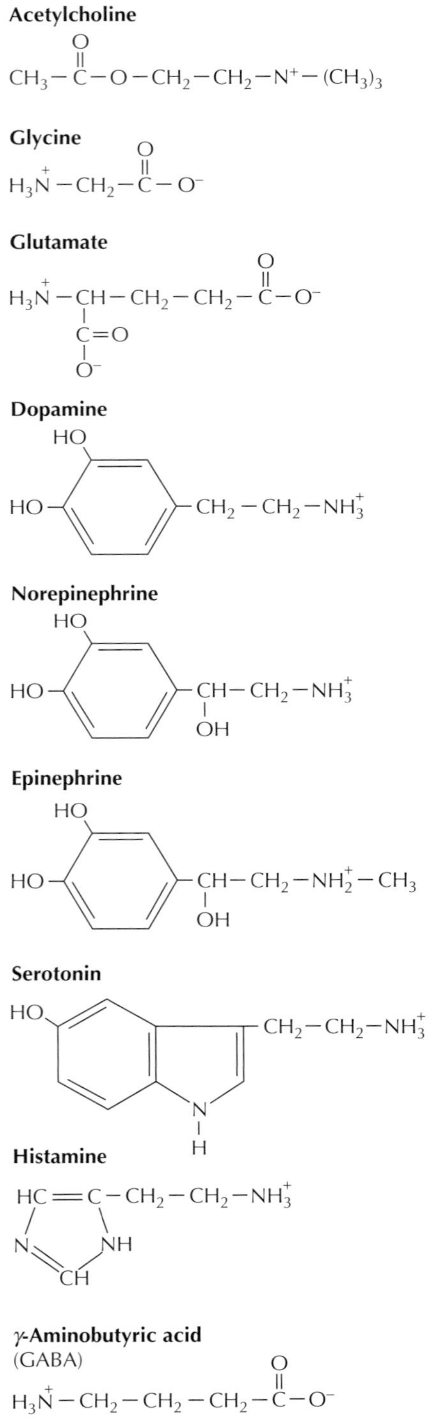

FIGURE 15.6 Structure of representative neurotransmitters The neurotransmitters are hydrophilic molecules that bind to cell surface receptors.

Because the neurotransmitters are hydrophilic molecules they are unable to cross the plasma membrane of their target cells. Therefore, in contrast to steroid hormones and NO or CO, the neurotransmitters act by binding to cell surface receptors. Many neurotransmitter receptors are ligand-gated ion channels, such as the acetylcholine receptor discussed in Chapter 13 (see Figure 13.23). Neurotransmitter binding to these receptors induces a conformational change that opens ion channels, directly resulting in changes in ion flux in the target cell. Other neurotransmitter receptors are coupled to G proteins—a major group of signaling molecules (discussed later in this chapter) that link cell surface receptors to a variety of intracellular responses. In the case of neurotransmitter receptors, the associated G proteins frequently act to indirectly regulate ion channel activity.

Peptide Hormones and Growth Factors

The widest variety of signaling molecules in animals are peptides, ranging in size from only a few to more than a hundred amino acids. This group of signaling molecules includes peptide hormones, neuropeptides, and a diverse array of polypeptide growth factors (Table 15.1). Well-known examples of **peptide hormones** include insulin, glucagon, and the hormones produced by the pituitary gland (growth hormone, follicle-stimulating hormone, prolactin, and others).

Neuropeptides are secreted by some neurons instead of the small-molecule neurotransmitters discussed in the previous section. Some of these

TABLE 15.1 Representative Peptide Hormones, Neuropeptides, and Growth Factors

Signaling molecule	Size[a]	Activities[b]
Peptide hormones		
Insulin	A = 21, B = 30	Regulation of glucose uptake; stimulation of cell proliferation
Glucagon	29	Stimulation of glucose synthesis
Growth hormone	191	General stimulation of growth
Follicle-stimulating hormone (FSH)	$\alpha = 92, \beta = 118$	Stimulation of the growth of oocytes and ovarian follicles
Prolactin	198	Stimulation of milk production
Neuropeptides and neurohormones		
Substance P	11	Sensory synaptic transmission
Oxytocin	9	Stimulation of smooth muscle contraction
Vasopressin	9	Stimulation of water reabsorption in the kidney
Enkephalin	5	Analgesic
β-Endorphin	31	Analgesic
Growth factors		
Nerve growth factor (NGF)	118	Differentiation and survival of neurons
Epidermal growth factor (EGF)	53	Proliferation of many types of cells
Platelet-derived growth factor (PDGF)	A = 125, B = 109	Proliferation of fibroblasts and other cell types
Interleukin-2	133	Proliferation of T lymphocytes
Erythropoietin	166	Development of red blood cells

[a] Size is indicated in number of amino acids. Some hormones and growth factors consist of two different polypeptide chains, which are designated either A and B or α and β.

[b] Most of these hormones and growth factors possess other activities in addition to those indicated.

FIGURE 15.7 Structure of epidermal growth factor (EGF)   EGF is a single polypeptide chain of 53 amino acids. Disulfide bonds between cysteine residues are indicated. (After G. Carpenter and S. Cohen, 1979. *Ann. Rev. Biochem.* 48: 193.)

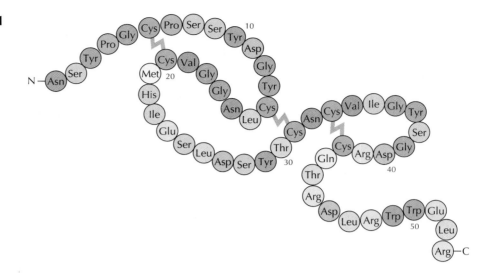

peptides, such as the **enkephalins** and **endorphins**, function not only as neurotransmitters at synapses but also as **neurohormones** that act on distant cells. The enkephalins and endorphins have been widely studied because of their activity as natural analgesics that decrease pain responses in the central nervous system. Discovered during studies of drug addiction, they are naturally occurring compounds that bind to the same receptors on the surface of brain cells as morphine does.

The polypeptide **growth factors** include a wide variety of signaling molecules that control animal cell growth and differentiation. The first of these factors (**nerve growth factor**, or **NGF**) was discovered by Rita Levi-Montalcini in the 1950s. NGF is a member of a family of polypeptides (called **neurotrophins**) that regulate the development and survival of neurons. During the course of experiments on NGF, Stanley Cohen serendipitously discovered an unrelated factor (called **epidermal growth factor**, or **EGF**) that stimulates cell proliferation. EGF, a 53-amino-acid polypeptide (Figure 15.7), has served as the prototype of a large array of growth factors that play critical roles in controlling animal cell proliferation, both during embryonic development and in adult organisms.

A good example of growth factor action is provided by the activity of **platelet-derived growth factor** (**PDGF**) in wound healing. PDGF is stored in blood platelets and released during blood clotting at the site of a wound. It then stimulates the proliferation of fibroblasts in the vicinity of the clot, thereby contributing to regrowth of the damaged tissue. Members of another large group of polypeptide growth factors (called **cytokines**) regulate the development and differentiation of blood cells and control the activities of lymphocytes during the immune response. Other polypeptide growth factors (**membrane-anchored growth factors**) remain associated with the plasma membrane rather than being secreted into extracellular fluids, therefore functioning specifically as signaling molecules during direct cell-cell interactions.

Peptide hormones, neuropeptides, and growth factors are unable to cross the plasma membrane of their target cells, so they act by binding to cell surface receptors, as discussed later in this chapter. As might be expected from the critical roles of polypeptide growth factors in controlling cell proliferation, abnormalities in growth factor signaling are the basis for a variety of diseases, including many kinds of cancer. For example, abnormal expression of the EGF receptor is an important factor in the development of many human cancers, and inhibitors of the EGF receptor appear to be promising agents for cancer treatment (see Chapter 18).

Eicosanoids

Several types of lipids serve as signaling molecules that, in contrast to the steroid hormones, act by binding to cell surface receptors. The most important of these molecules are members of a class of lipids called the **eicosanoids**, which includes **prostaglandins**, **prostacyclin**, **thromboxanes**, and **leukotrienes** (Figure 15.8). The eicosanoids are rapidly broken down and therefore act locally in autocrine or paracrine signaling pathways. They stimulate a variety of responses in their target cells, including blood platelet aggregation, inflammation, and smooth-muscle contraction.

All eicosanoids are synthesized from arachidonic acid, which is formed from phospholipids. The first step in the pathway leading to synthesis of either prostaglandins or thromboxanes is the conversion of arachidonic acid to prostaglandin H_2. Interestingly, the enzyme that catalyzes this reaction (cyclooxygenase) is the target of aspirin and other nonsteroidal anti-inflammatory drugs (NSAIDs). By inhibiting synthesis of the prostaglandins, aspirin reduces inflammation and pain. By inhibiting synthesis of thromboxane, aspirin also reduces platelet aggregation and blood clotting.

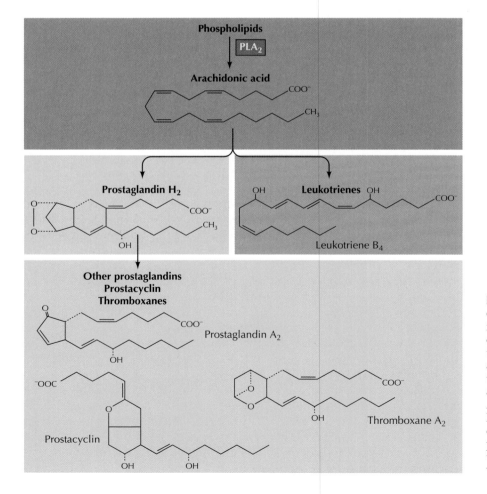

FIGURE 15.8 Synthesis and structure of eicosanoids   The eicosanoids include the prostaglandins, prostacyclin, thromboxanes, and leukotrienes. They are synthesized from arachidonic acid, which is formed by the hydrolysis of phospholipids catalyzed by phospholipase A_2 (PLA$_2$). Arachidonic acid can then be metabolized via two alternative pathways—one pathway leads to synthesis of prostaglandins, prostacyclin, and thromboxanes, while the other pathway leads to synthesis of leukotrienes.

Because of this activity, small daily doses of aspirin are frequently prescribed for prevention of strokes. In addition, aspirin and NSAIDs have been found to reduce the frequency of colon cancer in both animal models and humans, apparently by inhibiting the synthesis of prostaglandins that act to stimulate cell proliferation and promote cancer development. It is noteworthy that there are two forms of cyclooxygenase: COX-1 and COX-2. COX-1 is thought to be principally responsible for the normal physiological production of prostaglandins and COX-2 for the increased prostaglandin production associated with inflammation and disease. Consequently, selective inhibitors of COX-2 have been developed based on the rationale that such drugs would be more effective and have fewer side effects than aspirin or conventional NSAIDs, which inhibit both COX-1 and COX-2. However, selective COX-2 inhibitors may also be associated with serious side effects, including an increased risk of cardiovascular disease.

Plant Hormones

Plant growth and development are regulated by a group of small molecules called **plant hormones**. The levels of these molecules within the plant are typically modified by environmental factors, such as light or infection, so they coordinate the responses of tissues in different parts of the plant to environmental signals.

The plant hormones are classically divided into five major groups: **auxins**, **gibberellins**, **cytokinins**, **abscisic acid**, and **ethylene** (Figure 15.9), although several additional plant hormones (such as the plant steroid hormones) have recently been discovered. The first plant hormone to be identified was auxin, with the early experiments leading to its discovery having been performed by Charles Darwin in the 1880s. One of the effects of auxins

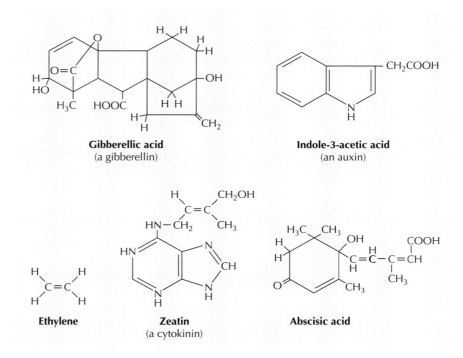

FIGURE 15.9 Structure of plant hormones

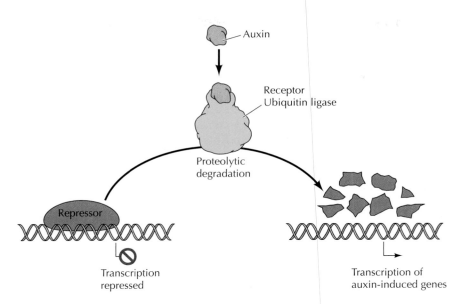

FIGURE 15.10 Auxin signaling Auxin binds to a receptor with ubiquitin ligase activity. This stimulates ubiquitination and proteolysis of a specific transcriptional repressor, leading to transcription of auxin-induced genes.

is to induce plant cell elongation by weakening the cell wall (see Figure 14.7). In addition, auxins regulate many other aspects of plant development, including cell division and differentiation. The other plant hormones likewise have multiple effects in their target tissues, including stem elongation (gibberellins), fruit ripening (ethylene), cell division (cytokinins), and the onset of dormancy (abscisic acid).

The signaling pathways triggered by these hormones in plants use a variety of mechanisms that are conserved in animal cells, as well as a number of elements that are unique to plants. For example, one well understood pathway, which has been elucidated by genetic analysis of *Arabidopsis thaliana*, signals the response of plant cells to ethylene. Elements of this pathway include an ethylene receptor on the plant cell surface, a protein kinase related to the Raf protein kinases of animal cells, and a novel transcription factor that regulates expression of ethylene-responsive genes. Recent studies have also elucidated the mechanism of auxin action (Figure 15.10). Auxin binds to a receptor associated with a ubiquitin ligase (see Figure 8.42), stimulating the ubiquitination and degradation of specific transcriptional repressors and resulting in auxin-mediated gene induction.

Functions of Cell Surface Receptors

As already reviewed, most ligands responsible for cell-cell signaling (including neurotransmitters, peptide hormones, and growth factors) bind to receptors on the surface of their target cells. Consequently, a major challenge in understanding cell-cell signaling is unraveling the mechanisms by which cell surface receptors transmit the signals initiated by ligand binding. As discussed in Chapter 13, some neurotransmitter receptors are ligand-gated ion channels that directly control ion flux across the plasma membrane. Other cell surface receptors, including the receptors for peptide hormones and growth factors, act instead by regulating the activity of

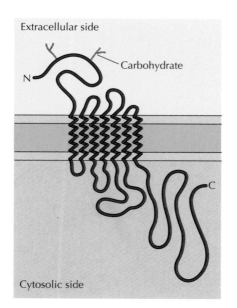

Extracellular side

Carbohydrate

N

C

Cytosolic side

FIGURE 15.11 Structure of a G protein-coupled receptor The G protein-coupled receptor is characterized by seven transmembrane α helices.

■ The G protein-coupled receptors responsible for our sense of smell (odorant receptors) are encoded by more than 500 genes in the human genome.

intracellular proteins. These proteins then transmit signals from the receptor to a series of additional intracellular targets, frequently including transcription factors. Ligand binding to a receptor on the surface of the cell thus initiates a chain of intracellular reactions, ultimately reaching the target cell nucleus and resulting in programmed changes in gene expression. The functions of the major classes of cell surface receptors are discussed here, with the pathways of intracellular signaling downstream of these receptors being considered in the next section of this chapter.

G Protein-Coupled Receptors

The largest family of cell surface receptors transmits signals to intracellular targets via the intermediary action of guanine nucleotide-binding proteins called **G proteins**. More than a thousand such **G protein-coupled receptors** have been identified, including the receptors for eicosanoids, many neurotransmitters, neuropeptides, and peptide hormones. In addition, the G protein-coupled receptor family includes a large number of receptors that are responsible for smell, sight, and taste.

The G protein-coupled receptors are structurally and functionally related proteins characterized by seven membrane-spanning α helices (Figure 15.11). The binding of ligands to the extracellular domain of these receptors induces a conformational change that allows the cytosolic domain of the receptor to bind to a G protein associated with the inner face of the plasma membrane. This interaction activates the G protein, which then dissociates from the receptor and carries the signal to an intracellular target, which may be either an enzyme or an ion channel.

The discovery of G proteins came from studies of hormones (such as epinephrine) that regulate the synthesis of cyclic AMP (cAMP) in their target cells. As discussed later in the chapter, cAMP is an important second messenger that mediates cellular responses to a variety of hormones. In the 1970s Martin Rodbell and his colleagues made the key observation that GTP is required for hormonal stimulation of adenylyl cyclase (the enzyme responsible for cAMP formation). This finding led to the discovery that a guanine nucleotide-binding protein (called a G protein) is an intermediary in adenylyl cyclase activation (Figure 15.12). Since then an array of G proteins have been found to act as physiological switches that regulate the activities of a variety of intracellular targets in response to extracellular signals.

FIGURE 15.12 Hormonal activation of adenylyl cyclase Binding of hormone promotes the interaction of the receptor with a G protein. The activated G protein α subunit then dissociates from the receptor and stimulates adenylyl cyclase, which catalyzes the conversion of ATP to cAMP.

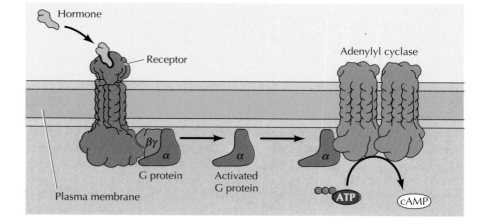

FIGURE 15.13 Regulation of G proteins

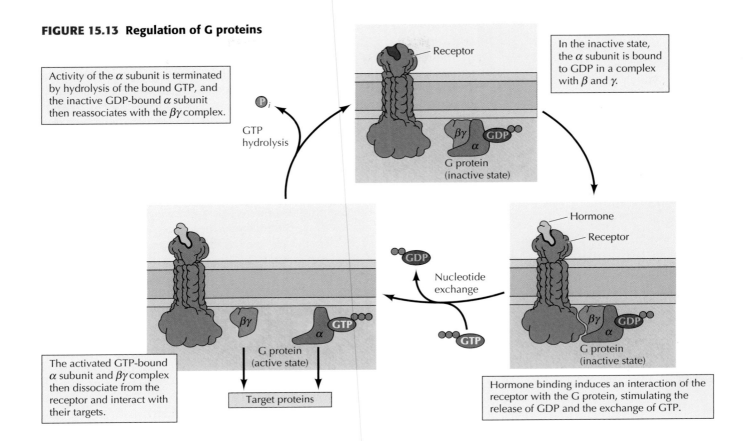

Activity of the α subunit is terminated by hydrolysis of the bound GTP, and the inactive GDP-bound α subunit then reassociates with the $\beta\gamma$ complex.

In the inactive state, the α subunit is bound to GDP in a complex with β and γ.

GTP hydrolysis

Receptor

$\beta\gamma$ GDP

α

G protein (inactive state)

The activated GTP-bound α subunit and $\beta\gamma$ complex then dissociate from the receptor and interact with their targets.

$\beta\gamma$ α GTP

G protein (active state)

Target proteins

GDP

Nucleotide exchange

GTP

Hormone

Receptor

$\beta\gamma$ GDP

α

G protein (inactive state)

Hormone binding induces an interaction of the receptor with the G protein, stimulating the release of GDP and the exchange of GTP.

G proteins consist of three subunits designated α, β, and γ (Figure 15.13). They are frequently called **heterotrimeric G proteins** to distinguish them from other guanine nucleotide-binding proteins, such as the Ras proteins discussed later in the chapter. The α subunit binds guanine nucleotides, which regulate G protein activity. In the resting state, α is bound to GDP in a complex with β and γ. Hormone binding induces a conformational change in the receptor, such that the cytosolic domain of the receptor interacts with the G protein and stimulates the release of bound GDP and its exchange for GTP. The activated GTP-bound α subunit then dissociates from β and γ, which remain together and function as a $\beta\gamma$ complex. Both the active GTP-bound α subunit and the $\beta\gamma$ complex then interact with their targets to elicit an intracellular response. The activity of the α subunit is terminated by hydrolysis of the bound GTP, and the inactive α subunit (now with GDP bound) then reassociates with the $\beta\gamma$ complex, ready for the cycle to start anew.

Mammalian genomes encode 20 different α subunits, 5 β subunits, and 12 γ subunits. Different G proteins associate with different receptors, so this array of G proteins couples receptors to distinct intracellular targets. For example, the G protein associated with the epinephrine receptor is called G_s because its α subunit stimulates adenylyl cyclase (see Figure 15.12). Other G protein α and $\beta\gamma$ subunits act instead to inhibit adenylyl cyclase or to regulate the activities of other target enzymes.

In addition to regulating target enzymes, both the α and $\beta\gamma$ subunits of some G proteins directly regulate ion channels. A good example is provided

15.2 WEBSITE ANIMATION

Signal Transduction
The largest family of cell surface receptors transmits signals inside the cell by activating G proteins, which bind GTP and then activate effector proteins.

by the action of the neurotransmitter acetylcholine on heart muscle, which is distinct from its effects on nerve and skeletal muscle. The nicotinic acetylcholine receptor on nerve and skeletal muscle cells is a ligand-gated ion channel (see Figure 13.23). Heart muscle cells have a different acetylcholine receptor, which is G protein-coupled. This G protein is designated G_i because its α subunit *inhibits* adenylyl cyclase. In addition, the G_i $\beta\gamma$ subunits act directly to open K^+ channels in the plasma membrane, which has the effect of slowing heart muscle contraction.

Receptor Protein-Tyrosine Kinases

In contrast to the G protein-coupled receptors, other cell surface receptors are directly linked to intracellular enzymes. The largest family of such enzyme-linked receptors is the **receptor protein-tyrosine kinases**, which phosphorylate their substrate proteins on tyrosine residues. This family includes the receptors for most polypeptide growth factors, so protein-tyrosine

---

KEY EXPERIMENT

The Src Protein-Tyrosine Kinase

Transforming Gene Product of Rous Sarcoma Virus Phosphorylates Tyrosine

Tony Hunter and Bartholomew M. Sefton

The Salk Institute, San Diego, CA

Proceedings of the National Academy of Science, USA, 1980, Volume 77, pages 1311–1315

Tony Hunter

Bartholomew Sefton

The Context

Following its isolation in 1911 Rous sarcoma virus (RSV) became the first virus that was generally accepted to cause tumors in animals (see the Molecular Medicine box in Chapter 1). Several features of RSV then made it an attractive model for studying the development of cancer. In particular, the small size of the RSV genome offered the hope of identifying specific viral genes responsible for inducing the abnormal proliferation that is characteristic of cancer cells. This goal was reached in the 1970s when it was established that a single RSV gene (called *src* for *sarcoma*) is required for tumor induction. Importantly, a closely related *src* gene was also found to be part of the normal genetic complement of a variety of vertebrates, including humans. Since the viral Src protein is responsible for driving the uncontrolled proliferation of cancer

cells, it appeared that understanding Src function would yield crucial insights into the molecular bases of both cancer induction and the regulation of normal cell proliferation.

In 1977 Ray Erikson and his colleagues identified the Src protein by immunoprecipitation (see Figure 4.30) with antisera from animals bearing RSV-induced tumors. Shortly thereafter, it was found that incubation of Src immunoprecipitates with radioactive ATP resulted in phosphorylation of the immunoglobulin molecules. Src therefore appeared to be a protein kinase, clearly implicating protein phosphorylation in the control of cell proliferation.

All previously studied protein kinases phosphorylated serine or threonine residues, which were also the only phosphoamino acids to have been detected in animal cells. However, Walter Eckhardt and Tony Hunter had

observed in 1979 that the oncogenic protein of another animal tumor virus (polyomavirus) was phosphorylated on a tyrosine residue. Hunter and Sefton therefore tested the possibility that Src might phosphorylate tyrosine, rather than serine/threonine, residues in its substrate proteins. Their experiments demonstrated that Src does indeed function as a protein-tyrosine kinase—an activity now recognized as playing a central role in cell signaling pathways.

phosphorylation has been particularly well studied as a signaling mechanism involved in the control of animal cell growth and differentiation. Indeed, the first protein-tyrosine kinase was discovered in 1980 during studies of the oncogenic proteins of animal tumor viruses—in particular, Rous sarcoma virus—by Tony Hunter and Bartholomew Sefton. The EGF receptor was then found to function as a protein-tyrosine kinase by Stanley Cohen and his colleagues, clearly establishing protein-tyrosine phosphorylation as a key signaling mechanism in the response of cells to growth factor stimulation.

The human genome encodes 59 receptor protein-tyrosine kinases, including the receptors for EGF, NGF, PDGF, insulin, and many other growth factors. All these receptors share a common structural organization: an N-terminal extracellular ligand-binding domain, a single transmembrane α helix, and a cytosolic C-terminal domain with protein-tyrosine kinase activity (Figure 15.14). Most of the receptor protein-tyrosine kinases consist of single polypeptides, although the insulin receptor and some

KEY EXPERIMENT

The Experiments

Hunter and Sefton identified the amino acid phosphorylated by Src by incubating Src immunoprecipitates with ^{32}P-labeled ATP. The amino acid that was phosphorylated by Src in the substrate protein (in this case, immunoglobulin) therefore became radioactively labeled. The immunoglobulin was then isolated and hydrolyzed to yield individual amino acids, which were analyzed by electrophoresis and chromatography methods that separated phosphotyrosine, phosphoserine, and phosphothreonine (see figure). The radioactive amino acid detected in these experiments was phosphotyrosine, indicating that Src specifically phosphorylates tyrosine residues.

Further experiments showed that the normal cell Src protein, as well as viral Src, functions as a protein-tyrosine kinase in immunoprecipitation assays. In addition, Hunter and Sefton extended these *in vitro* experiments by demonstrating the presence of phosphotyrosine in proteins extracted from whole cells. In normal cells, phosphotyrosine accounted for only about 0.03% of total phosphoamino acids (the rest being phosphoserine and phosphothreonine), explaining why it had previously escaped detection. However, phosphotyrosine was about

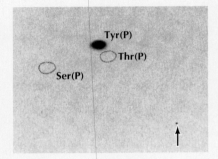

ten times more abundant in cells that were infected with RSV, suggesting that increased protein-tyrosine kinase activity of the viral Src protein was responsible for its ability to induce abnormal cell proliferation.

The Impact

The discovery that Src was a protein-tyrosine kinase both identified a new protein kinase activity and established this activity as being related to the control of cell proliferation. The results of Hunter and Sefton were followed by demonstrations that many other tumor virus proteins also function as protein-tyrosine kinases, generalizing the link between protein-tyrosine phosphorylation and the abnormal proliferation of cancer cells. Stanley Cohen and his colleagues further found that the EGF receptor is a protein-tyrosine kinase, directly implicating protein-tyrosine

Identification of phosphotyrosine in immunoglobulin phosphorylated by Src. An immunoprecipitate containing RSV Src was incubated with [^{32}P]-ATP. The immunoglobulin was then isolated and hydrolyzed. Amino acids in the hydrolysate were separated by electrophoresis and chromatography on a cellulose thin-layer plate. The positions of ^{32}P-labeled amino acids were determined by exposing the plate to X-ray film. Broken lines indicate the positions of unlabeled phosphoamino acids that were included as markers. Note that the principal ^{32}P-labeled amino acid is phosphotyrosine.

phosphorylation in the control of normal cell proliferation. Continuing studies have identified numerous additional receptor and nonreceptor protein-tyrosine kinases that function in a variety of cell signaling pathways. Studies of the mechanism by which a virus causes cancer in chickens thus revealed a previously unknown enzymatic activity that plays a central role in the signaling pathways that regulate animal cell growth, survival, and differentiation. Moreover, as discussed in Chapter 18, protein-tyrosine kinases encoded by oncogenes have provided the most promising targets to date for development of specific drugs against cancer cells.

FIGURE 15.14 Organization of receptor protein-tyrosine kinases
Each receptor consists of an N-terminal extracellular ligand-binding domain, a single transmembrane α helix, and a cytosolic C-terminal domain with protein-tyrosine kinase activity. The structures of three distinct subfamilies of receptor protein-tyrosine kinases are shown. The EGF receptor and insulin receptor both have cysteine-rich extracellular domains, whereas the PDGF receptor has immunoglobulin (Ig)-like domains. The PDGF receptor is also noteworthy in that its kinase domain is interrupted by an insert of approximately a hundred amino acids unrelated to those found in most other protein-tyrosine kinase catalytic domains. The insulin receptor is unusual in being a dimer of two pairs of polypeptide chains (designated α and β).

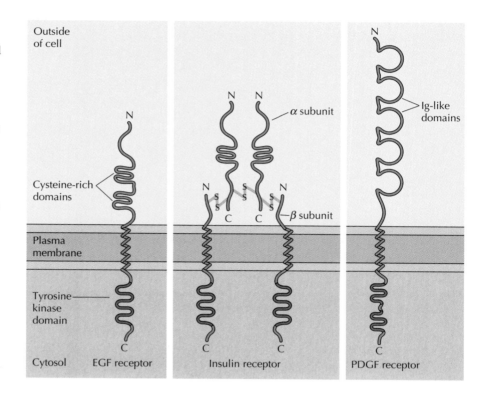

FIGURE 15.15 Dimerization and autophosphorylation of receptor protein-tyrosine kinases Growth factor binding induces receptor dimerization, which results in receptor autophosphorylation as the two polypeptide chains phosphorylate one another.

related receptors are dimers consisting of two polypeptide chains. The binding of ligands (e.g., growth factors) to the extracellular domains of these receptors activates their cytosolic kinase domains, resulting in phosphorylation of both the receptors themselves and intracellular target proteins that propagate the signal initiated by growth factor binding.

The first step in signaling from most receptor protein-tyrosine kinases is ligand-induced receptor dimerization (**Figure 15.15**). Some growth factors, such as PDGF and NGF, are themselves dimers consisting of two identical

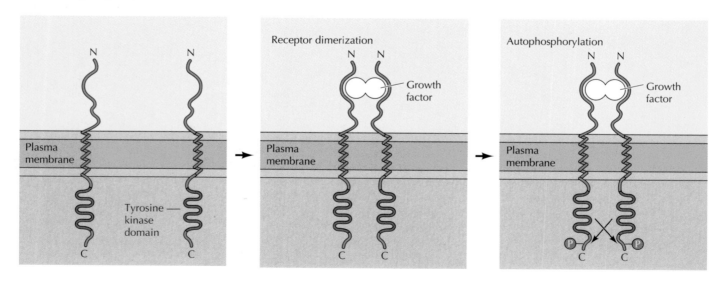

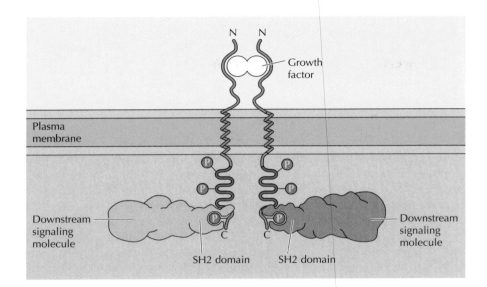

FIGURE 15.16 Association of downstream signaling molecules with receptor protein-tyrosine kinases SH2 domains bind to specific phosphotyrosine-containing peptides of the activated receptors.

polypeptide chains; these growth factors directly induce dimerization by simultaneously binding to two different receptor molecules. Other growth factors (such as EGF) are monomers but lead to receptor dimerization as a result of inducing conformational changes that promote protein-protein interactions between different receptor polypeptides.

Ligand-induced dimerization then leads to **autophosphorylation** of the receptor as the dimerized polypeptide chains cross-phosphorylate one another (see Figure 15.15). Such autophosphorylation plays two key roles in signaling from these receptors. First, phosphorylation of tyrosine residues within the catalytic domain increases protein kinase activity. Second, phosphorylation of tyrosine residues outside of the catalytic domain creates specific binding sites for additional proteins that transmit intracellular signals downstream of the activated receptors.

The association of these downstream signaling molecules with receptor protein-tyrosine kinases is mediated by protein domains that bind to specific phosphotyrosine-containing peptides (Figure 15.16). The first of these domains to be characterized are called **SH2 domains** (for *S*rc *h*omology 2) because they were initially recognized in protein-tyrosine kinases related to Src, the oncogenic protein of Rous sarcoma virus. SH2 domains consist of approximately 100 amino acids and bind to specific short peptide sequences containing phosphotyrosine residues (Figure 15.17). Other proteins bind to phosphotyrosine-containing peptides via **PTB domains** (for *p*hospho*t*yrosine-*b*inding). The resulting association of SH2- or PTB-containing proteins with activated receptor protein-tyrosine kinases can have several effects: It localizes these proteins to the plasma membrane, leads to their association with other proteins, promotes their phosphorylation, and stimulates their enzymatic activities. The association of these proteins with autophosphorylated receptors thus represents the first step in the intracellular transmission of signals initiated by the binding of growth factors to the cell surface.

FIGURE 15.17 Complex between an SH2 domain and a phosphotyrosine peptide The polypeptide chain of the Src SH2 domain is shown in red with its surface indicated by green dots. Purple spheres indicate a groove on the surface. The three amino acid residues that interact with the phosphotyrosine are shown in blue. The phosphotyrosine-containing peptide is shown as a space-filling model. Yellow and white spheres indicate the backbone and side-chain atoms, respectively, and the phosphate group is shown in red. (From G. Waksman and 13 others, 1992. *Nature* 358: 646.)

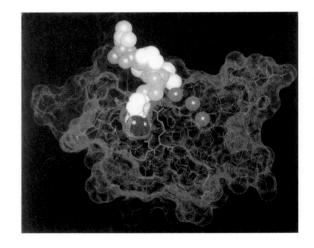

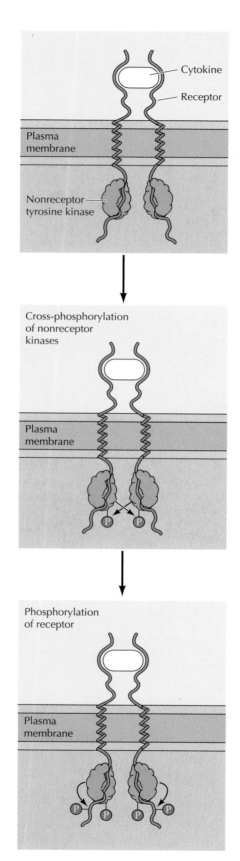

Cytokine

Receptor

Plasma membrane

Nonreceptor tyrosine kinase

Cross-phosphorylation of nonreceptor kinases

Plasma membrane

Phosphorylation of receptor

Plasma membrane

FIGURE 15.18 Signaling from cytokine receptors Ligand binding induces receptor dimerization and leads to the activation of associated nonreceptor protein-tyrosine kinases as a result of cross-phosphorylation. The activated kinases then phosphorylate tyrosine residues of the receptor, creating phosphotyrosine-binding sites for downstream signaling molecules.

Cytokine Receptors and Nonreceptor Protein-Tyrosine Kinases

Rather than possessing intrinsic enzymatic activity, many receptors act by stimulating intracellular protein-tyrosine kinases with which they are non-covalently associated. This family of receptors (called the **cytokine receptor superfamily**) includes the receptors for most cytokines (e.g., interleukin-2 and erythropoietin) and for some polypeptide hormones (e.g., growth hormone). Like receptor protein-tyrosine kinases, the cytokine receptors contain N-terminal extracellular ligand-binding domains, single transmembrane α helices, and C-terminal cytosolic domains. However, the cytosolic domains of the cytokine receptors are devoid of any known catalytic activity. Instead, the cytokine receptors function in association with **nonreceptor protein-tyrosine kinases**, which are activated as a result of ligand binding.

The first step in signaling from cytokine receptors is thought to be ligand-induced receptor dimerization and cross-phosphorylation of the associated nonreceptor protein-tyrosine kinases (Figure 15.18). These activated kinases then phosphorylate the receptor, providing phosphotyrosine-binding sites for the recruitment of downstream signaling molecules that contain SH2 domains. Combinations of cytokine receptors plus associated nonreceptor protein-tyrosine kinases thus function analogously to the receptor protein-tyrosine kinases discussed in the previous section.

The kinases associated with cytokine receptors belong to the **Janus kinase** (or **JAK**) family, which consists of four related nonreceptor protein-tyrosine kinases. Members of the JAK family appear to be universally required for signaling from cytokine receptors, indicating that JAK family kinases play a critical role in coupling these receptors to the tyrosine phosphorylation of intracellular targets.

Additional nonreceptor protein-tyrosine kinases belong to the **Src** family, which consists of Src and eight closely related proteins. As already noted, Src was initially identified as the oncogenic protein of Rous sarcoma virus and was the first protein shown to possess protein-tyrosine kinase activity, so it has played a pivotal role in experiments leading to our current understanding of cell signaling. Members of the Src family play key roles in signaling downstream of receptor protein-tyrosine kinases, from antigen receptors on B and T lymphocytes, and (as discussed later in this chapter) from integrins at sites of cell attachment to the extracellular matrix.

Receptors Linked to Other Enzymatic Activities

Although the vast majority of enzyme-linked receptors stimulate protein-tyrosine phosphorylation, some receptors are associated with other enzymatic activities. These receptors include protein-tyrosine phosphatases, protein-serine/threonine kinases, and guanylyl cyclases.

Protein-tyrosine phosphatases remove phosphate groups from phosphotyrosine residues, thus acting to counterbalance the effects of protein-tyrosine kinases. In many cases, protein-tyrosine phosphatases play negative regulatory roles in cell signaling pathways by terminating the signals initiated by protein-tyrosine phosphorylation. However, some protein-tyrosine phosphatases are cell surface receptors whose enzymatic activities play a positive role in cell signaling: 21 such receptor protein-tyrosine phosphatases are encoded in the human genome. A good example is provided by a receptor called CD45, which is expressed on the surface of T and B lymphocytes. Following antigen stimulation, CD45 dephosphorylates a specific phosphotyrosine that inhibits the enzymatic activity of Src family members. Thus the CD45 protein-tyrosine phosphatase acts (somewhat paradoxically) to stimulate nonreceptor protein-tyrosine kinases.

The receptors for **transforming growth factor β** (**TGF-β**) and related polypeptides are protein kinases that phosphorylate serine or threonine, rather than tyrosine, residues on their substrate proteins. TGF-β is the prototype of a family of polypeptide growth factors that control proliferation and differentiation of a variety of cell types. The cloning of the first receptor for a member of the TGF-β family in 1991 revealed that it is the prototype of a unique receptor family with a cytosolic **protein-serine/threonine kinase** domain. Since then, receptors for additional TGF-β family members have similarly been found to be protein-serine/threonine kinases. The binding of ligand to these receptors results in the association of two distinct types of polypeptide chains, which are encoded by different members of the TGF-β receptor family, to form heterodimers in which one of the receptor kinases phosphorylates the other. The activated TGF-β receptors then phosphorylate members of a family of transcription factors called Smads, which translocate to the nucleus and stimulate expression of target genes.

Some peptide ligands bind to receptors whose cytosolic domains are guanylyl cyclases, which catalyze formation of cyclic GMP. As discussed earlier, nitric oxide also acts by stimulating guanylyl cyclase, but the target of nitric oxide is an intracellular enzyme rather than a transmembrane receptor. The receptor **guanylyl cyclases** have an extracellular ligand-binding domain, a single transmembrane α helix, and a cytosolic domain with catalytic activity. Ligand binding stimulates cyclase activity, leading to the formation of cyclic GMP—a second messenger whose intracellular effects are discussed in the next section of this chapter.

Other receptors bind to cytoplasmic proteins with additional biochemical activities. For example, the cytokine tumor necrosis factor (TNF) induces cell death, perhaps (as discussed in Chapter 17) as a way of eliminating damaged or unwanted cells from tissues. The receptors for TNF and related death-signaling molecules are associated with specific proteases, which are activated in response to ligand binding. Activation of these receptor-associated proteases triggers the activation of additional downstream proteases, ultimately leading to degradation of a variety of intracellular proteins and death of the cell.

Pathways of Intracellular Signal Transduction

Many cell surface receptors stimulate intracellular target enzymes, which may be either directly linked or indirectly coupled to receptors by G proteins. These intracellular enzymes serve as downstream signaling elements that propagate and amplify the signal initiated by ligand binding. In most cases, a chain of reactions transmits signals from the cell surface to a variety of intracellular targets—a process called **intracellular signal transduction**. The targets of such signaling pathways frequently include transcription factors that function to regulate gene expression. Intracellular signaling pathways thus connect the cell surface to the nucleus, leading to changes in gene expression in response to extracellular stimuli.

■ Cytokine receptors are used by human immunodeficiency virus (HIV) as cell surface receptors for infection of immune cells.

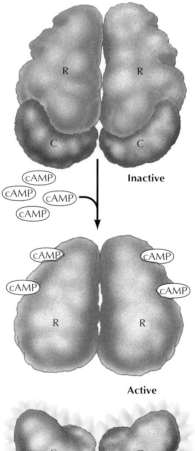

FIGURE 15.19 Synthesis and degradation of cAMP Cyclic AMP is synthesized from ATP by adenylyl cyclase and degraded to AMP by cAMP phosphodiesterase.

The cAMP Pathway: Second Messengers and Protein Phosphorylation

Intracellular signaling was first elucidated by studies of the action of hormones such as epinephrine, which signals the breakdown of glycogen to glucose in anticipation of muscular activity. In 1958 Earl Sutherland discovered that the action of epinephrine was mediated by an increase in the intracellular concentration of **cyclic AMP** (**cAMP**), leading to the concept that cAMP is a **second messenger** in hormonal signaling (the first messenger being the hormone itself). Cyclic AMP is formed from ATP by the action of **adenylyl cyclase** and degraded to AMP by **cAMP phosphodiesterase** (Figure 15.19). As discussed earlier, the epinephrine receptor is coupled to adenylyl cyclase via a G protein that stimulates enzymatic activity, thereby increasing the intracellular concentration of cAMP (see Figure 15.12).

How does cAMP then signal the breakdown of glycogen? This and most other effects of cAMP in animal cells are mediated by the action of **cAMP-dependent protein kinase**, or **protein kinase A**, an enzyme discovered by Donal Walsh and Ed Krebs in 1968. The inactive form of protein kinase A is a tetramer consisting of two catalytic and two regulatory subunits (Figure 15.20). Cyclic AMP binds to the regulatory subunits, leading to their dissociation from the catalytic subunits. The free catalytic subunits are then enzymatically active and able to phosphorylate serine residues on their target proteins.

In the regulation of glycogen metabolism, protein kinase A phosphorylates two key target enzymes (Figure 15.21). The first is another protein kinase, phosphorylase kinase, which is phosphorylated and activated by protein kinase A. Phosphorylase kinase in turn phosphorylates and activates glycogen phosphorylase, which catalyzes the breakdown of glycogen to glucose-1-phosphate. In addition, protein kinase A phosphorylates the enzyme glycogen synthase, which catalyzes glycogen synthesis. In this case, however, phosphorylation inhibits enzymatic activity. Elevation of cAMP and activation of protein kinase A thus blocks further glycogen synthesis at the same time as it stimulates glycogen breakdown.

The chain of reactions leading from the epinephrine receptor to glycogen phosphorylase provides a good illustration of signal amplification during intracellular signal transduction. Each molecule of epinephrine activates only a single receptor. However, each receptor may activate up to a hundred molecules of G_s. Each molecule of G_s then stimulates the enzymatic

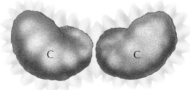

FIGURE 15.20 Regulation of protein kinase A The inactive form of protein kinase A consists of two regulatory (R) and two catalytic (C) subunits. Binding of cAMP to the regulatory subunits induces a conformational change that leads to dissociation of the catalytic subunits, which are then enzymatically active.

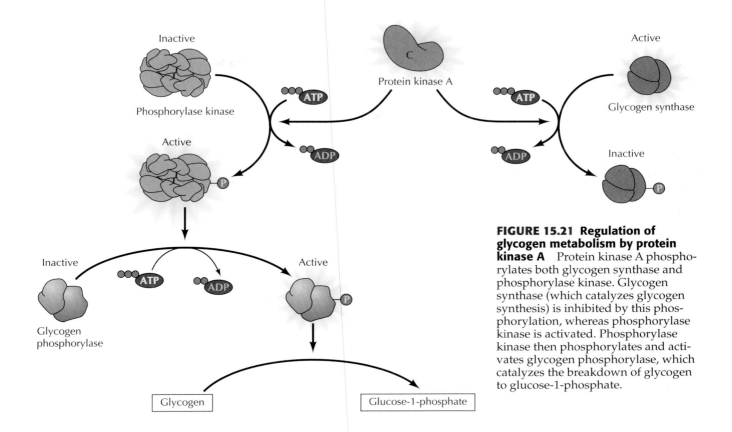

FIGURE 15.21 Regulation of glycogen metabolism by protein kinase A Protein kinase A phosphorylates both glycogen synthase and phosphorylase kinase. Glycogen synthase (which catalyzes glycogen synthesis) is inhibited by this phosphorylation, whereas phosphorylase kinase is activated. Phosphorylase kinase then phosphorylates and activates glycogen phosphorylase, which catalyzes the breakdown of glycogen to glucose-1-phosphate.

activity of adenylyl cyclase, which can catalyze the synthesis of many molecules of cAMP. Signal amplification continues as each molecule of protein kinase A phosphorylates many molecules of phosphorylase kinase, which in turn phosphorylates many molecules of glycogen phosphorylase. Hormone binding to a small number of receptors thus leads to activation of a much larger number of intracellular target enzymes.

In many animal cells, increases in cAMP activate the transcription of specific target genes that contain a regulatory sequence called the **cAMP response element**, or **CRE** (Figure 15.22). In this case, the signal is carried from the cytoplasm to the nucleus by the catalytic subunit of protein kinase A, which is able to enter the nucleus following its release from the regulatory subunit. Within the nucleus, protein kinase A phosphorylates a transcription factor called **CREB** (for CRE-*b*inding protein), leading to the recruitment of coactivators and transcription of cAMP-inducible genes. Such regulation of gene expression by cAMP plays important roles in controlling the proliferation, survival, and differentiation of a wide variety of animal cells, as well as being implicated in learning and memory.

It is important to recognize that protein kinases, such as protein kinase A, do not function in isolation within the cell. To the contrary, protein phosphorylation is rapidly reversed by the action of protein phosphatases. Some protein phosphatases are transmembrane receptors, as discussed in the preceding section. A number of others are cytosolic enzymes that remove phosphate groups from either phosphorylated tyrosine or serine/threonine residues in their substrate proteins. These protein phosphatases serve to terminate the responses initiated by receptor activation of protein kinases. For example, the serine residues of proteins that are phosphorylated by protein

15.3 WEBSITE ANIMATION

Signal Amplification
In a signal transduction cascade, each enzyme activated at a stage in the cascade may activate 100 molecules of the next enzyme, quickly amplifying the response to the receptor-bound ligand.

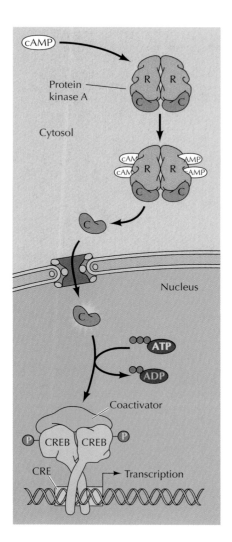

FIGURE 15.22 Cyclic AMP-inducible gene expression The free catalytic subunit of protein kinase A translocates to the nucleus and phosphorylates the transcription factor CREB (CRE-binding protein), leading to the recruitment of coactivators and expression of cAMP-inducible genes.

kinase A are usually dephosphorylated by the action of a phosphatase called protein phosphatase 1 (Figure 15.23). The levels of phosphorylation of protein kinase A substrates (such as phosphorylase kinase and CREB) are thus determined by a balance between the intracellular activities of protein kinase A and protein phosphatases.

Although most effects of cAMP are mediated by protein kinase A, cAMP can also directly regulate ion channels, independent of protein phosphorylation. Cyclic AMP functions in this way as a second messenger involved in sensing smells. Many of the odorant receptors in sensory neurons in the nose are G protein-coupled receptors that stimulate adenylyl cyclase, leading to an increase in intracellular cAMP. Rather than stimulating protein kinase A, cAMP in this system directly opens Na^+ channels in the plasma membrane, leading to membrane depolarization and initiation of a nerve impulse.

Cyclic GMP

Cyclic GMP (cGMP) is also an important second messenger in animal cells, although its roles are not as clearly understood as those of cAMP. Cyclic GMP is formed from GTP by guanylyl cyclases and degraded to GMP by a phosphodiesterase. As discussed earlier in this chapter, guanylyl cyclases are activated by nitric oxide and carbon monoxide as well as by peptide ligands. Stimulation of these guanylyl cyclases leads to elevated levels of cGMP, which then mediate biological responses, such as blood vessel dilation. The action of cGMP is frequently mediated by activation of cGMP-dependent protein kinases, although cGMP also regulates ion channels and phosphodiesterases.

One well-characterized role of cGMP is in the vertebrate eye, where it serves as the second messenger responsible for converting the visual signals received as light to nerve impulses. The photoreceptor in rod cells of the retina is a G protein-coupled receptor called **rhodopsin** (Figure 15.24). Rhodopsin is activated as a result of the absorption of light by the associated small molecule 11-*cis*-retinal, which then isomerizes to all-*trans*-retinal, inducing a conformational change in the rhodopsin protein. Rhodopsin

FIGURE 15.23 Regulation of protein phosphorylation by protein kinase A and protein phosphatase 1 The phosphorylation of target proteins by protein kinase A is reversed by the action of protein phosphatase 1.

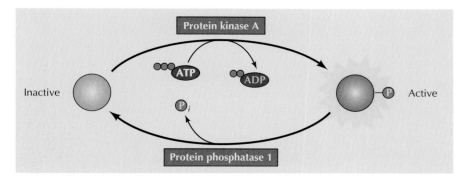

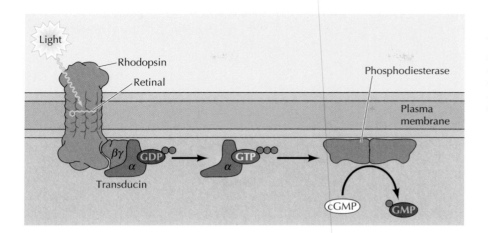

then activates the G protein **transducin**, and the α subunit of transducin stimulates the activity of **cGMP phosphodiesterase**, leading to a decrease in the intracellular level of cGMP. This change in cGMP level in retinal rod cells is translated to a nerve impulse by a direct effect of cGMP on ion channels in the plasma membrane, similar to the action of cAMP in sensing smells.

Phospholipids and Ca²⁺

One of the most widespread pathways of intracellular signaling is based on the use of second messengers derived from the membrane phospholipid **phosphatidylinositol 4,5-bisphosphate (PIP$_2$)**. PIP$_2$ is a minor component of the plasma membrane, localized to the inner leaflet of the phospholipid bilayer (see Figure 13.2). A variety of hormones and growth factors stimulate the hydrolysis of PIP$_2$ by **phospholipase C**—a reaction that produces two distinct second messengers, **diacylglycerol** and **inositol 1,4,5-trisphosphate (IP$_3$)** (Figure 15.25). Diacylglycerol and IP$_3$ stimulate distinct down-

FIGURE 15.25 Hydrolysis of PIP$_2$ Phospholipase C (PLC) catalyzes the hydrolysis of phosphatidylinositol 4,5-bisphosphate (PIP$_2$) to yield diacylglycerol (DAG) and inositol trisphosphate (IP$_3$). Diacylglycerol activates members of the protein kinase C family, and IP$_3$ signals the release of Ca²⁺ from intracellular stores.

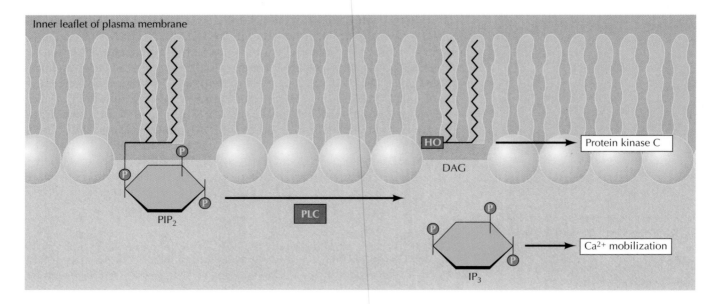

FIGURE 15.26 Activation of phospholipase C by protein-tyrosine kinases Phospholipase C-γ (PLC-γ) binds to activated receptor protein-tyrosine kinases via its SH2 domains. Tyrosine phosphorylation increases PLC-γ activity, stimulating the hydrolysis of PIP₂.

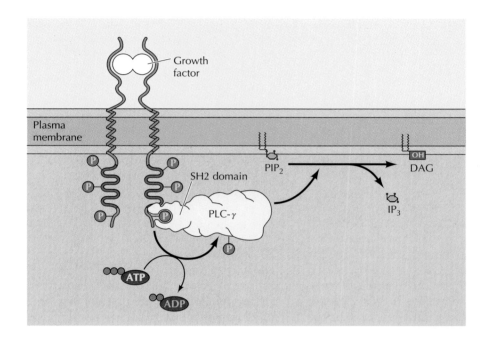

■ **Toxic snake venoms contain phospholipases. Hydrolysis of phospholipids by venom from rattlesnakes and cobras leads to the rupture of red blood cell membranes.**

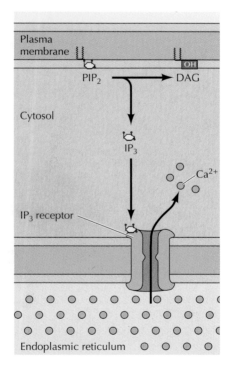

stream signaling pathways (protein kinase C and Ca^{2+} mobilization, respectively), so PIP₂ hydrolysis triggers a two-armed cascade of intracellular signaling.

It is noteworthy that the hydrolysis of PIP₂ is activated downstream of both G protein-coupled receptors and protein-tyrosine kinases. This occurs because one form of phospholipase C (PLC-β) is stimulated by G proteins, whereas a second form of phospholipase C (PLC-γ) contains SH2 domains that mediate its association with activated receptor protein-tyrosine kinases (Figure 15.26). This interaction localizes PLC-γ to the plasma membrane as well as leading to its tyrosine phosphorylation, which increases its catalytic activity.

The diacylglycerol produced by hydrolysis of PIP₂ remains associated with the plasma membrane and activates protein-serine/threonine kinases belonging to the **protein kinase C** family, many of which play important roles in the control of cell growth and differentiation. The other second messenger produced by PIP₂ cleavage, IP₃, is a small polar molecule that is released into the cytosol, where it acts to signal the release of Ca^{2+} from intracellular stores (Figure 15.27). As noted in Chapter 13, the cytosolic concentration of Ca^{2+} is maintained at an extremely low level (about 0.1 μM) as a result of Ca^{2+} pumps that actively export Ca^{2+} from the cell. Ca^{2+} is pumped not only across the plasma membrane but also into the endoplasmic reticulum, which therefore serves as an intracellular Ca^{2+} store. IP₃ acts to release Ca^{2+} from the endoplasmic reticulum by binding to receptors that are ligand-gated Ca^{2+} channels. As a result, cytosolic Ca^{2+} levels increase to about 1 μM, which affects the activities of a variety of target proteins, including

FIGURE 15.27 Ca^{2+} mobilization by IP₃ Ca^{2+} is pumped from the cytosol into the endoplasmic reticulum, which serves as an intracellular Ca^{2+} store. IP₃ binds to receptors that are ligand-gated Ca^{2+} channels in the endoplasmic reticulum membrane, thereby allowing the efflux of Ca^{2+} into the cytosol.

FIGURE 15.28 Function of calmodulin Calmodulin is a dumbbell-shaped protein with four Ca^{2+}-binding sites. The active Ca^{2+}/calmodulin complex binds to a variety of target proteins, including Ca^{2+}/calmodulin-dependent protein kinases.

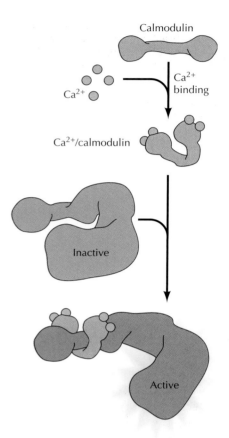

protein kinases and phosphatases. For example, some members of the protein kinase C family require Ca^{2+} as well as diacylglycerol for their activation, so these protein kinases are regulated jointly by both arms of the PIP_2 signaling pathway. In most cells, the transient increase in intracellular Ca^{2+} resulting from production of IP_3 triggers a more sustained increase caused by the entry of extracellular Ca^{2+} through channels in the plasma membrane. This entry of Ca^{2+} from outside the cell serves both to prolong the signal initiated by release of Ca^{2+} from the endoplasmic reticulum and to allow the stores of Ca^{2+} within the endoplasmic reticulum to be replenished.

Many of the effects of Ca^{2+} are mediated by the Ca^{2+}-binding protein **calmodulin**, which is activated when the concentration of cytosolic Ca^{2+} increases to about 0.5 μM (**Figure 15.28**). Ca^{2+}/calmodulin then binds to a variety of target proteins, including protein kinases. One example of such a Ca^{2+}/calmodulin-dependent protein kinase is myosin light-chain kinase, which signals actin-myosin contraction by phosphorylating one of the myosin light chains (see Figure 12.31). Other protein kinases that are activated by Ca^{2+}/calmodulin include members of the **CaM kinase** family, which phosphorylate a number of different proteins, including metabolic enzymes, ion channels, and transcription factors. One form of CaM kinase is particularly abundant in the nervous system where it regulates the synthesis and release of neurotransmitters. In addition, CaM kinases can regulate gene expression by phosphorylating transcription factors. Interestingly, one of the transcription factors phosphorylated by CaM kinase is CREB, which is phosphorylated at the same site by protein kinase A. This phosphorylation of CREB illustrates one of many intersections between the Ca^{2+} and cAMP signaling pathways. Other examples include the regulation of adenylyl cyclases and phosphodiesterases by Ca^{2+}/calmodulin, the regulation of Ca^{2+} channels by cAMP, and the phosphorylation of a number of target proteins by both protein kinase A and Ca^{2+}/calmodulin-dependent protein kinases. The cAMP and Ca^{2+} signaling pathways thus function coordinately to regulate many cellular responses.

The entry of extracellular Ca^{2+} is particularly important in the electrically excitable cells of nerve and muscle in which voltage-gated Ca^{2+} channels in the plasma membrane are opened by membrane depolarization (**Figure 15.29**). The resulting increases in intracellular Ca^{2+} then trigger the further release of Ca^{2+} from intracellular stores by activating distinct Ca^{2+} channels known as **ryanodine receptors**. One effect of increases in intracellular Ca^{2+} in neurons is to trigger the release of neurotransmitters, so Ca^{2+} plays a crit-

FIGURE 15.29 Regulation of intracellular Ca^{2+} in electrically excitable cells Membrane depolarization leads to the opening of voltage-gated Ca^{2+} channels in the plasma membrane causing the influx of Ca^{2+} from extracellular fluids. The resulting increase in intracellular Ca^{2+} then signals the further release of Ca^{2+} from intracellular stores by opening distinct Ca^{2+} channels (ryanodine receptors) in the endoplasmic reticulum membrane. In muscle cells, ryanodine receptors in the sarcoplasmic reticulum may also be opened directly in response to membrane depolarization.

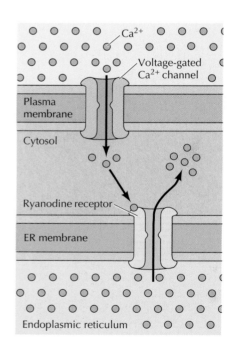

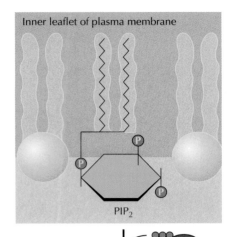

Inner leaflet of plasma membrane

PIP$_2$

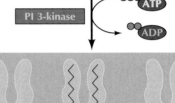

PI 3-kinase

ATP

ADP

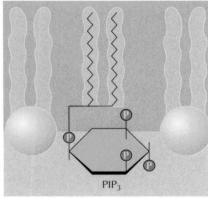

PIP$_3$

FIGURE 15.30 Activity of PI 3-kinase PI 3-kinase phosphorylates the 3 position of inositol, converting PIP$_2$ to PIP$_3$.

ical role in converting electric to chemical signals in the nervous system. In muscle cells Ca^{2+} is stored in the sarcoplasmic reticulum from which it is released by the opening of ryanodine receptors in response to changes in membrane potential. This release of stored Ca^{2+} leads to large increases in cytosolic Ca^{2+}, which trigger muscle contraction (see Chapter 12). Cells thus utilize a variety of mechanisms to regulate intracellular Ca^{2+} levels, making Ca^{2+} a versatile second messenger that controls a wide range of cellular processes.

The PI 3-Kinase/Akt and mTOR Pathways

PIP$_2$ not only serves as the source of diacylglycerol and IP$_3$ but is also the starting point of a distinct second messenger pathway that plays a key role in regulating cell growth and survival. In this pathway, PIP$_2$ is phosphorylated on the 3 position of inositol by the enzyme **phosphatidylinositide (PI) 3-kinase** (Figure 15.30). Like phospholipase C, one form of PI 3-kinase is activated by G proteins while a second form has SH2 domains and is activated by association with receptor protein-tyrosine kinases. Phosphorylation of PIP$_2$ yields the second messenger **phosphatidylinositol 3,4,5-trisphosphate (PIP$_3$)**.

A key target of PIP$_3$, which is critical for signaling cell proliferation and survival, is a protein-serine/threonine kinase called **Akt**. PIP$_3$ binds to a domain of Akt known as the pleckstrin homology domain (Figure 15.31). This interaction recruits Akt to the inner face of the plasma membrane where it is phosphorylated and activated by another protein kinase (called PDK1) that also contains a pleckstrin homology domain and binds PIP$_3$. The formation of PIP$_3$ thus results in the association of both Akt and PDK1 with the plasma membrane, leading to phosphorylation and activation of Akt. Activation of Akt also requires phosphorylation at a second site by a distinct protein kinase, which has recently been identified as a form of mTOR complexed with a protein called rictor. The mTOR/rictor complex is itself stimulated by growth factors, but its mechanism of activation remains to be understood.

Once activated, Akt phosphorylates a number of target proteins, including proteins that are direct regulators of cell proliferation and survival (discussed in Chapter 17), transcription factors, and other protein kinases. The critical transcription factors targeted by Akt include members of the Forkhead or FOXO family (Figure 15.32). Phosphorylation of FOXO by Akt creates a binding site for cytosolic chaperone proteins (14-3-3 proteins) that sequester FOXO in an inactive form in the cytoplasm. In the absence of growth factor signaling and Akt activity, FOXO is released from 14-3-3 and translocates to the nucleus, stimulating transcription of genes that inhibit cell proliferation or induce cell death. Another target of Akt is the protein kinase GSK-3β, which regulates metabolism as well as cell proliferation and survival. Like FOXO, GSK-3β is inhibited by Akt phosphorylation. The targets of GSK-3β include several transcription factors and the translation initiation factor eIF-2B. Phosphorylation of eIF-2B leads to a global downregulation of translation initiation (see Figure 8.20), so GSK-3β provides a link between growth factor signaling and control of cellular protein synthesis.

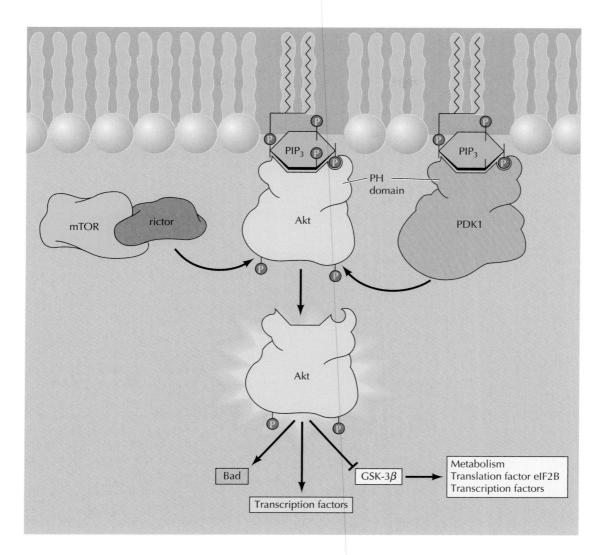

FIGURE 15.31 The PI 3-kinase/Akt pathway Akt is recruited to the plasma membrane by binding to PIP_3 via its pleckstrin homology (PH) domain. It is then activated as a result of phosphorylation by another protein kinase (PDK1) that also binds PIP_3, as well as by the mTOR/rictor complex. Akt then phosphorylates a number of target proteins, including direct regulators of cell survival (Bad, see Chapter 17), several transcription factors, and the protein kinase GSK-3β (which is inhibited by Akt phosphorylation). GSK-3β phosphorylates metabolic enzymes, transcription factors, and the translation initiation factor eIF-2B.

The **mTOR** pathway is a central regulator of cell growth that couples the control of protein synthesis to the availability of growth factors, nutrients, and energy (Figure 15.33). This is accomplished via the regulation of mTOR by multiple signals, including the PI 3-kinase/Akt pathway. The mTOR protein kinase exists in two distinct complexes in cells in which mTOR is associated with either rictor or raptor. As discussed above, the mTOR/rictor complex is one of the protein kinases that phosphorylates and activates Akt (see Figure 15.31). In contrast, the mTOR/raptor complex is activated downstream of Akt and functions to regulate cell size, at least in part by

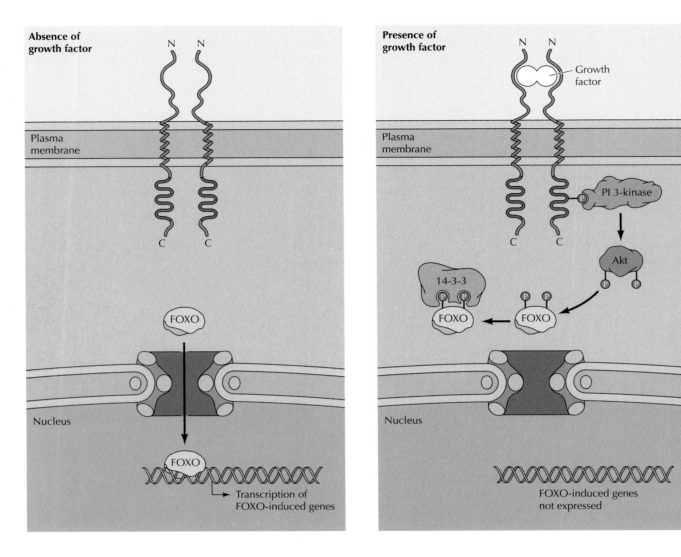

Absence of growth factor

Plasma membrane

Nucleus

FOXO

FOXO

Transcription of FOXO-induced genes

Presence of growth factor

Growth factor

Plasma membrane

PI 3-kinase

Akt

14-3-3

FOXO

FOXO

Nucleus

FOXO-induced genes not expressed

FIGURE 15.32 Regulation of FOXO In the absence of growth factor stimulation, the FOXO transcription factor translocates to the nucleus and induces target gene expression. Growth factor stimulation leads to activation of Akt, which phosphorylates FOXO. This creates binding sites for the cytosolic chaperone 14-3-3, which sequesters FOXO in an inactive form in the cytoplasm.

controlling protein synthesis. The mTOR/raptor complex is regulated by the Ras-related GTP-binding protein Rheb, which is in turn regulated by the GTPase-activating protein complex TSC1/2. Akt phosphorylates TSC2, leading to activation of mTOR/raptor in response to growth factor stimulation. In addition, TSC2 is regulated by another protein kinase called the AMP/activated kinase (AMPK). AMPK senses the energy state of the cell and is activated by a high ratio of AMP to ATP. Under these conditions, AMPK phosphorylates TSC2, leading to inhibition of mTOR/raptor when cellular energy stores are depleted. TSC2 is also regulated by the availability of amino acids, although the mechanism responsible remains to be established.

The mTOR/raptor complex phosphorylates at least two well-characterized targets that function to regulate protein synthesis: S6 kinase and eIF4E binding protein-1 (4E-BP1). S6 kinase controls translation by phosphorylating the ribosomal protein S6 as well as other proteins involved in translational regulation. The eIF4E binding protein controls translation by interacting with initiation factor eIF4E, which binds to the 5′ cap of mRNAs. In the absence of mTOR signaling, nonphosphorylated 4E-BPs bind to eIF4E and inhibit translation by interfering with the interaction of eIF4E with eIF4G (see Figure 8.11). Phosphorylation of 4E-BP1 by mTOR prevents its interaction with eIF4E, leading to increased rates of translation initiation.

MAP Kinase Pathways

The MAP kinase pathway refers to a cascade of protein kinases that are highly conserved in evolution and play central roles in signal transduction in all eukaryotic cells ranging from yeasts to humans. The central elements in the pathway are a family of protein-serine/threonine kinases called the **MAP kinases** (for *m*itogen-*a*ctivated *p*rotein *k*inases) that are activated in response to a variety of growth factors and other signaling molecules. In yeasts, MAP kinase pathways control a variety of cellular responses, including mating, cell shape, and sporulation. In higher eukaryotes (including *C. elegans*, *Drosophila*, frogs, and mammals), MAP kinases are ubiquitous regulators of cell growth and differentiation.

The MAP kinases that were initially characterized in mammalian cells belong to the **ERK** (*e*xtracellular signal-*r*egulated *k*inase) family. ERK activation plays a central role in signaling cell proliferation induced by growth factors that act through either protein-tyrosine kinase or G protein-coupled receptors. Protein kinase C can also activate the ERK pathway, and both the Ca^{2+} and cAMP pathways intersect with ERK signaling, either stimulating or inhibiting the ERK pathway in different types of cells.

Activation of ERK is mediated by two upstream protein kinases, which are coupled to growth factor receptors by the **Ras** GTP-binding protein (Figure 15.34). Activation of Ras leads to activation of the **Raf** protein-serine/threonine kinase, which phosphorylates and activates a second protein kinase called **MEK** (for *M*AP kinase/*E*RK *k*inase). MEK is a dual-specificity protein kinase that activates members of the ERK family by phosphorylation of both threonine and tyrosine residues separated by one amino acid (e.g., threonine-183 and tyrosine-185 of ERK2). Once activated, ERK phosphorylates a variety of targets, including other protein kinases and transcription factors.

The central role of the ERK pathway in mammalian cells emerged from studies of the Ras proteins, which were first identified as the oncogenic proteins of tumor viruses that cause sarcomas in rats (hence the name Ras, from *rat sarcoma* virus). Interest in Ras intensified considerably in 1982 when mutations in *ras* genes were first implicated in the development of

> ▪ **Rapamycin, an antibiotic produced by certain fungi, is a specific inhibitor of the mTOR/raptor complex and is used as an immunosuppressive drug in organ transplants.**

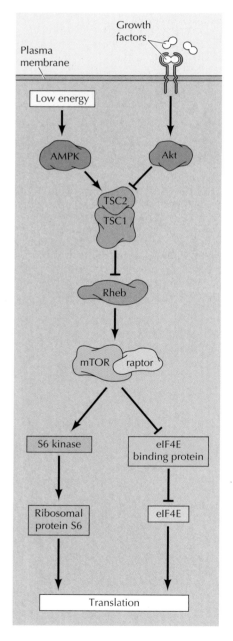

FIGURE 15.33 The mTOR pathway The mTOR/raptor protein kinase is activated by Rheb, which is inhibited by the TSC1/2 complex. Akt inhibits TSC1/2, leading to activation of Rheb and mTOR/raptor in response to growth factor stimulation. In contrast, AMPK activates TSC1/2, leading to inhibition of Rheb and mTOR/raptor if cellular energy stores are depleted. mTOR/raptor stimulates translation by phosphorylating S6 kinase (which phosphorylates ribosomal protein S6) and by phosphorylating eIF4E binding protein, relieving inhibition of translation initiation factor eIF4E.

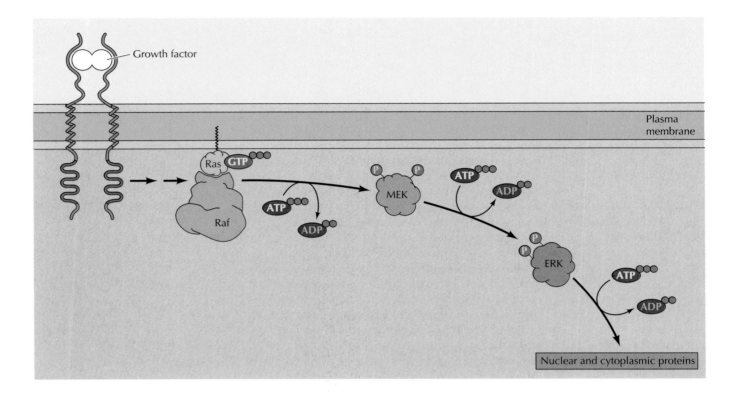

FIGURE 15.34 Activation of the ERK MAP kinases Stimulation of growth factor receptors leads to activation of the small GTP-binding protein Ras, which interacts with the Raf protein kinase. Raf phosphorylates and activates MEK, a dual-specificity protein kinase that activates ERK by phosphorylation on both threonine and tyrosine residues (Thr-183 and Tyr-185). ERK then phosphorylates a variety of nuclear and cytoplasmic target proteins.

human cancers (discussed in Chapter 18). The importance of Ras in intracellular signaling was then indicated by experiments showing that microinjection of active Ras protein directly induces proliferation of normal mammalian cells. Conversely, interference with Ras function by either microinjection of anti-Ras antibody or expression of a dominant negative Ras mutant blocks growth factor-induced cell proliferation. Thus Ras is not only capable of inducing the abnormal growth characteristic of cancer cells but also appears to be required for the response of normal cells to growth factor stimulation.

The Ras proteins are guanine nucleotide-binding proteins that function analogously to the α subunits of G proteins, alternating between inactive GDP-bound and active GTP-bound forms (**Figure 15.35**). In contrast to the G protein α subunits, however, Ras functions as a monomer rather than in association with $\beta\gamma$ subunits. Ras activation is mediated by **guanine nucleotide exchange factors** that stimulate the release of bound GDP and its exchange for GTP. Activity of the Ras-GTP complex is then terminated by GTP hydrolysis, which is stimulated by the interaction of Ras-GTP with **GTPase-activating proteins**. It is interesting to note that the mutations of *ras* genes in human cancers have the effect of inhibiting GTP hydrolysis by the Ras proteins. These mutated Ras proteins therefore remain continuously in the active GTP-bound form, driving the unregulated proliferation of cancer cells even in the absence of growth factor stimulation.

The Ras proteins are prototypes of a large family of approximately 50 related proteins frequently called small GTP-binding proteins because Ras and its relatives are about half the size of G protein α subunits. One member of this family is Rheb, which regulates mTOR signaling (see Figure 15.33). Other subfamilies of small GTP-binding proteins control a vast array

Cancer: Signal Transduction and the ras Oncogenes

The Disease

Cancer claims the lives of approximately one out of every four Americans, accounting for almost 600,000 deaths each year in the United States. There are more than a hundred different kinds of cancer, but some are more common than others. In this country the most common lethal cancers are those of the lung and colon/rectum, which together account for about 40% of all cancer deaths. Other major contributors to cancer mortality include cancers of the breast, pancreas, and prostate, which are responsible for approximately 7.2%, 5.6%, and 5.3% of U.S. cancer deaths, respectively.

The common feature of all cancers is the unrestrained proliferation of cancer cells, which eventually spread throughout the body, invading normal tissues and organs and leading to death of the patient. Surgery and radiotherapy are effective treatments for localized cancers but are unable to reach cancer cells that have spread to distant body sites. Treatment of these cancers therefore requires the use of chemotherapeutic drugs. Unfortunately, the commonly available chemotherapeutic agents are not specific for cancer cells. Most act by either damaging DNA or interfering with DNA synthesis, so they also kill rapidly dividing normal cells, such as the epithelial cells that line the digestive tract and the blood-forming cells of the bone marrow. The resulting toxicity of these drugs limits their effectiveness, and many cancers are not eliminated by doses of chemotherapy that can be tolerated by the patient. Consequently, although major progress has been made in cancer treatment, nearly half of all patients diagnosed with cancer ultimately die of their disease.

Molecular and Cellular Basis

The identification of viral genes that can convert normal cells to cancer cells, such as the src gene of RSV, provided the first demonstration that cancers can result from the action of specific genes (oncogenes). The subsequent discovery that viral oncogenes are related to genes of normal cells then engendered the hypothesis that non-virus-induced cancers (including most human cancers) might arise as a result of mutations in normal cell genes, giving rise to oncogenes of cellular rather than viral origin. Such cellular oncogenes were first identified in human cancers in 1981. The following year, human oncogenes of bladder, lung, and colon cancers were found to be related to the ras genes previously identified in rat sarcoma viruses.

Although many different genes are now known to play critical roles in cancer development, mutations of the ras genes remain one of the most common genetic abnormalities in human tumors. Mutated ras oncogenes are found in about 20% of all human cancers, including approximately 25% of lung cancers, 50% of colon cancers, and more than 90% of pancreatic cancers. Moreover, the action of ras oncogenes has clearly linked the development of human cancer to abnormalities in the signaling pathways that regulate cell proliferation. The mutations that convert normal ras genes to oncogenes substantially decrease GTP hydrolysis by the Ras proteins. Consequently, the mutated oncogenic Ras proteins remain locked in the active GTP-bound form, rather than alternating normally between inactive and active states in response to extracellular signals. The oncogenic Ras proteins thus continuously stimulate the ERK signaling pathway and drive cell proliferation, even in the absence of the growth factors that would be required to activate Ras and signal proliferation of normal cells.

Prevention and Treatment

The discovery of mutated oncogenes in human cancers raised the possibility of developing drugs specifically targeted against the oncogene proteins. In principle, such drugs might act selectively against cancer cells with less toxicity toward normal cells than that of conventional chemotherapeutic agents. Because ras is frequently mutated in human cancers,

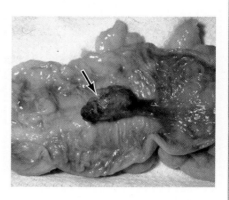

A human colon polyp (an early stage of colon cancer). The ras oncogenes contribute to the development of about half of all colon cancers. (E. P. Ewing, Jr., Centers for Disease Control.)

the Ras proteins and other elements of Ras signaling pathways have attracted considerable interest as potential drug targets.

As discussed in Chapter 18, effective drugs that selectively target cancer cells have been recently developed against some protein-tyrosine kinase oncogenes, including the EGF receptor, that act upstream of Ras. A variety of additional drugs are under investigation as potential cancer treatments, including drugs targeted against Ras itself and against protein kinases activated downstream of Ras, such as Raf. The identification of oncogenes in human tumors has thus opened new strategies to rational development of drugs that act effectively and selectively against human cancer cells by targeting the signaling pathways that are responsible for cancer development.

References

Der, C. J., T. G. Krontiris and G. M. Cooper. 1982. Transforming genes of human bladder and lung carcinoma cell lines are homologous to the ras genes of Harvey and Kirsten sarcoma viruses. Proc. Natl. Acad. Sci. USA 79: 3637–3640.

Parada, L. F., C. J. Tabin, C. Shih and R. A. Weinberg. 1982. Human EJ bladder carcinoma oncogene is homologue of Harvey sarcoma virus ras gene. Nature 297: 474–478.

Sawyers, C. 2004. Targeted cancer therapy. Nature 432: 294–297.

FIGURE 15.35 Regulation of Ras proteins
Ras proteins alternate between inactive
GDP-bound and active GTP-bound states.

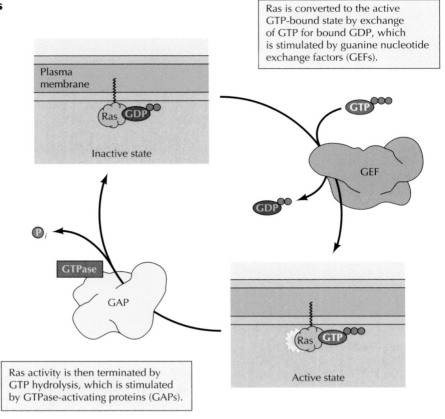

Ras is converted to the active
GTP-bound state by exchange
of GTP for bound GDP, which
is stimulated by guanine nucleotide
exchange factors (GEFs).

Ras activity is then terminated by
GTP hydrolysis, which is stimulated
by GTPase-activating proteins (GAPs).

of cellular activities. For example, the largest subfamily of small GTP-binding proteins (the Rab proteins) functions to regulate vesicle trafficking, as discussed in Chapter 10. Other small GTP-binding proteins are involved in the import and export of proteins from the nucleus (the Ran protein, discussed in Chapter 9) and organization of the cytoskeleton (the Rho subfamily, discussed later in this chapter).

The best understood mode of Ras activation is that mediated by receptor protein-tyrosine kinases (Figure 15.36). Autophosphorylation of these receptors results in their association with Ras guanine nucleotide exchange factors as a result of SH2-mediated protein interactions. One well-characterized example is provided by the guanine nucleotide exchange factor Sos, which is bound to the SH2-containing protein Grb2 in the cytosol of unstimulated cells. Tyrosine phosphorylation of receptors (or of other receptor-associated proteins) creates a binding site for the Grb2 SH2 domains. Association of Grb2 with activated receptors localizes Sos to the plasma membrane where it is able to interact with Ras proteins, which are anchored to the inner leaflet of the plasma membrane by lipids attached to the Ras C terminus (see Figure 13.9). Sos then stimulates guanine nucleotide exchange resulting in formation of the active Ras-GTP complex. In its active GTP-bound form Ras interacts with a number of effector proteins, including the Raf protein-serine/threonine kinase. This interaction with Ras recruits Raf from the cytosol to the plasma membrane and initiates Raf activation, which also involves phosphorylation of Raf by both protein-tyrosine and protein-serine/threonine kinases.

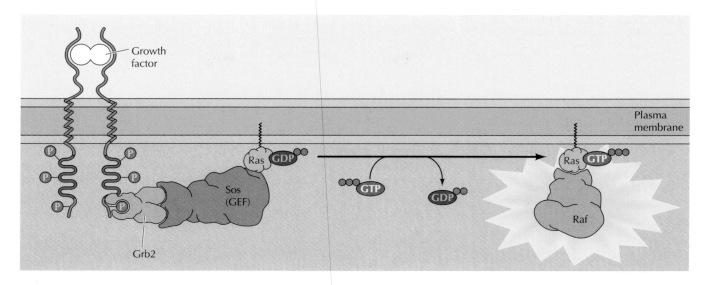

FIGURE 15.36 Ras activation downstream of receptor protein-tyrosine kinases A complex of Grb2 and the guanine nucleotide exchange factor Sos binds to a phosphotyrosine-containing sequence in the activated receptor via the Grb2 SH2 domain. This interaction recruits Sos to the plasma membrane where it can stimulate Ras GDP/GTP exchange. The activated Ras-GTP complex then binds to the Raf protein kinase.

As already noted, activation of Raf initiates a protein kinase cascade leading to ERK activation. ERK then phosphorylates a variety of target proteins, including other protein kinases. In addition, ERK regulates the mTOR pathway by phosphorylating TSC2 (see Figure 15.33). Importantly, a fraction of activated ERK translocates to the nucleus where it regulates transcription factors by phosphorylation (Figure 15.37). In this regard, it is notable that a primary response to growth factor stimulation is the rapid transcriptional induction of a family of approximately 100 genes called **immediate-early genes**. The induction of a number of immediate-early genes is mediated by a regulatory sequence called the **serum response element** (**SRE**), which is recognized by a complex of transcription factors including the **serum response factor** (**SRF**) and **Elk-1**. ERK phosphorylates and activates Elk-1, providing a direct link between the ERK family of MAP kinases and immediate-early gene induction. Many immediate-early genes themselves encode transcription factors, so their induction in response to growth factor stimulation leads to altered expression of a battery of other downstream genes called **secondary response genes**. As discussed in Chapter 16, these alterations in gene expression directly link ERK signaling to the stimulation of cell proliferation induced by growth factors.

Both yeasts and mammalian cells have multiple MAP kinase pathways that control distinct cellular responses. Each cascade consists of three protein kinases: a terminal MAP kinase and two upstream kinases (analogous to Raf and MEK) that regulate its activity. In the yeast *S. cerevisiae* five different MAP kinase cascades regulate mating, sporulation, filamentation, cell wall remodeling, and response to high osmolarity. In mammalian cells, three major groups of MAP kinases have been identified. In addition to

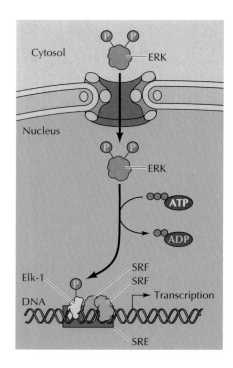

FIGURE 15.37 Induction of immediate-early genes by ERK Activated ERK translocates to the nucleus where it phosphorylates the transcription factor Elk-1. Elk-1 binds to the serum response element (SRE) in a complex with serum response factor (SRF). Phosphorylation stimulates the activity of Elk-1 as a transcriptional activator, leading to immediate-early gene induction.

FIGURE 15.38 Pathways of MAP kinase activation in mammalian cells
In addition to ERK, mammalian cells contain JNK and p38 MAP kinases. Activation of JNK and p38 is mediated by members of the Rho subfamily of small GTP-binding proteins (Rac, Rho, and Cdc42), which stimulate protein kinase cascades parallel to that responsible for ERK activation. The protein kinase cascades leading to JNK and p38 activation appear to be preferentially activated by inflammatory cytokines or cellular stress and generally lead to inflammation and cell death.

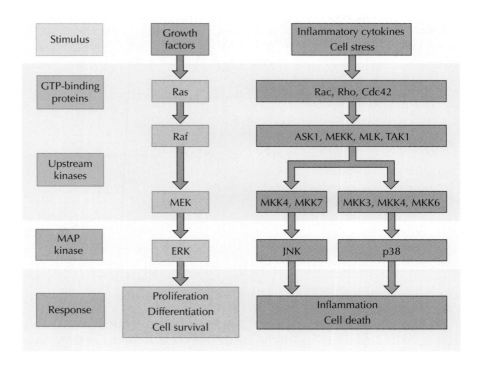

members of the ERK family these include the JNK and p38 MAP kinases, which are preferentially activated in response to inflammatory cytokines and cellular stress (e.g., ultraviolet irradiation) (Figure 15.38). The JNK and p38 MAP kinase cascades are activated by members of the Rho subfamily of small GTP-binding proteins (including Rac, Rho, and Cdc42) rather than by Ras. Whereas ERK signaling principally leads to cell proliferation, survival, and differentiation, the JNK and p38 MAP kinase pathways often lead to inflammation and cell death. Like ERK, the JNK and p38 MAP kinases can translocate to the nucleus and phosphorylate transcription factors that regulate gene expression. Multiple MAP kinase pathways thus function in all types of eukaryotic cells to control cellular responses to diverse environmental signals.

The specificity of MAP kinase signaling is maintained at least in part by the organization of the components of each MAP kinase cascade as complexes that are associated with **scaffold proteins**. For example, the JIP-1 scaffold protein organizes the JNK MAP kinase and its upstream activators MLK and MKK7 into a signaling cassette (Figure 15.39). As a result of the specific association of these protein kinases on the JIP-1 scaffold, activation of MLK by Rac leads to specific and efficient activation of MKK7, which in turn activates JNK. Distinct scaffold proteins are involved not only in the organization of other MAP kinase signaling cassettes but also in the association of other downstream signaling molecules with their receptors. The physical association of signaling pathway components as a result of their

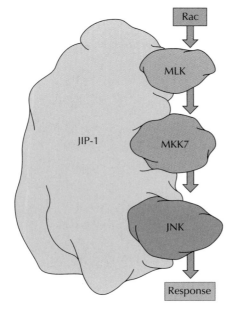

FIGURE 15.39 A scaffold protein for the JNK MAP kinase cascade The JIP-1 scaffold protein binds MLK, MKK7, and JNK, organizing these components of the JNK pathway into a signaling cassette.

FIGURE 15.40 The JAK/STAT pathway The STAT proteins are transcription factors that contain SH2 domains that mediate their binding to phosphotyrosine-containing sequences. In unstimulated cells, STAT proteins are inactive in the cytosol. Stimulation of cytokine receptors leads to the binding of STAT proteins where they are phosphorylated by the receptor-associated JAK protein-tyrosine kinases. The phosphorylated STAT proteins then dimerize and translocate to the nucleus where they activate the transcription of target genes.

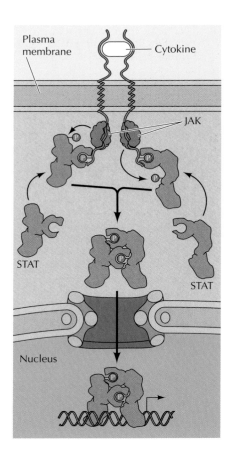

interaction with scaffold proteins is thought to play an important role in determining the specificity of signaling pathways within the cell.

The JAK/STAT and TGF-β/Smad Pathways

The PI 3-kinase and MAP kinase pathways are examples of indirect connections between the cell surface and the nucleus in which a cascade of protein kinases ultimately leads to transcription factor phosphorylation. The JAK/STAT and TGF-β/Smad pathways illustrate more direct connections between growth factor receptors and transcription factors in which the targeted transcription factors are phosphorylated directly by receptor-associated protein kinases.

The key elements in the **JAK/STAT pathway** are the **STAT proteins** (signal *t*ransducers and *a*ctivators of *t*ranscription), which were originally identified in studies of cytokine receptor signaling (Figure 15.40). The STAT proteins are a family of seven transcription factors that contain SH2 domains. They are inactive in unstimulated cells where they are localized to the cytoplasm. Stimulation of cytokine receptors leads to recruitment of STAT proteins, which bind via their SH2 domains to phosphotyrosine-containing sequences in the cytoplasmic domains of receptor polypeptides. Following their association with activated receptors the STAT proteins are phosphorylated by members of the JAK family of nonreceptor protein-tyrosine kinases, which are associated with cytokine receptors. Tyrosine phosphorylation promotes the dimerization of STAT proteins, which then translocate to the nucleus where they stimulate transcription of their target genes.

Further studies have shown that STAT proteins are also activated downstream of receptor protein-tyrosine kinases where their phosphorylation may be catalyzed either by the receptors themselves or by associated nonreceptor kinases. The STAT transcription factors thus serve as direct links between both cytokine and growth factor receptors on the cell surface and regulation of gene expression in the nucleus.

The receptors for members of the TGF-β family of growth factors are protein-serine/threonine kinases, which directly phosphorylate transcription factors of the **Smad** family (Figure 15.41). The receptors are composed of type I and type II polypeptides, which become associated following ligand binding. The type II receptor then phosphorylates the type I receptor, which in turn phosphorylates a Smad protein. The phosphorylated Smads then translocate to the nucleus and regulate gene expression. It should be noted that there are 42 different members of the TGF-β family in humans, which

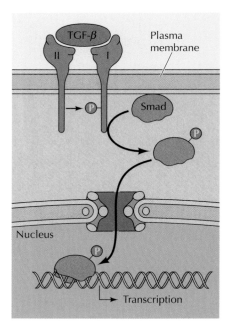

FIGURE 15.41 Signaling from TGF-β receptors TGF-β receptors are dimers of type I and II polypeptides. The type II receptor phosphorylates and activates type I, which then phosphorylates a Smad protein. Phosphorylated Smads translocate to the nucleus and activate transcription of target genes.

■ **Growth hormone functions by activating the JAK/STAT pathway.**

elicit different responses in their target cells. This is accomplished by combinatorial interactions of seven different type I receptors and five type II receptors, which lead to activation of different members of the Smad family (a total of 8 family members). Smad family members can also be phosphorylated by ERK, and this intersection between the TGFβ/Smad pathway and ERK plays a critical role in embryonic development.

NF-κB Signaling

NF-κB signaling is another example of a signaling pathway that directly targets a specific family of transcription factors. The **NF-κB** family consists of five transcription factors that play key roles in the immune system and in inflammation as well as in regulation of proliferation and survival of many types of animal cells. Members of this transcription factor family are activated in response to a variety of stimuli, including cytokines, growth factors, viral infection, and DNA damage. In unstimulated cells NF-κB proteins are bound to inhibitory **IκB** proteins that maintain NF-κB in an inactive state in the cytosol (Figure 15.42). Activation of NF-κB results from signals that activate the IκB kinase, which phosphorylates IκB. This phosphorylation targets IκB for ubiquitination and degradation by the proteasome, freeing NF-κB to translocate to the nucleus and induce expression of its target genes.

The Hedgehog, Wnt, and Notch Pathways

The **Hedgehog** and **Wnt** pathways are closely connected signaling systems that play key roles in determining cell fate during embryonic development. Both Hedgehog and Wnt pathways were first described in *Drosophila*, but members of the Hedgehog and Wnt families have been found to control a wide range of events that establish cell patterning during the development of both vertebrate and invertebrate embryos. Examples of the processes regulated by these signaling pathways include the determination of cell types and establishment of cell patterning during the development of limbs, the nervous system, the skeleton, lungs, hair, teeth, and gonads.

The *hedgehog* genes (one in *Drosophila* and three in mammals) encode secreted proteins that are modified by the addition of lipids. The functional receptor for Hedgehog consists of two transmembrane proteins, Patched and Smoothened (Figure 15.43). Hedgehog binds to Patched, which acts as a negative regulator of Smoothened. The binding of Hedgehog to Patched allows Smoothened to propagate an intracellular signal, leading to activation of a transcription factor called Cubitus interruptus (Ci) in *Drosophila* (Gli in mammals). In the absence of Hedgehog, Ci is maintained in a complex with a protein kinase called Fused and a kinesin-related protein called Coastal-2, which anchors the complex to microtubules. Within this complex, Ci is either completely degraded or cleaved to generate a transcriptional repressor (Ci75). Hedgehog signaling promotes the interaction of Smoothened with Coastal-2, leading to the release of full-length Ci (Ci155), which is then able to translocate to the nucleus and activate transcription of its target genes.

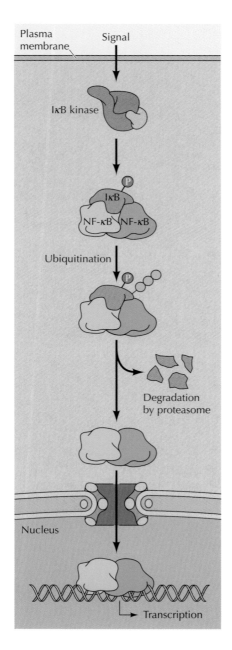

FIGURE 15.42 NF-κB signaling In the inactive state, homo- or heterodimers of NF-κB are bound to IκB in the cytoplasm. A variety of signals leads to activation of the IκB kinase, which phosphorylates IκB. This phosphorylation marks IκB for ubiquitination and degradation by the proteasome, allowing NF-κB to translocate to the nucleus and activate transcription of target genes.

(A) Absence of Hedgehog

(B) Presence of Hedgehog

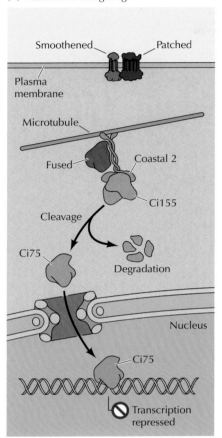

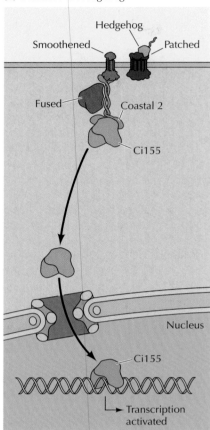

FIGURE 15.43 Hedgehog signaling In the absence of Hedgehog, the transcription factor Ci155 is anchored to microtubules by the kinesin-related protein Coastal-2 in association with the protein kinase Fused. Within this complex, Ci155 is degraded or cleaved to generate a transcriptional repressor (Ci75). The Hedgehog polypeptide binds to Patched on the surface of a target cell, relieving the inhibition of Smoothened by Patched and promoting the interaction of Smoothened with Coastal-2. This leads to the release of full-length Ci (Ci155), which translocates to the nucleus and activates transcription of its target genes.

The Wnt proteins are a family of secreted growth factors that bind to receptors of the LRP and Frizzled family, which are related to Smoothened (Figure 15.44). Signaling from LRP and Frizzled leads to activation of a cytoplasmic protein called Dishevelled, and inhibition of a complex of the proteins axin, APC, and the protein kinase GSK-3β. Within this complex, GSK-3β phosphorylates β-catenin, leading to its ubiquitination and degradation, so Wnt signaling results in increased β-catenin levels. β-catenin was discussed in Chapter 14 as a transmembrane protein that links cadherins to actin at adherens junctions (see Figure 14.22). Importantly, linking cadherins to actin is only one role of β-catenin. In Wnt signaling, β-catenin acts as a direct regulator of gene expression by forming a complex with members of the Tcf/LEF family of transcription factors. The association of β-catenin activates Tcf/LEF family members, leading to the expression of tar-

(A) Absence of Wnt

(B) Presence of Wnt

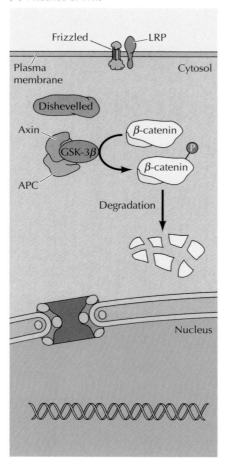
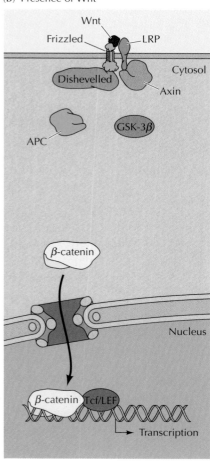

FIGURE 15.44 The Wnt pathway In the absence of Wnt, β-catenin is phosphorylated by GSK-3β in a complex with axin and APC, leading to its ubiquitination and degradation. Wnt polypeptides bind to Frizzled and LRP cell surface receptors, which activate Disheveled and lead to inhibition of the axin/APC/GSK-3β complex. This results in stabilization of β-catenin, which translocates to the nucleus and forms a complex with Tcf/LEF transcription factors.

get genes encoding other cell signaling molecules and a variety of transcription factors that control cell fate.

The **Notch** pathway is another highly conserved signaling pathway that controls cell fate during animal development. Notch signaling is an example of direct cell-cell interactions during development. Notch is a large protein with a single transmembrane domain that serves as a receptor for signaling by transmembrane proteins (e.g., Delta) on the surface of adjacent cells (Figure 15.45). Stimulation of Notch initiates a novel and direct pathway of transcriptional activation. In particular, ligand binding leads to proteolytic cleavage of Notch, and the intracellular domain of Notch is then translocated into the nucleus. The Notch intracellular domain then interacts with a transcription factor (called Su[H] in *Drosophila*, or CSL in mammals) and converts it from a repressor to an activator of its target genes. As in the

FIGURE 15.45 Notch signaling   Notch serves as a receptor for direct cell-cell signaling by transmembrane proteins (e.g., Delta) on neighboring cells. The binding of Delta leads to proteolytic cleavage of Notch by γ-secretase. This releases the Notch intracellular domain, which translocates to the nucleus and interacts with a transcription factor (Su[H] or CSL) to induce gene expression.

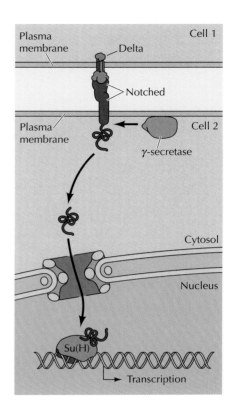

Wnt signaling pathway, the Notch target genes include genes encoding other transcriptional regulatory proteins, which act to determine cell fate.

Signal Transduction and the Cytoskeleton

The preceding sections focused on signaling pathways that regulate changes in metabolism, gene expression, and cell behavior in response to hormones and growth factors. However, the functions of most cells are also directly affected by cell adhesion and the organization of the cytoskeleton. The receptors responsible for cell adhesion thus act to initiate intracellular signaling pathways that regulate other aspects of cell behavior, including gene expression. Conversely, growth factors frequently act to induce cytoskeletal alterations resulting in cell movement or changes in cell shape. Components of the cytoskeleton thus act as both receptors and targets in cell signaling pathways, integrating cell shape and movement with other cellular responses.

Integrins and Signal Transduction

As discussed in Chapters 12 and 14, the integrins are the major receptors responsible for the attachment of cells to the extracellular matrix. At two types of cell-matrix junctions (focal adhesions and hemidesmosomes), the integrins also interact with components of the cytoskeleton to provide a stable linkage between the extracellular matrix and adherent cells (see Figure 14.19). In addition to this structural role, the integrins serve as receptors that activate intracellular signaling pathways thereby controlling cell movement and other aspects of cell behavior (including cell proliferation and survival) in response to adhesive interactions.

Like members of the cytokine receptor superfamily, the integrins have short cytoplasmic tails that lack any intrinsic enzymatic activity. However, protein-tyrosine phosphorylation is an early response to the interaction of integrins with extracellular matrix components, suggesting that the integrins are linked to nonreceptor protein-tyrosine kinases. One mode of signaling downstream of integrins involves activation of a nonreceptor protein-tyrosine kinase called **FAK** (*f*ocal *a*dhesion *k*inase) (Figure 15.46). As its name implies, FAK is localized to focal adhesions and rapidly becomes tyrosine-phosphorylated following the binding of integrin to extracellular matrix components, such as fibronectin. Like other protein-tyrosine kinases, the activation of FAK involves autophosphorylation induced by the clustering of integrins bound to the extracellular matrix as well as by their interactions with actin. Autophosphorylation of FAK creates docking sites for signaling molecules containing SH2 domains, including members of the Src family of nonreceptor protein-tyrosine kinases that phosphorylate additional sites on FAK. As discussed earlier for growth factor receptors, tyrosine phosphorylation of FAK creates binding sites for the SH2 domains of other downstream signaling molecules, including phospholipase C-γ, PI 3-kinase, and the Grb2-

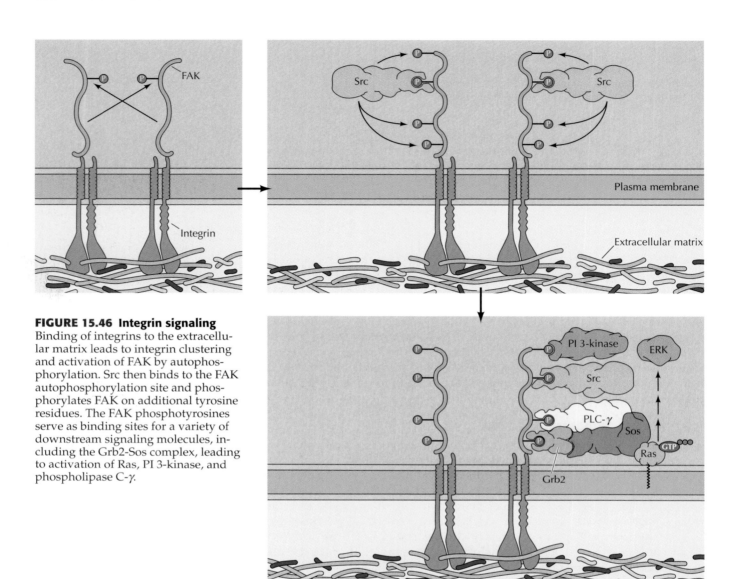

FIGURE 15.46 Integrin signaling
Binding of integrins to the extracellular matrix leads to integrin clustering and activation of FAK by autophosphorylation. Src then binds to the FAK autophosphorylation site and phosphorylates FAK on additional tyrosine residues. The FAK phosphotyrosines serve as binding sites for a variety of downstream signaling molecules, including the Grb2-Sos complex, leading to activation of Ras, PI 3-kinase, and phospholipase C-γ.

Sos complex. Recruitment of the Sos guanine nucleotide exchange factor leads to activation of Ras, which in turn couples integrins to activation of the ERK pathway. Integrin activation of the FAK and Src nonreceptor protein-tyrosine kinases thus links cell adhesion to the same downstream signaling pathways that regulate gene expression, cell proliferation, and cell survival downstream of growth factor receptors. In addition, integrins can interact with and stimulate the activities of receptor protein-tyrosine kinases, such as the EGF receptor, leading to a parallel activation of the signaling pathways stimulated by growth factors and by cell adhesion.

Regulation of the Actin Cytoskeleton

Signaling from integrins as well as from growth factor receptors also plays a central role in control of cell movement by regulating the dynamic behavior of the actin cytoskeleton. Cellular responses to growth factors as well as cell adhesion receptors frequently include changes in cell motility,

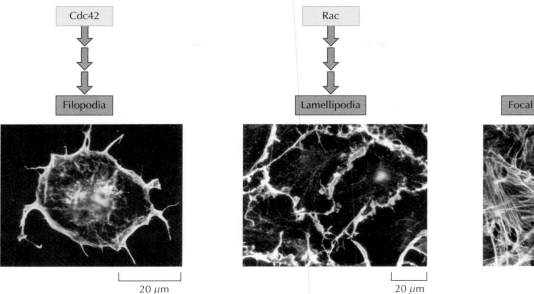

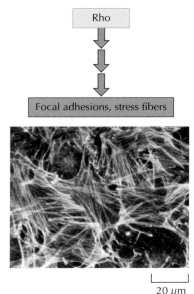

20 μm   20 μm   20 μm

which play critical roles in processes such as wound healing and embryonic development. As discussed in Chapter 12, these aspects of cell behavior are governed principally by the dynamic assembly and disassembly of actin filaments underlying the plasma membrane. Remodeling of the actin cytoskeleton therefore represents a key element of the response of many cells to extracellular stimuli.

Members of the Rho subfamily of small GTP-binding proteins (including **Rho**, **Rac**, and **Cdc42**) play central roles in regulating the organization of the actin cytoskeleton and thus control a variety of cell processes, including cell motility and cell adhesion. The role of Rho family members in cytoskeletal remodeling was first demonstrated by experiments showing that microinjection of fibroblasts with Rho proteins induced cytoskeletal alterations, including the production of cell surface protrusions (filopodia and lamellipodia) and the formation of focal adhesions and stress fibers (Figure 15.47). Microinjection of cells with specific mutants of different Rho family members has shown that Cdc42 induces the formation of filopodia, Rac mediates the formation of lamellipodia, and Rho is responsible for the formation of stress fibers. Further studies have demonstrated that the activities of Rho family members are not restricted to fibroblasts: They also play similar roles in regulating the actin cytoskeleton in all types of eukaryotic cells.

Rho family members are activated by integrin signaling as well as by growth factor receptors. A large number of proteins then serve to mediate the cytoskeletal changes induced by Rho, Rac, and Cdc42. For example, a key target of Rho in regulating cytoskeletal alterations is a protein-serine/threonine kinase called Rho kinase (Figure 15.48). Activation of Rho kinase increases the phosphorylation of the light chain of myosin II by two

FIGURE 15.47 Regulation of actin remodeling by Rho family proteins
Different members of the Rho family regulate the polymerization of actin to produce filopodia (Cdc42), lamellipodia (Rac), and stress fibers (Rho). Fluorescence micrographs illustrate the distribution of actin following microinjection of fibroblasts with Cdc42, Rac, and Rho. (From C. D. Nobes and A. Hall, 1995. *Cell* 81: 53.)

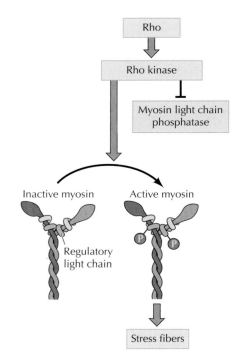

FIGURE 15.48 Regulation of myosin light chain phosphorylation by Rho
Rho activates the Rho kinase, which phosphorylates the regulatory light chain of myosin II and inhibits myosin light chain phosphatase. The resulting increase in phosphorylation of the light chain activates myosin II, leading to assembly of actin-myosin filaments and the formation of stress fibers.

mechanisms: Rho kinase not only directly phosphorylates the myosin light chain but also phosphorylates and inhibits myosin light chain phosphatase. The resulting increase in myosin light chain phosphorylation activates myosin and leads to the assembly of actin-myosin filaments, resulting in the formation of stress fibers and focal adhesions. Additional targets of Rho may also contribute to stress fiber formation by promoting actin polymerization. Both Rac and Cdc42 lead to the formation of cell surface protrusions (filopodia and lamellipodia) by stimulating actin polymerization, acting through several targets that associate with the Arp2/3 complex to induce actin filament formation (see Figure 12.35).

Signaling Networks

We have so far discussed signaling in terms of linear pathways that transmit information from the environment to intracellular targets. However, signaling within the cell is far more complicated. First, the activities of individual pathways are regulated by feedback loops that control the extent and duration of signaling activity. In addition, signaling pathways do not operate in isolation; rather, there is frequent crosstalk between different pathways, so that intracellular signal transduction ultimately needs to be understood as an integrated network of connected pathways. Computational modeling of such signaling networks is currently a major challenge in cell biology, which will be necessary to understand the dynamic response of cells to their environment.

Feedback and Crosstalk

The activity of signaling pathways is controlled by **feedback loops**, which are similar in principal to feedback regulation of metabolic pathways (see Figure 8.36). A good example of a negative feedback loop is provided by the NF-κB pathway (Figure 15.49). NF-κB is activated by signals that lead to proteolysis of the inhibitor IκB, allowing NF-κB to translocate to the nucleus and induce expression of its target genes. One of the target genes induced by NF-κB encodes IκB, so NF-κB signaling leads to the synthesis of new IκB, which inhibits continued NF-κB activity. This regulation is critical because the extent and duration of NF-κB activity can determine the transcriptional response of the cell. For example, some target genes are induced by transient NF-κB activity, persisting for only 30–60 minutes, whereas the induction of other genes requires several hours of sustained NF-κB signaling.

Signaling by the ERK MAP kinase provides another example of the importance of the duration of signaling. In a well-studied model of cell differentiation in response to nerve growth factor (NGF), ERK signaling can lead either to cell proliferation or to neuronal differentiation depending on the duration of ERK activity. In particular, transient activation of ERK (for 30–60 minutes) stimulates cell proliferation, but sustained activation of ERK for 2–3 hours induces differentiation of the NGF-treated cells into neurons. Although the mechanism by which these differences in the duration of ERK activity lead to such distinct biological outcomes remains to be understood, it is clear that quantitative considerations of signaling activity are critical to cell response.

Crosstalk refers to the interaction of one signaling pathway with another. Several examples have already been noted in this chapter, including junctions between Ca²⁺ and cAMP signaling, between the cAMP and ERK pathways, between the ERK and TGFβ/Smad pathways, and between integrin signaling and receptor protein-tyrosine kinases. A novel example

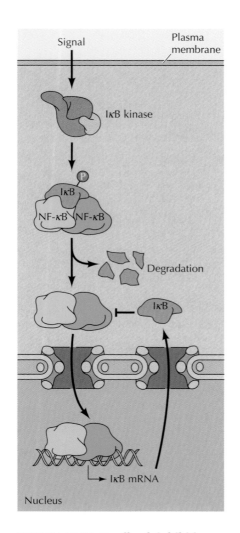

FIGURE 15.49 Feedback inhibition of NF-κB   NF-κB is activated as a result of phosphorylation and degradation of IκB, allowing NF-κB to translocate to the nucleus and activate transcription of target genes. One of the genes activated by NF-κB encodes IκB, generating a feedback loop that inhibits NF-κB activity.

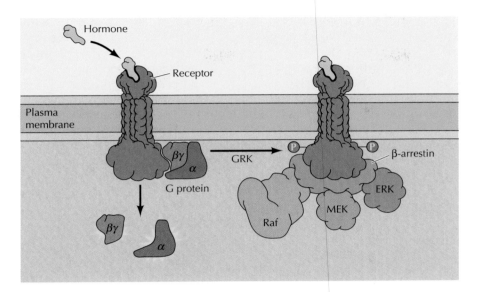

FIGURE 15.50 Crosstalk between G protein-coupled receptors and ERK signaling by β-arrestin Ligand binding stimulates G protein-coupled receptors, leading to activation of trimeric G proteins. The activity of the receptors is turned-off as a result of phosphorylation by GRKs and association of β-arrestin with the phosphorylated receptor. β-arrestin also acts as a scaffold protein for Raf, MEK, and ERK, linking G protein-coupled receptors to the ERK signaling pathway.

in which regulation of the duration of signaling is combined with crosstalk has come from recent studies of G protein-coupled receptors. These receptors are linked to MAP kinase signaling by **β-arrestins**, which were first identified as regulatory proteins that turn-off signaling from G protein-coupled receptors to G proteins (Figure 15.50). The signaling activity of these receptors is turned-off as a result of phosphorylation by a family of protein kinases called GRKs (*G* protein-coupled *r*eceptor *k*inases), followed by association of β-arrestin with the phosphorylated receptor. However, the β-arrestins do more than turn-off signaling to G proteins: They also act as signaling molecules themselves to stimulate additional downstream pathways, including nonreceptor protein-tyrosine kinases (e.g., Src family members) and MAP kinase pathways. In particular, β-arrestin-2 serves as a scaffold protein for Raf-MEK-ERK signaling, directly linking this MAP kinase pathway to G protein-coupled receptors.

Networks of Cellular Signal Transduction

The extensive crosstalk between individual signal transduction pathways means that multiple pathways interact with one another to form **signaling networks** within the cell. Some of the ways in which pathways can connect within a network are illustrated in Figure 15.51. The junctions between differ-

FIGURE 15.51 Elements of signaling networks In feedback loops, a downstream element of a pathway either inhibits (negative feedback) or stimulates (positive feedback) an upstream element. In feedforward relays, an upstream element of a pathway stimulates both its immediate target and another element further downstream. Crosstalk occurs when an element of one pathway either stimulates or inhibits an element of a second pathway.

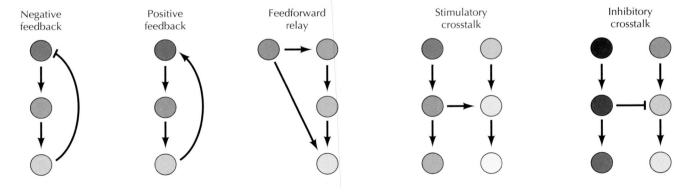

| Negative feedback | Positive feedback | Feedforward relay | Stimulatory crosstalk | Inhibitory crosstalk |

ent pathways can be either positive (where one pathway stimulates the other) or negative (where one pathway inhibits the other). In addition to negative feedback loops, signaling networks can contain positive feedback loops and **feedforward relays** in which the activity of one component of a pathway stimulates a distant downstream component.

A full understanding of cell signaling will require the development of network models that predict the dynamic behavior of the interconnected signaling pathways that ultimately result in a biological response. In this view of cell signaling as an integrated system, mathematical models and computer simulations will clearly be needed to deal with the complexity of the problem. For example, many signaling pathways involve receptors that stimulate cascades of protein kinases that ultimately affect gene expression by regulating transcription factors. The human genome encodes approximately 1500 different receptors, almost 700 protein kinases and phosphatases, and nearly 2000 transcription factors, so the potential for cross-regulation between pathways formed from combinations of these elements is enormous. A recently developed network model for signaling in a mammalian neuron provides a graphical representation of this complexity (Figure 15.52). Although clearly a daunting task, understanding cell signaling in quantitative terms using mathematical and computational approaches that view the cell as an integrated biological system is a critically important problem at the cutting edge of research in cell biology.

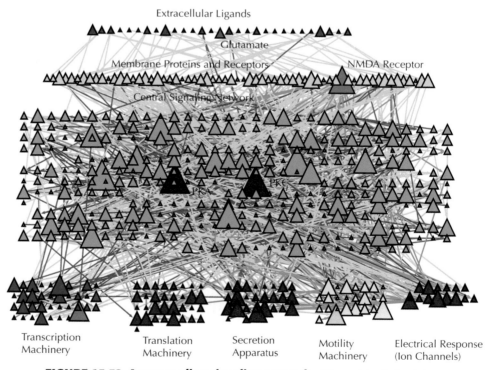

FIGURE 15.52 A mammalian signaling network The network depicts connections between 545 signaling elements in a mammalian neuron. Green arrows are stimulatory, red arrows are inhibitory, and blue arrows are neutral links between elements. (From A. Ma'ayan et al., 2005. *Science* 309: 1078.)

SUMMARY

KEY TERMS

SIGNALING MOLECULES AND THEIR RECEPTORS

Modes of Cell-Cell Signaling: Most signaling molecules are secreted by one cell and bind to receptors expressed by a target cell. Cell-cell signaling is divided into three general categories (endocrine, paracrine, and autocrine signaling) based on the distance over which signals are transmitted.

endocrine signaling, hormone, paracrine signaling, autocrine signaling

Steroid Hormones and the Nuclear Receptor Superfamily: The steroid hormones, thyroid hormone, vitamin D_3, and retinoic acid are small hydrophobic molecules that diffuse across the plasma membrane of their target cells and bind to intracellular receptors. Members of the nuclear receptor superfamily function as transcription factors to directly regulate gene expression in response to ligand binding.

steroid hormone, testosterone, estrogen, progesterone, corticosteroid, glucocorticoid, mineralocorticoid, ecdysone, brassinosteroid, thyroid hormone, vitamin D_3, retinoic acid, retinoid, nuclear receptor superfamily

Nitric Oxide and Carbon Monoxide: The simple gases nitric oxide and carbon monoxide are important paracrine signaling molecules in the nervous system and other cell types.

Neurotransmitters: Neurotransmitters are small hydrophilic molecules that carry signals between neurons or between neurons and other target cells at a synapse. Many neurotransmitters bind to ligand-gated ion channels.

neurotransmitter

Peptide Hormones and Growth Factors: The widest variety of signaling molecules in animals are peptides, ranging from only a few to more than a hundred amino acids. This group of molecules includes peptide hormones, neuropeptides, and growth factors.

peptide hormone, neuropeptide, enkephalin, endorphin, neuro-hormone, growth factor, nerve growth factor (NGF), neurotrophin, epidermal growth factor (EGF), platelet-derived growth factor (PDGF), cytokine, membrane-anchored growth factor

Eicosanoids: The eicosanoids are a class of lipids that function in paracrine and autocrine signaling.

eicosanoid, prostaglandin, prostacyclin, thromboxane, leukotriene

Plant Hormones: Small molecules known as plant hormones regulate plant growth and development.

plant hormone, auxin, gibberellin, cytokinin, abscisic acid, ethylene

FUNCTIONS OF CELL SURFACE RECEPTORS

G Protein-Coupled Receptors: The largest family of cell surface receptors, including the receptors for many hormones and neurotransmitters, transmit signals to intracellular targets via the intermediary action of G proteins.

G protein, G protein-coupled receptor, heterotrimeric G protein

Receptor Protein-Tyrosine Kinases: The receptors for most growth factors are protein-tyrosine kinases.

receptor protein-tyrosine kinase, autophosphorylation, SH2 domain, PTB domain

Cytokine Receptors and Nonreceptor Protein-Tyrosine Kinases: The receptors for many cytokines act in association with nonreceptor protein-tyrosine kinases.

cytokine receptor superfamily, nonreceptor protein-tyrosine kinase, Janus kinase (JAK)

KEY TERMS

protein-tyrosine phosphatase, transforming growth factor β (TGF-β), protein-serine/threonine kinase, guanylyl cyclase

intracellular signal transduction, cyclic AMP (cAMP), second messenger, adenylyl cyclase, cAMP phosphodiesterase, cAMP-dependent protein kinase, protein kinase A, cAMP response element (CRE), CREB

cyclic GMP (cGMP), rhodopsin, transducin, cGMP phosphodiesterase

phosphatidylinositol 4,5-bisphosphate (PIP$_2$), phospholipase C, diacylglycerol, inositol 1,4,5-trisphosphate (IP$_3$), protein kinase C, calmodulin, CaM kinase, ryanodine receptor

phosphatidylinositide (PI) 3-kinase, phosphatidylinositol 3,4,5-trisphosphate (PIP$_3$), Akt, mTOR

MAP kinase, ERK, Ras, Raf, MEK, guanine nucleotide exchange factor, GTPase-activating protein, immediate-early gene, serum response element (SRE), serum response factor (SRF), Elk-1, secondary response gene, scaffold protein

SUMMARY

Receptors Linked to Other Enzymatic Activities: Other kinds of cell surface receptors include protein-tyrosine phosphatases, protein-serine/threonine kinases, and guanylyl cyclases.

PATHWAYS OF INTRACELLULAR SIGNAL TRANSDUCTION

The cAMP Pathway: Second Messengers and Protein Phosphorylation: Cyclic AMP is an important second messenger in the response of animal cells to a variety of hormones and odorants. Most actions of cAMP are mediated by protein kinase A, which phosphorylates both metabolic enzymes and the transcription factor CREB.

Cyclic GMP: Cyclic GMP is also an important second messenger in animal cells. Its best-characterized role is in visual reception in the vertebrate eye.

Phospholipids and Ca^{2+}: Phospholipids and Ca^{2+} are common second messengers activated downstream of both G protein-coupled receptors and protein-tyrosine kinases. Hydrolysis of phosphatidylinositol 4,5-bisphosphate (PIP$_2$) yields diacylglycerol and inositol 1,4,5-trisphosphate (IP$_3$), which activate protein kinase C and mobilize Ca^{2+} from intracellular stores, respectively. Increased levels of cytosolic Ca^{2+} then activate a variety of target proteins, including Ca^{2+}/calmodulin-dependent protein kinases. In electrically excitable cells of nerve and muscle, levels of cytosolic Ca^{2+} are increased by the opening of voltage-gated Ca^{2+} channels in the plasma membrane and ryanodine receptors in the endoplasmic and sarcoplasmic reticula.

The PI 3-kinase/Akt and mTOR Pathways: In addition to being cleaved into diacylglycerol and IP$_3$, PIP$_2$ can be phosphorylated to the distinct second messenger PIP$_3$. This leads to activation of the protein-serine/threonine kinase Akt, which plays a key role in cell survival. One of the targets of Akt signaling is the protein kinase mTOR, which is a central regulator of cell growth and couples protein synthesis to the availability of growth factors, nutrients, and cellular energy.

MAP Kinase Pathways: The MAP kinase pathways are conserved chains of protein kinases activated downstream of a variety of extracellular signals. In animal cells, the best-characterized forms of MAP kinase are coupled to growth factor receptors by the small GTP-binding protein Ras, which initiates a protein kinase cascade leading to MAP kinase (ERK) activation. ERK then phosphorylates a variety of cytosolic and nuclear proteins, including transcription factors that mediate immediate-early gene induction. Other MAP kinase pathways mediate responses of mammalian cells to inflammation and stress. Components of MAP kinase pathways are organized by scaffold proteins, which play an important role in maintaining the specificity of MAP kinase signaling.

SUMMARY

The JAK/STAT and TGF-β/Smad Pathways: STAT proteins are transcription factors that contain SH2 domains and are activated directly by the JAK protein-tyrosine kinases associated with cytokine and growth factor receptors. Members of the TGF-β receptor family are protein-serine/threonine kinases that directly phosphorylate and activate Smad transcription factors.

NF-κB Signaling: Members of the NF-κB family of transcription factors are activated in response to cytokines, growth factors, and a variety of other stimuli. Their activation is mediated by phosphorylation and degradation of inhibitory IκB subunits.

The Hedgehog, Wnt, and Notch Pathways: The Hedgehog, Wnt, and Notch pathways play key roles in determination of cell fate and patterning during animal development. The Hedgehog and Wnt signaling pathways both act by preventing degradation of transcription factors in complexes in the cytoplasm. Notch signaling is mediated by direct cell-cell interactions, which induce proteolytic cleavage of Notch, followed by translocation of the Notch intracellular domain to the nucleus where it interacts with a transcription factor to affect expression of target genes.

SIGNAL TRANSDUCTION AND THE CYTOSKELETON

Integrins and Signal Transduction: Binding of integrins to the extracellular matrix stimulates the FAK and Src nonreceptor protein-tyrosine kinases, leading to activation of phospholipase C, PI 3-kinase, and Ras/Raf/ERK signaling pathways.

Regulation of the Actin Cytoskeleton: Signaling from integrins as well as growth factor receptors induces alterations in cell movement and cell shape by remodeling the actin cytoskeleton. These cytoskeletal alterations are mediated by members of the Rho subfamily of small GTP-binding proteins.

SIGNALING NETWORKS

Feedback and Crosstalk: The activity of signaling pathways within the cell is regulated by feedback loops that control the extent and duration of signaling. Different signaling pathways also interact to regulate each others activity.

Networks of Cellular Signal Transduction: The extensive crosstalk between individual pathways leads to the formation of complex signaling networks. A full understanding of signaling within the cell will require the development of quantitative network models.

KEY TERMS

JAK/STAT pathway, STAT protein, Smad

NF-κB, IκB

Hedgehog, Wnt, Notch

FAK

Rho, Rac, Cdc42

feedback loop, crosstalk, β-arrestin

signaling network, feedforward relay

Questions

1. What is the difference between paracrine and endocrine signaling?

2. How does signaling by hydrophobic molecules like steroid hormones differ from signaling by peptide hormones?

3. How does aspirin reduce inflammation and blood clotting?

4. Hormones that activate a receptor coupled to G_s stimulate the proliferation of thyroid cells. How would inhibitors of cAMP phosphodiesterase affect the proliferation of these cells?

5. The epinephrine receptor is coupled to G_s, whereas the acetylcholine receptor (on heart muscle cells) is coupled to G_i. Suppose you were to construct a recombinant molecule containing the extracellular sequences of the epinephrine receptor joined to the cytosolic sequences of the acetylcholine receptor. What effect would epinephrine have on cAMP levels in cells expressing such a recombinant receptor?

How would acetylcholine affect these cells?

6. Platelet-derived growth factor (PDGF) is a dimer of two polypeptide chains. How would PDGF monomers affect signaling by the PDGF receptor?

7. You have generated a truncated version of the EGF receptor that lacks the tyrosine kinase domain. Expression of this truncated receptor inhibits the response of cells to EGF. Why?

8. How would overexpression of protein phosphatase 1 affect the induction of cAMP-inducible genes in response to hormone stimulation of target cells? Would protein phosphatase 1 affect the function of cAMP-gated ion channels?

9. How does the PI 3-kinase/Akt pathway regulate cellular protein synthesis in response to growth factor stimulation?

10. You are studying an immediate early gene that is induced via the Ras/Raf/

MEK/ERK pathway in response to growth factor treatment of fibroblasts. How would expression of siRNA against Sos affect the induction of this gene?

11. A specific member of the STAT family induces certain liver genes in response to stimulation by a cytokine. How would the induction of these genes be affected if you overexpressed a dominant-negative mutant of JAK?

12. You are studying a gene that is induced by the Wnt signaling pathway. To determine the role played by β-catenin, you make different site-specific mutants of β-catenin. You find that changing a specific Lys residue to Arg leads to nuclear accumulation of β-catenin and constitutive expression of the gene (even in the absence of Wnt stimulation). What is the most likely mechanism by which this mutation alters β-catenin activity?

References and Further Reading

Signaling Molecules and Their Receptors

Alonso, J. M. and A. N. Stepanova. 2004. The ethylene signaling pathway. *Science* 306: 1513 1515. [R]

Arai, K., F. Lee, A. Miyajima, S. Miyatake, N. Arai and T. Yokota. 1990. Cytokines: Coordinators of immune and inflammatory responses. *Ann. Rev. Biochem.* 59: 783–836. [R]

Baranano, D. E. and S. H. Snyder. 2001. Neural roles for heme oxygenase: Contrasts to nitric oxide synthase. *Proc. Natl. Acad. Sci. U.S.A.* 98: 10996–11002. [R]

Burgess, W. H. and T. Maciag. 1989. The heparin-binding (fibroblast) growth factor family of proteins. *Ann. Rev. Biochem.* 58: 575–606. [R]

Callis, J. 2005. Auxin action. *Nature* 435: 436–437. [R]

Carpenter, G. and S. Cohen. 1990. Epidermal growth factor. *J. Biol. Chem.* 265: 7709–7712. [R]

Chawla, A., J. J. Repa, R. M. Evans and D. J. Mangelsdorf. 2001. Nuclear receptors and lipid physiology: Opening the X-files. *Science* 294: 1866–1870. [R]

Denninger, J. W. and M. A. Marletta. 1999. Guanylate cyclase and the NO/cGMP signaling pathway. *Biochem. Biophys. Acta* 1411: 334–350. [R]

Funk, C. D. 2001. Prostaglandins and leukotrienes: Advances in eicosanoid biology. *Science* 294: 1871–1875. [R]

Hess, D. T., A. Matsumoto, S.-O. Kim, H. E. Marshall and J. S. Stamler. 2005. Protein *S*-nitrosylation: Purview and parameters. *Nature Rev. Mol. Cell Biol.* 6: 150–166. [R]

Hull, M. A. 2005. Cyclooxygenase-2: How good is it as a target for cancer chemoprevention? *Eur. J. Cancer* 41: 1854–1863. [R]

Levi-Montalcini, R. 1987. The nerve growth factor 35 years later. *Science* 237: 1154–1162. [R]

Massagué, J. and A. Pandiella. 1993. Membrane-anchored growth factors. *Ann. Rev. Biochem.* 62: 515–541. [R]

Mayer, B. and B. Hemmens. 1997. Biosynthesis and action of nitric oxide in mammalian cells. *Trends Biochem. Sci.* 22: 477–481. [R]

McCarty, D. R. and J. Chory. 2000. Conservation and innovation in plant signaling pathways. *Cell* 103: 201–209. [R]

McDonnell, D. P. and J. D. Norris. 2002. Connections and regulation of the human estrogen receptor. *Science* 296: 1642–1644. [R]

McKenna, N. J. and B. W. O'Malley. 2002. Combinatorial control of gene expression by nuclear receptors and coregulators. *Cell* 108: 465–474. [R]

Minna, J. D., A. F. Gazdar, S. R. Sprang and J. Herz. 2004. A bull's eye for targeted lung cancer therapy. *Science* 304: 1458–1461. [R]

Ross, R., E. W. Raines and D. F. Bowen-Pope. 1986. The biology of platelet-derived growth factor. *Cell* 46: 155–169. [R]

Sheen, J. 2002. Phosphorelay and transcriptional control in cytokinin signal transduction. *Science* 296: 1650–1652. [R]

Smalley, W. and R. N. DuBois. 1997. Colorectal cancer and nonsteroidal anti-inflammatory drugs. *Adv. Pharmacol.* 39: 1–20. [R]

Snyder, S. H., S. R. Jaffrey and R. Zakhary. 1998. Nitric oxide and carbon monoxide: Parallel roles as neural messengers. *Brain Res. Rev.* 26: 167–175. [R]

Thummel, C. S. and J. Chory. 2002. Steroid signaling in plants and insects—common themes, different pathways. *Genes Dev.* 16: 3113–3129. [R]

Functions of Cell Surface Receptors

Alonso, A., J. Sasin, N. Bottini, I. Friedberg, I. Friedberg, A. Osterman, A. Godzik, T. Hunter, J. Dixon and T. Mustelin. 2004. Protein tyrosine phosphatases in the human genome. *Cell* 117: 699–711. [R]

Blume-Jensen, P. and T. Hunter. 2001. Oncogenic kinase signaling. *Nature* 411: 355–365. [R]

Hunter, T. and B. M. Sefton. 1980. Transforming gene product of Rous sarcoma virus phosphorylates tyrosine. *Proc. Natl. Acad. Sci. U.S.A.* 77: 1311–1315. [P]

Irvine, R. F. 2003. 20 years of Ins(1,4,5)P$_3$ and 40 years before. *Nature Rev. Mol. Cell Biol.* 4: 586–590. [R]

Lefkowitz, R. J. 2004. Historical review: A brief history and personal retrospective of seven-transmembrane receptors. *Trends Pharmacol. Sci.* 25: 413–422. [R]

Malbon, C. C. 2005. G proteins in development. *Nature Rev. Mol. Cell Biol.* 6: 689–701. [R]

Mishra, L., R. Derynck and B. Mishra. 2005. Transforming growth factor-β signaling in stem cells and cancer. *Science* 310: 68–71. [R]

Neves, S. R., P. T. Ram and R. Iyengar. 2002. G protein pathways. *Science* 296: 1636–1639. [R]

Parsons, S. J. and J. T. Parsons. 2004. Src family kinases, key regulators of signal transduction. *Oncogene* 23: 7906–7909. [R]

Schlessinger, J. 2000. Cell signaling by receptor tyrosine kinases. *Cell* 103: 211–225. [R]

Schlessinger, J. 2004. Common and distinct elements in cellular signaling via EGF and FGF receptors. *Science* 306: 1506–1507. [R]

Singer, A. L. and G. A. Koretsky. 2002. Control of T cell function by positive and negative regulators. *Science* 296: 1639–1640. [R]

Wedel, B. and D. Garbers. 2001. The guanylyl cyclase family at Y2K. *Ann. Rev. Physiol.* 63: 215–233. [R]

Pathways of Intracellular Signal Transduction

Aaronson, D. S. and C. M. Horvath. 2002. A road map for those who don't know JAK-STAT. *Science* 296: 1653–1655. [R]

Artavanis-Tsakonas, S., M. D. Rand and R. J. Lake. 1999. Notch signaling: Cell fate control and signal integration in development. *Science* 284: 770–776. [R]

Berridge, M. J., M. D. Bootman and H. L. Roderick. 2003. Calcium signaling: Dynamics, homeostasis and remodeling. *Nature Rev. Mol. Cell Biol.* 4: 517–529. [R]

Bradley, J., J. Reisert and S. Frings. 2005. Regulation of cyclic nucleotide-gated channels. *Curr. Opin. Neurobiol.* 15: 343–349. [R]

Brivanlou, A. H. and J. E. Darnell Jr. 2002. Signal transduction and the control of gene expression. *Science* 295: 813–818. [R]

Cantley, L. C. 2002. The phosphoinositide 3-kinase pathway. *Science* 296: 1655–1657. [R]

Chang, L. and M. Karin. 2001. Mammalian MAP kinase signalling cascades. *Nature* 410: 37–40. [R]

Chen, L.-F. and W. C. Greene. 2004. Shaping the nuclear action of NF-κB. *Nature Rev. Mol. Cell Biol.* 5: 392–401. [R]

Clapham, D. E. 2003. TRP channels as cellular sensors. *Nature* 426: 517–524. [R]

Cohen, P. T. W. 1997. Novel protein serine/threonine phosphatases: Variety is the spice of life. *Trends Biochem. Sci.* 22: 245–251. [R]

Coleman, M. L., C. J. Marshall and M. F. Olson. 2004. Ras and Rho GTPases in G1-phase cell-cycle regulation. *Nature Rev. Mol. Cell Biol.* 5: 355–366. [R]

Corcoran, E. E. and A. R. Means. 2001. Defining Ca^{2+}/calmodulin-dependent protein kinase cascades in transcriptional regulation. *J. Biol. Chem.* 276: 2975–2978. [R]

Datta, S. R., A. Brunet and M. E. Greenberg. 1999. Cellular survival: A play in three Akts. *Genes Dev.* 13: 290–2927. [R]

Davis, R. J. 2000. Signal transduction by the JNK group of MAP kinases. *Cell* 103: 239–252. [R]

Hay, N. and N. Sonenberg. 2004. Upstream and downstream of mTOR. *Genes Dev.* 18: 1926–1945. [R]

Hayden, M. S. and S. Ghosh. 2004. Signaling to NF-κB. *Genes Dev.* 18: 2195–2224. [R]

Hofmann, F. 2005. The biology of cyclic GMP-dependent protein kinases. *J. Biol. Chem.* 280: 1–4. [R]

Hunter, T. 1995. Protein kinases and phosphatases: The yin and yang of protein phosphorylation and signaling. *Cell* 80: 225–236. [R]

Hurley, J. H. 1999. Structure, mechanism, and regulation of mammalian adenylyl cyclase. *J. Biol. Chem.* 274: 7599–7602. [R]

Johnson, G. L. and R. Lapadat. 2002. Mitogen-activated protein kinase pathways mediated by ERK, JNK, and p38 protein kinases. *Science* 298: 1911–1912. [R]

Jope, R. S. and G. V. W. Johnson. 2004. The glamour and gloom of glycogen synthase kinase-3. *Trends Biochem. Sci.* 29: 95–102. [R]

Logan, C. Y. and R. Nusse. 2004. The Wnt signaling pathway in development and disease. *Ann. Rev. Cell Dev. Biol.* 20: 781–810. [R]

Lum, L. and P. A. Beachy. 2004. The Hedgehog response network: Sensors, switches, and routers. *Science* 304: 1755–1759. [R]

Manning, G., D. B. Whyte, R. Martinez, T. Hunter and S. Sudarsanam. 2002. The protein kinase complement of the human genome. *Science* 298: 1912–1934. [R]

Marinissen, M. J. and J. S. Gutkind. 2005. Scaffold proteins dictate Rho GTPase-signaling specificity. *Trends Biochem. Sci.* 30: 423–426. [R]

Martin, D. E. and M. N. Hall. 2005. The expanding TOR signaling network. *Curr. Opin. Cell Biol.* 17: 158–166. [R]

Massague, J., J. Seoane and D. Wotton. 2005. Smad transcription factors. *Genes Dev.* 19: 2783–2810. [R]

Mayr, B. and M. Montminy. 2001. Transcriptional regulation by the phosphorylation-dependent factor CREB. *Nature Rev. Mol. Cell Biol.* 2: 599–609. [R]

Morrison, D. K. and R. J. Davis. 2003. Regulation of MAP kinase signaling modules by scaffold proteins in mammals. *Ann. Rev. Cell Dev. Biol.* 19: 91–118. [R]

Neves, S. R., P. T. Ram and R. Iyengar. 2002. G protein pathways. *Science* 296: 1636–1639. [R]

O'Shea, J. J., M. Gadina and R. D. Schreiber. 2002. Cytokine signaling in 2002: New surprises in the Jak/Stat pathway. *Cell* 109: S121–S131. [R]

Sarbassov, D. D., S. M. Ali and D. M. Sabatini. 2005. Growing roles for the mTOR pathway. *Curr. Opin. Cell Biol.* 17: 596–603. [R]

Scheid, M. P. and J. R. Woodgett. 2001. PKB/Akt: Functional insights from genetic models. *Nature Rev. Mol. Cell Biol.* 2: 760–768. [R]

Soderling, T. R. 1999. The Ca^{2+}-calmodulin-dependent protein kinase cascade. *Trends Biochem. Sci.* 24: 232–236. [R]

Taylor, C. W. 2002. Controlling calcium entry. *Cell* 111: 767–769. [R]

Tran, H., A. Brunet, E. C. Griffith and M. E. Greenberg. 2003. The many forks in FOXO's road. *Science STKE.* re5. [R]

Venkatachalam, K., D. B. van Rossum, R. L. Patterson, H.-T. Ma and D. L. Gill. 2002. The cellular and molecular basis of store-operated calcium entry. *Nature Cell. Biol.* 4: E263–272. [R]

Wellbrock, C., M. Karasarides and R. Marais. 2004. The Raf proteins take centre stage. *Nature Rev. Mol. Cell Biol.* 5: 875–885. [R]

Yoon, K. and N. Gaiano. 2005. Notch signaling in the mammalian central nervous system: Insights from mouse mutants. *Nature Neurosci.* 8: 709–715. [R]

Signal Transduction and the Cytoskeleton

Burridge, K. and K. Wennerberg. 2004. Rho and Rac take center stage. *Cell* 116: 167–179. [R]

DeMali, K. A., K. Wennerberg and K. Burridge. 2003. Integrin signaling to the actin cytoskeleton. *Curr. Opin. Cell Biol.* 15: 572–582. [R]

Etienne-Manneville, S. and A. Hall. 2002. Rho GTPases in cell biology. *Nature* 420: 629–635. [R]

Giancotti, F. G. and E. Ruoslahti. 1999. Integrin signaling. *Science* 285: 1028–1032. [R]

Hynes, R. O. 2002. Integrins: Bidirectional, allosteric signaling machines. *Cell* 110: 673–687. [R]

Miranti, C. K. and J. S. Brugge. 2002. Sensing the environment: A historical perspective on integrin signal transduction. *Nature Cell Biol.* 4: E83–E90. [R]

Mitra, S. K., D. A. Hanson and D. D. Schlaepfer. 2005. Focal adhesion kinase: In command and control of cell motility. *Nature Rev. Mol. Cell Biol.* 6: 56–68. [R]

Nobles, C. D. and A. Hall. 1995. Rho, Rac, and Cdc42 GTPases regulate the assembly of multimolecular focal complexes associated with actin stress fibers, lamellipodia, and filopodia. *Cell* 81: 53–62. [P]

Ridley, A. J. 2001. Rho family proteins: Coordinating cellular responses. *Trends Cell Biol.* 11: 471–47. [R]

Schwartz, M. A. and M. H. Ginsberg. 2002. Networks and crosstalk: Integrin signalling spreads. *Nature Cell Biol.* 4: E65–E68. [R]

Signaling Networks

Eungdamrong, N. J. and R. Iyengar. 2004. Computational approaches for modeling regulatory cellular networks. *Trends Cell Biol.* 14: 661–669. [R]

Hoffmann, A., A. Levchenko, M. L. Scott and D. Baltimore. 2002. The IκB—NF-κB signaling module: Temporal control and selective gene activation. *Science* 298: 1241–1245. [P]

Lefkowitz, R. J. and S. K. Shenoy. 2005. Transduction of receptor signals by β-arrestins. *Science* 308: 512–517. [R]

Levchenko, A. 2003. Dynamic and integrative cell signaling: Challenges for the new biology. *Biotechnol. Bioeng.* 84: 773–782. [R]

Ma'ayan, A., S. L. Jenkins, S. Neves, A. Hasseldine, E. Grace, B. Dubin-Thaler, N. J. Eungdamrong, G. Weng, P. T. Ram, J. J. Rice, A. Kershenbaum, G. A. Stolovitzky, R. D. Blitzer and R. Iyengar. 2005. Formation of regulatory patterns during signal propagation in a mammalian cellular network. *Science* 309: 1078–1083. [P]

Marshall, C. J. 1995. Specificity of receptor tyrosine kinase signaling: Transient versus sustained extracellular signal-regulated kinase activation. *Cell* 80: 179–185. [R]

Papin, J. A., T. Hunter, B. O. Palsson and S. Subramaniam. 2005. Reconstruction of cellular signaling networks and analysis of their properties. *Nature Rev. Mol. Cell Biol.* 6: 99–111 [R]

CHAPTER

16

The Cell Cycle

■ **The Eukaryotic Cell Cycle 650**

■ **Regulators of Cell Cycle Progression 657**

■ **The Events of M Phase 669**

■ **Meiosis and Fertilization 678**

■ **KEY EXPERIMENT:**
The Discovery of MPF 658

■ **KEY EXPERIMENT:**
The Identification of Cyclin 662

SELF-REPRODUCTION IS PERHAPS THE MOST FUNDAMENTAL
characteristic of cells—as may be said for all living organisms. All cells reproduce by dividing in two, with each parental cell giving rise to two daughter cells on completion of each cycle of cell division. These newly formed daughter cells can themselves grow and divide, giving rise to a new cell population formed by the growth and division of a single parental cell and its progeny. In the simplest case, such cycles of growth and division allow a single bacterium to form a colony consisting of millions of progeny cells during overnight incubation on a plate of nutrient agar medium. In a more complex case, repeated cycles of cell growth and division result in the development of a single fertilized egg into the approximately 10^{14} cells that make up the human body.

The division of all cells must be carefully regulated and coordinated with both cell growth and DNA replication in order to ensure the formation of progeny cells containing intact genomes. In eukaryotic cells, progression through the cell cycle is controlled by a series of protein kinases that have been conserved from yeasts to mammals. In higher eukaryotes, this cell cycle machinery is itself regulated by the growth factors that control cell proliferation, allowing the division of individual cells to be coordinated with the needs of the organism as a whole. Not surprisingly, defects in cell cycle regulation are a common cause of the abnormal proliferation of cancer cells, so studies of the cell cycle and cancer have become closely interconnected, similar to the relationship between studies of cancer and the cell signaling pathways discussed in Chapter 15.

The Eukaryotic Cell Cycle

The division cycle of most cells consists of four coordinated processes: cell growth, DNA replication, distribution of the duplicated chromosomes to daughter cells, and cell division. In bacteria, cell growth and DNA replication take place throughout most of the cell cycle, and duplicated chromosomes are distributed to daughter cells in association with the plasma membrane. In eukaryotes, however, the cell cycle is more complex and consists of four discrete phases. Although cell growth is usually a continuous process, DNA is synthesized during only one phase of the cell cycle, and the replicated chromosomes are then distributed to daughter nuclei by a complex series of events preceding cell division. Progression between these stages of the cell cycle is controlled by a conserved regulatory apparatus, which not only coordinates the different events of the cell cycle but also links the cell cycle with extracellular signals that control cell proliferation.

Phases of the Cell Cycle

A typical eukaryotic cell cycle is illustrated by human cells in culture, which divide approximately every 24 hours. As viewed in the microscope, the cell cycle is divided into two basic parts: **mitosis** and **interphase**. Mitosis (nuclear division) is the most dramatic stage of the cell cycle, corresponding to the separation of daughter chromosomes and usually ending with cell division (**cytokinesis**). However, mitosis and cytokinesis last only about an hour, so approximately 95% of the cell cycle is spent in interphase—the period between mitoses. During interphase, the chromosomes are decondensed and distributed throughout the nucleus, so the nucleus appears morphologically uniform. At the molecular level, however, interphase is the time during which both cell growth and DNA replication occur in an orderly manner in preparation for cell division.

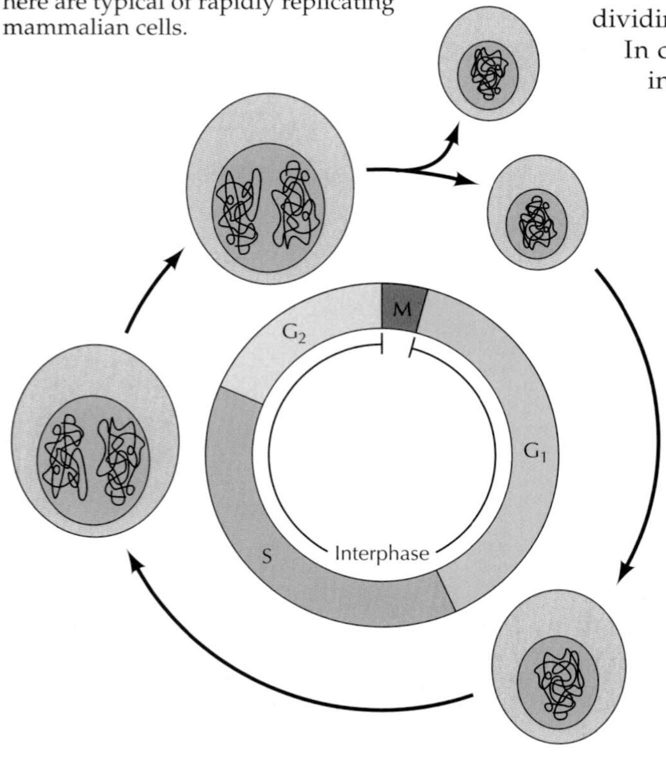

The cell grows at a steady rate throughout interphase, with most dividing cells doubling in size between one mitosis and the next. In contrast, DNA is synthesized during only a portion of interphase. The timing of DNA synthesis thus divides the cycle of eukaryotic cells into four discrete phases (Figure 16.1). The **M phase** of the cycle corresponds to mitosis, which is usually followed by cytokinesis. This phase is followed by the **G_1 phase** (gap 1), which corresponds to the interval (gap) between mitosis and initiation of DNA replication. During G_1, the cell is metabolically active and continuously grows but does not replicate its DNA. G_1 is followed by **S phase** (synthesis) during which DNA replication takes place. The completion of DNA synthesis is followed by the **G_2 phase** (gap 2) during which cell growth continues and proteins are synthesized in preparation for mitosis.

The duration of these cell cycle phases varies considerably in different kinds of cells. For a typical rapidly proliferating human cell with a total cycle time of 24 hours, the G_1 phase might last about 11 hours, S phase about 8 hours, G_2 about 4 hours, and M about 1 hour. Other types of cells, however, can divide much more rapidly. Budding yeasts, for example, can progress through all four stages of the cell cycle in only about 90 minutes. Even shorter cell cycles (30 minutes or

16.1 **WEBSITE ANIMATION**

Phases of the Cell Cycle
The cell cycle of most eukaryotic cells is divided into four discrete phases: M, G_1, S, and G_2.

FIGURE 16.1 Phases of the cell cycle
The division cycle of most eukaryotic cells is divided into four discrete phases: M, G_1, S, and G_2. M phase (mitosis) is usually followed by cytokinesis. S phase is the period during which DNA replication occurs. The cell grows throughout interphase, which includes G_1, S, and G_2. The relative lengths of the cell cycle phases shown here are typical of rapidly replicating mammalian cells.

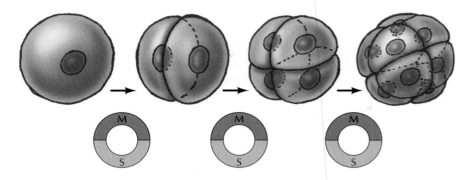

FIGURE 16.2 Embryonic cell cycles
Early embryonic cell cycles rapidly divide the cytoplasm of the egg into smaller cells. The cells do not grow during these cycles, which lack G_1 and G_2 and consist simply of short S phases alternating with M phases.

16.2 **WEBSITE ANIMATION**
Embryonic Cell Cycles
During early embryonic cell cycles, the cells do not grow, and instead divide into progressively smaller cells.

less) occur in early embryo cells shortly after fertilization of the egg (Figure 16.2). In this case, however, cell growth does not take place. Instead, these early embryonic cell cycles rapidly divide the egg cytoplasm into smaller cells. There is no G_1 or G_2 phase, and DNA replication occurs very rapidly in these early embryonic cell cycles, which therefore consist of very short S phases alternating with M phases.

In contrast to the rapid proliferation of embryonic cells, some cells in adult animals cease division altogether (e.g., nerve cells) and many other cells divide only occasionally, as needed to replace cells that have been lost because of injury or cell death. Cells of the latter type include skin fibroblasts, as well as the cells of some internal organs, such as the liver. As discussed further in the next section, these cells exit G_1 to enter a quiescent stage of the cycle called **G_0**, where they remain metabolically active but no longer proliferate unless called on to do so by appropriate extracellular signals.

Analysis of the cell cycle requires identification of cells at the different stages discussed above. Although mitotic cells can be distinguished microscopically, cells in other phases of the cycle (G_1, S, and G_2) must be identified by biochemical criteria. Cells in S phase can be readily identified because they incorporate radioactive thymidine, which is used exclusively for DNA synthesis (Figure 16.3). For example, if a population of rapidly pro-

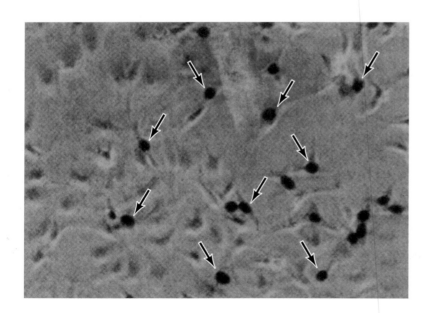

FIGURE 16.3 Identification of S phase cells by incorporation of radioactive thymidine
The cells were exposed to radioactive thymidine and analyzed by autoradiography. Labeled cells are indicated by arrows. (From D. W. Stacey et al., 1991. *Mol. Cell Biol.* 11: 4053.)

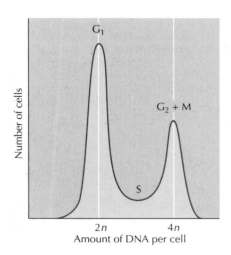

FIGURE 16.4 Determination of cellular DNA content A population of cells is labeled with a fluorescent dye that binds DNA. The cells are then passed through a flow cytometer, which measures the fluorescence intensity of individual cells. The data are plotted as cell number versus fluorescence intensity, which is proportional to DNA content. The distribution shows two peaks, corresponding to cells with DNA contents of $2n$ and $4n$; these cells are in the G_1 and G_2/M phases of the cycle, respectively. Cells in S phase have DNA contents between $2n$ and $4n$ and are distributed between these two peaks.

liferating human cells in culture is exposed to radioactive thymidine for a short period of time (e.g., 15 minutes) and then analyzed by autoradiography, about a third of the cells will be found to be radioactively labeled, corresponding to the fraction of cells in S phase.

Variations of such cell labeling experiments can also be used to determine the length of different stages of the cell cycle. For example, consider an experiment in which cells are exposed to radioactive thymidine for 15 minutes, after which the radioactive thymidine is removed and the cells are cultured for varying lengths of time prior to autoradiography. Radioactively labeled interphase cells that were in S phase during the time of exposure to radioactive thymidine will be observed for several hours as they progress through the remainder of S and G_2. In contrast, radioactively labeled mitotic cells will not be observed until 4 hours after labeling. This 4-hour lag time corresponds to the length of G_2—the minimum time required for a cell that incorporated radioactive thymidine at the end of S phase to enter mitosis.

Cells at different stages of the cell cycle can also be distinguished by their DNA content (Figure 16.4). For example, animal cells in G_1 are diploid (containing two copies of each chromosome), so their DNA content is referred to as $2n$ (n designates the haploid DNA content of the genome). During S phase, replication increases the DNA content of the cell from $2n$ to $4n$, so cells in S have DNA contents ranging from $2n$ to $4n$. DNA content then remains at $4n$ for cells in G_2 and M, decreasing to $2n$ after cytokinesis. Experimentally, cellular DNA content can be determined by incubation of cells with a fluorescent dye that binds to DNA, followed by analysis of the fluorescence intensity of individual cells in a **flow cytometer** or **fluorescence-activated cell sorter**, thereby distinguishing and allowing isolation of cells in the G_1, S, and G_2/M phases of the cell cycle.

Regulation of the Cell Cycle by Cell Growth and Extracellular Signals

The progression of cells through the division cycle is regulated by extracellular signals from the environment, as well as by internal signals that monitor and coordinate the various processes that take place during different cell cycle phases. An example of cell cycle regulation by extracellular signals is provided by the effect of growth factors on animal cell proliferation. In addition, different cellular processes, such as cell growth, DNA replication, and mitosis, all must be coordinated during cell cycle progression. This is accomplished by a series of control points that regulate progression through various phases of the cell cycle.

A major cell cycle regulatory point in many types of cells occurs late in G_1 and controls progression from G_1 to S. This regulatory point was first defined by studies of budding yeast (*Saccharomyces cerevisiae*), where it is known as **START** (Figure 16.5). Once cells have passed START, they are committed to entering S phase and undergoing one cell division cycle. However, passage through START is a highly regulated event in the yeast cell cycle where it is controlled by external signals, such as the availability of nutrients, as well as by cell size. For example, if yeasts are faced with a shortage of nutrients, they arrest their cell cycle at START and enter a resting state rather than proceeding to S phase. Thus START represents a decision point at which the cell determines whether sufficient nutrients are available to support progression through the rest of the division cycle. Polypeptide factors that signal yeast mating also arrest the cell cycle at

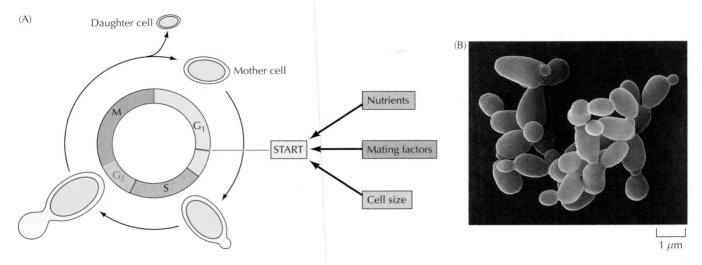

FIGURE 16.5 Regulation of the cell cycle of budding yeast (A) The cell cycle of *Saccharomyces cerevisiae* is regulated primarily at a point in late G_1 called START. Passage through START is controlled by the availability of nutrients, mating factors, and cell size. Note that these yeasts divide by budding. Buds form just after START and continue growing until they separate from the mother cell after mitosis. The daughter cell formed from the bud is smaller than the mother cell and therefore requires more time to grow during the G_1 phase of the next cell cycle. Although G_1 and S phases occur normally, the mitotic spindle begins to form during S phase, so the cell cycle of budding yeast lacks a distinct G_2 phase. (B) Scanning electron micrograph of *S. cerevisiae*. The size of the bud reflects the position of the cell in the cycle. (B, David M. Phillips/Visuals Unlimited.)

START, allowing haploid yeast cells to fuse with one another instead of progressing to S phase.

In addition to serving as a decision point for monitoring extracellular signals, START is the point at which cell growth is coordinated with DNA replication and cell division. The importance of this regulation is particularly evident in budding yeasts in which cell division produces progeny cells of very different sizes: a large mother cell and a small daughter cell. In order for yeast cells to maintain a constant size, the small daughter cell must grow more than the large mother cell does before they divide again. Thus cell size must be monitored in order to coordinate cell growth with other cell cycle events. This regulation is accomplished by a control mechanism that requires each cell to reach a minimum size before it can pass START. Consequently, the small daughter cell spends a longer time in G_1 and grows more than the mother cell.

The proliferation of most animal cells is similarly regulated in the G_1 phase of the cell cycle. In particular, a decision point in late G_1, called the **restriction point** in animal cells, functions analogously to START in yeasts (Figure 16.6). In contrast to yeasts, however, the passage of animal cells through the cell cycle is regulated primarily by the extracellular growth factors that signal cell proliferation, rather than by the availability of nutrients. In the presence of the appropriate growth factors, cells pass the restriction point and enter S phase. Once it has passed through the restriction point, the cell is committed to proceed through S phase and the rest of the cell

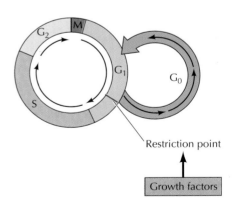

FIGURE 16.6 Regulation of animal cell cycles by growth factors The availability of growth factors controls the animal cell cycle at a point in late G_1 called the restriction point. If growth factors are not available during G_1, the cells enter a quiescent stage of the cycle called G_0.

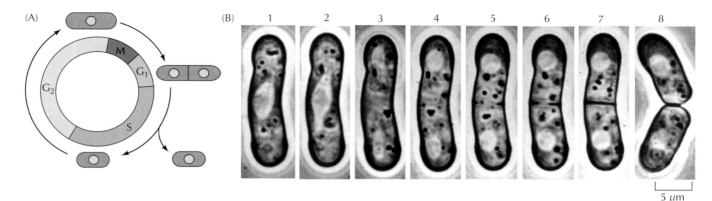

FIGURE 16.7 Cell cycle of fission yeast   (A) Fission yeasts grow by elongating at both ends and divide by forming a wall through the middle of the cell. In contrast to the cycle of budding yeasts, the cell cycle of fission yeasts has normal G_1, S, G_2, and M phases. Note that cytokinesis occurs in G_1. The length of the cell indicates its position in the cycle. (B) Light micrographs showing successive stages of mitosis and cytokinesis of *Schizosaccharomyces pombe*. (B, courtesy of C. F. Robinow, University of Western Ontario.)

cycle, even in the absence of further growth factor stimulation. On the other hand, if appropriate growth factors are not available in G_1, progression through the cell cycle stops at the restriction point. Such arrested cells then enter a quiescent stage of the cell cycle called G_0 in which they can remain for long periods of time without proliferating. G_0 cells are metabolically active, although they cease growth and have reduced rates of protein synthesis. As already noted, many cells in animals remain in G_0 unless called on to proliferate by appropriate growth factors or other extracellular signals. For example, skin fibroblasts are arrested in G_0 until they are stimulated to divide as required to repair damage resulting from a wound. The proliferation of these cells is triggered by platelet-derived growth factor, which is released from blood platelets during clotting and signals the proliferation of fibroblasts in the vicinity of the injured tissue.

Although the proliferation of most cells is regulated primarily in G_1, some cell cycles are instead controlled principally in G_2. One example is the cell cycle of the fission yeast *Schizosaccharomyces pombe* (Figure 16.7). In contrast to *Saccharomyces cerevisiae*, the cell cycle of *S. pombe* is regulated primarily by control of the transition from G_2 to M, which is the principal point at which cell size and nutrient availability are monitored. In animals, the primary example of cell cycle control in G_2 is provided by oocytes. Vertebrate oocytes can remain arrested in G_2 for long periods of time (several decades in humans) until their progression to M phase is triggered by hormonal stimulation. Extracellular signals can thus control cell proliferation by regulating progression from the G_2 to M as well as the G_1 to S phases of the cell cycle.

Cell Cycle Checkpoints

The controls discussed in the previous section regulate cell cycle progression in response to cell size and extracellular signals, such as nutrients and growth factors. In addition, the events that take place during different stages of the cell cycle must be coordinated with one another so that they occur in the appropriate order. For example, it is critically important that the cell not begin mitosis until replication of the genome has been completed. The alternative would be a catastrophic cell division in which the daughter cells failed to inherit complete copies of the genetic material. In most cells, this coordination between different phases of the cell cycle is dependent on a series of **cell cycle checkpoints** that prevent entry into the next phase of the cell cycle until the events of the preceding phase have been completed.

Several cell cycle checkpoints function to ensure that incomplete or damaged chromosomes are not replicated and passed on to daughter cells (Figure 16.8). These checkpoints sense unreplicated or damaged DNA and coordinate further cell cycle progression with the completion of DNA replication or repair. For example, the checkpoint in G_2 prevents the initiation of mitosis until DNA replication is completed. This G_2 checkpoint senses unreplicated DNA, which generates a signal that leads to cell cycle arrest. Operation of the G_2 checkpoint therefore prevents the initiation of M phase before completion of S phase, so cells remain in G_2 until the genome has been completely replicated. Only then is the inhibition of G_2 progression relieved, allowing the cell to initiate mitosis and distribute the completely replicated chromosomes to daughter cells. In addition to sensing unreplicated DNA, the G_2 checkpoint senses DNA damage, such as that resulting from irradiation. If DNA damage is detected, arrest at the checkpoint allows time for the damage to be repaired, rather than being passed on to daughter cells.

DNA damage not only arrests the cell cycle in G_2 but also at checkpoints in G_1 and S phase. Arrest at the G_1 checkpoint allows repair of the damage to take place before the cell enters S phase, where the damaged DNA would be replicated. The S-phase checkpoint provides continual monitoring of the integrity of DNA to ensure that damaged DNA is repaired before it is replicated. In addition, the S-phase checkpoint provides a quality control monitor to promote the repair of any errors that occur during DNA replication, such as the incorporation of incorrect bases or incomplete replication of segments of DNA.

Cell cycle arrest at the G_1, S, and G_2 checkpoints is mediated by two related protein kinases, designated **ATM** and **ATR**, that recognize damaged or unreplicated DNA and are activated in response to DNA damage. ATM and ATR then activate a signaling pathway that leads not only to cell cycle arrest, but also to the activation of DNA repair and, in some cases, programmed cell death. The importance of checkpoint regulation is emphasized by the fact that these proteins were initially identified because mutations in the gene encoding ATM are responsible for the disease ataxia telangiectasia, which results in defects in the nervous and immune systems as well as a high frequency of cancer in affected individuals.

Another important cell cycle checkpoint that maintains the integrity of the genome occurs toward the end of mitosis (see Figure 16.8). This checkpoint, called the spindle assembly checkpoint, monitors the alignment of chromosomes on the mitotic spindle, thus ensuring that a complete set of chromosomes is distributed accurately to the daughter cells. For example, the failure of one or more chromosomes to align properly on the spindle causes mitosis to arrest at metaphase, prior to the segregation of the newly replicated chromosomes to daughter nuclei. As a result of the spindle assembly checkpoint, the chromosomes do not separate until a complete complement of chromosomes has been organized for distribution to each daughter cell.

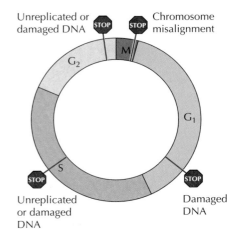

FIGURE 16.8 Cell cycle checkpoints Several checkpoints function to ensure that complete genomes are transmitted to daughter cells. DNA damage checkpoints in G_1, S, and G_2 lead to cell cycle arrest in response to damaged or unreplicated DNA. Another checkpoint, called the spindle assembly checkpoint, arrests mitosis if the chromosomes are not properly aligned on the mitotic spindle.

16.3 WEBSITE ANIMATION

Checkpoints in the Cell Cycle Several checkpoints function throughout the cell cycle to ensure that complete genomes are transmitted to daughter cells.

FIGURE 16.9 Restriction of DNA replication DNA replication is restricted to once per cell cycle by the MCM helicase proteins that bind to origins of replication together with ORC (origin recognition complex) proteins and are required for the initiation of DNA replication. MCM proteins are only able to bind to DNA in G_1, allowing DNA replication to initiate in S phase. Once initiation has occurred, the MCM proteins are displaced so that replication cannot initiate again until after mitosis.

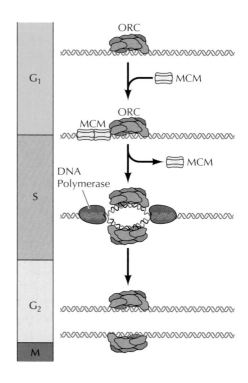

Restricting DNA Replication to Once per Cell Cycle

The G_2 checkpoint prevents the initiation of mitosis prior to the completion of S phase, thereby ensuring that incompletely replicated DNA is not distributed to daughter cells. It is equally important to ensure that the genome is replicated only once per cell cycle. Thus, once a segment of DNA has been replicated in S phase, control mechanisms must exist to prevent re-initiation of DNA replication until the cell cycle has been completed and the cell has passed through mitosis. As discussed in Chapter 6, mammalian cells use thousands of origins to replicate their DNA, so the initiation of replication at each of these origins must be carefully controlled so that each segment of the genome is replicated only once during the S phase of each cell cycle.

The molecular mechanism that restricts DNA replication to once per cell cycle involves the action of the MCM helicase proteins that bind to replication origins together with the origin recognition complex (ORC) proteins (see Figure 6.15). The MCM proteins act as "licensing factors" that allow replication to initiate (Figure 16.9). Their binding to DNA is regulated during the cell cycle such that the MCM proteins are only able to bind to replication origins during G_1, allowing DNA replication to initiate when the cell enters S phase. Once initiation has occurred, the MCM proteins are displaced from the origin, so replication cannot initiate again until the cell passes through mitosis and enters G_1 phase of the next cell cycle. The association of MCM proteins with DNA during the S, G_2 and M phases of the cell cycle is blocked by activity of the protein kinases that regulate cell cycle progression, as discussed in the next section of this chapter.

FIGURE 16.10 Identification of MPF Frog oocytes are arrested in the G_2 phase of the cell cycle, and entry into M phase of meiosis is triggered by the hormone progesterone. In the experiment diagrammed here, G_2-arrested oocytes were microinjected with cytoplasm extracted from oocytes that had undergone the transition from G_2 to M. Such cytoplasmic transfers induced the G_2 to M transition in the absence of hormonal stimulation, demonstrating that a cytoplasmic factor (MPF) is sufficient to induce entry into the M phase of meiosis.

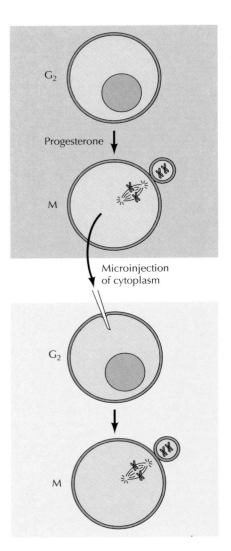

Regulators of Cell Cycle Progression

One of the most exciting developments in contemporary cell biology has been the elucidation of the molecular mechanisms that control the progression of eukaryotic cells through the division cycle. Our current understanding of cell cycle regulation has emerged from a convergence of results obtained through experiments on organisms as diverse as yeasts, sea urchins, frogs, and mammals. These studies have revealed that the cell cycle of all eukaryotes is controlled by a conserved set of protein kinases, which are responsible for triggering the major cell cycle transitions.

Protein Kinases and Cell Cycle Regulation

Three initially distinct experimental approaches contributed to identification of the key molecules responsible for cell cycle regulation. The first of these avenues of investigation originated with studies of frog oocytes (Figure 16.10). These oocytes are arrested in the G_2 phase of the cell cycle until hormonal stimulation triggers their entry into the M phase of meiosis (discussed later in this chapter). In 1971, two independent teams of researchers (Yoshio Masui and Clement Markert, as well as Dennis Smith and Robert Ecker) found that oocytes arrested in G_2 could be induced to enter M phase by microinjection of cytoplasm from oocytes that had been hormonally stimulated. It thus appeared that a cytoplasmic factor present in hormone-treated oocytes was sufficient to trigger the transition from G_2 to M in oocytes that had not been exposed to hormone. Because the entry of oocytes into meiosis is frequently referred to as oocyte maturation, this cytoplasmic factor was called **maturation promoting factor (MPF)**. Further studies showed, however, that the activity of MPF is not restricted to the entry of oocytes into meiosis. To the contrary, MPF is also present in somatic cells, where it induces entry into M phase of the mitotic cycle. Rather than being specific to oocytes, MPF thus appeared to act as a general regulator of the transition from G_2 to M.

The second approach to understanding cell cycle regulation was the genetic analysis of yeasts, pioneered by Lee Hartwell and his colleagues in the early 1970s. Studying the budding yeast *Saccharomyces cerevisiae*, these investigators identified temperature-sensitive mutants that were defective in cell cycle progression. The key characteristic of these mutants (called *cdc* for *cell division cycle* mutants) was that they underwent growth arrest at specific points in the cell cycle. For example, a particularly important mutant designated *cdc28* caused the cell cycle to arrest at START, indicating that the Cdc28 protein is required for passage through this critical regula-

KEY EXPERIMENT

The Discovery of MPF

Cytoplasmic Control of Nuclear Behavior during Meiotic Maturation of Frog Oocytes

Yoshio Masui and Clement L. Markert

Yale University, New Haven, CT

Journal of Experimental Zoology, 1971, Volume 177, Pages 129–146

Yoshio Masui

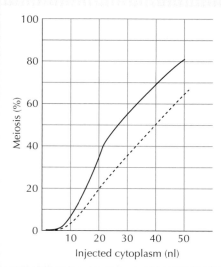

Clement Markert

The Context

Nuclear transplantation and cell fusion experiments performed in the 1960s indicated that nuclei transferred into cells at different stages of the mitotic cell cycle generally adopt the behavior of the host cell. Thus it appeared that mitotic activity of the nucleus was regulated by the cytoplasm. However, the existence of postulated cytoplasmic factors that control the mitotic activity of nuclei remained to be demonstrated by a direct experimental approach. This demonstration was provided by the studies of Masui and Markert, who investigated the role of cytoplasmic factors in regulating nuclear behavior during meiosis of frog oocytes.

Several features of frog oocyte meiosis suggested that it is controlled by cytoplasmic factors. In particular, meiosis of frog oocytes is arrested at the end of the prophase of meiosis I. Treatment with the hormone progesterone triggers the resumption of meiosis, which is equivalent to the G_2 to M transition of somatic cells. The oocytes then undergo a second arrest at the metaphase of meiosis II, where they remain until fertilization. Masui and Markert hypothesized that the effects of both hormone treatment and fertilization on meiosis were due to changes in the cytoplasm that secondarily controlled behavior of the nucleus. They tested this hypothesis directly by transferring cytoplasm from hormone-stimulated oocytes into unstimulated oocytes. These experiments demonstrated that a cytoplasmic factor, which Masui and Markert named maturation promoting factor (MPF), is responsible for the induction of meiosis following hormone treatment.

The Experiments

Because of both their large size and their ability to survive injection with glass micropipettes, frog oocytes appeared to provide a particularly suitable experimental system for testing the activity of cytoplasmic factors. The basic design of Masui and Markert's experiments was to remove cytoplasm from donor oocytes that had been treated with progesterone to induce the resumption of meiosis. Varying amounts of this cytoplasm were then injected into untreated recipient oocytes. The key result was that cytoplasm removed six or more hours after hormone treatment of donor oocytes induced the resumption of meiosis in injected recipients (see figure). In contrast, injection of cytoplasm from control oocytes that had not been exposed to progesterone had no effect on the recipients. It thus appeared that hormone-treated oocytes contained a cytoplasmic factor that could induce the resumption of meiosis in recipients that had never been exposed to progesterone.

Control experiments ruled out the possibility that progesterone itself is the meiosis-inducing factor in donor cytoplasm. In particular,

Recipient oocytes were injected with the indicated amounts of cytoplasm from oocytes that had been treated with progesterone. Donor cytoplasm was either withdrawn from the central region of oocytes with a micropipette (solid line) or prepared by homogenization of whole oocytes (dashed line). Results are presented as the percentage of injected oocytes that were induced to resume meiosis.

it was demonstrated that the injection of progesterone into recipient oocytes fails to induce meiosis. Only external application of the hormone is effective, indicating that progesterone acts on a cell surface receptor to activate a distinct cytoplasmic factor. Similar experiments performed independently by Dennis Smith and Robert Ecker (The interaction of steroids with *Rana pipiens* oocytes in the induction of maturation. *Dev. Biol.* 25: 232–247, 1971) led to the same conclusion.

[Figure: graph with y-axis "Meiosis (%)" from 0 to 100, x-axis "Injected cytoplasm (nl)" from 0 to 50]

KEY EXPERIMENT

Interestingly, the action of progesterone in this system is distinct from its action in most cells, where it diffuses across the plasma membrane and binds to an intracellular receptor (see Chapter 15). In oocytes, however, progesterone clearly acts on the cell surface to activate a distinct factor in the oocyte cytoplasm. Since the resumption of oocyte meiosis is commonly referred to as oocyte maturation, Masui and Markert coined the term "maturation promoting factor" for their newly discovered regulator of meiosis.

The Impact

Following its discovery in frog oocytes, MPF was also found to be present in somatic cells, where it induces the G_2 to M transition of mitosis. Thus MPF appeared to be a general regulator of entry into the M phase of both mitotic and meiotic cell cycles. The eventual purification of MPF from frog oocytes in 1988 then converged with yeast genetics and studies of sea urchin embryos to reveal the identity of this critical cell cycle regulator. Namely, MPF was

found to be a dimer of cyclin B and the Cdk1 protein kinase. Further studies have established that both cyclin B and Cdk1 are members of large families of proteins, with different cyclins and Cdk1-related protein kinases functioning analogously to MPF in the regulation of other cell cycle transitions. The discovery of MPF in frog oocytes thus paved the way to understanding a cell cycle regulatory apparatus that is conserved throughout all eukaryotes.

tory point in G_1 (Figure 16.11). A similar collection of cell cycle mutants was isolated in the fission yeast *Schizosaccharomyces pombe* by Paul Nurse and his collaborators. These mutants included *cdc2*, which arrests the *S. pombe* cell cycle both in G_1 and at the G_2 to M transition (the major regulatory point in fission yeast). Comparative analysis showed that *S. cerevisiae cdc28* and *S. pombe cdc2* are functionally homologous genes, which are required for passage through START as well as for entry into mitosis in both species of yeasts. Further studies demonstrated that *cdc2* and *cdc28* encoded a protein kinase—the first indication of the prominent role of protein phosphorylation in regulating the cell cycle. In addition, related genes were identified in other eukaryotes, including humans. The protein kinase encoded by the

FIGURE 16.11 Properties of *S. cerevisiae cdc28* mutants   The temperature-sensitive *cdc28* mutant replicates normally at the permissive temperature. At the nonpermissive temperature, however, progression through the cell cycle is blocked at START.

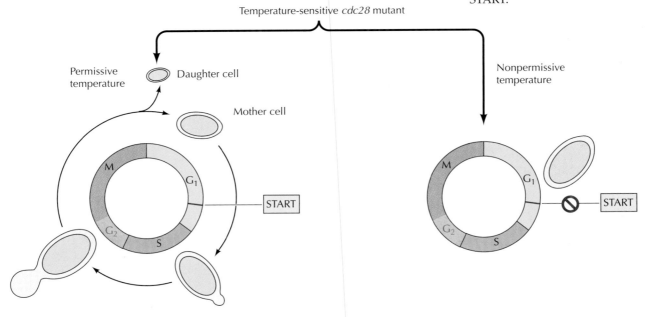

FIGURE 16.12 Accumulation and degradation of cyclins in sea urchin embryos The cyclins were identified as proteins that accumulate throughout interphase and are rapidly degraded toward the end of mitosis.

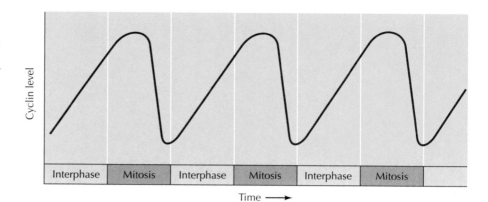

yeast *cdc2* and *cdc28* genes has since been shown to be a conserved cell cycle regulator in all eukaryotes, which is known as **Cdk1.**

The third line of investigation that eventually converged with the identification of MPF and yeast genetics emanated from studies of protein synthesis in early sea urchin embryos. Following fertilization, these embryos go through a series of rapid cell divisions. Intriguingly, studies with protein synthesis inhibitors had revealed that entry into M phase of these embryonic cell cycles requires new protein synthesis. In 1983, Tim Hunt and his colleagues identified two proteins that display a periodic pattern of accumulation and degradation in sea urchin and clam embryos. These proteins accumulate throughout interphase and are then rapidly degraded toward the end of each mitosis (Figure 16.12). Hunt called these proteins **cyclins** (the two proteins were designated cyclin A and cyclin B) and suggested that they might function to induce mitosis, with their periodic accumulation and destruction controlling entry and exit from M phase. Direct support for such a role of cyclins was provided in 1986, when Joan Ruderman and her colleagues showed that microinjection of cyclin A into frog oocytes is sufficient to trigger the G_2 to M transition.

These initially independent approaches converged dramatically in 1988 when MPF was purified from frog eggs in the laboratory of James Maller. Molecular characterization of MPF in several laboratories then showed that this conserved regulator of the cell cycle is composed of two key subunits: Cdk1 and cyclin B (Figure 16.13). Cyclin B is a regulatory subunit required for catalytic activity of the Cdk1 protein kinase, consistent with the notion that MPF activity is controlled by the periodic accumulation and degradation of cyclin B during cell cycle progression.

A variety of further studies have confirmed this role of cyclin B, as well as demonstrating the regulation of MPF by phosphorylation and dephosphorylation of Cdk1 (Figure 16.14). In mammalian cells, cyclin B is synthesized and forms complexes with Cdk1 during G_2. As these complexes form, Cdk1 is phosphorylated at two critical regulatory positions. One of these phosphorylations occurs on threonine-161 and is required for Cdk1 kinase activity. The second is a phosphorylation of tyrosine-15 and of the adjacent threonine-14 in vertebrates. Phosphorylation of tyrosine-15, catalyzed by a protein kinase called Wee1, inhibits Cdk1 activity and leads to the accumulation of inactive Cdk1/cyclin B complexes throughout G_2. The transition from G_2 to M is then brought about by activation of the Cdk1/cyclin B complex as a result of dephosphorylation of threonine-14 and tyrosine-15 by a protein phosphatase called Cdc25C.

MPF

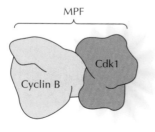

FIGURE 16.13 Structure of MPF MPF is a dimer consisting of cyclin B and the Cdk1 protein kinase.

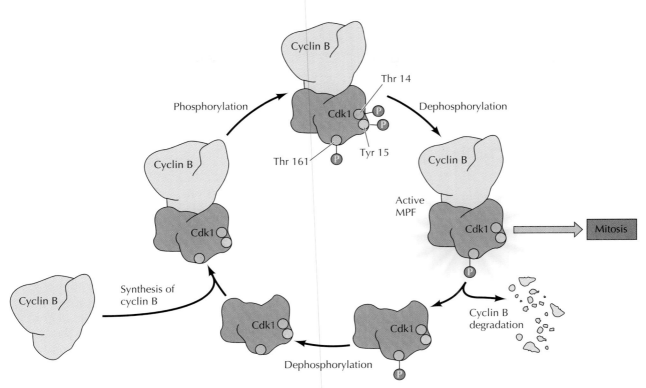

FIGURE 16.14 MPF regulation Cdk1 forms complexes with cyclin B during G_2. Cdk1 is then phosphorylated on threonine-161, which is required for Cdk1 activity, as well as on tyrosine-15 (and threonine-14 in vertebrate cells), which inhibits Cdk1 activity. Dephosphorylation of Thr14 and Tyr15 activates MPF at the G_2 to M transition. MPF activity is then terminated toward the end of mitosis by proteolytic degradation of cyclin B.

Once activated, the Cdk1 protein kinase phosphorylates a variety of target proteins that initiate the events of M phase, which are discussed later in this chapter. In addition, Cdk1 activity triggers the degradation of cyclin B, which occurs as a result of ubiquitin-mediated proteolysis. This proteolytic destruction of cyclin B then inactivates Cdk1, leading the cell to exit mitosis, undergo cytokinesis, and return to interphase.

16.4 WEBSITE ANIMATION

Cyclins, Cdks, and the Cell Cycle Maturation promoting factor (MPF)—consisting of cyclin B and Cdk1—regulates the transition from G_2 to M phase of the cell cycle.

Families of Cyclins and Cyclin-Dependent Kinases

The structure and function of MPF (Cdk1/cyclin B) provide not only a molecular basis for understanding entry and exit from M phase but also the foundation for elucidating the regulation of other cell cycle transitions. The insights provided by characterization of the Cdk1/cyclin B complex have thus had a sweeping impact on understanding cell cycle regulation. In particular, further research has established that both Cdk1 and cyclin B are members of families of related proteins, with different members of these families controlling progression through distinct phases of the cell cycle.

As discussed earlier, Cdk1 controls passage through START as well as entry into mitosis in yeasts. It does so, however, in association with distinct cyclins (Figure 16.15). In particular, the G_2 to M transition is driven by Cdk1 in association with the mitotic B-type cyclins (Clb1, Clb2, Clb3, and Clb4). Passage through START, however, is controlled by Cdk1 in association with a distinct class of cyclins called **G_1 cyclins** or **Cln's**. Cdk1 then associates

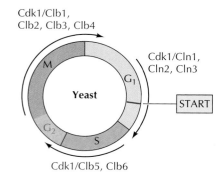

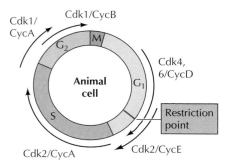

FIGURE 16.15 Complexes of cyclins and cyclin-dependent kinases In yeast, passage through START is controlled by Cdk1 in association with G_1 cyclins (Cln1, Cln2, and Cln3). Complexes of Cdk1 with distinct B-type cyclins (Clb's) then regulate progression through S phase and entry into mitosis. In animal cells, progression through the G_1 restriction point is controlled by complexes of Cdk4 and Cdk6 with D-type cyclins. Cdk2/cyclin E complexes function later in G_1 and are required for the G_1 to S transition. Cdk2/cyclin A complexes are then required for progression through S phase. Cdk1/cyclin A regulates progression to G_2, and Cdk1/cyclin B complexes drive the G_2 to M transition.

with different B-type cyclins (Clb5 and Clb6), which are required for progression through S phase. These associations of Cdk1 with distinct B-type and G_1 cyclins direct Cdk1 to phosphorylate different substrate proteins, as required for progression through specific phases of the cell cycle.

The cell cycles of higher eukaryotes are controlled not only by multiple cyclins but also by multiple Cdk1-related protein kinases, which are known as **Cdk's** for *cyclin-dependent kinases*. These multiple members of the Cdk family associate with specific cyclins to drive progression through the different stages of the cell cycle (see Figure 16.15). For example, progression from G_1 to S is regulated principally by Cdk2, Cdk4, and Cdk6 in association with cyclins D and E. Complexes of Cdk4 and Cdk6 with the D-type cyclins (cyclins D1, D2, and D3) play a critical role in progression through

---

KEY EXPERIMENT

The Identification of Cyclin

Cyclin: A Protein Specified by Maternal mRNA in Sea Urchin Eggs That Is Destroyed at Each Cleavage Division

Tom Evans, Eric T. Rosenthal, Jim Youngblom, Dan Distel, and Tim Hunt

Marine Biological Laboratory, Woods Hole, MA

Cell, 1983, Volume 33, Pages 389–396

Tim Hunt

The Context

Tim Hunt and his colleagues undertook an analysis of protein synthesis in sea urchin eggs following fertilization, with the goal of identifying newly synthesized proteins that might be required for cell division during embryo development. A variety of earlier experiments had shown that translation of maternal mRNAs, stored in unfertilized eggs, was activated at fertilization, leading to striking changes in the profile of protein synthesis following fertilization. In addition, studies with inhibitors of translation had established that protein synthesis was required for cell division during early embryonic

development. For example, inhibition of protein synthesis following fertilization of sea urchin eggs blocked nuclear envelope breakdown, chromosome condensation, and spindle formation. These results suggested that embryonic cell divisions required proteins that were synthesized from maternal mRNAs in fertilized eggs.

By studies of protein synthesis in fertilized eggs, Hunt and colleagues therefore hoped to identify newly synthesized proteins that were involved in cell division. Their experiments led to the identification of cyclins and elucidation of the fundamental mechanism that drives the division cycle of eukaryotic cells.

The Experiments

Unfertilized and fertilized sea urchin eggs were incubated in the presence of radioactive methionine, and cell extracts were analyzed by gel electrophoresis to identify newly synthesized proteins. Several prominent proteins were found to be synthesized in fertilized eggs, as had previously been seen in eggs of other organisms. Moreover, careful examination of the profile of radiolabeled proteins at different times after fertilization revealed a surprising

the restriction point in G_1. The E-type cyclins (E1 and E2) are expressed later in G_1, and Cdk2/cyclin E complexes are required for the G_1 to S transition and initiation of DNA synthesis. Complexes of Cdk2 with A-type cyclins (A1 and A2) function in the progression of cells through S phase. Cdk1 then regulates passage from S to G_2 and from G_2 to M in complexes with A- and B-type cyclins (B1, B2, and B3), respectively.

The roles of the Cdk/cyclin complexes summarized in Figure 16.15 are based on extensive research in many laboratories, principally using cultured cells as experimental models. However, recent studies of Cdk's and cyclins in genetically modified mice have yielded surprising results. In particular, embryonic cells from mice with mutated genes for the D-type and E-type cyclins as well as for Cdk2, Cdk4, and Cdk6 are still capable of proliferation. Moreover, mice lacking many of these genes are viable, although they may exhibit abnormalities in the development of specific cell types.

KEY EXPERIMENT

result. By analyzing the proteins present at 10 minute intervals after fertilization, Hunt and his colleagues identified a protein that was prominent in newly fertilized eggs, but then almost disappeared 85 minutes after fertilization. This protein again became abundant 95 minutes after fertilization, and declined again 30 minutes later. Because of these cyclic oscillations in its abundance, the protein was named "cyclin."

A further striking result was obtained when the oscillations in levels of cyclin were compared with the timing of cell divisions in the developing sea urchin embryos (see figure). The first division of the fertilized egg to a two-cell embryo took place 85 minutes after fertilization, coinciding with the decline of cyclin levels. The levels of cyclin then increased following this first cell division, and declined again 125 minutes after fertilization, coincident with the second cycle of embryonic cell division. Additional experiments demonstrated that cyclin was continuously synthesized after fertilization, but was rapidly degraded prior to each cell division.

The Impact

Based on its periodic degradation, coincident with the onset of embryonic cell division, Hunt and colleagues concluded that it was "difficult to believe that the behavior of the cyclins is not connected with the processes involved in cell division." However, in the

Sea urchin eggs were fertilized and radioactive methionine was added 5 minutes after fertilization. Samples were then harvested at 10 minute intervals and analyzed by gel electrophoresis to determine the levels of cyclin. In parallel, samples were examined microscopically to quantitate the percentage of cells undergoing cell division.

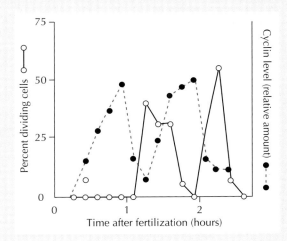

absence of experiments demonstrating the biological activity of cyclin, they had no direct evidence for this conclusion. Nonetheless, the activity of MPF was known to oscillate similarly during the cell cycle, and additional experiments had demonstrated that the formation of active MPF required protein synthesis. These similarities between MPF and cyclin led Hunt and his colleagues to suggest a direct relationship between MPF, defined as a biological activity, and cyclin, defined as a protein whose levels oscillated during cell cycle progression.

Further experiments of Joan Ruderman and her colleagues demonstrated that injection of cyclin was sufficient to induce the entry of frog oocytes into meiosis, showing that cyclin had the

biological activity of MPF. The biochemical relationship between cyclin and MPF was then established by the purification of MPF from frog eggs in the laboratory of James Maller and the subsequent characterization of MPF as a dimer of cyclin B and Cdk1. The periodic degradation of cyclin B, discovered by Tim Hunt and colleagues, controls the activity of Cdk1 during entry and exit from mitosis. Moreover, other cyclins play similar roles in regulating transitions between other phases of the cell cycle. The discovery of cyclin as an oscillating protein in sea urchin embryos was thus key to elucidating the molecular basis of cell cycle regulation.

FIGURE 16.16 Mechanisms of Cdk regulation The activities of Cdk's are regulated by four molecular mechanisms.

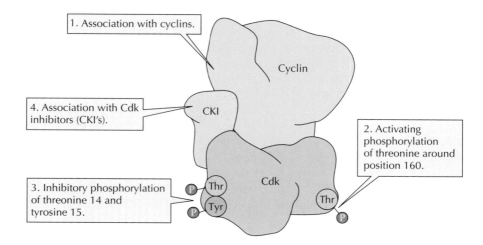

These results suggest that there may be substantial redundancy between different cyclins and Cdk's, so that one can compensate for the absence of another. For example, recent work suggests that Cdk1 may be able to substitute for Cdk2 (in complexes with both cyclin A and cyclin E). However, further studies are clearly necessary to understand the extent to which these cell cycle regulators can substitute for one another.

The activity of Cdk's during cell cycle progression is regulated by four molecular mechanisms (Figure 16.16). As already discussed for Cdk1, the first level of regulation involves the association of Cdk's with their cyclin partners. Thus the formation of specific Cdk/cyclin complexes is controlled by cyclin synthesis and degradation. Second, activation of Cdk/cyclin complexes requires phosphorylation of a conserved Cdk threonine residue around position 160. This activating phosphorylation of the Cdk's is catalyzed by an enzyme called CAK (for Cdk-*a*ctivating *k*inase), which is itself composed of a Cdk (Cdk7) complexed with cyclin H. Complexes of Cdk7 and cyclin H are also associated with the transcription factor TFIIH, which is required for initiation of transcription by RNA polymerase II (see Chapter 6), so this member of the Cdk family participates in transcription as well as cell cycle regulation.

In contrast to the activating phosphorylation by CAK, the third mechanism of Cdk regulation involves inhibitory phosphorylation of tyrosine residues near the Cdk amino terminus, catalyzed by the Wee1 protein kinase. In particular, both Cdk1 and Cdk2 are inhibited by phosphorylation of tyrosine-15, and the adjacent threonine-14 in vertebrates. These Cdk's are then activated by dephosphorylation of these residues by members of the Cdc25 family of protein phosphatases.

In addition to regulation of the Cdk's by phosphorylation, their activities are also controlled by the binding of inhibitory proteins (called **Cdk inhibitors** or **CKIs**). In mammalian cells, two families of Cdk inhibitors are responsible for regulating different Cdk's (Table 16.1). Members of the Ink4 family specifically bind to and inhibit monomeric Cdk4 and Cdk6, so the Ink4 CKIs act to inhibit progression through G_1. In contrast, members of the Cip/Kip family bind to and inhibit the protein kinase activity of both Cdk1 (complexed with either cyclin A or cyclin B) and Cdk2 (complexed with either cyclin A or cyclin E), thereby inhibiting progression through all phases of the cell cycle. The Cip/Kip CKI's also bind to complexes of Cdk4 and Cdk6 with cyclin D, but their role in regulation of these complexes is

■ Some of the cell cycle mutants of *S. pombe*, isolated by Paul Nurse and his colleagues at the University of Edinburgh, were identified because the cells divided when they were smaller in size than normal. These mutants were named *Wee*, the Scottish term for small.

TABLE 16.1 Cdk Inhibitors

Inhibitor	Cdk or Cdk/cyclin complex	Cell cycle phase affected
Ink4 family (p15, p16, p18, p19)	Cdk4 and Cdk6	G_1
Cip/Kip family (p21, p27, p57)	Cdk1/cyclin A	G_2
	Cdk1/cyclin B	G_2/M
	Cdk2/cyclin A	S
	Cdk2/cyclin E	G_1

unclear. Control of Cdk inhibitors thus provides an additional mechanism for regulating Cdk activity. The combined effects of these multiple modes of Cdk regulation are responsible for controlling cell cycle progression in response both to checkpoint controls and to the variety of extracellular stimuli that regulate cell proliferation.

Growth Factors and the Regulation of G_1 Cdk's

As discussed earlier, the proliferation of animal cells is regulated largely by a variety of extracellular growth factors that control the progression of cells through the restriction point in late G_1. In the absence of growth factors, cells are unable to pass the restriction point and become quiescent, frequently entering the resting state known as G_0 from which they can reenter the cell cycle in response to growth factor stimulation. This control of cell cycle progression by extracellular growth factors implies that the intracellular signaling pathways stimulated downstream of growth factor receptors (discussed in the preceding chapter) ultimately act to regulate components of the cell cycle machinery.

One critical link between growth factor signaling and cell cycle progression is provided by the D-type cyclins (Figure 16.17). Cyclin D1 synthesis is induced in response to growth factor stimulation in part as a result of signaling through the Ras/Raf/MEK/ERK pathway, and cyclin D1 continues to be synthesized as long as growth factors are present. However, cyclin D1 is also rapidly degraded, so its intracellular concentration rapidly falls if growth factors are removed. Thus, as long as growth factors are present through G_1, complexes of Cdk4, 6/cyclin D1 drive cells through the restriction point. On the other hand, if growth factors are removed prior to this key regulatory point in the cell cycle, the levels of cyclin D1 rapidly fall and cells are unable to progress through G_1 to S, instead becoming quiescent and entering G_0. The inducibility and rapid turnover of cyclin D1 thus integrates growth factor signaling with the cell cycle machinery, allowing the availability of extracellular growth factors to control the progression of cells through G_1.

Since cyclin D1 is a critical target of growth factor signaling, it might be expected that defects in cyclin D1 regulation could contribute to the loss of growth regulation characteristic of cancer cells. Consistent with this expectation, many human cancers have been found to arise as a result of defects in cell cycle regulation, just as many others result from abnormalities in the intracellular signaling pathways activated by growth factor receptors (see Chapter 15). For example, mutations resulting in continual unregulated expression of cyclin D1 contribute to the development of a variety of human cancers, including lymphomas and breast cancers. Similarly, mutations that inactivate the Ink4 Cdk inhibitors that bind to Cdk4 and Cdk6 are commonly found in human cancer cells.

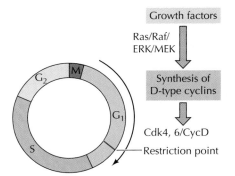

FIGURE 16.17 Induction of D-type cyclins   Growth factors regulate cell cycle progression through the G_1 restriction point by inducing synthesis of D-type cyclins via the Ras/Raf/MEK/ERK signaling pathway.

FIGURE 16.18 Cell cycle regulation of Rb and E2F   In its underphosphorylated form, Rb binds to members of the E2F family, repressing transcription of E2F-regulated genes. Phosphorylation of Rb by Cdk4, 6/cyclin D complexes results in its dissociation from E2F in late G_1. E2F then stimulates expression of its target genes, which encode proteins required for cell cycle progression.

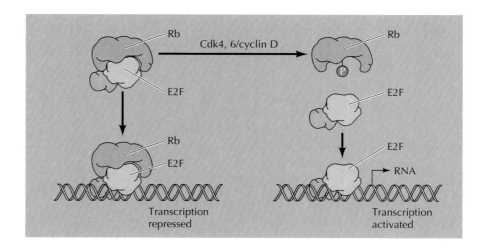

Transcription repressed

Transcription activated

The connection between cyclin D, growth control, and cancer is further fortified by the fact that a key substrate protein of Cdk4, 6/cyclin D complexes is itself frequently mutated in a wide array of human tumors. This protein, designated **Rb**, was first identified as the product of a gene responsible for retinoblastoma, a rare inherited childhood eye tumor (see Chapter 18). Further studies then showed that mutations resulting in the absence of functional Rb protein are not restricted to retinoblastoma but also contribute to a variety of common human cancers. Rb is the prototype of a **tumor suppressor gene**—a gene whose inactivation leads to tumor development. Whereas oncogene proteins such as Ras (see Chapter 15) and cyclin D drive cell proliferation, the proteins encoded by many tumor suppressor genes (including Rb and the Ink4 Cdk inhibitors) act as brakes that slow down cell cycle progression.

Further studies have revealed that Rb and related members of the Rb family play a key role in coupling the cell cycle machinery to the expression of genes required for cell cycle progression and DNA synthesis (Figure 16.18). The activity of Rb proteins is regulated by changes in phosphorylation as cells progress through the cycle. In particular, Rb becomes phosphorylated by Cdk4, 6/cyclin D complexes as cells pass through the restriction point in G_1. In its underphosphorylated form (present in G_0 or early G_1), Rb binds to members of the **E2F** family of transcription factors, which regulate expression of several genes involved in cell cycle progression, including the gene encoding cyclin E. E2F binds to its target sequences in either the presence or absence of Rb. However, Rb acts as a repressor, so the Rb/E2F complex suppresses transcription of E2F-regulated genes. Phosphorylation of Rb by Cdk4, 6/cyclin D complexes results in its dissociation from E2F, which then activates transcription of its target genes. Rb thus acts as a molecular switch that converts E2F from a repressor to an activator of genes required for cell cycle progression. The control of Rb by Cdk4, 6/cyclin D phosphorylation in turn couples this critical regulation of gene expression to the availability of growth factors in G_1.

Progression through the restriction point and entry into S phase is mediated by the activation of Cdk2/cyclin E complexes (Figure 16.19). This results in part from the synthesis of cyclin E, which is stimulated by E2F following phosphorylation of Rb. In addition, the activity of Cdk2/cyclin E is inhibited in G_0 or early G_1 by the Cdk inhibitor p27, which belongs to the Cip/Kip family (see Table 16.1). This inhibition of Cdk2 by p27 is relieved by multiple mechanisms as cells progress through G_1. First, growth factor

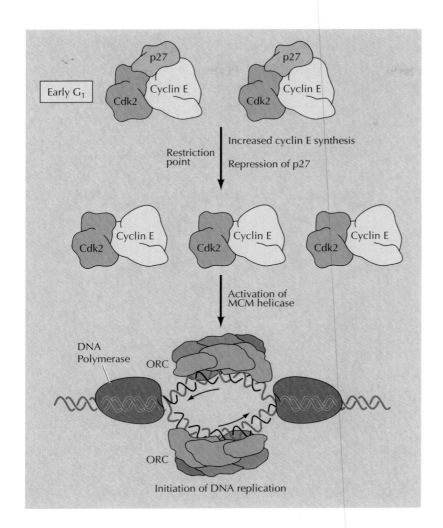

FIGURE 16.19 Cdk2/cyclin E and entry into S phase In early G_1, Cdk2/cyclin E complexes are inhibited by the Cdk inhibitor p27. Passage through the restriction point induces the synthesis of cyclin E via activation of E2F. In addition, growth factor signaling reduces the levels of p27 by inhibiting its transcription and translation. The resulting activation of Cdk2/cyclin E leads to activation of the MCM helicase and initiation of DNA replication.

signaling via both the Ras/Raf/MEK/ERK and PI 3-kinase/Akt pathways reduces the transcription and translation of p27, lowering the levels of p27 within the cell. In addition, increased synthesis of cyclin D leads to the binding of p27 to Cdk4, 6/cyclin D complexes, sequestering it from binding to Cdk2/cyclin E. Once Cdk2 becomes activated, it brings about the complete degradation of p27 by phosphorylating it and targeting it for ubiquitination. This positive autoregulation then results in full activation of Cdk2/cyclin E complexes. Cdk2 also phosphorylates Rb, completing its inactivation. Cdk2/cyclin E complexes then initiate S phase by activating the MCM helicase proteins at replication origins (see Figure 16.9), leading to the initiation of DNA synthesis.

DNA Damage Checkpoints

Cell proliferation is regulated not only by growth factors but also by a variety of signals that act to inhibit cell cycle progression. DNA damage checkpoints play a critical role in maintaining the integrity of the genome by arresting cell cycle progression in response to damaged or incompletely replicated DNA. These checkpoints, which are operative in G_1, S, and G_2 phases of the cell cycle, serve to halt cell cycle progression and allow time for the damage to be repaired before DNA replication or cell division proceeds (see Figure 16.8).

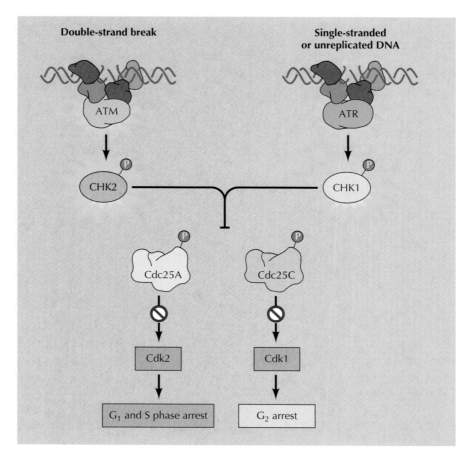

FIGURE 16.20 Cell cycle arrest at the DNA damage checkpoints The ATM and ATR protein kinases are activated in complexes of proteins that recognize unreplicated or damaged DNA. ATM is activated principally by double-strand breaks and ATR by single-stranded or unreplicated DNA. ATM and ATR then phosphorylate and activate the CHK2 and CHK1 protein kinases, respectively. CHK1 and CHK2 phosphorylate and inhibit the Cdc25A and Cdc25C protein phosphatases. Cdc25A and Cdc25C are required to activate Cdk2 and Cdk1, respectively, so their inhibition leads to arrest at the DNA damage checkpoints in G_1, S, and G_2.

Cell cycle arrest at the DNA damage checkpoints is initiated by the ATM or ATR protein kinases, which are components of protein complexes that recognize damaged or unreplicated DNA (Figure 16.20). ATM is activated principally by double-strand breaks, while ATR is activated by single-stranded or unreplicated DNA. Once activated by DNA damage, ATM and ATR phosphorylate and activate the **checkpoint kinases** CHK2 and CHK1, respectively, bringing about cell cycle arrest.

Both CHK1 and CHK2 phosphorylate and inhibit Cdc25 phosphatases, which are required to activate Cdk/cyclin complexes by removing inhibitory phosphorylations (see Figure 16.16). At the G_1 and S phase checkpoints, CHK1 and CHK2 phosphorylate Cdc25A, which is required to activate complexes of Cdk2 and cyclins A or E. Phosphorylation leads to the rapid degradation of Cdc25A, resulting in inhibition of Cdk2. At the G_2 checkpoint, CHK1 and CHK2 phosphorylate and inhibit Cdc25C, which is responsible for activating Cdk1/cyclin B complexes. In the absence of Cdk1 activation, progression to mitosis is blocked and the cell remains arrested in G_2.

In mammalian cells, arrest at the G_1 checkpoint is also mediated by the action of an additional protein known as **p53**, which is phosphorylated by both ATM and CHK2 (Figure 16.21). Phosphorylation stabilizes p53, which is otherwise rapidly degraded, resulting in a rapid increase in p53 levels in response to damaged DNA. The p53 protein is a transcription factor, and its increased expression leads to the induction of the Cip/Kip family Cdk inhibitor p21. The p21 protein inhibits Cdk2/cyclin E complexes, leading to cell cycle arrest in G_1.

Interestingly, the gene encoding p53 is frequently mutated in human cancers. Loss of p53 function as a result of these mutations prevents G_1 arrest in response to DNA damage, so the damaged DNA is replicated and passed on to daughter cells instead of being repaired. This inheritance of damaged DNA results in an increased frequency of mutations and general instability of the cellular genome, which contributes to cancer development. Mutations in the *p53* gene are among the most common genetic alterations in human cancers (see Chapter 18), illustrating the critical importance of cell cycle regulation in the life of multicellular organisms.

The Events of M Phase

M phase is the most dramatic period of the cell cycle, involving a major reorganization of virtually all cell components. During mitosis (nuclear division), the chromosomes condense, the nuclear envelope of most cells breaks down, the cytoskeleton reorganizes to form the mitotic spindle, and the chromosomes move to opposite poles. Chromosome segregation is then usually followed by cell division (cytokinesis). Although some of these events have been discussed in previous chapters, they are reviewed here in the context of a coordinated view of M phase and the action of MPF (Cdk1/cyclin B).

Stages of Mitosis

Although many of the details of mitosis vary among different organisms, the fundamental processes that ensure the faithful segregation of sister chromatids are conserved in all eukaryotes. These basic events of mitosis include chromosome condensation, formation of the mitotic spindle, and attachment of chromosomes to the spindle microtubules. Sister chromatids then separate from each other and move to opposite poles of the spindle, followed by the formation of daughter nuclei.

Mitosis is conventionally divided into four stages—**prophase, metaphase, anaphase,** and **telophase**—which are illustrated for an animal cell in Figures 16.22 and 16.23. The beginning of prophase is marked by the appearance of condensed chromosomes, each of which consists of two sister chromatids (the daughter DNA molecules produced in S phase). These newly replicated DNA molecules remain intertwined throughout S and G_2, becoming untangled during the process of chromatin condensation. The condensed sister chromatids are then held together at the **centromere,** which (as discussed in Chapter 5) is a DNA sequence to which proteins bind to form the **kinetochore**—the site of eventual attachment of the spindle microtubules. In addition to chromosome condensation, cytoplasmic changes leading to the development of the mitotic spindle initiate during prophase. The **centrosomes** (which had duplicated during interphase) separate and move to opposite sides of the nucleus. There they serve as the two poles of the **mitotic spindle,** which begins to form during late prophase.

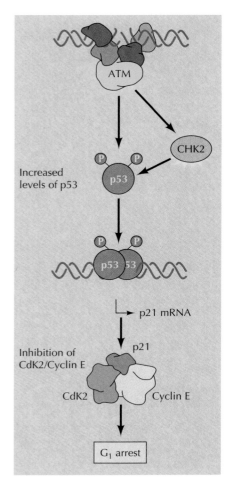

FIGURE 16.21 Role of p53 in G_1 arrest   The protein p53 plays a key role in cell cycle arrest at the G_1 checkpoint in mammalian cells. Phosphorylation by ATM and CHK2 stabilize p53, resulting in rapid increases in p53 levels in response to DNA damage. The protein p53 then activates transcription of the gene encoding the Cdk inhibitor p21, leading to inhibition of Cdk2/cyclin E complexes and cell cycle arrest.

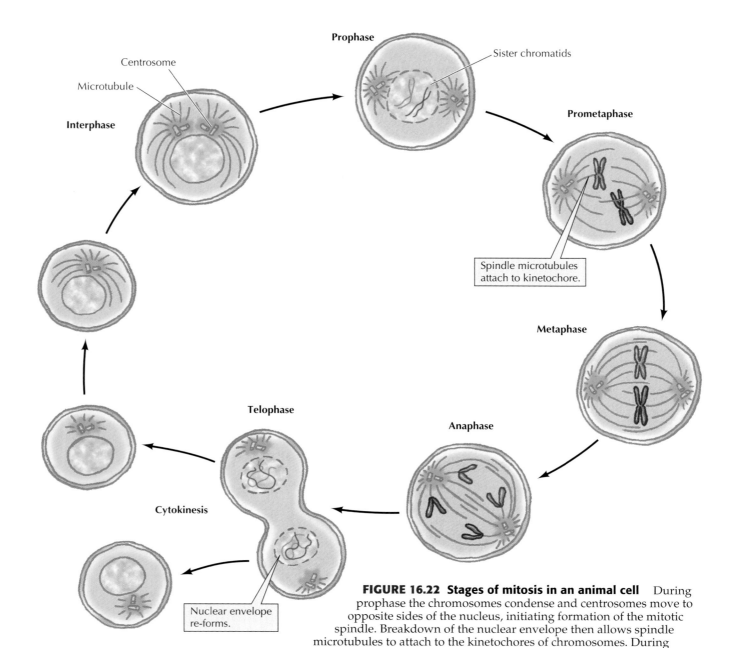

FIGURE 16.22 Stages of mitosis in an animal cell During prophase the chromosomes condense and centrosomes move to opposite sides of the nucleus, initiating formation of the mitotic spindle. Breakdown of the nuclear envelope then allows spindle microtubules to attach to the kinetochores of chromosomes. During prometaphase the chromosomes shuffle back and forth between the centrosomes and the center of the cell, eventually aligning in the center of the spindle (metaphase). At anaphase, the sister chromatids separate and move to opposite poles of the spindle. Mitosis then ends with re-formation of nuclear envelopes and chromosome decondensation during telophase, and cytokinesis yields two interphase daughter cells. Note that each daughter cell receives one centrosome, which duplicates prior to the next mitosis.

16.5 WEBSITE ANIMATION

Mitosis in an Animal Cell Mitosis, which is the division of the nucleus, consists of four phases—prophase, metaphase, anaphase, and telophase—and is followed by the division of the cytoplasm, called cytokinesis.

In higher eukaryotes the end of prophase corresponds to the breakdown of the nuclear envelope. However, this disassembly of the nucleus is not a universal feature of mitosis and does not occur in all cells. Some unicellular eukaryotes (e.g., yeasts) undergo so-called closed mitosis in which the nuclear envelope remains intact (Figure 16.24). In closed mitosis the daughter chromosomes migrate to opposite poles of the nucleus, which then divides

Mitosis

Interphase   Early prophase   Late prophase

Prometaphase   Metaphase   Early anaphase

Late anaphase   Telephase

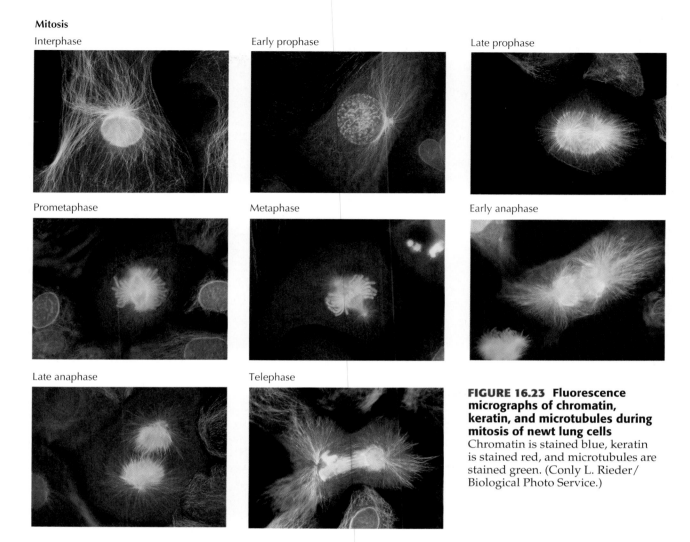

FIGURE 16.23 Fluorescence micrographs of chromatin, keratin, and microtubules during mitosis of newt lung cells
Chromatin is stained blue, keratin is stained red, and microtubules are stained green. (Conly L. Rieder/Biological Photo Service.)

in two. In these cells the spindle pole bodies are embedded within the nuclear envelope, and the nucleus divides in two following migration of daughter chromosomes to opposite poles of the spindle.

Following completion of prophase, the cell enters **prometaphase**—a transition period between prophase and metaphase. During prometaphase the microtubules of the mitotic spindle attach to the kinetochores of condensed chromosomes. The kinetochores of sister chromatids are oriented on opposite sides of the chromosome, so they attach to microtubules emanating from opposite poles of the spindle. The chromosomes shuffle back and forth until they eventually align on the metaphase plate in the center of the spindle. At this stage, the cell has reached metaphase.

Most cells remain only briefly at metaphase before proceeding to anaphase. The transition from metaphase to anaphase is triggered by breakage of the link between sister chromatids, which then separate and move to opposite poles of the spindle. Mitosis ends with telophase, during which nuclei re-form and the chromosomes decondense. Cytokinesis usually begins during late anaphase and is almost complete by the end of telophase, resulting in the formation of two interphase daughter cells.

FIGURE 16.24 Closed and open mitosis In closed mitosis, the nuclear envelope remains intact and chromosomes migrate to opposite poles of a spindle within the nucleus. In open mitosis, the nuclear envelope breaks down and then re-forms around the two sets of separated chromosomes.

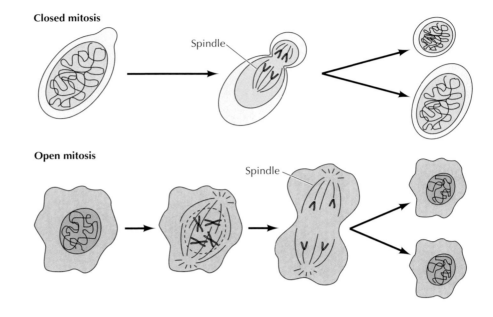

Cdk1/Cyclin B and Progression to Metaphase

Mitosis involves dramatic changes in multiple cellular components, leading to a major reorganization of the entire structure of the cell. As discussed earlier in this chapter, these events are initiated by activation of the Cdk1/cyclin B protein kinase (MPF). Cdk1/cyclin B not only acts as a master regulator of the M phase transition, phosphorylating and activating other downstream protein kinases but also acts directly by phosphorylating some of the structural proteins involved in this cellular reorganization (Figure 16.25).

The condensation of interphase chromatin to form the compact chromosomes of mitotic cells is a key event in mitosis, critical in enabling the chro-

FIGURE 16.25 Targets of Cdk1/cyclin B The Cdk1/cyclin B complex induces multiple nuclear and cytoplasmic changes at the onset of M phase both by activating other protein kinases and by phosphorylating proteins such as condensins, components of the nuclear envelope, Golgi matrix proteins, and proteins associated with centrosomes and microtubules.

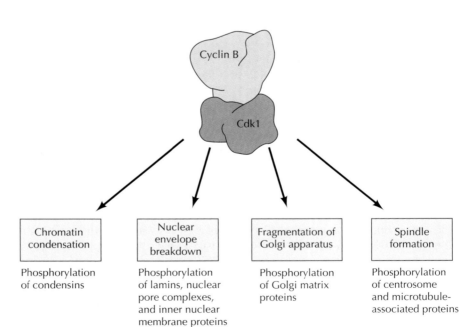

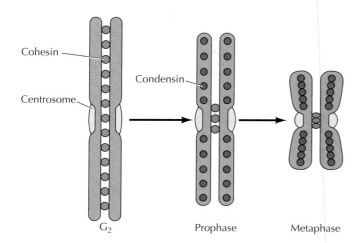

G₂　　Prophase　　Metaphase

FIGURE 16.26 The action of cohesins and condensins Cohesins bind to DNA during S phase and maintain the linkage between sister chromatids following DNA replication in S and G$_2$. As the cell enters M phase, the cohesins are replaced by condensins along most of the chromosome, remaining only at the centromere. Phosphorylation by Cdk1 activates the condensins, which drive chromatin condensation.

mosomes to move along the mitotic spindle without becoming broken or tangled with one another. As discussed in Chapter 5, the chromatin in interphase nuclei condenses nearly a thousandfold during the formation of metaphase chromosomes. Such highly condensed chromatin cannot be transcribed, so transcription ceases as chromatin condensation takes place. Despite the fundamental importance of this event, we do not fully understand either the structure of metaphase chromosomes or the molecular mechanism of chromatin condensation. However, it has been established that chromatin condensation is driven by protein complexes called **condensins**, which are members of a class of "structural maintenance of chromatin" (SMC) proteins that play key roles in the organization of eukaryotic chromosomes.

Both condensins and another family of SMC proteins, called **cohesins,** contribute to chromosome segregation during mitosis (**Figure 16.26**). Cohesins bind to DNA in S phase and maintain the linkage between sister chromatids following DNA replication. As the cell enters M phase, the condensins are activated by Cdk1/cyclin B phosphorylation. The condensins then replace the cohesins along most of the length of the chromosome, so that the sister chromatids remain linked only at the centromere. The condensins also induce chromatin condensation, leading to the formation of metaphase chromosomes.

Breakdown of the nuclear envelope, which is one of the most dramatic events of mitosis, involves changes in all of its components: the nuclear membranes fragment, the nuclear pore complexes dissociate, and the nuclear lamina depolymerizes. Depolymerization of the nuclear lamina (the meshwork of filaments underlying the nuclear membrane) results from phosphorylation of the lamins by Cdk1 (**Figure 16.27**). Phosphorylation causes the lamin filaments to break down into individual lamin dimers, leading directly to depolymerization of the nuclear lamina. Cdk1 also phosphorylates several proteins in the inner nuclear membrane and the nuclear pore complex, leading to disassembly of nuclear pore complexes and

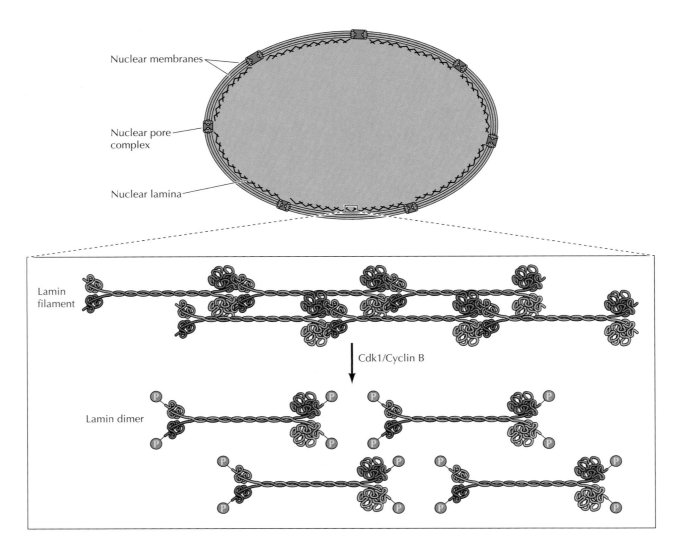

Nuclear membranes

Nuclear pore complex

Nuclear lamina

Lamin filament

Cdk1/Cyclin B

Lamin dimer

FIGURE 16.27 Breakdown of the nuclear envelope Cdk1/cyclin B phosphorylates the nuclear lamins as well as proteins of the nuclear pore complex and inner nuclear membrane. Phosphorylation of the lamins causes the filaments that form the nuclear lamina to dissociate into free lamin dimers.

detachment of the inner nuclear membrane from lamins and chromatin. Current evidence suggests that the nuclear membrane is then absorbed into the endoplasmic reticulum, which remains as an intact network and is distributed to daughter cells at mitosis.

The Golgi apparatus fragments into small vesicles at mitosis, which may either be absorbed into the endoplasmic reticulum or distributed directly to daughter cells at cytokinesis. The breakdown of these membranes is in part thought to be mediated by Cdk1 phosphorylation of Golgi matrix proteins (such as GM130 and GRASP-65), which are required for the docking of COPI-coated vesicles to the Golgi membrane. Phosphorylation of these proteins by Cdk1 inhibits vesicle docking and fusion, leading to fragmentation of the Golgi apparatus.

The reorganization of the cytoskeleton that culminates in formation of the mitotic spindle results from the dynamic instability of microtubules (see Chapter 12). At the beginning of prophase, activation of Cdk1 leads to separation of the centrosomes, which were duplicated during S phase. The centrosomes then move to opposite sides of the nucleus and undergo a process

FIGURE 16.28 Electron micrograph of microtubules attached to the kinetochore of a chromosome   (Conly L. Rieder/Biological Photo Service.)

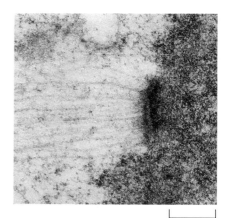

0.5 μm

of maturation during which they enlarge and recruit γ-tubulin and other proteins needed for spindle assembly. Centrosome maturation and spindle assembly involves the activity of protein kinases of the **Aurora** and **Polo-like kinase** families, which are located at the centrosome. Like Cdk1, Aurora and Polo-like kinases are activated in mitotic cells, and they play important roles in spindle formation and kinetochore function, as well as in cytokinesis. The rate of microtubule turnover increases five- to tenfold during mitosis, resulting in depolymerization and shrinkage of the interphase microtubules. This increased turnover is thought to result from phosphorylation of microtubule-associated proteins, either by Cdk1 or other mitotic protein kinases, such as the Aurora or Polo-like kinases. The number of microtubules emanating from the centrosomes also increases, so the interphase microtubules are replaced by large numbers of short microtubules radiating from the centrosomes.

The breakdown of the nuclear envelope then allows some of the spindle microtubules to attach to chromosomes at their kinetochores (Figure 16.28), initiating the process of chromosome movement that characterizes prometaphase. The proteins assembled at the kinetochore include microtubule motors that direct the movement of chromosomes toward the minus ends of the spindle microtubules, which are anchored in the centrosome. The action of these proteins, which draw chromosomes toward the centrosome, is opposed by plus-end directed motor proteins and by the growth of the spindle microtubules, which pushes the chromosomes away from the spindle poles. Consequently, the chromosomes in prometaphase shuffle back and forth between the centrosomes and the center of the spindle.

Microtubules from opposite poles of the spindle eventually attach to the two kinetochores of sister chromatids (which are located on opposite sides of the chromosome), and the balance of forces acting on the chromosomes leads to their alignment on the metaphase plate in the center of the spindle (Figure 16.29). As discussed in Chapter 12, the spindle consists of kinetochore and chromosomal microtubules, which are attached to the chromosomes, as well as polar microtubules, which overlap with one another in the center of the cell. In addition, short astral microtubules radiate outward from the centrosomes toward the cell periphery.

The Spindle Assembly Checkpoint and Progression to Anaphase

As discussed earlier in this chapter, the **spindle assembly checkpoint** monitors the alignment of chromosomes on the metaphase spindle. Once this has been accomplished, the cell proceeds to initiate anaphase and complete mitosis. The progression from metaphase to anaphase results from ubiquitin-mediated proteolysis of key regulatory proteins, triggered by acti-

(A)

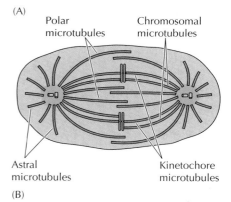

Polar microtubules

Chromosomal microtubules

Astral microtubules

Kinetochore microtubules

(B)

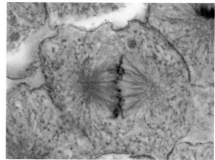

10 μm

FIGURE 16.29 The metaphase spindle   (A) The spindle consists of four kinds of microtubules. Kinetochore and chromosomal microtubules are attached to chromosomes; polar microtubules overlap in the center of the cell; and astral microtubules radiate from the centrosome to the cell periphery. (B) A whitefish cell at metaphase. (B, Michael Abbey/Photo Researchers, Inc.)

Unattached kinetochore

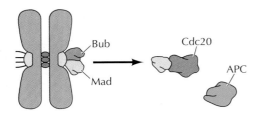

FIGURE 16.30 The spindle assembly checkpoint   Progression to anaphase is mediated by activation of the anaphase-promoting complex (APC) ubiquitin ligase. Unattached kinetochores lead to the assembly of a complex of Mad/Bub proteins in which Mad proteins are activated and prevent APC activation by inhibiting Cdc20. Once all chromosomes are aligned on the spindle, the Mad/Bub complex dissociates, relieving inhibition of Cdc20 and leading to APC activation. APC ubiquitinates cyclin B, leading to its degradation and inactivation of Cdk1. In addition, APC ubiquitinates securin, leading to activation of separase. Separase degrades a subunit of cohesin, breaking the link between sister chromatids and initiating anaphase.

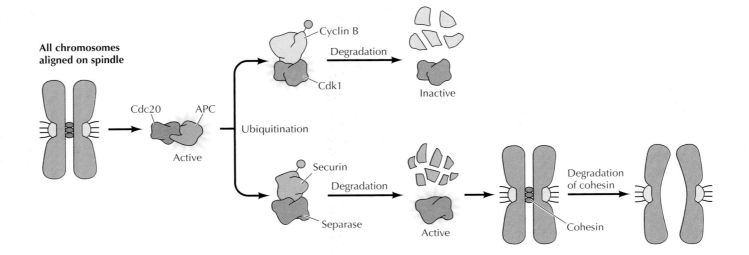

vation of an E3 ubiquitin ligase (see Figure 8.42) called the **anaphase-promoting complex**. Activation of the anaphase-promoting complex is induced at the beginning of mitosis, so the activation of Cdk1/cyclin B ultimately triggers its own destruction. The anaphase-promoting complex remains inhibited, however, until the cell passes the spindle assembly checkpoint, after which activation of the ubiquitin degradation system brings about the transition from metaphase to anaphase and progression through the rest of mitosis.

The spindle assembly checkpoint is remarkable in that the presence of even a single unaligned chromosome is sufficient to inhibit activation of the anaphase-promoting complex. The checkpoint is mediated by a complex of proteins, called the Mad/Bub proteins, that bind to Cdc20—a required component of the anaphase-promoting complex (**Figure 16.30**). The Mad/Bub proteins are assembled in a complex at unattached kinetochores. It appears that the Mad proteins are activated in this complex, and then released in an active form that inhibits Cdc20, maintaining the anaphase-promoting complex in an inactive state. Once microtubules have attached to the kinetochores, the Mad/Bub complex disassembles and inhibition of Cdc20 is relieved, leading to anaphase-promoting complex activation.

Activation of the anaphase-promoting complex results in ubiquitination and degradation of two key target proteins. The onset of anaphase results from proteolytic degradation of a component of the cohesins, which maintain the connection between sister chromatids while they are aligned on the metaphase plate (see Figure 16.26). Cohesin degradation is not catalyzed directly by the anaphase-promoting complex, which instead degrades a

■ **Abnormalities in chromosome segregation resulting from failures of the spindle-assembly checkpoint are common in cancer cells and are thought to play an important role in the development of many tumors.**

protein called securin that is a regulatory subunit of a protease called separase. Degradation of securin results in the activation of separase, which in turn degrades cohesin. Cleavage of cohesin breaks the linkage between sister chromatids, allowing them to segregate by moving to opposite poles of the spindle (Figure 16.31). The separation of chromosomes during anaphase then proceeds as a result of the action of several types of motor proteins associated with the spindle microtubules (see Figures 12.57 and 12.58).

The other key regulatory protein targeted for ubiquitination and degradation by the anaphase-promoting complex is cyclin B. Degradation of cyclin B leads to inactivation of Cdk1, which is required for the cell to exit mitosis and return to interphase. Many of the cellular changes involved in these transitions are simply the reversal of the events induced by Cdk1 during entry into mitosis. For example, reassembly of the nuclear envelope, chromatin decondensation, and the return of microtubules to an interphase state probably result directly from loss of Cdk1 activity and dephosphorylation of proteins that had been phosphorylated by Cdk1 at the beginning of mitosis. As discussed next, inactivation of Cdk1 also triggers cytokinesis.

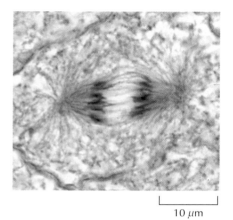

10 μm

FIGURE 16.31 A whitefish cell at anaphase   (Michael Abbey/Photo Researchers, Inc.)

Cytokinesis

The completion of mitosis is usually accompanied by cytokinesis, giving rise to two daughter cells. Cytokinesis usually initiates shortly after the onset of anaphase and is triggered by the inactivation of Cdk1, thereby coordinating nuclear and cytoplasmic division of the cell. As discussed in Chapter 12, cytokinesis of yeast and animal cells is mediated by a **contractile ring** of actin and myosin II filaments that forms beneath the plasma membrane (Figure 16.32). The location of this ring is determined by the position of the mitotic spindle, so the cell is eventually cleaved in a plane that passes through the metaphase plate perpendicular to the spindle. Cleavage proceeds as contraction of the actin-myosin filaments pulls the plasma membrane inward, eventually pinching the cell in half. The bridge between the two daughter cells is then broken, and the plasma membrane is resealed.

The mechanism of cytokinesis is different for higher plant cells. Rather than being pinched in half by a contractile ring, these cells divide by forming new cell walls and plasma membranes inside the cell (Figure 16.33). In

(A)

(B)

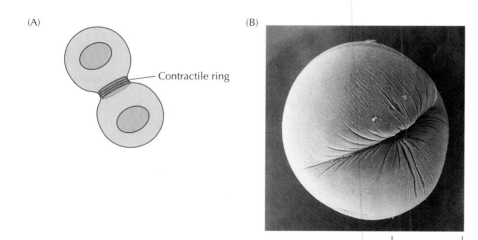

— Contractile ring

1 mm

FIGURE 16.32 Cytokinesis of animal cells   (A) Cytokinesis results from contraction of a ring of actin and myosin filaments, which pinches the cell in two. (B) Scanning electron micrograph of a frog egg undergoing cytokinesis. (B, David M. Phillips/ Visuals Unlimited.)

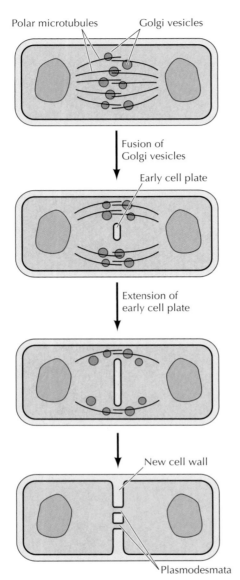

Polar microtubules   Golgi vesicles

Fusion of
Golgi vesicles

Early cell plate

Extension of
early cell plate

New cell wall

Plasmodesmata

FIGURE 16.33 Cytokinesis in higher plants Golgi vesicles carrying cell wall precursors associate with polar microtubules at the former site of the metaphase plate. Fusion of these vesicles yields a membrane-enclosed, disklike structure (the early cell plate) that expands outward and fuses with the parental plasma membrane. The daughter cells remain connected at plasmodesmata.

16.6 WEBSITE ANIMATION

Cytokinesis in Higher Plants During cell division, a plant cell divides its cytoplasm by depositing Golgi vesicles containing cell-wall precursors at the former site of the metaphase plate, building a larger and larger disclike structure that grows toward and fuses with the plasma membrane.

early telophase, vesicles carrying cell wall precursors from the Golgi apparatus associate with remnants of the spindle microtubules and accumulate at the former site of the metaphase plate. These vesicles then fuse to form a large, membrane-enclosed, disklike structure, and their polysaccharide contents assemble to form the matrix of a new cell wall (called a cell plate). The cell plate expands outward, perpendicular to the spindle, until it reaches the plasma membrane. The membrane surrounding the cell plate then fuses with the parental plasma membrane, dividing the cell in two. Connections between the daughter cells (plasmodesmata, see Figure 14.28) are formed as a result of incomplete vesicle fusion during cytokinesis.

Meiosis and Fertilization

The somatic cell cycles discussed so far in this chapter result in diploid daughter cells with identical genetic complements. Meiosis, in contrast, is a specialized kind of cell cycle that reduces the chromosome number by half, resulting in the production of haploid daughter cells. Unicellular eukaryotes, such as yeasts, can undergo meiosis as well as reproducing by mitosis. Diploid *Saccharomyces cerevisiae*, for example, undergo meiosis and produce spores when faced with unfavorable environmental conditions. In multicellular plants and animals, however, meiosis is restricted to the germ cells, where it is key to sexual reproduction. Whereas somatic cells undergo mitosis to proliferate, the germ cells undergo meiosis to produce haploid gametes (the sperm and the egg). The development of a new progeny organism is then initiated by the fusion of these gametes at fertilization.

The Process of Meiosis

In contrast to mitosis, **meiosis** results in the division of a diploid parental cell into haploid progeny, each containing only one member of the pair of homologous chromosomes that were present in the diploid parental cell (Figure 16.34). This reduction in chromosome number is accomplished by two sequential rounds of nuclear and cell division (called meiosis I and meiosis II), which follow a single round of DNA replication. Like mitosis, meiosis I initiates after S phase has been completed and the parental chromosomes have replicated to produce identical sister chromatids. The pattern of chromosome segregation in meiosis I, however, is dramatically different from that of mitosis. During meiosis I, homologous chromosomes first pair with one another and then segregate to different daughter cells. Sister chromatids remain together, so completion of meiosis I results in the formation of daughter cells containing a single member of each chromosome pair (consisting of two sister chromatids). Meiosis I is followed by meiosis II, which resembles mitosis in that the sister chromatids separate and segregate to different daughter cells. Completion of meiosis II thus results in the production of four haploid daughter cells, each of which contains only one copy of each chromosome.

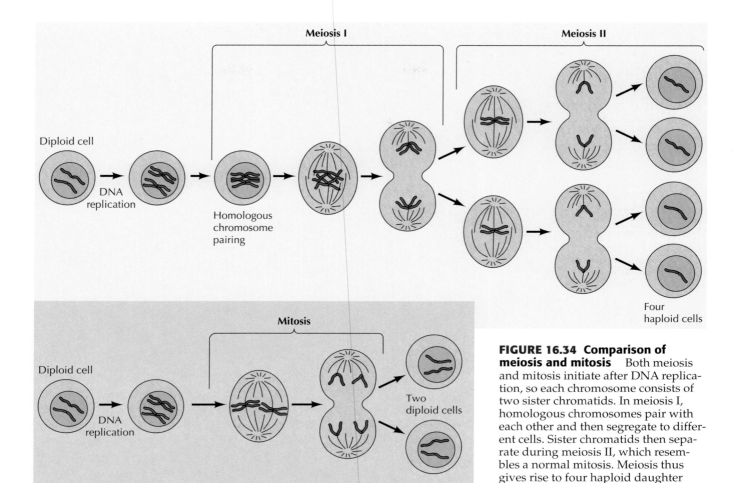

FIGURE 16.34 Comparison of meiosis and mitosis Both meiosis and mitosis initiate after DNA replication, so each chromosome consists of two sister chromatids. In meiosis I, homologous chromosomes pair with each other and then segregate to different cells. Sister chromatids then separate during meiosis II, which resembles a normal mitosis. Meiosis thus gives rise to four haploid daughter cells.

The pairing of homologous chromosomes after DNA replication results from recombination between chromosomes of paternal and maternal origin, so genetic recombination is critically linked to chromosome segregation during meiosis. Recombination between homologous chromosomes takes place during an extended prophase of meiosis I, which is divided into five stages (**leptotene**, **zygotene**, **pachytene**, **diplotene**, and **diakinesis**) on the basis of chromosome morphology (Figure 16.35). Recombination occurs at a high frequency during meiosis, and is initiated by double strand breaks that are induced early in meiotic prophase (leptotene) by a highly conserved endonuclease called Spo11. As discussed in Chapter 6, the formation of double strand breaks leads to the formation of single strand regions that invade a homologous chromosome by complementary base pairing (see Figure 6.32). The close association of homologous chromosomes (**synapsis**) begins during the zygotene stage. During this stage, a zipperlike protein structure, called the **synaptomenal complex**, forms along the length of the paired chromosomes (Figure 16.36). This complex keeps the homologous chromosomes closely associated and aligned with one another through the pachytene stage, which can persist for several days. Recombination between homologous chromosomes is completed by the end of pachytene,

16.7 **WEBSITE ANIMATION**
Meiosis In meiosis, a cell divides to produce daughter cells with half the number of chromosomes as the parent cell.

16.8 **WEBSITE ANIMATION**
Meiosis I and Mitosis Compared One difference between mitosis and meiosis can be seen at metaphase—in mitosis, homologous chromosomes line up separately on the metaphase plate, whereas in metaphase of meiosis I, homologous chromosomes line up in pairs at the metaphase plate.

Leptotene

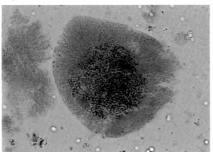

Zygotene

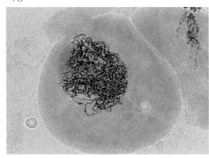

Pachytene

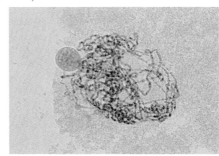

Diplotene

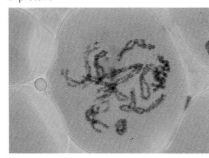

Diakinesis

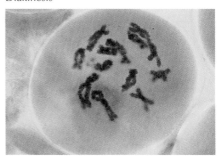

FIGURE 16.35 Stages of the prophase of meiosis I Micrographs illustrating the morphology of chromosomes of the lily. (C. Hasenkampf/Biological Photo Service.)

16.9 WEBSITE ANIMATION
Prophase I of Meiosis
Prophase I of meiosis consists of five stages, during which chromosomes condense and homologous chromosomes pair with each other and recombine.

leaving the chromosomes linked at the sites of crossing over (**chiasmata**; singular, chiasma). The synaptonemal complex disappears at the diplotene stage and the homologous chromosomes separate along their length. Importantly, however, they remain associated at the chiasmata, which is critical for their correct alignment at metaphase. At this stage, each chromosome pair (called a bivalent) consists of four chromatids with clearly evident chiasmata (Figure 16.37). Diakinesis, the final stage of prophase I, represents the transition to metaphase during which the chromosomes become fully condensed.

At metaphase I, the bivalent chromosomes align on the spindle. In contrast to mitosis (see Figure 16.29), the kinetochores of sister chromatids are adjacent to each other and oriented in the same direction, while the kinetochores of homologous chromosomes are pointed toward opposite spindle

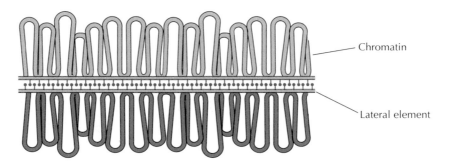

Chromatin

Lateral element

FIGURE 16.36 The synaptonemal complex Chromatin loops are attached to lateral elements, which are joined to each other by a zipperlike structure.

FIGURE 16.37 A bivalent chromosome at the diplotene stage   The bivalent chromosome consists of paired homologous chromosomes. Sister chromatids of each chromosome are joined at the centromere. Chromatids of homologous chromosomes are joined at chiasmata, which are the sites at which genetic recombination has occurred. (B. John/Visuals Unlimited.)

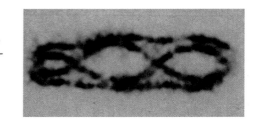

poles (Figure 16.38). Consequently, microtubules from the same pole of the spindle attach to sister chromatids, while microtubules from opposite poles attach to homologous chromosomes. Anaphase I is initiated by disruption of the chiasmata at which homologous chromosomes are joined. The homologous chromosomes then separate, while sister chromatids remain associated at their centromeres. At completion of meiosis I each daughter cell has therefore acquired one member of each homologous pair, consisting of two sister chromatids.

Meiosis II initiates immediately after cytokinesis, usually before the chromosomes have fully decondensed. In contrast to meiosis I, meiosis II resembles a normal mitosis. At metaphase II, the chromosomes align on the spindle with microtubules from opposite poles of the spindle attached to the kinetochores of sister chromatids. The link between the centromeres of sister chromatids is broken at anaphase II, and sister chromatids segregate to opposite poles. Cytokinesis then follows, giving rise to haploid daughter cells.

Regulation of Oocyte Meiosis

Vertebrate oocytes (developing eggs) have been particularly useful models for research on the cell cycle in part because of their large size and ease of manipulation in the laboratory. A notable example, discussed earlier in this chapter, is provided by the discovery and subsequent purification of MPF (Cdk1/cyclin B) from frog oocytes. Meiosis of these oocytes, like those of other species, is regulated at two unique points in the cell cycle, and studies of oocyte meiosis have illuminated novel mechanisms of cell cycle control.

The first regulatory point in oocyte meiosis is in the diplotene stage of the first meiotic division (Figure 16.39). Oocytes can remain arrested at this stage for long periods of time—up to 50 years in humans. During this diplotene arrest, the oocyte chromosomes decondense and are actively transcribed. This transcriptional activity is reflected in the tremendous growth of oocytes during this period. Human oocytes, for example, are about 100 μm in diameter (more than a hundred times the volume of a typical somatic cell). Frog oocytes are even larger, with diameters of approximately 1 mm. During this period of cell growth, the oocytes accumulate stockpiles of materials, including RNAs and proteins that are needed to support early development of the embryo. As noted earlier in this chapter, early embryonic cell cycles then occur in the absence of cell growth, rapidly dividing the fertilized egg into smaller cells (see Figure 16.2).

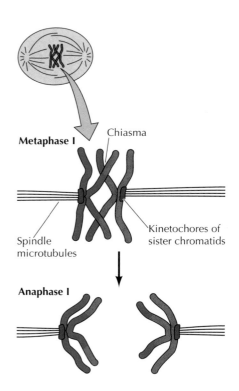

FIGURE 16.38 Chromosome segregation in meiosis I   At metaphase I, the kinetochores of sister chromatids are either fused or adjacent to one another. Microtubules from the same pole of the spindle therefore attach to the kinetochores of sister chromatids, while microtubules from opposite poles attach to the kinetochores of homologous chromosomes. Chiasmata are disrupted at anaphase I, and homologous chromosomes move to opposite poles of the spindle.

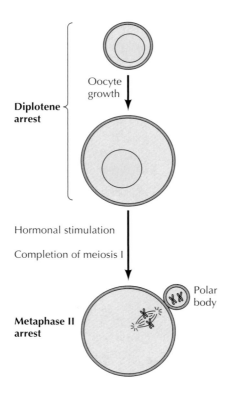

Diplotene arrest

Oocyte growth

Hormonal stimulation

Completion of meiosis I

Metaphase II arrest

Polar body

FIGURE 16.39 Meiosis of vertebrate oocytes Meiosis is arrested at the diplotene stage, during which oocytes grow to a large size. Oocytes then resume meiosis in response to hormonal stimulation and complete the first meiotic division, with asymmetric cytokinesis giving rise to a small polar body. Most vertebrate oocytes are then arrested again at metaphase II.

Oocytes of different species vary as to when meiosis resumes and fertilization takes place. In some animals, oocytes remain arrested at the diplotene stage until they are fertilized, only then proceeding to complete meiosis. However, the oocytes of most vertebrates (including frogs, mice, and humans) resume meiosis in response to hormonal stimulation and proceed through meiosis I prior to fertilization. Cell division following meiosis I is asymmetric, resulting in the production of a small **polar body** and an oocyte that retains its large size. The oocyte then proceeds to enter meiosis II without having re-formed a nucleus or decondensed its chromosomes. Most vertebrate oocytes are then arrested again at metaphase II, where they remain until fertilization.

Like the M phase of somatic cells, the meiosis of oocytes is controlled by the activity of Cdk1/cyclin B complexes. The regulation of Cdk1 during oocyte meiosis, however, displays unique features that are responsible for progression from meiosis I to meiosis II and for metaphase II arrest (Figure 16.40). Hormonal stimulation of diplotene-arrested oocytes initially triggers the resumption of meiosis by activating Cdk1, as at the G_2 to M transition of somatic cells. As in mitosis, Cdk1 then induces chromosome condensation, nuclear envelope breakdown, and formation of the spindle. Activation of

■ Animals can be cloned by the procedure of somatic cell nuclear transfer in which the nucleus of a somatic cell is transferred to a metaphase II oocyte from which the normal chromosomes have been removed. The oocyte is then stimulated to divide and, upon implantation into a surrogate mother, can give rise to an animal genetically identical to the donor of the somatic cell nucleus. Since the cloning of Dolly the sheep in 1997, this technology has been used to create cloned offspring of several mammalian species. It also offers the potential of therapeutic cloning for the treatment of a variety of human diseases.

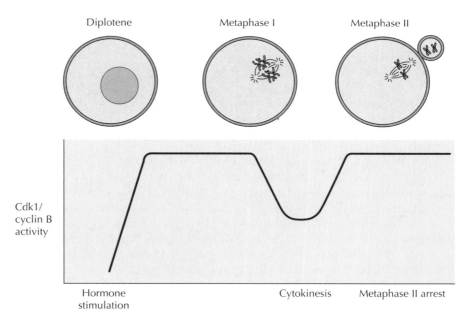

Diplotene

Metaphase I

Metaphase II

Cdk1/ cyclin B activity

Hormone stimulation

Cytokinesis

Metaphase II arrest

FIGURE 16.40 Activity of Cdk1/cyclin B during oocyte meiosis Hormonal stimulation of diplotene oocytes activates Cdk1/cyclin B, resulting in progression to metaphase I. Cdk1/cyclin B activity only partially falls at the transition from metaphase I to anaphase I, and the oocyte remains in M phase. Following completion of meiosis I, Cdk1/cyclin B activity again rises and remains high during metaphase II arrest.

FIGURE 16.41 Identification of cytostatic factor Cytoplasm from a metaphase II egg is microinjected into one cell of a two-cell embryo. The injected embryo cell arrests at metaphase, while the uninjected cell continues to divide. A factor in metaphase II egg cytoplasm (cytostatic factor) therefore has induced metaphase arrest of the injected embryo cell.

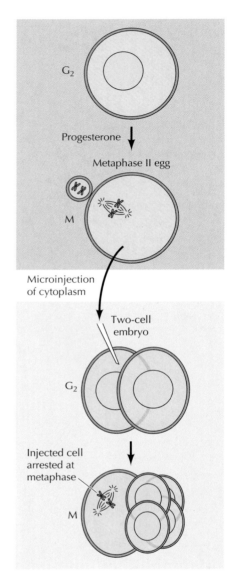

the anaphase-promoting complex then leads to the metaphase to anaphase transition of meiosis I, accompanied by a decrease in the activity of Cdk1. However, in contrast to mitosis, Cdk1 activity is only partially decreased, so the oocyte remains in M phase, chromatin remains condensed, and nuclear envelopes do not re-form. Following cytokinesis, Cdk1 activity again rises and remains high while the egg is arrested at metaphase II. A regulatory mechanism unique to oocytes thus acts to maintain Cdk1 activity during the metaphase to anaphase transition of meiosis I and subsequent metaphase II arrest, preventing the inactivation of Cdk1 that would result from cyclin B proteolysis during a normal M phase.

The factor responsible for metaphase II arrest was first identified by Yoshio Masui and Clement Markert in 1971, in the same series of experiments that led to the discovery of MPF. In this case, however, cytoplasm from an egg arrested at metaphase II was injected into an early embryo cell that was undergoing mitotic cell cycles (Figure 16.41). This injection of egg cytoplasm caused the embryonic cell to arrest at metaphase, indicating that metaphase arrest was induced by a cytoplasmic factor present in the egg. Because this factor acted to arrest mitosis, it was called **cytostatic factor** (**CSF**).

More recent experiments have identified a protein-serine/threonine kinase known as **Mos** as an essential component of CSF. Mos is specifically synthesized in oocytes around the time of completion of meiosis I and is required for the maintenance of Cdk1/cyclin B activity during the metaphase to anaphase transition of meiosis I as well during metaphase II arrest. The action of Mos results from activation of the ERK MAP kinase, which plays a central role in the cell signaling pathways discussed in the previous chapter. In oocytes, however, ERK plays a different role: It activates another protein kinase called Rsk, which maintains the activity of MPF both by stimulating cyclin B synthesis and by inhibiting cyclin B degradation (Figure 16.42). Inhibition of cyclin B degradation is mediated

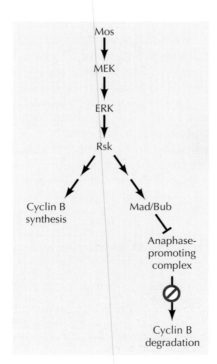

FIGURE 16.42 Maintenance of Cdk1/cyclin B activity by the Mos protein kinase The Mos protein kinase maintains Cdk1/cyclin B activity both by stimulating the synthesis of cyclin B and by inhibiting the degradation of cyclin B by the anaphase-promoting complex. The action of Mos is mediated by MEK, ERK, and Rsk protein kinases, and inhibition of the anaphase-promoting complex is mediated by Mad/Bub proteins.

16.10 WEBSITE ANIMATION
Polar Body Formation
During meiosis in female vertebrates, the meiotic divisions are often unequal, resulting in a single large egg and much smaller polar bodies.

FIGURE 16.43 Fertilization
Scanning electron micrograph of a human sperm fertilizing an egg. (David M. Philips/Visuals Unlimited.)

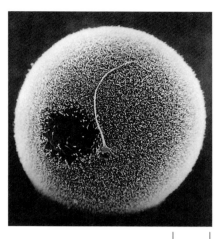

10 µm

by inhibition of the anaphase-promoting complex by the Mad/Bub proteins, similar to arrest at the spindle assembly checkpoint (see Figure 16.30). Mos thus maintains Cdk1/cyclin B activity during oocyte meiosis, leading to the arrest of oocytes at metaphase II. Oocytes can remain arrested at this point in the meiotic cell cycle for several days, awaiting fertilization.

Fertilization

At **fertilization**, the sperm binds to a receptor on the surface of the egg and fuses with the egg plasma membrane, initiating the development of a new diploid organism containing genetic information derived from both parents (Figure 16.43). Not only does fertilization lead to the mixing of paternal and maternal chromosomes, but it also induces a number of changes in the egg cytoplasm that are critical for further development. These alterations activate the egg, leading to the completion of oocyte meiosis and initiation of the mitotic cell cycles of the early embryo.

A key signal resulting from the binding of a sperm to its receptor on the plasma membrane of the egg is an increase in the level of Ca^{2+} in the egg cytoplasm, probably as a consequence of stimulation of the hydrolysis of phosphatidylinositol 4,5-bisphosphate (PIP_2) (see Figure 15.27). One effect of this elevation in intracellular Ca^{2+} is the induction of surface alterations that prevent additional sperm from entering the egg. Because eggs are usually exposed to large numbers of sperm at one time, this is a critical event in ensuring the formation of a normal diploid embryo. These surface alterations result from the Ca^{2+}-induced exocytosis of secretory vesicles that are present in large numbers beneath the egg plasma membrane. Release of the

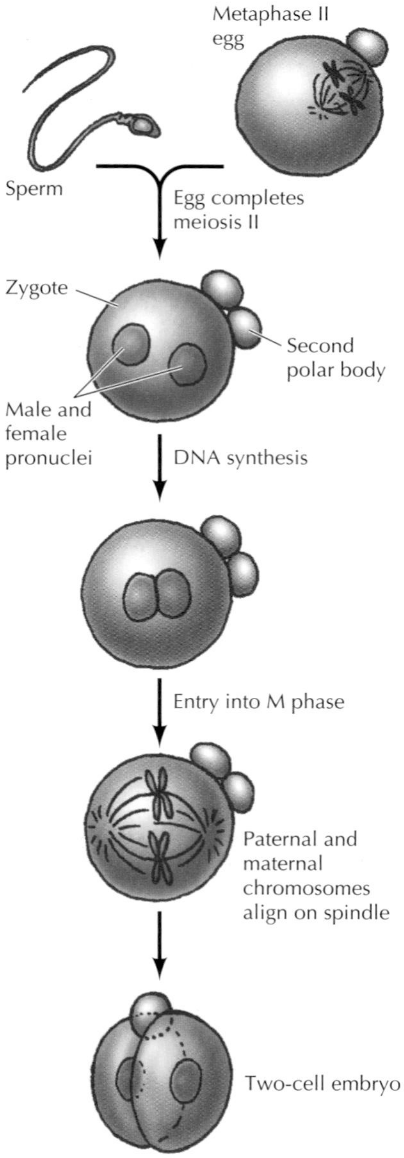

Metaphase II egg

Sperm

Egg completes meiosis II

Zygote

Second polar body

Male and female pronuclei

DNA synthesis

Entry into M phase

Paternal and maternal chromosomes align on spindle

Two-cell embryo

FIGURE 16.44 Fertilization and completion of meiosis Fertilization induces the transition from metaphase II to anaphase II, leading to completion of oocyte meiosis and emission of a second polar body (which usually degenerates). The sperm nucleus decondenses, so the fertilized egg (zygote) contains two haploid nuclei (male and female pronuclei). In mammals, the pronuclei replicate DNA as they migrate toward each other. They then initiate mitosis, with male and female chromosomes aligning on a common spindle. Completion of mitosis and cytokinesis thus gives rise to a two-cell embryo, with each cell containing a diploid genome.

contents of these vesicles alters the extracellular coat of the egg so as to block the entry of additional sperm.

The increase in cytosolic Ca^{2+} following fertilization also signals the completion of meiosis (Figure 16.44). In eggs arrested at metaphase II, the metaphase to anaphase transition is triggered by activation of the anaphase-promoting complex, resulting from Ca^{2+}-dependent phosphorylation and degradation of an inhibitory protein. The resultant degradation of cyclin B and condensin leads to completion of the second meiotic division, with asymmetric cytokinesis (as in meiosis I) giving rise to a second small polar body.

Following completion of oocyte meiosis, the fertilized egg (now called a **zygote**) contains two haploid nuclei (called **pronuclei**), one derived from each parent. In mammals, the two pronuclei then enter S phase and replicate their DNA as they migrate toward each other. As they meet, the zygote enters M phase of its first mitotic division. The two nuclear envelopes break down, and the condensed chromosomes of both paternal and maternal origin align on a common spindle. Completion of mitosis then gives rise to two embryonic cells, each containing a new diploid genome. These cells then commence the series of embryonic cell divisions that eventually lead to the development of a new organism.

■ *In vitro* fertilization (IVF) is widely used to help infertile couples. A variety of reproductive disorders can be dealt with by this procedure in which metaphase II eggs are recovered from the ovary, fertilized *in vitro* and then returned to the Fallopian tube or uterus of the mother. The first baby resulting from *in vitro* fertilization was born in 1978, and tens of thousands of infertile couples have since been treated by IVF.

COMPANION WEBSITE

Visit the website that accompanies **The Cell** (www.sinauer.com/cooper) for animations, videos, quizzes, problems, and other review material.

SUMMARY

KEY TERMS

THE EUKARYOTIC CELL CYCLE

Phases of the Cell Cycle: Eukaryotic cell cycles are divided into four discrete phases: M, G_1, S, and G_2. M phase consists of mitosis, which is usually followed by cytokinesis. S phase is the period of DNA replication.

mitosis, interphase, cytokinesis, M phase, G_1 phase, S phase, G_2 phase, flow cytometer, fluorescence-activated cell sorter

Regulation of the Cell Cycle by Cell Growth and Extracellular Signals: Extracellular signals and cell size regulate progression through specific control points in the cell cycle.

START, restriction point

Cell Cycle Checkpoints: Checkpoints and feedback controls coordinate the events that take place during different phases of the cell cycle and arrest cell cycle progression if DNA is damaged.

cell cycle checkpoint, ATM, ATR

Restricting DNA Replication to Once per Cell Cycle: Once DNA replication has taken place, initiation of a new S phase is prevented until the cell has passed through mitosis.

REGULATORS OF CELL CYCLE PROGRESSION

Protein Kinases and Cell Cycle Regulation: MPF is the key molecule responsible for regulating the G_2 to M transition in all eukaryotes. MPF is a dimer of cyclin B and the Cdk1 protein kinase.

maturation promoting factor (MPF), Cdk1, cyclin

Families of Cyclins and Cyclin-Dependent Kinases: Distinct pairs of cyclins and Cdk1-related protein kinases regulate progression through different stages of the cell cycle. The activity of Cdk's is regulated by association with cyclins, activating and inhibitory phosphorylations, and the binding of Cdk inhibitors.

G_1 cyclin, Cln, Cdk, Cdk inhibitor (CKI)

KEY TERMS

Rb, tumor suppressor gene, E2F

checkpoint kinase, p53

prophase, metaphase, anaphase, telophase, centromere, kinetochore, centrosome, mitotic spindle, prometaphase

condensin, cohesin, Aurora kinase, Polo-like kinase

spindle assembly checkpoint, anaphase-promoting complex

contractile ring

SUMMARY

Growth Factors and the Regulation of G_1 Cdk's: Growth factors stimulate animal cell proliferation by inducing synthesis of the D-type cyclins. Cdk4, 6/cyclin D complexes then act to drive cells through the restriction point in G_1. A key substrate of Cdk4, 6/cyclin D complexes is the tumor suppressor protein Rb, which regulates transcription of genes required for cell cycle progression, including cyclin E. Activation of Cdk2/cyclin E complexes is then responsible for entry into S phase.

DNA Damage Checkpoints: DNA damage or incompletely replicated DNA arrest cell cycle progression in G_1, S, and G_2. Cell cycle arrest is mediated by protein kinases that are activated by DNA damage and inhibit Cdc25 phosphatases, which are required for Cdk activation. In mammalian cells, arrest at the G_1 checkpoint is also mediated by p53, which induces synthesis of the Cdk inhibitor p21.

THE EVENTS OF M PHASE

Stages of Mitosis: Mitosis is conventionally divided into four stages: prophase, metaphase, anaphase, and telophase. The basic events of mitosis include chromosome condensation, formation of the mitotic spindle, nuclear envelope breakdown, and attachment of spindle microtubules to chromosomes at the kinetochore. Sister chromatids then separate and move to opposite poles of the spindle. Finally, nuclei re-form, the chromosomes decondense, and cytokinesis divides the cell in half.

Cdk1/Cyclin B and Progression to Metaphase: M phase is initiated by activation of Cdk1/cyclin B, which phosphorylates other protein kinases, as well as the nuclear lamins and other proteins of the nuclear envelope, condensins, and Golgi matrix proteins. Activation of Cdk1/cyclin B is responsible for chromatin condensation, nuclear envelope breakdown, fragmentation of the Golgi apparatus, and reorganization of microtubules to form the mitotic spindle. The attachment of spindle microtubules to the kinetochores of sister chromatids then leads to their alignment on the metaphase plate.

The Spindle Assembly Checkpoint and Progression to Anaphase: Activation of a ubiquitin ligase called the anaphase-promoting complex leads to degradation of key regulatory proteins at the metaphase to anaphase transition. The activity of the anaphase-promoting complex is inhibited until the cell passes the spindle assembly checkpoint and all chromosomes are properly aligned on the spindle. Ubiquitin-mediated proteolysis initiated by the anaphase-promoting complex then leads to the degradation of cohesin, breaking the link between sister chromatids at the onset of anaphase. The anaphase-promoting complex also ubiquitinates cyclin B, leading to inactivation of Cdk1 and exit from mitosis.

Cytokinesis: Inactivation of Cdk1/cyclin B also triggers cytokinesis. In yeast and animal cells, cytokinesis results from contraction of a ring of actin and myosin filaments. In higher plant cells, cytokinesis results from the formation of a new cell wall and plasma membrane inside the cell.

SUMMARY

MEIOSIS AND FERTILIZATION

The Process of Meiosis: Meiosis is a specialized cell cycle that gives rise to haploid daughter cells. A single round of DNA synthesis is followed by two sequential cell divisions. During meiosis I, homologous chromosomes first form pairs and then segregate to different daughter cells. Meiosis II then resembles a normal mitosis in which sister chromatids separate.

Regulation of Oocyte Meiosis: Meiosis of vertebrate oocytes is regulated at two unique points in the cell cycle: the diplotene stage of meiosis I and metaphase of meiosis II. Metaphase II arrest results from inhibition of the anaphase-promoting complex by a protein kinase expressed in oocytes.

Fertilization: Fertilization triggers the resumption of oocyte meiosis by Ca^{2+}-dependent activation of the anaphase-promoting complex. The fertilized egg then contains two haploid nuclei, which form a new diploid genome and initiate embryonic cell divisions.

KEY TERMS

meiosis, leptotene, zygotene, pachytene, diplotene, diakinesis, synapsis, synaptomenal complex, chiasma

polar body, cytostatic factor (CSF), Mos

fertilization, zygote, pronucleus

Questions

1. In what ways are cells in G_0 and G_1 similar? How do they differ?

2. Consider a mammalian cell line that divides every 30 hours. Microscopic observation indicates that 3.3% of the cells are in mitosis at any given time. Analysis in the flow cytometer establishes that 53.3% of the cells have DNA contents of $2n$, 16.7 % have DNA contents of $4n$, and 30% have DNA contents ranging between $2n$ and $4n$. What are the lengths of the G_1, S, G_2, and M phases of the cycle of these cells?

3. Radiation damages DNA and arrests cell cycle progression at checkpoints in G_1, S, and G_2. Why is this advantageous for the cell?

4. What are the mechanisms that regulate the activity of cyclin-dependent kinases?

5. The spindle assembly checkpoint delays the onset of anaphase until all chromosomes are properly aligned on the spindle. What would be the result if a failure of this checkpoint allowed anaphase to initiate while one chromosome was attached to microtubules from only a single centrosome?

6. What cellular processes would be affected by expression of siRNA targeted against Cdk7?

7. The Cdk inhibitor p16 binds specifically to Cdk4, 6/cyclin D complexes. What would be the predicted effect of overexpression of p16 on cell cycle progression? Would overexpression of p16 affect a tumor cell lacking functional Rb protein?

8. *In vitro* mutagenesis of cloned lamin cDNAs has been used to generate mutants that cannot be phosphorylated by Cdk1. How would expression of these mutant lamins affect nuclear envelope breakdown at the end of prophase?

9. What substrates are phosphorylated by Cdk1/cyclin B to initiate mitosis?

10. A mutant of cyclin B resistant to degradation by the anaphase-promoting complex has been generated. How would expression of this mutant cyclin B affect the events at the metaphase to anaphase transition?

11. How does the activity of anaphase-promoting complex lead to the separation of sister chromatids?

12. Homologous recombination has been used to inactivate the *mos* gene in mice. What effect would you expect this to have on oocyte meiosis?

13. How does the fertilized egg provide a long-lasting block to additional sperm entering the egg?

References and Further Reading

The Eukaryotic Cell Cycle

Abraham, R. T. and R. S. Tibbetts. 2005. Guiding ATM to broken DNA. *Science* 308: 510–511. [R]

Blow, J. J. and A. Dutta. 2005. Preventing re-replication of chromosomal DNA. *Nature Rev. Mol. Cell Biol.* 6: 476–486. [R]

Forsburg, S. L. and P. Nurse. 1991. Cell cycle regulation in the yeasts *Saccharomyces cerevisiae* and *Schizosaccharomyces pombe*. *Ann. Rev. Cell Biol.* 7: 227–256. [R]

Hartwell, L. H. and T. A. Weinert. 1989. Checkpoints: Controls that ensure the order of cell cycle events. *Science* 246: 629–634. [R]

Machida, Y. J., J. L. Hamlin and A. Dutta. 2005. Right place, right time, and only once: Replication initiation in metazoans. *Cell* 123: 13–24. [R]

McGowan, C. H. and P. Russell. 2004. The DNA damage response: Sensing and signaling. *Curr. Opin. Cell Biol.* 16: 629–633. [R]

Norbury, C. and P. Nurse. 1992. Animal cell cycles and their control. *Ann. Rev. Biochem.* 61: 441–470. [R]

Pardee, A. B. 1989. G₁ events and the regulation of cell proliferation. *Science* 246: 603–608. [R]

Russell, P. 1998. Checkpoints on the road to mitosis. *Trends Biochem. Sci.* 23: 399–402. [R]

Regulators of Cell Cycle Progression

Andrews, B. and V. Measday. 1998. The cyclin family of budding yeast: Abundant use of a good idea. *Trends Genet.* 14: 66–72. [R]

Bracken, A. P., M. Ciro, A. Cocito and K. Helin. 2004. E2F target genes: Unraveling the biology. *Trends Biochem. Sci.* 29: 409–417. [R]

Evans, T., E. T. Rosenthal, J. Youngbloom, D. Distel and T. Hunt. 1983. Cyclin: A protein specified by maternal mRNA in sea urchin eggs that is destroyed at each cleavage division. *Cell* 33: 389–396. [P]

Gottifredi, V. and C. Prives. 2005. The S phase checkpoint: When the crowd meets at the fork. *Sem. Cell Dev. Biol.* 16: 355–368. [R]

Hartwell, L. H., R. K. Mortimer, J. Culotti and M. Culotti. 1973. Genetic control of the cell division cycle in yeast: V. Genetic analysis of *cdc* mutants. *Genetics* 74: 267–287. [P]

Kastan, M. B. and J. Bartek. 2004. Cell-cycle checkpoints and cancer. *Nature* 432: 316–323. [R]

Lohka, M. J., M. K. Hayes and J. L. Maller. 1988. Purification of maturation-promoting factor, an intracellular regulator of early mitotic events. *Proc. Natl. Acad. Sci. U.S.A.* 85: 3009–3013. [P]

Malumbres, M. and M. Barbacid. 2005. Mammalian cyclin-dependent kinases. *Trends Biochem. Sci.* 30: 630–641. [R]

Massague, J. 2004. G₁ cell-cycle control and cancer. *Nature* 432: 298–306. [R]

Masui, Y. and C. L. Markert. 1971. Cytoplasmic control of nuclear behavior during meiotic maturation of frog oocytes. *J. Exp. Zool.* 177: 129–146. [P]

McGowan, C. H. and P. Russell. 2004. The DNA damage response: Sensing and signaling. *Curr. Opin. Cell Biol.* 16: 629–633. [R]

Murray, A. W. 2004. Recycling the cell cycle: Cyclins revisited. *Cell* 116: 221–234. [R]

Nyberg, K. A., R. J. Michelson, C. W. Putnam and T. A. Weinert. 2002. Toward maintaining the genome: DNA damage and replication checkpoints. *Ann. Rev. Genet.* 36: 617–656. [R]

Reed, S. I. 2003. Ratchets and clocks: The cell cycle, ubiquitylation and protein turnover. *Nature Rev. Mol. Cell Biol.* 4: 855–864. [R]

Sherr, C. J. and J. M. Roberts. 2004. Living with or without cyclins and cyclin-dependent kinases. *Genes Dev.* 18: 2699–2711. [R]

Smith, L. D. and R. E. Ecker. 1971. The interaction of steroids with *Rana pipiens* oocytes in the induction of maturation. *Dev. Biol.* 25: 232–247. [P]

Swenson, K. I., K. M. Farrell and J. V. Ruderman. 1986. The clam embryo protein cyclin A induces entry into M phase and the resumption of meiosis in *Xenopus* oocytes. *Cell* 47: 861–870. [P]

Vousden, K. H. 2002. Activation of the p53 tumor suppressor protein. *Biochim. Biophys. Acta* 1602: 47–59. [R]

The Events of M Phase

Barr, F. A. 2004. Golgi inheritance: Shaken but not stirred. *J. Cell Biol.* 164: 955–958. [R]

Barr, F. A., H. H. W. Sillje and E. A. Nigg. 2004. Polo-like kinases and the orchestration of cell division. *Nature Rev. Mol. Cell Biol.* 5: 429–440. [R]

Blagden, S. P. and D. M. Glover. 2003. Polar expeditions—provisioning the centrosome for mitosis. *Nature Cell Biol.* 5: 505–511. [R]

Burke, B. and J. Ellenberg. 2002. Remodeling the walls of the nucleus. *Nature Rev. Mol. Cell Biol.* 3: 487–497. [R]

Carmena, M. and W. C. Earnshaw. 2003. The cellular geography of Aurora kinases. *Nature Rev. Mol. Cell Biol.* 4: 842–854. [R]

De Gramont, A. and O. Cohen-Fix. 2005. The many phases of anaphase. *Trends Biochem. Sci.* 30: 559–568. [R]

Gadde, S. and R. Heald. 2004. Mechanisms and molecules of the mitotic spindle. *Curr. Biol.* 14: R797–R805. [R]

Glotzer, M. 2005. The molecular requirements for cytokinesis. *Science* 307: 1735–1739. [R]

Hetzer, M. W., T. C. Walther and I. W. Mattaj. 2005. Pushing the envelope: Structure, function, and dynamics of the nuclear periphery. *Ann. Rev. Cell Dev. Biol.* 21: 347–380. [R]

Jurgens, G. 2005. Plant cytokinesis: Fission by fusion. *Trends Cell Biol.* 15: 277–283. [R]

Kline-Smith, S. L. and C. E. Walczak. 2004. Mitotic spindle assembly and chromosome segregation: Refocusing on microtubule dynamics. *Mol. Cell* 15: 317–327. [R]

Kops, G. J. P. L., B. A. A. Weaver and D. W. Cleveland. 2005. On the road to cancer: Aneuploidy and the mitotic checkpoint. *Nature Rev. Cancer* 5: 773–785. [R]

Losada, A. and T. Hirano. 2005. Dynamic molecular linkers of the genome: The first decade of SMC proteins. *Genes Dev.* 19: 1269–1287. [R]

Nasmyth, K. 2002. Segregating sister genomes: The molecular biology of chromosome separation. *Science* 297: 559–565. [R]

Nasmyth, K. 2005. How do so few control so many? *Cell* 120: 739–746. [R]

Peters, J.-M. 2002. The anaphase-promoting complex: Proteolysis in mitosis and beyond. *Mol. Cell* 9: 931–943. [R]

Rieder, C. L. and A. Khodjakov. 2003. Mitosis through the microscope: Advances in seeing inside live dividing cells. *Science* 300: 91–96. [R]

Scholey, J. M., I. Brust-Mascher and A. Mogilner. 2003. Cell division. *Nature* 422: 746–752. [R]

Shorter, J. and G. Warren. 2002. Golgi architecture and inheritance. *Ann. Rev. Cell Dev. Biol.* 18: 379–420. [R]

Wolfe, B. A. and K. L. Gould. 2005. Split decisions: Coordinating cytokinesis in yeast. *Trends Cell Biol.* 15: 10–18. [R]

Meiosis and Fertilization

Evans, J. P. and H. M. Florman. 2002. The state of the union: The cell biology of fertilization. *Nature Cell Biol.* 4: S57–S63. [R]

Page, S. L. and R. S. Hawley. 2003. Chromosome choreography: The meiotic ballet. *Science* 301: 785–789. [R]

Pawlowski, W. P. and W. Z. Cande. 2005. Coordinating the events of the meiotic prophase. *Trends Cell Biol.* 15: 674–681. [R]

Petronczki, M., M. F. Siomos and K. Nasmyth. 2003. Un ménage a quatre: The molecular biology of chromosome segregation in meiosis. *Cell* 112: 423–440. [R]

Tunquist, B. J. and J. L. Maller. 2003. Under arrest: Cytostatic factor (CSF)-mediated metaphase arrest in vertebrate eggs. *Genes Dev.* 17: 683–710. [R]

Villeneuve, A. M. and K. J. Hillers. 2001. Whence meiosis? *Cell* 106: 647–650. [R]

Wassarman, P. M. 1999. Mammalian fertilization: Molecular aspects of gamete adhesion, exocytosis, and fusion. *Cell* 96: 175–183. [R]

CHAPTER 17

Cell Death and Cell Renewal

■ **Programmed Cell Death 690**

■ **Stem Cells and the Maintenance of Adult Tissues 700**

■ **Embryonic Stem Cells and Therapeutic Cloning 709**

■ **KEY EXPERIMENT:**
Identification of Genes Required for Programmed Cell Death 694

■ **KEY EXPERIMENT:**
Culture of Embryonic Stem Cells 710

CELL DEATH AND CELL PROLIFERATION are balanced throughout the life of multicellular organisms. Animal development begins with the rapid proliferation of embryonic cells, which then differentiate to produce the many specialized types of cells that make up adult tissues and organs. Whereas the nematode *C. elegans* consists of only 959 somatic cells, humans possess a total of approximately 10^{14} cells, consisting of more than 200 differentiated cell types. Starting from only a single cell—the fertilized egg—all the diverse cell types of the body are produced and organized into tissues and organs. This complex process of development involves not only cell proliferation and differentiation but also cell death. Although cells can die as a result of unpredictable traumatic events, such as exposure to toxic chemicals, most cell deaths in multicellular organisms occur by a normal physiological process of programmed cell death, which plays a key role both in embryonic development and in adult tissues.

In adult organisms, cell death must be balanced by cell renewal, and most tissues contain stem cells that are able to replace cells that have been lost. Abnormalities of cell death are associated with a wide variety of illnesses, including cancer, autoimmune disease, and neurodegenerative disorders, such as Parkinson's and Alzheimer's disease. Conversely, the ability of stem cells to proliferate and differentiate into a wide variety of cell types has generated enormous interest in the possible use of these cells, particularly embryonic stem cells, to replace damaged tissues. The mechanisms and regulation of cell death and cell renewal have therefore become areas of research at the forefront of biology and medicine.

Programmed Cell Death

Programmed cell death is carefully regulated so that the fate of individual cells meets the needs of the organism as a whole. In adults, programmed cell death is responsible for balancing cell proliferation and maintaining constant cell numbers in tissues undergoing cell turnover. For example, about 5×10^{11} blood cells are eliminated daily in humans by programmed cell death, balancing their continual production in the bone marrow. In addition, programmed cell death provides a defense mechanism by which damaged and potentially dangerous cells can be eliminated for the good of the organism as a whole. Virus-infected cells frequently undergo programmed cell death, thereby preventing the production of new virus particles and limiting spread of the virus through the host organism. Other types of cellular insults, such as DNA damage, also induce programmed cell death. In the case of DNA damage, programmed cell death may eliminate cells carrying potentially harmful mutations, including cells with mutations that might lead to the development of cancer.

During development, programmed cell death plays a key role by eliminating unwanted cells from a variety of tissues. For example, programmed cell death is responsible for the elimination of larval tissues during amphibian and insect metamorphosis, as well as for the elimination of tissue between the digits during the formation of fingers and toes. Another well-characterized example of programmed cell death is provided by development of the mammalian nervous system. Neurons are produced in excess, and up to 50% of developing neurons are eliminated by programmed cell death. Those that survive are selected for having made the correct connections with their target cells, which secrete growth factors that signal cell survival by blocking the neuronal cell death program. The survival of many other types of cells in animals is similarly dependent on growth factors or contacts with neighboring cells or the extracellular matrix, so programmed cell death is thought to play an important role in regulating the associations between cells in tissues.

The Events of Apoptosis

In contrast to the accidental death of cells that results from an acute injury, programmed cell death is an active process, which usually proceeds by a distinct series of cellular changes known as **apoptosis**, first described in 1972. During apoptosis, chromosomal DNA is usually fragmented as a result of cleavage between nucleosomes (Figure 17.1). The chromatin condenses and the nucleus then breaks up into small pieces. Finally, the cell itself shrinks and breaks up into membrane-enclosed fragments called apoptotic bodies.

Apoptotic cells and cell fragments are efficiently recognized and phagocytosed by both macrophages and neighboring cells, so cells that die by apoptosis are rapidly removed from tissues. In contrast, cells that die as a result of acute injury swell and lyse, releasing their contents into the extracellular space and causing inflammation. The removal of apoptotic cells is mediated by the expression of so-called "eat me" signals on the cell surface. These signals include phosphatidylserine, which is normally restricted to the inner leaflet of the plasma membrane (see Figure 13.2). During apoptosis, phosphatidylserine becomes expressed on the cell surface where it is recognized by receptors expressed by phagocytic cells (Figure 17.2).

Pioneering studies of programmed cell death during the development of *C. elegans* provided the critical initial insights that led to understanding the

> ■ The term apoptosis is derived from the Greek word describing the falling of leaves from a tree or petals from a flower. It was coined to differentiate this form of programmed cell death from the accidental cell deaths caused by inflammation or injury.

(A)

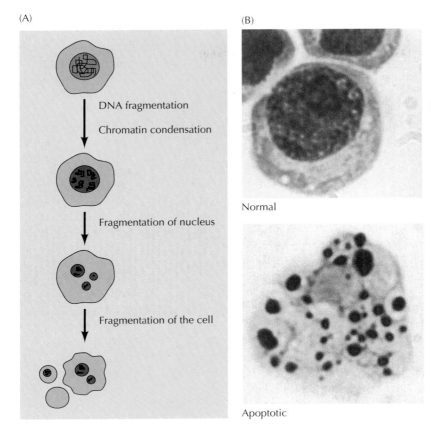

DNA fragmentation

Chromatin condensation

Fragmentation of nucleus

Fragmentation of the cell

(B)

Normal

Apoptotic

(C) Hours after induction of Apoptosis

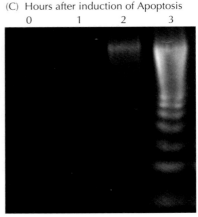

FIGURE 17.1 Apoptosis (A) Diagrammatic representation of the events of apoptosis. (B) Light micrographs of normal and apoptotic human leukemia cells illustrating chromatin condensation and nuclear fragmentation. (C) Gel electrophoresis of DNA from apoptotic cells, showing its degradation to fragments corresponding to multiples of 200 base pairs (the size of nucleosomes) at 1–4 hours following induction of apoptosis. (B, courtesy of D. R. Green/La Jolla Institute for Allergy and Immunology; C, courtesy of Ken Adams, Boston University.)

molecular mechanism of apoptosis. These studies in the laboratory of Robert Horvitz initially identified three genes that play key roles in regulating and executing apoptosis. During normal nematode development, 131 somatic cells out of a total of 1090 are eliminated by programmed cell death, yielding the 959 somatic cells in the adult worm. The death of these cells is highly specific, such that the same cells always die in developing embryos. Based on this developmental specificity, Robert Horvitz undertook a genetic analysis of cell death in *C. elegans* with the goal of identifying the genes responsible for these developmental cell deaths. In 1986 mutagenesis of *C. elegans* identified two genes, *ced-3* and *ced-4*, that were required for developmental *c*ell death. If either *ced-3* or *ced-4* was inactivated by mutation, the normal programmed cell deaths did not take place. A third gene, *ced-9*, functioned as a negative regulator of apoptosis. If *ced-9* was inactivated by mutation, the cells that would normally survive failed to do so. Instead, they also underwent apoptosis, leading to death of the developing animal. Conversely, if *ced-9* was expressed at an abnormally high level, the normal programmed cell deaths failed to occur. Further studies indicated that the proteins encoded by these genes acted in a pathway with Ced-4 acting to stimulate Ced-3, and Ced-9 inhibiting Ced-4 (**Figure 17.3**). Genes related to *ced-3*, *ced-4*, and *ced-9* have also been identified in *Drosophila* and mammals and found to encode proteins that represent conserved effectors and regulators of apoptosis induced by a variety of stimuli.

17.1 WEBSITE ANIMATION

Apoptosis During apoptosis, chromosomal DNA is usually fragmented, the chromatin condenses, the nucleus breaks up, and the cell shrinks and breaks into apoptotic bodies.

FIGURE 17.2 Phagocytosis of apoptotic cells Apoptotic cells and cell fragments are recognized and engulfed by phagocytic cells. One of the signals recognized by phagocytes is phosphatidylserine on the cell surface. In normal cells, phosphatidylserine is restricted to the inner leaflet of the plasma membrane, but it becomes expressed on the cell surface during apoptosis.

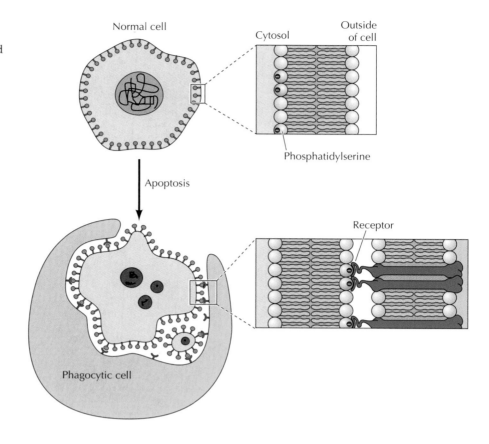

Caspases: The Executioners of Apoptosis

The molecular cloning and nucleotide sequencing of the *ced-3* gene indicated that it encoded a protease, providing the first insight into the molecular mechanism of apoptosis. We now know that Ced-3 is the prototype of a family of more than a dozen proteases, known as **caspases** because they have cysteine (C) residues at their active sites and cleave after aspartic acid (Asp) residues in their substrate proteins. The caspases are the ultimate effectors or executioners of programmed cell death, bringing about the events of apoptosis by cleaving nearly 100 different cell target proteins (Figure 17.4). Key targets of the caspases include an inhibitor of a DNase, which when activated is responsible for fragmentation of nuclear DNA. In addition, caspases cleave both nuclear lamins, leading to fragmentation of the nucleus, and cytoskeletal proteins, leading to disruption of the cytoskeleton, membrane blebbing, and cell fragmentation.

Ced-3 is the only caspase in *C. elegans*. However, *Drosophila* and mammals contain families of at least seven caspases that function in different

FIGURE 17.3 Programmed cell death in *C. elegans* Genetic analysis identified three genes that play key roles in programmed cell death during development of *C. elegans*. Two genes, *ced-3* and *ced-4*, are required for cell death, whereas *ced-9* inhibits cell death. The Ced-9 protein acts upstream of Ced-4, which activates Ced-3.

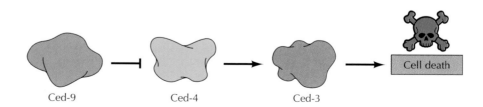

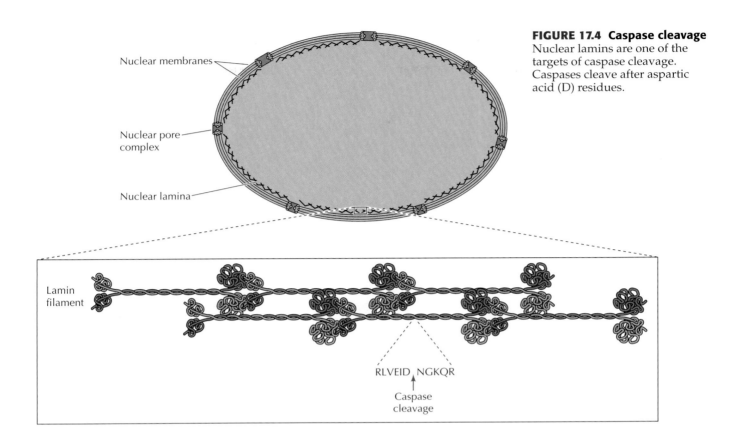

FIGURE 17.4 Caspase cleavage Nuclear lamins are one of the targets of caspase cleavage. Caspases cleave after aspartic acid (D) residues.

Nuclear membranes

Nuclear pore complex

Nuclear lamina

Lamin filament

RLVEID NGKQR

Caspase cleavage

aspects of apoptosis. These caspases are generally classified as either *initiator* or *effector* caspases, and they function in a cascade that ultimately brings about death of the cell. All caspases are synthesized as inactive precursors that can be converted to the active form by proteolytic cleavage, catalyzed by other caspases. The activation of an initiator caspase therefore starts off a chain reaction leading to activation of downstream effector caspases and death of the cell. Regulation of caspases is thus central to determining cell survival.

Genetic analysis in *C. elegans* initially suggested that Ced-4 functioned as an activator of the caspase Ced-3. Subsequent studies have shown that Ced-4 and its mammalian homolog (Apaf-1) bind to caspases and promote their activation. In mammalian cells, the key initiator caspase (caspase-9) is activated by binding to Apaf-1 in a multisubunit complex called the **apoptosome** (Figure 17.5). Formation of this complex in mammals also requires cytochrome *c*, which is released from mitochondria by stimuli that trigger apoptosis (discussed in the following section). Once activated in the apoptosome, cas-

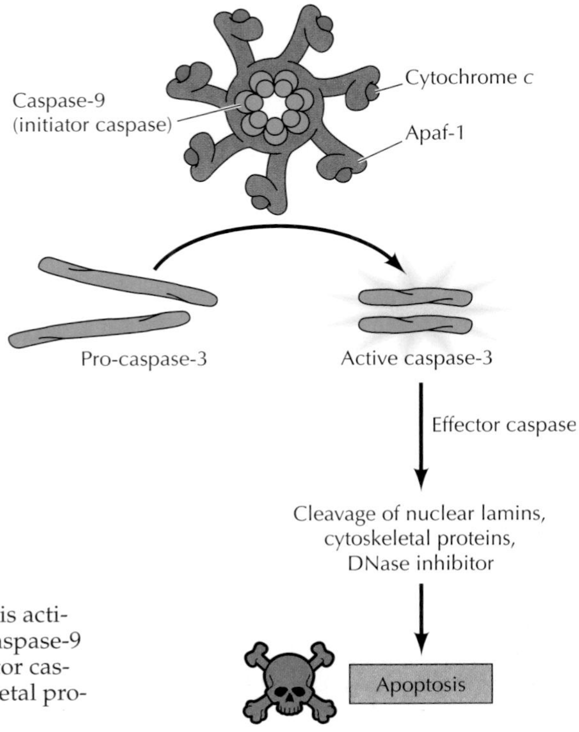

Caspase-9 (initiator caspase)

Cytochrome *c*

Apaf-1

Pro-caspase-3

Active caspase-3

Effector caspase

Cleavage of nuclear lamins, cytoskeletal proteins, DNase inhibitor

Apoptosis

FIGURE 17.5 Caspase activation The mammalian initiator caspase-9 is activated as a complex with Apaf-1 and cytochrome *c* in the apoptosome. Caspase-9 then cleaves and activates effector caspases, such as caspase-3. The effector caspases cleave a variety of cell proteins, including nuclear lamins, cytoskeletal proteins, and an inhibitor of DNase, leading to death of the cell.

Identification of Genes Required for Programmed Cell Death

Genetic Control of Programmed Cell Death in the Nematode *C. elegans*

Hilary M. Ellis and H. Robert Horvitz

Massachusetts Institute of Technology, Cambridge, MA

Cell, 1986, Volume 44, pages 817–829

H. Robert Horvitz

The Context

By the 1960s cell death was recognized as a normal event during animal development, implying that it was a carefully regulated process with specific cells destined to die. The simple nematode *C. elegans*, which has been a critically important model system in developmental biology, proved to be key to understanding both the regulation and the mechanism of such programmed cell deaths. Microscopic analysis in the 1970s established a complete map of *C. elegans* development so that the embryonic origin and fate of each cell was known. Importantly, *C. elegans* development included a very specific pattern of programmed cell deaths. In particular, John Sulston and H. Robert Horvitz reported in 1977 that the development of adult worms (consisting of 959 somatic cells) involved the programmed death of 131 cells out of 1090 that were initially produced. The same cells died in all embryos, indicating that the death of these cells was a normal event during development, with cell death being a specific developmental fate. It was also notable that all of these dying cells underwent a similar series of morphological changes, suggesting that these programmed cell deaths occurred by a common mechanism.

Based on these considerations, Horvitz undertook a genetic analysis with the goal of characterizing the mechanism and regulation of programmed cell death during *C. elegans* development. In the experiments reported in this 1986 paper, Hilary Ellis and Horvitz identified two genes that were required for all of the pro-

grammed cell deaths that took place during development of the nematode. The identification and characterization of these genes was a critical first step leading to our current understanding of the molecular biology of apoptosis.

The Experiments

Cells undergoing programmed cell death in *C. elegans* can readily be identified as highly refractile cells by microscopic examination, so Ellis and Horvitz were able to use this as an assay to screen for mutant animals in which the normal cell deaths did not occur. To isolate mutants that displayed abnormalities in cell death, they treated nematodes with the chemical mutagen ethyl methanesulfonate, which reacts with DNA. The progeny of approximately 4000 worms were examined to identify dying cells, and two mutant strains were found in which the expected cell deaths did not take place (see figure). Both of these mutant strains harbored recessive mutations of the same gene, which was called *ced-3*. Further studies indicated that the mutations in *ced-3* blocked all of the 131 programmed cell deaths that would normally occur during development.

Continuing studies identified an additional mutation that was similar to *ced-3* in preventing programmed cell death. However, this mutation was in a different gene, which was located on a different chromosome than *ced-3*. This

Photomicrographs of a normal worm (A) and a *ced-3* mutant (B). Dying cells are highly refractile and are indicated by arrows in panel A. These cells are not present in the mutant animal.

second gene was called *ced-4*. Similar to mutations in *ced-3*, recessive mutations in *ced-4* were found to block all programmed cell deaths in the worm.

The Impact

The isolation of *C. elegans* mutants by Ellis and Horvitz provided the first identification of genes that were involved in the process of programmed cell death. The proteins encoded by the *ced-3* and *ced-4* genes, as well as by the *ced-9* gene (which was subsequently identified by Horvitz and colleagues), proved to be prototypes of the central regulators

(A)

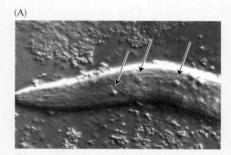

(B)

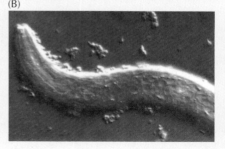

and effectors of apoptosis that are highly conserved in evolution. The cloning and sequencing of *ced-3* revealed that it was related to a protease that had been previously identified in mammalian cells, and which became the first member of the caspase family. The *C. elegans ced-9* gene was related to the *bcl-2* oncogene, first isolated from a human B cell lym-

phoma, which had the unusual property of inhibiting apoptosis rather than stimulating cell proliferation. And *ced-4* was found to encode an adaptor protein related to mammalian Apaf-1, which is required for caspase activation. The identification of these genes in *C. elegans* thus led the way to understanding the molecular basis of apoptosis, with broad

implications both for development and for the maintenance of normal adult tissues. Since abnormalities of apoptosis contribute to a wide variety of diseases, including cancer, autoimmune disease, and neurodegenerative disorders, the seminal findings of Ellis and Horvitz have impacted a wide range of areas in biology and medicine.

pase-9 cleaves and activates downstream effector caspases, such as caspase-3 and caspase-7, eventually resulting in cell death.

Central Regulators of Apoptosis: The Bcl-2 Family

The third gene identified as a key regulator of programmed cell death in *C. elegans, ced-9*, was found to be closely related to a mammalian gene called **bcl-2**, which was first identified in 1985 as an oncogene that contributed to the development of human B cell lymphomas (cancers of B lymphocytes). In contrast to other oncogenes proteins, such as Ras, that stimulate cell proliferation (see Molecular Medicine Chapter 15), Bcl-2 was found to inhibit apoptosis. Ced-9 and Bcl-2 were thus similar in function, and the role of Bcl-2 as a regulator of apoptosis first focused attention on the importance of cell survival in cancer development. As discussed further in the next chapter, we now recognize that cancer cells are generally defective in the normal process of programmed cell death and that their inability to undergo apoptosis is as important as their uncontrolled proliferation in the development of malignant tumors.

Mammals encode a family of approximately 20 proteins related to Bcl-2, which are divided into three functional groups (Figure 17.6). Some members of the Bcl-2 family (antiapoptotic family members)—like Bcl-2 itself—function as inhibitors of apoptosis and programmed cell death. Other members of the Bcl-2 family, however, are proapoptotic proteins that act to induce caspase activation and promote cell death. There are two groups of these proapoptotic proteins, which differ in function as well as in their extent of homology to Bcl-2. Bcl-2 and the other antiapoptotic family members share

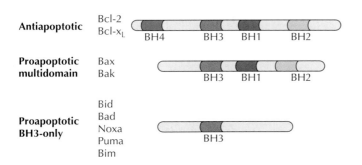

FIGURE 17.6 The Bcl-2 family The Bcl-2 family of proteins is divided into three functional groups. Antiapoptotic proteins (e.g., Bcl-2 and Bcl-x$_L$) have four Bcl-2 homology domains (BH1–BH4). The multidomain proapoptotic proteins (e.g., Bax and Bak) have three homology domains (BH1–BH3), whereas the BH3-only proapoptotic proteins (e.g., Bid, Bad, Noxa, Puma, and Bim) have only one homology domain (BH3).

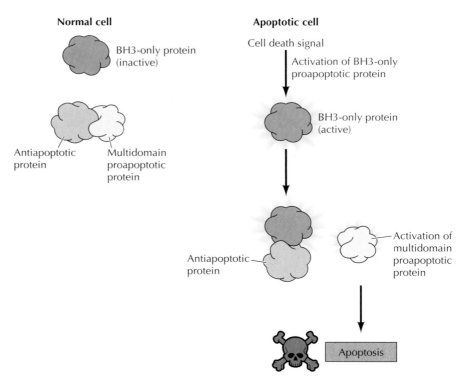

Normal cell

BH3-only protein (inactive)

Antiapoptotic protein

Multidomain proapoptotic protein

Apoptotic cell

Cell death signal

Activation of BH3-only proapoptotic protein

BH3-only protein (active)

Antiapoptotic protein

Activation of multidomain proapoptotic protein

Apoptosis

FIGURE 17.7 Regulatory interactions between Bcl-2 family members   In normal cells, the BH3-only proapoptotic proteins are inactive, and the multidomain proapoptotic proteins are inhibited by interaction with antiapoptotic proteins. Cell death signals activate the BH3-only proteins, which then interact with the antiapoptotic proteins, leading to activation of the multidomain proapoptotic proteins and cell death.

four conserved regions called Bcl-2 homology (BH) domains. One group of proapoptotic family members called the "multidomain" proapoptotic proteins have 3 BH domains (BH1, BH2, and BH3), whereas the second group, the "BH3-only" proteins, have only the BH3 domain.

The fate of the cell—life or death—is determined by the balance of activity of proapoptotic and antiapoptotic Bcl-2 family members, which act to regulate one another (Figure 17.7). The multidomain proapoptotic family members, such as Bax and Bak, are the downstream effectors that directly induce apoptosis. They are inhibited by interactions with the antiapoptotic family members, such as Bcl-2. The BH3-only family members are upstream members of the cascade, regulated by the signals that induce cell death (e.g., DNA damage) or cell survival (e.g., growth factors). When activated, the BH3-only family members antagonize the antiapoptotic Bcl-2 family members, activating the multidomain proapoptotic proteins and tipping the balance in favor of caspase activation and cell death.

In mammalian cells, members of the Bcl-2 family act at mitochondria, which play a central role in controlling programmed cell death (Figure 17.8). When activated, Bax and Bak form oligomers in the mitochondrial outer membrane. Formation of these Bax or Bak oligomers leads to the release of cytochrome *c* from the mitochondrial intermembrane space, either by forming pores or by interacting with other mitochondrial outer membrane pro-

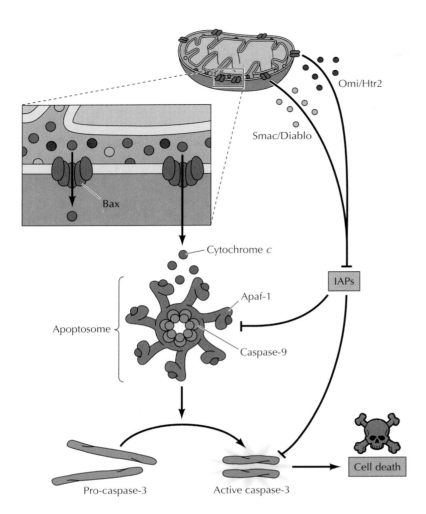

FIGURE 17.8 The mitochondrial pathway of apoptosis In mammalian cells, many cell death signals induce apoptosis as a result of damage to mitochondria. When active, the proapoptotic multidomain Bcl-2 family proteins (e.g., Bax) form oligomers in the outer membrane of mitochondria, resulting in the release of cytochrome *c* and other proapoptotic molecules from the intermembrane space. Release of cytochrome *c* leads to the formation of apoptosomes containing Apaf-1 and caspase-9 in which caspase-9 is activated. Caspase-9 then activates downstream caspases, such as caspase-3, by proteolytic cleavage. Smac/Diablo and Omi/Htr2 promote cell death by interfering with the action of IAPs, which are inhibitors of the caspases.

teins. The release of cytochrome *c* from mitochondria then triggers caspase activation. In particular, the key initiator caspase in mammalian cells (caspase-9) is activated by forming a complex with Apaf-1 in the apoptosome. In mammals, formation of this complex also requires cytochrome *c*. Under normal conditions of cell survival, cytochrome *c* is localized to the mitochondrial intermembrane space (see Figure 11.9) while Apaf-1 and caspase-9 are found in the cytosol, so caspase-9 remains inactive. Activation of Bax or Bak results in the release of cytochrome *c* to the cytosol, where it binds to Apaf-1 and triggers apoptosome formation and caspase-9 activation.

Caspases are also regulated by a family of proteins called the **IAP**, for *i*nhibitor of *ap*optosis, family. Members of the IAP family directly interact with caspases and suppress apoptosis by either inhibiting caspase activity or by targeting caspases for ubiquitination and degradation in the proteasome. IAPs are present in both *Drosophila* and mammals (but not *C. elegans*), and regulation of their activity or expression provides another mechanism for regulating apoptosis. In mammalian cells, the permeabilization of mitochondria by Bax or Bak results not only in the release of cytochrome *c* but also of other proteins that promote apoptosis. These include two proteins (called Smac/Diablo and Omi/Htr2) that stimulate caspase activity by interfering with the action of IAPs (see Figure 17.8). As discussed below,

■ **IAPs were first discovered in virus-infected insect cells as viral proteins that inhibited apoptosis of the host cell in response to viral infection.**

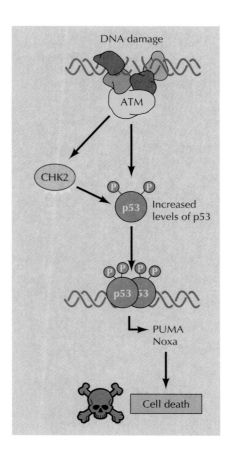

DNA damage

ATM

CHK2

p53 — Increased levels of p53

p53 53

PUMA
Noxa

Cell death

FIGURE 17.9 Role of p53 in DNA damage-induced apoptosis DNA damage leads to activation of the ATM and CHK2 protein kinases, which phosphorylate and stabilize p53 resulting in rapid increases in p53 levels. The protein p53 then activates transcription of genes encoding the proapoptotic BH3-only proteins PUMA and Noxa, leading to cell death.

IAPs as well as members of the Bcl-2 family are targets of the signaling pathways that control survival of mammalian cells.

Signaling Pathways that Regulate Apoptosis

Regulation of programmed cell death is mediated by the integrated activity of a variety of signaling pathways, some acting to induce cell death and others acting to promote cell survival. These signals control the fate of individual cells, so that cell survival or elimination is determined by the needs of the organism as a whole. As illustrated by the following examples, the multiple signaling pathways that regulate apoptosis converge on regulation of the BH3-only members of the Bcl-2 family.

One important role of apoptosis is the elimination of damaged cells. Cells with damaged genomes are particularly dangerous because of the possibility that they will have suffered mutations that may lead to the development of cancer. DNA damage is thus one of the principal triggers of programmed cell death, leading to the elimination of cells carrying potentially harmful mutations. As discussed in Chapter 16, several cell cycle checkpoints halt cell cycle progression in response to damaged DNA, allowing time for the damage to be repaired. In mammalian cells, a major pathway leading to cell cycle arrest in response to DNA damage is mediated by the transcription factor **p53**. The ATM and CHK protein kinases, which are activated by DNA damage, phosphorylate and stabilize p53. The resulting increase in p53 leads to transcriptional activation of p53 target genes. These include the Cdk inhibitor p21, which inhibits Cdk1/cyclin and Cdk2/cyclin complexes, halting cell cycle progression. However, activation of p53 by DNA damage can also lead to apoptosis (Figure 17.9). The induction of apoptosis by p53 results, at least in part, from transcriptional activation of genes encoding BH3-only propapoptotic Bcl-2 family members, such as PUMA and Noxa. Thus p53 mediates both cell cycle arrest and apoptosis in response to DNA damage. Whether DNA damage in a given cell leads to apoptosis or reversible cell cycle arrest may depend on the extent of damage and the resulting level of p53 induction, as well as the influence of other life/death signals being received by the cell.

Rather than inducing programmed cell death, many signaling pathways act to promote cell survival by inhibiting apoptosis. These signaling pathways control the fate of a wide variety of cells whose survival is dependent on extracellular growth factors or cell-cell interactions. As already noted, a well-characterized example of programmed cell death in development is provided by the vertebrate nervous system. About 50% of neurons die by apoptosis, with the survivors having received sufficient amounts of survival signals from their target cells. These survival signals are polypeptide growth factors related to nerve growth factor (NGF), which induces both neuronal survival and differentiation by activating a receptor protein-tyrosine kinase. Other types of cells are similarly dependent upon growth factors or cell contacts that activate nonreceptor protein-tyrosine kinases associated with integrins. Indeed, most cells in higher animals are programmed

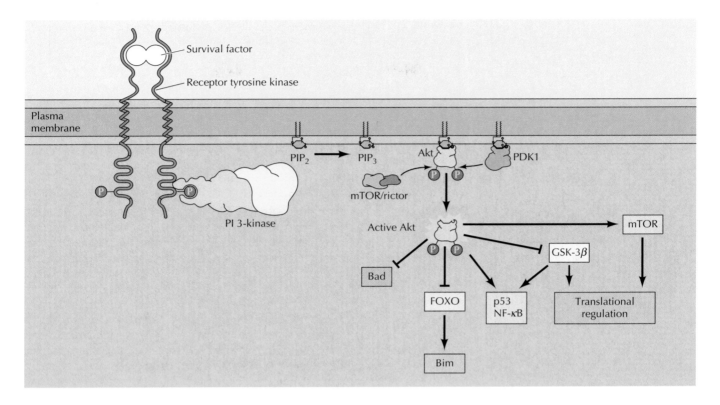

FIGURE 17.10 The PI 3-kinase pathway and cell survival Many growth factors that signal cell survival activate receptor protein-tyrosine kinases, leading to activation of PI 3-kinase, formation of PIP3, and activation of the protein kinase, Akt. Akt then phosphorylates a number of proteins that contribute to cell survival. Phosphorylation of the BH3-only protein Bad maintains it in an inactive state, as does phosphorylation of the FOXO transcription factor. In the absence of Akt signaling, activation of Bad promotes apoptosis and activation of FOXO stimulates transcription of another BH3-only protein, Bim. Additional targets of Akt that have been implicated in regulation of apoptosis include the protein kinase GSK-3β and additional transcription factors, such as p53 and NF-κB, both of which are regulated by Akt and GSK-3β phosphorylation. Translational regulation by GSK-3β and by the mTOR pathway (see Figure 15.33) may also affect cell survival.

to undergo apoptosis unless cell death is actively suppressed by survival signals from other cells.

One of the major intracellular signaling pathways responsible for promoting cell survival is initiated by the enzyme **PI 3-kinase**, which is activated by either protein-tyrosine kinases or G protein-coupled receptors. PI 3-kinase phosphorylates the membrane phospholipid PIP_2 to form PIP_3, which activates the protein-serine/threonine kinase **Akt** (see Figures 15.30 and 15.31). Akt then phosphorylates a number of proteins that regulate apoptosis (Figure 17.10). One key substrate for Akt is the proapoptotic BH3-only Bcl-2 family member called Bad. Phosphorylation of Bad by Akt creates a binding site for 14-3-3 chaperone proteins that sequester Bad in an inactive form, so phosphorylation of Bad by Akt inhibits apoptosis and promotes cell survival. Bad is similarly phosphorylated by protein kinases of other growth factor-induced signaling pathways, including the Ras/Raf/MEK/ERK pathway, so it serves as a convergent regulator of growth factor signaling in mediating cell survival.

Other targets of Akt, including the FOXO transcription factors, also play key roles in cell survival. Phosphorylation of FOXO by Akt creates a binding site for 14-3-3 proteins, which sequester FOXO in an inactive form in the cytoplasm (see Figure 15.32). In the absence of growth factor signaling and Akt activity, FOXO is released from 14-3-3 and translocates to the nucleus, stimulating transcription of proapoptotic genes, including the gene encoding the BH3-only protein, Bim. Akt and its downstream target GSK-3β also regulate other transcription factors with roles in cell survival, including p53 and NF-κB, which may control the expression of IAPs as well as Bcl-2 family members. Translational regulation by both GSK-3β and the mTOR pathway (see Figure 15.33) may also affect cell survival.

In contrast to the growth factor signaling pathways that promote cell survival, some secreted polypeptides signal programmed cell death by activating receptors that directly induce apoptosis of the target cell. These cell death signals are polypeptides belonging to the **tumor necrosis factor** (**TNF**) family. They bind to members of the TNF receptor family, which can signal apoptosis in a variety of cell types. One of the best characterized members of this family is the cell surface receptor called Fas, which plays important roles in controlling cell death in the immune system. For example, apoptosis induced by activation of Fas is responsible for killing target cells of the immune system, such as cancer cells or virus-infected cells, as well as for eliminating excess lymphocytes at the end of an immune response.

The cell death receptors signal apoptosis by directly activating an initiator caspase (Figure 17.11). TNF and related family members consist of three identical polypeptide chains, and their binding induces receptor trimerization. The cytoplasmic portions of the receptors bind adaptor molecules that in turn bind an initiator caspase called caspase-8. This leads to activation of caspase-8, which can then cleave and activate downstream effector caspases. In some cells, caspase-8 activation and subsequent activation of caspases-3 and -7 is sufficient to induce apoptosis directly. In other cells, however, amplification of the signal is needed. This results from caspase-8 cleavage of the proapoptotic BH3-only protein Bid, leading to Bid activation, permeabilization of mitochondria, and activation of caspase-9, thus amplifying the caspase cascade initiated by direct activation of caspase-8 at cell death receptors.

> ■ Therapies based on a member of the TNF family are under clinical trial for the treatment of certain cancers.

Stem Cells and the Maintenance of Adult Tissues

Early development is characterized by the rapid proliferation of embryonic cells, which then differentiate to form the specialized cells of adult tissues and organs. As cells differentiate, their rate of proliferation usually decreases, and most cells in adult animals are arrested in the G_0 stage of the cell cycle. However, cells are lost either due to injury or programmed cell death throughout life. In order to maintain a constant number of cells in adult tissues and organs, cell death must be balanced by cell proliferation. In order to maintain this balance, most tissues contain cells that are able to proliferate as required to replace cells that have died. Moreover, in some tissues a subpopulation of cells divide continuously throughout life to replace cells that have a high rate of turnover in adult animals. Cell death and cell renewal are thus carefully balanced to maintain properly sized and functioning adult tissues and organs.

Proliferation of Differentiated Cells

Most types of differentiated cells in adult animals are no longer capable of proliferation. If these cells are lost, they are replaced by the proliferation of less differentiated cells derived from self-renewing stem cells, as discussed in the following section. Other types of differentiated cells, however, retain the ability to proliferate as needed to repair damaged tissue throughout the life of the organism. These cells enter the G_0 stage of the cell cycle but resume proliferation as needed to replace cells that have been injured or have died.

Cells of this type include fibroblasts, which are dispersed in connective tissues where they secrete collagen (Figure 17.12). Skin fibroblasts are normally arrested in G_0 but rapidly proliferate if needed to repair damage

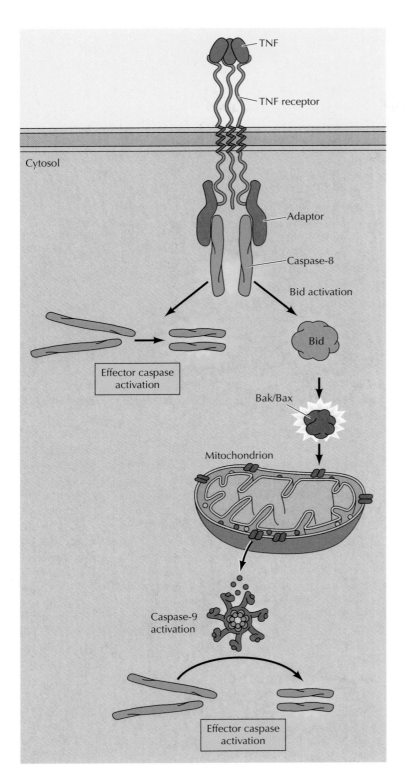

FIGURE 17.11 Cell death receptors TNF and other cell death receptor ligands consist of three polypeptide chains, so their binding to cell death receptors induces receptor trimerization. Caspase-8 is recruited to the receptor and activated via interaction with adaptor molecules. Once activated, caspase-8 can directly cleave and activate effector caspases. In addition, caspase-8 cleaves the BH3-only protein Bid, which activates the mitochondrial pathway of apoptosis, leading to caspase-9 activation.

FIGURE 17.12 Skin fibroblasts
Scanning electron micrograph of a fibroblast surrounded by collagen fibrils. (© CMEABG-UCBL/Photo Researchers Inc.)

resulting from a cut or wound. Blood clotting at the site of a wound leads to the release of platelet-derived growth factor (PDGF) from blood platelets. As discussed in Chapter 15, PDGF activates a receptor protein-tyrosine kinase, stimulating both the proliferation of fibroblasts and their migration into the wound where their proliferation and secretion of collagen contributes to repair and regrowth of the damaged tissue.

The endothelial cells that line blood vessels (Figure 17.13) are another type of fully differentiated cell that remains capable of proliferation. Prolifera-

FIGURE 17.13 Endothelial cells
Electron micrograph of a capillary. The capillary is lined by a single endothelial cell surrounded by a thin basal lamina. (© Dr. Don W. Fawcett/Visuals Unlimited.)

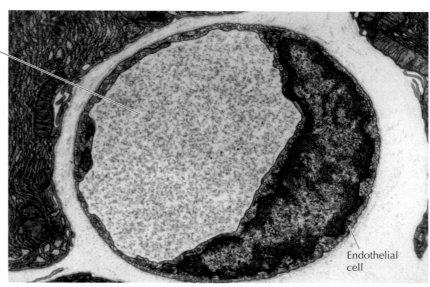

Capillary lumen

Endothelial cell

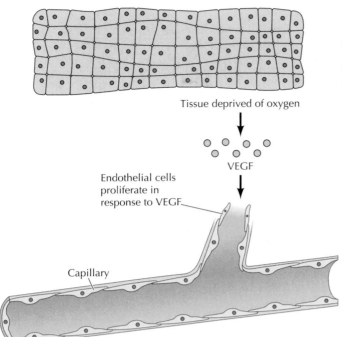

FIGURE 17.14 Proliferation of endothelial cells
Endothelial cells are stimulated to proliferate by vascular endothelial growth factor (VEGF). VEGF is secreted by cells deprived of oxygen, leading to the outgrowth of new capillaries into tissues lacking an adequate blood supply.

tion of endothelial cells allows them to form new blood vessels as needed for repair and regrowth of damaged tissue. Endothelial cell proliferation and the resulting formation of blood capillaries is triggered by a growth factor (vascular endothelial growth factor or VEGF) produced by cells of the tissue, which the new capillaries will invade. The production of VEGF is in turned triggered by a lack of oxygen, so the result is a regulatory system in which tissues that have a low oxygen supply resulting from insufficient circulation stimulate endothelial cell proliferation and recruit new capillaries (Figure 17.14). Smooth muscle cells, which form the walls of larger blood vessels (e.g., arteries) as well as the contractile portions of the digestive and respiratory tracts and other internal organs, are also capable of resuming proliferation in response to growth factor stimulation. In contrast, differentiated skeletal and cardiac muscle cells are no longer able to divide.

The epithelial cells of some internal organs, such as the liver and pancreas, are also able to proliferate to replace damaged tissue. A striking example is provided by liver cells, which are normally arrested in the G_0 phase of the cell cycle. However, if large numbers of liver cells are lost (e.g., by surgical removal of part of the liver), the remaining cells are stimulated to proliferate to replace the missing tissue (Figure 17.15). For example, surgical removal of two-thirds of the liver of a rat is followed by rapid proliferation of the remaining cells, leading to regeneration of the entire liver within a few days.

Stem Cells

Most fully differentiated cells in adult animals, however, are no longer capable of cell division. Nonetheless, they can be replaced by the proliferation of a subpopulation of less differentiated self-renewing cells called **stem cells** that are present

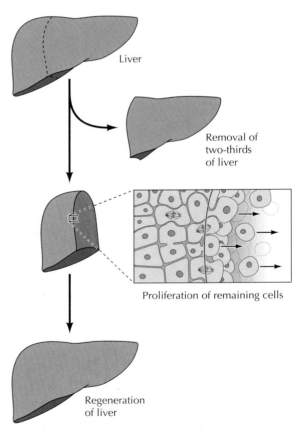

FIGURE 17.15 Liver regeneration Liver cells are normally arrested in G_0 but resume proliferation to replace damaged tissue. If two-thirds of the liver of a rat is surgically removed, the remaining cells proliferate to regenerate the entire liver in a few days.

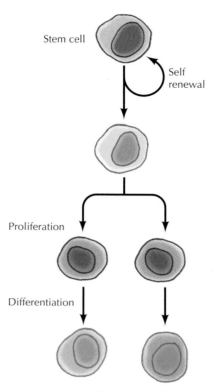

Stem cell

Self renewal

Proliferation

Differentiation

Differentiated cells

FIGURE 17.16 Stem cell proliferation Stem cells divide to form one daughter cell that remains a stem cell and a second that proliferates and then differentiates.

in most adult tissues. Because they retain the capacity to proliferate and replace differentiated cells throughout the lifetime of an animal, stem cells play a critical role in the maintenance of most tissues and organs.

The key property of stem cells is that they divide to produce one daughter cell that remains a stem cell and one that divides and differentiates (Figure 17.16). Because the division of stem cells produces new stem cells as well as differentiated daughter cells, stem cells are self-renewing populations that can serve as a source for the production of differentiated cells throughout life. The role of stem cells is particularly evident in the case of several types of differentiated cells, including blood cells, epithelial cells of the skin, and epithelial cells lining the digestive tract—all of which have short life spans and must be replaced by continual cell proliferation in adult animals. In all of these cases, the fully differentiated cells do not themselves proliferate; instead, they are continually renewed by the proliferation of stem cells that then differentiate to maintain a stable number of differentiated cells. Stem cells have also been identified in a variety of other adult tissues, including skeletal muscle and the nervous system, where they may function to replace damaged tissue.

Stem cells were first identified in the hematopoietic (blood-forming) system by Ernest McCulloch and James Till in 1961 in experiments showing that single cells derived from mouse bone marrow could proliferate and give rise to multiple differentiated types of blood cells. Hematopoietic stem cells are well-characterized and the production of blood cells provides a good example of the role of stem cells in maintaining differentiated cell populations. There are several distinct types of blood cells with specialized functions: erythrocytes (red blood cells) that transport O_2 and CO_2; granulocytes and macrophages, which are phagocytic cells; platelets (which are fragments of megakaryocytes) that function in blood coagulation; and lymphocytes that are responsible for the immune response. All these cells have limited life spans ranging from less than a day to a few months, and all are derived from the same population of hematopoietic stem cells. More than 100 billion blood cells are lost every day in humans, and must be continually produced from hematopoietic stem cells in the bone marrow (Figure 17.17). Descendants of the hematopoietic stem cell continue to proliferate and undergo several rounds of division as they become committed to specific differentiation pathways that are determined by growth factors that channel precursor cells along specific pathways of blood cell differentiation. Once they become fully differentiated, blood cells cease proliferation, so the maintenance of differentiated blood cell populations is dependent on continual division of the self-renewing hematopoietic stem cell.

The intestine provides an excellent example of stem cells in the self-renewal of an epithelial tissue. The intestine is lined by a single layer of epithelial cells that are responsible for the digestion of food and absorption of nutrients. These intestinal epithelial cells are exposed to an extraordinarily harsh environment and have a lifetime of only a few days before they die by apoptosis and are shed into the digestive tract. Renewal of the intestinal epithelium is therefore a continual process throughout life. New cells are derived from the continuous but slow division of stem cells at the bot-

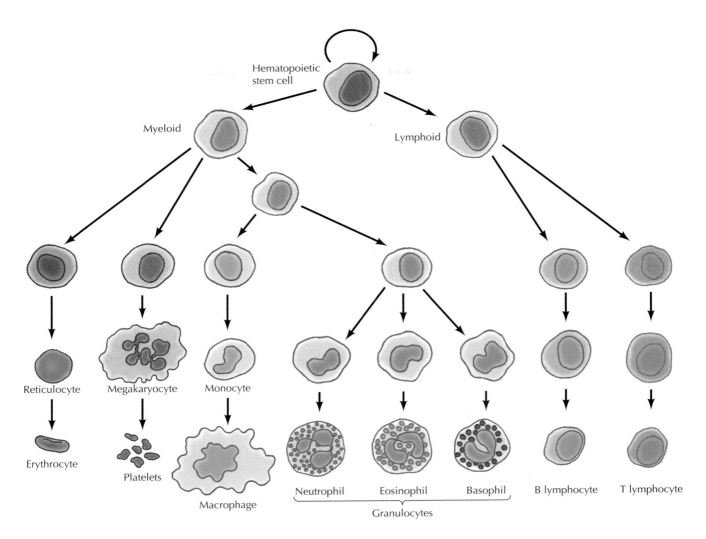

FIGURE 17.17 Formation of blood cells All of the different types of blood cells develop from a hematopoietic stem cell in the bone marrow. The precursors of differentiated cells undergo several rounds of cell division before they differentiate.

tom of intestinal crypts (Figure 17.18). The stem cells give rise to a population of transit-amplifying cells, which divide rapidly and occupy about two-thirds of the crypt. The transit-amplifying cells proliferate for 3–4 cell divisions and then differentiate into the three cell types of the colon surface epithelium: absorptive epithelial cells and two types of secretory cells, called goblet cells and enteroendocrine cells. The small intestine also contains a fourth cell type, Paneth cells, which secrete antibacterial agents. It is noteworthy that all of the cells in an adult crypt are derived from the continuous division of a single self-renewing stem cell.

Stem cells that are responsible for continuous renewal of the skin and hair are also well-characterized. Like the lining of the intestine, the skin and hair are exposed to a harsh external environment—including ultraviolet radiation from sunlight—and are continuously renewed throughout life. In humans, the epidermis turns over every two weeks, with cells being

(A)

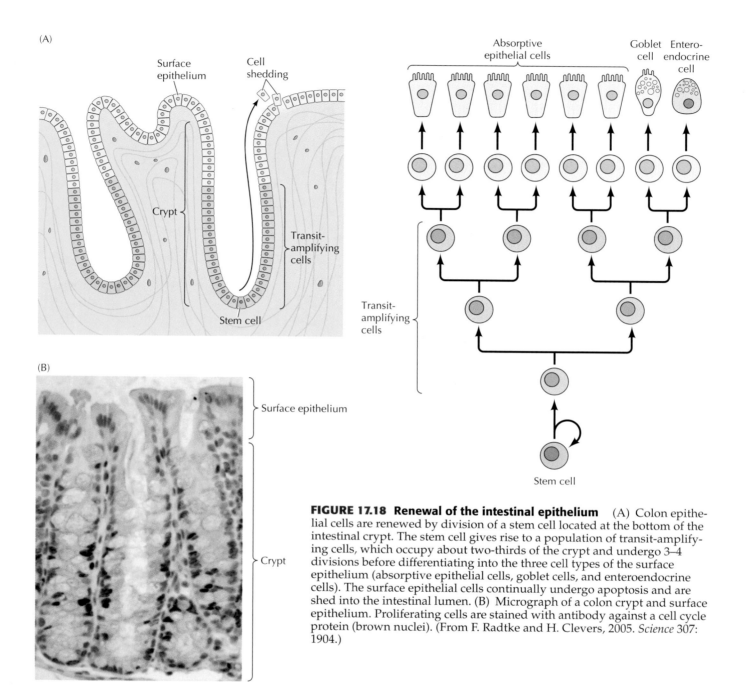

FIGURE 17.18 Renewal of the intestinal epithelium (A) Colon epithelial cells are renewed by division of a stem cell located at the bottom of the intestinal crypt. The stem cell gives rise to a population of transit-amplifying cells, which occupy about two-thirds of the crypt and undergo 3–4 divisions before differentiating into the three cell types of the surface epithelium (absorptive epithelial cells, goblet cells, and enteroendocrine cells). The surface epithelial cells continually undergo apoptosis and are shed into the intestinal lumen. (B) Micrograph of a colon crypt and surface epithelium. Proliferating cells are stained with antibody against a cell cycle protein (brown nuclei). (From F. Radtke and H. Clevers, 2005. *Science* 307: 1904.)

sloughed from the surface. The epidermis is a multi-layered epithelium, which is maintained by epidermal stem cells residing in a single basal layer (Figure 17.19). The epidermal stem cells give rise to transit-amplifying cells, which undergo 3–6 divisions before differentiating and moving outward to the surface of the skin. The stem cells responsible for producing hair reside in a region of the hair follicle called the bulge. The bulge stem cells give rise to transit-amplifying matrix cells, which proliferate and differentiate to form the hair shaft. If the skin is injured, stem cells of the bulge can also give rise to epidermis, demonstrating their activity as multipotent stem cells from which both skin and hair can be derived.

Epidermis

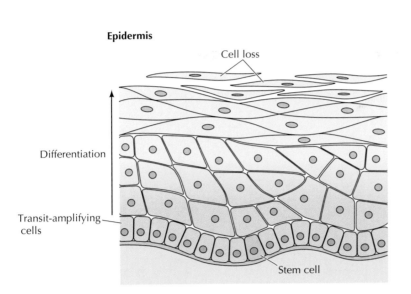

Cell loss

Differentiation

Transit-amplifying cells

Stem cell

Hair follicle

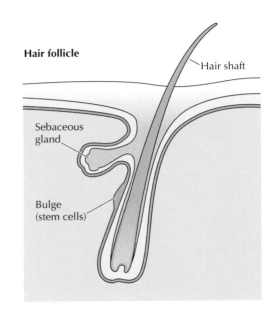

Hair shaft

Sebaceous gland

Bulge (stem cells)

FIGURE 17.19 Stem cells of the skin The epidermis consists of multiple layers of epithelial cells. Cells from the surface are continually lost and replaced by epidermal stem cells in the basal layer. The stem cell gives rise to transit-amplifying cells, which undergo several divisions in the basal layer before differentiating and moving to the surface of the skin. Stem cells of hair follicles reside in a region beneath the sebaceous gland called the bulge. The bulge stem cells not only give rise to cells that form the hair shaft but can also give rise to epidermis if the skin is injured.

Skeletal muscle provides an example of the role of stem cells in the repair of damaged tissue, in contrast to the continual cell renewal just described in the hematopoietic system, intestinal epithelium, and skin. Skeletal muscle is composed of large multinucleated cells (muscle fibers) formed by cell fusion during development (see Figure 12.21). Although skeletal muscle is normally a stable tissue with little cell turnover, it is able to regenerate rapidly in response to injury or exercise. This regeneration is mediated by proliferation of satellite cells, which are the stem cells of adult muscle. Satellite cells are located beneath the basal lamina of muscle fibers (Figure 17.20). They are normally quiescent, arrested in the G_0 phase of the cell cycle, but are activated to proliferate in response to injury or exercise. Once activated, the satellite cells give rise to progeny that undergo several divisions and then differentiate and fuse to form new muscle fibers. The continuing capacity of skeletal muscle to regenerate throughout life is due to self-renewal of the satellite stem cell population.

Stem cells have also been identified in many other adult tissues, including the brain, retina, heart, lung, kidney, liver, and pancreas, and it is possible that most—if not all—tissues contain stem cells with the potential of replacing cells that are lost during the lifetime of the organism. It is also possible that stem cells from one tissue might be able to differentiate into cells giving rise to other tissues—a phenomenon that has been called "developmental plasticity." For example, hematopoietic stem cells derived from bone marrow have been reported to give rise to nonhematopoietic tissues, such as skin or lung. However, the extent to which stem cells from

(A)

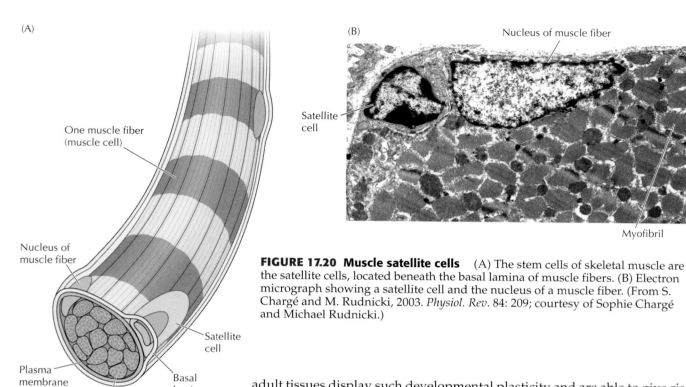

One muscle fiber
(muscle cell)

Nucleus of
muscle fiber

Plasma
membrane

Satellite
cell

Basal
lamina

Myofibril

(B)

Nucleus of muscle fiber

Satellite
cell

Myofibril

FIGURE 17.20 Muscle satellite cells (A) The stem cells of skeletal muscle are the satellite cells, located beneath the basal lamina of muscle fibers. (B) Electron micrograph showing a satellite cell and the nucleus of a muscle fiber. (From S. Chargé and M. Rudnicki, 2003. *Physiol. Rev.* 84: 209; courtesy of Sophie Chargé and Michael Rudnicki.)

adult tissues display such developmental plasticity and are able to give rise to other tissue types is an area of controversy that remains to be clarified by further research.

Medical Applications of Adult Stem Cells

The ability of adult stem cells to repair damaged tissue clearly suggests their potential utility in clinical medicine. If these stem cells could be isolated and propagated in culture, they could in principal be used to replace damaged tissue and treat a variety of disorders, such as diabetes or neurodegenerative diseases like Parkinson's or Alzheimer's disease. In some cases, the use of stem cells derived from adult tissues may be the optimal approach for such stem cell therapies, although the use of embryonic stem cells (discussed in the next section of this chapter) is likely to provide a more versatile approach to treatment of a wider variety of disorders.

A well-established clinical application of adult stem cells is **bone marrow transplantation**, which plays an important role in the treatment of a variety of cancers. As discussed in Chapter 18, most cancers are treated by chemotherapy with drugs that kill rapidly dividing cells by damaging DNA or inhibiting DNA replication. These drugs do not act selectively against cancer cells but are also toxic to those normal tissues that are dependent on continual renewal by stem cells, such as blood, skin, hair, and the intestinal epithelium. The hematopoietic stem cells are among the most rapidly dividing cells of the body, so the toxic effects of anticancer drugs on these cells frequently limit the effectiveness of chemotherapy in cancer treatment. Bone marrow transplantation provides an approach to bypassing this toxicity, thereby allowing the use of higher drug doses to treat the patient's cancer more effectively. In this procedure, the patient is treated with high doses of chemotherapy that would normally not be tolerated because of toxic effects on the hematopoietic system (Figure 17.21). The potentially lethal damage is repaired, however, by transferring new

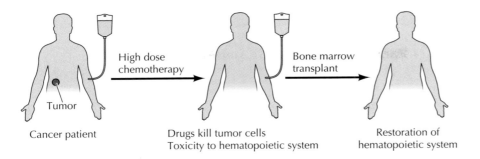

FIGURE 17.21 Bone marrow transplantation A cancer patient is treated with high doses of chemotherapy, which effectively kill tumor cells but normally would not be tolerated because of potentially lethal damage to the hematopoietic system. This damage is then repaired by transplantation of new hematopoietic stem cells (e.g., from bone marrow).

hematopoietic stem cells (obtained either from bone marrow or peripheral blood) to the patient following completion of chemotherapy, so that a normal hematopoietic system is restored. In some cases, the stem cells are obtained from the patient prior to chemotherapy, stored, and then returned to the patient once chemotherapy is completed. However, it is important to ensure that these cells are not contaminated with cancer cells. Alternatively, the stem cells to be transplanted can be obtained from a healthy donor (usually a close relative) whose tissue type closely matches the patient. In addition to their use in cancer treatment, hematopoietic stem cell transfers are used to treat patients with diseases of the hematopoietic system, such as aplastic anemia, hemoglobin disorders, and immune deficiencies.

Epithelial stem cells have also found clinical application in the form of skin grafts that are used to treat patients with burns, wounds, and ulcers. One approach to these procedures is to culture epidermal skin cells to form an epithelial sheet, which can then be transferred to the patient. Because the patient's own skin can be used for this procedure, it eliminates the potential complication of graft rejection by the immune system. The possibilities of using adult stem cells for similar replacement therapies of other diseases, including diabetes and Parkinson's disease, are being actively pursued. However, these clinical applications of adult stem cells are limited by the difficulties in isolating and culturing the appropriate stem cell populations, as well as by the fact that stem cells have not yet been identified in many adult tissues.

> ■ Stored umbilical cord blood from an unrelated donor can also be used as a source of hematopoietic stem cells for transplantation.

Embryonic Stem Cells and Therapeutic Cloning

While adult stem cells are difficult to isolate and culture, it is relatively straightforward to isolate and propagate the stem cells of early embryos (**embryonic stem cells**). These cells can be grown indefinitely as pure stem cell populations while maintaining the ability to give rise to all of the differentiated cell types of adult organisms. Consequently, there has been an enormous interest in embryonic stem cells from the standpoints of both basic science and clinical applications.

KEY EXPERIMENT

Culture of Embryonic Stem Cells

Isolation of a Pluripotent Cell Line from Early Mouse Embryos Cultured in Medium Conditioned by Teratocarcinoma Stem Cells

Gail R. Martin

University of California, San Francisco, CA

Proceedings of the National Academy of Science, USA, 1981, Volume 78, pages 7634–7638

Gail R. Martin

The Context

The cells of early embryos are unique in their ability to proliferate and differentiate into all of the types of cells that make up the tissues and organs of adult animals. In 1970 it was found that early mouse embryos frequently developed into tumors if they were removed from the uterus and transplanted to an abnormal site. These tumors, called teratocarcinomas, contained cells that were capable of forming an array of different tissues as they grew within the animal. In addition, cells from teratocarcinomas (called embryonal carcinoma cells) could be isolated and grown in tissue culture. These cells resembled normal embryo cells and could be induced to differentiate into a variety of cell types in culture. Some embryonal carcinoma cells could also participate in normal development of a mouse if they were injected into early mouse

embryos (blastocysts) that were then implanted into a foster mother.

The ability of embryonal carcinoma cells to differentiate into a variety of cell types and to participate in normal mouse development suggested that these tumor-derived cells might be closely related to normal embryonic stem cells. However, the events that occurred during the establishment of teratocarcinomas in mice were unknown. Gail Martin hypothesized that the embryonal carcinoma cells found in teratocarcinomas were essentially normal embryo cells that proliferated abnormally simply because, when they were removed from the uterus and transplanted to an abnormal site, they did not receive the appropriate signals to induce normal differentiation. Based on this hypothesis, she attempted to culture cells from mouse embryos with the goal of isolating normal embryonic stem cell lines.

Her experiments, together with similar work by Martin Evans and Matthew Kaufman (Establishment in culture of pluripotential cells from mouse embryos, *Nature*, 1981, 292: 154–156), demonstrated that stem cells could be cultured directly from normal mouse embryos. The isolation of these embryonic stem cell lines paved the way to genetic manipulation and analysis of mouse development, as well as to the possible use of human embryonic stem cells in transplantation therapy.

The Experiments

Based on the premise that embryonal carcinoma cells were derived from normal embryonic stem cells, Martin attempted to culture cells from normal

Embryonic Stem Cells

Embryonic stem cells were first cultured from mouse embryos in 1981 (Figure 17.22). They can be propagated indefinitely in culture and, if reintroduced into early embryos, are able to give rise to cells in all tissues of the mouse. Thus they retain the capacity to develop into all of the different types of cells in adult tissues and organs. In addition, they can be induced to differentiate into a variety of different types of cells in culture.

As discussed in Chapter 4, mouse embryonic stem cells have been an important experimental tool in cell biology because they can be used to introduce altered genes into mice (see Figure 4.36). Moreover, they provide an outstanding model system for studying the molecular and cellular events associated with embryonic cell differentiation, so embryonic stem cells have long been of considerable interest to cell and developmental biol-

KEY EXPERIMENT

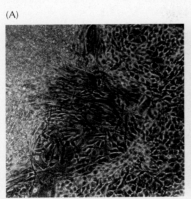

(A)

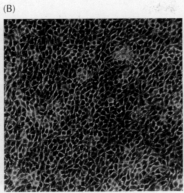

(B)

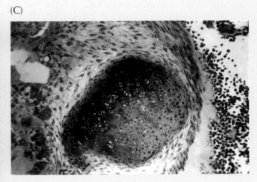

(C)

Embryonic stem cells differentiate in culture to a variety of cell types, including neuron-like cells (A), endodermal cells (B), and cartilage (C).

mouse blastocysts. Starting with cells from approximately 30 embryos, she initially isolated four colonies of growing cells after a week of culture. These cells could be repeatedly passaged into mass cultures, and new cell lines could be reproducibly derived when the experiment was repeated with additional mouse embryos.

The cell lines derived from normal embryos (embryonic stem cells) closely resembled the embryonal carcinoma cells derived from tumors. Most importantly, the embryonic stem cells could be induced to differentiate in culture into a variety of cell types, including endodermal cells, cartilage, and neuron-like cells (see figure). Moreover, if the embryonic stem cells were injected into a mouse, they

formed tumors containing multiple differentiated cell types. It thus appeared that embryonic stem cell lines, which retained the ability to differentiate into a wide array of cell types, could be established in culture from normal mouse embryos.

The Impact

The establishment of embryonic stem cell lines has had a major impact on studies of mouse genetics and development as well as opening new possibilities for the treatment of a variety of human diseases. Subsequent experiments demonstrated that embryonic stem cells could participate in normal mouse development following their injection into mouse embryos. Since gene transfer techniques could be used to introduce or mutate genes in cultured embryonic stem cells, these cells have been used to investigate the role of a variety of genes in mouse devel-

opment. As discussed in Chapter 4, any gene of interest can be inactivated in embryonic stem cells by homologous recombination with a cloned DNA, and the role of that gene in mouse development can then be determined by introducing the altered embryonic stem cells into mouse embryos.

In 1998 two groups of researchers developed the first lines of human embryonic stem cells. Because of the proliferative and differentiative capacity of these cells, they offer the hope of providing new therapies for the treatment of a variety of diseases. Although a number of technical problems and ethical concerns need to be addressed, transplantation therapies based on the use of embryonic stem cells may provide the best hope for eventual treatment of diseases such as Parkinson's, Alzheimer's, diabetes, and spinal cord injuries.

(A)

Embryo

Culture of embryonic stem cells

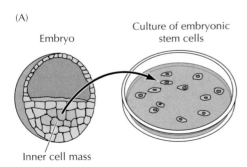

Inner cell mass

(B)

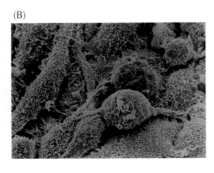

FIGURE 17.22 Culture of mammalian embryonic stem cells
(A) Embryonic stem cells are cultured from the inner cell mass of an early embryo (blastocyst). (B) Scanning electron micrograph of cultured embryonic stem cells. (Yorgos Nikas/Photo Researchers Inc.)

FIGURE 17.23 Differentiation of embryonic stem cells Mouse embryonic stem cells (ES cells) are maintained in the undifferentiated state in the presence of LIF. If LIF is removed from the culture medium, the cells aggregate to form embryoid bodies and then differentiate into a variety of cell types.

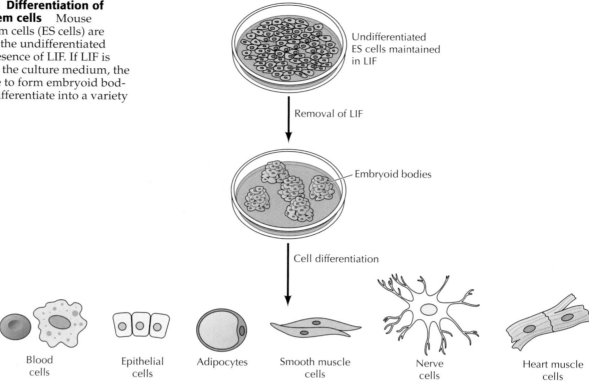

Undifferentiated ES cells maintained in LIF

Removal of LIF

Embryoid bodies

Cell differentiation

Blood cells   Epithelial cells   Adipocytes   Smooth muscle cells   Nerve cells   Heart muscle cells

ogists. Interest in these cells reached a new peak of intensity, however, in 1998 when two groups of researchers reported the isolation of stem cells from human embryos, raising the possibility of using embryonic stem cells in clinical transplantation therapies.

Mouse embryonic stem cells are grown in the presence of a growth factor called LIF (for *leukemia inhibitory factor*), which signals through the JAK/STAT pathway (see Figure 15.40) and is required to maintain these cells in their undifferentiated state (Figure 17.23). If LIF is removed from the medium, the cells aggregate into structures that resemble embryos (embryoid bodies) and then differentiate into a wide range of cell types, including neurons, adipocytes, blood cells, epithelial cells, vascular smooth muscle cells, and even beating heart muscle cells. Human embryonic stem cells do not require LIF but are similarly maintained in the undifferentiated state by other growth factors, which are not yet fully characterized.

Importantly, embryonic stem cells can be directed to differentiate along specific pathways by the addition of appropriate growth factors to the culture medium. It may thus be possible to derive populations of specific types of cells, such as heart cells or nerve cells, for transplantation therapy. For example, methods have been developed to direct the differentiation of mouse embryonic stem cells into neurons that have then been used for transplantation therapy in rodent models of myelin disease and Parkinson's disease. A great deal of current research is therefore focused on the development of culture conditions to promote the differentiation of embryonic stem cells along specific pathways, thereby producing populations of differentiated cells that can be used for transplantation therapy of a variety of diseases.

Somatic Cell Nuclear Transfer

The isolation of human embryonic stem cells in 1998 followed the first demonstration that the nucleus of an adult mammalian cell could give rise to a viable cloned animal. In 1997 Ian Wilmut and his colleagues initiated a new era of regenerative medicine with the cloning of Dolly the sheep (Figure 17.24). Dolly arose from the nucleus of a mammary epithelial cell that was transplanted into an unfertilized egg in place of the normal egg nucleus—a process called **somatic cell nuclear transfer**. It is interesting to note that this type of experiment was first carried out in frogs in the 1950s. The fact that it took over forty years before it was successfully performed in mammals attests to the technical difficulty of the procedure. Since the initial success of Wilmut and his colleagues, transfer of nuclei from adult somatic cells into enucleated eggs has been used to create cloned offspring of a variety of

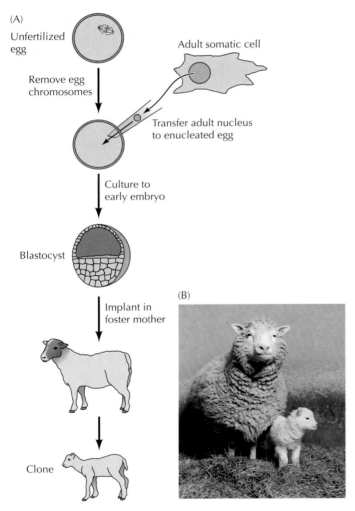

(A)
Unfertilized egg

Remove egg chromosomes

Adult somatic cell

Transfer adult nucleus to enucleated egg

Culture to early embryo

Blastocyst

Implant in foster mother

(B)

Clone

FIGURE 17.24 Cloning by somatic cell nuclear transfer (A) The nucleus of an adult somatic cell is transferred to an unfertilized egg from which the normal egg chromosomes have been removed (an enucleated egg). The egg is then cultured to an early embryo and transferred to a foster mother, who then gives birth to a clone of the donor of the adult nucleus. (B) Dolly (the adult sheep, left) was the first cloned mammal. She is shown with her lamb, Bonnie, who was produced by normal reproduction. (Photograph by Roddy Field; courtesy of T. Wakayama and R. Yanagimachi.)

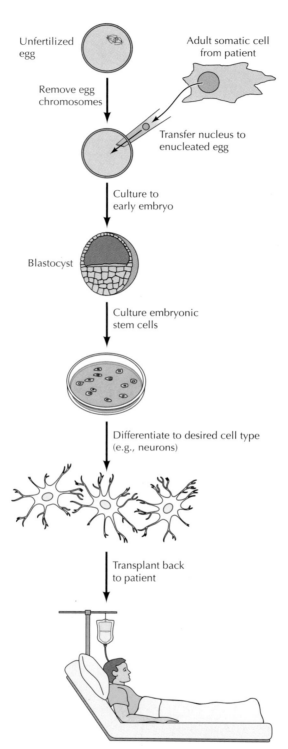

Unfertilized egg

Adult somatic cell from patient

Remove egg chromosomes

Transfer nucleus to enucleated egg

Culture to early embryo

Blastocyst

Culture embryonic stem cells

Differentiate to desired cell type (e.g., neurons)

Transplant back to patient

FIGURE 17.25 Therapeutic cloning In therapeutic cloning, the nucleus of a patient's cell would be transferred to an enucleated egg, which would be cultured to an early embryo. Embryonic stem cells would then be derived, differentiated into the desired cell type and transplanted back into the patient. The transplanted cells would be genetically identical to the recipient (who was the donor of the adult nucleus), so complications of immune rejection would be avoided.

mammalian species, including sheep, mice, pigs, cattle, goats, rabbits, and cats. However, cloning by somatic cell nuclear transfer in mammals remains an extremely inefficient procedure, such that only 1–2% of embryos generally give rise to live offspring.

Animal cloning by somatic cell nuclear transfer, together with the properties of embryonic stem cells, opens the possibility of **therapeutic cloning** (Figure 17.25). In therapeutic cloning, a nucleus from an adult human cell would be transferred to an enucleated egg, which would then be used to produce an early embryo in culture. Embryonic stem cells could then be cultured from the cloned embryo and used to generate appropriate types of differentiated cells for transplantation therapy. The major advantage provided by therapeutic cloning is that the embryonic stem cells derived by this procedure would be genetically identical to the recipient of the transplant, who was the donor of the adult somatic cell nucleus. This bypasses the barrier of the immune system in rejecting the transplanted tissue.

The application of therapeutic cloning combined with gene transfer has been demonstrated in mice by correction of an inherited immunodeficiency (Figure 17.26). The genetic defect treated in these experiments was the absence of a functional gene encoding the RAG2 protein, which is required for the gene rearrangements that produce functional antibodies and T cell receptors during the development of lymphocytes (see Figure 6.40). Nuclei from

FIGURE 17.26 Correction of an inherited immunodeficiency in mice by therapeutic cloning Fibroblasts from immunodeficient mice lacking RAG2 were used as donors for nuclear transfer and derivation of embryonic stem (ES) cells. The defect was then repaired by introduction of a cloned functional *RAG2* gene into the ES cells. The ES cells were then differentiated to hematopoietic stem cells and transferred back to the RAG2-deficient mice, at least partially correcting the immunodeficiency. (After W. M. Rideout et al., 2002. *Cell* 109: 17.)

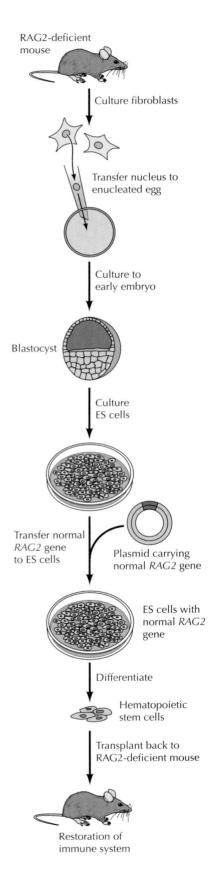

RAG2-deficient mouse

Culture fibroblasts

Transfer nucleus to enucleated egg

Culture to early embryo

Blastocyst

Culture ES cells

Transfer normal *RAG2* gene to ES cells

Plasmid carrying normal *RAG2* gene

ES cells with normal *RAG2* gene

Differentiate

Hematopoietic stem cells

Transplant back to RAG2-deficient mouse

Restoration of immune system

fibroblasts of RAG2-deficient mice were transferred to an enucleated egg, which was used as a source of embryonic stem cells. A functional *RAG2* gene was then introduced into the embryonic stem cells in culture, correcting the genetic defect. These corrected embryonic stem cells were then differentiated into hematopoietic stem cells, and transplantation of these cells into RAG2-deficient mice was found to at least partially rescue the immunodeficiency.

The possibility of therapeutic cloning provides the most general approach to treatment of the wide variety of devastating disorders for which stem cell transplantation therapy could be applied, including Parkinson's disease, Alzheimer's disease, diabetes, and spinal cord injuries. Although some success has been achieved in animal models, there remain major obstacles that need to be overcome before therapeutic cloning could be applied to humans. Substantial improvements are needed in the methods used to generate embryos by nuclear transfer, and it will be necessary to develop procedures that reliably differentiate embryonic stem cells into the appropriate cell types prior to transplantation. The development of therapeutic cloning also raises ethical concerns, not only with respect to the possibility of cloning human beings (**reproductive cloning**), but also with respect to the destruction of embryos that serve as the source of embryonic stem cells. While many challenges remain, continuing research in this area holds great promise of opening new approaches to the treatment of a broad array of human diseases.

COMPANION WEBSITE

Visit the website that accompanies **The Cell** (www.sinauer.com/cooper) for animations, videos, quizzes, problems, and other review material.

<table>
<tr><td>

</td><td>

</td></tr>
</table>

	### PROGRAMMED CELL DEATH
programmed cell death, apoptosis	*The Events of Apoptosis:* Programmed cell death plays a key role in both the maintenance of adult tissues and embryonic development. In contrast to the accidental death of cells from an acute injury, programmed cell death takes place by the active process of apoptosis. Apoptotic cells and cell fragments are then efficiently removed by phagocytosis. Genes responsible for the regulation and execution of apoptosis were initially identified by genetic analysis of *C. elegans.*
caspase, apoptosome	*Caspases: The Executioners of Apoptosis:* The caspases are a family of proteases that are the effectors of apoptosis. Caspases are classified as either initiator or effector caspases, and both function in a cascade leading to cell death. In mammalian cells, the major initiator caspase is activated in a complex called the apoptosome, which also requires cytochrome *c* released from mitochondria.
Bcl-2, IAP	*Central Regulators of Apoptosis: the Bcl-2 Family:* Members of the Bcl-2 family are central regulators of caspase activation and apoptosis. Some members of the Bcl-2 family function to inhibit apoptosis (antiapoptotic) whereas others act to promote apoptosis (proapoptotic). Signals that control programmed cell death alter the balance between proapoptotic and antiapoptotic Bcl-2 family members, which regulate one another. In mammalian cells, proapoptotic Bcl-2 family members act at mitochondria, where they promote the release of cytochrome *c*, leading to caspase activation. Caspases are also regulated directly by inhibitory IAP proteins.
p53, PI 3-kinase, Akt, tumor necrosis factor (TNF)	*Signaling Pathways that Regulate Apoptosis:* A variety of signaling pathways regulate apoptosis by controlling the expression or activity of proapoptotic members of the Bcl-2 family. These pathways include DNA damage-induced activation of the tumor suppressor p53, growth factor-stimulated activation of PI 3-kinase/Akt signaling, and activation of death receptors by polypeptides that induce programmed cell death.
	### STEM CELLS AND THE MAINTENANCE OF ADULT TISSUES
	Proliferation of Differentiated Cells: Most cells in adult animals are arrested in the G_0 stage of the cell cycle. A few types of differentiated cells, including skin fibroblasts, endothelial cells, smooth muscle cells, and liver cells are able to resume proliferation as required to replace cells that have been lost because of injury or cell death.
stem cell	*Stem Cells:* Most differentiated cells do not themselves proliferate but can be replaced via the proliferation of stem cells. Stem cells divide to produce one daughter cell that remains a stem cell and another that divides and differentiates. Stem cells have been identified in a wide variety of adult tissues, including the hematopoietic system, skin, intestine, skeletal muscle, brain, and heart.
bone marrow transplantation	*Medical Applications of Adult Stem Cells:* The ability of stem cells to repair damaged tissue suggests their potential use in clinical medicine. Adult stem cells are used to repair damage to the hematopoietic system in bone marrow transplantation, and epidermal stem cells can be used for skin grafts. However, clinical applications of adult stem cells are limited by difficulties in isolating and culturing these cells.

SUMMARY	KEY TERMS

EMBRYONIC STEM CELLS AND THERAPEUTIC CLONING

Embryonic Stem Cells: Embryonic stem cells are cultured from early embryos. They can be readily grown in the undifferentiated state in culture while retaining the ability to differentiate into a wide variety of cell types, so they may offer considerable advantages over adult stem cells for many clinical applications.

Somatic Cell Nuclear Transfer: Mammals have been cloned by somatic cell nuclear transfer in which the nucleus of an adult somatic cell is transplanted into an enucleated egg. This opens the possibility of therapeutic cloning in which embryonic stem cells would be derived from a cloned embryo and used for transplantation therapy of the donor of the adult nucleus. Although many obstacles need to be overcome, the possibility of therapeutic cloning holds great promise for the development of new treatments for a variety of devastating diseases.

embryonic stem cell

somatic cell nuclear transfer, therapeutic cloning, reproductive cloning

Questions

1. Why is cell death via apoptosis more advantageous to multicellular organisms than cell death via acute injury?

2. What molecular mechanisms regulate caspase activity?

3. You have expressed mutants of nuclear lamins in human fibroblasts. The Asp residue in the caspase cleavage site has been mutated to Glu in these lamins. How would these mutant lamins affect the progression of apoptosis?

4. How do Bcl-2 family proteins regulate apoptosis in mammalian cells?

5. How does p53 activation in response to DNA damage affect cell cycle progression and cell survival?

6. You have constructed a Bad mutant in which the Akt phosphorylation site has been mutated such that Akt no longer phosphorylates it. How would expression of this mutant affect cell survival?

7. How would expression of siRNA targeted against 14-3-3 proteins affect apoptosis?

8. You are considering treatment of a leukemic patient with TNF. Upon further analysis you determine that the leukemic cells have an inactivating mutation of caspase-8. Will treatment with TNF be an effective therapy for this patient?

9. How would siRNA against Ced-3 affect the development of *C. elegans*?

10. You have isolated a polypeptide from a toxic plant, which localizes to mitochondria after endocytosis by mammalian cells. The polypeptide aggregates and forms large channels in the mitochondrial outer membrane, releasing proteins from the intermembrane space into the cytoplasm. How will treatment with this polypeptide affect mammalian cells in culture?

11. Many adult tissues contain terminally differentiated cells that are incapable of proliferation. However, these tissues can regenerate following damage. What gives these tissues their regenerative capabilities?

12. What is the critical property of stem cells?

13. What are the potential advantages of embryonic stem cells as compared to adult stem cells for therapeutic applications?

References and Further Reading

Programmed Cell Death

Adams, J. M. 2003. Ways of dying: Multiple pathways to apoptosis. *Genes Dev.* 17: 2481–2495. [R]

Brunet, A., S. R. Datta and M. E. Greenberg. 2001. Transcription-dependent and -independent control of neuronal survival by the PI3K-Akt pathway. *Curr. Opin. Neurobiol.* 11: 297–305. [R]

Chen, G. and D. V. Goeddel. 2002. TNF-R1 signaling: A beautiful pathway. *Science* 296: 1634–1635. [R]

Danial, N. K. and S. J. Korsmeyer. 2004. Cell death: Critical control points. *Cell* 116: 205–219. [R]

Datta, S. R., A. Brunet and M. E. Greenberg. 1999. Cellular survival: A play in three Akts. *Genes Dev.* 13: 290–2927. [R]

Ellis, H. M. and H. R. Horvitz. 1986. Genetic control of programmed cell death in the nematode *C. elegans*. *Cell* 44: 817–829. [P]

Hengartner, M. O. 2000. The biochemistry of apoptosis. *Nature* 407: 770–776. [R]

Holcik, M. and N. Sonenberg. 2005. Translational control in stress and apoptosis. *Nature Rev. Mol. Cell Biol.* 6: 318–327. [R]

Jacobson, M. D., M. Weil and M. C. Raff. 1997. Programmed cell death in animal development. *Cell* 88: 347–354. [R]

Jope, R. S. and G. V. W. Johnson. 2004. The glamour and gloom of glycogen synthase kinase-3. *Trends Biochem. Sci.* 29: 95–102. [R]

Lauber, K., S. G. Blumenthal, M. Waibel and S. Wesselborg. 2004. Clearance of apoptotic cells: Getting rid of the corpses. *Mol. Cell* 14: 277–287. [R]

Riedl, S. J. and Y. Shi. 2004. Molecular mechanisms of caspase regulation during apoptosis. *Nature Rev. Mol. Cell Biol.* 5: 897–907. [R]

Samejima, K. and W. C. Earnshaw. 2005. Trashing the genome: The role of nucleases during apoptosis. *Nature Rev. Mol. Cell Biol.* 6: 677–688. [R]

Spierings, D., G. McStay, M. Saleh, C. Bender, J. Chipuk, U. Maurer and D. R. Green. 2005. Connected to death: The (unexpurgated) mitochondrial pathway of apoptosis. *Science* 310: 66–67. [R]

Vaux, D. L. and J. Silke. 2005. IAPs, RINGs and ubiquitylation. *Nature Rev. Mol. Cell Biol.* 6: 287–297. [R]

Vousden, K. H. and X. Lu. 2002. Live or let die: The cells response to p53. *Nature Rev. Cancer* 2: 594–604. [R]

Wajant, H. 2002. The Fas signaling pathway: More than a paradigm. *Science* 296: 1635–1636. [R]

Wang, X. 2001. The expanding role of mitochondria in apoptosis. *Genes Dev.* 15: 2922–2933. [R]

Yu, J. and L. Zhang. 2005. The transcriptional targets of p53 in apoptosis control. *Biochem. Biophys. Res. Comm.* 331: 851–858. [R]

Stem Cells and the Maintenance of Adult Tissues

Alonso, L. and E. Fuchs. 2003. Stem cells of the skin epithelium. *Proc. Natl. Acad. Sci. USA* 100: 11830–11835. [R]

Carmeliet, P. 2003. Angiogenesis in health and disease. *Nature Med.* 9: 653–660. [R]

Chargé, S. B. P. and M. A. Rudnicki. 2004. Cellular and molecular regulation of muscle regeneration. *Physiol. Rev.* 84: 209–238. [R]

Dhawan, J. and T. A. Rando. 2005. Stem cells in postnatal myogenesis: Molecular mechanisms of satellite cell quiescence, activation and replenishment. *Trends Cell Biol.* 15: 666–672. [R]

Dor, Y., J. Brown, O. I. Martinez and D. A. Melton. 2004. Adult pancreatic β-cells are formed by self-duplication rather than stem-cell differentiation. *Nature* 429: 41–46. [P]

Fuchs, E., T. Tumbar and G. Guasch. 2004. Socializing with their neighbors: Stem cells and their niche. *Cell* 116: 769–778. [R]

Griffith, L. G. and G. Naughton. 2002. Tissue engineering—Current challenges and expanding opportunities. *Science* 295: 1009–1014. [R]

Heldin, C.-H. and B. Westermark. 1999. Mechanism of action and the *in vivo* role of platelet-derived growth factor. *Physiol. Rev.* 79: 1283–1316. [R]

Mayhall, E. A., N. Paffett-Lugassy and L. I. Zon. 2004. The clinical potential of stem cells. *Curr. Opin. Cell Biol.* 16: 713–720. [R]

McCulloch, E. A. and J. E. Till. 2005. Perspectives on the properties of stem cells. *Nature Med.* 11: 1026–1028. [R]

Radtke, F. and H. Clevers. 2005. Self-renewal and cancer of the gut: Two sides of a coin. *Science* 307: 1904–1909. [R]

Sell, S., ed. 2004. *Stem Cells Handbook.* Totowa: New Jersey: Humana Press.

Suda, T., F. Arai and A. Hirao. 2005. Hematopoietic stem cells and their niche. *Trends Immunol.* 26: 426–433. [R]

Taub, R. 2004. Liver regeneration: From myth to mechanism. *Nature Rev. Mol. Cell Biol.* 5: 836–847. [R]

Wagers, A. J. and I. M. Conboy. 2005. Cellular and molecular signatures of muscle regeneration: Current concepts and controversies in adult myogenesis. *Cell* 122: 659–667. [R]

Wagers, A. J. and I. L. Weissman. 2004. Plasticity of adult stem cells. *Cell* 116: 639–648. [R]

Wang, D.-Z. and E. N. Olson. 2004. Control of smooth muscle development by the myocardin family of transcriptional coactivators. *Curr. Opin. Genet. Dev.* 14: 558–566. [R]

Weissman, I. L. 2000. Stem cells: Units of development, units of regeneration, and units in evolution. *Cell* 100: 157–168. [R]

Embryonic Stem Cells and Therapeutic Cloning

Boiani, M. and H. R. Scholer. 2005. Regulatory networks in embryo-derived pluripotent stem cells. *Nature Rev. Mol. Cell Biol.* 6: 872–884. [R]

Guasch, G. and E. Fuchs. 2005. Mice in the world of stem cell biology. *Nature Genet.* 37: 1201–1206. [R]

Keller, G. 2005. Embryonic stem cell differentiation: Emergence of a new era in biology and medicine. *Genes Dev.* 19: 1129–1155. [R]

Mayhall, E. A., N. Paffett-Lugassy and L. I. Zon. 2004. The clinical potential of stem cells. *Curr. Opin. Cell Biol.* 16: 713–720. [R]

Rhind, S. M., J. E. Taylor, P. A. DeSousa, T. J. King, M. McGarry and I. Wilmut. 2003. Human cloning: Can it be made safe? *Nature Genet.* 4: 855–864. [R]

Rideout, W. M. III, K. Hochedlinger, M. Kyba, G. Q. Daley and R. Jaenisch. 2002. Correction of a genetic defect by nuclear transplantation and combined cell and gene therapy. *Cell* 109: 17–27. [P]

Shamblott, M. J., J. Axelman, S. Wang, E. M. Bugg, J. W. Littlefield, P. J. Donovan, P. D. Blumenthal, G. R. Huggins and J. D. Gearhart. 1998. Derivation of pluripotent stem cells from cultured human primordial germ cells. *Proc. Natl. Acad. Sci. USA* 95: 13726–13731. [P]

Thompson, J. A., J. Itskovitz-Eldor, S. S. Shapiro, M. A. Waknotz, J. J. Swiergiel, V. S. Marshall and J. M. Jones. 1998. Embryonic stem cell lines derived from human blastocysts. *Science* 282: 1145–1147. [P]

Wilmut, I., N. Beaujean, P. A. de Sousa, A. Dinnyes, T. J. King, L. A. Paterson, D. N. Wells and L. E. Young. 2002. Somatic cell nuclear transfer. *Nature* 419: 583–586. [R]

Wilmut, I., A. E. Schnieke, J. McWhir, A. J. Kind and K. H. S. Campbell. 1997. Viable offspring derived from fetal and adult mammalian cells. *Nature* 385: 810–813. [P]

CHAPTER 18

Cancer

■ **The Development and Causes of Cancer** *719*

■ **Tumor Viruses** *729*

■ **Oncogenes** *733*

■ **Tumor Suppressor Genes** *746*

■ **Molecular Approaches to Cancer Treatment** *755*

■ **KEY EXPERIMENT:**
The Discovery of Proto-Oncogenes *737*

■ **MOLECULAR MEDICINE:**
STI-571: Cancer Treatment Targeted against the *bcr/abl* Oncogene *759*

CANCER IS A PARTICULARLY APPROPRIATE TOPIC for the concluding chapter of this book because it results from a breakdown of the regulatory mechanisms that govern normal cell behavior. As discussed in preceding chapters, the proliferation, differentiation, and survival of individual cells in multicellular organisms are carefully regulated to meet the needs of the organism as a whole. This regulation is lost in cancer cells, which grow and divide in an uncontrolled manner, ultimately spreading throughout the body and interfering with the function of normal tissues and organs.

Because cancer results from defects in fundamental cell regulatory mechanisms, it is a disease that ultimately has to be understood at the molecular and cellular levels. Indeed, understanding cancer has been an objective of molecular and cellular biologists for many years. Importantly, studies of cancer cells have also illuminated the mechanisms that regulate normal cell behavior. In fact, many of the proteins that play key roles in cell signaling, regulation of the cell cycle, and control of programmed cell death were first identified because abnormalities in their activities led to the uncontrolled proliferation of cancer cells. The study of cancer has thus contributed significantly to our understanding of normal cell regulation, as well as vice versa.

The Development and Causes of Cancer

The fundamental abnormality resulting in the development of cancer is the continual unregulated proliferation of cancer cells. Rather than responding appropriately to the signals that control normal cell behavior, cancer cells grow and divide in an uncontrolled manner, invading normal tissues and organs and eventually spreading throughout the body. The generalized loss of growth control exhibited by cancer cells is the net result of accumulated abnormalities in multiple cell regulatory systems and is reflected in several aspects of cell behavior that distinguish cancer cells from their normal counterparts.

FIGURE 18.1 A cancer of the pancreas   Light micrograph of a section through a pancreas showing pancreatic cancer. The cancer cells have dark purple nuclei and are invading the normal tissue (pink). (Astrid and Hanns-Frieder Michler/SPL/ Photo Researchers, Inc.)

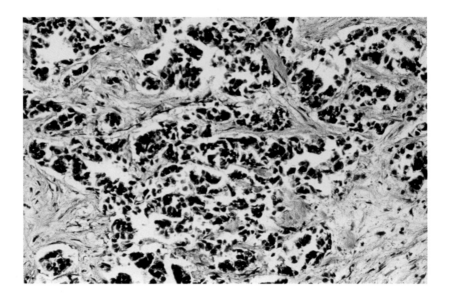

■ The earliest recorded description of cancer is found in an Egyptian papyrus dating back to around 3000 BC. Hippocrates is believed to be the originator of the word cancer. It is thought that the shape of tumors reminded him of crabs and he used the words *carcinos*, *carcinoma* and *cancer*, which refer to crabs in Greek, to describe the tumors.

Types of Cancer

Cancer can result from abnormal proliferation of any of the different kinds of cells in the body, so there are more than a hundred distinct types of cancer, which can vary substantially in their behavior and response to treatment. The most important issue in cancer pathology is the distinction between benign and malignant tumors (Figure 18.1). A **tumor** is any abnormal proliferation of cells, which may be either benign or malignant. A **benign tumor**, such as a common skin wart, remains confined to its original location, neither invading surrounding normal tissue nor spreading to distant body sites. A **malignant tumor**, however, is capable of both invading surrounding normal tissue and spreading throughout the body via the circulatory or lymphatic systems (**metastasis**). Only malignant tumors are properly referred to as cancers, and it is their ability to invade and metasta-

TABLE 18.1 Most Frequent Cancers in the United States

Cancer site	Cases per year	Deaths per year
Prostate	232,100 (16.9%)	30,400 (5.3%)
Breast	212,900 (15.5%)	40,900 (7.2%)
Lung	172,600 (12.6%)	163,500 (28.7%)
Colon/rectum	145,300 (10.6%)	56,300 (9.9%)
Lymphomas	63,700 (4.6%)	20,600 (3.6%)
Bladder	63,200 (4.6%)	13,200 (2.3%)
Skin (melanoma)	59,600 (4.3%)	7800 (1.4%)
Uterus	51,200 (3.7%)	11,000 (1.9%)
Kidney	36,200 (2.6%)	12,700 (2.2%)
Leukemias	34,800 (2.5%)	22,600 (4.0%)
Pancreas	32,200 (2.3%)	31,800 (5.6%)
Subtotal	1,103,800 (80.4%)	410,800 (72.1%)
All sites	1,373,000 (100%)	570,000 (100%)

Source: American Cancer Society, *Cancer Facts and Figures—2005.*

size that makes cancer so dangerous. Whereas benign tumors can usually be removed surgically, the spread of malignant tumors to distant body sites frequently makes them resistant to such localized treatment.

Both benign and malignant tumors are classified according to the type of cell from which they arise. Most cancers fall into one of three main groups: carcinomas, sarcomas, and leukemias or lymphomas. **Carcinomas**, which include approximately 90% of human cancers, are malignancies of epithelial cells. **Sarcomas**, which are rare in humans, are solid tumors of connective tissues, such as muscle, bone, cartilage, and fibrous tissue. **Leukemias** and **lymphomas**, which account for approximately 7% of human malignancies, arise from the blood-forming cells and from cells of the immune system, respectively. Tumors are further classified according to tissue of origin (e.g., lung or breast carcinomas) and the type of cell involved. For example, fibrosarcomas arise from fibroblasts, and erythroid leukemias from precursors of erythrocytes (red blood cells).

Although there are many kinds of cancer, only a few occur frequently (Table 18.1). More than a million cases of cancer are diagnosed annually in the United States, and more than 500,000 Americans die of cancer each year. Cancers of 11 different body sites account for more than 80% of this total cancer incidence. The four most common cancers accounting for more than half of all cancer cases are those of the prostate, breast, lung, and colon/rectum. Lung cancer, by far the most lethal, is responsible for nearly 30% of all cancer deaths.

The Development of Cancer

One of the fundamental features of cancer is tumor clonality—the development of tumors from single cells that begin to proliferate abnormally. The single-cell origin of many tumors has been demonstrated by analysis of X chromosome inactivation (Figure 18.2). As discussed in Chapter 7, one member of the X chromosome pair is inactivated by being converted to heterochromatin in female cells. X inactivation occurs randomly during embryonic development, so one X chromosome is inactivated in some cells, while the other X chromosome is inactivated in other cells. Thus, if a female is heterozygous for an X chromosome gene, different alleles will be expressed in different cells. Normal tissues are composed of mixtures of cells with different inactive X chromosomes, so expression of both alleles is detected in normal tissues of heterozygous females. In contrast, tumor tissues generally express only one allele of a heterozygous X chromosome gene. The implication is that all of the cells constituting such a tumor were derived from a single cell of origin in which the pattern of X inactivation was fixed before the tumor began to develop.

The clonal origin of tumors does not, however, imply that the original progenitor cell that gives rise to a tumor has initially acquired all of the characteristics of a cancer cell. On the contrary, the development of cancer is a multistep process in which cells gradually become malignant through a progressive series of alterations. One indication of the multistep development of cancer is that most cancers develop late in life. The incidence of

> ■ Although cancer has been with us throughout the history of mankind, it has become a leading cause of death only in the last century. Prior to 1900, most deaths were due to infectious diseases, such as pneumonia and tuberculosis, and life expectancy was less than 50 years. Cancer was a rare disease that accounted for only a small percentage of deaths.

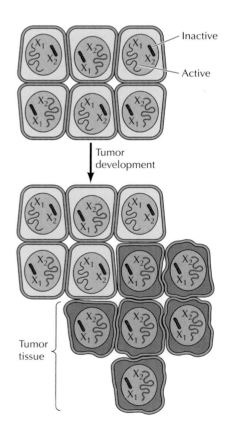

FIGURE 18.2 Tumor clonality   Normal tissue is a mosaic of cells in which different X chromosomes (X_1 and X_2) have been inactivated. Tumors develop from a single initially altered cell, so each tumor cell displays the same pattern of X inactivation (X_1 inactive, X_2 active).

FIGURE 18.3 Increased rate of colon cancer with age Annual death rates from colon cancer in the United States. (Data from J. Cairns, 1978. *Cancer: Science and Society*, New York: W. H. Freeman.)

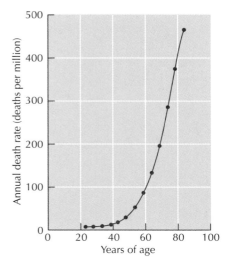

Initiation

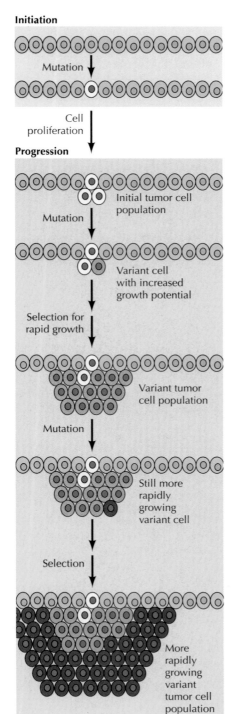

Mutation

Cell proliferation

Progression

Initial tumor cell population

Mutation

Variant cell with increased growth potential

Selection for rapid growth

Variant tumor cell population

Mutation

Still more rapidly growing variant cell

Selection

More rapidly growing variant tumor cell population

colon cancer, for example, increases more than tenfold between the ages of 30 and 50, and another tenfold between the ages of 50 and 70 (Figure 18.3). Such a dramatic increase of cancer incidence with age suggests that most cancers develop as a consequence of multiple abnormalities, which accumulate over periods of many years.

At the cellular level, the development of cancer is viewed as a multistep process involving mutation and selection for cells with progressively increasing capacity for proliferation, survival, invasion, and metastasis (Figure 18.4). The first step in the process, **tumor initiation**, is thought to be the result of a genetic alteration leading to abnormal proliferation of a single cell. Cell proliferation then leads to the outgrowth of a population of clonally derived tumor cells. **Tumor progression** continues as additional mutations occur within cells of the tumor population. Some of these mutations confer a selective advantage to the cell, such as more rapid growth, and the descendants of a cell bearing such a mutation will consequently become dominant within the tumor population. The process is called clonal selection, since a new clone of tumor cells has evolved on the basis of its increased growth rate or other properties (such as survival, invasion, or metastasis) that confer a selective advantage. Clonal selection continues throughout tumor development, so tumors continuously become more rapid-growing and increasingly malignant.

Studies of colon carcinomas have provided a clear example of tumor progression during the development of a common human malignancy (Figure 18.5). The earliest stage in tumor development is increased proliferation of colon epithelial cells. One of the cells within this proliferative cell population is then thought to give rise to a small benign neoplasm (an **adenoma** or **polyp**). Further rounds of clonal selection lead to the growth of adenomas of increasing size and proliferative potential. Malignant carcinomas then

FIGURE 18.4 Stages of tumor development The development of cancer initiates when a single mutated cell begins to proliferate abnormally. Additional mutations followed by selection for more rapidly growing cells within the population then result in progression of the tumor to increasingly rapid growth and malignancy.

FIGURE 18.5 Development of colon carcinomas   A single initially altered cell gives rise to a proliferative cell population, which progresses first to benign adenomas of increasing size and then to malignant carcinoma. The cancer cells invade the underlying connective tissue and penetrate blood and lymphatic vessels, thereby spreading throughout the body.

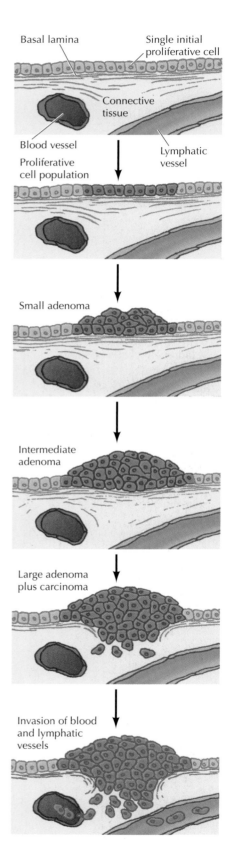

arise from the benign adenomas, indicated by invasion of the tumor cells through the basal lamina into underlying connective tissue. The cancer cells then continue to proliferate and spread through the connective tissues of the colon wall. Eventually the cancer cells penetrate the wall of the colon and invade other abdominal organs, such as the bladder or small intestine. In addition, the cancer cells invade blood and lymphatic vessels, allowing them to metastasize throughout the body.

Causes of Cancer

Substances that cause cancer, called **carcinogens**, have been identified both by studies in experimental animals and by epidemiological analysis of cancer frequencies in human populations (e.g., the high incidence of lung cancer among cigarette smokers). Since the development of malignancy is a complex multistep process, many factors may affect the likelihood that cancer will develop, and it is overly simplistic to speak of single causes of most cancers. Nonetheless, many agents, including radiation, chemicals, and viruses, have been found to induce cancer in both experimental animals and humans.

Radiation and many chemical carcinogens (Figure 18.6) act by damaging DNA and inducing mutations. Some of the carcinogens that contribute to human cancers include solar ultraviolet radiation (the major cause of skin cancer), carcinogenic chemicals in tobacco smoke, and aflatoxin (a potent liver carcinogen produced by some molds that contaminate improperly stored supplies of peanuts and other grains). The carcinogens in tobacco smoke (including benzo(*a*)pyrene, dimethylnitrosamine, and nickel compounds) are the major identified causes of human cancer. Smoking is the undisputed cause of 80 to 90% of lung cancers, as well as being implicated in cancers of the oral cavity, pharynx, larynx, esophagus, and other sites. In total, it is estimated that smoking is responsible for nearly one-third of all cancer deaths—an impressive toll for a single carcinogenic agent.

Other carcinogens contribute to cancer development by stimulating cell proliferation, rather than by inducing mutations. Such compounds are referred to as **tumor promoters**, since the increased cell division they induce facilitates the outgrowth of a proliferative cell population during early stages of tumor development. Hormones, particularly estrogens, are important as tumor promoters in the development of some human cancers. The proliferation of cells of the uterine endometrium, for example, is stimulated by estrogen, and exposure to excess estrogen significantly increases the likelihood that a woman will develop endometrial cancer. The risk of endometrial cancer is therefore substantially increased by long-term postmenopausal estrogen replacement therapy with high doses of estrogen alone. Fortunately, this risk is minimized by administration of progesterone to counteract the stimulatory effect of estrogen on endometrial cell proliferation. However, oral contraceptives and hormone replacement therapy with combinations of estrogen and progesterone still lead to an increased risk of breast cancer.

FIGURE 18.6 Structure of representative chemical carcinogens

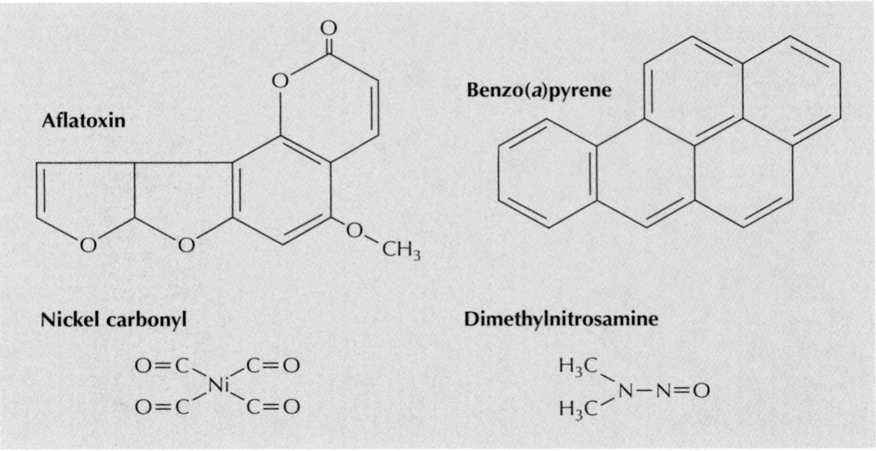

Aflatoxin

Benzo(a)pyrene

Nickel carbonyl

Dimethylnitrosamine

18.1 WEBSITE ANIMATION

Metastasis of a Cancer
A cancer begins when one altered cell gives rise to a proliferative cell population, forming a primary tumor from which cells can break away and form secondary tumors elsewhere in the body.

■ Barry Marshall and Robin Warren were the first to establish a link between *Heliobacter pylori* and stomach ulcers. To prove that *H. pylori* is the causative agent for stomach ulcers, Marshall deliberately infected himself with a pure culture of the microbe and studied the course of the infection. Thankfully, the illness resolved itself spontaneously.

Some viruses also cause cancer both in experimental animals and in humans, and infection with the bacterium *Heliobacter pylori* causes stomach cancer. The common human cancers caused by viruses include liver cancer and cervical carcinoma, which together account for 10 to 20% of worldwide cancer incidence. These viruses are important not only as causes of human cancer but, as discussed later in this chapter, studies of tumor viruses have played a key role in elucidating the molecular events responsible for the development of cancers induced by both viral and nonviral carcinogens.

Properties of Cancer Cells

The uncontrolled growth of cancer cells results from accumulated abnormalities affecting many of the cell regulatory mechanisms that have been discussed in preceding chapters. This relationship is reflected in several aspects of cell behavior that distinguish cancer cells from their normal counterparts. Cancer cells typically display abnormalities in the mechanisms that regulate normal cell proliferation, differentiation, and survival. Taken together, these characteristic properties of cancer cells provide a description of malignancy at the cellular level.

The uncontrolled proliferation of cancer cells *in vivo* is mimicked by their behavior in cell culture. A primary distinction between cancer cells and normal cells in culture is that normal cells display **density-dependent inhibition** of cell proliferation (Figure 18.7). Normal cells proliferate until they reach a finite cell density, which is determined in part by the availability of growth factors added to the culture medium (usually in the form of serum). They then cease proliferating and become quiescent, arrested in the G_0 stage of the cell cycle (see Figure 16.6). The proliferation of most cancer cells, however, is not sensitive to density-dependent inhibition. Rather than responding to the signals that cause normal cells to cease proliferation and enter G_0, tumor cells generally continue growing to high cell densities in culture, mimicking their uncontrolled proliferation *in vivo*.

A related difference between normal cells and cancer cells is that many cancer cells have reduced requirements for extracellular growth factors. As discussed in Chapter 15, the proliferation of most cells is controlled, at least in part, by polypeptide growth factors. For some cell types, particularly fibroblasts, the availability of serum growth factors is the principal determinant of their proliferative capacity in culture. The growth factor require-

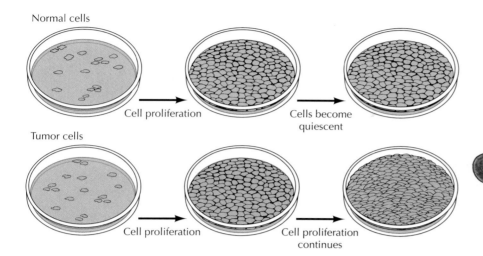

Normal cells

Cell proliferation → Cells become quiescent

Tumor cells

Cell proliferation → Cell proliferation continues

FIGURE 18.7 Density-dependent inhibition Normal cells proliferate in culture until they reach a finite cell density at which point they become quiescent. Tumor cells, however, continue to proliferate independent of cell density.

18.2 WEBSITE ANIMATION
Density-Dependent Inhibition In culture, normal cells proliferate until they reach a certain cell density, but tumor cells continue to proliferate independent of cell density.

ments of these cells are closely related to the phenomenon of density-dependent inhibition, since the density at which normal fibroblasts become quiescent is proportional to the concentration of serum growth factors in the culture medium.

The growth factor requirements of many tumor cells are reduced compared to their normal counterparts, contributing to the unregulated proliferation of tumor cells both *in vitro* and *in vivo*. In some cases, cancer cells produce growth factors that stimulate their own proliferation (Figure 18.8). Such abnormal production of a growth factor by a responsive cell leads to continuous autostimulation of cell division (**autocrine growth stimulation**), and the cancer cells are therefore less dependent on growth factors from other, physiologically normal sources. In other cases, the reduced growth factor dependence of cancer cells results from abnormalities in intracellular signaling systems, such as unregulated activity of growth factor receptors or other proteins (e.g., Ras proteins or protein kinases) that were discussed in Chapter 15 as elements of signal transduction pathways leading to cell proliferation.

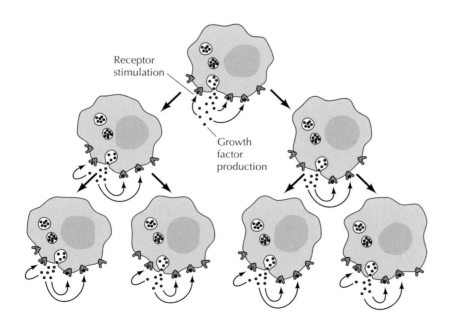

Receptor stimulation

Growth factor production

FIGURE 18.8 Autocrine growth stimulation A cell produces a growth factor to which it also responds, resulting in continuous stimulation of cell proliferation.

FIGURE 18.9 Contact inhibition
Light micrographs (left) and scanning electron micrographs (right) of normal fibroblasts and tumor cells. The migration of normal fibroblasts is inhibited by cell contact, so they form an orderly side-by-side array on the surface of a culture dish. Tumor cells, however, are not inhibited by cell contact, so they migrate over one another and grow in a disordered, multilayered pattern. (Courtesy of Lan Bo Chen, Dana-Farber Cancer Institute.)

Normal cells

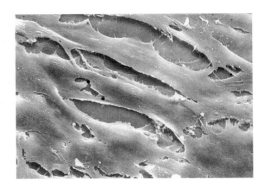

Tumor cells

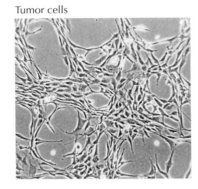
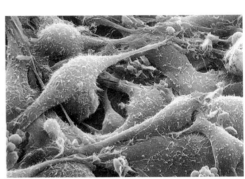

Cancer cells are also less stringently regulated than normal cells by cell-cell and cell-matrix interactions. Most cancer cells are less adhesive than normal cells, often as a result of reduced expression of cell surface adhesion molecules. For example, loss of E-cadherin, the principal adhesion molecule of epithelial cells, is important in the development of carcinomas (epithelial cancers). As a result of reduced expression of cell adhesion molecules, cancer cells are comparatively unrestrained by interactions with other cells and tissue components, contributing to the ability of malignant cells to invade and metastasize. The reduced adhesiveness of cancer cells also results in morphological and cytoskeletal alterations: Many tumor cells are rounder than normal, in part because they are less firmly attached to either the extracellular matrix or neighboring cells.

A striking difference in the cell-cell interactions displayed by normal cells and those of cancer cells is illustrated by the phenomenon of **contact inhibition** (Figure 18.9). Normal fibroblasts migrate across the surface of a culture dish until they make contact with a neighboring cell. Further cell migration is then inhibited, and normal cells adhere to each other, forming an orderly array of cells on the culture dish surface. Tumor cells, in contrast, continue moving after contact with their neighbors, migrating over adjacent cells, and growing in disordered, multilayered patterns. Not only the movement but also the proliferation of many normal cells is inhibited by cell-cell contact, and cancer cells are characteristically insensitive to such contact inhibition of growth.

Two additional properties of cancer cells affect their interactions with other tissue components, thereby playing important roles in invasion and metastasis. First, malignant cells generally secrete proteases that digest extracellular matrix components, allowing the cancer cells to invade adjacent normal tissues. Secretion of collagenase, for example, appears to be an important determinant of the ability of carcinomas to digest and penetrate through basal laminae to invade underlying connective tissue (see Figure 18.5). Second, cancer cells secrete growth factors that promote the formation of new blood vessels (**angiogenesis**). Angiogenesis is needed to support the growth of a tumor beyond the size of about a million cells, at which point new blood vessels are required to supply oxygen and nutrients to the proliferating tumor cells. Such blood vessels are formed in response to growth factors, secreted by the tumor cells, that stimulate proliferation of endothelial cells in the walls of capillaries in surrounding tissue, resulting in the outgrowth of new capillaries into the tumor (see Figure 17.14). The formation of such new blood vessels is important not only in supporting tumor growth, but also in metastasis. The actively growing new capillaries formed in response to angiogenic stimulation are easily penetrated by the tumor cells, providing a ready opportunity for cancer cells to enter the circulatory system and begin the metastatic process.

Another general characteristic of most cancer cells is that they fail to differentiate normally. Such defective differentiation is closely coupled to abnormal proliferation since, as discussed in Chapter 17, most fully differentiated cells cease cell division. Rather than carrying out their normal differentiation program, cancer cells are usually blocked at an early stage of differentiation, consistent with their continued active proliferation.

The leukemias provide a particularly good example of the relationship between defective differentiation and malignancy. All of the different types of blood cells are derived from a common hematopoietic stem cell in the bone marrow (see Figure 17.17). Descendants of these cells then become committed to specific differentiation pathways. Some cells, for example, differentiate to form erythrocytes whereas others differentiate to form lymphocytes, granulocytes, or macrophages. Progenitor cells of each of these types undergo several rounds of division as they differentiate, but once they become fully differentiated, cell division ceases. Leukemic cells, in contrast, fail to undergo terminal differentiation (Figure 18.10). Instead, they become arrested at early stages

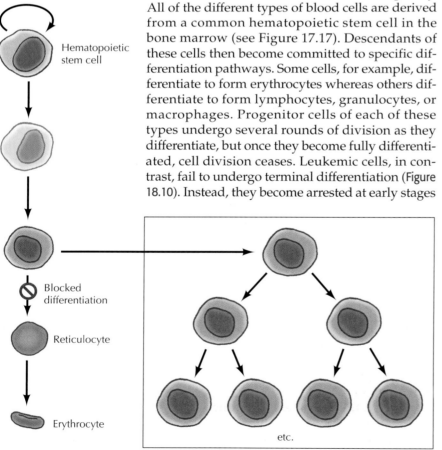

Hematopoietic stem cell

Blocked differentiation

Reticulocyte

Erythrocyte

etc.

Leukemic cells fail to differentiate and continue to divide

FIGURE 18.10 Defective differentiation and leukemia Different types of blood cells develop from the hematopoietic stem cell in the bone marrow. The precursors of differentiated cells undergo several rounds of cell division as they mature, but cell division ceases at the terminal stages of differentiation. The differentiation of leukemic cells is blocked at early stages of maturation, consistent with their continued proliferation.

of maturation at which they retain their capacity for proliferation and continue to reproduce.

As discussed in Chapter 17, **programmed cell death** or **apoptosis** is an integral part of the differentiation program of many cell types, including blood cells. Many cancer cells fail to undergo apoptosis and therefore exhibit increased life spans compared to their normal counterparts. This failure of cancer cells to undergo programmed cell death contributes substantially to tumor development. For example, the survival of many normal cells is dependent on signals from either growth factors or from the extracellular matrix that prevent apoptosis. In contrast, tumor cells are often able to survive in the absence of growth factors required by their normal counterparts. Such a failure of tumor cells to undergo apoptosis when deprived of normal environmental signals may be important not only in primary tumor development but also in the survival and growth of metastatic cells in abnormal tissue sites. Normal cells also undergo apoptosis following DNA damage, while many cancer cells fail to do so. In this case, the failure to undergo apoptosis contributes to the resistance of cancer cells to irradiation and many chemotherapeutic drugs, which act by damaging DNA. In addition to evading apoptosis, cancer cells generally acquire the capacity for unlimited replication as a result of expression of telomerase, which is required to maintain the ends of eukaryotic chromosomes (see Figure 6.16). Abnormal cell survival, as well as cell proliferation, thus plays a major role in the unrelenting growth of cancer cells in an animal.

Transformation of Cells in Culture

The study of tumor induction by radiation, chemicals, or viruses requires experimental systems in which the effects of a carcinogenic agent can be reproducibly observed and quantitated. Although the activity of carcinogens can be assayed in intact animals, such experiments are difficult to quantitate and control. The development of *in vitro* assays to detect the conversion of normal cells to tumor cells in culture, a process called **cell transformation**, therefore represented a major advance in cancer research. Such assays are designed to detect transformed cells, which display the *in vitro* growth properties of tumor cells, following exposure of a culture of normal cells to a carcinogenic agent. Their application has allowed experimental analysis of cell transformation to reach a level of sophistication that could not have been attained by studies in whole animals alone.

The first and most widely used assay of cell transformation is the focus assay, which was developed by Howard Temin and Harry Rubin in 1958. The focus assay is based on the ability to recognize a group of transformed cells as a morphologically distinct "focus" against a background of normal cells on the surface of a culture dish (Figure 18.11). The focus assay takes advantage of three properties of transformed cells: altered morphology, loss of contact inhibition, and loss of density-dependent inhibition of growth. The result is the formation of a colony of morphologically altered transformed cells that overgrow the background of normal cells in the culture. Such foci of transformed cells can usually be detected and quantified within a week or two after exposure to a carcinogenic agent. In general, cells transformed *in vitro* are able to form tumors following inoculation into suscepti-

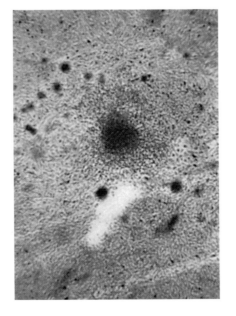

FIGURE 18.11 The focus assay   A focus of chicken embryo fibroblasts induced by Rous sarcoma virus. (From H. M. Temin and H. Rubin, 1958. *Virology* 6: 669.)

ble animals, supporting *in vitro* transformation as a valid indicator of the formation of cancer cells.

Tumor Viruses

Members of several families of animal viruses, called **tumor viruses**, are capable of directly causing cancer in either experimental animals or humans (Table 18.2). The viruses that cause human cancer include hepatitis B and C viruses (liver cancer), papillomaviruses (cervical and other anogenital cancers), Epstein-Barr virus (Burkitt's lymphoma and nasopharyngeal carcinoma), Kaposi's sarcoma-associated herpesvirus (Kaposi's sarcoma), and human T-cell lymphotropic virus (adult T-cell leukemia). In addition, HIV is indirectly responsible for the cancers that develop in AIDS patients as a result of immunodeficiency.

As already noted, tumor viruses not only are important as causes of human disease but have also played a critical role in cancer research by serving as models for cellular and molecular studies of cell transformation. The small size of their genomes has made tumor viruses readily amenable to molecular analysis, leading to the identification of viral genes responsible for cancer induction and paving the way to our current understanding of cancer at the molecular level.

Hepatitis B and C Viruses

The **hepatitis B** and **C viruses** are the principal causes of liver cancer, which is the third leading cause of cancer deaths worldwide. Both hepatitis B and C viruses specifically infect liver cells and can lead to long term chronic infections of the liver. Such chronic infections are associated with a high risk of liver cancer, which eventually develops in 10–20% of people chronically infected with hepatitis B and in about 5% of people chronically infected with hepatitis C virus.

The molecular mechanisms by which the hepatitis viruses cause liver cancer are not yet understood. Hepatitis B virus is a DNA virus with a genome of only 3 kb. Cell transformation by hepatitis B virus may be mediated by a viral gene (called the *X* gene) that affects expression of several cellular genes that drive cell proliferation. In addition, the development of cancers induced by hepatitis B virus is driven by the continual proliferation of liver cells that results from chronic tissue damage and inflammation.

TABLE 18.2 Tumor Viruses

Virus family	Human tumors	Genome size (kb)
DNA Genomes		
Hepatitis B viruses	Liver cancer	3
SV40 and polyomavirus	None	5
Papillomaviruses	Cervical carcinoma	8
Adenoviruses	None	35
Herpesviruses	Burkitt's lymphoma, nasopharyngeal carcinoma, Kaposi's sarcoma	100–200
RNA Genomes		
Hepatitis C virus	Liver cancer	10
Retroviruses	Adult T-cell leukemia	9–10

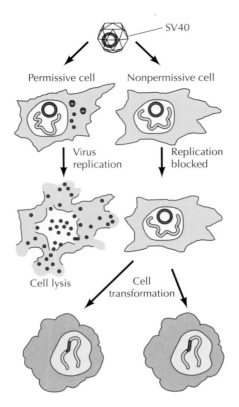

Permissive cell

Nonpermissive cell

Virus replication

Replication blocked

Cell lysis

Cell transformation

SV40

FIGURE 18.12 SV40 replication and transformation Infection of a permissive cell results in virus replication, cell lysis, and release of progeny virus particles. In a nonpermissive cell, virus replication is blocked, allowing some cells to become permanently transformed.

Hepatitis C virus is an RNA virus with a genome of approximately 10 kb. Cell proliferation in response to chronic inflammation is a major contributor to cancer development in people chronically infected with hepatitis C virus, although it is also possible that some hepatitis C virus proteins directly stimulate the proliferation of infected liver cells.

SV40 and Polyomavirus

Although neither **simian virus 40 (SV40)** nor **polyomavirus** are associated with human cancer, they have been critically important as models for understanding the molecular basis of cell transformation. The utility of these viruses in cancer research has stemmed from the availability of good cell culture assays for both virus replication and transformation, as well as from the small size of their genomes (approximately 5 kb).

SV40 and polyomavirus do not induce tumors or transform the cells of their natural host species—monkeys and mice, respectively. In cells of their natural hosts (permissive cells), infection leads to virus replication, cell lysis, and release of progeny virus particles (Figure 18.12). Since a permissive cell is killed as a consequence of virus replication, it cannot become transformed. The transforming potential of these viruses is revealed, however, by infection of nonpermissive cells in which virus replication is blocked. In this case, the viral genome sometimes integrates into cellular DNA, and expression of specific viral genes results in transformation of the infected cell.

The SV40 and polyomavirus genes that lead to cell transformation are the same viral genes that function in early stages of lytic infection. The genomes of SV40 and polyomavirus are divided into early and late regions. The early region is expressed immediately after infection and is required for synthesis of viral DNA. The late region is not expressed until after viral DNA replication has begun and includes genes encoding structural components of the virus particle. The early region of SV40 encodes two proteins (called small and large T antigens) of about 17 kd and 94 kd, respectively (Figure 18.13). Their mRNAs are generated by alternative splicing of a single early-region primary transcript. Polyomavirus likewise encodes small and large T antigens, as well as a third early-region protein of about 55 kd designated middle T. Transfection of cells with cDNAs for individual early-region proteins has shown that SV40 large T is sufficient to induce transformation, whereas middle T is primarily responsible for transformation by polyomavirus.

During lytic infection these early-region proteins fulfill multiple functions required for virus replication. SV40 T antigen, for example, binds to the SV40 origin and initiates viral DNA replication (see Chapter 6). In addition, the early-region proteins of SV40 and polyomavirus stimulate host cell gene expression and DNA synthesis. Since virus replication is dependent on host cell enzymes (e.g., DNA polymerase), such stimulation of the host cell is a critical event in the viral life cycle. Most cells in an animal are nonproliferating and therefore must be stimulated to divide in order to induce the enzymes needed for viral DNA replication. This stimulation of cell pro-

FIGURE 18.13 The SV40 genome   The genome is divided into early and late regions. Large and small T antigens are produced by alternative splicing of early-region pre-mRNA.

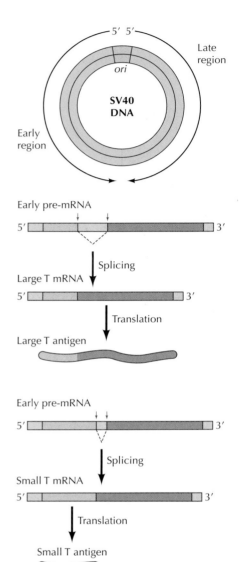

liferation by the early gene products can lead to transformation if the viral DNA becomes stably integrated and expressed in a nonpermissive cell.

As discussed later in this chapter, both SV40 and polyomavirus early-region proteins induce transformation by interacting with host proteins that regulate cell proliferation. For example, SV40 T antigen binds to and inactivates the host cell tumor suppressor proteins Rb and p53, which are key regulators of cell proliferation and survival.

Papillomaviruses

The **papillomaviruses** are small DNA viruses (genomes of approximately 8 kb) that induce both benign and malignant tumors in humans and a variety of other animal species. Approximately 60 different types of human papillomaviruses, which infect epithelial cells of several tissues, have been identified. Some of these viruses cause only benign tumors (such as warts), whereas others are causative agents of malignant carcinomas, particularly cervical and other anogenital cancers. The mortality from cervical cancer is relatively low in the United States, in large part as a result of early detection and curative treatment made possible by the Pap smear. In other parts of the world, however, cervical cancer remains common; it is responsible for 5 to 10% of worldwide cancer incidence.

Cell transformation by human papillomaviruses results from expression of two early-region genes, *E6* and *E7* (**Figure 18.14**). Like SV40 T antigen, the E6 and E7 proteins induce transformation by interacting with host cell proteins that control cell proliferation and survival, including Rb and p53. In particular, E7 binds to Rb, and E6 stimulates the degradation of p53 by ubiquitin-mediated proteolysis.

Adenoviruses

The **adenoviruses** are a large family of DNA viruses with genomes of about 35 kb. In contrast to the papillomaviruses, the adenoviruses are not associated with naturally occurring cancers in either humans or other animals. However, they are widely studied and important models in experimental cancer biology.

■ **The Pap smear was developed by George Papanicolaou in the 1930s. Uterine cells are collected and smeared on a microscope slide. Since cancer cells have altered morphology they can be readily detected upon microscopic examination.**

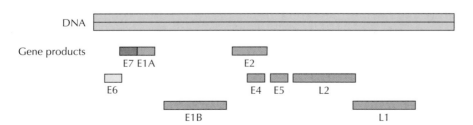

FIGURE 18.14 The genome of a human papillomavirus   Gene products are designated E (early) or L (late). Transformation results from the action of E6 and E7.

FIGURE 18.15 The adenovirus genome Two early-region genes, *E1A* and *E1B*, are responsible for induction of transformation.

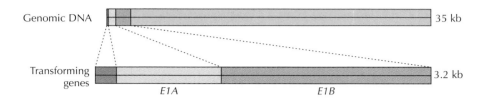

Like SV40 and polyomaviruses, the adenoviruses are lytic in cells of their natural host species but can induce transformation in nonpermissive hosts. Transformation by the adenoviruses results from expression of two early-region genes, *E1A* and *E1B*, which are required for virus replication in permissive cells (Figure 18.15). These transforming proteins inactivate the Rb and p53 tumor suppressor proteins, with E1A binding to Rb and E1B binding to p53. It thus appears that SV40, papillomaviruses, and adenoviruses all induce transformation by related pathways in which interfering with the activities of Rb and p53 plays a central role.

Herpesviruses

The **herpesviruses** are among the most complex animal viruses, with genomes of 100 to 200 kb. Several herpesviruses induce tumors in animal species, including frogs, chickens, and monkeys. In addition, two members of the herpesvirus family, **Kaposi's sarcoma-associated herpesvirus** and **Epstein-Barr virus**, cause human cancers. Kaposi's sarcoma-associated herpesvirus plays a critical role in the development of Kaposi's sarcomas, and Epstein-Barr virus has been implicated in several human malignancies, including Burkitt's lymphoma in some regions of Africa, B-cell lymphomas in AIDS patients and other immunosuppressed individuals, and nasopharyngeal carcinoma in China.

In addition to its association with these human malignancies, Epstein-Barr virus is able to transform human B lymphocytes in culture. Partly because of the complexity of the genome, however, the molecular biology of Epstein-Barr virus replication and transformation remains to be fully understood. The main Epstein-Barr virus transforming protein (LMP1) mimics a cell surface receptor on B lymphocytes and functions by activating signaling pathways that stimulate cell proliferation and inhibit apoptosis. Several other viral genes may also contribute to transformation of lymphocytes, but their functions in the transformation process have not been established.

Kaposi's sarcoma-associated herpesvirus DNA is regularly found in Kaposi's sarcoma cells, and several viral genes that affect cell proliferation and survival are expressed in these tumors. A noteworthy feature of Kaposi's sarcoma cells is that they secrete a variety of growth factors that drive their own proliferation. Interestingly, the transforming proteins of Kaposi's sarcoma-associated herpesvirus appear to act at least in part by stimulating growth factor secretion.

Retroviruses

Retroviruses cause cancer in a variety of animal species, including humans. One human retrovirus, human T-cell lymphotropic virus type I (HTLV-I), is the causative agent of adult T-cell leukemia, which is common in parts of Japan, the Caribbean, and Africa. Transformation of T lymphocytes by HTLV-I results from expression of the viral gene *tax*, which encodes a regu-

■ Denis Burkitt was a surgeon in Uganda when he came across several cases of patients with tumors of the head and neck and identified the disease as a previously uncharacterized lymphoma. Michael Epstein and Yvonne Barr subsequently isolated a virus from Burkitt's lymphoma tissue, providing one of the first links between viruses and human cancer.

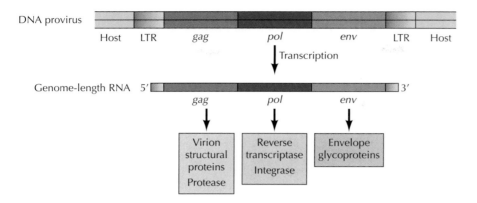

FIGURE 18.16 A typical retrovirus genome   The DNA provirus, integrated into cellular DNA, is transcribed to yield genome-length RNA. This primary transcript serves as the genomic RNA for progeny virus particles and as mRNA for the *gag* and *pol* genes. In addition, the full-length RNA is spliced to yield mRNA for *env*. The *gag* gene encodes the viral protease and structural proteins of the virus particle, *pol* encodes reverse transcriptase and integrase, and *env* encodes envelope glycoproteins.

latory protein affecting expression of several cellular growth control genes. AIDS is caused by another retrovirus, HIV. In contrast to HTLV-I, HIV does not cause cancer by directly converting a normal cell into a tumor cell. However, AIDS patients suffer a high incidence of some malignancies, particularly lymphomas and Kaposi's sarcoma. These cancers, which are also common among other immunosuppressed individuals, are associated with infection by other viruses (e.g., Epstein-Barr virus and Kaposi's sarcoma-associated herpesvirus) and apparently develop as a secondary consequence of immunosuppression in AIDS patients.

Different retroviruses differ substantially in their oncogenic potential. Most retroviruses contain only three genes (*gag*, *pol*, and *env*) that are required for virus replication but play no role in cell transformation (Figure 18.16). Retroviruses of this type induce tumors only rarely, if at all, as a consequence of mutations resulting from the integration of proviral DNA within or adjacent to cellular genes.

Other retroviruses, however, contain specific genes responsible for induction of cell transformation and are potent carcinogens. The prototype of these highly oncogenic retroviruses is **Rous sarcoma virus** (**RSV**), first isolated from a chicken sarcoma by Peyton Rous in 1911. More than 50 years later, studies of RSV led to identification of the first viral oncogene, which has provided a model for understanding many aspects of tumor development at the molecular level.

Oncogenes

Cancer results from alterations in critical regulatory genes that control cell proliferation, differentiation, and survival. Studies of tumor viruses revealed that specific genes (called **oncogenes**) are capable of inducing cell transformation, thereby providing the first insights into the molecular basis of cancer. However, the majority (approximately 80%) of human cancers are not induced by viruses and apparently arise from other causes, such as

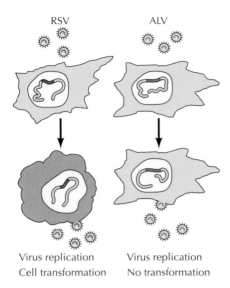

FIGURE 18.17 Cell transformation by RSV and ALV Both RSV and ALV infect and replicate in chicken embryo fibroblasts, but only RSV induces cell transformation.

radiation and chemical carcinogens. Therefore, in terms of our overall understanding of cancer, it has been critically important that studies of viral oncogenes also led to the identification of cellular oncogenes, which are involved in the development of non-virus-induced cancers. The key link between viral and cellular oncogenes was provided by studies of the highly oncogenic retroviruses.

Retroviral Oncogenes

Viral oncogenes were first defined in RSV, which transforms chicken embryo fibroblasts in culture and induces large sarcomas within 1 to 2 weeks after inoculation into chickens (Figure 18.17). In contrast, the closely related avian leukosis virus (ALV) replicates in the same cells as RSV without inducing transformation. This difference in transforming potential suggested the possibility that RSV contains specific genetic information responsible for transformation of infected cells. A direct comparison of the genomes of RSV and ALV was consistent with this hypothesis: The genomic RNA of RSV is about 10 kb, whereas that of ALV is smaller, about 8.5 kb.

In the early 1970s, Peter Vogt and Steven Martin isolated both deletion mutants and temperature-sensitive mutants of RSV that were unable to induce transformation. Importantly, these mutants still replicated normally in infected cells, indicating that RSV contains genetic information that is required for transformation but not for virus replication. Further analysis demonstrated that both the deletion and the temperature-sensitive RSV mutants define a single gene responsible for the ability of RSV to induce tumors in birds and transform fibroblasts in culture. Because RSV causes sarcomas, its oncogene was called **src**. The *src* gene is an addition to the genome of RSV; it is not present in ALV (Figure 18.18). It encodes a 60-kd protein that was the first protein-tyrosine kinase to be identified (see the Key Experiment in Chapter 15).

More than 40 different highly oncogenic retroviruses have been isolated from a variety of animals, including chickens, turkeys, mice, rats, cats, and monkeys. All of these viruses, like RSV, contain at least one oncogene (in some cases two) that is not required for virus replication but is responsible for cell transformation. In some cases, different viruses contain the same oncogenes, but more than two dozen distinct oncogenes have been identi-

FIGURE 18.18 The RSV genome RSV contains an additional gene, *src*, that is not present in ALV and encodes the Src protein-tyrosine kinase.

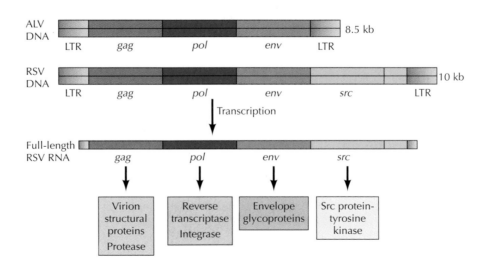

TABLE 18.3 Retroviral Oncogenes

Oncogene	Virus	Species
abl	Abelson leukemia	Mouse
akt	AKT8	Mouse
cbl	Cas NS-1	Mouse
crk	CT10 sarcoma	Chicken
erbA	Avian erythroblastosis-ES4	Chicken
erbB	Avian erythroblastosis-ES4	Chicken
ets	Avian erythroblastosis-E26	Chicken
fes	Gardner-Arnstein feline sarcoma	Cat
fgr	Gardner-Rasheed feline sarcoma	Cat
fms	McDonough feline sarcoma	Cat
fos	FBJ murine osteogenic sarcoma	Mouse
fps	Fujinami sarcoma	Chicken
jun	Avian sarcoma-17	Chicken
kit	Hardy-Zuckerman feline sarcoma	Cat
maf	Avian sarcoma AS42	Chicken
mos	Moloney sarcoma	Mouse
mpl	Myeloproliferative leukemia	Mouse
myb	Avian myeloblastosis	Chicken
myc	Avian myelocytomatosis	Chicken
p3k	Avian sarcoma-16	Chicken
qin	Avian sarcoma-31	Chicken
raf	3611 murine sarcoma	Mouse
*ras*H	Harvey sarcoma	Rat
*ras*K	Kirsten sarcoma	Rat
rel	Reticuloendotheliosis	Turkey
ros	UR2 sarcoma	Chicken
sea	Avian erythroblastosis-S13	Chicken
sis	Simian sarcoma	Monkey
ski	Avian SK	Chicken
src	Rous sarcoma	Chicken
yes	Y73 sarcoma	Chicken

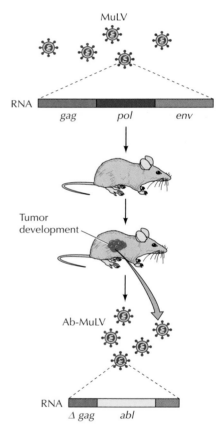

FIGURE 18.19 Isolation of Abelson leukemia virus The highly oncogenic virus Ab-MuLV was isolated from a rare tumor that developed in a mouse that had been inoculated with a nontransforming virus (Moloney murine leukemia virus, or MuLV). MuLV contains only the *gag*, *pol*, and *env* genes required for virus replication. In contrast, Ab-MuLV has acquired a new oncogene (*abl*), which is responsible for its transforming activity. The *abl* oncogene replaced some of the viral replicative genes and is fused with a partially deleted *gag* gene (designated Δ *gag*) in the Ab-MuLV genome.

fied among this group of viruses (Table 18.3). Like *src*, many of these genes (such as **ras** and **raf**) encode proteins that are now recognized as key components of signaling pathways that stimulate cell proliferation (see Figure 15.34).

Proto-Oncogenes

An unexpected feature of retroviral oncogenes is their lack of involvement in virus replication. Since most viruses are streamlined to replicate as efficiently as possible, the existence of viral oncogenes that are not an integral part of the virus life cycle seems paradoxical. Scientists were thus led to question where the retroviral oncogenes had originated and how they had become incorporated into viral genomes—a line of investigation that ultimately led to the identification of cellular oncogenes in human cancers.

The first clue to the origin of oncogenes came from the way in which the highly oncogenic retroviruses were isolated. The isolation of Abelson leukemia virus is a typical example (Figure 18.19). More than 150 mice were inoculated with a nontransforming virus containing only the *gag*, *pol*, and *env* genes required for virus replication. One of these mice developed a lymphoma from which a new, highly oncogenic virus (Abelson leukemia virus), which now contained an oncogene (*abl*), was isolated. The scenario suggested the hypothesis that the retroviral oncogenes are derived from genes of the host cell, and that occasionally such a host cell gene becomes incorporated into a viral genome, yielding a new, highly oncogenic virus as the product of a virus-host recombination event.

The critical prediction of this hypothesis was that normal cells contain genes that are closely related to the retroviral oncogenes. This was definitively demonstrated in 1976 by Harold Varmus, J. Michael Bishop, and their colleagues, who showed that a cDNA probe for the *src* oncogene of RSV hybridized to closely related sequences in the DNA of normal chicken cells. Moreover, *src*-related sequences were also found in normal DNAs of a wide range of other vertebrates (including humans) and thus appeared to be highly conserved in evolution. Similar experiments with probes for the oncogenes of other highly oncogenic retroviruses have yielded comparable results, and it is now firmly established that the retroviral oncogenes were derived from closely related genes of normal cells.

The normal-cell genes from which the retroviral oncogenes originated are called **proto-oncogenes**. They are important cell regulatory genes, in many cases encoding proteins that function in the signal transduction pathways controlling normal cell proliferation (e.g., *src*, *ras*, and *raf*). The oncogenes are abnormally expressed or mutated forms of the corresponding proto-oncogenes. As a consequence of such alterations, the oncogenes induce abnormal cell proliferation and tumor development.

An oncogene incorporated into a retroviral genome differs in several respects from the corresponding proto-oncogene. First, the viral oncogene is transcribed under the control of viral promoter and enhancer sequences, rather than being controlled by the normal transcriptional regulatory sequences of the proto-oncogene. Consequently, oncogenes are usually expressed at much higher levels than the proto-oncogenes and are sometimes transcribed in inappropriate cell types. In some cases, such abnormalities of gene expression are sufficient to convert a normally functioning proto-oncogene into an oncogene that drives cell transformation.

In addition to such alterations in gene expression, oncogenes frequently encode proteins that differ in structure and function from those encoded by their normal homologs. Many oncogenes, such as *raf*, are expressed as fusion proteins with viral sequences at the amino terminus (Figure 18.20). Recombination events leading to the generation of such fusion proteins often occur during the capture of proto-oncogenes by retroviruses, and sequences from both the amino and carboxy termini of proto-oncogenes are frequently deleted during the process. Such deletions may result in the loss of regulatory domains that control the activity of the proto-oncogene proteins thereby generating oncogene proteins that function in an unregulated manner. For example, the viral *raf* oncogene encodes a fusion protein in which amino-terminal sequences of the normal Raf protein have been deleted. These amino-terminal sequences are critical to the normal regulation of Raf protein kinase activity, and their deletion results in unregulated constitutive activity of the oncogene-encoded Raf protein. This unregulated Raf activity drives cell proliferation, resulting in transformation.

Many other oncogenes differ from the corresponding proto-oncogenes by point mutations, resulting in single amino acid substitutions in the oncogene products. In some cases, such amino acid substitutions (like the deletions already discussed) lead to unregulated activity of the oncogene proteins. An important example of such point mutations is provided by the *ras*

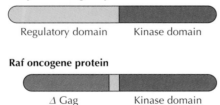

FIGURE 18.20 The Raf oncogene protein The Raf proto-oncogene protein consists of an amino-terminal regulatory domain and a carboxy-terminal protein kinase domain. In the viral Raf oncogene protein, the regulatory domain has been deleted and replaced by partially deleted viral Gag sequences (Δ Gag). As a result, the Raf kinase domain is constitutively active, causing cell transformation.

The Discovery of Proto-Oncogenes

DNA Related to the Transforming Gene(s) of Avian Sarcoma Viruses Is Present in Normal Avian DNA

Dominique Stehelin, Harold E. Varmus, J. Michael Bishop and Peter K. Vogt

Department of Microbiology, University of California, San Francisco (DS, HEV, and JMB) and Department of Microbiology, University of California, Los Angeles (PKV)

Nature, Volume 260, 1976, pages 170–173

J. Michael Bishop

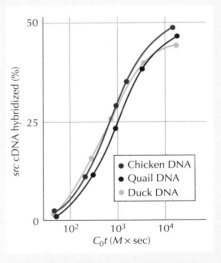

Harold Varmus

The Context

Genetic analysis of RSV defined the first viral oncogene (*src*) as a gene that was specifically responsible for cell transformation but was not required for virus replication. The origin of highly oncogenic retroviruses from tumors of infected animals then suggested the hypothesis that retroviral oncogenes are derived from related genes of host cells. Consistent with this suggestion, normal cells of several species were found to contain retrovirus-related DNA sequences that could be detected by nucleic acid hybridization. However, it was unclear whether these sequences were related to the retroviral oncogenes or to the genes required for virus replication.

Harold Varmus, J. Michael Bishop, and their colleagues resolved this critical issue by exploiting the genetic characterization of the *src* oncogene. In particular, Peter Vogt had previously isolated transformation-defective mutants of RSV that had sustained deletions of approximately 1.5 kb, corresponding to most or all of the *src* gene. Stehelin and collaborators used these mutants to prepare a cDNA probe that specifically represented *src* sequences. The use of this defined probe in nucleic acid hybridization experiments allowed them to definitively demonstrate that normal cells contain *src*-related DNA sequences.

The Experiments

First reverse transcriptase was used to synthesize a radioactive cDNA probe composed of short single-stranded DNA fragments complementary to the entire genomic RNA of RSV. This probe was then hybridized to an excess of RNA isolated from a transformation-defective deletion mutant.

Fragments of cDNA that were complementary to the viral replication genes hybridized to the transformation-defective RSV RNA, forming RNA-DNA duplexes. In contrast, cDNA fragments that were complementary to *src* were unable to hybridize and remained single-stranded. This single-stranded DNA was then isolated to provide a specific probe for *src* oncogene sequences. As predicted from the size of deletions in the transformation-defective RSV mutants, the *src*-specific probe was homologous to about 1.5 kb of RSV RNA.

The radioactive *src* cDNA was then used as a hybridization probe to attempt to detect related DNA sequences in normal avian cells. Strikingly, the *src* cDNA hybridized extensively to normal chicken DNA, as well as to DNA of other avian species (see figure). These experiments thus demonstrated that normal cells contain DNA sequences that are closely related to the *src* oncogene, supporting the hypothesis that retroviral oncogenes originated from cellular genes that became incorporated into viral genomes.

The Impact

Stehelin and colleagues concluded their 1976 paper by raising the possibility that the cellular *src* sequences are "involved in the normal regulation of cell growth and development or the transformation of cell behavior by physical, chemical or viral agents." This prediction has been strikingly substantiated, and the discovery of cellular *src* sequences has opened new doors to understanding both the regulation of normal cell proliferation and the molecular basis of human cancer. Studies of oncogene and proto-oncogene proteins, including Src itself, have proven critical in unraveling the signaling pathways that control the proliferation and differentiation of normal cells. The discovery of the *src* proto-oncogene further suggested that non-virus-induced tumors could arise as a result of mutations in related cellular genes, leading directly to the discovery of oncogenes in human tumors. By unifying studies of tumor viruses, normal cells, and non-virus-induced tumors, the results of Varmus, Bishop, and their colleagues have had an impact on virtually all aspects of studies of cell regulation and cancer research.

Hybridization of *src*-specific cDNA to normal chicken, quail, and duck DNA.

Human bladder carcinoma

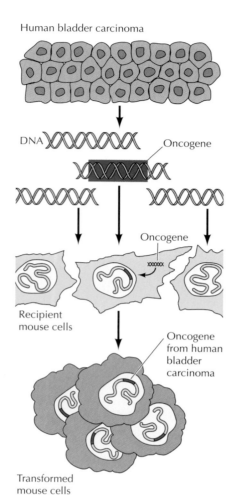

Recipient mouse cells

Oncogene from human bladder carcinoma

Transformed mouse cells

FIGURE 18.21 Detection of a human tumor oncogene by gene transfer DNA extracted from a human bladder carcinoma induced transformation of recipient mouse cells in culture. Transformation resulted from integration and expression of an oncogene derived from the human tumor.

oncogenes, which are discussed in the next section in terms of their role in human cancers.

Oncogenes in Human Cancer

Understanding the origin of retroviral oncogenes raised the question as to whether non-virus-induced tumors contain cellular oncogenes that were generated from proto-oncogenes by mutations or by DNA rearrangements during tumor development. Direct evidence for the involvement of cellular oncogenes in human tumors was first obtained by gene transfer experiments in the laboratories of Robert Weinberg and of one of the authors (GMC) in 1981. DNA of a human bladder carcinoma was found to efficiently induce transformation of recipient mouse cells in culture, indicating that the human tumor contained a biologically active cellular oncogene (Figure 18.21). Both gene transfer assays and alternative experimental approaches have since led to the detection of active cellular oncogenes in human tumors of many different types (Table 18.4).

Some of the oncogenes identified in human tumors are cellular homologs of oncogenes that were previously characterized in retroviruses, whereas others are new oncogenes first discovered in human cancers. The first human oncogene identified in gene transfer assays was subsequently identified as the human homolog of the *ras*H oncogene of Harvey sarcoma virus (see Table 18.3). Three closely related members of the *ras* gene family (*ras*H, *ras*K, and *ras*N) are the oncogenes most frequently encountered in human tumors. These genes are involved in approximately 20% of all human malignancies, including about 50% of colon and 25% of lung carcinomas.

The *ras* oncogenes are not present in normal cells; rather, they are generated in tumor cells as a consequence of mutations that occur during tumor development. The *ras* oncogenes differ from their proto-oncogenes by point mutations resulting in single amino acid substitutions at critical positions. The first such mutation discovered was the substitution of valine for glycine at position 12 (Figure 18.22). Other amino acid substitutions at position 12 (as well as at positions 13 and 61) are also frequently encountered in *ras* oncogenes in human tumors. In animal models, it has been shown that mutations that convert *ras* proto-oncogenes to oncogenes are caused by chemical carcinogens, providing a direct link between the mutagenic action of carcinogens and cell transformation.

FIGURE 18.22 Point mutations in *ras* oncogenes A single nucleotide change, which alters codon 12 from GGC (Gly) to GTC (Val), is responsible for the transforming activity of the *ras*H oncogene detected in bladder carcinoma DNA.

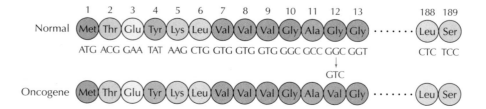

TABLE 18.4 Representative Oncogenes of Human Tumors

Oncogene	Type of cancer	Activation mechanism
abl	Chronic myeloid leukemia, acute lymphocytic leukemia	Translocation
akt	Breast, ovarian and pancreatic carcinomas	Amplification
bcl-2	Follicular B-cell lymphoma	Translocation
CCND1	Parathyroid adenoma, B-cell lymphoma	Translocation
CCND1	Squamous cell, bladder, breast, esophageal, liver, and lung carcinomas	Amplification
cdk4	Melanomas	Point mutation
CTNNB1 (β-catenin)	Colon carcinoma	Point mutation
*erb*B	Gliomas, many carcinomas	Amplification
*erb*B	Lung carcinomas	Point mutation
*erb*B-2	Breast and ovarian carcinomas	Amplification
gli	Glioblastoma	Amplification
kit	Gastrointestinal stromal tumors	Point mutation
c-*myc*	Burkitt's lymphoma	Translocation
c-*myc*	Breast and lung carcinomas	Amplification
L-*myc*	Lung carcinoma	Amplification
N-*myc*	Neuroblastoma, lung carcinoma	Amplification
PDGFR	Chronic myelomonocytic leukemia	Translocation
PDGFR	Gastrointestinal stromal tumors	Point mutation
PI3K	Breast carcinoma	Point mutation
	Ovarian, gastric, lung carcinoma	Amplification
PML/RARα	Acute promyelocytic leukemia	Translocation
B-*raf*	Melanoma, colon carcinoma	Point mutation
*ras*H	Thyroid carcinoma	Point mutation
*ras*K	Colon, lung, pancreatic, and thyroid carcinomas	Point mutation
*ras*N	Acute myeloid and lymphocytic leukemias, thyroid carcinoma	Point mutation
ret	Multiple endocrine neoplasia types 2A and 2B	Point mutation
ret	Thyroid carcinoma	DNA rearrangement
SMO	Basal cell carcinoma	Point mutation

As discussed in Chapter 15, the *ras* genes encode guanine nucleotide-binding proteins that function in transduction of mitogenic signals from a variety of growth factor receptors. The activity of the Ras proteins is controlled by GTP or GDP binding, such that they alternate between active (GTP-bound) and inactive (GDP-bound) states (see Figure 15.35). The mutations characteristic of *ras* oncogenes have the effect of maintaining the Ras proteins constitutively in the active GTP-bound conformation. In large part, this effect is a result of nullifying the response of oncogenic Ras proteins to GAP (GTPase-activating protein), which stimulates hydrolysis of bound GTP by normal Ras. Because of the resulting decrease in their intracellular GTPase activity, the oncogenic Ras proteins remain in the active GTP-bound state and drive unregulated cell proliferation.

Point mutations are only one of the ways in which proto-oncogenes are converted to oncogenes in human tumors. Many cancer cells display abnor-

FIGURE 18.23 Translocation of c-myc The c-*myc* proto-oncogene is translocated from chromosome 8 to the immunoglobulin heavy-chain locus (IgH) on chromosome 14 in Burkitt's lymphomas, resulting in abnormal c-*myc* expression.

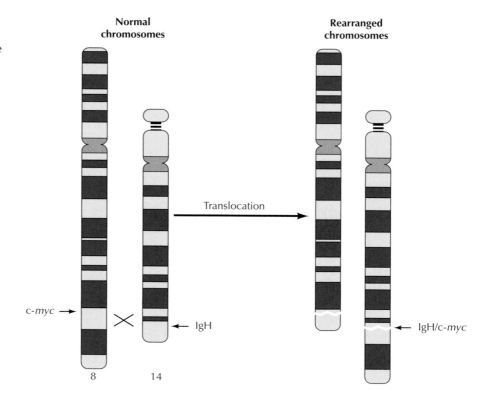

malities in chromosome structure, including translocations, duplications, and deletions. The gene rearrangements resulting from chromosome translocations frequently lead to the generation of oncogenes. In some cases, analysis of these rearrangements has implicated already known oncogenes in tumor development. In other cases, novel oncogenes have been discovered by molecular cloning and analysis of rearranged DNA sequences.

The first characterized example of oncogene activation by chromosome translocation was the involvement of the **c-myc** oncogene in human Burkitt's lymphomas and mouse plasmacytomas, which are malignancies of antibody-producing B lymphocytes (Figure 18.23). Both of these tumors are characterized by chromosome translocations involving the genes that encode immunoglobulins. For example, virtually all Burkitt's lymphomas have translocations of a fragment of chromosome 8 to one of the immunoglobulin gene loci, which reside on chromosomes 2 (κ light chain), 14 (heavy chain), and 22 (λ light chain). The fact that the immunoglobulin genes are actively expressed in these tumors suggested that the translocations activate a proto-oncogene from chromosome 8 by inserting it into the immunoglobulin loci. This possibility was investigated by analysis of tumor DNAs with probes for known oncogenes, leading to the finding that the c-*myc* proto-oncogene was the chromosome 8 translocation break point in Burkitt's lymphomas. These translocations inserted c-*myc* into an immunoglobulin locus, where it was expressed in an unregulated manner. Such uncontrolled expression of the c-*myc* gene, which encodes a transcription factor normally induced in response to growth factor stimulation, is sufficient to drive cell proliferation and contribute to tumor development.

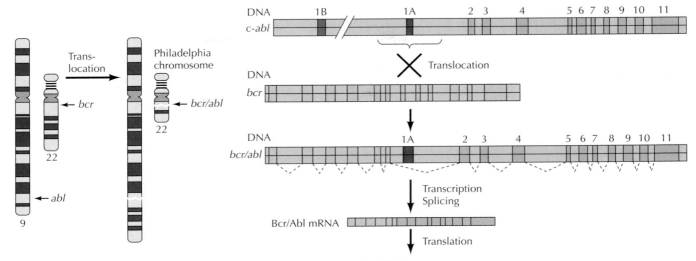

FIGURE 18.24 Translocation of *abl*
The *abl* oncogene is translocated from chromosome 9 to chromosome 22, forming the Philadelphia chromosome in chronic myeloid leukemias. The *abl* proto-oncogene, which contains two alternative first exons (1A and 1B), is joined to the middle of the *bcr* gene on chromosome 22. Exon 1B is deleted as a result of the translocation. Transcription of the fused gene initiates at the *bcr* promoter and continues through *abl*. Splicing then generates a fused Bcr/Abl mRNA in which *abl* exon 1A sequences are also deleted and *bcr* sequences are joined to *abl* exon 2. The Bcr/Abl mRNA is translated to yield a recombinant Bcr/Abl fusion protein.

Translocations of other proto-oncogenes frequently result in rearrangements of coding sequences, leading to the formation of abnormal gene products. The prototype is translocation of the ***abl*** proto-oncogene from chromosome 9 to chromosome 22 in chronic myeloid leukemia (**Figure 18.24**). This translocation leads to fusion of *abl* with its translocation partner, a gene called *bcr*, on chromosome 22. The result is production of a Bcr/Abl fusion protein in which the normal amino terminus of the Abl proto-oncogene protein has been replaced by Bcr amino acid sequences. The fusion of Bcr sequences results in unregulated activity of the Abl protein-tyrosine kinase, leading to cell transformation.

A distinct mechanism by which oncogenes are activated in human tumors is gene amplification, which results in elevated gene expression. DNA amplification (see Figure 6.50) is common in tumor cells, occurring more than a thousand times more frequently than in normal cells, and amplification of oncogenes may play a role in the progression of many tumors to more rapid growth and increasing malignancy. Indeed, novel oncogenes have been identified by molecular cloning and characterization of DNA sequences that are amplified in tumors.

A prominent example of oncogene amplification is the involvement of the **N-*myc*** gene, which is related to c-*myc*, in neuroblastoma (a childhood tumor of neuronal cells). Amplified copies of N-*myc* are frequently present in rapidly growing aggressive tumors, indicating that N-*myc* amplification is associated with the progression of neuroblastomas to increasing malignancy. Amplification of another oncogene, ***erb*B-2**, which encodes a receptor protein-tyrosine kinase, is similarly related to progression of breast and ovarian carcinomas.

Functions of Oncogene Products

The viral and cellular oncogenes have defined a large group of genes (about 100 in total) that can contribute to the abnormal behavior of malignant cells. As already noted, many of the proteins encoded by proto-oncogenes regulate normal cell proliferation; in these cases, the elevated expression or activity of the corresponding oncogene proteins drives the uncontrolled

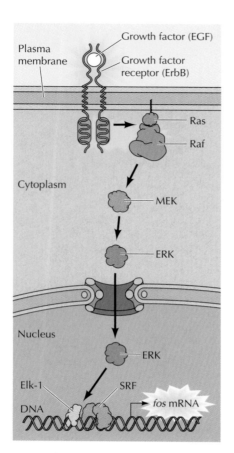

FIGURE 18.25 Oncogenes and the ERK signaling pathway Oncogene proteins act as growth factors (e.g., EGF), growth factor receptors (e.g., ErbB), and intracellular signaling molecules (Ras and Raf). Ras and Raf activate the ERK MAP kinase pathway (see Figures 15.34 and 15.37), leading to the induction of additional genes (e.g., *fos*) that encode potentially oncogenic transcriptional regulatory proteins. Proteins with known oncogenic potential are highlighted with a yellow glow.

proliferation of cancer cells. Other oncogene products contribute to other aspects of the behavior of cancer cells, such as failure to undergo programmed cell death or defective differentiation.

The function of oncogene proteins in regulation of cell proliferation is illustrated by their activities in growth factor-stimulated pathways of signal transduction, such as the activation of ERK signaling downstream of receptor protein-tyrosine kinases (**Figure 18.25**). The oncogene proteins within this pathway include polypeptide growth factors, growth factor receptors, intracellular signaling proteins, transcription factors, and the cell cycle regulator cyclin D1.

The action of growth factors as oncogene proteins results from their abnormal expression, leading to a situation in which a tumor cell produces a growth factor to which it also responds. The result is autocrine stimulation of the growth factor-producing cell (see Figure 18.9), which drives abnormal cell proliferation and contributes to the development of a wide variety of human tumors.

A large group of oncogenes encode growth factor receptors, most of which are protein-tyrosine kinases. These receptors can be converted to oncogene proteins by alterations of their amino-terminal domains, which would normally bind extracellular growth factors. For example, the receptor for platelet-derived growth factor (PDGF) is converted to an oncogene in some human leukemias by a chromosome translocation in which the normal amino terminus of the PDGF receptor (PDGFR) is replaced by the amino terminal sequences of a transcription factor called Tel (Figure 18.26). The Tel sequences of the resulting Tel/PDGFR fusion protein dimerize in the absence of growth factor binding, resulting in constitutive activity of the intracellular kinase domain and unregulated production of a proliferative signal from the oncogene protein. Alternatively, genes that encode receptor protein-tyrosine kinases can be activated by gene amplification or by point mutations that result in unregulated kinase activity. Other oncogenes (including *src* and *abl*) encode nonreceptor protein-tyrosine kinases that are constitutively activated by deletions or mutations of regulatory sequences.

The Ras proteins play a key role in mitogenic signaling by coupling growth factor receptors to activation of the Raf protein-serine/threonine kinase, which initiates a protein kinase cascade leading to activation of the ERK MAP kinase (see Figure 15.34). As discussed earlier, the mutations that convert *ras* proto-oncogenes to oncogenes result in constitutive Ras activity, which leads to activation of the ERK pathway. The *raf* gene can similarly be converted to an oncogene by deletions that result in loss of the amino-terminal regulatory domain of the Raf protein (see Figure 18.20). These deletions result in unregulated activity of the Raf protein kinase, which also leads to constitutive ERK activation. Alternatively, *raf* proto-oncogenes can be converted to oncogenes by point mutations that result in elevated Raf kinase activity.

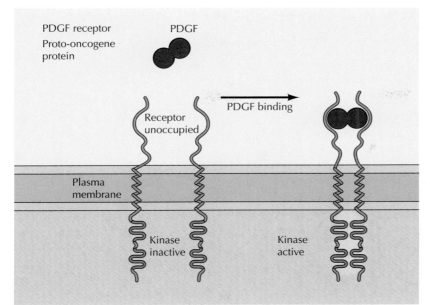

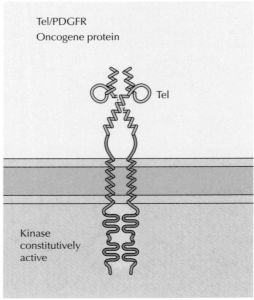

FIGURE 18.26 Mechanism of Tel/PDGFR oncogene activation
The normal PDGF receptor (PDGFR) is activated by dimerization induced by PDGF binding. The Tel/PDGFR oncogene encodes a fusion protein in which the normal extracellular domain of PDGFR is replaced by the amino terminal sequences of the Tel transcription factor, which include its helix-loop-helix dimerization domain (see Figure 7.29). These sequences dimerize in the absence of PDGF, leading to constitutive activation of the oncogene protein kinase.

The ERK pathway ultimately leads to the phosphorylation of transcription factors and alterations in gene expression. As might therefore be expected, many oncogenes encode transcriptional regulatory proteins that are normally induced in response to growth factor stimulation. For example, transcription of the *fos* proto-oncogene is induced as a result of phosphorylation of Elk-1 by ERK (see Figure 18.25). **Fos** and the product of another proto-oncogene, **Jun**, are components of the AP-1 transcription factor, which activates transcription of a number of target genes, including cyclin D1, in growth factor–stimulated cells (Figure 18.27). Constitutive activity of AP-1, resulting from unregulated expression of either the Fos or Jun oncogene proteins, is sufficient to drive abnormal cell proliferation, leading to cell transformation. The Myc proteins similarly function as transcription factors regulated by mitogenic stimuli, and abnormal expression of *myc* oncogenes contributes to the development of a variety of human tumors. Other transcription factors are frequently activated as oncogenes by chromosome translocations in human leukemias and lymphomas.

The signaling pathways activated by growth factor stimulation ultimately regulate components of the cell cycle machinery that promote progression through the restriction point in G_1. The D-type cyclins are induced in response to growth factor stimulation (at least in part via activation of the AP-1 transcription factor) and play a key role in coupling growth factor signaling to cell cycle progression (see Figure 18.27). Perhaps not surprisingly, the gene encoding cyclin D1 is a proto-oncogene, which can be activated as an oncogene (called *CCND*1) by chromosome translocation or gene amplifi-

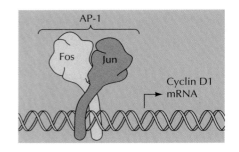

FIGURE 18.27 The AP-1 transcription factor   Fos and Jun dimerize to form AP-1, which activates transcription of cyclin D1 and other growth factor–inducible genes.

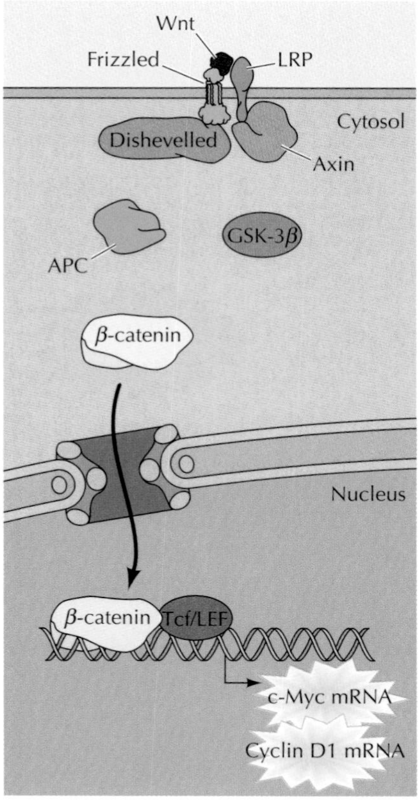

FIGURE 18.28 Oncogenic activity of the Wnt pathway   Wnt polypeptides bind to Frizzled and LRP receptors, which activate Dishevelled and lead to inhibition of the axin/APC/GSK-3β complex. This results in stabilization of β-catenin, which translocates to the nucleus and forms a complex with Tcf/LEF transcription factors, leading to activation of target genes, including the genes encoding c-Myc and cyclin D1. The genes encoding both Wnt and β-catenin can act as oncogenes.

cation. These alterations lead to constitutive expression of cyclin D1, which then drives cell proliferation in the absence of normal growth factor stimulation. The catalytic partner of cyclin D1, Cdk4, is also activated as an oncogene by point mutations in melanomas.

Components of other signaling pathways discussed in Chapter 15, including G protein-coupled signaling pathways, the NF-κB pathway, and the Hedgehog, Wnt, and Notch pathways, can also act as oncogenes. For example, Wnt proteins were identified as oncogenes in mouse breast cancers, and activating mutations frequently convert the downstream target of Wnt signaling, β-catenin, to an oncogene (*CTNNB1*) in human colon cancers (Figure 18.28). These activating mutations stabilize β-catenin, which then forms a complex with Tcf and stimulates transcription of target genes. The targets of β-catenin/Tcf include the genes encoding c-Myc and cyclin D1, leading to unregulated cell proliferation. It is noteworthy that Wnt signaling normally promotes the proliferation of stem cells and their progeny during continuing cell renewal in the colon (see Figure 17.18), indicating that colon cancer results from abnormal activity of the same pathway that signals physiologically normal proliferation of colon epithelial cells.

Although many oncogenes stimulate cell proliferation, the oncogenic activity of some transcription factors instead results from inhibition of cell differentiation. As noted in Chapter 15, thyroid hormone and retinoic acid induce differentiation of a variety of cell types. These hormones diffuse through the plasma membrane and bind to intracellular receptors that act as transcriptional regulatory molecules. Mutated forms of both the thyroid hormone receptor (**ErbA**) and the retinoic acid receptor (**PML/RARα**) act as oncogene proteins in chicken erythroleukemia and human acute promyelocytic leukemia, respectively. In both cases, the mutated oncogene receptors appear to interfere with the action of their normal homologs, thereby blocking cell differentiation and maintaining the leukemic cells in an actively proliferating state (Figure 18.29). In the case of acute promyelocytic

FIGURE 18.29 Action of the PML/RARα oncogene protein   The PML/RARα fusion protein blocks the differentiation of promyelocytes to granulocytes.

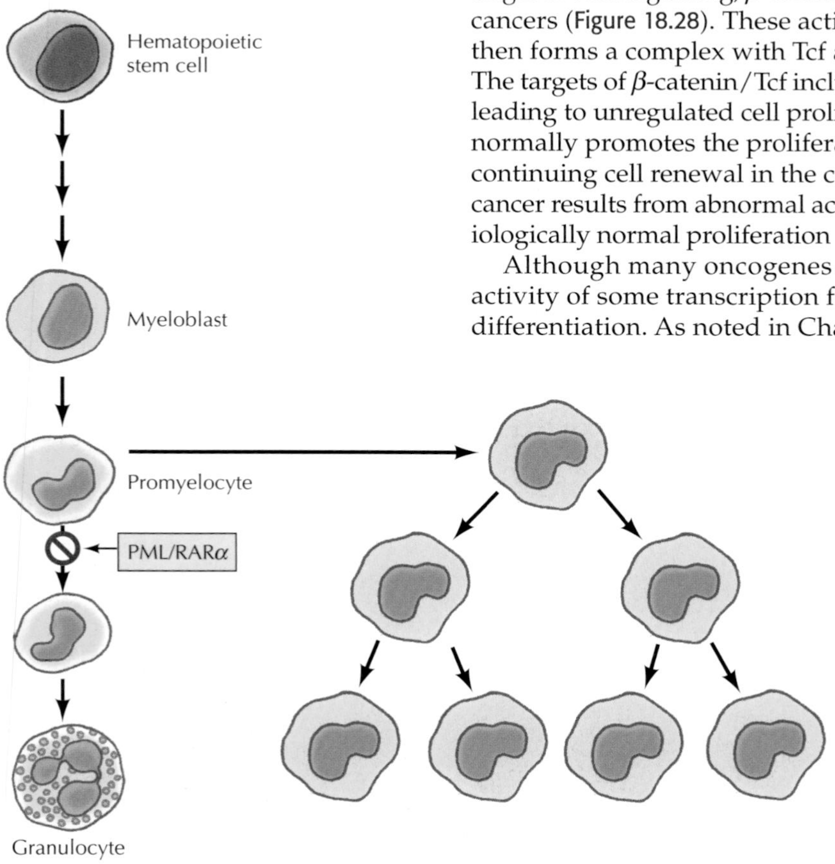

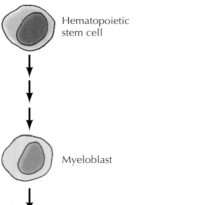

leukemia, high doses of retinoic acid can overcome the effect of the PML/RARα oncogene protein and induce differentiation of the leukemic cells. This biological observation has a direct clinical correlate: Patients with acute promyelocytic leukemia can be treated effectively by administration of retinoic acid, which induces differentiation and blocks continued cell proliferation.

As discussed earlier in this chapter, the failure of cancer cells to undergo programmed cell death (or apoptosis) is a critical factor in tumor development, and several oncogenes encode proteins that act to promote cell survival (Figure 18.30). The survival of most animal cells is dependent on growth factor stimulation, so those oncogenes that encode growth factors, growth factor receptors, and signaling proteins such as Ras act not only to stimulate cell proliferation but also to prevent cell death. As discussed in Chapter 16, the PI 3-kinase/Akt signaling pathway plays a key role in preventing apoptosis of many growth factor–dependent cells, and the genes encoding **PI 3-kinase** and **Akt** act as oncogenes in both retroviruses and

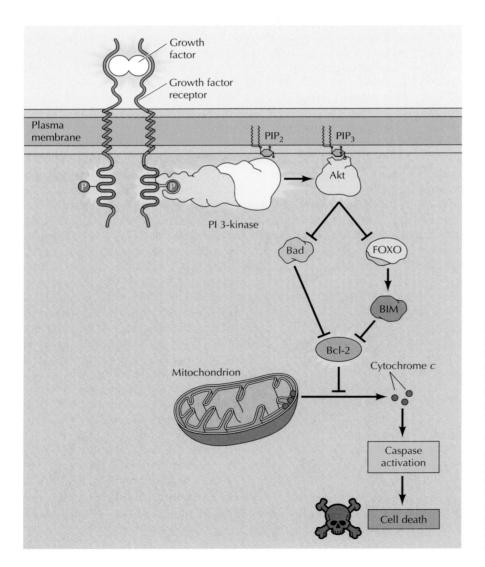

FIGURE 18.30 Oncogenes and cell survival   The oncogene proteins that signal cell survival include growth factors, growth factor receptors, PI 3-kinase, Akt, and Bcl-2. The targets of Akt include the proapoptotic Bcl-2 family member Bad and the FOXO transcription factor, which stimulates transcription of another proapoptotic Bcl-2 family member, Bim. Phosphorylation by Akt inhibits both Bad and FOXO, promoting cell survival. The antiapoptotic protein Bcl-2 also functions as an oncogene by promoting cell survival and inhibiting the release of cytochrome *c* from mitochondria. Proteins with oncogenic potential are highlighted with a yellow glow.

human tumors. The downstream targets of PI 3-kinase/Akt signaling include a proapoptotic member of the Bcl-2 family, Bad, which is inactivated as a result of phosphorylation by Akt, as well as the FOXO transcription factor, which regulates expression of the proapoptotic Bcl-2 family member, Bim. In addition, it is notable that **Bcl-2** itself was first discovered as the product of an oncogene in human lymphomas. The *bcl*-2 oncogene is generated by a chromosome translocation that results in elevated expression of Bcl-2, which blocks apoptosis and maintains cell survival under conditions that normally induce cell death. The identification of *bcl*-2 as an oncogene not only provided the first demonstration of the significance of programmed cell death in the development of cancer but also led to the discovery of the role of Bcl-2 and related genes as central regulators of apoptosis in organisms ranging from *C. elegans* to humans.

Tumor Suppressor Genes

The activation of cellular oncogenes represents only one of two distinct types of genetic alterations involved in tumor development; the other is inactivation of tumor suppressor genes. Oncogenes drive abnormal cell proliferation as a consequence of genetic alterations that either increase gene expression or lead to uncontrolled activity of the oncogene-encoded proteins. **Tumor suppressor genes** represent the opposite side of cell growth control, normally acting to inhibit cell proliferation and tumor development. In many tumors, these genes are lost or inactivated, thereby removing negative regulators of cell proliferation and contributing to the abnormal proliferation of tumor cells.

Identification of Tumor Suppressor Genes

The first insight into the activity of tumor suppressor genes came from somatic cell hybridization experiments initiated by Henry Harris and his colleagues in 1969. The fusion of normal cells with tumor cells yielded hybrid cells containing chromosomes from both parents (Figure 18.31). In

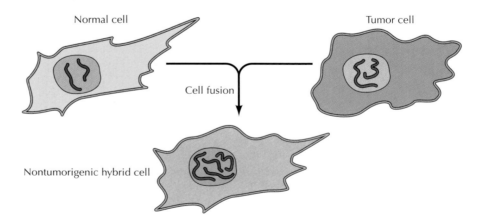

FIGURE 18.31 Suppression of tumorigenicity by cell fusion   Fusion of tumor cells with normal cells yields hybrids that contain chromosomes from both parents. Such hybrids are usually nontumorigenic.

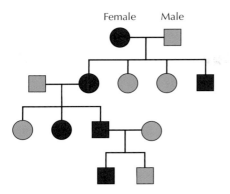

Female   Male

FIGURE 18.32 Inheritance of retinoblastoma Susceptibility to retinoblastoma is transmitted to approximately 50% of offspring. Affected and normal individuals are indicated by purple and green symbols, respectively.

most cases, such hybrid cells were not capable of forming tumors in animals. Therefore it appeared that genes derived from the normal cell parent acted to inhibit (or suppress) tumor development. Definition of these genes at the molecular level came, however, from a different approach—the analysis of rare inherited forms of human cancer.

The first tumor suppressor gene was identified by studies of retinoblastoma, a rare childhood eye tumor. Provided that the disease is detected early, retinoblastoma can be successfully treated and many patients survive to have families. Consequently, it was recognized that some cases of retinoblastoma are inherited. In these cases, approximately 50% of the children of an affected parent develop retinoblastoma, consistent with Mendelian transmission of a single dominant gene that confers susceptibility to tumor development (Figure 18.32).

Although susceptibility to retinoblastoma is transmitted as a dominant trait, inheritance of the susceptibility gene is not sufficient to transform a normal retinal cell into a tumor cell. All retinal cells in a patient inherit the susceptibility gene, but only a small fraction of these cells give rise to tumors. Thus tumor development requires additional events beyond inheritance of tumor susceptibility. In 1971 Alfred Knudson proposed that the development of retinoblastoma requires two mutations, which are now known to correspond to the loss of both of the functional copies of the tumor susceptibility gene (the *Rb* tumor suppressor gene) that would be present on homologous chromosomes of a normal diploid cell (Figure 18.33). In inherited retinoblastoma, one defective copy of *Rb* is genetically transmitted. The loss of this single *Rb* copy is not by itself sufficient to trigger tumor development, but retinoblastoma almost always develops in these individuals as a result of a second somatic mutation leading to the loss of the remaining normal *Rb* allele. Noninherited retinoblastoma, in contrast, is rare, since its development requires two independent somatic mutations to inactivate both normal copies of *Rb* in the same cell.

The functional nature of the *Rb* gene as a negative regulator of tumorigenesis was initially indicated by observations of chromosome morphology. Visible deletions of chromosome 13q14 were found in some retinoblastomas, suggesting that loss (rather than activation) of the *Rb* gene led to tumor development (Figure 18.34). Gene-mapping studies further indicated that tumor development resulted from loss of normal *Rb* alleles in the tumor cells, consistent with the function of *Rb* as a tumor suppressor gene. Isolation of the *Rb* gene as a molecular clone in 1986 then firmly established

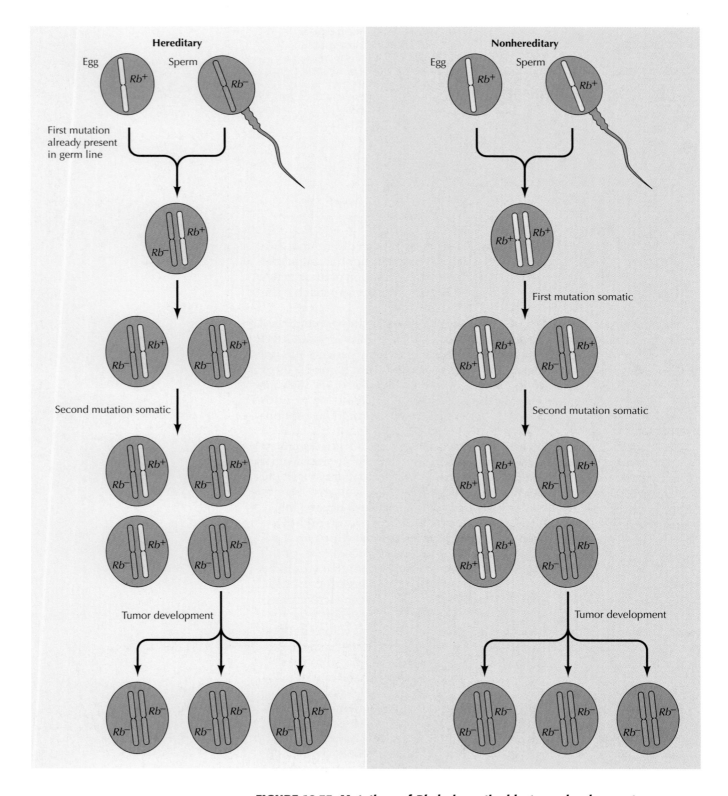

FIGURE 18.33 Mutations of *Rb* during retinoblastoma development
In hereditary retinoblastoma, a defective copy of the *Rb* gene (*Rb⁻*) is inherited from the affected parent. A second somatic mutation, which inactivates the single normal *Rb⁺* copy in a retinal cell, then leads to the development of retinoblastoma. In non-hereditary cases, two normal *Rb⁺* genes are inherited, and retinoblastoma develops only if two somatic mutations inactivate both copies of *Rb* in the same cell.

FIGURE 18.34 *Rb* deletions in retinoblastoma Many retinoblastomas have deletions of the chromosomal locus (13q14) that contains the *Rb* gene.

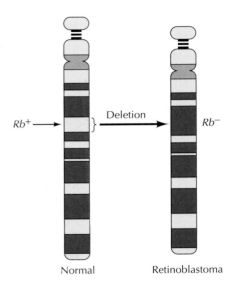

that *Rb* is consistently lost or mutated in retinoblastomas. Gene transfer experiments also demonstrated that introduction of a normal *Rb* gene into retinoblastoma cells reverses their tumorigenicity, providing direct evidence for the activity of *Rb* as a tumor suppressor.

Although *Rb* was identified in a rare childhood cancer, it is also involved in more common tumors of adults. In particular, studies of the cloned gene have established that *Rb* is lost or inactivated in many bladder, breast, and lung carcinomas. The significance of the *Rb* tumor suppressor gene thus extends beyond retinoblastoma, and mutations of the *Rb* gene contribute to development of a substantial fraction of human cancers. In addition, as noted earlier in this chapter, the Rb protein is a key target for the oncogene proteins of several DNA tumor viruses, including SV40, adenoviruses, and human papillomaviruses, which bind to Rb and inhibit its activity (Figure 18.35). Transformation by these viruses thus results, at least in part, from inactivation of Rb at the protein level rather than from mutational inactivation of the *Rb* gene.

Characterization of *Rb* as a tumor suppressor gene served as the prototype for the identification of additional tumor suppressor genes that contribute to the development of many different human cancers (Table 18.5). Some of these genes were identified as the causes of rare inherited cancers, playing a role similar to that of *Rb* in hereditary retinoblastoma. Other tumor suppressor genes have been identified as genes that are frequently deleted or mutated in common noninherited cancers of adults, such as colon carcinoma. In either case, it appears that most tumor suppressor genes are involved in the development of both inherited and noninherited forms of cancer. Indeed, mutations of some tumor suppressor genes appear to be the most common molecular alterations leading to human tumor development.

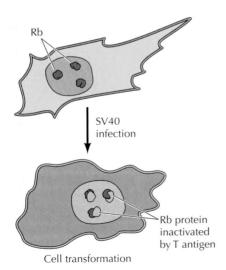

FIGURE 18.35 Interaction of Rb with oncogene proteins of DNA tumor viruses The oncogene proteins of several DNA tumor viruses (e.g., SV40 T antigen) induce transformation by binding to and inactivating Rb protein.

TABLE 18.5 Representative Tumor Suppressor Genes

Gene	Type of cancer
APC	Colon/rectum carcinoma
BRCA1	Breast and ovarian carcinomas
BRCA2	Breast carcinoma
INK4	Melanoma, lung carcinoma, brain tumors, leukemias, lymphomas
NF1	Neurofibrosarcoma
NF2	Meningioma
p53	Brain tumors; breast, colon/rectum, esophageal, liver, and lung carcinomas; sarcomas; leukemias and lymphomas
PTCH	Basal cell carcinoma
PTEN	Brain tumors; melanoma; prostate, endometrial, kidney, and lung carcinomas
Rb	Retinoblastoma; sarcomas; bladder, breast, and lung carcinomas
Smad2	Colon/rectum carcinoma
Smad4	Colon/rectum carcinoma, pancreatic carcinoma
TβRII	Colon/rectum carcinoma, gastric carcinoma
VHL	Renal cell carcinoma
WT1	Wilms' tumor

The second tumor suppressor gene to have been identified is *p53*, which is frequently inactivated in a wide variety of human cancers, including leukemias, lymphomas, sarcomas, brain tumors, and carcinomas of many tissues, including breast, colon, and lung. In total, mutations of *p53* play a role in about 50% of all cancers, making it the most common target of genetic alterations in human malignancies. It is also of interest that inherited mutations of *p53* are responsible for genetic transmission of a rare hereditary cancer syndrome in which affected individuals develop any of several different types of cancer. In addition, the p53 protein (like Rb) is a target for the oncogene proteins of SV40, adenoviruses, and human papillomaviruses.

Like *p53*, the *INK4* and *PTEN* tumor suppressor genes are very frequently mutated in several common cancers, including lung cancer, prostate cancer, and melanoma. Other tumor suppressor genes (including *APC*, *TβRII*, *Smad2*, and *Smad4*) are frequently inactivated in colon cancers. In addition to being involved in noninherited cases of this common adult cancer, inherited mutations of the *APC* gene are responsible for a rare hereditary form of colon cancer, called familial adenomatous polyposis. Individuals with this condition develop hundreds of benign colon adenomas (polyps), some of which inevitably progress to malignancy. Inherited mutations of two other tumor suppressor genes, *BRCA1* and *BRCA2*, are responsible for hereditary cases of breast cancer, which account for about 5% of total breast cancer incidence. Additional tumor suppressor genes have been implicated in the development of brain tumors, pancreatic cancers, and basal cell skin carcinomas, as well as in several rare inherited cancer syndromes, such as Wilms' tumor.

Functions of Tumor Suppressor Gene Products

In contrast to proto-oncogene and oncogene proteins, the proteins encoded by most tumor suppressor genes inhibit cell proliferation or survival. Inacti-

■ *p53* was initially thought to be an oncogene because mutated *p53* genes found in many cancer cells induced transformation in gene transfer assays. Subsequent studies revealed that *p53* was actually a tumor suppressor and that the mutant *p53* genes found in many tumors acted as dominant negatives that induced transformation by interfering with normal *p53* function.

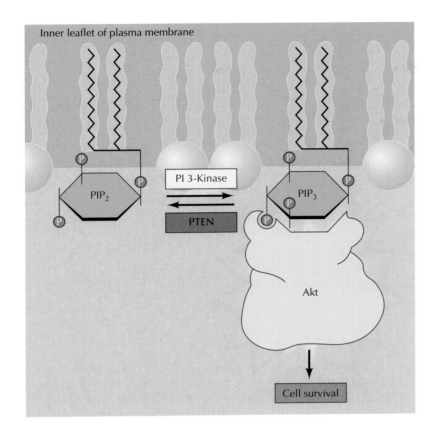

FIGURE 18.36 Suppression of cell survival by PTEN The tumor suppressor protein PTEN is a lipid phosphatase that dephosphorylates PIP$_3$ at the 3 position of inositol, yielding PIP$_2$. PTEN thus counters the action of the oncogene proteins PI 3-kinase and Akt, which promote cell survival.

vation of tumor suppressor genes therefore leads to tumor development by eliminating negative regulatory proteins. In many cases, tumor suppressor proteins inhibit the same cell regulatory pathways that are stimulated by the products of oncogenes.

The protein encoded by the *PTEN* tumor suppressor gene is an interesting example of antagonism between oncogene and tumor suppressor gene products (Figure 18.36). The PTEN protein is a lipid phosphatase that dephosphorylates the 3 position of phosphatidylinositides, such as phosphatidylinositol 3,4,5-bisphosphate (PIP$_3$). By dephosphorylating PIP$_3$, PTEN antagonizes the activities of PI 3-kinase and Akt, both of which can act as oncogenes by promoting cell survival. Conversely, inactivation or loss of the PTEN tumor suppressor protein can contribute to tumor development as a result of increased levels of PIP$_3$, activation of Akt, and inhibition of programmed cell death.

Proteins encoded by both oncogenes and tumor suppressor genes also function in the Hedgehog signaling pathway (see Figure 15.43). The receptor Smoothened is an oncogene in basal cell carcinomas, whereas Patched (the negative regulator of Smoothened) is a tumor suppressor gene. In addition, the Gli proteins (the mammalian homologs of the *Drosophila* Ci transcription factor activated by Smoothened) were first identified as the products of an amplified oncogene.

Several tumor suppressor genes encode transcriptional regulatory proteins. A good example is provided by the product of *WT1*, which is frequently inactivated in Wilms' tumors (a childhood kidney tumor). The WT1 protein is a repressor that appears to suppress transcription of a number of

FIGURE 18.37 Inhibition of cell cycle progression by Rb and p16 Rb inhibits progression past the restriction point in G₁. Cdk4, 6/cyclin D complexes promote passage through the restriction point by phosphorylating and inactivating Rb. The activity of Cdk4, 6/cyclin D is inhibited by p16. Rb and p16 are tumor suppressors, whereas cyclin D1 and Cdk4 are oncogenes.

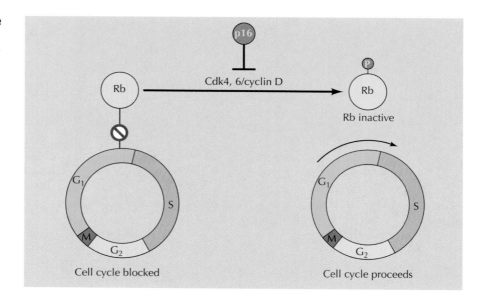

growth factor–inducible genes. One of the targets of WT1 is thought to be the gene that encodes insulin-like growth factor II, which is overexpressed in Wilms' tumors and may contribute to tumor development by acting as an autocrine growth factor. Inactivation of WT1 may thus lead to abnormal growth factor expression, which in turn drives tumor cell proliferation. Two other tumor suppressor genes, *Smad2* and *Smad4,* encode transcription factors that are activated by TGF-β signaling and lead to inhibition of cell proliferation (see Figure 15.41). Consistent with the activity of the TGF-β pathway in inhibiting cell proliferation, the TGF-β receptor is also encoded by a tumor suppressor gene (*TβRII*).

The products of the *Rb* and *INK4* tumor suppressor genes regulate cell cycle progression at the same point as that affected by cyclin D1 and Cdk4, both of which can act as oncogenes (Figure 18.37). Rb inhibits passage through the restriction point in G₁ by repressing transcription of a number of genes involved in cell cycle progression and DNA synthesis (see Figure 16.18). In normal cells, passage through the restriction point is regulated by Cdk4, 6/cyclin D complexes, which phosphorylate and inactivate Rb. Mutational inactivation of *Rb* in tumors thus removes a key negative regulator of cell cycle progression. The *INK4* tumor suppressor gene, which encodes the Cdk inhibitor p16, also regulates passage through the restriction point. As discussed in Chapter 16, p16 inhibits Cdk4, 6/cyclin D activity. Inactivation of *INK4* therefore leads to elevated activity of Cdk4, 6/cyclin D complexes, resulting in uncontrolled phosphorylation of Rb.

The *p53* gene product regulates both cell cycle progression and apoptosis. DNA damage leads to rapid induction of p53, which activates transcription of both proapoptotic and cell cycle inhibitory genes (Figure 18.38). The effects of p53 on apoptosis are mediated in part by activating the transcription of proapoptotic members of the Bcl-2 family (PUMA and Noxa) that induce programmed cell death. Unrepaired DNA damage normally induces apoptosis of mammalian cells, a response that is presumably advantageous to the organism because it eliminates cells carrying potentially deleterious mutations (e.g., cells that might develop into cancer cells).

FIGURE 18.38 Action of p53 Wild-type p53 is required for both cell cycle arrest and apoptosis induced by DNA damage. Cell cycle arrest is mediated by induction of the Cdk inhibitor p21 and apoptosis by induction of the proapoptotic Bcl-2 family members PUMA and Noxa.

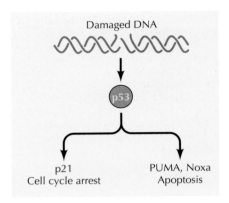

Cells lacking p53 fail to undergo apoptosis in response to agents that damage DNA, including radiation and many of the drugs used in cancer chemotherapy. This failure to undergo apoptosis in response to DNA damage contributes to the resistance of many tumors to chemotherapy. In addition, loss of p53 appears to interfere with apoptosis induced by other stimuli, such as growth factor deprivation and oxygen deprivation. These effects of p53 on cell survival are thought to account for the high frequency of *p53* mutations in human cancers.

In addition to inducing apoptosis, p53 blocks cell cycle progression in response to DNA damage by inducing the Cdk inhibitor p21 (see Figure 16.21). The p21 protein blocks cell cycle progression by acting as a general inhibitor of Cdk/cyclin complexes, and the resulting cell cycle arrest presumably allows time for damaged DNA to be repaired before it is replicated. Loss of p53 prevents this damage-induced cell cycle arrest, leading to increased mutation frequencies and a general instability of the cell genome. Such genetic instability is a common property of cancer cells, and it may contribute to further alterations in oncogenes and tumor suppressor genes during tumor progression.

Although their function remains to be fully understood, the products of the *BRCA1* and *BRCA2* genes, which are responsible for some inherited breast and ovarian cancers, also appear to be involved in checkpoint control of cell cycle progression and repair of double-strand breaks in DNA (see Figure 6.27). *BRCA1* and *BRCA2* thus function as **stability genes**, which act to maintain the integrity of the genome. Mutations in genes of this type lead to the development of cancer not as a result of direct effects on cell proliferation or survival but because their inactivation leads to a high frequency of mutations in oncogenes or tumor suppressor genes. Other stability genes whose loss contributes to the development of human cancers include the *ATM* gene, which acts at the DNA damage checkpoint (see Figure 16.20), the mismatch repair genes that are defective in some inherited colorectal cancers (see Chapter 6, Molecular Medicine), and the nucleotide excision repair genes that are mutated in xeroderma pigmentosum (see Figure 6.23).

Roles of Oncogenes and Tumor Suppressor Genes in Tumor Development

As discussed earlier, the development of cancer is a multistep process in which normal cells gradually progress to malignancy. Mutations resulting in both the activation of oncogenes and the inactivation of tumor suppressor genes are critical steps in tumor initiation and progression. Accumulated damage to multiple genes eventually results in the increased proliferation, survival, invasiveness, and metastatic potential that are characteristic of cancer cells.

The role of multiple genetic defects is best understood in the case of colon carcinomas, which have been studied extensively by Bert Vogelstein and his colleagues. These tumors frequently involve mutations of oncogenes or tumor suppressor genes with four distinct activities: 1) *ras* or *raf* oncogenes affecting the ERK pathway, 2) tumor suppressor or oncogene

Normal cells

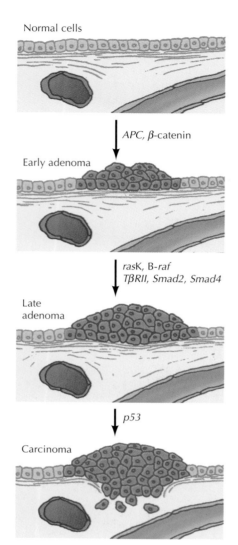

APC, β-catenin

Early adenoma

rasK, B-raf
TβRII, Smad2, Smad4

Late
adenoma

p53

Carcinoma

FIGURE 18.39 Genetic alterations in colon carcinomas Inactivation of APC (a component of the Wnt signaling pathway) is an early event in tumor development, giving rise to a proliferative cell population. In some cases, mutations of the gene encoding β-catenin (which is downstream of APC in Wnt signaling) occur instead. Development of adenomas is then associated with mutations in *ras*K or B-*raf* genes (which stimulate ERK signaling) and with genes encoding components of the TGF-β pathway (TGF-β receptor, Smad2 or Smad4). Mutations of *p53* are associated with later stages of tumor progression.

components of the Wnt pathway, 3) tumor suppressor proteins involved in TGF-β signaling, and 4) *p53*. Lesions representing multiple stages of colon cancer development are regularly obtained as surgical specimens, so it has been possible to correlate these genetic alterations with discrete stages of tumor progression (Figure 18.39).

These studies indicate that inactivation of *APC* (a component of the Wnt signaling pathway; see Figure 18.28) is an early event in tumor development. Genetic transmission of mutant *APC* genes in patients with familial adenomatous polyposis results in abnormal colon cell proliferation, leading to the outgrowth of multiple adenomas in the colons of affected patients. Mutations of *APC* also occur frequently in patients with noninherited colon carcinomas and are generally detected at early stages of the disease process. In some cases, activation of Wnt signaling results from mutations in the gene encoding β-catenin (which is downstream of APC in the Wnt pathway), rather than from mutations in *APC*. Wnt signaling is a major pathway responsible for proliferation of colon epithelial stem cells, and it appears that the constitutive unregulated activity of this pathway initiates the development of colon cancers.

Mutations of *ras*K genes then appear to occur, and *ras*K oncogenes are frequently present in colon adenomas. Some colon tumors have mutations leading to activation of a member of the *raf* oncogene family (B-*raf*), rather than mutations of *ras*K. Since Raf is immediately downstream of Ras, activation of either the *ras*K or B-*raf* oncogene leads to stimulation of ERK signaling in the tumor cells.

Almost all colon cancers have mutations affecting the TFG-β signaling pathway, which inhibits cell proliferation by inducing the synthesis of p15, a member of the Ink4 family of Cdk inhibitors. Mutations affecting TGF-β signaling also appear to occur relatively early in the development of colon cancers, and are frequently found in adenomas. In some tumors, mutations inactivate the *TβRII* tumor suppressor gene, which encodes the TGF-β receptor. In other cases, mutations inactivate tumor suppressor genes encoding the Smad2 or Smad4 transcription factors, which are targets of TFG-β signaling. Finally, the *p53* tumor suppressor gene is inactivated, usually at a later stage of tumor progression.

Accumulated damage to multiple oncogenes and tumor suppressor genes, affecting distinct pathways that regulate cell proliferation and survival, thus appears to be responsible for the multistage development of colon cancer. Accumulated damage to both oncogenes and tumor suppressor genes affecting distinct pathways of cell regulation similarly appears to be responsible for the development of other types of cancer, including breast and lung carcinomas. The progressive loss of growth control that is characteristic of cancer cells is thus thought to be the end result of abnormalities in the products of multiple genes that normally regulate cell proliferation and survival.

Molecular Approaches to Cancer Treatment

A great deal has been learned about the molecular defects responsible for the development of human cancers. However, cancer is more than a topic of scientific interest. It is a dread disease that claims the lives of nearly one out of every four Americans. Translating our growing understanding of cancer into practical improvements in cancer prevention and treatment therefore represents a major challenge for current and future research. Fortunately, advances in elucidating the molecular biology of cancer have begun to contribute to the development of new approaches to its prevention and treatment, which promise to ultimately yield major advances in dealing with this disease.

Prevention and Early Detection

The most effective way to deal with cancer would be to prevent development of the disease. A second-best, but still effective, alternative would be to reliably detect early premalignant stages of tumor development that can be easily treated. Many cancers can be cured by localized treatments, such as surgery or radiation, if they are detected before they spread throughout the body by metastasis. For example, early premalignant stages of colon cancer (adenomas) are usually completely curable by relatively minor surgical procedures (Figure 18.40). The cure rate for early carcinomas that remain localized to their site of origin is also high, about 90%. However, survival rates drop to about 50% for patients whose cancers have spread to adjacent tissues and lymph nodes, and to less than 10% for patients with metastatic colon cancer. Early detection of cancer can thus be a critical determinant of the outcome of the disease.

The major application of molecular biology to prevention and early detection may lie in the identification of individuals with inherited susceptibilities to cancer development. Such inherited cancer susceptibilities can result from mutations in tumor suppressor genes, in at least two oncogenes (*ret* and *cdk4*), and in stability genes, such as the mismatch repair genes responsible for development of hereditary nonpolyposis colon cancer (see Molecular Medicine in Chapter 6). Mutations in these genes can be detected by genetic testing, allowing the identification of high-risk individuals before disease develops.

In addition to contributing to family planning decisions, careful monitoring of high-risk individuals may allow early detection and more effective treatment of some types of cancer. For example, colon adenomas can be detected by colonoscopy and removed prior to the development of malignancy. Patients with familial adenomatous polyposis (resulting from inherited mutations of the *APC* tumor suppressor gene) typically develop hundreds of adenomas within the first 20 years of life, so the colons of these patients are usually removed before the inevitable progression of some of these polyps to malignancy. However, patients with hereditary nonpolyposis colon cancer develop a smaller number of polyps later in life and may therefore benefit from routine colonoscopy and drugs, such as nonsteroidal anti-inflammatory drugs (NSAIDs), that inhibit colon cancer development.

The direct inheritance of cancers resulting from defects in known genes is a rare event, constituting about 5% of total cancer incidence. The most common inherited cancer susceptibility is hereditary nonpolyposis colon cancer, which accounts for about 15% of colon cancers and 1 to 2% of all cancers in the United States. Mutations in the *BRCA1* and *BRCA2* tumor suppressor genes are also relatively common, accounting for approximately

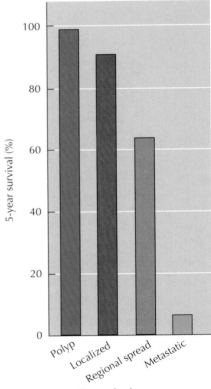

FIGURE 18.40 Survival rates of patients with colon carcinoma Five-year survival rates are shown for patients diagnosed with adenomas (polyps), with carcinoma still localized to its site of origin, with carcinoma that has spread regionally to adjacent tissues and lymph nodes, and with metastatic carcinoma.

5% of all breast cancers. However, still-unidentified genes conferring cancer susceptibility may contribute to the development of a larger fraction of common adult malignancies, such as breast, colon, and lung cancers. The continuing isolation of cancer susceptibility genes is thus an important undertaking with clear practical implications. Individuals with such inherited susceptibility genes could be appropriately advised to avoid exposure to relevant carcinogens (e.g., tobacco smoke in the case of lung cancer) and carefully monitored to detect tumors at early stages that are more readily treated. The reliable identification of susceptible individuals, if it were followed by appropriate preventive and early detection measures, might ultimately make a significant impact on cancer mortality.

Molecular Diagnosis

Molecular analysis of the oncogenes and tumor suppressor genes involved in particular types of tumors has the potential of providing information that is useful in the diagnosis of cancer and in monitoring the effects of treatment. Indeed, several applications of such molecular diagnosis have already been put into clinical practice. In some cases, mutations in oncogenes have provided useful molecular markers for monitoring the course of disease during treatment. The translocation of *abl* in chronic myeloid leukemia is a good example. As previously discussed, this translocation results in the fusion of *abl* with the *bcr* gene, leading to expression of the Bcr/Abl oncogene protein (see Figure 18.24). The polymerase chain reaction (see Figure 4.23) provides a sensitive method of detecting the recombinant *bcr/abl* oncogene in leukemic cells and is therefore used to monitor the response of patients to treatment.

In other cases, the detection of mutations in specific oncogenes or tumor suppressor genes may provide information pertinent to choosing between different therapeutic options. For example, amplification of N-*myc* in neuroblastomas and *erb*B-2 in breast and ovarian carcinomas predicts rapid disease progression. Therefore, it might be appropriate to treat patients with such amplified oncogenes more aggressively. As discussed in the following section, mutations in specific oncogenes, such as *erb*B, may also dictate the response of tumors to oncogene-targeted drugs.

In addition to analysis of individual genes, important diagnostic information may be gleaned from global analysis of gene expression in cancers. The use of DNA microarrays allows the expression of tens of thousands of genes to be analyzed simultaneously (see Figure 4.27), so it is possible to develop a molecular classification of cancers by comparing the gene expression profiles of different tumors. Studies of this type suggest that gene expression profiling can distinguish between otherwise similar tumors and provide information that predicts clinical outcome and response to treatment. Characterization of tumors by gene expression profiling may thus become a useful tool in cancer diagnosis.

Treatment

The most critical question, however, is whether the discovery of oncogenes and tumor suppressor genes will allow the development of new drugs that act selectively against cancer cells. Most of the drugs currently used in cancer treatment either damage DNA or inhibit DNA replication. Consequently, these drugs are toxic not only to cancer cells but also to normal cells, particularly those normal cells that are continually replaced by the division of stem cells (e.g., hematopoietic cells, epithelial cells of the gastrointestinal tract, and hair follicle cells). The action of anticancer drugs against these normal cell populations accounts for most of the toxicity associated with these drugs and limits their effective use in cancer treatment.

One promising new approach to cancer therapy is the use of drugs that inhibit tumor growth by interfering with angiogenesis (blood vessel formation) or disrupting tumor blood vessels, rather than acting directly against cancer cells. As noted earlier in this chapter, the formation of new blood vessels is needed to supply the oxygen and nutrients required for tumor growth. Promoting angiogenesis is thus critical to tumor development, and tumor cells secrete a number of growth factors, including VEGF, that stimulate the proliferation of capillary endothelial cells, resulting in the outgrowth of new capillaries into the tumor (see Figure 17.14). The importance of angiogenesis was first recognized by Judah Folkman in 1971, and continuing research by Folkman and his colleagues has led to the development of new drugs that inhibit angiogenesis by blocking the proliferation of endothelial cells. Because these drugs act specifically to inhibit the formation of new blood vessels, they are less toxic to normal cells than standard anticancer agents. Angiogenesis inhibitors have shown promising results in animal tests and are currently being tested in clinical trials to evaluate their effectiveness against human cancers. In 2004, positive clinical results led the US Food and Drug Administration (FDA) to approve the use of the first angiogenesis inhibitor, a monoclonal antibody against VEGF, for treatment of colon cancer.

An alternative strategy for achieving more selective cancer treatment is the development of drugs targeted specifically against the oncogenes that drive tumor growth. Unfortunately, from the standpoint of cancer treatment, oncogenes are not unique to tumor cells. Since proto-oncogenes play important roles in normal cells, general inhibitors of oncogene expression or function are likely to act against normal cells as well as tumor cells. The exploitation of oncogenes as targets for anticancer drugs is therefore not a straightforward proposition, but several promising advances indicate that it is possible to develop selective oncogene targeted therapies (Table 18.6).

The first therapeutic regimen targeted against a specific oncogene is used for the treatment of acute promyelocytic leukemia. This leukemia is characterized by a chromosome translocation in which the gene that encodes the retinoic acid receptor ($RAR\alpha$) is fused to another gene (PML) to form the $PML/RAR\alpha$ oncogene. The PML/RARα protein is thought to function as a transcriptional repressor that blocks cell differentiation. These leukemic cells, however, differentiate in response to treatment with high doses of

TABLE 18.6 Oncogene Targeted Therapies Approved for Clinical Use

Drug	Oncogene	Tumor Types
Retinoic acid	$PML/RAR\alpha$	Acute promyelocytic leukemia
Herceptin	erbB-2	Breast cancer
Erbitux	erbB	Colorectal cancer
STI-571	abl	Chronic myelogenous leukemia
	kit	Gastrointestinal stromal tumors
	PDGFR	Gastrointestinal stromal tumors, chronic myelomonocytic leukemia, hypereosinophilic syndrome, dermatofibrosarcoma protuberans
Gefitinib	erbB	Lung cancer

retinoic acid, which binds to and inactivates the PML/RARα oncogene protein. Such treatment with retinoic acid results in remission of the leukemia in most patients, although this favorable response is temporary and patients eventually relapse. However, combined treatment with retinoic acid and standard chemotherapeutic agents significantly reduces the incidence of relapse, so the use of retinoic acid is of substantial benefit in the treatment of acute promyelocytic leukemia. The therapeutic activity of retinoic acid was observed prior to identification of the *PML/RARα* oncogene, so its effectiveness against leukemic cells expressing this oncogene protein was discovered by chance rather than by rational drug design. Nonetheless, the use of retinoic acid for treatment of acute promyelocytic leukemia provides the first example of a clinically useful drug targeted against an oncogene protein.

Herceptin, a monoclonal antibody against the ErbB-2 oncogene protein, was the first drug developed against a specific oncogene to achieve FDA approval for clinical use in cancer treatment. The ErbB-2 protein is overexpressed in about 30% of breast cancers as a result of amplification of the *erb*B-2 gene. It was first found that an antibody against the extracellular domain of ErbB-2 (a receptor protein-tyrosine kinase) inhibited the proliferation of tumor cells in which ErbB-2 was overexpressed. These results led to the development and clinical testing of Herceptin, which was found to significantly reduce tumor growth and prolong patient survival in clinical trials involving over 600 women with metastatic breast cancers that overexpressed the ErbB-2 protein. Based on these results, Herceptin was approved by the FDA in 1998 for treatment of metastatic breast cancers that express elevated levels of ErbB-2. Erbitux, a monoclonal antibody against the EGF receptor (the ErbB oncogene protein), was also approved by the FDA in 2004 for use in treatment of advanced colorectal cancer.

The therapeutic use of monoclonal antibodies is limited to extracellular targets, such as growth factors or cell surface receptors. A more widely applicable area of drug development is the identification of small-molecule inhibitors of oncogene proteins, including the protein kinases that play key roles in signaling the proliferation and survival of cancer cells. The pioneering advance in this area was the development of a selective inhibitor of the Bcr/Abl protein-tyrosine kinase, which is generated by the Philadelphia chromosome translocation in chronic myeloid leukemia (see Figure 18.24). Brian Druker and his colleagues developed a potent and specific inhibitor of the Bcr/Abl protein kinase and showed that this compound (called STI-571 or Gleevec) effectively blocks proliferation of chronic myeloid leukemia cells. Based on these results, a clinical trial of STI-571 was initiated in 1998. The responses to STI-571 were remarkable, and the drug had minimal side effects. The striking success of STI-571 in these clinical studies served as the basis for its rapid approval by the FDA in 2001 for use in the treatment of chronic myeloid leukemia. Although some patients relapse and develop resistance against the drug, STI-571 is unquestionably a highly effective therapy for this leukemia. Interestingly, resistance to STI-571 most often results from mutations of the Bcr/Abl protein kinase domain that prevent STI-571 binding. By analysis of these resistant mutants, it has been possible to design new inhibitors, which are currently being tested in clinical trials to determine their effectiveness against leukemias that have become resistant to STI-571.

STI-571 is also a potent inhibitor of the PDGF receptor and Kit protein-tyrosine kinases, and it has proven to be an effective therapy for tumors in

MOLECULAR MEDICINE

STI-571: Cancer Treatment Targeted against the bcr/abl Oncogene

The Disease

Chronic myeloid leukemia (CML) accounts for approximately 20% of leukemias in adults. In 2005, it was estimated that about 4600 cases of CML would be diagnosed in the United States, and that there would be about 850 deaths from this disease.

CML originates from the hematopoietic stem cell of the bone marrow. It is a slowly progressing disease, which is clinically divided into two stages: chronic phase and blast crisis. The chronic phase of CML can persist for years and is associated with minimal symptoms. Eventually, however, patients progress to an acute life-threatening stage of the disease known as blast crisis. Blast crisis is characterized by the accumulation of large numbers of rapidly proliferating leukemic cells, called blasts. Patients in blast crisis are treated with standard chemotherapeutic drugs, which may induce remission to the chronic phase of the disease. Chemotherapy may also be used during the chronic phase of CML but does not usually succeed in eliminating the leukemic cells. CML can also be treated by transplantation of bone marrow stem cells, which can be curative for about half of patients in the chronic phase of the disease.

Molecular and Cellular Basis

Activation of the *abl* oncogene by translocation from its normal locus on chromosome 9 to chromosome 22 is a highly reproducible event in CML, occurring in about 95% of these leukemias. This translocation results in formation of a fusion between *abl* and the *bcr* gene on chromosome 22, yielding the *bcr/abl* oncogene. The oncogene expresses a Bcr/Abl fusion protein in which Bcr amino acid sequences replace the first exon of Abl (see Figure 18.24). The Bcr/Abl protein is a constitutively active protein-tyrosine kinase, which leads to leukemia by activating a variety of downstream signaling pathways.

Because activation of the *bcr/abl* oncogene is such a reproducible event in the development of CML, it appeared to be a good candidate against which to develop a selective tyrosine kinase inhibitor that might be of clinical use. These studies led to the development of STI-571 (or Gleevec) as the first therapeutic drug successfully designed as a selective inhibitor of an oncogene protein.

Prevention and Treatment

The development of STI-571 started with the identification of 2-phenylaminopyrimidine as a nonspecific inhibitor of protein kinases. A series of related compounds were then synthesized and optimized for activity against different targets, including the Abl tyrosine kinase. Among many compounds screened in these investigations, STI-571 was found to be a potent and specific inhibitor of Abl and two other protein-tyrosine kinases: the platelet-derived growth factor receptor and c-Kit. Further studies demonstrated that STI-571 specifically inhibited the proliferation of cells transformed by *bcr/abl* oncogenes, including cells from CML patients in culture. In addition, STI-571 prevented tumor formation by *bcr/abl*-transformed cells in mice. In contrast, normal cells or cells transformed by other oncogenes were not affected by STI-571, demonstrating its specificity against the Bcr/Abl tyrosine kinase.

Based on these results, an initial phase I clinical study of STI-571 was initiated in June, 1998. The results were strikingly positive. Of 54 chronic phase patients treated with STI-571, 53 responded to the drug. Moreover, responses to STI-571 were seen in more than half of treated blast crisis patients. The success of these initial studies prompted expanded Phase II studies involving over 1000 patients. These studies confirmed the promising results of the phase I study, with 95% of chronic phase patients and approximately 50% of blast crisis patients responding to STI-571. Moreover, in contrast to conventional chemotherapeutic drugs, STI-571 had minimal side effects. These clinical studies clearly demonstrated that STI-571 is a highly effective therapy for CML, and led to accelerated FDA approval of STI-571 in May 2001—a milestone in the translation of basic science to clinical practice.

Reference

Druker, B. J. 2002. Inhibition of the Bcr-Abl tyrosine kinase as a therapeutic strategy for CML. *Oncogene* 21: 8541–8546.

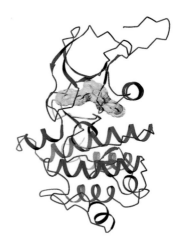

Crystal structure of the catalytic domain of Abl complexed with a derivative of STI-571. (From T. Schindler, W. Bornmann, P. Pellicena, W. T. Miller, B. Clarkson and J. Kuriyan. 2000. *Science* 289: 1938.)

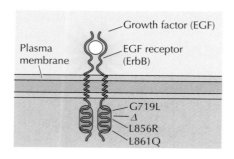

Growth factor (EGF)

Plasma membrane

EGF receptor (ErbB)

G719L
Δ
L856R
L861Q

FIGURE 18.41 EGF receptor mutations associated with sensitivity to gefitinib Lung cancers that respond to gefitinib have activating mutations within the EGF receptor kinase domain. These mutations include point mutations (G719L, L856R, and L861Q) and small deletions (Δ).

which the genes encoding these protein kinases are mutationally activated oncogenes. *Kit* is activated as an oncogene by point mutations resulting in constitutive protein kinase activity in approximately 90% of gastrointestinal stromal tumors, which are tumors of stromal connective tissue of the stomach and small intestine. Many of the gastrointestinal stromal tumors that do not have activating mutations of Kit instead have activating mutations of the PDGF receptor. Consequently, gastrointestinal stromal tumors are highly responsive to STI-571. In addition, STI-571 is active against three other types of tumors in which the PDGF receptor is activated as an oncogene, including chronic myelomonocytic leukemia in which it is activated by fusion with the Tel transcription factor (see Figure 18.26).

A small molecule inhibitor of the EGF receptor (gefitinib) has recently shown striking activity against a subset of lung cancers in which the EGF receptor is activated by point mutations. It is noteworthy that the rationale for treatment of lung cancers with gefitinib was the fact that EGF receptors are overexpressed in most lung cancers, rather than the fact that they can be mutationally activated as oncogenes. In contrast to the general effectiveness of STI-571 against chronic myeloid leukemia, clinical studies indicated that gefitinib was effective in only about 10% of lung cancer patients, although the responses in these patients were quite dramatic. A major advance was thus made in 2004 when two groups of researchers found that the subset of lung cancers that responded to gefitinib were those in which point mutations resulted in constitutive activation of the EGF receptor tyrosine kinase (Figure 18.41). These results indicated that inhibition of the EGF receptor was an effective treatment for tumors in which it had been mutated to act as an oncogene but not for tumors expressing the normal protein.

The finding that gefitinib is active against only those lung cancers with mutationally activated EGF receptors supports the hypothesis that tumors with activated oncogenes are particularly susceptible to inhibitors of those oncogenes. This is also consistent with the activity of STI-571 against tumors with mutationally activated Bcr/Abl, Kit, and PDGF receptor oncogene proteins, as well as with a variety of experiments in animal model systems. The sensitivity of tumors to inhibition of activated oncogenes has been referred to as **oncogene addiction**. It is thought that an activated oncogene becomes a major driving force in the tumor cell, such that other signaling pathways in the tumor cell become secondary in importance. Consequently, the proliferation and survival of a tumor cell may become dependent on continuing activity of the oncogene, whereas normal cells have alternative signaling pathways that can compensate if any one pathway is blocked.

The examples of STI-571 and gefitinib clearly suggest that the continuing exploitation of oncogenes as targets for drug development has the potential of leading to a new generation of drugs that act selectively against cancer cells. Indeed, a wide variety of drugs targeted against both oncogene proteins (including B-Raf, PI-3 kinase, and Akt) and downstream components of oncogenic signaling pathways (such as MEK and mTOR) are being tested to evaluate their potential in the treatment of human cancer. The apparent dependence of cancer cells on mutationally activated oncogenes further offers the promise that the use of oncogene targeted drugs combined with genetic analysis of the tumors of individual patients may lead to major advances in cancer treatment. Although the eventual impact of molecular biology on the treatment of cancer remains to be seen, it is clear that the rational design of drugs targeted against specific oncogene proteins will play an important role.

<table>
<tr><td>

SUMMARY

THE DEVELOPMENT AND CAUSES OF CANCER

Types of Cancer: Cancer can result from the abnormal proliferation of any type of cell. The most important distinction for the patient is between benign tumors, which remain confined to their site of origin, and malignant tumors, which can invade normal tissues and spread throughout the body.

The Development of Cancer: Tumors develop from single altered cells that begin to proliferate abnormally. Additional mutations lead to the selection of cells with progressively increasing capacities for proliferation, survival, invasion, and metastasis.

Causes of Cancer: Radiation and many chemical carcinogens act by damaging DNA and inducing mutations. Other chemical carcinogens contribute to the development of cancer by stimulating cell proliferation. Viruses also cause cancer in both humans and other species.

Properties of Cancer Cells: The uncontrolled proliferation of cancer cells is reflected in reduced requirements for extracellular growth factors and lack of inhibition by cell-cell contact. Many cancer cells are also defective in differentiation, consistent with their continued proliferation *in vivo*. The characteristic failure of cancer cells to undergo apoptosis also contributes substantially to tumor development.

Transformation of Cells in Culture: The development of *in vitro* assays for cell transformation has allowed the conversion of normal cells into tumor cells to be studied in cell culture.

TUMOR VIRUSES

Hepatitis B and C Viruses: The hepatitis B and C viruses cause liver cancer in humans.

SV40 and Polyomavirus: Although neither SV40 nor polyomavirus causes human cancer, they are important models for studying the molecular biology of cell transformation. SV40 T antigen induces transformation by interacting with the cellular Rb and p53 tumor suppressor proteins.

Papillomaviruses: Papillomaviruses induce tumors in a variety of animals, including cervical carcinoma in humans. Like SV40 T antigen, the transforming proteins of papillomaviruses interact with Rb and p53.

Adenoviruses: The adenoviruses do not cause naturally occurring cancers in either humans or other species but are important models in cancer research. Their transforming proteins also interact with Rb and p53.

Herpesviruses: The herpesviruses, which are among the most complex animal viruses, cause cancer in several species, including humans.

Retroviruses: Retroviruses cause cancer in humans and a variety of other animals. Some retroviruses contain specific genes responsible for inducing cell transformation, and studies of these highly oncogenic retroviruses have led to the characterization of both viral and cellular oncogenes.

</td><td>

KEY TERMS

cancer, tumor, benign tumor, malignant tumor, metastasis, carcinoma, sarcoma, leukemia, lymphoma

tumor initiation, tumor progression, adenoma, polyp

carcinogen, tumor promoter

density-dependent inhibition, autocrine growth stimulation, contact inhibition, angiogenesis, programmed cell death, apoptosis

cell transformation

tumor virus, hepatitis B virus, hepatitis C virus

simian virus 40 (SV40), polyomavirus

papillomavirus

adenovirus

herpesvirus, Kaposi's sarcoma-associated herpesvirus, Epstein-Barr virus

retrovirus, Rous sarcoma virus (RSV)

</td></tr>
</table>

KEY TERMS

SUMMARY

ONCOGENES

oncogene, *src, ras, raf*

Retroviral Oncogenes: The first oncogene to be identified was the *src* gene of RSV. Subsequent studies have identified more than two dozen distinct oncogenes in different retroviruses.

proto-oncogene

Proto-Oncogenes: Retroviral oncogenes originated from closely related genes of normal cells, called proto-oncogenes. The oncogenes are abnormally expressed or mutated forms of the corresponding proto-oncogenes.

c-*myc*, *abl*, N-*myc*, *erb*B-2

Oncogenes in Human Cancer: A variety of oncogenes are activated by point mutations, DNA rearrangements, and gene amplification in human cancers. Some of these human tumor oncogenes, such as the *ras* genes, are cellular homologs of oncogenes that were first described in retroviruses.

Fos, Jun, *CCND*1, ErbA, PML/RAR*α*, PI 3-kinase, Akt, Bcl-2

Functions of Oncogene Products: Many oncogene proteins function as elements of signaling pathways that stimulate cell proliferation. The genes that encode cyclin D1 and Cdk4 can also act as oncogenes by stimulating cell cycle progression. Other oncogene proteins interfere with cell differentiation, and oncogenes encoding PI 3-kinase, Akt, and Bcl-2 inhibit apoptosis.

TUMOR SUPPRESSOR GENES

tumor suppressor gene, *Rb, p53*

Identification of Tumor Suppressor Genes: In contrast to oncogenes, tumor suppressor genes inhibit tumor development. The prototype tumor suppressor gene, *Rb*, was identified by studies of inheritance of retinoblastoma. Loss or mutational inactivation of *Rb* and other tumor suppressor genes, including *p53*, contributes to the development of a wide variety of human cancers.

stability gene

Functions of Tumor Suppressor Gene Products: The proteins encoded by most tumor suppressor genes act as inhibitors of cell proliferation or survival. The Rb, INK4, and p53 proteins are negative regulators of cell cycle progression. In addition, p53 is required for apoptosis induced by DNA damage and other stimuli, so its inactivation contributes to enhanced tumor cell survival. Some genes, such as *BRCA1* and *BRCA2*, act to maintain genomic stability rather than directly influencing cell proliferation.

Roles of Oncogenes and Tumor Suppressor Genes in Tumor Development: Mutations in both oncogenes and tumor suppressor genes contribute to the progressive development of human cancers. Accumulated damage to multiple such genes results in the abnormalities of cell proliferation, differentiation, and survival that characterize the cancer cell.

MOLECULAR APPROACHES TO CANCER TREATMENT

Prevention and Early Detection: Many cancers can be cured if they are detected at early stages of tumor development. Genetic testing to identify individuals with inherited cancer susceptibilities may allow early detection and more effective treatment of high-risk patients.

SUMMARY

Molecular Diagnosis: Detection of mutations in oncogenes and tumor suppressor genes may be useful in diagnosis and in monitoring response to treatment. Global analysis of gene expression may distinguish subclasses of cancers with differing clinical prognosis or response to treatment.

Treatment: The development of drugs targeted against specific oncogenes is beginning to lead to the discovery of new therapeutic agents that act selectively against cancer cells.

KEY TERMS

oncogene addiction

Questions

1. How does a benign tumor differ from a malignant tumor?

2. What is the role of clonal selection in the development of cancer?

3. How do estrogens increase the risk of cancer?

4. How does autocrine growth stimulation contribute to the progression of tumors?

5. What properties of cancer cells give them the ability to metastasize?

6. You have constructed a mutant SV40 T antigen that fails to induce cell transfor-

mation because it no longer binds Rb. Would this mutant T antigen induce transformation if introduced into cells together with a papillomavirus cDNA that encodes E6? How about one that encodes E7?

7. Why do AIDS patients have a high incidence of some types of cancer?

8. How can a proto-oncogene be converted to an oncogene without a change or mutation in its coding sequence? Explain two ways by which this can occur.

9. What effect would overexpression of the INK4 tumor suppressor have on tumor cells that express inactive Rb?

10. Which would you expect to be more sensitive to treatment with radiation – tumors with wild-type *p53* genes or tumors with mutated *p53* genes?

11. What is the mode of action of STI-571? How do some tumors develop resistance to this drug?

12. What is 'oncogene addiction' and why is this concept important for selecting molecular targets for cancer therapy?

References and Further Reading

General References

Cooper, G. M. 1992. *Elements of Human Cancer.* Boston: Jones and Bartlett.

Cooper, G. M. 1995. *Oncogenes.* 2nd ed. Boston: Jones and Bartlett.

Varmus, H. and R. A. Weinberg. 1993. *Genes and the Biology of Cancer.* New York: Scientific American Library.

Vogelstein, B. and K. W. Kinzler. 2004. Cancer genes and the pathways they control. *Nature Med.* 10: 789–799. [R]

The Development and Causes of Cancer

Chambers, A. F., A. C. Groom and I. C. MacDonald. 2002. Dissemination and growth of cancer cells in metastatic sites. *Nature Rev. Cancer* 2: 563–572. [R]

Christofori, G. and H. Semb. 1999. The role of the cell-adhesion molecule E-cadherin as a tumour-suppressor gene. *Trends Biochem. Sci.* 24: 73–76. [R]

Colditz, G. A., T. A. Sellers and E. Trapido. 2006. Epidemiology—Identifying the caus-

es and preventability of cancer? *Nature Rev. Cancer* 6: 75–83. [R]

Fialkow, P. J. 1979. Clonal origin of human tumors. *Ann. Rev. Med.* 30: 135–143. [R]

Hanahan, D. and R. A. Weinberg. 2000. The hallmarks of cancer. *Cell* 100: 57–70. [R]

Kolonel, L. N., D. Altshuler and B. E. Henderson. 2004. The multiethnic cohort study: Exploring genes, lifestyle and cancer risk. *Nature Rev. Cancer* 4: 1–9. [R]

Liotta, L. A. and E. C. Kohn. 2001. The microenvironment of the tumour-host interface. *Nature* 411: 375–379. [R]

Nowell, P. C. 1986. Mechanisms of tumor progression. *Cancer Res.* 46: 2203–2207. [R]

Peto, J. 2001. Cancer epidemiology in the last century and the next decade. *Nature* 411: 390–395. [R]

Raff, M. C. 1992. Social controls on cell survival and cell death. *Nature* 356: 397–400. [R]

Sporn, M. B. and A. B. Roberts. 1985. Autocrine growth factors and cancer. *Science* 313: 745–747. [R]

Temin, H. M. and H. Rubin. 1958. Characteristics of an assay for Rous sarcoma virus and Rous sarcoma cells in culture. *Virology* 6: 669–688. [P]

Tenen, D. G. 2003. Disruption of differentiation in human cancer: AML shows the way. *Nature Rev. Cancer* 3: 89–101. [R]

Thompson, C. B. 1995. Apoptosis in the pathogenesis and treatment of disease. *Science* 267: 1456–1462. [R]

Tumor Viruses

Block, T. M., A. S. Mehta, C. J. Fimmel and R. Jordan. 2003. Molecular viral oncology of hepatocellular carcinoma. *Oncogene* 22: 5093–5107. [R]

Boshoff, C. and R. Weiss, 2002. AIDS-related malignancies. *Nature Rev. Cancer* 2: 373–382. [R]

Bouchard, M. J. and R. J. Schneider. 2004. The enigmatic X gene of hepatitis B virus. *J. Virol.* 78: 12725–12734. [R]

Coffin, J. M., S. H. Hughes and H. E. Varmus, eds. 1997. *Retroviruses.* New York: Cold Spring Harbor Laboratory Press.

Damania, B. 2004. Oncogenic γ-herpesviruses: Comparison of viral proteins involved in tumorigenesis. *Nature Rev. Microbiol.* 2: 656–668. [R]

Knipe, D. M., P. M. Howley, D. E. Griffin, R. A. Lamb, M. A. Martin, B. Roizman and S. E. Straus. 2001. *Fundamental Virology.* 4th ed. New York: Lippincott Williams and Wilkins.

Flint, S. J., L. W. Enquist and A. M. Skalka. 2003. *Principles of Virology: Molecular Biology, Pathogenesis, and Control of Animal Viruses.* 2nd ed. Washington, DC: ASM Press.

Grassmann, R., M. Aboud and K.-T. Jeang. 2005. Molecular mechanisms of cellular transformation by HTLV-1 Tax. *Oncogene* 24: 5976–5985. [R]

Helt, A.-M. and D. A. Galloway. 2003. Mechanisms by which DNA tumor virus oncoproteins target the Rb family of pocket proteins. *Carcinogenesis* 24: 159–169. [R]

Saenz-Robles, M. T., C. S. Sullivan and J. M. Pipas. 2001. Transforming functions of simian virus 40. *Oncogene* 20: 7899–7907. [R]

Wang, H.-W. and C. Boshoff. 2005. Linking Kaposi virus to cancer-associated cytokines. *Trends Mol. Med.* 11: 309–312. [R]

Young, L. S. and A. B. Rickinson. 2004. Epstein-Barr virus: 40 years on. *Nature Rev. Cancer.* 4: 757–768. [R]

Zur Hausen, H. 2002. Papillomaviruses and cancer: From basic studies to clinical application. *Nature Rev. Cancer* 2: 342–350. [R]

Oncogenes

Aaronson, S. A. 1991. Growth factors and cancer. *Science* 254: 1146–1153. [R]

Adhikary, S. and M. Eilers. 2005. Transcriptional regulation and transformation by Myc proteins. *Nature Rev. Mol. Cell Biol.* 6: 635–645. [R]

Altomare, D. A. and J. R. Testa. 2005. Perturbations of the AKT signaling pathway in human cancer. *Oncogene* 24: 7455–7464. [R]

Blume-Jensen, P. and T. Hunter. 2001. Oncogenic kinase signaling. *Nature* 411: 355–365. [R]

Danial, N. K. and S. J. Korsmeyer. 2004. Cell death: Critical control points. *Cell* 116: 205–219. [R]

Datta, S. R., Brunet, A. and M. E. Greenberg. 1999. Cellular survival: A play in three *Akts. Genes Dev.* 13: 2905–2927. [R]

Davies, H. and 51 others. 2002. Mutations of the *BRAF* gene in human cancer. *Nature* 417: 949–954 [P]

Der, C. J., T. G. Krontiris and G. M. Cooper. 1982. Transforming genes of human bladder and lung carcinoma cell lines are homologous to the *ras* genes of Harvey and Kirsten sarcoma viruses. *Proc. Natl. Acad. Sci. USA* 79: 3637–3640. [P]

Garnett, M. J. and R. Marais. 2004. Guilty as charged: B-RAF is a human oncogene. *Cancer Cell* 6: 313–319. [R]

Gregorieff, A. and H. Clevers. 2005. Wnt signaling in the intestinal epithelium: From endoderm to cancer. *Genes Dev.* 19: 877–890. [R]

Kranenburg, O. 2005. The *KRAS* oncogene: Past, present, and future. *Biochim. Biophys. Acta* 1756: 81–82. [R]

Krontiris, T. G. and G. M. Cooper. 1981. Transforming activity of human tumor DNAs. *Proc. Natl. Acad. Sci. USA* 78: 1181–1184. [P]

Leder, P., J. Battey, G. Lenoir, C. Moulding, W. Murphy, H. Potter, T. Stewart and R. Taub. 1983. Translocations among antibody genes in human cancer. *Science* 222: 765–771. [R]

Legauer, C., K. W. Kinzler and B. Vogelstein. 1998. Genetic instabilities in human cancers. *Nature* 396: 643–649. [R]

Look, A. T. 1997. Oncogenic transcription factors in the human acute leukemias. *Science* 278: 1059–1064. [R]

Martin, G. S. 1970. Rous sarcoma virus: A function required for the maintenance of the transformed state. *Nature* 227: 1021–1023. [P]

Massagué, J. 2004. G1 cell-cycle control and cancer. *Nature* 432: 298–306. [R]

Shih, C., L. C. Padhy, M. Murray and R. A. Weinberg. 1981. Transforming genes of carcinomas and neuroblastomas introduced into mouse fibroblasts. *Nature* 300: 539–542. [P]

Stehelin, D., H. E. Varmus, J. M. Bishop and P. K. Vogt. 1976. DNA related to the transforming gene(s) of avian sarcoma viruses is present in normal avian DNA. *Nature* 260: 170–173. [P]

Tabin, C. J., S. M. Bradley, C. I. Bargmann, R. A. Weinberg, A. G. Papageorge, E. M. Scolnick, R. Dhar, D. R. Lowy and E. H. Chang. 1982. Mechanism of activation of a human oncogene. *Nature* 300: 143–149. [P]

Taipale, J. and P. A. Beachy. 2001. The Hedgehog and Wnt signaling pathways in cancer. *Nature* 411: 349–354. [R]

Thompson, C. B. 1995. Apoptosis in the pathogenesis and treatment of disease. *Science* 267: 1456–1462. [R]

Vogelstein, B. and K. W. Kinzler. 2004. Cancer genes and the pathways they control. *Nature Med.* 10: 789–799. [R]

Vogt, P. K. 1971. Spontaneous segregation of nontransforming viruses from cloned sarcoma viruses. *Virology* 46: 939–946. [P]

Tumor Suppressor Genes

Cantley, L. C. and B. G. Neel. 1999. New insights into tumor suppression: *PTEN* suppresses tumor formation by restraining the phosphoinositide 3-kinase/AKT pathway. *Proc. Natl. Acad. Sci. USA* 96: 4240–4245. [R]

Classon, M. and E. Harlow. 2002. The retinoblastoma tumour suppressor in development and cancer. *Nature Rev. Cancer* 2: 910–917. [R]

Danial, N. N. and S. J. Korsmeyer. 2004. Cell death: Critical control points. *Cell* 116: 205–219. [R]

Friend, S. H., R. Bernards, S. Rogelj, R. A. Weinberg, J. M. Rapaport, D. M. Albert and T. P. Dryja. 1986. A human DNA segment with properties of the gene that predisposes to retinoblastoma and osteosarcoma. *Nature* 323: 643–646. [P]

Gregorieff, A. and H. Clevers. 2005. Wnt signaling in the intestinal epithelium: From endoderm to cancer. *Genes Dev.* 19: 877–890. [R]

Harris, H., O. J. Miller, G. Klein, P. Worst and T. Tachibana. 1969. Suppression of malignancy by cell fusion. *Nature* 223: 363–368. [P]

Kastan, M. B. and J. Bartek. 2004. Cell-cycle checkpoints and cancer. *Nature* 432: 316–323. [R]

Kinzler, K. W. and B. Vogelstein. 1996. Lessons from hereditary colorectal cancer. *Cell* 87: 159–170. [R]

Knudson, A. G. 1976. Mutation and cancer: A statistical study of retinoblastoma. *Proc. Natl. Acad. Sci. USA* 68: 820–823. [P]

Lee, W.-H., R. Bookstein, F. Hong, L.-J. Young, J.-Y. Shew and E. Y.-H. P. Lee. 1987. Human retinoblastoma susceptibility gene: Cloning, identification, and sequence. *Science* 235: 1394–1399. [P]

Massagué, J., S. W. Blain and R. S. Lo. 2000. TGFβ signaling in growth control, cancer, and heritable disorders. *Cell* 103: 295–309. [R]

Narod, S. A. and W. D. Foulkes. 2004. *BRCA1* and *BRCA2*: 1994 and beyond. *Nature Rev. Cancer* 4: 665–676. [R]

Sherr, C. J. 2004. Principles of tumor suppression. *Cell* 116: 235–246. [R]

Taipale, J. and P. A. Beachy. 2001. The Hedgehog and Wnt signaling pathways in cancer. *Nature* 411: 349–354. [R]

Venkitaraman, A. R. 2002. Cancer susceptibility and the functions of BRCA1 and BRCA2. *Cell* 108: 171–182. [R]

Vousden, K. H. and X. Lu. 2002. Live or let die: The cell's response to p53. *Nature Rev. Cancer* 2: 594–604. [R]

Molecular Approaches to Cancer Treatment

Adams, G. P. and L. M. Weiner. 2005. Monoclonal antibody therapy of cancer. *Nature Biotech.* 23: 1147–1157. [R]

Balmain, A. 2002. Cancer as a complex genetic trait: Tumor susceptibility in humans and mouse models. *Cell* 108: 145–152. [R]

Downward, J. 2002. Targeting Ras signaling pathways in cancer therapy. *Nature Rev. Cancer* 3: 11–22. [R]

Ferrara, N. and R. S. Kerbel. 2005. Angiogenesis as a therapeutic target. *Nature* 438: 967–974. [R]

Gschwind, A., O. M. Fischer and A. Ullrich. 2004. The discovery of receptor tyrosine kinases: Targets for cancer therapy. *Nature Rev. Cancer* 4: 361–370. [R]

Hampton, G. M. and H. F. Frierson, Jr. 2003. Classifying human cancer by analysis of gene expression. *Trends Mol. Med.* 9: 5–10. [R]

Hennessy, B. T., D. L. Smith, P. T. Ram, Y. Lu and G. B. Mills. 2005. Exploiting the PI3K/Akt pathway for cancer drug discovery. *Nature Rev. Drug Discovery* 4: 988–1003. [R]

Herbst, R. S., M. Fukuoka and J. Baselga. 2004. Gefitinib—A novel targeted approach to treating cancer. *Nature Rev. Cancer* 4: 956–965. [R]

Lallemand-Breitenbach, V., J. Zhu, S. Kogan, Z. Chen and H. de The'. 2005. How patients have benefited from mouse models of acute promyelocytic leukemia. *Nature* 5: 821–827. [R]

O'Hare, T., A. S. Corbin and B. J. Druker. 2006. Targeted CML therapy: Controlling drug resistance, seeking cure. *Curr. Opin. Gen. Dev.* 16: 92–99. [R]

Ponder, B. 2001. Cancer genetics. *Nature* 411: 336–341. [R]

Ramaswamy, S. and T. R. Golub. 2002. DNA microarrays in clinical oncology. *J. Clin. Oncol.* 20: 1932–1941. [R]

Sawyers, C. L. 2003. Opportunities and challenges in the development of kinase inhibitor therapy for cancer. *Genes Dev.* 17: 2998–3010. [R]

Sawyers, C. L. 2004. Targeted cancer therapy. *Nature* 432: 294–297. [R]

Thun, M. J., S. J. Henley and C. Patrono. 2002. Nonsteroidal anti-inflammatory drugs as anticancer agents: Mechanistic, pharmacologic, and clinical issues. *J. Natl. Cancer Inst.* 94: 252–266. [R]

Weinstein, I. B. 2002. Addiction to oncogenes—The Achilles heal of cancer. *Science* 297: 63–64. [R]

Answers to Questions

CHAPTER 1

1. That organic molecules, including several amino acids, can be formed spontaneously from a mixture of reducing gases.

2. RNA is uniquely capable of both serving as a template and catalyzing the chemical reactions required for its own replication.

3. Mitochondria and chloroplasts contain their own DNA and ribosomes, are similar in size to bacteria, and divide like bacteria. The ribosomal proteins and RNAs of these organelles are also more closely related to those of bacteria than to those encoded by eukaryotic nuclear genomes.

4. Because O_2 became abundant in Earth's atmosphere as a result of photosynthesis.

5. The volume of a sphere is given by the formula $4/3\pi r^3$, where r = the radius. Since the volume of the cells is proportional to the cube of their radii, the relative volume of a macrophage compared to *S. aureus* is the ratio of the cubes of their radii: $25^3/0.5^3$ = 125,000.

6. Yeast.

7. Mice.

8. The refractive index of air is 1.0; the refractive index of oil is approximately 1.4. Since resolution = $0.61/\eta \sin \alpha$ (η is the refractive index), viewing a specimen through air rather than through oil changes the limit of resolution from about 0.2 μm to about 0.3 μm.

9. Resolution = $0.61\lambda/NA$. Taking λ = 0.5 μm for visible light, the objective with an NA=1.3 will resolve objects about 0.23 μm apart, while the objective of NA =1.1 will resolve objects about 0.27 μm apart. Magnification of the images can be obtained by other means (such as projecting on a screen or using an ocular lens with higher magnification). Thus the bet-ter choice would be the objective with 60× magnification and the numerical aperture of 1.3.

10. GFP-tagged proteins can be expressed in live cells, so they can be seen without the need to fix and kill the cells for staining with a fluorescent antibody. Movement of GFP-tagged proteins can therefore be followed in living cells.

11. Velocity centrifugation separates organelles on the basis of size and shape, whereas equilibrium centrifugation separates them on the basis of density (independent of size and shape).

12. Serum contains growth factors that are required to stimulate division of most animal cells in culture.

13. Primary cell cultures are cells grown directly from an organism or tissue. Immortal cell lines have the ability to proliferate indefinitely in culture.

14. These cells retain the ability to differentiate into all of the different cell types of adult organisms, so they offer the potential of being used in transplantation therapies for treatment of a variety of diseases.

CHAPTER 2

1. Water is a polar molecule whose hydrogens and oxygen can form hydrogen bonds with each other (to make water a liquid at most of Earth's temperatures), as well as with polar organic molecules and inorganic ions. In contrast, nonpolar molecules or parts of molecules cannot interact with water, forcing them to associate with each other to form important structures like cell and organelle membranes.

2. Glycogen is a polymer of glucose residues connected primarily by $\alpha(1\rightarrow4)$ glycosidic bonds, and it has occasional $\alpha(1\rightarrow6)$ bonds that produce branches. Cellulose is a straight, unbranched polymer with $\beta(1\rightarrow4)$ glycosidic bonds between the glucose residues.

3. Fats accumulate in fat droplets of cells and are an efficient form of energy storage. Phospholipids function as the major components of cell membranes.

4. Nucleotides function as carriers of chemical energy (e.g., ATP) and as intracellular signaling molecules (e.g., cyclic AMP).

5. No. The information in nucleic acids is conveyed by the sequence of its bases. If the bases were removed, we would be left with a linear sugar phosphate chain with little informational content.

6. Christian Anfinsen and his colleagues showed that denatured ribonuclease can spontaneously refold into the active enzyme in the absence of other cellular constituents. This showed that the information required for folding was contained in the primary sequence of ribonuclease.

7. The side chain of cysteine contains a sulfhydryl group that can form covalent disulfide bonds with another cysteine residue, stabilizing the structure of cell surface and secreted proteins.

8. Cholesterol can be incorporated into membranes, where it regulates membrane fluidity. Additionally, cholesterol is used in the synthesis of steroid hormones.

9. The α-helical structure allows the CO and NH groups of peptide bonds to form hydrogen bonds with each other, thereby neutralizing their polar character. β-barrels are also capable of traversing lipid bilayers.

10. d. Alanine is the only amino acid on the list with a hydrophobic side chain capable of interacting with the fatty acids of membrane lipids.

11. In the first dimension, proteins are separated in a pH gradient according to their overall charge. In the second dimension, the proteins are separated on the basis of their mass.

12. In one approach, an organelle is purified by subcellular fractionation and the proteins contained within it are identified by mass spectrometry. Another approach, which has been used in the global analysis of yeast proteins, is to express a large number of proteins fused to GFP. The subcellular location of the fusion proteins can then be determined by fluorescence microscopy.

CHAPTER 3

1. Aspartate is an acidic amino acid that interacts with basic amino acids in the substrates of trypsin. Substitution of aspartate with lysine (a basic amino acid) would interfere with substrate binding and catalysis.

2. The side chain of histidine can be either uncharged or positively charged at physiological pH, thus allowing it to be used for exchange of hydrogen ions.

3. Enzyme E1 is probably regulated by feedback inhibition.

4. An energetically unfavorable reaction can be coupled to an energetically favorable reaction with a high negative free energy change (often the hydrolysis of ATP), such that the combined reaction is energetically favorable.

5. The reaction catalyzed by phosphofructokinase is fructose-6-phosphate + ATP $\leftrightarrow$ fructose-1,6-bisphosphate + ADP. The standard free energy change can be calculated as the sum of the standard free energy change of fructose-1,6-bisphosphate formation from fructose-6-phosphate and phosphate ($\Delta G^{o'}$ = +4 kcal/mol) and the standard free energy change of ATP hydrolysis (–7.3 kcal/mol), giving a standard free energy change of –3.3 kcal/mol for the coupled reaction.

6. Substituting the given values into the equation:

$$\Delta G = \Delta G^{o} + RT \ln \frac{[B][C]}{[A]}$$

gives the following result: ΔG = –1.93 kcal/mol. The reaction will therefore proceed from left to right, with A being converted to B plus C within the cell.

7. Phosphofructokinase is inhibited by high concentrations of ATP. Thus, glycolysis will be inhibited in response to an increased cellular concentration of ATP.

8. Under anaerobic conditions, glucose is metabolized only through glycolysis, with a net production of 2 ATP molecules per glucose molecule. Under aerobic conditions, glucose is completely oxidized to produce 36 – 38 molecules of ATP.

9. Under anaerobic conditions the NADH produced during glycolysis is used to reduce pyruvate to ethanol or lactate, thereby regenerating NAD$^+$.

10. Since lipids are more reduced than fatty acids, their oxidation yields much more energy per molecule.

11. In the light reactions, energy derived from the absorption of light by chlorophylls is used to split water to $2H^+ + 1/2\ O_2 + 2e^-$. The electrons enter an electron transport chain that results in synthesis of ATP and NADPH. In the dark reactions, ATP and NADPH drive the synthesis of glucose from CO_2 and H_2O.

12. Some reactions in glycolysis involve a large decrease in free energy and are not easily reversible. Gluconeogenesis bypasses these reactions by other reactions that are driven by the expenditure of ATP and NADH.

CHAPTER 4

1. You would breed flies with mutations in the two different genes. If mutations in the two genes are frequently inherited together, the genes would be linked and reside on the same chromosome.

2. Half the DNA will be intermediate density and half will be light.

3. Addition or deletion of one or two nucleotides would alter the reading frame of the gene, resulting in a completely different amino acid sequence of the encoded protein.

4. Addition or deletion of three nucleotides adds or removes only one amino acid, often yielding a protein that functions normally.

5. To clone this piece of human DNA, a yeast artificial chromosome would need telomeres and a centromere in order to replicate as a linear chromosome-like molecule, as well as an origin of replication and a cleavage site for *Eco*RI.

6. UGC also encodes cysteine, so this mutation would have no effect on enzyme function. UGA, however, is a stop codon, so this mutation would lead to a truncated protein that would be inactive.

7. The restriction map is

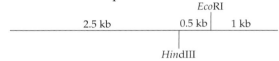

8. The haploid sperm contains a single starting copy of the DNA sequence. Each cycle of PCR amplifies the starting material twofold, so 10 cycles yields 2^{10}, or approximately a thousand, copies. Amplification for 30 cycles yields 2^{30}, or more than a billion, copies.

9. From Table 4.3, cosmids carry inserts of 30–45 kb. From Table 4.2, the recognition sequence of *Bam*HI is 6 bp long, which occurs with a random frequency of once every 4000 base pairs of DNA ($4^6 = 4096$). Dividing 30,000–45,000 by 4000, we expect about 10 *Bam*HI fragments in a cosmid insert.

10. The human genome is approximately 3×10^6 kb, and the size of an insert for a BAC vector is 120–300 kb (Table 4.3). Dividing the size of the genome by the size of an insert indicates that a minimum of 10,000 to 25,000 BAC clones would be needed to cover the genome.

11. Influenza virus replicates via RNA-directed RNA synthesis, so actinomycin D (which inhibits DNA-directed RNA synthesis) will not affect influenza virus replication.

12. A selectable marker, such as drug resistance, that allows you to select stably transfected cells.

13. Proteins are dissolved in a solution containing the negatively charged detergent SDS. Each protein binds many molecules of SDS, which imparts a net negative charge to the protein. The proteins can then be separated by gel electrophoresis according to their size.

CHAPTER 5

1. Organisms with larger genomes than their complexity have larger amounts of noncoding DNA.

2. Sharp and coworkers made hybrids between adenovirus hexon mRNA and a single strand of viral DNA. Upon observation under an electron microscope, they saw that the hexon mRNA hybridized with separate regions of viral DNA and the intervening unhybridized DNA formed loops. The loops of single stranded DNA corresponded to introns that were spliced out of the long primary transcript.

3. Through alternative splicing more than one protein can be synthesized from the same gene, thereby increasing the diversity of proteins expressed from a limited number of genes.

4. Because of their distinct base composition, simple sequence repeats (e.g., ACAAACT) differ in density from the average AT/GC ratio of bulk nuclear DNA. Sequences containing tandem copies of these repeats will separate upon CsCl density-gradient centrifugation into satellite bands that are distinct from the main band of genomic DNA.

5. A centromere of *S. cerevisiae* is small (125 base pairs) and presumably binds a limited number of kinetochore proteins to form a single microtubule-binding site. Centromeres of animal chromosomes are much larger, with repeated sequences that form more extensive kinetochores and provide attachment sites for multiple microtubules.

6. Cutting the plasmid with a restriction endonuclease creates a linear chromosome with two free ends but no telomeres. The plasmid genes are quickly lost because of the instability of the chromosome ends. To test this, you could attach telomere sequences to the ends of the linearized plasmid and see if this allows the plasmid to be stably inherited.

7. The answer can be obtained by dividing the length of the human genome (3×10^6 kb) by the number of genes, after subtracting the length of DNA encompassed by the genes themselves (calculated as the number of genes multiplied by the average length of a gene [30 kb]). Thus the average distance between genes is about 100 kb.

8. One molecule of histone H1 binds genomic DNA approximately every 200 bp, so the number of histone H1 molecules bound to the yeast genome can be determined by dividing the size of the yeast genome (12 Mb) by 200 bp to yield a value of 60,000.

9. From Table 5.1, we can calculate the average length of a human intron by dividing the total length of intron sequence by the number of introns in an average human gene: approximately 3400 bp.

10. From Table 5.1, the exon sequence of an average human gene totals 2500 bp. This would be the expected average length of complete cDNA inserts in your library.

11. The International Human Genome Sequencing Consortium sequenced BAC clones that had already been mapped to distinct regions of human chromosomes. Celera Genomics used a whole-genome shotgun approach in which genomic DNA fragments were sequenced at random and overlaps between fragments were then used to reassemble a complete genome sequence.

12. Unlike protein coding sequences, regulatory sequences are short, poorly defined sequences that occur frequently by chance in large genomes, making it difficult to identify functional regulatory sequences. Approaches used to identify functional regulatory sequences include looking for clusters of regulatory elements, for sequences that are conserved in evolution, and for sequences that are present in coordinately regulated genes.

13. Single nucleotide polymorphisms (SNPs) are differences in single base pairs between the genomes of individuals. By studying SNPs, we may be able to identify specific genes associated with susceptibility to different diseases.

CHAPTER 6

1. At high temperature the mutant would have defects in replacing RNA primers with DNA in Okazaki fragments and in filling gaps in DNA following excision repair.

2. Topoisomerase I cuts one strand of the DNA double helix and allows it to swivel around the other strand to relieve twist tension. Topoisomerase II cuts both strands of the DNA double helix and can pass another DNA molecule through the cut to untangle intertwined molecules (e.g., during mitosis).

3. From Table 5.6, the size of the yeast genome is 12 Mb (12 million base pairs), while Okazaki fragments are approximately 1–3 kb (1,000–3,000 base pairs) in length. We can calculate the number of Okazaki fragments synthesized during replication by dividing the size of the genome by the length of Okazaki fragments, giving an answer of 4,000–12,000.

4. DNA polymerases are unable to initiate DNA synthesis *de novo:* they can only add nucleotides to a pre-existing primer strand. Primase initiates *de novo* and synthesizes short RNA primers, which can then be extended by a DNA polymerase.

5. 3′ to 5′ exonuclease activity is required for the excision of mismatched bases in newly synthesized DNA during proofreading. A mutant *E. coli* with a DNA polymerase III lacking this activity would have a high frequency of mutations each time the DNA is replicated.

6. You would insert different DNA sequences into plasmids that lack origins of replication and determine whether the plasmids are able to transform mutant yeast that require a plasmid gene for their growth and division. Only plasmids with a functional origin will yield a high frequency of transformed yeast colonies.

7. DNA polymerases are unable to replicate the extreme 5′ ends of linear DNA molecules; yeast cells have evolved telomerase to maintain the ends of their linear chromosomes. Since the *E. coli* genome is a circular DNA molecule and has no ends, special mechanisms to replicate the ends of linear DNA are not required.

8. Double strand breaks can be repaired by recombinational repair, in which the missing portion of one chromosome can be recovered from a homologous chromosome. Single stranded breaks can be repaired by excision repair, since an undamaged complementary strand is available for use as template to direct synthesis of the excised portion of the damaged strand.

9. The high frequency of skin cancer results from DNA damage induced by solar UV irradiation, which is subject to repair by the nucleotide-excision repair system. The lack of elevated incidence of other cancers may suggest that similar types of damage are not frequent in internal organs, and that most cancers of these organs result from other types of mutations (e.g., the incorporation of mismatched bases during DNA replication).

10. The cellular processes that might be affected by these drugs include maintenance of telomeres by telomerase and the transposition of retrotransposons.

11. Instead of using methylation of parental strands, the mismatch repair system in humans uses single-strand breaks to identify newly replicated DNA. Thus, a homolog of *Mut H* is not required in human cells.

12. The mouse would be immunodeficient, lacking both B and T lymphocytes, as a result of being unable to rearrange its immunoglobulin and T cell receptor genes.

CHAPTER 7

1. The DNA sequence is radiolabeled at one end. It is then incubated with the protein and subjected to partial digestion with DNase. The site at which the protein binds will be protected from digestion, so no

labeled pieces of that length will be seen after electrophoresis of the digested DNA.

2. Sigma factors bind to sequences upstream of the transcription start site, bringing RNA polymerase specifically to the promoter region to initiate transcription.

3. The most common type of termination involves transcription of a GC-rich inverted repeat that forms a stable stem-loop structure by complementary base pairing. Formation of this structure disrupts association of the mRNA with the DNA and terminates transcription.

4. Since the diploid contains a functional repressor, it will bind the operator of the wild-type gene, in *trans*, and it will be regulated normally. The temperature-sensitive gene's operator cannot bind the repressor, so it will be expressed constitutively. β-galactosidase will therefore be produced at the permissive but not the nonpermissive temperature.

5. The promoter containing the TATA box can be transcribed *in vitro* in the presence of either TBP or TFIID. However, the Inr promoter requires TFIID, since the Inr sequence is recognized by TAFs rather than by TBP.

6. The activity of enhancers depends neither on their distance nor their orientation with respect to the transcription start site. Promoters are defined as being near the transcription start site.

7. The sequence element would be a potential binding site for a tissue-specific repressor.

8. They divide chromosomes into individual domains of chromatin structure that can be euchromatin or heterochromatin but cannot spread beyond an insulator. Also, they prevent an enhancer in one domain from acting on a promoter in the next domain.

9. One of the two X chromosomes is inactivated early in female development (or in XXY males). An RNA called *Xist* is produced by the *Xist* gene on one of the two X chromosomes, and binds to most of the genes on that chromosome. *Xist* RNA recruits proteins that induce chromatin condensation and conversion of most of the inactive X to heterochromatin.

10. The specific high-affinity binding of Sp1 to the GC box DNA sequence is exploited for its biochemical purification by DNA-affinity chromatography. To demonstrate that the purified protein is Sp1, you could carry out in vitro transcription assays.

11. The anti-Sm antiserum would bind to the snRNPs and prevent them from binding splice sites, thereby inhibiting splicing.

12. Noncoding RNAs can repress the transcription of a target gene via the RITS complex and they can induce the degradation of a target mRNA via the RISC complex. Thus noncoding RNAs can regulate both the synthesis and degradation of target transcripts.

13. Splicing factors direct snRNPs to the correct splice sites by binding to specific sequences in the pre-mRNA.

14. Apo-B100 is synthesized in the liver by translation of the unedited mRNA. The shorter Apo-B48 is synthesized in the intestine by translation of an edited mRNA in which the editing reaction has generated a stop codon.

15. Nonsense-mediated mRNA decay is a quality control process by which mRNAs that lack complete open-reading frames are degraded. This process eliminates defective mRNAs that would have given rise to abnormal truncated proteins.

CHAPTER 8

1. Many tRNAs are able to recognize more than one codon because the third base of their anticodon can "wobble" or form hydrogen bonds in nonstandard ways with bases other than the usual complementary pairs.

2. A Shine-Dalgarno sequence is needed.

3. Ribosomes depleted of most of their proteins can still synthesize polypeptides, whereas treatment with RNase completely abolishes translational activity. In addition, high-resolution structural analysis of the 50S ribosomal subunit showed that the site at which the peptidyl transferase reaction occurs is composed of rRNA, not ribosomal protein.

4. Polyadenylation is an important translational regulatory mechanism in early development. Its inhibition would block the translation of many oocyte mRNAs following fertilization.

5. Chaperones are proteins that aid the proper folding of other proteins. Elevated temperatures can denature proteins. Heat-shock proteins are chaperones that aid in the refolding of these denatured proteins, thereby restoring protein function.

6. Phospholipase treatment would release a GPI-anchored protein but not a transmembrane protein, from the cell surface.

7. The degradation of cyclin B, which allows cells to exit mitosis, requires a specific 9-amino-acid sequence called the destruction box. Mutations in this sequence prevent ubiquitination and degradation of cyclin B and block exit from mitosis.

8. No. In addition to targeting proteins for degradation, ubiquitination of a protein can have other functions, including targeting proteins for endocytosis and serving as part of the histone code.

9. miRNAs can target specific mRNAs via the RISC complex, leading to either mRNA degradation or inhibition of translation.

10. Sequences in the 3′ UTRs can regulate mRNA stability, localization, and translation by serving as binding sites for regulatory factors and miRNAs.

11. Since *T. aquaticus* grows at high temperatures, its rRNA is more stable than *E. coli* rRNA and better able to withstand the vigorous protein extraction procedures used in Noller's experiments.

12. Signal sequences that target proteins to specific cellular compartments are often removed by the action of signal peptidases. Additionally, many proteins are synthesized as larger precursor proteins and are proteolytically processed to yield the mature protein.

13. A "decoding center" in the small ribosomal subunit recognizes correct codon-anticodon base pairs and discriminates against mismatches. Insertion of a correct aminoacyl tRNA into the A site induces a conformational change that causes the hydrolysis of GTP bound to eEF-1α and the release of the elongation factor bound to GDP.

14. PDI allows rapid exchanges between paired disulfides, yielding a pattern of disulfide bonds that is compatible with the stably folded conformation of a protein. Inhibiting PDI with the siRNA will interfere with the correct folding of RNase, so you will detect much lower RNase activity in the medium of cells expressing the siRNA.

CHAPTER 9

1. Prokaryotic mRNAs are translated as they are transcribed. Separation of the site of transcription from the site of translation in eukaryotes allows regulation of mRNAs by posttranscriptional processes, such as alternative splicing, polyadenylation, and regulated transport to the cytoplasm. In addition, transcription can be regulated by modulating the nuclear localization of transcription factors.

2. Lamins form a filamentous network that supports and stabilizes the nuclear envelope. Lamins also provide binding sites for chromatin to attach to the inside of the nuclear envelope. In addition, many proteins that participate in transcription, DNA replication, and chromatin modification interact with lamins.

3. The 15 kd but not the 100 kd protein will be able to enter the nucleus, since proteins smaller than approximately 20 kd can pass freely through the nuclear pore complex.

4. The distribution of Ran/GTP across the nuclear envelope determines the directionality of nuclear transport.

5. One example is NF-κB. In unstimulated cells, IκB binds to NF-κB, and the complex cannot be imported to the nucleus. Upon stimulation, IκB is phosphorylated and degraded by ubiquitin-mediated proteolysis. This exposes a nuclear localization signal on NF-κB, allowing it to enter the nucleus and activate transcription of target genes.

6. The transcription factor could no longer be phosphorylated at these sites, so it would be constitutively imported to the nucleus and activate target gene expression.

7. Inactivating the nuclear export signal would result in retention of the protein in the nucleus.

8. You would label newly replicated DNA with bromodeoxyuridine, which is incorporated into DNA in place of thymidine. The bromodeoxyuridine-labeled DNA can be localized with antibodies against bromodeoxyuridine and fluorescence microscopy.

9. Kalderon and colleagues fused the T antigen amino acid sequence 126 to 132 to normally cytoplasmic proteins, β-galactosidase and pyruvate kinase. These fusion proteins accumulated in the nucleus.

10. Nuclear speckles contain concentrated populations of snRNPs and are thought to be storage sites for splicing components.

11. Small nucleolar RNAs (snoRNAs) localize to the nucleolus where they participate in the cleavage and modification of rRNAs.

12. Exportin-t is required for the export of tRNAs from the nucleus, so inhibiting exportin-t function would prevent tRNA export and lead to inhibition of translation.

CHAPTER 10

1. Palade and coworkers labeled pancreatic acinar cells with a pulse of radioactive amino acids, which were incorporated into proteins. Autoradiography showed that the labeled proteins were first detected in the rough ER. After a short "chase" with nonradioactive amino acids, the labeled proteins had moved to the Golgi apparatus and after longer periods, the labeled proteins were found in secretory vesicles and then outside of cells.

2. When an mRNA encoding a secreted protein is translated *in vitro* on free ribosomes, a larger protein results than when the same mRNA is translated in the presence of microsomes from the rough ER. In the latter case, the signal sequence is cleaved off by a signal peptidase in the rough ER vesicles.

3. Cotranslational translocation involves binding of the nascent polypeptide to the signal recognition particle and translocation through a translocon driven by the process of protein synthesis. Posttranslational

translocation targets a polypeptide to the ER after synthesis is complete and does not require SRP. A Sec62/63 complex recognizes a polypeptide to be incorporated, and inserts it into a translocon. The polypeptide is pulled through into the ER lumen by the chaperone BiP.

4. Proteins bound for the Golgi apparatus are unable to enter the ER in a Sec61 mutant and therefore remain in the cytosol.

5. Carbohydrate groups are added within the lumen of the ER and Golgi apparatus, both of which are topologically equivalent to the exterior of the cell.

6. Mutation of the KDEL sequence would inhibit the return of this ER-resident protein from the Golgi to the ER, so it would be secreted from the cell. Inactivation of the KDEL receptor protein would result in secretion of all KDEL-containing ER proteins.

7. Initially, a signal sequence targets the nascent lysosomal polypeptide to the rough ER. As it enters the ER, an *N*-linked oligosaccharide is added to the protein. After it moves to the Golgi, the three mannoses that are normally removed are not, but are instead converted to mannose-6-phosphate. These mannose-6-phosphates are recognized by a receptor in the *trans* Golgi network, which directs the transport of these proteins to lysosomes.

8. The normally cytosolic protein lacks a signal sequence and does not enter the ER. Therefore addition of a lysosome-targeting signal would have no effect. In contrast, such an addition would direct a normally secreted protein to lysosomes from the Golgi apparatus.

9. In the absence of mannose-6-phosphate formation in the Golgi apparatus, lysosomal proteins would be secreted.

10. Glycolipids and sphingomyelin are produced by addition of sugars or phosphorylcholine to ceramide on the cytosolic and lumenal surfaces, respectively, of the Golgi apparatus. Glucosylceramide is then flipped to the lumenal surface. After vesicular transport and fusion with the plasma membrane, these lipids are located on the outer half of the plasma membrane.

11. A cell-free transport system was used by Rothman and his colleagues to demonstrate vesicular transport of proteins between Golgi cisternae.

12. The diagnosis is Gaucher disease. Since this disease involves a mutation in the lysosomal hydrolase glucocerebrosidase, you might suggest enzyme replacement therapy.

13. The enzymes destined for lysosomes are acid hydrolases that are not active at the neutral pH of the cyto-

plasm, ER, or Golgi apparatus. The acid hydrolases are activated by the acid pH of lysosomes, which is maintained by lysosomal proton pumps.

14. The strong interaction between the coiled-coil domains of v-SNAREs and t-SNAREs places the two membranes nearly in direct contact. This produces membrane instability and causes the membranes to fuse.

15. Rab proteins are kept in the GDP bound form in the cytosol by association with GDP dissociation inhibitors (GDIs). At membranes, the GDIs are removed by GDI displacement factors and membrane-localized guanine nucleotide exchange factors stimulate the exchange of GTP for GDP.

CHAPTER 11

1. The surface area of the mitochondrial inner membrane is high due to the formation of cristae. Also, the protein content of the membrane is unusually high (>70 %) and includes many enzymes and electron carriers.

2. Mitochondrial tRNAs are able to do an extreme form of wobble in which U in the anticodon can pair with any of the four bases in the third codon position of mRNA, allowing four codons to be recognized by a single tRNA.

3. The proton gradient of 1 pH unit corresponds to a tenfold higher $[H^+]_o$. By substituting the above values in the equation

$$\Delta G = RT \ln [H^+]_i/[H^+]_o$$

We get an answer of approximately –1.4 kcal/mol

4. Hsp70 proteins in the cytosol maintain newly synthesized mitochondrial proteins in an unfolded state so they can be inserted into the TOM complex and be imported as an unfolded chain. Mitochondrial matrix Hsp70 chaperones bind to the polypeptide chain as it emerges from the TIM complex and use ATP hydrolysis to pull the polypeptide into the matrix. In some cases, an Hsp60 chaperonin complex binds the polypeptide and facilitates folding into its proper tertiary structure.

5. The first positively charged import signal leads to import into the mitochondrial matrix, where this presequence is removed; the second, hydrophobic sequence targets the polypeptide to Oxa1, a translocase in the inner membrane that passes cytochrome *b2* into the intermembrane space.

6. Coenzyme Q accepts a pair of electrons from either complex I or complex II and passes them to complex III. Coenzyme Q also binds two protons from the mitochondrial matrix, carries them across the inner membrane, and releases them to the intermembrane space, contributing to the generation of a proton gra-

dient. Cytochrome c is a small protein in the inter-membrane space that picks up a pair of electrons from complex III and transfers them to complex IV.

7. F_0 spans the inner membrane of mitochondria and the thylakoid membrane of chloroplasts. F_1 is the stalk and knob that extends from F_0 into the matrix of mitochondria and into the stroma of chloroplasts. The F_0 complex provides a channel through which protons can flow back toward the matrix or stroma. Mechanical coupling by the stalk drives a rotation within the F_1 complex that catalyzes the synthesis of ATP.

8. In contrast to mitochondria, there is no electric potential across the chloroplast membrane. Therefore the charge of transit peptides does not contribute to protein translocation.

9. Two high-energy electrons are required to split each molecule of H_2O, so 24 high-energy electrons are required for the synthesis of each molecule of glucose. The passage of these electrons through the two photosystems generates 12 molecules of NADPH and between 12 and 18 molecules of ATP, depending on the stoichiometry of proton pumping at the cytochrome bf complex. Since 18 molecules of ATP are required for the Calvin cycle, the synthesis of glucose may require additional ATPs produced by cyclic electron flow.

10. Three out of four carbon atoms converted to glycolate are returned to chloroplasts and re-enter the Calvin cycle.

11. Most peroxisomal proteins are synthesized on free ribosomes in the cytosol and targeted to peroxisomes by either a Ser-Lys-Leu signal sequence on their carboxy terminus or a nine-amino acid sequence on their amino terminus. These are recognized by receptors and imported through transporters in the peroxisomal membrane.

12. While the thylakoid membrane is impermeable to protons, it is freely permeable to other ions, which can neutralize the voltage component of the proton gradient.

CHAPTER 12

1. The asymmetrical actin monomers associate in a head-to-tail fashion to form actin filaments. Since all the monomers are oriented in the same direction, the filament has a distinct polarity. The polarity of actin filaments defines the direction of myosin movement. If actin filaments were not polar, the unidirectional movement of myosin that results in the sliding of actin and myosin filaments could not take place.

2. Treadmilling is a dynamic behavior of actin filaments (or microtubules) in which they maintain a near-constant length by adding ATP-actins (or GTP-tubulins) to the plus end and dissociating an equal number of ADP-actins or GDP-tubulins from the minus end. During this steady-state behavior, subunits hydrolyze their nucleoside triphosphates after assembly, flux through the filament, and exit from the minus end. Treadmilling occurs at a monomer concentration between the critical concentration for the plus end and the critical concentration for the minus end.

3. Cytochalasin binds to the plus ends of actin filaments and blocks their elongation, so it would lead to depolymerization of treadmilling filaments. Phalloidin binds to actin filaments and inhibits their depolymerization, so it would cause filaments to stop treadmilling but remain present and grow longer.

4. ADF/cofilin binds to actin filaments and increases the dissociation of actin monomers. Profilin binds actin/ADP and stimulates the exchange of ADP for ATP, forming actin/ATP monomers that can join growing filaments. The Arp2/3 complex initiates the formation of branches.

5. The I band and the H zone shorten during contraction. The A band doesn't shorten because it is occupied by thick myosin filaments.

6. The contraction of smooth muscle cells is regulated by the phosphorylation of the myosin regulatory light chain by myosin light-chain kinase, which in turn is regulated by association with the calcium-binding protein calmodulin. An increase in the cytosolic concentration of calcium leads to the binding of calmodulin to myosin light-chain kinase.

7. Intermediate filaments are not required for the growth of cells in culture, so siRNA against vimentin would have no effect.

8. Dimers of cytoskeletal intermediate filaments assemble in a staggered antiparallel manner to form tetramers, which can then assemble end-to-end to form protofilaments. Since they are assembled from antiparallel tetramers, intermediate filaments do not have distinct ends and are apolar.

9. The *in vitro* movement of microtubules required ATP and was inhibited by the nonhydrolyzable ATP analog AMP-PNP. Importantly, organelles remained attached to microtubules in the presence of AMP-PNP, suggesting that the motor proteins responsible for organelle movement might also remain attached.

10. By linking the microtubule doublets in cilia together, nexin converts the sliding of individual microtubules to a bending motion that leads to the beating of cilia. If it were eliminated, the microtubules would simply slide past one another.

11. Colcemid inhibits the polymerization of microtubules and would inhibit the transport of secretory vesicles along microtubules.

12. γ-tubulin plays a key role in the formation of microtubule organizing centers. γ-tubulin associates with other proteins to form the γ-tubulin ring complex, which functions as a seed for the nucleation of new microtubules.

CHAPTER 13

1. At high temperatures cholesterol's ring structure inhibits the movement of phospholipids in the bilayer, thereby reducing membrane fluidity and increasing stability. At low temperatures cholesterol interferes with fatty acid chain interactions and maintains membrane fluidity.

2. Peripheral membrane proteins can be removed from a membrane by a high-salt wash or by solutions of extreme pH that do not disrupt the phospholipid bilayer. Integral membrane proteins can only be extracted from membranes by detergents that disrupt the phospholipid bilayer.

3. Frye and Edidin fused mouse and human cells and examined the distribution of membrane proteins after staining with anti-mouse and anti-human antibodies labeled with different fluorescent dyes. Immediately after fusion, the proteins were located in different halves of the fused cell surface, but after a brief incubation at 37°C the proteins were intermixed. This demonstrated that the proteins could diffuse laterally in a fluid membrane. If incubated at 2°C, the proteins remained separated because the membrane is not fluid at this temperature.

4. Lipid rafts are discrete membrane domains that are enriched in cholesterol and sphingolipids. Lipid rafts are believed to play important roles in cell movement, endocytosis, and cell signaling.

5. The glycocalyx protects the cell surface and is involved in cell-cell interactions.

6. Given $C_o/C_i = 10$, the K$^+$ equilibrium potential calculated from the Nernst equation is about –59 mV. The actual resting membrane potential differs from the K$^+$ equilibrium potential because resting squid axons are more permeable to K$^+$ than to other ions.

7. The opening of nicotinic acetylcholine receptors is required for membrane depolarization in muscle cells. Thus curare blocks the contraction of muscle cells in response to acetylcholine.

8. The K$^+$ channel contains a selectivity filter lined with carbonyl oxygen atoms. The pore is just wide enough to allow passage of dehydrated K$^+$ ions from which all water molecules have been displaced as a result of association with the carbonyl oxygen atoms. Hydrated Na$^+$ ions are too small to interact with the carbonyl oxygen atoms and remain associated with water molecules. This complex is too large to pass through the channel pore.

9. The uptake of glucose against its concentration gradient is coupled to the transport of Na$^+$ ions in the energetically favorable direction.

10. The *mdr* gene encodes an ABC transporter that is frequently overexpressed in cancer cells. The transporter can recognize a variety of drugs and pump them out of the cell, conferring resistance to chemotherapeutic drugs.

11. The addition of excess unlabeled LDL reduced the binding of labeled LDL to the surface of normal cells. This indicated that the labeled and unlabeled LDL were competing for a limited number of specific binding sites on the surface of normal cells.

12. Two types of mutations in the LDL receptor resulting in the inability to take up LDL were identified in FH patients. Cells from most FH patients failed to bind LDL, demonstrating that a specific receptor was required for LDL uptake. Other mutant receptors bound LDL, but failed to cluster in coated pits, demonstrating the role of coated pits in receptor-mediated endocytosis.

CHAPTER 14

1. Gram-positive bacteria have a single membrane—the plasma membrane—surrounded by a thick cell wall. Gram-negative bacteria have both a plasma membrane and an outer membrane, separated by a thin cell wall.

2. The rigid plant cell wall prevents cell swelling and allows the buildup of turgor pressure.

3. Hemicelluloses are branched polysaccharides that hydrogen bond to the surface of cellulose microfibrils. This interaction stabilizes the cellulose microfibrils into tough fibers.

4. The correct localization of these glucose transporters is necessary for the polarized function of intestinal epithelial cells in transferring glucose from the intestinal lumen to the blood supply. Tight junctions prevent the diffusion of these transporters between domains of the plasma membrane, as well as sealing the spaces between cells of the epithelium.

5. Hydroxylysine stabilizes the collagen triple helix by forming hydrogen bonds between the polypeptide chains. Inhibition of lysyl hydroxylase would therefore decrease the stability of collagen fibrils.

6. Fibril-forming collagens are synthesized as soluble precursors known as procollagens. Procollagens have nonhelical segments on their ends, which prevent the assembly of fibrils. Only after procollagen is secreted outside the cell are the non-helical segments removed and the fibrils assembled.

7. GAGs contain acidic sugar residues, which are modified by the addition of sulfate groups. This imparts

a high negative charge to GAGs, so they bind positively charged ions and trap water molecules to form hydrated gels.

8. The peptide most likely disrupts tight junctions, thereby allowing free diffusion of the transporter between the apical and basolateral domains of the plasma membrane.

9. Since E-cadherins mediate selective adhesion of epithelial cells, overexpression of a dominant negative version would disrupt interactions between neighboring cells.

10. The cytoplasmic domain of $\alpha_6\beta_4$ integrin is required for its interaction with the intermediate filament cytoskeleton through plectin. The mutation would therefore disrupt hemidesmosome formation and lead to decreased cell-matrix interaction.

11. Electrical synapses are specialized gap junctions found in nerve cells. They allow the rapid passage of ions between cells, thereby conducting a nerve impulse.

12. Gap junctions and plasmodesmata are similar in that they both provide channels between the cytoplasms of adjacent cells. They are likely to be analogous rather than homologous structures in animals and plants, since their structures are extremely different.

CHAPTER 15

1. In paracrine signaling, molecules released by one or a few cells affect nearby cells. In endocrine signaling, hormones are carried throughout the body to act on any target cell that has a receptor for that hormone.

2. Hydrophobic molecules like steroid hormones can diffuse through the plasma membrane and bind to cytosolic or nuclear receptors. Hydrophilic molecules (like peptide hormones) cannot cross the plasma membrane, so they act by binding to receptors on the cell surface.

3. Aspirin inhibits the enzyme cyclooxygenase, which catalyzes the first step in the synthesis of prostaglandin and thromboxanes from arachidonic acid.

4. Inhibition of cAMP phosphodiesterase would result in elevated levels of cAMP, which would stimulate cell proliferation.

5. The recombinant molecule would function as an epinephrine receptor coupled to G_i. Epinephrine would therefore inhibit adenylyl cyclase, lowering intracellular cAMP levels. Acetylcholine would have no effect, since it would not bind to the recombinant receptor.

6. PDGF monomers would not induce receptor dimerization. Since this is the first critical step in signaling from receptor protein-tyrosine kinases, they would be unable to stimulate the PDGF receptor.

7. The truncated receptor would act as a dominant-negative mutant because it would dimerize with the normal receptor. The dimers would be inactive because they would be unable to cross-phosphorylate.

8. Protein phosphatase 1 dephosphorylates serine residues that are phosphorylated by protein kinase A. Cyclic AMP-inducible genes are activated by CREB, which is phosphorylated by protein kinase A, so overexpression of protein phosphatase 1 would inhibit their induction. However, protein phosphatase 1 would not affect the activity of cAMP-gated ligand channels, since these channels are opened directly by cAMP binding rather than by protein phosphorylation.

9. Akt phosphorylates the protein kinase GSK-3β which regulates the translation factor eIF-2B. In addition, Akt regulates the mTOR/raptor protein kinase, which regulates translation by phosphorylating S6 kinase and the eIF-4E binding protein 4E-BP1.

10. Sos is required for the activation of Ras, so siRNA against Sos would inhibit induction of the immediate early gene.

11. JAKs phosphorylate and activate STATs in response to cytokine signaling. Expression of a dominant-negative JAK would inhibit STAT activation and block gene induction.

12. The mutation most likely alters a ubiquitination site on β-catenin, leading to its stabilization and nuclear accumulation. In the nucleus, β-catenin forms a complex with Tcf/LEF transcription factors and induces gene expression.

CHAPTER 16

1. G_0 and G_1 cells both have the same amount of DNA ($2n$) and both are metabolically active. G_0 cells differ from G_1 cells in that they are in a quiescent state and do not proliferate unless stimulated to re-enter the cell cycle.

2. The percentage of cells in a specific stage of the cell cycle can be used to calculate the duration of that phase. Since the cell takes 30 hours to complete all stages and 53.3% of the cells have a DNA content of $2n$ (G_1 cells), the duration of G_1 is 16 hours (53.3% of 30 hours). Cells in S phase have DNA contents between $2n$ and $4n$: these correspond to 30% of the cells, so the duration of S is 9 hours. 16.7% of the cells have DNA contents of $4n$, corresponding to cells in G_2 and M phases. Since 3.3% of the cells are in M phase, 13.4% are in G_2. The duration of G_2 is therefore 4 hours and that of M is 1 hour.

3. Cell cycle arrest at the G_1 and S phase checkpoints allows time for damaged DNA to be repaired before it is replicated. Arrest at the G_2 checkpoint allows time for DNA breaks or other damage to be repaired

before mitosis occurs, preventing the damaged DNA from being passed on to daughter cells.

4. Cdk's are regulated by four different mechanisms: association with cyclins, activating phosphorylations, inhibitory phosphorylations, and association with Cdk inhibitors.

5. One daughter cell would receive two copies of the misaligned chromosome; the other daughter cell would receive none.

6. The Cdk7/cyclin H complex is a Cdk-activating kinase and is also required for initiation of transcription by RNA polymerase II. Thus inhibiting Cdk7 would lead to cell cycle arrest and the inhibition of transcription.

7. In a normal cell, overexpression of p16 would inhibit cell cycle progression at the restriction point in G_1. Because Rb is the principal target of Cdk4, 6/cyclin D complexes, a tumor cell lacking functional Rb would be unaffected by p16 overexpression.

8. The phosphorylation of nuclear lamins is required for nuclear lamina breakdown. Expression of mutant lamins lacking a phosphorylation site for Cdk1 would prevent breakdown of the nuclear envelope.

9. Cdk1/cyclin B phosphorylates several structural proteins directly to alter their properties and initiate mitosis. Among them are condensins, nuclear lamins, Golgi matrix proteins, and microtubule-associated proteins. In addition, Cdk1/cyclin B activates other protein kinases.

10. Anaphase would initiate normally. However, Cdk1/cyclin B would remain active, so re-formation of nuclei, chromosome decondensation, and cytokinesis would not occur.

11. The anaphase-promoting complex degrades securin, leading to activation of the protease separase. Separase then degrades cohesins, breaking the link between sister chromatids.

12. Oocytes of these mice would fail to arrest at metaphase II.

13. Binding of the first sperm to its receptor releases calcium within the egg, probably by cleavage of PIP_2 to release IP_3, which opens IP_3-gated Ca^{2+} channels in the ER. This calcium induces exocytosis of secretory granules whose contents alter the extracellular coat of the egg to block entry of additional sperm.

CHAPTER 17

1. Apoptotic cells are efficiently removed from tissues by phagocytosis, whereas cells that die by acute injury release their contents into the extracellular space and cause inflammation.

2. Caspases are synthesized as long inactive precursors that are activated in complexes (e.g., the apoptosome) or converted to active enzymes by proteolytic cleavage. In addition, cells contain IAPs that associate with caspases and inhibit their activity.

3. The cleavage of lamins by caspases is required for nuclear fragmentation during apoptosis. The mutated lamins will not be cleaved by caspases, so their expression will block nuclear fragmentation.

4. Proapoptotic multidomain members of the Bcl-2 family induce apoptosis by promoting the release of cytochrome c from mitochondria, which leads to caspase activation. The activity of the proapoptotic multidomain proteins is regulated by antiapoptic and BH3-only members of the Bcl-2 family.

5. Activation of p53 in response to DNA damage leads to the expression of its target genes, which include the Cdk inhibitor p21 and the BH3-only Bcl-2 family members PUMA and Noxa. p21 induces cell cycle arrest and the BH3-only Bcl-2 family members induce apoptosis.

6. The mutant Bad would no longer be maintained in an inactive state by 14-3-3 protein, so it will act to induce apoptosis.

7. 14-3-3 proteins sequester proapoptotic proteins, such as Bad and FOXO transcription factors, in an inactive state. Cells expressing siRNA against 14-3-3 proteins will therefore have an increased rate of apoptosis.

8. Caspase-8 is the initiator caspase downstream of TNF receptors, so cells with inactive caspase-8 will not undergo apoptosis upon treatment with TNF. Thus TNF therapy would not be effective for this patient.

9. Ced-3 is the only caspase in *C. elegans*. Mutating it leads to the survival of all the cells that would normally die by apoptosis during development, and RNAi against Ced-3 would have the same effect.

10. The polypeptide will lead to the release of the proapoptotic proteins cytochrome c, Smac/Diablo, and Omi/Htr2 from the mitochondria and induce apoptosis in treated cells.

11. These tissues contain stem cells that retain the ability to proliferate and replace differentiated cells.

12. The critical characteristic of stem cells is their capacity for self-renewal. They divide to produce one daughter cell that remains a stem cell and one that divides and differentiates.

13. Embryonic stem cells are easier to isolate and culture and are capable of giving rise to all of the differentiated cell types in an adult organism.

CHAPTER 18

1. Benign tumors remain confined to their original location, whereas malignant tumors can invade surrounding normal tissue and metastasize to other parts of the body.

2. As a tumor progresses, mutations occur within cells of the tumor population. Some of these mutations confer a selective advantage to the cells in which they occur and allow them to outgrow other cells in the tumor. This process is called clonal selection, and it leads to the continuing development of more rapidly growing and increasingly malignant tumors.

3. Estrogens act as tumor promoters by stimulating the proliferation of estrogen responsive cells, such as breast and endometrial cells.

4. Autocrine growth stimulation is a positive feedback system in which tumor cells produce growth factors that stimulate their own proliferation.

5. Cancer cells are less adhesive that normal cells and are not as strictly regulated by cell-cell and cell-matrix interactions. In addition, cancer cells secrete proteases that degrade components of the extracellular matrix, facilitating invasion of adjacent tissues. Finally, cancer cells secrete angiogenic factors that promote the formation of new blood vessels that supply tumors with oxygen and nutrients and facilitate metastasis by allowing the cancer cells easy access to the circulatory system.

6. E6 interacts with p53 but not with Rb, so it would not induce transformation in combination with the mutant T antigen. However, E7 binds and inactivates Rb, so it will induce transformation in conjunction with the mutant T antigen.

7. AIDS patients are immunosuppressed and are therefore susceptible to infection by oncogenic viruses.

8. A proto-oncogene may be expressed at abnormal levels or in abnormal cell types. These changes in expression can convert a proto-oncogene to an oncogene even though a structurally normal protein is produced. A proto-oncogene can be activated in this manner either by a translocation that puts it under the control of an active promoter or by gene amplification.

9. INK4 encodes the Cdk inhibitor p16, which inhibits Cdk4, 6/cyclin D complexes. Since Rb is the critical target of Cdk4, 6/cyclin D, overexpression of p16 would not affect the proliferation of cells with inactivated Rb.

10. p53 is required for cell cycle arrest and apoptosis in response to DNA damage induced by ionizing radiation. Thus tumors with wild-type p53 genes will be more sensitive to radiation.

11. STI-571 is a specific inhibitor of the Bcr/Abl protein-tyrosine kinase expressed in chronic myeloid leukemia cells. Inhibition of Bcr/Abl blocks proliferation of these tumor cells. Resistance to STI-571 is most often caused by mutations in Bcr/Abl that prevent binding of the drug.

12. It is believed that the proliferation and survival of tumor cells become dependent on activated oncogenes, with other signaling pathways becoming secondary in importance. This dependence of tumor cells on activated oncogenes has been termed oncogene addiction. It suggests that drugs against an activated oncogene would selectively target tumor cells, while normal cells would be able to compensate by using alternative signaling pathways.

Glossary

α-actinin An actin-binding protein that crosslinks actin filaments into contractile bundles.

α helix A coiled secondary structure of a polypeptide chain formed by hydrogen bonding between amino acids separated by four residues.

ABC transporters A large family of membrane transport proteins characterized by a highly conserved ATP binding domain.

abl A proto-oncogene that encodes a protein-tyrosine kinase and is activated by chromosome translocation in chronic myeloid leukemia.

abscisic acid A plant hormone.

actin An abundant 43-kd protein that polymerizes to form cytoskeletal filaments.

actin-binding proteins Proteins that bind actin and regulate the assembly, disassembly, and organization of actin filaments.

actin bundle Actin filaments that are crosslinked into closely packed arrays.

actin-bundling proteins Proteins that crosslink actin filaments into bundles.

actin network Actin filaments that are crosslinked into loose three-dimensional meshworks.

action potential Nerve impulses that travel along axons.

activation energy The energy required to raise a molecule to its transition state to undergo a chemical reaction.

activation-induced deaminase (AID) An enzyme expressed in B lymphocytes that deaminates cytosine in DNA to form uracil in the variable regions of immunoglobulin genes. AID is required for both class switch recombination and somatic hypermutation.

active site The region of an enzyme that binds substrates and catalyzes an enzymatic reaction.

active transport The transport of molecules in an energetically unfavorable direction across a membrane coupled to the hydrolysis of ATP or other source of energy.

adaptin A protein that binds to membrane receptors and mediates the formation of clathrin-coated vesicles.

adenine A purine that base-pairs with either thymine or uracil.

adenoma A benign tumor arising from glandular epithelium.

adenylyl cyclase An enzyme that catalyzes the formation of cyclic AMP from ATP.

ADF/cofilin A family of actin-binding proteins that disassemble actin filaments.

adherens junction A region of cell-cell adhesion at which the actin cytoskeleton is anchored to the plasma membrane.

adhesion belt A beltlike structure around epithelial cells in which a contractile bundle of actin filaments is linked to the plasma membrane.

Akt A protein-serine/threonine kinase that is activated by PIP_3 and plays a key role in signaling cell survival.

allele One copy of a gene.

allosteric regulation The regulation of enzymes by small molecules that bind to a site distinct from the active site, changing the conformation and catalytic activity of the enzyme.

alternative splicing The generation of different mRNAs by varying the pattern of pre-mRNA splicing.

amino acid Monomeric building blocks of proteins, consisting of a carbon atom bound to a carboxyl group, an amino group, a hydrogen atom, and a distinctive side chain.

aminoacyl tRNA synthetase An enzyme that joins a specific amino acid to a tRNA molecule carrying the correct anticodon sequence.

amphipathic A molecule that has both hydrophobic and hydrophilic regions.

amyloplast A plastid that stores starch.

anaphase The phase of mitosis during which sister chromatids separate and move to opposite poles of the spindle.

anaphase A The movement of daughter chromosomes toward the spindle poles during mitosis.

anaphase B The separation of the spindle poles during mitosis.

anaphase-promoting complex A ubiquitin ligase that triggers progression from metaphase to anaphase by signaling the degradation of cyclin B and cohesins.

angiogenesis The formation of new blood vessels.

ankyrin A protein that binds spectrin and links the actin cytoskeleton to the plasma membrane.

antibody A protein produced by B lymphocytes that binds to a foreign molecule.

anticodon The nucleotide sequence of transfer RNA that forms complementary base pairs with a codon sequence on messenger RNA.

antigen A molecule against which an antibody is directed.

antiport The transport of two molecules in opposite directions across a membrane.

antisense nucleic acids Nucleic acids (either RNA or DNA) that are complementary to an mRNA of interest and are used to block gene expression.

AP endonuclease A DNA repair enzyme that cleaves next to apyrimidinic or apurinic sites in DNA.

apical domain The exposed free surface of a polarized epithelial cell.

apoptosis An active process of programmed cell death, characterized by cleavage of chromosomal DNA, chromatin condensation, and fragmentation of both the nucleus and the cell.

apoptosome A protein complex in which caspase-9 is activated to initiate apoptosis following the release of cytochrome *c* from mitochondria.

Arabidopsis thaliana A small flowering plant used as a model for plant molecular biology and development.

archaebacteria One of two major groups of prokaryotes; many species of archaebacteria live in extreme conditions similar to those prevalent on primitive Earth.

ARF A GTP-binding protein required for vesicle budding from the *trans*-Golgi network.

armadillo protein family A family of proteins, including β-catenin, that link cadherins to the cytoskeleton at stable cell-cell junctions.

ARP 2/3 complex A protein complex that binds to actin filaments and initiates the formation of branches.

astral microtubules Microtubules of the mitotic spindle that extend to the cell periphery.

ATM A protein kinase that recognizes damaged DNA and leads to cell cycle arrest.

ATP (adenosine 5′-triphosphate) An adenine-containing nucleoside triphosphate that serves as a store of free energy in the cell.

ATP synthase A membrane spanning protein complex that couples the energetically favorable transport of protons across a membrane to the synthesis of ATP.

ATR A protein kinase related to ATM that leads to cell cycle arrest in response to DNA damage.

Aurora kinase A protein kinase family involved in mitotic spindle formation, kinetochore function, and cytokinesis.

autocrine growth stimulation Stimulation of cell proliferation as a result of growth factor production by a responsive cell.

autocrine signaling A type of cell signaling in which a cell produces a growth factor to which it also responds.

autonomously replicating sequence (ARS) An origin of DNA replication in yeast.

autophagosome A vesicle containing internal organelles enclosed by fragments of the endoplasmic reticulum membrane that fuses with lysosomes.

autophagy The degradation of cytoplasmic proteins and organelles by their enclosure in vesicles from the endoplasmic reticulum that fuse with lysosomes.

autophosphorylation A reaction in which a protein kinase catalyzes its own phosphorylation.

autoradiography The detection of radioisotopically labeled molecules by exposure to X-ray film.

auxin A plant hormone that controls many aspects of plant development.

axonemal dynein The type of dynein found in cilia and flagella.

axoneme The fundamental structure of cilia and flagella composed of a central pair of microtubules surrounded by nine microtubule doublets.

β-arrestin A regulatory protein that terminates signaling from G protein-coupled receptors, as well as stimulating other downstream signaling pathways.

β-barrel A transmembrane domain formed by the folding of β sheets into a barrel-like structure.

β sheet A sheetlike secondary structure of a polypeptide chain, formed by hydrogen bonding between amino acids located in different regions of the polypeptide.

bacterial artificial chromosome (BAC) A type of vector used for cloning large fragments of DNA in bacteria.

bacteriophage A bacterial virus.

baculovirus A virus commonly used as an expression vector for production of eukaryotic proteins in insect cells.

barrier element See insulator.

basal body A structure similar to a centriole that initiates the growth of axonemal microtubules and anchors cilia and flagella to the surface of the cell.

basal lamina A sheetlike extracellular matrix that supports epithelial cells and surrounds muscle cells, adipose cells, and peripheral nerves.

base-excision repair A mechanism of DNA repair in which single damaged bases are removed from a DNA molecule.

basement membrane See basal lamina.

basolateral domain The surface region of a polarized epithelial cell that is in contact with adjacent cells or the extracellular matrix.

Bcl-2 A member of a family of proteins that regulate programmed cell death.

benign tumor A tumor that remains confined to its site of origin.

bioinformatics The use of computational methods to analyze large amounts of biological data, such as genome sequences.

bone marrow transplantation A clinical procedure in which transplantation of bone marrow stem cells is used in the treatment of cancer and diseases of the hematopoietic system.

brassinosteroid A plant steroid hormone.

bright-field microscopy The simplest form of light microscopy in which light passes directly through a cell.

brush border The surface of a cell (e.g., an intestinal epithelial cell) containing a layer of microvilli.

cadherins A group of cell adhesion molecules that form stable cell-cell junctions at adherens junctions and desmosomes.

Caenorhabditis elegans A nematode used as a simple multicellular model for development.

callus An undifferentiated mass of plant cells in culture.

calmodulin A calcium-binding protein.

Calvin cycle A series of reactions by which six molecules of CO_2 are converted into glucose.

CaM kinase A member of a family of protein kinases that are activated by the binding of Ca^{2+}/calmodulin.

cAMP-dependent protein kinase See protein kinase A.

cAMP phosphodiesterase An enzyme that degrades cyclic AMP.

cAMP-response element (CRE) A regulatory sequence that mediates the transcriptional response of target genes to cAMP.

cancer A malignant tumor.

carbohydrate A molecule with the formula $(CH_2O)_n$. Carbohydrates include both simple sugars and polysaccharides.

carcinogen A cancer-inducing agent.

carcinoma A cancer of epithelial cells.

cardiolipin A phospholipid containing four hydrocarbon chains.

carrier proteins Proteins that selectively bind and transport small molecules across a membrane.

caspases A family of proteases that bring about programmed cell death.

catalase An enzyme that decomposes hydrogen peroxide.

caveolae Small invaginations of the plasma membrane that may be involved in endocytosis.

caveolin A protein that interacts with lipid rafts and forms caveolae.

CCND1 The gene encoding cyclin D1, which is an oncogene in a variety of human cancers.

Cdc42 A member of the Rho subfamily of small GTP-binding proteins.

Cdk1 A protein-serine/threonine kinase that is a key regulator of mitosis in eukaryotic cells.

Cdk inhibitor (CKI) A family of proteins that bind Cdks and inhibit their activity.

Cdks Cyclin dependent protein kinases that control the cell cycle of eukaryotes.

cDNA library A collection of recombinant cDNA clones.

cell adhesion molecules Transmembrane proteins that mediate cell-cell interactions.

cell cortex The actin network underlying the plasma membrane.

cell cycle checkpoints Regulatory mechanisms that prevent entry into the next phase of the cell cycle until the events of the preceding phase have been completed.

cell lines Cells that can proliferate indefinitely in culture.

cell plate A membrane-enclosed disclike structure that forms new cell walls during cytokinesis of higher plants.

cell transformation The conversion of normal cells to tumor cells in culture.

cell wall A rigid, porous structure forming an external layer that provides structural support to bacteria, fungi, and plant cells.

cellulose The principal structural component of the plant cell wall, a linear polymer of glucose residues linked by $\beta(1\rightarrow4)$ glycosidic bonds.

cellulose microfibrils Fibers in plant cell walls that are formed by the association of several dozen parallel chains of cellulose.

cellulose synthase An enzyme that catalyzes the synthesis of cellulose.

central dogma The concept that genetic information flows from DNA to RNA to proteins.

centriole A cylindrical structure consisting of nine triplets of microtubules in the centrosomes of most animal cells.

centromere A specialized chromosomal region that connects sister chromatids and attaches them to the mitotic spindle.

centrosome The microtubule-organizing center in animal cells.

cGMP phosphodiesterase An enzyme that degrades cGMP.

channel proteins Proteins that form pores through a membrane.

chaperone A protein that facilitates the correct folding or assembly of other proteins.

chaperonin A family of heat-shock proteins within which protein folding takes place.

checkpoint kinase (CHK1 and CHK2) A protein kinase that brings about cell cycle arrest in response to damaged DNA. CHK1 and CHK2 are activated by the ATM and ATR protein kinases.

chemiosmotic coupling The generation of ATP from energy stored in a proton gradient across a membrane.

chiasmata Sites of recombination that link homologous chromosomes during meiosis.

chitin A polymer of *N*-acetylglucosamine residues that is the principal component of fungal cell walls.

chlorophyll The major photosynthetic pigment of plant cells.

chloroplast The organelle responsible for photosynthesis in the cells of plants and green algae.

cholesterol A lipid consisting of four hydrocarbon rings. Cholesterol is a major constituent of animal cell plasma membranes and the precursor of steroid hormones.

chromatin The fibrous complex of eukaryotic DNA and histone proteins. See histones, nucleosome, and chromatosome.

chromatin immunoprecipitation A method for determining regions of DNA that bind transcription factors within a cell.

chromatosome A chromatin subunit consisting of 166 base pairs of DNA wrapped around a histone core and held in place by a linker histone.

chromoplast A plastid that contains carotenoids.

chromosomal microtubules Microtubules of the mitotic spindle that attach to the ends of condensed chromosomes.

chromosomes The carriers of genes, consisting of long DNA molecules and associated proteins.

cilium A microtubule-based projection of the plasma membrane that moves a cell through fluid or fluid over a cell.

cis-**acting control element** A regulatory DNA sequence that serves as a protein binding site and controls the transcription of adjacent genes.

cis-**Golgi network** The region of the Golgi apparatus at which proteins enter from the endoplasmic reticulum.

citric acid cycle A series of reactions in which acetyl CoA is oxidized to CO_2. The central pathway of oxidative metabolism.

class switch recombination A type of region specific recombination responsible for the association of rearranged immunoglobulin V(D)J regions with different heavy chain constant regions.

clathrin A protein that coats the cytoplasmic surface of cell membranes and assembles into basketlike lattices that drive vesicle budding.

clathrin-coated pit A specialized region of the plasma membrane that contains receptors for macromolecules to be taken up by endocytosis.

clathrin-coated vesicle A transport vesicle coated with clathrin.

c-**myc** A proto-oncogene that encodes a transcription factor and is frequently activated by chromosome translocation or gene amplification in human tumors.

codon The basic unit of the genetic code; one of the 64 nucleotide triplets that code for an amino acid or stop sequence.

coenzyme A (CoA) A coenzyme that functions as a carrier of acyl groups in metabolic reactions.

coenzyme Q A small lipid-soluble molecule that carries electrons between protein complexes in the mitochondrial electron transport chain.

coenzymes Low-molecular-weight organic molecules that work together with enzymes to catalyze biological reactions.

cohesins A complex of proteins that maintain the connection between sister chromatids.

colcemid A drug that inhibits the polymerization of microtubules.

colchicine A drug that inhibits the polymerization of microtubules.

collagen The major structural protein of the extracellular matrix.

collagen fibrils Fibrils formed by the assembly of collagen molecules in a regularly staggered array.

collenchyma Plant cells characterized by thick cell walls; they provide structural support to the plant.

complementary DNA (cDNA) A DNA molecule that is complementary to an mRNA molecule, synthesized *in vitro* by reverse transcriptase.

condensin A protein complex that drives metaphase chromosome condensation.

confocal microscopy A form of microscopy in which fluorescence microscopy is combined with electronic image analysis to obtain images with increased contrast and detail.

connexin A member of a family of transmembrane proteins that form gap junctions.

connexon A cylinder formed by six connexins in the plasma membrane.

contact inhibition The inhibition of movement or proliferation of normal cells that results from cell-cell contact.

contractile bundles Bundles of actin filaments that interact with myosin II and are capable of contraction.

contractile ring A structure of actin and myosin II that forms beneath the plasma membrane during mitosis and mediates cytokinesis.

COP I and **COP II** The two proteins other than clathrin that coat transport vesicles (COP indicates coat protein).

COP-coated vesicle Transport vesicles coated with COP I or COP II.

corepressor A protein that associates with repressors to inhibit gene expression, often by modifying chromatin structure.

corticosteroids Steroid hormones produced by the adrenal gland.

cosmid A vector that contains bacteriophage λ sequences, antibiotic resistance sequences, and an origin of replication. It can accomodate large DNA inserts of up to 45 kb.

CREB Cyclic AMP response element-binding protein. A transcription factor that is activated by cAMP-dependent protein kinase.

crista A fold in the inner mitochondrial membrane extending into the matrix.

crosstalk A regulatory mechanism in which one signaling pathway controls the activity of another.

cyanobacteria The largest and most complex prokaryotes in which photosynthesis is believed to have evolved.

cyclic AMP (cAMP) Adenosine monophosphate in which the phosphate group is covalently bound to both the 3′ and 5′ carbon atoms, forming a cyclic structure; an important second messenger in the response of cells to a variety of hormones.

cyclic electron flow An electron transport pathway associated with photosystem I that produces ATP without the synthesis of NADPH.

cyclic GMP (cGMP) Guanosine monophosphate in which the phosphate group is covalently bound to both the 3′ and 5′ carbon atoms, forming a cyclic structure; an important second messenger in the response of cells to a variety of hormones, and in vision.

cyclins A family of proteins that regulate the activity of Cdks and control progression through the cell cycle.

cytochalasin A drug that blocks the elongation of actin filaments.

cytochrome *bf* complex A protein complex in the thylakoid membrane that carries electrons during photosynthesis.

cytochrome *c* A mitochondrial peripheral membrane protein that carries electrons during oxidative phosphorylation.

cytochrome oxidase A protein complex in the electron transport chain that accepts electrons from cytochrome *c* and transfers them to O_2.

cytokine receptor superfamily A family of cell surface receptors that act by stimulating the activity of intracellular protein-tyrosine kinases.

cytokines Growth factors that regulate blood cells and lymphocytes.

cytokinesis Division of a cell following mitosis or meiosis.

cytokinin A plant hormone that regulates cell division.

cytoplasmic dynein The form of dynein associated with microtubules in the cytoplasm.

cytosine A pyrimidine that base-pairs with guanine.

cytoskeleton A network of protein filaments that extends throughout the cytoplasm of eukaryotic cells. It provides the structural framework of the cell and is responsible for cell movements.

cytostatic factor (CSF) A cytoplasmic factor that arrests oocyte meiosis at metaphase II.

dark reactions The series of reactions that convert carbon dioxide and water to carbohydrates during photosynthesis. See Calvin cycle.

density-dependent inhibition The cessation of the proliferation of normal cells in culture at a finite cell density.

density gradient centrifugation A method of separating particles by centrifugation through a gradient of a dense substance, such as sucrose or cesium chloride.

deoxyribonucleic acid (DNA) The genetic material of the cell.

2′-deoxyribose The five-carbon sugar found in DNA.

desmin An intermediate filament protein expressed in muscle cells.

desmosome A region of contact between epithelial cells at which keratin filaments are anchored to the plasma membrane. See also hemidesmosome.

diacylglycerol A second messenger formed from the hydrolysis of PIP_2 that activates protein kinase C.

diakinesis The final stage of the prophase of meiosis I during which the chromosomes fully condense and the cell progresses to metaphase.

dideoxynucleotides Nucleotides that lack the normal 3′ hydroxyl group of deoxyribose and are used as chain-terminating nucleotides in DNA sequencing.

differential interference-contrast microscopy A type of microscopy in which variations in density or thickness between parts of the cell are converted to differences in contrast in the final image.

differential centrifugation A method used to separate the components of cells on the basis of their size and density.

diploid An organism or cell that carries two copies of each chromosome.

diplotene The stage of mieosis I during which homologous chromosomes separate along their length but remain associated at chiasmata.

DNA-affinity chromatography A method used to isolate DNA-binding proteins based on their binding to specific DNA sequences.

DNA glycosylase A DNA repair enzyme that cleaves the bond linking a purine or pyrimidine to the deoxyribose of the backbone of a DNA molecule.

DNA ligase An enzyme that seals breaks in DNA strands.

DNA microarray A glass slide or membrane filter onto which oligonucleotides or fragments of cDNAs are printed at a high density, allowing simultaneous analysis of thousands of genes by hybridization of the microarray with fluorescent probes.

DNA polymerase An enzyme catalyzing the synthesis of DNA.

DNA transposons Transposable elements that move via DNA intermediates.

dolichol phosphate A lipid molecule in the endoplasmic reticulum upon which oligosaccharides are assembled for the glycosylation of proteins.

domains Compact, globular regions of proteins that are the basic units of tertiary structure.

dominant The allele that determines the phenotype of an organism when more than one allele is present.

dominant inhibitory mutant A mutant that interferes with the function of the normal allele of the gene.

Drosophila melanogaster A species of fruit fly commonly used for studies of animal genetics and development.

dynactin A protein that acts with cytoplasmic dynein to move cargo along microtubules.

dynamic instability The alternation of microtubules between cycles of growth and shrinkage.

dynamin A membrane-associated GTPase involved in vesicle budding.

dynein A motor protein that moves along microtubules towards the minus end.

dystrophin A cytoskeletal protein of muscle cells.

E2F A family of transcription factors that regulate the expression of genes involved in cell cycle progression and DNA replication.

ecdysone An insect steroid hormone that triggers metamorphosis.

ectoderm The outer germ layer; gives rise to tissues that include the skin and nervous system.

eicosanoid A class of lipids, including prostaglandins, prostacyclins, thromboxanes, and leukotrienes, that act in autocrine and paracrine signaling.

elaioplasts Plastids that store lipids.

elastic fibers Protein fibers that are present in the extracellular matrix of connective tissues in organs that stretch and then return to their original shape.

elastin The principal component of elastic fibers.

electrical synapse Specialized assemblies of gap junctions that allow the rapid passage of ions between nerve cells.

electrochemical gradient A difference in chemical concentration and electric potential across a membrane.

electron microscopy A type of microscopy that uses an electron beam to form an image. In transmission electron microscopy, a beam of electrons is passed through a specimen stained with heavy metals. In scanning electron microscopy, electrons scattered from the surface of a specimen are analyzed to generate a three-dimensional image.

electron tomography A method used to generate three-dimensional images by computer analysis of multiple two-dimensional images obtained by electron microscopy.

electron transport chain A series of carriers through which electrons are transported from a higher to a lower energy state.

electrophoretic-mobility shift assay An assay for the binding of a protein to a specific DNA sequence.

electroporation The introduction of DNA into cells by exposure to a brief electric pulse.

Elk-1 A transcription factor that is activated by ERK phosphorylation and induces expression of immediate-early genes.

elongation factor A protein involved in the elongation phase of transcription or translation.

embryonic stem (ES) cells Stem cells cultured from early embryos.

endocrine signaling A type of cell-cell signaling in which endocrine cells secrete hormones that are carried by the circulation to distant target cells.

endocytosis The uptake of extracellular material in vesicles formed from the plasma membrane.

endoderm The inner germ layer; gives rise to internal organs.

endoplasmic reticulum (ER) An extensive network of membrane-enclosed tubules and sacs involved in protein sorting and processing as well as in lipid synthesis.

endorphin A neuropeptide that acts as a natural analgesic.

endosome A vesicular compartment involved in the sorting and transport to lysosomes of material taken up by endocytosis.

endosymbiosis A symbiotic relationship in which one cell resides within a larger cell.

enhancer A transcriptional regulatory sequence that can be located at a site distant from the promoter.

enkephalin A neuropeptide that acts as a natural analgesic.

entactin An extracellular matrix protein that interacts with laminins and type IV collagen in basal laminae.

enzymes Proteins or RNAs that catalyze biological reactions.

epidermal cells Cells forming a protective layer on the surfaces of plants and animals.

epidermal growth factor (EGF) A growth factor that stimulates cell proliferation.

epithelial cells Cells forming sheets (epithelial tissue) that cover the surface of the body and line internal organs.

Epstein-Barr virus A human herpesvirus that causes B-cell lymphomas.

equilibrium centrifugation The separation of particles on the basis of density by centrifugation to equilibrium in a gradient of a dense substance.

*erb*A A proto-oncogene that encodes thyroid hormone receptor.

*erb*B-2 A proto-oncogene encoding a receptor protein-tyrosine kinase that is frequently amplified in breast and ovarian carcinomas.

ERK A member of the MAP kinase family that plays a central role in growth factor-induced cell proliferation.

ERM proteins A family of proteins that link actin filaments to the plasma membranes of many kinds of cells.

erythrocytes Red blood cells.

Escherichia coli (E. coli) A species of bacteria that has been extensively used as a model system for molecular biology.

estrogen A steroid hormone produced by the ovaries.

ethylene A plant hormone responsible for fruit ripening.

etioplast An intermediate stage of chloroplast development in which chlorophyll has not been synthesized.

eubacteria One of two major groups of prokaryotes, including most common species of bacteria.

euchromatin Decondensed, transcriptionally active interphase chromatin.

eukaryotic cells Cells that have a nuclear envelope, cytoplasmic organelles, and a cytoskeleton.

excinuclease The protein complex that excises damaged DNA during nucleotide-excision repair in bacteria.

exocyst A protein complex on the plasma membrane at which exocytosis occurs.

exon A segment of a gene that contains a coding sequence.

exonuclease An enzyme that hydrolyzes DNA molecules in either the 5′ to 3′ or 3′ to 5′ direction.

exportin A karyopherin that recognizes nuclear export signals and directs transport from the nucleus to the cytosol.

expression vector A vector used to direct expression of a cloned DNA fragment in a host cell.

extracellular matrix Secreted proteins and polysaccharides that fill spaces between cells and bind cells and tissues together.

facilitated diffusion The transport of molecules across a membrane by carrier or channel proteins.

FAK (focal adhesion kinase) A nonreceptor protein-tyrosine kinase that plays a key role in integrin signaling.

fats See triacylglycerols.

fatty acids Long hydrocarbon chains usually linked to a carboxyl group (COO^-).

feedback inhibition A type of allosteric regulation in which the product of a metabolic pathway inhibits the activity of an enzyme involved in its synthesis.

feedback loop A regulatory mechanism in which a downstream element of a signaling pathway controls the activity of an upstream component of the pathway.

feedforward relay A regulatory mechanism in which one element of a signaling pathway stimulates a downstream component.

fibroblast A cell type found in connective tissue.

fibronectin The principal adhesion protein of the extracellular matrix.

filamentous [F] actin Actin monomers polymerized into filaments.

filamin An actin-binding protein that crosslinks actin filaments into networks.

filopodium A thin projection of the plasma membrane supported by actin bundles.

flagellum A microtubule-based projection of the plasma membrane that is responsible for cell movement.

flavin adenine dinucleotide ($FADH_2$) A coenzyme that functions as an electron carrier in oxidation/reduction reactions.

flow cytometer An instrument that measures the fluorescence intensity of individual cells.

fluid mosaic model A model of membrane structure in which proteins are inserted in a fluid phospholipid bilayer.

fluid-phase endocytosis The nonselective uptake of extracellular fluids during endocytosis.

fluorescence-activated cell sorter An instrument that sorts individual cells on the basis of their fluorescence intensity.

fluorescence *in situ* hybridization (FISH) A method used to localize genes on chromosomes or RNAs within cells using fluorescent probes.

fluorescence microscopy Type of microscopy in which molecules are detected based on the emission of flourescent light.

fluorescence recovery after photobleaching (FRAP) A method used to study the movement of proteins within living cells.

fluorescence resonance energy transfer (FRET) A method used to study protein interactions within living cells.

focal adhesion A site of attachment of cells to the extracellular matrix at which integrins are linked to bundles of actin filaments.

focal complex A small cluster of integrins binding to the extracellular matrix that initiates the formation of a focal adhesion.

fodrin Nonerythroid spectrin.

formin An actin-binding protein that nucleates and polymerizes actin filaments.

Fos A transcription factor, encoded by a proto-oncogene, that is induced in response to growth factor stimulation.

freeze fracture Method of electron microscopy in which specimens are frozen in liquid nitrogen and then fractured to split the lipid bilayer, revealing the interior faces of cell membranes.

γ-tubulin ring complex A protein complex that nucleates the formation of microtubules.

G protein A family of cell signaling proteins regulated by guanine nucleotide binding.

G protein-coupled receptor A receptor characterized by seven membrane-spanning α helices. Ligand binding causes a conformational change that activates a G protein.

G_0 A quiescent state in which cells remain metabolically active but do not proliferate.

G_1 cyclins (Clns) Yeast cyclins that control passage through START.

G_1 phase The phase of the cell cycle between the end of mitosis and the begining of DNA synthesis.

G_2 phase The phase of the cell cycle between the end of S phase and the begining of mitosis.

gap junction A plasma membrane channel forming a direct cytoplasmic connection between adjacent cells.

gel electrophoresis A method in which molecules are separated based on their migration in an electric field.

gene A segment of DNA that encodes a polypeptide chain or an RNA molecule.

gene amplification An increase in the number of copies of a gene resulting from the repeated replication of a region of DNA.

gene family A group of related genes that have arisen by duplication of a common ancestor.

gene transfer The introduction of foreign DNA into a cell.

general transcription factors Transcription factors that are part of the general transcription machinery.

genetic code The correspondence between nucleotide triplets and amino acids in proteins.

genomic imprinting The regulation of genes whose expression depends on whether they are maternally or paternally inherited, apparently controlled by DNA methylation.

genomic library A collection of recombinant DNA clones that collectively contain the genome of an organism.

genomics The systematic analysis of entire cell genomes.

genotype The genetic composition of an organism.

gibberellin A plant hormone.

Gibbs free energy (G) The thermodynamic function that combines the effects of enthalpy and entropy to predict the energetically favorable direction of a chemical reaction.

globular[G] actin Monomers of actin that have not been assembled into filaments.

glucocorticoid A steroid produced by the adrenal gland that acts to stimulate production of glucose.

gluconeogenesis The synthesis of glucose.

glycerol phospholipids Phospholipids consisting of two fatty acids bound to a glycerol molecule.

glycocalyx A carbohydrate coat covering the cell surface.

glycogen A polymer of glucose residues that is the principal storage form of carbohydrates in animals.

glycolipid A lipid consisting of two hydrocarbon chains linked to a polar head group containing carbohydrates.

glycolysis The anaerobic breakdown of glucose.

glycoprotein A protein linked to oligosaccharides.

glycosaminoglycan (GAG) A gel-forming polysaccharide of the extracellular matrix.

glycosidase An enzyme that removes sugar residues from its substrate.

glycosidic bond The bond formed between sugar residues in oligosaccharides or polysaccharides.

glycosylation The addition of carbohydrates to proteins.

glycosylphosphatidylinositol (GPI) anchor Glycolipids containing phosphatidylinositol that anchor proteins to the external face of the plasma membrane.

glycosyltransferase An enzyme that adds sugar residues to its substrate.

glyoxylate cycle The conversion of fatty acids to carbohydrates in plants.

glyoxysome Peroxisomes in which the reactions of the glyoxylate cycle take place.

Golgi apparatus A cytoplasmic organelle involved in the processing and sorting of proteins and lipids. In plant cells, it is also the site of the synthesis of cell wall polysaccharides.

Golgi stack The compartments of the Golgi apparatus within which most metabolic activities take place.

granulocytes Blood cells that are involved in inflammatory reactions.

green fluorescent protein (GFP) A protein from jellyfish that is commonly used as a marker for fluorescence microscopy.

growth factors Polypeptides that control animal cell growth and differentiation.

GTPase-activating proteins Proteins that stimulate GTP hydrolysis by the small GTP-binding proteins.

guanine A purine that base-pairs with cytosine.

guanylyl cyclase An enzyme that catalyzes the formation of cyclic GMP from GTP.

haploid An organism or cell that has one copy of each chromosome.

hard keratin A keratin used for production of structures such as hair, nails, and horns.

heat-shock proteins A highly conserved group of chaperone proteins expressed in cells exposed to elevated temperatures or other forms of environmental stress.

helicase An enzyme that catalyzes the unwinding of DNA.

helix-loop-helix A transcription factor DNA-binding domain formed by the dimerization of two polypeptide chains. The dimerization domains of these proteins consist of two helical regions separated by a loop.

helix-turn-helix A transcription factor DNA-binding domain in which three or four helical regions contact DNA.

hemicellulose A polysaccharide that crosslinks cellulose microfibrils in plant cell walls.

hemidesmosome A region of contact between cells and the extracellular matrix at which keratin filaments are attached to integrin.

hepatitis B viruses A family of DNA viruses that infect liver cells and can lead to the development of liver cancer.

hepatitis C viruses A family of RNA viruses that infect liver cells and can lead to the development of liver cancer.

herpesviruses A family of DNA viruses, some members of which induce cancer.

heterochromatin Condensed, transcriptionally inactive chromatin.

heterophilic interaction An interaction between two different types of cell adhesion molecules.

heterotrimeric G protein A guanine nucleotide-binding protein consisting of three subunits.

high-energy bonds Chemical bonds that release a large amount of free energy when they are hydrolyzed.

histone acetylation The modification of histones by the addition of acetyl groups to specific lysine residues.

histone code Combinations of specific histone modifications that are thought to regulate the transcriptional activity of chromatin.

histones Proteins that package DNA in eukaryotic chromosomes.

HMGN proteins Nonhistone chromosomal proteins associated with decondensed transcriptionally active chromatin.

Holliday junction The central intermediate in recombination, consisting of a crossed-strand structure formed by homologous base pairing between strands of two DNA molecules.

Holliday model A molecular model of genetic recombination involving the formation of heteroduplex regions.

homeobox Conserved DNA sequences of 180 base pairs that encode homeodomains.

homeodomain A type of DNA binding domain found in transcription factors that regulate gene expression during embryonic development.

homologous recombination Recombination between segments of DNA with homologous nucleotide sequences.

homophilic interaction An interaction between cell adhesion molecules of the same type.

hormones Signaling molecules produced by endocrine glands that act on cells at distant body sites.

hydrophilic Soluble in water.

hydrophobic Not soluble in water.

IκB An inhibitory subunit of NF-κB transcription factors.

immediate-early genes A family of genes whose transcription is rapidly induced in response to growth factor stimulation.

immunoblotting A method that uses antibodies to detect proteins separated by SDS-polyacrylamide gel electrophoresis.

immunoglobulin See antibody.

immunoglobulin (Ig) superfamily A family of cell adhesion molecules containing structural domains similar to immunoglobulins.

immunoprecipitation The use of antibodies to isolate proteins.

importin A karyopherin that recognizes nuclear localization signals and directs nuclear import.

induced fit A model of enzyme action in which the configurations of both the enzyme and the substrate are altered by substrate binding.

in situ **hybridization** The use of radioactive or flourescent probes to detect RNA or DNA sequences in chromosomes or intact cells.

in vitro **mutagenesis** The introduction of mutations into cloned DNA *in vitro*.

in vitro **translation** Protein synthesis in a cell-free extract.

inositol 1,4,5-trisphosphate (IP$_3$) A second messenger, formed from the hydrolysis of PIP$_2$, that signals the release of calcium ions from the endoplasmic reticulum.

insulator A sequence that divides chromatin into independent domains and prevents an enhancer from acting on a promoter in a separate domain.

integral membrane proteins Proteins embedded within the lipid bilayer of cell membranes.

integrin A transmembrane protein that mediates the adhesion of cells to the extracellular matrix.

intermediate filament A cytoskeletal filament about 10 nm in diameter that provides mechanical strength to cells in tissues. See also keratins and neurofilaments.

interphase The period of the cell cycle between mitoses that includes G_1, S, and G_2 phases.

intracellular signal transduction A chain of reactions that transmits chemical signals from the cell surface to their intracellular targets.

intron A noncoding sequence that interrupts exons in a gene.

ion channel A protein that mediates the rapid passage of ions across a membrane by forming open pores through the phospholipid bilayer.

ion pump A protein that couples ATP hydrolysis to the transport of ions across a membrane.

JAK/STAT pathway A signaling pathway in which STAT transcription factors are activated as a result of phosphorylation by members of the JAK family of protein kinases.

Janus kinase (JAK) A family of nonreceptor protein-tyrosine kinases associated with cytokine receptors.

Jun A transcription factor, encoded by a proto-oncogene, that is activated in response to growth factor stimulation.

junctional complex A region of cell-cell contact containing a tight junction, an adherens junction, and a desmosome.

Kaposi's sarcoma-associated herpesvirus A human herpesvirus that causes Kaposi's sarcoma.

karyopherin A nuclear transport receptor.

keratin A type of intermediate filament protein of epithelial cells.

kilobase (kb) One thousand nucleotides or nucleotide base pairs.

kinesin A motor protein that moves along microtubules toward the plus end.

kinetochore A specialized structure consisting of proteins attached to a centromere that mediates the attachment and movement of chromosomes along the mitotic spindle.

kinetochore microtubules Microtubules of the mitotic spindle that attach to condensed chromosomes at their centromeres.

knockout Inactivation of a chromosomal gene by homologous recombination with a cloned mutant allele.

Krebs cycle See citric acid cycle.

lagging strand The strand of DNA synthesized opposite to the direction of movement of the replication fork by ligation of Okazaki fragments.

lamellipodium A broad, actin-based extension of the plasma membrane involved in the movement of fibroblasts.

laminin The principal adhesion protein of basal laminae.

lamins Intermediate filament proteins that form the nuclear lamina.

leading strand The strand of DNA synthesized continuously in the direction of movement of the replication fork.

leptotene The initial stage of the extended prophase of meiosis I during which homologous chromosomes pair before condensation.

leucine zipper A protein dimerization domain containing repeated leucine residues; found in many transcription factors.

leucoplast A plastid that stores energy sources in nonphotosynthetic plant tissues.

leukemia Cancer arising from the precursors of circulating blood cells.

leukotriene An eicosanoid synthesized from arachodonic acid.

ligand A molecule that binds to a receptor.

ligand-gated channels Ion channels that open in response to the binding of signaling molecules.

light reactions The reactions of photosynthesis in which solar energy drives the synthesis of ATP and NADPH.

lignin A polymer of phenolic residues that strengthens secondary cell walls.

LINEs (long interspersed elements) A family of highly repeated retrotransposons in mammalian genomes.

lipids Hydrophobic molecules that function as energy storage molecules, signaling molecules, and the major components of cell membranes.

liposome A lipid vesicle used to introduce DNA into mammalian cells.

lock-and-key model A model of enzyme action in which the substrate fits precisely into the enzyme active site.

long terminal repeat (LTR) Sequences found at the ends of retroviral DNA that are direct repeats of several hundred nucleotides resulting from reverse transcriptase activity.

low-density lipoprotein (LDL) A lipoprotein particle that transports cholesterol in the circulation.

lymphocyte A blood cell that functions in the immune response. B lymphocytes produce antibodies and T lymphocytes are responsible for cell mediated immunity.

lymphoma A cancer of lymphoid cells.

lysosomal storage diseases A family of diseases characterized by the accumulation of undegraded material in the lysosomes of affected individuals.

lysosome A cytoplasmic organelle containing enzymes that break down biological polymers.

M phase The mitotic phase of the cell cycle.

macrophage A type of white blood cell specialized for phagocytosis.

macropinocytosis The uptake of fluids in large vesicles.

malignant tumor A tumor that invades normal tissue and spreads throughout the body.

MAP kinases A family of mitogen-activated protein-serine/threonine kinases that are ubiquitous regulators of cell growth and differentiation.

mass spectrometry A method for identifying compounds based on accurate determination of their mass. Mass spectrometry is commonly used for protein identification.

matrix The inner mitochondrial space.

matrix processing peptidase (MPP) The protease that cleaves presequences from proteins imported to the matrix of mitochondria.

maturation promoting factor (MPF) A complex of Cdk1 and cyclin B that promotes entry into the M phase of either mitosis or meiosis.

Mediator A complex of proteins that allows eukaryotic protein-coding genes to respond to gene-specific regulatory factors.

megabase (Mb) One million nucleotides or nucleotide base pairs.

meiosis The division of diploid cells to haploid progeny, consisting of two sequential rounds of nuclear and cellular division.

membrane-anchored growth factors Growth factors associated with the plasma membrane that function as signaling molecules during cell-cell contact.

mesoderm The middle germ layer; gives rise to connective tissues and the hematopoietic system.

messenger RNA (mRNA) An RNA molecule that serves as a template for protein synthesis.

metal shadowing An electron microscopic technique in which the surface of a specimen is coated with a thin layer of evaporated metal.

metaphase The phase of mitosis during which the chromosomes are aligned on a metaphase plate in the center of the cell.

metastasis Spread of cancer cells through the blood or lymphatic system to other organ sites.

7-methylguanosine cap A structure consisting of GTP and methylated sugars that is added to the 5' ends of eukaryotic mRNAs.

microfilament A cytoskeleton filament composed of actin.

microRNA (miRNA) A naturally-occurring short noncoding RNA that acts to regulate gene expression.

microsome A small vesicle formed from the endoplasmic reticulum when cells are disrupted.

microspike See filopodium.

microtubule A cytoskeletal component formed by the polymerization of tubulin into rigid, hollow rods about 25 nm in diameter.

microtubule-associated proteins (MAPs) Proteins that bind to microtubules and modify their stability.

microtubule-organizing center An anchoring point near the center of the cell from which most microtubules extend outward.

microvillus An actin-based protrusion of the plasma membrane, abundant on the surfaces of cells involved in absorption.

middle lamella A region of the plant cell wall that acts as a glue to hold adjacent cells together.

mineralocorticoids Steroid hormones produced by the adrenal gland that act on the kidney to regulate salt and water balance.

mismatch repair A repair system that removes mismatched bases from newly synthesized DNA strands.

mitochondria Cytoplasmic organelles responsible for synthesis of most of the ATP in eukaryotic cells by oxidative phosphorylation.

mitosis Nuclear division.

mitotic spindle An array of microtubules extending from the spindle poles that is responsible for separating daughter chromosomes during mitosis. See also kinetochore microtubules, polar microtubules, chromosomal microtubules, and astral microtubules.

molecular clone See recombinant molecule.

molecular cloning The insertion of a DNA fragment of interest into a DNA molecule (vector) that is capable of independent replication in a host cell.

molecular motor A protein that generates force and movement by converting chemical energy to mechanical energy.

monocistronic Messenger RNAs that encode a single polypeptide chain.

monoclonal antibody An antibody produced by a clonal line of B lymphocytes.

monocyte A type of blood cell involved in inflammatory reactions.

monosaccharides Simple sugars with the basic formula of $(CH_2O)_n$.

Mos A protein kinase that is required for progression from meiosis I to meiosis II and maintenance of metaphase II arrest in vertebrate oocytes.

mTOR A protein kinase involved in regulation of protein synthesis in response to growth factors, nutrients, and energy availability.

multi-photon excitation microscopy A form of fluorescence microscopy in which the specimen is illuminated with a wavelength of light such that excitation of the fluorescent dye requires the simultaneous absorption of two or more photons.

muscle fibers The large cells of skeletal muscle, which are formed by the fusion of many individual cells during developmenht.

mutagen A chemical that induces a high frequency of mutations.

mutation A genetic alteration.

myofibril A bundle of actin and myosin filaments in muscle cells.

myosin A protein that interacts with actin as a molecular motor.

myosin I A type of myosin that acts to transport cargo along actin filaments.

myosin II The type of myosin that produces contraction by sliding actin filaments.

myosin light–chain kinase A protein kinase that activates myosin II by phosphorylating its regulatory light chain.

Na⁺-K⁺ ATPase See Na⁺-K⁺ pump.

Na⁺-K⁺ pump An ion pump that transports Na⁺ out of the cell and K⁺ into the cell.

NADP reductase An enzyme that transfers electrons from ferrodoxin to NADP⁺, yielding NADPH.

nebulin A protein that regulates the length of actin filaments in muscle cells.

nectin A cell adhesion molecule involved in the formation of adherens junctions.

Nernst equation The relationship between ion concentration and membrane potential.

nerve growth factor (NGF) A polypeptide growth factor that regulates the development and survival of neurons.

neurofilament A type of intermediate filament that supports the axons of nerve cells.

neurofilament (NF) proteins The major intermediate filament proteins of many types of mature nerve cells.

neurohormone Peptides that are secreted by neurons and act on distant cells.

neuron A nerve cell specialized to receive and transmit signals throughout the body.

neuropeptides Peptide signaling molecules secreted by neurons.

neurotransmitter A small, hydrophilic molecule that carries a signal from a stimulated neuron to a target cell at a synapse.

neurotrophin A member of a family of polypeptides that regulates neuron development and survival.

nexin A protein that links microtubule doublets to each other in the axoneme.

NF-κB A family of transcription factors that are activated in response to a variety of stimuli.

nicotinamide-adenine dinucleotide (NAD⁺) A coenzyme that functions as an electron carrier in oxidation/reduction reactions.

nitrogen fixation The reduction of atmospheric nitrogen (N_2) to NH_3.

nitrosylation Protein modification by addition of NO groups to the side chains of cysteine residues.

N-*myc* A proto-oncogene that encodes a transcription factor and is frequently activated by amplification in neuroblastomas.

N-myristoylation The addition of myristic acid (a 14-carbon fatty acid) to the N-terminal glycine residue of a polypeptide chain.

nonreceptor protein-tyrosine kinase An intracellular protein-tyrosine kinase.

nonsense-mediated mRNA decay Degradation of mRNAs that lack complete open-reading frames.

Northern blotting A method in which mRNAs are separated by gel electrophoresis and detected by hybridization with specific probes.

nuclear envelope The barrier separating the nucleus from the cytoplasm, composed of an inner and outer membrane, a nuclear lamina, and nuclear pore complexes.

nuclear export signal An amino acid sequence that targets proteins for transport from the nucleus to the cytosol.

nuclear lamina A meshwork of lamin filaments providing structural support to the nucleus.

nuclear localization signal An amino acid sequence that targets proteins for transportation from the cytoplasm to the nucleus.

nuclear membranes Membranes forming the nuclear envelope; the outer nuclear membrane is continuous with the endoplasmic reticulum and the inner nuclear membrane is adjacent to the nuclear lamina.

nuclear pore complex A large structure forming a transport channel through the nuclear envelope.

nuclear receptor superfamily A family of transcription factors that includes the receptors for steroid hormones, thyroid hormone, retinoic acid, and vitamin D_3.

nuclear transport receptor A protein that recognizes nuclear localization signals and mediates transport across the nuclear envelope.

nucleic acid hybridization The formation of double stranded DNA and/or RNA molecules by complementary base pairing.

nucleolar organizing regions The chromosomal regions containing the genes for ribosomal RNAs.

nucleolus The nuclear site of rRNA transcription, processing, and ribosome assembly.

nucleoside A purine or pyrimidine base linked to a sugar (ribose or deoxyribose).

nucleosome The basic structural unit of chromatin consisting of DNA wrapped around a histone core.

nucleosome core particles Particles containing 146 base pairs of DNA wrapped around an octamer consisting of two molecules each of histones H2A, H2B, H3, and H4.

nucleosome remodeling factors Proteins that disrupt chromatin structure, allowing transcription factors to bind nucleosomal DNA.

nucleotide A phosphorylated nucleoside.

nucleotide excision repair A mechanism of DNA repair in which oligonucleotides containing damaged bases are removed from a DNA molecule.

nucleus The most prominent organelle of eukaryotic cells; contains the genetic material.

Okazaki fragments Short DNA fragments that are joined to form the lagging strand of DNA.

oligonucleotide A short polymer of only a few nucleotides.

oligosaccharide A short polymer of only a few sugars.

oncogene A gene capable of inducing one or more characteristics of cancer cells.

oncogene addiction The dependence of cancer cells on the continuing activity of oncogenes.

open-reading frame A stretch of nucleotide sequence that does not contain stop codons and can encode a polypeptide.

operator A regulatory sequence of DNA that controls transcription of an operon.

operon A group of adjacent genes transcribed as a single mRNA.

origin of replication A specific DNA sequence that serves as a binding site for proteins that initiate replication.

origin recognition complex (ORC) A protein complex that initiates DNA replication at eukaryotic origins.

oxidative metabolism The use of molecular oxygen as an electron acceptor in the breakdown of organic molecules.

oxidative phosphorylation The synthesis of ATP from ADP coupled to the energetically favorable transfer of electrons to molecular oxygen as the final acceptor in an electron transport chain.

P1 artificial chromosome (PAC) A vector used for cloning large fragments of DNA in *E. coli*.

p53 A transcription factor (encoded by the *p53* tumor suppressor gene) that arrests the cell cycle in G_1 in response to damaged DNA and is required for apoptosis induced by a variety of stimuli.

pachytene The stage of meiosis I during which recombination takes place between homologous chromosomes.

palmitoylation The addition of palmitic acid (a 16-carbon fatty acid) to cysteine residues of a polypeptide chain.

papillomavirus A member of a family of DNA viruses, some of which cause cervical and other anogenital cancers in humans.

paracrine signaling Local cell-cell signaling in which a molecule released by one cell acts on a neighboring target cell.

parenchyma cell A type of plant cell responsible for most metabolic activities.

passive diffusion The diffusion of small hydrophobic molecules through a phospholipid bilayer.

passive transport The transport of molecules across a membrane in the energetically favorable direction.

patch clamp technique A method used to isolate and study the activity of single ion channels.

pectin A gel-forming polysaccharide in plant cell walls.

peptide bond The bond joining amino acids in polypeptide chains.

peptide hormone A signaling molecule composed of amino acids.

peptidoglycan The principal component of bacterial cell walls consisting of linear polysaccharide chains crosslinked by short peptides.

peptidyl prolyl isomerase An enzyme that facilitates protein folding by catalyzing the *cis-trans* isomerization of prolyl peptide bonds.

pericentriolar material The material in the centrosome that initiates microtubule assembly.

peripheral membrane proteins Proteins indirectly associated with cell membranes by protein-protein interactions.

peroxin A protein present in peroxisomes.

peroxisome A cytoplasmic organelle specialized for carrying out oxidative reactions.

phagocytosis The uptake of large particles, such as bacteria, by a cell.

phagolysosomes A lysosome that has fused with a phagosome or autophagosome.

phagosome A vacuole containing a particle taken up by phagocytosis.

phalloidin A drug that binds to actin filaments and prevents their disassembly.

phase-contrast microscopy A type of microscopy in which variations in density or thickness between parts of the cell are converted to differences in contrast in the final image.

phenotype The physical appearance of an organism.

phosphatidylcholine A glycerol phospholipid with a head group formed from choline.

phosphatidylethanolamine A glycerol phospholipid with a head group formed from ethanolamine.

phosphatidylinositide 3-kinase (PI 3-kinase) An enzyme that phosphorylates PIP_2, yielding the second messenger phosphatidylinositol 3,4,5-trisphosphate (PIP_3).

phosphatidylinositol 4,5-bisphosphate (PIP_2) A minor phospholipid component of the inner leaflet of the plasma membrane. Hormones and growth factors stimulate its hydrolysis by phospholipase C, yielding the second messengers diacylglycerol and inositol trisphosphate.

phosphatidylinositol 3,4,5-triphosphate (PIP_3) A second messenger formed by phosphorylation of PIP_2.

phosphatidylserine A glycerol phospholipid with a head group formed from serine.

phosphodiester bond A bond between the 5'-phosphate of one nucleotide and the 3'-hydroxyl of another.

phospholipase C An enzyme that hydrolyzes PIP_2 to form the second messengers diacylglycerol and inositol trisphosphate.

phospholipid bilayer The basic structure of biological membranes in which the hydrophobic tails of phospholipids are buried in the interior of the membrane and their polar head groups are exposed to the aqueous solution on either side.

phospholipid transfer protein A protein that transports phospholipid molecules between cell membranes.

phospholipids The principal components of cell membranes, consisting of two hydrocarbon chains (usually fatty acids) joined to a polar head group containing phosphate.

phosphorylation The addition of a phosphate group to a molecule.

photoreactivation A mechanism of DNA repair in which solar energy is used to split pyrimidine dimers.

photosynthesis The process by which cells harness energy from sunlight and synthesize glucose from CO_2 and water.

photosynthetic pigments Molecules that capture energy from sunlight by absorbing photons.

photosystem I A protein complex in the thylakoid membrane that uses energy absorbed from sunlight to synthesize NADPH.

photosystem II A protein complex in the thylakoid membrane that uses energy absorbed from sunlight to synthesize ATP.

pinocytosis The uptake of fluids or molecules into a cell by small vesicles.

plakin A member of a family of proteins that link intermediate filaments to other cellular structures.

plant hormones A group of small molecules that coordinate the responses of plant tissues to environmental signals.

plasma membrane A phospholipid bilayer with associated proteins that surrounds the cell.

plasmalogens A family of phospholipids that have an ether bond and an ester bond.

plasmid A small, circular DNA molecule capable of independent replication in a host cell.

plasmodesma A cytoplasmic connection between adjacent plant cells formed by a continuous region of the plasma membrane.

plastids A family of plant organelles including chloroplasts, chromoplasts, leucoplasts, amyloplasts, and elaioplasts.

platelet-derived growth factor (PDGF) A growth factor released by platelets during blood clotting to stimulate the proliferation of fibroblasts.

PML/RARα An oncogene formed by translocation of the retinoic acid receptor in acute promyelocytic leukemia.

polar body A small cell formed by asymmetric cell division following meiosis of oocytes.

polar microtubules Microtubules of the mitotic spindle that overlap in the center of the cell and push the spindle poles apart.

Polo-like kinase A protein kinase involved in mitotic spindle formation, kinetochore function, and cytokinesis.

poly-A tail A tract of about 200 adenine nucleotides added to the 3' ends of eukaryotic mRNAs.

polyadenylation The process of adding a poly-A tail to a pre-mRNA.

polycistronic Messenger RNAs that encode multiple polypeptide chains.

polymerase chain reaction (PCR) A method for amplifying a region of DNA by repeated cycles of DNA synthesis *in vitro*.

polynucleotide A polymer containing up to millions of nucleotides.

polyomavirus A widely-studied DNA tumor virus.

polyp A benign tumor projecting from an epithelial surface.

polypeptide A polymer of amino acids.

polysaccharide A polymer containing hundreds or thousands of sugars.

polysome A series of ribosomes translating a messenger RNA.

polytene chromosome A giant chromosome found in some tissues of *Drosophila* that arises from repeated replication of DNA strands that fail to separate from each other.

porin A member of a class of proteins that cross membranes as β-barrels and form channels in the outer membranes of some bacteria, mitochondria, and chloroplasts.

pre-mRNA The primary transcript, which is processed to form messenger RNA in eukaryotic cells.

pre-rRNA The primary transcript, which is cleaved to form individual ribosomal RNAs (the 28S, 18S, and 5.8S rRNAs of higher eukaryotic cells).

pre-tRNA The primary transcript, which is cleaved to form transfer RNAs.

prenylation The addition of specific types of lipids (prenyl groups) to C terminal cysteine residues of a polypeptide chain.

primary cell walls The walls of growing plant cells.

primary cultures Cell cultures established from a tissue.

primary structure The sequence of amino acids in a polypeptide chain.

primase An RNA polymerase used to intiate DNA synthesis.

processed pseudogene A pseudogene that has arisen by reverse transcription of mRNA.

procollagens Soluble precursors to the fibril-forming collagens.

product A compound formed as a result of an enzymatic reaction.

profilin An actin-binding protein that stimulates the assembly of actin monomers into filaments.

progesterone A steroid hormone produced by the ovaries.

programmed cell death A normal physiological form of cell death characterized by apoptosis.

prokaryotic cells Cells lacking a nuclear envelope, cytoplasmic organelles, and a cytoskeleton (bacteria).

prometaphase A transition period between prophase and metaphase during which the microtubules of the mitotic spindle attach to the kinetochores and the chromosomes shuffle until they align in the center of the cell.

promoter A DNA sequence at which RNA polymerase binds to initiate transcription.

pronuclei Two haploid nuclei in a newly fertilized egg.

proofreading The selective removal of mismatched bases by DNA polymerase.

prophase The beginning phase of mitosis, marked by the appearance of condensed chromosomes and the development of the mitotic spindle.

prostacyclin An eicosanoid formed from prostaglandin H_2.

prostaglandin A family of eicosanoid lipids involved in signaling inflammation.

prosthetic groups Small molecules bound to proteins.

proteasome A large protease complex that degrades proteins tagged by ubiquitin.

protein disulfide isomerase An enzyme that catalyzes the formation and breakage of disulfide (S–S) linkages.

protein kinase An enzyme that phosphorylates proteins by transferring a phosphate group from ATP.

protein kinase A A protein kinase regulated by cyclic AMP.

protein kinase C A family of protein-serine/threonine kinases that are activated by diacylglycerol and Ca^{2+} and function in intracellular signal transduction.

protein phosphatase An enzyme that reverses the action of protein kinases by removing phosphate groups from phosphorylated amino acid residues.

protein-serine/threonine kinase A protein kinase that phosphorylates serine and threonine residues.

protein-tyrosine kinase A protein kinase that phosphorylates tyrosine residues.

protein-tyrosine phosphatase An enzyme that removes the phosphate groups from phosphotyrosine residues.

proteins Polypeptides with a unique amino acid sequence.

proteoglycan A protein linked to glycosaminoglycans.

proteolysis Degradation of polypeptide chains.

proteome All of the proteins expressed in a given cell.

proteomics Large scale analysis of cell proteins.

proto-oncogene A normal cell gene that can be converted into an oncogene.

pseudogene A nonfunctional gene copy.

pseudopodium An actin-based extension of the plasma membrane responsible for phagocytosis and amoeboid movement.

PTB domain A protein domain that binds phosphotyrosine-containing peptides.

PTEN A lipid phosphatase that dephosphorylates PIP_3 and acts as a tumor suppressor.

pyrimidine dimer A common form of DNA damage caused by UV light in which adjacent pyrimidines are joined to form a dimer.

quaternary structure The interactions between polypeptide chains in proteins consisting of more than one polypeptide.

Rab A family of small GTP-binding proteins that play key roles in vesicular transport.

Rac A small GTP-binding protein involved in regulation of the actin cytoskeleton.

Rad51 A eukaryotic protein that functions similarly to RecA in homologous recombination.

Raf A protein-serine/threonine kinase (encoded by the *raf* oncogene) that is activated by Ras and leads to activation of the ERK MAP kinase.

Ran A small GTP-binding protein involved in nuclear import and export.

Ras A family of small GTP binding proteins (encoded by the *ras* oncogenes) that couple growth factor receptors to intracellular targets, including the Raf protein-serine/threonine kinase and the ERK MAP kinase pathway.

Rb A transcriptional regulatory protein that controls cell cycle progression and is encoded by a tumor suppressor gene that was identified by the genetic analysis of retinoblastoma.

RecA A protein that promotes the exchange of strands between homologous DNA molelcules during recombination.

receptor downregulation The loss of receptors from the cell surface as a result of their internalization by endocytosis following ligand binding.

receptor-mediated endocytosis The selective uptake of macromolecules that bind to cell surface receptors that concentrate in clathrin-coated pits.

receptor protein-tyrosine kinase Membrane-spanning protein-tyrosine kinases that are receptors for extracellular ligands.

recessive An allele that is masked by a dominant allele.

recombinant DNA library A collection of genomic or cDNA clones.

recombinant molecule A DNA insert joined to a vector.

recombination The exchange of genetic material.

recombinational repair The repair of damaged DNA by recombination with an undamaged homologous DNA molecule.

release factor A protein that recognizes stop codons and terminates translation of mRNA.

replication fork The region of DNA synthesis where the parental strands separate and two new daughter strands elongate.

repressor A regulatory molecule that blocks transcription.

reproductive cloning The use of nuclear transfer to create a cloned organism.

resolution The ability of a microscope to distinguish objects separated by small distances.

restriction endonuclease An enzyme that cleaves DNA at a specific sequence.

restriction map The locations of restriction endonuclease cleavage sites on a DNA molecule.

restriction point A regulatory point in animal cell cycles that occurs late in G_1. After this point, a cell is committed to entering S and undergoing one cell division cycle.

retinoic acid A signaling molecule synthesized from vitamin A.

retinoid A molecule related to retinoic acid.

retrotransposon A transposable element that moves via reverse transcription of an RNA intermediate.

retrovirus A virus that replicates by making a DNA copy of its RNA genome by reverse transcription.

retrovirus-like element A retrotransposon that is structurally similar to a retrovirus.

reverse genetics Analysis of gene function by introducing mutations into a cloned gene.

reverse transcriptase A DNA polymerase that uses an RNA template.

reverse transcription Synthesis of DNA from an RNA template.

Rho A family of small GTP-binding proteins involved in regulation of the cytoskeleton.

rhodopsin A G protein-coupled photoreceptor in retinal rod cells that activates transducin in response to light absorption.

ribonucleic acid (RNA) A polymer of ribonucleotides.

ribose The five-carbon sugar found in RNA.

ribosomal RNA (rRNA) The RNA component of ribosomes.

ribosomes Particles composed of RNA and proteins that are the sites of protein synthesis.

ribozyme An RNA enzyme.

RNA editing RNA processing events other than splicing that alter the protein coding sequences of mRNAs.

RNA interference (RNAi) The degradation of mRNAs by short complementary double-stranded RNA molecules.

RNA polymerase An enzyme that catalyzes the synthesis of RNA.

RNA splicing The joining of exons in a precursor RNA molecule.

RNase H An enzyme that degrades the RNA strand of RNA-DNA hybrid molecules.

RNase P A ribozyme that cleaves the 5′ end of pre-tRNAs.

RNA world An early stage of evolution based on self-replicating RNA molecules.

rough endoplasmic reticulum (ER) The region of the endoplasmic reticulum covered with ribosomes and involved in protein metabolism.

Rous sarcoma virus (RSV) An acutely transforming retrovirus in which the first oncogene was identified.

ryanodine receptors Calcium channels in muscle and nerve cells that open in response to changes in membrane potential.

S phase The phase of the cell cycle during which DNA replication occurs.

Saccharomyces cerevisiae A frequently studied budding yeast.

sarcoma A cancer of cells of connective tissue.

sarcomere The contractile unit of muscle cells composed of interacting myosin and actin filaments.

sarcoplasmic reticulum A specialized network of membranes in muscle cells that stores a high concentration of Ca^{2+}.

satellite DNA Simple-sequence repetitive DNA with a buoyant density differing from the bulk of genomic DNA.

scaffold proteins Proteins that bind to components of signaling pathways, leading to their organization in specific signaling cassettes.

scanning electron microscopy See electron microscopy.

sclerenchyma cells Plant cells characterized by thick cell walls that provide structural support to the plant.

SDS-polyacrylamide gel electrophoresis (SDS-PAGE) A commonly used method to separate proteins by gel electrophoresis on the basis of size.

second messenger A compound whose metabolism is modified as a result of a ligand-receptor interaction; it functions as a signal transducer by regulating other intracellular processes.

secondary cell wall A thick cell wall laid down between the plasma membrane and the primary cell wall of plant cells that have ceased growth.

secondary response gene A gene whose induction following growth factor stimulation of a cell requires protein synthesis.

secondary structure The regular arrangement of amino acids within localized regions of a polypeptide chain. See *α* helix and *β* sheet.

secretory pathway The movement of secreted proteins from the endoplasmic reticulum to the Golgi apparatus and then, within secretory vesicles, to the cell surface.

secretory vesicles Membrane-enclosed sacs that transport proteins from the Golgi apparatus to the cell surface.

selectins Cell adhesion molecules that recognize oligosaccharides exposed on the cell surface.

self-splicing The ability of some RNAs to catalyze the removal of their own introns.

semiconservative-replication The process of DNA replication in which the two parental strands separate and serve as templates for the synthesis of new progeny strands.

serum response element (SRE) A regulatory sequence that is recognized by the serum response factor and mediates the transcriptional induction of many immediate-early genes in response to growth factor stimulation.

serum response factor (SRF) A transcription factor that binds to the serum response element.

SH2 domain A protein domain of approximately 100 amino acids that binds phosphotyrosine-containing peptides.

Shine-Dalgarno sequence The sequence prior to the initiation site that correctly aligns bacterial mRNAs on ribosomes.

signaling network The interconnected network formed by the interactions of multiple signaling pathways within a cell.

signal patch A recognition determinant formed by the three-dimensional folding of a polypeptide chain.

signal peptidase An enzyme that removes the signal sequence of a polypeptide chain by proteolysis.

signal recognition particle (SRP) A particle composed of proteins and srpRNA that binds to signal sequences and targets polypeptide chains to the endoplasmic reticulum.

signal sequence A hydrophobic sequence at the amino terminus of a polypeptide chain that targets it for secretion in bacteria or incorporation into the endoplasmic reticulum in eukaryotic cells.

simian virus 40 (SV40) A widely-studied DNA tumor virus.

simple-sequence repeats A class of repeated DNA sequences consisting of tandem arrays of thousands of copies of short sequences.

SINEs (short interspersed elements) A family of highly repeated retrotransposons in mammalian genomes.

single-stranded DNA-binding proteins Proteins that stabilize unwound DNA by binding to single-stranded regions.

site-specific recombination Recombination mediated by proteins that recognize specific DNA sequences.

sliding filament model The model of muscle contraction in which contraction results from the sliding of actin and myosin filaments relative to each other.

Smad A family of transcription factors activated by TGF-*β* receptors.

small GTP-binding proteins A large family of monomeric GTP-binding proteins, including the Ras, Rab, Rho, and Ran proteins.

small nuclear RNAs (snRNAs) Nuclear RNAs ranging in size from 50 to 200 bases.

small nuclear ribonucleoprotein particles (snRNPs) Complexes of snRNAs with proteins.

small nucleolar RNAs (snoRNAs) Small RNAs present in the nucleolus that function in pre-rRNA processing.

smooth endoplasmic reticulum The major site of lipid synthesis in eukaryotic cells.

SNARE hypothesis The hypothesis that vesicle fusion is mediated by pairs of transmembrane proteins (SNAREs) on the vesicle and target membranes.

soft keratin The keratins found in the cytoplasm of epithelial cells.

somatic cell nuclear transfer The basic procedure of animal cloning in which the nucleus of an adult somatic cell is transferred to an enucleated egg.

somatic hypermutation The introduction of multiple mutations within rearranged immunoglobulin variable regions to increase antibody diversity.

Southern blotting A method in which radioactive probes are used to detect specific DNA fragments that have been separated by gel electrophoresis.

spacer sequences The DNA sequences between genes.

spectrin A major actin-binding protein of the cell cortex.

sphingomyelin A phospholipid consisting of two hydrocarbon chains bound to a polar head group containing serine.

spindle assembly checkpoint A cell cycle checkpoint that monitors the alignment of chromosomes on the metaphase spindle.

spliceosomes Large complexes of snRNAs and proteins that catalyze the splicing of pre-mRNAs.

Src A nonreceptor protein-tyrosine kinase encoded by the oncogene (*src*) of Rous sarcoma virus.

SRP receptor A protein on the membrane of the endoplasmic reticulum that binds the signal recognition particle (SRP).

srpRNA The small cytoplasmic RNA component of SRP.

stability gene A gene that acts to maintain the integrity of the genome and whose loss can lead to the development of cancer.

starch A polymer of glucose residues that is the principal storage form of carbohydrates in plants.

START A regulatory point in the yeast cell cycle that occurs late in G_1. After this point a cell is committed to entering S and undergoing one cell division cycle.

STAT proteins Transcription factors that have an SH2 domain and are activated by tyrosine phosphorylation, which promotes their translocation from the cytoplasm to the nucleus.

stem cell A cell that divides to produce daughter cells that can either differentiate or remain as stem cells.

stereocilium A specialized microvillus of auditory hair cells.

steroid hormones A group of hydrophobic hormones that are derivatives of cholesterol.

stroma The compartment of chloroplasts that lies between the envelope and the thylakoid membrane.

stromal processing peptidase (SPP) The protease that cleaves transit peptides from proteins imported to the chloroplast stroma.

substrate A molecule acted upon by an enzyme.

symport The transport of two molecules in the same direction across a membrane.

synapse The junction between a neuron and another cell, across which information is carried by neurotransmitters.

synapsis The association of homologous chromosomes during meiosis.

synaptic vesicle A secretory vesicle that releases neurotransmitters at a synapse.

synaptomenal complex A zipperlike protein structure that forms along the length of paired homologous chromosomes during meiosis.

systems biology A new field of biology in which large-scale experimental approaches are combined with quantitative analysis and modeling to study complex biological systems.

T cell receptor A T lymphocyte surface protein that recognizes antigens expressed on the surface of other cells.

talin A protein that mediates the association of actin filaments with integrins at focal adhesions.

TATA box A regulatory DNA sequence found in the promoters of many eukaryotic genes transcribed by RNA polymerase II.

TATA-binding protein (TBP) A basal transcription factor that binds directly to the TATA box.

taxol A drug that binds to and stabilizes microtubules.

TBP-associated factors (TAFs) Polypeptides associated with TBP in the general transcription factor TFIID.

telomerase A reverse transcriptase that synthesizes telomeric repeat sequences at the ends of chromosomes from its own RNA template.

telomeres Repeats of simple-sequence DNA that maintain the ends of linear chromosomes.

telophase The final phase of mitosis, during which the nuclei re-form and chromosomes decondense.

temperature-sensitive mutant A cell expressing a protein that is functional at one temperature but not at another, whereas the normal protein is functional at both temperatures.

tertiary structure The three-dimensional folding of a polypeptide chain that gives the protein its functional form.

testosterone A steroid hormone produced by the testis.

therapeutic cloning A procedure in which nuclear transfer into oocytes could be used to produce embryonic stem cells for use in transplantation therapy.

thylakoid membrane The innermost membrane of chloroplasts that is the site of electron transport and ATP synthesis.

thymine A pyrimidine found in DNA that base-pairs with adenine.

thyroid hormone A hormone synthesized from tyrosine in the thyroid gland.

thromboxane An eicosanoid involved in blood clotting.

Tic complex The protein translocation complex of the chloroplast inner membrane.

tight junction A continuous network of protein strands around the circumference of epithelial cells, sealing the space between cells and forming a barrier between the apical and basolateral domains.

Tim complex The protein translocation complex of the mitochondrial inner membrane.

Ti plasmid A plasmid used for gene transfer in plants.

titin A large protein that acts as a spring to keep myosin filaments centered in the muscle sarcomere.

Toc complex The protein translocation complex of the chloroplast outer membrane.

Tom complex The protein translocation complex of the mitochondrial outer membrane.

topoisomerase An enzyme that catalyzes the reversible breakage and rejoining of DNA strands.

trans-acting factors Transcriptional regulatory proteins.

trans Golgi network The Golgi compartment within which proteins are sorted and packaged to exit the Golgi apparatus.

transcription-coupled repair The preferential repair of damage to transcribed strands of DNA.

transcription factor A protein that regulates the activity of RNA polymerase.

transcription The synthesis of an RNA molecule from a DNA template.

transcriptional activators Transcription factors that stimulate transcription.

transcytosis The sorting and transport of proteins to different domains of the plasma membrane following endocytosis.

transducin A G protein that stimulates cGMP phosphodiesterase when it is activated by rhodopsin.

transfection The introduction of a foreign gene into eukaryotic cells.

transfer RNA (tRNA) RNA molecules that function as adaptors between amino acids and mRNA during protein synthesis.

transformation The transfer of DNA between genetically distinct bacteria. See also cell transformation.

transforming growth factor β (TGF-β) A polypeptide growth factor that generally inhibits animal cell proliferation.

transgenic mouse A mouse that carries foreign genes incorporated into the germ line.

transient expression The expression of unintegrated plasmid DNAs that have been introduced into cultured cells.

transitional ER The region of the ER from which proteins exit for the Golgi apparatus.

transition state A high energy state through which substrates must pass during the course of an enzymatic reaction.

transit peptides N-terminal sequences that target proteins for import into chloroplasts.

translation The synthesis of a polypeptide chain from an mRNA template.

translesion DNA synthesis A form of repair in which specialized DNA polymerases replicate across a site of DNA damage.

translocon The membrane channel through which polypeptide chains are transported into the endoplasmic reticulum.

transmembrane proteins Integral membrane proteins that span the lipid bilayer and have portions exposed on both sides of the membrane.

transmission electron microscopy See electron microscopy.

transposable element See transposon.

transposon A DNA sequence that can move to different positions in the genome.

treadmilling A dynamic behavior of actin filaments and microtubules in which the loss of subunits from one end of the filament is balanced by their addition to the other end.

triacylglycerol Three fatty acids linked to a glycerol molecule.

tropomyosin A fibrous protein that binds actin filaments and regulates contraction by blocking the interaction of actin and myosin.

troponin A complex of proteins that binds to actin filaments and regulates skeletal muscle contraction.

tubulin A cytoskeletal protein that polymerizes to form microtubules.

tumor Any abnormal proliferation of cells.

tumor initiation The first step in tumor development, resulting from abnormal proliferation of a single cell.

tumor necrosis factor (TNF) A polypeptide growth factor that induces programmed cell death.

tumor progression The accumulation of mutations within cells of a tumor population, resulting in increasingly rapid growth and malignancy.

tumor promoter A compound that leads to tumor development by stimulating cell proliferation.

tumor suppressor gene A gene whose inactivation leads to tumor development.

tumor virus A virus capable of causing cancer in animals or humans.

turgor pressure The internal hydrostatic pressure within plant cells.

twinfilin An actin-binding protein that stimulates the assembly of actin monomers into filaments.

two-dimensional gel electrophoresis A method for separating cell proteins based on both charge and size.

ubiquinone See coenzyme Q.

ubiquitin A highly conserved protein that acts as a marker to target other cellular proteins for rapid degradation.

ultracentrifuge A centrifuge that rotates samples at high speeds.

unfolded protein response A cellular stress response in which an excess of unfolded proteins in the endoplasmic reticulum leads to general inhibition of protein synthesis, increased expression of chaperones, and increased proteasome activity.

uniport The transport of a single molecule across a membrane.

3′ untranslated region A noncoding region at the 3′ end of mRNA.

5′ untranslated region A noncoding region at the 5′ end of mRNA.

uracil A pyrimidine found in RNA that base-pairs with adenine.

vacuole A large membrane-enclosed sac in the cytoplasm of eukaryotic cells. In plant cells, vacuoles function to store nutrients and waste products, to degrade macromolecules, and to maintain turgor pressure.

vector A DNA molecule used to direct the replication of a cloned DNA fragment in a host cell.

velocity centrifugation The separation of particles based on their rates of sedimentation.

video-enhanced microscopy The combined use of video cameras with the light microscope to allow the visualization of small objects.

villin The major actin-bundling protein of intestinal microvilli.

vimentin An intermediate filament protein found in a variety of different kinds of cells.

vinblastine A drug that inhibits microtubule polymerization.

vincristine A drug that inhibits microtubule polymerization.

vinculin A protein that mediates the association of actin filaments with integrins at focal adhesions.

voltage-gated channels Ion channels that open in response to changes in electric potential.

WASP/Scar complex A protein complex that stimulates actin filament branching.

Western blotting See immunoblotting.

X-chromosome inactivation A dosage compensation mechanism in which most of the genes on one X chromosome are inactivated in female cells.

X-ray crystallography A method in which the diffraction pattern of X rays is used to determine the arrangement of individual atoms within a molecule.

Xenopus laevis An African clawed frog used as a model system for developmental biology.

yeast artificial chromosome (YAC) A vector that can replicate as a chromosome in yeast cells and can accomodate very large DNA inserts (hundreds of kb).

yeast two-hybrid A genetic method for detecting protein interactions in yeast cells.

yeasts The simplest unicellular eukaryotes. Yeasts are important models for studies of eukaryotic cells.

zebrafish A species of small fish used for genetic studies of vertebrate development.

zinc finger domain A type of DNA binding domain consisting of loops containing cysteine and histidine residues that bind zinc ions.

zygote A fertilized egg.

zygotene The stage of meiosis I during which homologous chromosomes become closely associated.

Index

Note: Page numbers in *italics* indicate that the information will be found in figures or tables.

A bands, 488–489, *490*
ABC transporters, 553, *555*
Abelson leukemia virus, *735*, 735–736
abl oncogene, *735, 736, 739,* 742, 756, *757, 759*
abl proto-oncogene, 741
Abscisic acid, 608, *609*
Acetyl CoA
  in lipid synthesis, 93–94
  in oxidative metabolism, 86, *87,* 88, 89, *434,* 435
Acetylcholine, 547, 548, 604, *604,* 612
N-Acetylgalactosamine, 335, 413, 578
N-Acetylglucosamine, 335, 336, 411, 412, *413, 570, 571,* 578
N-Acetylglucosamine phosphates, 412
N-Acetylmuramic acid, *571*
Acidic activation domains, 279
Acidic keratins, 497
Actin, 473, 474, 534. *See also* Actin filaments
Actin-binding proteins, 476–479, 480, 482–483, 495, 580. *See also specific proteins*
Actin bundles, 480–481
Actin-bundling proteins, 480–481, 496
Actin cross-linking proteins, 480, 481
Actin cytoskeleton, regulation of, 638–640
Actin filaments
  assembly and disassembly, 474–480
  association with plasma membrane, 482–485
  cell movement and, 495–497
  cell surface protrusions, 485–486, *487*
  in contractile assemblies in nonmuscle cells, 491–493
  contractile ring, 677
  intermediate filaments and, 500
  in muscle contraction, 487–491, *492*
  organization of, 480–481
  overview of, 473–474
Actin networks, 480, 481
Actin-related proteins, 480
α-Actinin, *477,* 481, 484, 489, 580, 581
Actinomycin D, 117
Action potentials, *546,* 547
Activation energy, 74
Activation-induced deaminase (AID), 238

Active site, of enzymes, 74
Active transport, 64, *65,* 550–553, *555*
Acute lymphocytic leukemia, *739*
Acute myeloid leukemia, *739*
Acute promyelocytic leukemia, 374, *739,* 744–745, 757–758
Acyclovir, 99
Adaptor proteins, 419, 420
Adenine, *50, 51,* 108, *109*
Adenine nucleotide translocator, 450
Adenomas, 722, 750, 754, 755
Adenosine 5'-diphosphate (ADP)
  ATP hydrolysis and, 83, 84
  in glycolysis, 86
  transport across inner mitochondrial membrane, 450
Adenosine 5'-monophosphate (AMP), 83, 84
Adenosine 5'-triphosphate (ATP). *See also* ATP hydrolysis; ATP synthesis
  in actin polymerization, 474, 475–476
  in active transport, 64, *65*
  amino acid synthesis and, 95
  in anabolic pathways, 91
  in attachment of amino acids to tRNA, 311
  evolution of metabolism and, 7, 8
  free energy changes and, 83–84
  glucose synthesis and, 92, 93
  lipid synthesis and, 94
  as nucleotide, 51
  protein import into mitochondria and, 441
  protein synthesis and, 96, *97*
  transport across inner mitochondrial membrane, 450
  used by Na⁺-K⁺ pump, 550
Adenoviruses, *38*
  cancer studies and, 731–732
  introns discovered in, 157, 158–159
  tumors caused by, 729
Adenylyl cyclase, 261, 610, 611, 612, 618, 619, 620
Adenylyl imidodiphosphate (AMP-PNP), 514, 515
ADF/cofilin, *477,* 479, 496

Adherens junctions, 484–485, 584–588
Adhesion belts, 484, *485,* 491–492
Adipocytes, 14
ADP. *See* Adenosine 5'-diphosphate
ADP-ribosylation factors, 419
Adrenaline, 80
Adult stem cells
  developmental plasticity and, 707–708
  examples of, 704–707
  medical applications, 708–709
  properties of, 703–704
Adult T-cell leukemia, 729, 732
Aequoria victoria, 25
Aerobic bacteria, electron transport chain in, 89
Aflatoxin, 723, *724*
Aggrecan, 579
Aging syndromes, 216
Agrobacterium tumefaciens, 141, *142*
AID. *See* Activation-induced deaminase
AIDS, 99, 120, 733
akt oncogene, *735, 739*
Akt protein kinase, 624, 625, *626, 627,* 699, 745
AKT8 virus, *735*
Alanine, 52, *53*
Albinism, 422
Albuterol, 487
Aldosterone, *601*
Alkylation, 218–219
Allen, Robert, 512, 514
Allosteric regulation, 80, 340
ALS. *See* Amyotrophic lateral sclerosis
Alternative splicing, 160, *161,* 189, 299–300
Altman, Sidney, 6, 289, 316
ALV. *See* Avian leukosis virus
Alzheimer's disease, 331, 510
Amanita phalloides, 476
α-Amanitin, 476
Amino acids
  attachment to tRNAs, 310–311, *312*
  biosynthesis, 94–95
  dietary requirements, *96*
  as energy source, 89
  genetic code and, 113–115
  polymerization, 95–96

pyrrolysine, 325
selenocysteine, 325
structure, 52–53
Aminoacyl tRNA synthetases, 114, 310–311, 317
α-Aminobutyric acid (GABA), 604
Ammonia, 94, 95
Amoeba proteus, 13
Amoebas
phagocytosis, 557
pseudopodia, 486, 487
AMP. *See* Adenosine 5'-monophosphate
AMP/activated kinase (AMPK), 626, 627
AMP-PNP, 514, 515
Amphibians, DNA amplification in oocytes, 247–248
Amphipathic molecules, 6, 49, 533
AMPK. *See* AMP/activated kinase
Amyloid fibers, 331
Amylopectin, 45, 46
Amyloplasts, 457
Amylose, 45
Amyotrophic lateral sclerosis (ALS), 504–505
Anabolic pathways, 91
Analgesics, 606
Anaphase
meiosis I, 681
meiosis II, 681
mitosis, 522–523, 669, 670, 671, 676–677
Anaphase A, 522–523
Anaphase B, 522, 523
Anaphase-promoting complex, 676–677, 683, 685
Anderson, Richard, 560
Anfinsen, Christian, 54–55, 56, 330
Angiogenesis, 727, 757
Animal cell culture, 33–36
Animal cells
cell cycle regulation in, 653
structure of, 10
transfection, 139–140
Animal viruses
cancer and, 37
initiation of DNA synthesis, 211
protein cleavage and, 334–335
RNA tumor viruses, 115–117
structure, 36
as vectors for transfection, 140
Animals, tissue systems and cell types, 14–15
Ankyrin, 483, 534, 538
Anogenital cancers, 731
Antennapedia mutant, 277–278, 279
Anti-codon loop, 310
Antibiotic resistance, 320
Antibiotics, actions on bacteria, 570
Antibodies
inhibition of protein function, 147
as probes for proteins, 134–136
transport through transcytosis, 565
Anticancer drugs, 756–760
Antigens, 134
Antimalarial drugs, 424
Antimetabolites, 98–99
Antiport, 556
Antisense DNA, 145
Antisense nucleic acids, 145
Antisense RNA, 145
AP-1 transcription factor, 743
AP endonuclease, 220

AP sites. *See* Apurinic sites; Apyrimidinic sites
Apaf-1 protein, 693, 695, 697
APC, 635, 636
APC gene, 750, 754
Apical membrane domain, 415–416, 538–539
Apo-B48 protein, 301
Apo-B100 protein, 300–301
Apolipoprotein B, 300–301
Apoptosis
Bcl-2 family proteins and, 695–698
cancer cells and, 728
caspases and, 692–693, 695
events of, 691–692, 693
genetic analysis in *C. elegans*, 690–691, 692, 694–695
oncogenes and, 745
p53 protein and, 752–753
signaling pathways that regulate, 698–700, 701
tumor necrosis factor and, 617
Apoptosome, 693, 697
Apurinic (AP) sites, 220
Apyrimidinic (AP) sites, 220
Arabidopsis thaliana
centromeres, 174
complete genome sequence data, 176, 183–184
as experimental model, 19
genome size and composition, 16, 166, 167
mitochondrial genome, 436
telomeric repeat sequences, 175
Arachidonic acid, 607
Archaebacteria, 11–12, 178, 325
ARF protein family, 419, 420, 496
Arginine, 52, 53, 76
Armadillo protein family, 586–587
Arp2/3 complex, 477, 478, 479, 495, 640
Arp4–8 protein, 480
β-Arrestins, 641
ARSs. *See* Autonomously replicating sequences
Asparagine, 52, 53
Aspartate, 52–53, 76
Aspartic acid, 52–53
Aspirin, 607–608
Asthma, 487
Astral microtubules, 521, 522, 523, 675
Astrobiology, 8
ATM gene, 753
ATM protein kinase, 655, 668, 669, 698
Atmosphere, effects of photosynthesis on, 8
ATP-binding cassettes, 553
ATP-binding domains, 553
ATP hydrolysis, 83–84
in active transport, 550–553, 555
muscle contraction and, 490, 491
ATP synthase, 448–449, 461
ATP synthesis
in chemiosmotic coupling, 445–449, 453
electron transport chain and, 444–445, 446
in glucose oxidation, 84–89
in glycolysis, 7
in lipid oxidation, 89–90
in mitochondria, 435
in oxidative metabolism, 7, 8
in oxidative phosphorylation, 443
in photosynthesis, 90, 91, 459, 460, 461, 462, 462

ATR protein kinase, 655, 668, 669
Aurora kinase family, 675
Autocrine growth stimulation, 725
Autocrine signaling, 600
Autonomously replicating sequences (ARSs), 212, 213, 214
Autophagosomes, 347, 348, 428
Autophagy, 347–348, 428
Autophosphorylation, 614, 615
Auxins, 573, 608–609
Avery, Oswald, 108
Avian erythroblastosis viruses, 735
Avian leukosis virus (ALV), 734
Avian myeloblastosis virus, 735
Avian myelocytomatosis virus, 735
Avian sarcoma viruses, 735
Avian SK virus, 735
Axin, 635, 636
Axonemal dynein, 512, 513, 518, 519, 520
Axonemes, 518, 519, 520
Axons
action potentials, 546, 547
ion channels and membrane potential, 544–545
microtubules and, 510–511
transport of membrane vesicles and organelles, 514–515
AZT, 99

B-cell lymphomas, 695, 732, 739
B lymphocytes, 705
activation-induced deaminase, 238
antibodies and, 134, 135, 234
cancers and, 740
CD45 receptor, 617
Epstein-Barr virus and, 732
immunoglobulin enhancer, 271–272
myelomas, 390
B-*raf* oncogene, 739
BAC. *See* Bacterial artificial chromosome
Bacteria. *See also Escherichia coli*
ABC transporters, 553
antibiotic resistance, 320
antimetabolite experiments, 98–99
cancers and, 724
cell wall, 569, 570
complete genome sequence data, 176
dual membrane system and porins, 535–536
expression of cloned genes in, 127
flagella, 517
genetic transformation experiments, 107–108
genome size and composition, 16, 165
gram-positive and gram-negative, 570
initiation of translation in, 319, 321, 323
insertion sequences, 239
pathogenic, 495
restriction endonucleases, 118
Bacterial artificial chromosome (BAC), 123, 125, 183, 187, 188
Bacteriophage λ, 118, 123
Bacteriophage P1, 123
Bacteriophages, 38, 118, 123
Baculovirus vectors, 128
Bad protein, 699, 746
Bak protein, 696, 697
Baltimore, David, 116, 117
*Bam*HI enzyme, 118
Band 3 protein, 483, 534, 535, 538
Band 4.1 protein, 534
Barnacles, 58

Barr, Yvonne, 732
β-Barrels, 61, 63, 535
Barrier elements, 272
Basal bodies, 519–520
Basal cell carcinomas, 739, 750, 751
Basal laminae, 575, 577, 578, 580
Base-excision repair, 220
Base pairing
    codon-anticodon, 311, 312
    complementary, 51, 109, 314
Basement membranes, 575. *See also* Basal
    laminae
Basic keratins, 497
Basolateral membrane domain, 415, 416,
    538–539
Basophil, 705
Bax protein, 696, 697
Bcl-2 family proteins, 695–698, 746
bcl-2 oncogene, 695, 739, 746
bcr/abl oncogene, 759
Bcr/Abl protein, 741, 756, 758, 759
bcr gene, 741, 756
Beadle, George, 107
Becker's muscular dystrophy, 483
Benign tumors, 720
Benzo(*a*)pyrene, 723, 724
Berget, Susan, 158
Bid protein, 700, 701
Bile acids, 464
Bim protein, 699, 746
Bioinformatics, 192
Biosynthetic pathways
    carbohydrate synthesis, 92–93, 94
    lipid synthesis, 93–94
    nucleic acid synthesis, 97–98, 99
    overview of, 91–92
    protein synthesis, 94–96, 97
Biotin, 78
BiP protein, 393, 399, 401–402, 408
Bishop, J. Michael, 736, 737
Bivalents, 680, 681
Blackburn, Elizabeth, 214
Bladder cancers/carcinomas, 720, 738, 739,
    750
Blindness, 438
Blobel, Günter, 364, 389, 390, 391
Blood cells, types and formation of, 704,
    705
Blood platelets, 483, 702
Blood testing, 67
Blood tissue, 14
Blood vessels
    anticancer drugs and, 757
    cancer-related angiogenesis, 727
    endothelial cell proliferation, 702–703
    nitric oxide induced dilation, 604
Bone, 575
Bone marrow transplantation, 708–709
Boveri, Theodor, 507
BP180 protein, 581
Brady, Scott, 512, 514
Brain tumors, 750
BRCA1 gene, 750, 753, 754
BRCA2 gene, 227, 750, 753, 754
BRCA1 protein, 753
BRCA2 protein, 753
Breast cancers/carcinomas
    anticancer drug therapies, 757, 758
    BRCA genes and, 227
    *erb*B-2 gene and, 248, 741
    hormones and, 723

incidence and mortality, 720
molecular diagnosis, 756
oncogenes and, 739
tumor suppressor genes and, 750
Brenner, Sidney, 113
Bright-field microscopy, 23
Britten, Roy, 161
Broker, Tom, 158
Bromodeoxyuridine, 373
Brown, Michael, 559–560
Brown fat, 450
Brush border, 485
Buck, Clayton, 582
Buoyant density, 32
Burkitt, Denis, 732
Burkitt's lymphoma, 729, 732, 739
Butel, Janet, 364

C-terminal domain (CTD), of RNA poly-
    merase II, 265, 266, 290
Cadherins, 485, 584, 585–586, 600
Caenorhabditis elegans
    anatomy, 18
    complete genome sequence data, 176,
        180–181
    as experimental system, 17–18
    genes encoding ion channels, 549
    genome size and composition, 16, 18,
        166, 167
    kinesins in, 513
    programmed cell death in, 690–691, 692,
        694–695
Cajal bodies, 374
CAK enzyme, 664
Calcium (Ca^{2+})
    cytosolic concentrations, 622
    dynamics during fertilization, 684–685
    endoplasmic reticulum as repository for,
        386, 622
    extracellular and intracellular concentra-
        tions, 545
    mobilization by IP$_3$, 622–623
    muscle contraction and, 491, 492
    Na$^+$-Ca^{2+} antiporter, 556
    as second messenger, 622–624
Calcium (Ca^{2+}) channels, 544, 548, 549, 550,
    623
Calcium (Ca^{2+}) pump, 552
Callus, 36
Calmodulin, 485, 493, 623
Calnexin, 401, 402
Calponin protein family, 482–483
Calreticulin, 401
Calvin cycle, 90–91, 92, 465
CaM kinase family, 623
cAMP. *See* Cyclic AMP
cAMP-dependent protein kinase, 343, 344,
    618. *See also* Protein kinase A
cAMP pathway, 618–620, 623
cAMP phosphodiesterase, 618
cAMP response element (CRE), 619
Cancer
    alternative splicing and, 299
    causes of, 723–724
    common lethal forms, 629
    development of, 721–723
    E-cadherin and, 585
    general characteristics of, 719
    molecular diagnosis, 756
    oncogenes and, 738–741
    origin of term, 721

polypeptide growth factors and, 606–607
prevention and early detection, 755–756
proteasome inhibitors and, 347
ras oncogenes and, 629
treatment, 756–760
types of, 720–721
viruses and, 37
Cancer cells
    autocrine signaling in, 601
    cell cycle mutations and, 669
    cell transformation assays, 728–729
    defects in cell cycle regulation, 665–666
    farnesylation inhibitors and, 338
    gene amplification and, 248
    MDR transporters and, 553
    properties of, 724–728
    proteases and, 575, 726–727
    protein-tyrosine kinases and, 612–613
    Ras protein and, 628, 629
    telomerase and, 176
    telomeres and, 216
Cancer treatment, 755–760
CAP. *See* Catabolite activator protein
Capillaries, 703
CapZ, 477
Carbohydrates
    addition to proteins, 413
    biosynthesis, 92–93, 94
    glycosylation of proteins, 335–337
    lipid metabolism and, 413–414
    structures and functions, 44–46
Carbon dioxide (CO$_2$)
    conversion to carbohydrates, 464–465
    oxidative metabolism and, 434, 435
    photosynthesis and, 90, 91
Carbon monoxide (CO), 604
Carcinogens, 723, 724
Carcinomas, 721, 726, 755. *See also specific
    types*
Cardiac muscle, 487
Cardiolipin, 443, 444
Carins, John, 202
Carotenoids, 457
Carrier proteins, 64, 542–543
Cartilage, 575, 579
Cas NS-1 virus, 735
Caspases, 692–693, 695, 697, 700, 701
Catabolism, 91
Catabolite activator protein (CAP), 261–262,
    277
Catabolite repression, 261–262
Catalase, 463
Cataracts, 591–592
α-Catenin, 477, 586, 587
β-Catenin, 485, 586, 635, 636, 744, 754
Catenins, 485
Caveolae, 540, 562
Caveolin, 540, 562
cbl oncogene, 735
CCND1 oncogene, 739, 743
CD45 receptor, 617
Cdc20, 676
*cdc*2 mutant, 659, 660
*cdc*28 mutant, 659, 660
cdc mutants, 657, 659–660
Cdc25 phosphatase, 668
Cdc29 protein, 657
Cdc42 protein, 639
Cdc25C protein, 660
Cdk1/cyclin B complex. *See also* Maturation
    promoting factor

anaphase promoting complex and, 676, 677
Cdk inhibitor, *665*
in cell cycle regulation, 661, 662, 663, 668
control of meiosis in oocytes, 682–684
Cdk/cyclin complexes, 661–665, 666–667
Cdk4/cyclin D complex, 752
Cdk4,6/cyclin D1 complex, 665–666, 667
Cdk2/cyclin E complex, 666–667, 668, 669
cdk4 gene, *739*, 755
Cdk inhibitors (CKIs), 664, *665*
Cdk1 protein kinase. *See also* Cdk1/cyclin B complex
in cell cycle regulation, 660, 661, 662, 664
cyclin B and, 346
in cytokinesis, 677
MPF and, 659, 663
Cdk2 protein kinase, 663, 664, 668
Cdk4 protein kinase, 744
Cdk7 protein kinase, 664
Cdk's. *See* Cyclin-dependent kinases
cDNA, 122
cloning in plasmids, 123, *124*
DNA microarrays, 134
Cech, Tom, 6, 295, 316
ced-3 gene, 691, 692, 694, 695
ced-4 gene, 691, 694, 695
ced-9 gene, 691, 694, 695
Ced-3 protein, 691, 692, 693
Ced-4 protein, 691, 693
Ced-9 protein, 691, 695
Celera Genomics, 187, 188
Cell adhesion molecules, 584–588
Cell-cell interactions
adherens junctions, 484–485, 584–588
cancer cells and, 726
gap junctions, 589–590
intermediate filaments and, 500, *501*
plasmodesmata, 592, *593*
tight junctions, 588–589
types of, 584
Cell-cell signaling
carbon monoxide, 604
cell surface receptors, 609–617
eicosanoids, 607–608
modes of, 600–601
neurotransmitters, 604–605
nitric oxide, 603–604
overview of, 599
peptide hormones and growth factors, 605–607
steroid hormones, 601–603
Cell cortex, 482
Cell culture, 33–36
Cell cycle. *See also* Cell cycle regulation
cyclin B and, 346
cytokinesis, 677–678
mitosis, 669–678
phases of, 650–652
tumor suppressor genes and, 752, 753
Cell cycle checkpoints, 654–655
Cell cycle regulation
cancer cells and, 665–666
checkpoints, 654–655
cyclins and cyclin-dependent kinases in, 661–665
DNA damage checkpoints, 667–669
G_2 phase control, 652–653
growth factors and Cdk's at G_1 phase, 665–667
of oocytes in meiosis, 681–684
protein kinases and, 657–661

restriction of DNA replication, 656
restriction points and growth factors, 653–654
START mechanism, 652–653
Cell death. *See* Apoptosis; Programmed cell death
Cell death receptors, 700, *701*
Cell differentiation, cancer cells and, 727–728
Cell lines, 34
Cell-matrix adhesion
adhesion proteins, 579–580
cancer cells and, 726
integrins, 582–583
Cell membranes. *See also* Plasma membrane; *specific membranes*
fluid mosaic model, 59–50, 61, *62*
fluidity, 61
introduction to, 58
membrane lipids, 60–61, 403
membrane proteins, 61–63
phospholipid components, 47–49
polarized, protein transport to, 415–416
transport across, 63–64, *65*
vesicle fusion, 420–423
Cell movement/migration, 495–497
Cell plate, 678
Cell proliferation
density-dependent inhibition, 724, *725*
differentiated cells, 700, 702–703
stem cells, 703–708
stimulation by SV40 and polyomaviruses, 730–731
Cell proteins. *See also* Proteins
global analysis of localization, 68–69
identification of, 65–68
identifying protein-protein interactions, 69
Cell surface receptors
cytokine receptors, 616
G protein coupled, 610–612
neurotransmitters and, 605
overview of, 609–610
receptor protein-tyrosine kinases, 612–615
Cell survival
oncogenes and, 745–746
p53 protein and, 752–753
tumor suppressor proteins and, 751, 752–753
Cell theory, 21–22
Cell transformation
adenoviruses and, 731–732
retroviral oncogenes and, 734
by SV40 and polyomaviruses, 730–731
Cell transformation assays, 728–729
Cell walls
in bacteria, 569, 570
in eukaryotes, 569, 570–575
functions and types of, 569
in plants, *11*, 569, 570–575
in prokaryotes, 8, *9*
Cells
origin of life, 4–7
prokaryotic and eukaryotic compared, 4
Cellulose, 45–46, 414, 570, *572*, 573, 574–575
Cellulose synthase, 574–575
Cement proteins, 58
CenH3, 174–175
Centrifugation techniques, 30–33
Centrin, 510
Centrioles, *10*, 509, 510, 519

Centromeres, 171–175, 669
Centrosomes, 507–509, 520, 521, 669
Ceramide, 406, 413, 426
Cervical cancer/carcinoma, 724, *729*, 731
CFTR. *See* Cystic fibrosis transmembrane conductance regulator
cGMP. *See* Cyclic GMP
cGMP phosphodiesterase, 621
Channel proteins, 542
ion channels, 543–549, *550*
overview of, 63–64
porins, 535–536, 543–544
Chaperones, 399
posttranslational translocation and, 393
in protein degradation, 401
protein folding and, 330–332
protein import into mitochondria and, 440–441
Chaperonins, 331–332, 441
Charcot-Marie-Tooth disease (CMT), 359, 591
Chargaff, Erwin, 109
"Chase" experiments, 386, *387*
Checkpoint kinases, 668, 669
Chemical carcinogens, 723, *724*
Chemiosmotic coupling, 445–449, *453*
Chemotherapy, 629
anticancer drugs, 756–760
antimetabolites and, 98–99
Chiasmata, 680, 681
Chickens
complete genome sequence data, *176*
genome size and composition, *16*, 167
whole genome sequencing and, 191
Chimpanzees, 192
Chitin, 570, 574
CHK1 checkpoint kinase, 668
CHK2 checkpoint kinase, 668, 669
CHK protein kinase, 698
Chloramphenicol, 320
Chloride (Cl⁻), 462, 552
cystic fibrosis and, 553, 554, 555
extracellular and intracellular concentrations, *545*
gradient across plasma membrane, *553*
Chloride (Cl⁻) channels, 553, 554, 555
Chlorophyll special pair, 460
Chlorophylls, 90, *91*, 459, 460
Chloroplast envelope, 451, 452
Chloroplasts
chemiosmotic generation of ATP, 452, *453*
chromoplasts and, 457
compared to mitochondria, 451, *453*
development of, 457, *458*
electron transport system in, 452
etioplasts and, 458
function of, 9
genome, 452–454
H⁺ pumps, 553
inner membrane, 451, *452*, 454, *455*
origin through endosymbiosis, 10–11
outer membrane, 451, *452*, 454, *455*
protein import, 454–455, *456*
RNA editing, 300
structure and function, 451–452, 453
Chloroquine, 424
Cholera toxin, 564
Cholesterol, 49, 403
in lipid rafts, 539
in membranes, 60, 61
peroxisomes and, 464

in plasma membranes, 531–532
synthesis, 406
uptake by mammalian cells, 559–561
Choline, 47, *48*
Chondrocytes, 14
Chondroiton sulfate, *578, 579*
Chow, Louise, 158
Chromatin. *See also* Chromosomes
  Arp4–8 protein and, 480
  condensation, 672–673
  modifications during transcription,
    281–284
  nuclear lamina and, 360
  organization in the nucleus, 371–375
  structure and function, 166–171, 281
Chromatin assembly factor, 209
Chromatin immunoprecipitation, 273, *274*
Chromatosome, 168
Chromokinesin, 521
Chromoplasts, 457
Chromosomal microtubules, 521, 522, 523,
  675
Chromosomal theory of inheritance,
  104–105, *106*
Chromosomal translocations
  Bcr/Abl protein, 758
  oncogenes and, 740–741
Chromosomes. *See also* Chromatin
  association with microtubules during
    mitosis, 520–523
  centromeres, 171–175
  genes and, 104–105
  homologous, 679–681
  human, *186, 187*
  inheritance and, 104–105, *106*
  looped-domain organization, 373
  at M phase transition, 672–675
  in meiosis, 678–681
  in mitosis, 171, *172,* 669–671
  numbers in eukaryotic cells, *167*
  organization in the nucleus, 371–375
  polytene, *183, 184,* 281, *282*
  telomeres, 175–176
Chronic myelogenous leukemia, *757*
Chronic myeloid leukemia, *739, 741,* 756,
  758, 759
Chronic myelomonocytic leukemia, *739*
Chymotrypsin, 75–78
Ci155 transcription factor, 634, *635*
Ci75 transcriptional repressor, 634
Cilia, 517–520
Ciliated protozoa, 372
Cip/Kip protein family, 664, *665*
Cisternae, 409, *410*
Citric acid cycle, 86–87, 89, 434, 435
Clamp-loading proteins, 207
Class switch recombination, 236–237, 238
Clathrin, 419, *420*
Clathrin-coated pits, 558, 559, 560, 561–562
Clathrin-coated vesicles, 419, *420,* 425, 558,
  565
Clathrin-independent endocytosis, 562–563
Claude, Albert, 27, 30
Claudin, 589
Cleaver, James, 221
Clinical depression, 547
Cln's. *See* G$_1$ cyclins
Clonal selection, 722
Cloned genes
  expression, 126–128
  gene transfer techniques, 139–142
  mutagenesis, 142, *143*

yeast and, 137–139
Cloning, 682. *See also* Cloned genes;
    Molecular cloning
  Dolly (cloned sheep), 682, 713
  reproductive, 715
  somatic cell nuclear transfer, 713–715
  therapeutic, 714–715
Closed mitosis, 670–671
Closed-promoter complex, 257
CMT. *See* Charcot-Marie-Tooth disease
c-myc oncogene, *739, 740, 744*
CoA-SH, *86, 87*
Coactivators, 279–280
Coastal-2 protein, 634, *635*
Coat proteins, 419, 420, *421*
Coated vesicles, 419–420, *421*
Cobras, 622
Cockayne's syndrome, 222, 223
Codons
  described, 114–115
  initiating translation, 317, 318, 319
  nonstandard base pairing, 311, *312*
  redundancy, 310
Coenzyme A (CoA), *78,* 86, 89, *404,* 405
Coenzyme Q (CoQ), *88,* 444, *445,* 446, 460
Coenzymes, 78–79
Cohen, Stanley, 606, 613
Cohesins, 673, 677–678
Coilin, *375*
Colcemid, 506–507, 508
Colchicine, 506–507
Collagen fibrils, 576–577
Collagenase, 727
Collagens, 575–578, 580
Collenchyma, 13, *14*
Collins, Frances, 188
Colon adenomas, 722, 750, 754, 755
Colon cancers/carcinomas
  development of, 722–723
  incidence and mortality, *720,* 722
  mismatch DNA repair and, 223, 224
  prostaglandins and, 608
  role of oncogenes and tumor suppressor
    genes in, *739, 744, 750, 753–754*
Colonial green algae, 13
Colonoscopy, 224
Colorectal cancers, 223, 224, 753, *757*
Complementary base pairing, 51, 109, 314
Condensins, 673, 685
Confocal microscopy, 26–27
Connective tissues, 14, 575
Connexins, 590, 591–592
Consensus sequences, 255
Constitutive heterochromatin, 371
Constitutive secretory pathway, 414, 416
Contact inhibition, 726
Contractile bundles, 481
Contractile ring, 492–493, 650, 677
"Conventional" kinesin, 512
COP-coated vesicles, 419
COPI-coated vesicles, 419, 420, 674
COPII-coated vesicles, 419, 420
CoQ. *See* Coenzyme Q
Core histones, 168
Corey, Robert, 57
Corn, *167*
Cortical microtubules, 574
Corticosteroids, 601
Cortisol, *601*
Cosmid vectors, 123, *125*
Cotranslational translocation, 387–392
Cows, *167*

COX-1, 608
COX-2, 608
CRE. *See* cAMP response element
CREB transcription factor, 619, 623
Crick, Francis, 108, 109
Cristae, 434, 435
crk oncogene, *735*
Crosstalk, 6540–641
CSA protein, 223
CSB protein, 223
CsCl density gradients, 162
CSF. *See* Cytostatic factor
CSL transcription factor, 636
CT10 virus, *735*
CTD. *See* C-terminal domain
CTNNB1 oncogene, *739, 744*
Cubitus interruptus (Ci) protein, 634
Culture cells, as experimental model, 19
Curare, 548
Cyanobacteria, 8, 452
Cyclic AMP (cAMP)
  G proteins and, 610
  in glucose repression, 261
  in glycogen metabolism, 342–343
  regulation of cAMP-dependent protein
    kinase, 344
  synthesis and degradation, 618
Cyclic electron flow, 461, *462*
Cyclic GMP (cGMP), 604, 617, 620–621
Cyclin A, 660, 663, 668
Cyclin B, 346, 659, 660, 661, 662, 663. *See
    also* Cdk1/cyclin B complex
Cyclin D, 662, 665–666, 743–744
Cyclin-dependent kinases (Cdk's), 661–665
Cyclin E, 662, 663, 668
Cyclin G$_1$, 661, 662
Cyclin H, 664
Cyclins, 346, 660, 661–665
Cyclobutane, 217, 218
Cycloheximide, 320
Cyclooxygenase, 607, 608
Cysteine, 52, *53*
Cystic fibrosis, 553–555
Cystic fibrosis transmembrane conductance
    regulator (CFTR), 553, 554, 555
Cytochalasins, 476, 495
Cytochrome *b,* 88
Cytochrome *bc* complex, 460
Cytochrome *bf* complex, 460, 461, 462
Cytochrome *c, 88,* 444, *445,* 460, 693, 697
Cytochrome oxidase, 444
Cytokine receptors, 616, 617
Cytokines, 606
Cytokinesis, 492–493, 650, *670,* 671, 677–678
Cytokinins, 608, 609
Cytoplasmic dynein, 513
Cytosine, *50,* 51, 108, *109*
Cytoskeleton
  actin filaments (*see* Actin filaments)
  adherens junctions, 586–587
  in animal cells, *10*
  functions of, 9–10
  integrins and, 580
  intermediate filaments, 497–505
  microtubules, 505–511
  overview of, 473
  in plant cells, *11*
  regulation of, 638–640
  signal transduction and, 637–640
Cytostatic factor (CSF), 683

Dalgarno, Lynn, 318
Dark reactions, 90, 459
Deafness, 591
Deamination, *217*
Deathcap mushroom, 476
deDuve, Christian, 30, 426, 557
Dehydration reactions, 93, 95–96, *97*
Deinococcus radiodurans, 227
Deisenhofer, Johann, 459
DeLisi, Charles, 188
Delta protein, 636, *637*
Denaturation, 56
Dendrites, 510, 511
Density-dependent inhibition, 724, *725*
Density-gradient centrifugation, 31
Deoxyribonucleic acid (DNA). *See also*
cDNA; DNA content; DNA repair;
DNA replication; Recombinant DNA
amplification, 247–248, 741
antisense, 145
central dogma and, 113
chemical composition, 50–51
chromatin, 166–171
complementary base pairing, 51, 109
damage checkpoints during the cell
cycle, 667–669
damage leading to cell death, 698
determination of cell content, 652
difference from RNA, 112
of *E. coli*, 8
gene transfer techniques, 139–142
homologous recombination, 227–233
identified as the genetic material,
107–108
λ, 118–119
methylation, 286, *287*
mitochondrial, 436
nucleic acid hybridization techniques,
131–134
PCR amplification, 129–131
reassociation experiments, 161–162
repetitive sequences, 161–164
sequencing, 125, *126*
site-specific recombination, 233–238,
240–241
structure of, 108–109
telomeric sequences, 175–176
transposition with DNA intermediates,
239, 241–242
transposition with RNA intermediates,
242–247
Deoxyribonucleoside 5'-triphosphates
(dNTPs), 202, 210
Deoxyribose, *50*, 51
Depression, 547
Depurination, *217*
Dermal tissue, 13
Dermatan sulfate, *578*, 579
Dermatofibrosarcoma protuberans, *757*
Desmin, *497*, 498, 500
Desmocollin, *501*, 587
Desmoglein, *501*, 587
Desmoplakin, 500, *501*, 586, 587, 588
Desmosomes, 500, *501*, 586, 587–588
Destruction box, 346
Detergents, 533
Developmental plasticity, 707–708
Diacylglycerol, 621–622, 623
Diakinesis, 679, *680*
2,6-Diaminopurine, 98
Dicer enzyme, 145, *146*
Dictyostelium, *167*, *175*

Dideoxynucleotides, 125, *126*
Differential centrifugation, 30
Differential interference-contrast
microscopy, 23
Differentiated cells, proliferation and, 700,
702–703
Digestive enzymes, 415
Dihydrofolate reductase, 248
Dihydrouridine, *289*
Dihydroxyacetone, *45*
Dimethylnitrosamine, 723, *724*
Diploids/Diploidy, 105, 652
Diplotene, 679, 680, 681
Disaccharides, 578
Dishevelled protein, 635, *636*
Disulfide bonds, 332–333, 399
λ-DNA, 118–119
DNA-affinity chromatography, 275, 276
DNA amplification, 247–248, 741
DNA content
C. elegans, 18
Drosophila melanogaster, 18
E. coli, 16
humans, 16, 19
representative organisms, *16*
yeast, 17
DNA footprinting, 255–257
DNA glycosylase, 220
DNA ligase, 121, *122*, 203, 205
DNA looping, 271, 272
DNA methylation, 286, *287*
DNA microarrays, 133–134, 194, 756
DNA polymerase λ, 202, 205
DNA polymerase δ, 202, 205–206, 207, 222
DNA polymerase ε, 202, 206, 207, 222
DNA polymerase γ, 202
DNA polymerase I, 205
DNA polymerase II, 225
DNA polymerase III, 202, 205, *206*, 207
DNA polymerase IV, 225
DNA polymerase V (pol V), 225
DNA polymerases, 110, 129
accuracy of DNA replication and,
210–211
in DNA replication, 202, *203*, 204,
205–207, 209
telomeres and, 176
in translesion DNA synthesis, 225
types of, 202
DNA provirus, *733*
DNA provirus hypothesis, 116–117
DNA reassociation, 161–162
DNA repair
colon cancer and, 223, 224
direct reversal of damage, 216–219
of double strand breaks, 226–227
excision, 219–223
recombinational repair, 225–227
translesion DNA synthesis, 223–225
DNA repair genes, 220, 221–222
DNA replication, 110, *111*
DNA polymerases and, 202, *203*, 204,
205–207, 209
fidelity of, 209–211
localization in the nucleus, 373
origins of replication, 211–214
replication fork, 202–209
restricting during the cell cycle, 656
telomeres and telomerase, 214–216
DNA transposons, *162*, 164, 239, 241–242,
247

dNTPs. *See* Deoxyribonucleoside 5'-
triphosphates
Dobberstein, Bernhard, 390, 391
Dogs, *167*, 192–193
Dolichol, 399, 464
Dolichol phosphate, 336
Dolly (cloned sheep), 682, 713
Dominant genes, 104
Dominant inhibitory mutants, 147
Dopamine, *604*
Dosage compensation, 285
Downstream promoter element (DPE), 263,
265
DPE. *See* Downstream promoter element
Drosophila melanogaster, 368
alternative splicing in, 299, 300
centromeres, 174
chromosome number, *167*
complete genome sequence data, *176*,
182–183
evolutionary split in species, 368
as experimental model, 18
gene amplification in, 248
genetic experiments on heredity, 105
genome size and composition, *16*, 18,
166, *167*
Hedgehog pathway, 634
homeodomain proteins and homeotic
mutants, 277–278, *279*
origin recognition complex, 213–214
polytene chromosomes, *183*, 184, 281,
282, 372
protein interaction map, *69*
simple-sequence repeats in, 162
Drosophila simulans, 368
Drug testing, 67
Druker, Brian, 758
Dscam gene, 300
Dscam protein, 300
dTTP, 248
Duchenne's muscular dystrophy, 483
Dunnigan-type partial lipodystrophy, 359
Dyes, 23
Dynactin, 513
Dynamic instability, 506, *507*
Dynamin, 558, 561
Dynan, William, 276
Dyneins
in chromosome movement, 523
in cilia and flagella, 518, *519*, 520
microtubule motor protein, 511, 512,
513–514, 515, 516
Dystrophin, *477*, 483–484

E6/7 proteins, 731
4E-BPs, 328–329
4E-BP1, 627
E-cadherin, 585, 726
E1 enzyme, 345
E2 enzyme, 345
E3 enzyme, 345
E-selectin, 540, *541*
E1A/B genes, 732
E1A/B proteins, 732
Eagle, Harry, 33, 35
EBS. *See* Epidermolysis bullosa simplex
Ecdysone, 602
Ecker, Robert, 657, 658
Eckhardt, Walter, 612
*Eco*RI enzyme, 118, 119, *122*
Edidin, Michael, 60, 537
eEF1α elongation factor, 323, 324, 341

eEF2 elongation factor, 323
eEF1 α/GDP complex, 324
eEF1βγ elongation factor, 324
EF-G elongation factor, 323
E2F transcription factors, 666
EF-Ts elongation factor, 324
Ef-Tu elongation factor, 323, 324
Effector caspases, 693
EGF. *See* Epidermal growth factor
EGF receptor, 607, 613, *614*, 629, 638, 758, 760
Eggs
   DNA amplification and, 247–248
   in fertilization, 684–685
   of *Xenopus,* 20
Eicosanoids, 607–608
eIF2 initiation factor, 321, 327–328, *329*
eIF2B initiation factor, 328, *329*, 624
eIF4E binding protein-1 (4E-BP1), 627
eIF4E initiation factor, 323, 326, 328–329, 627
eIF4G initiation factor, 627
eIFs (eukaryotic initiation factors), 321–323, 327–329
Elaioplasts, 457
Elastase, 76
Elastic fibers, 578
Elastin, 578
Electric potential, 545
Electrical synapse, 590
Electrochemical gradients
   in chloroplasts, 452
   in mitochondria, 447–448, 450–451
Electron microscopy, 27–30
Electron tomography, 28
Electron transport chain, 88–89
   in chloroplasts, 452
   Leber's hereditary optic neuropathy and, 438
   in mitochondria, 435, 444–445, *446*
   in photosynthesis, 459, 460–461
Electrophoresis. *See also* Gel electrophoresis
   SDS-PAGE, 135
   two-dimensional gel, 66, 67
Electrophoretic-mobility shift assay, 272, *273*
Electroporation, 139
Elion, Gertrude, 98, 99
Elk-1 transcription factor, 631, 743
Ellis, Hilary, 694
Elongation factors, 284, *319*, 323–324
Embryonal carcinoma cells, 710
Embryonic stem (ES) cells
   culturing, 33
   described, 709–712
   gene transfer in mice, 141
   production of mutants through homologous recombination, 143–144
   therapeutic cloning and, 714
Embryos, cell cycle time, 651
Emerin, 358, 359
Emery, Alan, 359
Emery-Dreyfuss muscular dystrophy, 359, 360
Endocrine glands, 600
Endocrine neoplasias, *739*
Endocrine signaling, 600
Endocytic vesicles, 515
Endocytosis
   fluid phase, 561
   lysosome formation and, 424–425, *427*
   overview of, 556–557

phagocytosis, 557–558
protein trafficking in, 563–565
Rab family proteins and, *422*
receptor-mediated, 558–563
Endometrial cancer/carcinoma, 723, *750*
Endoplasmic reticulum
   in animal cells, *10*
   association with microtubules, 516, *517*
   calcium and, 386, 622
   continuity with inner nuclear membrane, 398
   degradation of misfolded proteins, 401
   export of proteins and lipids from, 406–408, *409*
   functions of, 9
   glycosylation and, 325
   key features of, 386
   lipid synthesis in, 403–406
   lumen, 394
   membrane characteristics, *61*
   nuclear envelope and, *356, 357*
   in plant cells, *11*
   in plasmodesmata, 592, *593*
   protein disulfide isomerase in, 333
   protein folding and processing in, 398–400, I401
   protein insertion into membrane, 393–398
   protein secretion and, 386–387, *395*
   targeting proteins to, 387–393
   transport of proteins into, 334
   unfolded protein response, 402, *403*
Endorphins, *605, 606*
Endosomes, 425, 553, 563–565
Endosymbiosis, 10–11, 435
Endothelial cells
   anticancer drugs and, 757
   cell-cell adhesion and, 584–585
   proliferation, 702–703
   transport of antibodies, 565
Endotoxin, 570
Energy diagrams, *74*
Enhancers, 270–272
Enkephalins, *605, 606*
Entactin, 580
Enteroendocrine cells, 705, *706*
env gene, 733, 736
Enzyme replacement therapy, 426
Enzyme-substrate interactions, 75
Enzymes
   catalytic activity and mechanisms, 73–78
   one gene-one enzyme hypothesis, 107
   regulation of, 79–80, 340
Eosinophil, *705*
Epidermal cells, 13, *14*
Epidermal growth factor (EGF), *605, 606,* 615
Epidermis, self-renewal with stem cells, 705–706, *707*
Epidermolysis bullosa simplex (EBS), 503, 504
Epinephrine, 80, 342, *343*, 604, 618
Epinephrine receptor, 618
Epithelial cancers, 726
Epithelial cells, 14, *15. See also* Intestinal epithelial cells; Skin epithelial cells
   adhesion belt, 485
   ciliated, 517–518
   intermediate filament proteins, 497
   keratin filaments in, 500
   microvilli, 485–486
   plasma membrane polarity, 538–539

protein transfer with endosomes, 563
protein transport to, 415–416
stem cells, 709
tight junctions, 588
transcytosis in, 565
Epithelial stem cells, 709
Epithelial tissue, 14
Epstein, Michael, 732
Epstein-Barr virus, 37, 188, 732, 733
Equilibrium centrifugation, 32–33, 162
ER-Golgi intermediate compartment, 407, 408, *409,* 410, *411,* 420
*erb*A/B oncogenes, *735*
ErbA protein, 744
erbB oncogene, *735, 739, 757,* 758
*erb*B-2 oncogene, 248, 741, *757,* 758
ErbB-2 protein, 758
Erbitux, 757, 758
ERCC1 repair protein, 222
eRF1,eRF2, eRF3 release factors, 325
Erikson, Ray, 612
ERK kinase family, 627, 634, 683, 743
ERK signaling pathway, 638
   duration of signaling and, 640
   oncogene proteins and, 742, 743
   overview of, 627–628, 630–631
   tumor development and, 754
ERM proteins, 483
Erythrocytes, 14, *15,* 704, *705*
   cortical cytoskeleton, 482–483
   daily loss of, 558
   facilitated diffusion of glucose, 542
   membrane characteristics, *61*
   membrane proteins, 534
   membrane studies and, 530
   mobility of membrane proteins, 538
Erythrokeratoderma variabilis, 591
Erythroleukemia, 744
Erythromycin, 320
Erythropoietin, *605*
Escherichia coli
   cell wall characteristics, 570, *571*
   complete genome sequence data, 178
   DNA cloning with plasmid vectors, 123, *124*
   DNA polymerase III, 202, 205, *206,* 207
   DNA replication studies, 110, *111*
   dual membrane system, 535
   as experimental system, 16–17
   expression of cloned genes in, 127
   genome size and composition, 16, 165
   homologous recombination enzymes, 230, 231–232
   length of time for DNA replication, 212
   membrane characteristics, 60, *61*
   mismatch repair in, 223
   molecular cloning and, 121
   negative regulation of transcription, 258–261
   nucleotide-excision repair in, 220
   origins of replication, 211
   positive control of transcription, 261–262
   recombinant DNA vectors and, 123
   replication fork, 209
   ribosomes in, 311, 313
   RNA experiments and, 113
   RNA polymerase and transcription, 254–258
   structure of, 8, *9*
   T4 bacteriophage, 38
   transcription-coupled repair in, 222–223
   translesion DNA synthesis in, 225

tRNAs in, 311
Esophageal carcinomas, *739, 750*
Estradiol, *49, 601*
Estrogen receptor, *602, 603*
Estrogens, 49, 600, 601, *602, 603,* 723
Ethanol, 86
Ethanolamine, 47, *48*
Ethylene, 608, 609
Etioplasts, 457–458
ets oncogene, *735*
Eubacteria, 10–11, 11–12
Euchromatin, 169, 371
Euglena, 16
Eukaryotes
    ABC transporters, 553
    cell wall, 569, 570–575
    characteristics of, 4, 9–10
    chromosome numbers, *167*
    closed mitosis, 670–671
    compared to prokaryotes, 4
    elongation factors, 323, 324
    expression of cloned genes in, 127–128
    homologous recombination enzymes, 231
    initiation of translation, 317–319
    mismatch repair system, 223
    mRNA degradation, 302–303
    mRNA processing, 290–292
    origin and evolution of, 10–12
    origins of DNA replication, 212–214
    protein disulfide isomerase in, 333
    protein folding by heat shock proteins, 331–332
    ribosome structure, 313
    RNA polymerases and general transcription factors, 262–268
    RNase H, 205
    rRNA processing, 288
    species of rRNA in, 287–288
    structure of, *10, 11*
    translation process, 321–325
    translocation of proteins into endoplasmic reticulum, 334
Eukaryotic genomes
    chromatin, 166–171
    composition of, 165–166
    gene duplication, 164–165
    introns and exons, 157–161
    paradox of genome size, 155–157
    pseudogenes, *164,* 165
    repetitive DNA sequences, 161–164
    telomeres, 175–176
Eukaryotic initiation factors (eIFs), 321–323, 327–329
Evans, Martin, 710
Evolution, rRNA and, 314, 317
Excision repair, 219–223
Excinuclease, 220
Exobiology, 8
Exocysts, 423
Exocytosis, *422,* 423
Exon shuffling, 161
Exons, 157–161
Exonucleases, 210
Expansins, 573
Experimental models
    Arabidopsis thaliana, 19
    C. elegans, 17–18
    Drosophila melanogaster, 18
    E. coli, 16–17
    vertebrate, 19–20
    yeasts, 17

Exportins, 363, *364,* 367, *368,* 370
Expression vectors, 127
Extracellular matrix
    adhesion proteins, 579–580
    cell-matrix interactions, 580–582
    integrins and cell adhesion, 582–583
    polysaccharides, 578–579
    structural proteins, 575–578
    types of, 575
Eye, photoreception, 620–621

F actin, 474, *475*
F factor, 123
Facilitated diffusion, 542–543
Facultative heterochromatin, 371–372, 373
FAD, *78*
FADH₂. *See* Flavin adenine dinucleotide
FAK protein kinase, 637, 638
Familial adenomatous polyposis, 224, 750, 754, 755
Familial hypercholesterolemia (FH), 559, 560–561
Farnesyl transferase, 338
Farnesylation, 338, *339*
Fas cell surface receptor, 700
Fats, 47
    oxidation, 89–90
Fatty acids
    attachment to proteins, 338–339
    biosynthesis, 93–94
    overview of, 46–47
    oxidation, 434–435
    oxidation in peroxisomes, 463–464
Fatty acyl-CoA, 89
FBJ murine osteogenic sarcoma virus, *735*
Feedback inhibition, 79–80, 340
Feedback loops, 640, *641*
Feedforward relay, *641*
Ferritin, 325–326
Ferrodoxin, 460, *461*
Fertilization, 105, 684–685
fes oncogene, *735*
fgr oncogene, *735*
FH. *See* Familial hypercholesterolemia
Fibrillarin, *375*
Fibroblasts, 14, *15*
    actin cytoskeleton and Rho proteins, 639
    culturing, 33
    extracellular matrix and, 575
    intermediate filament proteins, *497*
    lamellipodia, 486
    proliferation, 700, 702
Fibronectin, 579–580, *580*
Filamentous [F] actin, 474, *475*
Filamin, *477,* 481, 483
Filopodia, 486, *487,* 496, 639, 640
Fimbrin, *477,* 480–481, 485
Fire, Andrew, 145
Fischer, Ed, 342
FISH. *See* Fluorescence *in situ* hybridization
Fish, *176*
Fission yeast. *See Schizosaccharomyces pombe*
Flagella, 517, 518–520
Flavin adenine dinucleotide (FADH₂)
    ATP generated through oxidation, 449
    citric acid cycle, 87, 88
    electron transport chain, *88,* 89, 444, *445, 446*
    glycolysis, 89
    oxidation of fatty acids, *89,* 90
    oxidative metabolism and, 435
Flavin mononucleotide (FMN), *88*

Flaviviruses, *38*
Flippase, *405,* 406
Flow cytometer, 652
Fluid mosaic model, 59–50, 61, *62,* 532
Fluid phase endocytosis, 561
Fluorescence-activated cell sorter, 652
Fluorescence *in situ* hybridization (FISH), *134,* 185, 187
Fluorescence microscopy, 24–25
Fluorescence recovery after photobleaching (FRAP), 25, *26*
Fluorescence resonance energy transfer (FRET), 25–26
fms oncogene, *735*
Focal adhesions, 484, 496, 580, 581, *582*
Focal complexes, 581
Focus assay, 728–729
Fodrin, 483
Folate, *78*
Folkman, Judah, 757
Follicle-stimulating hormone (FSH), 605
Follicular B-cell lymphoma, *739*
Footprinting, 255–257
Forkhead transcription factor, 624, *626*
Formins, 477, 478, 479, 581
N-Formylmethionyl, 318, 319, *323*
fos oncogene, *735*
Fos protein, 743
fos proto-oncogene, 743
14-3-3 protein, 624, *626,* 699
FOXO transcription factor, 624, *626,* 699, 746
fps oncogene, *735*
Fractionation. *See* Subcellular fractionation
Franklin, Rosalind, 108
FRAP. *See* Fluorescence recovery after photobleaching
Free energy, 81–84
Freeze etching, 30
Freeze fracture, 29, *30*
FRET. *See* Fluorescence resonance energy transfer
Frizzled protein, 635, *636*
Frogs
    cell cycle regulation in oocytes, 657, 658–659
    oocyte size, 681
Frye, Larry, 60, 537
FSH. *See* Follicle-stimulating hormone
FtsZ protein, 505
Fuchs, Elaine, 502–503
Fuginami sarcoma virus, *735*
Fugu rubripes, 190. *See also* Pufferfish
Fungi, cell walls, 570
Fused protein kinase, 634, *635*

G actin, 474, *475*
G₁ cyclins, 661, 662
G₀ phase, 651, 654
G₁ phase, 650, 652–653, 655
G₂ phase, 650, 654, 655
G protein-coupled receptors, 610–612, 641
G proteins, 605, 610–612, 617, 622
GABA (γ-Aminobutyric acid), *604*
gag gene, 733, 735
GAGs. *See* Glycosaminoglycans
α-Galactosidase, 258, *259,* 260
Gallo, Robert, 120
Gap junctions, 544, 589–590
GAPs. *See* GTPase-activating proteins
Gardner-Arnstein feline sarcoma virus, *735*
Gardner-Rasheed feline sarcoma virus, *735*

Gas constant, 83
Gastric carcinoma, *739, 750*
Gastrointestinal stromal tumors, *739, 757,* 760
Gaucher disease, 424, 426
GC box, 274, 275, 276
Gcn5p protein, 282
GDIs. *See* GDP-dissociation inhibitors
GDP, in regulation of proteins, 341
GDP-dissociation inhibitors (GDIs), 422
Gefitinib, *757, 760*
Gel electrophoresis, 119, 131, 135
Gelinas, Richard, 158
Gelsolin, *477*
Gene amplification, 247–248, 741
Gene cloning. *See* Cloned genes
Gene evolution, 247
Gene expression
    chromatin condensation and, 169
    of cloned genes, 126–128
    cyclic-AMP induced, *620*
    ERK pathway and, 631
    gene-protein colinearity, 111–112
    genetic code, 113–115
    large-scale analysis of, 194–195
    role of mRNA in, 112–113
    techniques for inhibiting, 145–147
Gene families, 164–165
Gene-specific transcription factors, 263
Gene therapy, cystic fibrosis and, 554
Gene transfer, 139–142
General transcription factors
    for RNA polymerase I, 266–267
    for RNA polymerase II, 262–266
    for RNA polymerase III, 267–268
Genes
    amplification, 247–248, 741
    chromosomal theory of inheritance and, 104–105, *106*
    cloned (*see* Cloned genes)
    defined, 157
    dominant and recessive, 104
    duplication, 164–165
    evolution, 247
    functional analyses, 137–147
    gene-protein colinearity, 111–112
    gene transfer, 139–142
    imprinted, 286, 287
    introducing mutations into, 142–144
    introns and exons, 157–161
    large-scale systematic analysis of, 193–194
    one gene-one enzyme hypothesis, 107
    segregation and linkage, 105, *106*
    transfer in animal and plants, 138–142
Genetic analyses
    gene transfer, 139–142
    inhibiting cellular gene expression, 145–147
    introducing mutations in cellular genes, 142–144
    mutagenesis of cloned genes, 142, *143*
    yeast systems, 137–139
Genetic code, 113–115, 436–437
Genetic transformation, 107
Genetically modified plants, 142
Genome sequence data
    C. elegans, 180–181
    Drosophila, 182–183
    global studies of gene expression and, 194–195
    human, 185–190

medical implications, 195
plants, 183–185
prokaryotes, 177–178
pufferfish, 190
significance of, 176–177
systematic studies of gene function and, 193–194
Genomes. *See also* Eukaryotic genomes; Genome sequence data; Human genome
    E. coli, 16
    paradox of genome size, 155–157
    T4 bacteriophage, 38
Genomic imprinting, 286, 287
Genomics, 65
Genotypes, 104
Germ line cells, 140, 678
Germ-line mutations, 437
GFP. *See* Green fluorescent protein
Gibberellins, 608, 609
Gibbons, Ian, 512
Gibbs, Josiah Willard, 81
Gibbs free energy, 81–84
Gilbert, Walter, 261
Gleevec, 37, 758–760
gli oncogene, *739*
Gli proteins, 751
Glial cells, *497,* 498
Glial fibrillary acidic protein, *497*
Glioblastoma, *739*
Gliomas, *739*
β-Globin gene, 159, *160*
Globin genes, 164, 165
Globins, fetal and adult, 164–165
Globular [G] actin, 474, *475*
Glucagon, 605
Glucocerebrosidase, 426
Glucocorticoids, 601
Gluconeogenesis, 92–93
Glucose
    amino acid synthesis and, 95
    biosynthesis, 92–93
    in cellulose, 46
    control of *lac* operon, 261–262
    in glycogen and starch, 45
    movement across membranes, 63, 542–543
    oxidative breakdown, 84–89
    structure and function, 44, *45*
Glucose repression, 261–262
Glucose transporter, 542–543
Glucuronic acid, 578
Glutamate, 53, 94, *604*
Glutamic acid, 52–53
Glutamine, 52, *53,* 94
Glyceraldehyde, *45*
Glycerate, 465
Glycerol, *47,* 405
Glycerol phospholipids, 47
Glycine, 52, *53,* 465, *604*
Glycocalyx, 540, 575
Glycogen, 80
Glycogen metabolism, 342–343, 618
Glycogen phosphorylase, 80, *81,* 342, 343, 618, *619*
Glycogen synthase, 618, *619*
Glycolipids
    cell surface proteins and, 337, 339
    glycocalyx and, 540
    in lipid rafts, 539
    localization in Golgi apparatus, 414
    in membranes, 49, 60, *61,* 403, 531, 532

synthesis in the Golgi apparatus, 413
Glycolysis
    ATP generated, 7
    described, 84–86
    evolution of metabolism and, 7–8
    in mitochondria, 434
Glycophorin, 483, 534, *535*
Glycoproteins, 335–337, 534, 540
Glycosaminoglycans (GAGs), 578–579
Glycosidases, 411
Glycosidic bonds, 45, 93
Glycosylation, 335–337, 399–400, 410–413
Glycosylphosphatidylinositol, 339
Glycosylphosphatidylinositol (GPI) anchors, 339, *340,* 400, *401,* 536–537
Glycosyltransferases, 411
Glyoxylate cycle, 464
Glyoxysomes, 464
Glypicans, 579
GM130 protein, 674
Goblet cells, 705, *706*
Golden orb spider, 58
Goldstein, Joseph, 559–560
Golgi apparatus
    in animal cells, *10*
    association with microtubules, 516
    coated vesicles and, 419, *420*
    functions of, 9, 408–409
    glycosylation of proteins, 336, 337, 410–413
    lipid metabolism in, 413–414
    at M phase transition, 674
    movement of proteins through, 410, *411*
    organization of, 409–410, *411*
    in plant cells, *11*
    polysaccharide metabolism in, 414
    protein secretory pathway and, 386, 387, *395*
    protein sorting and export from, 414–416
    resident proteins, 414
    return of resident ER proteins and, 408
    vesicular transport from ER and, 407
cis-Golgi apparatus, 410, *411*
trans-Golgi apparatus, 410, *411,* 414–415, 416, 419, *420,* 425
Golgi complex. *See* Golgi apparatus
Golgi stack, 410, *411*
Gorter, Edwin, 530
GPI anchored proteins, 407, *408,* 536–537, 539
GPI anchors. *See* Glycosylphosphatidylinositol anchors
Gram, Christian, 570
Gram-negative bacteria, 570
Gram-positive bacteria, 570
Gram stain, 570
Grana, 451
Granulocytes, 14, *15,* 704, *705*
GRASP-65 protein, 674
Grb2 protein, 630, *631*
Grb2-Sos complex, 637–638
Green algae, 13
Green fluorescent protein (GFP), 25, 68, 418
Greider, Carol, 214
Grendel, F., 530
Griscella syndrome, 422
GRK protein kinases, 641
Ground substance, 575
Ground tissue, 13
Growth factors, 33
    cancer cells and, 724–725

and Cdk's at G_1 phase regulation, 665–667
cell cycle regulation and, 653–654
as oncogene proteins, 742
overview of, *605*, 606–607
receptor protein-tyrosine kinases and, 614–615
Growth hormone, 605
JAK/STAT pathway and, 634
Growth media, 33
GSK-3β protein kinase, 624, 635, *636*, 699
GTP. *See* Guanosine triphosphate
GTP-binding proteins, 419, 422–423
GTPase-activating proteins (GAPs), 628, 739
Guanine, *50*, 51, 108, *109*, 219
Guanine nucleotide-binding proteins. *See* G proteins
Guanine nucleotide exchange factors, 628
Guanosine triphosphate (GTP)
citric acid cycle and, 87
cotranslational translocation and, 392
eIF2 initiation factor and, 328, *329*
in G protein regulation, 611
glucose synthesis and, 93
in microtubule polymerization, 505–506, *507*
protein synthesis and, 96, *97*
Ran protein and nuclear transport of proteins, 365–366, *367*, *368*
Ras proteins and, 628
in regulation of proteins, 341
in translation process, 319, 321, *322*, 323, 324
Guanylate cyclase, 604
Guanylyl cyclases, 604, 617, 620
Guidance complex, 454, *455*

H19 gene, 286, *287*
H zone, *488*, 489, 490
H2A histone, 360
*Hae*III enzyme, *118*
Haemophilus influenzae, *176*, 177–178, 188
Hair, 706, *707*
Hair follicle, 706, *707*
Hair shaft, 706, *707*
Hanson, Jean, 489
Haploids/Haploidy, 105, 652
Hard keratins, 497
Hardy-Zuckerman feline sarcoma virus, *735*
Harris, Henry, 746
Hartwell, Lee, 657
Harvey sarcoma virus, *735*, 738
Hawking, Stephen, 504
H2B histone, 360
HDL. *See* High-density lipoprotein
Heart muscle cells, 590, 612
Heat shock proteins (Hsp), 331–332
hedgehog genes, 634
Hedgehog protein, 634, *635*
Hedgehog signaling pathway, 634, *635*, 751
HeLa cells, 35
Helicases, 207, *208*, 211
α-Helix, 57
glucose transporter and, 542
ion channels and, 549
transmembrane proteins and, 533, 535
Helix-loop-helix protein, 278–279
Helix-turn-helix motif, 277–278
Hematopoietic stem cells, 704, *705*, 708–709
Hemicelluloses, 414, 570, *572*, 573

Hemidesmosomes, 500, *501*, 580–581, 583
Hemoglobin, 164
Hepadnaviruses, *38*
Heparan sulfate, *578*, 579
Hepatitis B virus, 37, *38*, 729
Hepatitis C virus, 37, 729, 730
Herbicides, 460
Herceptin, *757*, 758
Hereditary nonpolyposis colorectal cancer (HNPCC), 223, 224, 755
Heredity, chromosomal basis, 104–105, *106*
Herpes simplex virus, *38*
Herpes simplex virus gene, 269
Herpesviruses, *38*, *729*, 732
Heterochromatin, 169, 371–372, 373
Heteroduplexes, 228, 229, 230, *232*
Heterophilic interaction, 585
Heterotrimeric G proteins, 611
Hexokinase, 84
Hexon, 158
High-density lipoprotein (HDL), 562
High-energy bonds, 83
*Hind*III enzyme, *118*
Hip dysplasia, 192
Hippocrates, 721
Histamine, *604*
Histidine, 52, *53*, 76
Histone acetylation, 282, *283*
Histone acetyltransferase, 282, *283*, *602*, 603
Histone code, 283
Histone deacetylases, 282, *283*, *603*
Histones
in centromeres, 174–175
in chromatin, 166, *167*, 168, 169, 281
DNA replication and, 209
modifications affecting transcription of chromatin, 282–284
nuclear lamina and, 360
ubiquitination and, 347
Hitchings, George, 98, 99
HIV, 733
AZT and, 99
cytokine receptors and, 617
medical overview, 120
protease inhibitors and, 335
proteases and, 76, 335
protein cleavage and, 334–335
HIV protease, 335
HMG-CoA reductase, 559, 560
HMGN proteins, 282
HNPCC. *See* Hereditary nonpolyposis colorectal cancer
Hoagland, Mahlon, 310
Hodgkin, Alan, 544
Holliday, Robin, 228
Holliday junctions, 228–229, 230, *231–232*
Holliday model, 228–229
Holmes, Kenneth, 474
Homeoboxes, 278
Homeodomain proteins, 277–278
Homeotic mutants, 277–278, *279*
Homogenate, 31
Homologous chromosomes, 679–681
Homologous recombination
enzymes involved in, 229–232
gene inactivation and, 143–144, 193
importance of, 227–228
models of, 228–229
Homophilic interactions, 585
Honjo, Tasuku, 238
Hooke, Robert, 21
Hormone replacement therapy, 723

Hormones. *See also* Steroid hormones; *individual hormones*
endocrine signaling, 600
of plants, 608–609
Horvitz, H. Robert, 691, 694
Horwitz, Alan, 582
Hozumi, Nobumichi, 240–241
*Hpa*I enzyme, *118*
*Hpa*II enzyme, *118*
Hsp. *See* Heat shock proteins
Hsp60 chaperones, 441
Hsp70 chaperones, 331–332, 393, 399, 440, *441*, 454, *455*
Hsp90 chaperones, 441, *602*, 603
Hsp100 chaperones, 454, *455*
Hsp60 chaperonin, 454
HTLV-I, 732
Huber, Robert, 459
Human genome
alternative splicing in, 189
centromeres, 174
chimpanzees and, 192
chromosomes of, *171*, *186*, *187*
complete sequence data, *176*, 185–190
composition of, 166
DNA repair genes, 221–222
DNA transposons, 241
encoded number of cell signaling network proteins, 642
encoding of receptor protein-tyrosine kinases, 613
genome size and composition, 19, *167*, 189
intron-exon structure of genes, 160, 161
LINEs, *245*
miRNAs and gene regulation, 327
pseudogenes, *164*, 165
repetitive sequences, *162*, 163–164, 189
retroposons and, 244, 247
reverse transcription and, 117
rRNA genes, 376
telomeric repeat sequences, *175*
transcription factors in, 263, 277
transposable elements and, 247
Human Genome Initiative, 188
Human immunodeficiency virus, *38*
Human-mouse cell hybrids, 537–538
Human papillomavirus virus, *38*
Human T-cell lymphotropic virus type I (HTLV-I), 732
Humans
Apo-B100 protein, 300–301
kinesins in, 513
mitochondrial genome, 436, 437
oocyte size, 681
tissue systems and cell types, 14–15
Hunt, Tim, 660, 662–663
Hunter, Tony, 612, 613
Hutchinson-Gilford progeria syndrome, 359
Huxley, Andrew, 489, 544
Huxley, Hugh, 489
Hyaluronan, 574, *578*, 579
Hydrogen bonds, 43, *44*
Hydrogen (H+) pumps, 553
Hydrogen ion (H+), in active transport, *65*
Hydrogen peroxide, 463
Hydrophilic molecules, 6, 44
Hydrophobic molecules, 6, 44
Hydroxylysines, 576
Hydroxyproline, 576
Hylobacter pylori, 724

Hynes, Richard, 582
Hypercholesterolemia, 560. *See also* Familial hypercholesterolemia
Hyperosinophilic syndrome, *757*

I bands, *488*, 489
I-cell disease, 424
IAP protein family, 697–698
IκB proteins, 369, 634, 640
ICAMs. *See* Intercellular adhesion molecules
Iduronic acid, 578
IgA antibodies, 236
IgE antibodies, 236
IgG antibodies, 236
IgM antibodies, 236
Immediate-early genes, 631
Immune responses, 234
Immune system, site-specific recombination and, 234–238, 240–241
Immunoblotting, 135–136
Immunodeficiency diseases, 714–715
Immunofluorescence, 136, *137*
Immunoglobulin genes, 740
    class switch recombination, 236–237
    enhancers and, 270–271
    site-specific recombination and, 234–235, 240–241
    somatic hypermutation, 237
Immunoglobulin (Ig) superfamily, 584, 585, 587
Immunoglobulins, 234–235, 390
Immunoprecipitation, 135, 136, 273, *274*
Importins, 363, *364*, 365–366
In situ hybridization, 134, 185
In vitro fertilization (IVF), 685
In vitro mutagenesis, 142
In vitro translation, 114
Indole-3-acetic acid, *608*
Induced fit, 75
Influenza virus, *38*, 115
Inheritance, chromosomal basis, 104–105, *106*
Inherited breast cancer, 227
Inherited colorectal cancers, 753
Inherited diseases, 299
Inherited immunodeficiency, 714–715
Inhibitory crosstalk, *641*
Initiation codons, 317, 318, 319
Initiation factors
    in translation, 319, 321, *322*
    in translational regulation, 326, 327–329
Initiation sites, 317–319
Initiator caspases, 693, 700, *701*
Initiator (Inr) element, 263, *264*
Ink4 protein family, 664, *665*
INK4 tumor suppressor gene, 750, 752
Inner chloroplast membrane, 451, *452*, 454, *455*
Inner mitochondrial membrane, 434, 435
    characteristics, 60
    chemiosmotic coupling, 445–449
    electron transport chain, 444–445, *446*
    protein translocation, 439–442, *443*
    transport of metabolites across, 450–451
Inner nuclear membrane, 398
Inosine, *289*, 311, *312*
Inositol, 47, *48*
Inositol 1,4,5-triphosphate (IP₃), 621–622, 623
Inr element. *See* Initiator element
Insecticides, 444

Insertion sequences, 239, 241
The Institute for Genomic Research, 188
Insulators, 272
Insulin, 605
    proteolytic processing, 334, *335*
    structure, 54, *55*
Insulin receptor, 613–614, *614*
Integral membrane proteins, 61, *62*, 532, 533, 534, *535*. *See also* Transmembrane proteins
Integrase, retroviral, 244
Integrins
    as cell adhesion molecules, 584
    in cell-matrix interactions, 580–582
    extracellular matrix and, 579, 580
    focal adhesions and, 484, 496
    hemidesmosomes and, 500, *501*
    overview of, 582–583
    signal transduction and, 600, 637–638
    structure of, *580*
Intercellular adhesion molecules (ICAMs), 584, 585
Interleukin-2, *605*
Intermediate filament proteins, 497–498
Intermediate filaments
    assembly and disassembly, 498–499
    characteristics of, 497
    component proteins, 497–498
    functions of, 502–505
    intracellular organization of, 499–502
The International Human Genome Sequencing Consortium, 187, 188
α-Internexin, *497*, 498
Interphase, *670*
    chromatin in, 169, *170*, 371, 372, 373
    defined, 650
    microtubule cytoskeleton, 520, *521*
Interspersed repetitive elements, 163–164
Intestinal crypts, 705, *706*
Intestinal epithelial cells
    active transport of glucose, 555–556
    microvilli, 485–486
    plasma membrane polarity, 415–416, 538–539
    self-renewal with stem cells, 704–705
Introns
    C. elegans, 181
    degradation, 301
    human genome, 189
    overview of, 157–161
    pufferfish, 190
    splicing mechanisms, 292–293, 295–298
    tRNAs, 289–290
    yeasts, 179, 180
Ion channels. *See also specific channels*
    cAMP regulation, 620
    described, 543–549, *550*
    functions of, 63–64
    G protein regulation, 611–612
Ion concentrations
    extracellular and intracellular, *545*
    membrane potential and, 545–547
Ion gradients
    active transport across membranes and, 555–556
    ion flow and, 544
    ion pumps and, 550–551
    membrane potential and, 544
    typical values, *553*
Ion pumps, 545, 550–553, 555
Ionic bonds, 533
IP₃. *See* Inositol 1,4,5-triphosphate

IRE. *See* Iron response element
Iron, 302–303, 326
Iron regulatory protein (IRP), 302–303
Iron response element (IRE), 302–303, 326
IRP. *See* Iron regulatory protein
Isoleucine, 52, *53*, 79–80
IVF. *See In vitro* fertilization

Jacob, François, 113, 258
JAK family kinases, 616, 633
JAK/STAT pathway, 633
JAMs. *See* Junctional adhesion molecules
Janus kinase (JAK) family, 616, 633
JIP-1 scaffold protein, 632
JNK kinase pathway, 632
jun oncogene, *735*
Jun protein, 743
Junctional adhesion molecules (JAMs), 589
Junctional complexes, 588

Kabsch, Wolfgang, 474
Kadonaga, James, 276
Kapα, 363, *364*
Kapβ1, 363, *364*
Kaposi's sarcoma, *729*, 732, 733
Kaposi's sarcoma-associated herpesvirus, 732, 733
Karyopherins, 363, *364*, 365–366, 367
Kaufman, Matthew, 710
Kb. *See* Kilobases
KDEL sequence, 408, *409*, 420
Kendrew, John, 56
Keratan sulfate, *578*, 579
Keratins, 497, 500, *501*
α-Ketobutyrate, *80*
α-Ketoglutarate, 94
Kevlar, 58
Kidney cancer/carcinoma, *720*, *750*
Kilobases (Kb), 160
Kinesin I, 512–513, 516
Kinesin II, 516
Kinesin-related proteins, 513
Kinesins, 511, 512–513, 514–516, 523
Kinetochore microtubules, 521, 522, 523, 675
Kinetochores, 171–172, 174, 521, 669, 671, 675, 680
Kirschner, Marc, 506, 521
Kit gene, 760
kit oncogene, *735, 739, 757*
KKXX sequence, 408, 420
Knockout mutation libraries, 144, 193
Knudson, Alfred, 747
Kohne, David, 161
Kornberg, Arthur, 202
Krebs, Ed, 342
Krebs cycle, 86–87, 89
Kristen sarcoma virus, *735*

L cells, 35
L-myc oncogene, *739*
L-selectin, 540
lac operon, *260*, 261–262, 341
Lactate, 86
Lactobacillus casei, 98
Lactose, 341
Lactose metabolism, 258–261
Lactose permease, 259, 260
Lagging strand, 203
Lamellipodia, 486, *487*, 495, 639, 640
Lamin B receptor, 358, 359
Lamin filaments, 673, *674*

Lamina-related diseases, 359–360, 371
Laminins, 580
Laminopathies, 360
Lamins, 357–360, 371
Lander, Eric, 188
Lasek, Ray, 512, 514
Laskey, Ron, 330
LDL. *See* Low-density lipoprotein
LDL receptor, 559–561, 563
Lead citrate, 28
Leading strand, 203
Leber's hereditary optic neuropathy, 437, 438–439
Leeuwenhoek, Anton van, 21
Leptotene, 679, *680*
Lerner, Michael, 294–295
Leucine, 52, *53*
Leucine zipper, 278–279
Leucoplasts, 457
Leukemias, 99, 721
    defective cell differentiation and, 727–728
    incidence and mortality, *720*
    tumor suppressor genes, *750*
Leukocytes, 584–585
Leukotrienes, 607
Levi-Montalcini, Rita, 606
Lewis, Ed, 278
Licensing factors, 656
LIF growth factor, 712
Ligand-gated channels, 544, 547–548, 605
Ligands, 558
Light reactions, 90, *91*, 459–461
Lignin, 573
Lily, *167*
LINEs. *See* Long interspersed elements
O-Linked glycosylation, 413
N-Linked oligosaccharides, 410–411, 412
Linker DNA, 168, 169
Lipid rafts, 532, 539–540, 564
Lipids
    attachment to proteins, 337–339
    biosynthesis, 93–94
    of cell membranes, 60–61, 403
    eicosanoids, 607–608
    export from the ER, 406–407
    lamins and, 358
    oxidation, 89–90
    peroxisomes and, 464
    in phospholipid bilayers, 533–531
    structures and functions, 46–49
    synthesis in smooth ER, 403–406
    synthesis in the Golgi apparatus, 413–414
Liposomes, 139
Liver, 406
    elimination of red blood cells, 558
    Gaucher disease and, 426
    regeneration, 603
Liver cancers/carcinomas, 724
    hepatitis viruses and, 37, 729–730
    oncogenes, *739*
    tumor suppressor genes, *750*
Liver cells, 543, 703
LMNA gene, 359, 360
LMP1 protein, 732
Lock-and-key model, 75
Long interspersed elements (LINEs), *162*, 163, 244–245, 246–247
Long terminal repeats (LTRs), 242, *243*
Lou Gehrig's disease, 504–505
Low-density lipoprotein (LDL), 559–561, 563

LRP protein, 635, *636*
LTR retrotransposons, 244
Lumen (endoplasmic reticulum), 394
Lumen (thylakoid). *See* Thylakoid lumen
Lung cancers/carcinomas, 721
    anticancer drug therapies, *757, 760*
    incidence and mortality, *720*
    oncogenes, *739*
    tumor suppressor genes, *750*
Lungfish, *167*
Lungs, 578
Lymphocytes, 14, *15*, 704, *705*. *See also* B lymphocytes; T lymphocytes
    antibodies and, 134–135
    HIV and, 120
Lymphoid cell, *705*
Lymphomas, 37, 695, *720, 721, 750*
Lysate, 31
Lysine, 52, *53*, 76, 464
Lysosomal acid hydrolases, 424
Lysosomal hydrolases, 424
Lysosomal proteolysis, 347–348
Lysosomal storage diseases, 424, 426
Lysosomes, 9
    acid hydrolases and, 424
    in animal cells, *10*
    endosomes mature into, 564
    H^+ pumps, 553
    phagocytosis and, 557
    positioning in the cell, 516
    protein transport to, 416
    proteolysis and, 347–348

M line, *488*, 489, 490, *491*
M phase, 650
    Cdk1/cyclin B complex and, 672–675
    cytokinesis, 677–678
    progression to anaphase, 676–677
    spindle assembly checkpoint, 675–676
    stages of mitosis, 669–671
MacLeod, Colin, 108
Macrophages, 14, 426, *487*, 558, 704, *705*
Macropinocytosis, 562
Mad/Bub proteins, 676, 684
maf oncogene, *735*
Magnesium (Mg^{2+}), 462
Magnification, 22
Malignant tumors, 720–721
Maller, James, 660, 663
Mammalian cells
    actin genes in, 474
    genes encoding G proteins, 611
    LINEs and SINEs, 244–247
    MAP kinases in, 631–632
    nuclear subcompartments, 373
    number of ribosomes in, 311
    origin recognition complex, 213–214
    origins of DNA replication, 212
    phagocytosis and, 558
    plasma membrane characteristics, 60
    RNA editing, 300–301
    transition-coupled repair in, 223
Mannose-6-phosphate, 412, *413*, 416, *421*, 424
Mannose-6-phosphate receptors, 425
MAP-1C protein, 513
MAP kinase pathways, 627–628, 630–633, 641
MAP kinases, 627
MAPs. *See* Microtubule-associated proteins
Markert, Clement, 657, 658, 683
Marshall, Barry, 724

Martin, Gail, 710
Martin, Steven, 734
Mass spectrometry, 67–68
Masui, Yoshio, 657, 658, 683
Matrix (mitochondria), 434, 439–441
Matrix polysaccharides, 578–579
Matrix processing peptidase (MPP), 440
Matrix proteins
    adhesion, 579–580
    structural, 575–578
Matthaei, Heinrich, 114
Maturation promoting factor (MPF), 657, 658–659, 660–661, 663. *See also* Cdk1/cyclin B complex
Mb. *See* Megabases
*Mbo*I enzyme, *118*
McCarty, Maclyn, 108
McClintock, Barbara, 233
McCulloch, Ernest, 704
McDonough feline sarcoma, *735*
McKnight, Steven, 276
MCM helicase proteins, 213, *214*, 656
mdr gene, 553
MDR transporters, 553
Measles virus, *38*
Mediator, 279
Mediator complex, 266
Megabases (Mb), 177
Megakaryocytes, 704, *705*
Meiosis, 105
    process of, 679–681
    regulation in oocytes, 681–684
Meiosis I, 678, 679–681
    in oocytes, 682, 683
Meiosis II, 678, *679*, 681
MEK protein kinase, 627, *628*
Melanomas, *720, 739, 750*
Melanosomes, 422
Mello, Craig, 145
Membrane-anchored growth factor, 606
Membrane potential
    dynamics of, 545–547
    ion channels and, 544, 545
    ion pumps and, 545
    Na^+-K^+ pump and, 551–552
Membrane proteins
    fluid mosaic model and, 532
    insertion into ER membrane, 393–398
    mobility of, 537–540
    receptor down-regulation, 564
    recycling, 563–564
    types of, 533–537
Membranes. *See* Cell membranes; Phospholipid membranes; Plasma membrane; *other specific membranes*
Meningioma, *750*
Mental retardation, 96
6-Mercaptopurine, 98, 99
Meselson, Matthew, 110, 113
Messenger RNA (mRNA)
    alternative splicing, 299–300
    functions of, 50
    in gene expression and protein synthesis, 112–113
    half-lives and degradation, 302–303
    initiation sites, 317–319
    internal binding sites, 319
    localization, 327
    miRNAs and, 285
    nuclear transport, 369–370
    poly- and monocistronic, 317, *318*
    processing in eukaryotes, 290–292

protein synthesis and, 96
RNA polymerase II and, 262
techniques for disrupting, 145–146
viral, 319
Metabolic energy
  evolution of, 7–8
  free energy and ATP, 81–84
  glucose oxidation, 84–89
  nucleotides and amino acids as energy
    sources, 89
  oxidation of lipids, 89–90
  photosynthesis, 90–91
  polysaccharides as energy sources, 89
  thermodynamics and, 81
Metal shadowing, 29
Metaphase
  Cdk1/cyclin B complex and, 672–675
  chromatin in, 170–171
  chromosome centromeres, 171–172
  human chromosomes, *171*
  meiosis I, 681–682
  meiosis II, 681, 682–684
  microtubule cytoskeleton during,
    521–522
  mitosis, 669, *670*, 671
  spindle assembly checkpoint, 675–677
Metaphase II arrest, 682–684
Metaphase plate, 521, 522
Metastasis, 720, 726
Methanococcus janaschii, 178
Methionine, 52, *53*
Methylation, damage to DNA, 219
O^6-Methylguanine methyltransferase, 219
Methylguanosine, *289*
7-Methylguanosine cap, *290*, 291
Mice
  Abelson leukemia virus, 735–736
  embryonic stem cell studies, 710–711, 712
  as experimental model, 20
  genome size and composition, *16, 167*
  β-globin gene intron-exon structure, 159,
    160
  human-mouse cell hybrids, 537–538
  immunoglobulin light chains in, 234
  knockout mutation library, 144
  transgenic, 140, 141, 143–144, 502–504
  whole genome sequencing and, 191–192
Michel, Hartmut, 459
Microfilaments, 473. *See also* Actin filaments
MicroRNAs (miRNAs)
  gene regulation and, 285
  regulation of transcription, 327, *328*
  in RNA degradation, 302
  RNA polymerases transcribed by, 262
Microscopy, 21–27. *See also* Electron
  microscopy
Microsomes, *388*, 389
Microspikes, 486
Microtubule-associated proteins (MAPs),
  510, 511
Microtubule motors
  in cargo transport, 514–516
  in chromosome movement, 522–523
  in cilia and flagella, 518, 520
  in intracellular organization, 516–517
  motor proteins, 511–514
Microtubules
  assembly of, 507–510
  in cilia and flagella, 517–520
  functions within cells, 514–517
  intermediate filaments and, 500
  at M phase transition, 675

mitosis and, 506–508, 520–523
mitotic spindle and, 171–172
organization within cells, 510–511
plant cell wall growth and, 574
structure and dynamic organization of,
  505–507
Microvilli, 485–486, 539
Middle lamella, 592, *593*
Milk, antibodies and, 565
Miller, Stanley, 5
Milstein, Cesar, 390
Mineralocorticoids, 601–602
Mismatch repair, 223
Mitchell, Peter, 445, 448–449
Mitchison, Tim, 506, 521
Mitochondria
  in animal cells, *10*
  apoptosis and, 696–697
  chemiosmotic coupling, 445–449, *453*
  compared to chloroplasts, 451, *453*
  DNA polymerase γ, 202
  electron transport chain, 88–89, 444–445,
    446
  function of, 9
  genetic system of, 435–437
  H$^+$ pumps, 553
  inner membrane (*see* Inner mitochondrial
    membrane)
  key features of, 434
  kinesin-mediated movement, 516
  Leber's hereditary optic neuropathy and,
    437, 438–439
  organization and function of, 434–435
  origin through endosymbiosis, 10–11
  outer membrane characteristics (*see*
    Outer mitochondrial membrane)
  phospholipid import, 442–443
  in plant cells, *11*
  protein import, 330, *331*, 437, 439–442,
    443
  proteins in, 437
  proteome analysis, 68
  RNA editing, 300
  transport of metabolites across inner
    membrane, 450–451
Mitochondrial genetic code, 437
Mitosis. *See also* M phase
  Cdk1/cyclin B complex and, 672–675
  chromatin and, 169–170
  chromosomes in, 171, *172*
  defined, 650
  microtubules and, 506–508, 520–523
  stages of, 669–671
Mitotic spindles, 171–172, *172*, 520, 521,
  669, 671, 674–675
Mizutani, Satoshi, 117
MKK7 protein kinase, 632
MLK protein kinase, 632
Model systems. *See* Experimental models
Molecular cloning, 121–122, 131
Moloney murine leukemia virus (MuLV),
  735
Moloney sarcoma virus, *735*
Monocistronic mRNA, 317, *318*
Monoclonal antibodies, 135, 758
Monocytes, 14, *15, 705*
Monod, Jacques, 258
Monosaccharides, 44–45
Montagnier, Luc, 120
Moore, Claire, 158
Moore, Peter, 315, 316
Morphine, 606

mos oncogene, *735*
Mos protein, 683
Motor neurons, diseases of, 504–505
MPF. *See* Maturation promoting factor
mpl oncogene, *735*
MPP. *See* Matrix processing peptidase
MreB protein, 474
mTOR pathway, 625–627, 631
mTOR protein kinase, 624, 625–627
mTOR/raptor complex, 625–627
mTOR/rictor complex, 624, *625*
Mullis, Kary, 129
Multi-photon excitation microscopy, 27
Multicellularity, development of, 12–15
Multiple myeloma, 347
3611 Murine sarcoma virus, *735*
Mus81 endonuclease, 232
Muscle cells, 15
  actin in, 474
  Ca^{2+} activity and, 623–624
  contraction, 487–491, *492*
  as experimental model, 19
  intermediate fibers in, 500
  intermediate filament proteins, *497*
  satellite stem cells, 707, *708*
Muscle tissue, 14
Muscles
  satellite stem cells and, 707, *708*
  types of, 487
Muscular dystrophy, 359, 483
Mutagenesis, of cloned DNA, 142, *143*
Mutations, 105
  gene-protein colinearity and, 111–112
  introducing into cellular genes, 142–144
MutH protein, 223
MutL gene, 223, 224
MutL protein, 223
MutS gene, 223, 224
MutS protein, 223
myb oncogene, *735*
myc oncogenes, *735*, 743
Myc proteins, 743
Mycoplasma, 16
  M. genitalium, 176, 178
Mycoplasmas, 178
Myelin disease, 712
Myelinated nerves, 591
Myeloid cell, *705*
Myelomas, 390
Myeloproliferative leukemia virus, *735*
Myofibrils, 487
Myoglobin, 56
Myosin filaments, 487–493
Myosin I, 486, 494
Myosin II, 490, 491–493, 497, 677
Myosin light chain, 639–640
Myosin light-chain kinase, 493, 623
Myosin V, 494, 496
Myosin VI, 494
Myosins, 491–495, 513
Myristic acid, 338, 536
N-Myristoylation, 337, 338

N-cadherin, 585–586
N-CAMs. *See* Nerve cell adhesion mole-
  cules
N-*myc* gene, *739, 741*, 756
Na$^+$-Ca^{2+} antiporter, 556
Na$^+$-H$^+$ antiporter, 556
Na$^+$-K$^+$ ATPase, 550–551, *552*
Na$^+$-K$^+$ pump, 550–551, *552*
NAD$^+$, 78, *79*, 86, *87*

NADH, 78, *79*
 ATP generated through oxidation, 449
 electron transport chain and, *88, 89,* 444, *445*
 glucose synthesis and, *92, 93*
 nitrogen fixation and, 95
 oxidative breakdown of glucose and, *85, 86, 87, 87, 88,* 89
 oxidative breakdown of lipids and, 89, *90*
 oxidative metabolism and, 435
NADH dehydrogenase, 438
NADP⁺, *78,* 90
NADP reductase, 460–461
NADPH
 amino acid synthesis and, 95
 in anabolic pathways, 91
 lipid synthesis and, 94
 nitrogen fixation and, 95
 photosynthesis and, 90, 91, 459, 460–461
Nanocrystals, 25
Narcolepsy, 192
Nasopharyngeal carcinoma, *729, 732*
National Institutes of Health, 188
National Research Council, 188
Nebulin, *477,* 489
Nectins, 587, 589
Negative feedback loops, 640, *641*
Negative staining, 28–29
Neher, Erwin, 544
Nematodes. *See Caenorhabditis elegans*
Nernst equation, 546
Nerve cell adhesion molecules (N-CAMs), 585
Nerve cells. *See also* Glial cells; Neurons
 cell-cell adhesion and, 585
Nerve growth factor (NGF), *605,* 606, 614–615, 640, 698
Nerve impulses. *See* Action potentials
Nervous system
 programmed cell death and, 690, 698
 tissues, 14
Nestins, *497,* 498
Neural cadherin, 585–586
Neuroblastomas, *739,* 741, 756
Neurofibrosarcoma, *750*
Neurofilament (NF) proteins, *497,* 498
Neurofilaments, 500, 502
Neurohormones, 606
Neurons, 14, 418
 action potentials, *546,* 547
 Ca²⁺ activity and, 623–624
 connexin mutations and, 591
 as experimental model, 19–20
 intermediate filament proteins, *497*
 ion channels and membrane potential, 544–545
 microtubules in axons and dendrites, 510–511
 neurofilaments, 500, 502
 neuropeptide signaling, 605–606
 programmed cell death and, 690, 698
 signaling at synapses, 547–548
 synaptic recycling and neurotransmitter release, 564–565
Neuropeptides, 605–606
Neurospora crassa, 107
Neurotoxins, 548
Neurotransmitters, 418, 549
 actions of, 547–548
 Ca²⁺ and, 623–624
 carbon monoxide synthesis and, 604

nitric oxide synthesis and, 604
 structure and actions of, 604–605
 synaptic vesicles and, 564, 565
Neurotrophins, 606
Neutral keratins, 497
Neutrophils, 558, *705*
Nexin, 518, *519,* 520
NF-κB pathway, 634, 640
NF-κB transcription factor family, 368–369, 634, 699
NF proteins, *497,* 498
NF1 tumor suppressor gene, *750*
NF2 tumor suppressor gene, *750*
NGF. *See* Nerve growth factor
Niacin, *78*
Nickel carbonyl, *724*
Nickel compounds, 723, *724*
Nicolson, Garth, 59–60, 61, 532
Nicotinamide adenine dinucleotide (NAD⁺), 78, *79,* 86, *87*
Nicotine, 548
Nicotinic acetylcholine receptor, 547–548, 612
Niedergerke, Ralph, 489
Nirenberg, Marshall, 114
Nitrate, 95
Nitric oxide (NO), 603–604, 617
Nitric oxide synthase, 603
Nitrogen, incorporation in organic compounds, 94, *95*
Nitrogen fixation, 94
Nitroglycerin, 604
Nitrosylation, 343
Noller, Harry, 315
Nomura, Masayasu, 314
Non-LTR retrotransposons, 244–247
Noncoding RNAs, 285–286, 327
Nonerythroid spectrin, 483
Nonreceptor protein-tyrosine kinases, 616, 742
Nonsteroidal anti-inflammatory drugs (NSAIDs), 607, 608, 755
Norepinephrine, *604*
Northern blotting, 131
Notch pathway, 636–637
*Not*I enzyme, *118*
Noxa protein, 698, 752, *753*
NSAIDs. *See* Nonsteroidal anti-inflammatory drugs
NTF2 protein, 366, 369
NTPs. *See* Ribonucleoside 5′-triphosphates
Nuclear envelope
 breakdown at M phase transition, 673–674
 key features of, 355–356
 during mitosis, 670
 structure of, 356–358, 360
Nuclear export signals, 367, *368*
Nuclear lamina, 371
 depolymerization at M phase transition, 673, *674*
 genetic diseases and, 359–360, 371
 intermediate filament proteins, *497*
 structure and function, *356,* 357–358, 360
Nuclear lamins, 371, *497,* 498, 499
Nuclear localization signals, 362–363, 364–365
Nuclear membranes
 continuity with ER membrane, 398
 structure and function, 356–357
Nuclear pore complexes, *356,* 357, 361, 362–362

Nuclear receptor superfamily, 602–603
Nuclear speckles, 373
Nuclear transport receptors, 363, *364,* 365–366, 367
Nucleation, in actin assembly, 474, 476–477
Nucleic acid bases, 108, 109. *See also specific bases*
Nucleic acid hybridization, 131–134
Nucleic acids
 biosynthesis, 97–98, *99*
 methods of detecting, 129–134
 overview of, 50–51
Nucleolar organizing regions, 377
Nucleolus
 in animal cells, *10*
 functions of, 375–376
 in plant cells, *11*
 ribosome assembly, 379, *380*
 structure, 376–377
 transcription and processing of rRNA, 377–379
Nucleoplasmin, 330, 363
Nucleoporins, 361
Nucleosome core particles, 168, *169*
Nucleosome remodeling factors, 284
Nucleosomes
 in chromatin structure, 167, *168, 169,* 281
 during DNA replication, 209
 HMGN proteins and, 282
 modification during chromatin transcription, 284
 nucleoplasmin and, 330
Nucleotide-excision repair, 220–222
Nucleotides
 biosynthesis, 97–98, *99*
 in DNA and RNA, 50–51
 as energy source, 89
 genetic code and, 113–115
Nucleus
 actin and, 480
 in animal cells, *10*
 in eukaryotic cells, 9
 intermediate filaments and, 500
 organization of chromosomes and chromatin in, 371–373, *371*–375
 in plant cells, *11*
 protein transport to and from, 362–367
 regulation of protein import, 368–369
 sub-compartments, 374–375
 transport of RNAs, 369–370
Numerical aperture, 22, 27
Nurse, Paul, 659, 664

Occludin, 589
Octyl glucoside, *533*
Odorant receptors, 620
Oil-immersion, 23
Okazaki, Reiji, 203
Okazaki fragments, 203, 204, 205, *206*
Oligonucleotides, 51
Oligosaccharides, 45, 46
 glycocalyx and binding of selectins, 540, *541*
 glycosylation and, 336–337, 399, *400*
Oligosaccharyl transferase, 399
Omi/Htr2 protein, 697
Oncogene addiction, 760
Oncogene proteins, 741–746
Oncogenes, 37, 629
 anticancer drugs and, 757–758
 benefits of molecular analysis, 756
 functions of oncogene proteins, 741–746

in human cancer, 738–741
proto-oncogenes, 735–738
retroviral, 734–735
in tumor development, 753–754
viral and cellular, 733–734
One gene-one enzyme hypothesis, 107
Oocytes
cell cycle regulation in frogs, 657, 658–659
DNA amplification and, 247–248
G$_2$ phase arrest, 654
growth in size, 681
maturation promoting factor, 657
regulation of meiosis in, 681–684
rRNA genes and, 376
translational regulation in, 327
Open-reading frames, 177
Operators, 260, 261
Operons, 260, 261
Oral contraceptives, 723
ORC. *See* Origin recognition complex
Organelles. *See also individual kinds*
in eukaryotic cells, 9
origin through endosymbiosis, 10–11
transport along microtubules, 515–516
Organic molecules, 44. *See also*
Carbohydrates; Lipids; Nucleic acids;
Proteins
Origin of life
emergence of cells, 4–7
evolution of metabolism, 7–8
RNAs and, 317
Origin recognition complex (ORC), 213–214, 656
Origins of replication, 123, 211–214
Orthomyxoviruses, *38*
Osmium tetroxide, 28
Osmotic pressure, ion gradients and, 553
Osteoblasts, 14
Outer chloroplast membrane, 451, *452, 454, 455*
Outer mitochondrial membrane, 434, 435
characteristics of, *61*
protein translocation, 439–442, *443*
Ovarian carcinomas, *739, 741, 750, 756*
Ovaries, 406
Oxa1 translocase, 442
Oxidation-reduction reactions, 78, *79*
Oxidative decarboxylation, 86
Oxidative metabolism. *See also* Metabolic
energy
ATP generated, *7, 8*
evolution of life and, 8
glucose oxidation, 84–89
lipid oxidation, 89–90
in mitochondria, 434–435
Oxidative phosphorylation, 88
ATP generated by, 443
in cell metabolism, 443
chemiosmotic coupling, 445–449
electron transport chain, 444–445, *446*
Leber's hereditary optic neuropathy and, 438
in mitochondria, 435
Oxidative reactions, peroxisomes and, 463–464
Oxygen
electron transport chain and, 444
photosynthesis and, 90, 91
Oxygen (atmospheric), photosynthesis and, 8
Oxytocin, 487, *605*

P1 artificial chromosome (PAC), 123, *125*
P-cadherin, 586
p27 Cdk inhibitor, 666–667
p38 MAP kinase pathway, 632
p15 protein, 754
p16 protein, 752
p21 protein, 669, 698, 753
p53 protein, 669, 698, 699, 731, 732, 752–753
p120 protein, 586
P-selectin, 540
p53 tumor suppressor gene, 754
PABP. *See* Poly-A binding protein
PAC. *See* P1 artificial chromosome
Pachytene, 679, *680*
Pacific yew, 507
Palade, George, 27, 386
Palmitate, *90, 94*
Palmitic acid, 339, 536
Palmitoylation, 337, 339
Pancreas
acinar cells, 386, *387*, 415
cystic fibrosis and, 554
Pancreatic cancer/carcinoma, *720, 739, 750*
Paneth cells, 705
Pantothenate, *78*
Pap smear, 731
Papanicolaou, George, 731
Papillomaviruses, 37, *729*, 731
Papovaviruses, *38*
Paracrine signaling, 600
Paramecium, 517, *581*
Paramyxoviruses, *38*
Parathyroid adenoma, *739*
Parenchyma, 13, *14*
Parkinson's disease, 331, 712
Passive diffusion, 541–542
Passive transport, 64, 541–542
Patch clamp technique, 544
Patched protein, 634, *635*, 751
Pauling, Linus, 57, 108
PCNA. *See* Proliferating cell nuclear antigen
PCR. *See* Polymerase chain reaction
PDGF. *See* Platelet-derived growth factor
PDGF receptor, *614*, 742, 760
PDGFR oncogene, *739, 757*
PDI. *See* Protein disulfide isomerase
PDK1 protein kinase, 624
Pectins, 414, 570–571, *572, 573*
Pelger-Huët anomaly, 359
Penicillin, 320, 570
Peptide bonds
catalyzed by rRNA, 315–317
in polypeptide formation, 53, 95–96, *97*
protein folding and, 333
Peptide hormones, 605–606
Peptides, 605
Peptidoglycan, 570, *571*
Peptidyl prolyl isomerase, 333
Peptidyl transferase reaction, 315–317
Pericentriolar material, 509
Peripheral membrane proteins, 61, *62, 532,* 533, 534, 540
Peripheral neurons, *497*
Peripherin, *497*
Permeability, of phospholipid bilayers, 541–542
Permissive cells, 730
Peroxins, 463, 467–468

Peroxisome targeting signal 1 (PTS1), 466, 467
Peroxisome targeting signal 2 (PTS2), 467
Peroxisomes, 9
in animal cells, *10*
assembly, 465–467
characteristics of, 462–463
functions of, 463–465
human diseases and, 467
in plant cells, *11*
Pex3 protein, 467–468
Pex19 protein, 467–468
pH
of endosomes, 563, 564
Na$^+$ -H$^+$ antiporter and, 556
in photosynthesis, 462
Phagocytic vacuoles, 428
Phagocytosis, 428, 486, *487*, 557–558, 690, *692*
Phagolysosomes, *347*, 428, 557
Phagosomes, 428, 557
Phalloidin, 476
Phase-contrast microscopy, 23
Phenotypes, 104
Phenylalanine, 52, *53*, 76, 107
Phenylalanine hydroxylase, 96
Phenylketonuria (PKU), 96, 107
Phenylpyruvate, 96
Pheophytins, *459*, 460
Philadelphia chromosome translocation, 758
Phloem, 14
Pho4 transcription factor, 369
Phosphate (P*i*)
ATP hydrolysis and, 83
in phospholipids, *48*
Phosphatidic acid, *48, 404*, 405
Phosphatidylcholine, *48*, 60, *61, 404*, 405, 442, 530
Phosphatidylethanolamine, *48*, 60, *61, 404, 405*, 442, 443, 530
Phosphatidylinositol, *48*, 339, *404*, 405, 531
Phosphatidylinositide 3-kinase (PI 3-kinase), *624, 625, 626, 638*, 673, 699, 745
Phosphatidylinositol 4,5-bisphosphate (PIP$_{2\gamma}$ *621, 622, 623, 624*, 684, 699)
Phosphatidylinositol 3,4,5-triphosphate (PIP$_3$), *624*, 699, 751
Phosphatidylserine, *48*, 60, *61, 404, 405*, 443, 530, 531, 536, 690, *692*
Phosphodiester bonds, 51
Phosphofructokinase, 84
Phosphoglycolate, 465
Phospholipase C, *621, 622*
Phospholipase C-γ, *638*, 673
Phospholipases, 622
Phospholipid bilayers, 6. *See also*
Phospholipid membranes
in cell membranes, 58, 60–61
permeability, 541–542
structure and characteristics of, 530–532
Phospholipid membranes
components, 47–49
origin of cells and, 6
Phospholipid transfer proteins, 442
Phospholipids
characteristics of, 6
export from the ER, 406–407
import to mitochondria, 442–443
in intracellular signal transduction, 621–622

structures and functions, 47–49
synthesis, 403–406
Phosphorylase kinase, 343, 618, *619*
Phosphorylation, 80, *81*
actin-myosin contraction in nonmuscle cells, 493
modification of intermediate filaments, 499
of proteins, 341–343
of transcription factors, 369
Phosphorylcholine, 413
Phosphotyrosine, 613
Phosphotyrosine-containing peptides, 615
Photocenters, 459
Photoreactivation, 218, *219*
Photoreception, 620–621
Photorespiration, 464–465
Photosynthesis
described, 458–462
energy generated by, 90–91
evolution of life and, 8
light reactions, 90, *91*
photorespiration and, 464–465
Photosynthetic cyanobacteria, 452
Photosynthetic pigments, 90, *91*, 459
Photosynthetic reaction centers, 459–460, 535
Photosystem I, 459–461, *462*
Photosystem II, 459–461
PI 3-kinase. *See* Phosphatidylinositide 3-kinase
PI 3-kinase/Akt signaling pathway, 624, *625*, 745–746
Picornaviruses, *38*
Piebaldism, 21
Pigments, photosynthetic, 90, *91*
PI3K oncogene, *739*
Pinocytosis, 557. *See also* Receptor-mediated endocytosis
PIP₂. *See* Phosphatidylinositol 4,5-bisphosphate
PIP₃. *See* Phosphatidylinositol 3,4,5-triphosphate
PIP₂ pathway, 621–622
Pituitary gland, 605
p3k oncogene, *735*
PKU. *See* Phenylketonuria
Placental cadherin, 586
Plakin protein family, 500, 581
Plakoglobin, *501*, 587, 588
Plakophilin, *501*, 587, 588
Plant cell culture, 36
Plant cells
cell wall, 569, 570–575
cytokinesis in, 677–678
gene transfer, 141
microtubules and, 508, 509
plasmodesmata, 592, *593*
polysaccharide synthesis in, 414
structure of, *11*
vacuole in, 416, *417*
Plant hormones, 608–609
Plant viruses, 115
Plants
plasma membrane fluidity and, *61*
tissue systems and cell types, 13–14
whole genome sequence data, 183–185
Plasma membrane. *See also* Cell membrane
active transport by ATP hydrolysis, 550–553, 555
active transport by ion gradients, 555–556

in animal cells, *10*
association of actin filaments with, 482–485
characteristics of, 529
dual membrane system in bacteria, 535–536
electron transport chain in bacteria and, 89
enodcytosis, 556–565
facilitated diffusion across, 542–543
fluid mosaic model, 532
ion channels, 543–549, *550*
membrane proteins, 532–537
mobility of membrane proteins, 537–540
passive diffusion of small molecules, 541–542
phospholipid bilayer, 530–532
in plant cells, *11*
polarized, 538–539
polarized, protein transport to, 415–416
in prokaryotes, 8, *9*
"railroad track" appearance, 530
structure of, 529–540
Plasmalogens, 464
Plasmid vectors, 123, *124*
Plasmids, 121
Plasmodesmata, 592, *593*, 678
Plasmodium falciparum, 436
Plastids, 456–458. *See also* Chloroplasts
Plastocyanin, 460, 461, *462*
Plastoquinone, 460, *461*
Platelet-derived growth factor (PDGF), *605*, 606, 614–615, 702
Platelets, 704, *705*
Pleckstrin homology domain, 624
Plectin, 500, *502*, 580, 581
Ploidy, 652
PML bodies, 374
PML gene, 757
PML/RARα oncogene, *739*, 757
PML/RARα protein, 744, 745, 757, 758
Pneumococcus, 107, *108*
Point mutations, proto-oncogenes and, 736, *738*
pol gene, 733, 735
Pol V. *See* DNA polymerase V
Polar bodies, 682, 685
Polar microtubules, 521, *522*, 523, 675
Polarized cells, 563, 565
Polarized membranes, 415–416, 538–539
Poliovirus, *38*, 115
Polo-like kinase family, 675
Poly-A binding protein (PABP), 327
Poly-A tails, 291, 327
Polyadenylation, *290*, 291, 327
Polycistronic mRNA, 317, *318*
Polymerase chain reaction (PCR), 129–131
Polymerization, of actin, 474–476
Polynucleotides, 51, 98, *99*
Polyomaviruses, 612, 729, 730–731
Polypeptide growth factors, 606–607, 617
Polypeptides/Polypeptide chains. *See also* Translation
attachment of lipids, 337–339
chaperone proteins and, 330–332
cleavage, 333–335
folding, 54–55, 330–333
glycosylation, 335–337
interactions between, 344
peptide bonds, 53, 95–96, *97*
structure, 53
Polyps, 224, 722, 750

Polyribosomes, 325
Polysaccharides, 45, 46
biosynthesis, 93, *94*
of cell walls, 570–571, *572*, 573 (*see also* Cellulose)
as energy sources, 89
of the extracellular matrix, 578–579
structure of, *46*
synthesis in the Golgi apparatus, 414
Polysomes, 325
Polytene chromosomes, 182, *183*, 281, *282*, 372
Porins, 435, 451, 535–536, 543–544
Porphyrin ring, 91
Porter, Keith, 27
Positive feedback, *641*
Positive staining, 28
Posttranslational translocation, 387, 392–393
Potassium (K⁺)
extracellular and intracellular concentrations, 545
gradient across plasma membrane, 553
membrane potential and, 545
Na⁺-K⁺ pump, 550–551, *552*
Potassium (K⁺) channels, *546*, 547, 548–549, *550*, 612
Poxviruses, *38*
Pre-mRNAs
alternative splicing, 299–300
processing, 290–292
splicing and splicing mechanisms, 292–298
Pre-rRNAs, 288
processing, 378–379
splicing, 295
Pre-tRNAs, 288–290
Prenyl groups, 536
Prenylation, 337, 338, *339*, 358
Preproinsulin, 334, *335*
Presequences, 439, 440
Primary cell walls, 572–573
Primary cultures, 33
Primary protein structure, 56
Primase, 204
Processed pseudogenes, 246
Procollagens, 576–577
Profilin, 477, 479, 496
Progesterone, 601, 658, 723
Programmed cell death, 690, 728, 745. *See also* Apoptosis
Proinsulin, 334, *335*
Prokaryotes. *See also* Bacteria; *Escherichia coli*
ancestor of actin, 474
archaebacteria and eubacteria, 8
bacterial electron transport chain, 89
characteristics of, *4*
compared to eukaryotic cells, 4
complete genome sequence data, *176*, 177–178
DNA polymerase I, 205
elongation factors, 323, 324
initiation of translation, 317–318
introns and, 160
mRNA half-life, 302
protein folding by heat shock proteins, 331–332
protein secretion and, 392
rRNA processing, 288
species of rRNA in, 288
transcription in, 254–262

typical structure of, 8, *9*
Prolactin, 605
Proliferating cell nuclear antigen (PCNA), 207
Proliferation. *See* Cell proliferation
Proline, 52, *53,* 333
Prolyl hydroxylase, 576
Prolyl-peptide bonds, 333
Prometaphase, *670, 671,* 675
Promoters
    for genes transcribed by RNA polymerase II, 263–265, 269
    for genes transcribed by RNA polymerase III, 267–268
    in prokaryotes, 254–257
    for ribosomal RNA genes, 266–267
Pronuclei, 685
Proofreading, by DNA polymerase, 210–211
Prophase
    meiosis, 679–680
    microtubule cytoskeleton during, *521*
    mitosis, 669, *670*
Proplastids, 457, *458*
Prostacyclin, 607
Prostaglandins, 607, *608*
Prostate cancer/carcinoma, *720, 750*
Prosthetic groups, 78
Protease inhibitors, 335
Proteases, 67
    cancer cells and, 575, 726–727
    in cell death, 617
    HIV and, 76
    in lysosomes proteolysis, 347
Proteasome, 345
Proteasome inhibitors, 347
14-3-3 protein, 624, *626*
Protein 4.1, 483
Protein-coding sequences
    Arabidopsis thaliana, 183, 184
    C. elegans, 181
    Drosophila, 183
    human genome, 189
    open-reading frames and, 177
    prokaryotes, 177–178
    pufferfish, 190
    representative organisms, *176*
    yeasts, 179, 180
Protein disulfide isomerase (PDI), 333, 399, 401, 408
Protein kinase A, 618, 619, 620, 623
Protein kinase C family, 622, 623
Protein kinases, 342–343, 612
Protein phosphatase 1, 620
Protein phosphatases, 342–343, 619–620
Protein-protein interactions, 344
    detecting with immunoprecipitation, 136
    identifying, 69
Protein secretion
    conservation in, 392
    export from ER, 406–408
    export from Golgi apparatus, 414–416
    protein folding and processing in the ER, 398–400
    secretory pathway, 386–387, *395*
    translocation of proteins into ER, 387–393
    transmembrane ER proteins and, 398
    vesicular transport, 417–423
Protein-serine/threonine kinases, 342–343, 617
Protein-tyrosine kinases, 342–343, 612–613, 616, 622, 742

Protein-tyrosine phosphatases, 616–617
Protein-tyrosine phosphorylation, 612–613
Proteins. *See also* Protein secretion
    addition of carbohydrates, 413
    amino acid components, 52–53
    antibody probes, 134–136
    attachment of lipids, 337–339
    biosynthesis, 94–96, *97,* 112–113 (*see also* Transcription; Translation)
    of cell membranes, 61–63
    chaperone proteins and, 330–332
    cleavage, 333–335
    degradation, 344–348, 401
    direct inhibition of function, 147
    export from the Golgi apparatus, 414–416
    folding and processing, 54–55, 330–333, 398–400
    gene-protein colinearity, 111–112
    glycosylation, 335–337, 399–400, 410–413
    import to chloroplasts, 454–455, *456*
    import to mitochondria, 437, 439–442, *443*
    misfolding, 331
    in mitochondria, 437
    movement through Golgi apparatus, 410, *411*
    phosphorylation, 341–343
    polypeptides, 53–54
    protein-protein interactions, 344
    proteomics, 65–69
    regulation of function, 339–344
    regulation of nuclear import, 368–369
    secretion (*see* Protein secretion)
    sorting process, 387, *388*
    structure, 54–58
    trafficking in endocytosis, 563–565
    translocation across membranes, 333–334
    transport to and from the nucleus, 362–367
    unfolded protein response, 402, *403*
α-Proteobacteria, 435–436
Proteoglycans, 579, 580
Proteolysis, 333–335
Proteomics, 65–69, 193
Proto-oncogenes, 735–738, 757
Proton (H⁺) gradients
    in active transport, 555, 556
    in chemiosmotic coupling, 445–449
    in photosynthesis, 462
    transport of metabolites across inner mitochondrial membrane and, 450–451
Proton (H⁺) pumps, 424
Protozoans, 175
    cilia, 517, *581*
    nuclei, 372
    transposable elements, 241–242
Prozac, 547
Pseudogenes, *164,* 246
Pseudopodia, 13, 486, *487,* 557
Pseudouridine, *289*
PTB domains, 615
PTCH tumor suppressor gene, *750*
PTEN protein, 751
PTEN tumor suppressor gene, 750, 751
PTS1. *See* Peroxisome targeting signal 1
PTS2. *See* Peroxisome targeting signal 2
Pufferfish, *176,* 190
PUMA protein, 698, 752, *753*
Purine bases, 50–51
    antimetabolites, 98–99

biosynthesis, 97
    in DNA, 108, *109*
    rate of loss in DNA, 220
Puromycin, 316, 320
Pyridoxal, *78*
Pyridoxal phosphate, *78*
Pyrimidine dimers, 217–218
Pyrimidines, 50–51
    biosynthesis, 97
    in DNA, 108, *109*
Pyrophosphate (PP*i*), 83, 98
Pyrrolysine, 325
Pyruvate, 88
    in glucose synthesis, 92, 93
    oxidative decarboxylation, *86*
    in oxidative metabolism, 434
    transport across inner mitochondrial membrane, 450–451

qin oncogene, *735*
Quantum dots, 25
Quaternary protein structure, 58
Quinones, *459,* 460

Rab family proteins, 419, 422–423, 630
Rab GTP-binding proteins, 563
Rabl, Carl, 372
Rac protein, 632, 639
RAD genes, 221, 222
Rad51 protein, 231, *233*
Radioactive thymidine, 651–652
raf oncogenes, *735,* 736, 742, 754
Raf protein, 627, *628,* 630, *631,* 736, 742
RAG1 protein, 236, 237
RAG2 protein, 236, 237, 714–715
Ran/GTP complex, 365, 366, 367, *368*
Ran GTPase-activating protein (Ran GAP), *366, 368*
Ran guanine nucleotide exchange factor (Ran GEF), *366*
Ran protein, 365–366, 367
Rapamycin, 627
Raptor protein, 625
RARα gene, 757
ras oncogenes, 629, *735,* 736, 738, 739
Ras proteins, *537*
    as oncogenic proteins, 739, 742
    prenylation and, 338, *339*
    regulation of, 341
    in signal transduction, 627–628, 629, 630, *631,* 638, 742
ras proto-oncogenes, 742
*ras*H oncogene, *735*
*ras*K oncogene, *735,* 754
Rayment, Ivan, 491
Rb protein, 666, 667, 731, 732, 749, 752
Rb tumor suppressor gene, 747–749, *750*
Reaction centers
    in photosynthesis, 459–460
    transmembrane proteins, 535
RecA protein, 230, 231, *232, 233*
Receptor down-regulation, 564
Receptor guanylyl cyclases, 617
Receptor-mediated endocytosis, 558–563
Receptor protein-tyrosine kinases
    described, 612–615
    integrins and, 638
    as oncogene proteins, 742
    Ras proteins and, 630, *631*
Recessive genes, 104
Reclinomonas americana, 436
Recombinant DNA

cloning, 121–122, 131
expression of cloned genes, 126–128
generation of, 121–122
restriction endonucleases and, 118–119,
121
sequencing, 125, *126*
vectors for, 122–123, *124, 125*
Recombinant DNA libraries, 132, *133*
Recombinant molecules, 121–122
Recombination, 679
Recombination signal (RS) sequence, 236
Recombinational repair, 225–227
Rectum cancers/carcinomas, *720, 750*
Red blood cells. *See* Erythrocytes
Reese, Thomas, 512, 514
Refractive index, 22
Regulated secretion, 414–415
rel oncogene, *735*
Release factors, 325
Renal cell carcinoma, *750*
Repetitive DNA, 161–164, 189, 190
Replication factor C (RFC), 207
Replication fork, 202–209
Replication protein A (RPA), 207, 222
Repressor proteins, 260–261, 325–326
Reproductive cloning, 715
Resolution, 22, 23
Respiratory diseases, 554
Restriction endonucleases, 118–119, 121
Restriction maps, 119
Restriction point, 653
ret oncogene, *739, 755*
Reticulocyte, *705*
Reticuloendotheliosis virus, *735*
Retinal, 620, *621*
Retinoblastoma, 666, 747–749, *750*
Retinoic acid, 601, 602, 744, 745, 758
Retinoic acid receptor, 744
Retinoic acid receptor gene, 757
Retinoids, 602
Retrotransposons, *162, 163,* 242, 244–247
Retroviral oncogenes, 734–735
Retrovirus-like elements, *162, 163,* 244
Retroviruses
cancers and, *729,* 732–733
overview of, *38,* 39
replication and reverse transcriptase,
116–117, 242–244
transposition and, 242–244
as vectors for transfection, 140
Reverse genetics, 142
Reverse transcriptase, 117, 122, 163, 214,
242–244. *See also* Telomerase
Reverse transcription
interspersed repetitive sequences and,
163–164
with LINEs, 245–246
overview of, 115–117
pseudogenes and, 165
retroviruses, 242–244
RFC. *See* Replication factor C
RFs. *See* Release factors
Rheb protein, 626, *627,* 628
Rheumatoid arthritis, 192
Rho protein family, 258, 496, 632, 639–640
Rhodobacter sphaeroides, 459
Rhodopseudomonas viridis, 459–460, 535
Rhodopsin, 620–621
Riboflavin, *78*
Ribonuclease, 54–55, *57*
Ribonucleic acid (RNA)
antisense, 145

catalytic properties, 6
central dogma and, 113
chemical composition, 50–51
cloning, 122
complementary base pairing, 51
degradation, 301–303
difference from DNA, 112
double-stranded, 145, 146
editing, 300–301
nuclear transport, 369–370
nucleic acid hybridization techniques,
131–134
origin of life and, 6–7, 317
PCR amplification, 129, 130
primers in DNA replication, 204, 205, *206*
self-replication, 5
snoRNPs and, 379
splicing (*see* Splicing)
telomerase and, 214–216
transposable elements and, 242–247
viruses and, 38–39
Ribonucleoprotein complexes (RNPs), 294,
369, *370*
Ribonucleoside 5′-triphosphates (NTPs),
254, 257
Ribose, *45, 50, 51*
Ribosomal proteins
chloroplast genome and, 454
ribosome assembly and, 379, *380*
Ribosomal RNA genes, 266–267, 376
Ribosomal RNA (rRNA), 113
amplification in amphibian oocytes,
247–248
catalytic activity of, 315–317
encoded by chloroplast DNA, 453
encoded in mitochondria, 436
functions of, 50
genes, 266–267, 376
nuclear transport, 369
nucleolus and, 375–376
in ribosomes, 311, 313–315
RNA polymerases transcribed by, 262
splicing, 295
stability of, 302
transcription and processing, 287–288,
377–379
types in eukaryotes, 287–288
types in prokaryotes, 288
Ribosomes
in animal cells, *10*
assembly, 379, *380*
component RNAs, 376
cotranslational translocation and,
387–392
in *E. coli,* 8
initiation of translation and, 318–319
in plant cells, *11*
polysomes, 325
structure of, 311, 313–135
subunits, 376
in translation, 319, 321, *322,* 323, 324, 325
in vitro assembly, 314
Ribothymidine, *289*
Ribozyme, 289
Ribulose bisphosphate carboxylase
(Rubisco), 454, 465
Rice, *176,* 185
Richardson, Timothy, 169
Rickettsia prowazekii, 435–436
Rictor protein, 624, 625
rII gene, 113–114
RISC. *See* RNA-induced silencing complex

RITS complex. *See* RNA-induced transcrip-
tional silencing complex
RNA-DNA hybrids, 145
RNA editing, 300–301
RNA-induced silencing complex (RISC),
145, *146,* 285, 327, *328*
RNA-induced transcriptional silencing
(RITS) complex, 285, 327
RNA interference (RNAi), 145–146, 192,
193–194, 285
RNA polymerase I, 262, 266–267, 376, 377
RNA polymerase II
α-amanitin and, 476
C-terminal domain, 265, 266, 290
general transcription factors, 262–266
mRNA processing and, 290
promoters, 269
proteins transcribed by, 262
RNA splicing and, 298
transcription of ribosomal protein genes,
379
RNA polymerase III
proteins transcribed by, 262
transcription of RNA genes, 267–268
transcription of rRNA genes, 376, 379
RNA polymerases, 113
chloroplast genome and, 454
in eukaryotes, 262, *263*
in prokaryotes, 254–258
RNA primers, 204, 205, *206*
RNA processing
editing, 300–301
mRNA in eukaryotes, 290–292
rRNAs, 288
splicing, 157, 292–300, 373
tRNAs, 288–290
RNA splicing. *See* Splicing
RNA transposable elements, 242–247
RNA tumor viruses, 115–117
RNA viruses, 115–117
RNA world, 6
RNase, 330
RNase H, 205, 242, *243,* 244
RNase P, 288–289, 317
RNPs. *See* Ribonucleoprotein complexes
Roberts, Richard, 157, 158
Rod cells, 620–621
Rodbell, Martin, 610
Roeder, Robert, 263
ros oncogene, *735*
Rotenone, 444
Rothman, James, 418, 421
Rough endoplasmic reticulum, 386
in animal cells, *10*
isolation of, 388–389
membrane characteristics, *61*
peroxisome assembly and, 467–468
in plant cells, *11*
signal sequence hypothesis and, 389
Rous, Peyton, 37, 39, 733
Rous sarcoma virus (RSV), 37
DNA provirus theory and, 116–117
oncogenes and, 733, 734, *735,* 737
protein-tyrosine kinase and, 612, 613
RPA. *See* Replication protein A
Rsk protein kinase, 683
RS sequence. *See* Recombination signal
sequence
RSV. *See* Rous sarcoma virus
Rubella virus, *38*
Rubin, Harry, 728
Ruderman, Joan, 660, 663

Ruv proteins, 231–232, *233*
Ryanodine receptors, 623, 624

S6 kinase, 627
S phase, 650, 655
Sabatini, David, 389, 390
Saccharomyces cerevisiae, 12, *13*
　cell cycle regulation in, 652–653, 657, 659–660
　centromeres, 172–173, *174*
　chromosome number, *167*
　complete genome sequence data, *176*, 178–180
　exocysts and, 423
　as experimental system, 17
　genome size and composition, *16, 17*, 137, 165, *167*
　MAP kinases in, 631
　origins of DNA replication, 212
　spores, 678
　telomeric repeat sequences, *175*
Sakmann, Bert, 544
Sanger, Frederick, 54
Sarcomas, 721, *750*
Sarcomeres, 487, 488–489, 490
Sarcoplasmic reticulum, 491, 624
Satellite cells, 707, *708*
Satellite DNAs, 162–163
Saturated fatty acids, 47
Scaffold proteins, 632–633
Scanning electron microscopy, 30
Schatz, Gottfried, 439
Schekman, Randy, 418
Schizosaccharomyces pombe
　cell cycle, 654
　cell cycle mutants, 659, 664
　centromeres, 173–174
　complete genome sequence data, *176*
　origin of recognition sequences, 213
　telomeric repeat sequences, *175*
Schleiden, Matthias, 21
Schwann, Theodor, 21
SCID. *See* Severe combined immunodeficiency
Sclerenchyma, 13
scRNAs, 262
Scurvy, 576
SDS. *See* Sodium dodecyl sulfate
SDS-polyacrylamide gel electrophoresis (SDS-PAGE), 135
sea oncogene, *735*
Sea urchin embryos, 660
Sec62/63 protein complex, 393
Sec pathway, 455, *456*
Sec61 proteins, 392, 418
Sec translocon, 455, *456*
SecA protein, 455, *456*
Second messengers
　Ca²⁺, 622–624
　cAMP pathway, 618–620
　cGMP, 620–621
　mTOR pathway, 625–627
　PI 3-kinase/Akt pathway, 624, *625*
　PIP₂, 620–621
Secondary cell walls, 572–573
Secondary protein structure, 56–57
Secondary response genes, 631
γ-Secretase, *637*
Secretory pathway, 386–387
Secretory vesicles, 386, 414–415
Securin protein, 677
Sedimentation, 31

Seeds, glyoxysomes in, 464
Sefton, Bartholomew, 613
Sela, Michael, 54–55
Selectins, 540, *541*, 584
Selectivity factor 1, 267
Selenocysteine, 325
Self-splicing, 295, 297, *298*
"Selfish DNA elements," 164
Semiconservative replication, 110, *111*
Separase protein, 677
Separation anxiety, 192
Serine, 47, *48*, 52, *53, 81*, 465
Serine proteases, 76
Serotonin, 547, *604*
Serotonin reuptake inhibitors (SSRIs), 547
Serum response element (SRE), 631
Severe combined immunodeficiency (SCID), 237
Sex determination, in *Drosophila*, 299
Sex steroids, 601
*Sfi*I enzyme, *118*
SH2 domains, 615, 622, 630, 633, 637
Sharp, Phillip, 157, 158, 276
β-Sheets, 57
Sheetz, Michael, 491, 512, 514
Shine, John, 318
Shine-Dalgarno sequence, 318
Shinoué, Shinya, 512
Short-interfering RNAs (siRNAs), 145, 146, 302
Short interspersed elements (SINEs), *162*, 163, 246–247
Shotgun mass spectrometry, 67–68
Shotgun sequencing technique, 187, 188
Signal amplification, 618–619
Signal hypothesis, 389, 390–391
Signal patches, 412
Signal peptidase, 334, 392, 408
Signal recognition particles (SRPs), 391–392, 455, *456*
Signal sequences, 334
　protein insertion into ER membrane, 396–397, *398*
　protein targeting to the ER, 388–392
Signal transduction pathways
　in apoptosis, 698–700, *701*
　cAMP pathway, 618–620, 623
　cancer cells and, 725
　cGMP and, 620–621
　crosstalk, 640–641
　cytoskeleton and, 637–640
　feedback loops, 640
　Hedgehog pathway, 634, *635*, 751
　JAK/STAT pathway, 633
　MAP kinase pathways, 627–628, 630–633, 641
　mTOR pathway, 625–627, 631
　NF-κB pathway, 634, 640
　Notch pathway, 636–637
　overview of, 617
　phospholipids in, 621–622
　PI 3-kinase/Akt pathway, 624, *625*, 745–746
　protein kinases and, 342–343
　signal amplification, 618–619
　signaling networks, 641–642
　TGF-β/Smad pathway, 633–634, 752, 754
　Wnt pathway, 634, 635–636, 744, 754
Signaling networks, 640–642
Silk, 58
Simian sarcoma virus, *735*
Simian virus 40 (SV40), *729*

cancer studies and, 730–731
enhancers, 270
origin of replication, 211
Sp1 transcription factor, 274, 276
T antigen, 362–363, 364–365
Simple-sequence repeats, 162
SINEs. *See* Short interspersed elements
Singer, Jonathan, 59–60, 61, 532
Single nucleotide polymorphisms (SNPs), 195
Single-stranded DNA-binding proteins, 207, *208*
siRNAs. *See* Short-interfering RNAs
sis oncogene, *735*
Site-specific recombination, 233–238, 240–241
Skeletal muscle
　muscle cell contraction, 487–491, *492*
　muscular dystrophy, 483
　stem cells and, 707, *708*
ski oncogene, *735*
Skin abnormalities, 502–504
Skin cancers, 217, 221, 222, *720*
Skin diseases, 591
Skin epithelial cells, keratin mutants, 502–504
Skin fibroblasts
　cell cycle regulation and, 654
　proliferation, 700, 702
Skin grafts, 709
SL1 transcription factor, 267
Sliding-clamp proteins, 207
Sliding filament model, 489–490
Slime molds, *167*
Sm antigen, 294–295
Smac/Diablo protein, 697
Smad2 transcription factor, 754
Smad4 transcription factor, 754
Smad transcription factor family, 617, 633, 634, 754. *See also* TGF-β/SMAD pathway
Smad2 tumor suppressor gene, *750, 752*
Smad4 tumor suppressor gene, *750, 752*
Small GTP-binding proteins, 365, 628, 630
Small intestine, 538–539
Small nuclear ribonucleotide particles (snRNPs), 293, 294–295, *298*
Small nuclear RNAs (snRNAs)
　nuclear transport, 369, 370
　RNA polymerases transcribed by, 262
　snRNPs and, 294–295
　spliceosomes and, 293, 295, 297
Small nucleolar RNAs (snoRNAs), 370, 378–379
Small ubiquitin-related modifier (SUMO), 347
SMC proteins, 673
Smith, Alan, 362, 364
Smith, Dennis, 657, 658
SMO oncogene, *739*
Smooth endoplasmic reticulum, 386
　in animal cells, *10*
　lipid synthesis and, 403–406
　in plant cells, *11*
Smooth muscle, 487, 703
Smoothened protein, 634, *635*, 751
Snake venom, 548, 622
SNAP transcription factor, 267
SNARE hypothesis, 421–422
SNARE proteins, 421–422, 423, 563
snoRNAs. *See* Small nucleolar RNAs
snoRNPs (snoRNA protein complexes), 379

SNPs. *See* Single nucleotide polymorphisms
snRNAs. *See* Small nuclear RNAs
snRNPs. *See* Small nuclear ribonucleotide particles
Sodium dodecyl sulfate (SDS), 135
Sodium (Na⁺)
    active transport and, 555–556
    extracellular and intracellular concentrations, *545*
    gradient across plasma membrane, *553*
    membrane potential and, 545
    Na⁺ -Ca²⁺ antiporter, 556
    Na⁺ -H⁺ antiporter, 556
    Na⁺-K⁺ pump, 550–551, *552*
Sodium (Na⁺) channels, 544, *546, 547, 548, 549, 550,* 620
Soft keratins, 497
Somatic cell nuclear transfer, 682, 713–715
Somatic hypermutation, 237, 238
Sos guanine nucleotide exchange factor, 630, *631,* 638
Southern, E. M., 131
Southern blotting, 131, *132*
Sp1 transcription factor, 274–275, 276, 277
Spacer sequences, 157
Specificity protein 1, 274–275, 276, 277
Spectrin, *477,* 482–483, 534, 538
Sperm
    in fertilization, 684
    flagella, *518*
Sphingolipids, 532, 539
Sphingomyelin, 47, *48,* 60, *61,* 406, 413, 531, 532, 539
Spindle assembly checkpoint, 655
Spindle microtubules, 522
Spindle pole bodies, 671
Spindle poles, 522, 523
Spleen, 426, 558
Spliceosomes, 293, 295, *296,* 297–298
Splicing, 157, 373
    alternative, 160, *161,* 189, 299–300
    mechanisms, 292–298
    in pre-tRNAs, 290
    in vitro, 292–293
Splicing factors, 297–298, 373
Spo11 endonuclease, 679
Spores, 678
SPP. *See* Stromal processing peptidase
Spudich, James, 491
Squamous cell carcinoma, *739*
Squid giant axon, 514–515, 544–545
SR protein family, 298, 300
Src nonreceptor protein-tyrosine kinase, 637, 638
src oncogene, 734, *735,* 736, 737, 742
Src protein family, *537,* 612, 613, 616, 617
SRE. *See* Serum response element
SRP. *See* Signal recognition particles
SRP pathway, 455, *456*
SRP receptor, 391, 392
srpRNA, 391–392
SSRIs. *See* Serotonin reuptake inhibitors
Stability genes, 753, 755
Stahl, Frank, 110
Staining, positive and negative, 28–29
Stains, 23, 570
Staphylococcus aureus, 570
Starch, 45, *46*
START, 652–653, 659, 661
STAT proteins, 633
Statins, 560
Stehelin, Dominique, 737

Steitz, Joan, 294–295
Steitz, Thomas, 315, 316
Stem cells. *See also* Adult stem cells; Embryonic stem cells
    hematopoietic, 704, *705,* 708–709
    intermediate filament proteins, *497*
Stereocilia, 485
Steroid hormone receptors, 277, 602–603
Steroid hormones, 49. *See also individual hormones*
    lipid metabolism and, 406
    nuclear receptor superfamily and, 602–603
    regulatory role of, 341
    types of, 601–602
Sterols, 532
STI-571 (drug), 37, *757, 758*–760
Stimulatory crosstalk, *641*
Stomach cancer, 724
Stomach ulcers, 724
Stomata, *14*
Stop codons, 115, 325
Stop-transfer sequence, 394, 396, 397, *398*
Streptomycin, 320
Stress fibers, 484, 491, 639
Stroma, 451, *452,* 454, *455*
Stromal processing peptidase (SPP), 454, *455*
Structural maintenance of chromatin (SMC) proteins, 673
Subcellular fractionation, 30–33, 68
Substance P, *605*
Substrates, 74
Succinate, 444, *446*
Sugars. *See also* Carbohydrates; Glucose
    lipid metabolism and, 413
    in nucleic acids, 50–51
    overview of, 44–45
Su[H] transcription factor, 636, *637*
Sulston, John, 694
SUMO, 347, 374
Supernatant, 31
Sutherland, Earl, 618
SV40. *See* Simian virus 40
Sweat glands, 554
Swinging-cross-bridge model, 490
SXL protein, 299, 300
Symport, 556
Synapses, 327, 418
    dynamics of, 547–548
    recycling of synaptic vesicles, 564–565
Synapsis, 679
Synaptic vesicles, 417–418, 564–565
Synaptomenal complex, 679
Syndecans, 579
Synthetic oligonucleotides, 142, *143*
Systemic lupus erythematosus, 294
Systems biology, 192–195

T antigens, 362–363, 364–365, 730, 731
T4 bacteriophage, 38, 113–114
T-cell lymphotropic virus, 37
T cell receptor genes, 236, 241
T cell receptors, 236
T lymphocytes, 234, 601, 617, *705*
T4 lymphocytes, 120
TAFs. *See* TBP-associated factors
Talin, *477,* 484, 496, 580, 581
Tamkun, John, 582, 583
Tandem mass spectrometry, 67
Taq polymerase, 129, *130*
*Taq*I enzyme, *118*

TAT pathway, 455, *456*
TATA-binding protein (TBP), *264,* 265, 266, 267, 268
TATA box, 263, *264,* 265, 267, 268, 269
Tatum, Edward, 107
Tau protein, 510, 511
tax gene, 732
Taxol, 507
TBP. *See* TATA-binding protein
TBP-associated factors (TAFs), *264,* 265
Tcf/LEF transcription factors, 635, *636*
Tel/PDGFR fusion protein, 742, *743*
Tel transcription factor, 760
Telomerase, 176, 214–216, 728
Telomeres, 175–176, 214–216
Telophase, 669, *670,* 671
Temin, Howard, 116–117, 728
Temperature-sensitive mutants, 138
Tendons, 575
Teratocarcinomas, 710
Terminal deoxynucleotide transferase, 236
Terminal web, 486
Termination factors, *319*
Tertiary protein structure, 57–58
Testes, 406
Testosterone, 49, 601
Tetracycline, 320
Tetrahydrofolate, *78*
Tetrahymena, 175, 214, 282, 295, 316, 317
Tetrodotoxin, 190
TFIIA, 267, 277
TFIIB, *264,* 265, 267–268, 279
TFIIC, 267
TFIID, *264,* 265, 279, 282
TFIIE, *264,* 265
TFIIF, *264,* 265
TFIIH, *264,* 265, 664
TGF-β. *See* Transforming growth factor β
TGF-β receptors, 633–634
TGF-β/SMAD signaling pathway, 633–634, 752, 754
Therapeutic cloning, 714–715
Thermoacidophiles, 8
Thermocyclers, 129
Thermodynamics, 81
Thermogenin, 450
Thermoregulation, 450
Thermus aquaticus, 129, 316
Thiamine, *78*
Thiamine pyrophosphate, *78*
6-Thioguanine, 98, 99
Threonine, 52, *53,* 80
Threonine deaminase, 80
Thrombin, 76
Thromboxanes, 607
Thy-1 protein, *537*
Thylakoid lumen, 451, 452, 462
    import of proteins, 455, *456*
    in photosynthesis, 460, *461*
Thylakoid membrane, 457
    functions of, 451–452
    photocenters, 459
    photosystems I and II, 460–461
    proton gradient across, 462
    structure, 451
    translocation across, 455, *456*
Thymine, *50, 51, 108, 109*
Thymine dimers, *219, 221*
Thymosin, 477
Thyroid carcinoma, *739*
Thyroid hormone, 601, 602, 744
Thyroid hormone receptor, 603, 744

Ti plasmid, 141, *142*
Tic complex, 454, *455*
Tight junctions, 539, 588–589
Till, James, 704
Tim complexes, 439, 440, 441, 442, *443*
"Tiny Tim" proteins, 441, *442*
Titin, 489
Tjian, Robert, 274, 276
TNF. *See* Tumor necrosis factor
TNF receptor, 700, *701*
Tobacco, cancer and, 723
Tobacco mosaic virus, 115
Toc complex, 454, *455*
Toc34 receptor protein, 454, *455*
Toc159 receptor protein, 454, *455*
Togaviruses, *38*
Tom complex, 439, *443*
Tom proteins, 439, 441, *443*
Tonegawa, Susumu, 234, 240–241
Topiosomerases, 208
Torpedo rays, 548
Tracheids, *14*
Transacetylase, 259, 260
Transcription, 113. *See also* Transcription
    factors; Transcriptional regulation
    chromatin condensation and, 169, 170
    differences between prokaryotes and
       eukaryotes, 262
    eukaryote RNA polymerases, 262, *263*
    in prokaryotes, 254–262
    RNA polymerase I, 266–267
    RNA polymerase II, 262–266
    RNA polymerase III, 267–268
Transcription-coupled repair, 222–223
Transcription factors
    ERK pathway regulation, 631
    identifying binding sites, 272–273
    intracellular signal transduction and, 617
    isolation of, 273–275, 276
    nuclear import, 368–369
    as oncogene products, 743
    regulation by small molecules, 340–341
    for RNA polymerase I, 266–267
    for RNA polymerase II, 262–266
    for RNA polymerase III, 267–268
Transcriptional activators, 277–280
Transcriptional regulation
    chromatin modifications and, 281–284
    cis-acting regulatory sequences, 269–272
    DNA methylation, 286, *287*
    key features of, 268–269
    by noncoding RNAs, 285–286
    positive and negative in prokaryotes,
       258–262
    transcriptional activators, 277–280
Transcytosis, 565
Transducin, 621
Transfection, 139–141
Transfer RNA genes, 267
Transfer RNA (tRNA)
    attachment of amino acids, 310–311, *312*
    encoded by chloroplast DNA, 453
    encoded in mitochondria, 436, 437
    functions of, 50, 113
    genetic codes and, 436–437
    nuclear transport, 369, 370
    processing, 288–290
    protein synthesis and, 96, *97*
    in reverse transcription by retroviruses,
       242, *243*
    RNA polymerases transcribed by, 262
    stability of, 302

structure, 310
    in translation, 319, 321, *322*, 323, 324
Transferrin receptor, 302–303
Transformation, genetic, 107
transformer gene, 299
Transforming growth factor β (TGF-β), 617
Transgenic mice, 140, 141, 143–144, 502–504
Transgenic plants, 141, *142*
Transient expression, 139–140
Transit-amplifying cells, 705, *706*
Transit peptides, 454
Transition state, 74
Transitional endoplasmic reticulum, 386,
    395, 407, 420
Translation, 113
    chaperone proteins and, 330, *331*
    initiation and mRNA organization,
       317–319
    key features of, 309–310
    process of, 319, 321–325
    regulation, 325–329
    ribosomes and rRNA, 311, 313–317
    tRNAs in, 310–311, *312*
Translation factors, 319, 321, *322*, 323–324,
    325
Translesion DNA synthesis, 223–225
Translocations
    Bcr/Abl protein and, 758
    oncogenes and, 740–741
Translocon, 392, 393, 398
Transmembrane proteins, 61, *62*
    characteristics of, *532*, 533–534, 535–536
    export from the ER, 406–407, *408*
    insertion in ER membrane, 393–398
Transmission electron microscopy, 28–30
Transport. *See also* Transport across plasma
       membranes; Vesicular transport
    active, 64, *65*, 550–553, 555
    passive, 64, 541–542
Transport across plasma membranes
    by ATP hydrolysis, 64, *65*, 550–553, 555
    channel proteins and, 63–64
    ion channels, 543–549, *550*
    by ion gradients, 555–556
    passive diffusion, 64, 541–542
Transport vesicles. *See also* Vesicular trans-
       port
    coated, 419–420, *421*
    fusion with target membrane, 420–423
    in secretory pathways, 398, 407–408, 410,
       411, 414, 415
Transporters, 63–64, *65*
Transposable elements, 163, 239. *See also*
    Transposons
Transposase, 241
Transposons. *See also* Transposable ele-
    ments
    DNA, 239, 241–242
    DNA methylation and, 286
    RNA, 242–247
Transvection, 271
Treadmilling, 476, 506
Triacylglycerols, 47, 89–90
Trichothiodystrophy, 222
Triplet code, 113–115
Tropomodulin, *477*
Tropomyosin, 477, 484, 491, *492*
Troponin, 491, *492*
Trypsin, 67, 76
Tryptophan, 52, *53*, 76
Tβ RII tumor suppressor gene, *750*, 752, 754
TSC2, 631

TSC1/2 protein complex, 626, *627*
Tubulin, 505, 509. *See also* Microtubules
δ-Tubulin, 510
γ-Tubulin, 509
γ-Tubulin ring complex, 509
Tumor clonality, 721
Tumor initiation, 722
Tumor necrosis factor (TNF), 617, 700, *701*
Tumor progression, 722
Tumor promoters, 723
Tumor suppressor genes, 666
    benefits of molecular analysis, 756
    functions of protein products, 750–753
    identification of, 746–750
    in tumor development, 753–754
Tumor suppressor proteins, 750–753
Tumor viruses
    adenovirus models, 731–732
    cancers caused by, 724, 729
    hepatitis B, 729
    hepatitis C, 729, 730
    herpesviruses, 732
    papillomaviruses, 731
    protein-tyrosine kinases and, 612–613
    retroviruses, 732–733
    RNA tumor viruses, 115–117
    SV40 and polyomavirus models, 730–731
Tumors, 720, 753–754
Turgor pressure, 573–574
Twinfilin, *477*, 496
Two-dimensional gel electrophoresis, 66, 67
Tyrosine, 52, *53*

U2AF splicing factor, 298, 299, 300
UBF transcription factor, 267
Ubiquinone, 444, 460
Ubiquitin, 344–347
Ubiquitin-activating enzyme (E1), 345
Ubiquitin-conjugating enzyme (E2), 345
Ubiquitin ligase (E3), 345
Ubiquitin-proteasome pathway, 344–347
Ubiquitination, 345–347
UDP-glucose, 93, *94*
Ultraviolet radiation (UV)
    cancer and, 723
    DNA damage and, 217, *218*
Umbilical cord blood, 709
UMP. *See* Uridine 5'-monophosphate
Unfolded protein response, 402, *403*
Unicellular eukaryotes, closed mitosis,
    670–671
Uniport, 556
Unsaturated fatty acids, 46–47
3'-Untranslated region, 317, *318*
5'-Untranslated region, 317, *318*
Upstream binding factor, 267
UR2 sarcoma virus, *735*
Uracil, *50*, 51, 220
Uranyl acetate, 28
Uridine, *50*
Uridine 5'-monophosphate (UMP), *50*
Uridine triphosphate (UTP), 93, *94*
U3snoRNA, 379
U8snoRNA, 379
U22snoRNA, 379
Uterine cancer, *720*
UTP. *See* Uridine triphosphate
UV radiation. *See* Ultraviolet radiation
uvr genes, 220
Uvr proteins, 220
UvrABC complex, 220, 223

Vaccinia virus, *38*
Vacuoles, *417*
    functions of, 9, 416
    in plant cells, *11*
Vale, Ronald, 512, 514
Valine, 52, *53*
Varmus, Harold, 736, 737
Vascular endothelial growth factor (VEGF),
    703, 757
Vascular system, in plants, 14
Vasopressin, *605*
Vector DNA, 121–122
Vectors
    expression vectors, 127
    for recombinant DNA, 122–123, *124, 125*
    for transfection, 140
Velocity centrifugation, 31–32
Venter, Craig, 177, 187, 188
Vertebrate eye, photoreception, 620–621
Vesicles. *See also* Transport vesicles
    clathrin-coated, 419, *420*
    coated, 419–420, *421*
    fusion with target membrane, 420–423
    secretory, 386, 414–415
    synaptic, 417–418, 564–565
    transport along microtubules, 515
Vesicular transport. *See also* Transport vesi-
        cles
    coated vesicles, 419–420, *421*
    experimental approaches, 417–418
    significance of, 417
    vesicle fusion, 420–423
Vessel elements, 14
VHL tumor suppressor gene, *750*
Video-enhanced differential interference-
        contrast microscopy, 24
Video-enhanced microscopy, 512
Villin, 477, 485
Vimentin, *497*, 498, 499, 500, 502
Vinblastine, 507
Vincristine, 507
Vinculin, 477, 484, 496, 580, 581, 587
Viral mRNA, 319
Viruses, 36–39, 724. *See also* Adenoviruses;
        Animal viruses; Retroviruses; Simian
        virus 40; Tumor viruses
Vision, photoreception, 620–621
Vitamin B_1, *78*
Vitamin B_2, *78*
Vitamin B_6, *78*
Vitamin C deficiency, 576
Vitamin D_3, 601, 602

Vitamins, coenzymes and, *78*, 79
Vogetstein, Bert, 753
Vogt, Peter, 734, 737
Vohwinkel's syndrome, 591
Voltage-gated Ca^{2+} channels, 623
Voltage-gated channels, 544, *546*, 547,
    548–549, *550*
Volvox, 13

Wallace, Douglas, 438
Warren, Robin, 724
WASP/Scar complex, 495
Water
    in photosynthesis, 90
    properties of, 43–44
Watson, Crick, 108
Wee mutant, 664
Wee1 protein kinase, 660, 664
Weinberg, Robert, 738
Western blotting, 135–136
White, Fred, 54–55
White blood cells, 14, *497*, 558. *See also spe-
    cific cell types*
Wilkins, Maurice, 108
Wilms' tumor, *750*, 751–752
Wilmut, Ian, 713
Wnt proteins, 635
Wnt signaling pathway, 634, 635–636, 744,
    754
Wobble, 311, 436
WT1 protein, 751–752
WT1 tumor suppressor gene, *750*

X chromosome inactivation, 285–286, 721
X-linked Emery-Dreyfuss muscular dystro-
    phy, 359
X-ray crystallography, 56, 535
Xenopus laevis, 20
    genome size and composition, *167*
    mRNA localization in oocytes, *327*
    oocyte rRNA genes, 376
Xeroderma pigmentosa (XP), 221, 222, 753
Xist gene, 285
Xist RNA, 285–286
XP. *See* Xeroderma pigmentosa
XP DNA repair genes, 222
XPB protein, 265
XPD protein, 265
XPF/ERCC1 endonuclease, 222
XPG endonuclease, 222
Xylem, 14

Y73 sarcoma virus, *735*
YAC. *See* Yeast artificial chromosome
Yeast, 12. *See also Saccharomyces cerevisiae;
    Schizosaccharomyces pombe*
    actin in, 474
    autonomously replicating sequences,
        212, 213, *214*
    cell cycle regulation in, 652–653, 657,
        659–660
    cell cycle time, 650
    cell signaling in, 599
    centromeres, 172–174
    closed mitosis, 670–671
    complete genome sequence data, *176*,
        178–180
    exocysts and, 423
    as experimental system, 17
    expression of cloned genes in, 128
    genetic analyses in, 137–139
    genome size and composition, *16, 17*,
        165, *167*
    knockout mutation library, 144
    MAP kinases in, 631
    mitochondrial genome, 436
    nucleotide-excision repair, 221
    origins of DNA replication, 212, 213
    posttranslational translocation in,
        392–393
    subcellular localization of proteins, 68–69
    telomeres, 175
    transposable elements, 241–242
    vacuole in, 416
    vesicular transport studies, 418
Yeast artificial chromosome (YAC), 123,
    125, 180–181
Yeast two-hybrid method, 69, 128
Yellow fever virus, *38*
yes oncogene, *735*

Z disc, 488, 489, 490
Zamecnik, Paul, 310
Zea mays, 16
Zeatin, *608*
Zebrafish, *16*, 20
Zellweger syndrome, 467
Zinc finger domain, 277, *278*
Zonula occludens proteins, 589
Zygote, 685
Zygotene, 679, *680*